W9-CIC-028

HANDBOOK OF PLANT AND CROP PHYSIOLOGY

BOOKS IN SOILS, PLANTS, AND THE ENVIRONMENT

Soil Biochemistry, Volume 1, edited by A. D. McLaren and G. H. Peterson
Soil Biochemistry, Volume 2, edited by A. D. McLaren and J. Skujiņš
Soil Biochemistry, Volume 3, edited by E. A. Paul and A. D. McLaren
Soil Biochemistry, Volume 4, edited by E. A. Paul and A. D. McLaren
Soil Biochemistry, Volume 5, edited by E. A. Paul and J. N. Ladd
Soil Biochemistry, Volume 6, edited by Jean-Marc Bollag and G. Stotzky
Soil Biochemistry, Volume 7, edited by G. Stotzky and Jean-Marc Bollag
Soil Biochemistry, Volume 8, edited by Jean-Marc Bollag and G. Stotzky

Organic Chemicals in the Soil Environment, Volumes 1 and 2, edited by C. A. I. Goring and J. W. Hamaker
Humic Substances in the Environment, M. Schnitzer and S. U. Khan
Microbial Life in the Soil: An Introduction, T. Hattori
Principles of Soil Chemistry, Kim H. Tan
Soil Analysis: Instrumental Techniques and Related Procedures, edited by Keith A. Smith
Soil Reclamation Processes: Microbiological Analyses and Applications, edited by Robert L. Tate III and Donald A. Klein
Symbiotic Nitrogen Fixation Technology, edited by Gerald H. Elkan
Soil-Water Interactions: Mechanisms and Applications, edited by Shingo Iwata, Toshio Tabuchi, and Benno P. Warkentin
Soil Analysis: Modern Instrumental Techniques, Second Edition, edited by Keith A. Smith
Soil Analysis: Physical Methods, edited by Keith A. Smith and Chris E. Mullins
Growth and Mineral Nutrition of Field Crops, N. K. Fageria, V. C. Baligar, and Charles Allan Jones
Semiarid Lands and Deserts: Soil Resource and Reclamation, edited by J. Skujiņš
Plant Roots: The Hidden Half, edited by Yoav Waisel, Amram Eshel, and Uzi Kafkafi
Plant Biochemical Regulators, edited by Harold W. Gausman
Maximizing Crop Yields, N. K. Fageria
Transgenic Plants: Fundamentals and Applications, edited by Andrew Hiatt
Soil Microbial Ecology: Applications in Agricultural and Environmental Management, edited by F. Blaine Metting, Jr.
Principles of Soil Chemistry: Second Edition, Kim H. Tan
Water Flow in Soils, edited by Tsuyoshi Miyazaki
Handbook of Plant and Crop Stress, edited by Mohammad Pessarakli
Genetic Improvement of Field Crops, edited by Gustavo A. Slafer
Agricultural Field Experiments: Design and Analysis, Roger G. Petersen
Environmental Soil Science, Kim H. Tan
Mechanisms of Plant Growth and Improved Productivity: Modern Approaches, edited by Amarjit S. Basra
Selenium in the Environment, edited by W. T. Frankenberger, Jr., and Sally Benson
Plant–Environment Interactions, edited by Robert E. Wilkinson

HANDBOOK OF PLANT AND CROP PHYSIOLOGY

edited by
MOHAMMAD PESSARAKLI

The University of Arizona
Tucson, Arizona

Marcel Dekker, Inc.   New York • Basel • Hong Kong

Library of Congress Cataloging-in-Publication Data

Handbook of plant and crop physiology / edited by Mohammad Pessarakli.
     p.   cm. – (Books in soils, plants, and the environment)
     Includes bibliographical references and index.
     ISBN 0-8247-9250-5 (acid-free paper)
     1. Crops–Physiology. 2. Plant physiology. I. Pessarakli,
Mohammad. II. Series.
SB112.5.H36   1994
581.1–dc20
                                          94-32077
                                            CIP

The publisher offers discounts on this book when ordered in bulk quantities. For more information, write to Special Sales/Professional Marketing at the address below.

This book is printed on acid-free paper.

Marcel Dekker, Inc.
270 Madison Avenue, New York, New York 10016

Current printing (last digit):
10 9 8 7 6 5 4 3 2 1

PRINTED IN THE UNITED STATES OF AMERICA

In memory of my beloved parents, Fatima and Vahab, who regretfully did not live to see this work, which in no small part resulted from their gift of many years of love.

Preface

After the successful completion of the *Handbook of Plant and Crop Stress*, I felt that the *Handbook of Plant and Crop Physiology* was needed to fill the gaps left by other handbooks dealing with plants and agronomy. In addition, it has long been recognized that physiological processes control plant growth and crop yields. This handbook is therefore intended to serve as a comprehensive resource and up-to-date reference book, covering the information relevant to plant physiology that is scattered among plant/crop physiology books and journals.

Several problems are encountered when editing a handbook, such as what material to include and exclude, how deeply to cover specific topics, and the best way to organize it all. In attempting to solve these problems, I have chosen to include information that will be beneficial to students, instructors and educators, agricultural researchers and practitioners, field specialists, and any others interested in plant and crop physiology. In order to plan, implement, and evaluate comprehensive and specific strategies for dealing with the problems in plant and crop physiology, strategies must be based on a firm understanding of the facts and principles.

The topics selected for discussion are those that I believe are relevant, common, or complex, and in which physiology plays the dominant role. The concepts of plant and crop physiology have been presented to allow both beginning students and specialists an opportunity to expand and refine their knowledge and practice. Certain conclusions have been provided throughout the text. These are related to the more significant and multifaceted problems of plant and crop physiology and are presented to provide a concise guide for students and specialists alike.

This practical guide has been prepared by 65 contributors from 15 countries, who are among the most competent and knowledgeable scientists, specialists, and researchers in agriculture. To facilitate the accessibility of the wealth of information covered in this collection, the volume has been divided into several sections, each of which discusses as many physiological factors as possible. Although the sections are interrelated, each can also stand alone as a self-contained, comprehensive resource for its specific topic.

Section I, on plants, crops, and growth environment, addresses nutrient uptake by plants, plant–water relationships, water absorption and loss by plants, and the role of temperature in the physiology of crop plants. Section II deals with the physiology of plant/crop growth and developmental stages, comprehensively covering plant physiological stages from seed germination to plant senescence and abscission. "Plant Growth Regulators: The Natural Hormones (Growth Promotors and Inhibitors)" (Section III) addresses roles of several growth promotors as well as growth inhibitor hormones in plants and crops.

Since plants and crops, like other living things, encounter stressful conditions at one time or another during their life cycle, Section IV is devoted to the physiological responses of plants and crops to stress and addresses plant and crop physiological responses to salt, drought, and/or other environmental stresses. Several examples of empirical investigations of specific plants and crops grown under stressful conditions are presented.

Physiology of plant genetics is presented in Sections V, VI, and VII. Section V, "Physiology of Lower-Plant Genetics and Development," consists of one chapter, which discusses development genetics in lower plants. "Physiology of Higher-Plant/Crop Genetics and Development" (Section VI) contains two chapters that present information on transpiration efficiency and the physiological mechanisms relevant to genetic improvement of salinity tolerance in crop plants. Section VII, "Whole Plant vs. Reductive Research on Physiological Genetics of Crop Physiology," addresses the whole-system research as complements of reductive research. The section supports this paradigm by presenting statistical models and analysis of several factors related to plant growth and development as well as the interactive effects of these factors on crop yields.

Finally, to link plant/crop physiology to production, Section VIII, "Physiological Aspects of Sustainable Plant/Crop Production," presents information on production and green crop fractionation and discusses the effects of species, growth conditions, and physiological development on fractionation products of plants.

As with other fields, the field of plant/crop physiology has been growing so rapidly that all plant/crop physiologists are faced with the problem of constantly updating their knowledge. In order to grow with their profession, plant/crop physiologists will need to extend their interests and skills. In this regard, even a casual reading of the material included in this handbook will help to set them in the right direction.

I would like to express my appreciation for the secretarial assistance that I received from Mrs. Elenor R. Loya, College of Agriculture, The University of Arizona. In addition, my sincere gratitude is extended to Mr. Russell Dekker, Vice President and Editor-in-Chief, Marcel Dekker, Inc., who supported this project from its initiation to its completion. Certainly, this job would not have been completed as smoothly and rapidly without Mr. Dekker's most valuable support and sincere efforts. Also, as the Production Editor, Ms. Christine Dunn's patience and work in the careful and professional handling of this volume are greatly appreciated.

The invaluable effort of each and every one of the authors who responded to my request for contributions to this volume is deeply appreciated. Their proficiency and knowledge in their area of expertise made this significant task possible.

My wife Vinca's support in the completion of this work is unforgettable. Last, but not least, I would like to admire my nine-year-old son, Mahdi, who had great patience and allowed me much time that would have otherwise been spent entertaining him during the course of the completion of this book.

Mohammad Pessarakli

Contents

Contents

Contributors

Muhammad Akbar Pakistan Agricultural Research Council, National Agricultural Research Centre, Islamabad, Pakistan

G. N. Amzallag Department of Botany, The Hebrew University of Jerusalem, Jerusalem, Israel

Roger A. Andersen Agricultural Research Service, U.S. Department of Agriculture, and Department of Agronomy, University of Kentucky, Lexington, Kentucky

Timothy S. Artlip Department of Soil, Crop, and Atmospheric Sciences, Cornell University, Ithaca, New York

J. P. Baudoin Faculte des Sciences Agronomiques, Phytotechnie des Regions, Gembloux, Belgium

J. Beaver Department of Agronomy and Soils, University of Puerto Rico, Mayaguez, Puerto Rico

Fred E. Below Department of Agronomy, University of Illinois, Urbana, Illinois

Bernard B. Bible Department of Plant Science, University of Connecticut, Storrs, Connecticut

Elizabeth A. Bray Department of Botany and Plant Sciences, University of California, Riverside, California

Donald P. Briskin Department of Agronomy, University of Illinois, Urbana, Illinois

Rolf Carlsson Department of Plant Physiology, Institute of Plant Biology, Lund University, Lund, Sweden

Calvin Chong   Horticultural Research Institute of Ontario, Ontario Ministry of Agriculture, Food and Rural Affairs, Vineland Station, Ontario, Canada

Gary R. Cline   Community Research, Kentucky State University, Frankfort, Kentucky

D. P. Coyne   Department of Horticulture, University of Nebraska, Lincoln, Nebraska

Frank G. Dennis, Jr.   Department of Horticulture, Michigan State University, East Lansing, Michigan

Francisco Domingo   Estación Experimentál de Zonas Aridas, Consejo Superior de Investigaciones Cientificas, Almeria, Spain

R. S. Dubey   Department of Biochemistry, Faculty of Science, Banaras Hindu University, Varanasi, India

V. P. Evangelou   Department of Agronomy, University of Kentucky, Lexington, Kentucky

Edward A. Funkhouser   Department of Biochemistry and Biophysics, Texas A&M University, College Station, Texas

Larry J. Grabau   Department of Agronomy, University of Kentucky, Lexington, Kentucky

William Grierson   Citrus Research and Education Center, University of Florida, Lake Alfred, Florida

Beverley A. Hale   Department of Horticultural Science, University of Guelph, Guelph, Ontario, Canada

Thomas R. Hamilton-Kemp   Department of Horticulture, University of Kentucky, Lexington, Kentucky

John E. Hendrix   Department of Plant Pathology and Weed Science, Colorado State University, Fort Collins, Colorado

David Hildebrand   Department of Agronomy, University of Kentucky, Lexington, Kentucky

Benjamin Jacoby   Department of Agricultural Botany, The Hebrew University of Jerusalem, Rehovot, Israel

Chris Johansen   Agronomy Division, International Crops Research Institute for the Semi-Arid Tropics (ICRISAT), Patancheru, Andhra Pradesh, India

Clarence J. Kaiser   Department of Agronomy, University of Illinois at Urbana-Champaign, Urbana, Illinois

Arvind Kumar   Department of Plant Physiology, S.K.N. College of Agriculture, Jobner, Rajasthan, India

H. R. Lerner   Department of Botany, The Hebrew University of Jerusalem, Jerusalem, Israel

Monica A. Madore   Department of Botany and Plant Sciences, University of California, Riverside, California

Porfirio N. Masaya   Regional Program for Support to Agronomic Research for Food Grain Crops in Central America, Interamerican Institute for Cooperation for Agriculture, San Jose, Costa Rica

Seja Mmopi   Instituto de Ciencia y Tecnologia Agricolas, Guatemala City, Guatemala

Joaquín Moreno Departmente de Bioquimica i Biologia Molecular, Facultat de Ciencies Biologiques, Universitat de Valencia, Burjassot, Valencia, Spain

Syed Shamshad Mehdi Naqvi Atomic Energy Agricultural Research Center, Tando Jam, Pakistan

James W. O'Leary Department of Plant Sciences, University of Arizona, Tucson, Arizona

D. P. Ormrod Department of Horticultural Science, University of Guelph, Guelph, Ontario, Canada

José A. Pardos Departamento de Silvopascicultura, E.T.S.I. Montes, Universidad Politécnica de Madrid, Madrid, Spain

Lola Peñarrubia Facultat de Ciencies Biologiques, Departamente de Bioquimica i Biologia Molecular, Universitat de Valencia, Burjassot, Valencia, Spain

Mohammad Pessarakli College of Agriculture, The University of Arizona, Tucson, Arizona

Zvi Plaut Agricultural Research Organization, Institute of Soils and Water, Bet Dagan, Israel

Francisco I. Pugnaire Department of Pure and Applied Biology, University of Leeds, Leeds, England

Pramila Rajput Department of Botany, University of Jodhpur, Jodhpur, India

R. C. Nageswara Rao Agronomy Division, International Crops Research Institute for the Semi-Arid Tropics (ICRISAT), Patancheru, Andhra Pradesh, India

A. S. N. Reddy Department of Biology, Colorado State University, Fort Collins, Colorado

Rafael Rodríguez Instituto de Ciencia y Tecnologia Agricolas, Guatemala City, Guatemala

Muhammad Salim Pakistan Agricultural Research Council, National Agricultural Research Centre, Islamabad, Pakistan

David N. Sen Department of Botany, University of Jodhpur, Jodhpur, India

Luis Serrano Departamento de Silvopascicultura, E.T.S.I. Montes, Universidad Politécnica de Madrid, Madrid, Spain

Roy Sexton Department of Biological and Molecular Sciences, Stirling University, Stirling, Scotland

J. P. Srivastava Department of Plant Physiology, Institute of Agricultural Sciences, Banaras Hindu University, Varanasi, Uttar Pradesh, India

G. V. Subbarao Agronomy Division, International Crops Research Institute for the Semi-Arid Tropics (ICRISAT), Patancheru, Andhra Pradesh, India

Ranga R. Velagaleti Batelle Memorial Institute, Columbus, Ohio

Karl M. Volkmar Land Resource Sciences, Agriculture Canada Research Station, Lethbridge, Alberta, Canada

Donald H. Wallace Department of Plant Breeding, New York State College of Agriculture and Life Sciences, Cornell University, Ithaca, New York

John C. Wallace Department of Biology, Bucknell University, Lewisburg, Pennsylvania

Jian Wang   Department of Agronomy, University of Kentucky, Lexington, Kentucky

J. W. White   Centro Internacional de Agricultura Tropical, Cah, Colombia

Kanapathipillai Wignarajah   Controlled Ecological Life Support Program, The Bionetics Corporation, NASA-AMES Research Center, Moffett Field, California

David W. Wolfe   Department of Fruit and Vegetable Science, Cornell University, Ithaca, New York

William Woodbury   Plant Science Department, University of Manitoba, Winnipeg, Manitoba, Canada

Graeme C. Wright   Field Crops Division, Queensland Department of Primary Industries, Kingaroy, Queensland, Australia

K. S. Yourstone   Pioneer Seed Company, Urbandale, Iowa

Hong Zhuang   Department of Nutrition and Food Science, University of Kentucky, Lexington, Kentucky

Richard W. Zobel   Agricultural Research Service, U.S. Department of Agriculture, Ithaca, New York

Nutrient Uptake by Plants

Benjamin Jacoby

The Hebrew University of Jerusalem
Rehovot, Israel

I. INTRODUCTION

This chapter deals with the absorption and accumulation by plant cells of mineral nutrient ions, organic acids, amino acids, and sugar molecules and with their translocation in the plant. The permeability of the phospholipid bilayer of biological membranes to all these hydrophilic compounds is very low. Their transport across the membranes is facilitated by transport proteins—carriers and channels—embedded in the phospholipid bilayer.

Plant cells accumulate all essential mineral ions, to higher concentrations than are present in their environment (Table 1). This accumulation is selective, as evidenced by the different accumulation ratios of the ions shown in Table 1. Some questions arising are:

How is passage through the impermeable liquid layer accomplished?
How is accumulation against the concentration gradient accomplished?
How is metabolic energy coupled to such transport?
What is the mechanism of selectivity?
How is vectorial transport accomplished?

These questions are dealt with in the sections that follow.

II. DEFINITIONS

At the outset, let us define some basic terms used in this chapter.

Electrochemical potential of solute $j - \overline{\mu}_j$ (J mol^{-1}). This is the Gibbs free energy [1] of j:

$$\overline{\mu}_j = \overline{\mu}_j^* + 2.3RT \log a_j + z_jF\Psi + P\overline{V}_j \tag{1}$$

where $\overline{\mu}_j^*$ is the electrochemical potential of j in the standard state ($a = 1.0$, $\Psi = 0$, $P = 0$); R is the gas constant (8.314 J mol^{-1}); T is temperature in degrees kelvin; a_j is the chemical activity of j ($a_j = \gamma_j c_j$, where γ_j is the activity coefficient and c_j the chemical concentration

Table 1 Composition of Pond Water and of the Sap of the Alga *Nitella clavata*, Growing in the Pond

Ion	Pond (mol m^{-3})	Sap (mol m^{-3})	Ratio of sap to pond
Mg^{2+}	1.5	5.5	3.36
Ca^{2+}	0.7	7.0	10.0
Na^+	1.2	49	41.0
K^+	0.5	49	97
$H_2PO_4^-$	0.008	1.7	212
Cl^-	1.0	101	101
SO_4^{2-}	0.34	6.5	20

of *j*); z_j is the electrical charge (ionic charge) of *j*; *F* is the Faraday constant (9.649×10^4 J mol^{-1} V^{-1}), Ψ is the electrical potential (V), *P* is the pressure in excess of atmospheric (MPa), and $\overline{V}_j$ is the partial molal volume of *j* (m^3 mol^{-1}).

Electrochemical potential difference. The driving force for transport of solutes across plant–cell membranes. It is the electrochemical potential difference across the membrane ($\Delta\overline{\mu}_j = \overline{\mu}_j^i - \overline{\mu}_j^o$, where i is inside and o is outside). Practically all biological experiments are performed at atmospheric pressure; hence the pressure difference is negligible, and so is $\Delta P\overline{V}$. The electrochemical potential difference across a membrane then is:

$$\Delta\overline{\mu}_j^{(i-o)} = 2.3RT \ \log \frac{a_j^i}{a_j^o} \ + \ z_jF(\psi^i - \psi^o) \tag{2}$$

Flux of solute *j* – J_j (mol s^{-1} m^{-2}). The unidirectional rate of solute movement across unit membrane area. The net flux J^{net}, or *uptake*, is the difference between the influx ($J^{o \rightarrow i}$) and efflux ($J^{i \rightarrow o}$): $J^{net} = J^{o \rightarrow i} - J^{i \rightarrow o}$.

Accumulation of a solute specifies a higher concentration (not necessarily higher electrochemical potential) of the solute inside.

Active transport. Transport of a solute against its electrochemical potential gradient [2]. Such transport always needs energy input.

Passive transport. Transport of a solute with its electrochemical potential gradient.

Metabolic transport. Any transport that depends on metabolic energy supply. Metabolic transport can be active or passive; it is inhibited by inhibitors of energy metabolism [3].

Electrogenic transport. Transport of an ion unaccompanied by equal opposite charge, thus creating an electrical potential difference.

Electrophoretic transport. Transport of an ion in response to a preexisting electrical potential difference.

III. FREE SPACE AND OSMOTIC VOLUME

Two phases are revealed in the time course of salt uptake by plant tissues: a rapid initial phase that is completed within a few minutes, and a slower uptake that may proceed for several hours at constant rate. The initial, rapid uptake is into the *free space* [4], namely, the extramembranal space of the plant tissue. The free space consists of the cell walls and the intercellular spaces. This uptake is reversible and nonmetabolic. All the anions and part of the cations that are absorbed in the first uptake phase can be washed out with water, and the remaining cations can be exchanged with another cation. Uptake in the second phase is into the *osmotic volume* [4], namely, the space that is surrounded by plasma membranes. This is a metabolic process and not easily reversible.

Cation exchange in the free space results from the presence of immobile negative charges in the cell walls. Dissociated carboxylic groups, in particular those of polygalacturonic acid, are responsible for these charges [4]. The presence of immobile negative charges in the cell wall, adjacent to the external aqueous phase, results in an electrical potential difference, the *Donnan potential* [5].

After equilibration of the external solution with the free space, electrochemical potential differences of cations and anions in the free space (FS) and external solution (sol) are zero. Hence, from Eq. (2):

$$0 = 2.3RT \; \log \; \frac{a_j^{FS}}{a_j^{sol}} \; + \; z_j F(\psi^{FS} - \psi^{sol})$$

The Donnan potential E_D is the electrical potential difference $(E_D = \Psi^{FS} - \Psi^{sol})$, and:

$$E_D = \frac{2.3RT}{z_j F} \; \log \; \frac{a_j^{sol}}{a_j^{FS}} \tag{3}$$

The Donnan potential in cell walls is negative. Equation (3) then shows that cations (z, positive) will accumulate in the negatively charged cell wall (Donnan phase) and that the anion concentration in the latter phase will be lower than in the (adjacent) aqueous phase. Donnan potentials from -7 mV to -289 mV have been calculated for various cell wall–solution systems [6]. The Donnan potential changes with the dissociation of the charged sites; it decreases with salt concentration and increases with the dissociation constants of the various cations.

IV. METABOLIC TRANSPORT

A. Proton Gradients: Uniport and Cotransport

Metabolic solute transport in plant cells is energized by an electrochemical potential gradient of protons ($\Delta\mu_{H^+}$) across the membranes and is facilitated by channels and carriers. The proton electrochemical potential difference is formed by active proton transport, from the cytoplasm to the free space and to the vacuoles. This proton transport is catalyzed by membrane-embedded electrogenic proton pumps that catalyze the transformation of chemical energy in adenosine 5′-phosphatase (ATP) and pyrophosphate, to an electrochemical potential gradient.

Metabolic transport in plant cells that is driven the electrochemical proton gradient is termed *uniport* [7] (also see Section V.A) or *cotransport* [7] (also see Section V.B.3). Uniport is passive and occurs via channels in the direction of the electrical potential gradient of the solute. Cotransport of solutes is active and, in plants, derives its energy from concomitant passive transport of protons back into the cytoplasm.

B. The Membrane Potential

The membrane potential E_M is the electrical potential difference across a membrane; it consists of a diffusion potential and a potential difference resulting from the action of electrogenic pumps. Diffusion potentials result from different diffusion velocities of anions and cations across a membrane.

Membrane potentials of plant cells are measured with reference to the cytoplasm ($E_M = \Psi^i - \Psi^o$), where inside (i) is always the cytoplasm; and outside (o) is the free space (with reference to the plasma membrane) and the vacuole (with reference to the tonoplast). These conventions will be maintained throughout this chapter. Accordingly, under physiological conditions, the membrane potential is negative at both membranes (positive charges in the free space as well as vacuole). An increase in the electrical potential difference, or *hyperpolarization* is synonymous with a decrease of E_M (to more negative values), and *depolarization* is synonymous with an increase of E_M.

C. Diffusion Potential

The unidirectional flux (J_j) of a solute j across a membrane depends on the driving force ($\Delta\bar{\mu}_j$) and the membrane permeability (P_j) of the solute: $J_j = P_j\Delta\bar{\mu}_j$ (mol s^{-1} m^{-2}). Thus, if a salt with different anion and cation permeabilities diffuses across the membrane, an excess charge of the more permeable ion is transported and a diffusion potential E_M^D results. This potential will then retard the diffusion of the more permeable ion. Practically, the same amount of ions of both kinds will pass through the membrane, because a very small anion/cation concentration difference creates a rather large diffusion potential (see Section VI).

The diffusion potential depends on the relative permeabilities of all the cations and anions in the system. Equation (4), the Goldman–Hodgkin–Katz equation [8,9], calculates the diffusion potential for Na$^+$, K$^+$, and Cl$^-$. These are often the quantitatively most important ions in biological systems, and they determine the diffusion potential:

$$E_M^D = 2.3\frac{RT}{F} \log \frac{P_K[K^+]^o + P_{Na}[Na^+]^o + P_{Cl}[Cl^-]^i}{P_K[K^+]^i + P_{Na}[Na^+]^i + P_{Cl}[Cl^-]^o} \qquad (4)$$

where P_K, P_{Na}, and P_{cl} are the membrane permeabilities of K$^+$, Na$^+$, and Cl$^-$, respectively and brackets [ ] designate concentration (mol m^{-3}). The diffusion potential across a membrane is the membrane potential that can be measured when electrogenic transport is inhibited.

Let us calculate the diffusion potential across the plasma membrane for the following situation: $[KCl]^o = 10$ mM, $[NaCl]^o = 10$ mM, $[K^+]^i = 100$ mM, $[Na^+]^i = 10$ mM, and $[Cl^-]^i = 110$ mM; the relative permeabilities of K$^+$, Na$^+$ and Cl$^-$ are 1, 0.2, and 0.01, respectively; the temperature is 30°C (303 K) and $2.3RT/F = 60$ mV. Then:

$$E_M^D = 60 \log \frac{10 + (0.2 \times 10) + (0.01 \times 110)}{100 + (0.2 \times 10) + (0.01 \times 20)} = -53.5 \text{ mV}$$

D. Protonmotive Force

A proton gradient across a membrane consists of an electrical component and a chemical proton concentration gradient—the pH difference across the membrane. The relations of these two components can be defined by replacing j in Eq. (2) with H$^+$ and a_j with [H$^+$]:

$$\Delta\bar{\mu}_{H^+} + 2.3RT \log\frac{[H^+]^i}{[H^+]^o} + zF(\psi^i - \psi^o) \qquad (5)$$

or, when ($\Psi^i - \Psi^o$) is replaced by E_M, log([H$^+$]^i/[H$^+$]^o) by $-\Delta$pH, and 1 is substituted for z (protons):

$$\Delta\bar{\mu}_{H^+} = FE_M - 2.3RT\Delta pH \text{ (J mol}^{-1}\text{)} \qquad (6)$$

The proton gradient can be expressed in electrical units (V), instead of energy units (J mol^{-1}), by division of Eq. (6) with F; $\Delta\mu_{H^+}/F$ then is the *protonmotive force* (pmf) [10,11], defined as follows:

$$\text{pmf} = E_M - \frac{2.3RT}{F}\Delta pH \qquad (7)$$

By substitution of values for R and F (R = 8.3 J mol^{-1} K^{-1}, F = 96.49 J mol^{-1} mV^{-1}) and calculation of the protonmotive force at 30°C (T = 303 K) Eq. (7) becomes

$$pmf = E_M - 60\Delta pH \text{ (mV)}$$

Thus E_M = −120 mV, pH^i = 7, and pH^o = 6, constitute a pmf of −180 mV.

E. Differentiation of Active and Passive Transport

Nonelectrolytes, such as sugars, are actively transported whenever they are accumulated in the cell to a higher concentration than outside. For nonelectrolytes, the electrical component of the electrochemical potential (Eq. 2) nullifies and becomes a function of the concentration ratio only. The electrical component of Eq. (2) is important for ions, and they can passively accumulate in response to an electrical potential difference. Suitable transport proteins may facilitate such passive accumulation. Cations may passively accumulate in the negatively charged cytoplasm, and anions in the positively charged vacuole. Passive ion accumulation is metabolic, because energy metabolism is needed to maintain the necessary membrane potential. The possible passive accumulation ratio (a^i/a^o) depends on the ionic charge of the accumulated ion and on the membrane potential. This ratio can be derived from Eq. (2). By assuming equilibrium, namely $\Delta\bar{\mu}_j$ = 0; and substituting E_M for $\Psi^i - \Psi^o$, Eq. (2) becomes:

$$0 = 2.3RT \ \log\frac{a_j^i}{a_j^o} + z_j F \ E_M$$

or

$$\log \frac{a_j^o}{a_j^i} = \frac{z_j F}{2.3RT} E_M \tag{8}$$

and at 30°C ($F/2.3RT$ = 1/60 mV):

$$\log \frac{a_j^o}{a_j^i} = \frac{z_j}{60} E_M$$

This is one form of the *Nernst equation* [12]; it gives the relation between the membrane potential and the expected ion accumulation ratio at equilibrium. Another form of this equation gives the Nernst potential (E^N), at 30°C:

$$E^N = \frac{60}{z_j} \log \frac{a_j^o}{a_j^i} \tag{9}$$

This is the membrane potential needed to sustain equilibrium at a certain ion accumulation ratio. Thus, at −120 mV a^o/a^i would be 10^{-2} for K$^+$, 10^{-4} for Ca^{2+}, and 10^2 for Cl$^-$.

The Nernst equation can be employed to determine whether specific ions have been transported passively. Such analysis must be performed for tissues in the steady state, namely, when net transport has ceased ($J^{i\to o} = J^{o\to i}$). For this analysis, the membrane potential and the concentration ratio of the ion must be known. Active transport is assumed when E_M differs from the expected E^N at the measured concentration ratio (a^o/a^i).

Whether transport is active can also be determined for the non-steady state, but the flux ratio must be known. Ussing [2] and Theorell [13] have shown that the ratio of passive fluxes ($J^{i\to o}/J^{o\to i}$) of a solute is proportional to the electrochemical potential difference. The *Ussing–Theorell equation* is:

$$\Delta\bar{\mu}_j = 2.3RT \log \frac{J_j^{i\to o}}{J_j^{o\to i}} \tag{10}$$

Substituting Eq. (2) for $\Delta\bar{\mu}_j$ and E_M for $\Psi^i - \Psi^o$ in Eq. (2), gives:

$$2.3RT \log \frac{a_j^i}{a_j^o} + z_j F E_M = 2.3RT \log \frac{J_j^{i \to o}}{J_j^{o \to i}}$$

Divising by $2.3RT$ and writing the equation for 30°C results in:

$$\log \frac{J_j^{i \to o}}{J_j^{o \to i}} = \log \frac{a_j^i}{a_j^o} + \frac{z_j}{60} E_M \tag{11}$$

If a flux is larger than expected by this relation, transport is assumed to be active.

V. TRANSPORT PROTEINS

Membrane-bound proteins that facilitate solute transport across the membrane are of two principal types: channels and carriers.

A. Channels

Channels are multisubunit proteins that span the membrane. They form hydrophilic pores, facilitating the passive passage of specific solutes. Such passive and electrophoretic transport of ions is termed uniport [7]. Channels are solute selective. The selectivity apparently depends on the hydrated radius of the solute and its charge [14]. Transport rates across channels are relatively high, about $10^7 \, s^{-1}$ [15], and 100–1000 times higher, per transport unit, than those of carriers [16]. Transport across channels is also controlled by "gating"; that is, various external and internal conditions regulate the conformation of channel proteins, thus affecting the probability that they are in the open state. Gating can be affected by voltage (membrane potential) [17], second messengers such as inositol 1,4,5-trisphosphate (IP$_3$) [15], Ca^{2+} [15], nucleotides [18], plant hormones [18], and physical membrane stretching [15,19].

1. Voltage Gating of Channels

A current–voltage (I/V) curve is obtained when the ion current through a membrane is plotted against the applied potential difference. Linear I/V curves indicate constant channel conductivity, while nonlinear curves indicate voltage-dependent changes of the conductivity (Figure 1). Nonlinear I/V curves result from voltage gating of ion channels. For certain voltage gated channels, the probability of being in the open state increases upon hyperpolarization; for others it increases upon depolarization [17]. The "patch clamping" technique is usually employed for measuring channel conductivity [20].

   The direction of the net ion flux that is facilitated by opening of voltage-gated channels depends on a number of factors. These include the charge of the transported ion and the prevailing Nernst potential for the specific ion. Let us consider the effect of depolarization and hyperpolarization on K$^+$ and Cl$^-$ fluxes, via voltage-gated channels in the plasma membrane of a plant cell. We assume that K$^+$ has been passively accumulated in the cytoplasm and is at equilibrium across the plasma membrane ($E_M = E^N$ for K$^+$) and that Cl$^-$ has actively accumulated in the cytoplasm. Activation of K$^+$ channels by depolarization would then induce K$^+$ efflux from the cytoplasm to the free space. Net K$^+$ efflux occurs because depolarization increased E_M above E^N for K$^+$.

   Depolarization-activated opening of a Cl$^-$ channel would also induce Cl$^-$ efflux. This is because Cl$^-$ has been accumulated, and even at complete depolarization ($E_M = 0$), the electrochemical potential of Cl$^-$ inside is larger than outside. Let us now consider hyperpolarization-activated K$^+$ and Cl$^-$ channels. Hyperpolarization creates an inward-directed electrochemical potential gradient for K$^+$ and results in net K$^+$ influx. Activation of a Cl$^-$ channel by

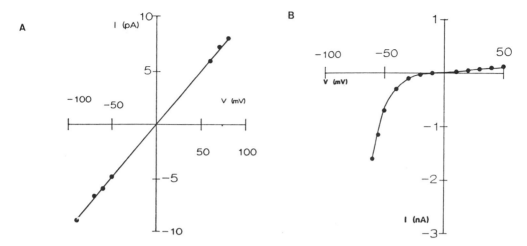

Figure 1 (A) Linear current–voltage (*I/V*) plot for an open tonoplast channel from *Vigna unguiculata* stems. (B) Whole vacuole *I/V* plot of voltage-gated channels (voltage-dependent channel opening). A large increase of negative current is evident at about –20 mV, the result of an increased number of open channels. In both plots the reversal potential (potential at which current direction changes) is zero, since the solutions on both sides of the membrane were the same and $E^N = 0$. (From Ref. 17.)

hyperpolarization would, however, again result in net Cl⁻ efflux through the open channel. This is because the outward-directed electrochemical potential gradient of Cl⁻ would increase with decreases in membrane potential.

2. Plasma Membrane Channels

Tester [15] listed two kinds of K^+ channel in the plasma membrane of plant cells. One kind is activated by hyperpolarization and conducts K^+ into the cells. The second type is activated by depolarization and facilitates K^+ efflux. Both these channels appear, as well, to conduct other cations, such as Na^+ and Ca^+ [21], with different permeabilities. For the first kind, a K^+/Na^+ selectivity of 5–10 is reported [17]. For the depolarization-activated channel, a K^+/Na^+ selectivity of 20–60 is reported [15]. When the external Na^+ concentration is high (saline conditions) and the electrochemical potential gradient of Na^+ is inward, depolarization-activated cation channels may conduct Na^+ inward. Sodium permeability of *Chara* plasma membranes was drastically decreased by Ca^{2+} [22] when ATP or other nucleotides were present in the cytosol [23]. A depolarization-activated channel, apparently, conducts Ca^{2+} as well, and such Ca^{2+} conductance was enhanced by abscisic acid [24].

Cytosolic Ca^{2+} activated Cl⁻ channels have been described [15,16] in the plasma membranes of Charophytes and higher plants. The necessary elevation of cytosolic Ca^{2+} concentration appeared to result from increased Ca^{2+} conductivity of the depolarization-activated cation channel [15]. In *Vicia faba* guard cell plasma membranes, an anion channels that conducts Cl⁻ and malate was found [25]. This channel was activated by depolarization, with maximal activity at a membrane potential of –40 mV. Another investigation [26] showed that the activation potential of an anion channel, in the plasma membrane of *Vicia faba* guard cells, could be shifted by application of auxin.

3. Tonoplast Channels

The tonoplasts from a variety of plant cells were found [27] to contain a rather unspecific depolarization-activated ion channel. In tonoplasts of beet root vacuoles, the permeability sequence of this channel was $K^+ > Na^+ > NO_3^-$ malate$^{2-} > Cl^-$. Calcium channels, also found in the tonoplast, were shown to be activated by cytosolic Cl^- [28], by voltage [29], and by IP$_3$ [30]. The latter compound is a second messenger, produced by the phosphatidylinositol cascade [31]. The IP$_3$-activated channel facilitates Ca^{2+} release from the vacuole to the cytosol.

B. Carriers

1. Transport Kinetics

Carriers catalyze the transport of specific solutes across the membrane. This transport is often vectorial (unidirectional). The relation between solute concentration and unidirectional flux in plant tissues was described by Epstein and Hagen [32]. They used terms introduced by Michaelis and Menten [33] for enzyme kinetics (the relation between the chemical activity of the substrate and the velocity of enzyme-catalyzed reactions). Carrier-catalyzed transport (or enzyme-catalyzed reaction) is modeled as follows:

$$E + S^o \underset{k_2}{\overset{k_1}{\rightleftarrows}} ES \overset{k_3}{\rightleftarrows} E + S^i \tag{12}$$

where S is the transported solute (or substrate), E the carrier (or enzyme), ES the carrier–solute (or enzyme–substrate) complex, and k_1, k_2, k_3 the rate constants of the reactions. Assuming that k_3 is much less than k_1, the transport velocity (v), or influx, is given by:

$$v = k_3[ES] \tag{13}$$

Maximal transport velocity V_{max} should be attained at saturating solute concentration, when all carrier molecules are occupied by solute:

$$V_{max} = k_3[E_T] \tag{14}$$

where $[E_T]$ is the sum of occupied and unoccupied carriers, or the total carrier concentration: $[E_T] = [E] + [ES]$. The affinity of the carrier for the solute is the reciprocal of the dissociation constant of the carrier–solute complex, or the reciprocal of the Michaelis–Menten constant (K_M):

$$K_M = \frac{k_2}{k_1} = \frac{[E][S]}{[ES]} \tag{15}$$

Transport kinetics are customarily defined by K_M and V_{max}. For this, [E] in Eq. (15) is replaced by $[E_T] - [ES]$, and the equation is rearranged:

$$K_M = \frac{([E_T] - [ES])[S]}{[ES]} \quad \text{and} \quad [ES] = \frac{[E_T][S]}{K_M + [S]}$$

The latter equation can be substituted for [ES] in Eq. (13), that is, $v = (k_3[E_T][S])/(K_M + [S])$ and by replacing $k_3[E_T]$ with V_{max} (Eq. 15), the Michaelis–Menten equation is obtained:

$$v = \frac{V_{max}[S]}{K_M + [S]} \tag{16}$$

This equation can be transformed into a liner function of the reciprocals of v (same as $J^{o \rightarrow i}$) and S (same as a_j^o):

$$\frac{1}{v} = \frac{K_M}{V_{max}} \frac{1}{[S]} + \frac{1}{V_{max}} \tag{17}$$

where K_M/V_{max} is the slope of the line and $1/V_{max}$ is the Y-intercept. By substitution of $V_{max}/2$ for v, it can be shown that K_M is the solute activity when the velocity is half-maximal. Various graphic analyses [34] show that competitive inhibition (inhibition by competition for carrier sites) of transport increases the apparent K_M (decreases apparent affinity) but does not affect V_{max}.

Strict compliance with Michaelis–Menten kinetics usually is obtained only in a narrow solute concentration range and, in particular, in the low concentration range, up to about 0.7 mM. In broader concentration ranges multiphasic uptake kinetic [35] is ordinarily encountered. The reason for multiphasic kinetics has not been resolved; various possible explanations have been discussed [36–41].

2. Primary Active Transport

The term "primary active transport" [10] is reserved for active transport that is directly driven by energy-rich metabolites, such as ATP, pyrophosphate, or electron donors. Cotransport is classified as secondary active, because it derives its energy from the protonmotive force that is produced by primary active transport of protons.

Two types of ATPase, at the plasma membrane and the tonoplast, respectively, and an inorganic pyrophosphatase (PP$_i$ase) at the tonoplast, are known to generate the protonmotive force at these membranes [18]. In addition, a protonmotive force is generated by vectorial electron transport across the inner membranes of mitochondria and chloroplasts. The latter protonmotive force is primarily used for ATP synthesis, catalyzed by a third type of ATPase [42].

ATPases. Protonmotive force generating ATPase, at the plasma membrane and tonoplast, transport protons actively from the cytoplasm to the free space and vacuole, respectively. They are classified as H^+-ATPases, and they need Mg^{2+} to function; their substrate, indeed, is Mg-ATP. These ATPases differ in their evolutional history, in their homology with bacterial and animal ATPases, and in some of their characteristics [43].

The plasma membrane H^+-ATPase forms a phosphorylated intermediate during ATP hydrolysis and belongs to the P-ATPase family [43]. Similar intermediates are formed by other ion-transporting ATPases. These ATPases are specifically inhibited by orthovanadate ions. The plasma membrane H^+-ATPase has a functional molecular weight of 200 kD and is apparently composed of two 100 kD subunits, each forming at least eight, and possibly ten, transmembrane helices [44]. Optimal pH for this enzyme is 6.5; its activity is enhanced (maximally doubled) by K^+, but the enzyme is not directly involved in K^+ transport [45]. H^+-ATPase is a major component of the plasma membrane. In cells highly active in transport, such as root cells, there are about 10^6 molecules of the enzyme per cell, with turnover numbers of 20–100 s^{-1}; this results in proton fluxes of 10–100 pmol s^{-1} cm^{-2} [46].

The tonoplast H^+-ATPase of plant cells does not form a phosphorylated intermediate and is not inhibited by vanadate. It belongs to the family of V-ATPases, operating as proton pumps at endomembranes of eukaryotic cells [42]. The vacuolar H^+-ATPase is inhibited by the antibiotic bafilomycin [47] and by NO_3^-. It is stimulated by Cl^-, not by monovalent cations; the pH optimum is 7.9 [43]. A functional molecular weight of 500 kD was ascribed to the enzyme, which is composed of seven [43] or ten [48] subunits and exhibits four potential transmembrane helices [43].

Another primary active carrier is the enzyme Ca^{2+}-ATPase, located in the plasma membrane [49] and in the endoplasmic reticulum [50]. The Ca^{2+}-ATPases form phosphorylated intermediates, are inhibited by vanadate [51,52], and belong to the P-ATPase family [53]. They transport Ca^{2+} out of the cytoplasm to the free space and into endoplasmic reticulum vesicles (Ca^{2+} transport into vacuole is performed by Ca^{2+}/H^+ antiport; see "Mineral Cations" under

Subsection 3). All these Ca^{2+} transporters are involved in Ca^{2+} homeostasis of the cytosol and regulate the Ca^{2+} activity there at about 0.1 μM [54]. Cytosolic Ca^{2+} activity rises only transiently, in response to certain stimuli [55].

The plasma membrane Ca^{2+}-ATPase seems to have a molecular weight of 270 kD and to function as a polypeptide dimer of about 140 kD [56]. Its preferred substrate is Mg-ATP; with Mg-ITP or Mg-GTP the activity was 30–40% of that with Mg-ATP. Optimal pH is 7.4–7.8, and the K_M for Ca^{2+} is in the micromolar range [57]. Calmodulin [58], a calcium-modulated protein involved in many Ca^{2+}-regulated processes, stimulated calcium uptake by inside-out plasma membrane vesicles [52]. The contribution of the Ca^{2+}-ATPase to the electrical potential difference across the plasma membrane is insignificant because its activity is two orders of magnitude lower than that of the H^+-ATPase [46].

The endoplasmic reticulum Ca^{2+}-ATPase differs in some characteristics from the plasma membrane enzyme. Apparently composed of 116 kD subunits, this enzyme seems not to be activated by calmodulin [53] and uses only Mg-ATP as substrate [51].

PYROPHOSPHATASE. A pyrophosphatase operates as a second proton pump at the tonoplast of plant cells [59]; like the V-ATPase, it secretes protons into the vacuole. This enzyme has not been identified in endomembranes of animal cells. The molecular weight of the H^+-PP_iase is approximately 81 kD, and it appears to comprise a single polypeptide [18]. Potassium ions greatly stimulate the H^+-PP_iase, but only when the cation is applied to the cytosolic face of tonoplast vesicles [60]. Recent evidence [61] indicated that the PP_iase directly translocates K^+ into the vacuoles, in addition to protons. The substrates of this enzyme are $MgHPP_i$ or $MgPP_i$, and it is stimulated by additional free Mg^{2+} [62]. Optimum pH for the H^+-PP_iase is 8.5–9.0, depending on the Mg^{2+} concentration [63]. The enzyme is not inhibited by vanadate or NO_3^- [64].

3. Cotransport

Cotransport is secondary active transport [65] of specific solute. It is coupled to transmembrane ion gradients, formed by primary active transport. Plant cells employ the protonmotive force as the energy source for active solute cotransport. The general principle of this transport is similar to that of a pulley: protons are carried "downhill" (passively) across the membrane, while the other solute is carried "uphill" (actively). The direction of passive proton transport is from the free space or vacuole to the cytoplasm. The cotransport of protons and the other solute in the same direction is called *symport*; when the two solutes are cotransported in opposite directions, the phenomenon is termed *antiport* [7].

Cotransporters are conceived as membrane-embedded transport proteins. Conformation changes are supposed to expose the solute binding sites alternatively to the inside or outside. When a proton binding site is exposed to the outside, where the electrochemical potential of protons is high, a proton binds. This supposedly induces a conformation change, resulting in increased affinity of the binding site for the cotransported solute. The increased affinity enables binding even at a low electrochemical potential of the solute. A symported solute binds on the same side of the membrane as the proton, while in antiporting the binding occurs on the opposite side. Binding of the cotransported solute is supposed to induce another conformation change that exposes the proton to the inside. It is then released inside, where the proton electrochemical potential is lower. A cotransported solute is now also exposed to the inside, and an antiported solute to the outside. The release of the proton decreases the affinity for the other solute and it is also released, at the side of its higher electrochemical potential (inside or outside for symport and antiport, respectively).

The stoichiometry of protons and cotransported solute differs in the various cases and is not always known. In some cotransport systems, the number of cotransported protons equals the negative charges of symported anions, or positive charges of antiported cations. This results in electroneutral transport, exclusively driven by the ΔpH component of the protonmotive force. In other cotransport systems, an excess of protons is cotransported and electrogenic transport results. The latter kind of transport is driven by both components of the protonmotive force, namely ΔpH and E_M.

For a known accumulation ratio of a specific cotransported solute, the minimal required number of cotransported protons per solute molecule or ion can be estimated when E_M and ΔpH are known [11]. At the steady state, the following relation between the electrochemical potential difference of the solute j and the number (n) of symported protons should apply [11]:

$$\Delta\bar{\mu}_j = -n\Delta\bar{\mu}_{H^+} \qquad \text{or} \qquad \Delta\bar{\mu}_j = -nF \text{ pmf} \tag{18}$$

Substitution of Eq. (2) for $\Delta\bar{\mu}_j$, replacing $(\Psi^i - \Psi^o)$ with E_M, and substitution of Eq. (7) for pmf gives:

$$2.3RT \log \frac{a_j^i}{a_j^o} + z_jFE_M = -nF(E_M - 60\Delta\text{pH})$$

$$\Delta pj = -\log \frac{a_j^i}{a_j^o}$$

when the $-\log$ quantity is replaced by p. By further division with F, and rewriting the equation for 30°C, we then have:

$$-60\Delta pj + z_jE_M = -\text{n}(E_M - 60\Delta\text{pH}) \qquad \text{and} \qquad \Delta pj = \frac{n + z_j}{60} E_M - n\Delta\text{pH} \tag{19}$$

For antiport, the steady state situation is given by $\Delta\bar{\mu}_j = nF$ pmf, and at 30°C,

$$\Delta pj = \frac{z - n}{60} E_M + n\Delta\text{pH} \tag{20}$$

Equation (19) can be employed to analyze Cl^--H^+ symport. If electroneutral transport is assumed ($n = 1$, $z = -1$), $-\Delta\text{pCl}$ will equal ΔpH, and such transport would not depend on the membrane potential. If a 100-fold Cl^- accumulation is found ($\Delta\text{pCl} = -2$) under such circumstances, a relatively large ΔpH of 2 would be needed. However, if $n = 2$ is assumed, and $E_M = -120$ mV, no ΔpH would be needed for a ΔpCl^- of -2.

SUCROSE. The transport of sucrose across the plasma membrane into sink cells and into the phloem of source tissues is critical for plant growth. Evidence for the involvement of sucrose-H^+ symport in phloem loading of source tissue and of sucrose accumulation in sink tissues was recently reviewed [66]. Sucrose-H^+ symport activity, with an apparent K_M of 1 mM for sucrose and 0.7 µM for H^+, was shown in isolated plasma membrane vesicles [67,68]. The system exhibited an acid pH optimum at the outside and an H^+/sucrose stoichiometry of 1. Sucrose-H^+ symport was sucrose specific in *Beta vulgaris* plasma membrane vesicles [67], but in *Ricinus communis* plasma membrane vesicles it was inhibited by maltose [68]. A 120 kD polypeptide was recently separated from *B. vulgaris* leaves and reconstituted into proteoliposomes [69]. The reconstituted system transported sucrose when a protonmotive force was applied.

Sucrose is the storage carbohydrate in the vacuoles of many plant cells. It is accumulated to high concentrations, in particular in the vacuoles of parenchyma cells in sugarcane stalks and in the storage tissue of sugar beet [70]. Recent investigations indicate that sucrose accumulation in vacuoles may be accomplished by sucrose/H^+ antiport. This was demonstrated in tonoplast vesicles from sugarcane [71], beet roots [72], and Japanese artichoke tubers [73]. The apparent K_M for antiported sucrose was 1.7 mM in red beet and 30–50 mM in Japanese artichoke tonoplast vesicles. The sucrose/H^+ antiporter transported palatinose, stachyose, raffinose, and fructose in addition to sucrose [71, 73]; it seemed to recognize a terminal fructosyl residue [73]. Optimal pH, at the outer tonoplast surface, was 7.5 for sucrose/H^+ antiport [71].

HEXOSES. Although oligosaccharides, in particular sucrose, present the major long-distance transported form of carbohydrate, a sucrose uptake system is not expressed in the plasma membrane of all cells. In the apoplast of many carbohydrate sinks, invertase catalyzes sucrose hydrolysis, and hexoses are then transported into the cells by specific hexose carriers [70].

The first cotransport system discovered in plants was, indeed, the hexose-H^+ symporter of *Chlorella* [74,75]. A functionally very similar hexose transporter was found in the plasma membranes of higher plant cells. Rausch [76] reviewed the characteristics of this system. The data originate from experiments with plant cells and protoplasts of different kinds. The apparent K_M of glucose-H^+ symporters for glucose varies from 7 μM to 1 Mm; it changes with the tissues and experimental conditions.

It is still controversial whether the hexose-H^+ symporter transports both glucose and fructose. Most of the evidence indicates, at least, a significantly larger affinity for glucose. Zamski and Wyse [77] concluded that distinctly separate proton symport systems transport glucose and fructose in sugar beet suspension cells. They ascribed the transport competition of glucose and fructose, found in starved cells, to a contest for available energy, not for carrier sites. In purified plasma membrane vesicles from *Nicotiana tabacum*, the proton symport of the glucose analogue 3-*O*-methyl-D-glucose, from a 16 μM solution, was strongly inhibited by 500 μM D-glucose and D-galactose, but only slightly by D-fructose [78].

The hexose-H^+ symporter from *Chlorella* has been cloned [79]; the clones were used to isolate genomic and complementary cDNA clones from *Arabidopsis thaliana*. The latter were expressed in a yeast and catalyzed active hexose transport [80].

Hexose transport at the tonoplast was also reviewed by Rausch [76]. The affinity sequence of different hexoses for transport into vacuoles was D-glucose = 2-deoxy D-glucose > D-fructose > D-galactose [78]. The K_M of this transport for D-glucose was 4.5 –20 Mm under different conditions and in different plants. The transport is saturable and is inhibited by SH reagents [78]. Some data support the proposal that hexose/H^+ antiport functions at the tonoplast, but this issue is unresolved [76].

AMINO ACIDS. Reinhold and Kaplan (81) reviewed much of the information about amino acid transport, derived from work with plant tissues, cells, and protoplasts. Here we summarize recent information about work with plasma membrane and tonoplast vesicles. Investigations with plasma membrane vesicles from *Cucurbita pepo* hypocotyls [82], *B. vulgaris* leaves [83,84], *B. vulgaris* roots, and *R. communis* cotyledons and roots [68] indicate symport of protons and 12 amino acids.

Li and Bush [83] resolved four different amino acid symport systems at the plasma membrane of *B. vulgaris* leaves, two for transport of neutral amino acids, one for acidic amino acids, and one for basic ones. One of the neutral amino acid symporters was active for alanine, methionine, glutamine, and leucine and the other for isoleucine, valine, and threonine [84].

In tonoplast vesicles three amino acid transport systems are reported [85,86]. One system is specific for aromatic amino acids and depends on protonmotive force, generated either by the V-ATPase or the PP_iase [87]. It apparently performs amino acid/H^+ antiport. A second system transports positively charged amino acids [84]. The latter and a third system may constitute an amino acid channel, with rather broad specificity. This channel seems to be gated by ATP but does not require ATP or PP_i as an energy source [85,86,88].

MINERAL CATIONS. Plant cells actively excrete Ca^{2+} and Na^+ from the cytosol at the plasma membrane and tonoplast. At the plasma membrane Ca^{2+} excretion is catalyzed by the Ca^{2+}-ATPase (see discussion above of ATPases) and at the tonoplast by Ca^{2+}/H^+ antiport. Primary active Na^+ transport exists in animal cells [89] but has not been found in plant cells. Evidence exists for Na^+/H^+ antiport at the plasma membrane and at the tonoplast.

The first evidence in higher plants for cotransport energized by protonmotive force was for Na^+/H^+ antiport in barley roots [90]. Much of the information about Na^+/H^+ antiport at the plasma membrane of higher plants still comes from experiments with intact plant tissues (for references, see Ref 91). At the level of plasma membrane vesicles, this antiporter was investigated in the halotolerant unicellular alga *Dunaliella salina* [92,93], in *Atriplex nummularia* [94,95], in *Gossypium hirsutum* [95], and in *B. vulgaris* (unpublished). The Na^+/H^+ antiporter in *Dunaliella* plasma membrane vesicles had a K_M for Na^+ of about 16 mM and was inhibited by amiloride, an inhibitor of Na^+/H^+ antiport in animal cells [92]. The V_{max} of the antiport increased when the cells had been adapted to a high NaCl concentration, or to ammonia at high pH; it decreased in LiCl-adapted cells [96]. This increase of V_{max} was interpreted as over production of the Na^+/H^+ antiporter and was correlated with over production of polypeptides of 20 and 50 kD. The Na^+/H^+ antiport in *Dunaliella* was Na^+ specific in comparison to K^+, Cs^+, and Li^+ [92]. However, in plasma membrane vesicles of *Atriplex* [94,95] and *Gossypium* [95], similar dissipation of pH gradients was found with Na^+ and K^+; and high concentrations of both ions were needed for activity. An additive effect was obtained when Na^+ was added to saturating K^+ concentrations, and vice versa. It was suggested that separate antiporters for K^+ and Na^+ may operate. Sodium proton antiport in plasma membrane vesicles from *B. vulgaris* storage tissue (L. Naveh and B. Jacoby, unpublished) was Na^+ specific, had an apparent K_M for Na^+ of about 12 mM, and was amiloride sensitive.

Evidence for Na^+/H^+ antiport at the tonoplast emanated first from experiments of Blumwald and Poole [97] with tonoplast vesicles. These investigators demonstrated amiloride-sensitive Na^+/H^+ antiport in tonoplast vesicles from *B. vulgaris* storage tissue. At a constant ΔpH the apparent K_M for Na^+ increased from 7.5 mM to 26.6 mM, with internal pH decrease in the range of 7.5–6.5. In tonoplast vesicles from suspension cultured cells of *B. vulgaris* [98] and from *Plantago maritima* roots [99], Na^+/H^+ antiport activity increased in response to NaCl in the growth medium. In *B. vulgaris* [98] this increased activity resulted from increased V_{max}, without change of K_M. Sodium antiport activity was also increased by cultivation of the *B. vulgaris* cells in the presence of amiloride [100]. In both cases the increase of Na^+/H^+ antiport activity was accompanied by synthesis of a 170 kD polypeptide. Polyclonal antibodies against this polypeptide almost completely inhibited the Na^+/H^+ antiport activity [100]. Garbarino and DuPont [101] induced Na^+/H^+ exchange in barley roots with sodium salts. The half-time of induction was 15 minutes, and this result was attributed to the activation of an existing protein.

Electroneutral K^+/H^+ exchange was found in tonoplast vesicles from *Brassica napus* [102]. This exchange was very specific for K^+ and was construed to be K^+/H^+ antiport. It operated at micromolar K^+ concentration, was saturated at 10 mM K^+, and was inhibited by K^+ concentrations above 25 mM.

The consensus regarding Ca^{2+} transport from the cytosol to the vacuole is that it is accomplished by Ca^{2+}/H^+ antiport. It depends on the V-ATPase and PP_iase generated protonmotive force [103,104]. Kasai and Muto (105) suggested that Ca^{2+}/H^+ antiport operates at the plasma membrane, in addition to the Ca^{2+}-ATPase. The apparent K_M values, for Ca^{2+} of the Ca^{2+}/H^+ antiporter in tonoplast vesicles from oat roots [104,106] and *B. vulgaris* [107] were 10–14 mM and 42–200 μM, respectively. In the latter case the K_M depended on the pH inside the tonoplast vesicles (homologous to inside the vacuole). ATP-driven Ca^{2+}/H^+ antiport created an 800–2000-fold Ca^{2+} gradient in *B. vulgaris* tonoplast vesicles [103]; an H^+/Ca^{2+} stoichiometry of 3 was indicated [108].

MINERAL ANIONS. Accumulation of anions in the negatively charged cytosol should be an active process, whether they are transported across the plasma membrane from the free space or across the tonoplast from the vacuole. Anion–proton symport is expected for such transport. Hitherto, evidence for proton symport of Cl^-, $H_2PO_4^-$, SO_4^{2-}, and NO_3^- comes almost exclusively from experiments with intact plant tissues and cells. Very few data are available at the membrane vesicle level.

Proton–chloride symport was shown [109] for barley roots that had been de-energized by anaerobic pretreatment. In such roots Cl^- influx could be induced by application of an artificial pH gradient (acid outside), but not by a low pH when the pH gradient was dissipated. Sanders [110] reviewed Cl^- transport in plants and concluded that the H^+/Cl^- stoichiometry of symport should be at least 2. Such stoichiometry is consistent with electrogenic Cl^- transport and with transient plasma membrane depolarization upon application of Cl^-.

The best available evidence for symport of protons with NO_3^- [111,112], $H_2PO_4^-$ [113,114], and SO_4^{2-} [115] in plant tissues and cells is the transient depolarization of the plasma membrane [110,111,113,115] or transient alkalinization of the medium [114] upon addition of these anions. In all these investigations the degree of transient depolarization was related to the transport rate of the anions. Transient depolarization indicates electrogenic transport with a stoichiometry exceeding 1 for H^+/anionic charge transported. Depolarization is transient because the membrane potential is restored by increased proton pump activity. Sakano [114] measured $H_2PO_4^-$ and H^+ uptake (transient external alkalinization) in *Caranthus roseus* cell cultures and arrived at an $H^+/H_2PO_4^-$ stoichiometry of 4. Recently $H_2PO_4^-$-H^+ symport was also shown in outside-out *B. vulgaris* plasma membrane vesicles (S. R. Stutz and B. Jacoby, unpublished). When ^{32}P-labeled $H_2PO_4^-$ was added to these vesicles, in the presence of a pH gradient, ^{32}P uptake and dissipation of the pH gradient could be measured concomitantly.

Evidence for anion–proton symport across the tonoplast, from the vacuole to the cytosol, is very limited. Blumwald and Poole [116] formed an ATPase-dependent protonmotive force in *B. vulgaris* tonoplast vesicles. Subsequent addition of Cl^- or NO_3^- dissipated the electrical component of the protonmotive force (the E_M), but only NO_3^- dissipated the ΔpH as well. Blumwald and Poole concluded that both anions entered the vesicles by uniport, dissipating the E_M, and that only NO_3^- was excreted again by symport with protons, thus dissipating the ΔpH. In similar experiments of Schumaker and Sze [117], with oat root tonoplast vesicles, both Cl^- and NO_3^- dissipated E_M and ΔpH.

VI. CHARGE BALANCE AND STOICHIOMETRY OF TRANSPORT

Uniport, cotransport, and primary active transport may all transport unbalanced charges, namely, they all may be electrogenic. A very small amount of electrogenic transport results in a large change of membrane potential. Primary active transport generates membrane potential, while electrogenic cotransport and uniport dissipate it. Continued electrogenic cotransport and uniport are sustained by persistent protonmotive force turnover (dissipation, followed by regeneration by primary active transport). The ion concentration difference that sustains the membrane potential of plant cells (up to about –240 mV at the plasma membrane) can be calculated from its relation to the capacitance C and the electrical charge Q_M of the membrane: $Q_M = CE_M$ [12]. For a spherical cell with a radius of 50 μm, such calculation shows that an uncompensated ion concentration difference of about 1.5 μM is needed to sustain a membrane potential of –240 mV. The total cytosolic salt concentration may be assumed to be about 50 mM. An anion concentration difference of 1.5 μM then constitutes an anion excess of only 0.003% and is practically equivalent to charge balance.

In spite of charge balance, different amounts of anion and cation of a salt are absorbed by plant cells. The difference is compensated by proton fluxes, resulting in transient pH shifts in the cells. Plant cells regulate the cytosolic pH at about 7.0 [118]. Such regulation, in response to nonstoichiometric salt absorption is accomplished by synthesis or decomposition of organic acids.

Plants usually absorb an excess of K^+ over SO_4^{2-} when presented with K_2SO_4. The excess of positive charge in the cytosol is compensated by proton excretion, which results in a transient elevation of cytosolic pH. Elevated pH induces respiratory CO_2 to form bicarbonate; this then enhances phosphoenolpyruvate carboxylase activity and malate synthesis. The consumption of phosphoenolpyruvate enhances its replenishment, via glycolytic synthesis of phosphoglyceric acid from triose phosphate. The elevated cytosolic pH is, thus, regulated by replacing KOH with K_2-malate [119] (Figure 2A).

Excess uptake of anionic charge may occur when plants are presented with $Ca(NO_3)_2$. The transient acidification of the cytoplasm by cotransported protons is regulated by activation of malic enzyme and malate decomposition [118] [Figure 2B].

VII. TRANSPORT TO THE SHOOT

Since the now classical experiments of Stout and Hoagland [120], it is accepted that mineral ions are transported from roots to shoots in the xylem. This pathway also applies to organic nitrogen compounds that are products of ammonium fixation and symbiotic N_2 fixation [121]. The mechanism of this transport is mass flow in the aqueous solution. The control of mineral ion transport to the shoots occurs during their passage across the roots to the xylem.

A. Transport Across the Root

Water can move across the root in the symplast, in the apoplast, or in both. The root *symplast* [122] comprises the plasmodesmata-connected cytoplasmic continuum of the root cells; it does not include the vacuoles. Symplastic mass flow across the root traverses two membranes: one upon entering the symplast and another upon exiting to the xylem. Biological membranes, which do not constitute a serious barrier to the diffusion of water, do not, however, allow free diffusion of the mineral ions dissolved in the water. Thus symplastic movement of solutes across roots can be regulated by metabolically controlled transport across these membranes [123].

Apoplastic water flow across the root occurs in the root–cell wall continuum. The *apoplast* [122] is the free space continuum throughout the plant. This continuum is interrupted by the Casparian bands, which are hydrophobic incrustations in the radial and transverse walls of the endodermis [124]. Water bypasses the Casparian bands by movement across the membrane into the endodermal symplast, and out again. Mineral ions that are dissolved in the water and move by mass flow also migrate into the symplast to bypass the Casparian bands. The symplastic and apoplastic flows of solutes are, thus, compelled to combine in the endodermal symplast, at least for a short distance.

Whatever path across the root mineral ions choose, they traverse two membranes, where control may occur. Solutes moving to the xylem must be absorbed into the symplast, from the medium at the root surface, or from the apoplast, by either cortical cells or the cortical face of the endodermis. The solutes move out again into the stelar apoplast or the xylem, at the stelar face of the endodermis or from the stelar parenchyma respectively. These membrane transport processes are qualitatively and quantitatively controlled.

A

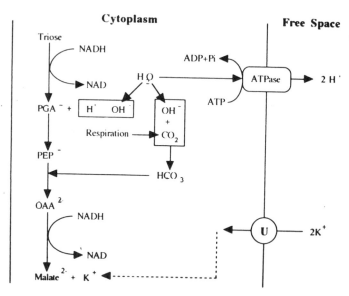

B

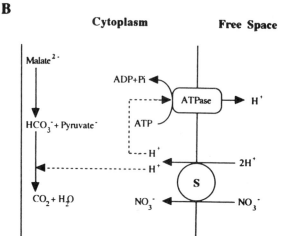

Figure 2 Charge compensation and pH regulation during counterion unaccompanied cation (A) and anion (B) uptake by plant tissues (U = uniport, S = symport). Details in text.

In contrast to earlier hypotheses that solutes leak into the xylem from adjacent stelar cells [125,126], in recent years evidence has increased for metabolic solute secretion into the xylem. Such evidence emanates from experiments with some inhibitors of protein synthesis that inhibit solute transport to the xylem, but not uptake by the roots [127,128]. This differential inhibition indicated that a separate metabolic process is involved in xylem loading (transport from the root symplast to the xylem). Additional evidence for separate regulation of xylem loading and uptake of K^+ comes from recent work in which root and shoot base temperatues were differentially manipulated [129]. Solute secretion into the xylem is further supported by evidence for the activity of proton pumps that maintain the pH in the xylem between 5.5 and 6.5 [130–132]. The protonmotive force generated by these proton pumps may provide the energy needed for solute secretion.

If solutes are secreted into the xylem, different transport processes should operate at the plasma membranes of the absorbing cortical cells and the secreting stelar ones. Similar primary active transport takes place at the plasma membranes of cells of both kinds; protons are pumped out of the cells at both sites. Hanson [133] suggested that the difference may emanate from disparate activities of the secondary transport systems. Whereas K^+ uniporter and anion (NO_3^-, $H_2PO_4^-$, SO_4^{2-}, etc.) symporters should be active at the plasma membranes of absorbing cells, the K^+ antiporters and anion uniporter should prevail at the plasma membranes of secreting cells.

Vacuoles of root cells are an additional site for selective regulation of solute transport across the root. They accumulate various solutes, removing them from the symplastic stream. In particular, most of the cellular Ca^{2+} is sequestered in the vacuoles, while the cytosolic Ca^{2+} concentration is maintained very low [54]. Under saline conditions, much of the Cl^- is also sequestered in the vacuoles, where it serves for turgor regulation. The same applies to Na^+ in plants that have conserved the Na^+/H^+ antiporter that is needed for Na^+ transport to the vacuole [90]. Other solutes are stored in the vacuoles when available in excess of requirement. They may then be released again upon demand [134].

B. Effects of Transpiration

Haberlandt [135] concluded more than a century ago that transpiration is not of major importance for plant mineral nutrition. This has since been reconfirmed repeatedly (see, e.g., Ref. 136 and, most recently, Ref. 137). Indeed, if membrane transport dependent delivery of solutes to the xylem is the rate-limiting process, the amount of solutes transported to the tops of plants should not be affected by the rate of water flow. A low solute concentration would be expected in the xylem sap at high transpiration rates, and a high concentration when transpiration is low. This seems essentially to be the situation under conditions of low external salt concentration and low salt status of the roots [136], and for solutes that are recognized by the transport proteins (not xenobiotics).

Broyer and Hoagland [136] stated in 1943 that delivery of ions to the xylem is metabolically controlled, while upward movement in the xylem is passive mass flow. They concluded that in plants of high salt status, the latter process may be rate limiting and, consequently, variation in the rate of transpiration may affect salt transport to the shoots. Good correlations between transpiration and transport to the shoot were also found for such nonessential elements as cadmium [138] and silicon [139] and for some xenobiotic organic compounds [140]. These solutes may not be recognized by the transport systems of the root. They are apparently transported in a fraction of the apoplastic mass flow that bypasses the Casparian strips [141,142]. It was suggested that such bypassing occurs when solutes enter at sites of secondary root emergence [141,142], or through the apical region of the root [143]. The latter pathway may also be utilized by Ca^{2+} that reaches the xylem sap. Most of the Ca^{2+} that enter the symplast via Ca^{2+} channels [21] is sequestered in the vacuoles and prevented from moving to the xylem. The relative amount of water bypassing the Casparian strips varies between a few percent [144] and more than 10% [138]. It seems to increase under conditions of stress damage [144]. For metabolically transported solutes at low external concentration, the amount transported in the bypass flow is an insignificant fraction of the total transport to the xylem. It may, however, comprise the whole amount of a xenobiotic solute that is transported to the tops [138].

C. Resorption and Exchange Binding

Solute transport in xylem vessels is driven strictly by mass flow. Nevertheless, the composition of xylem sap changes along its path. Two processes, resorption by adjacent living cells and binding by xylem walls, are responsible for composition changes.

The walls of xylem vessels, like other cell walls, contain immobile charges. These charges constitute a Donnan phase that retains cations moving in the vessels. In particular, polyvalent cations, such as free Ca^{2+} [145], Zn^{2+} [146], and Fe^{3+} [147] are retained. Cation retention can be diminished, or prevented, by chelating agents [147,148] and by displacement with similar or other cations [148].

Resorption by surrounding tissues of solutes from the xylem was demonstrated by Stout and Hoagland [120]. Resorption of Na^+ is rather pronounced and selective in some plants [149] and has been shown to depend on energy metabolism [150,151]. Xylem–parenchyma transfer cells are apparently involved in Na^+ resorption and its transfer to the phloem [152,153].

VIII. SUMMARY

Mineral nutrients in the root–cortex apoplast equilibrate with those in the root medium. These mineral ions are transported into the root symplast, moving across the plasma membranes of epidermis, cortex, or endodermis cells. Depending on the specific ion, this transport is facilitated by passive uniport through channels or by carrier-aided cotransport with protons.

Once in the root symplast, the ions may be transported into the vacuole, across the tonoplast, or excreted to the xylem, across the plasma membranes of xylem parenchyma cells. Some mineral ions, such as Na^+ and Ca^{2+}, may also be reexcreted to the apoplast.

In the xylem, mineral ions and other solutes move by mass flow, primarily to the leaf apoplast. There the ions are again absorbed into the symplast and vacuoles of leaf cells. From the leaves, secondary transport occurs in the phloem, together with the products of photosynthesis.

Metabolic transport across the roots, to the xylem, regulates the amount of mineral ions conveyed to the tops. Normally, this amount is very little affected by the velocity of xylem sap flow.

All the membrane transport processes mentioned depend on cellular energy metabolism. They are driven by the protonmotive force, which is composed of an electrical potential difference and a pH difference across the membranes. At the plasma membrane, the protonmotive force is generated by a P-type ATPase. At the tonoplast it is generated by two pumps that function in parallel, a V-type ATPase and a pyrophosphatase. The latter may also pump potassium ions into the vacuole.

Beyond the absorption of mineral nutrients from the environment, other transport processes occur in plants. Mineral ions are released and reexcreted from the vacuoles to the cytosol and from the cytosol to the apoplast. Organic nutrients, such as sugars and amino acids, are transported from the cytosol to the vacuole as well as to the apoplast, and vice versa.

Anions and cations composing a mineral salt are absorbed at different rates. The consequent charge imbalance is compensated by proton transport and results in pH shifts. The cytosolic pH is regulated, around 7.0, by organic acid synthesis and decomposition.

ACKNOWLEDGMENT

During the composition of this chapter, this laboratory was supported by the Endowment Fund for Basic Sciences: Charles H. Revson Foundation, administered by the Israel Academy of Sciences and Humanities.

REFERENCES

1. D. G. Nicholls, *Bioenergetics*, Academic Press, London, p. 189 (1982).
2. H. H. Ussing, *Acta Physiol. Scand.*, *36*: 43 (1949).
3. U. Luttge, *Stofftransport der Pflanzen*, Springer Verlag, Berlin, p. 280 (1973).

4. G. E. Briggs, *Annu. Rev. Plant Physiol.*, *8*: 11 (1957).
5. F. G. Donnan, *Z. Elektrochem.*, *17*: 572 (1911).
6. G. E. Briggs, A. B. Hope, and R. N. Robertson, *Electrolytes and Plant Cells*, Blackwell, Oxford, p. 217 (1961).
7. P. Mitchell, *Biol. Rev.*, *41*: 445 (1966).
8. D. E. Goldman, *J. Gen. Physiol.*, *27*: 37 (1943).
9. A. L. Hodgkin and B. Katz, *J. Physiol.*, *108*: 37 (1949).
10. P. Mitchell, *Membranes and Ion Transport*, Vol. 1 (E. E. Bittar, ed.), Wiley-Interscience, London, pp. 192–266 (1970).
11. R. Serrano, *Plasma Membrane ATPase of Plants and Fungi*, CRC Press, Boca Raton, FL, p. 174 (1985).
12. P. S. Nobel, *Physiochemical and Environmental Plant Physiology*, Academic Press, San Diego, CA, p. 635 (1991).
13. T. Theorell, *Arch. Sci. Physiolog.*, *3*: 205 (1949).
14. K. Imoto, *FEBS Lett.*, *325*: 100 (1993).
15. M. Tester, *New Phytol.*, *114*: 305–340 (1990).
16. S. D. Tyerman, *Annu. Rev. Plant Physiol. Plant Mol. Biol.*, *43*: 351 (1992).
17. F. J. M. Maathuis and H. B. A. Prins, *Acta Bot. Neerl.*, *40*: 197 (1991).
18. F. J. M. Maathuis and D. Sanders, *Curr. Opinion Cell Biol.*, *4*: 661 (1992).
19. D. Cosgrove and R. Hedrich, *Planta*, *186*: 143 (1991).
20. R. L. Satter and N. Moran, *Physiol. Plants*, *72*: 816 (1988).
21. D. P. Schachtman, S. D. Tyermann, and B. R. Terry, *Plant Physiol.*, *97*: 598 (1991).
22. R. Hoffman, J. Tufariello, and M. Bisson, *J. Exp. Bot.*, *40*: 875 (1989).
23. M. Katsuhara and M. Tazawa, *Plant Cell Environ.*, *13*: 179 (1990).
24. J. I. Schroeder and P. Thuleau, *Plant Cell*, *3*: 555 (1991).
25. B. U. Keller, R. Heidrich, and K. Raschke, *Nature*, *341*: 450 (1989).
26. I. Marten, G. Lohse, and R. Hedrich, *Nature*, *353*: 758 (1991).
27. R. Hedrich, H. Barbier-Brygoo, H. Felle, U. I. Flugge, U. Luttge, F. J. M. Maathuis, S. Marx, H. B. A. Prins, K. Raschke, H. Schnabl, J. I. Schroeder, I. Struve, L. Taiz, and P. Ziegler, *Bot. Acta*, *101*: 7 (1988).
28. O. Pantoja, J. Dainty, and E. Blumwald, *J. Membrane Biol.*, *125*: 219 (1992).
29. O. Pantoja, A. Gelli, and E. Blumwald, *Science*, *255*: 1567 (1992).
30. J. Alexandre, J. P. Lassalles, and R. T. Kado, *Nature*, *342*: 567 (1990).
31. M. J. Berridge, *Nature*, *361*: 315 (1993).
32. E. Epstein and C. E. Hagen, *Plant Physiol.*, *27*: 457 (1952).
33. L. Michaelis and M. L. Menten, *Biochem. Z.*, *49*: 333 (1913).
34. D. A. Baker and J. L. Hall, *Solute Transport in Plant Cells and Tissues* (D. A. Baker and J. L. Hall, eds.), Longman, Harlow, pp. 1–27 (1988).
35. P. Nissen, *Annu. Rev. Plant Physiol.*, *25*: 53 (1974).
36. A. D. M. Glass and J. Dunlop, *Planta*, *141*: 117 (1978).
37. G. G. J. Bange, *Z. Pflanzenphysiol.*, *91*: 75 (1979).
38. F. N. Dalton, *J. Exp. Bot.*, *35*: 1723 (1984).
39. J. C. Vincent and M. Thellier, *Biochem. J.*, *41*: 23 (1983).
40. A. C. Borstlapp, *Plant Cell Environ.*, *6*: 407 (1983).
41. P. Nissen, *Plant Cell Environ.*, *10*: 475 (1987).
42. N. Nelson, *Curr. Opinion Cell. Biol.*, *4*: 654 (1992).
43. N. Nelson and L. Taiz, *Trends Biochem. Sci.*, *14*: 113 (1989).
44. D. P. Briskin and J. B. Hanson, *J. Exp. Bot.*, *43*: 269 (1992).
45. D. P. Briskin, *Biochim. Biophys. Acta*, *1019*: 95 (1990).
46. R. Serrano, *Biochim. Biophys. Acta*, *947*: 1 (1988).
47. E. J. Bowman, A. Siebers, and K. Altendorf, *Proc. Natl. Acad. Sci. U.S.A.*, *85*: 7972 (1988).
48. C. Matsuura-Endo, M. Maeshima, and S. Yoshida, *Eur. J. Biochem.*, *287*: 745 (1990).
49. J. L. Giannini, J. L. Ruiz-Cristin, and D. P. Briskin, *Plant Physiol.*, *85*: 1137 (1988).
50. T. J. Buckhout, *Planta*, *159*: 962 (1983).

51. J. L. Giannini, L. H. Gildensoph, I. Reynolds-Niesman, and D. P. Briskin, *Plant Physiol.*, *85*: 1129 (1988).
52. L. E. Williams, S. B. Schueler, and D. P. Briskin, *Plant Physiol.*, *92*: 747 (1990).
53. L. E. Wimmer, N. N. Ewing, and A. B. Bennett, *Proc. Natl. Acad. Sci. U.S.A.*, *89*: 9205 (1992).
54. S. Gilroy and A. Trewavas, *The Plant Plasma Membrane: Structure, Function and Molecular Biology* (C. Larsson and I. M. Moller, eds.), Springer-Verlag, Berlin, pp. 203–232 (1990).
55. B. W. Poovaiah and A. S. N. Reddy, *CRC Crit. Rev. Plant Sci.*, *6*: 47 (1987).
56. F. Rasi-Caldagno, M. C. Pugliarello, C. Olivari, and M. I. De Michelis, *Bot. Acta*, *103*: 39 (1990).
57. M. Olbe and M. Sommarin, *Physiol. Plants*, *83*: 535 (1991).
58. D. M. Roberts, T. J. Lucas, and D. M. Watterson, *CRC Crit. Rev. Plant Sci.*, *4*: 311 (1986).
59. P. A. Rea and R. J. Poole, *Annu. Rev. Plant Physiol. Plant Mol. Biol.*, *44*: 157 (1993).
60. J. M. Davies, P. A. Rea, and D. Sanders, *FEBS Lett.*, *278*: 66 (1991).
61. J. M. Davies, R. J. Poole, P. A. Rea, and D. Sanders, *Proc. Natl. Acad. Sci. U.S.A.*, *89*: 11701 (1992).
62. E. Johannes and H. H. Felle, *Physiol. Plants*, *77*: 326 (1989).
63. P. Malsowski and H. Malsowska, *Biochim. Physiol. Pflanzen*, *182*: 73–84 (1987).
64. A. P. Rea, C. J. Griffith, and D. Sanders, *Plant Vacuoles* (B. Marin, ed.), Plenum Press, New York, pp. 157 ff (1987).
65. F. M. Harold, *Annu. Rev. Microbiol.*, *31*: 181 (1977).
66. D. R. Bush, *Photosynth. Res.*, *32*: 155 (1992).
67. T. J. Buckhout, *Planta*, *178*: 393 (1989).
68. L. A. Williams, S. J. Nelson, and J. L. Hall, *Planta*, *186*: 541 (1992).
69. Z.-S. Li, O. Gallet, C. Gaillard, R. Lemoine, and S. Delrot, *Biochim. Biophys. Acta*, *1103*: 259 (1992).
70. E. Zamski, *Photoassimilate Distribution in Plants and Crops: Source–Sink Relationships* (E. Zamski and A. A. Schaffer, eds.), Dekker, New York, 1994.
71. H. P. Getz, M. Thom, and A. Maretzki, *Physiol. Plants*, *83*: 401 (1991).
72. H. P. Getz, *Planta*, *185*: 261 (1991).
73. F. Keller, *Plant Physiol.*, *98*: 442 (1992).
74. E. Komor, *FEBS Lett.*, *38*: 16 (1973).
75. E. Komor and W. Tanner, *Eur. J. Biochem.*, *44*: 219 (1974).
76. T. Rausch, *Physiol. Plants*, *82*: 134 (1991).
77. E. Zamski and R. E. Wyse, *Plant Physiol.*, *78*: 291 (1985).
78. R. Verstappen, S. Ranostaj, and T. Rausch, *Biochim. Biophys. Acta*, *1073*: 366 (1991).
79. N. Sauer and W. Tanner, *FEBS Lett.*, *259*: 43 (1989).
80. N. Sauer, K. Friedlander, and U. Graml-Wicke, *EMBO J.*, *9*: 3045 (1990).
81. L. Reinhold and A. Kaplan, *Annu. Rev. Plant Physiol.*, *35*: 45 (1984).
82. D. R. Bush and P. J. Langston-Unkefer, *Plant Physiol.*, *88*: 486 (1988).
83. Z.-C. Li and D. R. Bush, *Plant Physiol.*, *94*: 268 (1990).
84. Z.-C. Li and D. R. Bush, *Plant Physiol.*, *96*: 1338 (1991).
85. E. Martinoia, M. Thume, E. Vogt, D. Rentsch, and K.-J. Dietz, *Plant Physiol.*, *97*: 644 (1991).
86. J. Goerlach and I. Willms-Hoff, *Plant Physiol.*, *99*: 134 (1992).
87. U. Homeyer and G. Schultz, *Planta*, *176*: 378 (1988).
88. K.-J. Dietz, R. Jager, G. Kaiser, and E. Martinoia, *Plant Physiol.*, *92*: 123 (1990).
89. J. C. Skou, *Methods Enzymol.*, *156*: 1 (1988).
90. A. Ratner and B. Jacoby, *J. Exp. Bot.*, *27*: 843 (1976).
91. H. Mennen, B. Jacoby, and H. Marschner, *J. Plant Physiol.*, *137*: 180 (1990).
92. A. Katz, H. R. Kaback, and M. Avron, *FEBS Lett.*, *202*: 141 (1986).
93. A. Katz, U. Pick, and M. Avron, *Biochim. Biophys. Acta*, *983*: 9 (1989).
94. Y. Braun, M. Hassidim, H. R. Lerner, and L. Reinhold, *Plant Physiol.*, *87*: 104 (1988).
95. M. Hassidim, H. R. Lerner, and L. Reinhold, *Plant Physiol.*, *94*: 1795 (1990).
96. A. Katz, U. Pick, and M. Avron, *Plant Physiol.*, *100*: 1224 (1992).
97. E. Blumwald and R. J. Poole, *Plant Physiol.*, *78*: 163 (1985).
98. E. Blumwald and R. J. Poole, *Plant Physiol.*, *83*: 884 (1987).

99. M. Staal, F. J. M. Maathuis, T. M. Elzenga, J. H. M. Overbeek, and H. B. A. Prins, *Physiol. Plant.*, *82*: 179 (1991).
100. B. J. Barkla and E. Blumwald, *Proc. Natl. Acad. Sci. U.S.A.*, *88*: 11177 (1991).
101. J. Garbarino and M. DuPont, *Plant Physiol.*, *89*: 1 (1989).
102. S. Cooper, H. R. Lerner, and L. Reinhold, *Plant Physiol.*, *97*: 1212 (1991).
103. F. M. DuPont, D. S. Bush, J. J. Windle, and R. L. Jones, *Plant Physiol.*, *94*: 179 (1990).
104. K. S. Schumaker and H. Sze, *Plant Physiol.*, *79*: 1111 (1985).
105. M. Kasai and S. Muto, *J. Membrane Biol.*, *114*: 133 (1990).
106. K. Schumaker and H. Sze, *J. Biol. Chem.*, *261*: 12171 (1986).
107. E. Blumwald and R. J. Poole, *Plant Physiol.*, *80*: 727 (1986).
108. S. Blackford, P. A. Rea, and D. Sanders, *J. Biol. Chem.*, *265*: 9617 (1990).
109. B. Jacoby and B. Rudich, *Ann. Bot.*, *46*: 493 (1980).
110. D. Sanders, *Chloride Transport Coupling in Biological Membranes and Epitelia* (G. A. Gerencser, ed.), Elsevier, Amsterdam, pp. 64–119 (1984).
111. W. R. Ulrich and A. Novacky, *Plant Sci. Lett.*, *22*: 211 (1981).
112. A. D. M. Glass, J. E. Shaff, and L. V. Kochian, *Plant Physiol.*, *99*: 456 (1992).
113. C. I. Ulrich-Eberius, A. Novack, and A. J. E. van Bel, *Planta*, *161*: 46 (1984).
114. K. Sakano, *Plant Physiol.*, *93*: 479 (1990).
115. B. Lass and C. I. Ulrich-Eberius, *Planta*, *161*: 53 (1984).
116. E. Blumwald and R. J. Poole, *Proc. Natl. Acad. Sci. U.S.A.*, *82*: 3683 (1985).
117. K. S. Schumaker and H. Sze, *Plant Physiol.*, *83*: 490 (1987).
118. A. F. Smith and J. A. Raven, *Annu. Rev. Plant Physiol.*, *30*: 289 (1979).
119. B. Jacoby and G. G. Laties, *Plant Physiol.*, *47*: 525 (1971).
120. P. R. Stout and D. R. Hoagland, *Am. J. Bot.*, *26*: 320 (1939).
121. J. S. Pate, *Annu. Rev. Plant Physiol.*, *31*: 313 (1980).
122. E. Munch, *Ber. Dtsch. Bot. Ges.*, *45*: 340 (1927).
123. M. G. Pitman, *Annu. Rev. Plant Physiol.*, *28*: 71 (1977).
124. D. T. Clarkson and A. W. Robards, *The Development and Function of Roots* (J. G. Torrey and D. T. Clarkson, eds.), Academic Press, London, pp. 415–446 (1975).
125. A. S. Crafts and T. C. Broyer, *Am. J. Bot.*, *24*: 415 (1938).
126. G. G. Laties and K. Budd, *Proc. Natl. Acad. Sci. U.S.A.*, *52*: 462 (1964).
127. M. G. Pitman, R. A. Wilds, N. Schaefer, and D. Welfare, *Plant Physiol.*, *60*: 240 (1977).
128. A. Lauchli, D. Kramer, and E. Ball, *Plant Cell Environ.*, *1*: 217 (1978).
129. C. Engels and H. Marschner, *Physiol. Plants*, *86*: 263 (1992).
130. A. H. De Boer and H. B. A. Prins, *Plant Cell Environ.*, *8*: 587 (1985).
131. D. T. Clarkson, L. Williams, and J. B. Hanson, *Planta*, *16*: 361 (1984).
132. D. T. Clarkson and J. B. Hanson, *J. Exp. Bot.*, *37*: 1136 (1986).
133. J. B. Hanson, *Plant Physiol.*, *62*: 402 (1978).
134. D. T. Clarkson, *Solute Transport in Plant Cells* (D. A. Baker and J. L. Hall, Eds.), Longman, Harlow, pp. 251–304 (1988).
135. G. Haberlandt, *Sitzungsber. Akad. Wissensch.*, Wien, I, *101*: 785 (1892).
136. T. C. Broyer and D. R. Hoagland, *Am. J. Bot.*, *30*: 261 (1943).
137. W. Tanner and H. Beevers, *Plant Cell Environ.*, *13*: 745 (1990).
138. R. T. Hardiman and B. Jacoby, *Physiol. Plants*, *61*: 670 (1984).
139. L. H. P. Jones and K. A. Handrek, *Plant Soil*, *23*: 79 (1965).
140. M. G. T. Shone, D. T. Clarkson, J. Sanderson, and A. V. Wood, *Ion Transport in Plants* (W. P. Anderson, ed.), Academic Press, New York, pp. 571–582 (1973).
141. E. B. Dumbroff and D. R. Peirson, *Can. J. Bot.*, *49*: 35 (1971).
142. C. A. Peterson, M. F. Emanuel, and G. B. Humphreys, *Can. J. Bot.*, *59*: 618 (1981).
143. D. T. Clarkson and J. B. Hanson, *Annu. Rev. Plant Physiol.*, *31*: 239 (1980).
144. A. R. Yeo, M. A. Yeo, and T. J. Flowers, *J. Exp. Bot.*, *38*: 1141 (1987).
145. C. W. Bell and O. Biddulph, *Plant Physiol.*, *38*: 968 (1963).
146. J. F. McGrath and A. D. Robson, *Ann. Bot.*, *54*: 231 (1984).
147. J. C. Brown and L. O. Tiffin, *Plant Physiol.*, *40*: 395 (1965).

148. B. Jacoby, *Ann. Bot.*, *31*: 725 (1967).
149. B. Jacoby, *Handbook of Plant and Crop Stress*, Vol. 1 (M. Pessarakli, ed.), Dekker, New York, pp. 97–123 (1993).
150. B. Jacoby, *Plant Physiol.*, *39*: 445 (1964).
151. B. Jacoby, *Physiol. Plants*,*18*: 730 (1965).
152. D. Kramer, A. Lauchli, A. R. Yeo, and J. Gullasch, *Ann. Bot.*, *41*: 1031 (1977).
153. B. Jacoby, *Ann. Bot.*, *43*: 741 (1979).

2

Plant–Water Relationships

Karl M. Volkmar
Agriculture Canada Research Station, Lethbridge, Alberta, Canada

William Woodbury
University of Manitoba, Winnipeg, Manitoba, Canada

I. INTRODUCTION

Water is essential to life on earth. The human body is about 85% water, and vegetative plant material is closer to 90% water. A typical crop or grassland will transpire about 500 kg of water per kilogram of dry matter produced. Water has a tremendous effect on the environment near the earth's surface, which on a regional basis we call "climate."

II. WATER AND ENVIRONMENT

About 70% of the earth's surface is covered by the oceans, which contain about 97% of the total water on earth (Table 1) [1]. The atmosphere contains about 0.00001 of the planet's total water. If we compare the total annual evaporation (Table 2) with the amount of water in the atmosphere

$$\frac{518 \times 10^{12} \text{ m}^3}{0.013 \times 10^{15} \text{ m}^3} = \text{about } 40$$

we find that the atmospheric water turnover occurs about 40 times per year or roughly every 9 days [1]. Because the latent heat of evaporation of water is high (2466 kJ kg^{-1}) [2], the annual cycling of water between the surface and the atmosphere transfers a large amount of energy to the atmosphere. This results in lower temperatures at the earth's surface.

Sunlight passes through water, whereas rocks and soil absorb it strongly. So, on land, solar energy is absorbed at the earth's surface, which can result in rapid warming of the surface. Sunlight is able to penetrate some distance into lakes and oceans before it is absorbed. As a result, large bodies of water tend to remain cooler than adjacent land surfaces. On the surface of the land, evaporation of water from a moist surface results in slower warming. Where the

23

Table 1 Stocks of Water on Earth

Stock	Value ($m^3 \times 10^{15}$)
Oceans	1350
Ice	29
Groundwater	8.3
Freshwater lakes	0.1
Saline lakes	0.1
Soil water	0.067
Atmosphere	0.013
Living biomass	0.003
Rivers and streams	0.001

Source: Ref. 1.

surface is covered with vegetation, which is able to transpire, warming of the land will be slower.

The water vapor in the atmosphere has another important effect on the temperature at the surface. Other atmospheric gases (CO_2, NO_2, CH_4) also contribute to this so-called greenhouse effect. The average temperature of the earth's surface is 15°C. Any physical body at this temperature will radiate infrared energy in all directions at a rate of [3]

$$W/m^2 = 315 + 5T = 315 + 75 = 390 \text{ W}$$

Where T is ambient air temperature (°C) at 1.4 m above the surface. Water molecules in the atmosphere absorb some of the IR and reradiate it toward the surface. The downward radiation can be calculated [3] as follows:

$$208 + 6T = 208 + 90 = 298 \text{ W/m}^2$$

The net energy exchange toward space is then about 100 W m^{-2}: considerably less than 390 W, which would occur without the greenhouse effect. Since this IR exchange occurs day and night, the earth would be a very cold place indeed.

Consider a lake that is cooling during the fall. The maximum density of water occurs at 4°C (Table 3) [4]. This means that as the lake cools toward 4°C, the colder, denser water will descend toward the bottom of the lake. Once the temperature has gone below 4°C, the cooler water will tend to remain at the surface. When the surface temperature goes below zero, ice will be formed. Since ice is less dense than water, it remains at the surface. In the process of freezing, heat (344 kJ kg^{-1}) is released [5]. This heat will delay the further cooling of the water in the lake. The presence of solid ice on the surface eliminates the mixing effect of surface

Table 2 Mean Annual Flow of Water on Earth ($m^3 \times 10^{12}$)

Precipitation on land	108
Precipitation on the seas	410
Evaporation from land	62
Evaporation from the seas	456
Runoff from land	46

Source: Ref. 2.

Table 3 Some Physical Properties of Water That Influence the Environment

Heat of evaporation	2466 kJ kg^{-1} (590 cal g^{-1})
Heat of fusion	344 kJ kg^{-1} (80 cal g^{-1})
Density	
at 20°C	997 kg m^{-3}
at 4°C	1000 kg m^{-3}
at 0°C	999 kg m^{-3}
Density of ice (0°C)	900 kg m^{-3}
Heat capacity	
water	4.18 kJ kg · °C^{-1} (1 cal g · °C^{-1})
ice	2.10 kJ kg · °C^{-1} (0.5 cal g · °C^{-1})
soil	0.80–1.6 kJ · °C^{-1} (0.2–0.4 cal g · °C^{-1})

Sources: Refs. 1, 2, 4, and 5.

waves, which further slows the cooling process. In fairly deep lakes, temperature at the bottom is thermostated at about 4°C.

III. WATER IN LIVING ORGANISMS

One of the peculiarities of water is that it is a liquid at ordinary temperatures, whereas substances with similar molecular weights, such as methane and ammonia, become gases at temperatures well below zero (−161 and −33°C, respectively). Water is a fairly good solvent for molecules that are biologically important, including salts, sugars, amino acids, and metabolic intermediates. The association of water molecules with polar groups on the surface of membranes and proteins contributes to their structure and to their biological activity. Water molecules participate in a number of important biochemical reactions, for example:

$$H_2O \rightarrow H^+ + OH^-$$
$$CO_2 + H_2O \rightarrow H_2CO_3 \rightarrow H^+ + HCO_3^-$$
$$2 H_2O \rightarrow O_2 + 4e^- + 4H^+$$

All these reactions are reversible under proper conditions. The dissociation of water molecules gives rise to our concept of pH. At pH 7.0, the concentration of hydrogen ions in pure water is 10^{-7} M. A change of one pH unit involves a tenfold change in concentration of hydrogen ions. The hydrogen and hydroxyl ions may participate in biochemical reactions.

The reaction of CO_2 with water to form carbonic acid and bicarbonate ions increases the total amount of CO_2 in solution. Although H_2CO_3 is a weak acid, its presence in rainwater and in the soil solution around respiring roots and microorganisms is important in the weathering of rocks and soil.

In photosynthesis, light provides the energy to split water to oxygen and hydrogen ions. The electrons released provide energy to reduce CO_2 to carbohydrate. In mitochondrial respiration, the reaction runs in the other direction to consume oxygen and electrons and produce all-important adenosine 5′-triphosphate (ATP).

It should be understood that while these biochemical reactions of water are very important to the functioning of living organisms, the amount of water involved in the reactions is only a small part of the total amount of water passing through the organisms during their lifetime.

A. Properties of Water

From the preceding discussion, we can conclude that water has a number of properties that distinguish it from similar materials and that these properties are important to the effects of water on the environment and as a component of biological materials. Ultimately these "weird and wonderful" properties of water derive from the structure of individual water molecules and the ways in which they can interact with other water molecules and with other materials.

B. The Water Molecule

The chemical formula of water is H_2O. The central oxygen atom is bonded via covalent bonds to two hydrogen atoms. The outer electron shell of the oxygen atom accommodates eight electrons, which group into four pairs equally spaced around the atom. The O—H bonds account for two of the pairs. The electrons in these bonds are located closer to the oxygen atom than to the hydrogen atoms. The oxygen then carries a partial negative charge and the hydrogen atoms carry partial positive charges. The O—H bonds are 0.99 Å long, and oriented at an angle of 105°. The electrical attraction between the negatively charged oxygen on one water molecule and positively charged hydrogen atoms on nearby water molecules results in the so-called hydrogen bond.

Because the water molecule is bent and because there is a distribution of positive and negative charges within the molecule, the molecule is said to be dipolar and water is called a polar solvent.

Ice is a crystalline solid. In the crystal, six water molecules are H-bonded to form a bent hexagon. The hexagons are in turn H-bonded to neighboring hexagons. In the process of melting ice, heat energy is absorbed at the rate of 344 kJ per kilogram of water formed, though the temperature remains constant at zero [5]. The heat absorbed is called *latent heat of fusion*. If the same amount of heat is added to a kilo of liquid water, the temperature will rise by 344/4.18 $\approx$ 80°C. This is called *sensible heat*, since it can be measured or "sensed" by a thermometer. Notice that a large amount of energy (2456 kJ kg^{-1}) is required to evaporate liquid water [4]. This is also a latent heat, because if the evaporation rate is slow and heat energy is supplied from the surrounding air, evaporation can occur without a change in temperature.

C. Water as a Solvent

To produce enough energy as ATP and metabolites (sugars, amino acids) to allow for the synthesis of cell walls, proteins, and membranes, a growing plant cell must be able to maintain a high metabolic rate. Enzymes act to speed up particular biochemical reactions, but the rate also depends on the concentration of the substrates in solution, which can be quite high. Water is an effective solvent for the wide variety of metabolites that occur in cells.

Consider a crystal of sodium chloride. Within the crystal, each positive sodium ion is surrounded by six neighboring, negatively charged chloride ions, and vice versa. Ionic bonds are relatively weak, but because the ions within the crystal are very close together, the attractive forces are very large. Sodium chloride melts at 804°C and boils at 810°C. But, NaCl dissolves readily in water: 1 g dissolves in 2.8 mL of water. When salt crystals are added to water, the polar water molecules are attracted to the charged ions on the surface of the crystal. The attractive force is very high, and the ions come to be surrounded with a shell of water molecules. The shells move the ions apart to distances great enough to ensure that electrical attraction between the ions is near zero.

A crystal of sucrose is held together by hydrogen bonds between hydrogen atoms on —OH groups and oxygen atoms in other —OH groups, in the glucose–fructose bond, and in the sugar rings. When water is present, water molecules displace these intermolecular H-bonds and sugar dissolves readily: 1 g dissolves in 0.5 mL water [6].

Amino acids can form ionic bonds between amino and carboxyl groups on the α_2 (second) carbon and amino and carboxyl groups on the side chains. Water dissociates these ionic bonds in the same way as described for salt. In the presence of water, nonpolar —CH_2— groups in the side chains of some of the amino acids tend to cluster together. Such hydrophobic interactions tend to limit the solubility of these amino acids.

D. Water and the Plant Cell Wall

Cell walls contribute to the physical properties of a plant. Secondary thickenings in the epidermal and fiber cells of a stem allow the plant to deploy an array of leaves above the ground in a way that favors the interception of solar energy and the exchange of CO_2. While the cell is growing, the wall must be able to stretch laterally to allow for cell enlargement. At the same time the wall must be strong enough to withstand the turgor pressure that develops within the cell.

The mechanical strength of the growing wall depends on the presence of cellulose fibrils. Cellulose is a linear polymer of glucose units linked through α_{1-4} bonds. The α linkages force individual glucose residues to orient 180° from the glucose units to which they are linked. This favors hydrogen bonding between adjacent glucose residues along each chain and also between adjacent chains within the cellulose fibril. Cellulose, a strong crystalline polymer, will swell a bit in boiling water, but its basic structure remains intact. Most other polysaccharides will dissolve in boiling water because the high temperature disrupts the hydrogen-bonding interaction.

E. Water and Protein Structure

Proteins are composed of linear chains of amino acids linked together through peptide bonds involving the amino and carboxyl groups on the α_2-carbons. In globular proteins the polypeptide chain winds itself into a right-handed helix, which is stabilized by hydrogen bonds between peptide groups on adjacent turns of the helix. The helix then folds back on itself in such a way that hydrophobic (nonpolar) amino acids are located within the core of the protein and the polar amino acids are located through hydrogen bonds. At higher temperatures the various bonding interactions that hold a protein in its correct functional configuration begin to break down and enzymatic activity is lost.

F. Water and Membrane Structure

The earliest model for the structure of biological membranes was the phospholipid bilayer proposed by Danieli and Davson [7]. In the phospholipid bilayer model, the polar groups (phosphates, carboxyls, and glycerol) occupy the surface, where they interact with water. The nonpolar fatty acid chains are buried within the bilayer. A membrane with this structure should have low permeability to water, ions, and polar solutes like sugar. This simple model explained how cells could retain high concentrations of solutes, but it did not explain how membranes could be selectively permeable to some of the solutes. By about 1965, the electron microscope began to provide evidence that some proteins were located within the membrane, while others were located in the surfaces. Some of the proteins, which span the thickness of the membrane, form channels through which selected solutes may pass, perhaps with the assistance of energy provided by the hydrolysis of ATP. Notice that water must be present if the membrane is to maintain its structure and function.

IV. FORCES DRIVING THE MOVEMENT OF WATER

A. The Vapor Phase: Movement of Water from the Plant to the Atmosphere

1. *Transpiration*

When a leaf is exposed to light, the stomatal pores open to allow the movement of CO_2 into the leaf for photosynthesis. When the stomates are open, water vapor will be lost from the leaf because of transpiration. This water loss can eventually lead to wilting of the leaf. To avoid wilting, the plant must be able to move water from the soil, through the roots and stems, and finally into the leaf. This pathway for movement of water is referred to as the soil–plant–air continuum (SPAC).

A plant exposed to air that is not at 100% relative humidity (RH) will lose water through evaporation from the internal surfaces of the leaves. The water vapor diffuses through the stomatal pores toward drier air in the surrounding atmosphere. The rate of water vapor movement will be proportional to the difference in concentration of water vapor between air spaces in the leaf and the surrounding air. Table 4 gives values for vapor pressure and concentration of water vapor in saturated air at different temperatures. Notice that the values almost double with each 10°C rise in temperature.

Assume we have exposed the same leaf to air at 50% RH but at 20 or 25°C. Leaf temperature is the same as air temperature, and air within the leaf is at 100% RH.

If transpiration rate is proportional to the difference in vapor concentration, the rate would be

	Concentration		
Temperature	In leaf (100% RH)	In air (50% RH)	Difference
20°C	17.3	8.6	8.7
25°C	23.0	11.5	11.5

higher by a factor of $11.5/8.7 = 1.32$ at 25°C than at 20°C. If the leaf is exposed to full sunlight, its temperature might rise to 25°C while the air remains at 20°C and 50% RH. The difference in vapor concentration is now $23.0 - 8.7 = 14.3$ g m^{-3}, which would increase the transpiration rate again.

ROLES OF TRANSPIRATION. Transpiration is an inevitable consequence of the requirement that the stomatal pores of the leaf open to allow photosynthesis. But, transpiration is also

Table 4 Effect of Temperature on Water Vapor (WV) in Air

Temperature (°C)	Concentration (g WV m^{-3})	Vapor pressure (mm Hg)
0	4.85	4.6
10	9.41	9.2
20	17.30	17.5
25	23.00	23.8
30	30.40	31.8
40	51.10	55.3

Source: Ref. 4.

essential to plants in several other ways. The global mean for solar radiation reaching the surface of the earth is 700 W m^{-2} [2]. This is simply the mean value over the entire illuminated surface at any time. When plant leaves absorb solar energy, most of the absorbed energy is converted to heat, and the leaves will tend to warm up. If the plants can obtain water from the soil, the evaporation of water can remove most of the heat. If the water supply is restricted, leaf temperatures may be higher than the air temperature. Leaf temperature affects rates of photosynthesis, respiration, and growth. At high temperatures, leaf function may be damaged.

Plants must also obtain inorganic nutrients from the soil. For nutrients such as nitrate and calcium, which can move readily in the soil, the flow of water toward the root of a transpiring plant may increase the nutrient supply. For other important ions (potassium and phosphate) that diffuse slowly in the soil, this mass flow effect is unimportant [8].

To reach the shoot, nutrient ions must be transprted along the length of the root and the stem in the transpiration stream that moves in the xylem vessels. In some plants, nitrate may be converted to organic nitrogen in the root. In legumes, root nodules convert nitrogen gas from the atmosphere to amino acids. The organic nitrogen is moved to the shoot in the transpiration stream. Movement of hormones (cytokinins, abscisic acid) from root to shoot also occurs in the xylem and is important in integrating growth and function of root and shoot.

2. Resistances to Movement of Water Vapor

The loss of water vapor from a leaf is a diffusion process and it obeys Fick's law:

$$\text{rate} = -D\,A\frac{dC}{dZ}$$

where

D = diffusion coefficient (m^2 s^{-1})
A = area (m^2)
dC = difference in concentration of water vapor (g m^{-3})
dZ = length of the diffusion path (m)

and rate is in grams per second.

The negative sign on the right-hand side indicates that the direction of movement is from higher to lower concentration. The diffusion coefficient for water vapor in air is 0.21 m^2 s^{-1} at 20°C [9].

Plant leaves are covered with waxy cuticles, which offer a large resistance to movement. Water vapor must, therefore, diffuse through the stomatal pores. It must also diffuse across a thin layer of still air located at the surface of the leaf. In most situations, we do not know the length of this diffusion path (dZ in the equation). We can eliminate dZ from the equation by dividing the diffusion coefficient D by dZ to get a new coefficient g, having units of meters per second [i.e., $g = D/dZ = (\text{m}^2\ \text{s}^{-1}\ \text{m}^{-1} = \text{m s}^{-1})$]. This is called the leaf conductance. Now if we know the value of dC, and measure the rate of water loss from a particular leaf of area A, we can calculate a value for g for that leaf. Once we know g, we can predict the rate of transpiration from our leaf and from similar leaves (if we know or can calculate the value of dC). Later, we will encounter a unit called the hydraulic conductivity for liquid water in the soil. (Conductance and conductivity differ in that with conductivity, all three lengths of the system must be known.)

In the real world, the situation becomes a little more complicated, because g is not a constant number. If the leaf is losing water rapidly, the stomates may begin to close, which restricts the rate of water loss and causes g to decrease. Also the thickness of the laminar boundary layer varies with the reciprocal of the square root of the wind speed, and g increases as wind speed increases.

The ability of the stomates to regulate transpiration may be an important part of a plant's response to water stress. We can set up the plant with dC known and control the wind speed, while measuring transpiration. Then, the values of g will give an indication of how the stomates have responded to water stress.

B. The Liquid Phase: Water Movement in the Plant and Soil

1. Water Potential

Movement of water vapor during transpiration is most easily evaluated using Fick's law and, as the driving force, either the difference in concentration or the difference in vapor pressure from leaf to air.

In the soil and in the plant we are concerned with the movement of liquid water. The interaction of water with solutes and with surfaces and capillaries modifies the properties of the water, so the "concentration" of water cannot be used in analyzing the flow process. Instead, the difference in potential energy of water is used as the driving force. The equation used, Darcy's law, is formally the same as for Fick's law, but the proportionality coefficient becomes K, the hydraulic conductivity.

$$\text{rate} = -KA \frac{dP}{dZ}$$

For practical purposes, Darcy expressed dP/dZ as meters of water column per meters depth of sand. But, since the pressure under a 10 m column of water is 1.0 kg cm^{-2}, a hydrostatic pressure is involved. This could also be expressed in units of pressure (0.1 MPa) or of energy (100 kJ kg^{-1}).

A simple definition is that water potential ψ_w is a free energy that is available to do the work. The work of interest is the movement of water from one location to another. Water will move from a region of higher ψ_w to a region of lower or more negative ψ. The reference point for ψ_w is a large mass of pure water at the same temperature and atmospheric pressure, which is assigned a ψ_w of zero.

2. Osmotic Potential: Effect of Solutes

The addition of one mole of solute to a liter of pure water lowers the ψ_π of the solution by a definite amount, which can be calculated according to [10]

$$\psi\pi = -CRT$$

where C is the molal concentration (moles per liter of water), R is the gas constant (8.31 m^3 · Pa mole · K^{-1}), and T is the temperature (K).

For pure water: $\psi_\pi = T^- 0 \times 8.31 \times 273 = 0.0$ MPa
For 1.0 molal solution: $\psi_\pi = T^- 1 \times 8.31 \times 273 = -2.27$ MPa

If a membrane separates the solution from pure water, water will move from the region of higher potential (pure water = 0.0 MPa) to the region of lower potential (−2.27 MPa). The rate of water movement will depend on the permeability of the membrane to water and the difference in potentials.

The addition of solutes to water changes a number of its colligative, or tied together, properties in a predictable way (Table 5) [11]. These values are easily determined if the proper equipment is available. A well-designed psychrometer allows the measurement of vapor pressure

Table 5 Colligative Properties of Water

Freezing point depression	−1.86°C
Boiling point increase	+0.51°C
Vapor pressure decrease (20°C)	−0.31 mm Hg
Osmotic potential (one mole of solute in molal water)	−2.27 MPa; −22.7 bar

Source: Ref. 11.

above samples, which may be solutions, plant sap, plant tissue, or soil. The colligative properties bear a constant relation to one another, so if vapor pressure is measured, water potential can be calculated.

Raoult's law states that the vapor pressure of a solvent above a solution is proportional to the ratio of moles of water to moles of solute + moles of water. One liter of water contains $1000/18 = 55.55$ moles of water. Comparing the osmotic potential of pure water and a one-molal solution, we get the following results if temperature is 0°C:

Pure water	One molal
$\psi_\pi = (RT/v) \times \ln (55.55/55.55)$	$\psi_\pi = (RT/v) \ln [55.55/(1.0 + 55.55)]$
$\psi_\pi = 0.0$ MPa	$\psi_\pi = -2.27$ MPa

Where

R = gas constant = 8.31 $m^3 \cdot$ Pa (mole $\cdot$ K^{-1})

T = temperature (K)

V = volume of one mole of water (0.018 L).

In soil, the ψ_π is affected by the association of water with surfaces and by the retention of water in capillary spaces. Dissolved salts have only a small effect in soils that are not saline. In the plant, the effects of solutes are most important, but water is also interacting with surfaces and pores in the cell walls. In either case, we cannot measure the ψ_π directly, but with a thermocouple psychrometer we can measure vapor pressure of water in air that is equilibrated with a sample of soil or leaf tissue or a solution. The psychometer itself must be calibrated against solutions of known molal concentration. This follows from Raoult's law. In a one-molal solution, the mole ratio is $55.5/(55.55 + 1.0) = 0.982$. The vapor pressure above the solution would than be 0.982×17.5 mm Hg, or 17.19 mm if the temperature is 20°C. To put this in more familiar terms, the relative humidity over a one-molal solution at 20°C would be 98.2%.

In soil, the lower limit of water availability to plants is the permanent wilting point, which is generally −15 bar (−1.5 MPa). If the sample is at 25°C,

$$\psi_\pi = -15 \text{ bar} = (0.0831 \times 0.018) \ln \frac{p}{p_0}$$

and the ratio p/p_0 comes out to 0.989. Some of the newer psychrometers contain a microcomputer, which will report the output in osmotic potential, and the researcher no longer needs to do the calculations.

It is useful to look at the numbers for p/p_0 in the last two calculations. These tell us that the relative humidity in the soil and in a plant are in the range of 98–100%. If the surrounding air is at 50% RH, the effects of decreasing ψ in the plant or in the soil as water is lost will have almost no effect on the difference in vapor pressure that drives the movement of water during evaporation from the soil and transpiration by plants. The plant can exercise some control over transpiration by adjusting the size of the stomatal pores.

3. Matric Potential: Effect of Porous Solids on Water

Soil consists of variable mixtures of sand, silt, and clay. In temperate soils, these materials are silicates and their surfaces are covered with an array of —OH groups. There are also ionic groups associated with the surfaces. Dipolar water molecules are attracted to the surfaces, where they form thin films of tightly bound water. In moist soil, the films are thicker and there is also liquid water retained in capillary spaces between the particles.

If the lower end of a glass capillary tube is immersed in water, water will rise to some definite height in the tube. While the water is rising, work is being done to move the mass of water upward against the force of gravity. After the tube has been filled, the water may be removed by applying air pressure at the top of the tube. Work must be done to empty the tube.

When the water column has become stationary, two equal, but opposite, forces are acting on the water. The upward force is due to the surface tension of water acting at the edge of the meniscus:

force up $= S2\pi r$

Notice that $2\pi Pr$ gives the circumference of the tube; S is the surface tension.

The downward force is due to gravity, acting on the mass of water in the tube.

force down $= (\pi r^2 * h)\rho g$

Here, $\pi r^2 * h$ is the volume of water, ρ is the density of water, and g is the gravitational constant. At equilibrium,

force up $=$ force down
$S2\pi r = (\pi r^2 h)\rho g$

After canceling and rearranging, we get

$$h = \frac{2S}{r\rho g}$$

To simplify this a bit, note that $2S/\rho g$ is a constant, so the height of the water column is proportional to reciprocal of the radius ($1/r$) of the capillary. If r is decreased by half, h will double. Using SI values and 20°C: $S = 0.0728$ N m^{-1}, $\rho = 998$ kg m^{-3}, and g $= 980$ m s^{-2} gives $h = (1.49 \times 10^{-5}$ m$^2)/r$, where r is expressed in meters.

A typical xylem vessel, which has a radius of 20 μm, has a capillary rise of 0.75 m. If the radius is reduced to 2 μm, the water column will rise to 7.5 m. Now, recall that a pressure of 1 bar or 0.1 MPa will support a column of water 10 m high. Thus, the downward, negative force acting in the 2 μm capillary is about –0.75 bar.

The "pores" in the polysaccharide matrix of plant cell walls may have radii about 5 nm. Surface tension acting in these pores would be able to support a water column of about 3 k. This far exceeds the height of very tall trees. Nobel [12] presents calculations showing that the tensile strength of water is great enough to maintain the continuity of such a column of water even at heights of more than a mile.

In soil, there are pores between the solid particles. These pores may contain air or water, depending on the moisture content of the soil. After a soil has been wetted by a rain, water will continue to move slowly downward for some time. When the rate of movement approaches zero, the soil is said to be at field capacity, which is usually taken as –0.3 bar (–0.03 MPa).

Pores larger than about 50 μm will be filled with air, while smaller pores will be filled with water. If the soil is allowed to dry by evaporation, the larger pores will empty first.

C. Movement of Water in Capillaries

Water is a viscous liquid, and its viscosity is due to the hydrogen bonding between the individual molecules. If water is moving in a capillary, some of the water molecules are held to the glass surface. The mathematics is a bit complicated. According to the Poiseuille equation, the rate of volume flow through a single tube is given by

$$\text{rate per tube} = \frac{-\pi r^4}{8v}\frac{dP}{dZ}$$

Where v is the viscosity of water and dP/dZ is the pressure difference divided by the length of the tube.

We can simplify this a bit because $P/8v$ is a constant. Then, if we compare flow rates through capillaries of different radii and hold dP/dZ constant, the flow rate will vary with r^4. A small change in radius of the tube will produce a large change in flow rate.

When dealing with water movement in the soil, we are more interested in flow rate per unit than per capillary. If we divide both sides of the equation by area $= \pi r^2$, we get

$$\text{rate per area} = \frac{r^2}{8v}\frac{dP}{dZ}$$

At 20°C, viscosity of water is 1.002×10^{-3} Pa · s. In a clay soil, the average pore radius is about 1 μm, so

$$\frac{r^2}{8v} = \frac{(1 \times 10^{-6}\ m)^2}{8(1.002 \times 10^3\ Pa \cdot s)}$$

$$= 1.2 \times 10^{-10}\ m^2\ s^{-1}\ Pa^{-1}$$

If we multiply this result by a realistic pressure gradient dP/dZ in pascals per meter, we may predict the actual flow per unit area. This idealized model assumes that the pores occupy the entire soil volume, whereas in real soil they occupy about half the volume. Also, in soil the pores are not shaped like cylinders all aligned in the same direction. Flow rates would be about ten fold smaller than predicted. In a very moist soil, pores smaller than about 50 μm would be filled with water. Assume that the soil now dries until the largest water-filled pores are 40 μm. Since the flow rate per unit area of pores varies with r^2, we would expect the flow rate to decrease by almost 40%. The decrease will be greater than this, however, because the area of water-filled pores per area of soil would decrease as well.

D. Water Movement in the Plant

1. Resistance to Water Uptake

If the resistance to water flow into and through the plant is higher than the rate of transpirational water loss, the plant will begin to dehydrate. It is important, therefore, that the two processes, transpiration and water uptake, be tightly controlled.

Traditionally, only a fairly narrow band of cells up to 10 cm behind the root tip is considered responsible for most of the radial water movement from the root surface to the xylem [13]. Within this zone, in a process called *apoplastic transport*, water passes mainly extracellularly (i.e., via cell walls) into the stele. Behind that region, an endodermal barrier to extracellular water movement gradually develops through deposition of a water-resistant suberin lamella. Movement of water into the stele within this mature region is possible only by first traversing an intracellular, symplasmic pathway, in a mode termed symplasmic transport. This account firmly implies that axial resistance—that is, the resistance to water flow in the xylem—is much smaller than a radial resistance.

Under conditions of low transpiration, the high resistance intracellular pathway of water movement probably dominates. At high transpiration rates, an increasing amount of water may travel extracellularly by moving through hydrophilic micropores embedded in the suberized cell wall [14], thereby enabling the uptake rate to match that of transpiration. If this model is accurate, mature regions of the root may be increasingly involved in water uptake. In practice, this would contribute little to total water uptake if the root system were following a descending water profile.

Recent evidence suggests an even greater role for mature roots in water uptake. The cross-wall of large water-conducting xylem elements, called *late metaxylem elements*, appears to persist for more than 25 cm behind the root tip [15,16]. McCully and Canny [15] suggested that the intervening rhizosheath region between the mature metaxylem and the root tip, characterized by encasement of aggregated soil tightly bound to the younger regions of grass roots, may be relatively inactive in water uptake. The much smaller, earlier opening xylem elements presumably could not account for reported flow rates. While this issue has not been resolved, there appears to be substantial apoplastic water movement in the stele in mature regions of the root at sites of lateral branching [17,18].

Episodes of soil dying can in some cases induce increased resistance to water uptake by root systems [19]. This is in part due to increased lignification and suberization of the endodermis. Axial resistance may also be affected by the persistence in water-stressed plants of late metaxylem cross-walls, which normally break down relatively quickly in unstressed plants [20].

2. Coupling Transpiration with Absorption

In sunlight, plants will open the stomates to admit CO_2 to the leaves for photosynthesis. The open stomates present a pathway for the efflux of water vapor. The rate of transpiration depends on the difference in vapor pressure between the leaf and the ambient air, which can be affected by leaf and air temperature. If supply of water from the soil is slower than the transpiration rate, water will be lost from plant cells. As the cells lose water, they will decrease in volume. This results in an increase in solute concentration and a reduction in ψ of the cells and tissues. Eventually the decrease in ψ is transmitted to the root. Because the potential of the soil is greater than the potential of the plant, water will move into the root. Much of this water will be removed from a layer of soil immediately adjacent to the root. Drying reduces the hydraulic conductivity of this soil, so to maintain a flow of water, ψ_{plant} must become more negative. At night, the stomates close, which limits the rate of water loss from the plant. Movement of water from the soil during the night will restore the ψ in the leaf to a value close to or equal to the value in the soil. Water movement in the soil brings the ψ of soil near the root to equilibrium with that of the bulk soil.

The diurnal variation in plant ψ is depicted in Figure 1, showing the effect of transpiration on leaf and soil ψ as a function of time over six days [21]. Transpiration during the first day removed some water from the soil. This results in a small decrease in ψ_{soil} and a larger decrease in hydraulic conductivity (K in the Darcy equation; $r^2/8v$ in the Poiseuille equation). On day 2, the transpiration cycle is repeated, but now K of the soil is lower and a larger potential difference ($\psi_{soil} - \psi_{plant}$) is needed to maintain the supply of water to the leaves. As a result, leaf ψ drops further than on the first day. By day 5, hydraulic conductivity of the soil is low

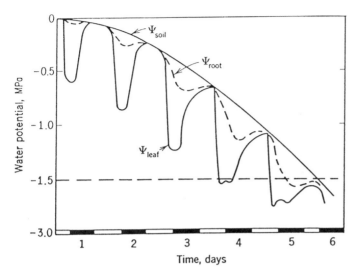

Figure 1 Schematic representation of the changes in leaf, root surface, and bulk soil water potentials associated with the exhaustion of the available water in a soil by a transpiring plant. Dark bars indicate darkness. (From Ref. 21.)

enough to cause the ψ in the leaf and the soil to drop below the permanent wilting point at -1.5 MPa.

The situation described in Figure 1 is somewhat artificial: the soil dried from field capacity to the permanent wilting point in only 5 days, suggesting a large plant growing in a fairly small volume of soil. A growing plant may be able to extend its roots into moister soil, deeper in the profile, however, and some plants are able to divert the use of photosynthates from shoot growth to root growth when water supply is limited [22]. This point is discussed further in a later section.

Deep-rooted plants have also been shown to carry water up to plants with shallower roots. The process, termed *hydraulic life*, occurs when roots tapped into a deep source of water leak the water into the shallow soil at night [23]. The ability of some species to adjust root development according to the availability of water in the soil profile and the transpiration demand of the local atmosphere is important to plant adaptation in semiarid grasslands.

V. WATER AND PLANT GROWTH

Growth of a plant begins with cell divisions, which occur in meristems located in the tip of the root and the apex of the shoot. As the cells age, they become larger and undergo structural differentiation according to their ultimate function. Since a cell is perhaps 90% water, an increase in cell size requires the movement of water into the cell. While the cell is enlarging, it must be able to develop an internal osmotic pressure or turgor pressure that is sufficient to stretch the wall.

The water immediately available to a cell is that located in the cell wall and adjacent intercellular spaces. Since solute concentrations normally are very low, water is retained by physical forces of cohesion and surface tension: that is, $\psi_{wall} = \psi_{matric}$. If the plant is well supplied with water and is not actively transpiring, as at night, ψ_{matric} will be high (> -0.1 MPa).

If water is moving into the cell, ψ_{cell} must be lower than ψ_{wall}. Inside the cell, water is held in the vacuole and in the cytoplasm. In the vacuole, simple solutes (sugars, ions, organic acids) contribute to ψ_π. In the cytoplasm, metabolic intermediates also contribute to ψ_π. Water bound to colloids (proteins) and to membranes contributes to a ψ_{matric} that is usually unknown and lumped with the ψ_π. Both ψ_π and ψ_{matric} will be negative. The turgor pressure P within the cell, created by the entry of water to balance ψ_π, is positive. Thus,

$$\psi_{cell} = \psi_\pi + P$$

A Höffler diagram (Figure 2) allows us to depict the effects of change in volume due to water uptake or loss on the components of water potential in a hypothetical leaf cell [24]. On the

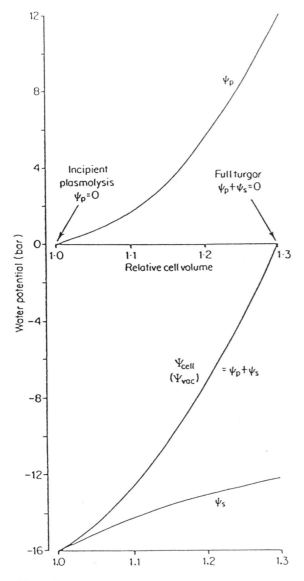

Figure 2 Höffler diagram for an idealized leaf cell showing the relationship between vacuolar water potential (ψ_{vac}) and its components, pressure potential (ψ_p) and solute potential (ψ_s), at different cell volumes. (From Ref 24.)

right, the cell is fully turgid and ψ_p is +12 bars (1.2 MPa). This is numerically equal to ψ_π (= −1.2 MPa), so ψ_{cell} = 0.0 MPa. As the cell loses water to transpiration, the value of ψ_p decreases. The value of ψ_π increases because the decrease in volume results in an increase in solute concentration. On the left, the cell, having lost a third of its water (ψ_p = 0), is flaccid, and the leaf would begin to wilt. Further water loss may result in mechanical disruption of intercellular contacts. As water is lost from the cells, solute concentrations increase, perhaps becoming high enough to precipitate or inactivate enzymes.

Lockhart [25] found that the relation between rate of cell enlargement associated with water uptake and ψ_p within the cell could be expressed as follows:

$$\frac{W}{dV/dt} = \epsilon(\psi_p - Y)$$

Where V is volume of the cell, ϵ is the coefficient of extensibility of the cell wall, and Y is the "yield pressure" (i.e., the minimum value of P needed to expand the cell wall). By monitoring the rate of cell elongation along with direct measurements of ψ_p either directly, using a micropressure probe [26], or indirectly, by measuring ψ_w and ψ_π [27], ϵ and Y can be estimated (Figure 3). Such measurements have revealed that ϵ and Y regulate cell expansion in a way that maintains a roughly constant rate of growth even when ψ_p changes [28]. Recently, attempts have been made to explain Lockhart's relationship between cell growth and the wall, yielding properties of the cell in terms of the breakage and re-formation of the hemicellulose molecules

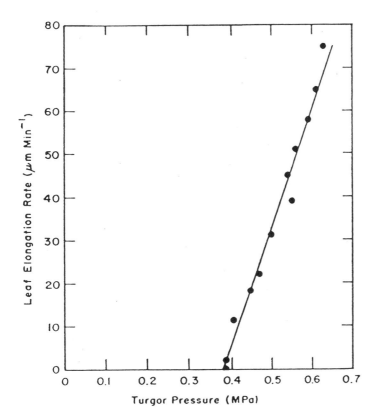

Figure 3 Elongation rate of fifth leaf of maize versus turgor pressure (P). (From Ref. 27.)

that connect transversely oriented cellulose microfibrils [29]. Hormones (auxins, cytokinins, and gibberellins) affect the rate of cell growth regulating the activity of enzymes that modify the properties of the cell wall. In any case, continuous growth requires that solute concentrations be regulated to maintain a turgor pressure sufficient to drive the expansion process.

The effects of water stress on the physiology of plants summarized by Hsiao et al [30] more than 15 years ago remain relevant (Figure 4). Most of their literature review dealt with plants grown in controlled environments. Cell enlargement appeared to be very sensitive; a drop in ψ_{cell} below -0.2 to 0.4 MPa slows or stops cell enlargement. Gallagher and Biscoe (31) used automated equipment to follow the rate of growth of barley leaves in the field for 24-hour periods and to measure air temperature and ψ_{leaf} (Figure 5). In the early morning, transpiration increased as a result of increasing temperature and stomatal opening. This resulted in a decrease in ψ_{leaf} to about -0.5 MPa at 0700 h, and the rate of leaf extension began to decline. However, by 1200 h, when ψ_{leaf} reached -1.5 MPa, the rate of leaf extension was still 60–70% of the rate predicted from the temperature response curve. Evidently plants growing in the field are considerably less sensitive than plants grown in controlled environments to the effects of water stress on cell enlargement. Similarly, when the effects of water stress on rates of photosynthesis and leaf conductance are compared, field-grown plants are usually found to be more tolerant [32].

The growth of the shoot and root also appear to respond differently to stress due to water deficit. A shift toward an increased ratio of root to shoot has often been observed in plants exposed to drought [33]. The adaptive advantage is clear. Allocating a greater portion of the photoassimilate to the root system as the soil dries ensures the reproductive security of the plant, albeit at some cost to yield. The physiological basis for this shift in carbon allocation is

process affected	sensitivity to stress			references
	very sensitive		insensitive	
	reduction in tissue Ψ_w required to affect the process			
	0	1	2MPa	
cell growth (−)	—— ·· ▪			Acevedo *et al.*,1971 ; Boyer,1968
wall synthesis[†] (−)	——			Cleland,1967
protein synthesis[†] (−)	——			Hsiao,1970
protochlorophyll formation[‡] (−)	——			Virgin,1965
nitrate reductase level (−)	——			Huffaker *et al.*,1970
ABA synthesis (+)	▪·· ——			Zabadal,1974; Beardsell &
stomatal opening (−):				Cohen,1974
(a) mesophytes		——		reviewed by Hsiao,1973
(b) some xerophytes		—————— ▪ ▪▪▪		Van den Driesche *et al.*,1971
CO₂ assimilation (−):				
(a) mesophytes		——		reviewed by Hsiao,1973
(b) some xerophytes		—————— ▪ ▪▪▪		Van den Driesche *et al.*,1971
respiration (−)	▪ ▪ ———			
xylem conductance[§] (−)	▪▪ ·· ▪ ——			Boyer,1971; Milburn,1966
proline accumulation (+)	▪▪ ··· ——			
sugar level (+)		——		

[†] fast growing tissue
[‡] etiolated leaves
[§] should depend on xylem dimension

Figure 4 The influence of water stress on the physiology of mesophyllic plants. Continuous horizontal bars indicate the range of stress levels within which a process is first affected; broken bars refer to effects that have not been firmly established. (From Ref. 30.)

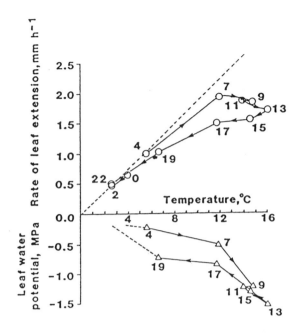

Figure 5 Relationship between the extension of the eighth leaf of a barley plant and air temperature during a single day in the field, and the corresponding time course of leaf water potential. Upper dash indicates the expected response to temperature in the absence of water stress. (From Ref. 31.)

the capacity of roots to continue growth at ψ_w that completely inhibits shoot growth [34]. Early investigations looked for evidence of sustained turgor in water-stressed roots. However, neither turgor nor solute concentrations appear to differ in the growing regions of water-stressed roots and shoots [35], suggesting that properties of the cell wall (i.e., ϵ and Y) may be responsible for the differential response of roots and shoots to water stress [36]. How these biophysical properties are regulated is an area of current interest.

A. Control of Transpiration by Stomata

To ensure the maximazation of photosynthesis, land plants deploy a large area of leaf material. They must be able to intercept solar radiation as efficiently as possible. The large leaf area facilitates gas exchange with the atmosphere. The evolution of an epidermis covered with a waxy cuticle restricted exchange of both water vapor and CO_2. When water-stressed, many plants increase the thickness of the cuticle. The guard cells surrounding the stomatal openings are able to respond to the divergent needs of the plant to conserve water while permitting photosynthesis. Raschke [37] provided a diagram (Figure 6) that summarizes the control mechanisms. Since all these regulatory loops may operate simultaneously, but with different time constants, it is not surprising that experimental results are often contradictory. Studies designed to examine changes within the guard cells themselves can be frustrated by the small proportion of the volume of the epidermis ($\approx 3\%$) that consists of guard cells.

In the light, the guard cells accumulate potassium ions and organic acids. The decrease in ψ_π results in movement of water into guard cells, which swell and enlarge the stomatal pore. As the leaf loses water to the transpiration stream, water will also be lost from the guard cells, which may then shrink to reduce the size of the stomatal pores.

Leaves exposed to water stress may produce abscisic acid (ABA), which can promote

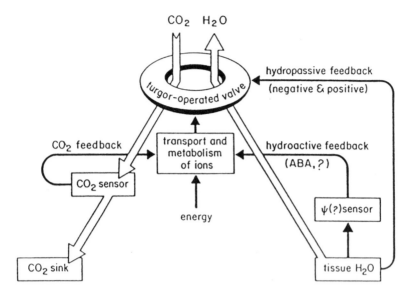

Figure 6 Simplified and hybrid schema of the stomatal feedback system. The guard cells function as a turgor-operated valve modulating uptake of CO_2 from the atmosphere and water loss from the mesophyll to the atmosphere. The mechanism sensing CO_2 is located in the guard cells. "Hydropassive" refers to changes in turgor caused by changes in the water potential in the leaf; the solute content of the guard cells remains unaffected. Hydropassive responses include effects of changes in epidermal pressure on stomatal aperture (positive feedback). The hydroactive feedback is probably between mesophyll and the guard cells, with ABA serving as one of possibly several messengers. (From Ref. 37.)

stomatal closure by causing the efflux of solutes and water from the guard cells. In some cases there is clear evidence that the ABA is produced in leaf chloroplasts, from which it is released under water stress conditions [38]. The concentration of ABA is often proportional to the severity of the stress and to the extent of stomatal closure. There is also some recent evidence that roots of water-stressed plants may produce ABA or other compounds that promote closure of the stomates. In addition, the guard cells are able to respond to the CO_2 requirements of the leaf. During photosynthesis the CO_2 concentration within the leaf decreases, and a feedback mechanism operates to enlarge the stomatal pore. The CO_2 sensor appears to be located in the guard cells themselves, since the stomates in epidermal strips are able to respond to changes in CO_2 [39]. The CO_2 level may regulate the synthesis of malate, which is the major organic anion in the guard cells [40].

B. Osmotic Adjustment

Field-grown plants often are less sensitive to water stress than plants grown in controlled environments. It is also known that the rate at which plants are dried can affect their sensitivity to stress. Figure 7a plots the decline in turgor pressure in sorghum plants subjected to a slow or fast drying cycle starting from a well-watered condition [41]. With slow drying, the drop in turgor potential (ψ_p) lags the decline in leaf water potential (ψ_w). This outcome can be traced to the accumulation of solutes (sugars, amino acids, organic acids) that lower the osmotic potential (ψ_π) and contribute to the decline in ψ_w according to the relationship

$$\psi_w = \psi_\pi + \psi_p$$

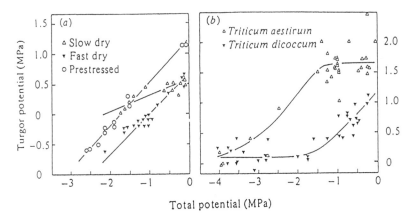

Figure 7 Examples of relationships between turgor potential and leaf water potential for (a) sorghum plants and (b) two wheat genotypes: ▼, previously well-watered and dried rapidly; △, previously well-watered and dried slowly; ○, previously stressed to –1.6 MPa and dried rapidly. (From Ref. 41.)

This is often termed "osmotic adjustment" or "osmoregulation."

Figure 7a also shows that sorghum plants that had been exposed to water stress maintained a higher ψ_p throughout the drying cycle. The values for ψ_p are for the total leaf, and they are related to the ability of leaves to continue growth. But since guard cells open and close by regulating their own ψ_p, different drying treatments can produce similar differential responses if rates of transpiration are measured. Figure 7b compares the changes in ψ_p and ψ_w in two species of wheat. In the *Triticum aestivum* plants, ψ_p did not begin to fall until ψ_w reached –1.5 MPa. This species was capable of osmotic adaptation, whereas *T. dicoccum* was not. From this distribution of properties, we can conclude that osmotic adaptation is not a universal property of plants.

The importance of osmotic adjustment as a mechanism of drought resistance remains controversial in spite of the attention it has received over two decades. The issue is not so much whether it occurs, but whether it is a valuable agronomic trait, worth incorporating into breeding programs. Failure to find a consistent positive relationship between osmotic adjustment, leaf photosynthetic rate, and stomatal conductance (see, e.g., Ref. 42), along with an apparent dissociation between cell turgor and growth, are reasons to question the value of the proposed mechanism.

VI. SUMMARY AND CONCLUSIONS

The fundamental processes of living systems are inextricably dependent on water's distinctive physical and chemical properties. The high melting and boiling points, surface tension, heats of fusion, and vaporization are all consequences of the strong hydrogen bonding between water molecules. The ready dissolution of many compounds in water enhances their reactivity; hence water is a catalyst for countless chemical reactions. The polarity of water strongly influences the configuration, and therefore the function, of proteins and membranes. Plant cell expansion itself is driven by the force exerted by hydraulic forces. A continuous column of water moves upward from the root to the leaf in response to transpirational water loss, driven by a water potential gradient between the leaf and the surrounding air. Transpiration helps cool the plant and also facilitates nutrient transport from the soil to the growing points of the shoot. The equilibrium between the components of water potential—essentially the matric, osmotic, and pressure potentials in the plant and the soil—governs the direction of water flow, with water moving in the direction of more negative potential. Because the gradient in water potential between the leaf and atmosphere normally far exceeds that between the root and soil, water moves from the soil into the plant and into the atmosphere. Movement of water across the root cortex to the xylem occurs largely intercellularly, though it may become increasingly extracellular with increasing transpiration rate. As soil water is depleted, a larger gradient in water potential is required to extract water from the increasingly larger proportion of small pores that still retain water. A plant's metabolic dependency on water sets a lower limit at which plant water potential may decrease before key metabolic processes are affected. Cell expansion is affected long before there has been appreciable dehydration, suggesting that enlargement is mediated by signals triggered by drying of soil around the root. Plants avoid dehydration by extending their roots beyond water-depleted zones. Differences in the cell wall yielding or extensibility coefficients in the root and shoot may be responsible for the preferential growth of root systems under conditions of soil drying. Osmotic adjustment may also serve to limit the effect of soil drying on growth.

Our understanding of plant–water relations has progressed rapidly with the advent of a universally adopted terminology and with improved methods of quantification. As we go deeper into the nature of plant–water relations, we derive an ever greater appreciation for the intricate complexity of the action of water in biological systems.

REFERENCES

1. J. Hart, *Consider a Spherical Cow. A Course in Environmental Problem Solving*, Kaufmann, Los Altos, CA, p. 238 (1986).
2. P. S. Nobel, *Physicochemical and Environmental Plant Physiology*, Academic Press, New York, p. 544 (1991).
3. J. L. Monteith, *Principles of Environmental Physics*, American Elsevier, New York, p. 35 (1973).
4. P. S. Nobel, *Physicochemical and Environmental Plant Physiology*, Academic press, New York, p. 547 (1991).
5. P. J. Kramer, *Water Relations of Plants*, Academic Press, New York, p. 8 (1983).
6. *CRC Handbook of Chemistry and Physics*, 60th ed. (R. C. Weast, ed.), CRC Press, Boca Raton, FL, p. C-503 (1979–1980).
7. J. F. Danieli and H. Davson, *J. Cell. Comp. Physiol.*, 5: 495 (1934).
8. S. A. Barber, *Soil Nutrient Bioavailability*, Wiley, New York, pp. 201–258 (1984).
9. J. L. Monteith, *Principles of Environmental Physics*, American Elsevier, New York, p. 13 (1973).
10. P. S. Nobel, *Physicochemical and Environmental Plant Physiology*, Academic Press, New York, p. 72 (1991).
11. P. J. Kramer, *Water Relations of Plants*, Academic Press, New York, p. 16 (1983).
12. P. S. Nobel, *Physicochemical and Environmental Plant Physiology*, Academic Press, New York, p. 55 (1991).
13. D. T. Clarkson and A. W. Robards, *The Development and Function of Roots*, (J. C. Torrey and D. T. Clarkson, eds.), Academic Press, London, pp. 415–436 (1975).
14. A. W. Robards, D. T. Clarkson, and J. Sanderson, *Plant Cell Environ.*, *11*: 247 (1988).
15. M. E. McCully, and M. J. Canny, *Plant Soil*, *111*: 159 (1988).
16. J. Sanderson, F. C. Whitbread, and D. T. Clarkson, *Plant Cell Environ.*, *11*: 247 (1988).
17. C. A. Peterson, G. B. Emanuel, and G. B. Humphries, *Can. J. Bot.*, *5*: 618 (1981).
18. M. E. McCully and D. Mallet, *Ann. Bot.*, *71*: 327 (1993).
19. C. Ramos and M. R. Kaufmann, *Physiol. Plants*, *45*: 311 (1979).
20. R. T. Cruz, W. R. Jordan, and M. C. Drew, *Plant Physiol.*, *99*: 203 (1992).
21. R. O. Slatyer, *Plant Water Relationships*, Academic Press, New York (1967).

22. A. H. El Nadi, R. Brouer, and J. T. Locker, *Neth. J. Agric. Sci.*, *17*: 133 (1969).
23. T. E. Dawson, *Oecologia*, *95*: 565 (1993).
24. H. Meidner and D. W. Sheriff, *Water and Plants*, Blackie, Glascow and London, p. 257 (1976).
25. J. A. Lockhart, *J. Theor. Biol.*, *8*: 264 (1965).
26. D. J. Cosgrove, E. Van Volkenburgh, and R. E. Cleland, *Planta*, *162*: 46 (1984).
27. T. C. Hsiao and J. Jing, *Physiology of Cell Expansion During Plant Growth* (D. J. Cosgrove and D. P. Knievel, eds.), American Society of Plant Physiologists, Rockville, MD, p. 180 (1987).
28. W. G. Spollen and R. E. Sharp, *Plant Physiol.*, *96*: 438 (1991).
29. J. B. Passioura and S. C. Fry, *Aust. J. Plant Physiol.*, *19*: 565 (1992).
30. T. C. Hsiao, E. Acevedo, E. Fereres, and D. W. Henderson, *Phil. Trans. R. Soc. London Ser. B*, *274*: 479 (1976).
31. J. N. Gallagher and P. V. Biscoe, *J. Exp. Bot.*, *30*: 645 (1979).
32. J. W. Jones, B. Zur, and K. J. Boote, *Agron. J.*, *75*: 281 (1983).
33. R. S. Malik, J. S. Dhankar, and N. C. Turner, *Plant Soil*, *53*: 109 (1979).
34. R. E. Sharp and W. J. Davies, *Planta*, *147*: 43 (1979).
35. M. E. Westgate and J. S. Boyer, *Planta*, *164*: 540 (1985).
36. J. Pritchard, R. G. Wyn Jones, A. D. Tomos, *J. Exp. Bot.*, *41*: 669 (1990).
37. K. Raschke, *Annu. Rev. Plant Physiol.*, *26*: 309 (1975).
38. B. R. Loveys, *Physiol. Plants*, *40*: 79–84 (1977).
39. J. Zhang and W. J. Davies, *Plant Cell Environ.*, *13*: 277 (1990).
40. K. Wardle and K. C. Short, *J. Exp. Bot.*, *32*: 303 (1981).
41. H. G. Jones, "Plants and Microclimate," University of Cambridge Press, Cambridge, p. 220 (1983).
42. F. S. Girma and D. R. Krieg, *Plant Physiol.*, *99*: 583 (1992).

3

Current Perspectives in Water Loss from Plants and Stomatal Action

J. P. Srivastava

Institute of Agricultural Sciences, Banaras Hindu University, Varanasi, Uttar Pradesh, India

Arvind Kumar

S.K.N. College of Agriculture, Jobner, Rajasthan, India

I. INTRODUCTION

Water is among the most important constituents of plants. Life originated in water, and primeval plant life thrived in water. Plants evolved from an aquatic mode to an amphibious mode, and vascular plants ultimately started to colonize the terrestrial environment. In analogy to the regulation of temperature in animals, there are two essential categories regarding regulation of water loss in plants [1]. These are *poikilohydry* and *homeohydry*. Poikilohydric plants are nonregulating and lose water in proportion to ambient dryness. But, such plants also have a remarkable ability to withstand repeated cycles of dehydration and rehydration. Because of this characteristic, such plants are often called resurrection (reviviscent) plants. Examples of resurrection plants are confined to lichens, Thallophyta, Bryophyta, and Pteridophyta. But some families of angiosperms such as Myrothamnaceae and Poaceae have poikilohydric representatives. Homeohydric, or water-regulating, plants are equipped with specialized structures to prevent or reduce water loss in a dry habitat. However, such plants can withstand only one cycle of dehydration and rehydration, dehydration being mostly in the seed or perennating propagule. As there are fewer poikilohydric species than homeohydric types, and since almost all crop plants are homeohydric, our information on water loss from plants concerns homeohydric species [2].

Water loss from plants assumes additional importance inasmuch as this phenomenon affects plant life at the cellular and molecular levels. Vegetation and topography of agroclimatic biomes are also determined by the availability and conservation of water. Thus, water loss from plants has implications in the entire range of the time–space scale [3]. Of all the water absorbed by plants, only 5% or even less is utilized; the remaining 95% is lost. If the total amount of water absorbed by plants were utilized, a single irrigation or modest rain would be adequate for the entire period of growth and development. It has been calculated that a corn plant loses more than 200 L of water during its life, or 100 times its fresh weight [4].

II. GUTTATION AND TRANSPIRATION

Transpiration is the phenomenon most responsible for the excessive water loss from plants. Other processes involved are guttation (loss of water from nectaries and glands), secretion, and bleeding (water loss from wounds and cut ends). The amount of water loss through secretion and bleeding is negligible. Guttation, however, is strikingly different from transpiration. In this process, water is lost in the form of liquid as opposed to the vapor form in transpiration. The liquid is an aqueous solution containing inorganic elements, sugars, organic acids, and vitamins (in contrast to pure water vapors in transpiration). The water droplets are visible at leaf margins and vein endings, and the anatomical structure associated with guttation is the water stomates or hydathodes (Figure 1). Transpiration takes place mainly through stomates but also through lenticels and cuticle.

Moreover, environmental conditions favoring guttation are high humidity, low temperature, low vapor pressure gradient, low water potential gradient, and darkness. Transpiration is enhanced under low humidity, high temperature, high vapor pressure deficit (VPD), adequate water potential gradient, and high incident radiation [5]. Root pressure is supposed to be a major cause of guttation, while transpiration is caused by an interaction of physical (boundary layer resistance) and biological (leaf conductance, which is a sum of stomatal, mesophyll, and cuticular conductances) factors. Although guttation is an interesting phenomenon, it is of little importance to plants. Occasionally salts dissolved in guttation water may be deposited on leaf margins, causing injury. Pathologists suggest that guttation water creates conditions favorable for infection with fungi and bacteria. It is sometimes argued that transpiration is beneficial because it lowers the foliage temperature through evaporative cooling. Absorption and translocation of minerals and salts are also increased by transpiration, but many plants thrive in shaded, humid habitats. Thus, transpiration can at best be regarded as an unavoidable evil [6].

III. MODES OF TRANSPIRATION

In addition to stomatal transpiration, water is lost as a vapor, directly from foliar surfaces and herbaceous stems. The other modes of transpiration, though of minor significance, are cuticular

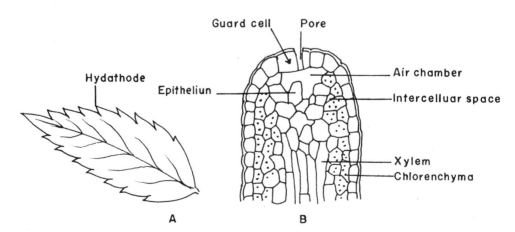

Figure 1   (A) Hydathode position on leaf. (B) V. S. of leaf through hydathode. (General adoption.)

and lenticular transpiration. Lenticels are small openings in the corky tissue covering stems and branches of the plants. Under dry conditions when stomates are closed, water loss through the cuticle and lenticels assumes importance. Lenticular transpiration may cause some desiccation in trees that shed leaves at the beginning of winter.

The plant cuticle, a waxy layer of cutin covering the surface of leaves, prevents water loss. However, the cuticle is now believed to be partially permeable to water vapors. Generally the cutin layer is thick in sun plants and in plants of arid regions. Recently, interest in cuticle has been revived, particularly because of new insights into its structural complexity.

IV. CUTICULAR TRANSPIRATION

The cuticle is a noncellular membrane covering all primary parts of higher plants, including substomatal cavities. In most of the crop plants, up to 20% water loss is through the cuticle, which is present on the outer surface of epidermal cells. Cuticular transpiration may exceed this value in xerophytes and in plants growing under moisture-stressed conditions.

A. Structural, Physical, and Chemical Properties of Cuticle

The cuticle, though very effective as a barrier for diffusion, is seldom complete and tight. Beside acting as a barrier to water vapor diffusion, the cuticle, being moderately permeable to water, is able to perform a number of other functions in plants. It protects the plants from injuries due to wind and physical abrasions, frost, and radiation. The nature of the cuticle influences the deposition and absorption of herbicides, pesticides, foliar nutrients, and other chemicals that contact plants. The cuticle also protects plants from attack by pathogenic fungi and insects [7].

The main constituent of cuticle is cutin, forming a matrix with deposited soluble lipids in and on its surface. Besides these dominating constituents, the cuticle contains trace amounts of various nonlipid components (e.g., cellulose, proteins, phenolic compounds).

The soluble cuticular lipid (SCL) makes up only a small percentage of the total mass of the cuticular membrane [8]. The SCL can be divided into two fractions on the basis of their arrangement within the cuticle. Intracuticular soluble lipids are embedded in the cutin matrix, and epicuticular soluble lipids are deposited on the surface of the cutin. The epicuticular soluble lipids are present as an amorphous layer covering the cuticle surface and partly as crystals [9]. These classifications are also based on the applicable extraction procedure. Epicuticular lipids can be extracted easily in organic solvents, and the lipid left on cuticle after extraction by organic solvent for a fixed time is termed intracuticular lipid. The term epicuticular lipids (waxes) also is used to denote surface crystals.

The SCL is a complex mixture of lipids containing mainly long hydrocarbon chains. Cyclic compounds like triterpenoids and flavones are also widely distributed among cuticular lipids.

The amount and the composition of epicuticular lipids vary in different species, plants, and organs, and even in the same plant.

The morphology of epicuticular crystal is closely related to the chemical structure of the individual compounds. The glaucous appearance of many leaves is due to the reflectance properties of SCL and is not necessarily a sign of large amounts of lipids on the leaf surface [10]. "Green" cultivars or varieties have surface lipids in the form of smooth films or platelets lying flat on the cuticular surface. The term "bloom" refers to the bluish coloration on the plant surfaces, caused by reflectance properties of the epicuticular lipids.

B. Mechanism of Water Loss Through Cuticle

Water loss through cuticular membranes takes place by a diffusion process, and the driving force is a vapor pressure gradient across the cuticle. Soluble cuticular lipids are the main barrier to water diffusion through cuticle. It is not clear, however, which properties of these lipids are most important in achieving low water permeability. It is also not known which lipids are the most important in determining the diffusion of water: the lipids embedded in the cutin matrix, the intracuticular lipids, or the lipids exuded to the outer surface of the cuticle and epicuticular lipids. It is, however, known that the removal of surface lipids by mechanical treatments (e.g., brushing) does increase water loss from plants [11].

C. Influence of Environmental Factors on Cuticular Transpiration

It is stated that the epicuticular soluble lipids play a dominant role in limiting the diffusion of water through the cuticle. Both amount and composition of epicuticular soluble lipids are very much influenced by factors such as season, growth temperature, light intensity, photoperiod, water availability, air humidity, salinity, air pollutants, herbicides, and insecticides. The cuticular transpiration rate was correlated with the amount of epicuticular soluble lipids obtained by dipping leaves in organic solvent [12]. However, it has been reported that it is not the quantity that predominantly regulates the diffusibility of cuticle to water vapors, but the composition and architecture of the crystals [13].

V. TRANSPIRATION: SOME LINKAGES

A. Water Loss Patterns in C_3, C_4, and Crassulacean Acid Metabolism (CAM) Plants

The water use efficiency (WUE)—that is, dry matter produced per unit of water used—of CAM plants is six times higher than that for plants exhibiting C_3 metabolism and three times more than C_4 plants [14]. Nocturnal stomatal opening of CAM plants is highly tactical, since radiant energy is at its minimum during the night, which means that stomatal opening is used mainly for the purpose of CO_2 uptake. Stomatal opening may totally be inhibited if the leaf water potential reaches -1.5 MPa [15]. However, higher WUE of C_4 plants than C_3 plants has been attributed to ability of C_4 plants to fix CO_2 under low CO_2 pressure, absence of temperature saturation, low photorespiratory losses, and recycling of internal CO_2. Stomates of C_3 plants are known to be more sensitive to water stress and, thus, to have a potential for efficient uptake of CO_2 under optimum conditions of the water supply. The C_3 plants are known to respond to photosynthetically active radiation (PAR) through the direct influence on guard cells, while in C_4 plants light acts via assimilation and concentration of CO_2.

B. Transpiration and Drought Tolerance

Levitt's scheme of the mechanism of drought resistance identifies low and high rates of transpiration as essential components of drought avoidance. Water savers conserve water by means of stomatal control or by employing other xerophytic adaptations. Water spenders are also expected to avoid drought through maintaining an optimum plant water status. This curious task is accomplished via efficient water absorption, particularly from the deeper strata of soil. However, plants develop and experience internal water deficit when transpiration rate exceeds that of absorption (absorption lag).

VI. TRANSPIRATION: PHYSICAL AND BIOLOGICAL COMPONENTS

Transpiration has two components: physical (e.g., VPD) and biological (e.g., stomatal conductance), and transpiration rate is an expression resulting from an interaction of these two components. It is usually believed that transpiration rate is proportional to stomatal aperture. This may be the case under most circumstances, but it is not always true. Experimental evidence indicates [16] that with increase in stomatal aperture from 0 to 20 μm, rate of transpiration increased very marginally in still air (boundary layer resistance r_a was high, around 2.0 s cm^{-1}), while in moving air (r_a low, around 0.1 s cm^{-1}), a similar increase in stomatal aperture greatly enhanced the rate of transpiration (Figure 2).

The transpiration rate of a particular plant organ varies considerably with time. It is determined not only by the taxonomic classification of plants but also by their ontogenetic age,

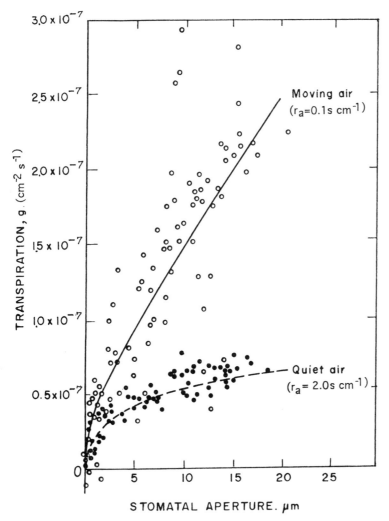

Figure 2 Interaction of physical (boundary layer resistance, r_a) and biological (stomatal aperture) components of transpiration. (From Ref. 6.)

as well as temporal oscillations of environmental variables. The energy balance equation of a leaf showed that the energy budget of a transpiring leaf is determined by radiation, convection, and evaporation [17]. However, it was pointed out that a change in a single component of the energy exchange system changes the entire balance of energy exchange with water vapors [18].

VII. METHODS OF MEASUREMENT AND QUANTIFICATION OF WATER LOSS FROM PLANTS

Methods used to measure transpiration vary widely, but essentially there is a progressive succession from gravimetry to porometry.

A. Gravimetry

Since the times of Stephen Hales in eighteenth century, the weighing of plants in containers at regular intervals has been a popular means of quantification.

Gravimetry has also been employed by several investigators on detached leaves or twigs weighed on a sensitive balance at intervals of 1, 2, 5, or 10 minutes. Stomatal closing and cuticular phases in transpiration decline curves were obtained by weighing the detached leaves [19]. There are some possibilities of error: for example, a transient increase in rate shortly after detachment (the Ivanov effect); moreover, the actual environment may differ from the location of measurement, and there may be large variations within replications. Nevertheless the cut shoot or rapid weighing method has been found to yield valuable information on comparative patterns of water loss [2]. Other methods, including the volumetric method, the measurement of water vapor loss, and the determination of the velocity of sap flow, have been discussed [6], and a comprehensive perspective of methods of measuring transpiration has been given by Slavik [20].

B. Porometry

A porometer is essentially a device for measuring stomatal or leaf resistance to water vapors diffusing from the leaf to the atmosphere. The instrument is calibrated in such a manner that readings can be converted into diffusion resistance or conductance of the leaf. With the adoption of new terminology in plant–water relations [21], the flow of water in the soil–plant–atmosphere continuum (SPAC) has been viewed as analogous to the flow of the electric current in a conducting systems. Transpiration as a function of driving force and resistance in diffusion pathway is measured as a flux.

1. Driving Force

The driving force for liquid movement in plant tissues is the water potential gradient, while the driving force for the movement of water vapors in transpiration is the vapor pressure gradient. The magnitude of the driving force and the resistances operative in the diffusional pathway determine the transpirational flux T:

$$T = \frac{\text{driving force}}{\text{resistance in the diffusional pathway}} \tag{1}$$

$$= \frac{C_{\text{leaf}} - C_{\text{air}}}{r_{\text{leaf}} + r_{\text{air}}} \tag{2}$$

where C_{air} and C_{leaf} are water vapor concentrations in air and at the evaporating surface of a leaf, respectively; r_{air} is the surface boundary layer resistance encountered by diffusing molecules, and r_{leaf} the diffusion resistance for water vapors inside the leaf.

Equation (2) can also be expressed as follows:

$$T = \frac{0.622 \, P_{air}}{\rho} \frac{e_{leaf} - e_{air}}{r_{leaf} + r_{air}}$$

where P is the density of the air, ρ is atmospheric pressure, e_{leaf} the vapor pressure at the evaporating surface inside the leaf, and e_{air} the vapor pressure in air.

Resistance r_{leaf} depends on resistance created by cuticular membrane (r_c) and stomata (r_s) to the diffusion of water vapor. Since r_c and r_s are parallel, we can write

$$\frac{1}{r_{leaf}} = \frac{1}{r_c} + \frac{1}{r_s} \tag{4}$$

Resistances of abaxial (r_{ab}) and adaxial (r_{ad}) leaf surfaces are also summed as reciprocals, and the leaf diffusive resistance (LDR) is calculated as follows:

$$LDR = \frac{r_{ad} \times r_{ab}}{r_{ad} + r_{ab}} \tag{5}$$

Stomatal resistance depends on the number of individual resistances operating in series, such as mesophyll cell walls (r_m), intercellular spaces (r_i), and the resistance created by stomatal pores (r_p). Stomatal resistance, therefore, may be given by following equation:

$$r_s = r_m + r_i + r_p \tag{6}$$

Figure 3 explains the relationships and relative importance of these resistances in determining r_{leaf}. Any factor, whether plant or environmental, that influences e_{leaf}, r_m, r_c, r_i, r_p, r_{air}, and/or e_{air} also influences water loss from plants.

2. Vapor Pressure

The major factors determining vapor pressure at the evaporating surface of a leaf (i.e., the cell walls of mesophylls and air) are the leaf and air temperatures, respectively. Considering that mesophyll cell walls are saturated with pure water, these may be considered to be at saturation

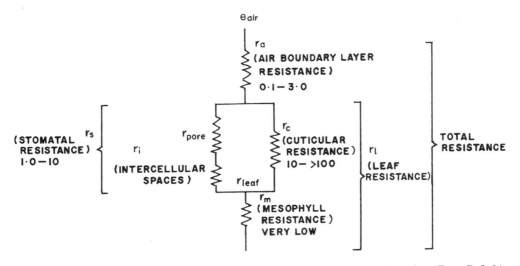

Figure 3 Diverse leaf and boundary layer resistances coupled in parallel and in series. (From Ref. 6.)

vapor pressure. A small increase in leaf temperature increases the vapor pressure inside the leaf. The dependence of e_{air} on air temperature is well known. A marginal increase in air temperature decreases e_{air}. Hence, a slight increase in leaf and air temperatures steepens the vapor pressure gradient and increases the water loss, provided the rest of the factors remain unchanged. Humidity determines e_{air}; therefore, a change in atmospheric humidity influences the water loss from plants.

3. Leaf Diffusive Resistances

Mesophyll cell walls and intercellular spaces are very small; thus the resistance they create is not of much significance. With the reduction in mesophyll cells size and intercellular spaces in xerophytes, however, resistance values increase and become important. Resistance created by stomatal pores is of much significance in determining the total water loss from plants. The value increases or decreases with closing and opening of stomata. Factors such as light, CO_2 concentration, temperature, and leaf water potential influence r_p significantly, and they regulate water loss from plants. Stomatal pore area is loss than 0.1% of total leaf area, but the water loss from the leaves may exceed even 50% of that from the free open surface of pure water. In fact, diffusion of gases through pores is roughly proportional to the circumference of the pores rather than to their area. Nevertheless, water loss from the plant also depends on stomatal frequency, the size of the stomata, and the extent of their opening.

4. Boundary Layer Resistance

Resistance created by air is due to boundary layer resistance. A layer of unstirred moist air surrounding transpiring leaves reduces the vapor pressure gradient and the water loss. Moving air tends to reduce the boundary layer resistance; hence with an increase in wind velocity, water loss is increased at low or moderate levels of radiation. At higher levels of radiation, the increase in wind velocity may decrease the rate of transpiration [17].

5. Porometers

Basically, there are two types of porometer: mass flow and diffusion devices. The mass flow porometer measures the rate at which air is forced across the thickness of leaf under pressure. The viscous or mass flow resistance (M_s, kg m^{-2} s^{-1}) is given by:

$$M_s = \frac{\Delta P}{f}$$

where ΔP is the pressure gradient (kg ms^{-2}) and f is the flow of air across the leaf (m^3 s^{-1}). Diffusion porometers are based on measurements of the rate of water vapor loss from a leaf or portion of a leaf enclosed in a porometer chamber. The diffusion resistance is determined from the rate of increase in humidity in a given period of transit time.

In modern null balance or steady state porometers, the transpiration rate is measured at a given flow rate of dry air to offset the increase in humidity.

If a flow rate of dry air f (cm^3 s^{-1}) is required to maintain a humidity H in a chamber enclosing a leaf area S (cm^2), the rate of transpiration E is

$$E = \frac{fe_a}{S}$$

where e_a is the vapor pressure of the air in the chamber. The details of calibrations and use have been presented by Slavik [20] and Coombs et al. [22]. A possible source of error in porometric determination of stomatal conductance has also been described by McDermitt [23].

VIII.   STOMATES: VALVES TO REGULATE WATER LOSS

Water loss from plants occurs mainly through the stomates, specialized apertures on foliar surfaces. These pores are surrounded by two kidney-shaped cells called guard cells. The guard cells fulfill two opposing priorities of plants: namely, the conservation of water and the photosynthetic uptake of carbon dioxide. The latter task is crucial inasmuch as the cuticle is almost totally impermeable to carbon dioxide.

A.   Morphology and Movement of Guard Cells

Stomates are distributed on both surfaces of leaves. They are found in nearly equal numbers on both sides of isobilateral leaves (e.g., members of Poaceae), but there are more stomates on the abaxial (lower) surface of dorsiventral leaves (e.g., most dicotylendons). In mono-cotyledonous plants, the guard cells are dumbbell shaped, and two cells just adjacent to them are strikingly different from the rest of the epidermal cells; these unusual cells are designated as subsidiary cells (Figure 4). Guard cells and subsidiary cells are often referred to as the stomatal complex (Figure 5a). The opening and closing of stomatal aperture is caused by

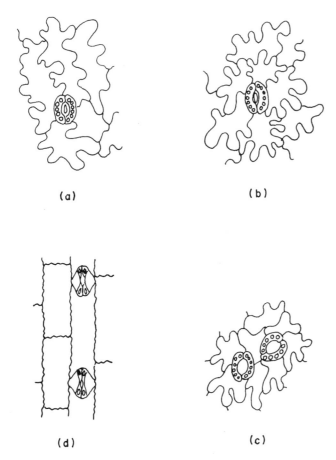

Figure 4   Stomates of various types: (a)–(c) dicotyledonous stomates and (d) monocotyledonous (dumbbell-shaped) stomates with subsidiary cells. (From Ref. 6.)

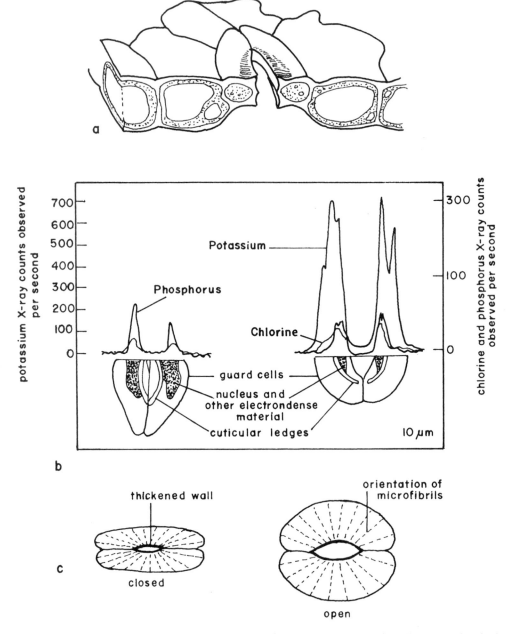

Figure 5   (a) Stomatal complex: note internal cuticle. (b) and (c) Physiological and microfibrillar changes in open and closed stomates. [From Irene Ridge, *Plant Physiology*, Hodden and Stoughton, Open University, Maynard Keynes, U.K. (1991).]

the expansion and contraction of guard cells. This sequence, in turn, depends on the turgor pressure of the guard cells. The movements of the guard cells are under the influence of myriad of endogenous and extraneous factors, but all culminate in a guard cell turgor pressure that must exceed that of neighboring epidermal (or subsidiary) cells before opening is elicited. The stomatal closure is invariably preceded by a decrease in the turgor pressure of the guard cells relative to that of the surrounding cells. The pressure-induced movement of the guard cells leads to an increase in pore diameter. This series of events has an anatomical explanation. The internal wall of the guard cells, surrounding the aperture, is thicker than the outer wall. When the turgor pressure of the guard cells increases, differential pressure is generated and the outer walls are stretched more than the inner walls (Figure 5c). Guard cells assume kidney or bean shapes, and pore diameter increases. In their closed condition, guard cells look like deflated cylindrical tubes (Figure 5c). Guard cells usually (but not always) differ from other epidermal cells in two respects. They are not linked by plasmodesmata to adjacent cells, and they have chloroplasts. Although guard cells possess nuclei, mitochondria, and other subcellular organelles like any other prokaryotic cell, the controversy about functional significance is unique to the chloroplasts. Contrasting pieces of evidence have been presented in the continuing debate as to whether guard cell chloroplasts participate in photosynthesis [24,25]. However, all guard cell chloroplasts of closed stomates contain starch (except onion guard cells, which are devoid of starch). The presence of starch in closed stomatal guard cells and its gradual disappearance during stomatal opening are among the first metabolic explanations of stomatal movements [26].

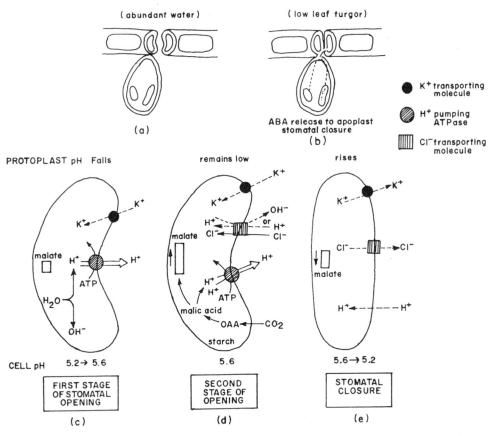

Figure 6 Summary of physiological events in guard cells leading to stomatal opening and closure. (a) ABA is not synthesized in open state. (b) ABA is synthesized in chloroplasts and transported to guard cells, causing stomatal closure. (c) and (d) pH, K^+, H^+, and malate changes during stomatal opening. (e) The same changes during stomatal closure. [From Irene Ridge, *Plant Physiology*, Hodden and Stoughton, Open University, Maynard Keynes, U.K. (1991).]

B. Stomatal Physiology: A Simplified Summary

Our ideas about the physiology and ecophysiology of stomatal behavior have undergone many revisions. Physiological events related to the movement of the guard cells (illustrated in Figures 5b and 6), are summarized as follows.

The excess turgor potential of the guard cells over the surrounding subsidiary/epidermal cells is responsible for stomatal opening.

Disappearance of starch is proportional to pore diameter.

Decrease in internal CO_2 (HCO_3^-) leads to stomatal opening.

Influx into guard cells of potassium and chloride ions from the surrounding cells leads to stomatal opening.

Malic acid synthesis is enhanced during stomatal opening.

Efflux of protons (H^+) from guard cells is associated with stomatal opening.

Stomatal closure is brought about by a decrease in guard cell turgor pressure. The metabolic reversions associated with stomatal closure are appearance of starch, increase in HCO_3^-, acidic pH, efflux of K^+ and Cl^-, decrease in malate synthesis, and influx of protons.

Among other endogenous factors, a low ratio of abscisic acid (ABA) to cytokinin (CK) is responsible for stomatal opening, while during closure a high ABA/CK ratio is commonly observed.

1. Environmental Control of Stomatal Movement

The environmental signals that cue stomatal movement are summarized as follows:

Increases in photosynthetically active radiation and blue light lead to stomatal opening in all photoactive stomates, whereas stomates of CAM plants are scotoactive and open at night.

Optimum soil water content and plant water status favor stomatal opening.

Increase in ambient humidity is known to cause rapid stomatal opening in many plants, although the response is not universal.

Stomates tend to open in response to increases in ambient temperature from 18°C to 35°C. However, ambient humidity and temperature usually are computed together as the vapor pressure deficit. The increase in VPD, the physical driving force, increases transpiration but also induces rapid stomatal closure [27].

2. Stomatal Movements: Complex Interplay of Factors

It is evident that stomatal movement is under the control of a myriad of biological factors and environmental signals. The interplay of these factors in the control of the stomatal aperture can be summarized in five broad categories.

Two signals—radiation and intercellular concentration of carbon dioxide [CO_2]—control stomatal aperture in relation to the plant's demand for CO_2 for photosynthesis. Low [CO_2] promotes stomatal opening, thus raising [CO_2] and leading to stomatal closure via a negative feedback loop, unless assimilation maintains a low [CO_2]. In addition, the degree of stomatal opening is associated with the supply of energy and assimilates through a positive feedback loop. Photosynthetically active radiation and blue light may elicit stomatal opening directly (feed-forward effect).

At least three signals—the VPD and the ABA levels in the leaf apoplast, and some unknown signal(s) from the roots—control stomatal aperture in relation to water supply. The VPD acts via cuticular (peristomatal) evaporation in a feed-forward manner. The levels of ABA, which serve as a monitoring system for leaf turgor, cause stomatal closure through a negative feedback loop. When stomates open, transpiration increases, leaf turgor falls, and ABA is synthesized in the chloroplasts and released. The ABA is then transported to the guard cells, where it inhibits the influx of K^+ to the guard cells and checks stomatal opening. When the soil is very dry and the water supply to roots does not keep pace with the demand by shoots, stomates open less widely in response to some negative feedback signal from the roots.

The role of PAR and low $[CO_2]$ is to cause a shift of guard cell pH toward alkalinity (5.2 → 5.6). Earlier, it was believed that this shift favored the enzymatic hydrolysis of starch into hexose phosphates, leading to a decline in the osmotic potential of the guard cells, followed by an influx of water from neighboring cells, which enhanced positive turgor in the guard cells. Now it is believed, however, that starch is converted to organic acids and sugar is replenished from the mesophyll cells. The lowering of guard cell turgor is also associated with an influx of K^+ and Cl^- ions. The presence of inorganic and organic anions (malate^{2-}), plus proton extrusion, account for the balance of charge in guard cells.

Even in well-known stomatal responses to PAR, $[CO_2]$, and humidity, certain aspects are unclear and remain to be understood. For example, the PAR, in addition to its effect through assimilation and low $[CO_2]$, may operate directly on the guard cells, as evidenced by its influence on isolated guard cells and CO_2-independent responses. It has been hypothesized that light stimulates proton secretion by providing ATP for the proton pump. Moreover, two separate light-absorbing systems may be involved: one consists of the usual chloroplast chlorophyll systems (photosystems I and II), and the other may be a membrane-bound pigment outside the chloroplasts which mainly absorbs blue light. The PAR photosystem is operative at medium to high irradiances and supplies ATP through photophosphorylation. The blue light receptor, which does not trigger wide stomatal opening, is present mostly in grasses and is responsible for rapid stomatal opening at dawn. In lady's slipper orchid (*Paphiopedalium* spp.), which grows in very shady habitats, the kidney-shaped guard cells lack chloroplasts and their response to light is apparently mediated through the blue light photosystem.

The effect of CO_2 clearly ties stomatal movement to assimilation and change in pH. But how CO_2 causes stomatal movement is unclear. According to one hypothesis, because CO_2 is fixed by phosphoenolpyruvate (PEP)-carboxylase (it is interesting that all guard cells, whether found in C_3 or C_4 plants, reveal high PEP-carboxylase activity) into malic acid in guard cells, the levels of one or both of the dissociation products—malate and H^+ ions (cytosolic pH)—reflect CO_2 levels and influence membrane permeability or proton gradient. Another hypothesis suggests that CO_2 and light affect the supply of ATP for proton pumping.

C. Stomatal Physiology: Some New Developments

1. Ion-Selective Channels in Guard Cell Protoplasts

Patch clamp techniques have been applied to guard cell protoplasts of *Vicia faba* to investigate dynamic membrane phenomena at high resolution. A potassium-selective single channel has been found which has the highest permeability for K^+ (followed by Rb^+, Na^+, Li^+, etc.), and this explains why stomatal opening occurs in the presence of these cations. Calcium ions have been shown to trigger stomatal closing and to inhibit stomatal opening. Patch clamp studies [28] revealed that an elevation of cytosolic free Ca^{2+} to the 1.5 μm level inhibited K^+ channels. This provides an explanation of the Ca^{2+}-dependent inhibition of stomatal opening and the induction of stomatal closure. These mechanisms have been summarized in a model for voltage- and signal-dependent plasma membrane ion transport during stomatal movement. The voltage-dependent, outward-conducting and inward-conducting K^+ channels have recently been found in guard cells of various types in higher plants [28].

2. Heterogeneity of Stomatal Conductance (Patchiness)

Studies on the effect of applying ABA to leaves initially suggested that this acid exerted no inhibitory action on photosynthesis per se, though it restricted uptake of CO_2. This situation arose because stomates on the same leaf were observed to behave differently [29]. This uneven behavior is said to be found in heterobaric leaves, where networks of vascular bundles cause the effective isolation of sections of intercellular spaces [30]. A fast method to detect patchiness was reported by Beyschlag and Pfanz [31].

3. Apoplastic and Cytosolic Calcium

The observations that calcium in the range of 50–250 μm can induce stomatal closure and that fluctuation in the concentration of calcium in the xylem sap significantly affects transpiration have triggered new interest in the role of calcium. Even external applications of calcium nitrate (1 mM) have been found to reduce stomatal aperture.

Abscisic acid, darkness, and cytokinins are now believed to elicit their characteristic response by employing calcium ions as second messengers. Interestingly, a stimulus-induced increase in cytosolic calcium may bring about both types of stomatal movement (i.e., opening and closing). The difference in the effects of apoplastic and cytosolic calcium is under critical investigation [30].

4. Stomatal Behavior in Vitro

Abnormal stomatal behavior has been found under certain conditions. For example, stomates on the floating leaves of many water plants and plants growing under conditions of permanent high humidity (e.g., the canyons of West Java) have been found to remain permanently open. Nonfunctional stomates also have been observed on leaves of plants cultured in vitro. Stomates of such plants do not respond, or respond to a smaller extent, to factors that are able to induce stomatal closure. In stomates of plants cultured in vitro, so-called vitreous plants, the abnormal response is predominant [32]. The conditions that lead to vitrification are liquid media, low VPD of the vessel, and application of cytokinin. It is important to determine the conditions that can offset vitrification.

IX. CONCLUSION: TRANSPIRATION VERSUS PHOTOSYNTHESIS

The dual role of stomates in transpiration and photosynthesis has evoked compassionate sentiments, particularly about plants surviving in unfavorable desert biomes. Professor Evenari has described the plight of desert plants as being killed by either Charybdis or Scylla. Professor Stocker found the life of desert plants analogous to navigation between two cliffs—the dangers of dehydration and starvation. The tragedy of desert plants indeed becomes explicit when it is realized that stomates must choose between food and water. These ideas have constituted the basis of modulation of water use efficiency and steady state stomatal responses in contrasting environments. To close on an optimistic note, we quote from Farquhar and Sharkey [33]:

> In general, the responses we have examined are qualitatively consistent with a role for stomata minimizing plant water loss, while only marginally limiting carbon gain. It appears that stomatal inhibition of photosynthesis is usually slight, whether the plants have C_3 or C_4 metabolism and whether or not they are stressed.

It is common knowledge that water resources are not infinite and that the planet is under the threat of thermal catastrophe due to the rise in atmospheric CO_2. If stomatal action can make a contribution in these areas via a fine-tuning of guard cell physiology respect to the environment of plants, the challenge is worth enduring.

REFERENCES

1. H. Walter, *Ecology of Tropical and Subtropical Vegetation*, Oliver and Boyd, Edinburgh (1971).
2. Arvind Kumar, Stomatal regulation and water economy in desert plants, Ph.D. dissertation, University of Jodhpur, Jodhpur, India (1982).
3. C. B. Osmond, O. Björkman, and D. J. Anderson, *Physiological Processes in Plant Ecology: A Synthesis with Atriplex*, Springer-Verlag, Berlin (1980).
4. H. L. Shantz and L. N. Piemeisel, *J. Agric. Res.*, 34: 1093 (1927).
5. J. P. Srivastava and S. N. Chaturvedi, *Ann. Arid Zone*, 28: 257 (1989).
6. P. J. Kramer, *Water Relations in Plants*, Academic Press, New York (1983).
7. J. T. Martin and B. E. Juniper, *The Cuticle of Plants*, Edward Arnold, London (1970).
8. K. Haas and J. Schonherr, *Planta*, *146*: 399 (1979).
9. E. A. Baker, *Plant Cuticle*, (D. F. Cutler, K. L. Alvin, and C. E. Price, eds.), Academic Press, London, p. 139 (1982).
10. D. J. Johnson, R. A. Richards, and N. C. Turner, *Crop Sci.*, *23*: 318 (1988).
11. C. E. R. Pitcairn, C. E. Jeffree, and J. Grace, *Plant Cell Environ.*, *9*: 191 (1986).
12. J. A. Clark and J. Levitt, *Physiol. Plant*, 9: 599 (1956).
13. K. Raja Reddy, J. V. S. Rao, and V. S. Rama Das, *Indian J. Plant Physiol.*, *25*: 55 (1982).
14. M. M. Ludlow, *Water and Plant Life* (O. L. Lange, L. Kappen, and E. D. Schulze, eds.), Springer-Verlag, Berlin, p. 364 (1976).
15. C. B. Osmond, *Annu. Rev. Plant Physiol.*, *29*: 379 (1978).
16. G. G. J. Bange, *Acta Bot. Neerl.*, 2: 255 (1953).
17. D. M. Gates, *Annu. Rev. Plant Physiol.*, *19*: 211 (1968).
18. R. O. Slatyer, *Plant Water Relationship*, Academic Press, New York (1967).
19. G. Hygen, *Physiol. Plants*, 6: 106 (1953).
20. B. Slavik, *Methods of Studying Plant Water Relations*, Springer-Verlag, Berlin (1974).
21. R. O. Slatyer and S. A. Taylor, *Nature*, *187*: 922 (1960).
22. J. Coombs, D. O. Hall, S. P. Long, and J. M. O. Scurlock, *Techniques in Bioproductivity and Photosynthesis*, Pergamon Press, Oxford (1985).
23. D. K. McDermitt, *Hortic. Sci.*, *25*: 1538 (1990).
24. S. Lurie, *Structure, Function and Ecology of Stomata* (D. N. Sen, D. D. Chawan, and R. P. Bansal, eds.), B. Singh and M. Singh, Dehradun, India, p. 61 (1979).
25. F. L. Milthorpe, N. Thorpe, and C. M. Willmen, *Structure, Function and Ecology of Stomata* (D. N. Sen, D. D. Chawan, and R. P. Bansal, eds.), B. Singh and M. Singh, Dehradun, India, p. 121 (1979).
26. G. W. Scarth, *Plant Physiol.*, 7: 481 (1932).
27. O. L. Lange, R. Losch, E. D. Schulze, and L. Kappen, *Planta*, *100*: 76 (1976).
28. J. I. Schroeder, *Stomata '89, No. 13*, Humboldt Universität, Berlin (1989).
29. A. Laisk, V. Oja, and K. Kull, *J. Exp. Bot.*, *38*: 126 (1980).
30. T. A. Mansfield, M. A. Hetherington, and C. J. Atkinson, *Annu. Rev. Plant Physiol. Plant Mol. Biol.*, *41*: 55 (1990).
31. W. Beyschlag, and H. Pfanz, *Oceologia*, *82*: 52 (1990).
32. S. Koshuchowa, K. Zoglauer, and H. Goring, *Stomata '89*, No. 13, Humboldt Universität, Berlin (1989).
33. G. D. Farquhar and T. D. Sharkey, *Annu. Rev. Plant Physiol.*, *33*: 317 (1983).

4

Role of Temperature in the Physiology of Crop Plants: Pre- and Postharvest

William Grierson
University of Florida, Lake Alfred, Florida

I. INTRODUCTION

A. Importance of Temperature

Temperature, like the poor, is always with us but, like the poor, it is only too often overlooked. This is unfortunate, as temperature is a major factor in all things biological.

To a physicist, temperature is simply a manifestation of the kinetic energy of component atoms, ions, and molecules.

To a chemist, the role of temperature is epitomized by the "Q 10 rule" whereby, over some reasonable range, the rate of a chemical reaction approximately doubles with every 10°C increase in temperature.

But to a biologist, temperature is the supreme conductor of the orchestra of life, initiating specific reactions and modulating, integrating, or suppressing them just as the conductor of a great orchestra calls upon, modulates, or dismisses the diverse instruments, whose discrete voices are thereby integrated into one harmonious whole. Regardless of the crop, or of the physiological response being monitored, consideration of the role of temperature can often be the sine qua non in interpreting the phenomena being investigated. It has been said that "The scientist shows his intelligence . . . by his ability to discriminate between the important and the negligible" [1]. Only too often, temperature may appear to be a negligible factor when, unnoticed, it plays some critical role.

B. Scope of This Chapter

Since every physiological and biochemical system of every crop plant is affected by temperature, it would be impossible to cover all its manifestations in a whole textbook, much less in a single chapter. Most aspects are, therefore, dealt with superficially. Specific examples are cited to indicate the types of relationship that invite further study and, when such study does not suffice, may inspire further research.

Citrus fruits, and most particularly the chilling injury (CI) syndrome, are represented in greater depth because the writer and his colleagues devoted many years to research on citrus, particularly the study of the basic mechanisms of chilling injury.

C. Definitions

1. Temperature

Temperature per se does not need definition. Not all the work cited, however, deals with the temperature of the actual plant tissues involved. Often "temperature" refers to that recorded for the immediate vicinity of the plant or organ.

2. Crop Plants

Crop plants are taken to be any grown for profit or pleasure, thus including ornamental plants grown for either indoor decoration or outdoor landscaping. No attempt has been made to include nondomesticated species.

II. ECOLOGICAL ROLE OF TEMPERATURE

Temperature obviously limits the geographical areas in which various crops can be grown. However, temperature per se is often not the only determinant: the effects of temperature extremes usually are associated with other factors such as availability of water, prevalence of high winds, and the duration and intensity of sunlight (insolation). An important aspect, as discussed below, is that limitations imposed by extremes of temperature differ sharply for annual versus perennial crops.

A. Extremes of Temperature

1. High Temperature Limitations

The limiting effect of high temperatures on crop production takes two principal forms: limitation of vegetative growth and adverse effects on fruit setting. Vegetable crops subject to very high transpiration losses, such as asparagus, lettuce, and all the *Brassica* species (cabbage, cauliflower, broccoli, brussels sprouts, etc.) are obviously limited by the excessive transpiration concurrent with exposure to extremely high temperatures. Tomato (*Lycopersicon esculentum* Mill.) is the quintessential example of a crop for which very high temperatures limit fruit setting. (In this regard, the small-fruited "cherry tomatoes" are more tolerant than the usual commercial varieties.) Plant breeders are having limited success in developing more heat-tolerant tomato varieties because heat and cold tolerance in fruit setting have only moderate heritability and such inheritance is complex [2].

A further complication is that the upper limit for fruit set can be correlated with humidity levels [3]. Successful breeding of truly heat-resistant tomatoes may well turn out to depend on the physiologists and biochemists more exactly defining the influence of temperature and humidity on the hormonal systems controlling anthesis, pollen tube activity, ovule receptivity and, in some instances [2], parthenocarpy. A recent press account [4] reports that a major U.S. seed company has developed both a tomato and a zucchini that set fruit in temperatures as high as 35.6°C (96°F). For commercial purposes, assuming that the report is correct, this "high temperature fruit set" will have to be incorporated into varieties having commercially acceptable yield and eating quality.

Very high temperatures can also limit fruit setting of citrus fruits. In this case, intensity of insolation appears to be another limiting factor, since flowers within the leafy canopy, protected from direct exposure to sunlight, will usually set some fruit [5]. A less subtle effect of extremely high temperatures on fruit set of citrus is the "burning" or "scorching" of blossoms, particularly on young trees, that is occasionally reported from desert areas such as southern California, Arizona, and the Negev of Israel. Even without such drastic effects, fruit set of navel oranges is reported to be sharply affected by temperatures during the bloom period [6].

A high temperature effect causing no visible symptoms is a cessation of growth even though nutrients and soil moisture are adequate, as reported for citrus trees during very hot weather in Arizona [7].

2. Low Temperature Limitations

The obvious limitation imposed by low temperature is killing of plant tissues by freezing. Most plant tissues can be destroyed by freezing temperatures suddenly imposed during a period of rapid growth. Some plants, given sufficient time under suitable conditions, can adapt themselves to freezing temperatures, and some cannot. This dichotomy is discussed in Section II. B.

3. Freezing of Plant Tissues

A more specific effect is the response to brief periods of freezing, or near-freezing, temperatures. The classic example, feared by fruit growers almost everywhere except in the tropics, is a freeze while the trees are in full bloom. This is much more drastic for deciduous fruit trees than for evergreen trees such as citrus. If the blossom-bearing wood is not damaged, such tropical or subtropical trees have a chance to replace fruit buds within the same bearing season, although yield and fruit quality may be impaired. As discussed below, this cannot happen with deciduous fruit trees.

A more subtle effect, to which green (English, garden) peas (*Pisum sativum*) are particularly susceptible, is low temperature stunting of young plants. When such peas and snap (wax) beans (*Phaseolus vulgaris*) are growing side by side, immature pea plants may be permanently stunted by a brief chilly period from which the beans usually recover.

4. Microclimates

It is apparent to even the most casual observer that on a frosty night, cold air can drain into hollows, thereby sometimes limiting damage to such small "microclimate" areas. In addition, vegetation can be markedly different on the north and south sides of a steep valley because the exposures to sunlight are very different. Foehn winds provide striking examples of rather larger microclimates utilized for the growing of specialized crops. A classic example is the chinook of the Rocky Mountains of Washington state and British Columbia. Strong winds off the Pacific Ocean are forced to rise on encountering the coastal range. As the air rises rapidly, moisture condenses, releasing great amounts of latent heat and forming a bank of clouds (the "foehn wall") that drenches the western slopes. This sequence of events provides a mild, moist area ideal for such crops as cane fruits, crucifers, and many ornamentals. By the time the air mass has crossed the coastal range, it is very dry, and on its leeward descent adiabatic compression warms it rapidly, providing a sudden spring. The resultant microclimate is (provided irrigation water is available) ideal for the growing of stone fruits. Apricots are particularly well served by this microclimate because they have a very short rest period, with consequent susceptibility to spring frosts, which are virtually unknown in inland chinook areas. The chinook occurs on such a grandiose scale as to almost exceed definition as "microclimate." But the eponymous foehn winds in the Austrian Alps, the ghibli in the Tripolitanian Mountains of Libya, and the zonda in the Argentine Andes produce the same effects on a much more local scale.

The writer's master's thesis [8], dated 1940, includes a map of a microclimate area once known as the "fruit bowl of Canada." Thirty-five miles (56 k) long at its maximum and varying in width from 5 to 14 miles (8–22 km), the fruit-growing area of the Niagara peninsula once produced most of Canada's peaches, plums, cherries, pears, and small fruits, and virtually all the wine grapes of eastern Canada. A high cliff (the Niagara escarpment) shelters this area on the south side. On the north, Lake Ontario moderates the temperature of the north winds in midwinter. In spring, the escarpment protects the orchards from unseasonably warm south winds that might induce too early a bloom, with consequent risk of a blossom freeze. Now, more than 50 years later, it is sad to return to the once overflowing "fruit bowl": this precious miracle of microclimate has been largely paved over with factories, shopping centers, and housing developments that could just as well have been located a few miles to the south, above the escarpment. Such squandering of invaluable microclimates is all too common everywhere.

But microclimate effects can also manifest themselves in far more subtle ways, often involving vertical as well as horizontal temperature differences. When air temperatures are favorable for growth, it is easy to forget that soil temperatures can also be limiting. Soil temperatures, both above and below optimum range, have been shown to limit uptake of soil water by citrus trees to the extent that visible wilting occurs even when soil moisture is adequate [9]. When water uptake is limited, obviously the uptake of water-soluble ions can be affected also. Iron deficiency chlorosis of citrus trees has been reported to be exacerbated by soil temperatures below 12.8°C [10]. Such ion uptake limitation also can be critical in nutrition experiments in which air temperatures are ignored. This relationship was confirmed in a controlled environment experiment with six varieties of spinach (*Spinacia oleracea*). The nitrate content of six cultivars of spinach grown at temperatures from 5 to 25°C varied significantly, not with whether nitrogenous fertilizer was applied, but with the growing temperature [11]!

Hazards from soil pathogens can depend directly on soil temperatures. All Florida citrus seemed to be doomed by a mysterious "spreading decline" until it was found that the cause was a nematode (*Radopholus similis*) that could be cultured only at subsoil temperatures. Because Florida laboratory temperatures normally exceed those of the soil below about 30 cm, cultures from diseased roots processed at ambient temperatures never indicated that *R. similis* was the causal agent [12,13].

5. Annual Versus Perennial Crops

Temperature limitations differ sharply for perennial and annual crops. For perennials (largely tree, vine, and bush crops, various grasses, and other pasture crops), ecological limits usually are set by *winter* temperatures. Few species are hardy enough to survive subarctic extremes of winter cold. In the tropics, the need for a cool winter rest period limits the cultivation of pome (e.g., apple and pear) and most drupe (e.g., peach, plum, cherry, apricot, almond, walnut, pecan, olive) fruits. Coconut (which botanically is a drupe with a desiccated mesocarp and liquid endosperm) is a conspicuous exception. Conversely, the lack of winter freeze hardiness limits the potential growing areas for purely tropical fruits (banana, mango, avocado, durian, mangosteen, etc.), tropical ornamentals, and purely tropical grasses, including sugarcane.

This set of limits is in sharp contrast to those applicable to purely annual crops such as almost all vegetables and grains, and annual flowers, for which *summer* temperatures are critical. All these annual crops require is about 3–5 months of suitable growing weather. Vegetables grow luxuriantly in the warm, long summer days in Alaska; the subarctic winters are of no consequence for them.

B. Various Interactions with Temperature

In the years immediately prior to World War II, the writer was a young graduate student in Canada working on storage and ripening of pears. At that time, it was customary for Canadian housewives to put fruits on sunny windowsills to ripen them. Since it seemed illogical that light should hasten ripening, I decided to put a row of unripe pears on the laboratory windowsill and cover half of them with a black cloth. Fortunately, I checked pulp temperatures: those under the black cloth were several degrees warmer. Then I tried shading with a white-painted board. Better, but still quite a difference. By the time the next year's pear crop came in, I was in uniform on the other side of the Atlantic. I never did return to the sunlight-pear-ripening problem, but have ever since been acutely aware that one way or another, temperature can be an interactant, wanted or not, in a great deal of plant research.

1. The "Day/Degrees" Concept

A very useful concept for expressing heat units is "total day degrees": that is, the accumulated number of days (or sometimes hours) above a certain base temperature. Another version is the accumulated sum of diurnal maximum temperatures times the number of days. For the reverse (cold units), the usual figure is the total number of hours below a given temperature, such as 40°F or 5°C. The usefulness of such methods is not helped by overreliance on statistical analysis of findings based on an initial arbitrary decision. In the United States, for example, 40°F (4.4°C) or 45°F (7.2°C) have been common baseline temperatures for determining chilling hours. As the Fahrenheit scale is abandoned in favor of Celsius, 5 and 7.5°C are more likely to be used. With such baseline variations, apparent fine statistical differences can be deceptive.

Peaches afford an excellent example of the use of such methods. Florida peach breeders have very successfully extended the southern limits for commercial production of peaches by breeding "200 hour" peaches and nectarines, in contrast to the 400, even 600, hour peaches grown in districts with cold winters [14]. In more northern states, versions of this day/degrees concept are used to forecast blossom freeze risks for varieties in a given area [15], and date of bloom in others [16]. Readers interested in a highly sophisticated discussion of the mathematics involved are referred to correspondence in a 1991 issue of *HortScience* [17].

2. Freezing of Plant Tissues

As depicted in older texts, freeze injury and freeze resistance were simply explained: in freeze-susceptible tissues, free water froze, forming crystals that disrupted cell membranes, whereas in freeze-resistant tissues the water was bound in the form of hydrophilic colloids. When this model was subjected to modern research, however, little if any of it turned out to be so simple. Interested readers are referred to two excellent reviews [18,19]. Freeze-hardy plants have hormonally controlled mechanisms enabling them to respond to gradual changes in temperature and day length in preparation for winter. Such changes are obvious with deciduous trees, vines, and shrubs, which shed their leaves, often after having displayed dramatic changes in leaf color. No such highly visible evidence is afforded by conifers, which, nevertheless, also need gradual autumnal climatic changes to induce similar hormone-controlled internal adaptation to prepare for winter [20]. But what of plants that survive a freeze without a prior hardening period? Expressed very briefly, water in certain woody plants can supercool to a surprising extent, although this protective mechanism often is negated by the presence of ice-nucleating bacteria [19]. Such bacteria are by no means ubiquitous, but they are very common and a real factor in freeze injury.

Exposure to freezing but nonlethal temperatures can cause various chemical changes in plant tissue. Only one is mentioned here. It is very common for oranges that survive a freeze to develop white, crystals clearly visible between the segment membranes. These are hesperidin, the principal flavone in citrus fruits and, although their presence sometimes causes alarm, they are completely nontoxic. Up to the 1950s, growers placed much credence on estimations of fruit damage as judged by the amount of hesperidin crystals. This mindset proved quite fallacious [21].

The once apparently simple field of tissue freezing is further complicated by work with detached plant parts. Celery pollen has been stored in viable condition at −10°C for as long as 9 months [22]. The use of "cryoprotectants" has made possible prolonged, very low temperature storage of living tissue for in vitro tissue culture and propagation. Using such cryoprotectants as polyethylene glycol + glucose, and dimethyl sulfoxide, such living material as apices of brussels sprouts [23] and *Rubus* [24] have been rapidly cooled, then held at −196°C until needed for tissue culture propagation.

3. Dormancy, Bud Initiation, and Fruit Setting

Obviously, it is well that autumnal climatic changes prepare perennials of the temperate zone for the rigors of winter. It might seem that if no winter was to be expected, such plants could grow happily in eternal summer. Or so thought the planners of the huge (> 1 hectare under glass) Devonian Gardens, located over a large shopping mall in Calgary, Alberta, Canada. Their concept had been to surround the clientele with familiar summer vegetation in the depths of Calgary's cold, snowy winter. It was a costly error. Deprived of their climate-induced cycle, the familiar native plants became spindly and unthrifty and soon began to die. The thousands of years of evolution that had fitted those plants for the rugged winter of the Rocky Mountain foothills had produced plants that could not do without it. Instead, the native plants had to be replaced with (as nearly as possible) "look alikes" imported from Florida and California [25].

The dormancy of winter-hardened plants is deceptive. Essential physiological and morphological changes are progressing and will do so only at the low temperatures to which evolution has adapted such plants. Spring bulbs (tulips, daffodils, narcissi, Easter lilies, etc.), brought indoors and kept in warm temperatures after flowering, will not bloom again. Such bulbs left in the winter ground (or held in correctly regulated cold storage) undergo histological changes clearly discernable under a dissecting microscope, or even a powerful hand lens. By the time the bulbs are ready to start growing again in the spring, each one contains all the necessary floral parts, minute but discernible. It is by use of a series of very exact storage temperatures that today's scientific flower producers are able to have spring bulbs in bloom timed for such occasions as Mothers' Day and Easter. Such imposed temperature regimes are very precise: there are sharp differences in temperature requirements, not only among genera, but even between individual cultivars [26].

The same thing happens (on a truly microscopic scale) within the fruit buds of deciduous fruit trees and shrubs. This is why, as horticultural students, we could cut apple boughs in late spring, place them in water in a warm building and, apparently miraculously, decorate our Easter dance with apple blossoms. The same phenomenon explains why a blossom freeze wipes out a deciduous tree fruit crop for a whole year. Those blossoms came from fruit buds initiated 10 or 11 months before, which had developed while dormant and apparently inactive during the winter months.

It is very different with citrus fruits. For one thing, fruit buds on deciduous fruit trees are clearly recognizable to anyone cognizant in such matters. Fruit and leaf buds are indistinguishable on citrus trees [27], however, and the initiation of fruit bud development takes place only a few weeks before bloom. The citrus industry, and the literature, usually speak of "dormant" citrus trees, but such dormancy is in no way comparable to that of deciduous fruit trees. "Quiescent" is a far better term. Blooming of quiescent citrus trees is usually initiated by the termination of a long cool spell or drought [28]. The best and most uniform blooms come when mild stresses from cool weather and drought are relieved simultaneously. A mild winter, followed by a warm, moist spring, tends to give a straggly bloom, spread over many weeks, or even months, with consequent poor yield, low fruit quality, and difficult harvesting.

When hormonal control of chilling injury was still a very new theory, a colleague and I sprayed a number of grapefruit trees with various combinations of growth regulators in

November. We definitely affected susceptibility to chilling injury of the fruit harvested in the following fall, although not in any clearly discernible pattern [29]. What was tantalizing about the test was that with one treatment we got a highly significant increase in yield, which we felt we could not publish. Temperatures were so mild that winter that bloom straggled on and on for many weeks—except on one of our growth regulator treatments, for which the bloom was a "snow bloom," on schedule in mid-March. The treatment would become useful only if long-range weather forecasts were so precise that each November they could forecast whether temperatures between November and March would be uniformly, and atypically, mild.

Obviously the occasional chilly spells so resented by winter tourists initiate the hormonal activity necessary for a desirably brief, early full bloom.

Even when fruit trees have bloomed satisfactorily, temperature can be a determinant of whether a good crop will be harvested. Most deciduous fruits need pollination, which is normally done by honey bees. It can be very difficult to get the attention of apple or pear growers whose trees are in full bloom if the temperature suddenly drops below that favored by the bees. If the temperature is not right, the bees just quit flying, and that can mean a very poor crop indeed. Even if the bees fly and pollen is spread, the pollen must germinate and the pollen tube grow down to the ovule, a process that can be severely restricted by unseasonably low temperatures [30]. And even when pollination has been successful, growth of individual grape berries (botanically, grapes are berries) can be restricted both by too high and too low temperatures [31]. Too high temperatures are more likely to affect fruit set of citrus than of deciduous fruits. In California, extremely high temperatures after fruit set can cause excessive fruit shedding of navel oranges [32]. In Florida, trouble is more apt to come from a combination of high temperature and high humidity, resulting in fungal invasion of the fruitlets [33].

Such problems are not limited to dessert fruits. The buying public having developed an unreasoning prejudice against seeds in fruits and vegetables of many types, parthenocarpy has become highly desirable. For some cucumber varieties, parthenocarpy can be induced with sprays of chlorfluorenol—unless the night temperatures are too high. Night temperatures between 16 and 21°C have been reported as favorable, with parthenocarpy very much reduced when the thermometer reaches 21°C [34].

4. Seed Dormancy and Germination

A very helpful specialist in seed science whom I consulted on the preparation of this chapter sent me, in addition to various published papers, a page-long list (which he considers incomplete) of textbooks, symposia, and so on dealing with the handling and storage of seeds. With temperature so often a critical factor in storage and germination of seeds, this account can be only the briefest of introductions for the nonspecialist.

An important temperature-related difference should be noted between seed-bearing plants of the temperate zone and those originating in the tropics or subtropics. In areas that experience killing winter freezes, seeds *must not* germinate until the following spring. Exceptions to this principle are seeds of plants that bloom early enough in the spring to be able to establish mature plants before the onset of winter. The dandelion (*Taraxacum officinale*) is a familiar, and usually unwelcome, example. Seeds of plants that evolved in tropical areas need no such protective device and so usually (but not always) can be germinated immediately on separation from the plant [35]. The lack of true seed dormancy severely limited the spread of many tropical species when they were first discovered by early European explorers. Even in modern times, dispersal of such crops as cocoa (*Theobroma cacao* L.) has been difficult because the seeds not only are adapted to immediate germination, but they are highly susceptible to chilling injury (see below) if held in cold storage.

A word on terminology: "dormancy" for seeds is used much as it is found in discussions of buds, bulbs, and so on. Seeds that will not respond to usually effective treatments are said (most appropriately) to be "recalcitrant." Some authorities designate as recalcitrant only seeds that do not survive dessication [35]. Such distinctions are, however, beyond the terms of reference of this chapter, which is limited to the effects, direct or indirect, of temperature. "Stratification" is used (not very logically) for chilling treatments to break dormancy. Perhaps this comes from the old custom of filling a box with alternate layers of sand and seeds from peaches (or other stone fruit) and setting it outside, exposed to the coldest possible weather.

For some seeds it has been demonstrated that dormancy is purely mechanical, being enforced as long as the tough impermeable testa is intact [36]. In this regard, it used to be argued that hard freezing only splits the peach pits, thus mechanically releasing the seed to germinate. Our pomology lecturer at the Ontario Agricultural College, Guelph, settled this for us more than 50 years ago. At his direction, we compared germination of "stratified" peach pits from the preceding year with that of fresh peach pits we had carefully cracked. The result was quite fascinating. The seedlings from the stratified seeds were normal. Those from the fresh, but mechanically cracked, seeds resembled tiny pineapple plants, producing leaves with no internodes. Prolonged cold temperatures (most effectively between 2 and 6°C) are definitely essential in such "stratification."

Various treatments (such as presoaking) to encourage emergence of seeds used to be called "vernalization," presumably because it hastened the effects of spring. The term was brought into disrepute by claims of permanent genetic changes by the Soviet charlatan Trofim Lysenko [37]. Today, "priming" is appropriately used for seed treatments (involving temperature, solutes, etc.) in wet or dry media to accelerate germination. But if seeds have been primed, subsequent permissible holding temperatures may be affected. Primed tomato seeds have been reported to retain viability at 4°C, but at 30°C they deteriorated within 6 months [38]. Similarly, primed tomato seed was reported to retain viability at storage temperatures as high as 20°C for 18 months. However, the seed degenerated at 30°C, particularly when primed with potassium nitrate rather than with polyethylene glycol (PEG) [39].

Priming does not necessarily overcome adverse weather conditions, as shown by 3 years of unsuccessful trials with primed sugar beet seed in cold Idaho spring weather [40]. Current research developments, however, promise to overcome these ill effects of too early sowing when they are due to a combination of moisture imbibition and too low temperature. A recent review article [41] reports success in such circumstances when seeds of table beet (*Beta vulgaris* L.) were primed with PEG.

Imbibitional chilling injury is of particular concern for seeds of plants of tropical origin, such as cotton, corn (maize, *Zea mays*), tomato, and many legumes, which are susceptible to chilling injury. For their seeds, the onset of CI is related to rate of water uptake [41]. Treatment with materials (such as PEG) that delay imbibition can be helpful but is not temperature specific. This problem appears to be surmountable by use of temperature-sensitive polymeric seed coatings that become permeable to water at specifically selected temperatures [41,42].

Too hot temperatures can also impede germination. Florida celery growers have been able to surmount this problem by using high temperature (30°C) priming in a solid matrix of calcined clay [43].

Recalcitrant seeds occur in all climates, and temperature can be a factor in achieving successful germination. Wild rice (*Zizania palustris*) is an excellent example. Deeply dormant at harvest, it will not germinate without prolonged cold treatment [44]. It is thus perfectly adapted to self-propagation in the Minnesota wetlands and as a food staple for Native Americans, who have depended on it over the centuries. Some of wild rice's reputation as a "recalcitrant seed" involves a supposed desiccation intolerance, but this misjudgment has been related to failure to understand the "novel relationship between seed viability, temperature, and moisture content" [45].

An interesting form of recalcitrance in tropical seeds is that some, such as kola (*Cola nitida*), must be aged for as long as 7–11 months, for which ambient temperatures are satisfactory [46]. This requirement accounts for how, for many centuries, the highly valued, but frail, caffeine-rich kola "nuts" (caffeine being a stimulant not prohibited to Muslims) were traded all over West Africa, wrapped in damp leaves and transported for weeks on the heads of slaves [47]. Such aging of tropical seeds is not necessarily completely temperature independent. Seeds of *Plantago ovata* (an annual herb grown in India), although completely recalcitrant at harvest, germinated freely after a single day at 15°C, plus treatment with gibberellic acid (GA3) [48].

A record for temperature-related recalcitrance is held by American ginseng (*Panax quinquefolius*). It is no wonder that this wild herb has been hard to domesticate: it is reported [49] to need cool–warm–cool stratification over a period as long as 540 days—18 months!

Temperature may or may not prove to be important in the storage and germination of a particular type of seed, but it can never be ignored as a possibly critical factor.

5. Temperature-Induced Ethylene Effects

Ethylene (C_2H_4) is the universal growth regulator. Until the advent of the gas chromatograph, it was believed that biosynthesis of ethylene was confined to certain plant tissues (e.g., apples) and was not present in others (e.g., oranges). As analytical equipment improved, it became apparent that under various forms of stress any plant tissue can produce ethylene, and the extent of this effect is temperature dependent [50]. Among the more striking temperature-induced effects of endogenous ethylene are the "fall colors" in deciduous woodlands, which result from the reaction of ethylene with plant pigments.

Bright colors are not only attractive but, as long as consumers insist on relying on their eyes rather than their taste buds, they can be very valuable. Thus, temperature-modulated ethylene effects become essential tools in the marketing of certain fruits and vegetables. Citrus fruits afford an excellent example. Citrus fruits grown at sea level in the humid tropics, where the species originated, are all green: no brilliant oranges or yellows gleaming amid the jungle foliage.

But for centuries, citrus fruits have been grown in cooler, usually more arid, areas, principally around the Mediterranean Sea. There, the considerable stress of cool nights on a tropical fruit forces production of minute amounts of ethylene, with consequent loss of chlorophyll and development of carotenoids. Thus, we have the obvious "fact" that oranges should be orange and lemons should be yellow. This consumer prejudice presents citrus growers in milder climates such as Florida and Brazil with a very real, temperature-induced problem. In such districts, early varieties may mature and pass their optimum maturity without ever developing "typical varietal color."

It has long been axiomatic among Florida citrus growers that their fruit would not change color without "a week of cool nights" (which in many years comes after the early varieties are over). A 1942 study confirmed this [51]. No significant color break was observed as long as night temperatures were above 55°F (12.8°C), and a week of nights below 50°F (10°C) resulted in good orange color on early varieties of oranges. Grapefruit, however, responded to the stress of low night temperatures much less predictably.

In California, an ingenious experiment studied the effect of temperature on the coloring of Valencia (late) oranges under controlled conditions. Fruit-bearing branches were grafted onto young potted rootstocks, and air and soil temperatures were controlled separately [52]. Both variables were found to affect fruit color, the best orange color being achieved with 7°C soil temperature and 20°C air temperature. Internal analyses found no correlation between fruit color and fruit maturity.

Various attempts by this author to reproduce such temperature-induced color changes with detached fruit have been unsuccessful. Once the fruit has been detached from the tree, exogenous ethylene must be supplied and the effect is, again, sharply temperature dependent, but with a relationship quite different from that observed for attached (nonpicked) fruit. In an early Florida study [53], we found a very sharply defined optimum for chlorophyll destruction in oranges at 85°F (29.4°C) and a very ill-defined optimum for grapefruit at approximately the same temperature. Such ethylene "degreening" had no apparent effect on carotenoids; the degreened oranges were pale yellow. California packinghouses that commonly degreened at 75°F (23.9°C) reported development of a deep orange color, but the process took 8–10 days, a prohibitive period in Florida because of endemic stem-end rot (caused by *Diplodia natalensis*), which is strongly stimulated by ethylene.

Nearly 20 years after the Florida work just described, the carotenoid development/chlorophyll destruction effect was studied in detail with very much more sophisticated equipment [54]. This time ethylene-induced carotenoid accumulation was shown to be (a) temperature sensitive and (b) inhibited at 30°C and above. The work was continued and showed that very high levels of specifically identified carotenoids could be achieved with concentrations of ethylene as low as 0.1 ppm. However, induced carotenoid development took weeks, rather than days, hence was commercially unacceptable in a stem-end rot district.

I still do not know why, prior to picking, cool (below ca. 12°C) temperatures are necessary to destroy chlorophyll in the peel of citrus fruits, but warm (ca. 30°C) temperatures maximize the rate of ethylene-mediated chlorophyll disappearance after picking. This paradox does, however, emphasize something that is too often ignored or forgotten: prior to picking, a fruit is an integral part of the physiology of the plant as a whole.

6. Temperature and Fruit Quality: Preharvest

There is no point in producing fruits commercially unless they are palatable, and in some instances palatability is strongly related to growing temperatures. Again, a citrus fruit, grapefruit (*Citrus paradisi*), will serve as a prime example, not so much because of its place in this writer's past research, but because the internal and external qualities of grapefruit have been extensively studied. All growing districts base their quality standards on what they do best [55], and since Florida's climate is so unsuited to the production of grapefruit with a bright, colorful exterior, standards have been developed largely based on internal quality. These are expressed in terms of sugar (as degrees Brix), acid (as ratio of citric acid to Brix), and juice volume (as cubic centimeters per fruit) [56,57]. Internal quality obviously varies widely among growing districts, leading to some totally unprofitable studies in day/degree relationships. But even the most casual observations make it apparent that districts famed for the high quality of their grapefruit (such as the Rio Grande Valley of Texas and the Indian River district of Florida) are areas with warm winter nights, during which growth of the tree and of the fruit can continue uninterrupted. A controlled climate experiment with Redblush grapefruit in Florida confirmed this. Maximum internal quality was found in fruit from little trees, which were grown where night temperatures were not allowed to drop below 21°C [58].

The shape of grapefruit is very sharply associated with internal quality: the flatter the fruit, the higher the internal quality. The influences of day and night temperatures, and of day length, were studied under controlled conditions [59]. A 32/7°C (day/night) temperature regime produced severely "sheep-nosed" fruit of very low internal quality. A 32/24°C temperature regime produced flat fruit (axis length < diameter) of high internal quality. No correlation between fruit quality and day length was found.

7. Wound Healing: Temperature × Humidity × Time

Some plant products have considerable ability to heal mechanical lesions after harvest. The ability depends on certain ranges of temperature and humidity, however, and the healing takes several days to complete. It has long been known that both sweet potatoes (*Ipomoea batatas*) and so-called Irish potatoes (*Solanum tuberosum*) can heal damage to their own tissue [60]; for this reason, it is advised that potatoes be harvested, then held for several days at ambient (or higher) temperature and very high humidity before being placed in cold storage, because such healing occurs only at high temperatures and humidities [61]. Similarly, when seed potatoes are cut into "planting pieces," they should be "cured" for several days prior to planting under the warmest conditions available. During this period of comparatively high temperature, a layer of suberized cells forms over the wounds.

A much more recent finding is that citrus fruits can heal shallow wounds into the flavedo (colored part of the peel), but only at very high humidity (ca 95% RH) and temperatures as high as 28–29°C (which, fortunately, are the conditions recommended within Florida citrus degreening rooms). An unusual aspect of this healing of citrus fruits is that it involves lignification, not suberization, and it is associated with sharp increases in phenolic compounds and of the enzyme phenylalanine ammonia-lyase (PAL) [62].

In both these types of healing, the role of comparatively high temperatures is critical. Such wound healing should not, however, be confused with drying treatments, which are essentially catabolic, rather than anabolic. The "curing" of onions prior to storage is an example of drying. The curing process aims at killing the outer layers of cells by heat and desiccation, a form of localized necrosis that would be disastrous with living products of most other types.

Attention is again drawn to the different physiological responses of plant organs on and off the mother plant. After a Florida hurricane, attached citrus fruits will heal severe wounds and continue to grow to maturity at normal ambient temperatures although badly scarred. Fruits with similar injuries that become detached from the tree promptly rot. Various forms of squash (*Curcubita* spp.) carved with a gardener's initials when immature will grow to maturity with the initials as prominent scars. Any such wounds inflicted on detached fruits would cause decay.

C. Temperature × Light Interactions

A factor that is easily overlooked in determining optimum temperature for a given response is light, which may play either a positive or a negative role.

Modern apple orchards are often based on clonal rootstocks rather than on seedling roots. The rootstocks must be rooted from cuttings, which is not always easy, and light can be a complicating factor. Rooting of M-26 clonal rootstock has been reported to be maximum at 25°C, but only in the absence of light, which may inhibit rooting [63].

The prospect of establishing life support systems in space has led to the prospect of crop production under controlled conditions not necessarily corresponding to those in terrestrial horticulture. One such study with lettuce (*Lactuca sativa*) found that maintaining a constant day/night temperature at 25°C maximized growth, but only with intensified light during the "day" period [64].

Many plants are known to respond sharply to photoperiod (a misnomer: it is the period of unbroken darkness, not of light, that is controlling). A study of the effect of photoperiod on the growth of West Indian mahogany (*Swietenia mahagoni*), grown in southern Florida as an ornamental, found that its typical response to photoperiod was inhibited by low temperatures atypical of its native tropics [65].

The relationship between temperature and photoperiod and flowering of traditional ornamentals such as *Chrysanthemum* is now well understood by both professional and amateur growers. But with the increasingly common introduction of exotic ornamentals, specific responses (to temperature, light, watering, etc.) must be established for the new arrivals. One such exotic is "kangaroo paw" (*Anigozanthos manglesii*), for which very sharp interactions between day and night temperatures and between temperature and day length control flowering and even mortality [66].

Individual species within a genus may respond quite differently to interactions of temperature and light. A *Peperomia* species imported to Indiana from the Andean highlands was unable to adapt to the double change, in summer, of temperature and photoperiod. Another *Peperomia* species from the lowlands of Ecuador made the transition successfully [67].

Temperature–light interactions are not limited to higher plants. For example, sporulation of some fungi, such as the citrus pathogen *Diplodia natalensis* (*Physalospora rhodina*), needs not only optimum temperature but also exposure to light of high intensity (G. Eldon Brown, personal communication).

A complicating role for light is always a possibility in the investigation of temperature relationships.

D. Temperature Control in Crop Production

1. Microclimate

Greenhouse (British "glasshouse," often a misnomer in this plastic age) production is the obvious example of microclimate temperature control. But greenhouse production has its own considerable expertise and literature. Thus, the examples of greenhouse research cited here are included only to illustrate specific situations in which individual control of air and soil temperatures is important.

Even outdoors, although climate (including temperature) is usually regarded as beyond the control of man, localized temperature control is sometimes effective on a microclimate scale. Vancouver, Canada, is a few miles north of the 49th parallel, about 60 miles farther north than Minot, North Dakota, with its legendary harsh winter temperatures. But constant foehn winds off the Pacific Ocean make Vancouver winters mild and wet, although sunshine is scanty. When I had a garden there in the late 1940s, a neighbor used to say that I "cheated God" to bring in my lettuce and tomatoes earlier than anyone else. The bed in which the vegetables grew was banked toward the south at approximately 50 degrees, and the area between the plants was covered with flat stones, gathered from the nearby beach, to maximize soil heating from the weak late winter–early spring sun.

This management was, of course, an extreme example of microclimate modification for crop production. Nevertheless, it was no more than ingenious growers have done to survive inhospitable climates throughout the ages, as with pre-Columbian Andean potato growers. In recent years, the native peoples of the Andean Altoplano have learned to revive the methods of their ancestors, growing potatoes on high narrow beds at the foot of mountain slopes. On freezing nights, the cold air settles between the raised beds without damaging the aerial parts of the plants, whose subterranean portions are protected by the latent heat of the water accumulated in the troughs between the beds; an ancient example of sophisticated microclimate control.

Poinsettia is typical of an ornamental grown for a specific date; unless the plants are marketable at Christmas, their value drops dramatically. Growth of the plants can be sharply reduced by too cool air temperatures. Maintaining temperatures in a greenhouse in very cold weather is very expensive. However, it has been found that raising soil temperature to 23°C (which is much cheaper to do) could counteract the adverse effects of air temperature as low as 11.5°C [68].

Sometimes the reverse modification is needed. Flowering of *Alstroemeria* (lily-of-the-Incas) was stimulated by cooling the root zone with 10°C circulating water. There was also an interaction with light, supplementary lighting being essential in winter but harmful in spring and summer [69]. A beneficial lowering of root zone temperature explained an anomalous result with azaleas pot-grown outside on either clamshell mulch or black polyethylene. Placing the pots close together increased growth of azaleas in black pots but not in white pots. The beneficial effect was traced to a decrease in root zone temperatures by shading when the plants in black pots were placed close together [70].

Another unexpected root zone temperature effect was traced to the chilling effect of cold greenhouse irrigation water in winter. The effect was noted with roses and chrysanthemums and was sufficient to affect turgidity, stomate opening, and flowering. Such unforseen deleterious temperature effects are particularly easy to overlook when they involve the temperatures of soil, rather than air [71].

Root zone heating usually involves use of expensive fuel. This potential cost was halved in an ingenious system of pumping comparatively warm water from a well 100 m deep, and circulating it through buried pipes [72].

Temperature, of course, affects more than plant growth. It sometimes is necessary to tread a fine line between temperatures optimum for growth and those that initiate or increase fungal attack. This can be a problem for Florida foliage growers in warm weather, as shown in a study of aerial blight (*Rhizoctonia solani*) infection of Boston fern (*Nephrolepis exalta*). Some plant quality had to be sacrificed if potting medium and air temperatures were to be regulated to restrict development of the pathogen [73].

Given sufficient irrigation water, many deserts will blossom as the rose. But sometimes the desert sun is too hot, with consequent potential for crop damage. An obvious remedy is to spray the crop with an overhead irrigation system. The cooling effect of such sprinkling is sharply dependent on initial air temperature. A California study [74], reported the following (the results have been converted from Fahrenheit to Celsius):

Macroclimate temperature	Lowered by
32°	2–3
38°C	≤ 5°C
39°C	≤ 7.5°C

In addition to other benefits, the water spray at 39°C was reported as being successful in reducing excessive "June drop" of small fruitlets. But such spraying of water in extremely hot weather can cause localized injury due to the "lens effect" of standing drops of water on the leaves [75]. "Lens effect" injury can be avoided, and better temperature reduction obtained, by using nozzles that emit a fine mist instead of streams of water [76].*

Microclimate is being modified on a very large scale. Whole hectares are commonly covered with plastic sheeting, which may be black, white, or transparent. Plastic covering may be spread over raised beds, with the plants inserted through holes in the plastic; it may lie over individual rows secured along the sides, with or without some form of framing [77]; or it may be used as "floating row covers," supported by the crop itself and rising as it grows [78]. Sometimes such plastic covering serves essentially for weed and soil moisture control. Often, however, some degree of temperature elevation is sought, and air and soil temperatures are commonly included in research reports. The elevation of temperature under plastic film will depend on both the

---

*A recent report indicator use of such "hot weather misting" to improve color of apples in Washington state.

climate and the type of plastic [79]. In sunny climates, temperature rise may be sufficient to provide effective disinfestation of pathogenic fungi [80].

2. Sunshading

Another form of large-scale microclimate control is by shading. A practice that started as "slat houses" for orchids and "cloth houses" for high quality tobacco has developed into very considerable industries, usually growing ornamentals. A high proportion are foliage plants, grown under coarse-woven plastic material developed to give certain "percentages" of shade. Obviously, any modification of insolation (irradiance) also modifies temperature. It is remarkable that although research reports commonly pay considerable attention to the expression of the exact degree of shade [81], temperature differences often are not mentioned. It can be very helpful to include temperature as a variable, as demonstrated in a study of disease intensity under different levels of shade [82]. Research workers in this field are urged to routinely measure and report the temperature variations that inevitably accompany any modification of irradiance.

Shade conditions can be expected not only to lower daytime temperatures, but also to raise night temperatures, particularly under cold night/clear sky conditions, in which ground-to-sky radiation can cause a very rapid, possibly harmful, drop in temperature near the ground.

3. Freeze Protection

The first, most obvious, and least expensive protection against freeze injury is to select a planting site where injurious freezing is unlikely to occur. Since this is often not possible, freeze protection measures may be necessary. Burning fossil fuels should be regarded as a last resort—the fuels themselves are very expensive, and their use is often environmentally questionable. Only too often, freeze protection methods are ineffective because of ignorance of the following basic thermodynamic and meteorological principles.

1. Cold air will roll down a slope until arrested by some physical barrier, which then forms a "frost pocket."

2. Hot air rises vertically. It cannot be made to move up a slope.

3. Radiated heat travels in all directions uniformly, but only in line-of-sight (straight) lines. Thus, to be warmed by irradiation from a heat source (such as an orchard heater), a plant must be able to "see" the heat source. Since radiated heat, like all forms of radiation, is subject to the inverse square law (i.e., intensity decreases proportionately to the square of the distance traveled), radiation warming decreases sharply with distance from the heat source.

4. The total heat content of a mass of air depends not only on its temperature (sensible heat), but also on its latent heat, the two together approximating its total energy content or enthalpy. Thus, total heat content can be very much greater for moist air than for dry air at the same temperature. Putting this in a different way: air masses at the same atmospheric pressure and conditions of 15°F (–9.4°C) and 100% RH, 20°F (–6.7°C) and 40% RH, and 25°F (–3.9°C) and 0% RH all have the same heat content of approximately 5.5 Btu per pound of dry air (ca. 3 kg · cal kg^{-1}) [83].

5. The latent heat of evaporation is approximately 7.5 times as great as the latent heat of freezing. Thus, when spraying irrigation water for freeze protection (a common practice for Florida strawberries and various other crops), it is essential to freeze at least 7.5 times as much water as is evaporated [84]. In calm or near-calm weather, this is no problem. Continuing to spray after the onset of a brisk breeze, however, can be disastrous. An ingenious application of this principle is to use such evaporative cooling to delay the blooming of fruit trees until the danger of a blossom freeze is over. The blooming of apple trees was delayed by as long as 17 days by use of thermostatically controlled sprinkling whenever prebloom temperatures exceeded 7°C (44.6°F) [85].

6. Smoke from burning oil or other fuel does NOT form a protective shield. It used to be

believed (particularly in California) that "smudge pots" could create a low cloud that reflected heat back to the crop below. It is now known that the smoke particles are not in a size range suited to reflect infrared emissions. It is, however, possible to generate a very fine water fog with droplets of the appropriate size. An added benefit is that any fog droplets that freeze give off latent heat to the surrounding atmosphere.

7. Freezes are classified as "convection freezes" or "advection freezes." Convection freezes occur with calm air and cloudless skies, conditions in which the earth is radiating heat to the sky, with consequent rapid cooling of the air near the ground. For orchard crops, it is beneficial to have bare ground to radiate ground heat into the trees. Weeds or cover crops trap such radiated heat at the expense of the trees. Convective conditions commonly result in atmospheric inversions, in which the lower air is colder than that at 10–30 m above the ground. In such conditions, "wind machines" mounted on tall towers or pylons can be beneficial. Helicopters have sometimes been used to achieve the same effect, particularly to prevent dangerously cold air from accumulating in the "frost pocket" hollows.

In an advective freeze, a wind strong enough to disrupt normal convection patterns freezes crops on the exposed higher ground, with much less freeze injury in the valleys and lowlands. Wind machines are worse than useless in an advective freeze, but rows of heaters placed at right angles to the wind direction can benefit crops for a considerable distance down wind.

A deadly interaction among temperature, humidity, and wind speed can occur in an advective freeze. Tender leaves and shoots can be killed, not by freezing, but by desiccation, if wind speed is high enough when the temperature approaches the freezing point of plant tissues under conditions of very low humidity (which frequently occur).

For further information on methods and principles of freeze protection, readers are referred to an extensive chapter on freeze protection [86].

E. Incidental Effects of Temperature

Old Ecclesiastes said, "Of the making of many books there is no end," and a number of them probably could be written on the incidental effects of temperature. However, only a very few examples can be cited here to indicate how often temperature is an unforeseen or unplanned-for variable.

Temperature can move in mysterious ways, its wonders to perform, through its subtle influence on the activity of growth regulators. As noted above, fruit setting in tomato plants is inhibited by too high temperatures. A role for growth regulators in this high temperature inhibition is indicated by a report [87] that relative levels of gibberellin and auxinlike growth regulators were sharply affected at high temperatures.

On a purely physical basis, temperature can be expected to affect gas diffusion rates, hence rates of photosynthesis and leaf respiration. However, not only can the physical effects of temperature be complicated by the metabolic effects of temperature on rates of photosynthesis and respiration, but such gas exchange is reported to be affected by an interaction between temperature and humidity [88]. Exact control of temperature is routine, but equivalent accuracy in control of humidity can be difficult, and exact simultaneous control of temperature and humidity can be very challenging indeed.

Vegetable transplants usually benefit from hardening by controlled temperature and/or moisture stress before being planted out in the field [89]. This does not appear to be the case for sweet potato transplants, which are vine cuttings rather than seedlings. Transplants held at 13–18°C were reported to have greatly increased vitality, and ultimately higher yields, compared with transplants held at ambient temperature of 26.7°C [90]. (That "26.7°C ambient" temperature is curiously exact and is possibly a translation from "ca. 80°F ambient.")

A particularly intriguing example of an unexpected temperature effect is reported in a study of male sterility in the common bean (*Phaseolus vulgaris* L.) [91]. When, in the course of an atypically cool summer, unexpected fertility was noted in supposedly male sterile plants, research was transferred to growth chambers. A day/night temperature regime of 30/18°C for an average of 12 days was sufficient to cause most unstable steriles to produce sterile buds. Day/night conditions of 18/7°C for an average of 14 days were effective in converting sterile to partially sterile phenotypes. Both temperature-stable and temperature-unstable genotypes were identified; this is an excellent example of valuable research findings achieved by following up on a temperature-related anomaly revealed in a field study.

The literature abounds in such examples. Many mysteries would be elucidated if research workers routinely reported temperatures (whether controlled or not) and included such data in their research reports. Subsequent research workers, if alert to the multitudinous roles of temperature, will then be in a position to carry the research further, perhaps with the advantages of better funding or instrumentation.

III. POSTHARVEST ROLE OF TEMPERATURE

A. Handling, Storage, and Shipping Temperatures

It is all too often forgotten that crops are still alive after harvest. No matter how meticulously grown, most horticultural crops will not realize their full economic or nutritional potential unless handled at suitable temperatures after harvest. How important this is depends both on the frailty of the crop and the time between harvest and consumption or processing. During this period, the importance of temperature and humidity depends very largely on the biological maturity of the plant part being harvested [92]. Temperature control is obviously of more consequence for asparagus than for coconuts. Only a very brief account of the principles involved can be given here. Attention is drawn to the U.S. Department of Agriculture handbook dealing with storage conditions for a very wide range of produce [93]. Most agronomic crops are far less sensitive to postharvest temperatures, but there are exceptions, such as potatoes (see Section III.D).

1. Fruits

Chapter 19 deals with the development and physiology of fruits, which, botanically, can mean any matured plant ovary from a grain of wheat to a watermelon. Thus, the comments here are very brief and are largely confined to temperature relationships of dessert fruits that sometimes are processed but more traditionally are eaten fresh. Bear in mind, however, that many products considered to be vegetables are botanically fruits: tomatoes, green (snap) beans, squash, bell peppers, and cucumbers are all botanically fruits.

Fruits can be classified according to their respiration pattern, as climacteric or nonclimacteric [94]. Soon after harvest, climacteric fruits (e.g., apples, pears, bananas) produce ethylene in quantities sufficient to overcome the antidoting effect of internal carbon dioxide [95]. The result is a rapid rise in respiration rate, at the conclusion of which the fruit is senescent, overripe, and unpalatable. The useful life of a climacteric-type fruit is typically ended by senescence, rather than by decay. Prompt refrigeration is thus critical for climacteric-type fruits. The more the climacteric rise in respiration can be suppressed, the longer the postharvest life of the fruit.

Nonclimacteric fruits (e.g., citrus and grapes) have no climacteric rise in postharvest respiration. At any constant temperature, their respiration rate remains constant. For such fruits, refrigeration functions more to prevent or delay the onset of decay than to lower respiration rate. For any type of fruit, one of the major functions of temperature regulation is to maintain fruit quality. This involves control of dessication, minimization of flavor and texture loss, and prevention of off-flavors.

Selection of optimum storage temperatures for some fruits can be conditioned by susceptibility to chilling injury (see Section III.C). Particularly for long-term storage, avoidance of chilling injury can override considerations of respiration rate or decay.

2. Seeds

Storage temperature, and thus potential storage life, are sharply conditioned by the tolerance of seeds to desiccation. "Orthodox" seeds that will survive desiccation (and often will desiccate on the plant) can be stored at very low (subfreezing) temperatures. "Recalcitrant" seeds that cannot survive desiccation are very difficult to store because they cannot survive low temperatures. These brief remarks oversimplify a complex situation. Readers needing to know more are referred to a very detailed review article by Ellis [35].

3. Other Plant Organs

The urgency of immediate postharvest temperature and humidity control is related to the maturity of the plant part involved [92]. Grain crops, mature root crops, and cabbage are typical of storage organs that enter a resting stage preparatory to winter. Their respiration rate is very low, and thus prompt postharvest refrigeration is of little consequence. Young actively growing tissues, such as asparagus, green peas, and sweet corn, have very high respiration rates that need to be reduced by refrigeration as soon as possible. The same is true of cut flowers, an intrinsically ephemeral product.

There is a tendency to forget the economic consequence of unrestricted respiration rate in crops for processing. Nevertheless, particularly when crops are paid for on the basis of sugar content, excessive respiration rates due to prolonged exposure to high temperature (as with truckloads of oranges waiting in the sun outside a Florida cannery) deplete sugar content, hence the cash value of the product. Even sugarcane stacked in the sun by the roadside after harvest is losing sugar for which the grower would otherwise be paid [96].

B. Prestorage "Curing": Temperature × Humidity × Time

Traditionally, those who handled horticultural crops for shipment or storage were advised to refrigerate as soon as possible after harvest. It is now known that there are marked exceptions to this general rule. One such exception is the group of products that need to be "cured" prior to storage to heal mechanical wounds (see above-Section II. B.7). The outstanding example is sweet potato, for which *Rhizopus* decay in cold storage was often calamitous until it was demonstrated that prior "curing" at ambient (or higher) temperature and very high humidity for several days healed wounds that otherwise would have been invasion sites for *Rhizopus* [97]. The same benefit can occur, although usually to a less marked extent, with other root and tuber crops.

C. The Chilling Injury Syndrome

Perhaps the most intriguing response of plants to temperature is the chilling injury syndrome exhibited by many plants of tropical origin (which include such familiar crops as cotton, soybeans, tomatoes, citrus, and cucumbers, commonly grown in the temperate zone). CI-susceptible plants (and their detached plant organs) are severely injured by temperatures well above freezing. Critical temperatures vary, but typically injury occurs at temperatures below 10°C. Preharvest chilling injury can occasionally be troublesome, particularly with cotton seedlings [98] and mature, but unripe, tomatoes [99]. But CI is particularly important after harvest, not only because of the products lost due to incorrect storage or transit temperatures, but (perhaps more significantly) because of severe limitations on marketing. If Florida grapefruit could be stored and shipped at the same temperatures as Florida oranges, markets for grapefruit growers would be enormously expanded.

The symptoms of CI can be either superficial or metabolic. Superficial effects are typically various forms of peel injury, which may be uniform as (e.g., the darkening of the peel of a banana held in a household refrigerator) or highly irregular (e.g., discrete, necrotic sunken areas of grapefruit or cucumbers, surrounded by healthy tissue).

The metabolic origin of CI is so profound that a remarkably precipient study demonstrated a parallel between behavior of mitochondria in CI-susceptible versus nonsusceptible plants and of mitochondria from poikilothermic (cold-blooded) versus homeothermic (warm-blooded) animals [100].

The tomato is an example of a climacteric-type fruit that is metabolically sensitive to CI. A mature green tomato that has been chilled will never ripen, even when treated with exogenous ethylene.

The literature on CI is dispersed among many types of plants and journals; moreover, research reports often deal solely with individual reactions or systems isolated from ecological considerations. Much of this literature up to 1986 has been reviewed [101].

Nevertheless, this account reviews the 25-year-long series of reports on grapefruit (and occasionally bananas, limes, and avocados, when grapefruit were out of season) at the University of Florida's Citrus Research and Education Center in Lake Alfred. There are several reasons for this duplication.

1. Grapefruit is uniquely suited for CI research in that fruit can be harvested from a single bloom on an individual tree for as long as 8 or 9 months (typically from September to May). Moreover, the same plant (tree) can be harvested year after year. In the last eight seasons (1974–75 to 1981–82), the same 28 trees were randomly picked (north, south, east, and west sides; upper and lower, inner and outer fruit) at 14-day intervals for a total of more than 100 pickings. We know of no comparable testbed material for CI research.

2. A reporting method was developed whereby the results of each individual picking were reduced to a single value, thus greatly facilitating statistical analysis of multiple experiments [102–104].

3. The program both sought immediate commercial results for the Florida citrus industry and provided training in basic research methods for a series of graduate students. Such training involved rigid adherence to the classical scientific method (i.e., constant testing and evaluation of hypotheses), evidence of which approach is singularly missing in many published reports on CI.

The initial hypothesis was that CI involved a breakdown of the respiratory system, resulting in toxic products of incomplete oxidation (typically acetaldehyde), which in turn caused the distinctive peel lesions. (Acetaldehyde was always detectable in the atmosphere around chilled fruit, and application of exogenous acetaldehyde caused superficially similar lesions.) A report that hypobaric (vacuum) storage greatly prolonged the useful lifetime of various products (at their usual recommended storage temperatures) attributed this effect to the continual removal of endogenous ethylene [105]. So we tried hypobaric storage of bananas at chilling temperatures. CI was completely controlled, which we attributed to continual removal of toxic acetaldehyde [106]. This same effect was soon confirmed for limes and mitochondrial respiration of CI-susceptible citrus fruits (limes and grapefruit) versus CI-resistant Florida-grown Valencia oranges [107].

The hypothesis of the breakdown of the respiratory mechanism appeared to be true. (It still does, but it is now regarded as a secondary effect.) In "micro" respiratory studies with 5 mm peel disks, the banana disks always chilled. In tissue culture, less than half the grapefruit peel disks chilled, which corresponded well to the curious pattern of CI-induced peel lesions [108].

An unsolved mystery is why, in fruits such as grapefruit and cucumber, the cells at the periphery of a necrotic lesion collapse and die while the immediately adjacent cells surrounding the lesion remain healthy. Carbon dioxide (a standard respiratory depressant) was found to minimize adenosine 5'-triphosphate (ATP) accumulation (apparent evidence for CI-induced impairment of the ATP/ADP energy transfer system). There was no correlation with CI and levels of three enzymes (pectinmethylesterase, polygalacturonase, and cellulase), which had been suspected of involvement in lesion formation [109,110].

Since "controlled atmosphere storage" has long been commercially used for other products, the effect of CO_2 in suppressing CI was investigated. Two treatments were tested: a prestorage treatment with very high levels (e.g., 25%) of CO_2 and also storage atmospheres developed under differentially permeable plastic films [111–113]. Success in suppressing CI was sometimes notable, but with three disconcerting caveats.

1. The early-season sensitivity to CI, which traditionally had been considered to decrease with increasing fruit maturity, was reappearing in late-season, very mature grapefruit. An alert graduate student, Kazuhide Kawada, found that such late-season susceptibility to CI had been reported in some detail for California grapefruit as long ago as 1936, but researchers had missed the paper because it had been given an inappropriate title [114].

2. Although extremely effective in early and midseason, CO_2 had absolutely no protective effect on grapefruit picked after the new bloom (ca. mid-March).

3. The length of delay between picking and postharvest treatments sometimes had more protective effect than the treatments being compared.

A new hypothesis was clearly called for, and the one produced was twofold: the tree and the fruit had to be considered as a whole (fruit off trees in full "growth flush" obviously behaved very differently from fruit from dormant trees), and the controlling mechanism between tree and fruit had to be growth regulators (GRs). A working hypothesis that CI was promoted by gibberellins and prevented by abscisic acid (ABA) was largely confirmed [115]. ABA, the protective "stress hormone," apparently can be developed either pre- or postharvest. Much of this material has been summarized elsewhere [116]. With this knowledge, it is easy to understand the protective effect of various prestorage treatments, not only for grapefruit [117], but for a wide variety of other products such as CI-sensitive Australian oranges [118] and zucchini squash [119].

D. Anomalous Chilling Injuries

Although the basic principles described above apply to a very wide range of CI-sensitive crops, there are other forms of low temperature injury. Apples are susceptible to a wide range of temperature-related storage diseases that comprise a field of study outside this discussion, with one exception. Apples grown in North America generally tolerate storage temperatures close to freezing point (1–2°C). Apples, even of the same variety, grown in Britain or Northern Europe cannot tolerate such low temperatures, however, and formerly this disadvantage sharply limited their marketing season. Thus "controlled atmosphere" (CA) storage (then called "gas storage") was developed in England in the early 1930s. Initially, CA relied on raising carbon dioxide levels to suppress the respiratory climacteric. Later practice favors lowering oxygen to just above a level that would induce anaerobiosis [120]. Such CA storage has made possible the year-round marketing of apples. I have seen no explanation of why apples from the two sides of the Atlantic should respond so differently to storage temperatures, but the effect is real. Similar differences in response to temperature exist for other products from widely dispersed growing areas. For example, Valencia oranges grown in California and Australia are susceptible to chilling injury during long-term storage and shipment, while those from Florida and Brazil are not.

Potatoes are subject to an important temperature-related storage disorder that can be very costly for manufacturers of such products as potato chips and frozen ready-to-cook french fries. At temperatures below about 5°C, potatoes undergo reversible starch–sugar hydrolysis, which causes potato products to darken when the sugar caramelizes upon exposure to high cooking temperatures. Such discolored products are discounted or are unsalable. If chilled potatoes are held at room temperature for several days, however, the reverse (condensation) reaction will convert the sugar back to starch.

Another anomalous postharvest "chilling" hazard is physical and pathological, rather than physiological. Some products, such as leafy vegetables, celery, and peaches, benefit from "hydro cooling" in refrigerated water. A marked exception is the tomato, which should never be immersed in water cooler than product temperature. The skin of a tomato is virtually impervious; gas exchange is through the porous stem scar. (A drop of molten wax on the stem scar of a green tomato will turn it into a self-contained "controlled atmosphere storage unit," thereby greatly delaying ripening.) When a warm tomato is immersed in cool water, contraction of its internal atmosphere draws nonsterile water in through the porous stem scar, with consequent greatly increased decay hazard [121]. The same problem obviously is possible with other products.

IV. CONCLUSION

With virtually any crop, from seed germination, bud sprouting, or anthesis to harvest, and after harvest to final consumption, temperature plays important, and sometimes unsuspected, roles.

REFERENCES

1. Hans Zinsser, *As I Remember Him*, cited from *Bartlett's Familiar Quotations*, 16th ed., Little, Brown, Boston (1992).
2. W. L. George, Jr., J. W. Scott, and W. E. Splitstoesser, *Hortic. Rev.*, *6*: 65–84 (1984).
3. D. W. Kretchman, *Ohio Agric. Res. Center Summ.*, *26*: 5–6 (1968).
4. Associated Press, *The Tampa Tribune*, Dec. 7, 1992.
5. M. Samedi and L. C. Cochran, *HortScience*, *10*(6): 593 (1976).
6. F. S. Davies, *Hortic. Rev.*, *8*: 129–120 (1986).
7. W. C. Cooper, R. H. Hilgeman, and G. E. Rasmussen, *Proc. Fla. State Hortic. Soc.*, *77*: 101–106 (1964).
8. W. R. F. Grierson-Jackson, The storage and ripening of Bartlett pears, M. Sc. Agriculture thesis, University of Toronto, Aug. 26, 1940.
9. Albert W. Marsh, *The Citrus Industry*, Vol. 3, University of California Press, Berkeley, pp. 235–236 (1973).
10. Homer D. Chapman, *The Citrus Industry*, Vol. 2, University of California Press, Berkeley, pp. 168–169 (1968).
11. Daniel J. Cantliffe, *J. Am. Soc. Hortic. Sci.*, *97*: 674–676 (1972).
12. R. F. Suit and E. P. DuCharme, *Plant Disease Rep.*, *37*: 379–383 (1953).
13. R. C. Baines, S. D. Van Gundy, and E. P. DuCharme, *The Citrus Industry* Vol. 4, University of California Press, Berkeley, pp. 321–345 (1978).
14. W. B. Sherman, J. Rodriguez, and E. P. Miller, *Proc. Fla. State Hortic. Soc.*, *97*: 320–322 (1984).
15. Joanne Logan, D. E. Deyton, and D. W. Lockwood, *HortScience*, *25*: 1382–1384 (1990).
16. Maria C. B. Raseira and J. N. Moore, *HortScience*, *22*: 216–218 (1987).
17. Robert K. Seagel, Steve Wiest, and Dale Linvil, correspondence in *HortScience*, *26*(2): 99–100 (1991).

18. M. J. Burke, L. V. Gusta, H. A. Quamme, C. J. Weiser, and P. H. Li, *Annu. Rev. Plant Physiol.*, *27*: 507–528 (1976).
19. Edward N. Ashworth, *HortScience*, *21*: 1325–1328 (1986).
20. R. Timmis, Pacific Forestry Research Center, Canadian Forestry Service Internal Report BC-35 (1972).
21. W. Grierson and F. W. Hayward, *Proc. Amr. Soc. Hortic. Sci.*, *73*: 278–288 (1959).
22. Vince D'Antonio and Carlos F. Quiros, *HortScience*, *22*: 479–481 (1987).
23. T. Harada, A. Inaba, T. Yakuwa, and T. Tamura, *HortScience*, *20*: 678–680 (1985).
24. Barbara M. Reed and E. B. Lagerstedt, *HortScience*, *22*(2): 302–303 (1987).
25. R. S. Benjamin, *The Tampa Tribune*, Jan. 10, (1993).
26. Jaap van Tuyl, *HortScience*, *18*: 754–756 (1983).
27. Charles E. Abbott, *Am. J. Bot.*, *22*: 476–485 (1933).
28. C. W. Coggins, Jr., and H. Z. Hield, *The Citrus Industry*, Vol. 2, University of California Press, Berkeley, pp. 383 (1968).
29. M. A. Ismail and W. Grierson, *HortScience*, *12*(2): 118–120 (1977).
30. M. Vasilakakis and I. C. Porlingis, *HortScience*, *20*: 733–735 (1985).
31. C. R. Hale and M. S. Buttrose, *J. Am. Soc. Hortic. Sci.*, *99*: 390–394 (1974).
32. Howard D. Frost and Robert K. Soost, *The Citrus Industry*, Vol. 2, University of California Press, Berkeley, pp. 299 (1968).
33. S. M. Southwick, F. S. Davies, N. E. El-Gholl, and C. I. Schoulties, *J. Am. Soc. Hortic. Sci.*, *107*(5): 800–904 (1982).
34. Bill B. Dean and L. R. Baker, *HortScience*, *18*(3): 349–351 (1983).
35. Richard H. Ellis, *HortScience*, *26*: 1119–1125 (1991).
36. Charles B. Heiser, Jr., *Of Plants and People*, University of Oklahoma Press, Norman, pp. 185–186 (1985).
37. Trofim Lysenko, *The Science of Biology Today*, International Publishers, New York (1948).
38. Cosme A. Argerich, K. J. Bradford, and Ana M. Tarquis, *J. Exp. Bot.*, *40*(214): 593–598 (1989).
39. Ana D. Alvarado and K. J. Bradford, *Seed Sci. Technol.*, *16*: 601–612 (1988).
40. Glen Murray, Jerry B. Swensen, and John J. Gallian, *HortScience*, *28*: 31–32 (1993).
41. A. G. Taylor, J. Prusinski, E. J. Hill, and M. D. Dickson, *HortTechnology*, *3*: 336–344 (1992).
42. Ray F. Stewart, U.S. Patent 5,129,180, assigned to Landec Laboratories, Inc., Menlo Park, CA (July 14, 1992).
43. Carlos A. Parera, Ping Qiao, and D. J. Cantliffe, *HortScience*, *28*: 20–22 (1993).
44. D. A. Kovach and K. J. Bradford, *Ann. Bot.*, *69*: 297–301 (1992).
45. D. A. Kovach and Kent J. Bradford, *J. Exp. Bot.*, *43*: 747–757 (1992).
46. Gholahan A. Ashiru, *J. Am. Soc. Hortic. Sci.*, *94*: 429–432 (1969).
47. Roland Oliver, *The African Experience*, Icon Editions, HarperCollins, New York, pp. 137–138 (1991).
48. D. L. McNeil and R. S. Duran, *Trop. Agric. (Trinidad)*, 229–234 (1992).
49. Leonard F. Stoltz and John C. Snyder, *HortScience*, *20*: 261–262 (1985).
50. Mikal E. Saltveit, Jr., and D. R. Dilley, *Plant Physiol.*, *61*: 675–679 (1978).
51. C. R. Stearns, Jr., and G. T. Young, *Proc. Fla. State Hortic. Soc.*, *55*: 59–61 (1942).
52. L. B. Young and L. C. Erickson, *Proc. Am. Soc. Hortic. Sci.*, *78*: 197–200 (1961).
53. W. Grierson and W. F. Newhall, *Proc. Fla. State Hortic. Soc.*, *66*: 42–46 (1953).
54. I. Stewart and T. A. Wheaton, *J. Agric. Food Chem.*, *20*: 448–449 (1972).
55. W. Grierson and S. V. Ting, *Proc. Int. Soc. Citriculture*, 21–27 (1978).
56. W. F. Wardowski, J. Soule, J. Whigham, and W. Grierson, Florida Agricultural Extension Service Special Publication 99 (1991).
57. W. Grierson, *Citrus Ind.* (Bartow, FL), *72*(1): 47–50 (1991).
58. Roger Young, Filmore Meredith, and Albert Purcell, *J. Am. Soc. Hortic. Sci.*, *94*: 672–674 (1969).
59. H. K. Wutscher, *J. Am. Soc. Hortic. Sci.*, *101*: 573–575 (1976).
60. J. W. Eckert, in *Postharvest Biology and Handling of Fruits and Vegetables*, AVI Publishing (Van Nostrand Reinhold, New York), pp. 81–117 (1975).

61. R. E. Hardenburg, A. E. Watada, and C. Y. Wang, U.S. Department of Agriculture Handbook 66 (rev. 1985).
62. M. A. Ismail and G. E. Brown, *J. Am. Soc. Hortic. Sci.*, *104*(1): 126–129 (1979).
63. C. L. Le, *HortScience*, *20*: 451–452 (1985).
64. Sharon L. Knight and Cary A. Miller, *HortScience*, *18*: 462–463 (1983).
65. Timothy K. Broschat and Henry M. Donselman, *HortScience*, *18*: 206–207 (1983).
66. A. Hagiladi, *HortScience*, *18*(3): 369–371 (1983).
67. Charles B. Heiser, Jr., *Of Plants and People*, University of Oklahoma Press, Norman, pp. 155–159 (1985).
68. Harry W. James and Richard McAvoy, *HortScience*, *18*(3): 363–364 (1983).
69. W. C. Lin, *HortScience*, *19*: 515–516 (1984).
70. G. J. Keever and C. S. Cobb, *HortScience*, *19*: 439–441 (1984).
71. William J. Carpenter and H. P. Rasmussen, *J. Am. Soc. Hortic. Sci.*, *95*(5): 578–582 (1970).
72. F. J. Regulski, *HortScience*, *18*: 476–428 (1983).
73. A. R. Chase and C. A. Conover, *HortScience*, *22*: 65–67 (1987).
74. Albert W. Marsh, *The Citrus Industry*, Vol. 3, University of California Press, Berkeley, pp. 272–274 (1973).
75. John W. Gerber, Jules Janick, David Martsolf, Charles Sacamano, Elden J. Stang, and Steven C. Wiest, *HortScience*, *18*: 402–404 (1983).
76. A. D. Mathias and W. E. Coates, *HortScience*, *21*: 1453–1455 (1986).
77. C. S. Wells and J. B. Loy, *HortTechnology*, *3*: 92–94 (1993).
78. C. S. Wells and J. B. Loy, *HortScience*, *20*: 800 (1985).
79. W. J. Lamont, Jr., *HortTechnology*, *3*: 35–39 (1993).
80. T. E. Hartz, J. E. DeVay, and C. L. Elmore, *HortScience*, *28*: 104–106 (1993).
81. James E. Barrett, *HortScience*, *20*: 812 (1985).
82. Ramsey L. Sealey, C. M. Kenerley, and E. L. McWilliams, *HortScience*, *25*: 293–294 (1990).
83. W. Grierson and W. F. Wardowski, *HortScience*, *10*: 356–360 (1975).
84. W. Grierson, *Proc. Fla. State Hortic. Soc.*, *77*: 87–93 (1964).
85. J. L. Anderson, G. L. Ashcroft, E. A. Richardson, J. F. Alfaro, R. E. Griffin, G. R. Hanson, and J. Keller, *J. Am. Soc. Hortic. Sci.*, *100*(3): 229–231 (1975).
86. F. M. Turrell, *The Citrus Industry*, Vol. 2, University of California Press, Berkeley, pp. 338–446 (1973).
87. C. G. Kuo and C. T. Tsai, *HortScience*, *19*: 870–872 (1984).
88. F. Lorenzo-Minguez, R. Ceulemans, R. Gabriels, I. Impens, and O. Verdonck, *HortScience*, *20*: 1060–1062 (1985).
89. W. Grierson, *Beneficial aspects of stress on plants*, *Handbook of Plant and Crop Stress* (M. Pessarakli, ed.), Dekker, New York, pp. 645–657 (1993).
90. Larry E. Hammett, *HortScience*, *20*: 198–200 (1985).
91. S. Estrada, M. A. Mutschler, and F. A. Bliss, *HortScience*, *19*: 401–402 (1984).
92. W. Grierson and W. F. Wardowski, *HortScience*, *13*(5): 570–574 (1978).
93. R. E. Hardenburg, A. E. Watada, and C. Y. Wang, U.S. Department of Agriculture Handbook 66 (rev. 1986).
94. Jacob B. Biale, *CSIRO Food Preserv. Q.* (*Aust.*) *22*(3): 57–62 (1962).
95. Stanley P. Burg and Ellen A. Burg, *Plant Physiol*, *42*(1): 144–152 (1967).
96. J. I. Lauritzen and R. T. Balch, U.S. Department of Agriculture Technical Bulletin 449 (1934).
97. J. I. Lauritzen and L. L. Harter, *J. Agric. Res.*, *33*: 527–539 (1926).
98. A. Rikin, D. Atsmon, and C. Gitler, *Plant Cell Physiol.*, *20*: 1537–1546 (1979).
99. L. L. Morris, Proceedings of the Association of American Railroads, Freight Loss Conference, Feb. 5–7, 1953, pp. 141–146 (1953).
100. James M. Lyons and John K. Raison, *Comp. Biochem. Physiol.*, *24*(1): 1–7 (1970).
101. Albert A. Markhart III, *HortScience*, *21*: 1329–1333 (1986).
102. W. Grierson, *Proc. Trop. Reg. Am. Soc. Hortic. Sci.*, *18*: 66–73 (1974). Reprinted in *Citrus Industry* (Bartow, FL), *56*(12): 15–17, 19, 21–22 (1975).
103. W. Grierson, *Proc. Trop. Reg. Am. Soc. Hortic. Sci.*, *23*: 290–294 (1979).

104. K. Kawada, W. Grierson, and J. Soule, *Proc. Fla. State Hortic. Soc.*, *91*: 128–130 (1978).
105. Stanley P. Burg and Ellen A. Burg, *Science*, *148*: 1190–1196 (1965), *153*: 314–315 (1966).
106. E. B. Pantastico, W. Grierson, and J. Soule, *Proc. Trop. Reg. Am. Soc. Hortic. Sci.*, *11*: 82–91 (1967).
107. E. B. Pantastico, J. Soule, and W. Grierson, *Proc. Trop. Reg. Am. Soc. Hortic. Sci.*, *12*: 171–183 (1968).
108. N. Vakis, W. Grierson, J. Soule, and L. G. Albrigo, *HortScience*, *5*(6): 472–473 (1971).
109. N. Vakis, W. Grierson, and J. Soule, *Proc. Trop. Reg. Am. Soc. Hortic. Sci.*, *14*: 89–100 (1970).
110. N. Vakis, J. Soule, R. H. Biggs, and W. Grierson, *Proc. Fla. State Hortic. Soc.*, *83*: 304–310 (1970).
111. W. Grierson, *Proc. Trop. Reg. Am. Soc. Hortic. Sci.*, *15*: 76–88 (1971).
112. W. F. Wardowski, W. Grierson, and G. J. Edwards, *HortScience*, *8*(3): 173–175 (1973).
113. W. F. Wardowski, L. G. Albrigo, W. Grierson, C. R. Barmore, and T. A. Wheaton, *HortScience*, *10*(4): 381–383 (1975).
114. E. M. Harvey and G. L. Rygg, *J. Agric. Res.*, *52*(10): 747–787 (1936).
115. Kazuhide Kawada, Some physiological and biochemical aspects of chilling injury of grapefruit (*Citrus paradisi* Macf.) with emphasis on growth regulators, Ph.D. thesis, University of Florida (1980).
116. W. Grierson, *Fresh Citrus Fruits*, (W. F. Wardowski, S. Nagy, and W. Grierson, eds.) AVI Publishing (Van Nostrand Reinhold, New York), pp. 371–373 (1986).
117. T. T. Hatton and R. H. Cubbedge, *HortScience*, *18*: 721–722 (1983).
118. B. L. Wild and C. W. Hood, *HortScience*, *24*(1): 109–110 (1989).
119. George F. Kramer and Chien Yi Wang, *HortScience*, *24*(6): 995–996 (1989).
120. Robert M. Smock and A. M. Neubert, *Apples and Apple Products*, Interscience, New York and London (1950).
121. R. K. Showalter, *HortTechnology*, *3*(1): 97–98 (1993).

5

Germination and Emergence

Calvin Chong

Horticultural Research Institute of Ontario, Ontario Ministry of Agriculture, Food and Rural Affairs, Vineland Station, Ontario, Canada

Bernard B. Bible

University of Connecticut, Storrs, Connecticut

I. INTRODUCTION

A seed (zygote) results from the fertilization or union of male and female gametes and is the reproductive structure of a plant. Thus, the regeneration or multiplication of plants from seed is termed sexual. Plants are also reproduced by asexual (vegetative) means from bulbs or pieces of stem, root, or other plant part [1].

A seed is essentially an embryo or young plant in the quiescent or dormant stage. In this state, the embryo has an extremely low metabolic rate, and most seeds can survive on their stored reserves for prolonged periods. The seed is the primary means by which a plant reproduces itself at a later time when conditions are suitable.

During fertilization, the genes controlling plant characteristics regenerate and recombine in many different ways, resulting in seeds that may or may not produce true to type. Seeds resulting from self-pollination may produce true-to-type specimens, while those resulting from cross-pollination usually do not. Cross-pollinated seeds provide genetic diversity for breeding and selection of new cultivars (cultivated varieties) and are often sources of novel plant material.

Seedling plants are commonly grown to be used as rootstocks for budding or grafting of cultivar fruit trees, nut trees, and woody landscape plants, or to produce superior landscape specimens that are difficult to propagate asexually or for which asexual methods of propagation are unknown [2]. Regeneration from seed is the most economical way to grow large numbers of plants. Most forestry species, vegetables, and flowering and other cultivated plants are grown from seeds.

Plants of a few kinds produce seeds without undergoing fertilization. This form of reproduction, called apomixis, is characteristic of species such as *Poa pratensis* (Kentucky bluegrass), which rarely or never set true seed. A plant grown from an apomictic seed is genetically identical to its parent.

II. SEED MORPHOLOGY

There are two major classes of seed-bearing plants: angiosperms (flower-bearing), in which seeds are borne in ovules enclosed within the ovary or fruit, and gymnosperms (cone-bearing), in which seeds are borne in pairs at the bases of scales of the cones. For an example of a seed from each of these classes of seed-bearing plants, see parts (A) and (B) of Figure 1 [3–6].

Most seeds consist of three parts: *embryo*, a miniature plant inside the seed; *endosperm*, stored food reserves for the growing embryo; and *seed coat* (testa), which encompasses and protects the embryo and endosperm from damage, excess water loss, and other unfavorable conditions.

Close examination of a seed embryo reveals one or more miniature seed leaves *(cotyledons)*, an embryonic stem *(plumule* or *epicotyl)*, and an embryonic root *(radicle)*. The *hypocotyl* is a transition zone between the embryonic stem and root. Among angiosperms, plants that have two cotyledons are classified as *dicotyledons* and those with a single cotyledon as *monocotyledons*. Gymnosperms may have as many as 15 cotyledons.

There are about quarter-million different seed-bearing plants in the world, each with its own kind of seed, which can be identified by size, shape, color, and other external features [7]. The endosperm contains stored food reserves consisting of carbohydrates, proteins, oils, and other biochemical substances. It usually makes up the larger part of the seed. All seeds contain stored food reserves, although in some the amount can be quite small. Generally, the larger the food reserve, the greater the vigor of the seedling. Plump seeds usually have more food reserves than small, shriveled seeds. Food reserves are also found in the cotyledons of some species.

The seed coat may appear dull, highly glossy, or smooth, wrinkled or pitted, hard or soft, thick or thin, or combinations of these characteristics. Many seeds also have attached wings or other appendages. In seeds having two seed coats, the inner one is usually thin, transparent, and physiologically active; that is, it restricts gaseous exchanges and movement of biochemical substances [8]. The outer layer is hard and thick. A nonviable seed may contain an empty seed coat without an embryo or one that is reduced and shrunken. Seed coverings play an important role in protecting the seed and in influencing viability.

III. SEED GERMINATION

Most seeds begin to germinate (resume activity) soon after being planted in a moist, warm soil or germinating medium. The germination process begins with a swelling of the seed as it takes up moisture. Usually the radicle emerges first from the softened or ruptured seed coat, grows downward, and develops into the primary root system. The plumule pushes up to form the shoot. During early growth, the young seedling derives its nourishment from the seed's cotyledons or endosperm. Cytokinins, which enhance cell division, promote the mobilization of the food reserves toward the developing shoot and to the root, which begins to function and take up nutrients from the soil.

In some instances, the cotyledon or cotyledons remain beneath the surface of the ground *(hypogeous* germination, Figure 1C), although in most species they push above the surface *(epigeous* germination, Figure 1D), turn green, and perform the functions of leaves. The food reserves continue to nourish the seedling until photosynthesis occurs at a rate capable of supporting the seedling, usually when the first true leaves are formed. At this stage, most seedlings are capable of independent existence, and germination is completed. Thus, germination is the first visual manifestation of the start of the life cycle of a plant.

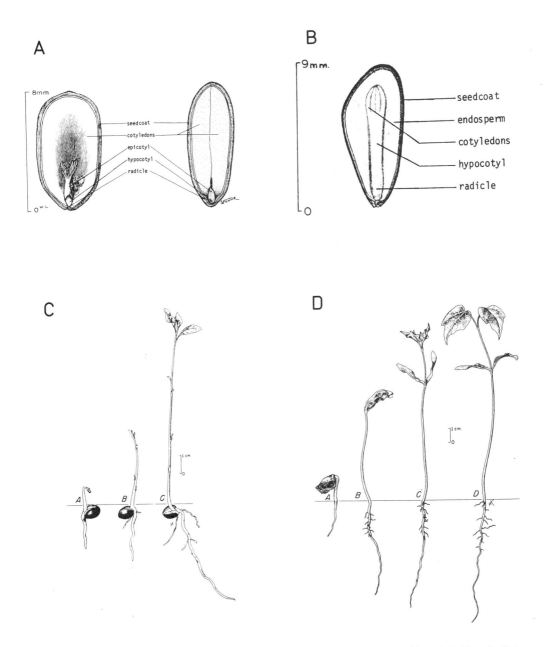

Figure 1 (A) Longitudinal section through seed of an angiosperm, *Albizia julibrissin* (silktree), 5×. (From Ref. 3.) (B) Longitudinal section through seed of a gymnosperm, *Pinus ponderosa* (ponderosa pine), 6×. (From Ref. 4.) (C) Hypogeous germination: development of *Lindera benzoin* (spicebush) 2, 3, and 10 days after germination. (From Ref. 5.) (D) Epigeous germination: development of *Acer platanoides* (Norway maple) 1, 3, 7, and 19 days after germination. (From Ref. 6.)

IV. PHYSIOLOGICAL AND ENVIRONMENTAL FACTORS

Each kind of seed has its own distinct set of requirements for moisture, temperature, air, light, and other factors.

A. Moisture

The need for moisture is perhaps the most important requisite for triggering germination. While some seeds require little moisture for germination, others, such as those from *Nymphaea* spp. (water lilies) and other aquatic plants, must be completely submerged in water. Seeds with hard or impermeable seed coats require special treatment (scarification) to allow water to seep in. Generally, water reaches the seed through contact with the soil or germinating medium. Once the germination process has begun, an adequate moisture level must be maintained, as temporary drying can result in death of the seed or seedling. Too much moisture can cause the soil or germinating medium to become waterlogged and the seeds to decay. Water uptake by seeds during germination has been described using a three-stage model [9].

1. Stage I (imbibition)

Dry seeds have a highly negative water potential (−100 to −200 MPa) as a result of the colloidal properties of the seed coat. The surfaces of proteins, cellulose, starch, and other substances must first become hydrated. The uptake or imbibition of water in stage I is purely physical, resulting in softening or rupturing of the seed coat and an increase in the volume of the seed. As the seed imbibes water, its internal water potential rises.

2. Stage II (active metabolism and hydrolysis)

Continuing water uptake activates stored enzymes and stimulates synthesis of new ones, which hydrolyze and transform some of the stored reserves into energy and smaller molecular weight, soluble compounds used for the production of more cells and tissues. These metabolic processes lower the water potential of the embryo and surrounding tissues.

Stage II appears to be a lag phase between imbibition and growth. During this stage, the driving force for water uptake is essentially the internal osmotic potential. Therefore, the duration of the lag phase may represent the time required to generate sufficient solutes to lower the osmotic potential to the degree necessary to induce further water uptake by the embryo and surrounding tissues. Exogenously applied gibberellins may assist with the intracellular generation of a lower water potential during stage II. The generally slow response to gibberellin treatment is probably related to the time lag for enzyme synthesis.

3. Stage III (visible germination)

In the last stage, visible germination appears, accompanied by rapid elongation of the radicle. For this process to occur, the water potential of the external solution should not be lower than −0.2 to −0.3 MPa.

B. Temperature

When moisture is adequate, the next most important requirement for germination is suitable temperature. Temperature affects metabolic processes such as the rate of water uptake, the translocation of nutrients and hormones, cell division and elongation, and other physiological and biochemical processes.

According to Hartmann et al. [8], temperature is the single most important factor in the regulation of the timing of germination, because of its role in dormancy control and/or release, or because it may necessitate climate adaptation. Generally, high temperatures induce or reinforce dormancy; low temperatures overcome dormancy.

Most seeds can tolerate prolonged hot weather if they are kept dry, and some can withstand even greater extremes of hot or cold. Seeds of some species with a very hard seed coat germinate only after exposure to intense heat, such as that from a brush fire [8,10]. The heat shock from the dry heat fractures the seed coat, allowing penetration of water, exchange of gases, or the freeing of the embryo from the physical constraint of the hard seed coat. Seeds are often placed in boiling water to control disease or to soften the seed coat without affecting seed viability.

Seeds of different species have been categorized according to temperature requirement: cool-temperature tolerant, cool-temperature requiring, warm-temperature requiring, and alternating temperature [8,11].

The optimum temperature requirement for germination may differ from that for early seedling growth. In the greenhouse, propagating nursery, or seed germinating laboratory, the usual practice is to shift the seedlings to a lower temperature regime, which makes them sturdy and more hardy for transplanting and growing-on [2].

C. Air

Most seeds require an abundant supply of oxygen during germination. Oxygen is required for respiration to oxidize starches, fats, and other food reserves. Its utilization is proportional to the amount of metabolic activity.

Although usually present in ample supply for germination, oxygen can become limited in soil or germinating medium if excessive water is present. Thus, a germinating medium or seedbed should be loose, friable, and well aerated. Seeds sown in heavy soils may germinate poorly, especially during wet seasons, when the soil becomes waterlogged and often lacks sufficient oxygen. Deep planting is unfavorable to germination because the oxygen supply may be restricted or seedlings may be unable to reach the surface, especially if the soil or medium is hard or compacted.

D. Light

Provided moisture and temperature are adequate, most seeds germinate equally well in darkness or light. Others are partially or completely inhibited, or require it to germinate. Some species, such as *Betula* (birch), respond to long photoperiods and are categorized as long-day seed. Light reaction (photosensitivity) is mediated through phytochrome, a pigment that absorbs either red or far-red light.

Germinating seeds that are favored by light need only low intensity and, thus, can be shaded to prevent excessive moisture loss from the germinating surface. Photosensitive seeds should be sown on or near the surface.

The *phytochrome* pigment system plays a key role in the photosensitivity of seeds and is of particular significance to many small-seeded species. Phytochrome is converted to Pfr (far-red light absorbing form) by 660 nm (red) light and, in turn, can be reconverted to Pr (red light absorbing form) by 730 nm (infrared) light.

Much of the phytochrome in quiescent seeds is in the Pfr form. However, within several hours after seeds have become fully hydrated, conversion of Pfr to Pr can occur in the absence of light [12]. Since Pfr actively promotes germination and lack of Pfr inhibits germination, photosensitive seeds germinate in response to exposure to the Pfr-forming 660 nm light.

The ratio of 655–665 nm light to 725–735 nm light varies significantly in nature. Sunlight transmitted through leaves has a low ratio of 655–665 nm to 725–735 nm light because chlorophyll selectively absorbs 655–665 nm light while transmitting 725–735 nm light [13].

The ecological significance of the seed phytochrome system involves allowing shaded, light-sensitive seeds on or near the soil surface to remain dormant until the leaf canopy above the seed disappears [14]. Buried, light-sensitive seeds, even though fully hydrated, would also remain dormant until returned to the soil surface by tillage or other soil disturbances.

Interactions between light and temperature are known for some kinds of seed. For example, photosensitivity may be overcome by alternating high and low temperatures. Externally applied chemicals can also interact with light and temperature. Many nitrogenous compounds, including cyanide, nitric acid, ammonium salts, urea, thiourea, and particularly potassium nitrate (10–20 mM solutions), have been found to stimulate the germination of photosensitive seeds [15].

V. ADAPTIVE FACTORS

A. Life Cycle

In the life cycle of every sexually reproduced plant, the seed germinates, the plant makes its vegetative growth, flowers and bears seeds (physiological maturity), and sooner or later dies. The duration of the cycle determines the three broad categories of plants: annuals, biennials, and perennials.

In its natural habitat, an annual plant usually lives for only one year or one season, a biennial for two years, and a perennial for more than two years. A perennial will continue to grow more or less indefinitely and, once physiologically mature, will continue to produce a succession of flowers, fruits, and seeds under suitable conditions.

The distinction between annual and the other categories is not absolute. A biennial that sets seed prematurely within the first growing season is converted to an annual by this process. An annual or biennial grown year-round in a greenhouse, or outdoors in a warmer climate, becomes a perennial. A perennial that would normally grow indefinitely in a warmer climate may be killed by frost in a colder climate.

Perennials are either *herbaceous* (having annual tops but perennial roots, crowns, or related underground structures) or *woody* (having biennial or perennial tops and perennial roots). Woody perennials consist mostly of trees, shrubs, and vines, which are readily distinguishable from herbaceous perennials, biennials, and annuals. Because of their longer juvenile periods, most woody species grown from seed do not start producing flowers and seeds for many years.

Thus, these categories describe the pattern of adaptation and cultural requirements of plants and, to a great extent also, their seed germination requirements. In general, but not always, seeds of woody perennials are more difficult to germinate or may possess more complex germination constraints than those of herbaceous types. Seeds of species within the same plant family and/or genera tend to have similar requirements for germination.

B. Dormancy

Dormancy is an all-inclusive term used to describe the state of a seed that will not germinate as a result of conditions associated with the seed itself (physical or physiological) or with external environmental factors [16]. Although often used synonymously with the term "rest period," the latter more often refers to delayed germination due to some internal or physiological seed condition [15,17]. Hartmann et al. [8] distinguish primary and secondary forms of dormancy. Primary dormancy is an adaptation of the seed to control the time and conditions for germination. Secondary dormancy is a further adaptation, which prevents germination of an imbibed seed if other environmental conditions are unfavorable.

Germination may be immediate or delayed. Seeds of many flowering garden annuals and trees, such as *Acer saccharinum* (silver maple), will germinate soon (within a month) after maturing, and some with almost no delay when removed from their protective fruits. Occasionally, germination in some species occurs on the parent plant, apparently because inhibiting chemicals are absent [17]. These seedlings are termed viviparous.

Readily germinable seeds should be stored promptly and properly until sown. This usually requires keeping dry or storing in a refrigerator in sealed plastic bags or other sealed containers. In unseasonal times, seeds should be sown in a greenhouse or other favorable environment. When the weather warms sufficiently, seedlings grown inside may be moved to a cold frame for further growth.

Viable seeds of many woody species may not germinate for considerable lengths of time, even when conditions are ideal. This condition, known as "rest," prevents seeds in their natural habitat from germinating in unseasonal times. Depending on the kind of seed, the rest period may last a few weeks, a few months, or even several years. By subjecting seeds to varying periods of low temperatures or repeated freezing and thawing (stratification), it is possible to hasten the afterripening process (release from rest) of seeds of many species.

Table 1 summarizes the types of seed dormancy, causes or factors associated with each type of dormancy, and the pretreatments required to overcome each type.

1. Dormancy Due to External Seed Coat and Associated Structures

In most seeds with delayed germination, dormancy is due to a hard seed coat that restricts water uptake and gaseous exchange, and/or to the actual mechanical constraint by the seed coat to the developing embryo [8]. Seeds from plants of the Cornaceae, Geraniaceae, Fabaceae, and Malvaceae are characterized by this condition [16]. The condition usually can be overcome by chemical or mechanical degradation *(scarification)*.

2. Dormancy Due to Internal and Physiological Factors

In many species, delayed germination results from internal conditions of the embryo and food storage tissues, or a portion of these tissues, which must undergo certain physiological changes or development before seeds germinate [8].

To overcome internal dormancy, an effective afterripening treatment is required. During this period, physiological and mechanical changes occur, hormones and enzymes are synthesized or activated, water is taken up, and gases exchanged. This rest period is often satisfied after a protracted exposure to low temperature *(cold stratification)*.

Immature or rudimentary embryos are characteristic of many species of seeds with double dormancy. Typical examples are *Ginkgo biloba* (maidenhair tree), *Ilex* spp. (holly), *Taxus* spp. (yew), and *Viburnum* spp. (viburnum). Seeds in this category will not germinate until dormancy as a result of both external and internal factors have been overcome sequentially.

In many other instances, the embryo may be fully developed and appears to be mature, but it may be in a state of rest because certain physiological and chemical changes have not been fulfilled. Specific regions of the embryos may be responsible, such as the seed coat, or a dormant radicle, hypocotyl, or epicotyl [8].

VI. SEED TREATMENTS

The most simple and practical approach to overcoming seed dormancy is to sow in outdoor seedbeds, allowing nature and its seasonal cycles to provide the appropriate conditions. As emphasized by Dirr and Heuser [18], cultural practices, including good seedbed preparation, appropriate seeding rate and depth, and protection from various diseases, insects, birds, and other pests, are all-important factors for success. Seedlings are usually allowed to grow from

Table 1 Types of Seed Dormancy, Location in the Seed, Causal Factors, and Treatments Required to Overcome Each Type

Types of dormancy	Location in seed	Cause	Treatments
Regulated by nonliving external covering			
Physical	Seed coat	Impervious to water	Soften, rupture, or remove; mechanical abrasion; alternate freezing and thawing; action of soil microbes; passage through intestines of animals
Mechanical	Seed coat	Hard; constrains the embryo	Rupture or remove; action of soil microbes
Chemical	Seed coat	Chemical inhibitors	Rupture or remove; soak or leach
Regulated by internal factors			
Morphological	Embryo	Rudimentary; underdeveloped	Moist chilling; warm exposure; alternating warm and cold temperatures; gibberellic acid; excise embryo and culture

Physiological	Inner seed coat or active membranes	Semipermeable; restrict gaseous exchange and inhibitor movement	Dry storage; short-term chilling; alternating temperatures; gibberellic acid; excise embryo and culture
thermodormancy	Active membranes	Thermosensitive	Specific temperature regime
photodormancy	Active membranes	Photosensitive	Expose to light Chilling hastens germination although seeds will eventually germinate
Intermediate[a]	Seed coverings	Endogenous conditions	
Embryo	Seed coverings and embryo	Endogenous conditions; deep dormancy; possible excess of abscisic acid; afterripening required	Moist chilling; excise embryo and culture; gibberellic acid
epicotyl		Separate afterripening required by epicotyl, hypocotyl, or radicle	Alternating cycles of warm–cold or cold–warm–cold
Double or complex	Combination of two or more types of dormancy		Eliminate all blocking conditions in sequence

[a]Occurring in coniferous species only.

one to three seasons in the beds and then transplanted to more permanent field locations or to containers [8]. However, good cultural practices may not always be reliable or successful.

Numerous interactions of the seed's heredity expression and the effect of environmental and other internal factors of the seed can render the germination of seeds of different kinds, or even of different seeds of one kind, extremely variable [17].

Pretreatment of seeds, which may hasten or induce more uniform and better germination, is more often required with seeds of woody trees and shrubs than with those of herbaceous species. In a survey of 400 species of woody plants, 33% had seeds that are commonly nondormant, 43% had seeds with internal dormancy, 7% had seeds with impermeable seed coat, and 17% had more than one kind of seed dormancy [19].

Without pretreatment, dormant seeds of many woody species may not germinate, or may do so sporadically over a prolonged period lasting 2 or 3 years, resulting in plants of irregular size and age in seedbeds or flats. Pretreatment procedures usually involve scarification or stratification. A combination of both procedures may be required for recalcitrant seeds with more complex dormancy problems. Pretreatment strategies for recalcitrant species have been described by various authors [2,8,18,20].

A. Scarification

Scarification is defined as any natural or artificial process of removing or altering the seed coat to make it permeable to water and gases. It can be accomplished physically by rupturing or abrading the seed coat, or chemically by soaking in acid, water, or other solvents.

1. Abrasion

Hard seed coats can be scratched or abraded with a file, sandpaper, or abrasive wheel, or cracked with a hammer or vise grip. A small mechanical tumbler lined with sandpaper or filled with sand or gravel may be more practical for larger amounts of seed [16]. The quantity of seeds should be sufficient to allow all the seeds to be abraded. For very large quantities of seeds, a concrete mixer containing coarse sand or gravel may be quite effective. The sand or gravel must be of a size that can easily be screened from the seeds [16].

Commercially designed machines are also available for scarifying large quantities of seeds. These scarifiers usually abrade or scar the seeds between two rubber-faced surfaces or impel seeds against roughened surfaces such as sandpaper. The severity of the abrasion or impact must be controlled to prevent damage to the seed [17,19].

2. Acid

Concentrated sulfuric acid (commercial grade, specific gravity 1.84) is often used because it is very effective. Caution is advised when using acid: NEVER ADD WATER TO ACID! Wear goggles and protective clothing, and handle the acid carefully to avoid spilling. If acid comes in contact with the skin, wash immediately under running cold water.

Using about twice the volume of acid over the seed in a glass container, stir gently with a glass rod during treatment. Fifteen minutes to 3 hours or more of exposure is required, depending on species. Carefully pour off the acid and rinse seeds several times under cold water to neutralize the acid. Stir the seeds carefully during rinsing. Pour off the water and spread the seeds uniformly on old newspaper and allow to dry at room temperature before sowing.

If properly treated, seeds remain firm, since little water is absorbed. The length of acid treatment, if unknown, must be determined empirically to prevent seed injury. While more suitable for a small amount of seed, acid treatment may not be practical for large quantities because of the danger of working with the concentrated acid. Treatment with nitric acid or with other chemicals, including potassium or sodium hydroxide, sodium hypochlorite, hydrogen peroxide, alcohol, acetone, and various growth-regulating substances may be effective for some seeds [2].

3. Hot Water

Soaking in hot water is a treatment commonly used with hard-seeded species. Soaking softens and sometimes ruptures the seed coat, and leaches from the seed coat naturally occurring substances that inhibit germination.

Pour about five times the volume of hot water (75–100°C) over the seeds and allow them to soak in the gradually cooling water for 6–24 hours. The amount of swelling of the seeds will indicate the degree of water absorption. Occasionally, seeds are actually boiled in water for 2–5 minutes, but the procedure is apparently injurious to seeds of most species. Seeds may also be soaked in running water or by exposing them to frequent changes of water to leach inhibitors.

Soaking in water may not yield as consistently good results as acid treatment. Soaking is easier to do, however, and needs no special equipment. Although damp or wet seeds are more difficult to sow, they should normally be sown immediately, since drying may make the treatment ineffective.

4. Other Considerations

In temperate climates, seeds planted in outdoor beds and subjected to natural abrasion by alternate freezing and thawing by the soil, or by movement of rock particles by water, may accomplish the desired result. In warm climates, seeds are ruptured primarily by the action of water. Organic acids in the soil or substances and enzymes excreted by soil microorganisms also soften or degrade the seed coat to some degree.

Removal of the fleshy seed coating or passage through the intestine of animals is implicated in overcoming dormancy of some species [8,10]. Dry heat may cause increased germination of some hard-seeded species by rupturing the seed coat or by denaturing seed coat inhibitors [10].

Commercial seed companies routinely pretreat seeds with chemical disinfectants to prevent infection by surface-borne fungi and bacteria, since most seeds appear to benefit from this pretreatment. Hot water appears to be a good disinfectant.

B. Stratification

1. Cold

The term *stratification* formerly applied to storing alternate layers of seeds and moist sand and subjecting the system to the cold, or more generally, freezing temperatures [21]. Nowadays, seeds are simply sown or mixed in the substrate rather than in layers, although the term is still used. The major requirements for cold stratification are adequate moisture, aeration, low temperature, and proper time of exposure. Research suggests that during stratification, the levels of growth-promoting substances in the seeds increase, while growth-inhibiting substances decrease, thus permitting germination [22].

In temperate climates, many types of seed sown directly in seedbeds, or in flats kept outdoors, undergo natural cold stratification during the winter and are ready to germinate in the spring. Seeds sown in flats or pots, or simply mixed with moist medium, may at any time be "stratified artificially" in refrigerators or coolers. For small lots, a plastic bag may be used. A moist medium such as sand, peat moss, or vermiculate, or a combination of these ingredients, is mixed with seed and placed in the bag, which is then sealed and placed in a refrigerator. The plastic allows gaseous exchange but holds moisture. Once stratification has begun, seeds should not be allowed to dry, since this may reverse the process. Therefore, the medium should be inspected periodically during the process. Seeds may also be chilled or frozen without being mixed in a substrate.

Freezing is apparently not essential but is sometimes recommended for certain herbaceous seeds. In general, best results occur with temperatures from just above freezing to 5°C. Although in milder climates, seeds may require higher stratification temperatures, from a practical viewpoint, lower temperatures will help to prevent germination while seeds are being stratified. During stratification, root radicles may emerge from the seeds, indicating readiness to germinate. Considerable exposure to freezing temperatures at this time may be injurious.

As shown in Tables 2 [23–30] and 3 [18,20,31–35], stratification, if required, varies with each species. The closer to optimum are the required temperature and duration, the better will be the outcome. Thus, knowing the proper temperatures and duration of stratification can result in more effective and efficient seedling production. Refrigeration (artificial stratification) is more predictable than outdoor sowing (natural stratification) and usually provides the best and most consistent results.

2. Warm

Some seeds require moist, warm stratification. Seeds with double dormancy, such as *Taxus* spp. and *Viburnum*, require both warm and cold stratification, while others with even more complex dormancy such as *Halesia carolina* (Carolina silverbell) require cold–warm–cold exposures in this sequence (Table 3).

During exposure to warm temperatures, usually between 20 and 30°C, immature or rudimentary embryos develop. If sufficient heat is gained after outdoor summer or fall seeding to satisfy the warm requirement, germination will occur during the first spring. Otherwise, germination will occur in the second spring. Dormancy of some seeds is also broken by storage in hot, dry conditions [10].

3. Embryo Culture

The technique of in vitro embryo culture, also referred to as embryo rescue, is routinely used by plant breeders and seed laboratories to obtain seedlings from otherwise nongerminable seeds that are not sufficiently mature when the fruit is ripe, or from recalcitrant seeds with very complex dormancies [2]. The procedure involves aseptically excising the embryo from the seed and culturing it on a suitable sterilized nutrient medium. When an immature or rudimentary embryo is cultured in this way, it may bypass the need for warm or cold stratification.

C. Growth Regulators

Seed dormancy and germination are believed to be controlled by the balance and interaction of substances that promote and inhibit growth. These regulatory substances accumulate in seeds during embryo development, although not necessarily in the embryo itself [8]. Gibberellic acid, in particular, appears to be essential for seed germination. It immobilizes food sources and stimulates growth of embryonic tissue. Differences in endogenous gibberellin concentrations of some cold-requiring seeds have been related to the amount of chilling exposure of the seeds [36]. Abscisic acid and related germination inhibitors are present in the seed coat, endosperm, or embryo.

Exogenously applied growth regulators sometimes influence seed germination. Some dormant seeds, particularly of wild plants that require light or cold for germination, may be induced to germinate by the application of gibberellins. Applied cytokinins also overcome dormancy in many species. Conversely, abscisic acid often inhibits germination when applied to nondormant seeds. In some instances, the abscisic acid induced germination inhibition can be reversed by cytokinin. Gibberellins do not usually reverse abscisic acid induced germination

Table 2 List of Selected Flowering Herbaceous Plants Showing Desired Medium Temperature Regimes for the Germinating and Posttransplant Stages and the Approximate Time for Germination Under Greenhouse Conditions; Need for Light or Seed Pretreatment Also Indicated

Definitions of plant type, column 2

A = **Annual.** A plant that normally completes its life cycle within one calendar year. Some may be considered short-lived perennials in areas of mild winters or with adequate protection (i.e., snow cover, mulch, sheltered location).

B = **Biennial.** A plant that requires 2 years to complete the life cycle, usually producing vegetative growth only in the first, then flowering and dying in the second.

HHP = **Half-hardy perennial.** A plant that may survive mild winters, or harsher winters with adequate protection (i.e., snow cover, mulch, sheltered location).

P = **Hardy herbaceous perennial.** A plant that lives longer than 2 years, normally producing flowers and seeds each year.

P-W = **Perennial wildflower.** A plant that is indigenous to North America or has become naturalized to the point of being considered so.

Species	Plant type	Medium temperature (°C)		Days to germinate	Light or dark	Pretreatment	Remarks [notes 1–6 follow last table entry]
		Germination	Posttransplant (night)				
Acanthus spinosus Bear's breeches Acanthaceae	P	10–12		21			Use fresh seed
Achillea filipendulina	P	20–24	15	7–10	L		Lower germinating medium night temperature by 3°C.
A. millefolium Yarrow Asteraceae	P-W	18	10	5–8	L		See note 6. Seeds may produce inferior plants; vegetative propagation may be more reliable.
Aconitum spp. Monkshood Ranunculaceae	P	12–15		30	D	Sow then freeze 3 weeks	Fresh seed will germinate much more quickly than older, ripened seed.
Actaea alba Baneberry Ranunculaceae	P-W	5–10	10	30	D	Stratify at 21°C 3 weeks, freeze 5 weeks, then sow	Drying of seed will delay germination.

Table 2 Continued

Species	Plant type	Medium temperature (°C) Germination	Posttransplant (night)	Days to germinate	Light or dark	Pretreatment	Remarks [notes 1–6 follow last table entry]
Adenophora confusa Lady bells Campanulaceae	P	21–23		14–21	L		
Adonis spp. Adonis Ranunculaceae	P	12–15		30	D	Sow, then freee 3 weeks	
Ageratum houstonianum Floss flower Asteraceae	A	21–24	12–16	7–10	L		Lower germinating medium night temperature by 3°C. After transplanting, gradually reduce night temperature to 10°C. See note 6.
Alcea rosea Hollyhock Malvaceae	B	18–21	10	12–18			
Alchemilla mollis Lady's mantle Rosaceae	P	10–15	10	7–21	D	Stratify at 21°C 3 weeks, freeze 5 weeks, then sow.	Fresh seed germinates readily at 21°C. Old or purchased seed needs pretreatment. Self-sows easily.
Amaranthus spp. Joseph's coat Amaranthaceae	A	21–24	15–18	8–10	D or L		Good seed, 80–85% germination.
Anacyclus depressus Rocky Mountain daisy Asteraceae	P	15–20	10–15	14–21			Fresh seed germinates best.
Anagalis monelli	P	>10		10–30	D		
A. linifolia	A	>10		10–30	D		
A. arvensis Pimpernel Primulaceae	P-W	21	16	21–28	D or L		Can be direct-sown to garden or container.

Plant / common name / Family	Type				Light	Treatment	Comments
Anaphalis margaritaceae Pearly everlasting Asteraceae	P-W	18–21	10–13	14–21	L		Reduce night temperature of medium by 3°C for best germination.
Anchusa capensis *A. azurea* Bugloss, alkanet Boraginaceae	A	20–25	15–18	14–21	D		Can be direct-sown to garden when soil has reached 20°C.
Anemone spp. Anemone Ranunculaceae	P	18–20	10–15	21			Rub seed with dry sand to remove cottony covering.
Antennaria spp. Pussytoes Asteraceae	P	22–24	15	14–21	D or L		Seed is very fine, cover lightly.
Anthemis spp. Golden marguerite, chamomile Asteraceae	P	20	8–14	7–21			
Antirrhinum majus Snapdragon Scrophulariaceae	HHP	18–21	8–12	10–14	L	Chill at 0°C 7 days	Sow 6–8 weeks before scheduled outdoor planting. Susceptible to damping off. Use a presow fungicide drench. Keep medium moist at all times during germination. Lower temperature after germination.
Aquilegia spp. Columbine Ranunculaceae	P	20–24	10	14–30	L	Sow then refrigerate 3 weeks	Reduce night temperature of medium by 7–8°C during germination.
Arabis spp. Rock cress Brassicaceae	P	21	10	15	L		
Arctotis stoechadifolia African daisy Asteraceae	A	15–20	10–15	21–35	D or L		Sow 6–8 weeks before scheduled outdoor planting. Use fresh seed. Viability decreases quickly.
Arenaria montana Sandwort Caryophyllaceae	P	12–17		15–20	L		

Table 2 Continued

Species	Plant type	Medium temperature (°C)		Days to germinate	Light or dark	Pretreatment	Remarks [notes 1–6 follow last table entry]
		Germination	Posttransplant (night)				
Armeria spp. Thrift Plumbaginaceae	P	15–18	10	14–21	D or L	Soak in warm water 6–8 hours	Germination is sometimes low or erratic
Arnica spp. Arnica Asteraceae	P	12		25–30			
Aruncus dioicus Goat's beard Rosaceae	P	22–24	10–15	14–21		If not sown fresh, stratify at 5°C, 4 weeks	Fresh seed can be sown immediately.
Asclepias tuberosa Butterfly weed Asclepidaceae	P-W	20–24		21–28	D	If not sown fresh, stratify at 5°C, 3 weeks	Germination is erratic. Fresh seed gives best results if sown immediately.
Aster spp. Aster Asteraceae	P	21	10	15			
Astilbe chinensis 'Pumila' A. × *arendsii* false spiraea Saxifragaceae	P	15–21		21–28		Sow then stratify at 21°C, 2 weeks, then at 5°C 4 weeks	
Astrantia major Masterwort Apiaceae	P	10–20	10	10–14		Stratify at 2–5°C 4 weeks	
Aubrieta sp. Rock cress Brassicaceae	P	18	10	20	D or L		
Aurinia saxitilis Basket-of-gold Brassicaceae	P	15	10	7–10	L		
Baptisia australis Wild indigo Fabaceae	P	21–24		5–10		Nick or file seed coat	When seeds turn black, collect and sow. Germination slow and uneven.

						Comments
Begonia semperflorens	A	21–24	16–18	15–20	L	Reduce night temperature of medium by 3°C for best germination. Fertilize with 100 ppm N immediately after germination. Never allow seed surface to dry.
B. × tuberhybrida Begonia Begoniaceae	A	21–24	16–18	15–20	L	
Belamcanda chinensis Blackberry lily Iridaceae	P	20–30		14–21		
Bellis perennis English daisy Asteraceae	B	20–24	10	6	L	Reduce night temperature of medium by 3°C for best germination.
Bergenia cordifolia Bergenia Saxifragaceae	P	13		15–20		
Boltonia asteroides Boltonia Asteraceae	P	21	10	15	D or L	
Brachycome iberidifolia Swan River daisy Asteraceae	A	21	10	10–18		
Brassica oleracea var. *acephala* Flowering cabbage, kale Brassicaceae	A	21–24	13–16	10	L	Sow outdoors 3 months before first frost to ensure cool weather for best color.
Browallia speciosa Browallia Solanaceae	A	24	18–21	7–14	L	Reduce night temperature of medium by 3°C for best germination. Predrench medium with fungicide to prevent damp-off; keep seedlings on the dry side.
Brunnera macrophylla Siberian bugloss Boraginaceae	P	21–24		14–21	D	

Table 2 Continued

Species	Plant type	Medium temperature (°C)		Days to germinate	Light or dark	Pretreatment	Remarks [notes 1–6 follow last table entry]
		Germination	Posttransplant (night)				
Buphthalmum speciosum Oxeye Asteraceae	P	21–24			D or L		
Calendula officinalis Pot marigold Asteraceae	A	21	10	7–10	D		
Callistephus chinensis Aster Asteraceae	A	21	16	6–14	D or L		
Campanula medium Bell flower Campanulaceae	B	20–24	10	6–16	L		Reduce night temperature of medium by 3°C for best germination.
C. carpatica	P	20–24	10	6–16	L		
C. rotundifolia	P-W	20–24	10	6–16	L		
Catananche caerula Cupid's dart Asteraceae	P	21		14–21	L		
Catharanthus roseus Annual periwinkle, vinca Apocynoceae	A	24–26	18–24	14–21	D		Reduce night temperature of medium by 3°C for best germination. Be careful not to overwater. Plants are sensitive to high moisture levels. Predrench medium with fungicide to prevent damp-off.
Celosia spp. Cockscomb Amaranthaceae	A	24–27	21	7–10	L		Reduce night temperature of medium by 3°C for best germination. Plant growth will be stunted if temperature falls below 18°C.

	HHP						
Centaurea cinerarea Dusty miller	HHP	18–21	10–13	5–10	D		Use fresh seed for best results.
C. cyanus	A	18–21	10–13	5–10	D		
C. montana Bachelor button Asteraceae	P	18–21	10–13	5–10	D		
Centranthus ruber Red valerian Valerianaceae	P	15	15	10–20			
Cerastium tomentosum Snow-in-Summer Caryophyllaceae	P	15	18	8–18	D or L		
Ceratostigma plumbaginoides Plumbago Plumbaginaceae	P	10–20		10–21		Stratify at 5°C 4–6 weeks	
Chelone glabra	P	15		10–14		Sow then stratify at 5°C 6 weeks	
C. lyonii Turtlehead Scrophulariaceae	P	15		10–14			
Chiastophyllum oppositifolium Cotyledon Crassulaceae	P	13		30	L	Stratify at 15–20°C 2–3 weeks then at 5°C 4–6 weeks	Seeds are tiny. Do not cover with medium.
Chieranthus chieri Wallflower Brassicaceae	B	21	10	7–10	D or L		Can be direct-sown to garden in summer.
Chrysanthemum parthenium Feverfew	A	18–21	10–13	5–10	L		Reduce night temperature of medium by 3°C for best germination.
C. coccineum Painted daisy	P	15–18	10	7–14	L		
C. ptarmiciflorum Dusty miller	P	15–18	10	7–14	D or L		
C. superbum Chrysanthemum Asteraceae	P	15–18	10	7–14	L		

Table 2 Continued

Species	Plant type	Medium temperature (°C)		Days to germinate	Light or dark	Pretreatment	Remarks [notes 1–6 follow last table entry]
		Germination	Posttransplant (night)				
Chrysogonum virginianum Golden star Asteraceae	P	22–24		14–21	D or L		Plants self-sow.
Chrysopsis sp. Golden aster Asteraceae	P	22–24		14–21	D or L		
Cimicifuga simplex Bugbane *C. racemosa* Snakeroot Ranunculaceae	P	10		7–30		Stratify at 20–25°C 6–10 weeks then at 1–5°C 6–8 weeks	Can be direct-sown to garden in summer. Use fresh seed. Germination is erratic.
Clarkia spp. Clarkia, godetia Onagaceae	A	20–23	12–15	5–10	D or L		Difficult to transplant. Best sown directly to garden or container.
Cleome hasslerana Spiderflower Capparaceae	A	28 (day) 21 (night)	21–24	5–14	L		Alternate day and night temperature as noted.
Coleus × *hybridus* Coleus Lamiaceae	A	21–24	16–18	10–12	L		Presow drench with fungicide to prevent damp-off.
Collinsia grandiflora Blue lips Scrophulariaceae	A	15–20	10–15	10–15			
Consolida ambigua Larkspur Ranunculaceae	A	15–20	13–16	15–21	D		
Convolvulus tricolor Dwarf morning glory Convolvulaceae	A	21–24	21	10–16	D or L		Difficult to transplant. Best sown directly to garden or container.

Coreopsis tinctoria	A	20–24	10	7–21	L		Reduce night temperature of medium by 3°C for best germination.
C. lanceolata	P-W	20–24	10	7–21	L		
C. verticillata	P	20–24	10	7–21	L		
Tickseed Asteraceae							
Coronilla varia	P	21–23		30	D or L	Nick hard seed coat or soak in water 6–9 h	
Crown vetch Fabaceae							
Corydalis lutea	P	22–25		7–14		Stratify at 22°C 6–8 weeks then at 0–2°C 6–8 weeks	Plants self-sow easily. Seed must be sown fresh. Vitality decreases rapidly with age.
Yellow corydalis Fumariaceae							
Cosmos sulphureus,	A	24	18	5–7	D or L		
C. bipinnatus	A	24	18	5–7	D or L		
Cosmos Asteraceae							
Cuphea ignea	A	21		8–10	L		
Cigar flower Lythraceae							
Cynaglossum amabile	B	18–23		5–10	D		
Chinese hound's tongue Boraginaceae							
Dahlia hybrids	A	21–24	13–16	7–10	D or L		
Dahlia Asteraceae							
Delphinium spp.	P	27 (day) 21 (night)	7–10	7–10	D	Sow then freeze 4–8 h or more	Use fresh seed. Viability decreases rapidly with age.
Delphinium Ranunculaceae							
Dianthus chinensis	P	21	10–13	8–10	D or L		
D. plumarius	P	21	10–13	8–10	D or L		
D. caryophyllus	P	21	10–13	8–10	D or L		
Pinks Caryophyllaceae							

Table 2 Continued

Species	Plant type	Medium temperature (°C) Germination	Posttransplant (night)	Days to germinate	Light or dark	Pretreatment	Remarks [notes 1–6 follow last table entry]
Dicentra spp. Bleeding heart Fumariaceae	P	10–12		21–28	D or L	Freeze or stratify at 15–18°C 4 weeks, then at 5°C 4 weeks	Self-sows easily. Fresh seed may germinate immediately. Older seed will need to be stratified.
Dictamnus albus Gas plant Rutaceae	P	12–15		30–40		Stratify at 3–5°C 6 weeks	If fall-sown outdoors, germination will occur over a 2-year period and may be more productive than "artificial" stratification.
Digitalis spp. Foxglove Scrophulariaceae	B	21	10	7–10	L		Reduce night temperature of medium by 3°C for best germination.
Dimorphotheca aurantiaca Cape marigold Asteraceae	A	16–18	16	7–10	D or L		Seeds are short-lived and should not be stored.
Disporum spp. Fairy bells Liliaceae	P	15–20	10	10–14	D or L		
Dodecatheon media Shooting star Primulaceae	P	15–22		30+		Freeze or stratify at 3–5°C 4–6 weeks	Germination may be erratic.
Doronicum spp. Leopard's bane Asteraceae	P	21	10	10–20	L		Reduce night temperature of medium by 3°C for best germination.
Dorotheanthus bellidiformis (*Mesembryanthemum criniflorum*) Livingston daisy Aizoaceae	A	18–22		15–20	D		Place in dark location but do not cover, since seed is fine.
Dryas octopetala Mountain avens Rosaceae	P	15–20		50+			

Species / Common name / Family		Day/Night temp (°C)		Germination (days)	Light	Treatment	Comments
Dyssodia tenuiloba Dahlberg daisy Asteraceae	A	18–21	15–18	10–15	D or L		
Echinacea purpurea Purple coneflower Asteraceae	P-W	21–24	13–16	10–12	L		Reduce night temperature of germinating medium by 3°C for best germination. Seeds may not come true to type and may take 2 years to flower.
Echinops ritro Globe thistle Asteraceae	P	20–24	10	8–20	L		
Erigeron spp. Fleabane Asteraceae	P	13	10	15–20	D or L		
Erinus alpinus Liver balsam Scrophulariaceae	P	18–24	10–12	20–25			
Eryngium Sea holly Apiaceae	P	18–23	15	5–10		Stratify at 20–24°C 2–4 weeks, then at 3–5°C 4–6 weeks	Use fresh seed. After 2 weeks, dormancy sets in and germination time may be extended over 1–2 years
Erysimum spp. Alpine erysimum Brassicaceae	B	13–18	10	7–10			
Eschscholzia californica California poppy Papaveraceae	A	16–18	10–12	14	D		Plants self-sow easily. Difficult to transplant. Best sown directly to garden or container.
Eupatorium spp. Boneset Asteraceae	P	20–24	15	2–3 weeks	D or L	Stratify at 5°C 4 weeks	
Euphorbia epithymoides	P	21–24	18–21	8–12	D or L	Chill 7 days then soak 4 h in water.	
E. marginata	A	21–24	18–21	8–12	D or L		
E. myrsinites Spurge Euphorbiaceae	P	27 (day) 21 (night)	10	12–18	L		

Table 2 Continued

Species	Plant type	Medium temperature (°C)		Days to germinate	Light or dark	Pretreatment	Remarks [notes 1–6 follow last table entry]
		Germination	Posttransplant (night)				
Filipendula spp. Meadowsweet Rosaceae	P	18–23	12–15				
Gaillardia pulchella	A	20–24	10	8–12	L		Reduce night temperature of medium by 3°C for best germination.
G. aristata Blanket flower Asteraceae	P-W	20–24	10	8–12	L		
Gaura lindheimeri Gaura Onagraceae	P	18–23		14–21			Difficult to transplant because of a long tap root.
Gazania spp. Gazania Asteraceae	A	15–17	13–16	10–14	D		
Gentiana acaulis Gentian Gentianaceae	P	20–25	10	14–28		Sow then freeze 3 weeks	Likes cool growing condition.
Geranium spp. Cranesbill Geraniaceae	P	20–22	10–12	7–28		Ripen fresh seeds at 22–24°C 2–4 weeks	Seeds are short-lived and should not be stored.
Gerbera jamesonii Transvaal daisy Asteraceae	A	20–22	16–18	7–14	D		Use fresh seed. Stored seed loses viability.
Geum spp. Avens Rosaceae	P	20–24	10	10–20	L		Reduce night temperature of medium by 3°C for best germination.
Globularia	P	13	10	10–12		Stratify at 5°C 21 days	
G. cordifolia Globe daisy Globulariaceae	P	13	10	10–12			

Godetia spp. Farewell-to-spring Onagraceae	A	21	10	10	D	
Gomphrena globosa Globe amaranth Amaranthaceae	A	21–24	21	10–14	D	Soak in water 24 h Germination may be erratic or slow. Predrench medium with fungicide to prevent damp-off. Keep seedlings relatively dry.
Gypsophila repens	P	21–25	13	10–14	L	
G. elegans Baby's breath Caryophyllaceae	A	21–25	13	10–14	L	
Helenium hoopesii	P-W	21	13–16	12–16	L	Reduce night temperature of medium by 3°C for best germination.
H. autumnale Sneezeweed Asteraceae	P-W	21	13–16	12–16	L	
Helianthus annuus Sunflower Asteraceae	A	20–30	15–20	14–21	D or L	Seeds best sown directly in garden, since they grow large and rapidly.
Helichrysum bracteatum Strawflower Asteraceae	A	24	16	7–10	L	
Heliopsis spp. Heliopsis Asteraceae	P	20	10–12	7–14		
Heliotropium arborescens Garden heliotrope Boraginaceae	A	21–24	16–18	14–21	L	Reduce night temperature of medium by 3°C for best germination.
Helleborus spp. Hellebore Ranunculaceae	P	20–23	5–10	30–60		Stratify at 20°C 8–10 weeks, then at 2–5°C 8–10 weeks—may need to repeat once

Table 2 Continued

Species	Plant type	Medium temperature (°C)		Days to germinate	Light or dark	Pretreatment	Remarks [notes 1–6 follow last table entry]
		Germination	Posttransplant (night)				
Hemerocallis spp. Daylily Liliaceae	P	15–22	10–12	21–50		Stratify at 3–5°C 6 weeks	
Herniaria glabra Rupture wort Illecebraceae	P	21	10–15	10–12			
Hesperis matronalis Dame's rocket Brassicaceae	P-W	21–24	10	7–10	L		Reduce night temperature of medium by 3°C for best germination.
Heuchera sanguinea Coral bells, alumroot Saxifragaceae	P	18–21	10	10–18	L		
Hibiscus trionum Mallow Malvaceae	A	21	10	14–20	D	Scarify in H₂SO₄ 30 min or soak in water 48 h	Hard seed coat may require scarification.
Hosta spp. Plantain lily Liliaceae	P	21	5–15	15–20			
Hunnemannia fumariifolia Mexican tulip poppy Papaveraceae	P	21–24	15	15–20	D		
Hypericum spp. St. John's wort Hypericaceae	P	21–23	10–15	25–30			

Hypoestes phyllostachya Polkadot plant Acanthaceae	A	21–24	15	10–12	L		
Iberis umbellata Candytuft Cruciferae	A	21	10	7–14			
I. sempervirens	P	16–18	7–10	16–20	L		Keep surface of medium moist at all times during germination.
Impatiens spp. Patience plant Balsaminaceae	A	21–24	16–18	7–18	L		
Incarvillea delavayi Hardy gloxinia Bignoniaceae	P	13–18	13–15	25–30			
Ipomoea tricolor Morning glory vine Convolvulaceae	A	21–28	15–20	7–10		Nick seed coat with a file or soak in water 24 h	
Iris spp. Iris Iridaceae	P	22	10–20	21–40			Use fresh seed. Germination may be slow and erratic.
Jasione perennis Sheep's bit Campanulaceae	P	21	15	10–15			
Kniphofia spp. Torch lily Liliaceae	P	20–24	10	10–18	L	Sow, then stratify at 5°C 6 weeks	Reduce night temperature of medium by 3°C for best germination.
Kochia scoparia Burning bush Chenopodiaceae	A	21–24	18	6–9	L	Soak in water 24 h	Reduce night temperature of medium by 3°C for best germination. Water sparingly after germination.
Lathyrus spp. Sweet pea Fabaceae	A or P	12–18	12–18	10–25	D	Nick seed coat with a file or soak in water for 24 h	Directly sow to garden or transplant container.
Lavandula spp. Lavender Lamiaceae	P	20	18	20–30	D of L	Stratify at 5°C, 30 days	

Table 2 Continued

Species	Plant type	Medium temperature (°C) Germination	Medium temperature (°C) Posttransplant (night)	Days to germinate	Light or dark	Pretreatment	Remarks [notes 1–6 follow last table entry]
Lavatera trimestris Tree mallow Malvaceae	A	21	18–20	15–20			Difficult to transplant. Best sown directly to garden or container.
Layia platyglossa Tidy tips Asteraceae	A	21–24	20	10–12			
Lewisia cotyledon Lewisia Portulacaceae	P	21		30		Stratify at 3–5°C 3 weeks	
Liatris Blazing star, gay feather Asteraceae	P	15–24	18–21	16–22			Reduce night temperature of medium by 3°C for best germination.
Limonium sinuatum Statice Plumbaginaceae	A	21	18	14–18	D		
Linaria maroccana *L. vulgaris* Toad flax Scrophulariaceae	A P-W	18–21 18–21	13 13	8–10 8–10			Reduce night temperature of medium by 3°C for best germination.
Linum grandiflorum *L. perenne* *L. flavum* Flax Linaceae	A P P	18–21 15–18 18–21	10 10 10	8–18 8–18 8–18	L L L		
Liriope spp. Lily-turf Liliaceae	P	18–21		30		Soak in water 24 h	

						Treatment	Remarks
Lisianthus russellianus (*Eustoma grandiflora*) Prairie gentian Gentianaceae	A	21–24		8–12	L		Seedlings grow slowly.
Lithodora diffusa Lithodora Boraginaceae	P	20		14–21		Scarify with sandpaper or file, or soak in warm water for 2 days	
Lobelia erinus	A	24	16	14–21	L	Stratify at 3–5°C 3 months	
L. cardinalis	P-W	21	10–13	10–14	L		
L. siphilitica	P-W	21	10–13	10–14	L		
Lobelia Lobeliaceae							
Lobularia maritima Sweet alyssum Brassicaceae	A	20–24	8–10	4–10	L		Can be direct-sown to garden or container. Predrench medium with fungicide to prevent damp-off.
Lunaria annua (*L. biennis*) Honesty Brassicaceae	B	15–18	10	16–20	D		
Lupinus spp. Lupine Fabaceae	P	21–24	10	16–20	D	Soak in warm water 18 h	
Lychnis chalcedonica Campion Caryophyllaceae	P	21	18	14–21	L		Reduce night temperature of medium by 3°C for best germination.
Machaeranthera tanacetifolia Tahoka daisy Asteraceae	A	21	10–12	25–30	L	Stratify at 3–5°C 2 weeks	

Table 2 Continued

Species	Plant type	Medium temperature (°C)		Days to germinate	Light or dark	Pretreatment	Remarks [notes 1–6 follow last table entry]
		Germination	Posttransplant (night)				
Malva alcea Mallow Malvaceae	B	18–21	10	12–18			
Mathiola incana Stocks Brassicaceae	A	20	10–13	7–10	L		Presow drench medium with fungicide to prevent damp-off.
Meconopsis spp. Welsh poppy Papaveraceae	P	18–22	10–12	20–25			
Mimulus × hybridus Monkey flower Scrophulariaceae	A	18–21	16	14–21	L		Can grow-on at low temps. (3–6° C) once established. Sow when days are at least 13 h to initiate flowering.
Mirabilis jalapa Four o'clock Nyctaginaceae	A	21	10–21	7–10			
Moluccella laevis Bells of Ireland Lamiaceae	A	25 (day) 10 (night)	16	25–30	L	Chill at 7°C 5 days	Difficult to transplant because of long tap root.
Monarda fistulosa Beebalm Lamiaceae	P-W	21	18–21	14–21	L		Reduce night temperature of medium by 3°C for best germination.
Myosotis spp. Forget-me-not Boraginaceae	P	18–21	10	8–12	D		
Nemesia strumosa Pouch nemesia Scrophulariaceae	A	16	13–16	5–10	D		Sensitive to medium temperature above 18°C.
Nemophila menziessii Baby blue eyes Hydrophyllaceae	A	12	13–15	7–12			
Nepeta × faassenii Catmint Lamiaceae	P	20–24	10	5– 7			

Nicotiana alata Flowering tobacco Solanaceae	A	21	13	7–12	L	Reduce night temperature of medium by 3°C for best germination. Transplants can be grown very cool (5–7°C).
Nierembergia hippomanica Cupflower Solanaceae	A	25°	18–20	14		
Nigella damascena Love-in-a-mist Ranunculaceae	A	20–22	15	10–15		Difficult to transplant. Best sown directly to garden or container.
Ocimum basilicum Basil Lamiaceae	A	22	15	10–14		
Oenothera argillicola	P-W	18–21	13–16	10–14	D or L	Reduce night temperature of medium by 3°C for best germination.
O. missouriensis Evening primrose Onagraceae	P-W	20–24	18–21	10–21	L	
Paeonia spp. Peony Paeoniaceae	P	22	20	30–60	Sow at 21°C, allow roots to emerge, then refrigerate at 3–5°C 8–10 weeks	Vegetative propagation is preferred, since it takes 5–7 years to produce a flowering plant from seed.
Papaver rhoeas	A	20–24	7–10	12–16	D	Reduce night temperature of medium by 3°C for best germination.
P. orientalis Poppy Papaveraceae	P	20–24	7–10	12–16	L	
Pelargonium × *hortorum* Geranium Geraniaceae	A	21–24	16–18	5–18	D or L	Soak in H_2SO_4 4 min; may induce uniform germination — Presow drench medium with fungicide to prevent damp-off.
Penstemon gloxinioides	A	16	10	7–14	L	
P. strictus Penstemon, beard tongue Scrophulariaceae	P-W	16	10	7–14	L	

Table 2 Continued

Species	Plant type	Medium temperature (°C)		Days to germinate	Light or dark	Pretreatment	Remarks [notes 1–6 follow last table entry]
		Germination	Posttransplant (night)				
Petunia × hybrida Petunia Solanaceae	A	21–24	13–16	7–12	L		Keep surface of medium constantly moist until germination is complete. Seedling vigor varies widely among cultivars.
Phacelia campanularia California bluebell Hydrophyllaceae	A	12–15	12–15	4–10			
Phlox drummondi	A	15–16	10–13	10	D	Stratify at 5°C 2 weeks	Presow drench medium with fungicide to prevent damp-off.
P. paniculata Phlox Polemoniaceae	P	21		25–30	D		
Physalis alkekengi Chinese lantern Solanaceae	P	21–24	18	12–24	L	Stratify at 5°C 4–6 weeks	Germination may be slow—up to 6 months.
Physostegia virginica Obedient plant Lamiaceae	P-W	24 (day) 13 (night)	10	12–14	D or L		
Platycodon grandiflorus Balloon flower Campanulaceae	P	18–22	10–15	14–21	L		
Polemonium spp. Jacob's ladder Polemoniaceae	P	20–24	13	10–20	D or L		Reduce night temperature of medium by 3°C for best germination.
Polygonatum spp. Solomon's seal Liliaceae	P-W	20–23	15–18	20–25		Stratify at 5°C 6 weeks	
Polygonum spp. Smart knotweed Polygonaceae	P	21	18	21–28	D or L		Germination may be erratic.

Species / Common name / Family	Type	Temp 1	Temp 2	Days	Light	Treatment	Comments
Portulaca grandiflora Rose moss Portulacaceae	A	21–24	16–18	7–10	L		Reduce night temperature of medium by 3°C for best germination. Seed can be direct-sown to garden or container.
Potentilla spp. Cinquefoil Rosaceae	P	21	10–12	15–20	D or L		Seeds lose viability quickly and should not be stored.
Primula spp. Primrose Primulaceae	P	16–18	3 weeks at 13 then 5–7	16–24	L	Stratify at 2–5°C 3–4 weeks	Reduce posttransplant temperature gradually. Sow seeds as soon as they ripen to enhance germination. Seed must be fresh or dormancy occurs.
Pulsatilla vulgaris Pasque flower Ranunculaceae	P	20–23	10–12	30–40			Seed must be fresh or dormancy occurs.
Ranunculus spp. Buttercup Ranunculaceae	P	15–16	10–12	40–56			Fresh seed can be sown immediately.
Ratibida columnifera Coneflower Asteraceae	P-W	20–24	13	17–20	D or L		Reduce night temperature of medium by 3°C for best germination.
Reseda odorata Mignonette Resedaceae	A	21	15	5–10	L		Difficult to transplant. Best sown directly to garden or container.
Ricinis communis Castor bean plant Euphorbiaceae	A	21–24	15–18	15–20	D	Nick seed coat or soak in warm water soak 24 h	
Rodgersia spp. Rodgers flower Saxifragaceae	P	21–23	10–15		D or L		Seeds are tiny. Do not cover.
Rudbeckia hirta Cone flower, gloriosa daisy Asteraceae	P-W	21–24	16	7–10	L		Reduce night temperature of medium by 3°C for best germination.
Salpiglossis sinuata Painted tongue Solanaceae	A	21–24	10–13	12–15	D		Seeds are very small. Do not cover. Place in an unlighted location to germinate.

Table 2 Continued

Species	Plant type	Medium temperature (°C) Germination	Posttransplant (night)	Days to germinate	Light or dark	Pretreatment	Remarks [notes 1–6 follow last table entry]
Salvia spp. Sage Lamiaceae	A	21–24	16	10–14	L		Seeds need light for only 48 h, then can be covered. Reduce night temperature of medium by 3°C for best germination. Harden-off seedlings at 13°C for 3–4 days before transplanting. Keep salt levels low in transplant medium.
Sanguinaria canadensis Bloodroot Papaveraceae	P-W	10–15	10			Keep at 20°C 2–4 weeks then 5°C 4–6 weeks	
Sanvitalia procumbens Creeping zinnia Asteraceae	A	21	16–18	10	L		Difficult to transplant. Best sown directly to garden or container. Reduce night temperature of medium by 3°C for best germination. Grow on the dry side after germination.
Saponaria ocymoides Soapwort Caryophyllaceae	P	15–20	8–12	14	D	Stratify at 5°C 2–4 weeks	
Saxifraga spp. Saxifrage Saxifragaceae	P	15	15	14–21	D or L	Stratify at 0–3°C 6 weeks	
Scabiosa spp. Pincushion flower Dipsacaceae	P	21	13	10–12	D or L		Reduce night temperature of medium by 3°C for best germination. Germination may be slow and erratic.
Sedum spp. Stonecrop Crassulaceae	P	15–25	10	10–14	L		
Sempervivum tectorum Hens and chickens Crassulaceae	P	21	10–13	10–14	L		Normally propagated vegetatively.

Species / Common name / Family	Type				Light	Pretreatment	Comments
Senecio cineraria Dusty miller Asteraceae	HHP	21–24	16–18	10	L		Predrench medium with fungicide to prevent damp-off.
Silene armeria Catchfly Caryophyllaceae	P	22	12–15	15–20			Seed is small. Cover lightly or not at all.
Sisyrinchium spp. Blue-eyed grass Iridaceae	P-W	21–23	15	21–28	D or L		
Stachys spp. Betony Lamiaceae	P	18–21	10	10–16	L		Reduce night temperature of medium by 3°C for best germination. Do not overwater. Germination may be erratic.
Stokesia laevis Stoke's aster Asteraceae	P	22–25	15–18	28–40		Stratify at 5°C 6 weeks	
Tagetes erecta Asteraceae	A	21–24	16–18	5–8	D or L		Reduce night temperature of medium by 3°C for best germination. *T. erecta* cultivars require 14 h or more of darkness per day at the germination and early seedling stages for proper flowering.
T. patula Marigold Asteraceae	A	21–24	16–18	5–8	D or L		
Teucrium chamaedrys Germander Lamiaceae	P	21	15–18	25–30			
Thalictrum rochebrunianum Meadow rue Ranunculaceae	P	22	15–18	28–40			
Thermopsis spp. False lupine Fabaceae	P	22	15–18	15–20		Nick with a file or soak in H_2SO_4 30 min	Seed vitality is quickly reduced as seed ages. Fresh seed may germinate without scarification.
Thunbergia alata Black-eyed Susan vine Acanthaceae	A	21–24	16	12	D or L		Can be direct-sown to garden or container.

Table 2 Continued

Species	Plant type	Medium temperature (°C)		Days to germinate	Light or dark	Pretreatment	Remarks [notes 1–6 follow last table entry]
		Germination	Posttransplant (night)				
Thymus spp. Thyme Lamiaceae	P	15–17	8–10	7–14	L		
Tiarella cordifolia Foamflower Saxifragaceae	P-W	10–15	8–10	21–28			
Tithonia rotundifolia Mexican sunflower Asteraceae	A	22	18–22	14–21	L		
Torenia fournieri Wishbone flower Scrophulanaceae	A	21–24	13–16	8–12	L		
Tradescantia spp. Spider wort Commelinaceae	P	15–22	13	16–21	D or L		Reduce night temperature of medium by 3°C for best germination.
Tropaeolum majus Nasturtium Tropaealaceae	A	18–21	16	10–14	D or L		Difficult to transplant. Best sown directly to garden or container.
Ursinia anethoides Ursinia Asteraceae	A	20	10	10			
Valeriana officinalis Valerian Valerianaceae	P	15–20	10–20	14–21			
Venidium fastuosum Monarch of the veldt Asteraceae	A	21–24		15–25	L		
Verbascum chaixii Mullein Scrophulariaceae	P	18–23	15–18	15–20			

	Type				Light	Treatment	Comments
Verbena × *hybrida* Garden verbena Verbenaceae	A	18	10–13	12–18	D	Chill at 5°C 7 days	Germination can be slow and erratic. Keep germinating medium on the dry side. Total darkness necessary until germination completed. Apply fungicide before sowing to prevent damp-off.
Veronica spp. Speedwell Scrophulariaceae	P	18–21	7–10	10–16	L		Reduce night temperature of medium by 3°C for best germination.
Viola spp. Pansy, violets Violaceae	P	18–21	10	8–15	D		Reduce night temperature of medium immediately after germination. Apply fungicide before sowing to prevent damp-off. Germination is erratic.
Waldsteinia spp. Barren-strawberry Rosaceae	P	21–24	15–16				Difficult to transplant. Best sown directly to garden or container.
Xeranthemum annuum Immortelle Asteraceae	A	22	18	10–15			Difficult to transplant. Best sown directly to garden or container.
Zinnia spp. Zinnia Asteraceae	A	24–27	21	5–10	D or L		

1. The list is limited to genera and species grown from seeds, primarily in North America, for ornamental and landscape purposes. References 23–30 were used in compiling the table.

2. If seeds require light for germination, maintain adequate moisture around the seeds during the germination period by a mist system, by careful watering with a fine spray, by use of a plastic cover over seed trays, or by watering from the bottom. Otherwise, cover the seed to a depth at least twice its size with vermiculite or sowing medium to ensure that adequate moisture is provided. *Caution*: Plastic covers cause heat buildup, which, if allowed to reach high levels, could be detrimental to seeds.

3. Low temperature, such as that in a refrigerator, increases the longevity of most species of seed. Unless otherwise indicated, store seeds dry in the original sealed packet or in a tightly closed plastic bag or sealed container. Silica gel or other hygroscopic substances may be added to resealed packages or containers to keep humidity low.

4. Freshly harvested (fresh) seeds of many herbaceous species will germinate if sown right away. The same seeds produced from a seed house or garden center have been stored for a period of time and may require chilling (stratification). Chilling temperature (near freezing to 5°C) in a refrigerator or cold room will usually suffice. However, freezing temperature (−10 to −20°C) such as in a household freezer is often recommended. Seeds may be frozen before or after sowing.

5. Fall sowing of herbaceous perennial seeds in outdoor beds, or in flats or containers kept outdoors during the winter, will satisfy their chilling (cold stratification) requirement to promote germination the following spring. This practice may also replace a need for warm stratification or for scarification. Alternately, germination can be scheduled to occur at any time after seeds have been chilled in a refrigerator.

6. Reducing the temperature of the germinating media at night by about 3°C may enhance germination of many herbaceous species. Gradually reducing the temperature during the posttransplant stage allows the development of short, sturdy plants, which are usually more desirable.

Table 3 List of Selected Woody Perennial Trees and Shrubs, Showing Recommended Dates for Seed Collection, Seed Longevity and Storage Requirements, and Pretreatments to Overcome Seed Dormancy

Definitions of plant type, column 2

BES = **Broadleaf evergreen shrub**. A multistemmed woody plant that retains its broad-leaved foliage throughout the year and attains a height usually less than 4 m.

DS = **Deciduous shrub**. A multistemmed woody plant that normally grows to less than 4 m and sheds its leaves at the end of each growing season.

DT = **Deciduous tree**. A woody plant consisting normally of a single trunk or stem with a leaf canopy more than 2 m high. Leaves are shed at the end of each growing season.

ES = **Evergreen shrub**. A multistemmed woody plant that retains its needles throughout the year and attains a height usually less than 4 m.

ET = **Evergreen tree**. A woody tree that retains its needle-shaped leaves throughout the year. Older needles may drop after the second or third year.

| | | | | | Pretreatment | | |
| | | | | | | Stratification (months) | |
Species	Plant type	Collection date	Storage	Scarification	Warm (20–30°C)	Cold (1–5°C)	Remarks [notes 1–6 follow last table entry]
Abies balsamea	ET	Aug–Sept	5 years at 1–5°C and 9–12% moisture content			2–3	Collect cones before they shatter.
A. concolor		Sept–Oct				1–2	Fresh seeds germinate best
A. fraseria		Sept–Oct				1–2	Light required for germination.
A. homolepis		Sept				1–2	See notes 2, 3, and 4.
A. nordmanniana		Sept–Oct				1–2	
A. procera		Sept				1–2	
A. veitchii		Sept–Oct				1–2	

Fir

Pinaceae

Species / common name / family		Collection	Storage	Pretreatment			Remarks
Acer campestre	DT	Oct–Nov	1–2 years at 1–5°C and 10–15% moisture content		1	3–6	Unless otherwise indicated, stratify or fall-sow outdoors before seeds turn completely brown; otherwise, germination may be delayed by up to one year. See notes 4 and 5.
A. ginnala		Oct–Nov			1–2	3–4	
A. negundo		Sept–Mar				2–3	
A. pennsylvanicum		Sept–Oct				3–4	
A. platanoides		Sept–Oct				3–4	
A. pseudoplatanus		Aug–Oct				3	
A. rubrum		May		None—sow immediately after collection.	None	None	
A. saccharinum		May		None—see *A. rubrum.*	None	None	
A. saccharum / Maple / Aceraceae	DT	Aug–Oct				2–3	
Aesculus glabra / Buckeye	DT	Sept–Oct	6–12 months at 5°C			4	Stratify or fall-sow outdoors immediately after harvest to prevent moisture loss from seed.
A. hippocastanum / Horse chestnut / Hippocastanaceae							
Ailanthus altissima / Tree of heaven / Simaroubaceae	DT	Oct–April	Store at 2–5°C in sealed containers	10-day soak may aid germination.		2	Fresh seeds may germinate without pretreatment.
Albizia julibrissin / Silktree / Fabaceae	DT	Fall	18 months at 4–8°C	Soak in H_2SO_4 30 min, or nick seed coat with a file.			Scarification not necessary if sown before seed coat hardens.
Alnus glutinosa / European alder / Betulaceae	DT	Fall–winter	1–2 years at 1–5°C			6 (dried seed only)	Fresh seeds will germinate without pretreatment.
Amelanchier alnifolia / Serviceberry / Rosaceae	DT	July–Aug		Soak in H_2SO_4, 15–30 min	Fresh, none; dried 3–6	3; 3–6	Collect fruits immediately upon ripening to avoid removal by birds; macerate to extract the seeds. Fall-sow fresh seeds outdoors to germinate the following spring. See note 5.

Table 3 Continued

Species	Plant type	Collection date	Storage	Scarification	Pretreatment Stratification (months) Warm (20–30°C)	Cold (1–5°C)	Remarks [notes 1–6 follow last table entry]
Aronia arbutifolia *A. melanocarpa* *A. prunifolia* Chokeberry Rosaceae	DS	Aug–Nov	2 years dry at 1–5°C			2–3	Macerate to extract seeds before sowing.
Asimina triloba Pawpaw Annonaceae	DT	Aug–Sept				2	Collect fruits when flesh is soft. Entire fruit can be sown or seeds can be extracted by maceration.
Berberis thunbergi Japanese barberry Berberidaceae	DS	May–Sept	4–5 years at 1–5°C			1–2	Collect fruits when red.
Betula nigra *B. papyrifera* *B. lenta* *B. pendula* Birch Betulaceae	DT	May–June Aug–Sept Aug–Sept July–Aug	1.5–2 years at room temperature if dried to 1–3% moisture content			None 1–2 1–2 1	Collect seeds while still green to avoid scattering by wind. Seeds are small and thin-coated; tend to lose viability quickly. Long-day photoperiod during germination eliminates need for stratification.
Carpinus betulus *C. caroliniana* Hornbean Betulaceae	DT	Sept	1+ years at 1–5°C and 10% moisture content		Fresh, none; dried, 2	Fresh, 3–4; dried, 2–4	Fall-sow fresh outdoors to germinate the following spring. Viability may be low; test by soaking 6 h in water and sowing only those that sink.

Species / Common name / Family	Type	Season	Storage	Pretreatment	No.	Remarks
Carya glabra Pignut *C. illinoensis* Pecan *C. ovata* Hickory *C. tomentosa* Mockernut Juglandaceae	DT	Sept–Dec	3–5 years at 5°C and 3–5% moisture content	Soak in H$_2$O 1–2 days at 20–25°C	2–4	Sow freshly harvested nuts immediately, or after 1–2 days of soaking at room temperature; germination may be erratic. Alternating day/night temperatures of 30/20°C may aid germination.
Castanea dentata C. mollisima Chestnut Fagaceae	DT	Aug–Oct	1 year at 1–3°C and 40–45% moisture content		1–3	Collect nuts as soon as burrs open; viability is lost if allowed to dry. Cure nuts 1–7 days at 15–22°C out of sunlight No pretreatment necessary.
Catalpa spp. Catalpa Bignoniaceae	DT	Oct–March	2+ years dry at 5°C			
Cedrus atlantica C. deodara Cedar Pinaceae	ET	Fall–winter	3–6 years at 0–5°C and 10% moisture content		1	Collect seeds when cones break open. Pretreatment not usually necessary but germination may be erratic and slow. Spring sowing recommended. Store immediately upon collection; seeds are oily and may deteriorate quickly upon drying.
Celtis occidentalis Hackberry Ulmaceae	DT	Oct–winter	5–6 years at 0–5°C		2–3	
Cercis canadensis Eastern redbud Fabaceae	DT	July–Aug		Soak in H$_2$SO$_4$ for 30 min before cold treatment	2–3	Fall planting without scarification will result in good germination if seeds are fresh and not dried.

Table 3 Continued

Species	Plant type	Collection date	Storage	Scarification	Stratification (months) Warm (20–30°C)	Cold (1–5°C)	Remarks [notes 1–6 follow last table entry]
Chaenomeles japonica Flowering quince Rosaceae	DS	Oct				2–3	
Chamaecyparis nootkatensis	ET	Sept–winter	2–5 years at 0°C and 10% moisture content		1	1	Seedlings are extremely variable; vegetative propagation is recommended for uniform plants. See notes 4 and 5.
C. lawsoniana					None	1	
C. obtusa False cypress Cupressaceae					None	1	
Chionanthus virginicus Fringetree Oleaceae	DT or DS	Oct	1–2 years at 0–5°C		1–3	2–3	Collect seeds when fruits turn purple. Fall sowing will produce seedlings the second spring. See note 5.
Cladrastis lutea Yellowwood Fabaceae	DT	Oct–Dec	3 years at 1–5°C	Soak in H2SO4 30 min before or instead of cold treatment		3	
Cornus alternifolia	DT	Aug–Sept			2–5	2–3	Collect fruits as soon as ripened to avoid removal by birds. Fleshy coverings may enhance dormancy and should be removed. See note 5.
C. canadensis	DS	Aug–Sept			3	3	
C. florida	DT	Oct				3	
C. kousa	DT	Sept–Oct.				3–4	
C. mas	DT	June–July			4–5	3	
C. racemosa	DS	Aug–Sept				3–4	
C. sericea Dogwood Cornaceae	DS	Aug–Oct				3	

Species / Common name / Family							Notes
Cotoneaster apiculata C. *horizontalis* Cotoneaster Rosaceae	BES	Oct	2 years dry at 5°C	Soak in H₂SO₄ 2 h Soak in H₂SO₄ 2 h		3 3–4	Warm stratification for 5–6 months before cold treatment may be used instead of acid scarification. Acid-treated seeds may be sown outdoors in fall.
Crataegus arnoldiana C. *crus-galli* C. *mollis* C. *phaenopyrum* Hawthorn Rosaceae	DT DT DT DT	Oct– winter	3 years at 5°C	4–5 h, H₂SO₄ 2–3 h	3	6 3 3 3	Macerate fruits in water to remove pulp from seeds Freshly harvested seeds may need less pretreatment time. Seed coat may become impermeable after drying. See notes 4, 5, and 6. Outdoor fall sowing may replace scarification.
Cytisus scoparius Scotch broom Fabaceae	BES	Sept	10+ years at 5°C	Repeated hot water soaks or 30 min in H₂SO₄		2–3	
Diospyros virginiana Persimmon Ebenaceae	DT	Oct–Nov					
Elaeagnus angustifolia Russian olive Elaeagnaceae	DS	Aug– winter	1–3 years at 5°C and 6–14% moisture content			2–3	
Euonymus alatus Winged euonymus Celastraceae	DS	Aug–Sept	Moist, cold storage			3–4	Collect fruits just as capsules start to split. Store moist. Drying reduces germination. Do not allow seeds to dry; sow outdoors immediately to germinate in spring.

Table 3 Continued

Species	Plant type	Collection date	Storage	Scarification	Pretreatment — Stratification (months) Warm (20–30°C)	Cold (1–5°C)	Remarks [notes 1–6 follow last table entry]
Fagus grandifolia F. sylvatica Beech Fagaceae	DT	Oct–Nov	3+ years at 1–5°C and 8–10% moisture content			3 3	Separate sound seeds by flotation method; seed should not dry out. Fall-sow outside or treat as soon as possible after collecting. See note 6.
Fraxinus americana F. excelsior F. ornus F. pennsylvanica Ash Oleaceae	DT	Oct–Nov	7+ years at 5°C and 7–10% moisture content		1 2–3 2 2	2 3 3 2–6	Collect and treat or sow as soon after ripening as possible. Older seed becomes more dormant. See note 5.
Ginkgo biloba Maidenhair tree Ginkgoaceae	DT	Late fall	1 year if stored moist		1–2	1–2	Collect fruits when dropped. Remove seeds by maceration in water before storing or sowing. Fresh seeds will germinate but at lower percentage. See note 5.
Gleditsia triacanthos Honeylocust Fabaceae	DT	Sept–winter	2+ years at 0–5°C	Soak in H_2SO_4 1–2 h; or hot water soak, leave overnight		none	
Gymnocladus dioicus Kentucky coffeetree Fabaceae	DT	Winter	5+ years dry	Soak in H_2SO_4, 2–4 h or nick seed coat with a file		none	Seeds are large.

Species / Common name / Family		Collection	Storage	Pretreatment	Stratification		Remarks
Halesia carolina Carolina silverbell Styracaceae	DS	Oct–Nov	1–2 years dry at 0–5°C		(2) 6 following first prechilling	(1) 3 initially (3) 4–5 after warm	Very complex dormancy; require cold–warm–cold in sequence shown in Stratification columns for stages (1)–(3), but outcome not dependable. Most reliable germination is obtained by fall sowing outdoors. Germination will occur the second spring. See note 5.
Hamamelis virginiana Witchhazel Hamamelidaceae	DS	Aug–Oct	1+ years at 0–5°C		3	3	Collect capsules while still yellowish and before open to avoid natural seed dispersal. Also, if fall-sown early, germination may occur the first spring. If seed is ripe and dry, a second season will be needed. See note 5.
Ilex aquifolium *I. glabra* *I. opaca* *I. verticillata* *I. vomitoria* Holly Aquifoliaceae	BES	Oct–winter	Dry, in sealed container	Soak in H₂SO₄ 30 min	None—fall-sow None—fall-sow 2 2 None—fall-sow	3 2	If fall-sown outdoors, *Ilex* seed may require up to 3 years in nature to complete germination. Fresh seeds sown immediately after collection may hasten germination by up to one year for *I. vomitoria*. Choose either acid treatment or fall sowing. See note 5.

Table 3 Continued

Species	Plant type	Collection date	Storage	Pretreatment Scarification	Stratification (months) Warm (20–30°C)	Cold (1–5°C)	Remarks [notes 1–6 follow last table entry]
Juglans nigra Black walnut Juglandaceae	DT	Oct–Nov	1+ years at 1–5°C and 20–50% moisture content			3–4	Protect against rodents, if fall-sown outdoors.
Juniperus communis *J. virginiana* Juniper Cupressaceae	ES	Sept.–winter	2+ years at 1–5°C and 10–12% moisture content		2–3	3 2–3	See note 5.
Kalmia latifolia Mountain laurel Ericaceae	BES	Oct–Nov	10+ years dry at room temperature				No special pretreatment required. Light required for germination.
Koelreuteria paniculata Golden rain tree Fabaceae	DT	Sept–Oct	1–2 years dry at 1–5°C	1–2 h in H$_2$SO$_4$ for fresh seed, or soak in hot water and let cool for 24 h		1	Collect before capsules open. Seeds may germinate after acid treatment only, depending on the state of dormancy. Fall-sown fresh seed gives best results.
Larix decidua *L. laricina* Larch, tamarack Pinaceae	DT (conifer)	Sept–winter	3+ years at 1–5°C			1–2	Fresh seeds may germinate without pretreatment if exposed to light after sowing.
Ligustrum amurense *L. japonicum* *L. lucidum* *L. vulgare* Privet Oleaceae	DS	Sept–Nov	Low temperature, sealed container			2–3	Fresh seeds may germinate without cold treatment. Stratification times refer to stored seed.

Species / Common name / Family	Type	Collection	Storage	Pretreatment			Remarks
Lindera benzoin Spicebush Lauraceae	DS	Sept–Oct	< 1 year: keep cool and sealed		1	3	Extract seeds by maceration. Sow outdoors in fall for best germination the following spring.
Liquidambar styraciflua Sweet gum Hamamelidaceae	DT	Sept–Nov	4+ years sealed at 5°C and 10–15% moisture content			1–3	Pick fruit when green color fades to avoid natural seed dispersal.
Liriodendron tulipifera Tulip tree Magnoliaceae	DT	Aug–Oct	Several years if stored dry at 1–5°C			2–3	Low percentage of viable seeds per cone (10%). Sow fresh, untreated seed in fall to germinate following spring.
Lonicera maackii *L. tatarica* Honeysuckle Caprifoliaceae	DS	Sept–Nov July–Aug	15+ years at 1–5°C			1–3	Low percentage of viable seeds per cone (10%). Sow fresh, untreated seed in fall to germinate following spring.
Maclura pomifera Osage orange Moraceae	DT	Sept–Oct	3+years at 5°C	Soak fresh seed in water for 2 days; replaces stratification		1	Collect fruits when dropped. Extract seeds by macerating fruits in water. Fruits may be allowed to ferment over winter in a heap. This makes seed separation easier and eliminates need for stratification.
Magnolia acuminata *M. grandiflora* *M. virginiana* Magnolia Magnoliaceae	DT	Aug–Sept	2+ years at 1–5°C			3–6	Collect when follicle opens, exposing the seeds. Fresh seeds with fleshy pulp provide best results. Macerate to remove pulp. Dried seeds lose viability.

Table 3 Continued

Species	Plant type	Collection date	Storage	Scarification	Pretreatment Stratification (months) Warm (20–30°C)	Cold (1–5°C)	Remarks [notes 1–6 follow last table entry]
Malus baccata *M. coronaria* *M. floribunda* *M. ioensis* Crabapple Rosaceae	DT	Sept–Oct	2+ years at 1–5°C and < 11% moisture content			1 4 2–4 2	Extract seeds from fruit by maceration before treating or sowing.
Metasequoia glyptostroboides Dawn redwood Taxodiaceae	DT (conifer)	Oct–Nov				1	Collect seeds when cones begin to open. Fresh seeds need no treatment.
Morus alba Mulberry Moraceae	DT	June–Aug	1+ years at −10 to −20°C			1–3	Harvest fruits as soon as ripened to avoid removal by birds. Macerate to separate pulp from seeds. Fresh seeds can be sown immediately for reasonable germination.
Myrica pennsylvanica Bayberry Myricaceae	BES	Oct–winter	10–15 years if dewaxed, dried, and stored at 1–5°C in sealed containers			1–3	Remove wax by rubbing over a screen before stratifying or sowing.
Nyssa sylvatica Tupelo Nyssaceae	DT	Sept–Nov	1+ year dry at 1–5°C			1–3	Pulp removal by maceration is often recommended but not necessary.

Ostrya virginiana Hop hornbeam Betulaceae	DT	Aug–Oct		3	3–5	Collect fruits when pale greenish brown and before seeds are naturally dispersed.
Parthenocissus quinquefolia Virginia creeper Vitaceae	DV	Sept–winter	2+ years at 1–5°C		2	Collect fruits when bluish black in color. Macerate to remove pulp.
Paulownia tomentosa Empress tree Bignoniaceae	DT	Sept–Oct				Collect fruits before they open and seeds are naturally dispersed. No pretreatment required. Light required for germination.
Phellodendron amurense Amur corktree Rutaceae	DT	Sept–Nov	Dry until sowing		2 (only for stored seed)	Macerate fruit to remove pulp. No pretreatment necessary for fresh seeds.
Picea abies	ET	Sept–Nov	5–20 years at 1–5°C and 4–8% moisture content		None	Collect cones before they shatter.
P. *engelmanni*		Sept–Oct			None	Generally no pretreatment required, so fall sowing is not recommended.
P. *glauca*		Sept			None	Cold stratification of 1 month may hasten and unify germination,
P. *omorika*		Oct			None	which takes place following spring sowing
P. *pungens*		Fall			None	See note 4.
P. *sitchensis* Spruce Pinaceae		Oct–spring			None	

Table 3 Continued

Species	Plant type	Collection date	Storage	Scarification	Pretreatment Stratification (months) Warm (20–30°C)	Cold (1–5°C)	Remarks [notes 1–6 follow last table entry]
Pinus cembra	ET	Aug–Oct	1+ years	For those requiring stratification first, soak in water for 1–2 days.		3–4	Seed provenance is an important consideration to future survival and growth characteristics.
P. densiflora	ET	Aug–Oct	2–5 years			0–1	
P. mugo	ES	Oct–Dec	5 years			None	
P. nigra	ET	Sept–Nov	10+ years			0–2	Collect cones of most species as soon as ripened and starting to crack open; otherwise seeds will be dispersed.
P. parviflora	ET	Sept–Nov	30 years			3	
P. resinosa	ET	Aug–Nov	10 years			2	
P. strobus	ET	Aug–Sept	15 years			2	
P. sylvestris	ET	Sept–March	11 years			1–3	Fresh seeds of some species requiring stratification may be sown without pretreatment with some success.
P. thunbergii	ET	Oct–Dec	5+ years			1–2	
P. virginiana	ET	Sept–Nov	5– years at 1–5°C and 5–10% moisture content			0–1	
Pine Pinaceae							
Platanus occidentalis	DT	Nov–Feb	1+ years at 1–5°C and 10–15% moisture content			None	Cold stratification for 2 months may improve germination.
P. orientalis Sycamore, planetree Platanaceae						None	
Populus deltoides	DT	May–June	2–3 years if air-dried 4 days and stored sealed at 5°C and 5–8% moisture content			None	Clean seeds before sowing Fresh seeds germinate rapidly.
P. tremuloides Poplar, aspen Salicaceae							

Species		Collection	Storage			Notes
Prunus armeniaca Apricot	DT	May–June	Many conditions tried; air-dried, several years at 0–5°C in sealed containers		1–5	Extract seeds by maceration and flotation Fall sowing eliminates need for prechilling, but seeds should be sown early enough to satisfy this requirement before freezing occurs. Mulching the seed beds delays freezing. See notes 4, 5, and 6.
P. avium Mazzard cherry	DT	June–July		1	3–4	
P. besseyi Western sand cherry	DS	July–Sept			3	
P. cerasifera Myrobolan plum	DT	July–Aug			3	
P. domestica Plum	DT	July–Oct			3	
P. persica Peach	DT	July–Oct			2	
P. pumila Sand cherry	DT	July–Sept			3	
P. serotina Black cherry	DT	Aug–Sept		1	4	
P. virginiana Choke cherry Rosaceae	DT	Aug–Sept			3	
Pseudotsuga menziesii Douglas fir Pinaceae	ET	Aug–Sept	3–4 months stored as cones dry and warm; 10–20 years at 0°C and 6–9% moisture content		1	Collect seeds when golden brown. Fall sowing not recommended as germination may begin too early. No pretreatment required but, if given, may enhance germination percentage and rate.
Pyrus communis Pear Rosaceae	DT	July–Oct	2–3 years at 1–5°C and 10% moisture content		3–4	Macerate fruit to extract seed. Presoaking seed in water for 24 h may improve stratification results.

Table 3 Continued

Species	Plant type	Collection date	Storage	Scarification	Stratification (months) Warm (20–30°C)	Cold (1–5°C)	Remarks [notes 1–6 follow last table entry]
Quercus spp.	DT	Aug–Dec	6 months or less for black oak group (B); white oak group (W) has almost no storage capabilities				With few exceptions, acorns of the white oak (W) group have little or no dormancy and will germinate immediately after dropping from the tree. Acorns of the black oak (B) group exhibit embryo dormancy. Fall-sow outdoors for germination the following spring. See note 4.
Q. alba (W) White oak						None	
Q. bicolor (W) Swamp white oak						None	
Q. coccinea (B) Scarlet oak						1–2	
Q. imbricaria (B) Shingle oak						1–2	
Q. macrocarpa (W) Bur oak						0–2	
Q. muhlenbergii (W) Chinkapin oak						None	
Q. nigra (B) Water oak						1–2	
Q. palustris (B) Pin oak						1–2	
Q. phellos (B) Willow oak						1–2	
Q. robur (W) English oak						None	
Q. rubra (B) Red oak						1–2	
Q. shumardii (B) Shumard oak						1–2	
Q. velutina (B) Black oak						1–2	
Q. virginiana (W)						None	

Species / Common name / Family	Method	Collection time	Storage	Treatment	Stratification (weeks)	Remarks
Live oak Fagaceae						
Rhamnus frangula Buckthorn Rhamnaceae	DS	July–Oct	2+ years at 1–5°C	Soak in H$_2$SO$_4$, 20 min	2–3	Collect seeds as soon as ripe before removal by birds.
Rhododendron maximum *R. catawbiense* Rhododendron Ericaceae	ES	Sept–Oct July–Oct	2 years at room temperature and 4–9% moisture content		None None	Collect seeds as soon as capsules begin to turn brown and before they open. Light is required for germination.
Ribes alpinum Alpine currant Saxifragaceae	ES	June–July			3–6	Macerate fruit in water to remove fleshy pulp.
Robinia pseudoacacia Black locust Fabaceae	DT	Aug–winter	Up to 10 years at 1–5°C	Nick with a file or soak 10–12 min in H$_2$SO$_4$.		Collect fruits before pods split open and seeds are dispersed. Seeds are large.
Rosa canina *R. multiflora* *R. rugosa* Rose Rosaceae	DS	Sept–Oct	2–4 years at 1–5°C	Soak 45 min in H$_2$SO$_4$ instead of warm stratification	3–5 3	Collect seeds fresh for best results. Hips should be just turning red. Macerate in water to remove pulp. Fresh seeds can be fall-sown immediately after cleaning. Alternative to warm stratification is scarification. See note 5.
Salix discolor *S. nigra* Willow Salicaceae	DT	April–May	4–6 weeks		None	Very short-lived. Viability 10 days at room temperature. Sow immediately after collection—no dormancy.

Table 3 Continued

Species	Plant type	Collection date	Storage	Scarification	Pretreatment Stratification (months) Warm (20–30°C)	Cold (1–5°C)	Remarks [notes 1–6 follow last table entry]
Sambucus canadensis Common elder Caprifoliaceae	DS	June–Aug	2 years dry at 5°C	Soak 10–20 min in H_2SO_4 instead of warm stratification	2	3–5	Collect fruits as soon as ripened to avoid removal by birds. Macerate in water to remove pulp. Alternative to warm stratification is scarification. See note 5.
Sassafras albidum Sassafras Lauraceae	DT	Sept	2+ years at 5°C			4	Collect seeds when color changes to dark blue. Remove pulp by maceration. Fall-sow as late as possible to prevent fall germination.
Sciadopitys verticillata Umbrella pine Taxodiaceae	ET	Fall	2 years at 5°C and 10% moisture content		3	3	See note 5.
Sorbus americana S. aucuparia Mountain ash Rosaceae	DT	Aug–Oct	2–8 years at 5°C and 6–8% moisture content			3	Collect fruit when just ripened to avoid removal by birds and to reduce length of pretreatment time. Remove pulp by maceration.

Species							Remarks
Symphoricarpos orbiculatus Coralberry Caprifoliaceae	DS	Sept–Oct	2 years at 5°C	Soak 20–60 min in H$_2$SO$_4$ in place of or in addition to warm stratification	3–4	4–6	Soak fruit several days then macerate to remove pulp. See note 4.
Syringa amurensis *S. persica* *S. vulgaris* Lilac Oleaceae	DS	Aug–Oct	2 years dry at 1–5°C			1–3	Seeds may germinate without pretreatment.
Taxodium distichum Bald cypress Taxodiaceae	DT	Oct–Dec	1+ years dry at 5°C			2–3	
Taxus baccata *T. canadensis* *T. cuspidata* Yew Taxaceae	ES	Aug–Oct July–Sept Oct–Nov	5–6 years dry at 5°C		5–7	2–4	Collect fruits as soon as ripened to avoid removal by birds. Macerate to remove pulp and do not allow to dry. Seeds sown after collecting in July will germinate naturally the second spring. Seeds allowed to dry after harvest will take longer to germinate or will need longer pretreatment times than fresh seed. See note 5.

Table 3 Continued

Species	Plant type	Collection date	Storage	Pretreatment Scarification	Stratification (months) Warm (20–30°C)	Stratification (months) Cold (1–5°C)	Remarks [notes 1–6 follow last table entry]
Thuja occidentalis *T. orientalis* Arborvitae Cuppressaceae	ES	Aug–Sept	5+ years at 1–5°C and 6–8% moisture content			2	Collect seeds as soon as cones turn light brown and before seeds are naturally dispersed. Some seed lots require no pretreatment. Seeds may require light to germinate.
Tilia americana *T. cordata* Linden Tiliaceae	DT	Sept–winter	2–3 years at 1–5° C with 10–12% moisture content	Soak in HNO_3 30 min, then H_2SO_4 for 30–60 min		3	For best results, harvest fruits as soon as ripened. Treat seeds immediately. Dried seed has deeper dormancy. Both fall and spring sowing result in germination over a 2–3 year period. Seeds have both hard seed coat and impermeable outer pericarp, hence the need for both acid scarification treatments. See note 5.

Species / Common name / Family	Code	Sowing	Storage			Comments
Tsuga canadensis	ET	Sept–winter	4 years at 1–5°C and 6–9% moisture content	1–4		Spring sowing is recommended. Fall sowing eliminates need for chilling but results in poor seedling survival. See note 4.
T. caroliniana						
Hemlock						
Pinaceae						
Ulmus americana	DT	March–June	15 years at –3°C and 3–4% moisture content	2–3		Sow seeds with wings attached.
U. parvifolia		Sept–Oct				
Elm						
Ulmaceae						
Viburnum acerifolium	DS	July–Oct	3+ years dry at 1–5°C	3	3–9	Macerate fruits in water to remove pulp. See note 5.
V. dentatum				1	10–12	
V. lantana				2	None	
V. lentago				2–4	5–9	
V. opulus				3	5	
Viburnum						
Caprifoliaceae						

Notes

1. The list is limited to genera and species grown in temperate North America primarily for landscape and ornamental purposes. References 18, 20, and 31–35 were used in compiling the table.

2. If seeds require light for germination, maintain adequate moisture around the seeds during the germination period by a mist system, by careful watering with a fine spray, by use of a plastic cover over seed trays, or by watering from the bottom. Otherwise, cover the seed to a depth at least twice its size with vermiculite or sowing medium to ensure that adequate moisture is provided. Very small seeds are usually sown on top of the germinating medium and are not covered. *Caution:* Plastic covers cause heat buildup, which, if allowed to reach high levels, could be detrimental to seeds.

3. Low temperature, such as that in a refrigerator, increases the longevity of most species of seed. Unless otherwise indicated, store seeds dry in the original sealed packet or in a tightly closed plastic bag or sealed container. Silica gel or other hygroscopic substances may be added to resealed packages or containers to keep humidity low.

4. Fall sowing of woody perennial seeds in outdoor beds, or in flats or containers kept outdoors during the winter, will satisfy the chilling (cold stratification) requirement to promote germination the following spring. This practice may also eliminate a need for warm stratification or for scarification. Alternately, germination can be scheduled to occur at any time after chilling seeds in a refrigerator.

5. Seeds with complex dormancy may require both warm and cold stratification in sequence. If sufficient heat is gained by unstratified seeds after a late summer or fall seeding to satisfy the warm requirement, germination will occur the following spring. Otherwise, germination will occur the second spring.

6. Seeds must usually be removed from fleshy fruits. Mechanical devices such as a fruit press can be used to macerate the fruit. Seeds can then be separated by flotation, screening, fermentation, or other means. Nonviable seeds are light and will usually float in water; viable seeds are heavier and will sink.

inhibition. However, the combined application of gibberellin and cytokinin can induce germination of dormant seeds in a wider range of species than either chemical administered separately.

Ethylene also breaks dormancy and initiates germination, but its effect is not as well documented as other growth regulators. According to Bell et al. [10], water-soluble chemical factors from charred wood or smoke may stimulate germination of some seeds. Interestingly, ethylene is a component of wood smoke. Ethylene released from seeds may also stimulate their germination [11].

D. Priming

Seed priming (osmoconditioning) has proven to be quite effective in improving germination, seedling emergence, and yield of many early-planted, small-seeded vegetable and flower crops. Priming refers to conditioning seeds in an aerated solution with a high solute content, which keeps the seed in a partially hydrated state [37]. Polyethylene glycol (PEG), an inert compound, is often used, although some systems use salt solutions of various compositions (Table 4) [38–43]. Primed seeds may be sown moist or dried, or even stored for later use.

Seeds treated with osmotic solutions of –1.0 to –1.7 MPa water potential may germinate more rapidly and uniformly under a wider range of temperatures than untreated seeds. The water potential of the priming solution, the priming temperature, and the duration of priming are all important if radicle elongation is to be prevented, while at the same time most other germination processes are allowed to proceed.

Osmoconditioning does not affect stage I water uptake (imbibition) because the priming solutions have much higher water potentials than the water potential of the colloidlike seed tissue. However, the stage II processes (active metabolism and hydrolysis) are triggered during the priming treatment. Thus, osmoconditioning enables the seed to imbibe enough water to become metabolically active and accumulate reserves of sugars, amino acids, proteins, and other substances required for germination. Since the osmotic potential of the priming solution is just negative enough to prevent the occurrence of the stage III phase of seed hydration (visible germination), radicle elongation is restricted.

The primed seeds germinate uniformly and rapidly once the osmotic stress has been relieved, and the final phase of seed hydration occurs. The water potential of the osmoconditioning solution varies among osmotica and species. Some examples are shown in Table 4.

A more recent improvement of the seed priming technique, referred to as matriconditioning, involves the use of a protective gel or colloidal agent that is high in water absorptive property instead of an osmotic solution [44,45]. Matriconditioning may be better suited than osmoconditioning for the treatment of large amounts of seeds.

Seed priming, either osmoconditioning or matriconditioning, may be integrated with fluid drilling (pregerminated seeds suspended in a protective gel) to improve plant emergence and performance under field conditions. Different gels are often tried, and growth regulators, fertilizers, and pesticides are incorporated into the gels in attempts to increase the effectiveness of the technique. Information on this type of seed technology is just unfolding.

E. Other Treatments

The pelleting of seeds has been used for a long time, with varied success. Although there are different ways of pelletizing seeds, in the simplest procedure, seeds are placed in a rotating drum and coated with a liquid binder and dust. The procedure results in uniform-sized, spherical pellets that facilitate more precise planting and often in increased and more uniform germination [2,46].

Table 4 Examples of Successful Seed Osmoconditioning Treatments for Selected Species

Species	Osmoconditioning treatments					Ref.
	Temperature (°C)	Duration (days)	Chemical	Amount (g Kg^{-1} H$_2$O)	Estimated water potential (MPa)[a]	
Beta vulgaris (sugar beet)	15	7	PEG 8000	302	-1.22	38
Daucus carota (carrot)	15	28	K$_3$PO$_4$ + KNO$_3$	21.65 20.6	-1.5	39
	15	14	K$_2$HPO$_4$ + KNO$_3$	18.28 21.1	-1.69	39
Allium cepa (onion)	15	14	PEG 8000	342	-1.55	40
Apium graveolens (celery)	15	14	PEG 8000	273	-1.0	41
Lycopersicum esculentum (tomato)	15	14	K$_2$HPO$_4$ + KNO$_3$	15.67 11.92	-1.0	39
Petroselinum crispum (parsley)	15	21	PEG 8000	296	-1.17	42

[a]Water potential estimates for PEG 8000 from Ref. 43.

Grass, vegetable, flowers, and sometimes seeds of woody species have been "seeded" in plastic rolls, in tapes, or in water-absorbent, fibrous mats. The seeds are held in position by water-soluble adhesives [46]. These procedures simplify planting and may result in more uniform germination and seedling establishment. The roll, tape, or mat serves as a mulch and provides a more uniform germinating environment. Fertilizers, inoculants, insecticides, fungicides, and other chemicals are often added to improve the effectiveness of these products [2].

VII. COLLECTION AND STORAGE

Seed companies regularly conduct germination tests in controlled laboratory conditions to determine the relative viability of seed lots and to maintain quality control. These tests also determine the temperature and moisture limits for successful storage of each type of seed [8].

Each type of seed must be collected, handled, and stored differently. In theory, seeds are ready to harvest when there is no further increase in weight. In practice, choosing the right time to harvest may be uncertain, since so many factors influence seed development [18]. Seeds from different species mature at different times of the year (Table 3). Some seeds that appear ripe may in fact contain undeveloped embryos. Fruits have many different shapes and sizes; they may be fleshy or dry, dehiscent or indehiscent. The fleshy coverings of some fruit may contain substances that inhibit germination and must be removed. Removing such coverings lessens the chance for bacterial or fungal growth, which may affect seed viability. Freshly harvested seeds of some species may require no pretreatment, or less stratification time, than those that have been dried and/or stored. The seed coat of *Crataegus* spp. (hawthorn), while not impermeable when freshly collected, becomes so after drying (Table 3).

Under normal conditions, many seeds are relatively short-lived or deteriorate with time. Seeds of *Acer saccharinum* remain viable for only a few days if they are not kept moist and cool (Table 3). *Salix* (willow) and *Populus* (poplar) seeds are viable for only 4 weeks, but many other seeds remain viable from several months to 15 years, and some longer. Since many woody species do not produce seed abundantly each year, commercial seed companies must collect and store seeds of these species for many years. Therefore, many different methods of collecting, handling, and storage are required. These methods have been described by other authors [2,8,18,20,47].

Under proper storage conditions, seeds of most species can be kept viable for a long time. Keeping them dry (usually 5–12% moisture content) and cool are the most important factors affecting longevity and viability. A temperature range of 0–5°C is usually adequate for most species, although lower temperatures may be acceptable for some. Freeze-drying at temperatures below 1°C with moisture control appears to offer the best storage conditions [11] but is not an economical way to store most seeds.

VIII. SUMMARY AND CONCLUSION

The geographic location or provenance of a seed can substantially influence its germinability. Seeds collected from different geographic sources may not germinate or perform uniformly under the same conditions [18]. Those from a more southerly location may require a shorter stratification period to overcome dormancy and may result in plants that are less winter-hardy in a more northerly location. Preharvest environmental conditions (which affect seed maturation) or seed handling procedures and humidity and temperature of storage (which affect the permeability of the seed coat) may cause seed treatments to yield different results between seed lots of the same species, or from year to year [16]. Therefore, treatments and other requirements as listed in Tables 2 and 3 should be considered as guides and may need to be modified to compensate for variations in seed condition.

Differences in germination requirements have evolved in response to species adaptation to changing environments or to selection pressure by cultivation and breeding. While seeds of most domesticated plants have been selected for ease and predictability of germination, the germination requirements of wild species, or those closer to their wild ancestry, appear to be more clued to ecological and environmental influences. Because of the complex interactions of the preharvest and postharvest history of seeds, and because of the large number of seeds for which germination requirements are unknown or not fully characterized, germination studies will continue to challenge seedsmen and plant propagators.

REFERENCES

1. C. Chong, *Plant Propagation*, Vol. I (B. R. Christie, ed.), CRC Press, Boca Raton, FL, p. 91 (1987).
2. B. Macdonald, *Practical Woody Plant Propagation for Nursery Growers*, Vol. I, Timber Press, Portland, OR (1986).
3. H. L. Wick and G. A. Walters, *Albizia*, Agriculture Handbook 450, U.S. Department of Agriculture Forest Service, Washington, DC, p. 203 (1974).
4. S. L. Krugman and J. L. Jenkinson, *Pinus*, Agriculture Handbook 450, U.S. Department of Agriculture Forest Service, Washington, DC, p. 598 (1974).
5. K. A. Brinkman and H. M. Phipps, *Lindera*, Agriculture Handbook 450, U.S. Department of Agriculture Forest Service, Washington, DC, p. 503 (1974).
6. D. F. Olson and W. J. Gabriel, *Acer*, Agriculture Handbook 450, U.S. Department of Agriculture Forest Service, Washington, DC, p. 187 (1974).
7. *What You Should Know About Seeds*, Ontario Factsheet AGDEX 500/010, Ontario Ministry of Agriculture and Food, Ontario, Canada (1987).
8. H. T. Hartmann, D. E. Kester, and F. T. Davies, *Plant Propagation: Principles and Practices*, 5th ed., Prentice-Hall, Englewood Cliffs, NJ (1990).
9. J. Bewley and M. Black, *Seeds: Physiology of Development and Germination*, Plenum Press, New York (1985).
10. D. T. Bell, J. A. Plummer, and S. K. Taylor, *Bot. Rev.*, *59*: 24 (1993).
11. R. P. Poincelot, *Horticulture: Principles and Practical Applications*, Prentice-Hall, Englewood Cliffs, NJ (1980).
12. H. A. Borthwick, S. B. Hendricks, M. W. Parker, E. H. Toole, and V. K. Toole, *Proc. Natl. Acad. Sci. U.S.A.*, *38*: 662 (1952).
13. R. B. Taylorson and H. A. Borthwick, *Weed Sci.*, *17*: 48 (1969).
14. B. Cumming, *Can. J. Bot.*, *41*: 1211 (1963).
15. S. B. Hendricks and R. B. Taylorson, *Nature*, *237*: 167 (1972).
16. J. P. Mahlstede and E. S. Haber, *Plant Propagation*, Wiley, New York (1957).
17. B. M. Pollock and V. K. Toole, *After Ripening, Rest Period, and Dormancy*, U.S. Department of Agriculture Yearbook of Agriculture, Government Printing Office, Washington, DC, p. 106 (1961).
18. M. A. Dirr and C. W. Heuser, Jr., *The Reference Manual of Woody Plant Propagation: From Seed to Tissue Culture*, Varsity Press, Athens, GA (1985).
19. P. O. Rudolf, *Collecting and Handling Seeds of Forest Trees*, U.S. Department of Agriculture Yearbook of Agriculture, Government Printing Office, Washington, DC, p. 221 (1961).
20. J. A. Young and C. G. Young, *Seeds of Woody Plants in North America*, Dioscorides Press, Portland, OR (1992).
21. M. C. Kains and L. M. McQueston, *Propagation of Plants*, Orange Judd Publishing, New York (1944).
22. H. T. Hartmann, W. J. Flocker, and A. M. Kofranek, *Plant Science: Growth, Development and Utilization of Cultivated Plants*, Prentice-Hall, Englewood Cliffs, NJ (1981).
23. *Seed and Plant Cultural Guide*, Jack Van Klaveren, St. Catharines, Ontario, Canada (1989).

24. A. M. Armitage, *Herbaceous Perennial Plants*, Varsity Press, Athens, GA (1989).
25. V. Ball (ed.), *Ball Red Book*, 14th ed., Reston Publishing, Reston, VA (1986).
26. J. Bennett and T. Forsyth, *The Harrowsmith Annual Garden*, Camden House Publishing, Ontario, Canada (1990).
27. B. Ferguson (ed.), *Color with Annuals*, Ortho Books, Chevron Chemical Co., San Ramon, CA (1987).
28. L. Jelitto and W. Schacht, *Hardy Herbaceous Perennials*, Vols. I and II, Timber Press, Portland, OR (1990).
29. H. R. Phillips, *Growing and Propagating Wildflowers*, University of North Carolina Press, Chapel Hill, (1985).
30. H. K. Tayama, and T. J. Roll (eds.), *Tips on Growing Bedding Plants*, Bulletin FP-763, Ohio Cooperative Extension Service (1989).
31. M. A. Dirr, *Manual of Woody Landscape Plants* (rev.), Stepes Publishing, Champaign, IL (1977).
32. T. A. Fretz, P. E. Read, and M. C. Peale, *Plant Propagation Lab Manual*, 3rd ed., Burgess Publishing, MN (1979).
33. R. C. Hosie, *Native Trees of Canada*, Queen's Printer, Ottawa, Ontario, Canada (1980).
34. P. D. A. McMillan-Browse, *Hardy Woody Plants from Seed*, Grower Books, London (1979).
35. J. S. Wells, *Plant Propagation Practices*, American Nurseryman Publishing, Chicago (1985).
36. J. D. Ross and J. W. Bradbeer, *Planta*, *100*: 288 (1971).
37. A. A. Khan, *Hortic. Rev.*, *13*: 131 (1992).
38. G. Murray, J. B. Swensen, and J. J. Gallian, *HortScience*, *28*: 31 (1993).
39. A. M. Haig, E. W. R. Barlow, F. L. Milthorpe, and P. J. Sinclair, *J. Am. Soc. Hortic. Sci.*, *111*: 660 (1986).
40. P. A. Brocklehurst and J. Dearman, *Ann. Appl. Biol.*, *105*: 391 (1984).
41. P. A. Brocklehurst and J. Dearman, *Ann. Appl. Biol.*, *102*: 577 (1983).
42. W. Heydecker and P. Coolbear, *Seed Sci. Technol.*, *5*: 353 (1977).
43. B. E. Michel, O. K. Wiggins, and W. H. Outlaw, *Plant Physiol.*, *72*: 60 (1983).
44. A. G. Taylor and G. E. Harman, *Annu. Rev. Phytopathol.*, *28*: 321 (1990).
45. R. Madakadze, E. M. Chirco, and A. A. Khan, *J. Am. Soc. Hortic. Sci.*, *118*: 330 (1993).
46. L. H. Purdy, J. E. Hamond, and G. B. Welch, *Special Processing and Treatment of Seeds*, U.S. Department of Agriculture Yearbook of Agriculture, Government Printing Office, Washington, DC, p. 322 (1961).
47. *Seeds of Woody Plants of the United States*, Agriculture Handbook 450, U.S. Department of Agriculture Forest Service, Washington, DC (1974).

6
Cell Cycle Control in Plants

A. S. N. Reddy
Colorado State University, Fort Collins, Colorado

I. INTRODUCTION

Cell division is one of the fundamental processes of growth and development of plants and animals. The time and place of cell division in an organism play a critical role in many developmental processes. The development of a complex organism with a defined form and structure requires tightly regulated cell growth and proliferation as well as transitions from cycling state to quiescent state and vice versa. To duplicate the genetic material and produce two daughter cells, the cell goes through a set of orderly events generally referred to as cell cycle. The cell cycle consists of four distinct phases called gap1 (G_1), synthetic phase (S), gap2 (G_2), and mitosis (M). In the G_1 phase cells prepare for S phase, during which DNA synthesis takes place and the cell replicates its chromosomes [1,2]. The completion of S phase leads into another gap phase (G_2). Upon completion of G_2, cells enter mitosis (M phase), where the duplicated chromosomes segregate into two daughter cells [3]. However, it should be pointed out that in some rare instances cycling cells have only two phases (M and S) without intervening gap phases (G_1 and G_2). For example, the first 13 nuclear division cycles during *Drosophila* embryo development do not have any gap phases [4]. Similarly, nuclear division cycles during early endosperm development in plants seem to lack gap phases [5].

Normal proliferating cells in G_1 can continue to cycle or revert to quiescent (G_0) state. The decision to undergo another round of DNA synthesis and continue to cycle, or to exit the cell cycle to enter into a quiescent state (G_0), is made during G_1 phase [1]. Cells in G_0 state either terminally differentiate or can be activated to reenter into cell cycle. These switches in and out of G_1 are primarily controlled by extracellular factors such as hormones and other mitogens [1]. However, once the cells have entered into S phase, the cell cycle events become independent of extracellular factors, leading to mitosis and production of two daughter cells. These events are mostly regulated by internal controls. Stringent control of decision points in the cell cycle is vital for normal growth and development of organisms [1,4,6,7]. Deregulation of the regulatory mechanisms that control decision points in the cell cycle results in uncontrolled cell division, leading to abnormal growth. The biochemical and molecular mechanisms that regulate the cell cycle are of great interest, not only to help us understand how cells divide during normal growth and development of organisms but also to get insight into abnormal growth processes such as cancer. Knowledge derived from cell cycle regulation in plants should enhance our ability to manipulate growth and developmental processes in plants and could have practical implications. For instance, regeneration of plants is very critical for crop improvement through genetic engineering [8,9]. However, the ability to regenerate a whole plant from differentiated somatic tissues varies considerably from species to species [10]. The induction of cell division in differentiated cells (G_0/G_1 transition) is the first critical step in the regeneration process. Hence, the analysis of cell cycle regulation and differentiated state of regenerable and nonregenerable (recalcitrant) tissues in terms of the cell cycle stage may provide some clues or correlation between cell cycle phase and cells' ability to divide [11,12].

In recent years, considerable advances with yeast and animal systems have led to our understanding of the control of different phases of the cell cycle [13–15]. The combination of genetic, biochemical, and molecular approaches has resulted in the identification of decision points in the cell cycle and of key regulatory proteins that control progression through the decision points of cell cycle. A number of excellent reviews describing the cell cycle regulation in fungi [14,16,17], insects [4], and mammalian cells [13,15,18] have appeared. Cell cycle research in plants is in its very early stages. However, research during the last 3 years shows that at least some of the key cell cycle regulatory proteins are structurally and functionally conserved among plants and other unicellular and multicellular eukaryotes. The goal here is to summarize what is known about cell cycle regulation in plants and some of the unique aspects of the cell cycle in plants. Because our information about plant systems is limited, and because there is considerable similarity in cell cycle regulation across phylogenetically divergent species, it is necessary to present an overview of cell cycle regulation in fungi and animal systems.

Largely based on genetic analysis in yeast, the eukaryotic cell cycle is believed to be regulated at two major decision points: a point late in G_1 called START, which marks a cell's commitment to DNA replication, and G_2/M phase transition [16]. Studies with fungi and animal systems indicate that both these transitions, as well as progression of cells through the S phase, are controlled by protein kinases whose activity is regulated in a very complex manner [14–16].

II. ONSET OF M PHASE

A. Key Proteins Involved in G2/M Phase Transition

1. p34 Protein Kinase

Research in the past several years with yeast and mammalian systems using various approaches indicates that the onset of M phase is regulated by a mechanism that is common to all eukaryotic cells [15]. Central to this mechanism is a serine/threonine protein kinase, $p34^{cdc2}$ (thereafter referred to as p34 protein kinase); in the active form, this kinase acts as a mitotic trigger. The p34 protein kinase was identified genetically in the fission yeast *(Saccharomyces pombe)* as the product of cell division cycle gene *(cdc2)* that encodes for a 34 kD protein [19–21]. Homologues of this gene have been found in budding yeast ($p34^{CDC28}$) [22–24], several vertebrates [25–27], invertebrates [28], and plants [29–35] and are shown to be highly conserved both structurally and functionally among all eukaryotes. Genes for p34 protein kinase from evolutionarily distant multicellular organisms, including vertebrates and plants, have been shown to complement yeast mutants in this gene [25,31–33]. Hence, it is considered to be a universal regulator of mitosis in eukaryotic cells [14,15].

2. Cyclins and Regulation of p34 Protein Kinase Activity

Genetic and biochemical studies indicate that the activity of the p34 kinase is required for entry into mitosis, and its inactivation is required for exit from mitosis [13,15]. The level of p34 protein kinase is fairly constant during the cell cycle of dividing cells in yeast [13,36]. However, the activity of this kinase increases significantly prior to the onset of M phase [13–15]. Extensive studies from various laboratories indicate that the activity of p34 kinase is regulated in a very complex manner by several positive and negative regulators. The p34 kinase by itself is inactive and is activated by its association with certain members of a family of proteins called cyclins, which are accumulated and degraded at very precise stages of the cell cycle [13]. Cyclins were first discovered in clams and sea urchins as a class of proteins that accumulate to high levels in interphase and are abruptly destroyed at the end of M phase [37–39]. Later it was found that the cyclins associate with p34 protein kinase and function as regulatory subunit of a maturation-promoting factor (MPF) [40–43]. First discovered in oocytes as a factor responsible for G_2/M transition, MPF activity was subsequently found in all dividing cells [13].

Cyclins associate with p34 kinase to form active MPF. Cyclins that associate with p34 protein kinase during G_2 phase are called mitotic cyclins and have been identified in a number of organisms ranging from yeast to man including plants [39,44–49]. Recent studies indicate that each major decision point in the cell cycle (START, G_2/M, and progression through S) is regulated by a distinct set of cyclins [50,51]. The activity of MPF is also regulated by phosphorylation and dephosphorylation of Thr14, Tyr15, and Thr161 of p34 protein kinase [15,52–56]. Two of the phosphorylation sites (Thr14 and Tyr15) are located in the ATP-binding region of the kinase [57]. In yeast, p34 protein kinase is inactivated by phosphorylation of a tyrosine (Tyr15) [53,54]. The products of *wee1* and *mik1* genes are implicated in phosphorylation of this residue [58–61]. Both these genes code for protein kinases and the product of *wee1* (p107[wee1]) has been shown to possess serine/tyrosine kinase activity in vitro [60,61].

During G_2/M transition, a phosphotyrosine phosphatase coded by a *cdc25* gene product (p80[cdc25]) in fission yeast and its homologues in other systems specifically dephosphorylates and activates MPF [61–65]. In animal cells, the p34 kinase is maintained in inactive state by phosphorylation of both tyrosine (Tyr15) and threonine (Thr14) residues. This phosphorylation is carried out by a protein kinase, probably a homologue of *wee1* kinase [56]. Dephosphorylation of Tyr15 alone is not sufficient to activate inactive MPF in animal cells. It has been shown that dephosphorylation of both residues is required for complete activation of p34 kinase and entry of cells into mitosis [15]. These studies indicate that the dephosphorylation of both tyrosine and threonine residues is important in animal cells. In animal cells, *cdc25* is capable of both serine/threonine and tyrosine dephosphorylation [65]. The genes *wee1* and *cdc25* in animal cells are unique in that they are capable of dual threonyl/tyrosyl phosphorylation and dephosphorylation, respectively [65]. Homologues of *cdc25* have been identified from a number of eukaryotic species and they are highly divergent outside catalytic domains [66–70]. However, in spite of this divergence, *Drosophila* and human *cdc25* genes have been shown to rescue yeast *cdc25* mutations [66,69,70]. A *cdc25* homologue has not yet been reported from plants.

Both in yeast and in vertebrates, phosphorylation of a threonine (Thr167 in fission yeast or Thr161 in *Xenopus*) residue of p34 protein kinase seems to be involved in the stabilization of p34 protein kinase and cyclin subunits, thereby activating the protein kinase complex [54–56]. The regulation of the activity of MPF by phosphorylation and dephosphorylation events in animal cells is shown schematically in Figure 1. Other modes of regulation of p34 kinase include proteins (e.g., p13[suc1] or p40) that bind to p34 kinase and inhibit its activity [71,72].

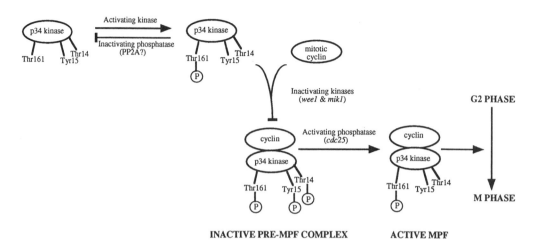

Figure 1 Regulation of p34 protein kinase activity during G_2/M transition in animal cells by protein phosphorylation/dephosphorylation of Thr14, Tyr14, and Thr161. Gene products that have a stimulatory effect on p34 kinase are indicated by pointed arrows; those that have an inhibitory effect are shown by blunt arrows. PP2A is protein phosphatase 2A; for other details, see text.

3. Mode of Action of p34 Kinase in G_2/M Transition and Exit from M Phase

A number of changes take place in the cell as it proceeds from G_2 to M phase [3,6,73]. These include chromosome condensation, nuclear envelope breakdown, reorganization of microtubules into a mitotic spindle, and transient inhibition of membrane traffic RNA and protein synthesis. In addition, the M phase is characterized by the appearance of newly phosphorylated proteins [14,74]. At the time of exit from M phase, these changes are reversed as the cell returns to interphase. Numerous studies indicate that in all eukaryotes, entry into M phase involves the activation of p34 protein kinase and exit from mitosis involves inactivation of p34 protein kinase [13,15]. As discussed above, the activation of p34 protein kinase prior to G_2/M transition is very complex, involving cyclins and phosphorylation and dephosphorylation events. Once the p34 protein kinase has been activated, it initiates various events of M phase such as disassembly of nucleus, cytoskeletal reorganization, and chromosome condensation. Understanding the mechanism(s) by which these M-phase events are initiated requires the identification of physiological substrates of p34 kinase and the role of these phosphorylated substrates in bringing about M-phase events. Although only a few proteins have been identified as in vivo substrates of p34 protein kinase [75–78], a number of substrates have been shown to be phosphorylated in vitro by p34 protein kinase (Table 1) [79–100]. At least some of the substrates listed in Table 1 (e.g., lamins, histone H_1, nucleolin, pp60[c-src], GTP-binding proteins, caldesmon) are among the physiological substrates of p34 protein kinase [79,81,82,93]. It has been shown that the nuclear disassembly during mitosis involves the phosphorylation of the lamin component of the nuclear envelope by p34 protein kinase [101,102]. Destruction of cyclins in M phase inactivates p34 protein kinase and is required for the transition from mitosis to interphase [13–15]. Sudden destruction of cyclins just prior to anaphase is mediated by the ubiquitin pathway of protein degradation [103–105]. In the amino-terminal region, mitotic cyclins have a conserved stretch of amino acids (RXXLXXIXN followed by a lysine-rich stretch) called the destruction box. In addition to inactivation of p34 protein kinase, reentry into the interphase requires dephosphorylation of proteins involving protein phosphatase action. Protein phosphatases that are required in late mitosis have been identified in yeast ("defective in sister chromatid disjoining," or *dis/*, and "bypass of *wee* suppression," or *bws1*) and *Aspergillus* ("blocked in mitosis," or *bimG*) [106].

Table 1 Substrates for p34 Kinase

Substrate	Nature of the protein	Possible function in M phase	Ref.
Nuclear lamins	Cytoskeletal protein	Nuclear lamin disassembly	79,80
Caldesmon	Cytoskeletal protein	Microfilament contraction	81
Vimentin	Cytoskeletal protein	Intermediate filament disassembly	82
Neurofilament H	Cytoskeletal protein	Unknown	83
Myosin regulatory light chain	Cytoskeletal protein	Contractile ring activation	84
Histone H_1	Chromatin-associate protein	Chromosome condensation	85
HMG I, Y, P_1	Chromatin-associate protein	Chromosome condensation	86
NO38, nucleolin	Chromatin-associate protein	Nuclear disassembly	75
SW 15	Transcription factor	Chromosome condensation	87
c-*myb*	Transcription factor	Chromosome condensation	88
$pp60^{c\text{-}src}$	Protein kinase	Cytoskeletal rearrangements	89
p150ab1	Protein kinase	Unknown	90
Casein kinase II (α/β)	Protein kinase	Chromosome condensation	91
RNA polymerase II	Transcription enzyme	Transcription inhibition	92
Rab1/Rab4	GTP-binding proteins	Inhibition of endomembrane traffic	93
p58 lamin B-receptor	Nuclear membrane protein	Dissociation of lamins	94
EF-1γ	Elongation factor	Translation inhibition	95
SV40 large T antigen	DNA binding protein	DNA replication	96
DNA polymerase α	Replication enzyme	DNA replication	97
pRb polymerases	Retinoblastoma gene product	Unknown	98
Cyclin B	Subunit of p34 kinase	Regulates p34 protein kinase	99
*Cdc*25	Protein phosphatase	Activates p34 protein kinase	100

B. M-Phase Regulatory Proteins in Plants

Cell cycle research in plants at the biochemical and molecular level started very recently and is greatly benefiting from the tools and information obtained with fungal and animal systems. The obvious first step was to find out which of the known cell cycle regulatory components are conserved in plants. Research in the 1990s has yielded some information indicating that at least some of the key cell cycle regulatory proteins (e.g., p34 protein kinase and cyclins, mitogen-activated protein kinase) are present and highly conserved, whereas the presence of various other proteins is yet to be explored. The availability of these genes will help in studying the detailed regulation of various components involved in cell cycle. The presence of the p34 protein kinase homologue in a number of plants was first reported by John et al. [29], using antipeptide antibodies made to a stretch of amino acids that is highly conserved in all yeast and animal p34 protein kinases and the antibodies raised against p34 protein kinase from yeast. A 34 kD protein was detected with these antibodies in different plant systems. In *Chlamydomonas*, it has been shown that the phosphorylation of p34 protein increases at the time the cells commit themselves to cell division [29]. The highly conserved nature of cell cycle regulatory proteins is enabling plant scientists to isolate plant counterparts. Primers made to conserved regions of key cell cycle regulatory proteins have helped in isolating partial sequences of p34 protein kinase by the polymerase chain reaction (PCR) [30–35,48]. Screening of libraries with such PCR generated probes resulted in isolation of full-length p34 protein kinase complementary DNA clones from *Arabidopsis* [32,34], rice [107,108], maize [31], soybean [35], and mothbean [109]. The isolated cDNAs showed extensive homology with yeast and animal p34 protein kinases (61–66% sequence identity at the amino acid level), indicating that the plant p34 protein kinase is structurally similar to other eukaryotic p34 protein kinases (Figure 2). It has been shown that p34 protein kinase cDNAs from *Arabidopsis*, corn, alfalfa, and soybean complement yeast *cdc2/CDC28* mutation [31,33,35]. These cross-species complementation studies indicate a high degree of functional conservation between plant and yeast p34 protein kinase. Currently very little is known about the regulation of the p34 protein kinase activity in plants. However, the amino acids that are known to be involved in the regulation of p34 protein kinase activity at the posttranslational level (viz., Thr14, Tyr15, and Thr161) are present in all known plant

```
                    ▼▼                              _____
athp34cdc2 :1 MDQYEKVEKIGEGTYGVVYKARDKVTNETIALKKIRLEQEDEGVPSTAIREISLLKEMQ.
mbeancdc2  :1 -E------------------R----------------------------------.
mzekinaa   :1 -E----------------L--A---------------------------------N.
ricecdc2   :1 -E----E-----------R-----------------------------------H.
yspcdc2    :1 -EN-Q---------------H-LSGRIV-M------D-S---------------VNd
hscdc2     :1 -ED-T-I------------G-H-T-GQVV-M------S-E--------------LR.

athp34cdc2 : ...HSNIVKLQDVVHSEKRLYLVFEYLDLDLKKHMDSTPDFSK...DLHMIKTYLYQILR
mbeancdc2  : ...-R---R------------------------S-E-V-...-PRQV-MF-----C
mzekinaa   : ...-G---R-H--------I-----------F---C-E-A-...NPTL--S------H
ricecdc2   : ...-G---R-H--I-----I-----------F---C-E-A-...NPTL--S-------
yspcdc2    : ennR--C-R-L-IL-A-SK------F--M----Y--RISETGAtsl-PRLVQKFT--LVN
hscdc2     : ...-P---S----LMQDS----I--F-SM----YL--I-PGQYm..-SSLV-S------Q
                                                          ▼
athp34cdc2 : GIAYCHSHRVLHRDLKPQNLLIDRRTNSLKLADFGLARAFGIPVRTFTHEVVTLWYRAPE
mbeancdc2  : ---------------------------------------------------------
mzekinaa   : -V-----------------------A-------------------------------
ricecdc2   : -V-----------------------A-------------------------------
yspcdc2    : -VNF---R-II-----------KEG-.----------S---V-L-NY---I---------
hscdc2     : --VF---R-------------DKGT.I--------------I-VY----------S--

athp34cdc2 : ILLGSHHYSTPVDIWSVGCIFAEMISQKPLFPGDSEIDQLFKIFRIMGTPYEDTWRGVTS
mbeancdc2  : -----R---------V----------VNRR---------E-------L---N-E--P---A
mzekinaa   : ----ARQ------V----------VN-----------E-------L---N-QS-P--SC
ricecdc2   : -----RQ------M----------VN-----------E------VL---N-QS-P--S-
yspcdc2    : V----R----G-------------RRS----------EI----QVL--N-EV-P---L
hscdc2     : V----AR---------I-T----LATK----H---------R---AL--NNEV-PE-E-

athp34cdc2 : LPDYKSAFPKWKPTDLETFVPNLDPDGVDLLSKMLLMDPTKRINARAALEHEYFKDLGGM
mbeancdc2  : ---F--T----P-K--A-V-----AA-LN---S--CL--S---T--I-V-------IKFV
mzekinaa   : ---F-T---R-QAQ--A-V------A-L-------RYE-S---T--Q---------EVV
ricecdc2   : ----------QAQ--A-I--T---A-L-------RYE-N---T--Q---------EMV
yspcdc2    : -Q----T--R--RM--HKV---GEE-AIE---A--VY--AH--S-KR--QQN-LR-FH..
hscdc2     : -Q---NT------GS-ASH-K---EN-L-------IY--A---SGKM--N-P--N--DNQ

athp34cdc2 : P...  294
mbeancdc2  : -...  294
mzekinaa   : Q...  294
ricecdc2   : Q...  294
yspcdc2    : ....  297
hscdc2     : IKKM  297
```

Figure 2 Amino acid sequence comparison of p34^{cdc2} protein kinase homologues from plants with yeast and human p34 protein kinase. Dashes indicate aligned identical amino acids, uppercase letters denote aligned nonidentical amino acids, and lowercase letters indicate unaligned amino acids. Gaps are denoted by dots. The solid line shows a highly conserved stretch of amino acids (PSTAIRE region), characteristic of p34 protein kinases. Arrows indicate the amino acids (Thr14, Tyr15, and Thr161) that have been shown to be involved in regulating the activity of p34 protein kinase by phosphorylation and dephosphorylation events in animal cells. The amino acid sequence of *Arabidospsis* p34^{cdc2} (athp34cdc2, Ref. 32) is aligned with mothbean (mbeancdc2, Ref. 109), corn (mzekinaa, Ref. 31), rice (ricecdc2, Ref. 108), fission yeast (yspcdc2, Ref. 21), and human (hscdc2, Ref. 25) p34^{cdc2} protein kinases.

p34 protein kinases (see Figure 2), indicating that they could be involved in regulating the activity of plant p34 protein kinase by protein phosphorylation/dephosphorylation. However, presence of the protein kinases (e.g., *wee1/mik1*) or protein phosphatases *(cdc25)* that are responsible for these phosphorylation/dephosphorylation events has not been shown in plants.

1. Expression of p34 Protein Kinase

In yeast, the level of protein kinase remains constant throughout the cell cycle, although the activity peaks at G_2/M transition [14,36]. In animal systems, noncycling cells (terminally differentiated cells or senescent cells) do not express p34 protein kinase in response to mitogen stimulation. Fluctuations in the level of the p34 protein kinase messenger RNA and protein are observed in synchronized human cells [110,111]. In other animal systems also p34 protein kinase mRNA level is correlated with the proliferative state of the cell [27,28]. In plants, the expression of p34 protein kinase at the mRNA level is detected in all the tissues including differentiated tissues [31,32]. However, the highest level of expression is found in meristematic tissues such as apical meristem, and immature leaf [31,112–115]. Furthermore, the expression of plant p34 protein kinase genes has been shown to be induced by external factors that are known to induce cell division, such as phytohormones [112,114] and *Rhizobium* [35] infection. Two distinct p34 protein kinase genes from soybean showed differential expression pattern [35]. One of the genes (*cdc2*-S5) is highly expressed in roots, whereas the other one (*cdc2*-S6) is active in aerial parts. Furthermore, infection of roots with *Rhizobium* resulted in enhanced expression of *cdc2*-S5 but not *cdc2*-S6, indicating that the two genes function in different developmental programs. Furthermore, these two genes differed in their response to auxin. These results suggest that in a multicellular organism, cell division in different tissues may be controlled by different p34 protein kinases and may respond to different signals. By in situ hybridization studies, Martinez et al. [112] have shown that p34 protein kinase expression at the mRNA level in the root and shoot apex parallels the pattern of mitotic activity. In addition, strong expression is observed in pericycle and perivascular parenchyma cells that are competent to divide but are not actively dividing. When the level of p34 protein kinase during cell differentiation in wheat leaf was studied using antibodies, it was found to be developmentally regulated [113]. There is about 20 times less p34 protein kinase in the differentiated cells that have ceased to divide than in cells in the meristematic region. In carrot, cessation of cell division during cotyledon development is shown to correlate with a decrease (16 times less than dividing cells) in the level of p34 protein kinase [114]. Furthermore, induction of cell division in explants from mature cotyledons by auxin resulted in a tenfold increase in the amount of p34 protein kinase [114]. This and other studies [35,112,113] with plant systems also indicate that the relative level of p34 kinase is important in determining whether a cell undergoes a division and its capacity to resume cell division.

In corn, two p34 protein kinase cDNAs (96% identical at the nucleotide sequence) have been isolated, and Southern analysis with PCR-generated probes under high stringency conditions suggests more *cdc2* genes [31]. Miao et al. [35] also isolated two cDNA clones that code for p34 protein kinase from soybean. These two clones shared 90% sequence identity in the coding region, whereas 5' and 3' untranslated sequences did not show significant sequence similarity.

Like yeast and mammalian p34 protein kinase, plant p34 kinase has affinity for p13^{suc1}. In yeast, p13 modulates p34 kinase activity and is shown to be necessary for mitosis [15,71]. Using antibodies raised to yeast p13, a plant homologue of identical size has been detected in wheat, pea, and *Chlamydomonas* [115]. Unlike p34 protein kinase, a high level of p13 has been found in differentiated cells of wheat.

2. Cyclins

Using yeast complementation and PCR approaches, several different types of cyclins have been isolated and characterized from mammalian systems [50,51,116,117]. Of the different animal cyclins, the A, B, and E types share considerable similarity in the cyclin box, whereas other cyclins (C, D, and E) do not have much homology among themselves or with the A, B, and E types [116,117]. Hata et al. [47] isolated plant cyclins from carrot and soybean by screening the cDNA libraries with degenerate oligonucleotides corresponding to conserved regions of mitotic cyclins. The carrot cDNA and one of the soybean cDNAs were partial, whereas the other soybean clones contained the entire coding region. The deduced amino acid sequence from the soybean clone is more similar to B-type cyclins, whereas carrot cyclin resembles A-type cyclin. However, because of the divergent amino acid sequence of plant cyclins, these cyclins could not be classified into any of the known cyclin sequences.

A full-length cyclin cDNA from *Arabidopsis* [48] and two partial cDNAs from alfalfa *cycMs1* and *cycMs2*, [49] have also been isolated. Alfalfa cyclins were found to be expressed in dividing suspension-cultured cells but not in the cells in stationary phase. Analysis of the expression of *cycMs1* and *cycMs2* during different stages of the cell cycle showed maximal expression of both these genes in the G_2 and M phases. However, *cycMs1* transcripts appeared early in G_2 whereas *cycMs2* expression was found only in late G2 and M phase cells [49]. Microinjection of synthetic mRNA from soybean and *Arabidopsis* into *Xenopus* oocytes has been shown to induce maturation, indicating the functional conservation of mitotic cyclins between plants and animals [47,48]. On the basis of the expression pattern of *cyc1*At and its homology with conserved regions of mitotic cyclins, it was concluded that *cyc1*At is a mitotic cyclin. We used primers designed to two stretches of conserved regions in A, B, and E type cyclins to amplify corresponding cyclin sequences from *Arabidopsis* using PCR. Among the 25 clones sequenced, three different cyclin sequences were detected. Southern analysis of genomic DNA from *Arabidopsis* with two of the PCR-generated probes indicated five to six hybridizing bands under high stringency conditions. Amplification of cyclin sequences with these primers using genomic DNA as a template yielded same-size product as cDNA, indicating the absence of an intron in this region [118]. Hence, each of the bands that is observed on Southern blots may represent a cyclin sequence, indicating the existence of several cyclin sequences. Screening of a cDNA library prepared from flower meristem with a mixture of two of the PCR-generated probes resulted in isolation of several cDNA clones. Restriction enzyme pattern and sequence analysis of 5′ and 3′ ends of these clones revealed the presence of four different cyclin sequences, indicating that there is a family of cyclins in *Arabidopsis* [118]. In maize also, four distinct cyclins have been isolated [119].

III. PROGRESSION THROUGH G_1 AND S PHASES

Compared to recent progress made in understanding the regulation of the onset of M phase, little is known about the mechanisms that control G_1/S transition [16,50,51]. In yeast, the same protein kinase (p34$^{cdc2/CDC28}$) is responsible for regulating START, a point in G_1 at which a cell commits itself to DNA synthesis. The G_1 cyclins (*CLN* genes) interact with p34$^{cdc2/CDC28}$ to drive the cell through this G_1 restriction point (START) to enter S phase [16]. In multicellular organisms, however, progression through G_1 and S phases seems to be much more complex and to be controlled by a family of protein kinases that are structurally related to p34 kinase [50,51,120–123]. These protein kinases, like p34 protein kinase, require specific cyclins to become activated; hence they are designated as cyclin-dependent protein kinases (cdks) [50,51] and are numbered in the order of their discovery. Several new cdks (e.g., *cdk2*, *cdk4*, *cdk5*—the p34 *cdc2* is sometimes referred to as *cdk1*) that are involved in cell cycle regulation have been identified in animal systems [120–123]. Each of the cdks seems to associate with a specific type of cyclin to be activated and to be involved in a specific phase of the cell cycle [50,51,116,122,124,125] (Table 2). Using different approaches such as yeast complementation

Table 2 p34 Protein Kinase Family Members and Their Cyclin Partners in Animal Systems

Type of protein kinase	Sequence in PSTAIRE motif	Cyclin partner	Ref.
cdkl (cdc2)	PSTAIRE	B types	25
cdk2	PSTAIRE	A, D, and E types	50, 122, 125
cdk3[a]	PSTAIRE	Not known	123
cdk4	PV/ISTVRE	D types	124
cdk5	PSSALRE	D types	124
cdk6	PLSTIRE	D type	50, 126
PCTAIRE	PCTAIRE	Unknown	124
KKIALRE	KKIALRE	Unknown	124
PITAIRE	PITAIRE	Unknown	123
PITSLRE	PITSLRE	Unknown	127
NRTALRE	NRTLARE	Unknown	128

[a]Cyclin partner for *cdk3* has not been identified. However, it was called *cdk3* because of its ability to complement *CDC28* in yeast.

and polymerase chain reaction, several different types of cyclins and cdks have been identified. There are seven cyclin types: A, B, C, D, E, F and G. Of these, types A, B, C, D, and E are implicated in regulating different control points in the cell cycle (Figure 3), whereas the functions of F and G are unknown [50]. About 10 different protein kinases that are related to the *cdc2* family have been reported. Some of these associate with a specific type of cyclin, whereas cyclin partners for some of the p34-related protein kinases are not identified [121–128]. Other *cdc2*-related kinases with no known cyclin partners are named after the amino acid sequences

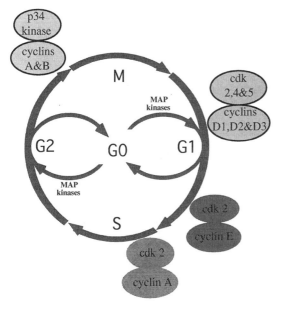

Figure 3 Schematic diagram of the cell cycle, indicating various cell cycle regulatory proteins and their involvement during different phases of the cell cycle in mammalian systems. MAP kinase, mitogen-activated protein kinase; cdk, cyclin-dependent protein kinase.

in the PSTAIRE motif of *cdc2*. Known *cdc2*-related protein kinases and their cyclin partners are summarized in Table 2.

In plants very little is known about progression through G_1 phase and G_1/S transition. Recent studies indicate that plants have several distinct p34-like protein kinases [129,130] and a family of cyclins [31,118]. Whether plants have G_1-specific cyclins and their corresponding protein kinase partners remains to be seen. Two *cdc2* homologues from alfalfa (*cdc2A* and *cdc2B*) appear to regulate different phases of cell cycle. The *cdc2A* could complement only G_2/M, transition whereas *cdc2B* complemented G_1/S function [130].

To study gene expression during the cell cycle in plants, Kodama et al. [131] analyzed phase-specific mRNAs and polypeptides using synchronized cultures. The rate of synthesis of 13 different polypeptides was found to change during the cell cycle. In addition, the levels of a small population of mRNAs (four) varied during the cell cycle, suggesting that the majority of the polypeptides whose synthesis varied during the cell cycle could be due to posttranscriptional and posttranslational regulation. By differential screening of a cDNA library prepared from the cell cycle S phase, two genes (*cyc02* and *cyc07*), which are preferentially expressed at the G_1/S boundary, have been isolated [132]. The deduced amino acid sequence of *cyc02*, which is 101 amino acids long, did not show homology with the sequences in the databases. Neither the identity nor the function of these genes is known.

Using auxin-dependent tobacco suspension cultures, seven different auxin-inducible cDNA clones have been isolated and characterized [133,134]. Messenger RNA corresponding to these clones is rapidly induced when quiescent cells are triggered to undergo cell division by exogenously supplied auxin. Takahashi *et al.* [135,136] isolated two auxin-induced cDNAs named *parA* and *parB* (protoplast auxin regulated) from tobacco mesophyll protoplasts. Addition of auxins and cytokinins can induce cell division in tobacco mesophyll protoplasts, which are differentiated cells that have ceased to divide. Expression of the *par* genes was not detected in differentiated cells, whereas it is expressed in protoplasts cultured in the presence of auxin. Both *parA* and *parB* genes are expressed during transition from G_0 to S phase of in vitro cultured protoplasts [135,136]. Furthermore, the expression of *par* genes was observed prior to initiation of DNA synthesis. The gene *parB* has been identified as glutathione *S*-transferase [136]. Although this enzyme is mostly known to be involved in the detoxification of xenobiotics, recent studies implicate its involvement in cell proliferation [137–139]. The role of a *parB*-coded enzyme in tobacco mesophyll protoplasts is not yet known.

Proliferating-cell nuclear antigen (PCNA)—an acidic, nonhistone nuclear protein and an auxiliary protein of DNA polymerase-δ—is shown to be present only in proliferating mammalian cells, not in nondividing cells [140]. A gene that codes for a plant homologue of PCNA has been cloned in rice and *Catharanthus* [141]. Like its animal counterpart, plant PCNA is preferentially expressed in proliferating cells and was not detectable in quiescent cells. In synchronized population of cells, PCNA was highly expressed in S phase [141].

It is somewhat intriguing that none of the auxin-regulated genes implicated in cell division nor any cell cycle phase specific genes showed any homology to known key cell cycle regulatory genes [132,133,135,136]. In recent studies using p34 protein kinase cDNAs and antibodies, it has been shown that auxin induces p34 protein kinase mRNA and protein [35,112,113–115]. However, it should be noted that the effect of auxin on p34 protein kinase mRNA and protein was studied a long time after the auxin treatment (the earliest time point is one day), whereas most of the auxin-regulated cDNAs that are implicated in cell division [133,135,136] have been isolated from the libraries that are made after several hours of auxin treatment. This and other factors, such as posttranscriptional regulation and abundance of mRNA corresponding to known key cell regulatory proteins in relation to other auxin-regulated genes, could account for the absence of known key cell cycle regulatory genes in the pool of auxin-induced cDNAs.

In animal cells, entry of quiescent state (G_0) into the cell cycle by mitogens or other external factors is accompanied by rapid changes in the status of phosphorylation of a number of cellular proteins. These protein phosphorylation changes are believed to be involved in changes in gene expression that trigger cells into cell cycle or a new developmental pathway. A key enzyme called mitogen-activated protein kinase (MAP kinase), which is involved in mitotic stimulation in animals [142,143], has been cloned from plant systems [144,145]. In pea, MAP kinase is expressed in dividing cells and quiescent cells; hence there is no correlation between the MAP kinase mRNA and cell proliferation [145]. However, it is likely that the activity of this enzyme could be regulated posttranslationally by protein phosphorylation [144,146].

IV. ROLE OF CALCIUM AND CALMODULIN IN CELL CYCLE REGULATION

Calcium, a key intracellular messenger in both plants and animals, has been shown to regulate many different processes in plants [147–149]. Calmodulin, a calcium-binding protein found in all eukaryotes, is one of the primary mediators of calcium action (see Chapter 34, Calcium as a Messenger in Stress Signals Transduction, for more information on calmodulin). For more than a decade, calcium and calmodulin have been implicated in controlling cell proliferation in eukaryotic cells, including plant cells [147,150–152]. Calcium is essential for the growth of all eukaryotic cells [153]. It has been shown that cells require the presence of millimolar level of extracellular calcium to proliferate [154,155].

Progression of normal cells through the cell cycle is found to be associated with transient changes in intracellular calcium concentration [150,151,156]. Neoplastic cells, which can proliferate in the absence of external calcium, contain higher levels of intracellular calcium than normal cells [157]. Manipulation of cytosolic calcium concentration has been shown to affect cell cycle events [158–160]. By determining the level of intracellular calcium during different stages of the cell cycle, it has been demonstrated that rapid and transient increases in intracellular calcium occur at specific stages of the cell cycle in plant and animal cells [161–164]. Calcium transients are observed at G_2/M transition, as the cells completed mitosis, and on both sides of the G_1/S boundary [151]. Mitotic events such as the breakdown of the nuclear envelope, chromatin condensation, and the onset of anaphase have been correlated with a transient increase in intracellular calcium [160,161]. Furthermore, these mitotic events could be induced prematurely by artificially elevating cytosolic calcium, whereas chelation of intracellular calcium by calcium chelating agents blocked the nuclear envelope breakdown and the metaphase/anaphase transition, suggesting that an increase in cytosolic calcium is required for these mitotic events to take place [158–160]. Blocking of intracellular calcium prior to the G_1/S boundary results in inhibition of DNA synthesis [151]. This finding suggests that calcium transients are critical for the progression of cells from G_1 to S phase of the cell cycle.

Studies with both plant and animal tissues have revealed higher levels of calmodulin in dividing cells than in nondividing cells [147,151,165]. Increased levels of calmodulin mRNA, protein, and activity are observed in meristematic tissues of the plants [147,166,167]. In vertebrates and lower eukaryotic cells, a twofold increase in the intracellular calmodulin concentration is observed at the G_1/S boundary [168–170]. Stimulation of quiescent cells to reenter the proliferative state elevated the amount of calmodulin. Furthermore, transformed mammalian cell lines have been shown to contain elevated levels of calmodulin [171,172]. To study the effect of altered levels of calmodulin on the cell cycle, Rasmussen and Means [173,174] manipulated the levels of calmodulin level by stably transforming the mouse cell lines with vectors that constitutively or inducibly express either calmodulin sense or antisense RNA. A transient increase in calmodulin resulted in acceleration of proliferation, while a decrease in calmodulin caused a transient cell cycle arrest. Constitutive elevation of intracellular calmodulin levels in these cells shortened G_1, and, in turn, the cell cycle. Calcium and calmodulin level determinations during different stages of the cell cycle and the data on the effect of elevated or reduced levels of calmodulin on the cell cycle indicate that three specific points in the cell cycle (G_1/S, G_2/M, and metaphase/anaphase) are sensitive to calcium and calmodulin (Figure 4). Overexpression of calmodulin in *Aspergillus nidulans* increased growth rate by decreasing cell cycle time, whereas reduced levels of calmodulin prevented entry into mitosis [151].

Calcium and calmodulin have multiple functions and regulate a variety of processes, including some housekeeping functions [147,165,175]. Hence, it has been argued that the observed effects of calcium and calmodulin manipulations on the cell cycle may not affect specific control points but could be due to the requirement for calcium and calmodulin to carry out many housekeeping functions. Recent studies with unicellular fungi (yeast and *Aspergillus nidulans*) that are amenable to genetic manipulations indicate that calcium and calmodulin regulate specific decision points during the cell cycle [151]. However, the mechanisms by which calcium and calmodulin control the cell cycle are beginning to be elucidated.

A. Mode of Calcium and Calmodulin Action in Regulating G_2/M Transition

Repression of calmodulin synthesis, and thereby calmodulin levels, or reduced extracellular calcium in *Aspergillus* cells, blocked entry into mitosis [176,177]. Under these conditions tyrosine dephosphorylation of p34 protein kinase, which is needed for its activation, is blocked; and the activity of NIMA protein kinase, a protein kinase required for G_2/M transition in *Aspergillus*, is also reduced. The effects of reduced calmodulin and calcium could be reversed by elevating their levels. The activation of p34 kinase and NIMA protein kinase by calcium and calmodulin could be due to direct interaction of NIMA protein kinase and the enzyme responsible for tyrosine dephosphorylation of p34 kinase with the calcium/calmodulin complex, or it could be indirect, through proteins that bind to this complex. The NIMA protein kinase and tyrosine phosphatase involved in p34 activation did not bind to calcium/calmodulin, and the activity of immunoprecipitated NIMA kinase was not affected by calcium and calmodulin.

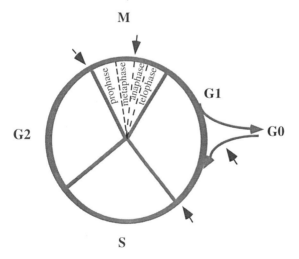

Figure 4 The phases of the cell cycle that require calcium and calmodulin; arrows indicate the control points that are regulated by calcium and calmodulin.

These results indicate that the activation of p34 and NIMA kinases could be mediated by the proteins that bind to the calcium/calmodulin complex. About 20 calmodulin-binding proteins have been identified in animal systems [178]. Some preliminary results suggest that a calmodulin-dependent protein kinase, a multifunctional enzyme that requires calcium and calmodulin for its activation, could be a likely candidate in mediating the calcium–calmodulin effect on NIMA protein kinase and NIMT (a *cdc25* homologue) of *Aspergillus* [151]. The purified calmodulin-dependent protein kinase has been shown to phosphorylate NIMA kinase and NIMT in vitro in a calcium/calmodulin-dependent manner. Furthermore, B-type cyclins that are known to associate with *cdc25* proteins and regulate their activity [70] have been found to act as substrates for calcium/calmodulin-dependent protein kinase in vitro [151]. However, the effect of such phosphorylation on the activity of these enzymes is not known.

Studies with plants indicate that there are a number of calmodulin-binding proteins in plants [148], although in many cases their identity and function are not known [179]. Studies using the antibodies raised to animal calcium/calmodulin-dependent protein kinase suggest that the plants contain a homologue of calcium/calmodulin-dependent protein kinase II [180]. A cDNA that encodes a calcium/calmodulin-dependent protein kinase has been recently isolated from plants [181].

In addition to calcium/calmodulin-dependent protein kinase, plants contain a unique calcium-regulated protein kinase that requires calcium but not calmodulin. This enzyme—calcium-dependent and calmodulin-independent protein kinase, also called calcium-dependent protein kinase (CDPK) [182,183]—appears to be present in all plants. Whether any of the calcium, calcium/calmodulin-regulated protein kinases, and calmodulin-binding proteins are involved in plant cell cycle regulation is not known.

B. Calcium/Calmodulin in Metaphase/Anaphase Transition

Several lines of evidence indicate that calcium and calmodulin are required for the metaphase/anaphase transition [161–164]. A transient increase in cytosolic free calcium at the onset of anaphase has been demonstrated. As indicated earlier, one of the critical events during metaphase/anaphase transition is inactivation of p34 kinase due to the degradation of cyclins, which according to recent studies involves calcium and calmodulin [151]. It has been demonstrated that micromolar concentrations of calcium induce cyclin B degradation in metaphase-arrested *Xenopus* egg extracts [184]. The addition of a synthetic peptide that binds to the calcium/calmodulin complex, prior to raising calcium level in the extract, blocked cyclin degradation and inactivation of p34 kinase [184]. The inhibition of cyclin degradation by micromolar concentrations of calcium with calcium/calmodulin-binding peptide could be reversed by adding calmodulin, suggesting that the calcium action is mediated by calmodulin. Furthermore, by using appropriate inhibitors, the involvement of calpain, a calcium-dependent protease, and protein kinase C was eliminated. These results indicate that calcium and calmodulin are involved in cyclin degradation in *Xenopus* eggs. It is known that cyclins are degraded by ubiquitin-dependent proteolysis [103]. The targets for calcium and calmodulin in cyclin degradation pathways remain to be identified. Also, further studies are needed to see whether calcium-induced cyclin degradation is a universal phenomenon.

V. PHYTOHORMONES AND CELL DIVISION

Phytohormones, especially auxins and cytokinins, have been shown to be intimately involved in the control of cell division [185]. In many plants these hormones, singly or in combination, induce cell division in dedifferentiated noncycling cells. It has been well established from plant tissue culture studies that auxin and cytokinin are necessary for inducing cell division. Also, apical meristems, which contain the cycling cells, contain high levels of auxin. Addition of these hormones to differentiated cells that have ceased to divide results in dedifferentiation and reentry of these cells into the cell cycle [186,187]. In a few plant systems that have been tested, the addition of auxin has been shown to induce the expression of p34 protein kinase, both at the mRNA and the protein level [35,112,115]. A severalfold increase in p34 protein kinase was observed during auxin-induced cell division in carrot cotyledons [115]. As described in Section III, suspension cultures or protoplasts that require an exogenous supply of auxin for cell division have been used to isolate auxin-regulated genes and eventually to understand the role of these genes in cell division. Several different auxin-inducible cDNA clones have been isolated and characterized from auxin-dependent tobacco suspension cultures and protoplasts that require auxin for cell division [133–135].

Other phytohormones such as abscisic acid (ABA) and gibberelic acid (GA) have been implicated in cell division control in certain plant systems [188–192]. In deepwater rice, GA induces growth, and part of this growth is found to be due to stimulation of cell division [188]. Abscisic acid is implicated in inhibiting cell division in endosperm of cultured maize kernels, maize root tips, pea buds, and in pollen mother cells [189–192].

VI. SYNCHRONIZATION OF PLANT CELLS

Synchronized cell populations are essential to study biochemical and molecular events that take place during different phases of the cell cycle. Much of our information about cell cycle regulatory proteins in animals was obtained by studying the level or activity of a given protein during different phases of the cell cycle. Cells in plant meristems have different cell cycle times and are highly asynchronous [5]. However, at a certain stage during the life cycle of a plant, cells divide synchronously for several cycles. For instance, microspore mother cells in anthers progress through meiosis synchronously. The first few divisions in the embryo and free nuclear divisions in endosperm are also synchronous. Natural synchrony, which occurs rarely, was found to be inappropriate for biochemical studies for various reasons [193]. Hence, several methods have been developed to obtain synchronized populations of cells in plant tissues and cultured cells. These methods include growing of cultured cells in the presence of DNA synthesis inhibitors (e.g., aphidicolin, hydroxyurea, 5-aminouracil, fluorodeoxyuridine) or in some nutrient-limiting medium [193,194]. However, only a few methods have been found to be effective in inducing synchronization in plant cells; in the majority of cases, either the methods were partially effective or the agents that cause synchrony were found to have toxic effects on cell metabolism.

Among the DNA synthesis inhibitors, aphidicolin is the most effective in inducing synchronous growth in suspension cultures, as well as in differentiated tissues. However, because of endogenous aphidicolin-inactivating activity in plant cells, which varies among cell types and plants, the concentration of aphidicolin and length of the incubation should be determined empirically in each case. Treatment of cells with aphidicolin, a mycotoxin that specifically blocks nuclear DNA replication by inhibiting DNA polymerase α [195], causes accumulation of cells at the G_1/S boundary of the cell cycle. The effect of this inhibitor is reversible; hence removal of aphidicolin from the medium results in synchronous resumption of DNA synthesis. In several plant cells, aphidicolin is shown to arrest about 80–95% of cells in G_1 that were found to move synchronously through the first round of mitosis after G_1/S arrest [49,194].

In suspension cultures of *Catharanthus roseus*, double phosphate starvation effectively induces synchrony [193]. This system is already helping to identify some of the phase-specific changes in mRNA and proteins [131,132]. In suspension cultures of *Datura*, hydroxyurea, another inhibitor of DNA synthesis, reversibly arrested the cells at the G_1/S boundary [196]. So far, only a few cell culture systems have been well characterized in terms of synchronization with either DNA synthesis inhibitors or nutrient-deficient media [193,196].

Synchronization of plant cells with the methods above coupled with flow cytometry should greatly expedite the progress in cell cycle research in plants [197]. During the last 10 years flow cytometry has been used increasingly in analyzing plant cells. Protoplasts and isolated nuclei are amenable to flow cytometry. When protoplasts are used, however, some modifications in methods and instrumentation are necessary because of their large (20–75 μm) size and fragility [197,198]. Recent developments in the use of flow cytometry for plant protoplasts have opened new avenues for the analysis of cell cycle regulatory proteins. Using multiparameter analysis, one could monitor the levels of two or more desired proteins during different phases of the cell cycle [199].

VII. CELL CYCLE IN PLANT DEVELOPMENT

Cell division is one of the primary determinants of development in multicellular eukaryotic organisms. The regulatory mechanisms that determine various aspects of the cell cycle (e.g., which cells in an organism should undergo cell division, the timing and the plane of cell division in these cells, which cells should remain quiescent and reenter into cell cycle) play a critical role in such plant developmental processes as embryogenesis, seed germination, and flowering. Hence investigating these regulatory mechanisms not only will help us to understand cell cycle regulation but also will enable us to elucidate developmental programming in plants. Various developmental processes that involve the cell cycle are unique to plants. Unlike animals, cell division in higher plants is restricted to meristematic regions (shoot apical meristem, root apical meristem, and lateral meristem). The primary meristems such as shoot and root apical meristems continuously divide and contribute to the production of new organs and growth of the plants. Furthermore, shoot apical meristem can lose its indeterminate vegetative growth to become determinate floral meristem. The transition from vegetative meristem to floral meristem involves the shortening of the cell cycle time as well as the synchronization of the cell cycle [5].

In plants, during the course of normal development, quiescent cells become proliferative. For instance, lateral meristems (pericycle and cambium), auxiliary buds, and cambium retain their ability to undergo cell division and enter into the cell cycle in response to developmental cues. The root apex in plants contains, in addition to dividing cells, a group of cells called the quiescent center, which do not normally undergo cell division. However, if the root meristem is damaged, cells in the quiescent center reenter the cell cycle and form new meristem. In addition, if cells from the quiescent center are cultured in vitro in the presence of hormones, they can undergo cell division and regenerate into whole plants. Pericycle cells retain the ability to divide and are responsible for the formation of lateral roots at vascular poles. Analysis of p34 protein kinase mRNA in roots has shown high levels of p34 protein kinase mRNA in meristem and in all pericycle cells, but not in the quiescent center. There is uniform expression of p34 protein kinase mRNA in all the pericycle cells, although lateral roots are initiated only at the vascular poles [112]. These results suggest that lateral root initiation opposite to vascular poles is controlled by a mechanism other than p34 protein kinase transcription. Studies on cell cycle regulation in plants should help us understand the mechanisms and factors that maintain the cells in root and shoot apical meristems in the proliferative state as well as the factors that cause withdrawal of cells from the cell cycle in other parts of the plant. Cell cycle studies should also help in elucidating the mechanisms by which differentiated cells dedifferentiate to enter the cell cycle by external factors such as hormones, wounding, and *Rhizobium* infection.

Unlike animal cells, plant cells are unique in the sense that they are totipotent. In several plant systems, terminally differentiated, nondividing somatic cells can dedifferentiate, divide, and regenerate into a whole plant. Reinitiation of cell division in differentiated and nondividing cells is a central feature in plant regeneration. Cytokinesis, a process by which cytoplasm is divided, is different in plants. In plant cells, cytokinesis is initiated by forming a phragmoplast (made of microtubules) between daughter cells, which is followed by deposition of cell wall material.

VIII. CONCLUSIONS

Several major themes are emerging from investigations on cell cycle regulatory mechanisms that used different model systems ranging from simple eukaryotes (yeast and *Aspergillus*) to complex metazoans including vertebrates, invertebrates, and plants. First, it is increasingly evident that a few key proteins are critical in controlling the decision points in the cell cycle, and these key proteins are highly conserved in all eukaryotes—indicating the universality of these key components. Second, the activity of certain protein kinases appears to play a key role in regulating the transition points between different phases of the cell cycle. Third, the mode of regulation of these key proteins may vary across phylogenetically divergent species. Finally, the regulatory mechanisms that control the cell cycle are far more complex in multicelluar organisms than in unicellular organisms.

Cell cycle research in plants is in its very early stages. Recent developments, and the tremendous progress made with fungi, vertebrate, and invertebrate systems, and the highly conserved nature of some of the key cell cycle regulatory proteins, should expedite progress in discovering similarities and differences in regulatory mechanisms among plants and other eukaryotic organisms. Some of the key proteins, known to be involved in yeast and mammalian systems, such as p34 protein kinase, cyclins, MAP kinases, PCNA, and calmodulin, have been identified in plants. Although the cell cycle is common to all eukaryotes, it is controlled by different hormones or growth factors in plants and animals. Hence, although some key proteins are highly conserved across phylogenetically divergent species, it is likely that regulatory mechanisms differ between plants and animals. Since cell division is so fundamental to growth and development, it is bound to be an exciting area of research in plant biology. Recent advances in molecular and cell biology offer new approaches to the investigation of this very complex process. Manipulation of cell cycle regulatory proteins in cultured cells and transgenic plants should provide more insights into cell cycle regulation in plants as well as in plant development.

ACKNOWLEDGMENTS

I thank Irene Day and Dr. Farida Chamberlain for reading the manuscript and for their help in preparing figures. Work in my laboratory is supported by Biomedical Research Support Grant, Plant Biotechnology Laboratory, Colorado Biotechnology Research Institute, Colorado RNA Center and Agricultural Experiment Station (Project no. 702).

REFERENCES

1. A. B. Pardee, *Science*, *246*: 603 (1989).
2. R. A. Laskey, M. P. Fairman, and J. J. Blow, *Science*, *246*: 609–614 (1989).
3. J. R. McIntosh, and M. P. Koonce, *Science*, *246*: 622 (1989).

4. P. H. O'Farrell, B. A. Edgar, D. Lakich, and C. F. Lehner, *Science, 246*: 635 (1989).
5. D. Francis, *New Phytol., 122*: 1 (1992).
6. L. H. Hartwell and T. A. Weinert, *Science, 246*: 629 (1989).
7. L. Hartwell, *Cell, 71*: 543 (1992).
8. R. Fraley, *Bio/Technology, 10*: 40 (1992).
9. C. S. Gasser and R. Fraley, *Science, 244*: 1293 (1989).
10. J. Marx, *Science, 235*: 31 (1987).
11. C. Bergounioux, C. Perennes, S. C. Brown, C. Sarda, and P. Gadal, *Protoplasma, 142*: 127 (1988).
12. S. Kartzke, H. Saedler, and P. Meyer, *Plant Sci., 67*: 63 (1990).
13. A. W. Murray and M. W. Kirschner, *Science, 246*: 614 (1989).
14. P. Nurse, *Nature, 344*: 503 (1990).
15. C. Norbury and P. Nurse, *Annu. Rev. Biochem., 61*: 441 (1992).
16. S. I. Reed, *Trends Genet., 7*: 95 (1991).
17. N. R. Morris, *Curr. Opinion Cell Biol., 2*: 252 (1990).
18. G. Draetta, *Trends Biochem. Sci., 15*: 378 (1990).
19. P. Nurse and P. Thuriax, *Genetics, 96*: 627 (1980).
20. V. Simanis and P. Nurse, *Cell, 45*: 261 (1986).
21. J. Hindley and G. Phear, *Gene, 31*: 129 (1984).
22. D. Beach, B. Durkacz, and P. Nurse, *Nature, 307*: 183 (1982).
23. A. Lorincz and S. Reed, *Nature, 307*: 183 (1984).
24. S. Reed, J. A. Hadwiger, and A. T. Lorincz, *Proc. Natl. Acad. Sci. U.S.A., 82*: 4055 (1985).
25. M. G. Lee and P. Nurse, *Nature, 327*: 31 (1987).
26. L. J. Cisek and J. L. Corden, *Nature, 339*: 679 (1989).
27. W. Krek and E. A. Nigg, *EMBO J., 8*: 3071 (1989).
28. C. F. Lehner and P. H. O'Farrell, *EMBO J., 9*: 3573 (1990).
29. P. C. L. John, F. J. Sek, and M. G. Lee, *Plant Cell, 1*: 1185 (1989).
30. H. S. Feiler and T. W. Jacobs, *Proc. Natl. Acad. Sci. U.S.A., 87*: 5397 (1990).
31. J. Colasanti, M. Tyers, and V. Sundaresan, *Proc. Natl. Acad. Sci. U.S.A., 88*: 3377 (1991).
32. P. C. G. Ferreira, A. S. Hemerly, R. Villarroel, M. Van Montagu, and D. Inzé, *Plant Cell, 3*: 531 (1991).
33. H. Hirt, A. Páy, J. Györgyey, L. Bakó, K. Németh, L. Bögre, R. J. Schweyen, E. Heberle-Bors, and D. Dudits, *Proc. Natl. Acad. Sci. U.S.A., 88*: 1636 (1991).
34. Y. Hirayama, T. Anai, M. Matsui, and A. Oka, *Gene, 105*: 159 (1991).
35. G.-H. Miao, Z. Hong, and D. P. S. Verma, *Proc. Natl. Acad. Sci. U.S.A., 90*: 943 (1993).
36. B. Durkacz, A. Carr, and P. Nurse, *EMBO J., 5*: 369 (1986).
37. T. Evans, E. T. Rosenthal, J. Youngblom, D. Distel, and T. Hunt, *Cell, 33*: 389 (1983).
38. K. I. Swenson, K. M. Farrel, and J. V. Ruderman, *Cell, 47*: 861 (1986).
39. C. Norbury and P. Nurse, *Curr. Biol., 1*: 23 (1991).
40. M. J. Lohka, M. K. Hayes, and J. L. Maller, *Proc. Natl. Acad. Sci. U.S.A., 85*: 3009 (1988).
41. J. Gautier, C. Norbury, M. Lohka, P. Nurse, and J. Maller, *Cell, 54*: 433 (1988).
42. G. Draetta, F. Luca, J. Westendorf, L. Brizuela, J. Ruderman, and D. Beach, *Cell, 56*: 829 (1989).
43. J.-C. Labbe, J.-P. Capony, D. Caput, J.-C. Cavadore, J. Derancourt, M. Kaghdad, J.-M. Lelias, A. Picard, and M. Doree, *EMBO. J., 8*: 3053 (1989).
44. W. G. F. Whitfield, C. Gonzalez, G. Maldonado-Codina, and D. M. Glover, *EMBO J., 9*: 2563 (1990).
45. C. F. Lehner and P. H. O'Farrell, *Cell, 56*: 957 (1989).
46. J. Pines and T. Hunter, *Cell, 58*: 833 (1989).
47. S. Hata, H. Kouchi, I. Suzuka, and T. Ishii, *EMBO J., 10*: 2681 (1991).
48. A. Hemerly, C. Bergounioux, M. Van Montagu, D. Inzé, and P. Ferreira, *Proc. Natl. Acad. Sci. U.S.A., 89*: 3295 (1992).
49. H. Hirt, M. Mink, M. Pfosser, L. Bögre, J. Gyorgyey, C. Jonak, A. Gartner, D. Dudits, and E. Heberle-Bors, *Plant Cell, 4*: 1531 (1992).
50. J. Pines, *Trends Biochem. Sci., 18*: 195 (1993).

51. C. J. Sherr, *Cell*, *73*: 1059 (1993).
52. J. Minshull, *BioEssays*, *15*: 149 (1993).
53. K. L. Gould, and P. Nurse, *Nature*, *342*: 39 (1989).
54. K. L. Gould, S. Moreno, D. J. Owen, S. Sazer, and P. Nurse, *EMBO J.*, *10*: 3297 (1991).
55. W. Krek and E. A. Nigg, *EMBO J.*, *10*: 305 (1991).
56. C. Norbury, J. Blow, and P. Nurse, *EMBO J.*, *10*:3321–3329 (1991).
57. S. K. Hanks, A. M. Quinn, and T. Hunter, *Science*, *241*: 42 (1988).
58. P. Russell and P. Nurse, *Cell*, *49*: 559 (1987).
59. K. Lundgren, N. Walworth, R. Booher, M. Dembski, M. Kirschner, and D. Beach, *Cell*, *64*: 1111 (1991).
60. C. Featherstone and P. Russell, *Nature*, *349*: 808 (1991).
61. L. L. Parker, S. Atherton-Fessler, M. S. Lee, S. Ogg, J. L. Falk, K. I. Swenson, and H. Piwnica-Worms, *EMBO J.*, *10*: 1255 (1991).
62. S. Moreno, J. Hayles, and P. Nurse, *Cell*, *58*: 361 (1989).
63. P. Fantes, *Nature*, *279*: 428 (1979).
64. P. Russell and P. Nurse, *Cell*, *45*: 145 (1986).
65. J. B. A. Millar and P. Russell, *Cell*, *68*: 407 (1992).
66. K. Sadhu, S. I. Reed, H. Richardson, and P. Russell, *Proc. Natl. Acad. Sci. U.S.A.*, *87*: 5139 (1990).
67. J. Jimenez, L. Alphey, P. Nurse, and D. M. Glover, *EMBO J.*, *9*: 3565 (1990).
68. B. A. Edgar and P. H. O'Farrell, *Cell*, *57*: 177 (1989).
69. P. Russell, S. Moreno, and S. Reed, *Cell*, *57*: 295 (1989).
70. K. Galaktionov and D. Beach, *Cell*, *67*: 1181 (1991).
71. W. G. Dunphy and J. W. Newport, *Cell*, *58*: 181 (1989).
72. M. D. Mendenhall, *Science*, *259*: 216 (1993).
73. S. Moreno and P. Nurse, *Cell*, *61*: 549 (1990).
74. J. L. Maller and D. S. Smith, *Dev. Biol.*, *109*: 150 (1985).
75. E. A. Nigg, *Curr. Opinion Cell Biol.*, *2*: 261 (1991).
76. E. A. Nigg, *Curr. Opinion Cell Biol.*, *4*: 105 (1992).
77. E. A. Nigg, *Curr. Opinion Cell Biol.*, *5*: 187 (1993).
78. R. Reeves, *Curr. Opinion Cell Biol.*, *4*: 413 (1992).
79. M. Peter, J. Nakagawa, M. Doree, and E. F. Nigg, *Cell*, *61*: 591 (1990).
80. G. E. Ward and M. W. Kirschner, *Cell*, *61*: 561 (1990).
81. Y. Yamakita, S. Yamashira, and F. Matsumura, *J. Biol. Chem.*, *267*: 12022 (1992).
82. J.-H. Chou, J. R. Bischoff, D. Beach, and R. D. Goldman, *Cell*, *62*: 1063 (1990).
83. S. Hisanaga, M. Kusubata, E. Okumura, and T. Kishimoto, *J. Biol. Chem.*, *266*: 21798 (1991).
84. L. L. Satterwhite, M. J. Lohka, K. L. Wilson, T. Y. Scherson, L. Cisek, J. L. Corden, and T. D. Pollard, *J. Cell Biol.*, *118*: 595 (1992).
85. A. Jerzmanowski and R. D. Cole, *J. Biol. Chem.*, *267*: 8514 (1992).
86. L. Meijer, A. C. Ostvold, S. I. Waiaas, T. Lund, and S. G. Laland, *Eur. J. Biochem.*, *196*: 557 (1991).
87. T. Moll, G. Tebb, U. Surana, H. Robitsch, and K. Nasmyth, *Cell*, *66*: 743 (1991).
88. B. Luscher and R. N. Eisenman, *J. Cell Biol.*, *118*: 775 (1992).
89. S. Shenoy, I. Chackalaparampil, S. Bagroda, P. H. Lin, and D. Shalloway, *Proc. Natl. Acad. Sci. U.S.A.*, *89*: 7237 (1992).
90. E. T. Kipreos and J. Y. J. Wang, *Science*, *56*: 382 (1992).
91. D. E. Litchfield, B. Luscher, F. J. Lozeman, R. N. Eisenman, and E. G. Krebs, *J. Biol. Chem.*, *267*: 13943 (1992).
92. L. J. Cisek and J. L. Corden, *Nature*, *339*: 679 (1989).
93. E. Bailly, M. McCaffrey, N. Touchot, A. Zahroui, B. Goud, and M. Bornens, *Nature*, *350*: 715 (1991).
94. J. C. Courvalin, N. Segil, G. Blobel, and H. J. Worman, *J. Biol. Chem.*, *267*: 19035 (1992).
95. O. Mulner-Lorillon, P. Cormier, J.-C. Cavadore, J. Morales, R. Poulhe, and R. Belle, *Exp. Cell Res.*, *202*: 549 (1992).

96. D. A. Jans, M. J. Ackermann, J. R. Bischoff, D. H. Beach, and R. Peters, *J. Cell Biol.*, *115*: 1203 (1991).

97. H.-P. Nasheuer, A. Moore, A. F. Wahl, and T. S.-F. Wang, *J. Biol. Chem.*, *266*: 7893 (1991).

98. B. T.-Y. Lin, S. Gruenwald, A. O. Moria, W.-H. Lee, and J. Y. J. Wang, *EMBO. J.*, *10*: 4279 (1991).

99. T. Izumi and J. L. Maller, *Mol. Cell Biol.*, *11*: 3860 (1991).

100. T. Izumi, D. H. Walker, and J. L. Maller, *Mol. Biol. Cell*, *3*: 927 (1992).

101. R. Heald and F. McKeon, *Cell*, *61*: 579 (1990).

102. T. Enoch, M. Peter, P. Nurse, and E. Nigg, *J. Cell Biol.*, *112*: 797 (1991).

103. M. Glotzer, A. W. Murray, and M. W. Kirschner, *Nature*, *349*: 132 (1991).

104. A. W. Murray, M. J. Solomon, and M. W. Kirschner, *Nature*, *339*: 280 (1989).

105. T. Hunt, *Nature*, *349*: 100 (1991).

106. M. S. Cyert and J. Thorner, *Cell*, *57*: 891 (1989).

107. S. Hata, *FEBS Lett.*, *279*: 149 (1991).

108. J. Hashimoto, T. Hirabayashi, Y. Hayano, S. Hata, Y. Ohashi, I. Suzuka, T. Utsugi, A. Toh-e, and Y. Kikuchi, *Mol. Gen. Genet.*, *233*: 10 (1992).

109. Z. Hong, G.-H. Miao, and D. P. S. Verma, *Plant Physiol.*, *101*: 1399 (1993).

110. C. H. McGowan, P. Russell, and S. I. Reed, *Mol. Cell Biol.*, *10*: 3847 (1990).

111. S. Dalton, *EMBO J.*, *11*: 1797 (1992).

112. M. C. Martinez, J.-E. Jorgensen, M. A. Lawton, C. J. Lamn, and P. W. Doerner, *Proc. Natl. Acad. Sci. U.S.A.*, *89*: 7360 (1993).

113. P. C. L. John, J. P. Carmichael, and D. W. McCurdy, *J. Cell Sci.*, *97*: 627 (1990).

114. J. R. Gorst, P. C. L. John, and F. J. Sek, *Planta*, *185*: 304 (1991).

115. P. C. L. John, F. J. Sek, and J. Hayles, *Protoplasma*, *161*: 70 (1991).

116. A. Koff, F. Cross, A. Fisher, J. Schumacher, K. Leguellec, M. Philippe, and J. M. Roberts, *Cell*, *66*: 1217 (1991).

117. D. J. Lew, V. Dulic, and S. I. Reed, *Cell*, *66*: 1197 (1991).

118. I. Day and A. S. N. Reddy, *Biochim. Biophys. Acta* in press (1994).

119. J. Colasanti, J.-P. Renaudin, S.-O. Cho, S. Wick, C. Jessus, H. Rime, Z. Yuan, and V. Sundaresan, "Regulation of cell-division in maize: When to divide and which way to divide," Proceedings of 12th Annual Missouri Plant Biochemistry, Molecular Biology and Physiology Symposium, University of Missouri, pp. 72–73 (1993).

120. S. J. Elledge, and M. R. Spottswood *EMBO J.*, *10*: 2653 (1991).

121. S. J. Elledge, R. Richman, F. L. Hall, R. T. Williams, N. Lodgson, and J. W. Harper, *Proc. Natl. Acad. Sci. U.S.A.*, *89*: 2907 (1992).

122. L.-H. Tsai, E. Harlow, and M. Meyerson, *Nature*, *353*: 174 (1991).

123. M. Meyerson, G. H. Enders, C.-L. Wu, L.-K. Su, C. Gorka, C. Nelson, E. Harlow, and L.-H. Tsai, *EMBO J.*, *11*: 2909 (1992).

124. H. Matsushime, M. E. Ewen, D. K. Strom, J.-Y. Kato, S. K. Hanks, M. F. Roussel, and C. J. Sherr, *Cell*, *71*: 323 (1992).

125. A. Koff, A. Giordano, D. Desai, K. Yamashita, J. W. Harper, S. Elledge, T. Nishimoto, D. O. Morgan, R. Franza, and J. M. Roberts, *Science*, *257*: 1689 (1992).

126. J. Lew, R. J. Winkfein, H. K. Paudel, and J. H. Wang, *J. Biol. Chem.*, *267*: 25922 (1992).

127. B. A. Bunnell, L. S. Heath, D. E. Adams, J. M. Lahti, and V. J. Kidd, *Proc. Natl. Acad. Sci. U.S.A.*, *87*: 7467 (1990).

128. J. Shuttleworth, R. Godrey, and A. Colman, *EMBO. J.*, *9*: 3233 (1990).

129. T. W. Jacobs, M. E. Prewett, B. K. Buerr, H. S. Feiler, J. Dunphy, H. Chen, and J. Poole, "Protein kinases in cell division control," Proceedings of 12th Annual Missouri Plant Biochemistry, Molecular Biology and Physiology Symposium University of Missouri, pp. 67–68 (1993).

130. L. Bögre, C. Jonak, D. T. Cam Ha, T. Murbacher, S. Kiegerl, L. Bako, C. Planck, M. Pfosser, A. Páy, I. Meskiene, M. Mink, J. Gyorgyey, K. Plame, E. Wagner, D. Dudits, E. Heberle-Bors, and H. Hirt, "Alfalfa cell cycle control elements: *cdc*2, cyclins, MAP kinases," Proceedings of 12th Annual Missouri Plant Biochemistry, Molecular Biology and Physiology Symposium University of Missouri, pp. 71–72 (1993).

131. H. Kodama, N. Kawakami, A. Watanabe, and A. Komamine, *Plant Physiol.*, *89*: 910 (1989).
132. H. Kodama, M. Ito, T. Hattori, K. Nakamura, and A. Komamine, *Plant Physiol.*, *95*: 406 (1991).
133. E. J. van der Zaal, J. Memelink, A. M. Mennes, A. Quinn, and K. R. Libbenga, *Plant Mol. Biol.*, *10*: 145 (1987).
134. E. J. van der Zaal, F. N. J. Droog, C. J. M. Boot, L. A. M. Hensgens, J. H. C. Hoge, R. A. Schilperoort, and K. R. Libbenga, *Plant Mol. Biol.*, *16*: 983 (1991).
135. Y. Takahashi, H. Kuroda, T. Tanaka, Y. Machida, I. Takebe, and T. Nagata, *Proc. Natl. Acad. Sci. U.S.A.*, *86*: 9279 (1989).
136. Y. Takahashi and T. Nagata, *Proc. Natl. Acad. Sci. U.S.A.*, *89*: 56 (1992).
137. M. Sakai, A. Okuda, and M. Muramatsu, *Proc. Natl. Acad. Sci. U.S.A.*, *85*: 9456 (1988).
138. K. Satoh, A. Kitahara, Y. Soma, Y. Inaba, I. Hatayama, and K. Sato, *Proc. Natl. Acad. Sci. U.S.A.*, *82*: 3964 (1985).
139. Y. Li, T. Seyama, A. K. Godwin, T. S. Winokur, R. M. Lebovitz, and M. W. Lieberman, *Proc. Natl. Acad. Sci. U.S.A.*, *85*: 344 (1988).
140. G. Prelich, C. Tan, M. Kostura, M. B. Mathews, A. G. So, K. M. Downey, and B. Stillman, *Nature*, *326*: 517 (1987).
141. H. Kodama, M. Ito, N. Ohnishi, I. Suzuka, and A. Komanine, *Eur. J. Biochem.*, *197*: 495 (1991).
142. S. L. Pelech and J. S. Sanghera, *Trends Biochem Sci.*, *17*: 233 (1992).
143. G. Thomas, *Cell*, *68*: 3 (1992).
144. B. Duerr, M. Gawienowski, R. Ropp, and T. Jacobs, *Plant Cell*, *5*: 87 (1993).
145. J. P. Stafstrom, M. Altschuler, and D. H. Anderson, *Plant Mol. Biol.*, *22*: 83 (1993).
146. D. M. Payne, A. J. Rossomando, P. Martino, A. K. Erikson, J. H. Her, J. Shabanowitz, D. F. Hunt, M. J. Weber, and T. W. Sturgill, *EMBO J.*, *10*: 885 (1991).
147. B. W. Poovaiah and A. S. N. Reddy, *CRC Crit. Rev. Plant Sci.*, *6*: 47 (1987).
148. B. W. Poovaiah and A. S. N. Reddy, *CRC Crit. Rev. Plant Sci.*, *12*: 185 (1993).
149. D. M. Roberts and A. Harmon, *Annu. Rev. Plant Physiol. Plant Mol. Biol.*, *43*: 375 (1992).
150. P. K. Hepler, *J. Cell Biol.*, *109*: 2567 (1989).
151. K. P. Lu and A. R. Means, *Endocr. Rev.*, *14*: 40 (1993).
152. M. Whitaker and R. Patel, *Development*, *108*: 525 (1990).
154. J. F. Whitfield, J. P. MacManus, R. H. Rixon, A. L. Boynton, T. Yoydale, and S. Swierenga, *In vitro*, *12*: 1 (1976).
155. J. F. Whitfield, A. L. Boynton, J. P. MacManus, R. H. Rixon, M. Sikorska, B. Tsang, and P. R. Walker, *Ann. N.Y Acad. Sci.*, *339*: 216 (1980).
156. R. B. Silver, *Ann. N.Y. Acad. Sci.*, *582*: 207 (1990).
157. M. L. Veigl, T. C. Vanaman, and W. D. Sedwick, *Biochim. Biophys. Acta*, *738*: 21 (1984).
158. J. P. Y. Kao, J. M. Alderton, R. Y. Tsien, and R. A. Steinhardt, *J. Cell Biol.*, *111*: 183 (1990).
159. J. G. Izant, *Chromosoma*, *88*: 1 (1983).
160. R. A. Steinhardt and J. Alderton, *Nature*, *332*: 364 (1988).
161. M. Poenie, J. Alderton, R. Y. Tsien, and R. A. Steinhardt, *Nature*, *315*: 147 (1985).
162. C. H. Keith, R. Ratan, F. R. Maxfield, A. Bajer, and M. L. Shelanski, *Nature*, *316*: 848 (1985).
163. M. Poenie, J. Alderton, R. A. Steinhardt, and R. Y. Tsien, *Science*, *233*: 886 (1986).
164. P. K. Hepler and D. A. Callaham, *J. Cell Biol.*, *105*: 2137 (1987).
165. D. M. Roberts, T. J. Lukas, and D. M. Watterson, *CRC Crit. Rev. Plant Sci.*, *4*: 311 (1986).
166. S. Muto and S. Miyachi, *Z. Pflanzenphysiol.*, *114*: 421 (1984).
167. P. K. Jena, A. S. N. Reddy, and B. W. Poovaiah, *Proc. Natl. Acad. Sci. U.S.A.*, *86*: 3644 (1989).
168. J. G. Chafouleas, W. E. Bolton, H. Hidaka, A. E. Boyd III, and A. R. Means, *Cell*, *28*: 41 (1982).
169. Y. Sasaki and H. Hidaka, *Biochem. Biophys. Res. Commun.*, *104*: 451 (1982).
170. J. G. Chafouleas, L. Legace, W. E. Bolton, A. E. Boyd III, and A. R. Means, *Cell*, *36*: 73 (1984).
171. J. G. Chafouleas, R. L. Pardue, B. R. Brinkely, J. R. Dedman, and A. R. Means, *Proc. Natl. Acad. Sci. U.S.A.*, *78*: 996 (1981).
172. D. C. LaPorte, S. Gidwitz, M. J. Weber, and D. R. Storm, *Biochem. Biophys. Res. Commun.*, *86*: 1169 (1979).
173. C. D. Rasmussen and A. R. Means, *EMBO. J.*, *6*: 3961 (1987).

174. C. D. Rasmussen and A. R. Means, *EMBO. J.*, *8*: 73 (1989).

175. A. R. Means, M. F. A. VanBerkum, I. C. Bagchi, K. P. Lu, and C. D. Rasmussen, *Pharmacol Ther.*, *50*: 255 (1991).

176. K. P. Lu, S. A. Osmani, and A. R. Means, *J. Cell Biol.*, *115*: 426 (1991).

177. K. P. Lu, C. D. Rasmussen, G. S. May, and A. R. Means, *Mol. Endocrinol.*, *6*: 365 (1992).

178. C. B. Klee *Neurochem. Res.*, *16*: 1059 (1991).

179. A. S. N. Reddy, D. Takezawa, H. Fromm, and B. W. Poovaiah, *Plant Sci.*, *94*: 109–117 (1993).

180. A. S. N. Reddy, Z. Q. Wang, Y. J. Choi, G. An, A. J. Czernik, and B. W. Poovaiah, *Proc. Cold. Spring Harbor Symp. Plant Signal Transduction*, 57 (1991).

181. B. Watillion, R. Kettmann, P. Boxus, and A. Burny, *Plant Physiol.*, *101*: 1381 (1993).

182. J. F. Harper, M. R. Sussman, G. E. Schaller, C. Putnam-Evans, H. Charbonneau, and A. C. Harmon, *Science*, *252*: 951 (1991).

183. J. H. Choi and K.-L. Suen, *Plant Mol. Biol.*, *17*: 581 (1991).

184. T. Lorca, S. Galas, D. Fesquet, A. Devault, J.-C. Cavadore, and M. Dorée, *EMBO J.*, *10*: 2087 (1991).

185. K. Lindsey (ed.), *Plant Tissue Culture: Fundamentals and Applications*, Kluwer Academic Publishers, Dordrecht (1991).

186. T. Nagata and I. Takebe, *Planta*, *92*: 201 (1970).

187. T. Nagata and I. Takebe, *Planta*, *99*: 12 (1971).

188. M. Sauter and H. Kende, *Planta*, *188*: 362 (1992).

189. P. N. Myers, T. L. Setter, J. T. Madison, and J. F. Thompson, *Plant Physiol.*, *94*: 1330 (1990).

190. E. S. Ober, T. L. Setter, J. T. Madison, J. F. Thompson, and P. S. Shapiro, *Plant Physiol.*, *97*: 154 (1991).

191. P. W. Barlow and P. E. Pilet, *Physiol. Plant.*, *62*: 125 (1984).

192. H. S. Saini and D. Aspinall, *Aust. J. Plant Physiol.*, *9*: 529 (1982).

193. S. Amino, R. Fujimura, and A. Komamine, *Physiol. Plant.*, *59*: 393 (1983).

194. F. Sala, M. G. Galli, G. Pedrali-Noy, and S. Spadari, *Methods Enzymol.*, *118*: 87 (1986).

195. S. Spadari, F. Sala, and G. Pedrali-Noy, *Trends Biochem. Sci.*, *7*: 29 (1982).

196. S. Amino, R. Fujimura, and A. Komamine, *Physiol. Plant.*, *59*: 393 (1983).

197. D. W. Galbraith, *Methods Cell Biol.*, *33*: 527 (1990).

198. M. H. Fox and D. W. Galbraith, *Flow Cytometry and Sorting* (M. R. Melamed, T. Lindmo, and M. L. Mendelsohn, eds.), Wiley-Liss, New York, p. 633 (1990).

199. S. Bruno, H. A. Crissman, K. D. Bauer, and Z. Darzynkiewic, *Exp. Cell Res.*, *196*: 99 (1991).

7

Vegetative and Elongation Growth Stages

Clarence J. Kaiser
University of Illinois at Urbana-Champaign, Urbana, Illinois

I. INTRODUCTION

Many attempts have been made to identify and quantify the stages of growth and development of crop plants. Accurate identification of growth stages in any population of crop plant species is critical to management decisions. Current systems describing and quantifying morphological development are available. This chapter selects economically important crop species as examples of the phenological and quantitative relationships adopted by many researchers to describe the vegetative and elongation stages of plant growth.

Each growth stage consists of a primary and secondary stage. A two-element code is assigned to each stage to assist in describing and memorizing substages. In each code, a capital letter (mnemonic) denotes the primary growth stage and a number (numerical index) refers to the substage within that primary stage. Substages of the vegetative and elongation stages describe specific events that occur similarly in most plant species. Vegetative and elongation substages are open-ended. Each substage event is equivalent to the number of morphological events that occur for that species and environment.

The vegetative growth stage of grasses is defined as leaves only, stems not elongated. The stem elongation stage follows the vegetative stage and terminates at the full boot stage (inflorescence enclosed in the flag leaf sheath and not visible). The vegetative growth stage of (upright-growing) legumes is best described as prebud. The elongation stage includes the bud formation and terminates prior to the first flower.

A system of describing primary growth stages of individual crop plant species covers the germination, vegetative, elongation, reproductive, and seed ripening phases. Only the vegetative and elongation stages are discussed in this chapter.

II. ANNUAL GRASSES

In this description of vegetative and elongation growth stages, the corn plant is our example of an annual grass. However, annual grass plants such as corn [1] commonly use only the

169

vegetative and reproductive stages (Table 1). The vegetative and elongation growth stages are included in stage number V1 through V(*n*). The notation *(n)* represents the last leaf stage before tasseling, VT. Each leaf stage is defined according to the uppermost leaf whose leaf collar is visible. The first part of the collar that is visible is the back of the leaf, which appears as a discolored line between the leaf blade and the leaf sheath. The characteristically oval first leaf is a reference point for counting upward to the top visible leaf collar. Beginning at about V6, increasing stalk growth tears the small lowest leaves from the plant. Lower leaves may be lost, making leaf count more difficult. The R stages (Table 1) of the corn plant are not discussed in this chapter. Leaf stages V2, V4, V7, V8, V11, V13, V14 and V16 are not discussed.

A. V1 Stage: First Leaf

The V1 stage occurs when the collar of the first leaf is visible above ground level: about 10 days after planting. The blades of leaves two and three are visible at the V1 stage. The nodal root system begins to elongate. Nearly all water and nutrients are supplied through the seminal root system. A set of nodal roots begins development at each progressively higher node on the stalk, up to V7–V10. The stem apex (growing point) is approximately 2.5–3.8 cm below the soil surface.

B. V3 Stage: Third Leaf

At the V3 stage (collar of the third leaf visible) the stem apex (growing point) is still below the soil surface. Very little stalk (stem) elongation has occurred. Root hairs are growing from the nodal root system. The seminal root system has virtually stopped growth. All leaf and ear shoots that the corn plant will eventually produce are being initiated at this stage.

C. V5 Stage: Fifth Leaf

Leaf and ear shoot initiation is nearly completed in the V5 stage. A microscopically small tassel is initiated in the stem apex tip. The stem apex at tassel initiation is just under or at the soil surface. The above-ground plant is about 20 cm high.

D. V6 Stage: Sixth Leaf

The nodal root system has become the major supplier of water and nutrients to the corn plant by the V6 stage. The growing point and tassel are above the soil surface. The stalk is beginning a period of greatly increased elongation. Some ear shoots or tillers are visible.

Table 1 Vegetative and Reproductive Stages of a Corn Plant

Vegetative (V) Stage		Reproductive (R) Stage	
Stage	Description	Stage	Description
VE	Emergence	R1	Silking
V1	First leaf	R2	Blister
V2	Second leaf	R3	Milk
V3	Third leaf	R4	Dough
		R5	Dent
		R6	Physiological
V(*n*)	*n*th leaf		
VT	Tasseling		

Source: Ref. 1.

E. V9 Stage: Ninth Leaf

Many ear shoots (potential ear) are visible in the V9 stage. The stalk continues rapid internode elongation. The lowest internode completes elongation first and, progressively, the next higher internodes in order. The tassel begins to develop rapidly.

F. V10 Stage: Tenth Leaf

The time between the appearance of new leaf stages begins to shorten in the V10 stage. New leaf stages appear every 2–3 days. The corn plant begins a steady increase in nutrient needs. Plant dry weight increases rapidly.

G. V12 Stage: Twelfth Leaf

The number of ovules (potential kernels) on each ear and the size of the ear are determined at the V12 stage. The top ear shoot is smaller than the lower ear shoots. Many of the upper ear shoots are close to the same size.

H. V15 Stage: Fifteenth Leaf

The V15 stage is the most crucial period in corn plant development in terms of seed yield determination. The upper ear shoot development has surpassed that of the lower ear shoots. New leaf stages are now occurring every 1–2 days. Silks are just beginning to grow from the upper ears.

I. V17 Stage: Seventeenth Leaf

Tips of the upper ear shoots are visible in the V17 stage. The tip of the tassel may also be visible.

J. V18 Stage: Eighteenth Leaf

Silks from the basal ear ovules are first and silks from the ear tip ovules are last to elongate over an 8–9 day period in the V18 stage. Brace roots (aerial nodal roots) are initiated.

K. VT Stage: Tasseling

The VT (vegetative tassel) stage is initiated when the last branch of the tassel is completely visible, but silks have not yet emerged. Silks will emerge in 2–3 days, the corn plant will attain its full height, and pollen shed will begin.

III. ANNUAL LEGUMES

The soybean plant is our example of an annual legume in this description of the vegetative and elongation growth stages. However, annual legume plants such as soybean [2] commonly use only the vegetative and reproductive stages (Table 2). The vegetative and elongation growth are included in the stages VC through V(*n*), where (*n*) represents the last leaf stage before R1 (beginning bloom). Each V stage following VC is defined according to the upper most fully developed leaf node. A fully developed leaf node is one that has a leaf above it with unrolled or unfolded leaflets. The V2 stage, for example, is defined when the leaflets on the first (unifoliolate) through the third node leaf are unrolled. The unifoliolate leaf node is the first node or reference point from which to begin counting up to identify upper leaf node numbers.

Table 2 Vegetative and Reproductive Stages of a Soybean Plant

Vegetative (V) stage		Reproductive (R) stage	
Stage	Description	Stage	Description
VE	Emergence	R1	Beginning bloom
VC	Cotyledon	R2	Full bloom
V1	First node	R3	Beginning pod
V2	Second node	R4	Full pod
V3	Third node	R5	Beginning seed
		R6	Full seed
V(n)	nth node	R7	Beginning maturity
		R8	Full maturity

Source: Ref. 2.

Note that the VC (cotyledons) arise opposite on the stem just below the unifoliolate node. The unifoliolate leaves are also produced on opposite sides of the stem. All other true leaves formed by the plant are trifoliolate leaves borne on long petioles and are produced singularly at each node. Petioles arise alternately (from side to side) on the stem. The R stages of the soybean plant are not discussed in this chapter.

A. VC Stage: Cotyledons

Shortly after emergence, the VE stage, the hook-shaped hypocotyl straightens out and discontinues growth as the cotyledons fold down. The subsequent expansion and unfolding of the unifoliolate leaves marks initiation of the VC stage. Nodule formation begin.

B. V1 Stage: First Node

At the V1 stage two nodes have leaves with completely unfolded leaflets (the unifoliolate node and the first trifoliolate leaf node). New vegetative (V) stages will appear about every 5 days through V5.

C. V2 Stage: Second Node

Plants at the V2 stage are 15–20 cm tall. Nitrogen fixation begins, continuing until midway between stages R5 and R6, when it decreases sharply.

D. V3 Stage: Third Node

Plants at the V3 stage are 18–23 cm tall and four nodes have leaves with completely unfolded leaflets. Leaf branching begins in the axil bud of the first trifoliolate leaf node. Axillary buds may develop a branch or a flower cluster, or they may remain dormant.

E. V5 Stage: Fifth Node

The total number of nodes that an indeterminate type soybean plant may potentially produce is set at the V5 stage. This total will be higher than the actual number of nodes that fully develop.

F. V6 Stage: Sixth Node

Plants at the V6 stage are 30–35 cm tall. Seven nodes have leaves with completely unfolded leaflets. Both the cotyledons and the unifoliolate leaves have senesced and fallen from the plant by this time. New vegetative (V) stages are now appearing every 3 days.

IV. PERENNIAL GRASSES

For perennial grasses, both cool (Kentucky bluegrass, orchardgrass, tall fescue, etc.) and warm season (switchgrass, big bluestem, indiangrass, etc.) growth stages vary very little among species [3]. An example of the vegetative and the elongation stage of the perennial grasses is given in Table 3. The numerical index may vary somewhat depending on species. The reproductive (R) and seed development (S) stages are not discussed in this chapter.

A. VE Stage: Emergence

The vegetative stage begins with the emergence of the first leaf from the coleoptile, or the prophyll in the case of tillers. Each successive substage refers to the number of fully emerged live leaves currently present. Leaves are considered fully emerged when collared.

B. V1 Stage: First Leaf

Only live leaves are counted in the V1 stage. In many grasses the rate of leaf senescence is nearly equal to the rate of new leaf appearance.

C. V(n) Stage: nth Leaf

The number of leaves present on a tiller becomes relatively constant once leaf senescence has begun to occur. In some cases only a small proportion of the total number of tillers will reach the highest observed vegetative stage V(n).

D. E0 Stage: Stem Elongation

Elongation is the stage during which culm (stem) elongation occurs. This stage is often referred to as jointing.

E. E1 Stage: First Node

By the E1 stage, the first node has become either visible or palpable as the result of stem elongation.

Table 3 Vegetative and Elongation Stages of Perennial Grasses

Vegetative (V) stage		Elongation (E) stage	
Stage	Description	Stage	Description
VE	Emergence of first leaf	E0	Onset of stem elongation
V1	First leaf collared	E1	First node visible/palpable
V(n)	nth leaf collared	E(n)	nth node visible/palpable

Source: Ref. 3.

F. E(n) Stage: nth Node

The elongation stage ceases when the inflorescence is enclosed in the uppermost leaf sheath (boot stage). Sterile culms are common in some perennial grass species. Regrowth of smooth bromegrass may elongate through several internodes, even though the terminal meristem remains vegetative.

V. PERENNIAL LEGUMES

The alfalfa plant [4] is selected as our example of a perennial legume to be used in describing the vegetative and elongation growth stages. These stages are combined into a single stage, elongation (E) (Table 4). The reproductive (R) stage is not discussed in this chapter.

A. E0 Stage: Early Vegetative

The early vegetative stage occurs for stem growth up to 15 cm. No buds, flowers, or seed pods are found at this stage.

B. E1 Stage: Midvegetative

The midvegetative stage occurs at a stem length of 16–30 cm. No buds, flowers, or seed pods are found.

C. E2 Stage: Late Vegetative

The late vegetative stage occurs when the stem length exceeds 31 cm of growth. No buds, flowers, or seed pods are found.

D. E3 Stage: Early Bud

The early bud stage occurs at one or two nodes with buds. No flowers or seed pods are found.

E. E4 Stage: Late Bud

The late bud stage occurs at three or more nodes with buds. No flowers or seed pods are found.

Table 4 Vegetative and Reproductive Stages of an Alfalfa Plant

Elongation (E) stage		Reproductive (R) stage	
Stage	Description	Stage	Description
E0	Early vegetative	R5	Early flower
E1	Mid vegetative	R6	Late flower
E2	Late vegetative		
E2	Late vegetative	R7	Early seed pod
E3	Early bud	R8	Late seed pod
E4	Late bud		

Source: Ref. 4; mnemonic code letter added.

VII. SUMMARY

This system for characterizing the morphological development of grasses and legumes emphasizes the vegetative and elongation growth stages. Accurate identification of the growth stage of economically important grasses and legume plants is critical in making many management decisions. Production practices involving establishment, grazing, harvesting, grain, and seed production require an assessment of development stages. The stages of morphological development described are necessary for making and communicating objective measurements relative to the growth and development of economically important plants.

REFERENCES

1. S. W. Ritchie, J. J. Hanway, and G. O. Benson, *How a Corn Plant Develops*, Special Report 48, Iowa State University of Science and Technology, Ames, p. 21 (1986).
2. S. W. Ritchie, J. J. Hanway, and H. E. Thompson, *How a Soybean Plant Develops*, Special Report 53, Iowa State University of Science and Technology, Ames, p. 20 (1982).
3. K. J. Moore, L. E Moser, K. P. Vogel, S. S. Waller, B. E. Johnson, and J. F. Pedersen, *Agron. J.*, *83*: 1073–1077 (1991).
4. B. A. Kalu and G. W. Fick, *Crop Sci.*, *21*: 267–271 (1981).

8

Ecophysiological Aspects of the Vegetative Propagation of Saltbush (*Atriplex* spp.) and Mulberry (*Morus* spp.)

David N. Sen and Pramila Rajput
University of Jodhpur, Jodhpur, India

I. INTRODUCTION

Nature has provided the phenomenon of reproduction to the living world in order to perpetuate species. The way in which reproduction in the plant kingdom is carried out may be broadly divided into two categories:

Vegetative propagation
Sexual reproduction

At times, under prevailing environmental conditions, some plants fail to complete their life cycle by means of seeds, yet they survive and perpetuate themselves. This is because nature has provided an alternative to sexual reproduction, that is, vegetative propagation.

The multiplication of species by vegetative means is practiced in forestry and horticulture to obtain plants of a desired genetic constitution for crossing in a breeding program for many reasons (to improve growth and yield, stem quality, wood quantity, or resistance to pests and diseases or other desirable characters, and also to maintain the purity of types so evolved for commercial exploitation). This process has been used for quick multiplication for a number of plant species, which is important for afforestation purposes in arid zones, where quick growth and development of plants is very much needed.

In easily rooting species, the ability of stem cuttings to root varies considerably with the season. In many cases, profuse rooting occurs when cuttings are taken from trees in an active season. The seasonal rooting response of stem cuttings is related to the disappearance of starch. The hydrolytic activity is high when rooting occurs, but is not detached when cuttings fail to root [1]. Nanda [2] revealed that the effectiveness of exogenously applied auxins varies with the season and that these differences may be ascribed to changes in a plant's nutritional and hormonal status during its annual cycle of growth.

In India, propagation through cuttings is the most common method of perpetuating vegetable species. It is restricted to varieties that are fully acclimatized to local conditions. Growers select plants that display the quantities chosen for multiplication, such as nutritious leaf, higher yield, quick growth, and resistance to disease, insect pests, and drought. Resistance to drought is an important property to be associated with other desirable characteristics in tropical and semiarid regions. In semiarid parts of India, rainfall is very low and the monsoon is erratic and unpredictable. Drought is frequent in semiarid tropical regions, hence, crop loss due to scarcity of water also is frequent. In addition to the development of more suitable farming technology, evolution of drought-resistant varieties would be desirable.

The arid ecosystem environment offers an adaptive challenge to the survival of plants: the only species that can survive possess adaptive mechanisms to adjust themselves under strong climatic fluctuations [3]. According to Sen [3] ecophysiological studies thus become important parameters for judging the capability of a particular species to adjust itself under prevailing climatic and edaphic conditions.

In arid regions of India, available soil moisture is used by the roots of annual and perennial plants from the end of the rainy season until early summer, by which time such moisture has been depleted. Later, a partial or total status quo is maintained in soil moisture, mainly in the open, with the result that water loss is eliminated by shedding or reduction of leaves by plants [3]. Rainwater is the only source of available moisture in the desert of northwest Rajasthan. Although the monsoon starts in mid-June and lasts through October, the rains are very erratic and scanty (Figure 1). The occurrence of rather long intervals between successive showers, sometimes ranging from a few days to a week, is not uncommon. Whatever rainwater is retained by the soil is used by the roots of annual and perennial species from June–July to November–December [4]. Physiological studies are helpful in determining the individual and collective influence of different factors on vegetative propagation.

A. Significance of Vegetative Propagation

Rooting in stem cuttings can be an important means of vegetative propagation for afforestation purposes. In arid zones, quick establishment of plants with ample root systems is a necessity.

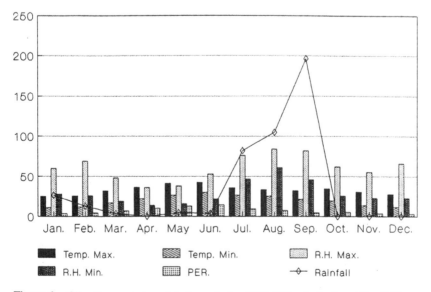

Figure 1 Climatic parameters at Jodhpur during 1992: RH, relative humidity; PER, potential evapotranspiration.

In arid Rajasthan, water in the form of precipitation is available only in the rainy season, and the plants must be established in suitable conditions of soil moisture. Therefore, the use of rooted stem cuttings is more useful than seed sowing, because rooted cuttings are far more able to survive in the stressful environment of the desert than delicate seedlings.

Thimann and Behnke-Rogers [5] showed that the rooting of cuttings of many tree species is stimulated by synthetic growth substances. Bose [6] has developed easier and better methods of vegetative propagation by the use of growth substances for ornamental and fruit plants. Bose and Mukherjee [7] used some growth substances to improve rooting in cuttings of *Legerstroemia indica*. Prasad and Dikshit [8] obtained maximum success in rooting with cuttings of essential oil yielding plants treated with growth regulators. Teaotia and Pandey [9] obtained better results in rooting guava stem cuttings with the assistance of growth substances.

B. Factors Affecting Vegetative Propagation

Saline and alkaline soils are the problems of individual localities, and their formation and causes of development must be considered before these soils are put to any economic use. Salt-tolerant plants have been used as forage in arid saline areas for millennia. The recognition of the value of certain salt-tolerant shrub and grass species is reflected in their incorporation into pasture improvement programs in many salt-affected regions throughout the world. However, reproduction, survival, and multiplication under the inhospitable conditions of arid saline areas are basic needs for any halophytic or glycophytic species. In many halophytes, germination of seeds is usually retarded by high concentrations of salt in the soil [10]. Germination is the most important stage in the life cycle of any species growing in an arid saline environment. Seed germination in saline environments occurs mostly with high precipitation, when soil salinity levels usually are reduced [11]. It is also known that when seeds are sown in a saline environment, there is a decrease in the rate of germination, delaying completion of germination; moreover, there is a water potential ψ below which germination will not occur [12–15]. In general, it is agreed that salinity affects germination by creating a ψ sufficiently low to inhibit water uptake (osmotic effect) and/or by providing conditions for the entry of ions that may be toxic to the embryo [16]. These constraints affect the different stages of seed germination and seed establishment to varying degrees.

Reduction in germination occurs when halophytes are subjected to salinities above 1% NaCl; increasing salt concentrations also delay germination [17]. In the Indian desert, salinity or alkalinity and water stress are the most important factors responsible for limiting seed germination and plant growth. To overcome the present environmental stress of inland saline areas, plants produce a variety of ecological adaptations. Propagation through vegetative means has been used as a method of multiplication for a number of plant species under arid saline conditions. Very little work on this aspect has been done in India, where the quick growth and development of plants is needed badly.

Among factors affecting rooting of cuttings, the position of the shoot plays an important role [18]. It is reported that without auxin treatment and without leaves, no roots were obtained in cuttings of red *Hibiscus* and *Allamenda cathartica* [19].

Vegetative reproduction substitutes for or at least contributes to the reproductive potential of many plants. This statement is more applicable to various halophytic species that are restricted to narrow ecological limits, either at the production of disseminules or by their germination [20]. Self-layering species of *Atriplex* are at an advantage in establishing themselves in salt-affected soil, which they accomplish faster than other species: the growth of developing roots results in rapid penetration through the upper salty soil layers. Furthermore, roots developing at different nodes are not dependent on a direct supply of water from the soil [21].

Being well supplied with water by the parent plant, roots can thus penetrate layers of extreme salinity.

C.   Vegetative Propagation in Saline Plants

The distribution of salinity varies spatially, temporally, qualitatively, and quantitatively. In addition, the responses of plants to salt stress vary during the life cycle of the individual [22,23].

Phenotype plasticity involving both morphological and physiological changes in response to episodic events is an important characteristic associated with the survival of long-lived plants under highly stressful environmental conditions. Transient reductions in yield in response to salinity may be the result of the adaptive reconstruction of the growth habits of a plant. The heterogeneity of saline habitats leads to considerable genetic differentiation among populations as a result of natural selection: an all-purpose genotype capable of growing in a wide range of saline habitats probably does not exist [20].

The growth and productivity of *Atriplex* under conditions of low and erratic rainfall is exceptional, and the adaptation of this species to high salinity makes its introduction very suitable [24]. Agronomic testing, feeding trials, and development of the best agronomic practices are necessary in the evaluation of suitable species for introduction and mass propagation [25].

Normal vegetation, except some halophytes, cannot survive on saline and sodic soils. Thus, the areas having soils of these types are of limited agricultural use, unless the salinity is quite mild. In Rajasthan, increased salinity has rendered many lands unfit for cultivation.

Plant species that are capable of accumulating large quantities of sodium in the body are the least sensitive to the presence of salt in the soil. The tolerance of a species to high amounts of absorbed or exchangeable sodium is modified by the pH of the soil and by the accumulation of CO_2.

Some saline lands provide fairly good pasture, unless the salinity is excessively high. The succession of grass species from highly saline to least saline conditions consists of *Sporobolus marginatus*, *Aristida* spp., *Eleusine compressa*, *Dactyloctenium sindicum*, *Eragrostis tremula*, and *Dichanthium annulatum* [26]. Among salt-tolerant trees in arid Rajasthan, the more important are *Prosopis juliflora*, *Salvadora persica*, and *Tamarix troupii*.

With increasing human and animal populations and the need for greater crop and fodder production, nonproductive salt-affected lands, such as barren saline tracts of Rajasthan, may be used to grow nonconventional crops of economic value and also such food crops as pearl millet. It is desirable to choose species well suited to saline habitats and to calculate the most economical means of reclamation to make the salt-affected soils productive. The essential ingredients of technology for meeting these problems consist of the use of tolerant species, special planting techniques, and aftercare.

Cultivation of salt-affected areas with palatable halophytes is one of the most promising and ecologically safe approaches in the reclamation process. It also helps cattle breeders and farmers to improve a chronically stagnant economy. Selection of the most suitable halophytic species for introduction into saline land needs extensive research. Malcolm [27,28] and Sen et al. [29] have produced a guide to the selection of salt-tolerant shrubs for forage production from saline lands in southwestern Australia. Important selection parameters include:

Growth and survival
Reproduction by seed or vegetative means
Production of biomass (in quantity and quality)
Growth habits
Ease in establishment
Persistence under a profitable management system

Many halophytic species appear to have significant economic potential for desert agriculture. In addition, the productivity of cultivated halophytes is high. Halophytic species of the genus *Atriplex* are widely used as fodder crops in otherwise unusable saline wastelands in many parts of the world. Many halophytic *Atriplex* species are promising in the reclamation of the salt-affected lands. Use of salt-affected soils for uncontrolled grazing, subsistence cropping, or intensive fuel gathering results in degradation of the natural vegetation cover. This process may take decades to reverse, and the land may never be returned to its original condition. To slow such deterioration, new economically useful exotic species can be introduced in these areas. Forage-yielding xerohalophytes like *Atriplex* can be suitable candidates for the management of saline wastelands, since these plants also can be irrigated with brackish water. Land reclamation and rehabilitation in arid zones can be achieved by using salt-tolerant plant species for a number of different purposes suited to the local conditions.

Many halophytic species (e.g., *Arthrocnemum* spp., *Nitraria retusa*, *Salicornia* spp.) are capable of forming adventitious roots on their twigs. This ability varies among species and according to the season of the year [30]. Vegetative propagation is of great advantage in revegetating salt-affected soils. It favors more assured establishment in the field than direct seeding or seedling transplantation. Rooted stem cuttings of *Atriplex* are also helpful in raising a large number of plants of such desired properties as favorable growth habits, regeneration capacity, leafiness, and palatability.

Vegetative propagation of desert shrubs is a means of producing genetically identical individuals in species whose sexually produced offspring normally exhibit higher variability. Reduced variability of plant materials can increase experimental precision, and many genetically identical individuals are necessary for varietal testing. Reproduction of desirable parent characteristics such as high seed yield would be valuable in the establishment of seed nurseries. Vegetative propagation is also a method of producing transplants of species whose seeds do not germinate readily.

II. VEGETATIVE PROPAGATION OF SALTBUSH (*Atriplex* spp.)

Rooted cuttings of *Atriplex* species are needed to establish a rapid plantation. Some *Atriplex* species are subdioecious, with at least three genders [31]. Moreover, rooted cuttings can be used to propagate superior individual plants for a variety of purposes, including breeding programs and provision of superior or uniform outplanting stock [32].

In *A. amnicola* Paul G. Wilson (river saltbush or swamp saltbush), vegetative propagation is the most suitable method. Observations made in the field have revealed that *A. amnicola* plants have natural ability to produce rooted cuttings. During the monsoon season, *A. amnicola* was found to produce nodal roots from the lateral branches wherever they touched the soil. This ability is of great importance in binding the loose topsoil. It also helps the plant to recover speedily from grazing pressure and enables the plant to spread rapidly and multiply. Vegetative propagation is much easier in *A. amnicola* because its nodal root formation helps in the production of a large number of rooted cuttings for field planting.

The effect of different growth regulators used on stem cuttings for root regulation and axillary shoot growth in different seasons of the year (Figures 2–5) is described in Sections A–D.

A. Indole Acetic Acid (IAA)

Observations regarding the effect of indole acetic acid on root and shoot growth are presented in Figure 2. Indole acetic acid did not produce much beneficial effect on root and shoot growth;

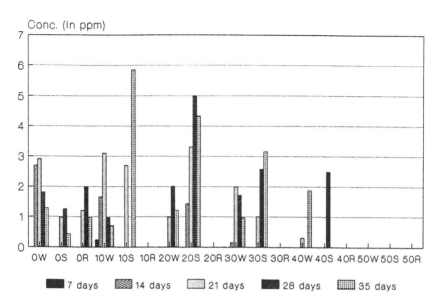

Figure 2 Effect of indole acetic acid (10–50 ppm) on rooting of *Atriplex amnicola* during winter (W), summer (S), and rainy season (R).

it promoted roots when administered in lower concentrations only. In higher concentrations (40 and 50 ppm) during winter, and at all concentrations in rainy seasons, root and shoot growth was affected severely: there was no root formation. Indole acetic acid favored root growth only in lower concentrations (10 and 20 ppm) during the winter and summer seasons, respectively. Slight yellowing and drying effects on leaves were seen at higher concentrations.

B. Naphthalene Acetic Acid (NAA)

Figure 3 shows that compared with other auxins, naphthalene acetic acid caused the maximum initiation of roots in cuttings. Root growth was affected more favorably only at lower concentrations (10 ppm) during winter; at higher concentrations the roots produced were thinner and had a minimum number of secondary roots. A distinct effect of NAA on root growth was seen by comparing results from winter and summer. In the rainy season, the length of the root was less than during the rest of the year. A drastic inhibition of root and axillary branch initiation and their growth in the rainy season was observed. Interestingly, in almost all concentrations, very large numbers of roots were also produced on the internodal region.

C. Indole Butyric Acid (IBA)

It is evident from Figure 4 that the effect on rooting of indole butyric acid is next to that of NAA: that is, IBA promotes root growth better in lower concentration (30 ppm) than in higher, and no distinct difference in the growth of axillary branches was observed. Root growth was maximum in winter at 30 ppm and with well-developed secondary roots. Very poor growth of roots and no initiation of axillary branches was observed in plants treated with IBA in summer.

D. Field Transfer and Establishment of Rooted Cuttings

The effect of growth regulators on root and shoot growth was observed by growing the cuttings in polyethylene bags for 35 days after treatment. It is clear from the results (Figure 5) that root growth was maximum at the higher concentration (20 ppm) of NAA, followed by IBA (10 ppm), and the least was from IAA (10 ppm) after 35 days. In the control set, the roots were

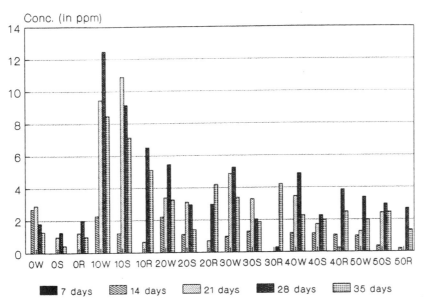

Figure 3 Effect of naphthalene acetic acid (10–50 ppm) on rooting of *Atriplex amnicola* during winter (W), summer (S), and rainy season (R).

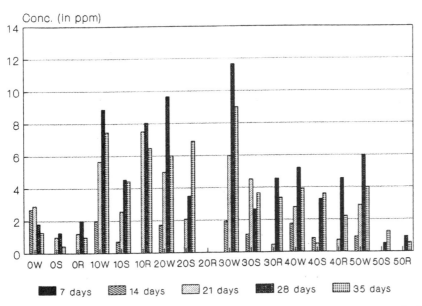

Figure 4 Effect of indole butyric acid (10–50 ppm) on rooting of *Atriplex amnicola* during winter (W), summer (S), and rainy season (R).

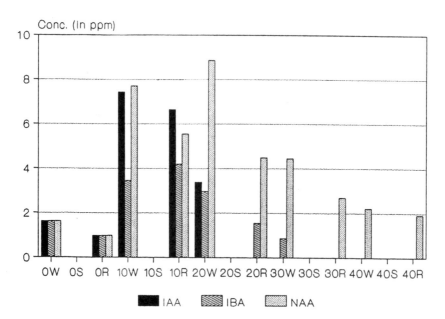

Figure 5 Effect of IAA, IBA, and NAA (10–50 ppm) on rooting of *Atriplex amnicola* in the field during winter (W), summer (S), and rainy season (R).

very much shorter than in the treated cuttings. The maximum development of roots with profuse secondary roots was observed in NAA and IBA. Whereas IAA suppressed the growth of root and axillary branches during summer, NAA and IBA enhanced the growth of axillary branches to a maximum, but the number and the length of the roots were diminished in comparison to NAA. The maximum number of axillary branches was observed in winter and rainy seasons, the least in summer.

The propagation of stem cuttings of several saltbush species and a few species from other salt desert shrub genera was studied by Nord and Goodin [33], Wieland et al. [34], Ellern [35], and Wiesner and Johnson [36]. Although Nord and Goodin [33] and Ellern [35] observed a general trend for better rooting of saltbush *(Atriplex)* species in spring than in fall, no data were available for summer and winter. Nord and Goodin [33] noted better rooting of green stem tips than ripe wood cuttings, but Ellern [35] failed to find any difference in rooting of soft, green cuttings and young woody stem cuttings.

Nanda et al. [37] used IAA, IBA, and NAA to enhance the rooting response of stem cuttings of forest trees and investigated the possibility that even seasonal changes in the effectiveness of different auxins are governed by morphophysiological factors. Auxins enhanced the rooting of stem cuttings of *Populus nigra* and *Hibiscus rosa-sinensis* even during December–February, but these hormones failed to cause rooting in *Ficus infectoria* cuttings during the same period. It was observed that auxins enhanced the rooting more in winter, followed by the rainy season, and least in summer.

Indole acetic acid has been one of the most commonly used auxins, but different workers have obtained varying results [8,9,37–40]. Chatterjee [39] found that *Pogostemon potehouli*, an essential oil yielding plant, responded more favorably to IAA than other auxins. Shanmugavelu [41] also obtained the maximum percentage of rooting in cuttings of certain shrubby plants with IAA. On the other hand, NAA gave favorable results in the induction of roots in cuttings of *Levendula* [39], *Ficus infectoria* [37], and *Hibiscus rosa-sinensis* [41]. The experimental results of our study showed that a large number of roots were produced at lower concentration of NAA, IAA, and IBA.

A number of saltbush species may be established from cuttings, including *A. amnicola, A. nummularia* [42], *A. canescens, A. halimus, A. lentiformis, A. paludosa,* and *A. polycarpa* [43]. The cuttings should be taken at the peak of spring growth or in the autumn of a Mediterranean climate [35]. The wood should be about 6 mm thick and 250 mm long, taken from young stems between two leaf axils. A rooting hormone (e.g., IBA) may be applied to encourage root growth before approximately half the stem is covered with a moist, sandy soil. The cuttings should root within 6 weeks and should be ready for transplanting in 10 weeks [44]. In our study, IBA also enhanced the rooting in *A. amnicola.*

According to Richardson et al. [32] fourwing saltbush cuttings could be rooted best in the summer, but *A. amnicola* rooted best in winter, followed by the rainy season and summer. According to Sharma and Sen [45] and Rajput and Sen [46], respectively, winter is the most suitable for the vegetative propagation of *Tamarix* and *Atriplex.* The present results also support these views.

The results of field experiments showed that NAA is more effective than IBA and IAA. The increased appearance of new leaves with the increase in percentage of rooting, also points to better rooting possibilities, with the emergence of more new leaves on the cuttings. The greater number of roots per cutting and the greater number of leaves may also help the cuttings to survive when sown in natural conditions.

III. VEGETATIVE PROPAGATION OF MULBERRY (*Morus* spp.)

Mulberry is propagated either through seeds or vegetatively. The latter is the most common method of propagation because of such advantages as maintenance of particular properties of the plant, relative speed in raising saplings in large numbers for plantation, adaptability to a particular habitat, and abilities to develop resistance to pests and diseases and to modify the growth of plants. Propagation through seeds has reached certain limitations. For example, triploid plants, which do not produce viable seeds, cannot be propagated. It is not possible to reproduce true to the type from a seed of biparental origin.

Mulberry is a highly heterozygous plant which is open for cross-fertilization. Therefore, the seeds that are formed through open pollination are natural hybrids. Seedling populations from such seeds provide wider chances for selection of superior types whose characteristics are perpetuated through vegetative propagation. Generally, the population thus obtained is a mixture of several clones. Each clone is heterozygous though homogeneous, and the same genotype is maintained because propagation is vegetative. Interclonal variations are due to heredity. Depending on climatic and soil conditions, different countries follow different modes of vegetative propagation. Hamada [47] described the methods used in Japan, which include (a) bark grafting (Fukurotsugi), (b) veneer grafting (Kiritsugi), (c) simple layers (Magedori), (d) continuous layers (Shumokudori), and (e) division (Shirodasmi), hardwood cuttings (Kojyosashiki), and softwood cuttings (Shinshosashiki). Generally grafting is used in places where the temperature is 6°C in March and more than 25°C in July, with rainfall of 175 mm. Shirodasmi cottage is popular in places having temperatures less than 4°C in March and less than 25°C in July, with rainfall lower than 175 mm. Propagation through hardwood and softwood cuttings is common in the northern districts and the southern region, respectively, of Japan [48]. In Italy [49] rooted grafting is a popular method of multiplying Japanese mulberry varieties.

In India the most common method of propagating mulberry is through cuttings in multivoltine regions (e.g., Karnataka and West Bengal). Exotic varieties that do not establish by cuttings are propagated through root grafts. Many of the indigenous varieties and well-acclimatized exotic varieties are propagated through cuttings. Bud grafting (budding) is used only when scion material is scarce. Whenever a large mulberry plant must be obtained in a shorter time than would be possible if started as a cutting, the method of layering is used. Layering allows the grower to fill in the gaps formed as a result of the failure to sprout of certain cuttings planted in pits of established plantations.

In univoltine areas, (e.g., Kashmir), the mulberry is propagated through seedlings and the exotic varieties through root grafts. In India, the field-scale propagation through cuttings of Japanese varieties of mulberry is still a problem.

Propagation through seeds is used mainly to bring about a varied population for the purpose of selection and hybridization. Since mulberry flowers are open for cross-pollination, the seeds thus collected serve mainly as sources of stock material for grafting.

The water requirement of mulberry does not differ greatly from species to species or from variety to variety. The plant must be capable of absorbing water from soils of low moisture regimes. Generally resistant plants should have well-developed root systems, hydrophilic colloids to absorb and hold water by imbibition, and adaptations to facilitate the lowering of transpiration. In this regard, certain Japanese varieties have thick cuticle, sometimes two-layered epidermis, palisade parenchyma, and other beneficial characteristics.

Although many tropical species root profusely through cuttings, certain temperate varieties do not ordinarily produce roots. Root induction has been successfully achieved in the latter varieties by the (artificial) application of the requisite quantity of root hormones. However, the efficacy of the substances varies from species to species and from variety to variety.

Development of root primordium depends on the relative amount of natural auxin present in the plant. Varieties that do not root apparently contain less auxin. The growth regulators act like auxins when applied in small quantities and move upward in mass translocation through the xylem when the bases of the cuttings are soaked in their solutions. The objective of treatment is to increase the percentage of cuttings that form roots, hasten the root initiation, and increase the number of roots per cutting. Indole butyric acid and naphthalene acetic acid appear to be better at producing roots than other agents [46]. The chemicals may be applied by various methods, including direct application of a powder, soaking the cuttings in dilute solutions, dipping the cuttings in concentrated solutions, and application as a paste in lanolin.

Table 1, which gives the results of experiments on rooting behavior in mulberry (cultivated variety), shows that with the addition of IAA and NAA in lower concentrations, more bud sprouting almost always was observed. Also the total leaves, generally increased, together with the amount of inflorescence. Increasing concentrations of hormones tended to decrease the values. Slightly higher values of these parameters were observed with 10 ppm than 20 ppm of IAA. Of the two auxins, IAA was more effective than NAA. It is also clear from Table 1 that in the case of NAA, also, lower concentration is more effective than higher.

From the observations of the rooting behavior in a wild variety of mulberry (Table 2), we see that the lower concentration of IAA is more effective than the higher concentration. The maximum number of sprouting buds was 11.33; afterward the values remained constant. However, in the case of NAA, 20 ppm was more effective than 10 ppm. Comparatively, IAA is more effective than NAA, and higher values were observed in the wild than in the cultivated variety.

Growth is as an irreversible increase in the weight, area, or length of a plant or a particular tissue, or organ of a plant, while development denotes the changing pattern of organization as growth progresses. Control over plant growth by the regulated exogenous supply of chemical substances may occur in different ways. In recent times, it has been made clear that total control of plants is vested not in a single hormonal type; rather, control is shared by a group of several specifically defined auxins, gibberellins, ethylene, and certain naturally occurring inhibitors

Table 1 Effect of Different Concentrations of Indole Acetic Acid and Naphthalene Acetic Acid on Bud Sprouting (BS), Initiation of Leaves (L), and Inflorescence (I) on Stem Cuttings of *M. alba* (cultivated variety)

Concentration	Total buds on cutting	Jan 25 BS	Jan 25 L	Jan 25 I	Feb 1 BS	Feb 1 L	Feb 1 I	Feb 3 BS	Feb 3 L	Feb 3 I	Feb 11 BS	Feb 11 L	Feb 11 I
Control	5	1			2	1		2.5	1	3	2.5	3.5	6
IAA, ppm													
10	5	3			4	1		4	2		5	4.5	7
20	8	4			5	1		5	1		5	3	4
NAA, ppm													
10	7	2			5			5			5	3	2
20	9				3			3			3	4	3

Table 2 Effect of Different Concentrations of Indole Acetic Acid and Naphthalene Acetic Acid on Bud Sprouting (BS), Initiation of Leaves (L), and Number of Inflorescences (I) on Stem Cuttings of *M. indica* (wild variety)

Concentration	Total buds on cuttings	Jan 25			Feb 1			Feb 3			Feb 11		
		BS	L	I	BS	L	I	BS	L	I	BS	L	I
Control	8	3.50			5			5	5.5	4	6.5	6	10
IAA, ppm													
10	15.66	9.33			11.33			11.33	8.5	26	11.33	8	28
20	16.50	6.50			10.50			11	3.3	3.66	11	4.33	6
NAA, ppm													
10	29.50	5.66			8.33			8.66	3.3	4	9	4	9
20	17.66	8.66			9.33			9.33	3.66	4	9.33	4.33	11

like phenols and abscisic acid. Thus, the plant growth regulators provide a very helpful tool for controlling physiological processes in plants.

Stem cuttings of *Ipomoea biloba* and species of *Morus* showed a large number of roots and buds in higher concentration, but with maximum suppression of growth, while lower concentrations showed only improvement in the growth of roots [50,51]. In our investigation also, the higher values were observed in connection with lower concentrations of the growth regulators.

Under favorable environmental conditions, during the period of root development, there forms at the basal end of a cutting a callus tissue: an irregular mass of parenchyma cells in various stages of lignification. Callus growth arises from cells and adjacent phloem, although various cortical and medullar cells also contribute. Because root development and callus formation occur simultaneously, it is believed that the formation of callus is essential for root development. In reality, these two are entirely different phenomena. Sometimes, roots develop even without callus from the nodes. Callus formation is sometimes beneficial in varieties that are slow to root, since it provides a protective layer, thereby preventing the cutting from becoming desiccated and decayed. Sometimes the callus interferes with the absorption of water by the cutting. In our investigation also, rooting did not start: instead, callus formation was observed after one week of treatment. The callus was creamy white and it had a granular texture.

The rate of sprouting of vegetative buds is of primary consideration in introducing a variety or species in an area. Mulberry varieties grown in Mysore and West Bengal sprout throughout the year, facilitating the attempts of sericulturists to rear the silkworms year-round. The axillary buds vary in size, shape, and position from variety to variety.

ACKNOWLEDGMENTS

Financial assistance received from DoEn (Dept. of Environment), CSB (Central Silk Board) and seeds from Texas Tech. University, Tucson, USA are gratefully acknowledged.

REFERENCES

1. A. Bala, V. K. Nanda, and K. K. Nanda, Seasonal changes in the rooting response of stem cuttings of *Dalbergia sisoo* and their relationship with biochemical changes, *Indian J. Plant Physiol.*, *12*: 154–165 (1969).

2. K. K. Nanda, Use of auxins in rooting stem cuttings of some forest tree species, Final Report PL-480 Project A 7-F5-11 (1971).

3. D. N. Sen, Ecology of saline areas of Rajasthan and exploitation of saline ecosystem for increased productivity, Department of Environment, Final Project Report, Government of India (1990).

4. D. N. Sen, "Water relations of psammophyte—*Convolvulus microphyllus* Sieb. ex. Spreng," Ecophysiological Foundation of Ecosystem Productivity in Arid Zone, International Symposium, USSR, pp. 79–83 (1972).

5. K. V. Thimann and J. Behnke-Rogers, The use of auxins in the rooting of wood cutting, Maria Moors Cabot Foundation Publication 1 (1950).

6. T. K. Bose, Improvement in the method of vegetative propagation of *Amherstia nobilis* and *Petrea arborea*, *Sci. Cult.*, *30*: 198–199 (1964).

7. T. K. Bose and D. Mukherjee, Use of root promoting chemicals on cuttings of *Legerstroemia indica*, *Sci. Cult.*, *34*: 217–218 (1968).

8. A. Prasad and A. P. Dikshit, Vegetative propagation of essential oil-yielding plants, *Sci. Cult.*, *29*: 460–461 (1963).

9. S. S. Teaotia and I. C. Pandey, Effect of growth substance on rooting of guava stem cuttings, *Sci. Cult.*, *27*: 442–444 (1961).

10. K. S. Rajpurohit and D. N. Sen, Soil salinity and seed germination under water stress, *Trans. Isdt. Ucds.*, *2*: 106–110 (1977).

11. A. McMahon and I. A. Ungar, Phenology, distribution and survival of *Atriplex triangularis* Willd. in an Ohio salt pan, *Am. Midl. Nat.*, *100*: 1–14 (1978).

12. R. Uhvits, Effect of osmotic pressure in water absorption and germination of alfalfa seeds, *Am. J. Bot.*, *33*: 278–285 (1946).

13. J. T. Prisco and J. W. O'Leary, Osmotic and "toxic" effects of salinity on germination of *Phaseolus vulgaris* L. seeds. *Turrialba*, *20*: 177–184 (1970).

14. I. A. Ungar, Halophyte seed germination, *Bot. Rev.*, *44*: 233–264 (1978).

15. D. N. Sen and S. Mohammed, General aspects of salinity and the biology of saline plants, *Handbook of Plant and Crop Stress* (M. Pessarakli, ed.), Dekker, New York, pp. 125–145 (1992).

16. J. D. Bewley and M. Black, *Physiology and Biochemistry of Seeds in Relation to Germination, Viability, Dormancy and Environmental Control*, Springer-Verlag, New York (1982).

17. V. J. Chapman, *Salt Marshes and Salt Deserts of the World*, J. Cramer, Bremerhaven, Germany (1974).

18. N. Yamdagni and D. N. Sen, Role of leaves present on the stem cuttings for vegetative propagation in *Portulaca grandiflora* L. *Biochem. Physiol. Pflanz.*, *164*: 447–449 (1973).

19. J. Van Overbeek, S. A. Gordon, and L. E. Gregory, Analysis of the function of the leaf in the process of root formation in cuttings, *Am. J. Bot.*, *33*: 100–107 (1946).

20. Y. Waisel, *Biology of Halophytes*, Acedemic Press, New York (1972).

21. Y. Waisel and G. Pollack, Estimation of water stresses in the active root zone of some native halophytes in Israel, *J. Ecol.*, *57*: 789–794 (1969).

22. D. N. Sen and K. S. Rajpurohit, *Contributions to the Ecology of Halophytes*, W. Junk, The Hague (1982).

23. E. Epstein and D. W. Rains, Advances in salt tolerance, *Plant Soil*, *99*: 17–29 (1987).

24. Le. Houreou, "Ecoclimatic and biogeographic comparison between the rangelands of the isoclimatic Mediterranean arid zone of Northern Africa and the Near East," Proceedings of the 2nd International Conference on Range Management in the Arabian Gulf, Kuwait, March 3–6, p. 20 (1990).

25. J. L. Gallagher, Halophytic crops for cultivation at seawater salinity, *Plant Soil*, *89*: 323–336 (1985).

26. P. C. Raheja, Salinity and aridity, new approaches to old problems, *Aridity and Salinity—A Survey of Soils and Land Use* (H. Boyko, ed.), W. Junk, The Hague, pp. 43–127 (1966).

27. C. V. Malcolm, *Wheatbelt Salinity—A Review of the Saltland Problem in South Western Australia*, Technical Bulletin 52, Department of Agriculture of Western Australia, Perth, p. 65 (1982).

28. C. V. Malcolm, Rehabilitation agronomy—Guidelines for revegetating degraded land. *Proc. Ecol. Soc. Aust.*, *16*: 551–556 (1990).

29. D. N. Sen, B. S. V. Prakash, and Thomas P. Thomas, Wasteland management with special reference in introduction of *Atriplex* spp, *Advances in Forestry Research in India*, (S. S. Negi, ed.), International Book Distributors, Booksellers and Publishers, Dehradun, India, pp. 221–240 (1987).

30. Y. Waisel and G. Pollack, Estimation of water stresses in the active root zone of some native halophytes in Israel, *J. Ecol.*, *57*: 789–794 (1969).

31. E. D. McArthur and D. C. Freeman, Sex expression in *Atriplex canescens*: Genetics and environment, *Bot. Gaz.*, *143*: 476–482 (1982).

32. S. G. Richardson, J. R. Barker, K. A. Crofts, and G. A. Van Epps, Factors affecting root of stem cuttings of salt desert shrubs, *J. Range Manage.*, *32*: 280–283 (1979).

33. E. C. Nord and J. R. Goodin, *Rooted Cuttings of Shrub Species for Planting in California Wildlands*, Research Note PSW 213, U.S. Department of Agriculture, Forest Service, Pacific Southwest Forest and Range Experiment Station, Berkeley, CA, 3 pp. (1970).

34. P. A. T. Wieland, E. F. Frolich, and A. Wallace, Vegetative propagation of woody shrub species from the northern Mojave and southern Great Basin deserts, *Madrono*, *21*: 149–152 (1971).

35. S. J. Ellern, Rooted cuttings of saltbush (*Atriplex halimus* L.), *J. Range Manage.*, *25*: 154–155 (1972).

36. L. E. Wiesner and W. J. Johnson, Fourwing saltbush *(Atriplex canescens)* propagation techniques, *J. Range Manage.*, *30*: 154–156 (1977).

37. K. K. Nanda, V. K. Anand, and P. Kumar, Some investigations of auxin effects on rooting of stem cuttings of forest plants, *Indian For.*, *96*: 171–187 (1970).

38. O. S. Jauhari and S. F. Rehman, Further investigation on rooting in cuttings of sweet lime (*Citrus limettoides* Tanaka), *Sci. Cult.*, *24*: 432–434 (1959).

39. S. K. Chatterjee, A note on the vegetative propagation of some of the essential oil yielding plants newly introduced at Mungpoo, Darjeeling, *Sci. Cult.*, *25*: 687–688 (1960).

40. K. K. Nanda, V. K. Kochhar, and S. Gupta, Effect of auxins, sucrose and morphactin in the rooting of hypocotyl cuttings of *Impatiens balsamina* during different seasons, *Biol. Land Plants*, *1*: 181–187 (1972).

41. K. G. Shanmugavelu, A note on the response of rooting of cuttings of the *Hibiscus rosa-sinensis* Linn. and *Allamanda cathartica* Linn. to the application of plant growth regulators, *Sci. Cult.*, *26*: 136–137 (1960).

42. R. L. Everett, R. O. Meeuwig, and J. H. Robertson, Propagation of Nevada shrubs by stem cuttings, *J. Range Manage.*, *31*: 426–429 (1978).

43. E. D. McArthur, A. P. Plummer, G. A. Van Epps, D. C. Freeman, K. R. Jorgensen, "Producing fourwing saltbush seed in seed orchards," Proceedings of the First International Rangeland Congress (D. N. Hyder, ed.), August 14–18, Denver, CO, Society for Range Management, pp. 406–410 (1978).

44. G. C. de Kock, "Drought resistant fodder shrub crops in South Africa," *Browse in Africa: The Current State of Knowledge* (H. Houeron, ed.), International Livestock Centre of Africa, Addis Ababa, Ethiopia, pp. 399–408 (1980).

45. T. P. Sharma and D. N. Sen, "Seasonal variation in the rooting capacity of *Tamarix* spp. in inland salines of Indian arid zone," 76th Indian Scientific Congress, Madurai, Abstr, p. 151 (1989).

46. P. Rajput and D. N. Sen, "Role of certain growth regulators on the vegetative propagation of an exotic species *Atriplex amnicola* Paul G. Wilson," Botanical Congress, Lucknow (1991).

47. S. Hamada, Propagation of mulberry tree in Japan, *J. Silkworm*, *10*: 273–278 (1958).

48. S. Taguchi, Distribution of the propagating methods of mulberry trees from Japan and the relation to the various climates, *J. Sericult. Sci. Japan*, *40*: 399–403 (1971).

49. P. L. Lombardi, Attuale stato della gelsicolture italiana, sistemazione per eventualli allevamenti successivi con rifermento anche ai tipi giapponesi impartati, *J. Silkworm*, *12*: 41–50 (1960).

50. S. S. Sharma and D. N. Sen, Role of certain growth regulators on the vegetative propagation of a desert species—*Ipomoea biloba* Forsk., *Indian For.*, *101*: 625–633 (1975).

51. P. Sehdev, Ecology and biology of mulberry in Indian desert, M. Phil. Dissertation, Jai Narain Vyas University, Jodhpur, India (1992).

9
Mineral Nutrition of Plants

Kanapathipillai Wignarajah

Controlled Ecological Life Support Program, The Bionetics Corporation, NASA-Ames Research Center, Moffett Field, California

I. NUTRIENT RESERVES ON EARTH

The ultimate source of nutrients for all living organisms consists of the inanimate nutrient reserves found on earth. Of the elements known to exist, seven are considered essential to plants in large amounts (macronutrients), and many others are required in smaller quantities (micronutrients). Essentiality of a nutrient is defined according to the concepts proposed by Arnon and Stout [1]:

(a) A deficiency of the element makes it impossible for the plant to complete the vegetative or reproductive stage of its cycle;
(b) such deficiency is specific to the element in question and can be prevented or corrected only by supplying this element;
(c) the element is directly involved in the nutrition of the plant quite apart from its possible effects in correcting some unfavorable microbiological or chemical condition of the soil or other culture medium. From that standpoint a favorable response from adding a given element to the culture medium does not constitute conclusive evidence of its indispensability in plant nutrition.

All the elements occurring in the outer part of the earth are in constant turnover among the different components of earth. This overall migration is referred to as geochemical cycling. When cycling includes a role for biological organisms, it is referred to as "biogeochemical cycling." Like most cyclical processes in nature, the biogeochemical cycling of elements is not continuous, nor does it proceed in a well-defined direction. At stages, it may be halted or short-circuited, or it may change. Any changes will eventually impact the survival, evolution, and development of biological species in the system. The relationship of the various systems is represented in a schematic manner in Figure 1 [2].

To assess the efficiency of operation of the biogeochemical cycles, it is important to include both natural and human activities. Often reliable values on use by man are difficult to obtain

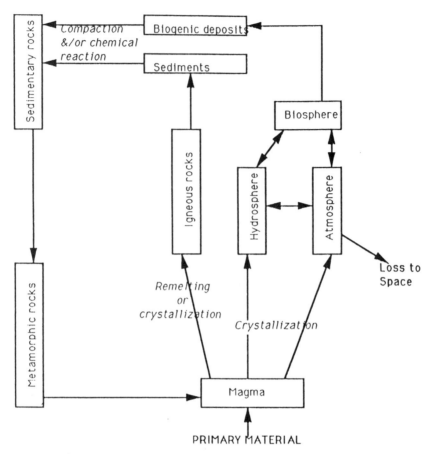

Figure 1   Simplified schematic representation of the biogeochemical cycle of an element, showing its potential relationship to the various components of earth. (From Ref. 2.)

for a number of reasons, such as lack of international cooperation, and lack of proper bookkeeping and auditing by individual nations. However, a general estimate of the annual world consumption of elements and their compounds is presented in Table 1 [2].

A.   Nutrients in the Lithosphere

The lithosphere is the solid crust of the earth or other celestial body, composed of rock essentially like that exposed at the surface. In the case of the earth, it is considered to be about 80 km thick. This crust represents only 0.6% of the total mass of earth, the mantle and core comprising the remaining 99.4%. Table 2 shows the commoner chemical elements in the earth's crust. The elements present in the lithosphere are the ones most readily available for the living matter on earth [3]. Table 3 compares the relative abundance of elements in the earth's crust, in lunar rock (regolith), and in meteorites [3].

A major difference between the earth's crust and the moon is the lack of potassium and hydrogen on the latter. Potassium is the major element of plant tissues, and its absence would be the major obstacle to the cultivation of plants on the lunar regolith in connection with NASA's contemplated manned lunar outpost [4]. Even though the percentage of mass contributed to the

Table 1 Annual World Consumption of the
Elements (and their compounds)

Element[a]	Annual tonnage
C	10^9-10^{10}
Na, Fe	10^8-10^9
N, O, S, K, Ca	10^7-10^8
H, F, **Mg**, Al, **P**, Cl, Cr, **Mn, Cu, Zn**, Ba, Pb	10^6-10^7
B, Ti, Ni, Zr, Sn	10^5-10^6
Ar, **Co**, As, **Mo**, Sb, W, U	10^4-10^5
Li, V, Se, Sr, Nb, Ag, Cd, I, rare earths, Au, Hg, Bi	10^3-10^4
He, Be, Te, Ta	10^2-10^3

[a]Elements of importance to plant growth are in bold type.

Source: Ref. 2.

earth's crust is only a very small fraction, it is evident that the elements added are very similar to that of earth, except that they are very low in K.

B. Nutrients in Aquatic Systems

The mass of the aquatic systems of earth (hydrosphere) is estimated to be 1.66×10^{24} g and the aquatic systems, with an area of 361×10^6 km^2, cover 70.8% of the earth's surface [3]. The nutrients in the aquatic systems make a substantial contribution to the earth's nutrient reserves. The nutrients available in aquatic systems are restricted mainly to the more soluble elements. Oxygen is present at a concentration of 8.57×10^5 mg L^{-1}, while phosphorus occurs at a concentration of 7×10^{-2} mg L^{-1}. The relative abundance of elements required by living organisms is shown in decreasing order of abundance:

$$O > H > Cl > Na > Mg > S > Ca > K > Br > C > B > Si > N > P > I > Se$$

C. Nutrients in the Atmosphere

The chemical composition of the atmosphere varies with altitude, both in density and in concentration of elements. All known living organisms are confined to the troposphere, and an average elemental composition of the troposphere is presented in Table 4 [4].

Table 2 The Chemical Elements of the Earth's
Crust

Element	Weight percent	Volume percent
O	46.60	91.7
Si	27.72	0.2
Al	8.13	0.5
Fe	5.00	0.5
Mg	2.09	0.4
Ca	3.63	1.5
Na	2.83	2.2
K	2.59	3.1

Source: Ref. 3.

Table 3 Relative Abundance (by weight) of
the Elements

Crust	Moon	Meteorites
O	O	O
Si	Si	Fe
Al	Mg	Si
Fe	Fe	Mg
Ca	Ca	S
Na	Al	Ni
K	Ni	Ca
Mg	S	Al
Ti	Ti	Na
H	Cr	Cr
P	Na	Mn
Mn	P	P
F	Mn	Co
Ba	V	K

Source: Ref. 3.

D. Nutrients from Traditional Waste Materials and Pollutants

Waste materials and pollutants have great potential to be major secondary resources of elements required for crop production [5]. Two major problems have limited our capacity for use of these materials:

Relative abundance in the present world of fairly pure supplies of the elements necessary for growing crops
Lack of the appropriate technology to process these wastes

With an increasing living biomass on earth, the demand for inorganic nutrients is likely to rapidly diminish the supply, thus necessitating the use of relatively less concentrated forms of

Table 4 Average Elemental Composition of the
Troposphere (exclusive of water vapor)

Element	Content (%) by volume	Content (ppm) by volume
N_2	78.08	
O_2	20.95	
CO_2	0.033	
Ar	0.934	
Ne		18.18
He		5.24
Kr		1.14
Xe		0.09
H_2		0.50
CH_4		2.00
N_2O		0.50

Source: Ref. 3.

these elements or scavenging them from the wastes that are generated. Advances in biotechnology and bioremediation are also enabling recovery of a number of inorganic elements, in particular carbon, hydrogen, nitrogen, and sulfur, and some of the cations and anions necessary for plant growth.

II. BIOGEOCHEMICAL CYCLING OF THE MAJOR PLANT NUTRIENTS

Figure 2 is a modified schematic representation of the biogeochemical cycling proposed by Delwiche [6]. The major reservoir of nitrogen, from which this element is made available to plants, is the atmosphere. Microorganisms and human industrial activity play a major role in converting the gaseous nitrogen to nitrate and ammonium salts, which then become available for plants and animals. The presence of a 1000-fold higher concentration in the atmosphere of nitrogen than in the rest of the earth's ecosystem thus ensures that this element is always readily available, if not by natural activities then through man-made fertilizers. In a natural ecosystem or a low fertilizer supplied agricultural system, problems of transferring nitrogen to biological organisms are the major limitation. These problems result from the limited capacity of nitrifying microorganisms relative to denitrifying microorganisms.

The carbon cycle contrasts sharply with the nitrogen cycle in that the capacity of microorganisms to mineralize organically bound carbon to atmospheric CO_2 is very limited. Hence, atmospheric CO_2 levels are in very low quantities and well below concentrations required for maximal photosynthesis by plants. In particular, they are below the capacity required by C_4-type plants. Much of the carbon of the earth is organically bound in living organisms in the

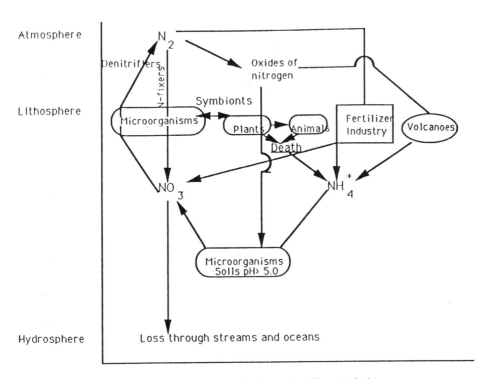

Figure 2 Generalized representation of the nitrogen cycle. (From Ref. 6.)

form of fossil deposits (e.g., oil and natural gases). Humans artificially increase the CO_2 generated into the atmosphere, and this activity has been advanced as a major cause of the greenhouse effect and global warming. At present, the only biological systems capable of refixing CO_2 are plants, which produce food and other economically important resources for humans; yet in some cases plants are operating at suboptimal CO_2 concentrations. The question of global warming can be remedied through effective balancing of the CO_2 cycle.

During the cycling of phosphorus, there occur great losses through movement to the oceans, and sedimentation at the bottom of the ocean is a major sink from which the element is only slowly regenerated. At present, the presence of large amounts of phosphorus does not threaten the supply available to man. However, large-scale mining processes are rapidly reducing the readily available surface deposits, and heavy use of phosphates is also resulting in the introduction of large amounts of these materials to the water supply, causing serious pollution and eutrophication problems.

The foregoing type of schematic of the biogeochemical cycle presents a qualitative picture only. For effective management of natural resources, it is important to arrive at detailed quantitative information of the individual elements. Ideally such studies should incorporate both spatial and temporal factors. Global-scale studies are not easy to perform, but major international organizations are working toward these goals. An important factor in their achievement is cooperation of governmental and national-level organizations, particularly from the developing countries. The availability of local and national inventories to incorporate into plans for the biogeochemical cycling of nutrients will also assist in controlled, managed use of resources.

III. PLANT GROWTH AND MINERAL ELEMENTS

Table 5 compares the average nutrient content of typical plants and man with that present in the soil solution [7,8]. The values for plants and man have been expressed as a percentage of

Table 5 Average Elemental Composition in a Typical nonfertilized Soil Solution, a Typical Plant Cell, and a Typical Animal Cell (human)

Element	Soil solution $(mg\ L^{-1})$	Plant (% of dry weight or ppm)	Humans (% of dry weight or ppm)
K	8–390	5–6.0%	0.64%
Mg	17-2400	0.1–0.3%	0.09%
Ca	20–1520	0.7–1.0%	5.36%
Na	10–3450	a	0.23%
P	0.003–30	0.2–0.7%	3.07%
S	2–4800	0.2–0.5%	1.07%
Total N	2.5–770	1–1.5%	7.38%
Fe	20,000	0.01	160.71 ppm
B	15–30	35–200 ppm	b
Mn	0.5–9	40–800 ppm	0.43 ppm
Zn	0.002–0.2	21–120 ppm	60.81 ppm
Cu	0.0006–0.04	2–20 ppm	3.69 ppm
Mo	0.6–3.5	0.5–0.8 ppm	0.32 ppm
Co	1–40	0.02–0.5 ppm	0.04 ppm

[a]Not found in mesophytic plants; present only in halophytic plants
[b]Not known to be present.
Source: Refs. 7 and 8.

dry matter. Some of the elements (e.g., iron), though present in abundance in the soil solution, are present only in trace amounts in plants or humans, while other elements are present in much higher concentrations in plants and humans. This imbalance provides support for the concept that even though plants and humans have only the elements that are present in the earth's system, they do show preference for some elements over others. They accommodate this preference by having present on their cell membranes specific carriers capable of selectively absorbing ions of importance to the organism.

A. Plant Macroelements

The elements required by plants in large quantities (C, H, O, N, S, K, Mg, Ca, and Fe) are referred to as the macroelements. Plants take up carbon as CO_2 from the atmosphere. Water and the other elements are generally taken up from the soil. Carbon, which constitutes between 40 and 45% of the dry weight of plant matter, can be released from the organic matter during burning as CO_2 or be eaten or colonized by bacteria, thus, after assimilation, becoming a source of energy for the organism. Hydrogen and oxygen constitute approximately 6 and 42%, respectively, of dry matter. Nitrogen can account for 1% of dry matter, while the rest of the elements constitutes 6–8% of dry matter.

B. Plant Microelements

The French scientist Gabriel Bertrand was the first to demonstrate the importance of manganese for growth of the fungus *Aspergillus* [9]. Since then, a number of other elements, including boron, zinc, copper, molybdenum, and chlorine, have been shown to be essential as microelements. Other elements often required by some plants, but not necessarily by others, include cobalt, selenium, and silicon. This is by no means a complete list, and research is still actively in progress to assess the nutrient requirements of crops as well as native plants. There are two major reasons for the incompleteness of the study to date:

1. The nutrient requirements show considerable differences at generic level, species level, and even within ecotypic differences. For example, cobalt is required as an essential element by legumes having nitrogen-fixing symbionts in their root systems [10,11]. On the other hand, the inclusion of cobalt was demonstrated to have beneficial effects on nonlegume crops, cacao [12], and wheat, as well as in legumes abundantly supplied with exogenous nitrogen, hence not dependent on fixed sources [13]. Thus, whether cobalt is a universally required micronutrient for all plants is a matter of debate. Until further research can resolve such conflicts, cobalt is categorized as a "potential" micronutrient.

2. Most of the microelements are found in trace levels in soils and often in most "purified" chemicals. The removal of these traces of the chemicals can be very time-consuming and laborious. Furthermore, such studies have little practical application to the practicing farmer, since the elements themselves are always found in trace levels in the soil and deficiency symptoms are rarely ever observed in the field.

The simplest method of investigating the effects of nutrients examines plant biomass production as a function of nutrient supplied to the plant. This method has a major limitation in that it is assumed that the nutrients supplied are readily available to the plant and that the leaf tissue concentration of the element bears a constant relationship to the nutrient supplied. Asher and colleagues developed a continuous flow hydroponic system for controlling the ionic environment of plant roots [14], which they used to study the relative growth of a number of plants in response to potassium [15] or phosphorus supply [16]. However, under field conditions (where the nutrient concentration in the leaf may not necessarily bear a relation to that supplied to the soil), it is preferable to analyze the nutrient content of the leaf tissue. Generalizations can be made on the response of plants to variations in nutrient content. Figure 3 represents a modification of the generalized responses proposed by Ulrich and Hills [17].

In Figure 3 it is assumed that nutrients are the only limiting factor to growth and that all the factors required for growth (light, water, CO_2, etc.) are not limiting. At low nutrient supply, growth is limited by the availability of a specific element, and this region is referred to the nutrient deficiency region. Often when plants in field conditions have been lacking in a nutrient of interest, characteristic visual symptoms of the deficiency are observed on the leaves. Early detection can be achieved by tissue analysis, in which the leaves are detached and the elemental composition analyzed by chemical procedures. Farmers can supply the nutrients needed by application either to the soil or to the leaves of the nutrient-deficient plants and thereby prevent a reduction in crop yield. The growth response plateaus on reaching 100% growth, and the minimal concentration required to ensure 100% growth is termed the optimal concentration. The term "critical concentration" is defined as the concentration of element in the tissue at which plants show 90% growth. To the farmer, the value of the critical concentration is the operational parameter, which ensures that crop yields are not significantly diminished by a reduction in the nutrient supply to the plant.

C.   Uptake and Functions of Elements

1.   Nitrogen

Most of the nitrogen taken up by plants is found in a reduced form as organic nitrogen, where it possesses an unshared pair of electrons. In this form, it can complex metals; in chlorophyll, for example, magnesium ions are chelated. Second, nitrogen plays a structural role by forming the peptide bonds during condensation of amino acids to form proteins. Third, it plays an important role in hydrogen bond formation, especially in nucleic acids and the secondary structures of proteins. Fourth, it is an essential element in the formation of heterocyclic compounds and secondary plant products. Nicotinamide adenine nucleotide is an important enzyme cofactor, while other heterocyclic compounds function as secondary plant products

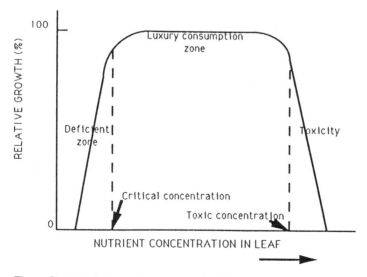

Figure 3   A typical growth response of a plant to added nutrients. (From Ref. 17.)

playing important functions such as the defense of plants against predation, pollination, and so on.

Plants can take up and metabolize nitrogen either as NO_3^- or NH_4^+ ions, and environmental factors and plant species play an important role in determining the species of nitrogen used by the plant. The pH value is the most critical environmental factor in the control of the N species to be supplied to the plant or to be present in the nutrient solution. When NO_3^- is supplied as the sole source of nitrogen, uptake results in an increase in the pH of the nutrient media, and the reverse is true for NH_4^+ ions. This commonly observed phenomenon has often been tested as a potential nutrient pH control system, but it has not proven to be very successful [18]. In general, it can be stated that at acidic pH values NO_3^- is taken up better by most plants, NH_4^+ ions are taken up better at high pH values, while at pH 6.8, the two ionic species are taken up equally [19]. This observation is important in making decisions on the type of nitrogen fertilizer to be supplied to plants showing a deficiency in nitrogen.

Nitrogen-deficient plants usually have yellow, pale leaves, and their growth rates are poor. Leaves are small, and stems are very spindly. Chloroplast development is very severely affected. Root growth is also affected, with the roots showing less branching than is seen in plants supplied with adequate levels of nitrogen. The ratio of root to shoot is increased during N deficiency, however, a circumstance that can be explained in terms of Brouwer's functional equilibrium hypothesis [20], namely, that plants subjected to N deficiency partition more of the available resources toward improving the physiological activity of the root—the site at which nitrogen is acquired. Other important symptoms of N deficiency are reduction of the vegetative growth phase, early onset of flowering, and early completion of the life cycle. Such a morphogenetic shift is common in most plants grown under environmental stress and has ecological as well as physiological significance.

2. Phosphorus

Phosphorus is required by plants in the oxidized form of orthophosphate. As the backbone of nucleic acid structure and as a component of membrane phospholipids, it plays an important structural role. It is also the major element involved in the conservation and transfer of energy in cells. During hydrolysis, the "high energy" bond compounds (e.g., ATP, ADP, phospho-creatine, etc.) release energy, which is made available for plant metabolic activity.

In any soil, only a very small fraction (0.1–1%) of the phosphate is available, the rest being adsorbed, or immobilized through precipitation. Two major ionic species, HPO_4^{2-} and $H_2PO_4^-$, are available to plants: the former species at neutral pH and the latter at a more acidic values. Plants take up phosphate very efficiently, and often phosphate levels in plants are 100–1000-fold higher than in the soil or nutrient solution. In some plants, phosphate uptake is enhanced by the presence of mycorrhizae [21].

3. Potassium

Potassium is the predominant inorganic element in plants, with cytoplasmic concentrations exceeding 100 mM. On a dry weight basis, potassium often exceeds 1%. The K^+ ions are known to be important in the activation of many enzymes of intermediary metabolism and biosynthesis. In addition, as the major cation, K^+ is an important contributor to the osmotic potential of cells. A reduced level of K^+ ions in the plant causes closure of the stomates [22,23] and reduces water uptake to the aerial parts of the plant through transpiration and root pressure [24–26]. Potassium-deficient plants have reduced turgor and wilt very easily, even with mild water stress. Potassium is very mobile, and the ion is transported during K deficiency from older leaves to younger leaves and meristematic regions, with the result that the deficiency symptoms are observed first in the older leaves of the plant.

4. Calcium

Calcium pectate is the major structural chemical of the middle lamella of the cell wall. Deficiency in Ca^{2+} results in reduced growth as well as browning of the roots. The growth reduction is the result of reductions in the rates of cell division and cell elongation [27]. The browning appears to be due to an increase in quinones (oxidized polyphenols). When plants have adequate calcium, the polyphenols are efficiently chelated and less susceptible to oxidation than when the ion is deficient. Calcium deficiency is most visible in acid soils, and liming has been shown to be effective in reducing calcium deficiency in plants. However, Ca^{2+} uptake by plants is dependent on both the supply of the ion to the soil solution and the rates of transpiration, since this ion is transported passively in the transpirational stream. Thus, environmental conditions favoring low transpiration reduce Ca^{2+} transport to the aerial parts of the plant [28,29], and application of Ca^{2+} must be timed to the proper conditions favoring Ca^{2+} uptake by plants. Liming is also a practice that has been used extensively in ameliorating saline soils, but in this instance the role of Ca^{2+} is based on its effect on the chemistry of soil rather than as a function in the plant system. When liming is necessary, the nitrogen should be supplied in the form of nitrate, not as ammonium fertilizers. The application of lime creates a basic environment in which hydroxyl ions predominate. These ions will react with the ammonium salts, producing gaseous ammonia, which will volatilize rapidly.

5. Magnesium

Magnesium uptake resembles calcium uptake in being mainly through the transpiration stream. Unlike Ca^{2+}, which is mobile only in the xylem, Mg^{2+} is mobile in both xylem and phloem. Its biochemical roles in plants include being a component in the chemical structure of the chlorophyll molecule. In addition, Mg^{2+} is a cofactor of a number of enzymes, including those involved in the phosphorylation processes. Thus, an absence of Mg^{2+} will stop all reactions involving energy production and transfer processes involving ATP and ADP. Such a step is tantamount to death, since it is the force of energy that distinguishes the animate from the inanimate state. It has been reported that Mn^{2+} sometimes may substitute for Mg^{2+} [30].

There is evidence that Mg^{2+}-deficient plants show decreased levels of CO_2 fixation as a result of decreased Rubisco activity. Rubisco—ribulose bisphosphate carboxylase–oxidase—is the major carboxylating enzyme in C_3-type plants and is found in the stroma of the chloroplast, and a decreased Mg^{2+} concentration in the stroma has been shown to increase the K_M and lower V_{max} values of this enzyme for CO_2, resulting in a reduction in the rate of photosynthetic production of carbohydrates [31,32]. This effect is distinct from the likely reduced level of chlorophyll and the production of yellow leaves, which would contribute further to decreased levels of photosynthesis.

Uptake of Mg^{2+} by plants shows a number of interactive effects, both synergistic and antagonistic, in the presence of other cations. NH_4^+ ions inhibit uptake of Mg^{2+} ions. Similar competitive effects have been observed in the uptake of K^+ and Mg^{2+}, resulting in increased uptake of Mg^{2+} ions in the presence of low K^+ concentration in the nutrient medium, and vice versa. Other ions that show competitive uptake effects are Ca^{2+} and Mn^{2+}. In the case of Ca^{2+}, the increased uptake ensures a concentration of divalent ion capacity sufficient to maintain cation/anion balance and proper functioning of physiological activity. The observed increase in Mn^{2+} ions during Mg deficiency prevents total failure of the biochemical processes of energy transfer, thereby forestalling collapse and death of the cell.

6. Iron

Iron occurs as the prosthetic group in a number of heme-type enzymes such as catalase and cytochrome oxidase. More than 75% of the iron in plants is located in the chloroplast, where it also forms an important component of a nonheme redox protein, ferrodoxin. A large proportion of the iron in plants is also stored as a ferric phosphoprotein, called phytoferritin.

Although iron is the most abundant of all the inorganic elements, only a small fraction of it appears in soluble form in the soil. If the nutrient environment is in an oxidized state, because of good aeration, the iron becomes available in the soluble forms [Fe^{3+}, $Fe(OH)_2^+$, $Fe(OH)^{2+}$]. If, however, an alkaline condition causes an abundance of OH^- ions, the solubility and availability of Fe^{3+} is reduced as a result of precipitation, as shown in Eq. (1).

$$Fe^{3+} + 3OH^- = Fe(OH)_3 \text{ (solid)} \tag{1}$$

The reduced availability of iron to plants growing on alkaline and calcareous soils results in the commonly observed conditions of "lime-induced chlorosis." The presence of carbonate and bicarbonate ions in the alkaline soils also immobilizes iron [33]. Lime-induced iron deficiency can be overcome by adjusting the pH of the nutrient solution to between 5.5 and 6.5, converting the element to its readily soluble ionic forms. In reduced and waterlogged environments, Fe^{2+} is the predominant form of iron. Consequently, the nutrient solution becomes alkaline, reducing the available iron to plants.

High levels of manganese in the soil induce iron chlorosis. This condition is the result of Mn^{2+} ions binding for the enzyme sites at which iron normally binds, or competitively reducing the uptake of iron. Iron deficiency chlorosis is also observed in the presence of high concentrations of heavy metals (e.g., Cu, Cr, Zn, Ni) at high soil moisture and high light intensity.

The availability of iron to plants was the first and one of the few instances in which uptake has been shown to be under genetic control. Oriental varieties of soybean (*Glycine max*) showed iron deficiency chlorosis, while the Hawkeye variety showed no deficiency when grown in a nutrient medium at low iron concentration [34]. Using grafting experiments, Brown and his colleagues have shown that the mechanism of efficient iron uptake resides at the root site. The Hawkeye cultivar, through the redox activity of its roots, makes the iron more soluble and available to the plant, while the oriental variety does not possess this competence [35,36].

In the early days, iron was supplied mainly as ferrous sulfate or complexed to organic acids such as citrate or tartrate. However, iron supplied in chelated form has been found to be more effective, and most commercial fertilizers use one of the three major chelators: ethylenediaminetetraacetic acid (EDTA), diethylenetriaminepentaacetic acid (DTPA), or ethylenediamine (di-*o*-hydroxyphenyl)acetic acid (EDDHA).

7. Sulfur

The most important function of sulfur in plants is in the structural role of proteins, where disulfide bonds (—S—S—) are formed. Sulfur is also a major component of secondary plant products (e.g., mustard oil). The biosynthesis of mustard oil requires amino acids, glucose, and sulfur. The sulfur in the mustard is present at two positions in the chemical structure of mustard oil—one as a sulfate anion attached and the other as the element linking glucose to the oxime form of the amino acid. Sulfur also occurs in the sulfoxide form in onion and garlic and is responsible for the smell and lachrymatory effects associated with these plants.

Sulfur is taken up by plants as SO_4^{2-} ions. The presence of selenate ions (SeO_4^{2-}) reduces the rate of SO_4^{2-} uptake by competing with the carrier. Furthermore, the production of biochemically active proteins is affected, since Se can substitute for S, causing distortions in the protein structure.

8. Manganese

Manganese is required as a cofactor in the IAA–oxidase enzyme system. In addition, it is known to be able to substitute for Mg^{2+} in enzymatic reactions involving, for example, membrane

ATPases and the enzymes of the Krebs cycle of respiration. It is also essential for photosystem II (PS II) activity in the electron transport chain of photosynthesis. The Mn^{2+} ion forms a manganometalloprotein, which is responsible for photolysis of water and production of electrons at the PS II sites [37].

Manganese deficiency affects chloroplast activity, and the plants show the interveinal chlorosis that is characteristic of Mg^{2+} deficiency, but different in that the symptoms in Mn^{2+}-deficient plants appear on younger leaves, instead of on the older leaves, as during Mg^{2+} deficiency. Tissue analysis of the young leaves of manganese-deficient plants show values in the range of 1.5–15 ppm [38].

In waterlogged and anaerobic soils, rice shows Mn^{2+} toxicity effects. The toxicity has been reported to accompany toxic affects of other toxins, such as Fe^{2+} and H_2S [39].

9. Copper

Plants have a very small range of tolerance to copper. Just over a 10- to 15-fold difference in concentration in copper content of plant tissues separates the deficiency range from the toxicity concentration level. Copper is a chelating element of the plastocyanin–protein complex, where it functions in the electron transport of photosynthesis. In addition, it is found to be cofactor for a number of other enzymes involved in electron transfer redox reactions or in enzymes responsible for oxidation–reduction reactions using molecular oxygen.

Cereal crops are particularly prone to problems of copper deficiency, and yield reductions are due to suppression of panicle formation. On the other hand, copper toxicity affects the membrane structure which, in turn, will cause damage to the plant through interference of the uptake of essential ions such as K^+ and NO_3^-.

10. Cobalt

Cobalt is an important micronutrient for plants such as the legumes, which rely for their supply of nitrogen on symbiotic root nodule bacteria [10]. This element is needed for the proper functioning of the bacteria. Reports have shown a requirement for cobalt by some nonleguminous plants, leaving the question of its universal importance to plants a matter of controversy.

Cobalt is taken up primarily through the roots and depends on transpiration for its mobility in the plant. Its strong chelating properties enable it to displace other ions from physiologically important binding sites and, thereby, to produce toxic effects in plants at high concentrations.

IV. AVAILABILITY OF NUTRIENTS FOR PLANTS

Carbon, which constitutes between 40 and 45% of the dry weight of any plant, is made available to plants from the atmosphere, through photosynthesis. In most plants, water is absorbed from the soil. It must be recognized, however, that the origin of the soil water depends on a rapid hydrological cycling involving the earth's atmosphere, hydrosphere, and lithosphere. Epiphytic orchids are an example of plants having the capacity to absorb the atmospheric water that condenses on special roots. All the other elements are taken up from the soil solution at the root site in natural systems. With human-controlled agricultural systems, mineral nutrients are also supplied by foliar application.

A. Soil Application of Fertilizers

The most common method of providing nutrients to a crop is by application at the site of the root, the organ responsible for uptake of water and minerals. Not all the nutrients applied in a soil system, however, are available to the plant. The crop takes up the nutrient from the soil solution, whose concentration, in turn, is dependent on the capability of the soil to supply the nutrient—that is, the "buffer capacity" for the particular element under consideration. A number of soil factors and environmental factors, as well as the chemical nature and solubility of the element, influence the buffering capacity. The concepts of nutrient intensity (I) and nutrient quantity (Q) were introduced by Schofield to explain these effects [40]. The intensity factor is a measure of the concentration of the nutrient in the soil solution and reflects the strength of retention of the nutrient by the soil solution. The quantity factor, which represents the amount of available nutrient in the soil, is a function of a number of events. The most significant of these is the large labile pool of nutrients in equilibrium with the soil solution. Second, there is a nonlabile pool, which holds nutrient adsorbed to the soil. This fraction includes nutrients released by plants during growth as well as nutrients added as fertilizers. The third component of quantity is the bulk mineral and organically bound reserves, which are released very slowly by weathering and the mineralization activity of soil microbes. Finally, the root/soil ratio is an important subsystem that will influence the quantity. To execute a well-developed strategy for fertilization, it is important to have knowledge of the buffering capacity of the soil at given times. Normally, the buffering capacity is calculated by plotting quantity (ion adsorbed) versus intensity (ion in solution). The slope of the curve represents the buffering capacity B for the particular element studied, for the particular soil studied, and under the conditions of interest.

$$B = \frac{\Delta Q}{\Delta T}$$

This information together, with a list of plant nutrient requirements, will allow the development of a plan for a fertilization strategy to optimize crop production.

The concept of buffering capacity applies well for potassium, phosphorus, and other elements. It does not hold for nitrogen, since NO_3^- ion is only weakly held in the soil, hence is not buffered. Furthermore, most of the nitrate released from the organic reserves is due to microbial activity, and this released nitrate is immediately taken up by the plant.

Application of soil fertilizers involves either broadcasting of the material or placing it in bands or rows. Broadcasting—the uniform distribution of fertilizer—is by far the most commonly used technique. However, placement is a superior technique when fertilizing soils of poor nutrient status or when applying nutrients such as phosphorus, which are known to adsorb very strongly to the soil.

The majority of the fertilizers in use are of the solid type, but a number of fluid fertilizers are also available. The liquid fertilizers have four major advantages over the solid type. They are:

easier to transport
easier to use (and relatively less labor intensive)
more concentrated
more evenly distributable in the field (since they are homogeneous)

The greatest drawback to the use of liquid by farmers is the need for expensive application equipment.

B. Foliar Application

Leaves can take up gaseous substances through the stomata. Any other nutrients must pass through the membrane before they can be made available to the plant. The leaf cell plasmalemma are very similar to those found in the roots and have the same carrier systems for ion uptake. However, leaf cells often have a cuticle made up cutin, which is hydrophobic and only weakly hydrophilic. Thus, any nutrient applied must first pass this water-impermeable layer of cutin.

Usually, fertilizers are applied together with dilute detergents and timed to ensure that the fertilizer has a long contact period with the leaf. The detergents have considerable hydrophobicity and assist in the passage through cutin. Furthermore, they enable the fertilizer to spend more time in contact with the leaf surface. Foliar applications are useful in supplying elements (e.g., Fe, Mn, Zn, Cu, B) that sometimes are not readily taken up because of changes in the redox state in the soil. Finally, foliar application has shown promise in deep-rooting crops such as fruit trees, which cannot readily take up nutrients applied to the soil.

V. MECHANISMS OF NUTRIENT UPTAKE BY PLANTS

In general, any living cell continuously exchanges water solutes, metabolites as well as gases, with the environment in which it lives. An understanding of nutrient uptake requires a knowledge of the nature of membranes. Typical biological membranes are lipoprotein structures. There are two layers of lipids with their hydrophobic ends facing each other in the center of the membrane. Two types of protein are found as components of any membrane: peripheral or extrinsic proteins, which exist on the two surfaces attached to the hydrophilic ends of the membrane, and "integral or intrinsic" proteins which extend through the bilipid layer and are exposed to both the inner and outer surfaces of the membrane (Figure 4) [41]. The membranes are freely permeable to water and other small molecular weight gases, such as CO_2 and O_2, but are intrinsically impermeable to electrically charged solutes (e.g., Na^+, K^+) and to polar uncharged molecules (e.g., carbohydrates) that are of importance to plants and animals. Integral proteins with hydophilic and hydrophobic features enable the transport of such charged solutes (e.g., Na^+) and uncharged polar molecules (e.g., carbohydrates). This mechanism is advantageous in that the solutes can be discriminated, enabling only solutes required by the plant to be taken up, while preventing potentially toxic solutes and those of no importance from entering the cell. However, all lipid-soluble molecules will readily pass through the membrane, since these substances can dissolve in the lipid fraction of the membrane. The mechanisms of exchange between the environment and a plant cell can be either passive or active.

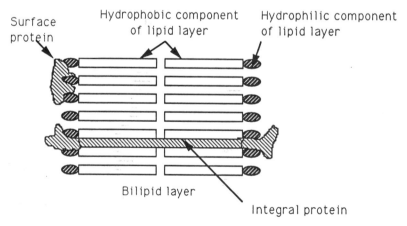

Figure 4 Structure of a typical cell membrane. The integral proteins traverse the full length of the membrane, with the very hydrophobic middle portion interacting with the hydrophobic region of the lipid molecules. The two ends of the integral proteins are more hydrophilic and associate with the hydrophilic regions of the lipids on the two exterior ends of the membranes. (From Ref. 41.)

A. Passive Transport

Passive transport involves diffusion of substances down a gradient of potential energy. A thermodynamic analogy can be drawn between this type of movement in the flow of heat from a warmer to a cooler body. Movement down a potential gradient is spontaneous and depends on the partial molar free energy G, which is identical to the chemical potential μ of the molecule.

For a noncharged molecule, this energy is given by

$$\mu_s = \mu_s^* + RT \ln a_s \tag{3}$$

where

μ_s = chemical potential of the solute (J mol^{-1})
μ_s^* = chemical potential of the solute in a standard state
R = ideal gas constant ($= 8.314$J K^{-1} mol^{-1})
T = absolute temperature (K)
$\ln a_s$ = natural logarithm of the activity of the solute (mol^{-1})

Very dilute solutions behave very much like ideal solutions and obey Raoult's law. Under these circumstances, the activity a_s must be the same as concentration C_s. Since C_s is easily measurable, it is substituted in our equation as follows:

$$\mu_s = \mu_s^* + RT \ln C_s \tag{4}$$

where C_s is concentration (mol^{-1}).

When the molecule is charged, the movement depends on the chemical as well as the electrical potential. Thus for a charged molecule, the new potential energy referred to as the electrochemical potential is given as follows:

$$\mu_s = \mu_s^* + RT \ln C_s + Z_j FE \tag{5}$$

where Z_j is an integer representing the charge number of species, F is the Faraday constant (96,487 J mol$\cdot$V^{-1}), and E is the electrical potential.

Most of the inorganic solutes of importance are charged molecules, and the electrochemical potential will be the important energy parameter involved in movement into and out of the cell through the membranes.

When substances move by diffusion through biological membranes, they may take one of the three pathways. Lipid-soluble solutes will move through the lipid component of the membrane. Small molecules, in particular gases, will move through the pores on the membranes. The third pathway involves the movement of the substance aided by a carrier, usually a transmembrane protein. This last type of movement is also referred to as "facilitated diffusion."

B. Active Transport

In general, plant nutrient uptake involves active mechanisms, and metabolic energy is utilized in the process. Active transport is also often referred to as "uphill transport." To determine whether transport across the membrane was active, it is necessary to know (a) the internal and external concentrations of the solute, and (b) the electrical potential difference across the membranes. In the case of noncharged solutes, a higher concentration of solutes inside the cell (accumulation) would indicate that transport of the solute was active. For charged solutes, it is important to consider both the concentration differences and the electrical potential of the membrane. Figure 5 shows schematically the differences between active and passive transport [42]. Measurement of cellular ion concentrations is relatively easy with giant cells, such as

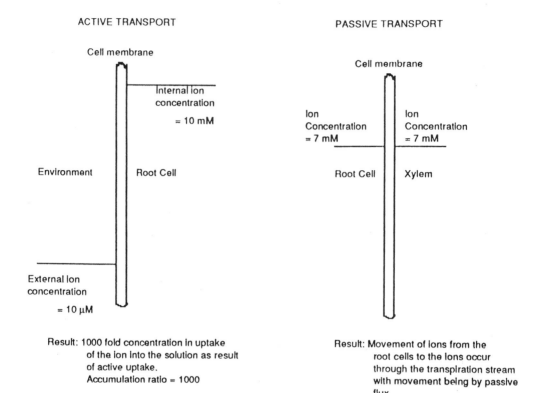

Figure 5 Schematization of some differences between active and passive transport of ions across plant cell membranes. The movement of solutes from the root environment to the root cortical cells is considered to illustrate active transport, and the movement from the root cells to the xylem during transpiration illustrates passive ion transport. (From Ref. 42.)

Nitella or *Valonia* [43]. Use of intracellular chemical probes is now making such measurements feasible with normal cells [44,45]. Such measurements permit a better understanding of the role of inorganic elements in plant growth.

VI. KINETICS OF ION TRANSPORT

A. The Free Space in Plants

Kinetic studies of solute uptake played an important role in elucidating the mechanism of uptake as well as providing an insight into the characteristics of specific carrier molecules. However, a good knowledge of the structure of plant cells is critical to an evaluation of kinetic-type studies. In a typical plant root, we can identify three major compartments in which solutes will be found. The outermost compartment and the one most readily accessible to solutes is that of apparent or outer free space (AFS or OFS) [46,47]. This compartment consists of two subcompartments. The first, which is in fact an extension of the soil solution, where the dissolved inorganic solutes move by diffusion, is referred to as the water free space (WFS). The second subcompartment is referred to as the Donnan free space (DFS) [48]. This second compartment is created as a result of the chemical composition of the cell wall and the membrane. Both these structures are made up of chemicals that have a large number of immobile negatively charged sites or electronegative sites (mainly—COO⁻ sites), which can bind cations. The net binding capacity is a measure of the cation exchange capacity or Donnan exchange capacity of the cell. Hence, whereas the WFS is a measure of the purely physical volume of intercellular spaces in the roots, the DFS is a measure of an exchange capacity (Figure 6).

The most compelling evidence for the presence of the compartments just described came from kinetic data obtained using root segments. Roots were loaded with radioactively labeled Sr^{2+} ions and then subjected to a "washout" or efflux system containing $Ca(NO_3)$ solution. When the rate of leakage or efflux of the radioactive ion into the external $Ca(NO_3)$ solution was measured, a triphasic rate of loss was reported. The first phase was very rapid and occurred within the first 3–5 minutes. The percentage of ions that efflux out of the tissue is representative of the volume of the AFS as a function of the plant root volume. The AFS may represent between 15 and 25% of the root volume. The second phase showed a slower rate of efflux, and the third phase is the slowest of the three [49].

A simplified scheme of the foregoing results is presented in Figure 7. The ions that leaked out during the first phase showed a rapid rate, since these were located in the AFS, where they

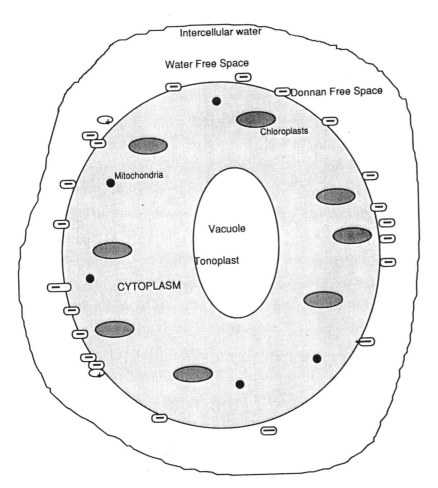

Figure 6 Representation of the water free space and the Donnan free space in relation to the plant cell structure. (Note the large numbers of negatively charged groups on the outside of the cell and constituting the Donnan free space, where mainly cations will bind preferentially.) (Modified from Refs. 46–48.)

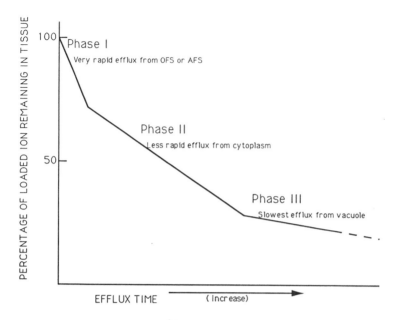

Figure 7 Efflux of labeled Sr^{2+} ions from a preloaded tissue during a washout or efflux experiment. (From Ref. 49.)

were in a compartment in free contact with the external medium, and movement between the outer free space and the external solution was by diffusion. The presence of $Ca(NO_3)$ solution was necessary to exchange ions that were bound to the DFS. The presence of a WFS separate from the DFS was demonstrated by washing out the tissue first in water and then in $Ca(NO_3)$ solution. The water was incapable of releasing the cations bound by electrovalent bonding to the negative charges, resulting in the cation content efflux in water being less than into a $Ca(NO_3)$ solution.

The efflux rate during phase II was slower, since the ions had to pass through a single membrane, the plasma membrane, which acts as a barrier to free diffusion. During the slowest phase, phase III, the ions could not reach the external solution until they had passed more than one barrier, either the tonoplast or some other organelle barrier, followed by the plasma membrane.

B. Carriers and Michaelis–Menten Kinetics

Uptake and transport of most ions across a membrane shows simple Michaelis–Menten kinetics, suggesting a substrate–enzyme kinetics such as is seen in uni-univalent enzyme reactions. This is hardly surprising, considering that ions are transported across a membrane by carriers which are proteins.

The Michaelis–Menten equation is given as:

$$v = \frac{V_{max}\ [S]}{K_M + [S]}$$

where

v = rate of absorption of the ion

V_{max} = maximal rate of absorption: also often called capacity factor

K_M = dissociation constant of the carrier–ion complex; a measure of affinity of the carrier for the ion

[S] = substrate concentration

In a few exceptional cases, the kinetics of the uptake mechanism deviates from Michaelis–Menten kinetics and exhibits the allosteric kinetic behavior shown by regulatory enzymes. Such behavior is representative of a situation that requires fine control of the uptake and features an ion carrier made up of a number of subunits [50].

The similarity in the behaviors of ion–carrier and enzyme–substrate complexes is a useful tool in predicting the effects of inhibitors on ion uptake as well as in studying the role of ion uptake in plant productivity. For example, based on known V_{max} values for a specific solute in a plant, one could calculate the amount of nutrients that would need to be supplied for all the applied elements to be taken up by the plant. In particular, if plants are grown in hydroponic systems, these calculations would show predictable behavior. For soil systems, other factors such as the cation exchange capacity and the immobilization of ions in the soil will also need to be considered. Other useful applications include selecting plants for efficiency in ion uptake. A plant having a carrier with a lower K_M value for a particular solute will be more efficient in taking up the solute present in low concentrations than a plant showing a higher K_M value. Thus, one could select for plants with better efficiency in ion uptake and use controlled fertilization, an approach that would minimize pollution and save on the cost and labor associated with fertilization.

VII. ENERGETICS OF ION TRANSPORT

Energy is required for the transport of ions across membranes, and two major hypotheses have been advanced to link ion transport and energy. The first hypothesis, which linked ion transport to respiratory energy, is referred to as the electrochemical hypothesis, while the second considered ion transport to be linked to ATP, the high energy intermediate associated with most energy-requiring processes.

A. Electrochemical Hypothesis

Early work of Lundegardh and Burstrom [51] linked anion uptake activity to respiration, and the term "anion respiration" was coined to signify that changes in respiration rate were directly parallel to anion uptake rates. Respiratory poisons such as cyanide inhibited anion respiration as well as anion uptake. However, addition of cyanide did not inhibit a basal rate of respiration, which was termed "ground respiration." The primary process of this hypothesis is the electron flow, where the electrons are accepted by oxygen, resulting in a counterflow of anions to maintain electrical neutrality. In this hypothesis, cations were assumed to move passively. This hypothesis, though no longer tenable in its original form, provided three key features that are still retained.

1. The process involves a separation of charges due to the oxidation of a substrate, which is the driving force for ion transport.
2. Only one ion of a salt moves during the primary transport process.
3. A stoichiometric relationship will exist between the rate of respiration and ion uptake.

In connection with the third feature above, it was predicted on the basis of Eq. (7), which shows the involvement of four electrons for each molecule of O_2 used, that no more than 4 moles of monovalent ions could be transported per mole of O_2 used in respiration [52, 53].

$$\underset{\substack{\text{(Respiratory} \\ \text{substrate)}}}{AH_2} \xrightarrow{\text{dehydrogenation}} \underset{\substack{+ \\ \text{4e (each electron can transport one anion)}}}{4H^+} \xrightarrow{\text{Oxidation by 1 mole oxygen}} = 2H_2O \tag{7}$$

Although the Lundegardh–Burstrom hypothesis supports Peter Mitchell's Nobel Prize–winning "chemiosmotic theory," it fails to explain two important aspects of ion transport across membranes.

1. Ion transport is directly linked to oxidation of substrates and the involvement of the respiratory chain, but the hypothesis does not explain how this system can operate at the tonoplast and the plasma membrane, where a respiratory chain is lacking and respiration does not occur. Thus, if such a mechanism is assumed to exist, it could operate only at the mitochondrial site or the chloroplastic sites, where electron transport chains are known to occur.

2. The dehydrogenation reaction that results in the release of electrons also cause a proton to be liberated into the solution. Thus electrical neutrality is maintained and there is no provision for a counterflow of anions. Further assumptions would have to be incorporated to account for such a movement of anions [54].

B. ATP-Mediated Transport

Oxidative phosphorylation involves the concomitant electron transport and production of the high energy intermediate adenosine 5–triphosphate (ATP) by phosphorylation of ADP. However, under certain conditions, such as the use of dinitrophenol or other inhibitors, it is possible to prevent production of ATP without inhibiting electron transport. A number of early workers showed that under conditions that inhibited ATP production, ion transport was inhibited, even though electron transport was not blocked. This finding led to the suggestion of ATP-mediated transport, where ATP was used as the fuel source. A simplified schematic representing this mechanism is presented in Figure 8 [55].

The enzyme ATPase is located on the membrane and is coupled to the carrier. The enzyme ATPase acts on ATP, in the presence of Mg^{2+} ions, and the energy released from the high energetic phosphate bonds is used to energize the ion–carrier protein. In the energetically

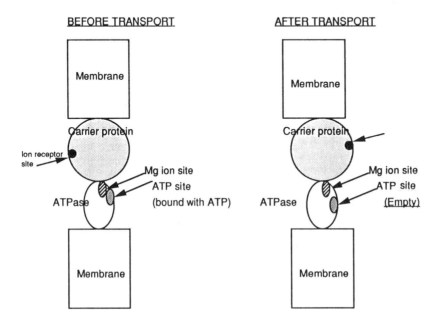

Figure 8 Hypothetical illustration of the relation between the energy source (ATP), the enzyme (ATPase), and the carrier protein in ion transport across cell membranes. (Redrawn from Ref. 55.)

activated form the carrier is capable of transporting the ion across the membrane. After the ion has been transported, the carrier must be reenergized before it can transport another ion.

VIII. PHYSIOLOGY AND BIOCHEMISTRY OF PROTON PUMPS

Proton pumps may be defined as pumps that extrude hydrogen ions. In plant systems, the activity of proton pumps is the major process responsible for transport of ions across membranes. Extrusion of protons is necessary in all cells to maintain a physiologically acceptable cytoplasmic pH (pH-stat mechanism). Proton pumps are also necessary for cell wall loosening to permit cell extension [56]. At the same time, if the proton pump were used to transport ions, it would permit the maintenance of the osmotic potential and turgor important for cell growth [57]. Two major types of proton pump have been identified based on the substrates used as the source of energy: ATPases use ATP as the source of energy, while PP_iases use inorganic pyrophosphatase.

Figure 9 shows the relation of the energy source to solute uptake through the proton pump. When the direction of movement is the same for the proton and the solute, the motion is called symport; antiport refers to movement of the proton and solute in opposite directions. Figure 10 shows the relationship of transport of a number of solutes to proton transport.

IX. MOLECULAR ASPECTS OF ION TRANSPORT

An understanding of the molecular aspects of proton pumps and ion transport systems is crucial to an understanding of ion uptake across membranes. In the past decade, many of the membrane ATPases and PP_iases have been purified and fractionated, and their subunits characterized. In addition, genes have been characterized, using nucleotide sequence methods on the comple-

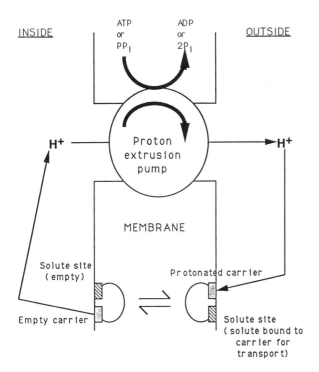

Figure 9 Schematization of the relation between solute uptake and the proton ATPases or proton PP_iases. (Redrawn from Ref. 55.)

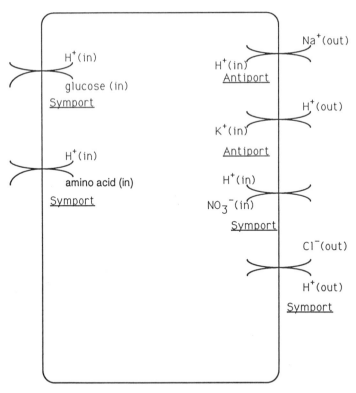

Figure 10 Relation of the uptake of some common solutes to the proton pump. (Wignarajah, Unpublished.)

mentary DNA. Future work in this direction will enable the production of genetically engineered plants (GEPs) having the appropriate characteristics for efficient uptake of ions important to plants (e.g., K^+, NO_3^- ions) or tonoplast membrane characteristics for the compartmentalization of potentially toxic ions into the vacuole. An example of the potential benefits of molecular biology was demonstrated recently by Sentenac et al. [58]. These workers produced a cDNA library of the high affinity K^+ transport system of a higher plant, *Arabidopsis thaliana*, cloned it into a yeast (*Saccharomyces cerevisiae*) mutant lacking a K^+ transport system, and transformed the yeast into an organism having the capacity to take up K^+ ions from low concentrations in the environment.

X. ROLE OF MICROORGANISMS IN NUTRIENT UPTAKE BY PLANTS

The roots of plants grown in soil are closely associated with the large diversity of microbes in the soils. These microbes can influence nutrient uptake by plants in a number of ways:

1. Microorganisms can degrade organic matter to simple inorganic molecules, which can be made available to plants.
2. Microbes assist in dissolving inorganic ions from insoluble ores.
3. Microbes can influence the uptake of nutrients by plants. The exact nature of the influence seems to be variable and clearly needs a great deal of research. For example, the presence of the microbes stimulated uptake of phosphate in young barley plants during short-term experimentation, but uptake was reduced in older plants in long-duration experiments [59].

4. Microbes could in some cases compete with plants for the limited nutrient supply, reducing the availability of nutrients to plants.
5. A number of microorganisms, in particular fungi, are known to degrade lignin, releasing phenolic acids. The presence of phenolic acid, in turn, inhibits phosphate uptake [60] and could interfere with the active transport of other essential elements.
6. A number of microorganisms are known to fix atmospheric nitrogen to ammonia. Such mechanisms can be important as pathways to the supply of nitrogen to plants.

The importance of soil microorganisms in terms of numbers and functions is best reflected by estimates that many agricultural soils may have as much as 3–5 Mg fresh weight of microorganisms per hectare of soil. The numbers can be estimated by means of simple microbial plating/counting techniques. Soil microbial activity can be estimated using techniques for measuring ATP or respiratory activity.

A. Rhizosphere Microorganisms

The microorganisms associated with the root are referred to as rhizosphere microorganisms [61]. The mass importance of the rhizosphere microorganisms is expressed by the ratio of root to soil (R/S), which expresses the number of microorganisms in the region of the rhizosphere relative to that in the rest of the soil. These numbers often are very high and can be used as a measure of the potential benefit derived by the plant from its rhizosphere microorganisms. Some of the most common rhizosphere microorganisms are *Pseudomonad*, *Azotobacter*, *Azospirillium*, and *Clostridium* species.

B. Mycorrhizal Associations

The many fungi known to form close associations with plant roots are referred to as mycorrhizae (*mycos* = fungus; *rhizos* = root). The degree of association between the plant species and the fungus is variable: from a simple external sheath of fungi around the root, the ectomycorrhizal type, to species that penetrate and invade the root cells—endomycorrhizal types. Orchid mycorrhizae and the vesicular–arbuscular (VA) mycorrhizae are examples of endomycorrhizae. Intermediate types, referred to as ectoendomycorrhizae, are also known to occur. Roots infected with mycorrhizae increase the uptake of phosphate ions. Three major postulates have been made to account for the increased uptake.

1. The ability of the fungus to take up sources of phosphate that the plants could not take up.
2. The ability of the fungus to explore parts of the soil not readily available to the plants, thereby making more P available to the plant.
3. The possibility that mycorrhizal roots have a lower K_M value than nonmycorrhizal roots, thus enabling a more efficient uptake by mycorrhizal roots at low concentrations of phosphorus in the soil solution. The subject of mycorrhizae has been extensively covered in a book by Harley and Smith [62].

XI. RELEVANCE OF ION UPTAKE TO CROP PRODUCTIVITY

The importance of ion uptake to crop productivity is best assessed using three major factors, all of which have economical basis.

1. An increase in the biologically important component of the crop must result from application of the nutrient. In a cereal crop, for example, the important component is the seed, and application of nutrients must be manipulated to increase seed production, not vegetative growth. On the other hand, with a vegetable crop such as lettuce, the application of nutrients

must be designed to ensure maximal leaf production, while keeping root production to a minimum without decreasing the ability, essential for rapid leaf production, to lower the uptake of nitrogen.

2. Nutrients that are not taken up by the plants often are leached away in the soil solution, decreasing output/input economy.

3. The loss of nutrients, particularly phosphate and nitrate, leads to serious environmental pollution. The European Community has established very stringent laws on the entry of these fertilizers from farms to streams, which could result in heavy penalties and fines to noncomplying farmers.

A. Problems of Mineral Deficiency and Toxicity

The characteristic response to varying concentrations of plant nutrients has been dealt with in Section III. At low concentrations, a plot of plant response to increasing concentrations is linear till the optimal concentration is reached in this linear range, plant growth will be limited by lack of nutrients. On reaching the optimal concentration, the curve plateaus (Figure 3). The concentration range of the plateau is variable. The plateau represents a phase referred to as the "luxury consumption" region, since at these concentrations the extra nutrients present do not contribute to any extra growth. For a number of elements (e.g., N, B, Mn, Zn), the curve will show a decline at higher concentrations, suggesting a toxicity effect, while for some elements, there is no toxicity effect. Despite this simple treatment of mineral deficiency and toxicity response, it must be borne in mind that very often interactions of nutrients in the growth medium can complicate the uptake of the nutrient into a plant system.

Morphological symptoms of mineral deficiency and toxicity are extremely useful diagnostic tools in the field. These symptoms do however have their limitations; hence there is a need to complement such visual or microscopic studies of plant samples with tissue analysis (i.e., study of the ions in the tissues). The first major limitation arises from the different morphological symptoms shown by different crops as the result of deficiency of a single ion. For example, zinc deficiency results in interveinal mottling on the older leaves of cucumber; in corn, there is a general chlorosis or blanching, which is particularly evident on the younger leaves—the symptom is often referred to as "white bud."

The second limitation results from the similarities in symptoms that are due to deficiencies of different elements. For example, plants lacking nitrogen or sulfur are unable to produce chlorophyll and the associated proteins, resulting in the leaves showing a general chlorotic appearance that even trained personnel will find difficult to distinguish based on visual symptoms alone. The third difficulty arises from multiple deficiency symptoms being interposed on one another. Finally, infections by pathogens may simulate symptoms of elemental deficiencies or toxicity, which may mislead field personnel.

Because of the complexities just enumerated, reliance on the use of morphological visual symptoms in diagnosing elemental problems requires a great deal of field experience. The usefulness of such studies as a first step, however, is very essential. A number of well-illustrated manuals with detailed descriptions are readily available [63–66] and must be part of the collection of any plant grower or field station. Unfortunately, these materials represent studies on crops grown mainly in the industrialized nations. There is a pressing need to collect such information for many of the lesser known crops, such as root crops, tubers, and other less familiar tropical crops grown in the rural parts of developing nations.

XII. HYDROPONICS AND NUTRIENT UPTAKE BY PLANTS

Hydroponics, sometimes referred to as "soil-less" culture, is the technique of growing plants without soil, in a liquid culture. This system has a number of advantages over soil culture.

1. All the nutrients supplied are readily available to the plant.
2. Lower concentrations of the nutrient can be used.
3. The pH of the nutrient solution can be controlled to ensure optimal nutrient uptake.
4. There are no losses of nutrients due to leaching.

A disadvantage of hydroponic systems is that any decline in the oxygen tension of the nutrient solution can create an anoxic condition, which inhibits ion uptake. To remedy this situation, a modification of the system called aeroponics is often used. Here, the nutrients are continuously run down along the roots, which have access to a ready supply of oxygen. The nutrient is continuously agitated and in motion, hence never becomes anoxic.

A. Types of Hydroponic System

A number of variations of the hydroponic system are currently in use. The type to be used is determined by:

The nature of the root system
The nature of the canopy architecture
The availability of water
The availability of space

Two major variants of the hydroponic system are the deep-feeding type and the nutrient film type (NFT). The deep-feeding type is used for crops with large root systems when water is available in large amounts. In the NFT system, the nutrient is fed as a thin film in a trough containing the roots. A comprehensive book by Molyneux describes use of NFT systems in crop production [67].

XIII. ENVIRONMENTAL CONTROL AND AUTOMATION

Nearly two centuries ago, Thomas Malthus published "An Essay on the Principle of Population," in which he predicted that while the human population tends to increase by geometric progression, food production increases only arithmetically, with the result that eventually the food produced would not be sufficient to meet the needs of the world's human population. However, through improvement in cultural practices, use of classical plant breeding techniques, development of rapidly growing plants, better use of nutrient supplies for plant productivity, and, in recent years, the use of genetic engineering tools, it has been possible to achieve plant productivity that increases exponentially. In environmentally controlled chambers equipped with higher light intensity, NASA scientists have achieved yields four times higher than world record yields in the field [68]. Key to success at NASA have been the abilities to control and manipulate the growing environment (light, temperature, and nutrient supply), to increase the CO_2 supply, and to use early maturing varieties.

Major advantages of crop production in controlled environments are as follows:

1. The ability to increase photosynthetic productivity through the use of increased light supply and higher CO_2 supply.
2. Minimization of water use by condensing the transpired water and reusing it in the production of the nutrient media for the plant.
3. Use of smaller space for production.
4. Grower can produce consistently predictable high yields with time.
5. Product is uniform in high quality—an important factor in the modern consumer market.

6. Through effective control of input and output requirements, a cleaner environment is ensured. Any potential pollutants are confined to the controlled environment, where they can be treated before further use or release.
7. Designs of controlled environment technology are continuously being upgraded. Examples include new lighting sources (e.g., use of LED technology) and new power sources for heating, ventilating, and air conditioning, which will continue to increase crop productivity levels. The best analogy of the relation of improved crop productivity to time is the manner in which improved controlled environment technology as applied to the earth has improved the quality of life for humans.

The disadvantages include, currently, the relatively higher production cost estimates compared to field production, and lack of sufficient knowledge and technology for crop production in controlled environment systems. The technology is now available for some of the major crops, such as lettuce, cucumber, and wheat, and agriculture of the twenty-first century will no doubt continue to bring more knowledge in this field. The costs of land space and water necessary for agriculture are at present unrealistically low and are likely to increase substantially into the 2000s, making the relative cost of food production in controlled environments economically attractive.

Automation provides an added feature that can help achieve increased crop production. The areas that require automated control are enumerated briefly.

1. *Nutrient media control.* A basic control system should monitor and control pH and electrical conductivity. Additional features can be added to allow the control of individual elements, particularly the major elements (K, Ca, Mg, and nitrate ions), using specific ion electrodes on-line in the nutrient delivery systems. Unfortunately, these on-line systems for monitoring specific ions are short-lived: the electrodes are fouled easily by microbes and chemicals in the system. Further refinement in the state of the art of specific ion electrode technology will offer promise for the future.

2. *Monitoring of water loss through evapotranspiration.* Automation can be achieved using basic devices, and the water in the nutrient system can be replaced.

3. *Control of the plant atmospheric system.* Monitoring of CO_2, temperature, humidity, and light will be necessary to ensure optimal crop production.

XIV. PRACTICAL CONCEPTS IN EFFICIENT FERTILIZER USAGE

Table 6 lists fertilizers in common usage and the elements supplied by each [69]. Any practical concept in fertilizer usage must be geared towards three major goals:

1. Enabling optimization of crop productivity.
2. Developing a capability for long-term maintenance of the field environmental conditions for optimal crop production. Clearly, controlled artificially created environments (e.g., crop growth in man-made controlled chambers) will be a step toward eliminating reliance on the fragile natural environment. This type of environment can be maintained over the long term without a drop in crop productivity. In fact, with refinements and advances in design, controlled environmental technology may continue to improve the potential of crop production.
3. Developing a continuing demand/supply economic model for world food production.

XV. ECOLOGICAL AND ENVIRONMENTAL ISSUES

With more than 80% of world food production located in ecological and environmental systems in which humans have undertaken major modifications to intensify production (e.g., chemical

Table 6 Commonly Used Fertilizers for Macroelement Fertilization

Element	Agricultural name	Formula	Availability
N	Ammonium sulfate	$(NH_4)_2SO_4$	21% N
	Ammonium chloride	NH_4Cl	26% N
	Ammonium nitrate	NH_4NO_3	35% N
	Nitrochalk	$NH_4NO_3 + CaCO_3$	21% N and 30% Ca
	Ammonium nitrate sulfate	$NH_4NO_3 (NH_4)_2SO_4$	26% N
	Urea	$CO(NH_2)_2$	46%
	Calcium cyanamide	$CaCN_2$	21%
	Anhydrous ammonia		82%
P	Superphosphate	$Ca(H_2PO_4)_2 + CaSO_4$	18–22% P_2O_5
	Triple	$Ca(H_2PO_4)_2$	46–47% P_2O_5
	Monoammonium phosphate	$NH_4H_2PO_4$	46–50% P_2O_5
	Diammonium phosphate	$(NH_4)_2H_2PO_4$	54% P_2O_5
	Superphosphoric acids	Mixtures of orthophosphoric and polyphosphoric acids	65–90% P_2O_5
	Sinterphosphate	$CaNaPO_4 + CaSiO_4$	18–22% P_2O_5
	Basic slag	$Ca_3P_2O_8)_2 \cdot CaO + CaO \cdot SiO_2$	10–22% P_2O_5
K	Muriate of potash	KCl	33–50% K depending on grade
	Potash of sulfate	K_2SO_4	approx. 43% K
	Potassium nitrate	KNO_3	37% K and 14% N
	Potassium metaphosphate	KPO_3	33% K and 13% N
	Sulfate of potash magnesia	$K_2SO_4 \cdot MgSO_4$	18% K and 11% Mg
	Magnesium kainite	$MgSO_4 + KCl + NaCl$	19% K + 3.6% Mg + 18% Na

Source: Ref. 69.

application, soil amendments, monocultural crop production), the question of the long-term stability of the ecosystem is always in question. At present, some of the major environmental issues are as follows.

1. The use of chemicals that have potential adverse effects on humans and other animals. Some progress is under way in using naturally occurring chemicals rather than the invariably more toxic synthetics.

2. A major concern is leaching of nitrates and phosphates into precious reservoirs of drinking water due to either excessive application of fertilizers or application to soils or at times that favor excessive leaching. A scientific approach to the problem requires an understanding of the uptake rates into the plant and application at levels that will minimize leaching. Such "point source" reduction is both cost-effective and environmentally sound.

3. Optimization of the plant resources not utilized by humans or animals: that is, the inedible biomass. Sound agricultural practices should be based on detailed input/output studies

and should attempt to increase the dependency and, thereby, close the loop of the chemical and biological resources on earth.

REFERENCES

1. D. I. Arnon and P. R. Stout, *Plant Physiol.*, *14*: 371 (1939).
2. B. H. Mason and C. B. Moore, *Principles of Geochemistry*. Wiley, New York (1982).
3. R. C. Weast (ed.), *CRC Handbook of Chemistry and Physics*, 70th ed., CRC Press, Boca Raton, FL (1990).
4. D. W. Ming and D. L. Henninger, *Lunar Base Agriculture: Soils for Plant Growth*, ASA-CSSA-SSSA Publications, Madison, WI (1989).
5. R. Upadhye, K. Wignarajah, and T. Wydeven, *Environ. Int.*, *19*(3): 81 (1993).
6. C. C. Delwiche, *Microbiology and Soil Fertility* (C. M. Gilmour and O. N. Allen, eds.), Oregon State University Press, Corvallis (1965).
7. M. L. Scott, *Nutrition of Humans and Selected Animal Species*, Wiley, New York (1986).
8. K. Wignarajah, "Mineral composition of crop residues," NASA-Ames Technical Memo, NASA-Ames Research Center, Moffett Field, CA (in preparation).
9. G. Bertrand, *Bull. Soc. Chim. Fr.*, *IV.*, *11*: 401 (1912).
10. S. Ahmed and H. J. Evans, *Biochem. Biophys. Res. Commun.*, *1*: 271 (1959).
11. H. M. Reisenauer, *Nature*, *186*: 375 (1960).
12. H. Evans, Proceedings of the Cocoa Conference, London, pp. 20 (1950).
13. S. B. Wilson and D. J. D. Nicholas, *Phytochemistry*, *6*: 1057 (1967).
14. C. J. Asher, P. G. Ozanne, and J. F. Loneragen, *Soil Sci.*, *100*: 149 (1965).
15. C. J. Asher and P. G. Ozanne, *Soil Sci.*, *103*: 155 (1967).
16. C. J. Asher and J. F. Loneragen, *Soil Sci.*, *103*: 225 (1967).
17. A. Ulrich and F. J. Hills, *Soil Testing and Plant Analysis*, Part II, *Plant Analysis*, Soil Science Society of America, Madison, WI (1967).
18. E. J. Hewitt, *Sand and Water Culture Methods Used in Plant Nutrition*, Commonwealth Agricultural Bureaux Press, Buckinghamshire, England (1952).
19. G. Michael, H. Schumacher, and H. Marschner, *Z. Pflanzennahr. Dueng. Bodenkd.*, *110*: 225 (1965).
20. R. Brouwer, *Jaarb. IBS*, *22*: 31 (1963).
21. F. E. Sanders and P. B. Tinker, *Pestic. Sci.*, *4*: 385 (1973).
22. R. A. Fischer and T. C. Hsiao, *Plant Physiol.*, *43*: 1953 (1968).
23. R. Raschke, *Annu. Rev. Plant Physiol.*, *26*: 309 (1975).
24. H. Brag, *Physiol. Plant*, *26*: 250 (1972).
25. D. A. Baker and P. E. Weatherly, *J. Exp. Bot.*, *20*: 485 (1969).
26. K. Mengel and R. Pfluger, *Physiol. Plant*, *22*: 840 (1969).
27. H. G. Burstrom, *Biol. Rev.*, *43*: 287 (1968).
28. G. Michael and H. Marschner, *Z. Pflanzennahr. Dueng. Bodenkd.*, *96*: 200 (1962).
29. N. Lazaroff and M. G. Pitman, *Aust. J. Biol. Sci.*, *19*: 991 (1966).
30. K. Wignarajah, T. Lundborg, T. Bjorkman, and A. Kylin, *Oikos*, *40*: 6 (1983).
31. D. E. Peaslee and D. N. Moss, *Soil Sci. Soc. Am.*, *30*: 220 (1966).
32. D. A. Walker, *International Review of Science*, Plant Biochemistry Series I, Vol. II (D. H. Northcote, ed.), Butterworths, London (1974).
33. J. C. Brown, *Adv. Agron.*, *13*: 329 (1961).
34. M. G. Weiss, *Genetics*, *28*: 253 (1943).
35. J. C. Brown, R. S. Holmes, and L. O. Tiffin, *Soil Sci.*, *86*: 75 (1958).
36. J. C. Brown, C. R. Weber, and B. E. Caldwell, *Agron. J.*, *59*: 459 (1962).
37. M. Spector and G. D. Winget, *Proc. Natl. Acad. Sci., U.S.A.*, *77*: 957 (1980).
38. R. F. Farley and A. P. Draycott, *J. Sci. Food Agric.*, *27*: 991 (1976).
39. F. N. Ponnamperuma, *Adv. Agron.*, *24*: 29–96 (1972).
40. R. K. Schofield, *Soils Fertil.*, *28*: 373 (1955).

41. R. W. Hendler, *Physiol. Rev.*, *51*: 66 (1971).
42. T. H. van den Honert, *Natur. Tijdschr. Ned. Ind.*, *97*: 150 (1937).
43. R. M. Spanswick, *J. Membrane Biol.*, *2*: 59 (1970).
44. G. D. Williams, T. J. Mosher, and M. B. Smith, *Anal. Biochem.*, *214*: 458 (1993).
45. R. P. Kraut, A. H. Greenberg, E. J. Cragoe, Jr., and R. Bose, *Anal. Biochem.*, *214*: 413 (1993).
46. G. E. Briggs, *New Phytol.*, *56*: 305 (1957).
47. G. E. Briggs and R. N. Robertson, *Annu. Rev. Plant Physiol.*, *8*: 11 (1957).
48. G. E. Briggs, A. B. Hope, and M. G. Pitman, *J. Exp. Bot.*, *9*: 128 (1958).
49. E. Epstein and J. E. Leggett, *Am. J. Bot.*, *41*: 785 (1962).
50. A. D. M. Glass, *Plant Physiol.*, *58*: 33 (1976).
51. H. Lundegardh and H. Burstrom, *Planta*, *18*: 683 (1935).
52. R. N. Robertson and M. J. Wilkins, *Nature*, *161*: 101 (1948).
53. R. N. Robertson and M. J. Wilkins, *Aust. J. Sci. Res.*, *B1*: 17 (1948).
54. G. G. Laties, *Survey Biol. Prog.*, *3*: 215 (1957).
55. D. T. Clarkson, *The Molecular Biology of Plant Cells* (H. Smith, ed.), Blackwell, Edinburgh (1977).
56. R. E. Cleland, "Integration of activity in the higher plant" (D. H. Jennings, ed.), *31st Symposium of the Society for Experimental Biology*, Cambridge Press, Cambridge (1977).
57. K. Wignarajah, T. Lundborg, T. Björkman, and A. Kylin, *Oikos*, *40*: 6 (1983).
58. H. Sentenac, N. Bonneaud, M. Minet, F. Lacroute, J.-M. Salmon, F. Gaymard, and C. Grignon, *Science*, *256*: 663 (1992).
59. D. A. Barber, G. D. Bowen, and A. D. Rovira, *Aust. J. Plant Physiol.*, *3*: 801 (1976).
60. A. D. M. Glass, *Plant Physiol.*, *51*: 1037 (1973).
61. R. L. Starkey, *Soil Sci.*, *27*: 319 (1927).
62. J. Harley and D. E. Smith, *Mycorrhizal Symbiosis*, Academic Press, London (1983).
63. E. Malavolta, F. A. Haag, F. A. F. Mello, and M. O. C. Brasil Sobra, *On the Mineral Nutrition of Some Tropical Crops*, International Potash Institute, Berne, Switzerland (1962).
64. H. B. Sprague (ed.), *Hunger Signs in Crops*, 3rd ed., McKay, New York (1964).
65. N. L. R. Van Eysinga and K. W. Smilde, *Nutritional Disorders in Glasshouse Lettuce*, Center for Agricultural Publishing and Documentation, Wageningen, Netherlands (1981).
66. A. Scaife and M. Turner, *Diagnosis of Mineral Disorders in Plants* (*A Colour Atlas and Guide*). Her Majesty's Stationery Office, London, England (1983).
67. C. J. Molyneux, *A Practical Guide to NFT*, Nutriculture Ltd., England (1988).
68. D. Bubenheim, *Waste Manage. Res.*, *9*: 435 (1991).
69. K. Mengel and E. A. Kirby, *Principles of Plant Nutrition*, International Potash Institute, Berne, Switzerland (1978).

10

Physiological and Growth Responses to Atmospheric Carbon Dioxide Concentration

David W. Wolfe
Cornell University, Ithaca, New York

I. ATMOSPHERIC CARBON DIOXIDE CONCENTRATION

A. Historical and Projected CO_2 Levels

During the past century atmospheric concentration of carbon dioxide has increased exponentially from about 280 μmol mol^{-1} in 1850 to present levels, which exceed 350 μmol mol^{-1} [1]. Analysis of air bubbles trapped in polar ice indicates that prior to this most recent period, CO_2 levels had been relatively stable between 270 and 280 μmol mol^{-1} for more than a thousand years (Figure 1), and had remained within the range of 200–300 μmol mol^{-1} for 150,000 years before that. Some of the initial increase in CO_2 that occurred in the early to mid-1800s was due to land clearing and destruction of temperate forest regions, but most of the dramatic rise since 1900 is attributed to the Industrial Revolution and increased burning of fossil fuels [1]. More recently, an unprecedented rate of deforestation tropical regions is contributing significantly to carbon loading of the atmosphere [2,3]. Deforestation is of particular concern because it not only leads to massive release of CO_2 into the atmosphere when vegetation is destroyed and subsequently burned or decomposed, but it also reduces the amount of terrestrial photosynthesis, a potential sink for excess atmospheric CO_2.

A measurement station utilizing infrared gas analyzers to accurately monitor atmospheric CO_2 concentration was established in 1958 at a site 11,200 feet (3414 m) above sea level in Mauna Loa, Hawaii. Data collected at this and other similar sites indicate that CO_2 levels are increasing at a rate of about 1.8 μmol mol^{-1}, or 0.5%, per year. The Intergovernment Panel on Climate Change (IPCC) sponsored by the United Nations has projected in their "business as usual" scenario that atmospheric CO_2 will double from preindustrial levels by the middle of the next century [3]. Their analysis also indicates that because of the long residence time of CO_2 in the atmosphere, even a halt to deforestation and implementation of worldwide emission controls and/or use of alternative energy sources may not lead to any significant reversal of the trend within the next century.

The focus of this chapter is the direct effects of CO_2 on plants, but the increase in CO_2

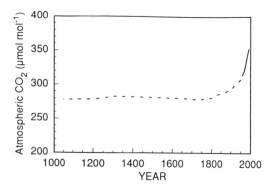

Figure 1 Atmospheric CO_2 concentration for the past thousand years based on ice core data (dashed line) and direct continual measurements made at the Mauna Loa Observatory in Hawaii since 1958 (solid line). (Adapted from Ref. 1.)

and other so-called greenhouse gases has received worldwide attention because it is likely to lead to a gradual warming of the earth's surface temperature. Such a trend could have substantial effects on regional weather patterns and global hydrology (e.g., rainfall, melting of polar ice), and it might affect natural vegetation and agriculture. The earth's surface temperature has increased by 0.6°C during the last 130 years [4], and global climate models predict that mean global temperatures will increase by somewhere between 1.5 and 4.5°C with an equivalent doubling of CO_2 [5].

B. The Role of Plants in the Global Carbon Balance

A simplified carbon cycle diagram, showing estimates of major active pools and fluxes, is presented in Figure 2. The annual net flux of carbon into the atmosphere from deforestation (and other land use changes) is currently estimated to be about 2 gigatons (Gt = 1 billion metric

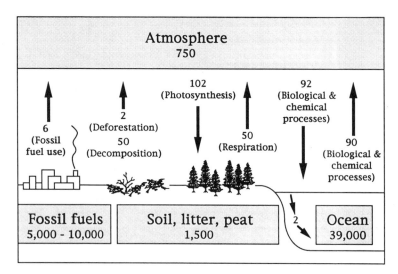

Figure 2 The global carbon cycle showing estimates of major carbon pools and annual carbon flux rates in gigatons (1 Gt = 1 billion metric tons). These values will shift with time and will also be modified as techniques for estimation are improved. (Adapted from Ref. 5.)

tons). Fossil fuel emissions contribute another 6 Gt of carbon per year. Plants play a major role in the global carbon balance. Photosynthetic activity of terrestrial and aquatic plants removes carbon from the atmosphere, and plant respiration and decomposition of organic matter release carbon back into the atmosphere. In the oceans, solid and dissolved carbonates equilibrate with CO_2, thereby buffering to some extent atmospheric CO_2 changes.

A comparison of flux rates in Figure 2 shows a carbon loading into the atmosphere of 3–4 Gt per year. This is more than can be accounted for by the current rates of increase in atmospheric CO_2 concentrations. In other words, our best estimates do not balance. Tans et al. [6] suggested that the unaccounted for carbon is being sequestered by terrestrial ecosystems, particularly in the Northern Hemisphere. A recent field study [7] provides evidence to support the notion that the net carbon uptake by decidous temperate forests has been underestimated in global carbon studies. Some have speculated that as the atmospheric CO_2 concentration increases, the carbon sink associated with photosynthesis will increase because of the "CO_2 fertilization effect" on plants [8]. However, as discussed in more depth later in this chapter, most evidence for the stimulatory effect of high CO_2 on net photosynthesis is based on short-term controlled environment experiments, the results of which are difficult to extrapolate to conditions of long-term high CO_2 exposure and a nonoptimum natural environment.

C. CO2 Gradients Within the Plant Canopy

A plant's photosynthetic and physiological response to CO_2 will be related to the CO_2 concentration of the air immediately surrounding the leaf surface, and this concentration is likely to differ from atmospheric concentrations measured outside the plant canopy. The depletion of CO_2 within plant canopies during the active daytime photosynthetic period is well documented [9,10]. The magnitude of CO_2 depletion is determined by many factors, including plant height, leaf density, light penetration into the canopy, wind speed, and photosynthetic capacity of individual leaves. Lemon [10] measured a 5–10% decrease in CO_2 concentration in the middle of a short canopy compared to the concentration above the canopy, and a 25% decrease has been observed in corn on a still day [9]. In the dark period, CO_2 concentrations within the canopy are similar to or slightly higher than those above the canopy, as a result of dark respiration by the plant and CO_2 efflux from the soil associated with root respiration and soil microbial activity.

In the absence of adequate mixing with outside air, or the addition of CO_2, plant photosynthetic activity during the daytime period in greenhouses can cause depletion of CO_2 to levels as low as 200–250 $\mu mol\ mol^{-1}$ [11].

II. STOMATAL RESPONSE TO CO2

A. CO2 Effects on Stomatal Aperture

That stomates respond to changes in CO_2 concentration was first reported almost 100 years ago [12], but the specific mechanisms of this response are still poorly understood. It has been well documented that in general, low CO_2 concentrations enhance stomatal opening, and increases in CO_2 concentration cause closure [13]. Recent evidence has confirmed the long-held assumption that the ambient CO_2 concentration (C_a) affects stomatal behavior only indirectly by its effect on the "intercellular" CO_2 concentration (C_i) within the substomatal cavity [14]. The three most important factors determining C_i at any point in time are C_a; the stomatal conductance (g_s), which controls the flux of CO_2 into the intercellular air space; and the rate of CO_2 assimilation by photosynthetic enzymes, which controls the flux of CO_2 from the intercellular air space into the mesophyll cells. Thus, C_i often increases as C_a increases, but the pattern of change in C_i depends on photosynthetic activity and the feedback effect associated with the stomatal closure response to increases in C_i, as well as C_a.

Various mechanisms have been proposed to explain how guard cells might perceive and respond to increases in C_i. One hypothesis is that the malate level in the guard cell cytoplasm reflects the level of CO_2 in the substomatal cavity, and that a high level of malate causes ion leakage from the vacuoles by reduction of the semipermeability of tonoplast and plasmalemma [13]. Ion leakage from the guard cells would then lead to loss of turgor and partial stomatal closure. This and other proposed mechanisms—reviewed by Mott [15]—have not been confirmed experimentally, however.

The magnitude of the sensitivity of the stomates to CO_2 varies with species and with environmental conditions. Typically, g_s will decrease by 30–40% when plants are exposed to a short-term doubling of CO_2 [15,16]. This has been observed in plant species with the C_4 and crassulacean acid metabolism (CAM) photosynthetic pathways, as well as the more prevalent C_3 species. There have been some notable exceptions. Complete insensitivity to CO_2 has been reported for Scotch pine [17] and other tree species [18].

It should be noted that most of our information is based on short-term experiments with plants grown at current CO_2 levels. We know relatively little about the stomatal behavior of plants acclimated to a high CO_2 environment, and so caution should be exercised when attempting to extrapolate to future CO_2 level scenarios. Stanghellini and Bunce [19] compared the stomatal response to CO_2 of plants raised at the current ambient CO_2 concentration (350 μmol mol^{-1}) to plants raised at high CO_2 (700 μmol mol^{-1}). The g_s values of the two groups of plants were similar when measured at 350 μmol mol^{-1} CO_2, but as the concentration was increased to 700 and 1000 μmol mol^{-1} CO_2, the g_s of plants raised at current CO_2 levels decreased significantly while the g_s of plants grown at high levels of CO_2 was not affected. These findings suggest that results based on the short-term response of plants grown at 350 μmol mol^{-1} CO_2 may overestimate the decline in g_s that would occur in a future environment high in CO_2. More research on acclimation of stomatal sensitivity to CO_2 is needed.

B. CO_2 Effects on Stomatal Density

Experiments have shown that CO_2 not only affects stomatal behavior, but can affect stomatal densities, with fewer stomates per unit leaf area in plants grown at increased CO_2 levels [20]. In addition to direct experimental evidence, these effects have been demonstrated by comparing stomate counts on given species in relation to elevation (CO_2 partial pressure decreases with increasing elevation), and by comparison of leaf specimens sampled and preserved before the Industrial Revolution with leaves of the same species growing under current ambient CO_2 concentrations [21]. Stomatal densities of the species examined have decreased by 40% over the past two centuries as CO_2 in the atmosphere has been increasing.

C.   Implications for Plant Water Use Efficiency

The stomatal closure response to high levels of CO_2 tends to increase the ratio of photosynthesis to transpiration, thus increasing water use efficiency (WUE) on a per unit leaf area basis. This is because the flux of water vapor from the leaf is relatively more sensitive than photosynthesis to changes in stomatal resistance. The rate-limiting factor for photosynthesis is more often the resistance associated with CO_2 assimilation into mesophyll cells rather than stomatal aperture. Sometimes an increase in atmospheric CO_2 concentration also causes an increase in photosynthetic rate—the "CO_2 fertilization effect" (Section III)—and this will increase WUE even further. In many short-term experiments in which other environmental conditions were maintained at optimum levels, a doubling of CO_2 decreased transpiration by 25–50% and increased WUE per unit leaf area as much as twofold [22,23].

The WUE response to CO_2 will vary among leaves of different ages and location within the plant canopy, and will change with time as the plant acclimates to changes in ambient CO_2. The effect of CO_2 on the short-term WUE response of individual leaves is usually much greater than the effect on WUE at the whole-plant level (defined as the ratio of dry weight produced to the amount of water transpired). One of the few well-documented examples of this was reported by Nijs et al. [24], who measured an 87% increase in WUE of individual leaves of CO_2-enriched plants, but found that WUE on a whole-plant basis was increased by only 25%.

D. Implications for Whole-Plant and Regional Water Requirements

Despite the many reports of increases in WUE in controlled environment experiments, it remains difficult to predict the effect of CO_2 on actual water use by plants in natural conditions. Leaf area production is often stimulated when plants are grown at increased CO_2 levels (Section IV.A), and this effect can completely counteract a reduction in transpiration *per unit leaf area,* with the result that water use by the plant is not reduced or is even increased at high levels of CO_2. Morison and Gifford [25], who compared 18 plant species grown at normal and twice-present atmospheric CO_2 concentrations, found that while WUE increased from 40 to 80% at high CO_2 levels, the rate of water use per plant was similar between treatments because of greater leaf area of plants grown at the higher CO_2 concentration.

Several factors associated with the physical dynamics of energy exchange between plants and the environment also create feedback effects that can counteract the reduction in transpiration when stomates close at high CO_2 levels. For example, lower g_s will cause an increase in leaf temperature because there is less water transpired at the leaf surface to cool it. This higher temperature increases the water vapor pressure of air within the leaf, causing an *increase* in transpiration rate. A second feedback effect occurs when transpiration decreases because of stomatal closure and the air surrounding the leaves becomes drier: both the leaf-to-air vapor pressure deficit and the subsequent transpiration potential increase. Jarvis and McNaughton [26] suggested that based on these factors alone, transpiration on a regional basis may not be significantly reduced by elevated CO_2. It is important to note that in many of the controlled environment experiments reporting substantial gains in WUE with increased CO_2, the temperature and the humidity of the air surrounding the leaves were maintained constant, so that these feedback effects were not operative.

A recent analysis of stomatal control of transpiration on a regional basis [27] concluded that scaling up from leaf to region "leads to an increase in number of negative feedback paths that stabilize the system and diminish the sensitivity of transpiration to change in stomatal conductance." This analysis showed that daily transpiration at the regional scale is particularly insensitive to changes in canopy conductance when that conductance is larger than 20 mm s^{-1}. This finding implies that CO_2 effects on g_s may be less important in well-watered agricultural canopies and pastures, where canopy conductance is often above this value, than for forest or rangeland regions, where canopy conductances are frequently less than 20 mm s^{-1}, even when well-supplied with water.

III. CO_2 EFFECTS ON PHOTOSYNTHESIS AND RESPIRATION

A. Short-Term Response

Most experiments in which plants were grown at current ambient CO_2 concentrations (e.g., 350 μmol mol^{-1}) and then measured for their instantaneous photosynthetic response to a doubling of CO_2 have found a significant positive effect from the increase in CO_2. It is important to emphasize that the vast majority of such studies have focused on the short-term response under optimum conditions (e.g., saturating light level, optimum temperature, water and nutrients nonlimiting). Within this experimental context, it is not unusual to observe photosynthetic increases of 30–50% or more in plants with the C_3 photosynthetic pathway when CO_2 is doubled, and increases of 5–15% in plants with the C_4 and CAM photosynthetic pathways. For a review, see Cure and Acock [22].

Atmospheric CO_2 concentration affects photosynthesis by several mechanisms. First, more CO_2 will tend to enter the leaves of plants exposed to high external CO_2 levels simply by mass action and the increased CO_2 gradient between leaf and air. This, however, is of relatively minor significance compared to the influence of CO_2 on the primary photosynthetic enzyme of C_3 plants, namely, ribulose bisphosphate carboxylase-oxygenase (Rubisco). Both CO_2 and O_2 are substrates of Rubisco, and so the two molecules compete for sites on the enzyme. The presence of CO_2 activates the enzyme, and CO_2 fixation initiates the photosynthetic carbon reduction cycle, whereas O_2 fixation initiates the photorespiratory carbon oxidation cycle. An increase in the atmospheric ratio of CO_2 to O_2 thus favors photosynthesis and inhibits photorespiration. The net carbon gain increases as CO_2 increases, not only because more carbon is assimilated, but also because photorespiratory carbon loss is reduced. Under normal ambient CO_2 levels, photorespiration of C_3 plants may cause the loss of as much as 50% of their recently fixed carbon [28].

Photorespiration losses are often negligible for C_4 and CAM plants because they have biochemical and morphological mechanisms for concentrating CO_2 internally. Therefore, these species shown relatively little stimulation of photosynthesis when external CO_2 concentrations are increased. Figure 3 illustrates the distinction between C_3 and C_4 plants with regard to their short-term photosynthetic response to CO_2. The C_4 plants have a greater photosynthetic efficiency at current CO_2 levels, but C_3 plants tend to be more responsive to a CO_2 doubling.

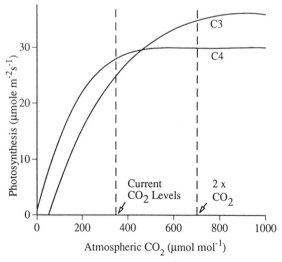

Figure 3 Net photosynthesis per unit leaf area in relation to atmospheric CO_2 concentration. Plants with the C_3 photosynthetic pathway typically show a greater relative benefit from a doubling of CO_2 than C_4 plants. These curves [103] illustrate the short-term response of plants grown under optimum conditions at an ambient CO_2 concentration of 350 μmol mol^{-1}. Figure 4 gives the long-term response of plants acclimated to high levels of CO_2.

B.   Photosynthetic Acclimation to Increased CO$_2$

The initial stimulatory effect of CO$_2$ enrichment on C$_3$ photosynthesis is seldom maintained, at least in container-grown plants, when plants are exposed to high CO$_2$ for a prolonged period (days or weeks). Cure and Acock [22] reviewed the published literature for several important C$_3$ agronomic species and found that an average 52% increase in net photosynthetic rate occurred with initial exposure to a CO$_2$ doubling; the average increase was less than 30%, however, when plants were acclimated to high levels of CO$_2$ for a week or more prior to measurement. Experiments with tomato [29] and monoecious cucumber [30] found that after acclimation to enhanced CO$_2$ for several weeks, an initial 30–40% stimulation declined to less than 10%.

It is not unusual to observe a reduced photosynthetic capacity in plants grown at high levels of CO$_2$ when comparisons are made at normal ambient CO$_2$ concentrations. This effect has been reported for bean [31], cotton [32,33], rice [34], and tropical and temperate tree species [20,35]. In some cases, photosynthetic rates of plants acclimated to high levels of CO$_2$ are lower than rates of control plants even when comparisons are made at the respective growth concentrations [35–38]. These and other photosynthetic acclimation responses reported in the literature are illustrated in Figure 4 as CO$_2$ assimilation versus C_i, or "A/C_i," curves. The most common responses for container-grown plants are those shown in Figures 4A and 4B, and to a lesser extent, 4C. No negative effect from prolonged exposure to high levels of CO$_2$ (Figure 4D) is based on data for potato [37], an anomaly that may be related to the large sink capacity of this species (see Section III.-B. 1, below).

When CO$_2$-enriched plants are grown in the field, with unrestricted rooting volume and adequate water and nutrients, photosynthetic capacity is more likely to be maintained, and sometimes increases, as illustrated in Figures 4E and 4F. Arp [39] and Sage [40] found evidence in the literature to suggest that the small containers used in most greenhouse and growth chamber studies create an experimental artifact that negatively affects the photosynthetic response to CO$_2$ enrichment by limiting root growth and activity. For example, photosynthetic enhancement was sustained in field-grown, CO$_2$-enriched cotton [41] and soybean [42], while photosynthetic capacity was reduced in the same species when grown in small pots at high CO$_2$ concentrations [32,43]. However, an experiment specifically designed to examine this issue [44] found that there was no "pot size effect" on CO$_2$-induced growth enhancement when nutrient concentrations were maintained at high levels.

1. Source–Sink Balance and Photosynthetic Acclimation

Recent reviews of the CO$_2$ enrichment literature [39,45] have concluded that the long-term response to exposure to high levels of CO$_2$ depends in large measure on a plant's ability to develop new sinks, accelerate the growth of existing sinks, and/or develop temporary storage capacity for excess carbohydrates. The extent to which a plant can expand its sink or storage capacity is dependent on both genetic factors (e.g., determinant vs. indeterminant growth habit) and environmental factors that may limit growth potential (e.g., temperature, or water and nutrient availability). Limitations in expansion of sink capacity lead to the frequently observed downregulation of photosynthesis that occurs at elevated CO$_2$ concentrations and may also explain the container size effect discussed above. Other mechanisms, however, were suggested by Sage [40], such as hormonal signals from roots growing in a restricted volume, or limited nutrient supply in small containers.

Perhaps the clearest evidence that the source–sink balance can determine the magnitude of photosyntheic acclimation to high levels of CO$_2$ comes from experiments in which this balance was artificially manipulated. Clough et al. [36] found that limiting the number of pods on soybean plants grown at high CO$_2$ levels exacerbated the reduction in photosynthetic capacity.

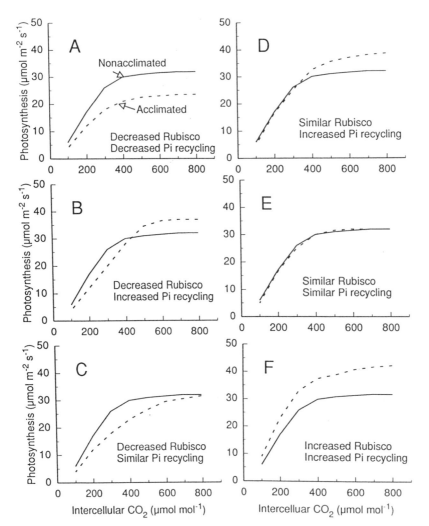

Figure 4 Representation of the range of photosynthetic responses to intercellular CO_2 ("A/C_i curves") that have been observed in C_3 plants acclimated to a CO_2-enriched atmosphere versus nonacclimated plants grown at normal CO_2 levels. For reference, atmospheric CO_2 concentrations of 350 and 700 μmol mol^{-1} typically result in C_i values of 200–250 and 500–550 μmol mol^{-1}, respectively. Shifts in photosynthetic capacity are attributed either to changes in Rubisco content or activity or to changes in P_i recycling, as indicated: (A) Decreased Rubisco, decreased P_i recycling; (B) decreased Rubisco, increased P_i recycling; (C) decreased Rubisco, similar P_i recycling; (D) similar Rubisco, increased P_i recycling; (E) similar Rubisco, similar P_i recycling; and (F) increased Rubisco, increased P_i recycling. (Adapted from Ref. 40.)

Peet [43] reduced the carbohydrate source by trimming leaves from soybean plants, and this alleviated the reduction in photosynthetic capacity of remaining leaves on plants grown in high levels of CO_2.

Kramer [46] suggested that indeterminate species, which continue producing new branches, leaves, and fruit throughout their life cycle, will tend to respond more positively to increased CO_2 than determinate types, which have a more limited sink capacity. The lack of downregulation of photosynthesis in citrus grown at high levels of CO_2 was attributed in part to an indeterminate growth habit [47]. Studies with cucumber [30] suggest that within this species, gyneocious varieties that have a higher fruit load (i.e., greater sink demand) than moneocious varieties show a more positive response to CO_2 enrichment.

2. Biochemical Aspects of Photosynthetic Acclimation

An accumulation of starch or other carbohydrates in the leaves, seen in almost all studies of plant growth at elevated levels of CO_2 [45,48], is reversible when plants are removed from the environment high in CO_2 [32,49]. In some extreme cases, enlargement of starch granules can cause physical damage within the chloroplasts and reduce subsequent photosynthetic capacity [50–52], but this is probably not a common mechanism of feedback inhibition at high CO_2 levels [39].

Photosynthetic acclimation to high CO_2 levels is associated in most cases with a reduction in the amount or activity of Rubisco [45,53]. The evidence for this includes not only direct measurements of Rubisco levels and kinetic properties, but also leaf gas exchange data. A decrease in the initial slope of an A/C_i curve reflects a Rubisco limitation to photosynthesis [54], and this effect is frequently observed in container-grown plants acclimated to a high CO_2 environment (Figures 4A–C).

A direct end-product inhibition of photosynthesis occurs when the early photosynthesis products, triose phosphates, are not utilized at the rate at which they are being produced in the Calvin cycle. Under these circumstances the supply of inorganic phosphate (P_i) for photophosphorylation becomes inadequate to meet the needs of the photosynthetic process. Such a limitation imposed by inadequate triose phosphate utilization and P_i recycling is identified experimentally when net assimilation becomes insensitive to changes in CO_2 (Figure 4A) or O_2 concentrations [55]. Such insensitivity has been reported for bean [56] and cotton [57] plants grown at elevated levels of CO_2. Sitt [45] suggested that P_i limitations may be an initial, transitory response to source–sink imbalance at high CO_2 levels, but reduced enzyme levels of reductions in enzyme activity are a more likely long-term acclimation mechanism.

The principal factor(s) limiting assimilation in plants grown at high CO_2 levels will depend on other environmental conditions. For example, P_i limitations become more significant at low temperatures because growth and sink demand for photosynthetic products are reduced [55,58]. In contrast, when environmental conditions that stimulate sink activity, such as increased temperature, occur in conjunction with elevated CO_2 levels, ribulose bisphosphate (RuBP) regeneration capacity may be the factor that limits assimilation [59].

We know that the overall process of photosynthesis is highly regulated and that the potential limiting factors are not mutually exclusive. As one process becomes limiting, other processes adjust and are brought into balance. We also know that Rubisco is not the only Calvin cycle enzyme affected by CO_2. Decreases of carbonic anhydrase [60] and NADP-glyceraldehyde 3-phosphatedehydrogenase [61] can occur in plants grown at high CO_2 levels. The activity of sucrose phosphate synthetase, a key enzyme in the rate control of sucrose synthesis, is negatively correlated with the increase in leaf starch content of CO_2-enriched bean [56] and cucumber [30].

C. Dark Respiration

Very few CO_2 enrichment studies have included measurements of dark respiration, despite the importance of this process on whole-plant carbon balance and growth response. Poorter et al. [62] evaluated several of the reports that are available and found that on average in plants grown at elevated levels of CO_2, leaf respiration per unit leaf area increased 16%, while leaf respiration per unit leaf weight decreased about 14%. More research will be needed before any generalizations can be made regarding CO_2 effects on dark respiration.

IV.   MORPHOLOGICAL AND GROWTH RESPONSES TO CO_2

Ultimately, CO_2 effects on plant processes at the biochemical and cellular levels manifest themselves as shifts in biomass accumulation, biomass partitioning, and plant morphology. These shifts have a substantial impact on subsequent growth and yield potential. For example, in terms of whole-plant growth, the amount of biomass allocated to new leaf area development can easily overwhelm any influence of CO_2 on photosynthetic rates per unit leaf area. A lack of correlation between photosynthesis and yield has long been recognized by plant breeders [63].

For agricultural species, increased CO_2 effects on the proportion of total biomass partitioned to the harvested portion is as important as the total amount of biomass produced. Ecologists studying the potential impact on natural vegetation will be interested in the influence of increased CO_2 on factors determining reproductive success and the ability of plants to survive under fluctuating and sometimes harsh environments.

A.   Leaf Area Development

An increase in leaf area, usually associated with increased branching or tillering, has been observed in many plant species when grown at high CO_2 levels and optimum conditions [64,65]. Since the increase in leaf area is often proportionally less than the increase in leaf weight, the ratio of the two—that is, specific leaf area (SLA)—decreases. The reduction in SLA is, in part, associated with an accumulation of starch and sugars in the leaves at high CO_2 levels, but increases in structural dry matter, as well as carbohydrates, contribute to the leaf weight increase [64]. Anatomical studies of soybean leaves from plants grown at high CO_2 showed that the leaves are thicker, have more densely packed mesophyll cells, and often have one or two extra layers of palisade cells [49].

B.   Root/Shoot Ratio and Harvest Index

Increased partitioning to roots at high CO_2 levels is sometimes observed in herbaceous species, but this response has been consistent and pronounced in only root crops, such as sugar/beet [66] and carrot and radish [67], where CO_2 enrichment increases the root/shoot ratio by 30–50% or more. This effect may be explained by the large root sink for carbon in these crops, and less end-product inhibition of photosynthesis. Cereal crops show relatively little root/shoot response to CO_2 [22]. Temperate tree saplings show little response or a reduction in root/shoot ratio at high CO_2 levels; when nutrients are limiting growth, however, the ratio may increase regardless of CO_2 concentration [18]. Arp [39] pointed out that the small container size used in many CO_2 enrichment studies tends to reduce the root/shoot ratio, possibly countering the increase in partitioning to roots that would occur at elevated levels of CO_2 under field conditions. More research with unrestricted rooting volume will be required to clarify this issue.

The harvest index (the proportion of total biomass allocated to the harvested portion of the plant) usually increases with a CO_2 doubling in plants with underground storage organs [66,67], but most other species show little response. Harvest index may increase or decrease slightly, but as with the root/shoot ratio, results are not always consistent within a single species because of differences in experimental conditions.

C.   Feedback Effects of Partitioning Response to CO_2

When considering the long-term growth response in the field, shifts in biomass partitioning within the plant can have either a beneficial or a negative effect, depending on environmental circumstances. For example, an increase in leaf area due to high levels of CO_2 would be of obvious benefit in terms of carbon- and energy-gaining ability and could possibly compensate for the downregulation of photosynthesis phenomenon discussed in Section III.B. However, large plants with greater leaf area will develop water and nutrient deficits more quickly if these inputs are in limited supply.

As a second example of feedback effects at the whole-plant level, an increase in root/shoot ratio may result in better ability to extract water and nutrients for growth, but if this improvement occurs at the expense of reducing the amount of biomass allocated to the harvested portion of the plant, economic yield may be much less. Also, a higher root/shoot ratio increases the proportion of nonphotosynthetic tissue of the plant, thus reducing subsequent net carbon and energy gain. Another feedback effect associated with root growth was reported by Körner and Arnone [68], working in a greenhouse-simulated tropical ecosystem. They found that an increase in fibrous root turnover and root exudation under conditions of enhanced CO_2 stimulated soil microbial activity and caused greater CO_2 losses from the soil. Such complex effects of atmospheric CO_2 concentration on processes at the whole-plant and ecosystem levels will be difficult to predict without more research under field conditions.

D. Yield Response to CO₂ Under Optimum Conditions

1. Controlled Environments

The vast majority of our quantitative information regarding yield response to enhanced CO_2 is based on experiments in which CO_2 concentration alone was varied, while water, nutrients, temperature, and pest pressure were maintained near optimum for growth. Kimball [69] and Cure and Acock [22] reviewed this scientific literature and concluded that under these circumstances, a doubling of CO_2 from about 350 μmol mol^{-1} to 700 μmol mol^{-1} increases the productivity of C_3 and C_4 crop plants about 33 and 10%, respectively, on average.

Perhaps the best testimony of a positive CO_2 effect in controlled environments is the widespread use of supplemental CO_2 for flower and vegetable crop production in greenhouses. Wittwer [70] estimated that more than half the commercial greenhouse growers worldwide use some type of CO_2 enrichment for at least some of their crops. Yield increases of 30–40% for crops such as tomato, cucumber, strawberry, and some ornamentals are common when CO_2 concentrations are maintained near 1000 μmol mol^{-1} [70]. This type of response requires maintaining other environmental factors near the optimum for growth. It has been recognized for some time that an increase in water and fertilizer applications is necessary to obtain maximum CO_2 benefits in greenhouses [71].

There are several reasons to suspect that extrapolation of the greenhouse experience to the field situation may overestimate the potential benefits of a CO_2 doubling. First, particularly for natural and nonintensive agricultural ecosystems, adverse environmental conditions and weed and pest pressure can limit the CO_2 response (Section V). Second, many greenhouse crops have an indeterminate growth habit, which may make them less "sink limited" and more responsive to CO_2 enrichment than some field crops (Section III.B). A third factor to consider in evaluating the greenhouse data is that often the CO_2 concentration in "control" greenhouses (i.e., greenhouses without supplemental CO_2) is depleted to levels as low as 200–250 μmol mol^{-1} [11], and at this low range the slope of the photosynthetic response to CO_2 is greater than at CO_2 concentrations in the range of current atmospheric CO_2 concentrations (350 μmol mol^{-1}: see Figure 3).

2. Field Studies

Numerous field CO_2 enrichment studies have been conducted in recent years using open-top canopy chambers. Most of these studies were conducted with water and nutrients nonlimiting and with minimal pressure from weeds, disease, and insects. Under these conditions, many C_3 crops show yield increases from a CO_2 doubling similar to those reported in controlled environments, although results are much more variable in the field. This literature has been reviewed by Lawlor and Mitchell [72].

Results of field investigations with natural plant communities have been less conclusive. Within a Chesapeake wetland, dominated by a C_3 sedge species, there were sustained increases in photosynthetic capacity and carbon accumulation over a 4-year period characterized by elevated CO_2 levels [73]. In contrast, sustained increases in carbon accumulation and productivity were not observed in a CO_2-enriched Arctic tundra ecosystem, also dominated by a C_3 sedge [74]. A study with young poplar trees [75] and a long-term greenhouse experiment with a simulated tropical ecosystem [68] both reported sustained stimulation of photosynthesis with CO_2 enrichment, but no significant increase in aboveground biomass production.

V. INTERACTIONS BETWEEN CO_2 AND OTHER ENVIRONMENTAL VARIABLES

Most of what we know about plant response to CO_2 is based on experiments conducted under optimum conditions. Under field conditions, however, there is often considerable variation in temperature, light, water, nutrients, weed competition, and pest pressure, and these variations will affect plant physiology, growth, and response to CO_2. Elevated CO_2 levels may partially compensate for or exacerbate plant stresses associated with these other factors. It is therefore important that we consider the potential interaction between CO_2 and some key environmental variables.

A. Temperature

The interaction between temperature and photosynthetic response to elevated levels of CO_2 has been well documented. Generally, C_3 plants show little or no benefit from CO_2 enrichment at low temperatures (e.g., $<15°C$) because electron transport and enzyme activities are slowed, perhaps in part because triose phosphate utilization and P_i recycling are reduced at low temperatures. As temperatures rise within the 20–35°C range, the proportionate stimulation of net photosynthesis by high CO_2 levels increases rapidly, and the greatest benefit from a high level of CO_2 is usually observed at warm temperatures (30–40°C). This is because a high level of CO_2 inhibits photorespiration, and at warmer temperatures, photorespiratory carbon losses in the absence of CO_2 enrichment increase (Rubisco oxygenase activity is favored over carboxylase activity) [76]. At temperatures above about 40°C, the response to CO_2 may again become minimal as photosynthetic activity is slowed by the damaging effects of high temperature on organelle membranes and enzyme function.

The interaction between CO_2 and temperature will shift as plants acclimate to prolonged periods of exposure to a new CO_2 concentration or temperature. Leaf morphology and composition are altered by changes in either CO_2 [49] or temperature [77], but the possible implications of such modifications for photosynthetic response to CO_2 at various temperature regimes have not been examined. When plants are grown at high levels of CO_2 for a prolonged period, Rubisco activity may be reduced (Section III.B), and when this effect is combined with a reduction in metabolism induced by low temperature, CO_2-enriched plants can have a *lower* photosynthetic capacity than plants raised at normal CO_2 concentrations.

The $CO_2 \times$ temperature interaction is illustrated in the simulations of Figure 5, based on the photosynthesis model of Farquhar et al. [54] and modified by Long [78]. The benefit from CO_2 enrichment becomes very small at low temperatures, even assuming no reduction in Rubisco activity. In simulations incorporating the assumption that acclimation to high levels of CO_2

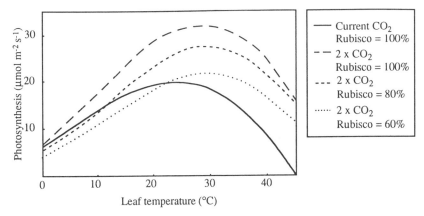

Figure 5 Computer simulations of net photosynthesis per unit leaf area in relation to leaf temperature, CO_2 concentration, and activity level of Rubisco. Reductions in Rubisco are sometimes observed after prolonged exposure to high CO_2. (Adapted from Ref. 78.)

reduces Rubisco activity by 20 and 40%, the photosynthetic rate of CO_2-enriched plants becomes lower than the control at temperatures of 12.5 and 22.5°C, respectively. It should also be noted that the temperature optimum for photosynthesis tends to increase at increased CO_2 concentrations. This effect may be beneficial in a future high CO_2 world, where air temperatures may be higher and leaf temperatures may increase because of the partial stomatal closure in response to increased CO_2.

Very few investigations have focused on the $CO_2 \times$ temperature interaction at the whole-plant level in terms of growth and yield. Little benefit from CO_2 enrichment was observed in natural tundra vegetation at a high latitude site with cool temperatures [74], whereas enrichment of warm wetland vegetation at a low latitude site produced significant increases in productivity [73]. Other factors, such as limited nutrient availability at the high latitude site, may also have been involved.

Field experiments by Idso et al. [79] used open-top canopy chambers to examine the effect of temperature on weekly growth rates of five plant species (carrot, cotton, radish, water fern, and water hyacinth) with and without a CO_2 doubling. The response to CO_2 enrichment ranged from a 60% *reduction* in growth at 12°C, to no effect at 18.5°C, to relative growth increases of 30% at 22°C and 100% at 30°C. When the study was repeated for carrot and radish [80], the threshold temperature for lack of positive CO_2 response was lower (about 12°C) for both crops, and the benefit at high temperatures was found to be less for radish than for carrot. Rawson [81] compiled data from several other sources and found less interaction between CO_2 and temperature. Some of these results are compared in Figure 6.

The specific slopes describing the growth response to CO_2 enrichment in relation to temperature regime will undoubtedly be modified as more field studies are completed, but our current understanding of the temperature $\times$ CO_2 interaction with regard to photosynthesis (Figure 5) suggests that growth response trends similar to those shown in Figure 6 are likely to be found for other plant species. Confirmations of the negative growth response to CO_2 enrichment at low temperatures by subsequent studies would have important implications for many temperate regions of the world. Even assuming that global climate warming will occur in conjunction with increases in atmospheric CO_2 concentration, the mean temperature in many temperate high latitude/high altitude locations will be less than 18°C for significant portions of the growing season.

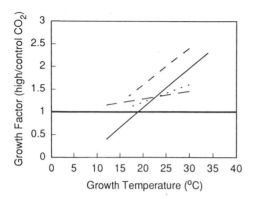

Figure 6 Growth response of plants to a doubling of atmospheric CO_2 concentration in relation to average air temperature. A growth factor of 1.0 indicates no response to elevated CO_2; values below or above 1.0 indicate negative or positive responses to CO_2, respectively. Results are based on combined data for five plant species (solid line) from Idso et al. [79]; data for carrot (medium dashes) and radish (dotted line) from Idso and Kimball [80]; and combined data for six plant species (long dashes) from Rawson [81].

B. Water

Major shifts in precipitation patterns are anticipated as the increases in CO_2 and other greenhouse gases cause warmer climates. For this reason, an understanding of the water $\times$ CO_2 interaction will be important. While most climate models predict a global annual increase in precipitation, conditions may be drier during the summer months for large portions of North and Central America, western Europe, central Asia, eastern Brazil, and north and western Africa [82]. Some climate researchers predict that droughts will be much more common in the next century [83].

The partial stomatal closure and improved water use efficiency of plants grown at elevated levels of CO_2 (Section II.C) should be beneficial under water-limited conditions. Experiments with bean and cotton [41,84] found that the stomates of CO_2-enriched plants were more sensitive to water deficits, closing more quickly or completely relative to photosynthetic capacity, than those of plants grown at normal CO_2 levels. However, as discussed in Section II.D, early leaf expansion is often stimulated in CO_2-enriched plants, and this effect can increase transpirational water losses and hasten the depletion of stored soil water despite a more efficient use of water per unit leaf area. When the increase in transpirational surface area exceeds the increase in water use efficiency, water deficits develop earlier or become more severe in environments that are high CO_2. This change in water deficit patterns has important implications for many natural ecosystems during the dry season, as well as for dryland farmers in developed and developing regions who rely completely on stored soil water without supplemental irrigation to raise their crops.

The results obtained by Gifford [85], working with wheat, showed that the relative yield increase from CO_2 enrichment increased with increasing water deficit, but in absolute terms, the beneficial CO_2 effect became negligible under severe drought because yields were very low regardless of CO_2 concentration. Kimball [86], who compared results from several water $\times$ CO_2 studies in which growth and/or yield were measured, found that most data corroborate the results of Gifford [85]. In absolute terms, the greatest benefit from CO_2 enrichment occurs when water is not limiting but, on a relative basis, the percent yield increase with CO_2 enrichment is as large or larger under mild to moderate water stress as it is under nonstress conditions.

C. Nutrients

A reduction in leaf nitrogen concentration combined with faster growth rates is often observed in CO_2-enriched plants raised under optimum conditions, and knowledge of this tendency has led to the suggestion that CO_2 enrichment improves nitrogen use efficiency [87]. However, Coleman et al. [88] pointed out that most nitrogen concentration comparisons of CO_2 treatments are made at a common time, and this would tend to highlight differences between bigger, more developed plants (grown at high CO_2 levels) and smaller, less developed plants (grown at low CO_2 levels). They found no effect of CO_2 enrichment on tissue nitrogen concentrations in *Amaranthus* and *Abutilon* species when comparisons were made at similar plant sizes. They also speculated that if nitrogen use efficiencies are not improved by CO_2 enrichment, plant growth in a future environment high in CO_2 may frequently be nitrogen limited.

Field experiments with cotton [89] and rice [90] found that on a relative basis, the response to CO_2 enrichment improved at low (vs. high) soil nitrogen levels. In contrast, a greenhouse study with cotton and maize [33] found that the response to CO_2 was reduced for both crops, on a relative basis and on an absolute basis, when soil nitrogen fertility was low. The growth response to CO_2 may be more sensitive to phosphorus and other nutrients than it is to nitrogen. Goudriaan and de Ruiter [91] did not observe any positive growth response to CO_2 enrichment in phosphorus-deficient plants, but some small beneficial effect from CO_2 was maintained when nitrogen alone was in short supply.

The mechanisms of interaction between CO_2 and plant nutrient status are not well understood, and we know almost nothing about the direct effects of CO_2 on soil microbial processes that affect nutrient availability. The field studies conducted by Allen [90] with rice found that CO_2 enrichment had a stimulatory effect on nitrogen-fixing bacteria in the soil, and this improved nitrogen nutrition and yield. In contrast, a greenhouse experiment that simulated a multispecies tropical ecosystem [68] found that increased soil microbial activity and increased turnover of fibrous roots in an elevated CO_2 environment led to a doubling of nitrogen, phosphorus, and potassium leached below the root zone, and a significant increase in CO_2 efflux from the soil. Clearly, more research on the effect of atmospheric CO_2 concentration on rhizosphere processes will be required to improve our estimates of the long-term effects on ecosystem carbon balance and productivity resulting from a CO_2 doubling.

D. Light

Light is frequently a limiting factor for photosynthesis and growth in the natural environment. Even for plants exposed to high solar radiation levels at midday, lower leaves are light-limited because of shading by upper leaves. It has been well documented that in absolute terms, the stimulatory effects of increased CO_2 concentrations on photosynthesis and dry weight gain are much less pronounced under low light than high light conditions [92]. Pearcy and Björkman [23] and Kimball [86] suggested that a decrease in the photosynthetic light compensation point, which may occur at elevated levels of CO_2, would result in a greater benefit from CO_2 at low light on a relative basis, even though the absolute size of the increase is very small at low light intensities.

Experiments examining the CO_2 × light interaction have had varying results. Acock et al. [93] did observe a decline in the light compensation point for CO_2-enriched tomatoes, and Gifford [94] showed a relatively greater stimulation of wheat growth from CO_2 enrichment at low compared to high light levels. However, Morgan [95] reported the reverse for tomato, with a greater percentage increase in yield at higher light intensities. Studies with soybeans [96] and bean [97] found a greater reduction in photosynthetic capacity of CO_2-enriched plants when they were grown at high light. Because of these conflicting results, and the paucity of data collected under field conditions, it is difficult to predict the growth response to CO_2 for specific light situations. Sage [98] developed a model describing some aspects of photosynthetic regulation under varying light and CO_2 regimes, but this model has not been tested rigorously or incorporated into whole-plant growth models used in climate change research.

E. Weed, Disease, and Insect Pests

An increase in atmospheric CO_2 concentrations will affect plant–plant interactions indirectly through its effects on plant growth and competition for other, more limiting, resources [99]. An increase in atmospheric CO_2 is as likely to stimulate the growth of weed species as it is crop species, but most agricultural CO_2 enrichment experiments have been conducted in a weed-free environment. Wittwer [65] optimistically speculated that rising CO_2 levels will generally favor crop production because the majority of our important food crops have the C_3 photosynthetic pathway, while a high percentage of the major weed species are C_4 plants that will likely benefit less from CO_2 enrichment. This simplistic analysis ignores the fact that not even a substantial increase in growth of C_3 crops would be sufficient to overcome the existing growth rate and competitive ability of some C_4 weeds. Wittwer's analysis also provides little comfort to the many commercial and subsistence farmers who grow C_4 crops, such as maize, millet, sorghum, and sugarcane. These farmers may suffer substantial yield losses from increased C_3 weed competition, particularly in areas where economic or environmental factors will limit the use of herbicides. The site-specific mix of weed and crop species, and the relative ability of these species to compete for limiting resources in the future CO_2-rich world, will determine the economic outcome for both farmers and consumers.

The interaction between CO_2 enrichment and insect pressure is another area in which our current research base is very small. Lincoln et al. [100] and Johnson and Lincoln [101] found that insect herbivores consumed greater amounts of leaf tissue from CO_2-enriched plants, presumably to compensate for the lower protein–nitrogen concentration in the leaves (Section V.C). Fajer et al. [102] found that lower nutritional quality of leaves from CO_2-enriched plants increased insect mortality rates but led to higher consumption rates of surviving insects. Wolfe and Eirckson [103] speculated that natural selection may tend to favor the evolution of insect genotypes that consume more plant material more rapidly.

Shifts in climate that will accompany the increase in CO_2 and other greenhouse gases will undoubtedly alter the geographic range of insect and disease pests. Crops and natural vegetation in some regions will suffer more frequent or more severe pest outbreaks, while other areas may benefit. Warmer temperatures at high latitudes will allow more insects to overwinter and populations to increase in these areas. Crop damage from fungal diseases could increase in regions of climate shifts from cool and dry to warm and humid.

VI. MODELING PLANT RESPONSE TO CO_2

The formulations describing the biochemistry of photosynthesis developed by Farquhar et al. [54] and Sage [98] provide a means of examining the direct and interactive effects of CO_2, light, and temperature on the photosynthesis of individual leaves. This mechanistic approach has not been incorporated into crop growth models currently being used in climate change research (for review, see Ref. 103). Instead, most crop modelers have used a CO_2 fertilizer effect "multiplier" that globally increases C_3 crop yields, often by as much as 30–35% with an equivalent CO_2 doubling, regardless of temperature, light, or other environmental conditions. The magnitude of the CO_2 fertilizer multiplier has been based almost exclusively on the publications by Kimball [69] and Cure and Acock [22], which consolidate results from a multitude of controlled environment experiments. Wolfe and Erickson [103] and Bazzaz and Fajer [104] argued that reliance on data from experiments conducted with water and nutrients nonlimiting, temperatures near optimum, and weed, disease, and insect pressures nonexistent may lead to overestimates of the benefits from CO$_2$ enrichment for many natural and agricultural ecosystems, where environmental conditions are seldom, if ever, optimum.

Two forest growth models, MAESTRO and BIOMASS [105], have been designed to incorporate the photosynthesis formulations of Farquhar et al. [54]. The simulations generated by these models show that for a clear day, with optimum water and nutrients, the magnitude of the increase in canopy photosynthesis when CO$_2$ is doubled is temperature dependent, with increases of 10, 45, and 70% at 10, 25, and 40°C, respectively. BIOMASS and MAESTRO have also been linked with a plant–soil model, G'DAY, to examine the feedbacks associated with rates of decomposition and nutrient cycling. Simulations show that on nutrient-limited sites, an initial 27% increase in productivity declines over a 10-year period to a sustained 8% increase in the long term. Expanded use of this nested modeling approach, which allows examination of CO$_2$ × environment interactive effects at a wide range of temporal and spatial scales, might improve our estimates of plant response to CO$_2$ and climate change.

VII. SUMMARY

Atmospheric CO$_2$ concentration has risen exponentially in the past century from about 280 μmol mol^{-1} in 1850 to the current level of about 350 μmol mol^{-1}. It is anticipated that this trend will continue, and atmospheric CO$_2$ concentration will likely double within the next century.

Most of our information regarding the physiological response of plants to CO$_2$ is based on short-term controlled environment experiments, where plants were well supplied with water and nutrients, temperatures were near optimum, and pressure from weeds, disease, and insect pests was nonexistent. Under these conditions, most C$_3$ plants, when raised at normal CO$_2$ levels and then exposed to a doubling of CO$_2$ (i.e., to 700 μmol mol^{-1}) show an initial increase in net photosynthesis of as much as 30–50%, a reduction in stomatal aperture, and improved WUE on a per unit leaf area basis. Photosynthesis of C$_4$ and CAM plants is much less affected by increases in atmospheric CO$_2$ because the CO$_2$ inhibition of photorespiration has relatively little impact on the net carbon gain of these species.

Longer term experiments have found that the initial stimulation of photosynthesis in C$_3$ plants is seldom maintained, and frequently when plants are grown at high CO$_2$ levels for prolonged periods, there is a reduction in Rubisco activity and photosynthetic capacity. Most evidence indicates that this lack of sustainability of increased photosynthetic rates is due to a negative feedback effect on photosynthesis that occurs when carbohydrate supply exceeds sink demand. A reduction in photosynthetic capacity with prolonged CO$_2$ exposure is less common in field-grown plants, and it has been suggested that the small container size used in many greenhouse and growth chamber CO$_2$ enrichment studies artificially limits root sink activity and photosynthetic response. Sink limitations on the long-term photosynthetic response to CO$_2$ may nevertheless be an important factor in some field situations: for example, when growth rate is constrained by low temperatures or nutrient deficiency, when environmental stress impairs reproductive development, or when dry or compacted soils limit root growth.

Reviews of the literature on the yield response to CO$_2$ indicate that under optimum growing conditions, a doubling of CO$_2$ increases the productivity of C$_3$ and C$_4$ crops about 33 and 10%, respectively. Maximum benefit from CO$_2$ enrichment usually requires an increase in fertilizer and other inputs. The improved WUE on a per unit leaf area basis in CO$_2$-enriched plants is often counterbalanced by a substantial increase in transpirational leaf area, with the result that water use per plant is unaffected or may be greater at higher CO_2 concentrations. At low temperatures (e.g., $<15°C$) there is little benefit from increased levels of CO_2, and some low temperature studies have reported slower growth of CO_2-enriched plants than control plants. An improvement in our estimates of the impact of increased atmospheric CO_2 concentration on the earth's plant life will require more research under field conditions on the interaction between plant response to CO_2 and other abiotic and biotic environmental factors.

REFERENCES

1. W. M. Post, T. H. Peng, W. R. Emanuel, A. W. King, V. H. Dale, and D. L. DeAngelis, *Am. Sci.*, *78*: 310 (1990).
2. D. D. Houghton, *Clim. Change*, *19*: 99 (1991).
3. R. T. Watson, L. G. Meira Filho, E. Sanhueza, and A. Janetos, *Climate Change: Supplementary Report to the IPCC Scientific Assessment* (J. T. Houghton, B. A. Callander, and S. K. Varney, eds.), Cambridge University Press, Cambridge, pp. 25–46 (1992).
4. J. D. Jones and T. M. Wigley, *Sci. Am.*, *263*: 84 (1990).
5. D. M. Gates, *Climate Change and Its Biological Consequences*, Sinauer, Sunderland, MA, p. 30 (1993).
6. P. P. Tans, I. Y. Fung, and T. Takahashi, *Science*, *247*: 1431 (1990).
7. S. C. Wofsy, M. L. Goulden, J. W. Munger, S.-M. Fan, P. S. Bakwin, B. C. Daube, S. L. Bassow, and F. A. Bazzaz, *Science*, *260*: 1314 (1993).
8. J. A. Taylor and J. Lloyd, *Aust. J. Bot.*, *40*: 407 (1992).
9. H. W. Chapman, L. S. Gleason, and W. E. Loomis, *Plant Physiol.*, *29*: 500 (1954).
10. E. Lemon, *Physiological Aspects of Crop Yield* (J. D. Eastin, ed.), American Society of Agronmy, Madison, WI, pp. 117–142 (1969).
11. K. L. Goldsberry, *Carbon Dioxide Enrichment of Greenhouse Crops* Vol. 2 (H. Z. Enoch and B. A. Kimball, eds.), CRC Press, Boca Raton, FL, Chapter 9 (1986).
12. F. Darwin, *Phil. Trans. R. Soc. London, B. Biol. Sci.*, *190*: 531–621 (1898).
13. K. Raschke, *Physiology of Movements*, Vol. 7, *Encyclopedia of Plant Physiology* (W. Haupt and M. E. Feinleib, eds.), Springer-Verlag, Berlin, p. 420 (1979).
14. K. A. Mott, *Plant Physiol.*, *86*: 200 (1988).
15. K. A. Mott, *Plant Cell Environ.*, *13*: 731 (1990).
16. J. L. Morison, *Stomatal Function* (E. Zeiger, G. D. Farquhar, and I. R. Cowan, eds.), Stanford University Press, Stanford, CA, pp. 229–252 (1987).
17. P. A. P. Ng, Ph.D. dissertation, University of Edinburgh (1978).
18. D. Eamus and P. G. Jarvis, *Adv. Ecol. Res.*, *19*: 1 (1989).
19. C. Stanghellini and J. A. Bunce, *Photosynthetica* (in press) (1994).
20. S. F. Oberbauer, B. R. Strain, and N. Fetcher, *Physiol. Plant.*, *65*: 352 (1985).
21. F. I. Woodward, *Nature*, *327*: 617 (1987).
22. J. D. Cure and B. Acock, *Agric. Forest Meteorol.*, *38*: 127 (1986).
23. R. W. Pearcy, and O. Björkman, *CO_2 and Plants* (E. R. Lemon, ed.), Westview Press, Boulder, CO, Chapter 4 (1983).
24. I. Nijs, I. Impens, and T. Behaeghe, *Planta*, *177*: 312 (1989).
25. J. L. Morison and R. M. Gifford, *Aust. J. Plant Physiol.*, *11*: 361 (1984).
26. P. G. Jarvis and K. G. McNaughton, *Adv. Ecol. Res.*, *15*: 1 (1986).
27. K. G. McNaughton and P. G. Jarvis, *Agric. Forest Meterol.*, *54*: 279–301 (1991).
28. N. E. Tolbert and I. Zelitch, *CO_2 and Plants* (E. R. Lemon, ed.), Westview Press, Boulder, CO, Chapter 3 (1983).
29. S. Yelle, R. C. Beeson, Jr., M. J. Trudel, and A. Gosselin, *Plant Physiol.*, *90*: 1473 (1989).
30. M. M. Peet, S. C. Huber, and D. T. Patterson, *Plant Physiol.*, *80*: 63 (1986).
31. F. X. Socias, H. Medrano, and T. D. Sharkey, *Plant Cell Environ.*, *16*: 81 (1993).
32. T. W. Sasek, E. H. DeLucia, and B. R. Strain, *Plant Physiol.*, *78*: 619 (1985).
33. S. C. Wong, *Oecologia*, *44*: 68 (1979).

34. J. T. Baker, L. H. Allen, K. J. Boote, P. Jones, and J. W. Jones, *Agron. J.*, *82*: 834 (1990).
35. P. Kaushal, J. M. Guehl, and G. Aussenac, *Can. J. Forest Res.*, *19*: 1351 (1989).
36. J. M. Clough, M. M. Peet, and P. J. Kramer, *Plant Physiol.*, *67*: 1007 (1981).
37. R. F. Sage, T. D. Sharkey, and J. R. Sieman, *Plant Physiol.*, *89*: 590 (1989).
38. T. F. Neales and A. O. Nicholls, *Aust. J. Plant Physiol.*, *5*: 45 (1978).
39. W. J. Arp, *Plant Cell Environ.*, *14*: 869 (1991).
40. R. F. Sage, *Photosynth. Res.*, (in press) (1994).
41. J. W. Radin, B. A. Kimball, D. L. Hendrix, and J. R. Mauney, *Photosynth. Res.*, *12*: 191 (1987).
42. W. J. Campbell, L. H. Allen, and G. Bowes, *Plant Physiol.*, *88*: 1300 (1988).
43. M. M. Peet, *Physiologia Plant.*, *60*: 38 (1984).
44. K. D. M. McConnaughay, G. M. Berntson, and F. A. Bazzaz, *Oecologia*, *94*: 550 (1993).
45. M. Stitt, *Plant Cell Environ.*, *14*: 741 (1991).
46. J. Kramer, *Bioscience*, *31*: 29 (1981).
47. S. B. Idso and B. A. Kimball, *Plant Physiol.*, *96*: 990 (1991).
48. J. F. Farrar and M. L. Williams, *Plant Cell Environ.*, *14*: 819 (1991).
49. G. Hofstra and J. D. Hesketh, *Environmental and Biological Control of Photosynthesis* (R. Marcelle, ed.), W. Junk, The Hague, pp. 71–80 (1975).
50. E. D. Nafziger and R. M. Koller, *Plant Physiol.*, *57*: 560 (1976).
51. E. H. DeLucia, T. W. Sasek, and B. R. Strain, *Photosynth. Res.*, *7*: 175 (1985).
52. A. Carmi and I. Shomer, *Ann. Bot.*, *44*: 497 (1979).
53. G. Bowes, *Plant Cell Environ.*, *14*: 795 (1991).
54. G. D. Farquhar, S. von Caemmerer, and J. A. Berry, *Planta*, *149*: 78 (1980).
55. T. D. Sharkey, *Plant Physiol.*, *78*: 71 (1985).
56. F. X. Socias, H. Medrano, and T. D. Sharkey, *Plant Cell Environ.*, *16*: 81 (1993).
57. P. C. Harley, R. B. Thomas, J. F. Reynolds, and B. R. Strain, *Plant Cell Environ.*, *15*: 271 (1992).
58. R. F. Sage and T. D. Sharkey, *Plant Physiol.*, *84*: 658 (1987).
59. C. A. Labate, M. D. Adock, and R. C. Leegood, *Planta*, *181*: 547 (1990).
60. M. Porter and b. Grodzinski, *Plant Physiol.*, *74*: 413 (1986).
61. R. T. Besford, *J. Plant Physiol.*, *136*: 458 (1990).
62. H. Poorter, R. M. Gifford, P. E. Kriedemann, and S. C. Wong, *Aust. J. Bot.*, *40*: 501 (1992).
63. C. D. Elmore, *Predicting Photosynthesis for Ecosystem Models* (J. D. Hesketh and J. W. Jones, eds.), CRC Press, Boca Raton, FL, Chapter 9 (1980).
64. B. Acock and D. Pasternak, *Carbon Dioxide Enrichment of Greenhouse Crops*, Vol. 2 (H. Z. Enoch and B. A. Kimball, eds.), CRC Press, Boca Raton, FL, pp. 41–52 (1986).
65. S. H. Wittwer, *HortScience*, *25*: 1560 (1990).
66. R. Wyse, *Crop Sci.*, *20*: 456 (1980).
67. S. B. Idso, B. A. Kimball, and J. R. Mauney, *Agric. Ecosyst. Environ.*, *21*: 293 (1988).
68. C. Körner and J. A. Arnone III, *Science*, *257*: 1672 (1992).
69. B. A. Kimball, *Agron. J.*, *75*: 779 (1983).
70. S. H. Wittwer, *Carbon Dioxide Enrichment of Greenhouse Crops*, Vol. 1 (H. Z. Enoch and B. A. Kimball, eds.), CRC Press, Boca Raton, FL, Chapter 1 (1986).
71. S. H. Wittwer and W. Robb, *Econ. Bot.*, *18*: 34 (1964).
72. D. W. Lawlor and A. C. Mitchell, *Plant Cell Environ.*, *14*: 807 (1991).
73. B. G. Drake and P. W. Leadley, *Plant Cell Environ.*, *14*, 853 (1991).
74. N. E. Grulke, G. H. Riechers, W. C. Oechel, U. Hjelm, and C. Jaeger, *Oecologia*, *83*: 485 (1990).
75. R. J. Norby, C. A. Gunderson, S. D. Wullschleger, E. G. O'Neill, and M. K. McCracken, *Nature*, *357*: 322 (1992).
76. D. B. Jordan and W. L. Ogren, *Planta*, *161*: 308 (1984).
77. D. W. Wolfe and M. O. Kelly, *Photosynthetica*, *26(3)*: 475 (1992).
78. S. P. Long, *Plant Cell Environ.*, *14*: 729 (1991).
79. S. B. Idso, B. A. Kimball, M. G. Anderson, and J. R. Mauney, *Agric. Ecosyst. Environ.*, *20*: 1 (1987).
80. S. B. Idso and B. A. Kimball, *Environ. Exp. Bot.*, *29*: 135 (1989).

81. H. M. Rawson, *Aust. J. Bot.*, *40*: 473 (1992).

82. M. Parry, *Climate Change and World Agriculture*, Earthscan, London, p. 53 (1990).

83. C. Rosenzweig, *Am. J. Agric. Econ.*, *71*: 265 (1989).

84. S. C. Wong, *Vegetatio*, *104/105*: 211 (1993).

85. R. M. Gifford, *Aust. J. Plant Physiol.*, *6*: 637 (1979).

86. B. A. Kimball, *Carbon Dioxide Enrichment of Greenhouse Crops*, Vol. 2 (H. Z. Enoch and B. A. Kimball, eds.), CRC Press, Boca Raton, FL, Chapter 5 (1986).

87. D. W. Hilbert, A. Larigauderie, and J. F. Reynolds, *Ann. Bot.*, *68*: 365 (1991).

88. J. S. Coleman, K. D. M. McConnaughay, and F. A. Bazzaz, *Oecologia*, *93*: 195 (1993).

89. B. A. Kimball, *Response of Vegetation to Carbon Dioxide*, Report 39, U.S. Water Conservation Laboratory, Phoenix, AZ (1986).

90. L. H. Allen, *Response of Vegetation to Carbon Dioxide*, Report 62, Plant Stress and Protection Unit, USDA-ARS, Gainesville, FL (1991).

91. J. Goudriaan and H. E. deRuiter, *Neth. J. Agric. Sci.*, *31*: 157 (1983).

92. B. Acock and L. H. Allen, Jr., *Direct Effects of Increasing Carbon Dioxide on Vegetation* (B. R. Strain and J. D. Cure, eds.), U. S. Department of Energy, DOE/ER-0238, Washington, DC, Chapter 4 (1985).

93. B. Acock, D. A. Charles-Edwards, D. J. Fitter, D. W. Hand, L. J. Ludwig, J. Warren Wilson, and A. C. Withers, *J. Exp. Bot.*, *29*: 815 (1978).

94. R. M. Gifford, *Aust. J. Plant Physiol.*, *4*: 99 (1977).

95. J. V. Morgan, *Acta Hortic.*, *22*: 187 (1971).

96. J. A. Bunce, *Physiol. Plant.*, *86*: 173 (1992).

97. D. L. Ehret and P. A. Joliffe, *Can. J. Bot.*, *63*: 2026 (1985).

98. R. F. Sage, *Plant Physiol.*, *94*: 1728 (1990).

99. F. A. Bazzaz and K. D. M. McConnaughay, *Aust. J. Bot.*, *40*: 547 (1992).

100. D. E. Lincoln, N. Sionit, and B. R. Strain, *Oecologia*, *69*: 556 (1986).

101. R. H. Johnson and D. E. Lincoln, *Oecologia*, *87*: 127 (1991).

102. E. D. Fajer, M. D. Bowers, and F. A. Bazzaz, *Science*, *243*: 1198 (1989).

103. D. W. Wolfe and J. Erickson, *Agricultural Dimensions of Climate Change* (H. Kaiser, and T. Drennen, eds.), St. Lucie Press, Delray Beach, FL, Chapter 8 (1993).

104. F. A. Bazzaz and E. D. Fajer, *Sci. Am.*, *266*: 68 (1992).

105. R. E. McMurtrie, H. N. Comins, M. U. F. Kirschbaum, and Y. P. Wang, *Aust. J. Bot.*, *40*: 657 (1992).

11

Absorption of Radiation, Photosynthesis, and Biomass Production in Plants

Luis Serrano and José A. Pardos
Universidad Politécnica de Madrid, Madrid, Spain

Francisco I. Pugnaire
University of Leeds, Leeds, England

Francisco Domingo
Consejo Superior de Investigaciones Científicas, Almería, Spain

I. INTRODUCTION

Photosynthesis is the conversion by plants of solar energy into several forms of chemical energy via a series of reactions which represents the largest synthetic process on earth and the main source of energy for living beings. The rate of dry matter production by plants depends on the interception of incident solar radiation by the leaf canopy and the conversion of this energy into carbohydrates.

Photosynthetic productivity, which may be described as the carbon balance of a plant over a time period [1], depends on internal and environmental factors [2]. The internal factors range from leaf age and chlorophyll content to osmotic adjustment, the presence of strong sinks, and photorespiration. The environmental factors include light availability, temperature (which influences photosynthesis but also dark respiration, which in turn may result in a substantial loss of carbohydrates), air humidity, water, and nutrient availability.

II. INTERCEPTION OF SOLAR RADIATION

There is a wide range of types of radiation characterized by wavelength (λ), expressed in nanometers (1 nm $= 10^{-9}$ m) or in angstroms (1 Å $= 10^{-10}$ m), from the most energetic γ-radiation (0.001 nm) to hertz waves or radio waves (several kilometers). The wavelengths of radiation that are of primary concern in environmental plant physiology lie between about 300 nm and 100 mm and include some of the ultraviolet (UV, < 360 nm), the photosynthetically active radiation (PAR, which is broadly similar to the visible spectrum, between 360 and 760 nm), and the infrared (IR, > 760 nm). Radiation has properties of both waves (it has a wavelength) and particles (energy is transferred as discrete units termed quanta or photons). The energy E of a photon is related to its wavelength λ or its frequency of oscillation ν by the equation:

$$E = \frac{hc}{\lambda} = h\nu$$

where h is Planck's constant (6.63×10^{-34} J·s) and c is the speed of the light (3×10^8 m s^{-1}). This energy is expressed in calories-gram or joules (1 cal·g = 4.1855 J). The amount of radiant energy emitted, transmitted, or received by a surface per unit time is called the radiant flux, which has units of power [J s^{-1} or watts (W)]. The net radiant flux through a unit area of surface is the radiant flux density (W m^{-2}). The component of the flux incident on a surface is termed irradiance (W m^{-2}), while that emitted by a surface is termed the emittance (or radiant excitance) (W m^{-2}).

Another term used in plant physiology is *radiation intensity* (or, an incorrect synonym, flux density), which is defined as a flux per unit solid angle emitted from a point source [watts per steradian (W sr^{-1})]. The term *fluence rate*, another incorrect synonym for flux density, in fact measures the flux per unit cross-sectional area incident from all directions on a spherical volume element. The ecological importance of radiation is based on the interaction between electromagnetic radiation and matter, which depends on the wave energy:

The energy associated with a shortwave radiation (UV) can alter the molecular organization.
PAR has adequate energy to be effectively captured by a live molecular structure.
Infrared radiation is composed of quanta with little energy (long-wavelength radiation) and its
    effect is, in general, confined to the acceleration of chemical reactions or the increase of
    molecular mobility.

A. Leaf Energy Balance

The energy balance of a leaf can be summarized as follows:

energy into the leaf (I) – energy out of the leaf (II) = energy storage by a leaf (III)

Term I is composed of absorbed solar irradiation and absorbed infrared irradiation from surroundings; term II is a combination of emitted infrared radiation, heat convection, heat conduction, and heat loss due to water evaporation; and term III represents photosynthesis, other metabolic processes, and leaf temperature changes. These terms differ greatly in magnitude, and in general the energy storage is relatively small. The efficiency of photosynthetic processes on a global scale is low, because the average solar irradiation at the soil level is 171 W m^{-2}, while average productivity in the main ecosystems is 0.35 g C m^{-2} day^{-1} (0.24 W m^{-2}). Photosynthetic efficiency is then 0.240/171 or 0.14%. Different approaches yield somewhat different results, but always show very low efficiencies. The contribution of photosynthesis to the energy balance of a leaf, therefore, can generally be ignored. Other metabolic processes in a leaf, such as respiration and photorespiration, are usually even less important than photosynthesis on an energy basis, so they too generally can be ignored [3]. The energy balance in the budget equation then is as follows:

energy into the leaf (I) – energy out of the leaf (II)

The energy balance of plant leaves has been investigated extensively, and some of the problems addressed included (a) establishment of quantitative relations between wind speed and leaf boundary–layer conductance (see, e.g., Refs 4–6), (b) determining influences of air turbulence and leaf flapping on these conductances (see, e.g., Refs. 7–9), and (c) establishing whether boundary layers on leaves are laminar or turbulent [10].

The leaf energy balance technique also has been used to examine bias in whole-tree transpiration rates, measured with large ventilated chambers [11], and to determine transpiration rates based on temperature differences [12–14].

III. PHOTOSYNTHESIS AND CANOPY ARCHITECTURE

Canopy structure is a central consideration in any description of plant–environment interactions, and especially of whole-plant photosynthesis, because it is related to the interception, scattering, and emission of radiation.

As the leaves of a plant develop, light interception by the canopy increases, increasing the plant net photosynthesis. Nevertheless, too many leaves will shade each other, decreasing the maximum potential photosynthesis by self-shading. The amount and spectral distribution of shortwave radiation plays an important role in the regulation of growth and development. For instance, leaves at the edge of the canopy are more sun-exposed than those within the canopy. The interception of light by the canopy is expressed by the Beer's law [15]:

$$I = I_0 \, e^{-kL}$$

where I_0 is the irradiance at the top of the plant canopy, I is the irradiance at a point in the canopy above which there is a leaf area index (LAI) of L, and k represents the extinction coefficient, which is determined empirically. Therefore, radiation interception and carbon gain are influenced by several architectural attributes, including vertical distribution of leaf area [16–19], leaf specific mass [20], leaf orientation [21–25] and diurnal leaf movements [26–28], and a gradient of leaf shape and size within the canopy associated with the light gradient.

Different plants have evolved different ways of acclimation to the light environment [29]: sun-adapted species show enhanced axis development in response to shade conditions, increased internode and petiole extension, strong apical dominance, and limited leaf development. On the other hand, shade-adapted plants tend to show greater leaf development and generally to have low relative growth rates, with low respiration, low photosynthetic rate, and low rates of leaf turnover [30].

The measurement and description of canopy structure is a complex task, and practical descriptions of canopy structure at present must use statistical tools or mathematical assumptions [31]. Foliage quantity is often normalized by ground area, or canopy volume. Leaf area index, the most commonly used canopy structure parameter, is defined as total leaf area per ground area. Branch area index or stem area index could be defined in the same way.

In some cases, other parameters are more useful than LAI, such as leaf area density (total leaf area per canopy volume), and drip-line LAI (DLLAI) [31]:

$$\text{DLLAI} = \frac{(\text{leaf area density} \times \text{canopy volume})}{\text{area within the plant's drip line}}$$

Because radiative transfer and canopy structure are coupled so tightly, one can often be used to predict the other. Thus, relatively simple measurements of radiation can be used to estimate various structural parameters, if a model is available to predict the influence of the canopy on the radiation. The scope of radiative techniques for estimating canopy structure is broad. Several techniques have recently been reviewed [31,32].

The amount of radiation absorbed by a plant canopy depends on many factors [33]: amount of leaf area, leaf angle distribution, clumping characteristics of the foliage, and angle of incident solar radiation, which consists of two components (direct radiation, as a beam of parallel rays, and diffuse radiation, scattered in the atmosphere and reaching the canopy from all directions).

A. Leaf Sizes and Shapes

Leaf sizes and shapes are very variable and have important consequences for leaf physiology and its energy balance [34,35]. For instance, in Mediterranean environments the upper, sun-exposed leaves in the canopy are intensely heated during the summer months, just when drought limits transpiration and radiation is maximum. These leaves may reach a very high temperature, exceeding that of surrounding air. Reduced leaf surface and increased number of lobes are examples of modifications that increase convective heat flux with the atmosphere, because a reduced boundary layer improves heat diffusion [36]. The heat convection coefficient increases with the depth of serrations on model leaves [37,38], so plants from dryland habitats and upper leaves in closed canopies are usually smaller and more dissected, reflecting adaption to a high light environment. In oak trees, for instance, where leaves almost double in size from top to bottom, the larger leaves are less deeply lobed and have fewer teeth.

B. Leaf Orientation

Foliage orientation is more completely described by a distribution of fractional areas in various inclination and azimuthal angle classes than by LAI, because the former is independent of leaf orientation. One effect of foliage orientation concerning radiation is that it changes radiation quality and penetration depending on canopy direction. Certain plants, especially those exposed to intense shortwave irradiation, have vertically oriented leaves; examples include willow, many *Eucalyptus* species, and certain chaparral, Mediterranean, and desert shrubs. Over the course of a day, vertical leaves often intercept nearly as much shortwave irradiation as do horizontal leaves, but they intercept less at midday, when air temperatures tend to be high [25]. As a consequence, these leaves avoid high leaf temperatures at midday, which would lead to higher transpiration for a given stomatal opening and possibly to temperatures above the optimal range for photosynthesis. On the other hand, reduced shortwave flux can result in lower foliage temperatures and increased water use efficiency [23,26,28,39]. Since leaves of most plants cannot effectively utilize the full midday irradiance, not only is the reduced flux associated with vertically oriented foliage more efficiently used, but the potential for photoinhibition due to excess energy is also lessened [40–42]. Leaves generally become more vertical upon wilting, thereby reducing their interception of shortwave irradiation for higher sun angles. Moreover, vertical leaves have different boundary layer conductance, hence affecting heat and water fluxes. Despite the high efficiency in light utilization, carbon gain may not always be improved. For instance, the steeply inclined foliage in *Lactuca serriola* may cause lower photosynthesis due to the reduced midday light interception [25]. For plants composing homogeneous canopies, daily carbon gain is relatively independent of leaf orientation [20].

C. Spatial Disposition

The rate of net photosynthesis of leaves increases with irradiance, but the slope of the response curve decreases steadily, giving an asymptote at the irradiance level corresponding to light saturation. In contrast, the photosynthetic efficiency (quantum yield) appears to decrease continuously with increasing irradiance [43], and above the point of saturation, further increments of PAR give no increase in the rate of net photosynthesis. The additional input of radiation absorbed by the leaf is converted into heat and used as energy for transpiration and convective heat exchanges with surrounding air [58].

The vertical distribution of canopy leaf area can be described by a normal curve, which indicates maximum leaf area at intermediate positions within the canopy. Maximum rates of photosynthesis are usually found in the upper part of the canopy [44]. But plant canopies can differ widely in overall height, in vertical distribution of LAI, in leaf angle, and in total LAI [58]. In ryegrass, for instance, the combination of low LAI in the top layers of the canopy and the vertical arrangement of the youngest leaves allows radiation to penetrate deeply into the plant stands, and canopy photosynthesis can be increased because the leaf area is larger. In contrast, a larger proportion of the LAI distribution is situated in the top layers of the clover canopy, and the leaves are disposed in a nearly horizontal plane; consequently, a much larger fraction of incoming PAR is intercepted at the top of the canopy, and photosynthesis tends to be distributed over a smaller area of leaf [45].

The pattern of mean leaf angle has been used to classify the plant canopies as erectophile, plagiophile, or planophile, depending on the mean leaf angle [46].

The LAI at which full interception is achieved determines how effectively PAR can be distributed through the canopy at irradiances below that required to saturate photosynthesis. An optimum canopy architecture with maximum LAI will lead to higher rates of canopy photosynthesis and plant growth [47].

D. Light and Biomass Production

The chlorophyll molecule in plants is specialized in radiation interception. Associated with the light gradient through the canopy, there is a gradient in chlorophyll concentration. In oak trees, for example, chlorophyll concentration is maximum just under the first layer of leaves, one meter under the canopy edge. Above this level, light saturates photosystems (about 30% of radiation in a clear day is enough to saturate a photosystem). Two meters below the canopy edge, light is no longer saturating, and a maximum in light-harvesting pigments is found. Further inside the canopy, concentration progressively decreases as light decreases [48].

Associated with the radiation gradient, there is a reduction in transpiration. Three factors tend to reduce the flux of water vapor out of a shaded leaf [3]: (a) the increase in stomatal resistance at the lower light levels, which is the main factor reducing transpiration; (b) a lower wind speed for protected leaves, which leads to a thicker air boundary layer; and (c) a concentration of water vapor in the turbulent air that is generally higher than at the top of the plant canopy. In the absence of differential permeability to water and CO_2, stomata open at times of high photosynthetic activity and abundant water and close when water is limiting or photosynthesis is not occurring, such as at night.

There is a marked coupling between net photosynthesis and stomatal conductance in the leaf [49]. In fact, a tight coupling between mesophyll photosynthesis and stomatal conductance has been demonstrated under many experimental conditions and constitutes a basic physiological property of leaves (Figure 1) [50]. This coupling poses an important question: If a change in photosynthetic rate in response to an environmental signal such as light is always accompanied by a change in stomatal conductance, could the observed stomatal response be a specific reaction to the environmental stimulus rather than an obligatory, direct tracking of photosynthesis?

Considerable progress has been made in understanding how variations in aerial factors such as evaporative demand and edaphic factors such as soil water availability are sensed and transduced into appropriate stomatal regulatory responses. Nevertheless, studies at multiple scales of observation are still needed to understand how external environmental factors and intrinsic plant properties interact to determine the role of stomata in regulating transpiration from different types of vegetation [51].

It is well established that plants growing in open habitats (sun plants) have much higher photosynthetic capacities and saturate at higher irradiance than plants growing in shaded habitats [52–54]. Low photosynthetic rates at light saturation would inevitably result in a large excess of excitation energy whenever a leaf became exposed to direct sunlight. Excessive excitation energy is known to be potentially injurious to the photosynthetic system, especially photosystem II (PS II) [55]. However, shade plants, which have lower dark respiration rates, use low irradiance more effectively for net photosynthetic CO_2 uptake than do sun plants.

Growth rate is often proportional to the amount of radiation intercepted by the canopy [56], and biomass *(W)* can be considered as the time-integrated product of three factors [57]:

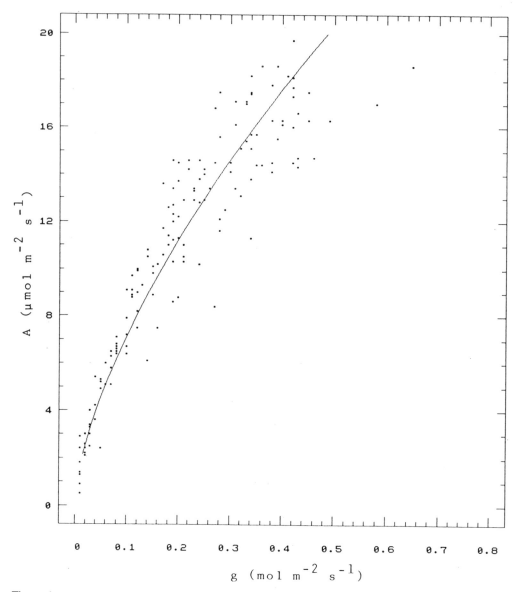

Figure 1 Relationship between net assimilation rate (*A*) and stomatal conductance (*g*) in field-grown *Eucalyptus globulus* trees. Measurements were taken between February and December 1991. Temperatures ranged from 21 to 40°C and dawn water potential from –0.21 to –2.06 MPa during this period. (From Ref. 50.)

$$W = \int \epsilon I Q \, dt$$

where ϵ is the amount of dry matter produced per unit of radiation intercepted (photosynthetic efficiency), Q is the daily radiation incident on the top of the canopy, I is the fraction of incident radiation intercepted by the canopy, which depends on its structure, and t is time measured in days. Variations in I account for most of the differences in yield, because ϵ and Q are relatively constant in the absence of severe stress. If the harvesting material of crops (stem, grain, or oil) is a constant proportion of the total biomass at maturity, this analysis can be extended. However, more complex relationships are needed when harvest index changes with treatment [57]. The partitioning of dry matter to the harvested parts is not a well-understood process, and little is known about the mechanisms controlling the allocation of dry matter among the various competing sinks (leaves, roots, stems, maintenance of mature tissues, etc.). However, the harvest index has been recognized as a crucial characteristic: for example, the steady increase in the grain yield of wheat and barley during this century is now seen in terms of a progressive increase in harvest index, with little or no increase in total plant biomass production [58].

IV. PHOTOSYNTHETIC POTENTIAL

Photosynthesis can be described as the process by which light energy is absorbed by chlorophyll and used to produce carbohydrates from carbon dioxide and water. The radiant energy is converted to chemical energy (ATP) and to reducing power ($NADPH_2$). These molecules are used to fix carbon dioxide in phosphoglyceric acid and reduce it to triose phosphate, from which glucose is synthesized. This process is known as the C_3 carbon fixation pathway. The carboxylation enzyme, ribulose bisphosphate carboxylase (Rubisco), also has oxygenase activity; hence oxygen is a competitive inhibitor of CO_2 fixation. In the C_4 carbon fixation pathway, malic and aspartic acids, each of which has four carbon atoms, are formed in the mesophyll. These acids are transported to the bundle sheath cells, where they are decarboxylated; the CO_2 then goes through the C_3 pathway. The initial carboxylating enzyme in the C_4 pathway is phosphoenolpyruvate carboxylase, which is not inhibited by oxygen. The major advantages of the C_4 plants in relation to C_3 plants are as follows: their compensation point is low; they are not light saturated; and photorespiration is minimal.

A. Rates of Photosynthesis

Rates of photosynthesis vary widely among species, between sun and shade leaves and during the day, and according to the growing season [33]. These differences are due to interactions among plant factors, such as leaf age, structure and exposition to light, canopy development, stomatal behavior, amount and activity of Rubisco, and environmental factors (e.g., light intensity, temperature, water deficit). It is well known [59] that woody plants generally have lower rates of photosynthesis than herbaceous plants, and evergreen trees have lower rates than deciduous trees. There are also differences among clones of the same species; for example, large differences have been observed among poplar clones with respect to the rate of CO_2 uptake and light compensation point [60].

B. Diurnal and Seasonal Patterns

Photosynthesis is limited to the daytime, and the rate is lower early in the morning because light intensity and temperature are low, and stomata are closed. As light intensity and temperature increase and stomata open, the rate increases to a maximum at midday and decreases later in the afternoon [33]. On hot, sunny days there is often a midday depression in the rate of photosynthesis which is attributed in part to stomatal closure and in part to inhibition of photosynthesis by water stress [61]. A seasonal variation of photosynthesis has been reported for many species. The measured total photosynthesis of a young apple tree throughout the entire year showed that early in the season respiration exceeded photosynthesis. As leaf area developed, however carbon fixation increased and remained high until the leaves senesced and temperature and daylength decreased [62]. Throughout the season, the most important factor affecting photosynthesis was radiation intensity. For evergreen species growing in regions with temperate winters and dry summers (Mediterranean climates), winter photosynthesis can account for as much as 65% of the total carbon fixed during the year [63,64].

V. PHOTOSYNTHETIC PRODUCTIVITY

A. Photosynthesis and Respiration

The rate of net photosynthesis of a plant is determined by the rates of gross photosynthesis, photorespiration, and dark respiration. These processes determine the efficiency with which intercepted PAR effects the conversion of CO_2 into plant dry matter [58]. As a result, net assimilation rate may be summarized as follows:

$$A = A_g - PR - R$$

where A is net photosynthesis, A_g is gross photosynthesis, PR is photorespiration, and R is dark respiration. It is assumed that the rate of dark respiration in the light is similar to that in the dark. It is now widely held that dark respiration continues in the light, although it is somewhat changed [65]. Photorespiration and respiration are two different processes, which produce CO_2 by oxidation of appropriate substrates. Photorespiration can be defined as the light-dependent oxidation of a metabolite of two carbons (mainly glycollate) derived from photosynthesis. The requirement for light is a consequence of the dependence of glycollate synthesis on the energy and reducing power generated by photosynthetic electron transport. Respiration is the oxidation of different substrates by glycolysis, the oxidative pentose phosphate pathway, and the tricarboxylic acid cycle. The reactions involved in photorespiration occur in the chloroplasts, peroxisomes, mitochondria, and cytosol, whereas those of respiration are restricted to the cytosol and mitochondria. Photorespiration can account for 25–50% of the net CO_2 assimilated by different C_3 species [66,67]. Thus if photorespiration were eliminated, net photosynthesis would increase by the same amount. For respiratory losses, values of 45% of the carbon fixed have been reported [68,69]. Respiration in the dark, when there is no CO_2 fixation, produces high energy compounds and provides carbon skeletons for the synthesis of new metabolites. In the light, however, ATP and $NADPH_2$ are produced by photosynthetic reactions, and respiration is necessary only as a source of carbon skeletons [65].

B. Photosynthetic Efficiency

The efficiency of photosynthetic processes can be considered from different points of view. At the molecular level, photosynthetic efficiency is expressed as the quantity of the absorbed light required for the fixation of one mole of CO_2, while at a higher level it is expressed as the dry matter production related to the available PAR intercepted over a season. Three molecules of ATP and two of $NADPH_2$ are required to reduce a molecule of CO_2 so in theory at least three mole quanta are required for the reduction of each mole of CO_2 [58].

However, not all the absorbed quanta are used for CO_2 reduction; energy losses (chlorophyll fluorescence, dissipation of heat) plus respiration and photorespiration account for some of the total. It has been proposed that the quantum requirement for photosynthesis in C_3 plants is 20 mole quanta per mole of net CO_2 assimilated [54]. The number of moles of CO_2 assimilated per mole quanta of PAR is called the quantum yield. With 20 mole quanta, the quantum yield obtained is 0.05; therefore the efficiency is approximately 13% (energy of reduced CO_2/energy of 20 mole quanta).

Actual efficiencies for crops in the field are much lower, since not all the PAR is available for photosynthesis (because of reflectance and transmittance of light through leaves) and because in C_3 plants exposed to high irradiances, the efficiency will be significantly diminished (radiation saturation). The potential photosynthetic efficiency in terms of dry matter related to radiation has been estimated as approximately 5 g MJ^{-1} (PAR) [57]. This value is much greater than estimates obtained for annual crops, which are in the range 2.4–3.4 g MJ^{-1} (PAR) [57], with an average value of 1.4 g MJ^{-1} (total solar radiation) [70].

These differences are explained in part by the reasons mentioned above and in part by noting that the plants are subjected to environmental stresses, such as low temperature and shortage of water. Periods of incomplete interception of radiation (seasonal development of leaf area) and different turnover rates also contribute to the explanation.

1. The Resistance Analogy: Carboxylation Efficiency

The movement of CO_2 from the air to the sites of carboxylation in the leaf can be treated as a diffusive process and analyzed in terms of Fick's law, which relates the rate of diffusion to the concentration gradient. In the same way, it can be expressed as the total resistance imposed by the leaf as a difference in concentration. In this case, Fick's law is similar to Ohm's law in an electrical circuit, and the rate of assimilation can be written as follows:

$$A = \frac{C_a - C_c}{r_a + r_s + r_m}$$

where C_a is the external CO_2 concentration, C_c is the concentration of CO_2 at the sites of carboxylation, and r_a, r_s, and r_m are the boundary layer, stomatal, and mesophyll resistances, respectively.

The first resistance to CO_2 flux is that imposed by the boundary layer. The movement of CO_2 across this laminar layer is by molecular diffusion only, because turbulent transfer of CO_2 is not possible in this unstirred air layer. The thickness of the boundary layer, hence the boundary layer resistance, depends on wind speed, leaf size, and leaf structure and shape. The lower the wind speed and the larger the leaf, the thicker will be the boundary layer [71]. The second resistance to CO_2 flux is offered by stomata. The stomatal resistance depends on the density of stomata and on the degree of closure, which in turn depends on light, air humidity, temperature, water potential, and hormone concentration. The closure of stomata has much less effect on net assimilation rate than on diffusion of water to the atmosphere [72].

The third resistance, known as mesophyll resistance, accounts for different factors, including diffusion of CO_2 from stomatal cavities through the cell walls, cell membranes, and chloroplasts to the sites of carboxylation, as well as the photosynthetic capacity of the leaf. The resistances to net assimilation are linked in series, and at the steady rate, the flux of CO_2 is the same through each portion of the pathway, so that it is possible to establish the relation of each one to net assimilation. Since under most circumstances the CO_2 concentration at the site of carboxylation is similar to that in the intercellular spaces (C_i), the net assimilation rate can be equated as

$$A = \frac{C_i - \Gamma}{r_m}$$

where Γ is the CO_2 compensation point and r_m is the reciprocal of the carboxylation efficiency. The response of net photosynthesis as a function of C_i (A/C_i curves) shows first a linear section with positive slope, which reflects the carboxylation efficiency of Rubisco under limiting CO_2 concentrations, followed by a curved part with slope near zero, in which net photosynthesis is limited by Rubisco regeneration, which depends, in turn, on the capacity for electron transport [73].

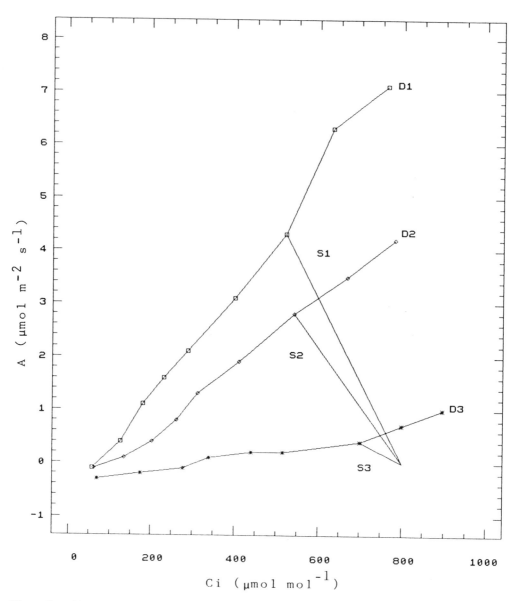

Figure 2 Effects of water stress on the relation between intercellular CO_2 concentration (C_i) on the net assimilation rate (A) in leaves of *Eucalyptus globulus*, measured at a photosynthetically active photon flux density of 1200 μmol m^{-2} s^{-1} and a leaf temperature of 25°C. Demand functions: D_1 measured at a water potential of –0.80 MPa, D_2 at –1.51 MPa, and D_3 at –2.70 MPa. Supply functions are also shown (S_1, S_2, S_3). (From Ref. 50.)

C. Effects of Water Stress on Photosynthesis

The effects of water stress can be divided into stomatal and nonstomatal limitations of net assimilation rate. The stomatal limitation is quantified from A/C_i curves by comparing the actual measured photosynthetic rate (A) and the photosynthetic rate (A_0) that could be maintained at an infinite stomatal conductance, when C_i is high and saturating for photosynthesis [74] and is expressed as follows:

$$I = \frac{A_0 - A}{A_0}$$

where I is stomatal limitation.

Reductions in the rate of net photosynthesis of water-stressed plants are caused by the closure of stomata, but under more severe stress there are also increases in the mesophyll resistance [75,76]. Other methods have been developed to calculate stomatal and nonstomatal limitations based on A/C_i curves at different water potentials by comparison of slopes of the demand and supply functions by path-dependent or state function methods [77]. Stomatal limitations imply a reduction in stomatal conductance (supply function, which has a slope of $-g$), hence, a reduction in the intercellular CO_2 concentration. But in parallel, there is a great increase in the nonstomatal limitations, because the slope of the demand functions is clearly reduced by water stress, indicating a decrease in the carboxylation efficiency and in the photosynthetic capacity of the leaf (Figure 2). It has been suggested that for a given level of stress, photosynthesis is maintained by an intercellular CO_2 concentration near the transition from Rubisco limitation to ribulose bisphosphate regeneration limitation [78]. In the field, water-stressed plants are often exposed to high irradiance and temperature and consequently exposed to overenergized conditions. Stomatal closure restricts the supply of CO_2, and under these circumstances there is a risk of damage to the photochemical system [79].

Photorespiration acts as a photoprotective mechanism and can dissipate photochemical energy generated in excess [80].

ACKNOWLEDGMENT

We thank P. Haase for comments on the manuscript.

REFERENCES

1. J. J. Landsberg, *Physiological Ecology of Forest Production*, Academic Press, London (1986).
2. M. Faust, *Physiology of Temperate Zone Fruit Trees*, Wiley, New York (1989).
3. P. S. Nobel, *Physicochemical and Environmental Plant Physiology*, Academic Press, San Diego, CA (1991).
4. K. Raschke, Micrometeorologically measured energy exchanges of an *Alocasia* leaf, *Arch. Meteorol. Geophys. Bioklim.*, *B7*: 240 (1956).
5. D. F. Parkhurst, P. R. Duncan, D. M. Gates, and F. Kreith, Convection heat transfer from broad leaves of plants, *J. Heat Transfer*, *C90*: 71 (1968).
6. G. I. Pearman, H. L. Weaver, and C. B. Tanner, Boundary layer heat transfer coefficients under field conditions, *Agric. Meteorol.*, *10*: 83 (1972).
7. J. Y. Parlange, P. E. Waggoner, and G. H. Heichel, Boundary layer resistance and temperature distribution on still and flapping leaves. I. Theory and laboratory experiments, *Plant Physiol.*, *48*: 437 (1971).
8. J. Y. Parlange and P. E. Waggoner, Boundary layer resistance and temperature distribution on still and flapping leaves. II. Field experiments, *Plant Physiol.*, *50*: 60 (1972).

9. D. F. Parkhurst and G. I. Pearman, Convective heat transfer from a semi-infinite flat plate to periodic flow at various angles of incidence, *Agric. Meteorol.*, *13*: 383 (1974).

10. J. Grace and J. Wilson, The boundary layer over a *Populus* leaf, *J. Exp. Bot.*, *97*: 231 (1976).

11. I. J. Foster and R. Leuning, *Comparison of Three Methods for Estimating Transpiration by Single Trees: Ventilated Chamber, Leaf Energy Budget and Penman–Monteith Equation*, Commonwealth Industrial and Research Organization, Division of Forest Research, User Series 7, Canberra, Australia (1987).

12. I. I. Impens, Leaf wetness, diffusion resistances and transpiration rates of bean leaves (*Phaseolus vulgaris* L.) through comparison of "wet" and "dry" leaf temperatures, *Oecol. Plant.*, *1*: 327 (1966).

13. E. T. Linacre, Leaf temperatures, diffusion resistances, and transpiration, *Agric. Meteorol.*, *10*: 365 (1972).

14. R. Leuning and I. J. Foster, Estimation of transpiration by single trees: Comparison of a ventilated chamber, leaf energy budgets and a combination equation, *Agric. For. Meteorol.*, *51*: 63 (1990).

15. M. Monsi and T. Saeki, Über der Lichtfaktor in den Pflanzengesellschaften und seine Bedeutung für die Stoffproduktion, *Jpn. J. Bot.*, *14*: 22 (1953).

16. T. W. Mulroy and P. W. Rundel, Annual plants: Adaptations to desert environments, *Bioscience*, *27*: 109 (1977).

17. S. W. Roberts and P. C. Miller, Interception of solar radiation as affected by canopy organization in two Mediterranean shrubs, *Oecol. Plant.*, *12*: 273 (1977).

18. P. W. Barnes, W. Beyschlag, R. J. Ryel, S. D. Flint, and M. M. Caldwell, Plant competition for light analyzed with a multispecies canopy model. III. Influence of canopy structure in mixtures and monocultures of wheat and wild oat, *Oecologia*, *82*: 560 (1990).

19. J. D. Tenhunen, A. Sala Serra, P. C. Harley, R. L. Dougherty, and J. F. Reynolds, Factors influencing carbon fixation and water use by Mediterranean sclerophyll shrubs during summer drought, *Oecologia*, *82*: 381 (1990).

20. V. P. Gutschick and F. W. Wiegel, Optimizing the canopy photosynthetic rate by patterns of investment in specific leaf mass, *Am. Natur.*, *132*: 67 (1988).

21. W. G. Duncan, Leaf angles, leaf area and canopy photosynthesis, *Crop Sci.*, *11*: 482 (1971).

22. M. J. McCree and M. E. Keener, Simulations of the photosynthetic rates of three selections of grain sorghum with extreme leaf angles, *Crop Sci.*, *14*: 584 (1974).

23. G. G. McMillen and J. H. McClendon, Leaf angle: An adaptative feature of sun and shade leaves, *Bot. Gaz.*, *140*: 437 (1979).

24. P. Oker-Blom and S. Kellomdki, Effect of angular distribution of foliage on light absorption and photosynthesis in the plant canopy: Theoretical computations, *Agric. Meteorol.*, *26*: 105 (1982).

25. K. S. Werk and J. R. Ehleringer, Non-random leaf orientation in *Lactuca serriola* L., *Plant Cell Environ.*, *7*: 81 (1984).

26. H. A. Mooney and J. R. Elheringer, The carbon gain benefits of solar tracking in a desert annual, *Plant Cell Environ.*, *1*: 307 (1978).

27. I. N. Forseth and J. R. Ehleringer, Ecophysiology of two solar tracking desert winter annuals. III. Gas exchange responses to light, CO_2 and VPD in relation to long-term drought, *Oecologia*, *57*: 344 (1983).

28. I. N. Forseth and J. R. Ehleringer, Ecophysiology of two solar tracking desert winter annuals. IV. Effect of leaf orientation on calculated daily carbon gain, *Oecologia*, *58*: 10 (1983).

29. J. W. Hart, Light and plant growth, Vol. 1, *Topics in Plant Physiology* (M. Black and J. Chapman, ser. eds.), Unwin Hyman, London (1988).

30. F. S. Chapin, The mineral nutrition of wild plants, *Annu. Rev. Eco. Sys.*, *11*: 233 (1980).

31. J. M. Welles, Some indirect methods of estimating canopy structure. Instrumentation for studying vegetation canopies for remote sensing in optical and thermal infrared regions (S. G. Narendra and J. M. Norman, eds.), *Remote Sensing Rev.*, *5*: 31 (1990).

32. J. M. Norman and G. S. Campbell, Canopy structure, *Plant Physiological Ecology: Field Methods and Instrumentation* (R. W. Pearcy, J. Ehleringer, H. A. Mooney, and P. W. Rundel, eds.), Chapman & Hall, London and New York, p. 301 (1989).

33. T. T. Kozlowski, P. J. Kramer, and S. G. Pallardy, *The Physiological Ecology of Woody Plants*, Academic Press, San Diego, CA (1991).

34. D. M. Gates, *Biophysical Ecology*, Springer-Verlag, Berlin (1980).
35. S. E. Taylor, Optimal leaf form, *Perspectives of Biophysical Ecology* (D. M. Gates and R. B. Schmerl, eds.), Springer-Verlag, Heidelberg, p. 73 (1975).
36. S. Vogel, Convective cooling at low airspeeds and the shapes of broad leaves, *J. Exp. Bot.*, *21*: 91 (1970).
37. D. E. Gottschlich and A. P. Smith, Convective heat transfer characteristics of toothed leaves, *Oecologia*, *53*: 418 (1982).
38. J. Grace, *Plant–Atmosphere Relationships*, Chapman & Hall, London (1983).
39. I. N. Forseth and J. R. Ehleringer, Ecophysiology of two solar tracking desert winter annuals. II. Leaf movements, water relations and microclimate, *Oecologia*, *54*: 41 (1982).
40. S. B. Powles, and O. Björkman, Leaf movement in the shade species *Oxalis oregano*. II. Role in protection against injury by intense light, *Carnege Inst. Washington Yearbook*, *81*: 63 (1981).
41. M. Ludlow and O. Björkman, Paraheliotropic leaf movements in *Siratro* as a protective mechanism against drought-induced damage to primary photosynthetic reactions: Damage by excessive light and heat, *Planta*, *161*: 505 (1984).
42. J. R. Ehleringer and I. N. Forseth, Diurnal leaf movements and productivity in canopies, *Plant Canopies: Their Growth, Form and Function* (G. Russell, B. Marshal, and P. G. Jarvis, eds.), Cambridge University Press, Cambridge, p. 129 (1989).
43. P. Gaastra, Light energy conversion in field crops in comparison with the photosynthetic efficiency under laboratory conditions, *Meded. Landbouwhogesch.*, *Wageningen*, *58*(4) (1958).
44. J. N. Woodman, Variation of net photosynthesis within the crown of a large forest-grown conifer, *Photosynthetica*, *5*: 50 (1971).
45. W. R. Stern and C. M. Donald, Light relationship in grass–clover swards, *Aust. J. Agric. Res.*, *13*: 599 (1962).
46. B. R. Trenbath and J. F. Angus, Leaf inclination and crop production, *Field Crop Abstr.*, *28*: 231 (1975).
47. A. F. Hawkins, Light interception, photosynthesis and crop productivity, *Outlook Agric.*, *11*: 104 (1982).
48. C. A. Gracia, La clorofila en los encinares del Montseny, Doctoral thesis, Universidad de Barcelona (1983).
49. E. Zeiger, C. Field, and H. A. Mooney, Stomatal opening at dawn: Possible roles of the blue-light response in nature, *Plants and the Daylight Spectrum* (H. Smith, ed.), Academic Press, New York, p. 391 (1981).
50. L. Serrano, Respuestas ecofisiológicas à la sequía en *Eucalyptus globulus* Labill.: Relaciones hídricas y parámetros de intercambio gaseoso, Doctoral thesis, Universidad Autónoma de Madrid (1992).
51. F. C. Meinzer, Somatal control of transpiration, *Trends Ecol. Evol.*, *8*: 289 (1993).
52. N. K. Boardman, Comparative photosynthesis of sun and shade plants, *Annu. Rev. Plant Physiol.*, *28*: 355 (1977).
53. A. Wild, Physiology of photosynthesis in higher plants. The adaptation of photosynthesis to light intensity and light quality, *Ber. Dtsch. Bot. Ges.*, *92*: 341 (1979).
54. O. Björkman, Responses to different quantum flux densities, *Physiological Plant Ecology*, *I: Responses to the Physical Environment* (O. L. Lange, P. S. Nobel, C. B. Osmond, and H. Ziegler, eds.), *Encyclopedia of Plant Physiology*, New Series, Vol. 12A, Springer-Verlag, Berlin, p. 57 (1981).
55. S. B. Powles, Photoinhibition of photosynthesis induced by visible light, *Annu. Rev. Plant Physiol.*, *35*: 15 (1984).
56. P. V. Biscoe and J. N. Gallagher, Weather, dry matter production and yield, *Environmental Effects on Crop Physiology* (J. J. Landsberg and C. V. Cutting, eds.), Academic Press, New York (1977).
57. G. Russell, P. G. Jarvis, and J. L. Monteith, Absorption of radiation by canopies and stand growth, *Plant Canopies: Their Growth, Form and Function* (G. Russell, B. Marshall, and P. G. Jarvis, eds.), Cambridge University Press, Cambridge (1989).
58. R. K. M. Hay and A. J. Walker, *An Introduction to the Physiology of Crop Yield*, Longman, London (1989).
59. W. Larcher, *Physiological Plant Ecology*, Springer-Verlag, Berlin (1980).

60. O. Luukkanen and T. T. Kozlowski, Gas exchange in six *Populus* clones, *Silvae Genet.*, *21*: 220 (1972).

61. B. Demming, K. Winter, A. Krüger, and F. C. Czygan, Zeaxanthin and heat dissipation of excess light energy in *Nerium oleander* exposed to a combination of high light and water stress, *Plant Physiol.*, *87*: 17 (1988).

62. D. R. Heinicke and N. F. Childers, The daily rate of photosynthesis during the growing season of 1935, of a young apple tree of bearing age. *Mem. N.Y., Agric. Exp. Stn. (Ithaca)*, *201* (1937).

63. W. H. Emmingham and R. H. Waring, An index of photosynthesis for comparing forest sites in western Oregon, *Can. J. For. Res.*, *7*: 165 (1977).

64. R. H. Waring and J. F. Franklin, The evergreen coniferous forest of the Pacific Northwest, *Science*, *204*: 1248 (1979).

65. D. Graham, Effects of light on dark respiration, *The Biochemistry of Plants*, Vol. 2 (D. D. Davis, ed.), Academic Press, New York (1980).

66. I. Zelitch, Basic research in biomass production: Scientific opportunities and organizational challenges, *Linking Research to Crop Production* (R. C. Staples and R. J. Kuhr eds.), Plenum Press, New York (1980).

67. A. J. Keys and C. P. Whittingham, Photorespiratory carbon dioxide loss, *Physiological Processes Limiting Plant Productivity* (C. B. Johnson ed.), Butterworths, London (1981).

68. M. J. Robson, The growth and development of simulated swards of perennial ryegrass. II. Carbon assimilation and respiration in a seedling sward, *Ann. Bot.*, *37*: 501 (1973).

69. C. L. Morgan and R. B. Austin, Respiratory loss of recently assimilated carbon in wheat, *Ann. Bot.*, *51*: 85 (1983).

70. J. L. Monteith, Climate and the efficiency of crop production in Britain, *Phil. Trans. R. Soc. London, B281*: 277 (1977).

71. H. G. Jones, *Plants and Microclimate*, Cambridge University Press, Cambridge (1983).

72. J. R. Cooke and R. H. Rand, Diffusion resistance models, *Predicting Photosynthesis for Ecosystem Models*, Vol. 1 (J. B. Hesketh, and J. W. Jones eds.), CRC Press, Boca Raton, FL (1980).

73. S. von Caemmerer and G. D. Farquhar, Some relationships between the biochemistry of photosynthesis and the gas exchange of leaves, *Planta*, *153*: 376 (1981).

74. G. D. Farquhar and T. D. Sharkey, Stomatal conductance and photosynthesis, *Ann. Rev. Plant Physiol.*, *33*: 317 (1982).

75. T. C. Hsiao, Plant response to water stress, *Ann. Rev. Plant Physiol.*, *24*: 519 (1973).

76. J. Levitt, *Responses of Plants to Environmental Stresses.* Vol. II, *Water, Radiation, Salt and Other Stresses* 2nd ed., Academic Press, New York (1980).

77. H. G. Jones, Partitioning stomatal and non-stomatal limitations to photosynthesis, *Plant Cell Environ.*, *8*: 95 (1985).

78. S. von Caemmerer and G. D. Farquhar, Effects of partial defoliation, changes of irradiance during growth, short-term water stress and growth at enhanced $p(CO_2)$ on the photosynthetic capacity of leaves of *Phaseolus vulgaris* L., *Planta*, *160*: 320 (1984).

79. B. Halliwell, *Chloroplasts Metabolism*, Oxford University Press, London (1981).

80. C. B. Osmond, O. Björkman, and D. J. Anderson, *Physiological Processes in Plant Ecology. Towards a Synthesis with Atriplex*, Springer-Verlag, Berlin (1980).

12
Carbohydrate Synthesis and Crop Metabolism

Monica A. Madore
University of California, Riverside, California

I. INTRODUCTION

Plants are capable of producing all organic materials required for growth, metabolism, and reproduction from very simple inorganic molecules obtained from the atmosphere and the soil. Using light energy trapped by chlorophyll in the process of photosynthesis, these inorganic molecules (principally CO_2, phosphate, and nitrate or ammonia) are incorporated within the chloroplasts of mature leaves into a number of relatively simple biomolecules (e.g., triose phosphates, amino acids), which are then used elsewhere in the cell for respiration or for the construction of the more complex biomolecules (e.g., carbohydrates, proteins, nucleic acids) required for growth and metabolism. In agronomic crop species, the incorporation of fixed carbon into carbohydrates is particularly important, for carbohydrate production largely determines the yield of crop plants.

Plant carbohydrates can be classified into two forms: structural and nonstructural. The form of carbohydrate in a particular plant part will also to a large extent determine its agronomic usage. Structural carbohydrates, as the name implies, are polymers that help to form the rigid plant cell wall and give support to the plant body. These carbohydrates are in the form of permanent, extracellular structures, and carbon incorporated into structural elements is in general not available for further metabolism by the plant. An exception may occur in some seeds, in which cell wall polysaccharides can be metabolized by the germinating seedling [1,2].

Structural carbohydrates, of which cellulose is a prime example, have agronomic importance as components of livestock feed and as sources of fiber for industrial purposes (e.g., cotton). In terms of human nutrition, they are of no direct nutritional value, as the enzymes for their metabolism are lacking in humans. The same is also true of many of the nonstructural carbohydrates. Indeed, of the many nonstructural carbohydrates characteristic of plant organs, only two forms, sucrose and starch, are directly metabolizable by humans.

Both structural and nonstructural carbohydrates form a large proportion of the dry weight of plant organs. The way in which carbon is partitioned among different carbohydrate types in a particular crop, therefore, becomes an important determinant of the agronomic value of that crop, particularly from a nutritional standpoint. In light of this, it is evident that a basic understanding of carbohydrate synthesis and its control in crop plants is central to our understanding of crop physiology.

II. CARBOHYDRATE FORMATION IN SOURCE LEAVES

Triose phosphates, the first stable products of the photosynthetic process, represent key metabolic intermediates, for they are the immediate precursors of all carbohydrates synthesized in the source leaves. The type of carbohydrate synthesized is, in turn, regulated by compartmentation of triose phosphates within chloroplastic and cytoplasmic pools within the source leaf. Since the major photosynthetic tissues are fully matured leaves, cell wall synthesis (i.e., the synthesis of structural carbohydrate) is of little importance in these tissues. Photosynthetic carbon is instead partitioned to nonstructural carbohydrates, which, in leaves of higher plants, may take the form of insoluble polymers (starch), soluble polymers (fructans), and soluble low molecular weight carbohydrates (sucrose, raffinose family oligosaccharides, simple monosaccharides, and polyols).

A. Starch

Starch is an insoluble glucan polymer that exists as granules within the chloroplast where it is formed. Starch consists of two molecular species: amylose, an essentially linear (α–1,4)-glucose polymer, and amylopectin, in which linear α–1,4-glucans are linked via α–1,6 linkages to form a highly branched structure. The proportion of amylose to amylopectin in starch grains varies in different plant species and in different cultivars of the same species [3–5].

In source leaves, starch is often called "assimilatory" starch because it is a major reserve of photosynthetically fixed carbon [4]. During nonphotosynthetic periods (i.e., at night), this starch is mobilized and utilized to support growth and maintenance of the plant. Starch granules form by apposition of newly formed polymers onto existing grains; they are degraded by the reverse process. Therefore, increases and decreases in the size of starch grains are seen during photosynthetic and nonphotosynthetic periods, respectively [3,6].

As indicated in Figure 1, starch formation in source leaves begins by the assimilation of CO_2 by photosynthesis, with the subsequent formation of triose phosphates (triose-P). Further operation of the photosynthetic carbon reduction (PCR) cycle results in the formation of fructose 6-phosphate (Fru-6-P). If conditions are right (i.e., if enough carbon exists in the PCR cycle to allow sufficient regeneration of ribulose-1,5-bisphosphate to maintain photosynthetic CO_2 fixation rates), some carbon may be diverted out of the PCR cycle for starch synthesis. Fru-6-P is converted to glucose 6-phosphate (Glu-6-P) via a chloroplastic form of the enzyme hexose phosphate isomerase (reaction 1 in Figure 1), and Glu-6-P is converted to glucose 1-phosphate (Glu-1-P) by chloroplastic phosphoglucomutase (reaction 2). Glu-1-P is then converted to a sugar nucleotide, adenosine diphosphoglucose (ADPG), via the enzyme ADPG pyrophosphatase (ADPGPPase, reaction 3). The sugar nucleotide ADPG then acts as the glucose donor for the reaction catalyzed by starch synthase (reaction 4), which lengthens the glucan chain by one α-1,4 linkage. A further enzyme, the branching enzyme (not shown) is responsible for creation of the α-1,6 linkages of amylopectin. There appear to be multiple enzyme forms of both starch synthase and branching enzyme, which may be related to the structural asymmetries associated with the starch molecule [3–5].

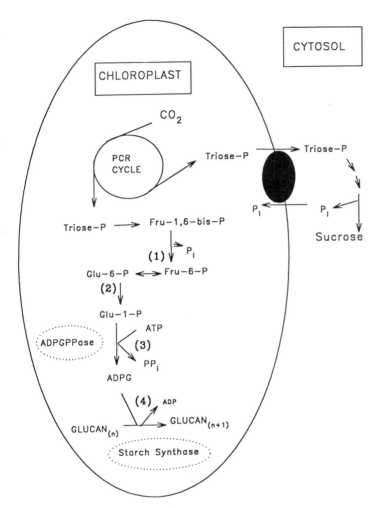

Figure 1 Pathway of starch synthesis in the chloroplasts of source leaves. Reaction 1, fructose 1,6-bisphosphatase; reaction 2, phosphoglucomutase; reaction 3, ADPG pyrophosphorylase (ADPGPPase); reaction 4, starch synthase.

1. Regulation of Starch Synthesis in Leaves

Regulation of starch synthesis in leaves is at the level of the enzyme ADPGPPase [4,5,7]. This enzyme is allosterically controlled by levels of phosphoglyceric acid (PGA), the initial product of CO_2 fixation, by the PCR cycle, which activates the enzyme, and inorganic phosphate (P_i), which inactivates it. The ratio of PGA to P_i in the chloroplast thus determines the activity of the ADPGPPase enzyme. Consequently, starch synthesis is promoted during periods of high photosynthetic rate, during which high levels of PGA are formed and P_i is rapidly incorporated into ATP and other phosphorylated intermediates of the PCR cycle.

Additionally, starch synthesis rate is coupled to sucrose synthesis rate through the export of triose-P out of the chloroplast. As indicated in Figure 1, this export occurs in strict exchange with the import of P_i via operation of the phosphate translocator of the chloroplast membrane [8]. Thus, conditions which favor triose-P export out of the chloroplast (i.e., high rates of cytosolic sucrose synthesis) will result in a low PGA/P_i ratio inside the chloroplast and will inhibit the formation of starch through inhibition of ADPGPPase activity [4,5,7]. Conversely, under conditions of reduced sucrose synthesis, cytosolic levels of P_i, a product of the sucrose synthetic pathway (Figure 1), will be low, preventing the export of triose-P from the chloroplast. The resulting reduction in import of P_i coupled with a reduced export of triose-P will raise the PGA/P_i ratio and activate the ADPGPPase [4,5,7].

B. Sucrose

As indicated above, the synthesis of sucrose and starch in photosynthesizing leaves is coupled with the operation of the phosphate translocator of the chloroplast membrane. Unlike starch synthesis, which occurs in the chloroplast, synthesis of the disaccharide sucrose (α-D-glucose-1,2-β-D-fructofuranoside, Figure 2) occurs in the cytosol of the photosynthetic cell, from triose-P that is exported to this compartment via the phosphate translocator [8–10].

As indicated in Figure 3, once in the cytosol, triose-P is converted to fructose 1,6-bisphosphate (Fru-1,6-bis-P), which is dephosphorylated to fructose 6-phosphate (Fru-6-P) via a specific fructose 1,6-bisphosphatase (FBPase, reaction 1 in Figure 3). In a series of reactions paralleling that seen in the chloroplast for starch synthesis, Fru-6-P can then be converted to glucose 1-phosphate (Glu-1-P) and then to a sugar nucleotide, in this case uridine-diphosphoglucose (UDPG), via the enzyme uridine diphosphoglucose pyrophosphorylase (UDPGPPase, reaction 2). This glucose residue of this sugar nucleotide is then transferred to Fru-6-P in the reaction catalyzed by sucrose-phosphate synthase (SPS; reaction 3). The sucrose phosphate produced in this reaction is finally converted to sucrose by sucrose phosphate phosphatase (reaction 4), resulting in the release of P_i to the cytosol.

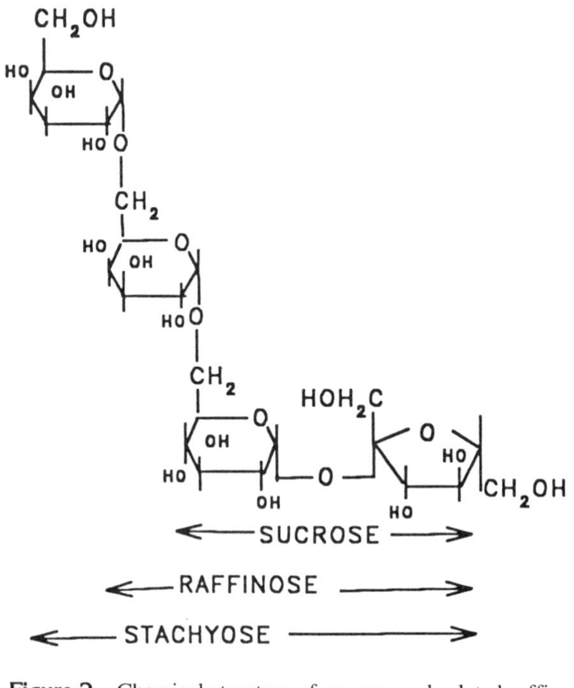

Figure 2   Chemical structure of sucrose and related raffinose family oligosaccharides.

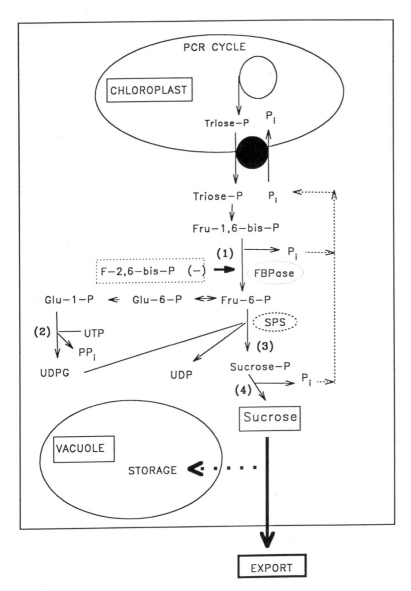

Figure 3 Pathway of sucrose synthesis in the cytoplasm of photosynthetic leaf cells. Reaction 1, cytoplasmic fructose 1,6-bisphosphatase (FBPase); reaction 2, UDPG pyrophosphorylase; reaction 3, sucrose phosphate synthase (SPS); reaction 4, sucrose phosphate phosphatase.

1. Regulation of Sucrose Formation in Leaves

To prevent inhibition of photosynthetic carbon fixation during sucrose synthesis, the export of triose-P, which is also required to run the PCR cycle, must be controlled [9]. Specifically, to maintain optimal CO_2 fixation rates, no more than one triose-P molecule out of six produced photosynthetically can leave the PCR cycle. As indicated in Figure 3, the synthesis of sucrose results in the formation of P_i in the cytosol. [Although not indicated in the diagram, the inorganic pyrophosphate (PP_i) released by reaction 1 can also be converted to P_i by the action of an inorganic pyrophosphatase [9].] Since this production of P_i could result in a large drain of triose-P from the chloroplast, the synthesis of sucrose in the cytosol (or perhaps more correctly the synthesis of cytosolic P_i) must be coordinated with ongoing photosynthetic rates in the chloroplast. Sucrose synthesis is, therefore, a tightly regulated metabolic reaction in photosynthetic cells [9].

At present, there appear to be at least two different strategies for regulation of sucrose production in green plant cells: one mechanism involving regulation of cytosolic FBPase [9,10], which provides hexose phosphates for UDPG and sucrose formation, and the other involving regulation of the SPS enzyme itself [11,12].

REGULATION OF CYTOSOLIC FBPase ACTIVITY. The cytosolic FBPase reaction (reaction 1, Figure 3) is the first irreversible step in carbon flow to sucrose and is subject to strong inhibition by a specific metabolite, fructose 1,2-bisphosphate (F-2,6-bis-P). Formation of F-2,6-bis-P in the cytosol is controlled by a specific Fru-6-P,2-kinase, while degradation is controlled by a Fru-2,6-bis-P phosphatase. The total concentration of F-2,6-bis-P is, therefore, a net result of the combined activities of these two enzymes. The amount of F-2,6-bis-P can, therefore, control the flow of carbon to sucrose by modulating the activity of the FBPase [9,10].

The kinase enzyme, which forms the inhibitor, is activated by Fru-6-P and P_i, the two products of the FBPase reaction. The same two products also inactivate the phosphatase enzyme, which degrades the inhibitor. Thus, the FBPase can indirectly inhibit its own activity, inasmuch as increased activity of the FBPase will lead eventually to increased levels of Fru-1,2-bis-P through increased synthesis and a slower rate of degradation. Conversely, the kinase, which forms the inhibitor, is inactivated by PGA and triose-P (in the form of dihydroxyacetone phosphate), so that when high levels of triose-P are being exported to the cytosol, the inhibition of FBPase by F-2,6-bis-P is relieved and carbon flow to sucrose can continue [9,10].

REGULATION OF SPS ACTIVITY. Regulation of sucrose phosphate synthase (SPS, reaction 3, Figure 3) occurs by a number of different mechanisms. One is the developmental control of the amount of SPS present in leaves as they mature. A second mechanism involves control by intermediates in the sucrose biosynthetic pathway of the SPS enzyme. For example, the SPS enzyme is activated by a high Glu-1-P/P_i ratio, a form of allosteric feed-forward control. The SPS enzyme may also be inactivated by high levels of sucrose, a form of feedback control [11,12].

There is now increasing evidence that in a number of plant species, including agronomically important ones such as maize, the SPS enzyme may also be directly controlled by covalent modification through phosphorylation. This process may modulate activity of SPS in the light, in which dephosphorylation of the enzyme causes its activation. Conversely, in the dark, rephosphorylation of the enzyme by a specific phosphatase results in inactivation. The phosphorylation status of the SPS enzyme, therefore, regulates its diurnal activity and is controlled by factors controlling the enzymes responsible for the phosphorylation/dephosphorylation steps, namely, protein kinase and phosphatase [11,12].

Species differences in SPS regulation are now being reported. Interestingly, the type of SPS regulation seen in a given plant appears to be correlated with the type of carbohydrate stored diurnally in its leaves. For example, in maize and spinach, which accumulate both sucrose and starch as temporary storage reserves, SPS is subject to allosteric control via Glu-1-P and P_i and also appears to be under phosphorylation control. In contrast, in soybean, in which starch alone is accumulated as a storage carbohydrate, there appears to be no such regulation of SPS [11,12].

C. Fructans

In some plant species, water-soluble polymers known as fructans accumulate as carbohydrate storage products. Fructans, as the name implies, are linear polymers of fructose; in leaves, they are derived from photosynthetically produced sucrose. Fructans have as their core starting component a single molecule of sucrose, to which chains of fructose residues are attached. The type of linkage between adjacent fructose residues, as well as the point of attachment of the fructose chains to the sucrose molecule, determines the type of fructan accumulated in a given plant [13–15].

1. Fructan Structure

There are three major classes of fructans found in agronomically important crop plants—the isokestose or inulin series, the kestose or phlein series, and the neokestose series, each of which is named for its characteristic trisaccharide sucrosyl–fructose [13–15]. In the isokestose series, which is synthesized in members of the Asteraceae such as Jerusalem artichoke (*Helianthus tuberosus* L.), fructose residues are attached to the fructosyl residue of sucrose in nonreducing β-2,1-linkages. Fructans of the isokestose series, therefore, have the general form:

$$Glu-1,2-Fru-1, (2-Fru-1)_n, 2-Fru$$
$$\uparrow$$
$$sucrose$$

where n_{max} is approximately 35.

Fructans of the kestose series, which are common in many temperate grass species including wheat and barley, consist of fructose residues joined by β-2,6 linkages and have the general form:

$$Glu-1,2-Fru-6, (2-Fru-6)_n, 2-Fru$$
$$\uparrow$$
$$sucrose$$

where n_{max} is approximately 250.

Fructans of the neokestose series, which have been isolated from asparagus (*Asparagus officinalis* L.), have fructose residues joined to both the glucose and the fructose residues of sucrose and have the general form:

$$Fru-2, (1-Fru-2)_m-1-Fru-2, 6-Glu-1,2-Fru-1, (2-Fru-1)_n, 2-Fru$$
$$\uparrow$$
$$sucrose$$

where m_{max} and n_{max} are each approximately 10.

Branched fructans also occur in nature, and frequently there is more than one fructan series in the same plant.

2. Fructan Synthesis in Leaves

The biosynthetic pathway leading to fructan synthesis differs substantially from that leading to sucrose or starch synthesis in that the fructosyl donor is not a sugar nucleotide. Instead, sucrose itself acts as the fructosyl donor to create the fructosylsucrose isokestose, by the reaction catalyzed by the enzyme sucrose:sucrosylfructose transferase (SST, reaction 1, Figure 4). The glucose released by the SST reaction is thought to reenter the general cytoplasmic hexose–phosphate pool following phosphorylation (Figure 4). Chain elongation then proceeds by the reaction catalyzed by another enzyme, fructan:fructosyltransferase (FFT, reaction 2, Figure 4) which utilizes the fructosylsucrose as a fructose donor to another fructosylsucrose. Distinct FFT enzymes can be isolated from plant tissues, which can form β-1,2 or β-2,6 linkages to fructose or glucose residues of fructosylsucroses formed by the SST reactions [13,14].

Both SST and FFT enzymes appear to be localized exclusively in plant vacuoles, where

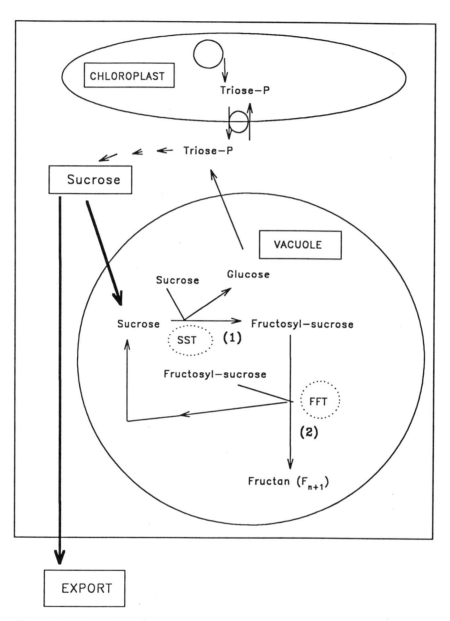

Figure 4 Pathway of fructan synthesis in the vacuole of photosynthetic leaf cells. Reaction 1, sucrose sucrosyl transferase (SST); reaction 2, fructan:fructan fructosyltransferase (FFT).

fructan accumulation occurs [13,14]. In leaves, fructan levels are usually low, but fructans can accumulate in response to environmental conditions, which serve to elevate carbohydrate levels—for example, in response to low temperatures [16]. Experimentally, cereal and grass leaves can be induced to form large quantities of fructan following excision, which eliminates phloem transport of sucrose, and continuous illumination, which promotes sucrose synthesis [13,17]. The genetic machinery for fructan synthesis, therefore, is present in leaves, although the growth conditions a plant experiences may not always result in activation of this machinery.

Fructans appear to be a form of readily accessible carbon and are degraded by the action of a fructan hydrolase, β-fructofuranosidase. It is thought that fructan metabolism in the vacuole of photosynthetic cells may serve to buffer chloroplasts from the adverse changes in cytosolic metabolites that occur when phloem transport is limited. Additionally, utilization of the vacuole provides a larger compartment for short-term carbohydrate storage than either the chloroplast or the cytoplasm, while polymerization avoids the osmotic problems that would occur if the large amounts of carbon partitioned into fructan were stored in the form of sucrose [13–15]. The mechanisms that control carbon partitioning into fructans are not yet established.

D. Polyols

Polyhydroxy alcohols, or polyols, are probably ubiquitous in all plant species, but only in relatively few plant families are these compounds found to be synthesized from photosynthetically fixed carbon in source leaves [18,19]. The most commonly occurring polyols are derivatives of hexose sugars in which the aldose or ketose group has been reduced to a hydroxyl group. Thus, mannitol, sorbitol, and dulcitol (Figure 5) are the polyol equivalents of the hexoses glucose, fructose, and galactose, respectively.

Formation of a polyol from a hexose sugar requires reduction of the aldehyde or ketone group. In higher plants, this reduction takes place through a hexose–phosphate intermediate, as indicated in Figure 6. In source leaves of celery [20,21] and privet [20], reduction of mannose-6-P to mannitol-1-P is catalyzed by the enzyme mannose-6-P reductase (M6PR), which utilizes NADPH as reductant. Similarly, in leaves of apple, peach, pear, apricot [22], and loquat [23], an aldose 6-phosphate reductase catalyzes the reduction of glucose-6-P to sorbitol-1-P, again using NADPH. A similar NADPH-dependent enzyme is also present in *Euonymus* leaves, producing dulcitol [24].

As indicated in Figure 6, synthesis of polyols always occurs in addition to sucrose synthesis, not in substitution for it. The regulatory mechanisms that control the allocation of carbon between sucrose and polyols are not yet known. Immunological evidence clearly indicates that polyol synthesis is a cytoplasmic event [25], but the regulation of this biosynthetic pathway has not been deciphered. Both sucrose and polyols are exported in the phloem and/or may also be stored in the vacuole for later export, but again, the regulation of compartmentation between storage and export pools is not fully understood.

Since polyols are not rapidly utilized by source leaf tissues, which lack enzymes to reconvert them to hexoses or hexose–phosphates, they are particularly useful as storage and transport forms of carbon in source leaves. Polyols may also serve a role as compatible solutes in source

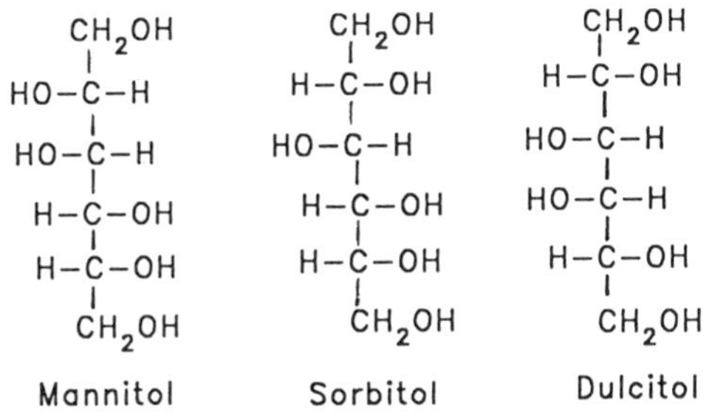

Figure 5 Structures of commonly occurring plant polyols.

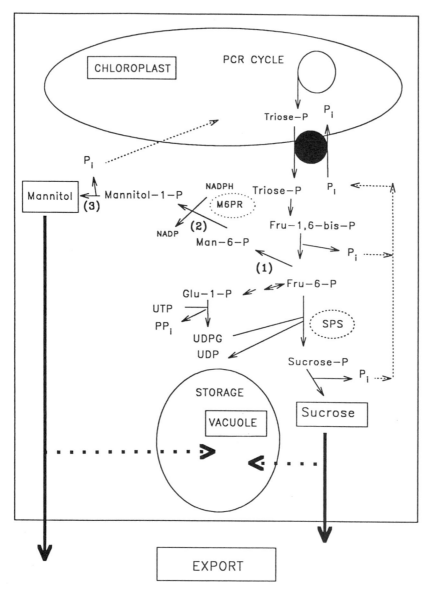

Figure 6 Pathway of polyol (mannitol) synthesis in the cytoplasm of photosynthetic leaf cells. Reaction 1, mannose 6-phosphate isomerase; reaction 2, mannose 6-phosphate reductase (M6PR); reaction 3, mannitol phosphate phosphatase.

leaves, allowing continuation of photosynthetic activity and carbon metabolism under adverse environmental conditions such as water stress. Additionally, the intriguing hypothesis has been put forward that the utilization of reductant in polyol synthesis allows recycling of NADPH between the chloroplast and cytosol, preventing photoinhibition under stress conditions. This possibility could also explain the unusually high photosynthetic rates commonly seen in polyol-synthesizing plants [18,19]. The underlying mechanisms that regulate the synthesis of polyols in source leaves have not as been established.

E. Raffinose Family Oligosaccharides

Like the polyols, raffinose family oligosaccharides are probably ubiquitous in the plant kingdom. There are a large number of plant families in which these oligosaccharides are synthesized in leaves and used as translocatable forms of carbon [26]. However, of the many plant species in which translocation of these sugars may occur, only a few, such as the cucurbit vine crops, are of major agronomic importance. As a result, this biochemical pathway of carbohydrate formation has been relatively neglected by crop physiologists. This is truly unfortunate, for recent evidence now clearly indicates that the synthesis of raffinose family oligosaccharides is quite different from that of other known soluble carbohydrates in a number of ways.

The raffinose family oligosaccharides, of which raffinose and stachyose (Figure 2) are the most common examples, are all simple galactosides of sucrose. The addition to the sucrose molecule of the galactose residues (which are linked by α-1,6 linkages to the glucose moiety of sucrose) occurs, as was seen with fructans, without the use of a sugar nucleotide. Instead, galactinol, a novel galactoside of myoinositol, is used as the galactose donor [27]. What is particularly unique about the raffinose family oligosaccharides is that the transfer of galactose residues to sucrose probably does not occur in the photosynthetic cell where sucrose is synthesized [28–30].

As indicated in Figure 7, the synthesis of raffinose family oligosaccharides is now believed to take place in two separate leaf cell types: the photosynthetic mesophyll cell and the modified phloem companion cell, or intermediary cell, which is characteristically found in leaves in which these oligosaccharides are synthesized [31,32]. As far as is known, production of sucrose in leaves in these plants occurs in much the same fashion as in other plant species. In the cucurbit vine crops, sucrose synthesis does not appear to be light-regulated [33], which suggests that SPS is not controlled by protein phosphorylation. The actual mechanisms controlling sucrose production in raffinose oligosaccharide synthesizing plants have not been elucidated. There is evidence, however, that sucrose synthesis occurs within the cytoplasm of the photosynthetic cells [28,34].

In plants that synthesize the raffinose family oligosaccharides, sucrose is used as a phloem-mobile and as a storage carbohydrate. However, it also must be used as the sucrose backbone for the synthesis of the raffinose saccharides. The way in which mobile or storage sucrose pools are kept separated from metabolizable sucrose pools is not clear, but compartmentation within the two different cell types involved in raffinose oligosaccharide biosynthesis may be occurring.

To further complicate carbon partitioning in these leaves, carbon must also be diverted *away* from the sucrose biosynthetic pathway, to allow formation of the galactose donor, galactinol. In some plants, this may occur within the photosynthetic cell, probably through the conversion of UDPglucose to uridine diphosphogalactose (UDPGal), the galactose donor used by galactinol synthase (GS, reaction 1, Figure 7). Galactinol and sucrose then cross into the intermediary cell, via the abundant plasmodesmata that interconnect these cells with the photosynthetic cells, where raffinose oligosaccharide synthesis takes place via the operation of raffinose synthase (RS, reaction 2) and stachyose synthase (SS, reaction 3).

In other plant species, there is evidence that galactinol synthesis may also take place within the intermediary cell [29]. In this case, as indicated in Figure 7, sucrose alone may leave the photosynthetic cell, to be used both as the sucrose moiety of the raffinose sugars and also for the synthesis of galactinol. Metabolism of sucrose may take place via sucrose synthase (reaction 4), which yields UDPG, from which UDPGal could be synthesized. In squash leaves, immunological data indicate the presence in the intermediary cells of both stachyose synthase (SS, reaction 3) and galactinol synthase (GS, reaction 1), but the complete details concerning

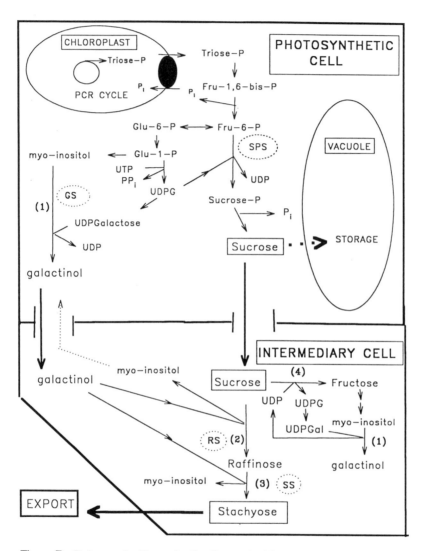

Figure 7 Pathway of raffinose family oligosaccharide biosynthesis in leaf tissues. Reaction 1, galactinol synthase (GS); reaction 2, raffinose synthase (RS); reaction 3, stachyose synthase (SS).

the location of the biosynthetic enzymes in this pathway remain to be established. In fact, since these oligosaccharides can also serve a storage function in plant tissues, it is likely to prove that the entire pathway leading to raffinose family oligosaccharide synthesis occurs both in the photosynthetic cells, where they are used for storage, and in the intermediary cell, where they are used for transport [29,35].

The regulation of the raffinose pathway in source leaves is not understood. From preliminary reports, the key regulating enzyme in the pathway would appear to be galactinol synthase (GS, reaction 1, Figure 7), which changes its activity in response to light treatments [36] and to changes in photoassimilate export rate. Additionally, stachyose synthase (SS, reaction 3, Figure 7), which has higher activity in fruiting than in vegetative plants [37], may also be under some form of metabolic control. During the dark period, export of stachyose in muskmelon declines and the plant becomes predominantly a sucrose transporter [38], an observation suggesting that some form of light regulation of these enzymes is a possibility. In this context, the common observation that GS requires a reducing environment for full activity suggests that this enzyme may be under some form of redox control. Certainly further study is warranted.

III. CARBOHYDRATE FORMATION IN NONPHOTOSYNTHETIC (SINK) TISSUES

The soluble and insoluble forms of carbohydrate listed above are all used as temporary storage reserves in the leaf. However, certain of the soluble forms, such as sucrose, the raffinose family oligosaccharides, and the polyols, are also phloem-mobile and can be delivered to nonphoto-synthetic tissues to support growth and development of these plant parts [24]. It is commonly found, therefore, that even nonphotosynthetic tissues will contain some or all of the same carbohydrates that commonly occur in phloem sap. However, since carbohydrates are also required for growth processes such as respiration or cell wall synthesis, sink tissues are also equipped with enzymes for breakdown, interconversion, and metabolism of whatever phloem-mobile carbohydrates are supplied to them. As a result, it is also not uncommon to find carbohydrates that are in fact quite different from those supplied to the sink by phloem transport.

Carbohydrates formed in sink tissues may act as storage reserves and, as occurs in source leaves, they are found to be compartmentalized in specialized cells or cellular compartments such as the plastids or vacuoles. In most agronomic crops, it is these storage reserves that are of economic importance—for example, the yield of seeds, grains, and storage roots or tubers is dictated principally by the size of their carbohydrate reserves at harvest. The enzymes for synthesis of common storage carbohydrates, including soluble and polymeric forms, are therefore found in a range of plant tissues, not just the mature leaves. Research into carbohydrate metabolism in nonphotosynthetic tissues is showing that the controlling factors in the regulation of carbohydrate synthesis are often surprisingly similar to those in photosynthetic cells.

A. Starch

Starch synthesis by sink tissues is probably one of the most important plant biochemical reactions in terms of human nutrition, since starch, particularly from grain crops (where it can make up 70% of the dry weight) is a major provider of nutritional calories in the human diet everywhere [15]. Despite the importance of starch biosynthesis in crop plants, we actually know very little concerning the biochemistry of starch deposition in sink tissues.

Starch deposition in sink tissues occurs at the expense of imported assimilates and appears to require the conversion of phloem-delivered solutes into a usable hexose–phosphate form [39]. For sucrose, there are at least two pathways by which this conversion occurs: by invertase hydrolysis to hexose followed by phosphorylation to hexose-P, and by reversal of the sucrose synthase reaction to provide fructose and UDPG:

Invertase: sucrose $\rightarrow$ glucose + fructose
Sucrose synthase: sucrose + UDP $\rightarrow$ UDPG + fructose

Depending on the sink tissue, there is evidence for operation of both these pathways in sinks. Formation of hexose-P takes place by direct phosphorylation or, in the case of UDPG, by reversal of the UDPG pyrophosphorylase reaction:

UDPPase: UDPG + PP$_i$ $\rightarrow$ Glu-1-P + UTP

It is now believed that synthesis of starch in the amyloplasts of sink tissues takes place from hexose-P imported from the cytoplasm [39]. The subsequent enzymatic steps appear to be similar to the series found in source leaves. Somewhat surprisingly, the regulation of

ADPGPPase by PGA and P_i appears to be a constitutive property of this enzyme from all sources studied so far [4,5,7], even though for most sinks PGA is not a major amyloplast metabolite.

There is relatively little information available concerning the conversion of raffinose family oligosaccharides or polyols to starch in sink tissues. Enzymes for raffinose oligosaccharide hydrolysis, the α-galactosidases, have been reported and characterized for a number of sink tissues [40], and there have been reports of enzymes for conversion to hexose of sorbitol [19] and mannitol [41] (sorbitol dehydrogenase and mannitol dehydrogenase, respectively). The remaining biochemistry remains to be clarified by further research, however.

B. Sucrose

Synthesis of sucrose is an important physiological function in some sink tissues—for example, in ripening fruits, where stored starch is metabolized to sugar during the ripening process. Similar metabolism of starch or fructans also occurs upon sprouting of perennating organs such as tubers and bulbs. During germination of seeds, conversion to sucrose of stored carbohydrates, and also other reserves such as wall materials and oils, occurs in endosperm tissues and cotyledons and is an essential process for growth and emergence of the embryo. The regulation of sucrose synthesis in sink tissues is under active investigation in many research laboratories, but comprehensive details of control of sucrose synthesis in sinks are still lacking.

C. Fructans

There are many reports of fructan accumulation in vegetative sinks of fructan-accumulating plants. Indeed, fructan accumulation is far more common in vegetative storage sinks than it is in source leaves, where experimental manipulation often must be used to induce fructan accumulation. Fructans accumulate naturally to particularly high levels in overwintering organs such as bulbs and tubers and also in stems of grasses, possibly in response to environmental cues. The exact regulatory mechanisms controlling fructan accumulation from imported solutes in sinks are not yet known.

D. Polyols

Although sink tissues such as seeds may contain polyols, these are usually only trace components, not major carbohydrates. In polyol translocating species, vegetative tissues such as petioles, stems, and roots may accumulate polyols, but there is no evidence that this occurs by direct synthesis rather than by simple import from the phloem. Polyol metabolism in sink tissues is, therefore, only very poorly understood and needs to be addressed in further research.

E. Raffinose Family Oligosaccharides

Raffinose oligosaccharides are prevalent in seeds of numerous plant species, even those that do not use these sugars as phloem-mobile compounds. Thus, although the biosynthetic pathway for raffinose oligosaccharide synthesis is not operative in the source leaves of many plants, it is encoded in the genome and is expressed in the developing seeds. It is thought that these oligosaccharides, which accumulate during seed drying, may allow maintenance of seed cellular membranes during dessication. Despite their prevalence, and the possibility that they perform an essential metabolic role in the dry seed, the regulatory mechanisms controlling synthesis of these sugars from imported assimilates in seeds has received relatively little attention. Indeed, current research appears to be focusing on genetic and molecular protocols for removal of this biochemical pathway from seeds, since the raffinose oligosaccharides are major antinutritional factors in many seed crops used for human consumption. It will be interesting to see what success these approaches achieve if indeed the raffinose pathway turns out to be necessary for seed viability.

Raffinose oligosaccharides also accumulate in some vegetative storage tissues, where they may simply reflect accumulation of imported stachyose [42]. However, there are numerous cases in which these sugars appear to be synthesized de novo in vegetative tissues including, interestingly, the sugar beet taproot. Although the first isolation of galactinol was from this tissue [43], and although raffinose produced in sugar beet is a major "contaminant," which interferes with sucrose crystallization, there has been little study of this *de novo* biochemical pathway in vegetative tissues. Certain environmental stresses, such as low temperature, can also result in accumulation of these oligosaccharides in vegetative tissues, including source leaves in which they would not be normally synthesized. The metabolic function and cellular location of these induced oligosaccharides is not known. Clearly more research is also needed in the area of raffinose family oligosaccharide synthesis in sink tissues.

F. Structural Carbohydrates

A key structural component of plant cells that differentiates them from animal cells is the presence of a rigid, highly structured cell wall. The plant cell wall consists largely of cellulose and related heteroglucan polymers [44] and can, therefore, be looked at as a very complex carbohydrate structure. The formation of the cell wall, which requires a tremendous input of carbon, is one of the most important uses of photosynthetically fixed carbon in the developing plant body. From an agronomic standpoint, the cell wall gives rise to many important fibers (e.g., cotton) with many industrial applications and also is an important fiber component of animal feeds.

The principal cell wall component is the carbohydrate cellulose, which is a linear glucose polymer containing β-1,4 linkages. Synthesis of cellulose occurs by chain elongation of a cellobiose primer via UDPG and the enzyme UDPG:glucan synthase [45]. We have considerable biochemical detail concerning cellulose biosynthesis, but relatively little is known concerning synthesis of any of the other numerous cell wall components. Again, the synthesis of structural components is an important area for future research.

IV. FUTURE PERSPECTIVES

With the advent of molecular biological techniques, we have entered a particularly exciting era in terms of carbohydrate biochemistry. Already many of the initial biochemical studies on regulation of sucrose and starch metabolism have been very elegantly corroborated in vivo through the direct manipulation of enzymes capable of synthesizing sucrose and starch in transgenic plants [46]. This ability to manipulate expression of carbohydrate-synthesizing enzymes by molecular approaches provides a radical new way to study carbohydrate formation in plant tissues and will certainly revolutionize our understanding of carbohydrate formation in crop plants.

REFERENCES

1. P. Hamer, *Physiol. Veg.*, *23*: 107 (1985).
2. D. A. DeMason, M. A. Madore, K. N. Chandra Sekhar, and M. J. Harris, *Protoplasma*, *166*: 177 (1992).
3. J. Preiss and C. Levi, *The Biochemistry of Plants*, Vol. 3, *Carbohydrates: Structure and Function* (J. Preiss, ed.), Academic Press, New York, pp. 371–423 (1980).

4. J. Preiss, *The Biochemistry of Plants*, Vol. 14, *Carbohydrates* (J. Preiss, ed.), Academic Press, New York, pp. 182–249 (1988).
5. P. John, *Biosynthesis of the Major Crop Products*, Wiley, Chichester, pp. 32–53 (1992).
6. M. Steup, *The Biochemistry of Plants*, Vol. 14, *Carbohydrates* (J. Preiss, ed.), Academic Press, New York, pp. 255–295 (1988).
7. D. M. Stark, K. P. Timmerman, G. G. Barry, J. Preiss, and G. M. Kishore, *Science*, *258*: 287–292 (1992).
8. U. Heber and H. W. Heldt, *Annu. Rev. Plant Physiol.*, *32*: 139–167 (1981).
9. M. Stitt, *Plant Physiol.*, *84*: 201–201 (1987).
10. M. Stitt, S. C. Huber, and P. Kerr, *The Biochemistry of Plants*, Vol. 10, *Photosynthesis* (M. D. Hatch and N. K. Boardman, eds.), Academic Press, New York, pp. 327–408 (1987).
11. S. C. Huber, J. L. A. Huber, and R. W. McMichael, *Carbon Partitioning Within and Between Organisms* (C. J. Pollock, J. F. Farrar, and A. J. Gordon, eds.), BIOS Scientific Publishers, Oxford, pp. 1–26 (1992).
12. S. C. Huber, *Plant Physiol.*, *99*: 1275–1278 (1992).
13. C. J. Pollock and A. J. Cairns, *Annu. Rev. Plant Physiol. Plant Mol. Biol.*, *42*: 77–101 (1991).
14. C. J. Pollock and N. J. Chatterton, *The Biochemistry of Plants*, Vol. 14, *Carbohydrates* (J. Preiss, ed.), Academic Press, New York, pp. 109–140 (1988).
15. P. John, *Biosynthesis of the Major Crop Products*, Wiley, Chichester, pp. 55–69 (1992).
16. P. Bancal and J. P. Gaudillere, *New Phytol.*, *124*: 375–379 (1993).
17. U. Simmen, D. Obenland, T. Boller, and A. Wiemken, *Plant Physiol.*, *101*: 459–468 (1993).
18. W. H. Loescher, J. K. Fellman, T. C. Fox, J. M. Davis, R. J. Redgwell, and R. A. Kennedy, *Regulation of Carbon Partitioning in Photosynthetic Tissue* (R. L. Heath and J. Preiss, eds.), American Society of Plant Physiologists, Rockville, MD, pp. 309–332 (1985).
19. W. H. Loescher, *Physiol. Plant.*, *70*: 553–557 (1987).
20. W. H. Loescher, R. H. Tyson, J. D. Everard, R. J. Redgwell, and R. L. Bieleski, *Plant Physiol.*, *98*: 1396–1402 (1992).
21. M. E. Rumpho, G. E. Edwards, and W. H. Loescher, *Plant Physiol.*, *73*: 869–873 (1983).
22. F. B. Negm and W. H. Loescher, *Plant Physiol.*, *67*: 139–142 (1981).
23. F. B. Negm, *Plant Physiol.*, *80*: 972–977 (1986).
24. M. Hirai, *Plant Physiol.*, *67*: 221–224 (1981).
25. J. D. Everard, V. R. Franceschi, and W. H. Loescher, *Plant Physiol.*, *102*: 345–356 (1993).
26. M. H. Zimmerman and H. Ziegler, *Transport in Plants*, Vol. I, *Phloem Transport* (M. H. Zimmerman and J. A. Milburn, eds.), Springer-Verlag, Berlin, pp. 480–503 (1975).
27. O. Kandler, *Harvesting the Sun* (A. San Pietro, F. A. Greer, and T. J. Army, eds.), Academic Press, New York, pp. 131–152 (1967).
28. K. Schmitz and U. Holthaus, *Planta*, *169*: 529–535 (1985).
29. D. U. Beebe and R. Turgeon, *Planta*, *188*: 354–361 (1992).
30. L. L. Flora and M. A. Madore, *Planta*, *189*: 484–490 (1993).
31. R. Turgeon, *Phloem Transport and Assimilate Compartmentation* (J. L. Bonnemain, S. Delrot, W. J. Lucas, and J. Dainty, eds.), Ouest Editions Presses Academique, Nantes, France, pp. 18–22 (1991).
32. A. J. E. van Bel, *Annu. Rev. Plant Physiol. Plant Mol. Biol.*, *44*: 253–281 (1993).
33. S. C. Huber, T. H. Neilsen, J. L. A. Huber, and D. M. Pharr, *Plant Cell Physiol.*, *30*: 277–285 (1989).
34. M. A. Madore and J. A. Webb, *Can. J. Bot.*, *60*: 126–130 (1982).
35. M. A. Madore, *Planta*, *187*: 537–541 (1992).
36. N. S. Robbins and D. M. Pharr, *Plant Physiol.*, *85*: 592–597 (1985).
37. U. Holthaus and K. Schmitz, *Planta*, *184*: 525–531 (1991).
38. D. E. Mitchell, M. V. Gadus, and M. A. Madore, *Plant Physiol.*, *99*: 959–965 (1992).
39. T. Ap Rees, *Carbon Partitioning Within and Between Organisms* (C. J. Pollock, J. F. Farrar, and A. J. Gordon, eds.), BIOS Scientific Publishers, Oxford, pp. 115–132 (1992).
40. P. M. Dey, *Methods in Plant Biochemistry*, Vol. 2, *Carbohydrates* (P. M. Dey and J. B. Harborne, eds.), Academic Press, New York, pp. 189–217 (1990).

41. J. H. Stoop and D. M. Pharr, *Arch. Biochem. Biophys.*, *298*: 612–619 (1992).

42. F. Keller, *Plant Physiol.*, *98*: 442–445 (1992).

43. R. J. Brown and R. F. Serro, *J. Am. Chem. Soc.*, *75*: 1040–1044 (1953).

44. A. Bacic, P. J. Harris, and B. A. Stone, *The Biochemistry of Plants*, Vol. 14, *Carbohydrates* (J. Preiss, ed.), Academic Press, New York, pp. 297–371 (1988).

45. D. P. Delmer and B. A. Stone, *The Biochemistry of Plants*, Vol. 14, *Carbohydrates* (J. Preiss, ed.), Academic Press, New York, pp. 373–419 (1988).

46. A. M. Smith, H. E. Neuhaus, and M. Stitt, *Planta*, *181*: 310–315 (1990).

13

Nitrogen Metabolism and Crop Productivity

Fred E. Below
University of Illinois, Urbana, Illinois

I. INTRODUCTION

Among the mineral nutrient elements, nitrogen (N) most often limits the growth and yield of nonleguminous crop plants, which require relatively large quantities of N (from 1.5 to 5% of the plant dry weight) for incorporation into numerous organic compounds. These compounds include proteins, nucleic acids, chlorophyll, and growth regulators, all of which have crucial roles in plant growth and development. The N composition of plant tissues also has important nutritional consequences, since plants are a major source of proteins in the diet of humans and animals. Because N deficiency can seriously decrease yield and crop quality, elaborate steps are often taken to assure that adequate N levels are available to plants.

Although plants can absorb small amounts of N from the atmosphere through their foliage, by far the greater part of it is acquired from specific forms in the soil such as nitrate (NO_3^-) or ammonium (NH_4^+). Most soils, however, do not have sufficient N in available form to support desired production levels. Therefore, addition of N from fertilizer is typically needed to maximize crop yields; this requirement has resulted in the development of a large N fertilizer industry. Some estimates suggest that N fertilizer accounts for 80% of all fertilization costs and 30% of all energy costs associated with crop production [1].

While it is well accepted that sufficient N is needed to obtain high yields, growers each year must determine how much fertilizer N to apply. This problem results from the complex cycle of N in the environment, which can allow for loss from the rooting zone. It is further complicated by mechanistic inconveniences associated with fertilizer N application, and by uncertainty related to weather conditions, especially water availability. Unused fertilizer N is economically wasteful and can become an environmental hazard if it is lost from the soil. Excessive use of fertilizer N has been implicated in the contamination of groundwater by NO_3 [2–5], which represents a potential health hazard to humans and animals [6,7]. As public awareness focuses on environmental quality, there are increasing pressures on growers to improve N management.

275

Additional knowledge regarding N use by crop plants is clearly one way to help improve N fertilizer management. Although complex, factors such as N use that limit or enhance crop productivity do so by affecting specific physiological processes within the plant. A better understanding of how N governs crop growth and yield will add to information required to improve N management, and will help to minimize the adverse environmental impact of N fertilizer use.

II. NITROGEN ACQUISITION BY CROP PLANTS

A. Nitrogen Availability

Under natural conditions, N enters the soil environment as the result of biological fixation and/or decomposition of animal or plant residues. Most (>90%) of the N in soils is contained in organic matter, which is relatively stable and not directly available to plants. Although a portion of the N in organic matter can be made available through mineralization by soil microorganisms, the amount released is variable depending on management practices and environmental conditions. In addition, the release is normally too slow to meet the needs of a growing crop, with only 2–3% of the N converted to available forms per year. As a result, addition of N from chemical fertilizers is usually required to optimize crop growth and yield.

Nitrogen is unique among the mineral nutrients in that it can be absorbed by plants in two distinct forms, either as the anion NO_3^- or the cation NH_4^+. Although numerous N fertilizer formulations are available that contain varying proportions of NO_3-N to NH_4-N, ammoniacal fertilizers are used more extensively because they are lower in cost [8]. However, NO_3 is the predominant form of N absorbed by plants, regardless of the source of applied N [9,10]. This preference is due to two groups of chemoautotropic soil bacteria, which rapidly oxidize NH_4 to NO_3 (nitrification) in warm, well-aerated soils that are favorable to crop growth.

The form of N (NH_4 or NO_3) can affect the availability of N to the plant as a result of differences in mobility of each form in the soil solution. In soil, the positively charged NH_4^+ ion is bound to negatively charged soil particles and is relatively immobile. In contrast, the negatively charged NO_3^- ion is repelled by soil particles, which aids in its movement to plant roots. Even though NO_3 is the N form most available to plants, however, it can be more readily lost from the rooting zone because it is susceptible to leaching and denitrification [11]. Both these economically and environmentally undesirable processes (i.e., leaching and denitrification) perpetuate a large amount of the uncertainty associated with N fertilizer management.

In the United States, N fertilizer recommendations are usually based on the past crop history of the field and expected yield goal and, to a lesser extent, on formulas calculated to estimate the soil's capacity for N mineralization [12,13]. Other factors (e.g., fertilizer cost and value of the crop) must also be considered [14]. While generally sound, problems with fertilizer recommendations can arise if the yield goal is unrealistic, or if growers fail to accurately assess the capacity of the soil to supply the crop with N. As a result, two main types of test have been developed to measure soil N: tests to determine the soil's potential to mineralize N from organic matter and direct measurement of residual inorganic N.

Several techniques have been developed to measure mineralization of soil N, which are collectively known as N availability indices [15,16]. These methods estimate the potential for organic N to be mineralized and involve either incubations [16–18] or some type of chemical extraction [19,20]. Some studies have shown that these tests can provide reasonable estimates of potentially mineralizable N [21,22]. They have not been widely used for making N fertilizer recommendations, however, because of difficulty in conducting the measurements and lack of supporting data to help interpret the results.

The other approach to assessing the soil N supply involves measuring the level of inorganic N in the soil profile and then adjusting the fertilizer N recommendation to account for N that is already present [15,16,19]. One such test for maize, known as the "late spring nitrate test," takes some of the uncertainty associated with N cycling into account by not removing soil samples until after the crop has been established (plants are in the early vegetative stage), when the potential for N loss is lessened. Based on soil analysis and yield response to applied N, a soil NO_3-N concentration in excess of 20–25 mg kg^{-1} (ppm) is considered adequate for maximum yield of maize; whereas lower values indicate the need for additional fertilizer N [23–26]. Although good at identifying situations in which no fertilizer N is required, the test does not work as well when the degree of responsiveness to fertilizer N application must be predicted or when a high percentage of the N is available as NH_4 [27]. In addition, this technique cannot be used if all the N is applied preplant, or if the N is knifed in as anhydrous ammonium. As a result, tests based on plant characters have also been developed as a way of assessing the soil N supply.

An advantage of plant measurements is that they integrate the effects of soil N availability and plant N uptake, regardless of the N source or application method. Additionally, because they are based on the plant, rather than the soil, plant measurements are more likely to reflect the direct impact of N availability on growth and yield. Tissue testing of plants to compare N concentrations with critical levels is a well-established procedure to document crop N status. These tests involve measuring organic N (also called reduced N) in the leaves [28,29] or inorganic N (NO_3) in the stems [30,31] and can be used to determine deficiencies as well as excessive applications of N. This technology, however, has typically been used in diagnostic work rather than as a management tool, because the measurements usually are made too late to permit corrective N applications.

Recently, leaf chlorophyll measurements have been advocated as a means of taking advantage of the close association between chlorophyll and leaf N concentration to assess soil N availability and plant N status [32,33]. The development of a handheld leaf chlorophyll meter (SPAD-502, Minolta Camera Co.) allows for rapid and nondestructive measurement of leaf greenness [34,35]; and some evidence suggests that this technique can be used as a management tool for making fertilizer N recommendations [36,37]. Other work, however, has shown that widespread calibration of chlorophyll meters to determine crop N status may not be practical, given differences in leaf greenness among cultivars and/or effects on the readings of growth stage, N form, and management practices [32,38]. As a result, normalization procedures may be necessary to standardize chlorophyll meter readings across cultivars, locations, and growth stages by comparing readings from well-fertilized rows with those from the test area [32].

B. Nitrogen Accumulation

1. Nitrogen Uptake

Plants acquire the vast bulk of their N from the soil via the root system. This process involves the movement of inorganic N (NO_3 and NH_4) across membranes, transport or storage within the plant, and ultimately assimilation into organic compounds. The uptake of both N forms is generally considered to require metabolic energy mediated by enzyme permeases located in or on the plasmalemma of external root cells. Absorption of both forms is affected by the ion's concentration in the external solution, with the uptake rate exhibiting diminishing returns in response to increasing internal concentrations. Absorption is also affected by external factors like temperature and pH (see Section II. B.2).

The consequences to plant metabolism from the uptake of NO_3 and NH_4 are vastly different because of differences in the charge of NO_3 and NH_4. With NH_4 nutrition, plants absorb cations in excess of anions, resulting in a net efflux of H^+ from the root and an acidification of the external medium [9,10]. Conversely, with NO_3 nutrition, plants absorb an excess of anions, which causes the medium to become more alkaline [9,10]. Also because of these differences in charge, the mechanisms for uptake by plant roots differ for NO_3 and NH_4.

In evaluations of N uptake, plants that have depleted their N supply (both in solution and in storage) are typically used to observe all phases of uptake and the influence of N in inducing the uptake system. For NO_3-depleted plants, the pattern of NO_3 uptake generally exhibits a two-phase pattern, with an initial lag period followed by an exponential increase in uptake [39–41]. The initial lag in NO_3 uptake is in contrast to that observed with many other ions [40,41] and suggests the induction of a specific NO_3 transporter by NO_3. The accelerated phase of NO_3 uptake is also indicative of induction because it is dependent on a critical NO_3 concentration in the root, in a manner similar to enzyme induction by its substrate [41,42]. In addition, the accelerated phase is restricted by inhibitors of protein or RNA synthesis, or by conditions that limit or inhibit respiration [40,43]. Collectively, these studies show that the NO_3 uptake system is dynamic and capable of adjusting to changes in the level of NO_3 in the root environment.

The uptake of NO_3 is an active process, which must overcome an unfavorable electrochemical gradient between the soil and the root. However, because of this gradient, NO_3 can also efflux (or leak) back out of the root. Efflux has been described as a passive diffusion process [44] or as a carrier-mediated process [45], but in either case dependent on the internal concentration of NO_3 in the root. As a result, the net accumulation of NO_3 is a function of the difference between influx and efflux. As might be expected, efflux is greatest when high concentrations of NO_3 have been accumulated by root tissues [46,47].

Unlike NO_3 uptake, the absorption of NH_4 does not exhibit a prolonged lag under N-depleted conditions [48]; although uptake can also be characterized by two main phases [40]. The initial phase of NH_4 is insensitive to low temperatures or metabolic inhibitors, hence is thought to occur passively [40,49]. In contrast, the second phase of NH_4 uptake involves metabolic energy and is sensitive to low temperatures and inhibitors [49]. In some plant species, the active phase of NH_4 uptake is also multiphasic, exhibiting uptake and growth rates associated with deficiency, luxury consumption, and toxicity [40,50,51].

Although the process of NH_4 uptake is not completely understood, it is clear that passive and active uptake must occur by different mechanisms. For passive uptake, the positively charged NH_4 ion may be absorbed by a uniport following the electrochemical gradient across the plasmalemma [52,53]. Conversely, since membrane permeability to NH_3 is greater than that of NH_4, passive uptake could also occur by nonspecific diffusion of NH_3 gas [54,55]. In the soil solution, the distribution of NH_4^+ and NH_3 is a function of an equilibrium relationship driven by pH. At neutral (or lower) soil pH values, more than 99% of the total ammoniacal N is in the protonated (NH_4) form, which would result in limited absorption of gaseous NH_3 by the root [54]. While aboveground plant parts can also absorb gaseous NH_3 through stomata, the amounts acquired are limited in unpolluted air [56,57]. In addition, because high concentrations of NH_3 are toxic to plant growth, especially roots [58], it seems unlikely that passive NH_3 absorption is a major source of N for plant growth. Therefore, the bulk of ammoniacal N absorbed by plants is likely the result of active uptake of NH_4. The mechanism of active NH_4 uptake, which has not been clearly established, appears to be carrier regulated, as indicated by saturation kinetics and the depression of uptake by factors that limit energy metabolism [39,52].

2. Factors Affecting Nitrogen Uptake

The uptake of NO_3 and NH_4 can be affected by internal factors, such as N and carbohydrate status, and by external factors, such as temperature, O_2 level, and rhizosphere pH. Plant species and stage of plant development can also influence N uptake. When the uptake of a specific N form is affected differentially by these factors, contrasting patterns of N uptake and growth can result, depending on the form of N available to the plant.

While NH_4 uptake does not appear to be affected by the presence of NO_3 [59,60], there are many reports of NH_4-induced inhibition of NO_3 uptake [40,43,61–63]. However, there are also cases in which NH_4 appeared to have little or no effect on NO_3 uptake [64,65], or even resulted in a stimulation in uptake [66]. Although the precise manner by which NH_4 inhibits NO_3 uptake is not clear, possibilities include (a) a decrease in NO_3 reduction, resulting in feedback inhibition of NO_3 uptake, (b) an alteration in the rate of activation or synthesis of the NO_3 uptake system, thereby restricting influx, and/or (c) an acceleration in NO_3 efflux. For a description of NO_3 reduction and nitrate reductase, see Section III. B.3. Various lines of evidence support each of these mechanisms.

Although some researchers have shown a decrease in the level of extractable nitrate reductase by NH_4 treatment [67–70], others have shown that NH_4 or products of NH_4 assimilation do not interfere with nitrate reductase [71–73]. In addition, the ability of NH_4 to inhibit NO_3 uptake in plants without detectable nitrate reductase activity [74] and the lack of proportional changes in activity and NO_3 uptake in response to NH_4 [63,75,76] further indicate that a change in nitrate reductase is not the main mechanism responsible for NH_4-induced inhibition in NO_3 uptake.

Alternatively, NH_4 or one of its assimilation products may interact with NO_3 transporters at either the external or internal surfaces of the plasmalemma and inhibit the activation or synthesis of the NO_3 absorption system [43,53,62]. One possibility is that NH_4 or the high acidity adjacent to the plasmalemma resulting from NH_4 uptake in excess of NO_3 uptake causes an alteration in membrane permeability, thereby restricting the capacity for NO_3 absorption [77]. Another possibility is that NH_4 may inhibit net NO_3 uptake by increasing NO_3 efflux [78]; yet others suggest that NO_3 influx, not efflux, is inhibited by NH_4 [79–81]. Although additional research is needed to elucidate the exact mechanism(s) involved in NH_4-induced inhibition of NO_3 uptake, the identification of genotypic variation for the extent of this inhibition [82–84] indicates that the process is under genetic control.

Many studies have shown that NO_3 uptake is more sensitive to low temperatures than is the uptake of NH_4 [85–90]. For example, at temperatures below 9°C, perennial ryegrass plants absorbed more than 85% of their total N as NH_4, while the proportion decreased to only 60% absorption as NH_4 at temperatures of 17°C or above [86]. Although the reason for the preferential uptake of NH_4 over NO_3 at low temperatures is unclear, physical changes in the membrane may be responsible, rather than differences in temperature sensitivity of the two transport systems [85]. Alternatively, because temperature has a strong influence on the rate of nitrification, it is reasonable to assume that the largest amounts of NH_4 will occur in cool soils. Thus, the greater uptake of NH_4 at low soil temperatures may partly be the result of more NH_4-induced inhibition of NO_3 uptake.

Another important difference between NO_3 and NH_4 uptake lies in the sensitivity to pH of these two N forms. The maximal uptake of NH_4 occurs at neutral pH values, and uptake is depressed as the pH falls [74,91,92]. The limitation in NH_4 uptake at low pH can lead to N stress and a decrease in growth when NH_4 is the only form of N supplied to the plant [92,93]. This problem is further exacerbated by the decrease in rhizosphere pH associated with the uptake of NH_4. Compared to NH_4 uptake, the opposite pH optimum occurs for NO_3, where more rapid uptake occurs at pH values of around 4–5, while uptake is depressed at higher pH values [9,74]. The reduction in NO_3 uptake at high pH may be related to a competitive effect of OH^- ions on the NO_3 uptake system [74]. Similar to NH_4, the alkalinity generated from NO_3 uptake could further restrict NO_3 uptake. Thus, the consequences of absorbing NO_3 or NH_4 can have rather detrimental effects to the subsequent uptake, as a result of differences in the optimum pH for uptake of the ion absorbed.

In addition to environmental and soil factors, the stage of plant development may also influence the relative proportions of uptake between NO_3 and NH_4. Some evidence suggests that plants absorb NH_4 more rapidly than NO_3 during early vegetative growth, while the reverse situation occurs and more NO_3 is absorbed than NH_4 as growth progresses [9,94,95]. Possibly, young plants may lack a completely functional systems for NO_3 uptake and assimilation [96]. Alternatively, changes in the carbohydrate status of the root during plant development could alter the N form that is preferentially absorbed [97,98].

3. Nitrogen Assimilation

Regardless of the form absorbed, the inorganic N must be assimilated into organic forms, typically amino acids, to be of use to the plant. Because NH_4 is toxic to plant tissues at relatively low levels, it is rapidly assimilated in the roots and the N translocated as organic compounds. In contrast, NO_3 can be assimilated in the root, stored in the vacuoles of root cells, or transported to the shoot, where it can also be stored or assimilated. Nitrate storage and translocation play important roles in N metabolism inasmuch as NO_3 in the vacuole can be made available for assimilation when external sources of N are depleted. However, relatively little is known about factors that regulate the entry and exit of NO_3 in the vacuole.

While NH_4 can be used directly for amino acid synthesis, NO_3 must first be reduced to NH_4. The reduction of NO_3 to NH_4 is an energy-requiring process occurring by two main partial reactions. The first step involves a two-electron reduction of NO_3^- to NO_2^- and is catalyzed by the enzyme nitrate reductase, while the second step involves a six-electron reduction of NO_2^- to NH_4^+ catalyzed by nitrite reductase. Of these two enzymes, nitrate reductase is considered to be the rate-limiting step in the assimilation of NO_3 because it initiates the reaction and is the logical point of control when NO_3 is available. Nitrate reductase is also induced by its substrate NO_3; it has a short half-life, and its activity varies diurnally and with environmental factors that affect the flux of NO_3 to the sites of induction and assimilation [61,99,100].

The reduction of NO_3 by nitrate reductase can occur in either the root or the shoot, and in both cases, the energy is derived from the oxidation of carbohydrates [61]. The extent to which NO_3 is reduced in roots and shoots varies widely with plant species and environmental conditions [101,102]. Based on the contribution of total NO_3 reduction by the roots, plants can be classified into three main groups:

Species in which the root is the major site for reduction
Species exhibiting NO_3 reduction in both the root and the shoot
Species in which the shoot is the primary site for reduction

These three classifications are roughly typified by woody plants, perennial herbs, and fast-growing annuals, respectively [101,102]. While many studies have indicated a cytosolic location for nitrate reductase [61,100,103], others have suggested that nitrate reductase is associated with chloroplasts, microbodies, or the plasmalemma [100,104–106].

Two main types of nitrate reductase, which differ in the electron donor have been identified in higher plants [61,99,100]. One nitrate reductase uses NADH (reduced nicotinamide dinucleotide), while another nitrate reductase uses NADH or NADPH (reduced nicotinamide dinucleotide phosphate). Essentially, all higher plants contain the NADH-specific nitrate reductase, and it is the only form of nitrate reductase in some species [99]. In contrast, other plant species contain both an NADH-specific and an NADPH-bispecific nitrate reductase [107]. In some plant species, the NADH-specific nitrate reductase is found in both leaves and roots and constitutes the majority of the total nitrate reductase activity, while the NADPH-bispecific form is found only in the roots [99].

Similar to NO_3 reduction, the reduction of NO_2^- to NH_4^+ can occur in either the root or the shoot; the cellular location and the electron donor, however, vary depending on the site of reduction. In the shoot, NO_2^- reduction occurs in the chloroplast and is coupled to the light reaction of photosynthesis by the use of reduced ferredoxin as the electron donor [61]. Nitrite reduction in the root occurs in a plastid, which is analogous to a chloroplast of the leaf, but the reaction differs from shoot reduction in the following respects: (a) the nitrite reductase of the root is similar but not identical to the leaf enzyme, (b) the electron donor is a ferredoxin-like protein that is not identical to the leaf protein, and (c) the root ferredoxin is reduced by NADPH and a corresponding enzyme, with the energy supplied from the oxidation of carbohydrates [108].

The NH_4^+ that results from both NO_3 assimilation and the NH_4^+ absorbed directly by the roots is assimilated by the glutamate synthase cycle, which involves two reactions operating in succession and catalyzed by the enzymes glutamine synthetase and glutamate synthase [109]. A characteristic of this pathway is the cyclic manner by which the amino acid glutamate acts as both acceptor and product of ammonia assimilation. In this cycle, NH_4^+ is incorporated into glutamine by glutamine synthetase, which attaches NH_3 to the carboxy group of glutamate, using energy supplied by ATP. In leaf cells, this reaction occurs in chloroplasts, while in roots it most likely occurs in plastids [110]. In the chloroplast, the light-trapping system provides the energy to regenerate ATP, while in root cells, other enzyme systems oxidize carbohydrates to provide the energy for ATP regeneration.

Another isoform of glutamine synthetase is found in the cytoplasm of both leaf and root cells which is not identical to the plastid enzyme [111,112]. The cytoplasmic enzyme can assimilate any free NH_3 or NH_4^+ regardless of its origin (from either deamination of amino acids or absorption from the soil). Thus, in addition to producing the key intermediate, glutamine, the glutamine synthetase reaction is also a detoxification process that avoids injury from the accumulation of NH_4^+ or NH_3.

Following the formation of glutamine, the amino group ($-NH_2$) is transferred to α-oxo-glutarate via glutamate synthase to form two molecules of glutamate. This reaction can occur in shoots or roots, and in both cases the enzyme is located in plastids [109,111,112]. There are three isoforms of glutamate synthase in plant cells, which utilize different electron donors [108]. In leaf chloroplasts, the electron donor is reduced ferredoxin derived directly from the trapping of light energy. Conversely, the electron donor in root cells is NADH or NADPH, where the energy to reduce the oxidized form of the pyridine-linked nucleotides is derived from oxidation of carbohydrates [109,111,112].

A further series of related reactions mediated by specific transaminases transfers the amino group from glutamate to the 2-oxy group (=O) of a 2-oxoacid. Biochemical modification of glutamate, glutamine, and the array of amino acids produced by the transaminase reactions generates the 20 amino acids required for protein synthesis. These amino acids can also be anabolized into a variety of complex nitrogenous compounds (e.g., chlorophyll, growth regulators, alkaloids, nucleic acids) that are involved in plant growth and metabolism.

C. Timing of Nitrogen Accumulation

Similar to dry matter production, the seasonal accumulation of N by crop plants can be divided into three main phases:

An initially slow accumulation due to limited crop biomass
A period of rapid, nearly linear accumulation that coincides with the onset of rapid plant growth
A cessation of N accumulation with advancing maturity

Examination on a daily rate basis generally reveals two periods of rapid N accumulation, corresponding to late vegetative growth and the onset of linear seed fill [113,114]. Although the maximum accumulation rate usually occurs during linear vegetative growth, it can be delayed by a delay in the availability of N [115]. The period of maximum N accumulation can also be affected by such other factors as planting date, irrigation, and climate [114,115].

Under most conditions, the majority of plant N accumulated by cereal plants is acquired during vegetative growth. Numerous reports in the literature for maize and wheat show cases of 75% or more of the total plant N accumulation having occurred by anthesis [116–120]. However, there is some indication that continued accumulation of N during grain fill can be a beneficial trait, especially for high yielding genotypes in good growing environments [121,122]. For example, the hybrid FS854, which holds the world record yield for maize, 23.2 Mg ha^{-1} [123], has been shown to accumulate a substantial proportion of its N during grain fill [119,124,125]. The proportion of plant N accumulated after anthesis, however, is highly influenced by growing season [117], soil N level [126], and cultivar [120,126,127].

More extensive early-season N uptake may also be the result of a larger supply of N in the soil, since available soil N often exhibits a marked decline coincident with rapid vegetative growth, with the lowest N levels occurring around anthesis [128]. However, while N applications made during the early stages of reproductive development can increase protein percentage [129,130], there is often little or no response in terms of grain yield [131,132]. Similarly, foliar N sprays have the least impact on increasing plant N levels when applied around anthesis, because the additional N interferes with the metabolism of indigenous N [124]. Collectively, these findings suggest there is a level of genetic control over N accumulation and distribution that is independent of the availability of N. Complicating the understanding of how N availability and genetics determine plant growth, however, is the inability to stringently control the supply of N in the soil.

In the few cases in the use of hydroponic culture to deprive maize plants of N at anthesis, yield either has been unaffected [133] or has decreased only modestly [134,135]. Similarly, grain yield in soils is generally affected more by the N supply before anthesis than after [136,137]. This evidence, and the lack of ability to increase yields of most maize cultivars with postanthesis foliar N sprays [138,139], suggests that N imparts its main impact on yield before anthesis. However, redistribution of previously accumulated N from vegetative to reproductive plant parts could minimize the need for postanthesis N uptake. For both maize and wheat, the grain typically contains about 70% of the total N in the plant at maturity, with more than half of it coming from remobilization from other plant parts [119,120,140,141]. Thus, because of extensive changes in the distribution of N among plant parts, it is difficult to separate the effects of the timing of N accumulation from the contribution of N to grain development and yield.

III. PHYSIOLOGICAL ROLES FOR NITROGEN IN CROP PRODUCTIVITY

A. Importance of Nitrogen to Plant Growth

Crop growth and productivity involve the integrated effect of a large number of components and metabolic processes that act, with variable intensity, throughout the life cycle of the crop. The interdependence of N and C metabolism creates additional problems in describing an independent role for N in achieving maximum crop productivity. Nevertheless, four major roles for N have been proposed for attaining high yields of rice [142] and maize [122], and these roles appear to be valid for many crops:

Establishment of photosynthetic capacity
Maintenance of photosynthetic capacity

Establishment of sink capacity (the number and potential size of seeds)
Maintenance of functional sinks throughout seed development

Each of these roles is discussed briefly with reference to the potential impact on crop productivity.

The objective in establishing photosynthetic capacity is to ensure that the supply of N does not limit development of the photosynthetic apparatus (enzymes, pigments, and other compounds needed for photosynthesis). Within limits, and if no other restrictive factors are present, an increase in N supply increases the growth, the composition of N and chlorophyll, and the photosynthetic capacity of leaves [143–145]. Nitrogen supply has also been shown to regulate the synthesis of photosynthetic carboxylating enzymes by affecting transcription and/or the stability of messenger RNA [146,147]. Collectively, these effects result in greater light interception, higher canopy photosynthesis, and higher yield. However, because little N is accumulated by the leaf after it has reached full expansion [148], a sufficient supply of N must be available throughout the development of each leaf, if the individual leaves are to attain their full genetic potential for photosynthetic capacity.

To achieve high yields, plants must not only establish photosynthetic capacity, but continue photosynthesis throughout the grain-filling period. Thus, once established, sufficient N must be available to maintain the photosynthetic apparatus. This role is particularly important, since dry matter accumulation in cereal grains is dependent on current photosynthesis [119,149,150]. Most of the N in the leaf is associated with proteins in the chloroplast—60% in C_4 plants and up to 75% in C_3 plants [122,151,152]—and these proteins are subject to breakdown and remobilization of the resultant amino acids [153,154]. Thus, as leaves age and senesce, their capacity for photosynthesis declines, with a correspondingly negative effect on assimilate supply and yield.

Many studies have shown a concurrent loss of photosynthetic activity and organic N from the leaves, especially during seed development [155–158]. An example of this relationship for maize leaves is shown in Figure 1: the losses in leaf N and photosynthesis were initiated at or near pollination and declined nearly linearly during the grain-filling period. Although it is clear that the loss of N from the leaf impairs photosynthetic activity, management practices that

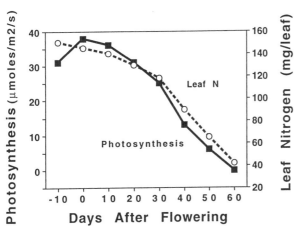

Figure 1 Changes in photosynthesis and N content of a selected leaf (the first leaf above the ear) of maize during the grain filling period. Values presented are for adequately fertilized plants (200 kg N ha^{-1}) averaged over two hybrids at the University of Illinois research farm in 1985 and 1986.

increase the N supply (such as supplementary side dressing or foliar sprays of N) do not automatically increase leaf N status and photosynthetic activity [159–162]. The absence of these effects is likely attributable to several key photosynthetic enzymes (the large subunit of RuBPCase) that are encoded for and synthesized by the chloroplast [163]. After full leaf expansion, the chloroplast loses much of its ability to synthesize these proteins, regardless of the availability of N [148,163]. This phenomenon indicates that the application of supplementary N to maintain photosynthetic activity may be of limited value until a technique is found that will reactivate protein synthesis in the chloroplast.

Another important role for N in assuring high productivity of crop plants is establishment of reproductive sink capacity. Sink capacity of a cereal plant is a function of the number and the potential size of grains. Grain number is dependent on the number of ears per unit area, the number of florets per ear, and the proportion of florets that develop into grain [149,164,165], while the potential size of individual grains depends on the number of endosperm cells and starch granules [166–169]. In either case, reproductive initials, like all growing tissues, are characterized by high concentrations of N and high metabolic activities. This need could indicate that sufficient amounts of both C and N assimilates are required for full expression of the genetic potential for initiation and early development of grains.

For cereal crops, grain number is usually more closely related to yield than other yield components [149,164,165]. Consequently, many studies have shown that N-induced yield increases are the result of more grains per plant [170–173]. For wheat, this enhancement is related to an increase in tiller production and survival [174,175], and to a lesser extent to a decrease in floret abortion [176]. In contrast, for maize, N supply affects kernel number primarily by decreasing kernel abortion [172,177]. An example of the effect of N supply on kernel number and kernel abortion of maize is shown in Figure 2: kernel number increases as the N supply is increased from a deficient to a sufficient level, which is associated with a decrease in kernel abortion. Other studies, however, have indicated that N supply can also affect individual grain weights [178,179], perhaps by means of a change in endosperm cell number [180].

Although the number of ears and grains is usually the yield component most affected by N supply, increases in kernel weight can also affect yield [149,164]. Because vegetative development in cereal crops is negligible after flowering, the N subsequently acquired, or

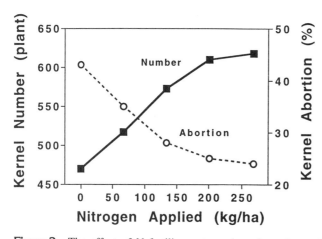

Figure 2 The effect of N fertilizer rate on kernel number and kernel abortion of maize. Values are averaged over two hybrids at the University of Illinois research farm in 1990.

remobilized from the vegetation, is used exclusively for grain development. This need for N is demonstrated by the fact that adequately fertilized cereal crops typically contain from 9 to 13% protein in the grain. Indeed, some workers have suggested that the deposition and/or accumulation of storage protein by the kernel is a factor regulating grain development [181–183]. This suggestion is based on the positive correlation between storage protein, kernel weight, and grain yield [182] and on genetic studies showing reduced levels of storage protein (zein) and starch in zein-deficient mutants of maize [184]. An alternative explanation, however, is that the availability of N within the plant and to the grain is positively associated with kernel development, and as such the amount of storage protein deposited is only an accurate reflection of the N supply [122].

Other needs for N by developing kernels could include embryo growth, and the initial and continued synthesis of enzymes needed for energy generation and the deposition of storage products in the kernel. Embryo development could affect the kernel's hormonal balance, since a large portion of kernel phytohormones are produced by the embryo [185,186]. Because several of the key classes of phytohormones either contain N (auxins, cytokinins, polyamines) or are synthesized from amino acids (auxins, ethylene, polyamines), an adequate supply of N may be needed for their production. With regard to storage product formation, provision of N to developing maize kernels has been shown to increase their capacity to synthesize proteins and to utilize sugars for the biosynthesis of starch [187]. Nitrogen supply also exerts a marked effect on endosperm enzymology and on the deposition of storage proteins in the endosperm [187,188]. Thus, it appears that at least a portion of the yield increase produced by N fertilization results from a modification of kernel metabolism in response to N supply.

B. Interactions of Carbon and Nitrogen

Grain yield of crops is primarily a function of the plant's ability to acquire, metabolize, and utilize C and N assimilates, and its genetic potential for maximum grain production. For cereal crops, the relative abundance of C versus N in the plant (approximately 44% C vs. 1.5% N) dictates a predominant role for photosynthesis in achieving maximum yields. However, as discussed in Section III. A, the metabolism of N plays a major role in the production of C assimilates and in their utilization for reproductive development. In addition, as evidenced by the use of reduced ferredoxin in NO_2^- reduction and NH_4^+ assimilation (see Section II. B.3), C and N interact at numerous points in plant metabolism [189]. This interdependence in C and N metabolism creates problems when one is attempting to describe an independent role for either C or N in achieving maximum productivity.

Grain composition offers a prime example of the complexities involved in understanding how C and N interact to affect productivity. A negative relationship between grain yield and protein percentage is widely noted in cereals, especially in cultivars selected for abnormally high or low percentages of grain protein [190,191]. The higher metabolic cost associated with the synthesis of protein than with carbohydrate has been proposed to explain this relationship [192,193]. However, evidence showing that carbohydrate supply does not normally limit kernel development [194–196] and progress toward identification and breeding of high protein, high-yielding cereals [197,198] make this explanation seem unlikely. In addition, source–sink alteration experiments have indicated that C- and N-storage processes in cereal kernels seem to operate autonomously [199–201].

Further complicating attempts to understand the relation between yield and protein concentration is the tendency for individual grain weight to vary with grain number [173]. A negative relationship between grain number and grain weight is often observed [202], which may be in part due to the relative supplies of C and N from the vegetative plant [203]. There is some indication that the composition of assimilates (C and N) channeled to the kernel by the mother plant controls the amounts of starch and protein accumulated in maize grain [204,205]. In other work, however, it is shown that the genotype of the kernel primarily dictates the range of grain composition, with external factors modulating the phenotype within this range either to a large or small degree [206]. Alternatively, zygotic factors and the source supply may interact to control grain composition [207]. Compensation phenomena also complicate our understanding of how C and N relationships control grain composition, since plants can make up for a lack of current assimilate (both C and N) with enhanced remobilization from the vegetation [172].

On a whole-plant basis, the supply of N often appears more limiting for grain development than does the supply of carbohydrate. For maize, the capacity of the plant to supply N to the ear was more limiting than the capacity to provide photosynthate, inasmuch as the net remobilization of vegetative N occurred earlier and was much more extensive than the remobilization of vegetative carbohydrate [119,140,208]. Although shading plants during grain fill decreased yield, and enhanced the remobilization of both dry matter and N, the availability of newly reduced N was still more limiting to grain fill than current photosynthate [209]. Similarly, while supplemental illumination to the lower two-thirds of the canopy increased carbohydrate status and yield, these effects could not be separated from an enhancement in the total accumulation, and the tissue concentration, of N [210]. Collectively, these data suggest that the availability of N to and within the plant is more variable than the availability of photosynthate, and at least as limiting to grain development.

IV. CROP RESPONSE TO APPLIED NITROGEN

A. Growth and Yield Response

Increases in crop productivity due to fertilizer N additions may be realized as dry matter yield, protein yield, or an improvement in quality factors. For cereal crops, grain yield and protein quality exhibit a typical pattern in response to N supply that can be divided into three main components [211]:

1. Grain yield and protein content (total amount present) and concentration (protein per unit weight) increase in unison with increasing N supply.
2. Grain yield reaches a plateau, but protein continues to increase with additional increments of N.
3. Grain protein content peaks, grain yield begins to decline, and protein concentration continues to rise with further increases in N supply.

Responses to applied N are affected by many environmental [212,213], cultural [214,215], and soil factors [14], with the result that the response curves can vary considerably at different locations. For example, in a fertile soil with a high residual N supply, applications of N may have no effect or may even decrease crop yields. Alternatively, if some factor other than N, such as soil moisture or another nutrient, is limiting, then applications of N fertilizer will not increase growth and yield even if the supply of soil N is low. The optimal economic N rate also depends on the soil type and the ratio of fertilizer N costs to the value of the crop. In general, the fertilizer N rate required for maximum yield and the economic optimal N rate are lower for soils with higher organic matter such as silt loams than for sandy soils [14].

Despite the variation associated with crop response to fertilizer N, an example of a general pattern for maize is presented in Table 1. In the absence of other limiting factors, addition of fertilizer N will increase maize yields in a curvilinear fashion. Similar to other growth inputs,

Table 1 General Effect of N Fertilizer Rate on Grain Yield and N Recovery of Maize Grown on Highly Fertile Silt Loam Soil in Illinois[a]

Nitrogen rate (kg ha^{-1})	Grain yield (Mg ha^{-1})	N in crop (kg ha^{-1})	N removed with grain (kg ha^{-1})
0	8.4	124	83
67	10.5	170	119
134	12.2	212	150
202	13.1	241	170
267	13.2	262	180

[a]Values are averaged over two hybrids grown at the University of Illinois research farm in 1990.

response to N fertilizer decreases as more and more fertilizer is added. As a result, plants are always the most efficient at utilizing fertilizer N when it is available at low levels (see Section IV. D for a discussion of N use efficiency). In this example, 202 kg of N per hectare increased yield by 4.8 Mg, compared to no N application, with nearly half (2.1 Mg) of this increase coming from the first 67 kg increment of N (Table 1). Although the second and third increments also increased yield, the size of these increases diminished successively (1.7 and 0.9 Mg, respectively). In contrast, the fourth 67 kg increment increased yield by only 0.1 Mg, which would not be considered economically (or environmentally) sound. Therefore, while it is obvious that fertilizer N was needed to maximize yield, the optimum rate required is much less clear.

In addition to demonstrating the need for fertilizer N, the foregoing data show that a reasonable yield (8.4 Mg ha^{-1}) was produced without adding any N to the soil (Table 1). The crop accumulated 124 kg ha^{-1} from the soil, of which 83 kg was removed with the grain. For heavily fertilized plants (267 kg ha^{-1} N), these values increased to 262 kg of plant N and 180 kg of grain N. When the level of fertilizer N limited grain yield (at N rates of 134 kg ha^{-1} or less), more N was removed with the grain than was provided by the fertilizer. Conversely, N levels in excess of those needed for maximum yields (202 kg ha^{-1} and up) resulted in the removal of less N than had been applied. This situation greatly increases the potential for accumulation of residual N (usually as NO$_3$) in the soil. Based on this example, and other published reports [216,217], it is suggested that the N level that just maximizes grain yield also results in the best balance between fertilizer N added and the amount removed with the grain.

Similar to grain yield, the largest increase in total plant N accumulation, and the greatest plant recovery, occurred with the first increment of N. For example, the first 67 kg of fertilizer N increased plant N accumulation by 46 kg, representing a plant recovery of 68% (calculated from data of Table 1). Of this 46 kg of plant N, 36 kg was removed with the grain for an N removal recovery of just over 50%. In contrast, plant N accumulation was increased by just 21 kg (32% recovery) for the fourth N increment, of which only 10 kg (15% recovery) was removed with the grain. These data demonstrate the inherent inefficiency with which fertilizer N is recovered by the maize plant, emphasizing the potential for environmental damage at excessive rates.

B. Genotypic Variation

Different cultivars grown at the same location can exhibit different response patterns to N fertilization, and such variation has been observed for wheat [218,219], rice [220], sorghum [221], and maize [126,127]. However, as might be expected, this variation is highly affected by the environment and growing conditions and is most apparent under controlled conditions (e.g., in hydroponics) [135,222]. Interest in identifying genetic differences in responsiveness to N fertilizer is intensifying, as producers and agricultural consultants see genotypic variation as one way to fine-tune N fertilizer management. There is also a desire to develop or identify genotypes that perform well under a low N supply or, conversely, to find genotypes that will respond to high fertility conditions.

From a botanical standpoint, plants can vary in their use of N in two major ways: in how much N the plant uses to produce maximum yield, or in when (i.e., what stage during the growing season) the plant acquires its N. An example of this type of variation is depicted for maize in Figure 3. In this example, the low N response type produced its maximum yield at 120 kg N ha^{-1} compared with an N requirement of 200 kg ha^{-1} for the high N type (Figure 3, left). Although high N types are usually capable of producing the highest yields, the low N types may outyield the high N types at low levels of soil N. Cultivars can also differ in their timing of N acquisition: some accumulate the majority of their N before flowering, while others may have a substantial requirement for N accumulation after flowering (Figure 3, right).

A cultivar's N acquisition pattern can affect N accumulation and productivity, since plants acquiring most of their N by flowering should be less subject to fluctuations in the N supply during grain fill. These types may be more consistent from year to year, because adverse growing conditions usually occur after flowering. However, these types might also be more susceptible to N deficiency early in the growing season, and less likely to recover from such setbacks. Further complicating the ability to characterize a cultivar's response to fertilizer N is the fact that these two strategies can interact to determine the final N use. In other words, the amount of N required for maximum yield may or may not be related to when the N is accumulated by the plant.

Because of its economic value and high requirement for fertilizer N, much of the recent effort in identifying genotypic variation for N use has been directed toward maize. Although differences in N use among maize genotypes have been reported for inbreds [223] and open-pollinated populations [121], there is considerable controversy regarding whether these differences can be used to improve N fertilizer management of hybrids. While some studies have reported large differences among maize hybrids for their response to fertilizer N [127,224], others have shown no or limited differences [117,225]. Similarly, a separate large-scale study did not observe hybrid × N rate interactions, although it was noted that hybrids within an individual

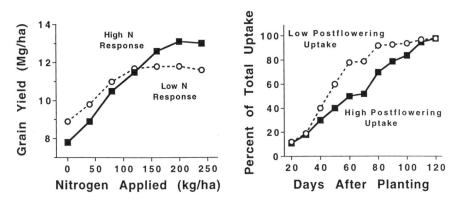

Figure 3 Representation of the two major ways in which maize cultivars can vary in their use of N. Cultivars can differ in the amount of fertilizer N required for maximum yield (high or low N response types; left) or in their timing of N acquisition (high or low postflowering uptake types; right).

location responded differently to the level of soil-applied N [226]. In an attempt to reconcile these differences, hybrids were divided into breeding groups based on their inbred parents, and differences in the N rate required for maximum yield were observed between, but not within, the groups [126]. Collectively, these studies suggest that maize hybrids do respond differently to the level of fertilizer-applied N but that the growing environment and the hybrid's genetic makeup can markedly influence the response.

C. Form of Nitrogen

As mentioned earlier, N can be utilized by plants as either NO_3 or NH_4, although under production conditions the greater amount is thought to be absorbed as NO_3. Thus, enhancing the supply of soil N as NH_4 is one way to improve N fertilizer management (i.e., by minimizing the potential for N losses). Increasing the supply of NH_4 in soils could also enhance plant performance: a survey of the literature shows numerous examples of improved vegetative growth and N accumulation when growing cereal plants are provided with mixtures of NO_3 and NH_4 compared to NO_3 alone [64,84,227–229]. Similar results have been reported for tomato [230], soybean [231], and sunflower [232]. These data imply that seedlings of many plant species cannot acquire sufficient N for maximum vegetative growth when N is supplied solely as NO_3.

While much of the earlier work has involved vegetative growth, recent studies show that cereal crops supplied with both NO_3 and NH_4 (mixed-N nutrition) also produce higher yields than those supplied with only NO_3 [233–238]. For a variety of maize hybrids grown in field hydroponics, an equal mixture of NO_3 and NH_4 increased the yield by 11–14% [217,235,239] compared to plants grown with only NO_3. Even greater mixed-N-induced yield increases (average of 21–43%) have been reported for hydroponically or pot-grown spring wheat [236,238,240,241].

Although these responses were obtained with hydroponics or pot culture, where a finer degree of control over the NO_3/NH_4 ratio is possible, there is evidence that enhancing the supply of NH_4 to cereal crops is beneficial under production conditions [242–247]. Several reports [243,246,247] have shown moderate yield increases of 6–11% when maize plants were grown under field conditions designed to provide mixed-N nutrition. However, not all environments [243] or hybrids [247] are responsive to mixed N, indicating that factors other than the availability of NH_4 can impact on the N use and productivity of maize. Other work shows that cultivars differ in their physiological strategy for achieving mixed-N-induced yield increases, and in the magnitude of response [235,238,247].

In most cases, mixed-N-induced yield increases are the result of more grains per plant [235,237,238,248], although increases in individual grain weight have also been reported [217,233]. For wheat, the additional kernels are primarily achieved by increasing the number of grain-bearing tillers [238,240,248], and to a lesser extent the number of grains per tiller [248]. Alteration of the N form at anthesis showed that mixed N supplied continuously, or during vegetative growth only, increased yield and tillering over all NO_3 plants, while mixed N during reproductive growth only did not [240]. Thus, for wheat, it appears that mixed-N-induced increases in yield potential occur during the early stages of plant development, when tillers are being formed.

In contrast to wheat, the main effect of enhanced NH_4 to maize is an increase in the number of grains per plant through more kernels per ear [235,239], although there is also a tendency for increased prolificacy [247,249]. Additional kernels per ear result primarily from a decrease in kernel abortion [246,250] and sometimes an increase in ovules per ear [247]. These findings suggest a direct physiological effect of N form on kernel development, inasmuch as all studies presumably supplied a more than adequate level of N (either as NO_3 or mixed N) to the plant.

These results also suggest that mixed-N-induced yield increases are associated with events that occur during ovule initiation and pollination rather than with processes occurring during the grain-filling period.

Support for a pre-grain-fill effect of mixed N on productivity has been obtained by transfer experiments in which N was supplied either as all NO_3 or as an equal mixture of NO_3 and NH_4 until anthesis, whereupon half the plants in each group were switched to the other N form [233,239]. In both sets of experiments, yield was increased over all NO_3 plants when mixed N was available continuously or only before anthesis, but not when it was available only after anthesis. Similarly, Reddy et al. [134] reported that the N form supplied before, but not after, anthesis affected growth and nutrient status of maize. These studies and other data [251] suggest that late vegetative and early reproductive development are the most crucial times to supply mixed N to the plant.

Although the physiological basis for improved productivity with mixed N is not understood, additional plant N accumulation has been implicated. Similar to reports for vegetative growth [64,228], cereal plants grown to maturity with mixed N typically contain more plant N (both content and concentration) than those grown with NO_3 alone. These results have been observed for plants grown hydroponically [233,235,238–240] and in soils [241,244,247,252]. Like the results of earlier work with seedlings, these data suggest that when N is supplied primarily as NO_3, cereal crops are unable to acquire sufficient N for maximal productivity.

While it is unclear exactly how this additional N (from mixed nutrition) enhances productivity, it is well known that N supply and plant N status affect tillering in wheat [174,175], and kernel abortion and prolificacy in maize [172,209,253,254]. Alternatively, a certain level of NH_4 may exert a direct effect on reproductive development, with a corresponding change in plant metabolism. For example, NH_4 nutrition has been reported to stimulate sucrose uptake by maize kernels, which in turn increased the production and translocation of assimilates from the leaves [182]. Similarly, NO_3-N was uniformly assimilated throughout the plant, while NH_4-N was assimilated in the root and preferentially exported as organic N to meristematic regions like the ear [255,256]. Indicative of an enhanced supply of N to the ear is an increase in grain protein concentration under mixed N, compared with plants grown on all, or predominantly, NO_3 [134,239,247,257].

In addition to enhanced N accumulation, mixed-N-induced increases in reproductive development and yield may be related to energy status. Because assimilation of NH_4 requires a third as many ATP equivalents as does NO_3 [258], plants acquiring a large percentage of their N as NH_4 may expend less total energy, especially if NO_3 is assimilated in the root [259]. Although the physiological impact of this energy saving is unclear, it seems possible that any effect would be largest for crops such as maize, which require high levels of N. However, based on cost estimates for NO_3 assimilation [173], Alexander et al. [233] recently concluded that mixed-N-induced increases could not be explained solely on the basis of energetics.

Alternatively, partitioning effects may play a role in altered growth: NH_4 must be assimilated immediately by the root, resulting in greater amounts of nitrogenous compounds in the roots and an altered partitioning of carbon between the root and the shoot [260]. Enhanced movement of sugars out of the leaves has been shown to relieve feedback inhibition of photosynthesis due to carbohydrate accumulation [261], which has been theorized to result in higher photosynthesis under mixed-N conditions [182]. However, while some studies have reported higher photosynthetic rates for NH_4-grown plants [262,263], others have shown greater photosynthesis for NO_3-grown plants [264,265]. In addition, field studies have shown equivalent (or greater) rates, and a similar duration, of canopy photosynthesis, but lower grain yields for maize plants supplied with predominantly NO_3 than with mixed N [247]. These findings, and the observation that mixed-N nutrition alters dry matter partitioning between shoots and roots

[228,246] and between vegetative and reproductive fractions [235,239,246,247], suggest that altered partitioning may be more important than photosynthesis in the enhanced productivity observed with mixed-N nutrition.

Other studies have shown that additional physiological processes are beneficially altered by mixed-N nutrition. For example, increasing the proportion of N used by the plant as NH_4 usually results in an increase in anion uptake, especially for P [258,266,267]. Because of its acidifying effect on the rhizosphere [268], enhanced uptake of NH_4 may also make trace elements like iron and zinc more available [170,269]. In addition, mixed-N nutrition has been shown to increase root branching [84,270] and the supply of cytokinins to the shoot [246], compared to NO_3-grown plants. It is also possible that by utilizing both N forms, plant cells are able to more tightly control their intracellular pH [271,272].

The experiments discussed in this section show that mixed-N nutrition can increase crop productivity as the result of alterations in several important physiological processes such as reproductive development, N acquisition, dry matter production, and assimilate partitioning. Indirect effects on other mineral nutrients and on endogenous phytohormone balance may also be important. For maximum yield enhancement, mixed N needs to be available during the period when reproductive potential is determined and set. Thus, although additional work is needed to further elucidate the physiological basis for mixed-N-induced increases in crop growth and yield, the prospect of using mixed N to improve fertilizer use efficiency is encouraging.

D. Nitrogen Use Efficiency

The efficient use of N is an important goal in maximizing yield in ways that have a minimal impact on the environment. Various methods have been used to define and characterize nitrogen use efficiency (NUE), so care must be taken to specify the method or definition that is used [13]. These methods can reflect agronomic, economic, or environmental perspectives, and they can be characterized on an incremental basis, on a cumulative basis, or as a yield efficiency index [13].

From an agronomic perspective, NUE refers to three main functions detailing the relationships between:

N availability and yield
N availability and N recovered
Yield and N recovered

To calculate these values requires measurements of grain yield, the total nitrogen in the plant, and the total available soil N. However, because the soil N availability and the total N recovered by the plant are difficult to determine in field experiments, the N content in the aboveground plant parts and the N rate supplied as fertilizer are typically used. In all cases, the most accurate estimates subtract the yield or plant N accumulated in unfertilized plots from those values obtained in fertilized plots.

The relationship between yield and N rate is most often referred to as "yield efficiency" or "agronomic efficiency" and is defined as the yield increase per unit of applied N for a specific portion of the yield response curve. The yield efficiency is a function of the efficiencies of N recovery and N utilization, which are known as "recovery efficiency" and "physiological efficiency," respectively. The recovery efficiency represents the N accumulated by the plant per unit of applied N, while the physiological efficiency is the grain produced per unit of N accumulated by the plant. Physiological efficiency integrates the effect of plant factors on N use and yield, while recovery efficiency is a measure of how much fertilizer N is absorbed by the plant. The yield efficiency for N use can be improved by increasing the recovery efficiency, or the physiological efficiency, or both.

From a soil standpoint, the overall NUE depends on the interaction of factors responsible for N loss (leaching, denitrification, volatilization, and immobilization) with such N management variables as N rate, N source, N placement, and timing of N application. In conjunction with soil factors, NUE from a plant standpoint depends on the processes associated with the absorption, translocation, assimilation, and redistribution of N. The NUE is greatest at low levels of N and is highly influenced by soil type, which determines the mineralization and N loss characteristics [13,173,214]. The NUE can also be influenced by plant characters such as tissue N concentration and the size and number of reproductive sinks [173,253].

Another measure of NUE uses data on yield and plant N content, without correcting for dry matter or N accumulation by unfertilized plants [273]. This procedure was developed to assess genotypic variation in response to N supply, where evaluation of a large number of genotypes by traditional methods is constrained by the size of the necessary field experiments. The procedure denotes dry weight and N values as a series of ratios, all expressed in the same unit—often grams per plant [273]. As with the traditional measures of NUE, this method defines NUE as grain production per unit of fertilizer N; there are two main components: the efficiency of N absorption (uptake efficiency), and the efficiency with which the N absorbed is utilized to produce grain (utilization efficiency). The uptake efficiency is denoted as the N in the plant divided by the fertilizer N applied, while the utilization efficiency is the grain produced divided by the N in the plant. Thus, the overall NUE can be expressed as a product function of uptake and utilization efficiencies.

Further subdivision of uptake and utilization efficiencies can be made to reflect more specific aspects of plant N use (e.g., translocation, remobilization, distribution, timing of N acquisition) [273]. In addition, converting the appropriate dry matter and N ratios to logarithms provides a means of partitioning variation in NUE into the proportion attributable to each of its components. Such data have shown that genotypic differences in NUE of eight maize hybrids were primarily the result of differences in N utilization efficiency when the crops were grown with a low supply of N, and differences in uptake efficiency at a high N supply [273]. The data also showed that either high or low values of NUE could be attained by different combinations of uptake and utilization efficiency. Similar cases of such variation have been noted for wheat [219]. The NUE and its components have also been shown to vary as a function of N fertilizer rate and the timing of N availability [126,173,274]. Collectively, these data emphasize that each of the plant traits involved in the acquisition and utilization of N is subject to genetic diversity, which may contribute to N use and crop productivity in different degrees under different environmental conditions.

V. ENVIRONMENTAL ASPECTS OF NITROGEN FERTILIZER USE

In addition to being removed with the crop, N can be lost from agricultural ecosystems in large amounts as the result of several processes. These include leaching, denitrification, volatilization, surface runoff, and soil erosion. Nitrogen can also be temporarily removed from the available soil pool because of adsorption, fixation, and microbial immobilization. The economic implications of these losses are self-evident, especially when they are large enough to limit crop productivity. These losses can also have environmental consequences with regard to water and air quality.

Losses of N are highly affected by which ionic N form (NO_3^- or NH_4^+) predominates in the soil [275,276]. Both forms are lost by soil erosion, but only NH_4 is lost directly to the atmosphere from volatilization. Ammonium can also be temporarily removed from the plant-available N pool by cation exchange with soil particles, fixation by clay lattices of the soil, fixation by organic matter, immobilization into microbial biomass, and conversion to NO_3. Conversely, NO_3-N is not readily used by soil microbes, nor does it bind to soil particles or organic matter. It is, however, subject to losses from leaching and denitrification.

The least controllable of these N losses, which are determined by soil type and rainfall, are the leaching and denitrification of NO_3. Leaching is a physical process that occurs because NO_3^- is repelled by negatively charged soil colloids and readily moves with soil water. However, if too much downward movement of water occurs, NO_3 can be leached below the plant's rooting zone, ultimately to accumulate in groundwater [2]. In denitrification, a separate microbial process that occurs under waterlogged or anaerobic conditions, NO_3 is converted to gaseous compounds, which are lost to the atmosphere. Both leaching and denitrification are economically and environmentally undesirable and add a large degree of uncertainty to N fertilizer management.

The problems of leaching and denitrification have stimulated the identification and development of nitrification inhibitors that block microbial conversion of NH_4 to NO_3 [277,278]. Research has shown that the use of these inhibitors can reduce losses of fertilizer N, especially under soil and weather conditions that favor N loss [276,279,280]. For several reasons, however, a consistent yield increase from the use of nitrification inhibitors is not always observed [276,280]. These reasons include:

A lack of opportunity for the nitrification inhibitor to express its potential for reducing N loss
Inadequate duration of the inhibitory effect
Inadequate experimental sensitivity to permit statistical detection of small benefits that may
occur
Adverse effects on other soil microorganisms
Genetic differences among cultivars to N level or to NH_4 nutrition

Although nitrification inhibitors were originally developed to minimize N losses, they have also been proposed as a means of altering the predominant form of N in the soil [245]. The use of ammoniacal fertilizers along with nitrification inhibitors may alter plant nutrition by supplying a greater proportion of the N to the plant as NH_4. Enhancing the supply and utilization of NH_4-N may also be beneficial to plant growth, since several crop species have been shown to absorb more N and to grow more rapidly when supplied with mixtures of NO_3 and NH_4 (see Section IV. C).

Urease inhibitors represent another approach to preventing fertilizer N loss [281,282]. When applied to the soil surface, urea [$(NH_2)_2CO$] is rapidly cleaved to NH_4^+ and CO_2 by the action of urease enzymes present in the soil and plant residue. This conversion gives rise both to high NH_4 levels and to elevated soil pH, two properties that are conducive to volatilization of N as NH_3. Urease inhibitors temporarily decrease the activity of urease enzymes, maintaining urea-applied N as urea for several days. Because the uncharged urea molecule is quite mobile in soil, rainfall can move surface-applied urea into the soil profile, where it can hydrolyze with less opportunity for N losses via volatilization. Similar to nitrification inhibitors, use of a urease inhibitor will generally be effective only when the crop can respond to the N conserved by the inhibitor, and when environmental conditions are conducive to large losses of surface-applied urea (such as warm soils with abundant plant residue). Conversely, urease inhibitors are of limited value when urea-based fertilizers can be easily and inexpensively incorporated into the soil during or immediately following their application [281,282]. They also require sufficient rainfall within a few days to facilitate urea movement into the soil.

In summary, although the use of fertilizer N has the potential for negative environmental consequences, several cultural practices can be used to minimize this possibility. These practices include:

Use of N rates appropriate for the historical productivity of the land and the yield of the crop being grown
Timing of N applications to better fit plant N needs
Specific placement of N-containing fertilizers
Use of appropriate N sources
Use of nitrification inhibitors to slow the breakdown of NH_4 to NO_3
Use of urease inhibitors to minimize volatilization of surface-applied urea
Taking into account the soil's capacity to supply the crop with N
Adequate fertilization with other mineral nutrients to maximize the plant's use of N.

VI. CONCLUSIONS

Because of the high requirement by crop plants for elemental N, and its numerous important roles in growth and development, N is the mineral nutrient element that most often limits crop productivity. Because N mineralization from the soil is normally too low to support desired production levels, soil N levels are typically increased through fertilization. However, the complex cycle of N in the environment causes uncertainty in N fertilizer management, increasing the chances for economic loss and environmental damage. Nitrogen use and productivity of crop plants is also complex, resulting from an interaction of biochemical, physiological, and morphological processes in the plant.

Nitrogen is unique among the mineral nutrients in that it can be absorbed by plants in two distinct forms, either as the anion NO_3^- or the cation NH_4^+. The form of N absorbed has a pronounced effect on the mechanisms for uptake, transport, assimilation, and storage, and, in some cases, on the physiology and morphology of crop productivity. The use of a specific N form also can be affected differentially by environmental or culture factors, such as plant N status, temperature, and pH. While N is usually applied as NH_4-fertilizer, the nitrification process renders NO_3-N the soil form most available to the crop. In addition, NO_3-N is the N form most susceptible to losses from the crop's rooting zone. Several relatively new techniques have been developed in an attempt to better assess the soil N supply; however, their usefulness is still being evaluated. Plant-based estimates of soil N supply are also receiving attention.

Although the maximum rate of N accumulation usually occurs during vegetative growth, the timing of N acquisition can be altered by cultural and environmental factors. Extensive redistribution of N among plant parts further confuses our understanding of when N has the greatest impact on crop productivity. Nevertheless, the major roles for N in crop productivity can be divided into four general areas:

Establishment of photosynthetic capacity
Maintenance of photosynthetic capacity
Establishment of sink capacity
Maintenance of sink capacity

Although the relative abundance of C and N in the plant dictates a predominant role for photosynthesis in the productivity of cereal crops, some evidence suggests that the availability of N to and within the plant is more variable than the availability of photosynthate, and at least as limiting to grain development.

Improved crop productivity from N fertilization can result from increases in dry matter yield and/or improvements in quality factors. In either case, increases with N supply follow the law of diminishing returns; thus N is used most efficiently when available at low levels. Cultivars grown at the same location may exhibit different responses to N supply that result from differences in how much N they need for maximum yield, or when in the life cycle they mainly acquire their N. Supplying N as mixtures of NO_3^- and NH_4^+ can also increase productivity as the result of alterations in important plant processes (e.g., reproductive development, N acquisition, dry matter production, assimilate partitioning). The efficient use of N is an important goal in strategies to maximize yield potential while minimizing negative effects of fertilizer N on the environment. Several methods have been used to assess N use efficiency, and its components, in crop plants.

The use of nitrogen by crop plants is dictated by a complex interaction of plant metabolism with cultural and environmental factors that alter the availability of N. Each of the plant processes involved in the acquisition and utilization of N is under genetic control, and each may contribute to varying degrees depending on the environmental conditions. A better understanding of these processes will undoubtedly help in developing strategies to improve the management of fertilizer nitrogen.

ACKNOWLEDGMENT

The author expresses his sincere gratitude to P. S. Brandau for providing critical comments on the text and for help with the graphics.

REFERENCES

1. P. J. Stangel, *Nitrogen in Crop Production* (R. D. Hauck, ed.), ASA, CSSA, SSSA, Madison, WI, p. 23 (1984).
2. S. R. Aldrich, *Nitrogen in Crop Production* (R. D. Hauck, ed.), ASA, CSSA, SSSA, Madison, WI, p. 663 (1984).
3. P. Newbould, *Plant Soil, 115*: 297 (1989).
4. J. S. Schepers, K. D. Frank, and C. Bourg, *J. Fertil. Issues, 3*: 133 (1986).
5. R. F. Spalding, M. E. Exner, C. W. Lindau, and D. W. Eaton, *Hydrology, 58*: 307 (1982).
6. C. A. Black, "Reducing American exposure to nitrate, nitrite, and nitroso compounds," Comments from CAST, Ames, IA, 16 pp. (1989).
7. H. I. Shuval and N. Gruener, *Am. J. Public Health, 62*: 1045 (1972).
8. D. A. Russel, *Nitrogen in Crop Production* (R. D. Hauck, ed.), ASA, CSSA, SSA, Madison, WI, p. 183 (1984).
9. R. H. Hageman, *Nitrogen in Crop Production* (R. D. Hauck, ed.), ASA, CSSA, SSA, Madison, WI, p. 67 (1984).
10. R. J. Haynes and K. M. Goh, *Biol. Rev., 53*: 465 (1978).
11. F. E. Allison, *Adv. Agron., 18*: 219 (1966).
12. S. A. Barber, *Agronomic Research for Food* (F. L. Patterson, ed.), ASA, Madison, WI, p. 13 (1976).
13. B. R. Bock, *Nitrogen in Crop Production* (R. D. Hauck, ed.), ASA, CSSA, SSA, Madison, WI, p. 273 (1984).
14. S. L. Oberle and D. R. Keeney, *J. Prod. Agric., 3*: 522 (1990).
15. D. R. Keeney, *Methods of Soil Analysis* (A. L. Page, ed.), ASA, Madison, WI, p. 711 (1982).
16. G. Stanford, *Nitrogen in Agricultural Soils*, ASA, Madison, WI, p. 651 (1982).
17. K. L. Sahrawat, *Adv. Agron., 36*: 415 (1983).
18. G. Stanford, J. N. Carter, and S. J. Smith, *Soil Sci. Soc. Am. Proc., 38*: 99 (1974).
19. R. H. Fox and W. P. Piekielek, *Soil Sci. Soc. Am. Proc., 42*: 751 (1978).
20. K. Németh, I. Q. Makhdum, K. Koch, and H. Beringer, *Plant Soil, 53*: 445 (1979).
21. R. Saint-Fort, K. D. Frank, and J. S. Schepers, *Commun. Soil Sci. Plant Anal., 21*: 1945 (1990).

22. A. Van Diest, *Plant Soil*, *64*: 115 (1985).
23. G. D. Binford, A. M. Blackmer, and M. E. Cerrato, *Agron. J.*, *84*: 53 (1992).
24. A. M. Blackmer, D. Pottker, M. E. Cerrato, and J. Webb, *J. Prod. Agric.*, 2: 103 (1989).
25. R. H. Fox, G. W. Roth, K. V. Iverson, and W. P. Piekielek, *Agron. J.*, *81*: 971 (1989).
26. F. R. Magdoff, W. E. Jokela, R. H. Fox, and G. F. Griffin, *Commun. Soil Sci. Plant Anal.*, *21*: 1103 (1990).
27. H. M. Brown, R. G. Hoeft, and E. D. Nafziger, "Evaluation of soil NO_3-N profile for prediction of N fertilizer requirement," Proceedings of Illinois Fertilizer Conference, 1992, Urbana, IL, pp. 21–30 (1992).
28. S. W. Melsted, H. L Motto, and T. R. Peck, *Agron. J.*, *61*: 17 (1969).
29. R. E. Voss, J. J. Hanway, and L. C. Dumenil, *Agron. J.*, *62*: 726 (1970).
30. G. D. Binford, A. M. Blackmer, and N. M. El-Hout, *Agron. J.*, *82*: 124 (1990).
31. T. C. Knowles, T. A. Doerge, and M. J. Ottman, *Agron. J.*, *83*: 353 (1991).
32. J. S. Schepers, D. D. Francis, M. Vigil, and F. E. Below, *Commun. Soil Sci. Plant Anal.*, *23*: 2173 (1992).
33. M. Takebe, T. Yoneyama, K. Inada, and T. Murakami, *Plant Soil*, *122*: 295 (1990).
34. L. M. Dwyer, M. Tollenaar, and L. Houwing, *Can. J. Plant Sci.*, *71*: 505 (1991).
35. U. L. Yadava, *HortScience*, *21*: 1449 (1986).
36. M. Takebe and T. Yoneyma, *Plant Soil*, *122*: 295 (1990).
37. C. W. Wood, D. W. Reeves, R. R. Duffield, and K. L. Edmisten, *J. Plant Nutr.*, *15*: 487 (1992).
38. W. P. Piekielek and R. H. Fox, *Agron. J.*, *84*: 59 (1992).
39. S. S. Goyal and R. C. Huffaker, *Plant Physiol.*, *82*: 1051 (1986).
40. R. J. Haynes, *Mineral Nitrogen in the Plant–Soil System*, Academic Press, Orlando, FL, p. 303 (1986).
41. W. A. Jackson, *Nitrogen in the Environment* (D. R. Neilsen and J. G. MacDonald, eds.), Academic Press, New York, p. 45 (1978).
42. W. A. Jackson, R. J. Volk, and T. C. Tucker, *Agron. J.*, *64*: 518 (1972).
43. D. T. Clarkson, *Fundamental, Ecological, and Agricultural Aspects of Nitrogen Metabolism in Higher Plants* (H. Lambers, J. J. Neeteson, and I. Stulen, eds.), Martinus Nijhoff, Dordrecht, Netherlands, p. 3 (1986).
44. M. A. Morgan, R. J. Volk, and W. A. Jackson, *Plant Physiol.*, *51*: 267 (1973).
45. C. E. Deane-Drummond and A. D. M. Glass, *Plant Physiol.*, *73*: 100 (1983).
46. W. A. Jackson, K. D. Kwik, R. J. Volk, and R. G. Butz, *Planta*, *132*: 149 (1976).
47. R. H. Teyker, W. A. Jackson, R. J. Volk, and R. H. Moll, *Plant Physiol.*, *86*: 778 (1988).
48. M. A. Morgan and W. A. Jackson, *Physiol. Plant.*, *73*: 38 (1988).
49. H. Sasakawa and Y. Yamamoto, *Plant Physiol.*, *62*: 665 (1978).
50. R. A. Joseph, T. van Hai, and J. Lambert, *Physiol. Plant.*, *34*: 321 (1975).
51. P. Nissen, N. K. Fageria, A. J. Rayar, M. M. Hassan, and T. van Hai, *Physiol. Plant.*, *49*: 222 (1980).
52. W. A. Ullrich, M. Larsson, C. M. Larsson, S. Lesch, and A. Novacky, *Physiol. Plant.*, *61*: 369 (1984).
53. W. A. Ullrich, *Inorganic Nitrogen Metabolism* (W. R. Ullrich, P. J. Aparicio, P. J. Syrtee, and F. Castillo, eds.), Springer-Verlag, Berlin, p. 32 (1987).
54. D. Kleiner, *Biochim. Biophys. Acta*, *639*: 41 (1981).
55. H. W. Scherer, C. T. MacKown, and J. E. Leggett, *J. Exp. Bot.*, *35*: 1060 (1984).
56. V. N. Faller, *J. Plant Nutr. Soil Sci.*, *131*: 120 (1972).
57. G. D. Farquhar, P. M. Firth, R. Wetselaar, and B. Weir, *Plant Physiol.*, *66*: 710 (1980).
58. A. C. Bennett, *The Plant Root and Its Environment* (E. W. Carson, ed.), University of Virgina, Charlottesville, p. 669 (1974).
59. K. Mengel and M. Viro, *Soil Sci. Plant Nutr.*, *24*: 407 (1978).
60. L. J. Youngdahl, R. Pacheco, J. J. Street, and G. L. P. Viek, *Plant Soil*, *69*: 225 (1982).
61. L. Beevers and R. H. Hageman, *Inorganic Plant Nutrition* (A. Läuchi and R. L. Bieleski, eds.), Springer-Verlag, Berlin, p. 351 (1983).

62. W. A. Jackson, W. L. Pan, R. H. Moll, and E. J. Kamprath, *Biochemical Basis of Plant Breeding*, Vol. II (C. A. Neyra, ed.), CRC Press, Boca Raton, FL, p. 73 (1986).
63. C. T. MacKown, W. A. Jackson, and R. J. Volk, *Plant Physiol.*, *69*: 353 (1982).
64. L. E. Schrader, D. Domska, P. E. Jung, and L. A. Peterson, *Agron. J.*, *64*: 690 (1972).
65. D. D. Warncke and S. A. Barber, *Agron. J.*, *65*: 950 (1973).
66. L. M. Bernardo, R. B. Clarck, and J. W. Maranville, *J. Plant Nutr.*, *7*: 1401 (1984).
67. J. C. Lycklama, *Acta Bot. Neerl.*, *12*: 361 (1963).
68. C. T. MacKown, R. J. Volk, and W. A. Jackson, *Plant Sci. Lett.*, *24*: 295 (1982).
69. T. O. Orebamjo and G. R. Stewart, *Planta*, *122*: 37 (1975).
70. J. W. Radin, *Plant Physiol.*, *55*: 178 (1975).
71. P. L. Minotti, D. C. Williams, and W. A. Jackson, *Planta*, *86*: 267 (1969).
72. A. Oaks, M. Aslam, and I. L. Boesel, *Plant Physiol.*, *59*: 391 (1977).
73. H. S. Srivastava, *Phytochemistry*, *31*: 2941 (1992).
74. K. P. Rao and D. W. Rains, *Plant Physiol.*, *57*: 55 (1976).
75. N. L. V. Datta, M. Rao, S. Guha-Mukherjee, and S. K. Sorory, *Plant Sci. Lett.*, *20*: 305 (1981).
76. P. Mehta and H. S. Srivastava, *Phytochemistry*, *19*: 2527 (1980).
77. P. L. Minotti, D. C. Williams, and W. A. Jackson, *Crop Sci.*, *9*: 9 (1969).
78. C. E. Deane-Drummond and A. D. M. Glass, *Plant Physiol.*, *73*: 105 (1983).
79. P. Oscarson, B. Ingemarson, M. Ugglas, and C. M. Larson, *Planta*, *170*: 550 (1987).
80. R. B. Lee and D. T. Clarkson, *J. Exp. Bot.*, *37*: 1753 (1986).
81. R. B. Lee and M. C. Drew, *J. Exp. Bot.*, *37*: 1768 (1986).
82. A. J. Bloom and J. Finazzo, *Plant Physiol.*, *81*: 67 (1986).
83. W. L. Pan, W. A. Jackson, and R. H. Moll, *Plant Physiol.*, *77*: 560 (1985).
84. X. T. Wang and F. E. Below, *Crop Sci.*, *32*: 997 (1992).
85. D. T. Clarkson and A. J. Warner, *Plant Physiol.*, *64*: 557 (1979).
86. D. T. Clarkson, M. J. Hooper, and L. J. P. Jones, *Plant Cell Environ.*, *6*: 535 (1986).
87. J. C. Lycklama, *Acta Bot. Neerl.*, *12*: 361 (1963).
88. J. H. MacDuff and M. J. Hooper, *Plant Soil*, *91*: 303 (1986).
89. H. Sasakawa and Y. Yamamoto, *Plant Physiol.*, *62*: 665 (1978).
90. T. Yoneyama, Y. Akiyama, and K. Kumazawa, *Soil Sci. Plant Nutr.*, *23*: 85 (1977).
91. J. R. Magalhaer and D. M. Huber, *J. Plant Nutr.*, *12*: 985 (1989).
92. J. K. Vessey, L. T. Henry, S. Chaillou, and C. D. Raper, *J. Plant Nutr.*, *13*: 95 (1990).
93. L. Tolley-Henry and C. D. Raper, *Plant Physiol.*, *82*: 54 (1986).
94. E. J. Hewitt, *Nitrogen Nutrition of Plants* (E. A. Kirkby, ed.), Waverly Press, Leeds, England, p. 68 (1970).
95. E. D. Spratt, *Agron. J.*, *66*: 57 (1974).
96. E. D. Spratt and J. K. R. Gasser, *Can. J. Soil Sci.*, *50*: 263 (1970).
97. G. Michael, P. Martin, and I. Owissia, *Nitrogen Nutrition of Plants* (E. A. Kirkby, ed.), Waverly Press, Leeds, England, p. 22 (1970).
98. H. M. Reisenauer, *Nitrogen in the Environment* (D. R. Neilsen and J. G. MacDonald, eds.), Academic Press, New York, p. 157 (1978).
99. T. Hoff, B. M. Stummann, and K. W. Henningsen, *Physiol. Plant.*, *84*: 616 (1992).
100. L. P. Solomonson and M. J. Barber, *Annu. Rev. Plant Physiol. Plant Mol. Biol.*, *41*: 225 (1990).
101. M. Andrews, *Plant Cell Environ.*, *9*: 511 (1986).
102. G. R. Stewart, N. Sumear, and M. Patel, *Inorganic Nitrogen Metabolism* (W. R. Ullrich, P. J. Aparicio, P. J. Syrtee, and F. Castillo, eds.), Springer-Verlag, Berlin, p. 39 (1987).
103. K. Vaughn and W. H. Campbell, *Plant Physiol.*, *88*: 1354 (1988).
104. K. Kamachi, Y. Amemiya, N. Ogura, and H. Nakagawa, *Plant Cell Physiol.*, *28*: 333 (1987).
105. S. H. Lips and Y. Avissar, *Eur. J. Biochem.*, *29*: 20 (1972).
106. M. R. Ward, H. D. Grimes, and R. C. Huffaker, *Planta*, *177*: 470 (1989).
107. M. G. Redinbaugh and W. H. Campbell, *Plant Physiol.*, *68*: 115 (1981).
108. A. Oaks and B. Hirel, *Annu. Rev. Plant Physiol.*, *36*: 345 (1985).
109. B. J. Miflin and P. J. Lea, *The Biochemistry of Plants* (B. J. Miflin, ed.), Academic Press, New York, p. 169 (1980).

110. R. B. Lee, *Plant Cell Environ.*, *3*: 65 (1980).
111. B. J. Miflin and P. J. Lea, *Nucleic Acids and Proteins in Plants* (I. D. Boulter and B. Parthiev, eds.), Springer-Verlag, New York, p. 5 (1982).
112. G. R. Stewart, A. F. Mann, and P. A. Fenten, *The Biochemistry of Plants* (B. J. Miflin, ed.), Academic Press, New York, p. 271 (1980).
113. D. L. Karlen, R. L. Flannery, and E. J. Sadler, *Agron. J.*, *80*: 232 (1988).
114. A. Olness, G. R. Benoit, K. Van Sickle, and J. Rinke, *J. Agron. Crop Sci.*, *164*: 42 (1990).
115. M. P. Russelle, R. D. Hauck, and R. A. Olson, *Agron. J.*, *75*: 293 (1983).
116. R. B. Austin, M. A. Ford, J. A. Edrich, and R. D. Blackwell, *J. Agric. Sci.*, *88*: 159 (1977).
117. L. G. Bundy and P. R. Carter, *J. Prod. Agric.*, *1*: 99 (1988).
118. J. J. Heitholt, L. I. Croy, N. O. Maness, and H. T. Nguyen, *Field Crops Res.*, *23*: 133 (1990).
119. J. C. Swank, F. E. Below, R. J. Lambert, and R. H. Hageman, *Plant Physiol.*, *70*: 1185 (1982).
120. D. A. Van Sanford and C. T. MacKown, *Crop Sci.*, *27*: 295 (1987).
121. B. I. Muruli and G. M. Paulsen, *Maydica*, *26*: 63 (1981).
122. R. H. Hageman and F. E. Below, *Nitrogen in Higher Plants* (Y. P. Abrol, ed.), Research Studies, Taunton, England, p. 313 (1990).
123. W. L. Nelson and H. F. Reetz, *Crops Soils*, *38*(8): 5 (1986).
124. F. E. Below, R. J. Lambert, and R. H. Hageman, *Agron. J.*, *76*: 777 (1984).
125. S. J. Crafts-Brandner, F. E. Below, J. E. Harper, and R. H. Hageman, *Plant Physiol.*, *74*: 360 (1984).
126. K. D. Smiciklas and F. E. Below, *Maydica*, *35*: 209 (1990).
127. C. Y. Tsai, D. M. Huber, D. V. Glover, and H. L. Warren, *Crop Sci.*, *24*: 277 (1984).
128. D. T. Walters and G. L. Malzer, *Soil Sci. Soc. Am. J.*, *54*: 115 (1990).
129. D. W. Altman, W. L. McCuistion, and W. E. Kronstad, *Agron. J.*, *75*: 87 (1983).
130. E. L. Deckard, R. J. Lambert, and R. H. Hageman, *Crop Sci.*, *13*: 343 (1973).
131. L. D. Maddux and P. L. Barnes, *J. Fertil. Issues*, *2*: 124 (1985).
132. F. M. Rhoads and A. Manning, *Soil Crop Sci. Soc. Fla. Proc.*, *45*: 50 (1986).
133. J. W. Friedrich, L. E. Schrader, and E. V. Nordheim, *Agron. J.*, *71*: 461 (1979).
134. K. S. Reddy, H. A. Mills, and J. B. Jones, *Agron. J.*, *83*: 201 (1991).
135. L. E. Gentry and F. E. Below, *Crop Sci.*, *33*: 491 (1993).
136. V. V. Rendig and T. W. Crawford, *J. Sci. Food Agric.*, *36*: 645 (1985).
137. C. J. Pearson and B. C. Jacobs, *Aust. J. Agric. Res.*, *38*: 1001 (1987).
138. F. E. Below, R. J. Lambert, and R. H. Hageman, *Agron. J.*, *76*: 773 (1984).
139. J. S. Tomar, A. F. MacKenzie, R. G. Mehuys, and I. Alli, *Agron. J.*, *80*: 802 (1988).
140. F. E. Below, L. E. Christensen, A. J. Reed, and R. H. Hageman, *Plant Physiol.*, *68*: 1186 (1981).
141. R. T. Weiland and T. C. Ta, *Aust. J. Plant Physiol.*, *19*: 77 (1992).
142. Y. Murata and S. Matsushima, *Crop Physiology: Some Case Histories* (L. T. Evans, ed.), Cambridge University Press, Cambridge, p. 73 (1975).
143. J. R. Evans, *Plant Physiol.*, *72*: 297 (1983).
144. S. C. Huber, T. Sugiyama, and R. S. Alberte, *Plant Cell Physiol.*, *30*: 1063 (1989).
145. S. C. Wong, I. R. Cowan, and G. D. Farquhar, *Plant Physiol.*, *78*: 821 (1985).
146. B. Sugiharto, K. Miyata, H. Nakomoto, H. Sasakawa, and T. Sugiyama, *Plant Physiol.*, *92*: 963 (1990).
147. B. Sugiharto and T. Sugiyama, *Plant Physiol.*, *98*: 1403 (1992).
148. T. Mae, A. Makino, and K. Ohira, *Plant Cell Physiol.*, *24*: 1079 (1983).
149. L. T. Evans, I. F. Wardlaw, and R. A. Fisher, *Crop Physiology: Some Case Histories* (L. T. Evans, ed.), Cambridge University Press, Cambridge, p. 101 (1975).
150. S. R. Simmons and R. J. Jones, *Crop Sci.*, *25*: 1004 (1985).
151. M. J. Dalling, *Exploitation of Physiological and Genetic Variability to Enhance Crop Productivity* (J. E. Harper, L. E. Schrader, and R. W. Howell, eds.), American Society of Plant Physiologists, Rockville, MD, p. 55 (1985).
152. K. Morita and M. Kono, *Soil Sci. Plant Nutr.*, *20*: 79 (1974).
153. J. A. Friedrich and R. C. Huffaker, *Plant Physiol.*, *65*: 1103 (1980).
154. K. Morita, *Ann. Bot.*, *46*: 297 (1980).

155. P. Boon-Long, D. B. Egli, and J. E. Leggett, *Crop Sci.*, *23*: 617 (1983).
156. S. J. Crafts-Brandner and C. G. Poneleit, *Plant Physiol.*, *84*: 255 (1987).
157. J. R. Evans, *Plant Physiol.*, *72*: 297 (1983).
158. A. Makino, T. Mae, and K. Ohira, *Plant Cell Physiol.*, *25*: 429 (1984).
159. F. E. Below, S. J. Crafts-Brandner, and R. H. Hageman, *Plant Physiol.*, *79*: 1077 (1985).
160. K. J. Boote, R. N. Gallaher, W. K. Robertson, K. Hinson, and L. C. Hammond, *Agron. J.*, *70*: 787 (1978).
161. H. J. Harder, R. E. Carlson, and R. H. Shaw, *Agron. J.*, *74*: 759 (1982).
162. R. Killorn and D. Zourarakis, *J. Prod. Agric.*, *5*: 142 (1992).
163. H. J. Thomas and L. Stoddart, *Annu. Rev. Plant Physiol.*, *31*: 83 (1980).
164. W. G. Duncan, *Crop Physiology: Some Case Histories* (L. T. Evans, ed.), Cambridge University Press, Cambridge, p. 23 (1975).
165. M. Tollenaar, *Maydica*, *22*: 49 (1977).
166. R. Capitanio, E. Gentinetta, and M. Motto, *Maydica*, *28*: 365 (1983).
167. V. M. Reedy and T. B. Daynard, *Maydica*, *28*: 339 (1983).
168. B. K. Singh and C. F. Jenner, *Aust. J. Plant Physiol.*, *9*: 83 (1982).
169. B. K. Singh and C. F. Jenner, *Aust. J. Plant Physiol.*, *11*: 151 (1984).
170. K. M. Goh and R. J. Haynes, *Mineral Nitrogen in the Plant–Soil System*, Academic Press, Orlando, FL, p. 379 (1986).
171. B. D. Jacobs and C. J. Pearson, *J. Exp. Bot.*, *43*: 557 (1992).
172. J. H. Lemcoff and R. S. Loomis, *Crop Sci.*, *26*: 1017 (1986).
173. R. Novoa and R. S. Loomis, *Plant Soil*, *58*: 177 (1981).
174. J. F. Power and J. Alessi, *J. Agric. Sci.*, *90*: 97 (1978).
175. J. H. Spiertz and N. M. DeVos, *Plant Soil*, *75*: 162 (1983).
176. S. M. Thomas, G. N. Thorne, and I. Pearman, *Ann. Bot.*, *42*: 827 (1978).
177. A. Mozafar, *Agron. J.*, *82*: 511 (1990).
178. H. V. Eck, *Agron. J.*, *76*: 421 (1984).
179. I. Pearman, S. M. Thomas, and G. N. Thorne, *Ann. Bot.*, *41*: 93 (1977).
180. J. R. Czyzewicz and F. E. Below, *Agron. Abstr.*, p. 110 (1993).
181. H. Doll and M. Kreis, *Crop Physiology and Cereal Breeding* (J. H. J. Spiertz and T. Kramer, eds.), Centre for Agricultural Publishing and Documentation, Wageningen, Netherlands, p. 173 (1979).
182. C. Y. Tsai, D. M. Huber, and H. L Warren, *Crop Sci.*, *18*: 399 (1978).
183. C. Y. Tsai, D. M. Huber, and H. L Warren, *Plant Physiol.*, *66*: 330 (1980).
184. C. Y. Tsai, B. A. Larkins, and D. V. Glover, *Biochem. Gen.*, *16*: 883 (1978).
185. A. A. Khan, R. Verbeck, E. C. Waters, and H. A. van Onckelen, *Plant Physiol.*, *51*: 641 (1973).
186. G. Michael and H. Beringer, "Physiological aspects of crop productivity," 15th Colloquium of the International Potash Institute, Berne, Switzerland, p. 85 (1980).
187. G. W. Singletary and F. E. Below, *Plant Physiol.*, *92*: 160 (1990).
188. G. W. Singletary, D. C. Doehlert, C. M. Wilson, M. J. Muhitch, and F. E. Below, *Plant Physiol.*, *94*: 858 (1990).
189. I. Stulen, *Fundamental, Ecological, and Agricultural Aspects of Nitrogen Metabolism in Higher Plants* (H. Lambers, J. J. Neeteson, and I. Stulen, eds.), Martinus Nijhoff, Dordrecht, Netherlands, p. 261 (1986).
190. J. W. Dudley, R. J. Lambert, and I. A. de la Roche, *Crop Sci.*, *17*: 111 (1977).
191. J. Mesdag, *Crop Physiology and Cereal Breeding* (J. H. J. Spiertz and T. Kramer, eds.), Centre for Agricultural Publishing and Documentation, Wageningen, Netherlands, p. 166 (1979).
192. F. W. T. Penning de Vries, A. H. M. Brunsting, and H. H. van Laar, *J. Theor. Biol.*, *45*: 399 (1974).
193. C. R. Bhatia and R. Rabson, *Science*, *194*: 1418 (1976).
194. P. Chevalier and S. E. Lingle, *Crop Sci.*, *23*: 272 (1983).
195. A. J. Reed and G. W. Singletary, *Plant Physiol.*, *91*: 986 (1989).
196. B. K. Singh and C. F. Jenner, *Aust. J. Plant Physiol.*, *11*: 151 (1984).
197. M. C. Cox, C. O. Qualset, and D. W. Rains, *Crop Sci.*, *25*: 430 (1985).

198. P. I. Payne, *Seed Proteins* (J. Daussant, I. Mosse, and J. Vaughan, eds.), Academic Press, New York, p. 223 (1983).
199. R. J. Jones and S. R. Simmons, *Crop Sci.*, *23*: 129 (1983).
200. Y. Z. Ma, C. T. MacKown, and D. A. Van Sanford, *Crop Sci.*, *30*: 1099 (1990).
201. F. Zink and G. Michael, *Z. Acker- Pflanzenbau*, *154*: 203 (1986).
202. L. T. Evans and I. F. Wardlaw, *Adv. Agron.*, *28*: 301 (1976).
203. D. T. Canvin, *Genetic Improvement of Seed Proteins*, U.S. National Academy of Sciences, Washington, DC, p. 172 (1976).
204. R. Reggiani, C. Soave, N. Di Fonzo, E. Gentinetta, and F. Salamini, *Genet. Agric.*, *39*: 221 (1985).
205. C. I. Tsai, I. Dweikat, and C. Y. Tsai, *Maydica*, *35*: 391 (1990).
206. C. S. Wyss, J. R. Czyzewicz, and F. E. Below, *Crop Sci.*, *31*: 761 (1991).
207. D. C. Doehlert and R. J. Lambert, *Crop Sci.*, *31*: 151 (1991).
208. J. B. Cliquet, E. Deléens, and A. Mariotti, *Plant Physiol.*, *94*: 1547 (1990).
209. A. J. Reed, G. W. Singletary, J. R. Schussler, D. E. Williamson, and A. L. Christy, *Crop Sci.*, *28*: 819 (1988).
210. M. J. Ottman and L. F. Welch, *Agron. J.*, *80*: 619 (1988).
211. D. H. Sander, W. H. Allaway, and R. A. Olsen, *Nutritional Quality of Cereal Grains: Genetic and Agronomic Improvement* (R. A. Olsen and K. J. Krey, eds.), ASA, Madison, WI, p. 45 (1987).
212. M. Asghari and R. G. Hanson, *Agron. J.*, *76*: 911 (1984).
213. S. E. Hollinger and R. G. Hoeft, *Agron. J.*, *78*: 818 (1986).
214. S. L. Oberle and D. R. Keeney, *J. Prod. Agric.*, *3*: 527 (1990).
215. E. D. Nafziger, R. L. Mulvaney, D. L. Mulvaney, and L. E. Paul, *J. Fertil. Issues*, *1*: 136 (1984).
216. F. E. Broadbent and A. B. Carlton, *Nitrogen in the Environment* (D. R. Nielsen and J. G. MacDonald, eds.), Academic Press, New York, p. 1 (1978).
217. J. S. Schepers and F. E. Below, "Influence of corn hybrids on nitrogen uptake and utilization efficiency," Proceedings of the 42nd Annual Corn and Sorghum Industry Research Conference, Chicago, pp. 172–186 (1987).
218. D. T. Gehl, L. D. Bailey, C. A. Grant, and J. M. Sadler, *Can. J. Plant Sci.*, *70*: 51 (1990).
219. D. A. Van Sanford and C. T. MacKown, *Theor. Appl. Genet.*, *72*: 158 (1986).
220. F. E. Broadbent, S. K. De Datta, and E. V. Laureles, *Agron. J.*, *79*: 786 (1987).
221. V. B. Ogunlela and P. N. Okoh, *Fertil. Res.*, *21*: 67 (1989).
222. P. Chevalier and L. E. Schrader, *Crop Sci.*, *17*: 897 (1977).
223. L. G. Balko and W. A. Russell, *Agron. J.*, *72*: 723 (1980).
224. R. J. Sabata and S. C. Mason, *J. Prod. Agric.*, *5*: 137 (1992).
225. M. B. Hatilitligil and W. A. Compton, *Fertil. Res.*, *5*: 321 (1984).
226. C. A. C. Gardner, P. L. Bax, D. J. Bailey, A. J. Cavalieri, C. R. Clausen, G. A. Luce, J. M Meece, P. A. Murphy, T. E. Piper, R. L. Segebart, O. S. Smith, C. W. Tiffany, M. W. Trimble, and B. N. Wilson, *J. Prod. Agric.*, *3*: 39 (1990).
227. L. Gashaw and L. M. Mugwira, *Agron. J.*, *73*: 47 (1981).
228. L. E. Gentry, X. T. Wang, and F. E. Below, *J. Plant Nutr.*, *12*: 363 (1989).
229. W. J. Cox and H. M. Reisenauer, *Plant Soil*, *38*: 363 (1973).
230. R. Ganmore-Neumann and U. Kafkafi, *Agron. J.*, *72*: 758 (1980).
231. T. W. Rufty, C. D. Raper, and W. A. Jackson, *Bot. Gaz.*, *144*: 466 (1983).
232. G. S. Weissman, *Plant Physiol.*, *39*: 947 (1964).
233. K. G. Alexander, H. M. Miller, and E. G. Beauchamp, *J. Plant Nutr.*, *14*: 31 (1991).
234. R. S. Antil, D. S. Yadav, V. Kumar, and M. Singh, *Trop. Plant Sci. Res.*, *1*: 353 (1983).
235. F. E. Below and L. E. Gentry, *J. Fertil. Issues*, *4*: 79 (1987).
236. B. R. Bock, *J. Fertil. Issues*, *4*: 68 (1987).
237. J. J. Camberto and B. R. Bock, *Plant Soil*, *113*: 79 (1989).
238. J. A. Heberer and F. E. Below, *Ann. Bot.*, *63*: 643 (1989).
239. F. E. Below and L. E. Gentry, *Crop Sci.*, *23*: 163 (1992).
240. F. E. Below and J. A. Heberer, *J. Plant Nutr.*, *13*: 667 (1990).
241. J. J. Camberto and B. R. Bock, *Agron. J.*, *82*: 463 (1990).

242. F. G. Adriaanse and J. J. Human, *S. Afr. J. Plant Soil*, *3*: 57 (1986).
243. K. L. Barber, L. D. Maddux, D. E. Kessel, G. M. Pierzynski, and B. R. Bock, *Soil Sci. Soc. Am. J.*, *56*: 1166 (1992).
244. J. Hagin, S. R. Olsen, and A. Shaviv, *J. Plant Nutr.*, *13*: 1211 (1990).
245. J. R. Huffman, *J. Agron. Educ.*, *18*: 93 (1989).
246. K. D. Smiciklas and F. E. Below, *Plant Soil*, *142*: 307 (1992).
247. K. D. Smiciklas and F. E. Below, *Crop Sci.*, *32*: 1220 (1992).
248. J. J. Camberto and B. R. Bock, *Agron. J.*, *82*: 467 (1990).
249. W. L. Pan, E. J. Kamprath, R. H. Moll, and W. A. Jackson, *Soil Sci. Soc. Am. J.*, *48*: 1101 (1984).
250. F. E. Below, L. E. Gentry, and K. D. Smiciklas, "Role of mixed N in enhancing productivity of maize," Proceedings of Symposium, Division S-8, Soil Science Society of America, Denver, CO, pp. 1–11 (1991).
251. L. E. Gentry and F. E. Below, *Agron. Abstr.*, p. 278 (1992).
252. P. E. Jung, L. A. Peterson, and L. E. Schrader, *Agron. J.*, *64*: 668 (1972).
253. E. L. Anderson, E. J. Kamprath, and R. H. Moll, *Agron. J.*, *76*: 397 (1984).
254. M. Motto and R. H. Moll, *Maydica*, *28*: 53 (1983).
255. G. S. McNaughton and M. R. Presland, *J. Exp. Bot.*, *34*: 880 (1983).
256. T. Yoneyama, Y. Akiyama, and K. Kumazawa, *Soil Sci. Plant Nutr.*, *23*: 85 (1977).
257. W. L. Pan, J. J. Camberato, W. A. Jackson, and R. H. Moll, *Plant Physiol.*, *82*: 247 (1986).
258. L. Salsac, S. Chillou, J. F. Morot-Gaudry, C. Lesaint, and E. Jolivet, *Plant Physiol. Biochem.*, *25*: 805 (1987).
259. J. A. Raven, *New Phytol.*, *101*: 25 (1985).
260. O. A. M. Lewis, B. Fulton, and A. A. A. van Zelewski, *Inorganic Nitrogen Metabolism* (W. R. Ulrich, P. J. Aparicio, P. J. Syrett, and F. Casrillo, eds.), Springer-Verlag, Berlin, p. 240 (1987).
261. T. F. Neales and L. D. Incoll, *Bot. Rev.*, *34*: 107 (1968).
262. N. P. Hall, R. Reggiani, J. Franklin, A. J. Keys, and P. J. Lea, *Photosynth. Res.*, *5*: 361 (1984).
263. G. F. Morot-Gaundry, F. Thuillier, C. Lesaint, S. Chaillou, and E. Jolivet, *Physiol. Veg.*, *23*: 257 (1985).
264. A. M. Amory and C. F. Cresswell, *Ann. Bot.*, *54*: 719 (1984).
265. O. A. M. Lewis, M. I. M. Soares, and S. H. Lips, *Fundamental, Ecological and Agricultural Aspects of Nitrogen Metabolism in Higher Plants* (H. Lambers, J. J. Neeteson, and I. Stulen, eds.), Martinus Nijhoff, Dordrecht, Netherlands, p. 295 (1986).
266. G. J. Blair, H. M. Miller, and W. A. Mitchell, *Agron. J.*, *62*: 530 (1970).
267. E. P. Papanicolaou, V. D. Skarlou, C. Nobeli, and N. S. Katranis, *J. Agric. Sci.*, *101*: 687 (1983).
268. R. W. Smiley, *Soil Sci. Soc. Am. Proc.*, *38*: 795 (1974).
269. E. Schnug and A. Fink, *Plant Nutrition 1982* (A. Scaife (ed.), Commonwealth Agriculture Bureau, Slough, England, p. 583 (1982).
270. R. H. Teyker and D. C. Hobbs, *Agron. J.*, *84*: 694 (1992).
271. J. Gerendas, R. G. Ratcliffe, and B. Sattelmacher, *J. Plant Physiol.*, *137*: 125 (1990).
272. J. A. Raven and F. A. Smith, *New Phytol.*, *76*: 415 (1976).
273. R. H. Moll, E. J. Kamprath, and W. A. Jackson, *Agron. J.*, *74*: 562 (1982).
274. E. L. Anderson, E. J. Kamprath, and R. H. Moll, *Crop Sci.*, *25*: 598 (1985).
275. K. C. Cameron and R. J. Haynes, *Mineral Nitrogen in the Plant–Soil System*, Academic Press, Orlando, FL, p. 166 (1986).
276. P. C. Scharf and M. M. Alley, *J. Fertil. Issues*, *5*: 109 (1988).
277. R. D. Hauck, *Nitrification Inhibitors—Potentials and Limitations* (J. J. Meisinger, G. W. Randall, and M. L. Vitosh, eds.), ASA and SSSA, Madison, WI, p. 19 (1980).
278. D. M. Huber, H. L. Warren, and C. Y. Tsai, *BioScience*, *27*: 523 (1977).
279. H. F. Chancy and E. J. Kamprath, *Agron. J.*, *74*: 656 (1982).
280. R. G. Hoeft, *Nitrogen in Crop Production* (R. D. Hauck, ed.), ASA, CSSA, SSA, Madison, WI, p. 561 (1984).
281. L. L. Hendrickson, *J. Prod. Agric.*, *5*: 131 (1992).
282. R. D. Voss, *Nitrogen in Crop Production* (R. D. Hauck, ed.), ASA, CSSA, SSA, Madison, WI, p. 571 (1984).

14

Biological Nitrogen Fixation in Legumes and Nitrogen Transfer in Crop Plants

Ranga R. Velagaleti
Battelle Memorial Institute, Columbus, Ohio

Gary R. Cline
Kentucky State University, Frankfort, Kentucky

I. INTRODUCTION

Nitrogen deficiency is often the most important factor limiting plant growth even though the atmosphere contains approximately 78% N_2, a form of nitrogen not available to plants. Chemically inert N_2 gas can be made available to plants by symbiotic N_2 fixation, which commonly occurs in nodules formed by rhizobia bacteria on the roots of leguminous plants. Plants supply the rhizobia with carbohydrates in return for N_2 fixed by the rhizobia.

Much of the research regarding nitrogen fixation has focused on grain and forage legumes. These legumes are able to fix significant amounts of nitrogen and thereby reduce requirements for inorganic nitrogen fertilizer. If the legumes are not harvested, they can be incorporated into the soil as green manure to provide nitrogen to subsequent crops.

Less attention has been given to examining the abilities of established legumes to supply nitrogen to neighboring plants when legumes occur in mixtures with nonleguminous plants. This process has been termed nitrogen transfer.

In this chapter, we have attempted to show the benefits of nitrogen fixation to legume yield and to associated nonlegume crops through nitrogen transfer from legumes. The evidence for nitrogen transfer and considerations of the mechanism by which it occurs are presented in detail through literature review and examples.

II. BENEFITS OF *RHIZOBIUM* INOCULATION ON GROWTH AND SEED YIELD

Benefits of *Rhizobium* inoculation have been demonstrated by a number of investigators [1,2]. In tropical western Nigeria, U.S. soybean cultivars (TGm 80 and TGm 294-4) responded to inoculation with *Bradyrhizobium japonicum*, showing significant increases in nodule dry weight, shoot dry weight, and seed yield. On the other hand, the Asiatic soybean variety Orba, with its ability to nodulate with indigenous cowpea rhizobia, did not respond to inoculation as well (Table 1). In the same experiment, *Bradyrhizobium* strains were shown to survive in the field

Table 1 Effects of Rhizobium Inoculation or Fertilizer N (100 kg ha^{-1}) on Shoot Dry Weight, Nodule Dry Weight, and Seed Yield of Three Soybean Cultivers, at 7 Weeks After Planting

Treatment	Shoot dry weight (g plant^{-1})			Nodule dry weight (mg)			Seed yield (kg ha^{-1})		
	TGm 80	TGm 294-4	Orba	TGm 80	TGm 294-4	Orba	TGm 80	TGm 294-4	Orba
Uninoculated	7.7	6.1	10.3	14	46	287	1526	1655	1924
Uninoculated + 150 kg N ha^{-1}	15.7	13.5	20.5	46	30	275	2314	2151	1848
IRj 2101	10.1	12.1	11.0	251	186	351	2429	2279	1915
2102	10.1	11.2	11.9	349	317	505	2422	2309	1772
2110	10.6	9.7	11.6	609	552	679	2387	2576	1798
2111	13.1	10.4	20.2	650	509	720	2451	2120	2046
2112	8.1	9.4	10.6	431	306	443	2736	2834	1789
2113	12.4	10.2	14.9	479	388	538	2934	2598	1950
2119	9.1	12.0	11.6	386	472	569	2331	1540	1670
2123	13.0	11.4	12.9	532	344	452	2433	1997	1471
2125	9.3	8.4	16.3	458	303	488	2271	1936	2044
2126	11.8	9.7	16.6	391	405	348	2362	2188	1857
2127	11.5	11.7	12.4[a]	565	410	495[a]	1847	1807	1758
Nitrogen commercial inoculum	9.4	10.6	13.0	645	424	648	2335	2335	2067
Least significant difference ($p = 0.05$)									
same treatment		4.1			185		3031	2935	
different treatments		4.2					508		

[a]Missing data; value was estimated.

Source: Adapted from Ref. 2.

over a 2-year fallow period (Table 2), and significant nodule and seed yields were reported in soybeans planted after the 2-year fallow.

Rhizobium inoculation has been found to be beneficial under conditions of environmental stresses. Selection of host genotypes and stress-tolerant *Rhizobium* strains and their appropriate matching assumes special significance in attempts to derive maximum benefits of inoculation under stress conditions. Velagaleti et al. [3–5] observed a wide diversity for salt tolerance in soybean host cultivars and *Bradyrhizobium* strains, as well as benefits of inoculation of salt-tolerant soybean cultivars with salt-tolerant *Bradyrhizobium* strains.

III. APPLICATIONS OF NITROGEN TRANSFER

Mixtures of perennial legumes and other plants often occur in pasture situations, where a number of different plant species may be required to achieve sustained productivity and fertility over an entire growing season. For example, legumes such as clover, alfalfa, and birdsfoot trefoil can be grown in combination with (and transfer nitrogen to) fescue, ryegrass, Bermuda grass, cocksfoot, reed canarygrass, and timothy [6–12]. Legumes may be needed to maintain adequate nitrogen availability via nitrogen fixation and transfer, whereas both warm-season and cool-season forage crops may be required to maintain adequate yields as temperature and precipitation patterns change throughout the growing season [13]. The ability of perennial legumes to transfer nitrogen to other plants may be just as, if not more important, than their ability to fix large amounts of nitrogen. This is because nitrogen fixed in their tissue is not recycled quickly, as would be the case for annual legumes or legumes in crop rotations. If no nitrogen transfer occurred, neighboring plants would obtain little in the way of nitrogen benefits from perennial legumes until the legumes expired and their tissue was recycled.

Intercropping agronomic crops is another agricultural cropping system in which nitrogen transfer is of importance. In these systems legume and nonlegume crops can be grown simultaneously, and the legume can provide nitrogen to the other crops via nitrogen fixation (Table 3) and subsequent transfer, especially when the available soil nitrogen is low [14,15]. Under moderate to high soil N conditions, nitrogen fixation by legumes may be reduced [Table 4]. Examples of such intercroppings include cowpea/corn [14,15] alfalfa/wheat and oats [16], and soybean/sorghum [17].

Table 2 Fraction of Nodules Formed by Introduced Rhizobium After a 2-Year Fallow Period

	% Occupancy[a] on		
Rhizobium	TGm 80	TGm 294-4	Orba
Pre fallow			
IRj 2110	81	98	72
2111	93	100	78
2112	98	88	16
2113	90	94	45
Post fallow			
IRj 2110	100	100	100

[a]Based on the examination of 192 nodules per cost.

Source: Adopted from Ref. 2.

Table 3 Net N Balances of Cowpea Cultivars ER-1, TVu 1190, Ife Brown, and TVu 4552 and Soybean Cultivars N59-5253 (Nodulating Isoline) and Williams, When Grown in a Soil Low in Available N*

| | N component (kg ha^{-1}) | | | | |
Cultivar	Mineral N uptake	Fixed N input[a]	Grain N removed	Residue N returned	N balance
ER-1	32	50	48	34	+2
TVu 1190	33	101	49	85	+52
Ife Brown	25	81	57	49	+24
TVu 4552	27	49	46	30	+3

**Overall mean of difference and A_N value data at 25 N.

***Calculation of N fixed.

[a]The estimates of amounts of N fixed (kg ha^{-1}) were made by the difference method and the A_N value method, with reference in turn to each nonfixing crop, nonnodulating soybean, maize, and celosia.

  Difference method

  N fixed = total N in fixing crop − total N in nonfixing crop

  A_N − value method.

  N fixed = (A_N of fixing crop − A_N of nonfixing crop) × %N-fertilizer utilization of fixing crop

  where $A_N (\text{kgha}^{-1}) = \dfrac{(100 - \%\text{NDFF})}{\%\text{NDFF}} \times$ rate N Applied

  %NDFF = %N derived from fertilizer = $\dfrac{\%^{15}\text{N atom excess of crop}}{\%^{15}\text{N atom excess of fertilizer}} \times 100$

  %N fertilizer utilization = $\dfrac{\%\text{NDFF} \times \text{total N in crop}}{\text{rate N applied}}$

Source: Adapted from Ref. 15.

Table 4 Net N Balances of Cowpea Cultivars ER-1, TVu 1190, Ife Brown, and TVu 4552, and Soybean Cultivars N59-5253 (Nodulating isoline) and Williams, When Grown in a Soil Intermediate in Available N*

| | N component (kg ha^{-1}) | | | | |
Cultivar	Mineral N uptake	Fixed N input[a]	Grain N removed	Residue N returned	N-balance
ER-1	66	28	54	40	−26
TVu 1190	69	49	49	69	0
Ife Brown	64	44	52	56	−8
TVu 4552	54	19	53	20	−34

[a]Overall mean of difference and A_N value data at 100 N (also see Table 3).

Source: Adapted from Ref. 15.

Another agricultural application of nitrogen transfer may be the use of legume/nonlegume mixtures as winter cover crops. In areas not subjected to severe winters (e.g., central and southern portions of the United States), legumes planted in late summer following the harvesting of cash crops can fix significant amounts of nitrogen before they are replaced with cash crops the following spring. Mixtures of legumes (for nitrogen fixation/transfer) and nonlegumes (biomass and scavenging of residual nitrogen fertilizer) may be more beneficial than monocultures of any particular species. Examples of such winter-cropped mixtures include various combinations of rye, ryegrass, and oats with vetch and clover [18].

Nitrogen transfer would also appear to be important regarding tree farms, natural ecosystems, and mineland reclamation. Most productive forest tree species are not legumes, but perennial legumes exist as forbs, shrubs, and small trees (e.g., redbud) in the forest understory. These ecosystems are normally not intensively managed, and nitrogen is often a factor limiting tree growth. Thus, nitrogen transfer from perennial legumes to trees may be extremely important, especially since these legumes may persist for many years before they die and nitrogen fixed in their tissue is recycled. Perennial legumes (e.g., birdsfoot trefoil and Sericea lespedeza) are planted on mine sites to improve nitrogen availability to other nonleguminous plant species [19]. Again, without nitrogen transfer, such perennial legumes may be of little use in supplying nitrogen to surrounding plants.

IV. EVIDENCE FOR NITROGEN TRANSFER

Although nitrogen transfer may be extremely wasteful regarding the "transferrer," and although it cannot always be demonstrated, it appears to be a common occurrence. Initial evidence came from comparisons of increases in yields and nitrogen contents of crops/grasses grown with legumes and grown alone [6,7,17,20]. Cowling and Lockyer [6] estimated that nitrogen transferred annually by clover to four grasses ranged from 26 to 52 kg of N per hectare with an overall mean of 39 kg ha^{-1} year^{-1} for all grasses. This amounted to an average of 29% of the aboveground clover nitrogen. In another study four tropical legumes fixed 51–140 kg ha^{-1} and transferred 12–17% of this nitrogen to a companion grass [21]. Under low soil nitrogen conditions, cowpeas contributed up to 52 kg/ha^{-1} (Table 3) of fixed nitrogen to soil [15].

Grasses and other nonlegumes normally compete favorably with legumes for nitrogen. Thus, when grown in combination, soil nitrogen may be spared by legumes, which then rely more on nitrogen fixation because of the competition [22]. Such sparing of nitrogen may not always be a significant factor [14]. However, its possibility makes it difficult to distinguish nitrogen sparing from actual nitrogen transfer between plants when results are based on tissue nitrogen contents and yields. Nitrogen sparing by legumes may have occurred in studies by Haystead and Lowe [23] and Vallis et al. [22]. In these field experiments nitrogen transfer was not detected using isotopic labeling methodology (see following discussion), although foliar nitrogen contents of neighboring grasses were increased in the presence of legumes.

Isotopic nitrogen has been used to document the transfer of nitrogen fixed from living legumes to neighboring plants. In these experiments soil is often labeled with ^{15}N, and the ratio of ^{15}N to unlabeled nitrogen is measured in plant shoots. Amounts of fixed nitrogen in plant shoots (legume or nonlegume) are calculated based on the knowledge that nitrogen fixed from the atmosphere contains less isotopic nitrogen. A nonleguminous plant grown in the absence of legumes is needed for comparison. Using isotopic nitrogen, Mallarino et al. [10] found that 20–60% of fescue nitrogen was transferred from three legumes to fescue. Annual uptake of transferred nitrogen averaged 18 and 34 kg ha^{-1}, respectively, for 2 years. Using similar methodology, Goh et al. [24] determined that 60–80% of the nitrogen present in ryegrass was originally fixed by clover. A value of 79% was reported by Broadbent et al. [25], also using clover and ryegrass. Brophy et al. [26], using isotopic nitrogen, reported that up to 68 and 79% of reed canarygrass nitrogen was derived from alfalfa and birdsfoot trefoil, respectively. This represented 13 and 17% of nitrogen fixed by birdsfoot trefoil and alfalfa, respectively. In other cases, the percentage of nonlegume derived from legumes has been considerably less than the percentages described above: 36% [27], <16% [28], and <12% [29,30].

Conditions favoring transfer of nitrogen have been difficult to define, and in some studies transfer has not been demonstrated [22,31,32]. In other cases transfer was not detected until late in the growing season after several harvests [30,33], whereas in still others it has not been demonstrated until several years after planting [34]. Amounts of nitrogen transferred have differed widely according to whether plants are grown under field or greenhouse conditions. Ryegrass obtained up to 80% of its nitrogen from clover under field conditions, but nitrogen transfer was undetectable in greenhouse experiments [25]. Similar results regarding pot versus field experiments were reported for ryegrass by Goh et al. [24]. However, nitrogen transfer from clover to ryegrass has also been reported to be greater under greenhouse conditions than in the field [30,35].

Some of the apparent inconsistencies in earlier studies may be due to differences in growth conditions. Prerequisites for nitrogen fixation and transfer include nitrogen deficient soil [14,30,36] and favorable conditions of moisture, light, and temperature [7,29,30,32]. Also differences in the proportions of legumes and nonlegumes in plant mixtures may explain variability in results among experiments [10].

V. MECHANISMS FOR NITROGEN TRANSFER

A number of different mechanisms have been suggested for nitrogen transfer. Legumes can lose fixed nitrogen to soil via

Excretion or leakage from living nodulated roots
Root cell sloughing
Senescence and degradation of nodules and roots
Mycorrhizal fungi
Leaching of foliage
Decomposition of dead herbage on the soil surface

Although direct excretion and particularly nodule/root degradation have received considerable attention, the actual mechanisms of transfer are unclear [37,38].

A. Excretion of Soluble Nitrogenous Compounds

Direct excretion of soluble nitrogenous compounds is known to occur [39–41], but the biological significance of such excretions remains unclear. The occurrence of such a direct mechanism is suggested by the work of Vest [36], who reported that only particular cultivars of soybean were able to transfer nitrogen, and only when in symbiosis with particular strains of rhizobia.

Using isotopic nitrogen in nutrient solution, Ta et al. [42] documented the excretion of ammonia and amino acids containing nitrogen fixed by alfalfa. However, this amounted to only 3% of the total nitrogen fixed by the plant. Ruschel et al. [43] measured rhizosphere deposition of labeled nitrogen fixed by bean and soybean 72 hours after exposure to an ^{15}N-enriched atmosphere. Again, only 3% of the total nitrogen fixed appeared in the soil rhizosphere, and it could have resulted from contamination. When isotopic nitrogen was used under field conditions, no fixed nitrogen was detected in the rhizosphere of alfalfa after 135 days and three harvests. Fixed nitrogen was measured in the rhizosphere in greenhouse experiments and detected in soil surrounding nodules in both greenhouse and field experiments. However, total amounts of fixed nitrogen surrounding nodulated alfalfa roots represented 1% or less of the amount fixed by the plant. The studies of Ta et al. [42] and Ruschel et al. [43] indicate that direct excretion and/or root cell sloughing may be significant but not necessarily primary pathways for nitrogen transfer; this conclusion is based on amounts of fixed nitrogen detected in the rhizosphere. However, a small pool of excreted nitrogen in the rhizosphere does not necessarily imply a small rate of transfer into and out of the pool.

B. Senescence and Degradation of Nodules and Roots

There is considerable evidence that decomposition of senescenced roots and nodules may contribute significantly to nitrogen transfer [44]. Root and nodule turnover has been observed throughout the growing season, with increasing disintegration occurring in autumn [45]. Mallarino et al. [10] noted that white clover had a greater propensity for nitrogen transfer than red clover or birdsfoot trefoil, and white clover also has a high rate of root and nodule turnover [46]. Large amounts of transfer by white clover also occurred during summer, when root and nodule turnover would be expected to be highest for this somewhat temperature- and drought-susceptible species [46]. Goh et al. [24] were unable to detect nitrogen transfer between clover/ryegrass in pot experiments or in the first two cuttings of a field experiment. They concluded that excretion of soluble compounds probably was not important (since it should have manifested itself much earlier) and that nitrogen transfer probably was due to the decomposition of dead root/nodule tissue. However, there is always the possibility that soluble nitrogenous compounds may have been excreted and initially immobilized by microorganisms. Such nitrogen might then be available to plants at a later date [30].

Factors that suppress the legume component of plant mixtures increase the rate at which fixed nitrogen is released to the soil and made available to nonlegumes [29,30]. Contributions of nitrogen from decaying nodules and roots are particularly evident following defoliation, shading, water stress, or high summer temperatures, since as root/nodule breakdown often occurs, releasing N to the rooting medium [7,29,47–49]. However, remobilization of nitrogen from senescing tissue by legumes is a factor that may act to decrease nitrogen transfer via tissue turnover [50,51].

C. Other Mechanisms

Bakhuis and Kleter [52] reported significant nitrogen transfer between clover and ryegrass via leaching of aboveground parts and loss of clover leaves. Leaf fall has also been concluded to be responsible for at least half the nitrogen transfer observed in tropical legume–grass associations [39].

There is evidence of mycorrizae-aided transfer of isotopic nitrogen from soybean to corn [53] and from clover to ryegrass [29,30]. However, Hamel et al. [54] found no effects of mycorrhizae on nitrogen transfer between alfalfa and timothy or bromegrass.

VI. PLANT INTERACTIONS

When legumes are grown in mixtures with nonlegumes, they acquire a higher percentage of nitrogen via fixation than legumes grown in monoculture. This common occurrence has been shown for clover [55,56], birdsfoot trefoil [9,26], and alfalfa [8,57]. Competition for nitrogen by associated nonlegumes is the usual explanation [46,58].

Although more nitrogen may be fixed per plant by legumes in mixtures with nonlegumes, total amounts of fixed nitrogen are normally directly related to the proportion of legume in the mixture [9,57]. That is, higher legume yields tend to offset the lower percentage of fixed nitrogen/legume as the ratio of legume to nonlegume increases. However there are exceptions. For example, Mallarino et al. [9], reported obtaining maximum nitrogen fixation when white clover comprised 50–70% of a sward with tall fescue. This combination allows for greater yield potentials and less risk of bloat. In contrast to red clover and birdsfoot trefoil in this study, stimulation of nitrogen fixation of white clover by competition from an associated nonlegume apparently was sufficient to compensate for nitrogen that would be fixed by additional white clover plants in a pure clover stand.

VII. CONCLUSIONS

Symbiotic nitrogen fixation by legumes is a unique way of converting atmospheric nitrogen into useful nitrogenous products beneficial to plant and human nutrition. Several examples of increased legume yields through *Rhizobium* inoculation exist, offering potential for the manipulation of symbiotic nitrogen fixation to further its benefits to plants and humans. The fixed nitrogen from some legumes could be transferred to soils and associated nonlegume crops (in a mixed cropping or rotaion cropping scenario) as a result of excretion of fixed nitrogen from nodules and/or contribution of legume residues that contain fixed nitrogen. The subject of nitrogen transfer from legumes to soil and nonlegumes and the mechanisms underlying this phenomenon have been investigated. As seen from the extensive literature reviewed in this chapter, the positive contributions of this subject have not been conclusively demonstrated and should be a focus of future research.

ACKNOWLEDGMENT

The authors thank Ms. Charlene Chenault for coordinating word-processing efforts during the preparation of the manuscript.

REFERENCES

1. B. E. Caldwell and G. West, Effect of *Rhizobium* insulation on soybean yields, *Crop Sci.*, *10*: 19–21 (1970).
2. V. Ranga Rao (R. R. Velagaleti), A. Ayanabea, A. R. J. Eaglesham, and G. Thottappilly, Effects of *Rhizobium* inoculation on field-grown soybeans in western Nigeria and assessment of inoculum persistence during a two-year follow, *Trop. Agric. (Trinidad)*, *62*: 125–130 (1985).
3. R. R. Velagaleti and S. Marsh, Influence of host cultivars and *Bradyrhizobium* strains on the growth and symbiotic performances of soybean under salt stress, *Plant Soil*, *119*: 133–138 (1989).
4. R. R. Velagaleti, S. Marsh, D. Kramer, D. Fleischman, and J. Corbin, Genotypic differences in growth and nitrogen fixation among soybean (*Glycine Max* (L.) Merr.) cultivars grown under salt stress, *Trop. Agric. (Trinidad)*, *67*: 169–177 (1990).
5. R. R. Velagaleti, D. Kramer, S. S. Marsh, N. G. Reichenbach, and D. E. Fleischman, Some approaches to rapid and pre-symptom diagnosis of chemical stress in plants, *Plants for Toxicity Assessment, ASTM STP 1091* (W. Wang, J. W. Gorsuch, and W. R. Lower, eds.), American Society for Testing and Materials, Philadelphia, pp. 333–355 (1990).
6. D. W. Cowling and D. R. Lockyer, A comparison of the reaction of different grass species to fertilizer nitrogen and to growth in association with white clover. II. Yield of nitrogen, *J. Br. Grassl. Soc.*, *22*: 53–61 (1967).
7. F. R. Simpson, Transfer of nitrogen from three pasture legumes under periodic defoliation in a field environment, *Aust. J. Agric. Animal Husb.*, *16*: 863–870 (1976).
8. T. C. Ta and M. A. Faris, Effects of alfalfa proportions and clipping frequencies on timothy–alfalfa mixtures. I. Competition and yield advantages, *Agron. J.*, *79*: 817–820 (1987).
9. A. P. Mallarino, W. F. Wedin, R. S. Goyenola, C. H. Perdomo, and C. P. West, Legume species and proportion effects on symbiotic dinitrogen fixation in legume–grass mixtures, *Agron. J.*, *82*: 785–790 (1990).

10. A. P. Mallarino, W. F. Wedin, C. H. Perdomo, R. S. Goyenola, and C. P. West, Nitrogen transfer from white clover, red clover, and birdsfoot trefoil to associated grass, *Agron. J.*, *82*: 790–795 (1990).

11. C. S. Hoveland and M. D. Richardson, Nitrogen fertilization of tall fescue–birdsfoot trefoil mixtures, *Agron. J.*, *84*: 621–627 (1992).

12. G. W. Burton and E. H. DeVane, Growing legumes with "coastal bermudagrass" in the lower coastal plain, *J. Prod. Agric.*, *5*: 278–281 (1992).

13. C. S. Hoveland, Grazing systems for humid tropic, *J. Prod. Agric.*, *5*: 23–27 (1992).

14. A. R. K. Eaglesham, A. Ayanaba, V. Ranga Rao (R. R. Velagaleti), and D. L. Eskew, Improving the nitrogen nutrition of maize by intercropping with cowpea, *Soil Biol. Biochem.*, *13*: 169–171 (1981).

15. A. R. K. Eaglesham, A. Ayanaba, V. Ranga Rao (R. R. Velagaleti), and D. L. Eskew, Mineral N effects on cowpea and soybean crops in a Nigerian soil. II. Amount of N fixed and accrual to the soil, *Plant Soil*, *68*: 183–192 (1982).

16. O. B. Hesterman, T. S. Griffin, P. T. Williams, G. H. Haris, and D. R. Christenson, Forage legume–small grain intercrops: Nitrogen production and response of subsequent corn, *J. Prod. Agric.*, *5*: 340–348 (1992).

17. R. W. Elmore and J. A. Jackobs, Yield and nitrogen yield of sorghum intercropped with nodulating and nonnodulating soybeans, *Agron. J.*, *78*: 780–782 (1986).

18. W. H. Mitchell and M. R. Teel, Winter–annual cover crops for no-tillage corn production, *Agron. J.*, *69*: 569–573 (1977).

19. W. G. Vogel, *A Guide for Revegetating Coal Minespoils*, U.S. Department of Agriculture Forest Service General Technical Report NE-68 (1981).

20. P. S. Nutman, IPB field experiments on nitrogen fixation by nodulated legumes, *Symbiotic Nitrogen Fixation in Plants* (P. S. Nutman, ed.). Cambridge University Press, Cambridge, pp 221–237 (1976).

21. C. Johansen and P. C. Kerridge, Nitrogen fixation and transfer in tropical legume–grass swards in south-eastern Queensland, *Trop. Grassl.*, *13*: 165–170 (1979).

22. I. Vallis, E. F. Henzell, and T. R. Evans, Isotopic studies on intake of nitrogen by pasture plants. III. The uptake of small additions of N-15 labelled fertilizer by Rhodes grass and Townsville Lucerne, *Aust. J. Agric. Res.*, *18*: 858–877 (1967).

23. A. Haystead and A. G. Lowe, Nitrogen fixation by white clover in hill pasture, *J. Br. Grassl. Soc.*, *32*: 57–63 (1977).

24. K. M. Goh, D. C. Edmeades, and B. W. Robinson, Field measurement of symbiotic nitrogen fixation in an established pasture using acetylene reduction and a ^{15}N method, *Soil Biol. Biochem.*, *10*: 13–20 (1978).

25. F. E. Broadbent, T. Nakashima, and G. Y. Chang, Estimation of nitrogen fixation by isotope dilution in field and greenhouse experiments, *Agron. J.*, *74*: 625–628 (1982).

26. L. S. Brophy, G. H. Heichel, and M. P. Russelle, Nitrogen transfer from forage legumes to grass in a systematic planting design. *Crop Sci.*, *27*: 753–758 (1987).

27. G. H. Heichel and K. I. Henjum, Dinitrogen fixation, nitrogen transfer, and productivity of forage legume–grass communities, *Crop Sci.*, *31*: 202–208 (1991).

28. G. Hardarson, S. K. A. Danso, and F. Zapata, Dinitrogen fixation measurements in alfalfa–ryegrass swards using nitrogen-15 and influence of the reference crop, *Crop Sci.*, *28*: 101–105 (1988).

29. A. Haystead and C. Marriott, Fixation and transfer of nitrogen in a clover-grass sward under hill conditions, *Ann. Appl. Biol.*, *88*: 453–457 (1978).

30. A. Haystead and C. Marriott, Transfer of legume nitrogen to associated grass, *Soil Biol. Biochem.*, *11*: 99–104 (1979).

31. T. A. T. Wahua and D. A. Miller, Effects of intercropping on soybean N_2 fixation and plant composition on associated and soybeans, *Agron. J.*, *70*: 292–295 (1978).

32. T. A. T. Wahua and D. A. Miller, Effects of shading on the N_2 fixation, yield, and plant composition of field-grown soybeans, *Agron. J.*, *70*: 387–392 (1978).

33. F. R. Simpson, The transference of nitrogen from pasture legumes to an associated grass under several systems of management in pot culture, *Aust. J. Agric. Res.*, *16*: 863–871 (1967).

34. B. F. Bland, The effect of cutting frequency and root segregation on the field from perennial ryegrass–white clover associations, *J. Agric. Sci. Cambridge*, *69*: 391–397 (1967).

35. S. F. Ledgard, J. R. Freney, and J. R. Simpson, Assessing nitrogen transfer from legumes to associated grasses, *Soil Biol. Biochem., 17*: 575–577 (1985).

36. G. Vest, Nitrogen increases in a non-nodulating soybean genotype grown with nodulating genotypes, *Agron. J., 63*: 356–359 (1971).

37. E. G. Mulder, T. A. Lie, and A. Houwers, The importance of legumes under temperate conditions, *A Treatise on Dinitrogen Fixation*, Section IV, *Agronomy and Ecology* (R. W. F. Hardy and A. H. Gibson, eds.), Wilet, Interscience, New York, pp. 221–242 (1977).

38. G. H. Heichel, Legumes as a source of nitrogen in conservation tillage systems, *The Role of Legumes in Conservation Tillage Systems* (J. F. Power, ed.), Soil Conservation Society of America, Washington, DC, pp. 29–35 (1987).

39. O. Wyss and P. W. Wilson, Factors influencing excretion nitrogen by legumes, *Soil Sci., 52*: 15–23 (1941).

40. A. S. Whitney and Y. Kanehiro, Pathways of nitrogen transfer in some tropical legume–grass associations, *Agron. J., 59*: 585–588 (1967).

41. M. Richter, W. Wilms, and F. Scheffer, Determination of root exudates in sterile continuous flow culture. II. Short-term and long-term variations of exudation intensity, *Plant Physiol., 43*: 1747–1754 (1968).

42. T. C. Ta, F. D. H. MacDowall, and M. A. Faris, Excretion of nitrogen assimilated from N_2 fixed by nodulated roots of alfalfa (*Medicago sativa* L.), *J. Plant Physiol., 132*: 239–244 (1986).

43. A. P. Ruschel, E. Salati, and P. B. Vose, Nitrogen enrichment of soil and plant by *Rhizobium phaseaoli–Phaseolus vulgaris* symbiosis, *Plant Soil, 51*: 425–429 (1979).

44. I. Vallis, Nitrogen relationships in grass/legume mixtures, *Plant Relationships in Pastures* (J. R. Wilson, ed.), Commonwealth Scientific and Industrial Research Organisation, Melbourne, Australia, pp. 190–201 (1978).

45. D. J. B. Young, A study on the influence of nitrogen on the root weight and nodulation of white clover in a mixed sward, *J. Br. Grassl. Soc., 13*: 106–114 (1958).

46. G. W. Butler, R. M. Greenwood, and R. M. Sopper, Effects of shading and defoliation on the turnover of root and nodule tissue of plants of *Trifolium repens, Trifolium pratense* and *Lotus uliginosus, N.Z. J. Agric. Res., 2*: 415–426 (1959).

47. J. S. Pate, Functional biology of dinitrogen fixation by legumes, *A Treatise on Dinitrogen Fixation* (R. W. Hardy and W. J. Silver, eds.), New York, pp. 473–517 (1977).

48. J. D. Wilson, The loss of nodules from legume roots and its significance, *J. Am. Soc. Agron., 34*: 460–471 (1942).

49. A. C. P. Chu and A. G. Robertson, The effect of shading and defoliation on nodulation and nitrogen fixation by white clover, *Plant Soil, 41*: 509–519 (1974).

50. R. W. Brougham, Leaf development in swards of white clover (*Trifolium repens*), *N.Z. J. Agric. Res., 1*: 707–718 (1958).

51. H. T. Craille and G. H. Heichel, Nitrogen fixation and vegetative regrowth of alfalfa and birdsfoot trefoil after successive harvests or floral debudding, *Plant Physiol., 67*: 898–905 (1981).

52. J. A. Bakhuis and H. J. Kleter, Some effects of associated growth on grass and clover under field conditions, *Neth. J. Agric. Sci., 13*: 280–310 (1965).

53. C. Hamel, C. Nesser, U. Borrantes-Cortin, and D. L. Smith, Endomycorrhizal fungi in nitrogen transfer from soybean to corn, *Plant Soil, 138*: 33–40 (1991).

54. C. Hamel, V. Furlan, and D. L. Smith, Mycorrhizal effects on interspecific plant competition and nitrogen transfer in legume–grass mixtures, *Crop Sci., 32*: 991–996 (1992).

55. D. C. Edmeades and K. M. Goh, Symbiotic nitrogen fixation in a sequence of pastures of increasing age measured by a ^{15}N dilution technique, *N.Z. J. Agric. Res., 21*: 623–628 (1978).

56. B. C. Boller and J. Nosberger, Symbiotically fixed nitrogen from field-grown white clover and red clover mixed with ryegrass at low levels of ^{15}N fertilization, *Plant Soil, 112*: 167–175 (1988).

57. C. P. West and W. F. Wedin, Dinitrogen fixation in alfalfa–orchard grass pastures, *Agron. J., 77*: 89–94 (1985).

58. I. Vallis, E. F. Henzell, and T. R. Evans, Uptake of soil nitrogen by legumes in mixed swards, *Aust J. Agric. Res., 28*: 413–425 (1977).

15

C6 Aldehydes in Plant Leaves and Preliminary Information on the in Vivo Control of Their Formation

David Hildebrand, Hong Zhuang, and Thomas R. Hamilton-Kemp
University of Kentucky, Lexington, Kentucky

Roger A. Andersen
Agricultural Research Service, U.S. Department of Agriculture, and University of Kentucky, Lexington, Kentucky

I. INTRODUCTION

In plant tissues, C_6-aldehydes, consisting of hexanal and hexenals (Figure 1), are formed via the lipoxygenase/hydroperoxide lyase (LOX/HPL) pathway [1–4]. When plant leaves are damaged by wounding or freezing, the polyunsaturated fatty acids, 18:2 and/or 18:3, can be rapidly dioxygenated by LOX into 13-hydroperoxy fatty acids. These hydroperoxides can be cleaved by HPL to form the C_6-aldehydes, hexanal and/or *cis*-3-hexenal. *cis*-3-Hexenal is usually isomerized to *trans*-2-hexenal both enzymically and nonenzymically. As shown in Figure 2, the C_6-aldehydes could be further reduced by alcohol dehydrogenase (ADH) into corresponding alcohols [5,6].

Along with the corresponding alcohols, C_6-aldehydes are the major volatile compounds responsible for the "green odor" of green leaves [2,7]. Because of their low odor threshold, C_6-aldehydes directly influence food quality. For example, *cis*-3-hexenal contributes to the fresh flavor of certain fruits and vegetables [8–10]. Hexanal may contribute an undesirable aroma, which has limited the use of soybean products in the United States [11]. Several investigations have shown that hexanal and hexenals are toxic to microorganisms in vitro, particularly fungal species [12–17]. Other work has shown that the C_6-aldehydes inhibited the germination of several seed species [14,18–20], as well as pollen germination [21], with the α,β-unsaturated C_6-aldehyde showing considerably more inhibitory action than the saturated aldehyde hexanal. Similar effects are seen on bacterial proliferation [22]. In addition, C_6-aldehydes are found to inhibit viruses [23].

The C_6-aldehydes were first isolated from green leaves more than 80 years ago [24]. Although biochemistry of in vivo C_6-compound production in plant leaves is well understood, the control of C_6-compound formation in leaves is not yet clear. Earlier results showed that

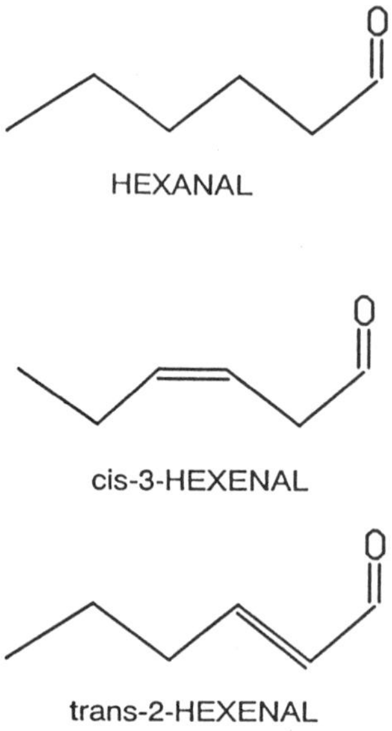

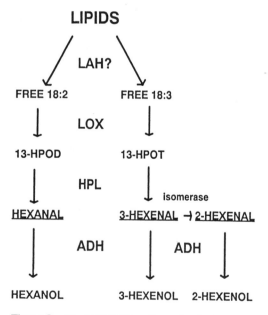

HEXANAL

cis-3-HEXENAL

trans-2-HEXENAL

Figure 1   Structures of C_6-aldehydes.

environmental factors such as temperature and solar radiation appeared to influence C_6-compound formation in leaves [25–28]. It was suggested that changes in LOX and lipolytic acylhydrolase (LAH) activities were responsible for the seasonal variation in C_6-aldehyde formation in tea leaves [6,26,28]. Different plant species had different capacities of forming C_6-aldehydes [29–32]. The investigation of enzyme activities of LOX and HPL in the green

LIPIDS

LAH?

FREE 18:2   FREE 18:3

LOX

13-HPOD   13-HPOT

HPL   isomerase

HEXANAL   3-HEXENAL → 2-HEXENAL

ADH   ADH

HEXANOL   3-HEXENOL   2-HEXENOL

Figure 2   The LOX/HPL pathway in plants.

leaves of different plant species showed that LOX activity rather than HPL activity affected overall C_6-aldehyde formation [30]. The formation of C_6-aldehydes increased with maturation of black tea leaves [33,34]. Young leaves had relatively higher in vitro activity for C_6-aldehyde formation than old leaves during development of kidney bean [35]. Our preliminary results showed that there were large differences in C_6-aldehyde formation by frozen soybean leaf disks taken from the same plant. These observations led us to hypothesize that the formation of C_6-aldehydes in plant leaves also is controlled by plants themselves. The objectives of our preliminary investigations were:

1. To determine the pattern of formation of in vivo C_6-compounds by soybean leaves at different positions on the plants and at different leaf ages.
2. To investigate the biochemical mechanism controlling the formation of C_6-aldehydes in plant leaves.

II. REVIEW OF THE LITERATURE

The C_6-aldehydes are widely distributed in plants. Mort plants and many plant organs are apparently able to produce C_6-aldehydes. The early studies on those compounds were carried out by phytochemists using plant leaves, which have a high level of volatile compounds. The first report of C_6-aldehydes can be dated to the last century (see Ref. 31). The first identified C_6-aldehyde, 2-hexenal, was isolated from the green leaves of 20 different species of bushes by Curtius and Frunzen [24]. The aldehyde 2-hexenal was proposed to be an intermediate step in the conversion of sugar to fats. Takei and coworkers [36] reported that hexenals and their corresponding alcohols were produced by leaves of 16 species of plants. They were first associated with the characteristic odor of freshly cut grass. Nye and Spoehr [31] tested the leaves of 15 species of trees and found a range of quantities of the compound. In the characterization of the ginkgo tree, Major et al. (27) found C_6-aldehyde formation from tree leaves. Evidence presented indicated that 2-hexenal was related to the pest defense of this species.

The identification of C_6-aldehydes produced by damaged plant tissues mainly came from food science investigations after 1960. Bengtsson and Bosund [37] found that stored frozen peas produced an off-flavor, and one of the three major volatile compounds responsible was identified as hexanal. Hexanal and 2-hexenal also were identified in macerated tomatoes [38–40]. Cutting or mechanical rupture of cucumber and cooked cabbage produced trace amounts of 2-hexenal, as well [41,42]. In addition, C_6-aldehydes were detected in various fruits such as strawberries, raspberries, apples, grapes, bananas, pears, and plums [10,23,43].

The studies of food uses of soybean proteins and the production of tea greatly enriched our knowledge of the formation of C_6-aldehydes. Although soybean is a potential sources of food protein because of its high levels of storage proteins, a very important factor limits the development of large Western food markets for the soybean: namely, the characteristic soybean flavor, often described as bitter, beany, and astringent [11]. Analysis of the composition of soybean flavor compounds showed that *n*-hexanal and hexanol were among the main aroma components of the flavor [44–48]. This C_6-aldehyde formation was thought to be due to the high levels of LOX activity in soybean seeds [46,49–52]. The formation of C_6-aldehyde during black tea manufacture was related to tea qualities [6,36,53]. In the 1930s, 2-hexenal and 3-hexenol were isolated from black tea during fermentation or from fresh tea leaves [36]. It was believed that these volatile compounds were the major contributors to the characteristic "green odor" of green leaves, and 2-hexenal and 3-hexenol were called leaf aldehyde and leaf alcohol, respectively. The analysis of the C_6-aldehydes produced by tea leaves showed that the geometrical structures of 2-hexenal was (*E*)- or trans form and 3-hexenol was (*Z*)- or cis form by comparison with synthetic specimens [54]. Further research showed that in addition to leaf alcohol and leaf aldehyde, *trans*-3-hexenol, *trans*-2-hexenol, *cis*-3-hexenal, and *trans*-3-hexenal were also present in fresh tea leaves [55–57].

The C_6-aldehydes are produced not only by plant leaves and fruits but also by buds, bark, stems, roots, cotyledons, embryo axes, pod walls, and testa [2,13,33,35,58].

In C_6-aldehyde formation, it is notable that large amounts of these compounds are observed only after plant tissues have been damaged by wounding or freezing. The extent of chopping and grinding leaves and freezing temperature affected the yield of C_6-aldehydes of plant leaves [31,37, and unpublished results]. 2-Hexenal did not seem to be present in significant amounts in intact, live tree leaves, but the hexenals were formed when the leaves were damaged in an atmosphere containing O_2 [27]. The mere physical disruption of cucumber tissue resulted in the production of volatile carbonyl compounds, which are responsible for the characteristic taste and aroma of this material [59]. *cis*-3-Hexenal was formed immediately after fresh tea leaves were macerated [60]. The same observation was made in fruits. Drawert et al. [10] reported that apples and grapes produced C_6-aldehydes only when homogenized. *cis*-3-Hexenal developed during the maceration of tomato tissue [8]. When soybean seeds were ground under high moisture, the C_6-aldehydes were formed in relatively large quantities. However, in intact seeds and under dry conditions only small amounts of C_6-aldehydes could be measured [46,61]. Freezing also dramatically increased the formation of C_6-aldehydes by plant leaves and fruits. Frozen soybean and wheat leaves produced at least 50 times more C_6-aldehydes than fresh, live leaves (unpublished data). In fresh peas, the hexanal level was very small, but the hexanal levels increased when peas had been maintained below −26°C [37]. Lee et al. [62] reported that frozen raw vegetables produced off-flavors. The freezing temperature also affected the yields of C_6-aldehydes formed by plants [37; our unpublished results].

Enzymic processes appear to be involved in C_6-aldehyde formation by macerated plant tissues. Nye and Spoehr [31] reported that tree leaves treated with hot water prior to chopping and grinding gave no aldehyde. However leaves treated with hot water after they had been chopped and ground produced a large amount of 2-hexenal. Yamanishi et al. [63] and Gonzalez et al. [64] showed that steamed green tea leaves could not produce hexanal and 2-hexenal. The same phenomenon also was observed for homogenized leaves of *Ginkgo biloba* [32]. Heat treatment of wheat plant homogenates caused a marked reduction in the production of C_6-aldehydes [65]. No C_6-aldehydes were detected from winter wheat leaves that had been microwaved for 1 minute, then treated with freezing temperature (unpublished results). The formation of C_6-aldehydes by plant tissues is also influenced by pH. Hatanaka and Harada [60] found that the pH optimum of C_6-aldehyde formation by fresh tea leaves is 5.5. Major et al. [27] found different pH ranges for C_6-aldehyde formation by leaves from different plants. The maximum C_6-aldehyde formation by *G. biloba* was in the pH range of 3.7–5.0, but the maximum by *A. glandulasa* was in the range of 6.0–7.4. At pH 6.0, which is the optimum pH of soybean seed LOX, the crude protein extract produced about four times more hexanal than was measured at pH 5.0 [46].

Molecular oxygen is required for the formation of C_6-aldehydes by plant tissues. When plant leaves were chopped and ground under CO_2 and N_2 instead of air, little or no 2-hexenal was demonstrated. Furthermore, the formation of hexenals that occurs when the ground leaves are exposed to oxygen can be inhibited by treating the ground leaves with hot water before such exposure. Extensive research showed that the antioxidant pyrogallol blocked the synthesis of C_6-aldehydes, but the addition of $MnCl_2$, which specifically enhances peroxidase activity, had no effect on the yield of the aldehyde [31]. Drawert et al. [10] found that apples and grapes produced hexanal and hexenals only when homogenized in air. Kazeniac and Hall [8] and

Jadhav et al. [66] reported that no C_6-aldehydes were produced when fresh tomatoes were blended under N_2 in the absence of O_2, or heated. Under an atmosphere of N_2 during tea fermentation, formation of hexanal and *trans*-2-hexenal decreased [63,64]. No C_6-aldehydes were obtained from the tree leaf homogenates in the absence of O_2 or under vacuum or N_2 [67]. Arai et al. [46] showed that N_2 reduced the formation of C_6-aldehydes by soybean preparations. When the coenzyme nicotinamide adenine dinucleotide (NAD) was ground with ginkgo leaves in the absence of oxygen, only traces of 2-hexenal were detected. The same result was obtained when methylene blue was substituted for NAD. The same experiment yielded little evidence of inhibition of the formation of hexenals in the presence of copper sulfate, hydrogen sulfide, carbon monoxide, and potassium ferricyanide, all of which would be expected to inhibit metalloporphyrin-containing enzymes [27]. The phenolic antioxidant pyrogallol, however, inhibited the formation of C_6-aldehydes, and peroxidase did not seem to be involved in the process [31]. At a later date, phenolic antioxidants were found to inhibit the LOX-catalyzed oxidation of linoleate [68].

The aliphatic nature of C_6-aldehydes led to the hypothesis that their biogenesis might entail oxidative degradation of fatty acids [31,66]. Nye and Spoehr [31] found that oleic acid increased the formation of hexenals by the leaves. The volatile carbonyl compounds including *n*-hexanal and 2-hexenals were produced by autoxidation of soybean oil [44,69,70]. Lee and Mattick [71] and Mattick and Lee [72] observed that enzymic hydrolysis of lipids and peroxide formation in peas took place in frozen material and suggested that successive reactions might be responsible for the development of off-flavors. Lipids isolated from fresh peas reacted with enzyme preparations also obtained from peas, to produce a wide range of carbonyl compounds including *n*-hexanal [73]. Sastry and Kates [74] showed that the disruption of leaves from runner bean caused extensive breakdown of galactolipids. Homogenization of potato tubers at 0°C resulted in rapid enzymic hydrolysis of the endogenous phospholipids and galactolipids to produce free fatty acids and fatty acid hydroperoxides [75]. A lethal frost caused rapid degradation of crown and root tissue of winter wheat and breakdown of polar lipids, with an increase first in diglycerides and later in unidentified fatty acids. Polyunsaturated polar lipids were preferentially degraded [76]. Analyses of fatty acids in tea leaves showed that the proportions of the fatty acids decreased during tea fermentation [77]. Drawert et al. [10] reported that in the presence of oxygen, the content of 18:3 in certain leaves decreased as hexenals were formed. Hexanal and 2-hexenal were produced enzymatically, with participation of atmospheric O_2, linoleic and linolenic acids from (18:2 and 18:3), respectively, in apples and bananas. It was found that 18:3 and 18:2 decreased markedly during blending of tea leaves. Without blending, 95% of the total 18:3 and 77% the total 18:2 were present in the neutral fat. After the summer leaves had been blended for 3 minutes, half the 18:3 and 18:2 disappeared from the neutral fat at the same time that a strong "green odor" was released from the homogenate of tea leaves [78].

Nye and Spoehr [31] suggested that 18:3 might be the precursor of 2-hexenal in leaves. Studies with isotope-labeled possible precursors showed that in the air, 2-hexenal was formed from 18:3 at pH 4.5 during the maceration of fresh ginkgo leaves. However, no radioactive 2-hexenal was formed when ^{14}C-labeled 18:3 in water was homogenized in the presence of air and steam-distilled leaves [32,79]. Addition of 18:2 and 18:3 to tomato and Japanese silver leaf homogenates increased the amounts of hexanal and hexenals [8,80]. When radioactive 18:2 and 18:3 were added during tea fermentation, the hexenals and hexanal were the main products in the volatile fractions [64,77]. Hatanaka and Harada [60] reported that homogenates of fresh tea leaves converted 18:2 and 18:3 into hexanal and hexenals, respectively. However, 18:2 or 18:3 incubated with a homogenate heated at 65°C for 30 minutes did not affect C_6-aldehyde formation. In further studies, when [14]C-18:3 was incubated with tea chloroplasts, two radioactive peaks, with the retention times of *cis*-3-hexenal and *trans*-2-hexenal, were detected in the headspace vapor by gas chromatography [81]. Grosch and Schwarz [82] also found that linoleic acid was a precursor of volatile and nonvolatile aldehydes in cucumber. The products identified were hexanal and *trans*-2-nonenal. Furthermore, the enzyme system that catalyzed the formation of C_6-aldehydes in the presence of 18:3 was detected in about 40 plant species. Among these plants, the green leaves of many dicotyledons had high enzyme activities, but edible leafy vegetables, fruits, and monocotyledonous plants had low activities. About 90% of the entire activities of plant tissues was localized in the green leaves [29]. Tomato enzyme extract did not catalyze the conversion of saturated fatty acids (C-6 to C-18) or even monounsaturated fatty acids (C-16, C-18). Only linoleic and linolenic acid yielded carbonyls [66].

Lipoxygenases are a group of enzymes catalyzing the oxygenation, by molecular oxygen, of fatty acids containing a *cis,cis*-1,4-pentadiene system to produce conjugated hydroperoxy derivatives. The fatty acids commonly found in plants that fit these requirements are linoleic acid and linolenic acids [1–3,83,84]. Enzymes of this type were formerly known as fat oxidase, carotene oxidase, and lipoxidase. LOXs are widely distributed in higher plants and in various plant organs [85–88]. Copper sulfate, basic lead acetate, sodium cyanide, thiourea, and sodium fluoride have no effect on soybean LOX. However, antioxidants such as nordihydroguaiaretic acid, propylgallate, hydroquinone, α-tocophenol, and H_2O_2 inhibit lipid-dependent O_2 uptake by LOXs [1,4,68,89,90]. It was suggested more than 50 years ago that LOX is involved in the formation of C_6-aldehydes [31]. During the growth of tomato fruit, LOX activity increases, consistent with the biosyntheses of hexanal, whereas polyunsaturated fatty acids decrease. When ^{14}C-18:3 and ^{14}C-18:2 were added to the enzyme extract of tomato fruits, radioactivity was incorporated into C_6-aldehydes. However, boiling of the extract did not produce C_6-aldehydes. Further studies on the inhibitors of oxidation showed that KCN, EDTA, and *p*-chloromercuricbenzoate did not inhibit the oxidation of polyunsaturated fatty acids; but propylgallate, phloridzin pyrocatechol, pyrogallol, and H_2O_2 reduced the uptake of O_2 by 18:2 and 18:3. In the presence of ATP, COA, and NADH, which are cofactors in the degradation of fatty acids to CO_2, no significant effects of these compounds were noted [66]. A similar effect of these compounds on the formation of C_6-aldehydes was observed using plant leaves [27,65]. Bengtsson and Bosund [37] reported that hexanal could be easily formed in reaction following the oxidation of 18:2 by LOX. In soy milk production, the development of off-flavor was attributed to LOX activity [52]. In defatted and raw soybean flour and tea leaves, C_6-aldehyde production showed the same pH optimum observed for LOXs [46,91]. The H_2O_2 oxidized LOX rapidly and inactivated it. The H_2O_2 also inhibited the development of hexenals [8]. When LOX was added to soybean oil, hexanal formation was increased [46]. Among the mutant lines of soybean embryo LOX, the seeds containing LOX 3 reduced the formation of C_6-aldehydes [92,93]. LOX 2 is the most effective isozyme in C_6-aldehyde production [93,94].

Hydroperoxides of 18:2 (HOO-18:2) and 18:3 (HOO-18:3) are the products of LOXs [95] (Figure 2). Kalbrener et al. [96] indicated that HOO-18:2 and HOO-18:3 from LOX oxidation decomposed to form a grassy–beany taste at 50 and 10 ppm, respectively. Tressl and Drawert [43] reported that radiolabeled 13- and 9-hydroperoxy 18:2 were transformed into C_6- and C_9-aldehydes and the corresponding oxo acids by an enzyme extract, indicating the presence of lyase in banana fruits. In tea chloroplasts, C_6-aldehyde formation increased linearly with increase in substrate concentration up to 0.4 mM of 18:3 or 18:2 or 0.6 mM of HOO-18:3 or -18:2 [97]. When HOO-acids were added to the homogenate of cucumber and tomato fruits, both oxo acids and C_6-aldehydes were formed. Only hexanal was produced from HOO-18:2 and hexenals from HOO-18:3. N_2 or anaerobic condition did not influence this process. When 1-^{14}C-18:2 and -18:3 were mixed with the enzyme extract, labeled HOO-18:2, 18:3, and oxo acids were detected. The major products were hydroperoxides. Furthermore, addition of

[14]C-18:2 formed labeled hexanal. With boiled homogenates or 18:1, relatively little conversion of fatty acid occurred.

Tressl and Drawert [43] found that a LOX preparation did not produce C_6- and C_9-aldehydes in the presence of 18:3. However, a crude enzyme preparation showed high activity, suggesting that there was an HPL in the banana fruits. Mixing LOX 1 and 18:3 (Sigma Chemical Co.) did not produce 2-hexenal, but the mixture of [14]C-18:3 with ginkgo leaf extract promoted 2-hexenal formation, and the optimum of pH for C_6-aldehyde formation was different from that for the LOX. Accordingly, Major and Thomas [32] concluded that LOX was not the only enzyme involved in the formation of C_6-aldehydes. Galliard et al. [98] reported that boiling the cucumber extract prevented production of aldehydes from HOO-18:2 and -18:3. The HPL was identified first in nongreen tissues, watermelon seedlings [99], and cucumber fruits [59,100]. Subsequently, the occurrence of this enzyme was reported in green tissues, such as peels of cucumber fruits [101], soybean pericarp and testa [58], leaves of kidney bean [35,102] and tea [97], and soybean [103] and other plant species [30], as well as in such nongreen tissues as etiolated seedlings of alfalfa and cucumber [104], soybean seedlings [105], soybean seed [106,107], cultured tobacco cells [108], plant roots [109], and fruits [110–112]. These studies indicate that HPL, like LOX and C_6-aldehyde formation, is also widely distributed in plants. Based on these results, the LOX/HPL pathway in plants was hypothesized [35,91,96,98,107, 111,113].

When 18:3 is used as substrate by the LOX/HPL pathway, hexenals, including both 3-hexenal and 2-hexenal, are formed as shown in Figures 1 and 2. Detailed analysis of the development of these two hexenals showed a transformation between the two compounds. Kazeniac and Hall [8] reported that the *cis*-3-hexenal was unstable both to heat and to the acidic juice. It was isomerized to the *trans*-2-hexenal upon standing or heating, suggesting that *cis*-3-hexenal was the source of *trans*-2-hexenal. Freshly picked tea leaves produced only *cis*-3-hexenal, which was observed immediately after fresh tea leaves were macerated. If the preparation stood for 50 minutes at 40°C, the aldehyde isomerized to *trans*-2-hexenal. This isomerization of *cis*-3-hexenal to *trans*-2-hexenal in macerated tea leaves was very fast in comparison with thermal isomerization, and it was not affected by the introduction of N_2. In the presence of tea leaf homogenate, *cis*-3-hexenal was isomerized to *trans*-2-hexenal, with pH 5.5 optimum [60]. By using [14]C-18:3, Hatanaka et al. [81] found that at the beginning *cis*-3-hexenal was the major hexenal. When incubation was prolonged, radioactivity in *cis*-3-hexenal decreased and that in *trans*-2-hexenal increased. *cis*-3-Hexenal in the headspace rapidly decreased to undetectable levels in the presence of a homogenate of soybean seed meal in about 300 seconds, with *trans*-2-hexenal dramatically increasing. No isomerization was observed if the soybean meal was heated before use [93]. *cis*-3-*trans*-2-Enal isomerase was reported to be present in cucumber [98] and tea [6].

The C_6-aldehydes can be further reduced to form C_6-alcohols. MacLeod and MacLeod [42] studying cabbage volatiles and Ralls et al. [115], studying green pea volatiles showed a close relation between C_6-aldehydes and C_6-alcohols. Schormuller and Grosch [38] observed the aldehyde–alcohol relation in tomato volatiles. Eriksson [116,117] found that ADHs were widely distributed in plants. The equilibrium constant for *trans*-2-hexenal and *trans*-2-hexenol very strongly favored the aldehyde over the alcohol in pea. When ground tomatoes were held without heat treatment, increase in the amounts of lipid-related alcohols, such as *cis*-3-hexen-1-ol and *n*-hexanol, at the expense of the corresponding aldehydes, was especially noticeable [8]. The reduction of hexenals to hexenols was also observed in tea leaves [60,117] and tea seeds [118].

The regulation of C_6-aldehyde production in plants was indicated from studies of the pathway by which C_6-aldehydes were synthesized. Major et al. [67] reported that two plant species yielded 2-hexenal in the presence of air from leaves collected late in summer, but not early in the summer.

Hatanaka et al. [25] found that the generation of C_6-aldehyde declined from summer to winter. The C_6-aldehyde levels in tea leaves increased rapidly when the temperature was changed from 7°C to 25°C in a growth chamber [28]. Hexanal concentration varies in tomato fruit with the stage of maturity, light conditions, nutrient availability, and enzymic activity during ripening [39]. Hexanal increased with the growth of tomato fruit [40], and C_6-aldehyde formation was greater in ripe fruits than in green fruit [66]. A similar change in C_6-aldehyde formation during ripening of banana fruit was also reported [43]. During the development of soybean fruit, the formation of C_6-aldehydes also changed [58,106]. A relation between HPL activity and C_6-aldehyde formation was observed [58]. In plant leaves, the levels of C_6-aldehydes quantitatively changed with plant species and seasons. Leaves had much higher enzymic activity than other organs in tea for C_6-aldehyde synthesis [2,33]. Leaf extract of dicotyledonous plants produced more C_6-aldehydes than those of monocotyledonous plants [29]. The formation of C_6-aldehydes by tea leaves increased with maturation [33]. In the presence of 18:2, young leaf extracts of kidney bean formed higher levels of hexanal than older leaf extracts [35]. These observations indicate that plants control the C_6-aldehyde formation in some way.

III.   PRELIMINARY STUDIES OF IN VIVO CONTROL OF C_6-ALDEHYDE FORMATION

A.   Materials and Methods

1. Chemicals and Plant Materials

Linoleic (18:2), linolenic (18:3), heptadecanoic (17:0), and nonadecanoic (19:0) acids, and phosphatidylcholine, (PC, 17:0), were purchased from Sigma Chemical company. Trichloroacetic acid (TCA), chloroform, methanol, and hexane were purchased from Fisher Chemical Company. 13-Hydroperoxyoctadecatrienoic acid (HPOT) and 13-Hydroperoxyoctadecadienoic acid (HPOD) were prepared from soybean LOX 1 (Sigma) following the procedure described by Gardner [120]. The concentration of hydroperoxide was measured spectrophotometerically at 235 nm with 25,000 M^{-1} cm^{-1} as the extinction coefficient.

The soybean genotype Century was used for these studies. *Arabidopsis thaliana* genotypes, designated *fad*B, *fad*C, *fad*D, *fad*2, *fad*3, and *act*1, recently identified by Browse and Somerville [121], were provided by Dr. Chris Somerville, Michigan State University. *Arabidopsis* was grown in a growth chamber with 16 hours of light and 8 hours of dark at 25°C for 30 days.

2. Measurement of C_6-Aldehydes

Soybean leaf disks were sampled by using a cork borer with a diameter of 1.5 cm. Whole *Arabidopsis* leaves or soybean disks were put into screwcap vials (1.8 mL). The tightly sealed vials were then frozen at –80°C for at least a day. Before measurement of compounds, the vials were placed in a water bath at 30°C for 20 minutes for soybean disks and at 30°C for 20 minutes and then at 80°C for 5 minutes for *Arabidopsis* leaves. A sample of vapor from the headspace of the vials was withdrawn with a gastight syringe and injected directly into a Varian 3700 gas chromatograph with a 30 m × 0.53 mm DB-Wax (polyethylene glycol) fused silica column operated under the following conditions: injector, 220°C; oven temperature, 50°C for 5 minutes then 3°C per minute to 150°C; flame ionization detection, 240°C; helium carrier, 6 mL/min^{-1}. For the wounding experiment, soybean leaves were ground in liquid N_2 and placed in vials. Before gas was injected into the chromatograph, the vials were incubated at 30°C for 30 minutes.

3. Lipid Analysis

Soybean and *Arabidopsis* leaves were collected and weighed when the plant materials were prepared for the C_6-aldehyde analysis. Some of the older leaves were frozen with liquid nitrogen and incubated at 30°C in the dark for 10 minutes to avoid photooxidation of the unsaturated fatty acids in cellular membranes by the low temperature [121,122]. The internal standards 17:0 or 19:0 and PC-17:0 were added to the leaves. Soybean leaves were homogenized on ice in 5% TCA to stop lipase action, and total lipids were extracted in chloroform:methanol (1:2) with 0.01% butylated hydroxytoluene [94]. Lipids of *Arabidopsis* leaves were extracted as described by Browse et al. [124]. Total lipid analysis was performed by a direct transmethylation method [124]. The isolation of different lipid fractions was accomplished by loading the total extract onto a 600 mg silica Sep-Pak cartridge (Waters). Sequential elution was performed following the method of Parkin and Kuo [123]. The efficacy in separating these fractions was verified by thin-layer chromatography. After the elution, 25 mg of 19:0 was added and the samples were dried under nitrogen. The fatty acid composition was analyzed and calculated as above.

B. Results

1. Time Course of C6-Aldehyde Formation by Soybean and Arabidopsis Leaves

The C_6-aldehydes are rapidly formed once plant leaves have been damaged by wounding and freezing. The levels of C_6-aldehydes in the headspace depends on the balance between their formation, reaction, and partitioning between air and leaf tissues. The time course of C_6-aldehyde formation by frozen soybean and *Arabidopsis* leaves was examined before the effect of leaf positions, leaf disk locations, and leaflets on C_6-aldehyde production was investigated. The results show that the maximum formation of C_6-aldehyde by both species was in the range of 5–30 minutes (Figure 3).

Hatanaka and Harada [60] found that maceration time influenced the formation of C_6-aldehydes by tea leaves. The C_6-aldehydes increased in the first 10 minutes. When fresh

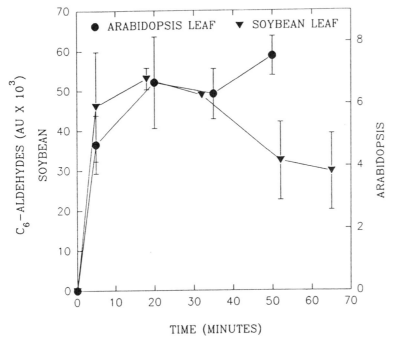

Figure 3 The time course of C_6-aldehyde formation by frozen *Arabidopsis* and soybean leaves incubated at 30°C. Data are mean of five replicates.

tea leaves were macerated with linolenic and linoleic acid, a 20-minute incubation produced more hexenals than a 10-minute incubation at pH 5.5. Kazeniac and Hall [8] reported that when blended tomatoes were held at 40°C, total hexenals increased at the beginning and reached maximum levels at 20 minutes, then decreased. Hexanal level was maximal at 30 minutes. A 20-minute incubation time was chosen for experiments on C_6-aldehyde formation by soybean and *Arabidopsis* leaves.

To determine the factors in plant leaves that might influence the levels of C_6-aldehydes, Century soybean leaves (with wild-type levels of embryo LOXs) were classified into three types based on the sources of leaf disks, leaf positions on soybean plants, leaflets of a trifoliate, and locations in a leaflet. Study of C_6-aldehyde formation showed that only the leaf positions had an obvious effect [125]. No significant difference was observed in the different leaflets of a trifoliate and different locations in a leaflet. Further analysis of the effect of leaf positions on C_6-aldehyde formation based on the positions and ages showed that the young leaves (near the top of the plant) produced relatively high amounts of total C_6-aldehydes. C_6-Aldehyde production showed a large decrease with subsequent leaf growth and expansion until the leaves reached almost full size. After the leaves reached full size, C_6-aldehyde formation rapidly increased.

2. The Effect of Maceration on C_6-Aldehyde Formation

It is not known whether freezing and wounding have the same mechanism to induce C_6-aldehyde formation. To look into this issue, soybean leaves of different ages and C_6-aldehyde production capacities were ground in the presence of liquid N_2 and packed into the vials. Measurement of C_6-aldehydes by ground soybean leaves showed trends in C_6-aldehyde formation similar to frozen leaves (Figure 4). The young leaves produced much more hexanal than did the older leaves (Figure 4A). The ratio of hexenals to hexanal increased during leaf aging (Figure 4D), whereas the ratio of 2-hexenal to 3-hexenal decreased (Figure 4C). The change in hexenals (Figure 4B) and total C_6-aldehydes formed by the ground leaves during leaf development was not exactly the same as that produced by the frozen leaves (i.e., no obvious increases in hexenals, therefore in total C_6-aldehydes, were observed in older leaves). A rapid depletion of hexenals in the headspace has been observed by Gardner et al. [103]. These preliminary results showed that the macerated leaves produced a much greater amount of C_6-alcohols than did the frozen leaves. In the presence of 18:3, younger leaves produced about two times more hexenals than old leaves. In both cases, HPOT resulted in the formation of considerably more hexenals than 18:3.

3. The Effect of Liquid Nitrogen Treatment on Fatty Acid Composition of Lipid Fractions in Older Soybean Leaves

The direct effect of freezing on fatty acid composition of different lipid fractions was investigated using older leaves. The results showed that freezing caused a large decrease in total 18:3 and a small decrease in total 18:2, but did not influence total 16:0, 18:0, and 18:1 levels. The large reduction in 18:3 accounts for more than 90% of the loss. A similar trend with glycolipid (GL) and triacylglycerol (TG) was observed (Figure 5). In the phospholipid fraction, almost all types of fatty acid decreased, with the main changes occurring in 16:0 and 18:3 (Figure 6). The decrease in the relative levels of C-18-polyunsaturated fatty acids was less than that seen with TG and GL. In the free fatty acid fraction, however, there were increases in all fatty acids except for 18:3. More than 70% of 18:3 in the total loss of fatty acids was derived from glycolipids.

4. C_6-Aldehyde Formation by Leaves of Fatty Acid Mutant Lines of Arabidopsis thaliana

Fatty acid mutant lines of *Arabidopsis* were used in experiments designed to further our understanding of the control of C_6-aldehyde formation by plant leaves. These mutants were: *fad*B and *act*1 with a much lower relative amount of 16:3 than wild type

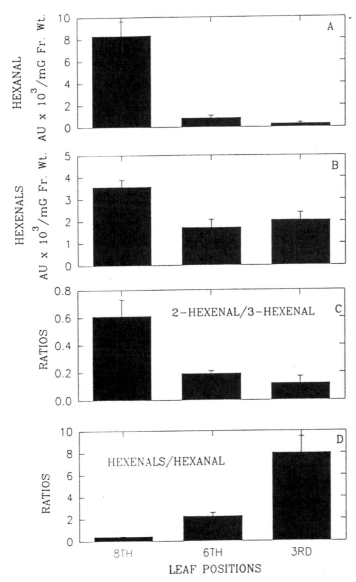

Figure 4 The formation of C_6-aldehydes from ground soybean leaves at different positions on plants. (A) hexanal, (B) hexenals, (C) 2-hexenal/3-hexenal, and (D) hexenals/hexanal. Data are mean of three replicates.

*fad*C with a much higher relative level of 18:1 and much lower relative levels of both 18:2 and 18:3 in chloroplasts

*fad*D with a much higher relative level (about two times) of 18:2 and a much lower relative level of 18:3 in chloroplasts

*fad*2 with a much lower relative amount of 18:2 in leaves

*fad*3 with a small decrease in 18:3 and higher level of 18:2 in leaves

Measurement of C_6-aldehydes showed that *fad*D produced much higher hexanal (about eight times) and *fad*C produced less than half that compared to the wild type. The *fad*3 line formed

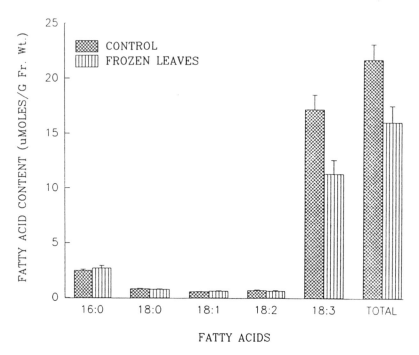

Figure 5 Effect of freezing on the fatty acid composition of glycolipids in soybean leaves (third position). Fatty acids were analyzed after soybean leaves had been treated with or without liquid nitrogen, then incubated at 30°C for 10 minutes. Data are mean of four replicates.

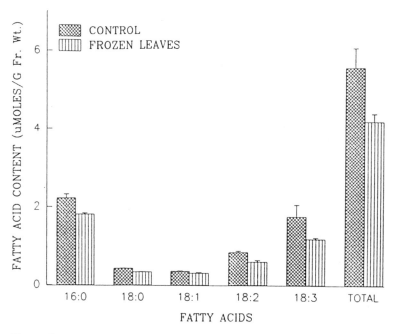

Figure 6 Effect of freezing on the fatty acid composition of phospholipids in soybean leaves (third position). Fatty acids were analyzed after soybean leaves had been treated with or without liquid nitrogen, then were incubated at 30°C for 10 minutes. Data are mean of four replicates.

about twofold more hexanal than the wild type, but no significant difference was observed among the other mutants, compared with the wild type (Figure 7). Mutants *fad*C and *fad*D produced about half the levels of hexenals, but *fad2* and *fad3* synthesized more than fourfold higher levels of hexenals, and *fad*B and *act*1 formed as much as the wild type (Figure 8C). The main component of hexenals produced by *fad2* and *fad3* was *trans*-2-hexenal, which was about 10 times more than that produced by the wild type (Figures 8A and 8B). Much greater amounts of total C_6-aldehydes were formed by *fad*D, *fad2*, and *fad3* than by the wild type. In contrast, *fad*C, developed much less total C_6-aldehydes.

5. The Contents of Polyunsaturated Fatty Acids and Total Fatty Acids in Leaves of Fatty Acid Mutant Lines of Arabidopsis

Total fatty acid and polyunsaturated fatty acid levels in leaves of the different mutant lines of *Arabidopsis* were measured: *fad*D and *fad3* leaves had higher 18:2 than the wild type (Figure 9A), while *fad*B and *fad2* had much higher levels of 18:3 (Figure 9B). The total fatty acid content also showed some differences among all the genotypes (Figure 9C).

C. Discussion

The investigation of the regulation of C_6-aldehyde formation by soybean leaves after freezing treatment was based on prior observations made in our laboratory. Those studies showed that freezing (in addition to maceration) rapidly stimulated C_6-aldehyde release from leaves, at least 50 times more than uninjured controls. Zhuang et al. (126) found that frozen soybean leaves of the same size collected from the same plant produced different levels of C_6-aldehydes. Based

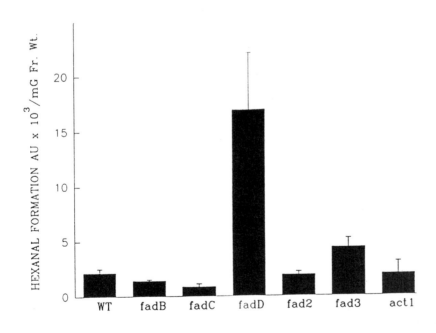

FATTY ACID MUTANT LINES OF ARABIDOPSIS

Figure 7 Hexanal formation by the leaves of wild-type (WT) and fatty acid mutant lines of *Arabidopsis*. Data are mean of three replicates.

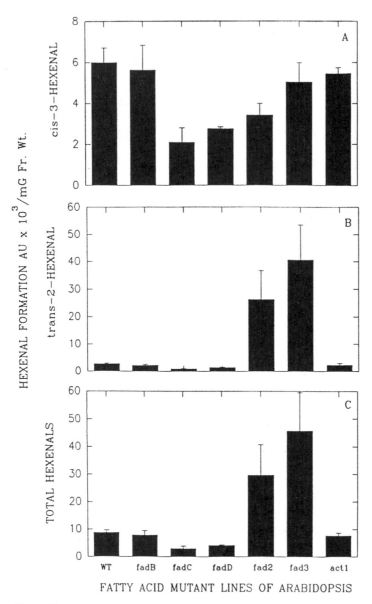

Figure 8 Hexenal formation by the leaves of wild-type (WT) and fatty acid mutant lines of *Arabidopsis*: (A) *cis*-3-hexenal, (B) *trans*-2-hexenal, and (C) total hexenals. Data are mean of three replicates.

on this observation, the investigators hypothesized that C_6-aldehyde formation was controlled by the plant leaves themselves, as well as being affected by environmental factors [2,39]. Soybean leaves were used as plant material because they could easily be classified according to their positions on the plant.

The current investigations showed that C_6-aldehyde formation appeared to be controlled by leaf ages (based on leaf positions on the plant). The locations in a leaflet and different leaflets of a trifoliate did not influence C_6-aldehyde formation. Detailed analysis showed that the formation of C_6-aldehydes during leaf development was regulated quantitatively and

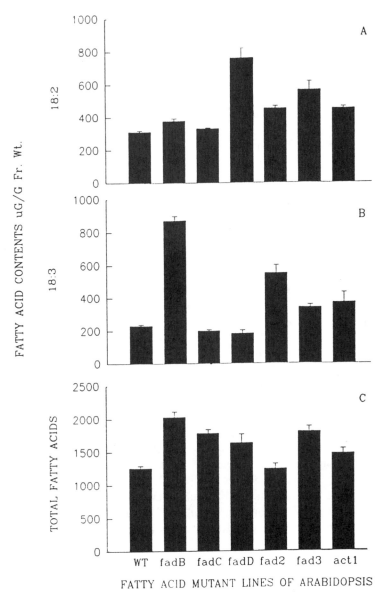

Figure 9 Contents of polyunsaturated fatty acids and total fatty acids in leaves of mutant lines of *Arabidopsis*: (A) 18:2, content, (B) 18:3 content, and (C) total fatty acids. Data are mean of four replicates.

qualitatively. The formation of C_6-aldehydes dramatically decreased and then recovered during leaf development [126]. A similar quantitative comparison of C_6-aldehyde formation by leaves at different positions on kidney bean plant has been made by Sekiya et al. [35]. However, these workers used young plants with only three or four trifoliates and studied extracts made from these leaves.

Attempts were made to gain insight into the in vivo C_6-aldehyde substrate pool by analyzing the fatty acid composition change after soybean leaves had been treated with freezing temperatures and by using fatty acid mutant lines of *Arabidopsis*. The results showed that freezing reduced only polyunsaturated fatty acids in total fatty acid of the old soybean leaves, suggesting the metabolism of lipids induced by freezing did not result from α- and β-oxidation. The same conclusion was made by Willenot [76], upon examining the change in fatty acid composition in lethal-frost-treated wheat crown and roots, and by Galliard [75], using cold-wounded potato tubers. Moreover, the main change happened to 18:3, namely, about 90% of total loss, consistent with the observation that old leaves produced much more hexenals than hexanal (about four times). Among the four lipid fractions, only phospholipids lost fatty acids in addition to polyunsaturated fatty acids, indicating that hydrolysis of phospholipids should have occurred during the incubation. This phenomenon obviously was responsible for the increase in the free fatty acid reaction (Figures 5 and 6). Phospholipid hydrolysis was also reported in macerated tea leaves [25], ground potato tubers [75], and cucumber fruits at low temperature [59]. In vitro, free fatty acids were reported to be the best substrates for C_6-aldehyde formation [91,93]. The possibility of the involvement of glycerides in C_6-aldehyde formation could not be excluded. In the presence of triacylglycerol, soybean LOX 2 and LOX 3 stimulated the uptake of O_2 [93,103]. Furthermore, LOX peroxidized, in vitro, glycolipids and phospholipids [126–129]. We found that MGD increased the C_6-aldehyde formation by soybean flour [93].

Further analysis of C_6-aldehyde formation by fatty acid mutant lines of *Arabidopsis* showed that the alteration in the metabolism of 18:2 and 18:3 directly impacted C_6-aldehyde biosynthesis (Figures 7 and 8). For example, *fad*C, which has very low levels of both 18:2 and 18:3 in chloroplast lipids [130], produced much lower levels of both hexanal and hexenals (Figures 7 and 8). It was found, as well, that *fad*D that has about two times more 18:2 in chloroplasts than the wild type and a low level of 18:3 [123] formed much higher levels of hexanal and total C_6-aldehydes (Figure 7), but much lower levels of hexenals than the wild type (Figure 8). The mutants *fad*2 and *fad*3 are deficient in the endoplasmic reticulum enzymes 18:1 desaturase (PC 18:1 desaturase) and the 18:2 desaturase (PC 18:2 desaturase), respectively [120,131]. Both largely increased the formation of hexenals and total C_6-aldehydes, however (Figure 8). The *fad*B and *act*1 mutants did not show any significant impact on C_6-aldehyde formation (Figures 7 and 8), although those two mutants had a much lower level of 16:3 compared with the wild type [132,133]. Measurement of total fatty acid levels in these mutants indicated an absence of correlation between C_6-aldehyde formation and total polyunsaturated fatty acid levels (Figure 9). Neither the increases in total C_6-aldehyde formation by *fad*D, *fad*2, and *fad*3 nor the decrease in the total C_6-aldehyde formation by *fad*C resulted from changes in the enzyme activities of the LOX/HPL pathway.

It has been suggested that LAH, LOXs, and HPL, which are needed for C_6-aldehyde production in vivo, are localized in the chloroplasts of plant leaves. In leaves, LAHs were found in chloroplasts of runner bean [74], spinach [134], and potato [135]. In wheat shoots [136] and pea shoots [137,138], LOXs appeared to be bound to chloroplastic lamella or to be soluble in the chloroplast stroma. In tea leaves most of the LOX [78,94,112,117] and HPL [81,91,97,138] activities were found in chloroplast fragments (lamella-rich fraction). In soybean leaves, HPL activity appeared to occur predominantly in chloroplasts [103]. Our experiments using *Arabidopsis* mutants, suggested that C_6-aldehydes are formed from chloroplast lipids, but the increases in total C_6-aldehyde formation by *fad*D, *fad*2, and *fad*3 remain to be further investigated.

IV. SUMMARY AND CONCLUSIONS

The control of C_6-aldehyde formation by plant leaves has never been investigated directly, although C_6-aldehydes were first identified more than 80 years ago. These volatile compounds have a low odor threshold, and their levels directly impact the quality of foods. In addition, C_6-aldehydes may be involved in plant pest defense. Although the pathway to the formation of C_6-aldehydes by plant tissues has been elucidated, the mechanism that regulates C_6-aldehyde formation is not yet known. Preliminary studies were conducted to gain insight into this issue by using soybean leaves and *Arabidopsis* mutants. The investigations began with the observation that soybean leaves had different capacities for forming C_6-aldehydes. An effect of leaf position on C_6-aldehyde production was found. The results showed that the regulation of C_6-aldehyde formation by plant leaves is a complicated process. In *Arabidopsis* leaves, this process appears to mostly use chloroplasts lipids.

In conclusion, these results show:

That C_6-aldehyde formation by soybean leaves is developmentally regulated.

That the regulation of C_6-aldehyde formation by soybean leaves may depend in part on substrate pool levels.

That chloroplast lipids may be the main substrates used by *Arabidopsis* leaves for the formation of C_6-aldehydes.

REFERENCES

1. B. A. Vick, D. C. Zimmerman, Oxidative systems for modification of fatty acids: The lipoxygenase pathway, *The Biochemistry of Plants: A Comprehensive Treatise*, Vol. 9, Academic Press, Orlando, FL, pp. 53–90 (1987).
2. A. Hatanaka, T. Kajiwara, and J. Sekiya, Biosynthetic pathway for C_6-aldehyde formation from linolenic acid in green leaves, *Chem. Phys. Lipids, 44*: 341–361 (1987).
3. D. F. Hildebrand, Lipoxygenase, *Physiol. Plant., 76*: 249–253 (1989).
4. H. W. Gardner, Recent investigations into the lipoxygenase pathway of plants, *Biochim. Biophys. Acta, 1084*: 221–239 (1991).
5. C. E. Eriksson, Aroma compounds derived from oxidized lipids. Some biochemical and analytical aspects (fresh, heat treated, and frozen vegetables), *J. Agric. Food Chem., 23*: 126–128 (1975).
6. J. Sekiya, T. Kajiwara, and A. Hatanaka, Seasonal changes in activities of enzymes responsible for the formation of C_6-aldehydes and C_6-alcohols in tea leaves and the effects of environmental temperature on the enzyme activity, *Plant Cell Physiol., 25*: 269–280 (1984).
7. H. W. Gardner, How the lipoxygenase pathway affects the organoleptic properties of fresh fruit and vegetables, *Flavor Chemistry of Lipid Foods* (D. B. Min and T. H. Smouse, eds.), American Oil Chemists Society, Champaign, IL, pp. 98–112 (1989).
8. S. J. Kazeniac and R. M. Hall, Flavor chemistry of tomato volatiles, *J. Food Sci., 35*: 519–530 (1970).
9. D. A. Forss, E. A. Dunstone, E. H. Ramshaw, and W. Stark, The flavor of cucumbers, *J. Food Sci., 27*: 90–93 (1962).
10. F. Drawert, W. Heimann, R. Emberger, and R. Tressl, On the biogenesis of aromatic substances in plants and [apple, banana] fruits. II. Enzymatic formation of hexen-(2)-al-(1), hexanal and their first steps, *Justus Liebigs Ann. Chem., 694*: 200–208 (1966).
11. W. J. Wolf, Lipoxygenase and flavor of soybean protein products, *J. Agric. Food Chem., 23*: 136–141 (1975).
12. R. T. Major, P. Marchini, and T. Sproston, Isolation from *Ginkgo biloba* L. of an inhibitor of fungus growth, *J. Biol. Chem., 235*: 3298–3299 (1960).
13. H. Schildknecht and G. Rauch, Report on the defensive substances of plants. II. The chemical nature of the volatile phytocides of leafy plants, particularly of *Robinia pseudacacia*, *Z. Naturforsch., 16B*: 422–429 (1961).
14. B. Nandi and N. Fries, Volatile aldehydes, ketones, esters and terpenoids as preservatives against storage fungi in wheat, *Z. Pflanzenkrankh. Pflanzenschutzber., 83*: 284–294 (1976).

15. R. C. Gueldner, D. M. Wilson, and A. R. Heidt, Volatile compounds inhibiting *Aspergillus flavus*, *J. Agric. Food Chem.*, *33*: 411–413 (1985).

16. I. Urbasch, Transformations of *trans*-2-hexenal by *Botrytis cinerea* Pers. as detoxification mechanism, *Z. Naturforsch.*, *42C*: 64–68 (1987).

17. T. R. Hamilton-Kemp, C. T. McCracken, J. H. Loughrin, R. A. Andersen, and D. F. Hildebrand, Effects of some natural volatile compounds on the pathogenic fungi *Alternaria alternata* and *Botrytis cinerea*, *J. Chem. Ecol.*, *18*: 1083–1091 (1992).

18. J. M. Bradow and W. J. Connick, Seed germination inhibition by volatile alcohols and other compounds associated with *Amaranthus palmeri* residues, *J. Chem. Ecol.*, *14*: 1633–1648 (1988).

19. R. C. French and G. R. Leather, Screening of nonanal and related volatile flavor compounds on the germination of 18 species of weed seed, *J. Agric. Food Chem.*, *27*: 828–832 (1979).

20. H. W. Gardner, D. L. Dornbos, Jr., and A. E. Desjardins, Hexanal, *trans*-2-hexenal, *trans*-2-nonenal inhibit soybean, *Glycine max* seed germination, *J. Agric. Food Chem.*, *38*: 1316–1320 (1990).

21. T. R. Hamilton-Kemp, J. H. Loughrin, D. D. Archbold, R. A. Andersen, and D. F. Hildebrand, Inhibition of pollen germination by volatile compounds including 2-hexenal and 3-hexenal, *J. Agric. Food Chem.*, *39*: 952–956 (1991).

22. W. Deng, T. R. Hamilton-Kemp, M. T. Nielsen, R. A. Andersen, G. B. Collins, and D. F. Hildebrand, Effects of six-carbon aldehydes and alcohols on bacterial proliferation. *J. Agric. Food Chem.*, *41*: 506–510 (1993).

23. E. Schauenstein, H. Esterbauer, and H. Zollner, Aldehydes in biological systems: Their natural occurrence and biological activities, Pion Limited, 1977.

24. T. Curtius and H. Frunzen, Chemical constituents of green plants. IV. Volatile aldehydes of hornbean leaves, *Justus Liebigs Ann. Chem.*, *390*: 89–121 (1912).

25. A. Hatanaka, T. Kajiwara, and J. Sekiya, Seasonal variations in *trans*-2-hexenal and linolenic acid in homogenates of the *Thea sinensis* leaves, *Phytochemistry*, *15*: 1889–1891 (1976).

26. J. Sekiya, T. Kajiwara, and A. Hatanaka, Seasonal changes in activity of the enzyme system producing *cis*-3-hexenal and *n*-hexanal from linolenic and linoleic acid in tea leaves, *Plant Cell Physiol.*, *18*: 283–286 (1977).

27. R. T. Major, O. D. Collins, P. Marchini, and H. W. Schnalsel, Formation of 2-hexenal by leaves, *Phytochemistry*, *11*: 607–610 (1972).

28. A. Kitamura, K. Matsui, T. Kajiwara, and A. Hatanaka, Changes in volatile C_6-aldehydes emitted from and accumulated in tea leaves, *Plant Cell Physiol.*, *33*: 493–496 (1992).

29. A. Hatanaka, J. Sekiya, and T. Kajiwara, Distribution of an enzyme system producing *cis*-3-hexenal, and *n*-hexanal from linolenic and linoleic acids in some plants, *Phytochemistry*, *17*: 869–872 (1978).

30. J. Sekiya, T. Kajiwara, T. Munechika, and A. Hatanaka, Distribution of lipoxygenase and hydroperoxide lyase in the leaves of various plant species, *Phytochemistry*, *22*: 1867–1869 (1983).

31. W. Nye and H. A. Spoehr, The isolation of hexenals from leaves, *J. Food Sci.*, *2*: 23–35 (1943).

32. R. T. Major and M. Thomas, Formation of 2-hexenal from linolenic acid by macerated *Ginkgo* leaves, *Phytochemistry*, *11*: 611–617 (1972).

33. R. R. Selvendran, J. Reymolds, and T. Galliard, Production of volatiles by degradation of lipids during manufacture of black tea, *Phytochemistry*, *17*: 233–236 (1978).

34. T. Takeo and T. Tsushida, Changes in lipoxygenase activity in relation to lipid degradation in plucked tea shoots, *Phytochemistry*, *19*: 2521–2522 (1980).

35. J. Sekiya, T. Kajiwara, and A. Hatanaka, Lipoxygenase, hydroperoxide lyase and volatile C_6-aldehyde formation from C_{18}-fatty acids during development of *Phaseolus vulgaris* L, *Plant Cell Physiol.*, *23*: 631–638 (1982).

36. S. Takei, Y. Sakato, M. Ono, and Y. Kuriowa, Leaf alcohol. I. Distribution of leaf alcohol in various plants, *Bull. Agric. Chem. Soc. Japan*, *14*: 709–716 (1938).

37. B. Bengtsson and I. Bosund, Gas chromatographic evaluation of the formation of volatile substances in stored pears, *Food Technol.*, *18*: 773–776 (1964).

38. J. Schormuller and W. Grosch, Aromatics in foods. I. Identification of carbonyl compounds in tomatoes, *Z. Lebensm.-Unters.-Forsch.*, *118*: 385–393 (1962).

39. B. M. Shah, D. K. Salunkhe, and L. E. Olson, Effect of ripening on chemistry of tomato volatiles, *J. Am. Soc. Hortic. Sci.*, *94*: 171–176 (1969).

40. K. B. Dalal, D. K. Salunkhe, L. E. Olson, J. Y. Do, and M. H. Yu, Volatile components of developing tomato fruit growing under field and green house condition, *Plant Cell Physiol.*, *9*: 389–400 (1968).

41. H. P. Fleming, W. Y. Cobb, J. L. Etchells, and T. A. Bell, The formation of carbonyl compounds in cucumbers, *J. Food Sci.*, *33*: 572–576 (1968).

42. A. J. MacLeod and G. MacLeod, Volatiles of cooked cabbage, *J. Sci. Food Agric.*, *19*: 273–277 (1968).

43. R. Tressl and F. Drawert, Biogenesis of banana volatiles, *J. Agric. Food Chem.*, *21*: 560–565 (1973).

44. M. Fujimaki, S. Arai, N. Kirigaya, and Y. Sakurai, Studies on flavor components in soybean. I. Aliphatic carbonyl compounds, *Agric. Biol. Chem.*, *29*: 855–863 (1965).

45. S. Arai, O. Koyanagi, and M. Fajimaki, Studies on flavor components in soybean. IV. Volatile neutral compounds, *Agric. Biol. Chem.*, *31*: 868–873 (1967).

46. S. Arai, M. Noguchi, M. Kaji, H. Kato, and M. Fajimaki, *n*-Hexanal and some volatile alcohols. Their distribution in raw soybean tissues and formation in crude soy protein concentrate by lipoxygenase, *Agric. Biol. Chem.*, *34*: 1420–1423 (1970).

47. J. A. Maga, A review of flavor investigations associated with the soy products raw soybeans defatted flakes and flours, and isolates, *J. Agric. Food Chem.*, *21*: 864–868 (1973).

48. W. F. Wilkens and F. M. Lin, Gas chromatographic and mass spectral analyses of soybean milk volatiles, *J. Agric. Food Chem.*, *18*: 333–336 (1970).

49. D. Arens, G. Laskany, and W. Grosch, Lipoxygenase aus erbsen bildung fluchtiger Aldehyde aus Linolsaure, *Z. Lebensm.-Unters.-forsch.*, *151*: 162–166 (1973).

50. H. W. Gardner, Decomposition of linoleic acid hydroperoxides. Enzymic reactions compared with nonenzymic, *J. Agric. Food Chem.*, *23*: 129–136 (1975).

51. G. C. Mustakas, W. J. Albrecht, J. E. McGhee, L. T. Black, G. N. Bookwalter, and E. L. Griffin, Jr., Lipoxidase deactivation to improve stability, odor and flavor of full fat soy flours, *J. Am. Oil Chem. Soc.*, *46*: 623–626 (1969).

52. W. F. Wilkens, L. R. Mattick, and D. B. Hand, Effect of processing method on oxidative off-flavors of soybean milk, *Food Technol.*, *21*: 1630–1633 (1967).

53. R. Saijyo and Y. Kuwabara, Volatile flavor of black tea. I. Formation of volatile components during black tea manufacture, *Agric. Biol. Chem.*, *31*: 389–396 (1967).

54. A. Hatanaka and M. Ohno, Leaf alcohol. IX. A simple synthesis and configuration of leaf aldehydes, *Agric. Biol. Chem.*, *25*: 7–9 (1961).

55. A. Hatanaka and M. Ohno, Occurrence of *trans* isomers in natural leaf alcohol fractions, *Bull. Agric. Chem. Soc. Japan*, *24*: 614–617 (1960).

56. A. Hatanaka and M. Ohno, Leaf alcohol. XIX. Chronic acid oxidation of isomeric *n*-hexenols, *Agric. Biol. Chem.*, *35*: 1044–1051 (1971).

57. T. Kajiwara, T. Harada, and A. Hatanaka, Isolation of Z-3-hexenal in tea leaves, *Thea sinensis*, and synthesis thereof, *Agric. Biol. Chem. Japan*, *39*: 243–247 (1975).

58. H. Zhuang and D. F. Hildebrand, Changes in lipoxygenase pathway enzyme activities during soybean fruit development, *Plant Physiol.* 99 (suppl.): 113 (1992).

59. T. Galliard and D. R. Phillips, The enzymic cleavage of linoleic acid to C_9-carbonyl fragments in extracts of cucumber (*Cucumis sativus*) fruit and the possible role of lipoxygenase, *Biochim, Biophys. Acta*, *431*: 278–287 (1976).

60. A. Hatanaka and T. Harada, Formation of *cis*-3-hexenal, *trans*-2-hexenal and *cis*-3-hexenol in macerated *Thea sinensis* leaves, *Phytochemistry*, *12*: 2341–2346 (1973).

61. A. I. Nelson, L. S. Wei, and M. P. Steinberg, Food products from whole soybeans, *Soybean Dig.*, *31*(3): 32–34 (1971).

62. E. A. Lee, A. C. Wagenknecht, and J. C. Hening, A chemical study of the progressive development of off-flavor in frozen raw vegetables, *Food Res.*, *20*: 289–297 (1955).

63. T. Yamanishi, A. Kobayashi, A. Uchida, and Y. Kawashima, Studies on the flavor of green tea. VII. Flavor components of manufactured green tea, *Agric. Biol. Chem.*, *30*: 1102–1105 (1966).

64. J. G. Gonzalez, P. Cogyon, and G. W. Sanderson, Biochemistry of tea fermenting formation of *trans*-2-hexenal from linolenic acid, *J. Food Sci.*, *37*: 797–798 (1972).

65. T. R. Hamilton-Kemp, R. A. Andersen, D. F. Hildebrand, J. H. Loughrin, and P. D. Fleming, Effects of LOX inhibitors on the formation of volatile compounds in wheat, *Phytochemistry*, *26*, 1273–1277 (1987).

66. S. Jadhav, B. Singh, and D. K. Salunkhe, Metabolism of unsaturated fatty acids in tomato fruit: Linoleic and linolenic acid as precursors of hexanal, *Plant Cell Physiol.*, *13*: 449–459 (1972).

67. R. T. Major, P. Marchini, and A. J. Boulton, Observation on the production of alpha-hexenal by leaves of certain plants, *J. Biol. Chem.*, *238*: 1813–1816 (1983).

68. A. L. Tappel, W. O. Lundberg, and P. D. Boyer, Effect of temperature and antioxidants upon the lipoxygenase-catalyzed oxidation of sodium linoleate, *Arch. Biochem. Biophys.*, *42*: 293–300 (1953).

69. H. van Duin and J. K. Poll, The identification of *cis*-2-enals in autoxidation products, *Neth. Milk Dairy J.*, *21*: 248–249 (1967).

70. E. Selke, H. A. Moser, and W. K. Rohwedder, Tandem gas chromotography–mass spectrometry analysis of volatiles from soybean oil, *J. Am. Oil Chem. Soc.*, *47*: 393–397 (1970).

71. E. A. Lee and L. R. Mattick, Fatty acids of the lipids of vegetables. I. Peas (*Pisum sativum*), *J. Food Sci.*, *26*: 273–275 (1961).

72. L. R. Mattick and E. A. Lee, The fatty acids of vegetables. II. Spinach, *J. Food Sci.*, *26*: 356–358 (1961).

73. W. Grosch, Enzymic formation of neutral carbonyl compounds from the lipids of peas (*Pisum sativum*; var. Goettinga), *Z. Lebensm.-Unters.-Forsch.*, *135*: 75–76 (1967).

74. P. S. Sastry and M. Kates, Hydrolysis of monogalactosyl and digalactosyl diglycerides by specific enzymes in runner-bean (*Phaseolus multiflornus*) leaves, *Biochemistry*, *3*: 1280–1287 (1964).

75. T. Galliard, The enzymic deacylations of phospholipids and galactolipids in plants, *Biochem. J.*, *121*: 379–390 (1970).

76. C. Willenot, Rapid degradation of polar lipids in frost damaged winter wheat crown and root tissue, *Phytochemistry*, *22*: 861–863 (1983).

77. R. Saijyo and T. Takeo, The importance of linoleic acid and linolenic acid as precursors of hexanal and *trans*-2-hexenal in black tea, *Plant Cell Physiol.*, *13*: 991–998 (1972).

78. A. Hatanaka, T. Kajiwara, and J. Sekiya, Biosynthesis of *trans*-2-hexenal in chloroplasts from *Thea sinensis*, *Phytochemistry*, *15*: 1125–1126 (1975).

79. R. T. Major, The *Ginkgo*, the most ancient living tree, *Science*, *157*: 1270–1273 (1967).

80. A. Hatanaka, J. Sekiya, and T. Kajiwara, Enzyme system catalyzing formation of *cis*-3-hexenal and *n*-hexanal from linolenic and linoleic acids in Japanese silver (*Farfugium iaponicum* Kitamura) leaves (drug plants), *Plant Cell Physiol.*, *18*: 107–116 (1977).

81. A. Hatanaka, T. Kajiwara, and J. Sekiya, Biosynthesis of *trans*-2-hexenal in chloroplasts from *Thea sinensis*, *Phytochemistry*, *15*: 1125–1126 (1976).

82. W. Grosch and J. Schwarz, Linoleic and linolenic acid as precursors of the cucumber flavor, *Lipids*, *6*: 351–352 (1971).

83. A. L. Tappel, Lipoxidase, *Methods Enzymol.*, *5*: 539–542 (1962).

84. D. F. Hildebrand, T. R. Hamilton-Kemp, and G. Bookjans, Plant lipoxygenases: Occurrence, properties and possible functions, *Curr. Top. Plant Biochem. Physiol.*, *7*: 201–219 (1988).

85. T. Galliard and H. W. S. Chan, Lipoxygenases, *The Biochemistry of Plants: A Comprehensive Treatise*, Vol. 4, *Lipids: Structure and Function* (P. K. Stumpf and E. E. Conn, eds.), Academic Press, New York, pp. 131–161 (1980).

86. A. Prinsky, S. Grossman, and M. Trop, Lipoxygenase content and antioxidant activity of some fruits and vegetables, *J. Food Sci.*, *36*: 571–572 (1971).

87. M. Holden, Lipoxidase activity of leaves, *Phytochemistry*, *9*: 507–512 (1970).

88. K. S. Rhee and B. M. Watts, Evaluation of lipid oxidation in plant tissues, *J. Food Sci.*, *31*: 664–668 (1966).

89. A. M. Siddiqi and A. L. Tappel, Catalysis of linoleate oxidation by pea lipoxidase, *Arch. Biochem. Biophys.*, *60*: 91–99 (1956).

90. A. L. Tappel, Lipoxidase, *The Enzymes*, 2nd ed., Vol. 8 (P. D. Boyer, H. Lardy, and K. Myrback, eds.), Academic Press, New York, pp. 275–283 (1963).

91. A. Hatanaka, T. Kajiwara, J. Sekiya, M. Imoto, and S. Inouye, Participation and properties of

lipoxygenase and hydroperoxide lyase in volatile C_6-aldehyde formation from C_{18}-unsaturated fatty acids in isolated tea chloroplasts, *Plant Cell Physiol.*, *23*: 91–99 (1982).

92. D. F. Hildebrand, T. R. Hamilton-Kemp, J. H. Loughrin, K. Ali, and R. A. Andersen, Lipoxygenase 3 reduces hexanal production from soybean seed homogenates, *J. Agric. Food Chem.*, *38*: 1934–1936 (1990).

93. H. Zhuang, D. F. Hildebrand, R. A. Andersen, and T. R. Hamilton-Kemp, Effects of polyunsaturated free fatty acids and esterified linoleoyl derivatives on oxygen consumption and C_6-aldehyde formation with soybean seed homogenates, *J. Agric. Food Chem.*, *39*: 1357–1364 (1991).

94. T. Matoba, H. Hidaka, H. Narita, K. Kitamura, N. Kaizuma, and M. Kito, Lipoxygenase-2 isozyme is responsible for generation of *n*-hexanal in soybean homogenate, *J. Agric. Food Chem.*, *33*: 852–855 (1985).

95. M. Hamberg and B. Samuelsson, On the specificity of the oxygenation of unsaturated fatty acids catalyzed by soybean lipoxidase, *J. Biol. Chem.*, *242*: 5329–5335 (1967).

96. J. I. Kalbrener, K. Warner, and A. C. Eldrighe, Flavors derived from linoleic and linolenic acid hydroperoxides (soybean), *Cereal Chem.*, *51*: 406–416 (1974).

97. A. Hatanaka, T. Kajiwara, J. Sekiya, and S. Inouye, Solubilization and properties of the enzyme cleaving 13-*S*-hydroperoxylinolenic acid in tea leaves, *Phytochemistry*, *21*: 13–17 (1982).

98. T. Galliard, D. R. Phillips, and J. Reynolds, The formation of *cis*-3-nonenal, *trans*-2-nonenal and hexanal from linoleic acid hydroperoxide isomers by a hydroperoxide cleavage enzyme system in cucumber (*Cucumis sativus*) fruits, *Biochim. Biophys. Acta*, *441*: 181–192 (1976).

99. B. A. Vick and D. C. Zimmerman, Lipoxygenase and hydroperoxide lyase in germinating water melon seedlings, *Plant Physiol.*, *57*: 780–788 (1976).

100. D. R. Phillips and T. Galliard, Flavor biogenesis: Partial purification and properties of a fatty acid hydroperoxide cleaving enzyme from fruits of cucumber, *Phytochemistry*, *17*: 233–236 (1978).

101. D. A. Wardale, E. A. Lambert, and T. Galliard, Localization of fatty acid hydroperoxide cleavage activity in membranes of cucumber fruit, *Phytochemistry*, *17*: 205–212 (1978).

102. J. A. Matthew and T. Galliard, Enzymic formation of carbonyls from linoleic acid in leaves of *Phaseolus vulgaris* (kidney beans), *Phytochemistry*, *17*: 1043–1044 (1978).

103. H. W. Gardner, D. Weisleder, and R. D. Plattner, Hydroperoxide lyase and other hydroperoxide-metabolizing activity in tissues of soybean, *Glycine max*, *Plant Physiol.*, *97*: 1059–1072 (1991).

104. J. Sekiya, T. Kajiwara, and A. Hatanaka, Volatile C_6-aldehyde formation via hydroperoxide from C_{18} carbon-unsaturated fatty acids in etiolated alfalfa and cucumber seedlings, *Agric. Biol. Chem.*, *43*: 969–980 (1979).

105. J. M. Olias, J. J. Rios, M. Valle, R. Zamore, L. C. Sanz, and B. Axelrod, Fatty acid hydroperoxide lyase in germination soybean seedlings, *J. Agric. Food Chem.*, *38*: 624–630 (1990).

106. J. Sekiya, T. Monma, T. Kajiwara, and A. Hatanaka, Changes in activities of lipoxygenase and hydroperoxide lyase during seed development of soybean, *Agric. Biol. Chem.*, *50*: 521–522 (1986).

107. T. Matoba, H. Hidaka, K. Kitamura, N. Kaizuma, and M. Kito, Contribution of hydroperoxide lyase activity to *n*-hexanal formation in soybean, *J. Agric. Food Chem.*, *33*: 856–858 (1985).

108. J. Sekiya, S. Tanigawa, T. Kajiwara, and A. Hatanaka, Fatty acid hydroperoxide lyase in tobacco cells cultured in vitro, *Phytochemistry*, *23*: 2439–2443 (1984).

109. A. Hatanaka, T. Kajiwara, and K. Matsui, Concentration of hydroperoxide lyase activities in root of cucumber seedlings, *Z. Naturforsch.*, *43c*: 308–310 (1988).

110. T. Galliard and J. A. Matthew, Lipoxygenase-mediated cleavage of fatty acids to carbonyl fragments in tomato fruits, *Phytochemistry*, *16*: 339–343 (1977).

111. L-S. Kim and W. Grosch, Partial purification and properties of a hydroperoxide lyase from fruits of pear, *J. Agric. Food Chem.*, *29*: 1220–1225 (1981).

112. P. Schreier and G. Lorenz, Separation, partial purification and characterization of a fatty acid hydroperoxide cleaving enzyme from apple and tomato fruits, *Z. Naturforsch.*, *37c*: 165–173 (1982).

113. T. Galliard, J. A. Matthew, J. Fishwick, and A. J. Wright, The enzymic degradation of lipids resulting from physical disruption of cucumber (*Cucumis sativus*) fruits, *Phytochemistry*, *15*: 1647–1650 (1976).

114. T. Galliard, J. A. Matthew, A. F. Wright, and J. Fishwick, The enzymic degradation of lipids to volatile and nonvolatile carbonyl fragments in disrupted tomato fruits, *J. Sci. Food Agric.*, *28*: 863–868 (1977).

115. J. W. Ralls, W. H. McFadden, R. M. Siefert, D. R. Black, and P. W. Kilpatrick, Volatiles from a commercial pea blancher. Mass spectra identification, *J. Food Sci.*, *30*: 228–232 (1965).

116. C. E. Eriksson, Alcohol:NAD oxidoreductase from peas, *Acta Chem. Scand.*, *21*: 304–308 (1967).

117. C. E. Eriksson, Alcohol:NAD oxidoreductase (E.C.1.1.1.1.) from peas, *J. Food Sci.*, *33*: 525–532 (1968).

118. J. Sekiya, S. Numa, T. Kajiwara, and A. Hatanaka, Biosynthesis of leaf alcohol formation of 3Z-hexenal from linolenic acid in chloroplasts of *Thea sinensis* (tea) leaves, *Agric. Biol. Chem.*, *40*: 185–190 (1976).

119. A. Hatanaka and T. Harada, Purification and properties of alcohol dehydrogenase from tea seeds, *Agric. Biol. Chem.*, *36*: 2033–2035 (1972).

120. H. W. Gardner, Isolation of a pure isomer of linoleic acid hydroperoxide, *Lipids*, *10*: 248–252 (1975).

121. J. Browse and C. Somerville, Glycerolipid synthesis: Biochemistry and regulation, *Annu. Rev. Plant Physiol. Plant. Mol. Biol.*, *42*: 467–506 (1991).

122. R. R. Wise and A. W. Naylor, Chilling-enhanced photooxidation. Evidence for the role of singlet oxygen and superoxide in the breakdown of pigments and antioxidants, *Plant Physiol.*, *83*: 278–282 (1987).

123. K. L. Parkin and S-J. Kuo, Chilling-induced lipid degradation in cucumber (*Cucumis sativa* L. cv hybrid C) fruit, *Plant Physiol.*, *90*: 1049–1056 (1989).

124. J. Browse, P. McCourt, and C. Somerville, A mutant of *Arabidopsis* deficient in $C_{18:3}$ and $C_{16:3}$ leaf lipids, *Plant Physiol.*, *81*: 859–864 (1986).

125. M. L. Dahmer, P. D. Fleming, G. B. Collins, and D. F. Hildebrand, A rapid screening technique for determining the lipid composition of soybean seeds, *J. Am. Oil Chem. Soc.*, *66*: 543–548 (1989).

126. H. Zhuang, D. F. Hildebrand, T. R. Hamilton-Kemp, and R. A. Andersen, Developmental change in C_6-aldehyde formation by soybean leaves, *Plant Physiol.*, *100*: 80–87 (1992).

127. R. Yanauchi, M. Kojima, K. Kato, and Y. Ueno, Lipoxygenase catalyzed oxygenation of monogalactosyl-dilinolenoyl-glycerol in dipalmitoyl-phosphatidylcholine liposomes, *Agric. Biol. Chem.*, *49*: 2475–2478 (1985).

128. P. L. Guss, T. Richardson, and M. A. Stahmann, Oxidation of various lipid substrates with unfractionated soybean and wheat lipoxidase, *J. Am. Oil Chem. Soc.*, *45*: 272–276 (1968).

129. A. R. Brash, C. D. Ingram, and T. M. Harris, Analyses of a specific oxygenation reaction of soybean lipoxygenase-1 with fatty acid esterified in phospholipids, *Biochemistry*, *26*: 5465–5471 (1987).

130. J. Eskola and S. Laakso, Bile salt-dependent oxygenation of polyunsaturated phosphatidylcholines by soybean lipoxygenase-1, *Biochim. Biophys. Acta*, *751*: 305–311 (1983).

131. J. Browse, L. Kunst, S. Anderson, S. Hugly, and C. Somerville, A mutant of *Arabidopsis* deficient in the chloroplast 16:1/18:1 desaturase, *Plant Physiol.*, *90*: 522–529 (1989).

132. B. Lemieux, M. Miquel, J. Browse, and C. Somerville, Mutants of *Arabidopsis* with alterations in seed lipid fatty acid composition, *Theor. Appl. Genet.*, *80*: 234–240 (1990).

133. L. Kunst, J. Browse, and C. Somerville, Altered regulation of lipid biosynthesis in a mutant of *Arabidopsis* deficient in the chloroplast glycerol-3-phosphate acyltransferase activity, *Proc. Natl. Acad. Sci. U.S.A.*, *85*: 4143–4147 (1988).

134. L. Kunst, J. Browse, and C. Somerville, A mutant of *Arabidopsis* deficient in desaturation of palmitic acid in leaf lipids, *Plant Physiol.*, *90*: 943–947 (1989).

135. M. M. Anderson, R. E. McCarty, and E. A. Zimmer, The role of galactolipids in spinach chloroplast lamellar membranes, *Plant Physiol.*, *53*: 699–704 (1974).

136. H. Matsuda and O. Hirayama, Purification and properties of a lipolytic acyl hydrolase from potato leaves, *Biochim. Biophys. Acta*, *573*: 155–165 (1979).

137. R. Douillard and E. Bergeron, Lipoxygenase activity of wheat shoot chloroplasts, *C. R. Hebd. Seances Acad. Sci. Ser. D*, *286*: 753–755 (1978).

138. R. Douillard and E. Bergeron, Chloroplastic localization of soluble lipoxygenase activity in young leaves, *Plant Sci. Lett.*, *22*: 263–268 (1981).

139. R. Douillard, E. Bergeron, and A. Scalbert, Some characteristics of soluble chloroplastic and nonchloroplastic lipoxygenases of pea leaves *Pisum sativum*, *Physiol. Veg.*, *20*: 377–384 (1982).

140. A. Hatanaka, T. Kajiwara, J. Sekiya, and Y. Kido, Formation of 12-oxo-*trans*-10-dodecenoic acid in chloroplasts from *Thea sinensis* (tea) leaves, *Phytochemistry*, *16*: 1828–1829 (1977).

16
Phloem Transport of Solutes in Crop Plants

Monica A. Madore
University of California, Riverside, California

I. INTRODUCTION

Many plant parts, including many flowers, fruits, and seeds, do not contain chlorophyll and are therefore not photosynthetically competent. Other plant parts, particularly underground roots, rhizomes, and tubers, are located on the plant in areas where light reception is insufficient to drive photosynthesis. In plant parts such as meristems, stems, and developing leaves, modification, incomplete development, or insufficient number of plastids also limits photosynthetic competency. Photosynthetic activity is therefore found to be largely confined to organs located in areas of maximal light interception and containing fully functional chloroplasts. In higher plants, these organs are represented by mature, fully expanded leaves.

The consequence of this separation of the plant body into photosynthetically competent and nonphotosynthetic organs is that photosynthesizing leaves become the sole "source" of photosynthetically produced biomolecules (photoassimilates) for the rest of the plant. Thus to supply the demands of nonphotosynthetic plant parts, which act as competing "sinks" for photosynthetic products, leaves must produce photoassimilates in amounts far in excess of what is required simply for maintenance of leaf metabolism. In higher plants, the delivery of photoassimilates from "source" to "sink" regions within the plant body is accomplished by translocation in the phloem tissues.

In crop plants, phloem transport is a particularly important physiological process, for with very few exceptions, the agronomically important plant parts that are harvested from our major agricultural crop plants are "sink" tissues. From a physiological standpoint, what this ultimately means is that the ability of a particular crop plant to carry out photosynthesis during a growth season will only partly determine the final harvestable yield of that crop. The phloem transport process will be of equal importance, for it is this process that determines just how efficiently photosynthetically produced nutrients are made available to the plant part to be harvested. A complete understanding of phloem transport and its regulation is therefore basic to our understanding of crop physiology.

II. PHLOEM STRUCTURE

It is beyond the scope of this chapter to provide more than a general description of the anatomy of the phloem transport system. Readers should consult a general plant anatomy textbook (e.g., Ref. 1) or reviews of phloem structure [2–4] for more details regarding the anatomy, morphology, and differentiation of vascular tissues. This chapter emphasizes the phloem structure of the minor veins of leaves, the key interface of the phloem transport system with the photosynthetic tissues.

A. General Feature of Phloem Tissues

1. Sieve Elements

Phloem tissues in general consist of several structurally distinct cell types: sieve elements, companion cells, parenchyma cells, and fibers [1]. The most characteristic cells are the sieve elements, which are linked end to end to form the conduit for the long-distance movement of solutes (Figure 1). Unlike xylem tracheids, which are dead at maturity, functional sieve elements are living cells. During maturation of the sieve element, the tonoplast and nucleus degenerate and all ribosomes disappear. Mitochondria and plastids assume a spherical shape, lose internal organization and, together with the endoplasmic reticulum, assume a parietal position next to the plasma membrane. Plastids accumulate either starch or protein inclusions. Proteinaceous strands (P-protein) may also be present in the cell lumen [1–4].

The end walls of the sieve element are modified to form the sieve plate (Figure 1). Contiguous sieve elements are interconnected to form a sieve tube through strands of protoplasm, which pass through the plasma membrane lined sieve plate pores. The side walls of adjacent sieve tubes may also contain sieve areas connecting the protoplasts of the neighboring sieve elements [1–4]. The cytoplasmic compartments of the sieve elements, therefore, form a continuum through which solutes can be moved.

2. Companion Cells

Companion cells are associated with sieve elements and arise concurrently with the sieve elements by division of a common mother cell. Unlike the sieve elements, companion cells retain their nuclei and vacuoles and are characterized by densely staining cytoplasm containing numerous free ribosomes and many highly differentiated mitochondria and plastids [1–4]. These structural features are indicative of high metabolic activity, and it is thought that the companion cells act to maintain the structural integrity of the sieve elements, which lose metabolic capability as a result of the structural changes that occur during maturation. The protoplasts of the companion cells are connected to sieve elements by numerous branched (on the companion cell side) plasmodesmata, providing a cytoplasmic connection for metabolite exchange between the two cell types. Because of the high degree of symplastic continuity between the companion cell and sieve element, these are often referred to as sieve element–companion cell (SE-CC) complexes [2–4].

B. Minor Vein Structure

The venation of source leaves is designed such that individual photosynthetic mesophyll cells are never more than a few cells away from a minor vein (Figure 2). This arrangement drastically reduces the distance that assimilates must travel from the sites of photosynthesis to the phloem transport system [5]. It is in the SE-CC complexes of the leaf minor veins that loading of the phloem transport system with photosynthetic products is initiated. Not surprisingly, the companion cells within the minor veins are very much larger than the sieve elements with which

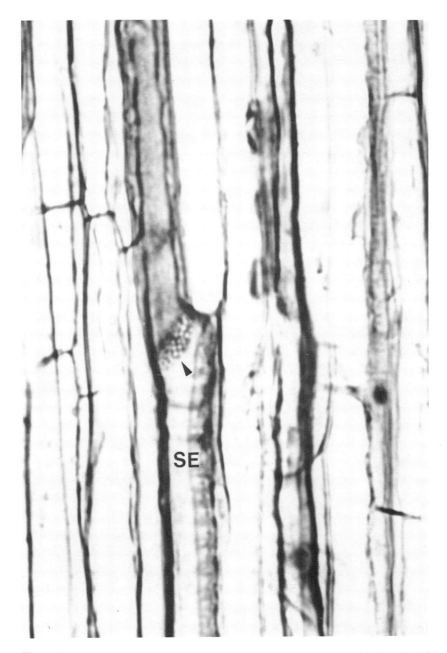

Figure 1 Longitudinal section of a squash (*Cucurbita pepo* L.) stem, showing the sieve elements (SE). Arrow indicates a sieve plate. (Paraffin section slide courtesy of D. A. DeMason.)

they are associated (Figure 3), which more than likely reflects the added metabolic activity imparted by the phloem loading process.

Minor vein companion cells form a key interface between the photosynthetic tissues of the leaf and the conduits of the phloem system. Based on ultrastructural differences, Gamalei [6], Van Bel and Gamalei [7], and Van Bel [8] have categorized three classes of minor vein companion cell within source leaves (Figure 4).

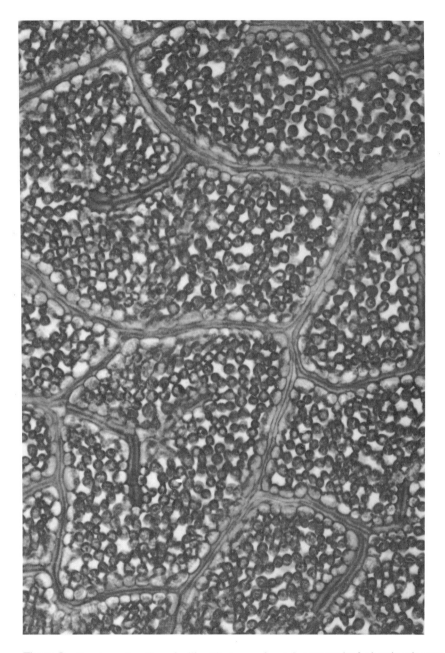

Figure 2 Paradermal section of a lilac (*Syringa vulgaris* L.) source leaf, showing the arrangement of the minor venation within the photosynthetic tissues. (Paraffin section slide courtesy of D. A. DeMason).

Type 1 companion cells (Figure 4A), referred to as "intermediary cells" [9–11], are characterized by large numbers of plasmodesmata, which link the cytoplasm of these cells to that of the adjacent photosynthetic cells. These cells are specialized for the symplastic transfer of assimilates from the photosynthetic cells to the sieve elements. Companion cells of this type are common in many species of horticultural importance, including tree species such as olive [12], most woody ornamental vines and shrubs [6–8], culinary herbs and ornamentals of the

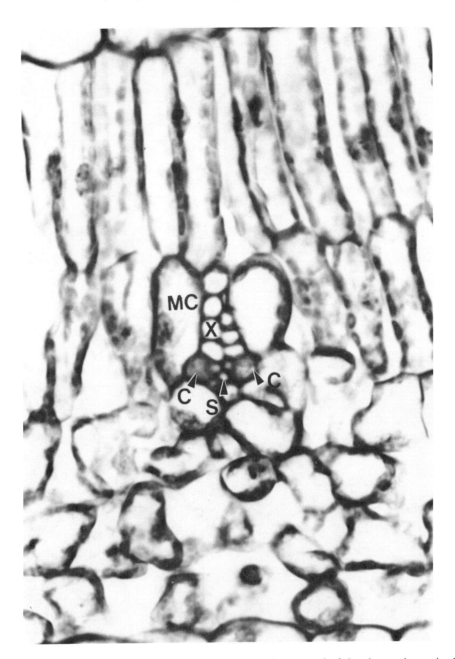

Figure 3 Cross section of a lilac (*Syringa vulgaris* L.) source leaf showing a minor vein. C, companion cell; S, sieve element; X, xylem; MC, mesophyll (photosynthetic) cell. The companion cell in this species is a type 1 (intermediary) cell. (Paraffin section slide courtesy of D. A. DeMason).

mint family such as coleus [13], and the cucurbit vine crops [14,15]. Crop species that have this companion cell type tend to be of tropical or subtropical origin [16].

Type 2a companion cells (Figure 4B) lack the extensive plasmodesmatal connections to the photosynthetic tissues that are typical of type 1 cells. Assimilates produced in leaves with this type of companion cell, therefore, do not have an elaborate symplastic pathway through

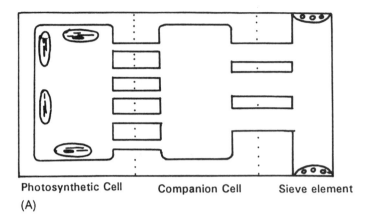

Photosynthetic Cell   Companion Cell   Sieve element

(A)

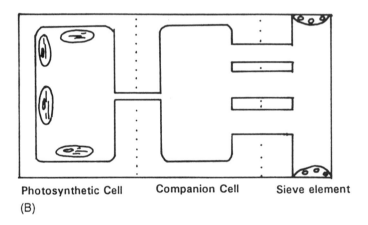

Photosynthetic Cell   Companion Cell   Sieve element

(B)

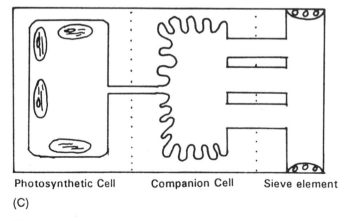

Photosynthetic Cell   Companion Cell   Sieve element

(C)

Figure 4 Diagrammatical representations of the classes of minor vein companion cells found in source leaves: (A) Type 1 companion cell (intermediary cell), (B) type 2a companion cell, and (C) type 2b companion cell (transfer cell).

which to travel into the SE-CC complex and must be released from the cytoplasm of the photosynthetic cells [6]. Assimilates are transferred into the cell wall space (the apoplast) of the SE-CC complex and are taken up across the plasma membrane of the SE-CC complex for export in the phloem [17,18]. Leaves with this type of minor vein configuration, therefore, use a membrane transport mechanism for phloem loading. Plant species with type 2a companion cells are almost exclusively herbaceous annuals of temperate origin [16]. Most of the major crop species of agronomic importance fall into type 2a.

Type 2b companion cells (Figure 4C) are similar to type 2a in that they lack extensive symplastic connections to the photosynthetic tissues [6–8]. However, the cell walls of type 2b companion cells are characterized by extensive wall ingrowths, which serve to greatly amplify the plasma membrane surface exposed to the apoplastic space. This type of companion cell is referred to as a "transfer cell" [18], for it has apparently been modified to facilitate the transfer of assimilates from the apoplast into the SE-CC complex. Type 2b companion cells are again typical of temperate herbaceous crops [16] and are a particularly common characteristic of legume species [19].

III. LONG-DISTANCE SOLUTE TRANSPORT

The vascular system of higher plants can be regarded as a series of parallel conduits of xylem and phloem tissue, which run the length of the plant body from root to shoot and permeate all major plant organs. Despite the co-occurence of xylem and phloem tissues in the vascular strands, these tissues are functionally quite distinct.

The main function of the xylem is the transport of water and dissolved mineral nutrients to the shoot following uptake from the soil by the roots. Xylem transport is unidirectional (upward from roots to shoots) and is driven by the water potential gradient created by evaporation of water from leaves (transpiration) [20]. Rates of transport can be of the order of 10 cm min^{-1} [20]. The transport pathway is formed by the cell walls of the xylem vessels and tracheids, which are dead at maturity [20]. Xylem transport therefore does not require cells with a living protoplast. Xylem sap consists primarily of water, with low levels of dissolved solutes (Table 1). Most noticeably, sugars are absent from xylem sap (Table 1).

Table 1 Typical Ranges for Components of Xylem and Phloem Saps in Higher Plants

Substance	Concentration (μg mL^{-1})	
	Xylem	Phloem
Sugars	Absent	140,000–210,000
Amino acids	200–1000	900–10,000
P	70–80	300–550
K	200–800	2800–4400
Ca	150–200	80–150
Mg	30–200	100–400
Mn	0.2–6.0	0.9–3.4
Zn	1.5–7.0	8–23
Cu	0.1–2.5	1.0–5.0
B	3.0–6.0	9–11
NO$_3^-$	1500–2000	Absent
NH$_4^+$	7–60	45–846

Sources: Data collated from Refs. 30 and 34.

In contrast, phloem transport occurs through living cells, the sieve elements [1]. The rates of transport are much lower than one sees in the xylem (in the order of 1 cm min^{-1}) and can occur in either an upward or downward direction [21]. Rate and direction of transport are dictated by differences in solute concentrations between sites of solute synthesis (sources) and solute consumption (sinks) within the plant body. It is these solute concentration differences that provide the driving force for phloem transport [22]. In contrast to xylem saps, phloem saps contain very high solute levels (Table 1), and particularly high levels of sugars, amino acids, and potassium.

A. The Munch Pressure Flow Mechanism

The pressure flow mechanism first postulated by Munch [22] provides the best explanation for the driving force for phloem transport presently available based on our knowledge of rates of transport and phloem structure. Phloem loading in source tissues leads to the very high solute concentrations characteristic of the phloem. The high solute levels create a water potential gradient within the sieve element, and water moves into the sieve element from the adjacent xylem tissues. Because the sieve element is a living cell and has a functional plasma membrane, this influx of water creates a very high hydrostatic pressure within the sieve element. At the sink end, the solutes are removed from the sieve element for use by the sink cells, and the hydrostatic pressure is reduced. This combination of solute loading at the source end and solute unloading at the sink end of the phloem system creates a strong hydrostatic pressure gradient. Because the sieve elements are linked end to end by open sieve plates, water containing the dissolved solutes passes through the pores of the sieve plates in response to the pressure gradient and solutes are moved by this bulk flow from source to sink.

B. Solutes Translocated in the Phloem

A mature, fully expanded leaf not only is the primary site of photosynthesis, it also has the highest rate of transpiration. As a result, a significant percentage of the dissolved mineral nutrients present in the xylem sap will end up in leaves, not in the agronomically important plant parts. The phloem of the minor veins of leaves is therefore very important, not only for the transport of photosynthate produced in the leaves but also for the redistribution of mineral elements delivered by the xylem. Additionally, phloem transport plays a major role in the transduction of developmental and environmental stimuli via the transport of growth regulators and systemic signal molecules.

1. *Carbohydrates*

Carbohydrates translocated in the phloem are all low molecular weight nonreducing sugars (Figure 5A) or sugar alcohols (Figure 5B). The disaccharide sucrose (Figure 5A) is ubiquitous in the phloem of crop plants. However, many important crop species transport sugars in addition to sucrose [23]. Plant species that possess type 1 companion cells (Figure 4a) all translocate the raffinose family oligosaccharides such as raffinose and stachyose [6–8,13–15,24], which are galactoside derivatives of sucrose (Figure 5A). Some members of the Rosaceae, including tree crops such as apples, cherries, plums, and apricots, also translocate significant quantities of the sugar alcohol sorbitol [23,25,26]. Members of the Apiaceae, such as celery, transport the sugar alcohol mannitol in addition to sucrose [25,27,28]. Still other plant species, such as olive [12] and euonymus [23], translocate both raffinose family sugars and a sugar alcohol (mannitol and dulcitol, respectively).

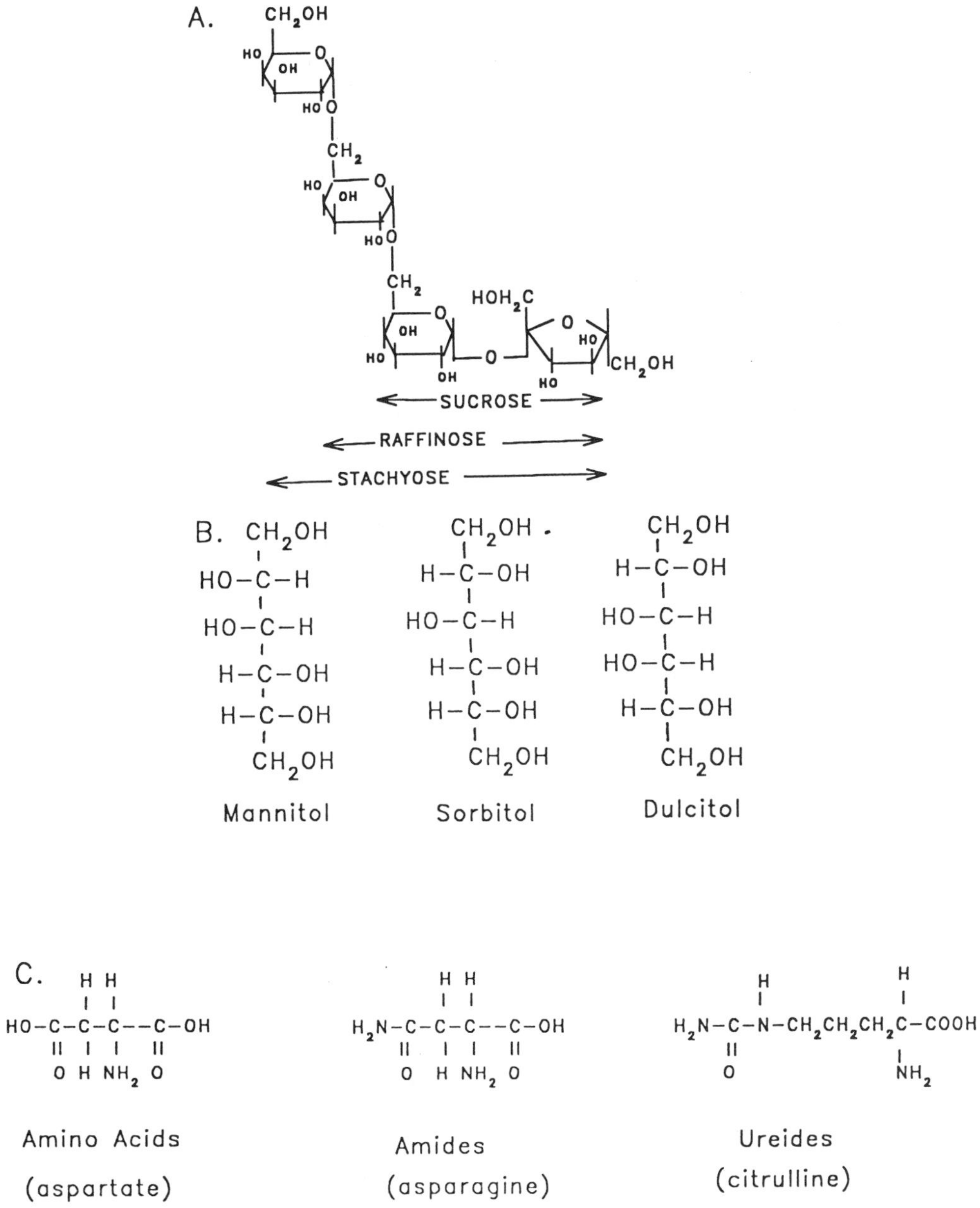
A.
SUCROSE
RAFFINOSE
STACHYOSE
B.
Mannitol
Sorbitol
Dulcitol
C.
Amino Acids
(aspartate)
Amides
(asparagine)
Ureides
(citrulline)

Figure 5 Chemical structures of (A) phloem-mobile sugars, (B) sugar alcohols, and (C) nitrogenous compounds found in phloem saps.

2. Nitrogen-Containing Compounds

Most protein amino acids are found in phloem saps (Table 2) [29–32]. The predominant amino acids tend to be those having a high ratio of nitrogen to carbon (Figure 5C), particularly the amides asparagine and glutamine [2]. The amino acid composition of the phloem sap can vary greatly depending on the species (Table 2) or environmental conditions [33]. In addition to amides, some species, particularly the nitrogen-fixing legumes, transport small quantities of

Table 2 Typical Range of Amino Acid
Composition of Phloem Sap

Amino acid	Phloem sap concentration (mM)
Aspartate	2–20
Glutamate	7–25
Asparagine	2–275
Glutamine	10–25
Serine	5–15
Glycine	Trace–6
Homoserine	0-trace
Citrulline	0–20
Histidine	0–trace
Arginine	Trace–5
Threonine	1–10
Alanine	1–8
Proline	5–15
Tyrosine	0.5–2.0
Valine	0–9
Methionine	0–trace
Cysteine	0–1
Isoleucine	2–6
Leucine	0–6
Phenylalanine	3–5
Tryptophan	0–trace
Ornithine	0–trace
Lysine	1–3

Sources: Data collated from Refs. 30–32.

the ureides allantoin and allantoic acid in the phloem [29]. Nonprotein amino acids, such as canavanine and ornithine, other ureides such as citrulline, and polyamines such as putrescine may also be found in limited quantities in phloem saps of many plant species [29].

3. Mineral Nutrients

Many of the same mineral ions found in xylem saps are also found in phloem saps (Table 2) [29,30,34,35], indicating that these nutrients can be removed from the xylem and loaded into the phloem transport system. Many sink tissues, being only poorly supplied with these nutrients by the xylem because of low transpiration rates, must depend on phloem transport for much of their mineral requirements [29,35].

In general, the relative mobility of mineral ions in the phloem can be determined by the site at which deficiency symptoms first appear. Some ions (e.g., boron and calcium) are only poorly loaded into the phloem [29,30,34,35]. In these cases, deficiency symptoms appear predominantly in sink tissues such as fruits and young leaves, which must depend on transpiration and xylem movement for a supply of these minerals [35]. In contrast, in the cases of minerals that are highly phloem-mobile (e.g., magnesium, potassium), deficiency symptoms appear first in the mature leaves [35]. This indicates a remobilization of minerals from the mature leaves and delivery of these elements to the sink leaves via phloem transport.

Mineral ions may be translocated as free elemental ionic forms (e.g., K^+; Cl^-;), but frequently may exist in other chemical forms (e.g., phosphate, sulfate, ammonium). Notably, although free nitrate is a common constituent of xylem saps, it is never found in phloem saps

(Table 2) [29,30,34]. Mineral elements may also be combined into organic complexes (e.g., ferric chelates, zinc peptides, phosphate esters, sulfur-containing amino acids, etc.) for transport in the phloem [29,35].

4. Growth Regulators

All classes of naturally occurring plant growth regulators (auxins, gibberellins, cytokinins, abscisic acid) can be recovered in phloem saps, indicating that these compounds are normally translocated in the phloem [29,36].

5. Systemic Signals

Grafting experiments using source leaves have indicated that other growth factors apart from the known growth regulators are also translocated in the phloem. These include floral initiation signals., cold-hardiness inducing signals, and pathogen resistance factors [29]. The chemical nature of these systemic signals is only beginning to be deciphered. Salicylic acid, which appears to be one prime candidate as a signaling molecule for these responses in some plant species, is thought to be phloem-mobile [37,38]. Additionally, a recently characterized phloem-mobile peptide, systemin, has been shown to induce pathogen resistance [39]. The plant growth regulator, abscisic acid, is a likely candidate as a phloem-mobile cold-hardiness inducing factor [40]. Phloem tissues may be capable of limited synthesis of phloem-specific proteins, whose function is unknown, but they may also be involved in signaling [41,42]. The recent localization of sucrose synthase within phloem tissues [43,44] may be involved in the signaling pathway that leads to callose synthesis [43] in response to wounding or pathgen invasion. It is likely that many more signaling mechanisms will be discovered in the phloem.

6. Xenobiotics

A number of man-made chemicals of agronomic importance, including many herbicides and pesticies [29], are also translocated in the phloem. The limitations to phloem mobility appear to be due mostly to failure of the applied chemical to cross cuticular barriers and plasma membranes: it is proposed that ionized groups or low lipophilicity prevent entry into the SE-CC complex [45]. One particularly good systemic herbicide is glyphosate (N-phosphonomethylglyc-ine), which is highly mobile in the phloem (Table 3).

IV. PHLOEM LOADING

As indicated by structural differences, there appear to be two pathways by which assimilates can be transferred from the photosynthetic cells to the minor vein SE-CC complexes in the source leaf. In species possessing the type 1 minor vein configuration, this transport can occur by a symplastic route through the numerous plasmodesmata that interconnect the photosynthetic cells and the phloem transport system. In plants with type 2 configurations, which lack a high degree of symplastic interconnection, transport can occur via a transmembrane route via the apoplast.

A. Apoplastic Phloem Loading

The textbook model of phloem loading in source leaves consists of a sequence of events starting with cell-to-cell transport of assimilates, primarily sucrose, through mesophyll cell plasmodes-mata to a site close to the SE-CC complex. At this point, sucrose is unloaded into the apoplast, where it actively accumulated into the SE-CC complex by a proton–sucrose symport mechanism (Figure 6A). The apoplastic proton symport model of phloem loading [17,18] affords a very

Table 3 Symplastically (phloem) Transported Herbicides

Herbicide class	Typical representative	Chemical structure
Phenoxy herbicides	2,4-D	2,4-Dichlorophenoxyacetic acid
	2,4,5-T	2,4,5-Trichlorophenoxyacetic acid
Benzoic acids	Dicamba[a]	3,6-Dichloro-2-methoxybenzoic acid
	2,3,6-TBA	2,3,6-Trichlorobenzoic acid
Picolinic acids	Picloram[a]	4-Amino-3,5,6-trichloropicolinic acid
	Triclopyr	[(3,5,6-Trichloro-2-pyridyl)oxy]acetic acid
Chlorinated aliphatics	Dalapon	2,2-Dichloropropionic acid
Triazoles	Amitrole[a]	3-Amino-*s*-triazole
Organic arsenicals	DSMA	Disodium methanoarsenate
Glyphosate		N-Phosphonomethylglycine
Sulfonylureas	Chlorsulfuron	2-Chloro-*N*-[(4-methoxy-6-methyl-1,3,5-triazin-2-yl)aminocar-bonyl]benzenesulfonamide

[a]Also transported apoplastically (xylem).

satisfactory mechanism for establishing the high concentration gradient required within the phloem to drive phloem transport by Munch pressure flow.

One of the key demonstrations of the apoplastic loading pathway is the inhibition of sucrose–proton cotransport by inhibitors such as *p*-chloromercuriphenylsulfonic acid (PCMBS). In isolated leaf plasma membrane systems [46,47] and leaf tissues [48], this compound has been shown to bind to the sucrose carrier [49] and to prevent transfer of sucrose. In many plants, this compound will also inhibit the delivery of photosynthetically produced sucrose from the photosynthetic cells to the minor veins of leaf tissues [48]. This finding lends significant support to the apoplastic loading theory.

Interestingly, though, only species possessing the type 2 minor vein configurations show this sensitivity to PCMBS [50]. In these species, which include most of the important agronomic crops, sucrose is the only sugar transported in the phloem. Therefore, apoplastic phloem loading probably best explains delivery of sucrose to the phloem in most agronomic species. However, although there is appreciable experimental evidence in support of the apoplastic pathway, it is now becoming apparent that this model may not hold for all crop plants.

B. Symplastic Phloem Loading

In plants with the type 1 minor vein configuration, where a symplastic route through plasmodesmata is available for the delivery of endogenously produced photoassimilates to the minor veins, phloem loading is not affected by PCMBS [12,50,51]. This is true despite the obvious sensitivity to inhibition by PCMBS of the uptake of exogenously supplied sugars into leaf tissues [12,52]. This observation suggests that although a PCMBS-sensitive proton symport mechanism may exist in type 1 plants, it is not utilized for loading of assimilates into the phloem.

If an apoplastic step is not involved in phloem loading in type 1 plants, some alternative mechanism must be invoked to create the high sieve element solute levels necessary for phloem transport. Plasmolysis studies clearly show that high solute levels do exist in the phloem of type 1 plants [53]. The answer to this dilemma may come from the observations that all type 1 plants export raffinose family oligosaccharides such as stachyose in the phloem [6–8] and

A.

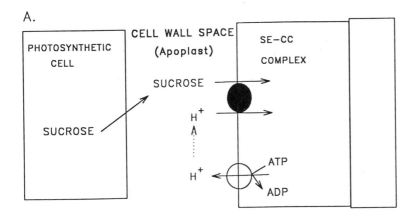

B.

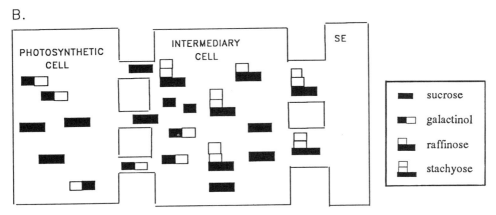

Figure 6 Diagrammatic representations of the processes involved in loading of the phloem via the apoplast or symplast. (A) In apoplastic loading, sucrose leaves the photosynthetic cell and enters the cell wall space. It is then taken up across the plasma membrane of the sieve element–companion cell (SE-CC) complex by a sucrose carrier (solid circle), which transports sucrose in conjunction with a proton (H^+). The proton gradient is established by proton extrusion via a plasma membrane ATPase (open circle).

(B) In symplastic loading, disaccharides (sucrose, galactinol) are passed through the plasmodesmata from the photosynthetic cells to the intermediary cells, where the galactose residues (□) of galactinol are transferred to sucrose to form the tri- (raffinose) and tetrasaccharide (stachyose) with the release of myo-inositol (■). The tri- and tetrasacchrides then pass into the sieve element (SE) but are prevented from passing back into the photosynthetic cell by the smaller diameter of the plasmodesmata connecting this cell to the intermediary cell.

that synthesis of the oligosaccharides destined for export most likely occurs within the intermediary cells characteristic of the type 1 morphology [54–57].

The biochemistry of raffinose sugar biosynthesis is somewhat anomalous compared to that of other sugars in that the galactose donor is not a sugar nucleotide but a simple disaccharide, galactinol [58]. (For more details of this biochemistry, readers are referred to Chapter 12 of this volume). A recently proposed hypothetical model reconciles both the odd ultrastructural features (i.e., the ubiquitous presence of the symplastic links between the photosynthetic cells and the minor vein intermediary cells) and the rather unusual carbohydrate biochemistry of type 1 plants [59].

The crux of this model (Figure 6B) is the hypothesis that the pore size of the plasmodesmata connecting the intermediary cells with the photosynthetic cells is wide enough to allow passage of disaccharides such as sucrose and galactinol only, not their oligosaccharide products, the tri- and tetrasaccharides raffinose and stachyose (Figure 6B). Thus, when stachyose and raffinose are synthesized within the intermediary cells, they cannot move anywhere, except into the adjacent sieve tubes. This "polymerization trap" model remains to be proven [60], and the details of compartmentation of the stachyose reactions within the intermediary cell need to be elucidated, but the model does give a feasible explanation of how symplastically linked cells might operate in establishing a solute gradient.

C.   Loading of Other Solutes

Because sugars are the predominant solutes translocated in the phloem, most of what is known about phloem loading concerns the movement of sugars into the phloem. Relatively little is known of either the pathways taken or the mechanisms used to load the other component solutes characteristically found in phloem saps. There is evidence for the operation of proton–amino acid transporters in plant tissues [47], but whether these are phloem tissue specific is not known. It is likely that active accumulation of the potassium ion takes place in exchange for protons, but the carrier(s) involved have not yet been characterized. How other ions enter the phloem is not yet clear. It is quite likely that further studies will reveal that the phloem sap composition in crop plants is determined by means of a combination of apoplastic and symplastic transport.

V.   REGULATION OF PHLOEM TRANSPORT

Despite the great numbers of different sink tissues and organs that comprise a typical plant, most plants tend to maintain a balanced ratio of shoot tissue to root tissue. This indicates that the plant has some means of regulating the amount of photoassimilate that is delivered to developing roots and shoots and that some metabolic control exists to control the direction of phloem transport. Since rate and direction of phloem transport are dictated by the solute gradient between sources and sinks, the regulatory mechanisms can exist either at the source end (where assimilates are loaded) or at the sink end (where assimilates are removed from the phloem).

A.   Regulation by Sources

Source leaves are the primary site of photoassimilate production, but the plant faces a dilemma with respect to allocation choices for photoassimilates. Since the photosynthetic period does not encompass the entire diurnal period but the demand for assimilates does, source tissues must conserve part of the carbon fixed during photosynthesis for use during nonphotosynthetic periods. The role of the source leaf in controlling phloem transport is therefore one of allocation, assigning fixed carbon to export or storage pools in such a manner that export can be maintained at some "set point" level throughout the diurnal period [61,62]. Additionally, the source must be able to accumulate enough reserve carbon to allow for environmental conditions (clouds, water stress, temperature fluctuations) that may interfere with photosynthetic processes even in the light [63]. How the plant decides which carbon is destined for export and which is to be stored is not fully understood.

Many tightly regulated metabolic steps control the accessibility of photosynthetically fixed carbon to the phloem transport system. Control at the source end is governed largely by rates of photosynthetic incorporation of CO_2, but for photosynthetic rate to have any direct effect on the rate of phloem transport, the carbon must be fixed into phloem-mobile intermediates (predominantly sucrose in most agronomically important crops). The flow of carbon into soluble sugars, which are synthesized in the cytoplasm of the photosynthetic cell, is regulated by complex biochemical interactions, which direct the export of fixed carbon out of the chloroplast [64]. Carbon not released from the chloroplast is stored as insoluble starch and may not be immediately available for phloem transport [65]. In addition, once synthesized, soluble sugars can be siphoned off into the vacuole for storage, and this carbon, also, would not be available for phloem transport [61].

The phloem loading process, which establishes the high solute level in the phloem, must therefore compete with storage processes also occurring in the chloroplast and vacuole, which can divert substantial amounts of photosynthetically fixed carbon from the phloem loading site. Control of phloem transport by source tissues is therefore exerted largely by control of the availability of phloem-mobile solutes and not directly by the rate of photosynthesis per se. Indeed, environmental factors that reduce rates of photosynthesis do not necessarily result directly in lowered rates of phloem transport. This is because stored carbon, either in the source tissues or in storage tissues along the pathway, can be mobilized to maintain the high solute levels in the phloem [61].

One key contribution of source leaf metabolism, arising from the combination of photosynthetic activity and membrane transport between cellular compartments, is therefore the control of amounts and probably the types (sugars or amino acids) of assimilates that have access to the loading sites. Source leaf metabolism, therefore, directly regulates the overall composition of the phloem sap [33], including both organic and inorganic constituents. In general, though, the "set point" for rates of phloem transport is established in the sink tissues, where these nutrients are utilized.

B. Regulation by Sinks

A typical higher plant has a myriad of sink tissues that depend on the source leaves for photoassimilates. Reproductive sinks (flowers, seeds, fruits) are of prime agronomic importance, and as a result most studies of sink regulation of phloem transport have tended to focus on carbon partitioning to these sinks. However, reproductive sinks represent only a small proportion of potential sinks on a plant, and we are now beginning to realize that during the growth period, carbon partitioning to other sinks, particularly temporary vegetative sinks, can be important in determining final crop yield.

1. Vegetative ("Buffering") Sinks
During the translocation process, carbon is continuously diverted from the phloem to surrounding parenchyma cells for temporary storage. Parenchyma tissues of leaves, petioles, stems, and roots can all act as sinks for assimilates, which usually are stored in the form of starch. These stored reserves can be drawn on and reloaded into the phloem under conditions of reduced photosynthesis [61–63,65] (e.g., during adverse environmental conditions) or when sink demand increases (e.g., during the reproductive phase of plant growth) [66]. An amplified version of this type of sink activity is seen in perenniating organs such as tubers and taproots and also in ray cells of woody species, in which large amounts of carbon are diverted to storage to allow for regrowth of vegetative tissues in the next growing season. The phenomenon of alternate bearing in perennial tree crops may also reflect this type of sink activity: that is, carbon diverted to vegetative storage sinks in nonbearing years may be utilized for crop production in the subsequent bearing year.

The vegetative "buffering" sinks, therefore, have the unique property of being able to act both as sinks for assimilates and as sources of assimilates for phloem transport, depending on the carbon needs of the plant at a particular growth phase or under the prevailing environmental conditions. Sinks of these types can, therefore, regulate phloem transport by coarse control of the assimilates available to the sieve elements along the phloem transport path.

Like minor vein phloem loading, sieve tube unloading into sinks can occur by either symplastic or apoplastic routes (Figute 7B). Unloading into "buffering" sinks such as the sugar beet taproot [67] and sugarcane stem [68] occurs via an apoplastic route. Unloading of sugar into the apoplast from the sieve tubes lowers the hydrostatic pressure and also promotes the flow of water out of the sieve element, thus allowing bulk flow to occur from source to sink. Unloaded solutes are then taken up into the sink cell, where compartmentation into the vacuole or conversion to insoluble starch (Figure 7A) can further dissipate the hydrostatic pressure between the phloem and the sink organ. In some cases, sucrose is hydrolyzed prior to uptake into the sink cell, and in other cases it may be taken up intact, then hydrolyzed in the vacuole (Figure 7A).

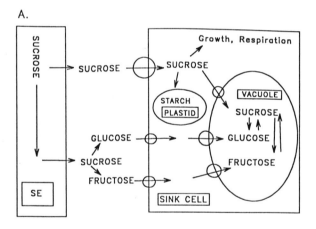

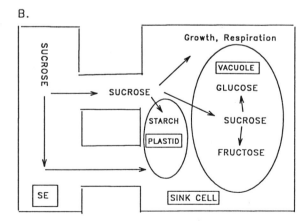

Figure 7   Pathways for phloem unloading in sink tissues. (A) apoplastic and (B) symplastic.

2. Terminal Sinks

Unlike the buffering sinks, terminal sinks, as the name implies, act as sinks only for assimilates. Carbon partitioned to terminal sinks is unavailable for remobilization out of those sinks, usually because it is incorporated into structural, as opposed to storage, components. Prime examples of terminal sinks are reproductive tissues, such as fruits and seeds, and rapidly growing meristems. Carbon partitioned to these sinks cannot be reaccessed by the plant, even if the carbon is stored in a conventional storage form such as starch. Sinks of these types, therefore, can exert a strong regulatory influence on phloem transport by controlling the low end of the hydrostatic pressure gradient created in the sieve tubes.

Phloem unloading in terminal sinks occurs either by apoplastic or symplastic routes (Figure 7). In rapidly growing meristematic organs such as developing roots [69] and leaves [70,71], unloading occurs via the symplast. The conversion of imported assimilates to insoluble structural components, principally cellulose, and other polymers (protein, nucleic acid, etc.) and their rapid utilization as respiratory substrates reduce the hydrostatic pressure in the phloem and allow continued phloem transport by bulk flow. Additionally, in expanding cells, some of the water for expansion may come from the phloem, allowing further dissipation of the phloem turgor pressure.

VI. FUTURE PERSPECTIVES

Our understanding of the physiology of carbon allocation, partitioning, and phloem transport in crop plants is still evolving. In recent years, the advent of molecular biological techniques has greatly facilitated our ability to answer once-difficult phloem transport questions. Already, these new techniques are rapidly advancing our understanding of phloem loading processes. Convincing evidence for the operation of the apoplastic pathway in members of the Solanaceae has developed from the use of transgenic plants in which apoplastic invertase has been overexpressed. In tobacco [72] and tomato [73], expression of apoplastic invertase results in the hydrolysis of apoplastic sucrose destined for phloem loading. Since glucose and fructose, the resulting reducing sugars, cannot be loaded into the phloem, plant growth and development are adversely affected. Similar experiments performed on plants with type 1 morphology would be instrumental in determining the role, if any, of apoplastic transport in these plants.

Recently, the genes coding for a sucrose–proton transporter from spinach leaves [49] and for a sucrose-binding protein from soybean tissues [74] have been isolated and characterized. The availability of these genes will greatly facilitate localization of the gene products within source leaves with either type 1 or type 2 companion cells and should allow confirmation of the role of sucrose–proton cotransport in phloem loading in different plant species. The techniques that have allowed isolation of sucrose carriers should also allow rapid characterization of other carriers that may be operating to load other solutes, such as K^+ and amino acids, into the SE-CC complex.

Similar experiments should also yield valuable information concerning phloem unloading in sink tissues. The recent demonstration of regulation of gene expression by sucrose in maize roots [75] clearly indicates the importance of molecular approaches in the study of sink tissue metabolism.

In short, the ability to genetically modify plants in highly specific ways through molecular approaches should continue to revolutionize the study of crop physiology. In combination with conventional physiology studies, these techniques should allow significant progress to be made in our understanding of assimilate transport processes in crop plants, and indeed there is still much to be learned.

REFERENCES

1. K., Esau, *Anatomy of Seed Plants*, Wiley, New York, pp. 157–182 (1977).
2. M. V. Parthasarathy, *Transport in Plants, Vol. I, Phloem Transport* (M. H. Zimmermann and J. A. Milburn, eds.), Springer-Verlag, Berlin, pp. 3–38 (1975).
3. R. F. Evert, *BioScience, 32*: 789 (1982).
4. H. D. Behnke, *Transport of Photoassimilates* (D. A. Baker and J. A. Milburn, eds.), Loughman Scientific and Technical, Harlow, Essex, pp. 79–137 (1989).
5. K. Esau, *Anatomy of Seed Plants*, Wiley, New York, pp. 321–332 (1977).
6. Y. V. Gamalei, *Trees, 5*: 50 (1991).
7. A. J. E. van Bel and Y. V. Gamalei, *Plant Cell Environ., 15*: 265 (1992).
8. A. J. E. van Bel, *Annu. Rev. Plant Physiol. Plant Mol. Biol., 44*: 253 (1993).
9. Y. V. Gamalei, *Trees, 3*: 96 (1989).
10. D. G. Fisher, *Planta, 169*: 141 (1986).
11. R. Turgeon, D. U. Beebe, and E. Gowan, *Planta*, in press (1993).
12. L. L. Flora and M. A. Madore, *Planta, 189*: 484 (1993).
13. D. G. Fisher, *Plant Cell Environ., 11*: 639 (1988).
14. R. Turgeon, J. A. Webb, and R. F. Evert, *Protoplasma, 83*: 217 (1975).
15. K. Schmitz, B. Cuypers, and M. Moll, *Planta, 171*: 19 (1987).
16. A. J. E. van Bel, *Acta Bot. Neerl., 41*: 121 (1992).
17. R. T. Giaquinta, *Annu. Rev. Plant Physiol., 34*: 347 (1983).
18. S. Delrot, *Transport of Photoassimilates* (D. A. Baker and J. A. Milburn, eds.), Loughman Scientific and Technical, Harlow, Essex, pp. 167–205 (1989).
19. B. E. S. Gunning, and J. S. Pate, *Dynamic Aspects of Plant Ultrastructure* (A. W. Robards, ed.), McGraw-Hill, New York, pp. 441–480 (1974).
20. M. H. Zimmermann, *Xylem Structure and the Ascent of Sap*, Springer-Verlag, Berlin, pp. 4–62 (1983).
21. T. G. Mason and C. J. Lewin, *Sci. Proc. R. Dublin Soc., 18*: 203 (1926).
22. E. Munch, *Die Stoffbewugungen in der Pflanze*, Gustav Fischer, Jena (1930).
23. M. H. Zimmermann and H. Ziegler, *Transport in Plants, Vol. I, Phloem Transport* (M. H. Zimmermann and J. A. Milburn, eds.), Springer-Verlag, Berlin, pp. 480–503 (1975).
24. M. A. Madore, *Plant Physiol., 93*: 617 (1990).
25. D. H. Lewis and D. C. Smith, *New Phytol., 66*: 143 (1967).
26. R. J. Redgwell and R. L. Bieleski, *Phytochemistry, 17*: 407 (1978).
27. M. E. Rumpho, G. E. Edwards, and W. H. Loescher, *Plant Physiol., 73*: 869 (1983).
28. W. H. Loescher, R. H. Tyson, J. D. Everard, R. J. Redgwell, and R. L. Bieleski, *Plant Physiol., 98*: 1396 (1992).
29. H. Ziegler, *Transport in Plants, Vol. I, Phloem Transport* (M. H. Zimmermann and J. A. Milburn, eds.), Springer-Verlag, Berlin, pp. 59–100 (1975).
30. P. J. Hocking, *Ann. Bot., 45*: 633 (1980).
31. C. Girousse, J. L. Bonnemain, S. Delrot, and R. Bournoville, *Plant Physiol. Biochem., 29*: 41 (1991).
32. D. E. Mitchell, M. V. Gadus, and M. A. Madore, *Plant Physiol., 99*: 959 (1992).
33. D. E. Mitchell and M. A. Madore, *Plant Physiol., 99*: 966 (1992).
34. B. J. Shelp, *Ann. Bot., 61*: 83 (1988).
35. K. Mengel and E. A. Kirkby, *Principles of Plant Nutrition*, International Potash Institute, Bern, Switzerland, pp. 207–219 (1982).
36. E. Komor, I. Liegl, and C. Schobert, *Planta, 191*: 252 (1993).
37. I. Raskin, *Annu. Rev. Plant Physiol. Plant Mol. Biol., 43*: 439 (1992).
38. J. Malemy and D. F. Klessig, *Plant J., 2*: 643 (1992).
39. G. Pearce, D. Strydom, S. Johnson, and C. A. Ryan, *Science, 253*: 895 (1991).
40. T. L. Setter, W. A. Brun, and M. L. Brenner, *Plant Physiol., 67*: 774 (1981).
41. D. B. Fisher, Y. Wu, and M. S. B. Ku, *Plant Physiol., 100*: 1433 (1992).

42. T. Sakuth, C. Schobert, A. Pecsvaradi, A. Eichholz, E. Komor, and G. Orlich, *Planta*, *191*: 207 (1993).
43. K. D. Nolte and K. E. Koch, *Plant Phyiol.*, *101*: 899 (1993).
44. T. Martin, W. Frommer, M. Salanoubat, and L. Willmitzer, *Plant J.*, *4*: 367 (1993).
45. M. A. Ross and C. A. Lmbi, *Applied Weed Science*, Burgess, Minneapolis, MN, pp. 157–198 (1985).
46. D. R. Bush, *Photosynth. Res.*, *32*: 155 (1992).
47. D. R. Bush, *Annu. Rev. Plant Physiol. Plant Mol. Biol.*, *44*: 513 (1993).
48. S. Bourquin, J. L. Bonnemain, and S. Delrot, *Plant Physiol.*, *92*: 97 (1990).
49. J. W. Riesmeier, L. Willmitzer, and W. B. Frommer, *EMBO J.*, *11*: 4705 (1992).
50. A. J. E. van Bel, Y. V. Gamalei, A. Ammerlaan, and L. P. M. Bik, *Planta*, *186*: 518 (1992).
51. L. A. Weisberg, L. E. Wimmers, and R. Turgeon, *Planta*, *175*: 1 (1988).
52. R. Turgeon and E. Gowan, *Plant Physiol.*, *94*: 1244 (1990).
53. R. Turgeon and P. K. Hepler, *Planta*, *179*: 24 (1989).
54. M. A. Madore, *Planta*, *187*: 537 (1992).
55. K. Schmitz and U. Holthaus, *Planta*, *169*: 529 (1986).
56. U. Holthaus and K. Schmitz, *Planta*, *185*: 479 (1991).
57. D. U. Beebe and R. Turgeon, *Planta*, *188*: 354 (1992).
58. O. Kandler, *Harvesting the Sun* (A. San Pietro, F. A. Greer, and T. J. Army, eds.), Academic Press, New York, pp. 131–152 (1967).
59. R. Turgeon, *Phloem Transport and Assimilate Compartmentation* (J. L. Bonnemain, S. Delrot, W. J. Lucas, and J. Dainty, eds.), Ouest Editions Presses Academique, Nantes, France, pp. 18–22 (1991).
60. R. Turgeon and D. U. Beebe, *Plant Physiol.*, *96*: 349 (1991).
61. D. R. Geiger and B. R. Fondy, *Phloem Transport and Assimilate Compartmentation* (J. L. Bonnemain, S. Delrot, W.J. Lucas, and J. Dainty, eds.), Ouest Editions Presses Academique, Nantes, France, pp. 1–9 (1991).
62. D. L. Hendrix and R. I. Grange, *Plant Physiol.*, *95*: 228 (1991).
63. D. R. Geiger and J. C. Servaites, *Response of Plants to Multiple Stresses* (H. A. Mooney, W. E. Winner, and E. J. Pell, eds.), Academic Press, New York, pp. 103–127 (1991).
64. M. Stitt, *Plant Physiol.*, *84*: 201 (1987).
65. J. C. Servaites, B. R. Fondy, B. Li, and D. R. Geiger, *Plant Physiol.*, *90*: 1168 (1989).
66. V. R. Franceschi, *Plant Biology, Vol. 1, Phloem Transport* (J. Cronshaw, W. J. Lucas, and R. T. Giaquinta, eds.), Liss, New York, pp. 399–409 (1986).
67. R. E. Wyse, *Plant Physiol.*, *63*: 828 (1978).
68. K. T. Glaziou and K. R. Gayler, *Plant Physiol.*, *49*: 912 (1972).
69. R. T. Giaquinta, W. Lin, N. L. Sadler, and V. R. Franceschi, *Plant Physiol.*, *72*: 362 (1983).
70. J. Gougler-Schmalstig and D. R. Geiger, *Plant Physiol.*, *79*: 237 (1985).
71. B. Ding, M. V. Parthasarathy, K. Niklas, and R. Turgeon, *Planta*, *176*: 307 (1988).
72. A. von Schaewen, M. Stitt, R. Schmidt, U. Sonnewald, and L. Willmitzer, *EMBO J.*, *9*: 3033 (1990).
73. C. D. Dickinson, T. Altabella, and M. J. Crispeels, *Plant Physiol.*, *95*: 420 (1991).
74. H. S. Grimes, P. J. Overvoorde, K. Ripp, V. R. Franceschi, and W. D. Hitz, *Plant Cell*, *4*: 1561 (1992).
75. K. E. Koch, K. D. Nolte, E. R. Duke, D. R. McCarty, and W. T. Avigne, *Plant Cell*, *4*: 59 (1992).

17
Assimilate Transport and Partitioning

John E. Hendrix

Colorado State University, Fort Collins, Colorado

I. INTRODUCTION

Typically, only part of a plant—roots, stems, leaves, fruit, or seeds—is harvested for its economic value. Therefore, it is important to maximize the proportion of the total assimilate pool that is partitioned into the plant part(s) having economic value. The prime assimilatory process is photosynthesis. The assimilation products must be transported to the growing plant part (sink) if it is to grow and attain economic value. This is true even if the economically important part of the plant is leaves, for young leaves are sinks for photosynthate until they are about half-expanded. Therefore, any procedure used to enhance the partitioning of assimilate into these sinks of economic value will serve to increase the economic gain. The processes involved are the production of assimilates, the loading of the assimilates into the phloem, their transport through the phloem (translocation), and their unloading from the phloem into the appropriate sinks.

If phloem function and assimilate partitioning are to be related to crop productivity, several questions must be addressed:

How much material is translocated and at what velocity?
What is the mechanism of translocation through the sieve tubes?
What materials are translocated, and why these and not others?
How are materials loaded into and unloaded from the sieve tubes?
How are materials partitioned among plant parts?
How are these processes controlled?
What is the mechanism of xenobiotic transport?

Then we must learn which components of genetics and environment influence partitioning.

These topics are not fully understood, but they have been studied extensively and discussed in several reviews and symposia publications in recent years (see, e.g., Refs. 1–9). The effort here is directed toward relating these topics to crop productivity.

II. CROP PRODUCTIVITY

The total biomass production of a plant is dependent on the balance between photosynthesis and respiration. Therefore, it might seem appropriate to develop genetic or cultural strategies to control these two processes. Cultural practices have been directed primarily toward increasing total biomass by increasing plant density and affording increased use of fertilizer and/or water, with the assumption that such an effort would result in greater economic productivity. Yet it has long been known that excessive use of nitrogen or excessively high plant densities often lowers production of economically important plant parts.

Environment strongly influences the development of economically important plant parts at certain critical stages in the life of a plant. Therefore, cultural practices have the potential of influencing harvest index (HI). The understanding of the timing of development gains in importance as certain inputs such as water, pesticides, and fertilizer become more expensive, difficult to obtain, or use-restricted through legislation and other regulations.

Genetic selection of crop plants over the past century, for the most part, has not resulted in an increase in biomass; rather, it has changed the proportion of that biomass partitioned into economically important plant parts. Altering the partitioning of biomass (i.e., changing HI) has been the pathway to increased economic production [10,11]. Plant breeding through selection or development of hybrids has been successful in changing HI.

III. PEST MANAGEMENT

With a few important exceptions, crop breeding programs have been devoted to improving crop production under the umbrella of chemical pest control. With growing concern about and control of pesticide use and increased resistance to pesticides that is developing among pests, one must be concerned about the crop varieties now in use. Even so, Dunan and Zimdahl [12] demonstrated that Moravian III barley was more competitive than wild oats in both greenhouse and field studies. However, Gressel [13] pointed out, "The use of herbicides . . . allowed breeders to develop less competitive crops with higher harvest indices." One must wonder whether the highly productive semidwarf grain cultivars are as competitive with weeds as are the taller cultivars.

Recently the emphasis of crop development has changed. Genetic research on the production of phytoalexins, materials toxic to pests, is being incorporated into the design of crop plants. However, the technique may not always be successful, for these materials can be toxic to the plant itself or to humans or farm animals [14]. Therefore we will likely need to continue to use chemical pest control, but the timing of application will be much more important, since it will be critical to obtain the most economic benefit with the smallest amount of material, the lowest possible rate of resistance development by the pests, and the lowest possible environmental impact. Timing of pesticide application has emphasized the vulnerable part of the life cycle of the pest. Timing of crop vulnerability to pests must also be considered, for the economically important plant parts are not equally vulnerable at all portions of the plant's life cycle. In addition, if we learn how pesticides are transported through plants, how to target plant parts, and how to use appropriate timing of application, we will be able to lower total pesticide use by avoiding many prophylactic applications.

IV. MECHANISM OF LONG-DISTANCE TRANSPORT

As early as 1900, one of the predominant hypotheses of assimilate translocation was pressure flow through the sieve tubes [1]. In 1927 Munch [15] proposed an osmotic model for the generation of pressure in this translocation system. This came to be known as the "Munch pressure flow hypothesis." Even so, textbooks published in the early 1930s [16,17] made no mention of a pressure flow mechanism.

Over the subsequent decades, several different mechanisms of long-distance transport through the phloem were proposed [18–21]. Arguments were put forth that the microanatomy of the sieve tubes could not support a flow mechanism because the holes in the sieve plates are blocked with "slime plugs" and, even if they are not blocked, sieve plates would create far too much resistance for flow to occur at observed rates. During the 1950s and 1960s, substantial amounts of physiological data were interpreted as refuting a flow mechanism. When plants were supplied with $^{14}CO_2$, the profile of ^{14}C in the transport system decreased from source to sink in a pattern that would be expected for diffusion, though the rates and velocities of movement were too great to be accounted for by diffusion. Furthermore, when plants were double-labeled with [^{14}C]sugar, ^3H_2O, and/or ^{32}P, it was found that no two nuclides moved together, as had been predicted for a flow system. Horwitz [22] and Biddulph [23] attempted to explain these data with a flow model representing differential exchange between the sieve tubes and adjacent, nontransporting cells. Nonetheless, several alternative proposals were put forth during this period. Included was a proposal for electro-osmotic pumping by the removal of K^+ from the downstream side of the sieve plates and cycling it to the upstream side [24–26]. The movement of K^+ would require metabolic energy, and the resultant potential gradient across the sieve plate would drive K^+ through small pores in the "slime plugs," causing a solution flow. Thaine and coworkers [27,28] reported observing "transcellular strands" of cytoplasm moving in both directions in the sieve tubes. Canny [29] developed a model of assimilate transport based on this system that would result in bidirectional movement within one sieve tube. Trip and Gorham [30] supplied different leaves of squash plants with $^{14}CO_2$ and ^3H [sugars] and found both nuclides in the same sieve tube. They interpreted their data as supporting bidirectional movement by transcellular strands.

Both the electro-osmotic and the transcellular strand models would require metabolic energy along the translocation pathway. For many plants, such as bean and squash, cooling of the translocation stream, but not the source or sink, inhibits translocation [31,32]. However, Swanson and Geiger [33] demonstrated that chilled (1°C) petioles of sugar beets translocated sugars at the nonchilled rate after only a few minutes of acclimation. In addition, Sij and Swanson (34, and personal observations) demonstrated that squash, a chilling-sensitive plant, translocated carbon through petioles exposed to an N_2 atmosphere even though such an environment eventually caused tissue death. Furthermore, Peterson and Currier [35], using a fluorescent dye, demonstrated that bidirectional movement within one sieve tube was unlikely.

Concurrently, data were accumulating regarding the concentration of solutes and the pressure in sieve tubes [36–38]. Many of the proposed mechanisms could be eliminated by considering the specific mass transfer [g dry wt. cm^{-2} (phloem) h^{-1}] and velocity of translocation (cm h^{-1}). Crafts and Crisp [1] compiled values for specific mass transport per unit cross section area of phloem that varied from 0.14 to 4.8 (average 3.6) g dry wt. cm^{-2} h^{-1}. The phloem is composed of many cells in addition to functional sieve tubes; therefore, the rates and velocities through sieve tubes must be two to four times those computed values. Assuming that the total solute in the sieve tubes is 18% w/v [39], likely a low estimate, using the minimum and maximum (and average) specific mass transfer rates, and assuming that the area of the sieve tubes is 0.5–0.25 the total phloem area [40], velocities of 1.6–106 cm h^{-1} (average 40–80) are obtained. Similar velocities have been obtained using labeled materials. At the average velocity, this would result in material moving from one end of a sieve tube element to the other in 2 seconds (sieve tube element of 0.03 cm measured from micrograph in Ref. 41).

Finally Cataldo et al. [42,43] explained the differential movement of [^{14}C]sucrose and ^3H_2O

by differential lateral exchange. Then Christy and Ferrier [44] presented a mathematical model of phloem translocation by flow that is in agreement with empirical observation. These efforts confirmed, in principle, the model developed by Horwitz [22] and elevated the pressure flow "hypothesis" to "theory" status.

Briefly, as understood today, assimilates are loaded in the sieve tube–companion cell (ST-CC) complex against a free energy gradient using metabolic energy. This causes the sieve tubes, sometimes called sieve elements, to have an osmotic potential more negative than other cells (except companion cells) in the source. Water follows osmotically, causing pressure to develop. In the sink, assimilates are unloaded and water follows. These processes in the source and sink generate a pressure drop from source to sink, and a flow results.

V. WHAT IS TRANSLOCATED AND WHY

A. Sugars and Sugar Alcohols

Sugars and sugar alcohols carry 60–95% of the translocated carbon and have a total concentration ranging up to 180 mg mL^{-1} [39]. Sucrose is the primary translocated carbohydrate in most plants [45–47]. In many plants, sucrose represents essentially all the carbohydrate that is translocated. Other nonreducing oligosaccharides, such as raffinose, stachyose [45,48–51], gentianose, and umbelliferose [46], are translocated.

Sugar alcohols, along with sucrose, are translocated in a few groups of plants. Specifically, sorbitol is translocated by several members of the Rosaceae and Oleaceae [45,52], while mannitol is translocated by celery [53]. Even the giant alga *Macrocystis* [54] and *Fucus* [55], translocate mannitol through "sieve cells" [56].

Occasionally, there are reports of glucose and fructose being isolated from phloem exudate [39]. However, these may be artifacts that resulted from hydrolysis of sucrose by enzymes released from damaged cells or from a portion of the sap supplied by damaged cells other than sieve tubes [57]. Sucrose hydrolysis should result in a 1:1 ratio between glucose and fructose; however, Glad et al. [58] reported a ratio of about 2:1 (glucose to fructose) from sieve tube exudate of grape. Nevertheless, Swanson and El-Shishiny [59] obtained data from grape that did not involve exudation and concluded that grape did not translocate hexoses.

In general, most, if not all of the sugars and sugar alcohols translocated by plant are nonreducing. Why should that be? Arnold [60] proposed that these are "protected" molecules. Molecules acted on by only a few enzymes would be favored for translocation because, to maintain the "protected" state, the activity of only these few enzymes would need to be suppressed within the sieve tubes.

Why do some plants translocate only sucrose while others translocate a mixture of sucrose, larger oligosaccharides, and sugar alcohols? Handley et al. [61] suggested that plants translocate sugars different from those accumulated in storage tissue, thereby maintaining a concentration gradient for the translocated sugar from sieve tubes to sink cells. Examples of such a system are represented for cucurbits [61] and legumes [1,46]. Even some parasites operate similarly, accumulating carbohydrates not found in appreciable concentrations in their hosts [62,63]. Obviously, this does not apply to all plants, for sugar beets and sugarcane translocate and store sucrose, but they store it within the vacuole. Recently it has been proposed that selection of the carbohydrate to be translocated is based on phloem loading mechanisms (discussed in Section VI, below).

In spite of the variability observed in translocated carbohydrates, all plants appear to translocate some carbon as sucrose. There is no known study that accounts for this observation. A possible explanation involves the observation that callose synthesis in sieve plate occurs rapidly [64]. This implies that the synthetic enzymes must be present at all times but are usually inactive. If injury were to activate sucrose synthetase, which Nolte and Koch [65] located in companion cells, UDP-glucose, the substrate of callose synthesis, would be provided. In addition, UDP, an inhibitor of callose synthesis, would be removed [66]. Other oligosaccharides or sugar alcohols could not provide appropriate conditions so easily.

B. Other Organic Compounds

Although most of the carbon is translocated through the phloem as carbohydrate, other organic compounds are translocated in significant quantities [47]. Ziegler [67] summarized the literature then available for nitrogenous compounds in sieve tube sap. The reported concentrations varied from 0.8 to 137 μmol mL^{-1}. Pate [68] reported that the concentration of amino acids and amides in the phloem of lupine varied from 14 mg mL^{-1} to 21 mg mL^{-1} and that asparagine accounted for more than half of that content, while glutamine was second most concentrated. In sieve tube exudate of yucca, glutamine and glutamic acid predominated [39], while in grape exudate glutamine predominated [58]. Interestingly, the same amino acids predominate in the translocation stream of giant algae [54]. Glad et al. [58] also reported that the molar quantities of amino acids approximated those of sugar. It should be noted, however, that on average, each translocated amino acid molecule is much smaller than sucrose; therefore, mass-based quantities strongly favored sucrose. As for sugars, the concentrations and ratios of concentrations of various amino acids in sieve tubes vary markedly from the values for the same compounds in mesophyll cells [69]. Other nitrogenous compounds (e.g., ureides and alkaloids) are translocated through the phloem in certain species [70].

During development, the ratio of amino acids to sugar in the translocation stream increased [58,68]. This would be expected, for as seeds developed, more nitrogenous compounds would be required by those sinks.

In a review of the literature, Ziegler [67] reported on several studies that found large amounts of protein in sieve tube exudate of members of the Cucurbitaceae. In the same volume, Eschrich and Heyser [71] concluded that the protein in the exudate was the result of surging induced by a sudden release of turgor when cuts were made. That conclusion is supported by ^{14}C labeling studies of selected members of this family, indicating that most of the carbon is carried as sucrose, raffinose, and stachyose [50,51]. More recently, a study using ^{35}S demonstrated that a small amount of protein was translocated through sieve tubes [72].

Reduction of nitrate and sulfate in mesophyll cells produces OH$^-$ [73,74], which is neutralized by the H$^+$ supplied as organic acids are synthesized. The acid anions are then loaded into the phloem. Although the concentration organic anions in the sieve tube sap of yucca plants was only 7 meq. mL^{-1} [39], these anions are exceedingly important in maintaining the pH of the sieve tube sap [73], for, together with the amino acids, they produce a salt of weak acids and strong bases with inorganic cations in the sieve tubes. This yields a pH near 8.0, a value that may be critical in the control of enzymes that might otherwise degrade the translocated saccharides.

Many other organic compounds are undoubtedly translocated through the phloem. Data indicate that auxins [75], cytokinins [76], and gibberellins [77] are phloem-mobile. Furthermore, there is strong evidence that abscisic acid [78], salicylic acid [79], and jasmonic acid [80] are phloem-mobile. These substances are present in such low concentrations in the sieve tubes that their osmotic influence on the transport process is minimal; therefore, they move passively with the flow through this system. However, they undoubtedly influence the translocation of assimilates indirectly by modifying the metabolic activity of phloem sinks [81], and possibly, even of the membrane processes of the sieve tubes and companion cells [82–85].

C. Inorganics

Most plant scientists consider the phloem as the conduit for organic materials. However, most are aware of the concept of phloem mobility of inorganic nutrients, and it is generally realized that deficiency symptoms of "phloem-mobile" nutrients first appear in older leaves, while symptoms of "phloem-immobile" nutrients first appear in young leaves. As noted above, the combination of organic anions and inorganic cations contributes to the control of sieve tube pH.

The concentration of inorganic materials for yucca [39] and lupine [86] sieve tube sap was about 2 mg mL^{-1}. This low concentration is nevertheless an extremely important component of the phloem transport system, for nutrients that enter mature leaves via the transpiration stream could not be repartitioned within the plant if they were not transported through the phloem. Therefore, organs such as expanding buds, young leaves, and developing flowers and fruit, which get most of their nutrients through the phloem, would receive inadequate supplies of nutrients if they were not translocated through the phloem. Therefore, the mobility of inorganic nutrients through the phloem is critical to plant growth and development.

Standard classroom deficiency experiments often used to determine the mobility of nutrients in the phloem are based on the assumption that the initial appearance of deficiency symptoms in older leaves indicates phloem mobility of the nutrient being tested, whereas initial appearance of symptoms in younger leaves indicates phloem immobility. Analysis of phloem exudates generally supports that conclusion. However, one must not use absolute concentrations of phloem exudate to determine mobility, because the requirements for different nutrients vary by orders of magnitude.

A more reasonable approach is to compare concentrations of phloem and xylem in plants that were adequately supplied with nutrients. Table 1 [86–90] offers such a comparison for studies in which concentrations of nutrients of both xylem and phloem exudate were obtained from the same plant in any one experiment. The following assumptions were used to construct Table 1:

Table 1 Relative Phloem Mobility[a,b]

Nutrient	Lupinus angustifola[c]	Lupinus alba[c]	Quercus rubra[d]	Ricinus cammus[e]	Lupinus alba[f]
K	100	100	100	100	100
Na	20	12	7	9	23
Mg	173	18	5	NA	49
Ca	9	7	2	4	0.4
Fe	69	32	NA	NA	NA
Mn	15	14	NA	NA	NA
Zn	78	85	NA	NA	NA

[a]Ratio of xylem to phloem concentration of several ions is divided by the same concentration for potassium, to set all values for potassium equal to 100, with all other values for other nutrients representing a percentage of the phloem mobility of potassium. See Eq. (1) in the text.

[b]NA, data not available.

[c]From Ref. 86.

[d]From Ref 87, compiled by Pate [70].

[e]From Refs. 88 and 89, compiled by Pate [70].

[f]From Ref. 90.

1. Over short periods of time, the same amount of potassium is translocated into a leaf through the xylem and out through the phloem.
2. Therefore, the mobility of any other nutrient is made relative to potassium by computing the ratio of phloem (P) to xylem (X) concentration (P_N/X_N) of any nutrient (N), dividing by same ratio for potassium (K), and multiplying by 100:

$$\text{mobility} = \frac{P_N/X_N}{P_K/X_K} \, 100 \tag{1}$$

To use Eq (1), we set potassium equal to 100, whereupon the phloem mobility of any other nutrient can be found as a percentage of potassium. When this procedure was applied to data from several studies (Table 1), the phloem mobilities of sodium, calcium, and manganese were found to be low; iron and zinc were intermediate; and magnesium was variable. A low value should reflect the accumulation of that nutrient in a leaf as it ages. Calcium does accumulate in leaves as they age [91,92].

One must wonder why this approach to the analysis of the phloem mobility of iron disagrees with the standard nutrient deficiency experiment. It may be that under deficiency conditions the iron is rapidly incorporated into cellular components as it enters a leaf and, therefore, is unavailable to the phloem, whereas under sufficient supply, some iron is available to be translocated through the phloem. In general support of this hypothesis, Loneragan et al. [93] reported that copper supplied to leaves of copper-deficient plants was retained in the leaves, but that supplied to copper-sufficient plants was exported rapidly. Certainly, the availability in the sieve tubes of organic anions, including amino acids, is critical to maintaining the solubility of iron, copper, and zinc [94a]. Stephen and Scholz [94b] reported that plants unable to produce the unusual amino acid nicotinamine (NA) are deficient in their ability to transport iron and other heavy metals through the phloem. These investigators report that NA is an excellent chelator of these metal ions and speculate about the mechanisms of action of NA in normal plants. Variability in the sufficiency of magnesium in the various experiments may explain the results for that nutrient (Table 1).

Data of Gorham et al. [95], reproduced in Table 2, support the foregoing conclusions regarding phloem mobility of inorganic ions, for ions with high phloem mobility were at higher concentrations in the phloem sink, while the ions of lower mobility were at higher concentrations in the xylem sink. It seems likely that this pattern has survival significance insofar as it protects the next generation (seeds) from enzyme-inhibiting ions [96]. It should be noted that the cells

Table 2 Chemical Composition of Leaves and Florets[a]

Nutrient	Leaves (mol m^{-3})	Florets (mol m^{-3})	Florets/leaves (K = 100)[b]
Potassium	72 ±6	133 ±21	100
Sodium	360 ±18	56 ±3	8
Calcium	35 ±9	25 ±2	39
Magnesium	37 ±5	45 ±2	66
Chloride	320 ±15	51 ±9	9

[a]Data for *Aster tipolium* on a plant water basis (± standard errors, $n = 3$).

[b]Ratio of florets to leaves adjusted such that potassium is equal to 100 and all other elements are a percentage of potassium.

Source: Ref. 95.

of the embryo have essentially no vacuole in which to sequester these ions. Other studies comparing accumulation of nutrients in phloem sinks (fruit) with accumulation in xylem sinks (leaves) lead to similar conclusions about the relative mobilities of various ions [97–100].

It should also be noted that an understanding of the relative phloem/xylem mobility of nutrients leads to an understanding of the best supplies of materials for human nutrition. In addition, toxic metal ions are generally more mobile in the xylem than in the phloem [101]. Therefore leafy vegetables grown on sites contaminated with such materials are more likely to be toxic to humans than are crops grown for their fruit [100].

Inorganic anions are not as concentrated in the phloem as are the cations, for much of the negative charge is accounted for by organic ions. Van Die and Tammes [39] did report chloride, phosphate, and sulfate in sieve tube exudate of yucca. About 75% of the phosphorus was combined into organic ions, however, and undoubtedly a significant portion of the sulfur was in amino acids. Nitrate is seldom reported to be a component of the phloem sap, but it is occasionally reported in low concentrations [57,102,103].

Wolterbeek and Van Die [104], using neutron activation analysis, were able to identify small amounts of several other inorganics in phloem exudate, including rubidium, copper, bromine, vanadium, and even gold. These data provide little information regarding the relative phloem/xylem mobility of these materials.

VI.   PHLOEM LOADING

Phloem loading refers to the transfer of assimilate into the ST-CC complex from photosynthetic cells or cells involved in temporary storage. It has been a difficult subject to study, for the cells and sites involved are not easily isolated from the portion of the system supplying the assimilates. Therefore, it is a relatively new field of study.

In discussing phloem loading, several characteristics of the system must be considered: (a) the loading process is selective, (b) concentrations of oligosaccharides and certain amino acids are higher in the ST-CC complex than in other leaf cells, (c) as a result, the ST-CC complex has an osmotic potential more negative than that of adjacent cells, and (d) the second and third conditions (b and c) result in a higher pressure in the ST-CC complex than in the other source cells.

The pathways available for phloem loading are apoplastic (cell wall) and symplastic (plasmodesmata). The conditions listed above indicate that metabolic energy must be involved in phloem loading. This energy requirement is most easily explained by invoking the apoplastic pathway. However, the mechanism of transport through plasmodesmata is not sufficiently understood to permit the elimination of energy input for the symplastic pathway. In 1987 Delrot [105] and Van Bel [106] separately published parallel papers in which they discussed the merits of each proposed pathway.

A.   Apoplastic Pathway

Serious early steps in developing an understanding of phloem loading were taken by Gunning and Pate [107] in their study of transfer cells associated with the phloem of minor veins in several families. Transfer cells have a large number of cell wall intrusions toward the interior of cells. This characteristic results in a large surface between the symplast and the apoplast. Other plant structures that transfer materials between apoplast and symplast contain transfer cells [108]; therefore, it seems reasonable to assume that phloem transfer cells are involved in phloem loading.

Sucrose supplied to the apoplast of sugar beet leaves was readily translocated through the phloem [109]. Later, Giaquinta [110] using sugar beet and Robinson and Hendrix [111] using wheat demonstrated that asymmetrically labeled sucrose did not have its label randomized, as would be expected if the supplied sucrose were in fact hydrolyzed, then resynthesized before phloem loading. When $^{14}CO_2$ was used to label sugars of squash plants, most of the translocated label was in stachyose [50]. When [^{14}C]sucrose was used, most of the label initially translocated was in sucrose; with time, however, an increasing proportion of the translocated label appeared in stachyose [112]. Furthermore that stachyose was most highly labeled in its galactose moieties [113]. These results indicated that a portion of the sucrose was not loaded directly into the phloem but moved into the metabolic pools of the plant and was hydrolyzed, whereupon the resultant hexoses were used in stachyose synthesis. Taken together, these data clearly indicate that sucrose can be absorbed from the apoplast directly into the translocation system, and that, at least in part, it can also be absorbed into other cells, where it is degraded before the translocated sugar is synthesized.

Giaquinta [114] demonstrated that [^{14}C]sucrose supplied to sugar beet leaf disks was accumulated in minor veins. He also demonstrated that the nonpenetrating sulfhydryl reagent *p*-chloromercuribenzenesulfonic acid (PCMBS) inhibited the absorption of sucrose without altering absorption of glucose, fructose, or 3-*O*-methylglucose. In addition, PCMBS did not alter the rate of photosynthesis [114,115].

It was also demonstrated that the optimum pH for apoplastic phloem loading is between 5.0 and 6.0 [115,116]. Furthermore, Giaquinta [115] demonstrated that changing the apoplastic pH from 5.0 to 8.0 more than doubled the K_M for sucrose absorption but did not change the V_{max}. On that basis, he proposed a model for phloem loading in which a sieve tube plasmalemma ATPase pumps protons out of the sieve tubes into the apoplast [117]. Sucrose secreted into the apoplast by mesophyll cells would then be loaded into the ST-CC complex by a sucrose–proton cotransporter against the sucrose concentration gradient, using the free energy gradient of protons that had been established by the ATPase. The charge gradient established by that proton-ATPase would also account for the high concentration of potassium in the sieve tubes (discussed above).

In the early 1980s mechanistic studies of apoplastic phloem loading were fully reviewed by Giaquinta [2,3]. Briefly, it was understood at that time that assimilates within mesophyll cells would move symplasticly from cell to cell via plasmodesmata until reaching a minor vein. The assimilates would then be transferred to the apoplast, where they would be absorbed by the ST-CC complex. It was seen as essential that the movement be symplastic up to the vein, for water moving from the vein to the point of transpiration in the apoplast would move assimilate away from the phloem.

B. Symplastic Pathway

The apoplastic pathway of phloem loading was favored by most phloem physiologists in the early 1980s. A few, however, held a different view. Lucas and coworkers repeated much of Giaquinta's work and extended it to include additional tests and additional plants. Their studies included the demonstration of a saturable and a nonsaturable (linear) component of sugar absorption by leaf cells [118,119], the demonstration of dye movement from mesophyll cell to veins through plasmodesmata [120], and the demonstration that modifying the apoplastic concentration of sucrose resulted in a modification of the rate at which sugars were absorbed by leaf tissues [121]. These data led to the conclusion that the absorption of exogenously supplied sugars was accomplished by a sugar retrieval system used normally to absorb sugars that leak from cells [122]. Some of this sugar would be loaded into the phloem. The investigators further concluded that the symplast is the prime pathway of phloem loading for many plants.

Furthermore they concluded that the partitioning of sucrose between the cytosol and vacuole of source cells may control the availability of assimilate to the phloem loading system [121].

Giaquinta [114] supplied sugar beet leaves with either $^{14}CO_2$ or [^{14}C]sucrose. He demonstrated that adding PCMBS to leaves inhibited vein loading of exogenously supplied [^{14}C]sucrose and carbon export of photosynthetically fixed ^{14}C, but PCMBS did not inhibit photosynthesis. Giaquinta [2,3] used these and other data to support the apoplastic model of phloem loading. However, Madore and Lucas [123] found that PCMBS inhibited vein loading of *Ipomoea tricolor* leaf disks at pH 5.0 only when [^{14}C]sucrose was supplied, not when the system was labeled via photosynthesis by supplying $^{14}CO_2$. From all these data, Madore and Lucas concluded that apoplastic phloem loading occurs in some species while symplastic phloem loading occurs in others.

Van Bel and coworkers then pointed out that the frequency of plasmodesmata connecting the ST-CC complex with other leaf cells varies by close to two orders of magnitude [124,125]. Therefore, they categorized vein types into those that have frequent plasmodesmata connections between ST-CC complex and other cells as type 1 veins and those with few such connections as type 2 veins. They also indicated that there are intermediate types, so they added subcategories (2a, 2b, 1/2). Type 1 vein companion cells have an intermediary cell structure, while the type 2b vein companion cells have a transfer cell structure. The transfer/companion cells have a typical transfer cell wall configuration, which results in a large plasmalemma surface, appropriate for loading material from the apoplast. Various types are represented in Figure 1.

Van Bel et al. [125] reported the pattern of vein loading correlated with vein anatomy. Plants in which PCMBS did not inhibit phloem loading of carbon from photosynthesis contain intermediary-type companion cells (type 1). Plants in which PCMBS inhibits phloem loading of photosynthetically fixed carbon have type 2b structure. These workers concluded that plants with the type 1 vein anatomy use the symplastic pathway as the prime pathway of phloem loading, while those with the type 2 anatomy use the apoplastic pathway. They reported that

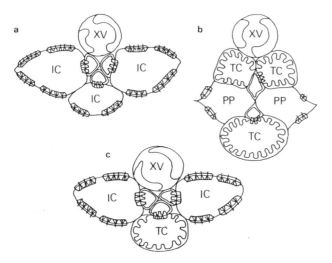

Figure 1 Typical minor vein structures of the species investigated: (a) open type (type 1) with intermediary cells (IC) as companion cells (*Hydrangea petiolaris*), (b) closed type (type 2b) with transfer cells (TC) as companion cells (*Pisum sativum*), and (c) composite type (type 1/2b) with intermediary and transfer cells in one single minor vein (*Acanthus mollis*). XV, xylem vessel; PP, phloem parenchyma; the central unmarked elements are sieve elements. (From Ref. 125.)

plants with intermediate anatomy likely use either pathway, or both. Physiological studies support this hypothesis of a dual nature of phloem loading [117,126].

Furthermore, Van Bel [127] proposed that the type 1 (symplastic loading) is evolutionarily primitive and type 2 is evolutionarily advanced. This seems questionable, for *Paulownia tomentera* is listed as a type 1 species but is a member of the Scrophulariaceae, a highly evolved family from the perspective of floral structure. In addition, Van Bel pointed out that type 1 is more prevalent in tropical rain forests, while type 2a (some plasmodesmata between ST-CC complexes and other cells) predominates in steppe and deciduous forest communities, and type 2b (containing transfer cells) predominates in cold deserts and arctic–alpine communities.

For plants using the symplastic pathway, one must be concerned over how the plasmodesmata can provide the selective control of phloem loading, as well as the osmotic and pressure gradients between the ST-CC complex and other leaf cells, as discussed above. To be able to understand the functioning of plasmodesmata, their structure must be understood. Plasmodesmata have been studied extensively, and the literature on that subject has been reviewed periodically (see, e.g., Refs. 128–130). Earlier papers were devoted to structure and distribution of plasmodesmata. Later papers also discussed function.

Briefly, plasmodesmata are tubes of cytoplasm that connect the symplast of adjacent cells through the cell wall with the plasmalemma surrounding these cytoplasmic tubes. On average, the diameters of plasmodesmata are about an order of magnitude more narrow than the connections between sieve tube member through sieve plates. Each plasmodesma varies in diameter across the cell wall, with the narrowest portion being near the point at which the ends connect with the main body of each cell (neck region). A desmotubule, which is an extension of (or attached to) the endoplasmic reticulum of each cell, extends through each plasmodesma. (Despite the name, desmotubules do not appear to be open tubes.) Between the desmotubule and the plasmalemma within the "cytoplasmic sleeve" is an extension of the cytosol. Surrounding the desmotubule, and apparently adhering to it, are cytoplasmic sleeve subunits that almost completely occlude the space between the desmotubule and the plasmalemma in the neck region.

Plasmodesmata vary in their degree of branching; many appear unbranched while others are highly branched [131]. The plasmodesmatal connections between sieve tubes and companion cells have a single pore into the sieve tube but are branched such that multiple pores enter the companion cells [120,132]. No suggestion has been put forth in an attempt to explain a functional basis for such branching.

The most likely pathway for materials moving through the plasmodesmata appears to be through the cytoplasmic sleeve. Electron microscopic measurements of the most restricted region, the neck, and experimental studies on movement of various sized dye molecules, support that hypothesis. Madore and coworkers [120,133] and Van Kesteren et al. [134] demonstrated the movement of fluorescent dyes from mesophyll into vascular bundles via plasmodesmata. Madore et al. [120] also showed that continuous plasmodesmotal connections exist from the mesophyll to the sieve tubes. However, Erwee and Goodwin [135] demonstrated that Ca^{2+} could induce blockage of cell-to-cell movement of dye. Robards and Lucas [130] pointed out that the exclusion limit for diffusion of dye through plasmodesmata varies from 376 D for roots and stems to 870 D between mesophyll and bundle sheath cells of plants using the C_4 photosynthetic pathway. However, they stated, ". . . dye-coupling results do not establish that . . . phloem loading occur(s) via this symplastic pathway. However, . . . modeling of the phloem system must incorporate the finding that plasmodesmata within the vascular bundle are not vestigial."

Another approach for studying phloem loading has been the introduction of yeast invertase genes into tomato [136] and tobacco and *Arabidopsis* [137]. This enzyme is secreted into the apoplast and there hydrolyzes any sucrose present. In tobacco and tomato this genetic alteration has resulted in accumulation of carbohydrate in mature leaves, major inhibition of carbohydrate translocation, stunted growth, and other malformations that are characteristic of some plant diseases. *Arabidopsis* was only slightly affected. These data indicate that the prime (exclusive?) pathway of phloem loading for tobacco and tomato is apoplastic, while *Arabidopsis* may have a symplastic alternative.

We are yet faced with the problem of proposing a mechanism of symplastic phloem loading that meets the criteria for phloem loading discussed above. However, some data suggest that not all those criteria need be met. Richardson et al. [138] found that sieve tube exudate of cucurbits had a rather low concentration of osmotically active substances. If these osmotic values were close to in vivo values, one could eliminate the criterion specifying large osmotic and pressure differences between ST-CC and mesophyll cells. But the same workers [57] did question the assumption that phloem exudate represents the in vivo contents of sieve tubes. Others [37,38], using a plasmolytic method, reported that *Cucurbita* and *Coleus* species, plants with type 1 anatomy, had quite negative osmotic potentials in their ST-CC complexes compared to mesophyll cells, thereby demonstrating that all the previously listed criteria for phloem loading indeed must be retained.

Van Bel [139] proposed a mechanism by which the plasmodesmata could act as an osmotic trap. In that scheme sucrose and galactinol, the precursors of raffinose and stachyose synthesis, would diffuse from mesophyll cells into intermediary cells (companion cells), where the larger oligosaccharides would be synthesized. The resultant oligosaccharides would be too large to return to the mesophyll cells through plasmodesmata, but they could penetrate the larger diameter plasmodesmata into the sieve tubes. This would provide the osmotic gradient reported for *Cucurbita* and *Coleus* species. However, no explanation was offered as to the fate of the myo-inositol that would result from synthesis of the oligosaccharides.

The mechanism proposed by Van Bel [139] would require that much of the carbon translocated by plants with type 1 anatomy be in the form of large oligosaccharides. Although the correlation between species that transport various sugars [45, and others] and the vascular bundle types [127] is quite strong, there is one striking exception: *Fraxinus ornus*, the most extreme example of the type 1 group, transports not only the higher oligosaccharides, but also significant amounts of mannitol. In addition, the mechanism of phloem loading suggested by Van Bel does not explain the observation that the ratio of various amino acids in sieve tube exudate fails to match that of mesophyll cells [69]. This lack of correspondence, however, could be accounted for by transfer of amino acids from xylem to phloem [68], using an apoplastic pathway.

We are left with the conclusion that both symplastic and apoplastic phloem loading occur, although only the mechanism of the apoplastic pathway is understood (see Ref. 140 for most recent review).

VII. PHLOEM UNLOADING

Assimilates from the phloem are unloaded either into vegetative cells of the plant producing the assimilate or into a developing embryo and/or endosperm (i.e., a separate plant). In special situations this process also includes transfer of assimilates to a parasite or symbiont. Clearly, the pathway must be apoplastic if the sink is a separate organism, for no plasmodesmata exist between parent and seed or between a plant and its parasite. However, if the sink is part of the same organism, the pathway may be symplastic or apoplastic. Phloem unloading has been recently reviewed [141–144].

Thorne [142] presented several mechanisms of phloem unloading.

1. Sucrose is passed into sink cells via plasmodesmata. Once in the sink cells, sucrose

is degraded into hexose in either the cytoplasm or vacuole of the sink cells. This type of unloading is reported to occur in sugar beet sink leaves, corn root tips, and bean endocarp.

2. Sucrose is unloaded from sieve tubes into the apoplast, where it is degraded into hexoses. The hexoses are transferred into the sink cell cytoplasm, where sucrose is resynthesized. The sucrose is then accumulated into the vacuole. The mechanism has been reported for assimilate movement into corn and sorghum kernels and into sugarcane stalks, and for movement into some parasites [63].

3. Sucrose is unloaded from sieve tubes into the apoplast, transferred into sink cell cytoplasm, then accumulated into vacuoles. This mechanism has been reported for sugar beet taproot, legume seeds, and wheat grain.

Eschrich [145] has added a fourth mechanism: carbohydrates follow any of these pathways, then are incorporated into starch. It could be argued that any metabolic removal of sugar would be analogous. One could visualize various combinations of steps of the listed processes to obtain even more possibilities.

From a thermodynamic standpoint, no energy should be required for transfer of sugar across the sieve tube or sink cell plasmalemma or through the plasmodesmata, since other processes maintain appropriate gradients along the pathways such that diffusion can account for movement. However, metabolic energy may be a mechanistic requirement.

As stated above, unloading of assimilates to a developing embryo must involve transfer of the assimilates into the apoplast. Thorne and Rainbird [146] developed a technique for studying this process. They removed the embryo from the seed coat while leaving the seed coat attached to the parent plant. They demonstrated that the unloading of assimilate into agar in the seed coat proceeded at a rate comparable to the unloading into an intact embryo. They also demonstrated that the nonpenetrating sulfhydryl reagent PCMBS inhibited unloading into the agar-filled seed coat cavity but not transfer of assimilates into the seed coat, thus indicating that transfer of assimilates into seed coat cells was symplastic. However, NaF and $NaAsO_2$ inhibited both processes, indicating an energy dependence of a portion of the unloading process. The technique of embryo removal has been used extensively to study the unloading process. Thorne [141] indicated that assimilates are transferred from sieve tubes symplastically through several cell layers before being deposited into the apoplast, either at the surface of the embryo/endosperm or several cell diameters from that surface. Once at the embryo/endosperm surface, the assimilates are absorbed into the developing tissue cells. Oparka [147] has referred to this as an "apoplastic unloading sink" rather than as "apoplastic unloading." He restricts the latter to unloading of assimilates into the apoplast directly from sieve tubes.

It appears that within the vegetative plant body, symplastic unloading is more common in immature tissue, whereas the apoplastic pathway is more prevalent in mature (nongrowing) tissue. This observation might be explained by differential growth rates of adjacent cells and the development of intercellular spaces. These growth processes would result in the disruption of plasmodesmata formed at cell division.

VIII. XENOBIOTIC TRANSPORT

The understanding of the transport of xenobiotics is essential to the judicious development and use of systemic pesticides. An example of the failure to use such information is provided by a program in which solutions of Benlate were injected into trunks of elm trees in an attempt to control Dutch elm disease. Benlate is soluble at the pH of the xylem (5.0–6.0) but essentially insoluble at the sieve tube pH (~8.0). As a result, the Benlate was translocated through the xylem to the leaves, providing some temporary protection to those structures. However, no protection was provided to any part of the plant supplied by the phloem (i.e., roots and even newly developed xylem).

The first extensive study of xenobiotic translocation was done by Crafts [148]. He was able to demonstrate whether a material was transported through xylem, phloem, neither, or both. It has been generally assumed that once a xenobiotic has entered the translocation system, it moves passively with the flow, which is driven by processes independent of the xenobiotic. In addition, the xenobiotic must not leak from the translocation stream too easily, or it will not travel far.

An important mechanism for phloem translocation of xenobiotics is ion trapping of weak acids [149, cited in Ref. 150]. The basis of ion trapping is as follows: a weak acid will be protonated in the lower pH of the apoplast and ionized in the higher pH of the sieve tubes. The protonated form readily penetrates the lipid portion of membranes. Once in the sieve tube, however the ionized form does not easily penetrate the plasmalemma of the sieve tubes (i.e. is ion trapped), and, therefore, is carried to the sink with the flow.

Tyree et al. [150] proposed a model for phloem mobility of substances that do not ionize. This model is based on idea that there is an optimum membrane permeability for translocation of a xenobiotic. If the permeability is too low, very small amounts of the material will enter the sieve tubes. If the permeability is too high, much of the xenobiotic will enter the sieve tubes but will leak out rapidly once it has moved away from the supply region. If the supply region consisted of leaves, the material would be carried back into the leaves in the xylem; therefore, the xenobiotic would never reach the sink (target). Figure 2 provides models of movement of two materials differing in membrane permeability. Kleier [151] combined the models for ion trapping and permeability to predict mobility of xenobiotics through the phloem.

For all these models, the required low membrane permeability of xenobiotics (for movement through the phloem) results in an accumulation of the xenobiotic in the sink region within the sieve tubes (see Figure 2). This buildup is due to the continuous flow of solution into the sink, which is driven by unloading of osmotically active materials and water following the movement of the osmoticum. Since some leakage of the xenobiotic does occur, its accumulation in the sink phloem results in significant delivery of the xenobiotic to the sink cells.

It might be possible to develop pesticides that attach to sites used by materials normally loaded into the phloem. These xenobiotics would then use the loading and unloading mechanisms used by metabolites. To date, that possibility seems not to have been exploited.

IX. ASSIMILATE PARTITIONING

A. General Considerations

Assimilate partitioning, recently reviewed by Wardlaw [9], includes the partitioning of all assimilated materials among plant parts. This topic is important for both theoretical and applied plant physiology. The incorporation of assimilates into the economically important components of a crop determines the potential for economic reward.

Processes that control assimilate partitioning are cell-to-cell transport, including transfer of materials between the xylem and phloem, loading and unloading of vascular tissues, long-distance translocation through the vascular tissues, and metabolic sequestration of materials such that they are either temporarily or permanently eliminated from transport processes. Several of these topics have been discussed, others are developed in this section, with emphasis on carbon partitioning. One of the most comprehensive models for carbon and nitrogen partitioning was developed by Pete et al. [152].

As plants develop, assimilates are partitioned differently at different times [153]. During

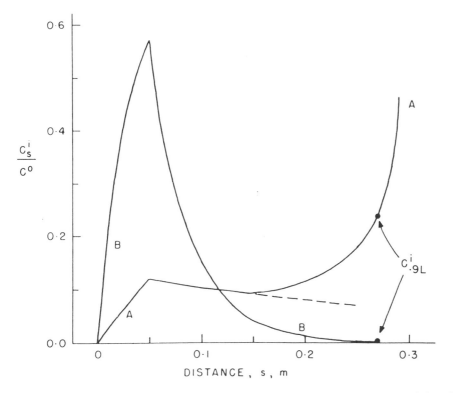

Figure 2 Theoretical distribution of xenobiotic in sieve tube of a "linearized" plant 0.3 m. Concentration in the sieve tube is plotted against distance, as a fraction of the concentration in the source leaf apoplast. Source is assumed to be 0.05 m long. In this calculation the sieve tube radius times sap velocity, rV, is 1.5×10^{-9} m s^{-1}. Curve A is for the optimum permeability, P^*, of about 2×10^{-9} m s^{-1}. In curve B, P^* is 10 times larger, 2×10^{-8} m s^{-1}. Dashed line extending from curve A shows how concentration would decline if V in the root remained constant instead of decreasing. (From Ref. 150.)

seed germination, the radical elongates first. A day or two later, plumule growth rate increases. As each leaf matures in sequence, it converts from sink to source [154]. In plants of determinate growth habit (e.g., corn, wheat, barley, sunflower), essentially all vegetative growth is completed at flowering. For a short time there are few growing sinks, so assimilate is stored in vegetative parts. As fruit and seeds enlarge, not only is current photosynthate partitioned into reproductive parts, but, depending on circumstances, stored materials are remobilized. This remobilization of assimilate from stems of wheat has been well illustrated (see, e.g., Refs. 155 and 156); however, tomato plants remobilized proportionately less assimilate from vegetative parts to fruit, and most of what they did remobilize was from leaves [157]. During these changes, the direction of translocation through the phloem often switches from primarily downward to primarily upward.

Also, the chemical mix of materials translocated is modified. Pate [70] reported increasing concentrations of nitrogenous compounds in phloem sap during seed growth of lupine. A significant portion of the nitrogenous compounds in the phloem was derived from the xylem, presumably transferred from the xylem via transfer cells in the stem and in minor veins in leaves. This system appears to provide control for partitioning of nitrogen compounds required for seed development. Amino acid concentration of sieve tube exudate increased from 14 µg mL^{-1} one week after anthesis to 21 µg mL^{-1} 6 weeks later, while the sucrose concentration fell from 112 to 80 μg mL^{-1}. Glad et al. [58] reported a changing pattern of sieve tube sap composition for grape similar to that for lupine.

Patterns for indeterminate annuals appear to be similar, though more complex, for most of their reproductive growth occurs with the first group of flower, yet they do have the capacity for continued vegetative growth and formation of later crops if the early crop is removed. In perennials, there are several variations of partitioning patterns. In many plants, however, flowering is followed by early rapid vegetative growth using assimilates from the preceding season. The major portion of assimilate is then used in reproductive growth. Finally, assimilate is stored [158]. In some plants such as elm, reproductive growth is completed in early spring before appreciable vegetative growth has occurred. Many other variations could be cited for perennial plants. For example, *Agave* develops much like a determinate annual plant, but the pattern extends over several years.

The mechanism by which an increase in economic productivity has been attained is as follows: increasing total dry matter production and increasing the proportion of assimilate partitioned into the economically important part of the plant (i.e., increasing HI). It should be pointed out that harvest index is somewhat misleading when total assimilate partitioning is considered, for HI usually accounts for only the aboveground parts of a crop. Dunan and Zimdahl [12] reported that roots accounted for 13.6 and 16.4% of the dry weight of barley and oats, respectively. Pate et al. [152] did the most complete carbon balance study known. Unfortunately, their study plant was *Lupinus albus*, which has a rather large taproot. Also, in considering their data, one must realize that gas exchange studies account for net photosynthesis of the shoot, yielding low values for actual photosynthesis and shoot respiration, while root gas exchange represents 24-hour root respiration. Considering those limitations on the data, Pate and coworkers reported that about 44% of the carbon fixed by net photosynthesis was used by roots, with about one-fourth of that going into growth. In addition, the root nodules used about 12% of the photosynthate in growth and respiration. Additional carbon was used in the formation of nitrogenous compounds that were supplied to the rest of the plant. These computations did not account for carbon that might have been lost as exudate, in sloughed cells, or because of use by other organisms. Buwalda [159], in his review of perennial crops, stated that mycorrhizal fungi account for 5–10% of total carbon acquired by photosynthesis. He also stated: "For mature plants, root growth is . . . a relatively small sink for carbon. . . ." Reports of high root/shoot ratios for plants that have perennial roots and annual shoots [160] do not negate Buwalda's statement.

The genetic component of increased crop productivity has not been assessed with regard to the proportion of assimilate partitioned to roots; rather, it seems to have resulted from an increase in HI based on analysis of aboveground parts only. These changes in HI have been associated with shorter plants for both small grains and soybeans [10,11,161], but not with increased rates of photosynthesis (carbon fixed per unit time and leaf area).

Economic yield also is increased is by increasing total dry matter yield without altering HI. Increase in total yield, as well as HI, is influenced by cultural practice, environment, and genotype. Gifford [11] compiled data for several crops to determine the basis of increased crop production over the years. Those data indicate that total shoot yield and HI have increased for all the crops he studied. HI was increased by genetic selection for high yield and by developing cultural practices that favored high yield.

It is also clear that an alteration in the pattern of assimilate partitioning occurs as a result of crop thinning or by removal of plant parts by pruning, herbivory, or violent weather. Various manipulative experiments have been performed to develop an understanding of the control of assimilate partitioning. One involved bean seedlings with nearly fully expanded primary leaves (the source) and a small, rapidly expanding first trifoliate (a sink). The experimenters removed the terminal leaflet, leaving the two lateral leaflets as sinks for each respective primary leaf [162]. Using, $^{14}CO_2$, they demonstrated that each leaflet received ~80% of its carbon from the nearest primary leaf. When one primary leaf was removed, the amount of assimilate translocated into the two leaflets did not decrease. The remaining primary leaf became the source for both, doubling its export to the leaflets without changing its rate of photosynthesis.

Loss of leaf area eaten by insects, destroyed by hail, or subjected to experimental desiccation [163] during grain or seed filling does not decrease production in proportion to the loss of leaf area because there is an increase in the utilization of stored carbohydrates. In addition, over longer time periods than in experiments cited above, the remaining leaf tissue will increase its rate of photosynthesis. Water stress can also influence partitioning [164].

How does this redirection of partitioning occur? The short-term response is as indicated by the work discussed above [162]. We should think of the vascular system of plants as a pipeline distribution system that runs vertically in the stem, with interconnections at nodes. Pressure will be greatest where loading is greatest and least where unloading is greatest. The flow follows the pressure gradients. In addition, the greatest resistance in the system is across the stem at the nodal interconnections. Therefore, with no perturbations in the system, most of the translocation will be vertical, with little movement across the nodal interconnections [165]. However, if the system is altered by removal of sources or sinks, cross-movement becomes significant. In the longer term, adjustments are made in photosynthetic rates and in utilization of stored assimilates.

Plants have an ability to adjust partitioning when exposed to different environments. By exposing tomato plants to photoperiods of 8, 16, or 20 hours each day, Logendra et al. [166] demonstrated that leaves of plants exposed to the shortest light periods accumulated carbohydrates the most rapidly during the light period available. Even though the total amount of carbohydrate accumulated in the light period by plants on the shortest photoperiod was less than that retained by the plants exposed to longer photoperiods, during the day, the former retained a higher proportion of their photosynthate for later export during the long dark period, thereby maintaining some supply for growth and maintenance during that dark period (see Table 3). Grange [167] obtained similar results for pepper.

B. Productivity

1. Small Grain

Shanahan et al. [168] demonstrated that grain number rather than grain size was the stronger indicator of winter wheat yields, even in the plains of eastern Colorado, where low moisture and high temperatures during grain filling often result in large variability in grain size. Bremner

Table 3 Rates of Net Photosynthesis (PS), Translocation, and Carbohydrate Accumulation of Tomato Leaves on Plants Exposed to 8, 16, or 20 hours of Light per Day[a]

Photoperiod (h)	Net PS	Translocation rates ($g\ m^{-2}\ h^{-1}$)		Carbohydrate accumulation rates ($g\ m^{-2}\ h^{-1}$)		
		Light	Dark	Hexose	Sugar	Starch
8	0.42b	0.0583	0.0744	0.062c	0.012b	0.123b
16	0.35b	0.1813	0.2700	0.046b	0.010b	0.115c
20	0.35b	0.1900	0.2700	0.050b	0.011b	0.054b

[a]Where available, significant differences at the 0.05 leval are indicated by b or c.

Source: Logendra et al. [166].

and Rawson [169] studied the influence of grain position and grain removal 9 days after anthesis on grain mass at maturity. They reported that position within the spike and within each spikelet had a large influence on grain mass in control (unthinned) plants, the largest grains within the spike being about one-third the distance from the base. Within each spikelet, the most basal grains were the largest. Thinning (removal of some grains) resulted in some increase in size, primarily at the ends of the spike. In no case did the most distal grain on the thinned spikelet attain the mass of those at the base. The investigators interpreted these results as indicating that the basal positions of each spikelet had a more adequate vascular system supplying assimilates, rather than suggesting a limiting supply of assimilates within the plants. That conclusion is supported by recent work by Natrova and Natr [170].

Further evidence that sink capacity is limiting to production was presented by several other workers. Blade and Baker [171] demonstrated that changing the source/sink ratio by lowering plant density, removal of developing grain, or removal of the flag leaf had little influence on mass of individual grains. Other work supports the conclusion that wheat [172] and oats [173] are sink-limited during grain filling.

To illustrate the impact of lowered photoassimilate production on productivity of wheat grain, Fischer [174] conducted several shading experiments. In one of the more severe treatments, he shaded plants for a single 21-day period, during which the plants received only 35% of the natural light. When the shading period was centered during vegetative development or early in floral development, the treatment had little impact on grain production compared with unshaded controls. The shading period centered near the midpoint of floral development had the most impact, lowering grain production to just over 40% of controls. The shading period centered at anthesis lowered yield to 80% of controls, whereas subsequent shading periods had progressively less impact, with the final period, which extended to maturity, having little impact.

Supporting data were provided by Kiniry [175] in the demonstration that shading sorghum plants during inflorescence development resulted in production of far fewer grains than were produced by controls. Removal of the shade at anthesis resulted in larger grains than in controls, but not nearly enough larger to compensate for loss in number. Caldiz and Sarandon [176] obtained similar results in that shading wheat during inflorescence development had a far greater negative impact on grain yield than did shading during grain filling. Furthermore, these workers showed that lowered yield resulted primarily from lower grain number.

High temperature during inflorescence development also had a greater negative impact on grain number than high temperature at any other developmental stage [177]. Fischer [178] also demonstrated that a combination of high temperature and low light during the 30-day period prior to anthesis resulted in low kernel number. Both these environmental impacts would lower assimilate accumulation during inflorescence development. High temperature during grain filling inhibited starch synthesis [179] and shortened the grain-filling period [180], thereby lowering production of wheat by lowering grain weight. Fischer and HilleRisLambers [181] concluded that grain production is source limited when they attained increases in grain size that were inadequate to compensate for lower grain numbers found when developing grains were removed. Shading the crop by 50% did lower grain mass, indicating source limitation under those conditions. Nonetheless, it appears that environmental impacts during grain/seed filling do not lower yield as much as comparable environmental pressures during inflorescence development. A further indication that assimilate supply during vegetative growth does not limit grain production was the observation that grazing of wheat prior to floral initiation did not lower grain yield [182].

2. Maize

Maize appears to be more closely balanced on the basis of source/sink capacity during grain filling. Jones and Simmons [183] demonstrated that removing a portion of the developing grains from an ear had little if any impact on grain size, whereas defoliation of the plants 12 days after mid silking resulted in lower grain number and mass per grain and defoliation 24 days after midsilking resulted only in lowered mass per grain. Defoliation at either date resulted in rapid depletion of carbohydrates from stems, while control plants had repartitioned about 40% of their maximum stem carbohydrates at maturity. Maize is more sensitive than wheat to loss of photosynthetic capacity during grain filling, for stem-stored carbohydrates are mostly sugars [183] rather than larger polymers. A lower storage capacity is the result. Shading corn plants during reproductive development lowered the quantity of harvested grain more than did shading at any other time [184]. Unlike wheat, shading during the vegetative stage caused significantly lower grain production. None of the shading treatments caused lower stover yield. Setter and Flannigan [185] demonstrated that decreasing source by shading early in grain development resulted in a correlation between lowered number of endosperm nuclei and lowered grain dry mass. Another interesting phenomenon in maize is the relation of the sugars in "sweet corn" to a lowered ability to accumulate starch [186,187]. This lowered ability to produce starch generally results in lower grain mass, which is then reflected in lower HI.

3. Legumes

It is more difficult to associate the pattern of carbohydrate partitioning and accumulation to reproductive activity in indeterminate plants. Aufhammer et al. [188] demonstrated that, unlike wheat, lowering number of seeds by removal of flower buds from *Vicia faba* did not lower production, for these plants were able to compensate by increasing seed size. In addition, they demonstrated that removal of basal flower buds increased the fruit set at more distal positions. It is likely that some minimum carbohydrate concentration must be met if buds are to develop into flowers and then into fruit. Sage and Webster [189] demonstrated that more distal buds, flowers, or fruit of a raceme of *Phaseolus* are more likely to abort than those at the more proximal positions. These studies imply that development at the more distal positions was limited by insufficient assimilate supply. Mauk and Breen [190] were able to support that hypothesis with data of ^{14}C-assimilation studies. This work appears to be parallel to the conclusion of Fischer and HilleRisLambers [181] that more distal reproductive sites are limited by vasularization. White et al. [191] reported that genetically determined seed size and yield (including HI) are negatively correlated in *Phaseolus vulgaris*. This suggests that the growth of basal fruit on a raceme or on a basal raceme of a stem results in less depletion of the carbohydrate supply if the seeds that developed were small, thus allowing more fruit to set. However, Stockman and Shibles [192] showed that neither increasing light intensity nor removing leaves altered flower and pod abscission. They concluded that carbohydrate supply was not immediately involved in abscission.

Wiles and Wilkerson [193] studied soybean production when plants were in competition with cocklebur. They reported that little production loss occurred if the cocklebur plants were removed before the fifth week of plant growth, about the time the soybean plants started to flower. Losses increased as the time of competition continued through the sixteenth week, when flowering was complete. No additional loss of seed production occurred in plants when cocklebur was allowed to compete to harvest. As with shading studies cited above, these data support the hypothesis that the period of flower development is the most critical portion of the life of crop plants if sexually produced parts are the economically important part of the plant.

A positive relationship between seed number and yield, as indicated above for grain, has also been demonstrated for soybeans [194]. Guffy et al. [195] also demonstrated a stronger correlation between nitrogen supply (nodulated vs. nonnodulated) than total assimilate. It should be pointed out that nitrogen fixation is an energy- (assimilate-) expensive process [196]; therefore, competition by weeds, cited above, could result in lowered N_2 fixation rates. The observation by Salado-Navarro et al. [197] that 80–90% of soybean's protein is in seeds at harvest further implicates the importance of nitrogen partitioning to soybean production. It should also be noted that much of this nitrogen, which accumulates in seed, is stored temporarily in specialized cells within the soybean leaves [198].

4. Tree Fruit

It is common for some fruit trees, especially certain apple varieties, to have a pattern of alternate years of heavy and light crop [199]. Floral initiation occurs in the summer at a time when a large crop would compete for carbohydrates. This would result in low flower bud formation and few flowers the following spring. During the subsequent summer of a small crop, there would be little competition for carbohydrates by the crop, so large numbers of flower buds would form. A common practice has been to break this alternate-year cycle by thinning the heavy crop. Unfortunately, a frost during flowering in one year can restart this cycle. This scenario is supported by Ryugo [200], but Westwood [199] seems to favor a concept of hormonal control. It is likely that there is an interaction between carbohydrate supply and hormones that provides control of floral development.

Treating apple or peach trees by shading (10% natural light) or with a photosynthetic inhibitor [201,202] induced abscission of fruit comparable to a "June drop" but markedly increased the proportion of fruit lost. This observation indicates that a similar mechanism based on assimilate supply is controlling sexual reproduction in trees as in annual plants. Miller and Walsh [203] compared partitioning of assimilate in peach trees where fruit had been either thinned or not thinned. This is a matter of economic importance, for thinning is a common cultural practice used for obtaining larger fruit, which have greater economic value. The unthinned trees had an HI of 0.50, while the thinned trees had an HI of 0.37. The point of interest is that the economically valuable part of the fruit, the fleshy mesocarp, is markedly increased by thinning, but the seed size is not increased nearly as much. In the closely related tree crop, almond (*Prunus amydalus*), from which the seeds are harvested, thinning would result in economic loss.

5. Floral Evocation

Studies cited above make it clear that plants are variable with respect to the stage of development at which reproduction is controlled. Some plants will continue to flower for a period of time after the fruit has set, but most of those later flowers will undergo abscission if the earlier set of fruit has not been removed. Ornamentalists have known for centuries that removal of old flowers and young fruit (carbohydrate sinks) markedly increases subsequent flowering. It is reasonable to conclude that assimilate supply is critical in determining degree of flowering, fruit, and seed set, although the control point might differ among different species.

Hendrix et al. [204] demonstrated that the content of fructan in wheat inflorescences 7 days before anthesis was highly correlated with grain number. Bodson [205] and Bodson and Outlaw [206] reported studies with *Sinapis* demonstrating that accumulation of carbohydrates in buds was associated with floral evocation. In their review of this topic, Bodson and Bernier [207] stated:

> Available evidence suggests that an early change in carbohydrate concentration in the apical bud is critical to floral initiation, but that this modification is not sufficient alone to trigger initiation. It is not possible to conclude whether assimilate accumulation in the reproductive structures is responsible for inflorescence development since the timing of events that are integral parts of reproductive development is generally very poorly known.

Furthermore, Lejeune et al. [208] demonstrated that floral induction of the long-day plant *Sinapsis alba* by one photoinductive cycle resulted in a threefold increase in sucrose concentration in sieve tube exudate. Ishioka et al. [209] were able to induce floral development on cultured *Pharbitis* apices by raising sucrose concentration and/or lowering ammonium concentration. Nitrate did not inhibit floral induction.

One might conclude that it is the environmental clues, such as day length, vernalization temperatures, or treatment (e.g., removal of assimilate sinks), that induce plants to alter partitioning patterns required for reproductive activity. One can further conclude that once floral development has been initiated, the degree of reproductive development is controlled by assimilate supply. It also appears that the importance of carbohydrate supply extends to shortly after anthesis, a period of potential abortion. Once the abortion period has passed, studies cited above indicate that capacity of sinks, rather than sources, is limiting.

It appears that in all crops, assimilate supply and partitioning patterns determine the degree of flowering and/or fruit set. The amount of fruit, seed, or grain that then develops determines the pattern of assimilate partitioning and may even exert a feedback effect on supply by impacting photosynthesis. From an evolutionary standpoint, this seems reasonable, for a plant will be more successful in producing progeny if it produces only the number of seeds that assuredly will be viable. So, at some stage, it must sense the number of seeds it can successfully supply even if a disaster, such as drought or loss of leaves to insects, should occur. Once a critical developmental stage has passed, the plant is committed to the seeds that are set, and the seed number is determined. One possible mechanism is suggested by the demonstrations [170,181] that the more basal sites on inflorescences are more adequately vascularized than the more distal sites. Thus the more distal location would receive adequate assimilate supply only if that supply were larger or the more proximal cites unoccupied. From the economic standpoint, the floral development period is critical, for it is during this period that the maximum potential sink capacity is established. Subsequent events may diminish that capacity, but it cannot be increased once that critical time has passed.

6. Vegetative Crops

Unlike the growth of seeds and fruit, growth of vegetative sinks may be limited by source capacity rather than by sink capacity. Working in a cool climate, Engels and Marschner [210–212] reported on a series of experiments with potato (*Solanum tuberosum*). They demonstrated that current photosynthate is rapidly used in tuber growth. In addition, tubers that initiated only 2–4 days after the first tubers had initiated were at a large disadvantage in accumulating assimilate. By the fourteenth day of the first tuber initiation, the tubers that had initiated only 4 days later were about one-tenth the size of the earlier tubers and were growing much more slowly.

When Engels and Marschner altered source–sink relationships by removing more than half of the tuber mass, the total tuber growth rate (cm^3 plant^{-1}) returned to the earlier rate within 4 days. When 50% of the leaves were removed with no reduction in tuber mass, the tuber growth rate was halved almost at once. Both these experiments support the investigators' conclusions that photosynthate is used immediately in tuber growth and that growth of potato tubers was source limited. However, Midmore et al. [213,214] reported that shading potato plants, especially if done early in development, enhanced tuber production. This would seem to contradict the above-mentioned study, but Midmore's work was done in Peru, where soil temperatures were reportedly as high as 32°C for the unshaded plots. It seems likely that lowering soil temperature was more important than loss of light in a region of high temperatures and high irradiance levels. Midmore and coworkers used this study to demonstrate the desirability of intercropping potatoes and corn.

In study of sweet potatoes, Nakatani et al. [215] demonstrated that productivity is limited by source. They made several reciprocal grafts between various clones to obtain potential differences of source and sink capacity. They reported that tuber dry weight was correlated with leaf area. Since there was no difference in rate of photosynthesis (mg CO_2 dm^{-2} h^{-1}), they concluded that growth rate of tubers was limited by the source.

C. Metabolic Consequences of Sink Limitation

When sinks are limiting, accumulation of assimilate can feed back and inhibit photosynthesis. Claussen and Lenz [216], working with eggplant, demonstrated that removal of fruit resulted in the accumulation of sucrose and starch in source leaves. They also reported that the treatment lowered the activity of sucrose phosphate synthetase and the rate of photosynthesis. These changes were observed over periods of days to weeks. Azcon-Bieto [217] demonstrated inhibition of photosynthesis in wheat leaves when export of assimilate was inhibited. That work demonstrated that photosynthesis was inhibited within hours as assimilate was accumulating in the leaves. The feedback mechanisms of assimilates that affect the enzymes of photosynthesis are quite complex and not completely understood. For reviews on this subject, see Stitt [218,219], Foyer [220], Stitt and Quick [221], and Sonnewald and Willmitzer [222]. In addition to the direct feedback on photosynthesis, Mandahar and Garg [223] reported that removal of sinks from barley plants resulted in loss of chlorophyll from flag leaves.

D. Environmental Control

Many factors in the environment influence plant productivity. This section does not cover these factors in detail; rather it relates these factors to material discussed earlier.

The realization that sink development occurs in crops that yield seed or fruit as their economic product is most critical: it indicates that adjustments in environment, genetics, or cropping cycle designed to enhance floral development and seed or fruit set will increase HI and, very likely, yield. The increasing CO_2 concentration in the atmosphere can increase crop yields only if the plants are genetically able to increase sinks in response to the increased photosynthetic opportunity. If they are not able to increase their sinks, assimilate accumulation will inhibit photosynthesis, and there will be no production increase. If the projected warming occurs, the temperatures may move out of the desired range for sink establishment, with lower productivity as a result. Such an environmental change may require changing crops or time of planting.

For factors that can be controlled, but have a cost, such as water supply or management, cost/benefit ratios, both economic and ecological, can be enhanced by awareness that the most critical time for economical development of a crop is the period of establishment of the sinks. Therefore, if some input is limited, emphasis should be placed on this critical phase of plant development. One must also be aware that the critical phase might be at very different calendar dates for different crops. For example, floral initiation in annual crops occurs in late spring or early summer, whereas floral initiation occurs in late summer for the next season's crop on woody perennials. No matter how favorable the conditions during seed and fruit growth, partitioning of assimilates into the economically important part of the plant cannot occur if those parts failed to form earlier in the season.

E. Extraterrestrial Agriculture

The problems discussed in this chapter also relate to problems important to NASA. If we are to inhabit a lunar base or travel to Mars, human needs must be met, at least in part, as they are here on earth: by plants. Interest in this topic is illustrated by support NASA has provided for research (see, e.g., Refs. 224–227, and references therein).

If food is to be provided most effectively for space travelers, a high HI is needed. A NASA-funded project on wheat is illustrative. In an attempt to achieve the highest possible food and O_2 production, temperature at or near the optimum for photosynthesis was used, together with high irradiance level. An HI of 0.25 was obtained, but when the temperature was lowered, an HI of near 0.50 was obtained (Salisbury, personal communication). It is also important that volume be used efficiently. To get maximum production per unit volume, high irradiance level must be provided to the plants. However, as irradiance level is increased, efficiency of light utilization in photosynthesis drops [224]. Since light (energy) is also expensive, a satisfactory compromise must be developed. If NASA's goals are to be attained, these and other problems (e.g., providing pure water from transpiration, recycling nutrients through plants, and removing CO_2 and providing O_2 to the crew) must be addressed in a way that most effectively uses both volume and energy.

X. SUMMARY

An understanding of the mechanisms of assimilate transport is useful, even essential, if crop inputs such as pesticides, fertilizer, and water are to be used economically. Understanding of these processes is also helpful in developing the most effective use of both traditional and molecular genetics.

Over the last several decades, the mechanism of long-distance translocation of assimilates through the sieve tubes has been established as an osmotically driven pressure flow system. The carbohydrates translocated by this system are nonreducing oligosaccharides such as sucrose, raffinose, and stachyose, as well as several sugar alcohols. For many plants, sucrose appears to be the only carbohydrate translocated. Other plants translocate a mixture of these sugars and sugar alcohols. Many different nitrogenous compounds are also translocated through the phloem. These include most, if not all, of the protein amino acids and amides as well as several ureides. The most common are aspartic and glutamic acids and their amides. Nitrate is of little, if any, significance in nitrogen translocation through the phloem.

Various inorganic materials are translocated through the phloem. By far the most concentrated, other than water, is potassium. Together with the anions of weak organic and amino acids, the inorganic cations form a salt that buffers the pH of sieve tube sap at about 8.0.

It now appears that there are two pathways of phloem loading. In the apoplastic route, the assimilates pass through the cell wall before entering the sieve tubes. The symplastic pathway follows the plasmodesmata from cell to cell and finally into the sieve tubes. Based on anatomy of the vascular bundles, it appears that many plants use primarily one or the other pathway, while some plants appear to be able to use either or both. Both mechanisms result in a higher concentration of sugars in the sieve tube–companion cell complex than in other cells. The mechanism of concentrating the carbohydrates is understood for the apoplastic pathway, but not for the symplastic pathway.

Phloem unloading also follows either pathway. Since there is no symplastic connection between a parent plant and the seed it is forming, transfer of assimilates into a developing seed must be apoplastic, although the assimilates may not enter the apoplast directly from sieve tubes. When assimilates are being transferred to cells within the same plant, they may follow either the symplastic or the apoplastic pathway. It appears that chemical modifications maintain free energy gradients for carbohydrates along the unloading pathway.

Assimilate partitioning is the most critical component of the translocation processes that determine economic productivity. When the economic product is the result of sexual reproduction, the establishment of the sinks is the critical step in productivity, for it is during floral development that maximum yield potential is established. Once established, that potential might be lowered, but it cannot be increased. Genetic and environmental factors interact to limit yield potential. It appears that sink formation is the limiting component of crop production in many situations, for experimental evidence supports the proposition that during grain/seed filling, the systems are sink-limited, that indeed a feedback of assimilate limits the rate of photosynthesis. For crops in which the economic product is vegetative, it appears that genetic and environmental determinants affect the establishment of sinks; once sinks are established, however, their growth is more likely to be source-limited.

ACKNOWLEDGMENT

I thank Drs. P. Bauer and C. Ross for reading and commenting on a draft manuscript. In addition, I express my appreciation to all the scientists whose work made this review possible.

REFERENCES

1. A. S. Crafts and C. E. Crisp, *Phloem Transport in Plants*, Freeman, San Francisco (1971).
2. R. T. Giaquinta, *Ber. Dtsch. Bot. Gest.*, *93*: 187 (1980).
3. R. T. Giaquinta, *Annu. Rev. Plant Physiol.*, *34*: 347 (1983).
4. R. L. Heath and J. Preiss (eds.), *Regulation of Carbon Partitioning in Photosynthetic Tissue*, American Society of Plant Physiologists, Rockville, MD (1985).
5. J. Cronshaw, *Annu. Rev. Plant Physiol.*, *32*: 465 (1981).
6. J. Cronshaw, W. J. Lucas, and R. T. Giaquinta (eds.), *Plant Biology*, Vol. I, *Phloem Transport*, Liss, New York (1986).
7. L. Ho, *Annu. Rev. Plant Physiol. Plant Mol. Biol.*, *39*: 355 (1988).
8. D. A. Baker and J. A. Milburn (eds.), *Transport of Assimilates*, Longman, New York (1989).
9. I. F. Wardlaw, *New Phytol.*, *116*: 341 (1990).
10. R. B. Austin, J. Bingham, R. D. Blackwell, L. T. Evans, M. A. Ford, C. L. Morgan, and M Taylor, *J. Agric. Sci.*, *94*: 675 (1980).
11. R. M. Gifford, *Plant Biology*, Vol. I, *Phloem Transport* (J. Cronshaw, W. J. Lucas, and R. T. Giaquinta, eds.), Liss, New York, p. 435 (1986).
12. C. M. Dunan and R. L. Zimdahl, *Weed Sci.*, *39*: 558 (1991).
13. J. Gressel, *Weed Technol.*, *6*: 509 (1992).
14. D. A. Smith, *Phytoalexins* (J. A. Bailey and J. W. Mansfield, eds.), Halsted Press, New York, p. 218 (1982).
15. E. Munch, *Ber. Dtsch. Bot. Ges.*, *44*: 68 (1927).
16. E. C. Miller, *Plant Physiology*, McGraw-Hill, New York, (1931).
17. O. Raber, *Principles of Plant Physiology*, Macmillan, New York (1933).
18. C. A. Swanson, *Plant Physiology: A Treatise*, Vol. II, *Plants in Relation to Water and Solutes* (F. C. Steward, ed.), Academic Press, New York, p. 481 (1959).
19. F. L. Milthorpe and J. Moorby, *Annu. Rev. Plant Physiol.*, *20*: 117 (1969).
20. A. J. Peel, *Transport of Nutrients in Plants*, Wiley, New York (1974).
21. S. Aronoff, J. Dainty, P. R Gorham, L. M. Srivastava, and C. A. Swanson (eds.), *Phloem Transport*, Plenum Press, New York (1975).
22. L. Horwitz, *Plant Physiol.*, *33*: 81 (1958).
23. O. Biddulph, *Plant Physiology: A Treatise*, Vol. II, *Plants in Relation to Water and Solutes* (F. C. Steward, ed.), Academic Press, New York, p. 553 (1959).
24. D. S. Fensom, *Can. J. Bot.*, *35*: 573 (1957).
25. D. C. Spanner, *J. Exp. Bot.*, *9*: 332 (1958).
26. D. J. F. Bowling, *Planta*, *80*: 21 (1968).
27. R. Thaine, *Nature*, *192*: 772 (1961).
28. R. Thaine, M. E. de Maria, and H. I. M. Sarisalo, *J. Exp. Bot.*, *26*: 91 (1975).

29. M. J. Canny, *Ann. Bot.*, *25*: 152 (1962).
30. P. Trip and P. R. Gorham, *Plant Physiol.*, *43*: 877 (1968).
31. J. A. Webb, *Can. J. Bot.*, *49*: 717 (1971).
32. R. T. Giaquinta and D. R. Geiger, *Plant Physiol.*, *51*: 372 (1973).
33. C. A. Swanson and D. R. Geiger, *Plant Physiol.*, *42*: 751 (1967).
34. J. W. Sij and C. A. Swanson, *Plant Physiol.*, *51*: 368 (1973).
35. C. A. Peterson and H. B. Currier, *Physiol. Plant.*, *22*: 1238 (1969).
36. J. P. Wright and D. B. Fisher, *Plant Physiol.*, *65*: 1133 (1980).
37. D. G. Fisher, *Planta*, *169*: 141 (1986).
38. R. Turgeon and P. K. Hepler, *Planta*, *179*: 24 (1989).
39. J. Van Die and P. M. L. Tammes. *Encyclopedia of Plant Physiology*, Vol. I, *Transport in Plants*, Part I (M. H. Zimmermann and J. A. Milburn, eds.), Springer-Verlag, New York, p. 196 (1975).
40. R. I. Grange and A. J. Peel, *Planta*, *124*: 191 (1975).
41. K. Esau, *Plant Anatomy*. Wiley, New York (1953).
42. D. A. Cataldo, A. L. Christy, C. L. Coulson, and J. M. Ferrier, *Plant Physiol.*, *49*: 685 (1972).
43. D. A. Cataldo, A. L. Christy, and C. L. Coulson, *Plant Physiol.*, *49*: 690 (1972).
44. A. L. Christy and J. M. Ferrier, *Plant Physiol.*, *52*: 531 (1973).
45. M. H. Zimmermann and H. Ziegler, *Encyclopedia of Plant Physiology*, Vol. I, *Transport in Plants*, Part I (M. H. Zimmermann and J. A. Milburn, eds.), Springer-Verlag, New York, p. 480 (1975).
46. O. Kandler and H. Hoff, *Encyclopedia of Plant Physiology*, Vol. 13(A) (F. A. Loewus and W. Tanner, eds.), Springer-Verlag, New York, p. 348 (1982).
47. J. A. Milburn and D. A. Baker, *Transport of Photoassimulates* (D. A. Baker and J. A. Milburn, eds.), Longman, New York, p. 345 (1989).
48. L. M. Cruz-Perez and D. Durkin, *Proc. Am. Soc. Hortic. Sci.*, *85*: 414 (1964).
49. J. A. Webb and P. R. Gorham, *Plant Physiol.*, *39*: 663 (1964).
50. J. E. Hendrix, *Plant Physiol.*, *43*: 1631 (1968).
51. J. E. Hendrix, *Plant Sci. Lett.*, *25*: 1 (1982).
52. K. L. Webb and J. W. A. Burley, *Science*, *137*: 766 (1962).
53. J. M. Davis and W. H. Loescher, *Physiol. Plant.*, *79*: 656 (1990).
54. K. Schmitz and L. M. Srivastava, *Plant Physiol.*, *63*: 995 (1979).
55. M. Diouris, *Phycologia*, *28*: 504 (1989).
56. B. C. Parker, *J. Phycol.*, *2*: 38 (1966).
57. P. T. Richardson, D. A. Baker, and L. C. Ho, *J. Exp. Bot.*, *33*: 1239 (1982).
58. C. Glad, J.-L. Regnard, Y. Querou, O. Brun, and Morol-Gaudry, *Vitis*, *31*: 131 (1992).
59. C. A. Swanson and E. D. H. El-Shishiny, *Plant Physiol.*, *33*: 33 (1958).
60. W. N. Arnold, *J. Theor. Biol.*, *21*: 13 (1968).
61. L. W. Handley, D. M. Pharr, and R. F. McFeeters, *Plant Physiol.*, *72*: 498 (1983).
62. J. M. Daly, *Encyclopedia of Plant Physiology*, Vol. 4 (R. Heilefuss and P. H. Williams, eds.), Springer-Verlag, New York, p. 450 (1976).
63. W. E. Seel, I. Cechin, G. A. Vincent, and M. C. Press, *Carbon Partitioning Within and Between Organisms* (C. J. Pollock, J. F. Farrar, and A. J. Gordon, eds.), BIOS Science Publishers, Oxford, p. 199 (1992).
64. D. H. Webster and H. B. Currier, *Can. J. Bot.*, *46*: 1215 (1968).
65. K. D. Nolte and K. E. Koch, *Plant Physiol.*, *101*: 899 (1993).
66. D. L. Morrow and W. J. Lucas, *Plant. Physiol.*, *84*: 565 (1987).
67. H. Ziegler, *Encyclopedia of Plant Physiol*, Vol. I, *Transport in Plants*, Part I (M. H. Zimmermann, and J. A. Milburn, eds.), Springer-Verlag, New York, p. 59 (1975).
68. J. S. Pate, *Transport and Transfer Processes in Plants* (I. F. Wardlaw and J. B. Passioura, eds.), Academic Press, New York, p. 447 (1976).
69. D. Schobert and E. Komer, *Planta*, *177*: 342 (1989).
70. J. S. Pate, *Transport and Transfer Processes in Plants* (I. F. Wardlaw and J. B. Passioura eds.), Academic Press, New York, p. 253 (1976).
71. W. Eschrich and W. Heyser, *Encyclopedia of Plant Physiology*, Vol. I, *Transport in Plants*, Part I (M. H. Zimmermann and J. A. Milburn, eds.), Springer-Verlag, New York, p. 101 (1975).

72.   D. B. Fisher, Y. Wu, and M. S. B. Ku, *Plant Physiol.*, *100*: 1433 (1992).
73.   J. A. Raven and F. A. Smith, *New Phytol.*, *76*: 415 (1976).
74.   E. A. Kirkby and A. H. Knight, *Plant Physiol.*, *60*: 349 (1977).
75.   M. H. M. Goldsmith, M., D. A. Cataldo, J. Karn, T. Brenneman, and P. Trip, *Planta*, *116*: 301 (1974).
76.   J. S. Taylor, B. Thompson, J. S. Pate, C. A. Atkins, and R. P. Pharis, *Plant Physiol.*, *94*: 1714 (1990).
77.   M. Katsumi, D. E. Foard, and B. O. Phinney, *Plant Cell Physiol.*, *24*: 379 (1983).
78.   E. S. Ober and T. L. Setter, *Plant Physiol.*, *98*: 353 (1992).
79.   I. Raskin, *Plant Physiol.*, *99*: 799 (1992).
80.   P. E. Staswick, *Plant Physiol.*, *99*: 804 (1992).
81.   U. Petzold, S. Peschel, I. Dahse, and G. Adam, *Acta Bot. Neerl.*, *41*: 469 (1992).
82.   J. W. Patrick, *Transport and Transfer Processes in Plants* (I. F. Wardlaw and J. B. Passioura, eds.), Academic Press, New York, p. 433 (1976).
83.   I. Sakeena and M. A. Salam, *Indian J. Plant Physiol.*, *31*: 428 (1988).
84.   S. Jahnke, D. Bier, J. J. Estruch, and J. P. Beltran, *Planta*, *180*: 53 (1989).
85.   V. Borkovec and S. Prochazka *J. Agron. Crop Sci.*, *169*: 229 (1992).
86.   J. S. Pate, *Encyclopedia of Plant Physiology*, Vol. I, *Transport in Plants*, Part I (M. H. Zimmermann and J. A. Milburn, eds.), Springer-Verlag, New York, p. 441 (1975).
87.   J. Van Die and P. C. M. Willense, *Acta Bot. Neerl.*, *24*: 237 (1975).
88.   S. M. Hall and D. A. Baker, *Planta*, *106*: 131 (1972).
89.   S. M. Hall, D. A. Baker, and J. A. Milburn, *Planta*, *100*: 200 (1971).
90.   W. D. Jeschke, C. A. Atkens, and J. S. Pate, *J. Plant Physiol.*, *117*: 319 (1985).
91.   D. Lamb, *Plant Soil*, *45*: 477 (1976).
92.   G. J. Waughman and D. J. Bellamy, *Ann. Bot.*, *47*: 141 (1981).
93.   J. F. Loneragan, K. Snowball, and A. D. Robson, *Transport and Transfer Processes in Plants* (I. F. Wardlaw and J. B. Passioura eds.), Academic Press. New York, p. 463 (1976).
94.   (a) G. L. Mullins, L. E. Sommers, and T. L. Housley, *Plant Soil*, *96*: 377 (1986). (b) V. W. Stephan and G. Scholz, *Physiol. Plant.*, *88*: 522 (1993).
95.   J. Gorham, L. L. Hughes, and R. G. WynJones, *Plant Cell Environ.*, *3*: 309 (1980).
96.   A. Pollard and R. G. WynJones, *Planta*, *144*: 291 (1979).
97.   G. A. Mitchell, F. T. Bingham, and A. L. Page, *J. Environ. Qual.*, *7*: 165 (1978).
98.   D. M. Maynard, O. A. Lorenz, and V. Magnifico, *J. Am. Soc. Hortic. Sci.*, *105*: 79 (1980).
99.   L. J. Sikora, R. L. Chaney, N. H. Frankos, and C. M. Murry, *J. Agric.*, *6*: 1281 (1980).
100. E. M. Romney, A. Wallace, R. K. Schulz, J. Kinnear, and R. A. Wood, *Soil Sci.*, *132*: 40 (1981).
101. G. S. Dollard and N. W. Lepp, *Z. Pflanzenphysiol.*, *97*: 409 (1980).
102. H. Hayashi and M. Chino, *Plant Cell Physiol.*, *26*: 325 (1985).
103. H. Hahashi and M. Chino, *Plant Cell Physiol.*, *27*: 1387 (1986).
104. B. Wolterbeek and J. Van Die, *Acta Bot. Neerl.*, *29*: 307 (1980).
105. S. Delrot, *Plant Physiol. Biochem.*, *25*: 667 (1987).
106. A. J. E. Van Bel, *Plant Physiol. Biochem.*, *25*: 677 (1987).
107. B. E. S. Gunning and J. S. Pate, *Protoplasma*, *68*: 107 (1969).
108. E. Schnepf and E. Proso, *Protoplasma*, *89*: 105 (1976).
109. D. R. Geiger, S. A. Sovonick, T. L. Shock, and R. J. Fellows, *Plant Physiol.*, *54*: 892 (1974).
110. R. T. Giaquinta, *Plant Physiol.*, *60*: 339 (1977).
111. N. L. Robinson and J. E. Hendrix, *Plant Physiol.*, *71*: 701 (1983).
112. J. E. Hendrix, *Plant Physiol.*, *52*: 688 (1973).
113. J. E. Hendrix, *Plant Physiol.*, *60*: 567 (1977).
114. R. T. Giaquinta, *Plant Physiol.*, *57*: 872 (1976).
115. R. T. Giaquinta, *Plant Physiol.*, *63*: 744 (1979).
116. S. Delrot and J.-L. Bonnemain, *Plant Physiol.*, *67*: 560 (1981).
117. R. T. Giaquinta, *Nature*, *267*: 369 (1977).
118. J. W. Maynard and W. J. Lucas, *Plant Physiol.*, *69*: 734 (1982).
119. J. W. Maynard and W. J. Lucas, *Plant Physiol.*, *70*: 1436 (1982).

120. M. A. Madore, J. W. Oross, and W. J. Lucas, *Plant Physiol.*, *82*: 432 (1986).
121. C. Wilson and W. J. Lucas, *Plant Physiol.*, *84*: 1088 (1987).
122. M. A. Madore and W. J. Lucas, *Transport of Photoassimilates* (D. A. Baker and J. A. Milburn, eds.), Wiley, New York, p. 49 (1989).
123. M. A. Madore and W. J. Lucas, *Plants*, *171*: 197 (1987).
124. C. E. J. Botha and A. J. E. Van Bel, *Planta*, *187*: 359 (1992).
125. A. J. E. Van Bel, Y. V. Gamalei, A. Ammerlaan, and L. P. M. Bik, *Planta*, *186*: 518 (1992).
126. D. Schmitz, B. Cuypers, and M. Moll, *Planta*, *171*: 19 (1987).
127. A. J. E. Van Bel, *Acta Bot. Neerl.*, *41*: 121 (1992).
128. A. W. Robards, *Annu. Rev. Plant Physiol.*, *26*: 13 (1975).
129. A. W. Robards, *Intercellular Communication in Plants: Studies on Plasmodesmata* (B. E. S. Gunning and A. W. Robards, eds.), Springer-Verlag, New York, p. 2 (1976).
130. A. W. Robards and W. J. Lucas, *Annu. Rev. Plant Physiol. Plant Mol. Biol.*, *41*: 369 (1990).
131. Y. V. Gamalei, *Sov. Plant Physiol.* (Eng. transl.), *28*: 649 (1981).
132. J. W. Oross, M. A. Grusak, and W. J. Lucas, *Plant Biology*, Vol. I, *Phloem Transport* (J. Cronshaw, W. J. Lucas, R. T. Giaquinta, eds.), Liss, New York, p. 477 (1986).
133. M. A. Madore and W. J. Lucas, *Plant Biology*, Vol. I, *Phloem Transport* (J. Cronshaw, W. J. Lucas, and R. T. Giaquinta, eds.), Liss, New York, p. 129 (1986).
134. W. J. P. Van Kesteren, C. Van der Schoot, and A. J. E. Van Bel, *Plant Physiol.*, *88*: 667 (1988).
135. M. G. Erwee and P. B. Goodwin, *122*: 162 (1984).
136. C. D. Dickinson, T. Allabella, and M. Chrispeels, *Plant Physiol.*, *95*: 420 (1991).
137. A. Von Schaewen, M. Stitt, R. Schmidt, U. Sonnewald, and L. Willmitzer, *EMBO J.*, *9*: 3033 (1990).
138. P. T. Richardson, D. A. Baker, and L. C. Ho, *J. Exp. Bot.*, *35*: 1575 (1984).
139. A. J. E. Van Bel, *Carbon Partitioning Within and Between Organisms* (C. J. Pollock, J. F. Farrar, and A. J. Gordon, eds.), BIOS Science Publishers, Oxford, p. 53 (1992).
140. A. J. E. Van Bel, *Annu. Rev. Plant Physiol. Plant Mol. Biol.*, *44*: 253 (1993).
141. J. H. Thorne, *Annu. Rev Plant Physiol.*, *36*: 317 (1985).
142. J. H. Thorne, *Plant Biology*, Vol. I, *Phloem Transport* (J. Cronshaw, W. J. Lucas, and R. T. Giaquinta, eds.), Liss, New York, p. 211 (1986).
143. W. Eschrich, *Transport of Photoassimilates* (D. A. Baker and J. A. Milburn, eds.), Wiley, New York, p. 206 (1989).
144. J. W. Patrick, *Physiol. Plant.*, *78*: 298 (1990).
145. W. Eschrich, *Plant Biology*, Vol. I, *Phloem Transport* (J. Cronshaw, W. J. Lucas, and R. T. Giaquinta, eds.), Liss, New York, p. 225 (1986).
146. J. H. Thorne and R. M. Rainbird, *Plant Physiol.*, *72*: 268 (1983).
147. K. J. Oparka, *Plant Physiol.*, *94*: 393 (1990).
148. A. S. Crafts, *Ann. N.Y. Acad. Sci.*, *144*: 357 (1967).
149. C. E. Crisp, "Insecticides," Proceedings of the 2nd IUPAC Congress on Pesticidal Chemicals, Vol. 1, Tel Aviv, Israel, p. 211 (1972).
150. M. T. Tyree, C. A. Peterson, and L. V. Edgington, *Plant Physiol.*, *63*: 367 (1979).
151. D. A. Kleier, *Plant Physiol.*, *86*: 803 (1988).
152. J. S. Pate, D. B. Layzell, and D. L. McNeil, *Plant Physiol.*, *63*: 730 (1979).
153. J. F. Farrar, *Carbon Partitioning Within and Between Organisms* (C. J. Pollock, J. F. Farrar and A. J. Gordon, eds.), BIOS Science Publishers, Oxford, p. 163 (1992).
154. R. Turgeon, *Annu. Rev. Plant Physiol. Plant Biochem.*, *40*: 119 (1989).
155. C. J. Bell and L. D. Incoll, *J. Exp. Bot.*, *41*: 949 (1990).
156. D. J. Davidson and P. M. Chevalier, *Crop Sci.*, *32*: 186 (1992).
157. J. D. Hewett and M. Murrush, *J. Am. Soc. Hortic. Sci.*, *111*: 142 (1986).
158. J. D. Keller and W. H. Loescher, *J. Am. Soc. Hortic. Sci.*, *114*: 969 (1989).
159. J. G. Buwalda, *Environ. Exp. Bot.*, *33*: 131 (1993).
160. M. M. Caldwell, *Root Development and Function—Effects on the Physical Environment* (P. J. Gregory, J. V. Lake, and D. A. Rose, eds.), Cambridge University Press, Cambridge, p. 167 (1987).

161. M. S. Lin and R. L. Nelson, *Crop Sci.*, *28*: 218 (1988).
162. C. Borchers-Zampini, A. B. Glamm, J. Hoddinott, and C. A. Swanson, *Plant Physiol.*, *65*: 1116 (1980).
163. R. H. Blum, Poiarkova, G. Golan, and J. Mayer, *Field Crop Res.*, *6*: 51 (1983).
164. E.-D. Schulze, K. Schilling, and S. Nagarajah, *Oecologia*, *58*: 169 (1983).
165. Y. Shishido and Y. Hori, *Tohoku J. Agric. Res.*, *28*: 82 (1977).
166. S. Logendra, J. D. Putman, and H. W. Janes, *Sci. Hortic.*, *42*: 75 (1990).
167. R. I. Grange, *J. Exp. Bot.*, *36*: 1749 (1985).
168. J. F. Shanahan, D. H. Smith, and J. R. Welsh, *Agron. J.*, *76*: 611 (1984).
169. P. M. Bremner and H. M. Rawson, *Aust. J. Plant Physiol.*, *5*: 61 (1978).
170. Z. Natrova and L. Natr, *Field Crop Res.*, *31*: 121 (1993).
171. S. F. Blade and R. J. Baker, *Crop Sci.*, *31*: 1117 (1991).
172. B. Borghi, M. Corbellini, M. Cattaneo, M. E. Fornasari, and L. Zucchelli, *J. Argon. Crop Sci.*, *157*: 245 (1986).
173. D. M. Peterson, *Field Crop Res.*, *7*: 41 (1983).
174. R. A. Fischer, *Crop Sci.*, *15*: 607 (1975).
175. J. R. Kiniry, *Agron. J.*, *80*: 221 (1988).
176. D. O. Caldiz and S. J. Sarandon, *Agronomie*, *8*: 327 (1988).
177. I. J. Warrington, R. L. Dunstone, and L. M. Green, *Aust. J. Agric. Res.*, *28*: 11 (1977).
178. R. A. Fischer, *J. Agric. Sci.*, *105*: 447 (1985).
179. S. S. Bhuller and C. F. Jenner, *Aust. J. Plant Physiol.*, *13*: 605 (1986).
180. I. Sofield, L. T. Evans, M. G. Cook, and I. F. Wardlaw, *Aust. J. Plant Physiol.*, *4*: 785 (1977).
181. R. A. Fischer and D. HilleRisLambers, *Aust. J. Agric. Res.*, *29*: 443 (1978).
182. S. C. Torbit, R. B. Gill, A. W. Alldredge, and J. C. Liever, *J. Wildl. Manag.*, *57*: 173 (1993).
183. R. J. Jones and S. R. Simmons, *Crop Sci.*, *23*: 129 (1983).
184. D. M. N. Mbewe and R. B. Hunter, *Can. J. Plant Sci.*, *66*: 53 (1986).
185. T. L. Setter and B. A. Flannigan, *Ann. Bot.*, *64*: 481 (1989).
186. J. W. Gonzales, A. M. Rhodes, and D. Dickinson, *Plant Physiol.*, *58*: 28 (1976).
187. D. B. Dickinson, C. D. Boyer, and J. G. Velu, *Phytochemistry*, *22*: 1371 (1983).
188. W. Aufhammer, E. Nalborezyk, B. Geyer, I. Gotz, C. Mack, and S. Paluck, *J. Agric. Sci.*, *112*: 419 (1989).
189. T. L. Sage and B. D. Webster, *Bot. Gaz.*, *148*: 35 (1987).
190. C. S. Mauk and P. J. Breen, *J. Am. Soc. Hortic. Sci.*, *111*: 416 (1986).
191. J. W. White, S. P. Sing, C. Pino, M. J. Rios, and I. Buddenhagen, *Field Crop Res.*, *28*: 295 (1992).
192. Y. M. Stockman and R. Shibles, *Iowa State J. Res.*, *61*: 35 (1986).
193. L. J. Wiles and G. G. Wilkerson, *Agric. Syst.*, *35*: 37 (1991).
194. R. D. Guffy, J. D. Hesketh, R. L. Nelson, and R. L. Bernard, *Biotronics*, *20*: 19 (1991).
195. R. D. Guffy, B. L. Vasilas, and J. D. Hesketh, *Biotronics*, *21*: 1 (1992).
196. A. J. Gordon, *Carbon Partitioning Within and Between Organisms* (C. J. Pollock, J. F. Farrar, and A. J. Gordon eds.), BIOS Science Publishers, Oxford, p. 133 (1992).
197. L. R. Salado-Navarro, K. Hinson, and T. S. Sinclair, *Crop Sci.*, *25*: 451 (1985).
198. V. R. Franceschi and R. T. Giaquinta, *Planta*, *159*: 415 (1983).
199. M. N. Westwood, *Temperate-Zone Pomology, Phynology and Culture*, 3rd ed., Timber Press, Portland, OR (1993).
200. K. Ryugo, *Fruit Culture: Its Science and Art*, Wiley, New York (1988).
201. R. E. Byers, J. A. Barden, R. F. Polomoki, R. W. Young, and D. H. Carbaugh, *J. Am. Soc. Hortic. Sci.*, *115*: 14 (1990).
202. R. E. Byers, C. G. Lyons, Jr., K. S. Yoder, J. A. Barden, and R. W. Young, *J. Hortic. Sci.*, *60*: 465 (1985).
203. A. N. Miller and C. S. Walsh, *J. Am. Soc. Hortic. Sci.*, *113*: 309 (1988).
204. J. E. Hendrix, J. C. Linden, D. H. Smith, C. W. Ross, and I. K. Park, *Aust. J. Plant. Physiol.*, *13*: 391 (1986).
205. M. Bodson, *Planta*, *135*: 19 (1977).

206. M. Bodson and W. H. Outlaw, Jr., *Plant Physiol.*, *79*: 420 (1985).
207. M. Bodson and G. Bernier, *Physiol. Veg.*, *23*: 491 (1985).
208. P. Lejeune, G. Bernier, and J.-M. Kinet, *Plant Physiol. Biochem.*, *29*: 153 (1991).
209. N. Ishioka, S. Tanimoto, and H. Harada, *J. Plant Physiol.*, *138*: 573 (1991).
210. C. H. Engels and H. Marschner, *J. Exp. Bot.*, *37*: 1804 (1986).
211. C. H. Engels and H. Marschner, *J. Exp. Bot.*, *37*: 1813 (1986).
212. C. H. Engles and H. Marschner, *Potato Res.*, *30*: 177 (1987).
213. D. J. Midmore, J. Roca, and D. Berrios, *Field Crop Res.*, *18*: 141 (1988).
214. J. D. Midmore, D. Berrios, and J. Roca, *Field Crop Res.*, *18*: 159 (1988).
215. M. Nakatani, A. Oyanagi, and Y. Watanabe, *Jpn. J. Crop Sci.*, *57*: 535 (1988).
216. W. Claussen and F. Lenz, *Z. Pflanzenphysiol.*, *109*: 459 (1993).
217. J. Azcon-Bieto, *Plant Physiol.*, *73*: 681 (1983).
218. M. Stitt, *Plant Biology*, Vol. I, *Phloem Transport* (J. Cronshaw, W. J. Lucas, and R. T. Giaquinta, eds.), Liss, New York, p. 331 (1986).
219. M. Stitt, *Annu. Rev. Plant Physiol. Plant Mol. Biol.*, *41*: 53 (1990).
220. C. H. Foyer, *Plant Physiol. Biochem.*, *26*: 483 (1988).
221. M. Stitt and W. P. Quick, *Physiol. Plant.*, *77*: 633 (1987).
222. U. Sonnewald and L. Willmitzer, *Plant Physiol.*, *99*: 1267 (1992).
223. C. L. Mandahar and I. D. Garg, *Photosynthetica*, *9*: 407 (1975).
224. B. G. Bugbee and F. B. Salisbury, *Plant Physiol.*, *88*: 869 (1988).
225. D. W. Ming and D. L. Henninger (eds.), *Lunar Base Agriculture: Soils for Plant Growth*, ASA, SSA, Madison, WI (1989).
226. B. Bugbee and O. Monje, *BioScience*, *42*: 494 (1992).
227. K. A. Corey and R. M. Wheeler, *BioScience*, *42*: 503 (1992).

18

Solute Transport Across Plant Cell Membranes: Biochemical and Biophysical Aspects

Donald P. Briskin
University of Illinois, Urbana, Illinois

I. INTRODUCTION

Like all other living organisms, plants have the capability to maintain within their cells conditions that may differ substantially from those present in the external environment. This capacity is dependent on the presence of membranes, which define not only the outer boundary of the living plant protoplast but also internal organelles such as the central vacuole, the endoplasmic reticulum, mitochondria, plastids, the Golgi apparatus, and the nucleus. When bounding the cytoplasm or an internal organelle, membranes serve as selective barriers to free solute movement; hence they regulate solute flux between the cytoplasm and the cell exterior or the lumen of organelles. This control of the intracellular chemical environment is essential not only for metabolic reactions taking place in the cytoplasm but also for the specialized biochemical functions associated with organelles. Such specialized functions often require metabolic reactions occurring in organelles to be isolated from those taking place in the cytoplasm and for metabolite movement into and out of the organelle to be tightly regulated (see Ref. 1 and references therein).

In addition to having a central role in the maintenance of intracellular conditions, the coordination of membrane transport events taking place at numerous cellular sites often represents the basis for a number of essential functions occurring at the tissue, organ, and whole-plant levels. Examples of plant functions that depend on membrane transport include mineral nutrient uptake and allocation [2,3], photoassimilate partitioning [4,5], stomatal movements [6], growth and morphogenesis [7,8], and various beneficial symbiotic associations with microorganisms [9–11]. Recent studies have also provided strong evidence for the involvement of membranes and transport phenomena in a variety of signal transduction events important for plant responses to environmental cues and possibly, growth regulators [12–14].

The ability of membranes to regulate solute movement is intimately related to their chemical structure and the presence of transport systems for physiologically relevant solutes. This chapter focuses on the biochemical and biophysical aspects of plant membrane structure and transport phenomena. The involvement of membranes and transport processes in plant signal transduction events also is considered.

II. STRUCTURE OF PLANT MEMBRANES

Because of their central importance in cell physiology and organismal function, biological membranes have been the subject of intensive research since the late 1800s. A major objective of this research has been to develop an understanding of membrane structure, since this forms the basis for our knowledge of functions such as selective solute transport.

A. Models of Membrane Structure: A Historical Perspective

For almost a century it has been known that lipids represent a substantial structural component of plant membranes. From studies conducted by Overton [15] with *Hydrocharis* species, it was shown that the degree to which a number of solutes could penetrate the cell protoplast depended on their solubility in organic solvents. From this observation, it was concluded that at least the plasma membrane and most likely all membranes of the plant cell were composed of a "lipoid substance," likely consisting of fatty acids and cholesterol. A structural arrangement for the lipids present in biological membranes was proposed 26 years later by Gorder and Grendel [16], based on measurements of erythrocyte membrane surface area and phospholipid content. By assuming an approximate volume occupied by a phospholipid molecule, these workers estimated that about twice as much phospholipid was present as was necessary to account for the estimated surface area. From this observation, it was concluded that the phospholipids were likely present in a bilayer arrangement (Figure 1A). Although the individual results of this study are now known to be in error as a result of problems with surface area estimation and the phospholipid extraction method, these errors apparently canceled out to yield the correct answer.

From comparative measurements of surface tension between pure lipid layers and natural membranes (e.g., red blood cell plasma membrane) conducted about 35 years after Overton's work, it was proposed that membranes also contained protein [17]. However, this protein was thought to occur as a thin layer over the outer surface of the membrane and to serve as a structural rather than functional component (Figure 1B). It was recognized that such a model could not account for solute movement across the membrane, however, and later versions included proteinacous pores. With the advent of electron microscopy and electron-dense staining technology, early results appeared to support the Danielli–Davson-type model in that images of membranes demonstrated a "dark–light–dark" or "railroad track" staining pattern [18]. This pattern was thought to reflect selective binding of electron-dense stain to the protein layer on the membrane surface, as opposed to the lipid interior of the membrane (Figure 1B). However by the late 1960s, it was apparent a Danielli–Davson-type model involving a protein-coated lipid bilayer could not account for many physical as well as physiological properties of biological membranes.

Our current view of membrane structure is strongly influenced by the "fluid mosaic" model proposed by Singer and Nicolson in 1972 (see Ref. 19 for review) and shown in Figure 1C. To date, this model best explains both the structural and functional aspects of most biological membranes. This model will represent the focus for our consideration of the major biochemical components of plant membranes.

B. Lipids of Plant Membranes

The major class of lipids associated with plant membranes is composed of phospholipids and sterols. As shown in Figure 2, membrane phospholipids consist of two fatty acyl side chains

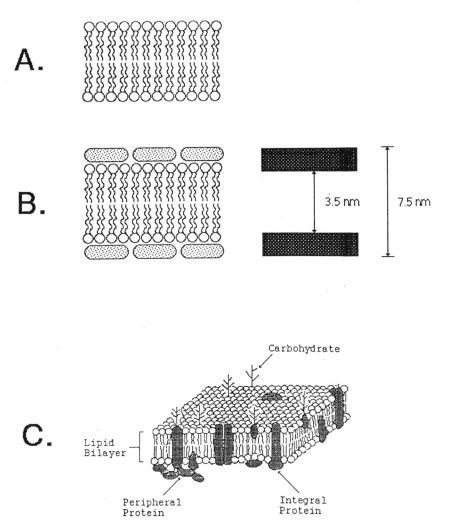

Figure 1 Models for the structure of cellular membranes. (A) Lipid bilayer model of Gorder and Grendel [16]. (B) Protein-coated lipid bilayer model of Danielli and Davson [17]. The pattern of electron-dense staining observed in electron microscopy and proposed by Robertson [18] to support this model is also shown, with molecular dimensions estimated. (C) The fluid mosaic model of Singer and Nicholson (described in Ref. 19), showing the association of integral and peripheral proteins with the lipid bilayer. Carbohydrate associated with membrane glycoprotein is also shown. For all models, membrane phospholipids (see Figure 2) are indicated by the "two-string balloon" symbol, and membrane proteins are indicated by the shaded globular structures.

esterified to the first and second hydroxyl groups of glycerol, while the third glycerol hydroxyl is linked to either phosphate alone (phosphatidic acid) or a phosphorylated "head group". The head groups generally observed consist of either phosphoserine, phosphoethanolamine, phosphocholine, phosphoglycerol, or phosphoinositol [20,21]. The fatty acids composing the acyl side chains esterified at positions 1 and 2 are generally 16–18 carbon units long and can include palmitic acid (16:0), oleic acid (18:1), linoleic acid (18:2), and linolenic acid (18:3) [20,22]. Since membrane phospholipids contain both hydrophobic (fatty acyl side chains) and hydrophilic (phospho head group) domains, these molecules are said to be "amphipathic." In the fluid

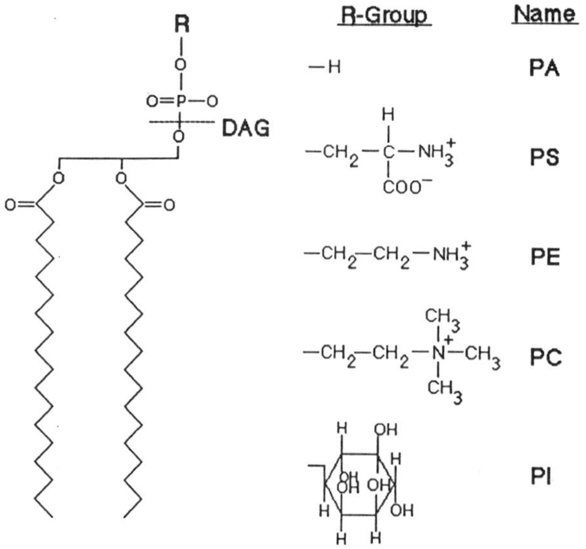

Figure 2 Chemical structure of common membrane phospholipids. The major phospholipids associated with plant membranes are shown with head groups (R) and nomenclature involving the following substances: PA, phosphatidic acid; PS, phosphatidylserine; PE, phosphatidylethanolamine; PC, phosphatidylcholine; PI, phosphatidylinositol. Plant membranes can also contain diacylglycerol (DAG) when a phosphorylated head group is absent.

mosaic model, membrane phospholipids are arranged as a lipid bilayer: the hydrophobic acyl side chains are present in the interior of the membrane, and the hydrophilic head groups are oriented to the membrane exterior [19]. This arrangement of the amphipathic phospholipids is most stable in an aqueous environment, which allows the hydrophobic acyl chains to cointeract and to be shielded from water molecules, while also allowing the hydrophilic head groups to strongly interact with water [19].

The sterols associated with plant membranes are also amphiphiles containing hydrophilic (hydroxyl) and hydrophobic (ring structure) domains. As such, these molecules orient themselves within the lipid bilayer of the membrane so that their hydroxyl group associates with the outer hydrophilic region and their complex ring structure associates with the hydrophobic interior region (Figure 3). While plant membranes contain only minor amounts of cholesterol, sterols such as sitosterol, stigmasterol, and campesterol are common [23].

C. Plant Membrane Proteins

Although earlier models of membrane structure considered protein to be in an extended conformation over the membrane surface [17], the more current view is that membrane proteins exist in a globular arrangement in association with the lipid bilayer (see Figure 1). This proposal was initially suggested after circular dichroism studies of membrane proteins indicated that substantial amounts of α-helix were present, similar to what was observed for soluble globular proteins [24,25]. In terms of their association or arrangement in the membrane, membrane proteins can be classified as either peripheral or integral [19]. Peripheral proteins though globular, are associated with the surface of the membrane by electrostatic interactions and/or hydrogen bonding either to lipid head groups or to other proteins. When dissociated from the membrane by treatment with salts or chelators, peripheral proteins retain their native conformation and enzyme activity in

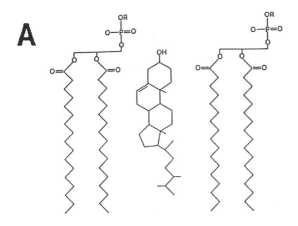

Figure 3 Orientation and chemical structure of the sterols associated with plant membranes. (A) Orientation of a sterol molecule adjacent to phospholipids in a upper bilayer half of a membrane. (B) Chemical structure of the major sterols present in plant membranes.

solution, similar to the behavior of soluble proteins [19]. Often, peripheral proteins may represent subunits of oligomeric enzymes associated with the membrane. In contrast, integral proteins are deeply embedded into the membrane and may even span the entire width of the lipid bilayer [19]. Integral proteins, like membrane lipids themselves, are amphipathic structures having both hydrophilic and hydrophobic domains. When present in the membrane, the hydrophobic domains of integral proteins associate with the nonpolar interior of the lipid bilayer, while the hydrophilic domains protrude from the surface of the membrane and interact with water. Unlike peripheral proteins, integral proteins cannot be removed from the membrane by treatment with salts or chelators. Instead, the removal or "solubilization" of integral proteins requires the use of detergent molecules to break the strong interaction between hydrophobic domains of the protein and the lipid interior [26]. Being amphipathic molecules, detergents also can interact with water, and hence stabilize the integral protein once it has been removed from the membrane. Although shown as vague globular structures in Figure 1, it should be remembered that both integral and peripheral proteins consist of a primary amino acid sequence twisted (α-helix) and folded back (β-pleated sheet or random coil) on itself to form the compact globular structure.

For membrane integral proteins, the peptide chain (primary amino acid sequence) can take a number of possible arrangements within the lipid bilayer. The arrangement of the integral protein peptide chain across the membrane is referred to as its "topology" [27], and the terminology developed by Blobel [28] is used in the following discussion (Figure 4). A single peptide portion of the protein might protrude only partway into the bilayer, then exit to the surface (monotopic protein) or extend all the way through the bilayer only once (bitopic protein). Alternatively, the peptide chain may cross the lipid bilayer more than once (polytopic protein) and have various arrangements of the N-terminal and C-terminal regions of the protein with respect to each other across the membrane. For proteins involved in the transmembrane

A

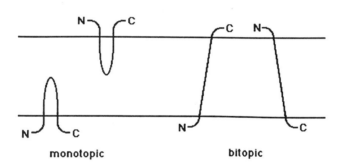

monotopic   bitopic

B

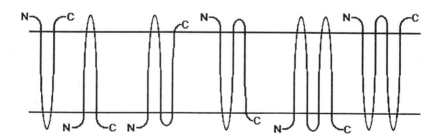

Figure 4 Possible arrangements for the peptides of integral membrane proteins according to the nomenclature of Blobel [28]. (A) Possible arrangements for monotopic and bitopic integral membrane proteins. Monotopic proteins associate with the membrane but do not cross the bilayer. Bitopic integral proteins traverse the bilayer only once. (B) Some possible arrangements for polytopic integral membrane proteins. Polytopic proteins traverse the membrane more than once and can have several possible arrangements of the N- and C-terminal regions based on both the origin of insertion in the membrane and whether the peptide crosses the membrane an even or odd number of times. (Adapted from Refs. 28 and 27.)

movement of solutes, the polytopic arrangement is most generally observed because it allows for the production of a protein-lined pathway for solute conduction. Based on theoretical arguments, it has been proposed that four to six transmembrane peptides may be required to form such a path for solute conductance across the membrane [29,30].

From both biochemical and molecular studies, it is apparent that there are particular requirements for the peptide regions of integral proteins that span the lipid bilayer (see Ref. 27 for review). These regions of integral proteins are α-helical and contain amino acids with nonpolar side chains. Since in the α-helical arrangement the side chains of amino acids point to the outside of the helix, the presence of nonpolar side chains would be in accordance with the nonpolar, hydrophobic environment of the membrane interior. The average lipid bilayer is about 6–8 nm thick, so it takes about 20 amino acids in an α-helix to cross the membrane from one side to another. These empirically determined "rules" for membrane-spanning regions of integral proteins have formed the basis for theoretical predictions of integral membrane protein topology that are often presented in the literature [27]. If the entire amino acid sequence can be determined—for example, by gene cloning and deduction of the amino acid sequence from the corresponding nucleic acid sequence—one of several computer programs can be used to predict membrane protein topology. In this approach, which is now quite common, putative membrane-spanning regions of the protein are denoted as portions of the sequence containing at least 20 nonpolar amino acids that promote formation of the α-helix. This approach must be viewed with some caution, however, because such sequence regions might instead be associated with non-membrane-spanning domains. For this reason, it has been pointed out that putative topological models must always be confirmed by direct biochemical methods [27].

D. Plant Membrane-Associated Carbohydrate

Depending on the cellular location, carbohydrate can represent either a relatively minor or a significant component of membranes relative to lipid and protein [22]. Most often, the carbohydrate represents the polysaccharide side chains of membrane glycoproteins. However, glycolipids can also be present to a minor extent. The carbohydrate associated with glycoproteins appears to take on an asymmetric arrangement with respect to the bilayer, being present on the membrane surface facing away from the cytoplasm. The best demonstration of this asymmetry for the plasma membrane and vacuolar membrane comes from cytochemical localization studies using labeled lectins where staining can be observed to occur on the outside surface of the protoplast [31] and the inside, luminal surface of the vacuole [21].

E. Asymmetry in Membrane Structural Organization

In the fluid mosaic model, membranes represent asymmetric structures with respect to protein composition on different sides of the membrane. This key property is absolutely essential for the functional activity of transport systems associated with membranes, since a defined orientation of the proteins mediating these processes is required (see Ref. 19 and references therein). Although this configuration is better documented in animal cell systems (see Ref. 32 and references therein), asymmetry with respect to the lipid composition of each bilayer half of the membrane also has been attributed, at least for the plant plasma membrane, to differences in isoelectric point between the cytoplasmic and apoplastic facing surfaces [33]. While two-dimensional lateral movement of proteins in the plane of the membrane was originally thought to be purely random, it is now apparent that the position and movement of specific proteins in a membrane may be controlled by elements of the cytoskeletal system (see Ref. 34 for discussion). In plants, there is some evidence that the plasma membrane associated glucan synthetase complexes (rosettes) responsible for cellulose synthesis may be controlled by microtubules present on the cytoplasmic side of the membrane [35].

F. Limitations of the Fluid Mosaic Model for Membrane Structure

The fluid mosaic model as often depicted (i.e., as in Figure 1) presents an idealized view of a membrane with wide regions of lipid and a small array of peripheral and integral proteins. However, it should be pointed out that membranes from different organelles can differ substantially in their relative protein and lipid content. For example, a membrane with low protein content, such as the vacuolar membrane, could be visualized as containing a few proteins in a sea of lipid [21]. Conversely, a membrane with high protein content, such as the mitochondrial inner membrane (about 85% protein), could be visualized as a proteinaceous membrane containing isolated patches of lipid (see Ref. 36 and references therein). Hence, the fluid mosaic model should be taken to represent a generalized view of membrane structure that may vary depending on cellular membrane origin.

III. NATURE OF SOLUTE MOVEMENT ACROSS PLANT MEMBRANES

For a solute to either enter or exit the plant cell or an internal organelle, it must cross at least one membrane during its transit. Depending on the chemical nature of the solute, two general routes for movement across membranes may be possible. If sufficiently lipid soluble, solutes may move directly through the bilayer without the intervention of a membrane protein. On the other hand, if the solute is relatively hydrophilic, as are most metabolites and mineral nutrients, the involvement of a membrane-associated transport system is necessary for translocation.

A. Movement of Lipid-Soluble Solutes

For direct movement through the lipid bilayer of the membrane, a lipid-soluble solute present in aqueous solution must dissolve into the hydrophobic interior of the membrane, move across the bilayer, and then re-dissolve into the aqueous solution on the other side of the membrane [37]. In the absence of an electrical or pH difference across the membrane, an uncharged solute without ionizable groups will move across the membrane by simple diffusion. However, the presence of an electrical potential or pH difference across the membrane can provide a driving force for solute movement through the bilayer, depending on whether the solute is charged or contains ionizable groups.

1. Simple Diffusion Across the Membrane

When solute movement across the membrane takes place by diffusion, translocation will occur from a higher concentration to a lower concentration at a rate according to:

$$\Phi = D_j K_j \frac{(C^{\text{out}} - C^{\text{i}})}{\Delta x}$$

where Φ is the solute flux across the membrane, C^{out} and C^{in} are the solute concentrations on the outside and inside surface of the membrane, D_j is the diffusion coefficient for the solute within the membrane, K_j is the partition coefficient, and Δx represents the thickness of the lipid bilayer. In this relationship, the solute concentration gradient across the membrane represents the driving force for movement and the partition coefficient represents the parameter that takes into consideration the tendency of the solute to dissolve into the lipid bilayer. From a practical standpoint, the partition coefficient represents the ratio of solute solubility in a nonpolar versus a polar solvent [37].

The nature of solute movement within the lipid bilayer is quite different from that occurring for solute movement in free solution. Although the interior of the bilayer is normally in a fluid state, solute movement differs from that occurring in free solution because the membrane interior as a "solvent" consists of a network of polymers (fatty acyl side chains) oriented parallel to the direction of transmembrane movement [37]. Because of this difference, solute movement across the bilayer shows a steeper dependence on solute size and temperature than that observed for diffusive movement in free solution [37]. When considering the rate of movement across the lipid bilayer, this difference explains the need to use a special solute diffusion coefficient (D_j) for the bilayer interior, as opposed to a diffusion coefficient for solute movement in free solution.

2. Solute Movement Driven by Electrical or pH Gradients Across the Membrane

In addition to a solute concentration gradient, the presence of an electrical and/or pH difference across plant membranes can serve to drive solute movement across the lipid bilayer. In contrast to solute movement driven by its own concentration gradient, these driving forces can lead to solute accumulation on one side of the membrane or within a membrane-bound compartment. For solutes known as "lipiphilic ions," which are charged, yet reasonably lipid soluble, the presence of an electrical difference across a membrane can result in an "electrophoretic" movement across the bilayer. At equilibrium, the extent of lipiphilic ion accumulation across the membrane can be described according to the Nernst equation, shown here in an approximate form for a temperature of 25°C:

$$\Delta\Psi = \frac{-59.2}{z_j} \ln \frac{C^{in}}{C^{out}}$$

where $\Delta\Psi$ is the membrane electrical difference (inside with respect to outside of membrane, in millivolts), z_j is the charge on the solute, and C^{out} and C^{in} represent the final equilibrium levels (concentration) of the solute on the respective sides of the membrane.

For solutes that are either weak acids and lipid soluble when protonated or weak bases and lipid soluble when deprotonated, the presence of a pH difference across a membrane can lead to solute movement by a process known as "ion trapping." Based on the magnitude of the pH difference across the membrane, the actual pH values on each side of the membrane, and the pK_a of the ionizable group, ion trapping will result in the accumulation of either a weak acid on the alkaline side of the membrane or a weak base on the acid side of the membrane (Figure 5). Quantitatively, the distribution of the weak acid anion (A^-) relative to the pH gradient across a membrane ($\Delta pH = pH^{in} - pH^{out}$) can be described according to:

$$\Delta pH = \ln \frac{C^{in}_{A^-}}{C^{out}_{A^-}}$$

while that for the weak base cation (BH^+) can be described according to:

$$\Delta pH = \ln \frac{C^{out}_{BH^+}}{C^{in}_{BH^+}}$$

Solute movement across membranes by ion trapping can have great physiological significance. For example, it is thought that ion trapping may represent a primary basis for the cytoplasmic accumulation of the plant growth regulator, abscissic acid [38]. The abscissic acid molecule is a weak acid that is membrane permeant in the protonated form. As such, this molecule would accumulate in the cytoplasm relative to more acid cellular compartments, such as the central vacuole or apoplast. Likewise, many herbicides are weak acids that are membrane permeant when protonated. This property is thought to be essential for their movement into the cytoplasm of plant cells following application [39].

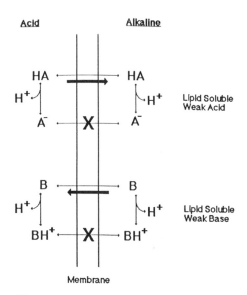

Figure 5 Mechanism of lipid-soluble weak acid and weak base movement across membranes by ion trapping. Bold arrows indicate the direction of net movement across the membrane. For weak acids, the protonated form of the acid is membrane permeable and will equilibrate across the membrane, while the deprotonated form is nonpermeant. Deprotonation of the weak acid on the alkaline side of the membrane drives further movement of the protonated form into the alkaline compartment. For weak bases, the deprotonated form is membrane permeable and equilibrates across the membrane, while the protonated form is nonpermeant. Protonation of the base on the acid side of the membrane then drives movement of the deprotonated form into the acidic compartment.

B.   Involvement of Membrane-Associated Transport Systems in Mineral Nutrient and Metabolite Movement Through Membranes

For ions and most metabolites that contain polar groups, the hydrophobic environment of the lipid bilayer represents a significant barrier to direct movement through the membrane. Instead, the transmembrane movement of these solutes will necessarily involve the activity of membrane proteins functioning as transport systems. The existence of transport systems for specific solutes represents the basis for the selectivity in solute accumulation observed for plant membranes and for linking transport to metabolic energy [40]. Clearly, the function of specific transport systems in plant membranes is essential for such processes as mineral nutrient acquisition by plant roots [2,3], phloem transport [5], and metabolite transport into and out of cells [41] and organelles [1]. To mediate the transmembrane movement of a solute, some components of these transport systems must consist of integral membrane proteins that span the lipid bilayer. Either large single peptide subunits or a series of smaller peptides that collectively cross the bilayer may be used. In spanning the membrane, the proteinaceous transport system provides a hydrophilic pathway for movement of polar solutes across the hydrophobic lipid bilayer. Furthermore, specific protein–solute interactions associated with binding and translocation events can allow for the precise specificity observed for transport processes in vivo.

The involvement of a proteinaceous transport system in mediating solute movement across plant membranes has often been viewed as analagous to an enzyme acting on its substrate. However, instead of conducting chemical work to transform the substrate into a product, the transport system conducts the physical work of moving the solute across the distance of the membrane. This enzyme analogy has been extended even further in the kinetic characterization of transport processes, with the observation that solute flux across a membrane measured as a function of solute concentration often displays saturation kinetics (Figure 6). Such a relationship for transport can be described by a modified version of the Michaelis–Menten equation used in steady state enzyme kinetics:

$$\Phi = \Phi_{max} \frac{C_s}{K_s + C_s}$$

where Φ represents the solute flux, C_s the solute concentration, Φ_{max} an asymptotic maximal flux (analogous to V_{max}), and K_s the substrate level at which Φ equals $0.5\,\Phi_{max}$ (i.e., analogous to K_M).

This kinetic approach has been especially useful in attempts to determine whether different solutes might be transported across a membrane by the same transport system. Again, by analogy to enzyme kinetics, the use of a single transport system by two different solutes should yield a mutual competitive inhibition of the flux of each solute by the other. In a kinetic analysis, this movement would be reflected by increased values of K_j with no effect on Φ_{max} when the flux of one solute is measured in the presence of the other.

While the enzyme kinetic analogy has proven useful in characterization of numerous transport processes (see Ref. 42 for examples), it does not have general applicability. There are examples of transport processes, often measured with tissues or whole cells, that substantially deviate from simple saturation. In the case of both K^+ [43,44] and sucrose uptake [45] into

A

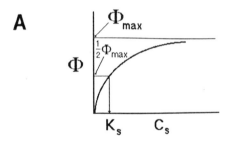

B

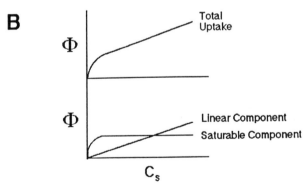

Figure 6 Kinetic analysis of solute transport. (A) Use of the enzyme kinetic analogy for transport processes displaying saturation kinetics. This hypothetical plot shows the solute flux into cells (Φ) as a function of external solute concentration (C_S). As in enzyme kinetics, a K_S value analogous to the K_M value can be estimated as the solute concentration at half the estimated maximum flux (Φ_{max}), analogous to V_{max}. (B) Hypothetical plot for solute uptake displaying nonsaturation. As discussed by Kochian and Lucas [43], this complex kinetic profile can be described as the sum of a saturable and linear nonsaturable uptake profiles.

plant cells, the kinetics of uptake has been described as representing the sum of a saturable process together with a linear nonsaturable process (Figure 6). While this description often considers the transport process to involve, most likely, the participation of multiple transport systems, attemps have also been made to describe complex kinetic phenomona in terms of single carriers displaying multiphasic kinetics [46] or negative cooperatively [47,48].

IV. MECHANISM OF ENERGY COUPLING TO SOLUTE TRANSPORT ACROSS PLANT MEMBRANES

The production and maintenance of selective solute gradients across plant membranes requires an input of energy generated by metabolic processes. This is clearly shown in experiments in which plant cell energy metabolism is inhibited by treatments such as low temperature, anoxia, and metabolic poisons [48,49]. Such an inhibition of cellular energy metabolism leads to a rapid loss in transport capacity for a number of solutes (see Ref. 49 and references therein). The basis for linking metabolic processes to solute transport consists of energy currency molecules such as ATP and PP_i. Free energy available from the hydrolysis of these compounds is utilized to energize solute transport in a stepwise process involving coordination of transport systems and generation of a proton electrochemical gradient across the membrane.

A. Definition of Primary and Secondary Transport Processes: "Pumps, Carriers, and Channels"

The membrane-associated transport systems of plant cells can be classified as "pumps," "carriers," or "channels" (see Refs. 32, 37, and 41, and references therein). Pumps are transport systems that couple an energy-releasing chemical reaction such as ATP hydrolysis to the transmembrane movement of a solute. Some examples of pumps associated with plant membranes include H^+-translocating ATPases associated with the plasma membrane [50] and vacuolar membrane [51], Ca^{2+}-translocating ATPases associated with the plasma membrane [52] and endoplasmic reticulum [53], and an H^+-translocating pyrophosphatase also associated with the vacuolar membrane [54].

Transport pumps are also often called primary transport systems, since the linkage of transport to a metabolic reaction is direct. In contrast, solute carriers are transport systems that mediate the movement of one or more solutes across the membrane without direct coupling to a chemical reaction. For this reason, carriers, unlike pumps, generally lack an associated enzyme activity. Carriers can mediate the transmembrane movement of a single solute alone (uniport), or the movement of one solute may be coupled to the movement of another solute in the same (symport) or the opposite direction (antiport). While channels, like uniport carriers, mediate the transport of a single solute across the membrane, the major difference between these two transport systems lies in the mechanism by which transmembrane movement takes place. In carriers, the transmembrane movement of a solute is achieved by a series of conformational changes in the carrier protein associated with solute binding, translocation across the membrane, and solute release [32,37]. With channels, on the other hand, transmembrane movement occurs through a proteinaceous pore whose opening and closing are closely regulated in response to chemical or environmental signals [32,55]. This difference in mechanism of solute transport used by carriers and channels explains why solute transport by channels is so much faster than that observed for carriers and why it requires the use of such special techniques for measurement as the patch clamp method [55]. While carriers can be involved in the transport of either ions or organic molecules, channels are typically involved in ion transport alone and represent an important component of cell signal transduction systems [32,55]. Because neither carriers nor channels directly link transport to a chemical reaction, they are termed secondary transport systems. As discussed next, transport processes are linked to the chemical energy generated by cell metabolism on the basis of the presence and coordination of primary and secondary transport systems in plant membranes.

B. Energy Coupling for Solute Transport Across Plant Membranes

From a number of studies conducted over the past two decades, it is apparent that the coupling of metabolic energy to membrane transport in plants [41,56] and other organisms depends on the integrated functioning of primary and secondary transport systems [32]. Figure 7 outlines this linking of primary and secondary transport system function. In plant membranes such as the plasma membrane, the vacuolar membrane, and very likely the Golgi membrane, a primary transport system utilizes a hydrolytic reaction for a chemical substrate to drive the transport of H^+ across the membrane. This results in the production of a proton electrochemical gradient across the membrane consisting of a pH difference and an electrical difference. This proton electrochemical gradient or "protonmotive force" can then be utilized by other secondary transport systems in the membrane as a driving force for solute movement. Because H^+-linked symports and antiports are associated with the membrane, the proton electrochemical gradient can be utilized to drive the transport of solutes in either direction across the membrane and to allow for the uphill transport of a solute against its own electrochemical gradient. Depending on solute charge and the stoichiometry of H^+/solute co- or countertransport, these H^+-linked carriers may also be able to utilize the electrical difference across the membrane as an additional driving force for transport (see Figure 8). For uniports mediating the transport of charged solutes, the electrical difference component of the proton electrochemical gradient can be utilized to drive solute movement and accumulation in an electrophoretic manner. In the case of channels mediating ion flux, the electrical component of the proton electrochemical gradient can also serve as a driving force for ion movement, when the channel is in the open-gated mode.

Although plant cells use the proton electrochemical gradient as a driving force for solute transport across cell membranes (i.e., plasma membrane, vacuolar membrane, Golgi membrane), the concept of a protonmotive force was initially conceived by Nobel laureate Peter Mitchell to explain energy transduction taking place in mitochondria, chloroplasts, and bacterial cells [57]. In the case of mitochondria and chloroplasts present in plant cells, a proton electrochemical gradient is involved in the indirect linking of primary transport processes associated with oxidation–reduction reactions and ATP synthesis. At the inner mitochondrial membrane, the oxidation of NADH and succinate by the electron transport chain results in the

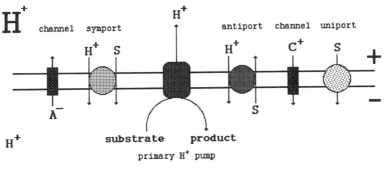

Figure 7 Energy coupling for solute transport through the action of primary and secondary transport systems. A primary transport system couples a chemical reaction such as ATP hydrolysis to the transport of protons across the membrane. This generates a pH and an electrical difference across the membrane which can be utilized by secondary transport systems such as uniports, symports, antiports, and channels.

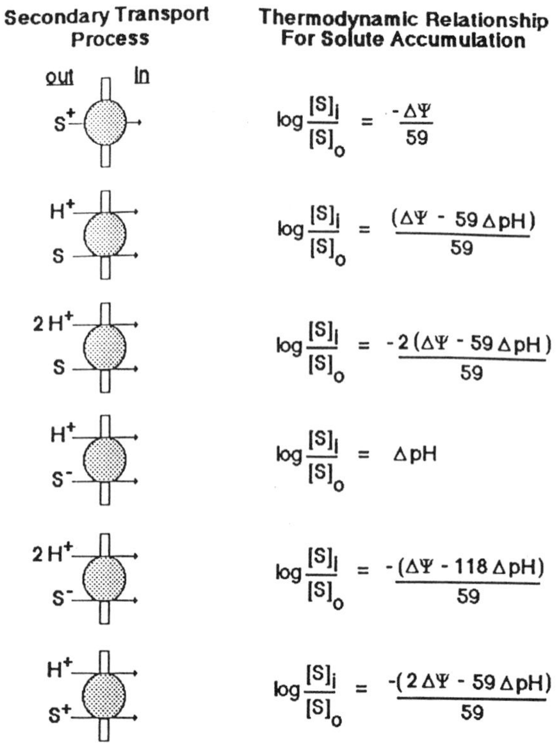

Secondary Transport Process　　**Thermodynamic Relationship For Solute Accumulation**

$$\log \frac{[S]_i}{[S]_o} = \frac{-\Delta\Psi}{59}$$

$$\log \frac{[S]_i}{[S]_o} = \frac{(\Delta\Psi - 59\,\Delta pH)}{59}$$

$$\log \frac{[S]_i}{[S]_o} = \frac{-2(\Delta\Psi - 59\,\Delta pH)}{59}$$

$$\log \frac{[S]_i}{[S]_o} = \Delta pH$$

$$\log \frac{[S]_i}{[S]_o} = \frac{-(\Delta\Psi - 118\,\Delta pH)}{59}$$

$$\log \frac{[S]_i}{[S]_o} = \frac{-(2\,\Delta\Psi - 59\,\Delta pH)}{59}$$

Figure 8 Thermodynamic relationships describing solute transport by various secondary transport systems. The relationships describe the distribution or accumulation ratio for a solute across a membrane at equilibrium. For these relationships, the membrane electrical potential ($\Delta\Psi$) is quantitated as the electrical difference across the membrane inside and outside ($\Delta\Psi = \Psi_{inside} - \Psi_{outside}$). For accumulation, an acid–exterior pH difference (ΔpH) is also assumed to be present, and these relationships are approximated for 25°C. (Adapted from F. M. Harold, *The Vital Force: A Study of Bioenergetics*, Freeman, New York, 1986.)

production of a transmembrane proton electrochemical gradient, which is then utilized by the F_1F_0 ATP synthase to produce ATP from ADP and P_i. Likewise, at the chloroplast thylakoid membranes, light-driven oxidation–reduction reactions result in the production of a proton electrochemical gradient that is then used by the CF_1CF_0 ATP synthase for ATP synthesis in a similar manner.

C. Proton Electrochemical Gradients as a Driving Force in Plant Membrane Transport: Thermodynamics and Quantification

Energization of solute transport across plant membranes by primary and secondary transport systems is possible because the proton electrochemical gradient linking these processes represents an interconvertable form of energy. As shown in Figure 7, primary transport systems (such as H^+-ATPases) couple chemical free energy available from substrate hydrolysis to the establishment of a proton electrochemical gradient across the membrane. Depending on the extent of buffering and charge compensation, this proton electrochemical gradient can consist of both a proton concentration ($[H^+]$) difference and an electrical potential difference ($\Delta\Psi$) across the membrane, where the total free energy available (kcal mol^{-1}) can be described by:

$$\Delta G_{H^+ \text{ electrochemical grad.}} = \Delta\mu H^+ = F\Delta\Psi + 2.3\ RT\ \log\frac{[H^+]_{\text{in}}}{[H^+]_{\text{out}}}$$

where R represents the gas constant and T the absolute temperature. The term "protonmotive force" ($\Delta\mu H^+$) has often been used to describe this free energy present in a proton electrochemical gradient. From the definition of both pH (i.e., $-\log[H^+]$) and the pH gradient across a membrane ($\Delta pH = pH^{\text{in}} - pH^{\text{out}}$), this equation can be simplified to:

$$\Delta\mu H^+ = F\Delta\Psi - 2.3\ RT\ \Delta pH$$

The magnitude of the protonmotive that can be established by a primary transport system is, in turn, related to the free energy available from substrate hydrolysis. At equilibrium, this can be described by:

$$\Delta G_{\text{substrate hydrolysis}} = n\Delta\mu H^+$$

where n is the number of protons transported with each catalytic cycle of substrate hydrolysis. This equation demonstrates that the stoichiometry of proton transport by a primary transport system is an important factor governing protonmotive force generation by primary H^+-transport systems. As proton transport stoichiometries exceed one, a smaller protonmotive force can be generated with a given amount of free energy available from substrate hydrolysis.

In the utilization of the protonmotive force by secondary transport systems, a second energy interconversion process takes place, involving conversion of the energy available in a proton electrochemical gradient to the work of moving a solute across the membrane. Figure 8 shows thermodynamic relationships describing the linkage of the protonmotive force to solute transport for several types of secondary transport. It is apparent that with a given proton electrochemical gradient present across a membrane, the extent to which a secondary transport system can establish a solute gradient is highly dependent on the H^+/solute stoichiometry of the coupled transport process and on whether transport involves a net movement of charge.

V. PRIMARY AND SECONDARY TRANSPORT SYSTEMS ASSOCIATED WITH PLANT CELL MEMBRANES: ORGANIZATION AND BIOCHEMICAL CHARACTERISTICS

From studies conducted over the past 20 years, it is apparent that the linking of primary and secondary transport systems through a transmembrane proton electrochemical gradient represents the basis for energy coupling to solute transport activity at the plasma membrane and tonoplast of plant cells. A number of the transport systems associated with these membranes have been well characterized using biochemical and, now, molecular techniques. Furthermore, the integration of primary and secondary transport has been demonstrated for these membranes using intact plant cells (for the plasma membrane), isolated vacuoles, and isolated membrane vesicles. At this point, only primary transport processes have been characterized for the endoplasmic reticulum and possibly the Golgi apparatus. As yet, there is no evidence that proton electrochemical gradients are used in transport energization at these two internal membranes.

A. Overview of Primary and Secondary Transport System Organization in Plant Cells

The primary and secondary transport systems associated with plant cellular membranes that have been identified and studied are shown in Figure 9. Each transport system shown has been directly characterized using either intact cells, intact organelles, or isolated membrane vesicle preparations generated by cell fractionation. At the plasma membrane, an H^+-translocating ATPase utilizes the free energy available from ATP hydrolysis to generate an inwardly directed

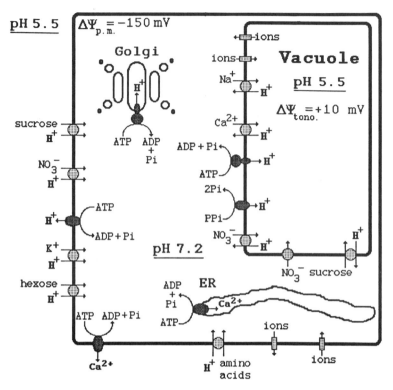

Figure 9 Summary of transport processes characterized in plant cells. Proton electrochemical gradients established by primary transport systems at the plasma membrane (H$^+$-transporting ATPase) and vacuolar membrane (H$^+$-transporting ATPase and pyrophosphatase) serve to drive solute movement across these membranes via secondary transport systrems (H$^+$-linked symports and antiports, uniports, and channels). Separate Ca^{2+}-transporting ATPases are also present at the plasma membrane and endoplasmic reticulum (ER), while an H$^+$-transporting ATPase similar to the vacuolar H$^+$-ATPase is present at the Golgi membrane.

proton electrochemical gradient [50]. This gradient consists of an acid–exterior pH difference of about 1.5–2.0 units and an interior–negative membrane electrical potential difference between about −100 and −150 mV [50,56].

Secondary transport carriers associated with the plant plasma membrane that have been characterized include an H$^+$/NO$_3^-$ symport [60], an H$^+$/K$^+$ symport [61], H$^+$/amino acid symports [62], and an array of ion channels demonstrated the use of the patch clamp method [55].

At the vacuolar membrane, an H$^+$-translocating ATPase, which differs biochemically from the plasma membrane H$^+$-ATPase, generates an outwardly directed proton electrochemical gradient consisting primarily of an acid–interior pH difference of about 2 units and a small interior–positive membrane potential of about 10–20 mV [51,56]. Interestingly, this membrane also contains an H$^+$-translocating pyrophosphatase as a primary transport system, operating in parallel with the H$^+$-ATPase for generation of the proton electrochemical gradient [54]. The pyrophosphate produced from a number of metabolic reactions taking place in the cytoplasm can be used by the pyrophosphatase as a chemical energy source for driving transport [54]. As at the plasma membrane, the proton electrochemical gradient across the vacuolar membrane is utilized by a number of secondary transport systems which have been identified, including an H$^+$/Na$^+$ antiport [63], an H$^+$/Ca^{2+} antiport [64], an H$^+$/sucrose antiport [65], an NO$_3^-$ uniport [66], an H$^+$/NO$_3^-$

antiport [67], and ion channels examined using patch clamp methods [55]. There is also some evidence for the presence of an H^+-translocating ATPase at the Golgi membrane that may be biochemically similar to the H^+-translocating ATPase associated with the vacuolar membrane [51].

While the transport of a number of solutes appears to be linked to the proton electrochemical gradient, it is now apparent that the transport of Ca^{2+} out of the cytoplasm at two cellular membranes utilizes a unique system of primary transport ATPases. These Ca^{2+}-translocating ATPases at the plasma membrane [54] and endoplasmic reticulum [53] have an important role in the maintenance of the low cytoplasmic concentration levels of Ca^{2+} (about 100 nM) required for the function of this divalent cation as a second messenger in cellular signal transduction [12,53]. This function is supplemented by the activity of the H^+/Ca^{2+} antiport present at the vacuolar membrane [64]. The roles of transport and membrane structural components in signal transduction are considered in Section VI.

B. Biochemical Characteristics of Primary and Secondary Transport Systems Associated with Plant Cell Membranes

The biochemical characteristics of only a limited number of plant membrane transport systems are known, based on studies conducted primarily with isolated membrane and purified protein preparations. In this respect, the bulk of available biochemical information is for primary transport systems, since these were more easily examined in isolated membrane vesicle fractions because of their associated enzyme activity. The measurement of this enzyme activity, using standard assay techniques, has allowed biochemical properties to be directly determined and has simplified quantification of the transport system during purification. In contrast, work with secondary transport systems has been more difficult, since these proteins have no associated enzyme activity. With the later development of methods for producing transport-competent membrane vesicles from plant cells, the transport characteristics of both primary and secondary transport systems could be evaluated. The more recent application of molecular methods has resulted thus far in the successful cloning of genes for only a few transport systems. For the transport systems that have been characterized in this manner, molecular approaches have allowed deduction of the amino acid sequence of the transport system protein(s) and the development of models for protein arrangement across the membrane (topology).

1. Plasma Membrane H^+-ATPase

Biochemical characterization of the plant plasma membrane H^+-ATPase began with early studies by Hodges and colleagues, who examined the ATP hydrolytic activity of this enzyme in plasma membrane fractions isolated from plant roots [47,48]. From this work, it was apparent that the enzyme demonstrated properties that distinguished it from other phosphohydrolyzing enzymes such as phosphatases, which might also be present in plant membrane fractions. These properties included a pH optimum at 6.5, magnesium dependence with further stimulation by K^+, and a pronounced substrate specificity for ATP [48].

Although it was reasoned in these early studies that the hydrolytic activity being measured most likely was involved in transport, no direct evidence for a transport role was shown. However, two properties of ATP hydrolytic activity appeared to be consistent with such a role in transport, namely, the apparent similarity in the complex kinetics for K^+ stimulation of ATP hydrolysis and ($^{86}Rb^+$) K^+ uptake by roots [47], and the general similarity between the specificity of monovalent cations (K^+, Rb^+, Cs^+, Na^+, Li^+) for stimulating ATP hydrolysis and their levels of uptake [68]. These results were interpreted to indicate a possible direct role in K^+ transport, where the ATPase would act as an H^+/K^+ exchange pump (see Ref. 69 and references therein).

With the subsequent development of methods for producing transport-competent plasma membrane vesicles (see Ref. 70 and references therein), direct evidence that the plasma membrane H^+-ATPase could couple ATP hydrolysis to H^+ extrusion was presented [71–73]. Using transport-competent plasma membrane vesicles, a stoichiometry of one proton transported per ATP molecule hydrolyzed was shown, indicating that the enzyme had the capability to produce a pH gradient of up to 7 units across the membrane, given typical values for the magnitude of the membrane electrical potential difference [74]. This gradient, being much greater than the 2–2.5 unit difference generally observed [75], demonstrated that the activity of the enzyme could account for the observed proton electrochemical difference. Although most researchers are of the view that the plasma membrane H^+-ATPase acts only as a primary H^+ extrusion pump and hence establishes the driving force for solute transport at the plasma membrane, the question of direct involvement in K^+ influx has not been fully resolved [50,69,76].

Both the ATP hydrolytic and H^+-transport activities of the plant plasma membrane H^+-ATPase are strongly inhibited by orthovanadate [76–79]. This indicates that the enzyme forms a covalent phosphorylated enzyme intermediate with the terminal phosphate group of ATP during the catalytic cycle of ATP hydrolysis and H^+ transport [50,77]. Enzyme assays conducted in the presence of γ-^{32}P-labeled ATP have directly demonstrated such an intermediate [80,81]. The phosphorylated intermediate is sufficiently stable if the reaction is stopped in ice-cold trichloroacetic acid so that the radiolabeled phosphate group can allow covalent labeling of the ATPase catalytic subunit [80–82]. Even before the enzyme was purified, covalent labeling, in turn, allowed identification of the ATPase catalytic subunit as a 100 kD polypeptide on electrophoretic gels. The formation of a phosphoenzyme has also proven useful in kinetic studies to elucidate the step-by-step reaction mechanism for ATP hydrolysis, using transient state kinetic methods [83,84]. These results, together with studies conducted on enzyme exchange reactions involving inorganic ^{32}P [85], have allowed the development of a fairly complete understanding of the enzyme mechanism for ATP hydrolysis (Figure 10) (see Refs. 50 and 77 and references therein).

It should be noted that transport ATPase, that form phosphorylated intermediates during their catalytic/transport cycles are often referred to as "P-type" transport enzymes [86]. This is in contrast to "V-type" ATPases (comprising the vacuolar ATPases) and "F-type" ATPases (comprising the ATP synthases of mitochondria, chloroplasts, and bacteria) [87]. The V- and

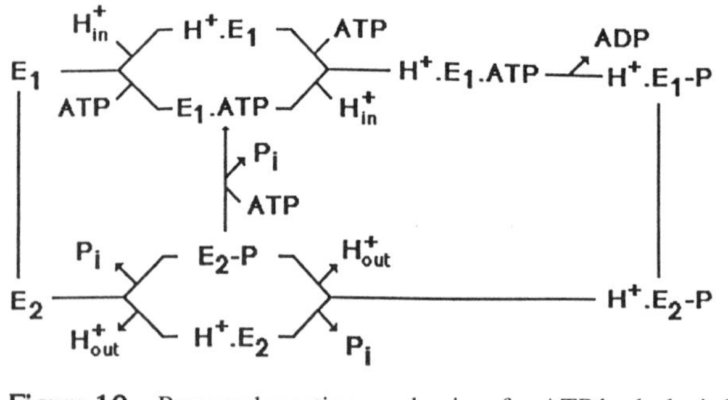

Figure 10 Proposed reaction mechanism for ATP hydrolysis by the plant plasma membrane H^+-ATPase. This reaction scheme is largely based on kinetic studies performed on the enzyme (see Refs. 50 and 77 for details), using radiolabeled substrate and products. The enzyme is present in two general reaction states E_1 and E_2, which include their phosphorylated forms E_1-P and E_2-P. This scheme also shows hypothetical coupling of the hydrolytic reaction to H^+ binding, translocation, and release. In this scheme, it is proposed that both H^+ and ATP binding and H^+ and P_i release occur as random ordered processes. In addition to a kinetic route around the outside of the pathway, a possible shortcut route by an ATP-mediated release of P_i is indicated. (Adapted from Ref. 77.)

F-type ATPases do not form phosphorylated intermediates, and their biochemical and structural characteristics are strikingly different.

The plasma membrane H^+-ATPase has been purified and reconstituted into phospholipid vesicles. Purification of the enzyme results in enrichment of a single 100 KD band on sodium dodecylsulfate gels (see Refs. 88 and 89 for examples) that is similar to what was observed from phosphoenzyme labeling [80–82]. When H^+ transport was measured for the enzyme purified and reconstituted into liposomes, properties similar to those originally observed for ATP hydrolytic activity were found with respect to substrate specificity, Mg^{2+} dependence, K^+ stimulation, pH optimum, and inhibitor sensitivity (see Refs. 50, 76, and 77 and references therein). From radiation inactivation analysis, a dimeric structure of 100 kD subunits for the native enzyme has been suggested [90,91]. Such a model for the native plasma membrane H^+-ATPase is shown in Figure 11.

Through molecular techniques, the gene for the plasma membrane H^+-ATPase was cloned [92–94]. This allowed deduction of the amino acid sequence for the enzyme and the development of a topological model for how the ATPase polypeptide is organized in the membrane (Figure 12). The predicted plasma membrane H^+-ATPase peptide sequence consisted of 999 amino acids, an amount consistent with the 100 kD size observed on electrophoretic gels [92]. With an N-terminus on the cytoplasmic side of the membrane, the enzyme is organized across the membrane with eight to ten transmembrane peptide regions. Based on similarities in regions of the amino acid sequence to those for much better characterized transport ATPases such as the animal cell Na,K-ATPase (see Refs. 32 and 37 and references therein), functional domains involved in ATP hydrolysis and transport have been proposed for this topological model [50,76,77]. The ATP binding region would represent the portion of the peptide at which ATP would bind during catalysis. The kinase region would represent the portion of the protein involved in formation of the phosphorylated intermediate, while the phosphatase region would represent the portion of the protein involved in phosphate release. The region labeled (energy) "transduction domain" represents the part of the peptide around the site, where the terminal phosphate group of ATP forms a covalent phosphointermediate at an essential aspartyl group. It is believed that collectively, the transmembrane domains, or at least four to six transmembrane peptides form a proteinaceous path involved in the transmembrane H^+-transport process [50]. From further molecular studies, it has become apparent that the plasma membrane H^+-ATPase is present in several isoforms as a multigenic family [95]. At present, 10 different isoforms have been demonstrated, some of which are expressed in different regions of the plant [95].

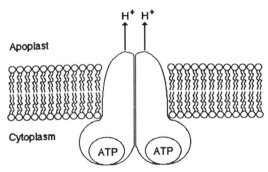

Figure 11   Model for the structure of the plasma membrane H^+-ATPase in the native membrane. The enzyme in the native membrane is proposed to exist in a dimer arrangement of 100 kD catalytic subunits, each with an ATP binding site facing the cytoplasm. From molecular studies and analysis, it is proposed that a substantial portion of the extramembrane peptide is located on the cytoplasmic face of the plasma membrane and constitutes the catalytic active site region of the protein.

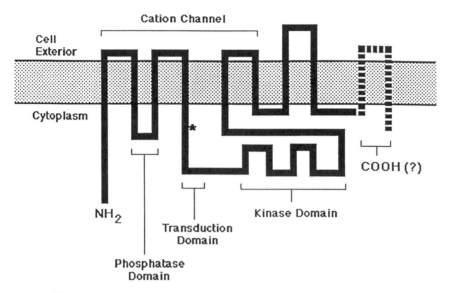

Figure 12 Model for the proposed transmembrane arrangement of the plant plasma membrane H^+-ATPase catalytic subunit peptide. Functional domains conserved between different P-type transport ATPases are indicated. The asterisk indicates the approximate location of an aspartic acid residue that becomes phosphorylated as an acylphosphate during formation of the enzyme-phosphorylated intermediate. Uncertainty regarding a cytoplasmic or apoplastic location for the carboxy terminus is indicated by the dashed segment of the peptide. (Adapted from Ref. 77.)

From studies conducted with tissues or whole plants, it is apparent that activity of the plasma membrane H^+-ATPase is modulated in accordance with environmental and chemical factors (see Ref. 96 and references therein). This implies some need for cellular control of the H^+- pump. There is evidence for phosphorylation of the plasma membrane H^+-ATPase by protein kinase activity [95,97], although only in the case of modulation by a phytotoxin has this effect been shown to have any direct effect on pump activity [98,99]. Recent studies have suggested that the C-terminal region of the enzyme itself may have an important regulatory function. Removal of this portion of the enzyme by limited proteolysis results in an increase in ATP hydrolytic activity and ATP-driven H^+-transport (see Ref. 96 and references therein). The characteristic stimulation of the plasma membrane H^+-ATPase by fusicoccin is also lost when this region of the polypeptide is cleaved off [100]. From this work, it has been suggested that this region of the enzyme not involved in catalysis may represent an "autoinhibitory" domain central to regulation of the H^+-ATPase [96]. Changes in conformation of this autoinhibitory domain could serve to modulate the H^+ pump. Such a hypothesis is intriguing in light of the observation that plasma membrane H^+-ATPase isoforms may differ in their C-terminal sequences. Perhaps, then, these different isoforms represent versions of the pump having differing control properties (see Ref. 96 for discussion).

2. Tonoplast H^+-ATPase

Although similar to the plasma membrane H^+-ATPase in coupling ATP hydrolysis to H^+ transport, the tonoplast H^+-ATPase demonstrates substantial differences in enzyme mechanism, transport mechanism, and protein structure. For this reason, this enzyme is considered to represent a different class of H^+-translocating ATPases known as the "V-type" transport ATPases (see Ref. 101 and references therein). When its activity was examined using vacuole preparations [102] or isolated tonoplast fractions [51], it was apparent that this enzyme represented a novel class of transport pump. Unlike the plant plasma membrane H^+-ATPase, the tonoplast H^+-ATPase was insensitive to orthovanadate, and hence did not form a covalent phosphorylated intermediate during ATP hydrolysis [51,101]. The tonoplast H^+-ATPase was similar to F-type mitochondrial ATP synthases in being inhibited by nitrate, but it differed from this enzyme class in being insensitive to azide and oligomycin [51,101,103]. Recent work has shown a characteristic and profound sensitivity of the tonoplast H^+-ATPase to bafilomycin A_1, a macrolide antibiotic [104]. Although originally identified in plants, the V-type ATPases have been shown to be present in animal cells and involved in H^+ transport in secretory vesicles and coated vesicles [105].

A further distinguishing feature of the tonoplast H^+-ATPase is its stimulation by anions other than nitrate, which is inhibitory. This is in contrast to the plasma membrane H^+-ATPase, which is stimulated by cations (see Ref. 77 and references therein). The stimulation of the tonoplast H^+-ATPase by anions such as Cl^- represents a direct effect on enzyme activity. This anion can stimulate ATP hydrolytic activity by the tonoplast H^+-ATPase in the absence of a membrane potential [106] and when the enzyme is solubilized from the membrane using detergent [107]. Furthermore, Cl^- stimulation of ATP hydrolytic activity is also observed for the purified enzyme [108].

Using intact vacuoles [109] or tonoplast vesicles [110,111], the H^+/ATP transport stoichiometry was shown to be 2 protons transported per molecule of ATP hydrolyzed. Again, this differs from the H^+/ATP stoichiometry of 1, observed for the plasma membrane H^+-ATPase [74]. Thermodynamic calculations indicate that this enzyme could generate a pH gradient of 2–3.5 units across the tonoplast, which is within the range of ΔpH values observed in vivo [109,111]

Structural studies have shown that the tonoplast H^+-ATPase is composed of several subunits arranged as a larger complex, which forms membrane-integral and membrane-peripheral protein regions. At present, 10 subunits have been identified. A model for the structure of the tonoplast H^+-ATPase is shown in Figure 13. This proposed model, based on the work of Sze and colleagues [51], indicates the molecular size of each peptide subunit. Identification of the

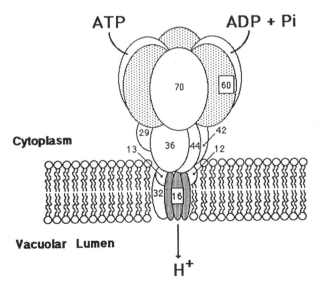

Figure 13 Model for the structure of the tonoplast H^+-ATPase. The various subunits for the enzyme are shown: numbers indicate molecular size in kilodaltons. (Adapted from Ref. 51.)

membrane-integral and -peripheral protein regions has been based on a determination of which subunits can be discharged from the membrane following treatment with high concentrations of KI [112,113]. Based on studies of this type, it was shown that the 70, 60, 44, 42, 36, and 29 kD subunits were released and thus most likely were not membrane-integral proteins [51]. The remaining peptides were integral and mostly likely function in mediating H^+ transport across the membrane. Based on covalent labeling with functional reagents, possible roles for at least the 70, 60, and 16 kD subunits have been proposed. Since both the 70 and 60 kD peptides are covalently labeled with reactive forms of ATP, it has been suggested that these subunits have ATP binding sites or collectively form ATP binding sites [51]. The observed labeling of the 16 kD peptide by radiolabeled N, N-dicyclohexylcarbodiimide (DCCD) has suggested a possible role for this subunit in H^+ translocation [114]. As yet, specific functions for the remaining subunits of the tonoplast H^+-ATPase complex have not been proposed.

Molecular studies have resulted in the cloning of genes for the 70, 60, and 16 kD subunits (see Ref. 51 and references therein). Based on sequence analysis for just the 16 kD subunit at this point, a preliminary suggestion for tonoplast isoforms has been made [51]. If confirmed, the model would be similar to what is observed for the plasma membrane H^+-ATPase and could imply functions in regulation and possibly tissue-specific gene expression.

3. Tonoplast H^+-Pyrophosphatase

In initial characterization of tonoplast-associated hydrolytic and H^+-transport activity using isolated vacuoles [115] or membrane fractions [116], it was often observed that significant pyrophosphate hydrolytic activity was present and that H^+ translocation could be driven by this substrate. This H^+ transport observed with PP_i was subsequently shown to be mediated by a separate H^+-translocating PP_iase that operates in parallel with the tonoplast H^+-ATPase at this membrane [116]. The presence of this enzyme at the tonoplast is widespread, being observed in most vascular plants [54].

Since its initial demonstration as a separate H^+-translocating entity, the tonoplast H^+-PP_iase has been extensively characterized [54] and purified [117,118], and its transport activity has been measured in a reconstituted system [119]. Recent molecular studies have resulted in the isolation and sequencing of complementary cDNA clones for the tonoplast H^+-PP_iase (see Ref. 54 and references therein).

These studies have shown that the structure of the tonoplast H^+-PP_iase is rather simple compared to the tonoplast H^+-ATPase. The former enzyme consists of a single 70 kD peptide, which is fully competent in mediating PP_i hydrolysis coupled to H^+ transport [119]. From radiation inactivation [120] and chromatographic [121] studies, it has been suggested that the native enzyme may be present as an oligomeric arrangement of 70 kD subunits. The deduction of the amino acid sequence from cDNA clones for the tonoplast H^+-PP_iase led to the prediction of a 770 amino acid peptide with a molecular mass (81 kD) close to that estimated by gel electrophoresis [54]. Analysis of these sequences by Rea and coworkers [54] resulted in a topological model for the enzyme (Figure 14). An N-terminus lies on the cytoplasmic side of the tonoplast membrane, and 13 transmembrane α-helices are proposed, with connecting regions of either α-helix or random coil. One large protein region on the cytoplasmic side of the membrane, which connects the fourth and fifth transmembrane peptides, is thought to contain catalytic domains involved in PP_i hydrolysis [54].

In isolated tonoplast vesicles, the H^+-PP_iase has been shown to have a transport stoichiometry of one proton pumped per molecule of PP_i hydrolyzed [111]. In consideration of the tonoplast H^+-ATPase stoichiometry of 2, the free energy available in both ATP and PP_i hydrolysis, and the magnitude of the proton electrochemical gradient present at the tonoplast, thermodynamic calculations have indicated that both pumps are likely to operate in the hydrolysis

Vacuolar
Lumen

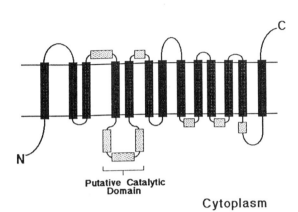

Cytoplasm

Figure 14 Model for the proposed transmembrane arrangement of the plant tonoplast H^+-pyrophosphatase catalytic subunit. In this model, based on the work of Rea et al. [54], the dark-shaded tubes represent transmembrane α-helical regions, the light-shaded tubes represent α-helical regions outside the membrane, and the lines connecting the tubes represent random coil regions of the H^+-PP$_i$ase peptide. The region of the peptide thought to represent the catalytic active site is also indicated. (Adapted from Ref. 54.)

direction and contribute jointly to generation of the protonmotive force at this membrane [111,116]. Such a function could prove useful in recovering free energy available from PP$_i$ generated in metabolic reactions involving ATP hydrolysis to AMP and PP$_i$ and for energizing transport under stress conditions where ATP might be limiting [54,116].

4. Plasma Membrane and Endoplasmic Reticulum Ca^{2+}-ATPases

The presence of both plasma membrane and endoplasmic reticulum Ca^{2+}-ATPases in plant cells as unique enzymes was only recently demonstrated through the use of isolated membrane vesicle preparations (see Refs. 52 and 53 and references therein). Although there were early reports in the literature describing plant Ca^{2+}-ATPases, such studies were not definitive, since only ATP hydrolytic activities were examined. This posed a particular problem in reports on "plasma membrane Ca^{2+}-ATPase," since it was generally uncertain whether the measured ATP hydrolytic activity actually reflected a separate Ca^{2+}-ATPase involved in Ca^{2+} transport or simply plasma membrane H^+-ATPase hydrolytic activity being examined in the presence of Ca^{2+} (see Ref. 52 for discussion). The use of transport-competent vesicles alleviated this problem, since measurements then focused on actual Ca^{2+} transport measured with either $^{45}Ca^{2+}$ (for examples, see Refs. 122–124) or optical probes such as chlorotetracycline [125,126].

From studies conducted with isolated vesicles, it became apparent that two fractions appeared to possess Ca^{2+} transport activity that was directly linked to ATP hydrolysis. Early studies by Buckhout [123,124], Lew et al. [125], and Bush and Sze [127] provided strong evidence for association of a Ca^{2+}-translocating ATPase with the endoplasmic reticulum. However, the membrane preparations used in these initial studies did not contain substantial amounts of transport-competent plasma membrane vesicles, and thus only a single component of ATP-dependent Ca^{2+} transport was observed. Only in subsequent work involving transport-competent plasma membrane vesicles was it possible to observe a seond component of ATP-dependent Ca^{2+} transport [122,128,129].

Using Ca^{2+} transport (as opposed to ATP hydrolysis) as an assay, the characteristics of the plasma membrane and endoplasmic reticulum Ca^{2+}-ATPases have been examined. These studies demonstrated that both enzymes appeared to display similarities with respect to pH optimum, in their dependence of activity on Ca^{2+} and ATP concentration, and in their sensitivity to orthovanadate [52,53]. However, there were some differences between these enzymes with respect to their sensitivity to the inhibitor erythrosin B and in their specificity for nucleotide substrates. While the endoplasmic reticulum Ca^{2+}-ATPase showed only a moderate sensitivity to erythrosin B, the plasma membrane Ca^{2+}-ATPase was almost fully inhibited at concentrations of this compound that were 1000-fold lower [52,130]. Furthermore, while the endoplasmic reticulum Ca^{2+}-ATPase appeared to be substrate specific for ATP, the plasma membrane Ca^{2+}-ATPase had the capacity to use either GTP or ITP as alternative substrates, supporting activity to levels of 50–70% of that observed with ATP (see Ref. 52 and references therein). This ability of the plasma membrane Ca^{2+}-ATPase to use alternative substrates that differ substantially in terms of chemical determinants on the purine ring would suggest major differences in active site structure between these enzymes (see Ref. 130 for discussion).

The sensitivity of both plasma membrane and endoplasmic reticulum Ca^{2+}-ATPases to orthovanadate would indicate that these enzymes are "P-type" ATPases, which form covalent phosphorylated intermediates during their catalytic/transport cycle. This was subsequently shown for both enzymes using radiolabeled substrates (see Refs. 126 and 131 for examples). When utilized to covalently label the enzyme during electrophoresis, a phosphorylated catalytic subunit of about 100 kD was observed for the endoplasmic reticulum Ca^{2+}-ATPase [126]. However, differences in catalytic subunit size ranging from 100 kD to 150 kD have been observed for the plasma membrane Ca^{2+}-ATPase (see Refs. 52 and 53 and references therein). Although the reason for this discrepancy is uncertain, it could result from use of different gel systems or from problems with proteolysis (see Refs. 52 and 53 for discussion). It should be noted that in work by Briars et al. [132], an antibody raised against purified animal erythrocyte plasma membrane Ca^{2+}-ATPase cross-reacted with the plant enzyme and demonstrated a molecular size of about 150 kD when subjected to Western blot analysis.

Although recent studies have tended to focus on characterization of plant Ca^{2+}-ATPases using transport-competent vesicles, purification of a Ca^{2+}-ATPase that is likely to be involved in transport had been reported earlier [133]. Since the starting material for this purification was a "microsomal" membrane fraction containing membranes derived from numerous cellular components (plasma membrane, tonoplast, endoplasmic reticulum, Golgi, etc.), the membrane origin of the enzyme was uncertain. However, the purification was based on calmodulin affinity chromatography of detergent-solubilized membranes, and it was reasoned that only a "plasma membrane-type" Ca^{2+}-ATPase should be calmodulin activated; therefore, it was assumed that this enzyme was derived from the plasma membrane [52,53]. Quite surprisingly, recent work has suggested that this calmodulin-affinity-purified enzyme most likely represents the endoplasmic reticulum rather than the plasma membrane Ca^{2+}-ATPase [134]. An origin for this enzyme in the endoplasmic reticulum would appear to be consistent with the observation that its activity was inhibited by erythrosin B in the micromolar rather than the nanomolar concentration range [134].

Since molecular approaches have only recently begun to be applied to plant Ca^{2+}-ATPases, a minimum of information is available. Both Wimmers et al. [135] and Parez-Prat et al. [136] have reported the isolation of cDNA clones for a putative Ca^{2+}-ATPase. Although the localization of this enzyme encoded by the nucleic acid sequence was uncertain, homologies in the sequence relative to those observed for the animal cell endoplasmic reticulum

Ca^{2+}-ATPase led investigators to assume that it represents the endoplasmic reticulum Ca^{2+}-ATPase [135].

5. Golgi H^+-ATPase

Although there is evidence that Golgi vesicles can mediate ATP-dependent H^+ transport and acidify their interior [137,138], the nature of the H^+-translocating ATPase mediating this function is not entirely clear. However, in many ways, the Golgi H^+-ATPase appears similar to the tonoplast H^+-ATPase. Both enzymes are inhibited by nitrate, yet insensitive to orthovanadate. However, the Golgi H^+-ATPase appears to be less sensitive to nitrate than the tonoplast H^+-ATPase [51]. Similarity to the tonoplast H^+-ATPase is also indicated by the observation that immunogold-labeled antibodies generated against the tonoplast H^+-ATPase 70 kD subunit also stained the Golgi membranes in thin sections from corn root tip cells [139]. This observation together with the finding that antibodies against the 70 kD subunit also cross-reacted with coated vesicles from zucchini hypocotyl cells [140] would suggest a possible localization of the "V-type" H^+-ATPase at different internal membranes of plant cell membranes.

6. Biochemical Characteristics of Secondary Transport Systems Associated with Plant Cell Membranes

Because of their lack of associated enzyme activity, secondary transport systems of plant cells have proven quite difficult to characterize. Significant advances in secondary transport system characterization in plants has been associated with the development of two major methodologies. The elucidation of transport-competent vesicle systems has allowed transport measurements to be conducted with a variety of vesicle fractions and solutes. Using radiolabeled solutes and applying the appropriate driving force for transport into the vesicles (pH gradient and/or membrane electrical potential) has allowed the identification of several H^+ or $\Delta\Psi$-linked secondary transport systems [70,141]. The recent application of the patch clamp method to plant cells has facilitated the identification of an array of ion channels associated with plant cells [55]. However, for only a very limited number of secondary transport systems has biochemical characterization progressed much beyond initial identification of the transport process mediated by the system. At present, extensive biochemical characterization has been conducted only on H^+-linked sugar and amino acid transport systems in plants (see Ref. 141 for review).

For the plasma membrane H^+/sucrose symport, the use of reactive radiolabeled probes has allowed putative identification of possible peptide components representing the carrier. Using radiolabeled *n*-ethylmaleimide (NEM), Gallet et al. [142] identified a 42 kD peptide as representing the plasma membrane sucrose carrier. In support of this proposal was the observation that antibodies generated against the 42 kD peptide inhibited sucrose transport in cells [143] and in plasma membrane vesicles [144].

Using antibodies produced against the 42 kD peptide to screen detergent-solubilized fractions, Li et al. [145] were able to solubilize and reconstitute ΔpH-dependent sucrose transport activity into liposomes. However, the idea of the 42 kD peptide having a role in mediating sucrose transport at the plasma membrane has been criticized, since inhibition of sucrose transport by NEM could not be demonstrated in sugar beet plasma membrane vesicles (see Ref. 141 for discussion).

Plasma membrane vesicle preparations have allowed further characterization of the nature of H^+/amino acid symport carriers [141]. From a kinetic analysis of the transport of various radiolabeled amino acids and the degree of competition by other amino acids, Li and Bush [146] concluded that four H^+/amino acid symports exist at the plasma membrane: an acidic amino acid symport, a basic amino acid symport, and two different symports for neutral amino acids. Although each symport appeared to be relatively specific for a given amino acid grouping, all four carriers demonstrated some "crossover" transport specificity for other amino acids outside the group [146].

Molecular approaches have been particularly helpful in the structural characterization of secondary transport systems associated with plant membranes. Sauer et al. [147] isolated a cDNA clone for a plasma membrane H^+/hexose carrier which predicted a sequence of 522 amino acids and a molecular weight of about 57.6 kD for the carrier protein. By an approach involving complementation of yeast mutants by a plant cDNA library, Riesmeier et al. [148] were able to identify a cDNA that is likely to represent the plant H^+/sucrose symport. When analyzed further, this cDNA encodes a sequence of 525 amino acids for a 55 kD peptide. Using a similar cDNA complementation approach, Hsu et al. [149] were able to identify a clone for neutral amino acid carrier most likely representing a plasma membrane H^+/neutral amino acid symport. This clone encoded a peptide consisting of 486 amino acids and predicted a molecular weight of about 52.9 kD for the carrier protein.

For both the hexose and sucrose carriers characterized in molecular studies, the deduced amino acid sequence predicts an arrangement of the peptide across the membrane involving 12 transmembrane α-helices and a central, nonspanning hydrophilic region [147,148]. Sequence deduction for the neutral amino acid carrier also predicts 12 transmembrane peptide regions [149]. Although the linear amino acid sequence for the hexose transport system shares significant homology with a variety of sugar carriers in diverse organisms (i.e., a superfamily of porters), the sequence for both the sucrose and neutral amino acid carriers appears to be unique (see Ref. 141 for discussion).

Molecular methods have also been useful in the cloning a secondary transport systems thought to represent ion channels. Using this approach, Anderson et al. [150] have recently reported the isolation and characterization of cDNA clones for a K^+ channel that most likely is associated with the plasma membrane.

VI. ROLE OF MEMBRANES AND CELLULAR TRANSPORT SYSTEMS IN SIGNAL TRANSDUCTION PHENOMENA

It is now apparent that in addition to their involvement in nutrient and metabolite transport in plants, membranes and transport systems have an important role in signal transduction [12,14]. As in animal cells, it is believed that response coupling involves events of Ca^{2+} transport as well as small molecules such as inositiol triphosphate (IP_3) and diacylglycerol (DAG) generated by the rapid metabolism of membrane components [151]. In this respect, Ca^{2+}, IP_3, and DAG appear to act as "second messengers," coupling a cellular response to chemical or environmental signals [14,151]. Such an involvement of Ca^{2+} in signal transduction could explain the need for separate transport systems, independent of the proton electrochemical gradient, for this cation. This would be particularly important if H^+-translocating ATPases themselves are found to be regulated by phosphorylation by protein kinases, which might, in turn, be modulated by cytoplasmic Ca^{2+} (see Ref. 130 for discussion).

An attempt to integrate concepts regarding putative signal transduction functions for Ca^{2+} and membrane components appears in Figure 15, which represents the current "paradigm" for plant cellular regulation (see Refs. 12, 14, and 151 for reviews).

Figure 15 shows several modes of response coupling, although only a single mode might operate in signal transduction associated with a particular stimulus. Two possible options are shown for the period after the interception of a signal by a receptor. Signal transduction might involve the transient opening of a Ca^{2+} channel at the plasma membrane, where upon Ca^{2+} would move downhill in electrochemical potential into the cytoplasm, elevate the cytoplasmic Ca^{2+} concentration, and activate a cellular response through enzymes such as Ca^{2+}- or

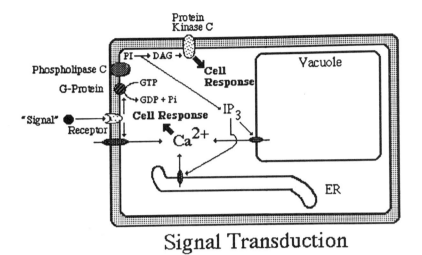

Signal Transduction

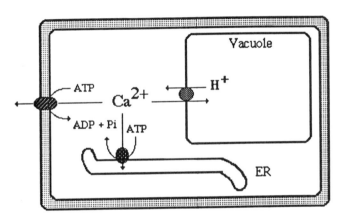

System Restoration

Figure 15 Model for signal transduction in plant cells involving membranes and Ca^{2+} transport systems. The reception of a "signal" (chemical or environmental) by a receptor leads to either the opening of a plasma membrane Ca^{2+} channel or, ultimately, the production of inositol triphosphate (IP_3) (i.e., by a G-protein-activated phospholipase C), which triggers the opening of Ca^{2+} channels associated with intracellular membranes. The increase in cytoplasmic Ca^{2+} due to channel opening results in a cell response by Ca^{2+}-regulated enzymes such as protein kinases. Likewise, the diacylglycerol produced by phospholipase attack on phosphatidylinositol (PI) can activate protein kinase C, which can also lead to a cellular response. To restore the signal transduction system to its prestimulus state, the cytoplasmic Ca^{2+} must be reduced by transport either out of the cell or into intracellular organelles.

Ca^{2+}/calmodulin-activated protein kinases [12,152]. Alternatively, the response of the receptor could involve an activation of G-proteins, which hydrolyze GTP and, in turn, activate a specific phospholipase C in the membrane. This phospholipase C acts on a membrane lipid, phosphatidylinositol diphosphate, to release IP_3 and produce DAG, which remains in the membrane [14,151]. The IP_3 so released could activate Ca^{2+} channels associated with intracellular membranes such as the endoplasmic reticulum and tonoplast, to release Ca^{2+} from internal stores and initiate a cellular response. Likewise, the DAG remaining in the membrane could then activate a protein kinase C associated with the plasma membrane, which would also lead to a cellular response by phosphorylation of proteins [14,151].

To be truly functional in signal transduction, the system must be capable of being restored to the prestimulus state following each event. This ability is highly dependent on the Ca^{2+} transport systems associated with the membranes that collectively transport Ca^{2+} out of the cytoplasm. Here, Ca^{2+} efflux from the cytoplasm can involve either transport out of the cell or sequestration within internal organelles. The observation that the plasma membrane and possibly the endoplasmic reticulum Ca^{2+}-ATPases are stimulated by calmodulin indicates that these enzymes are activated in response to elevated cytoplasmic Ca^{2+} levels to restore the original low cytoplasmic Ca^{2+} concentration [52,53].

This model is strongly analogous to what takes place in animal cells [151], and information for plant systems is only now emerging for many of the key aspects of the model. There is a substantial body of literature regarding the presence of protein kinases in plants, with more recent evidence being presented for enzymes specifically regulated by Ca^{2+} [152,153]. It has been shown that IP_3 is generated in plant cells (see Ref. 154 and references therein) and that this compound can cause Ca^{2+} release from internal membranes such as the tonoplast [155]. However, there is only preliminary evidence for the presence of Ca^{2+} channels associated with plant membranes [12]. Studies with GTP and GDP analogues have shown the presence of G-proteins in plant cells [156–158] and their involvement in IP_3 production via phospholipase C [159,160]. However, demonstrating the presence of DAG in the plasma membrane in a manner that correlates with signal transduction has proven to be much more difficult, given its rapid turnover and participation in multiple metabolic pathways (see Ref. 14 for discussion). Although it has been difficult to quantitate DAG, evidence for the association of protein kinase C activity with plant membranes has been presented [14,151]. Hence, while there is substantional information available on one or more processes in the model, we lack data showing true integration of processes during a signal transduction event. It is hoped that such data will be forthcoming.

VII. SUMMARY AND PERSPECTIVE

A key function of membranes is to allow selective solute transport, which is essential for most cellular functions, including metabolism. The ability of membranes to act in this manner depends on their chemical structure and the presence of membrane-associated proteins, which act as transport systems. Maintenance of selective solute gradients across plant membranes is an energy-requiring process dependent on cellular metabolism. Plant cells couple cellular energy to transport functions in a stepwise manner that entails primary and secondary transport processes. A central feature of this process is the involvement of proton electrochemical gradients in the interaction of primary and secondary transport processes. Studies conducted with cells, isolated organelles, membrane vesicles, and purified proteins have allowed the preliminary characterization of a number of the transport systems associated with plant membranes. However, detailed biochemical information is known for relatively few transport systems, most of which are involved in primary transport. The recent application of molecular methods involving gene cloning has had a significant impact in plant membrane research, allowing the elucidation of protein structural details for a limited number of primary as well as secondary transport systems. It is now apparent that membranes and transport systems may have an important function in cellular signal transduction. Since this possibility involves not only transport phenomena (Ca^{2+} ATPases, Ca^{2+} carriers, Ca^{2+} channels, etc.) but also membrane-associated enzymes (G-proteins, protein kinases, phospholipases) and membrane structural components (IP$_3$, DAG), this mode of information processing lies at the interface of the biochemistry of membrane structure and transport.

REFERENCES

1. W. Tanner and N. Sauer, *The Plant Plasma Membrane—Structure, Function and Molecular Biology* (C. Larsson and I. M. Møller, eds.), Springer-Verlag, Berlin, p. 295 (1989).
2. D. T. Clarkson, *Annu. Rev. Plant Physiol.*, *36*: 77 (1985).
3. D. T. Clarkson, *Solute Transport in Plant Cells and Tissues* (D. A. Baker and J. L. Hall, eds.), Loughman Scientific and Technical, Harlow, Essex, p. 251 (1988).
4. L. C. Ho, *Annu. Rev. Plant Physiol.*, *39*: 355 (1988).
5. T. H. Humphreys, *Solute Transport in Plant Cells and Tissues* (D. A. Baker and J. L. Hall, eds.), Loughman Scientific and Technical, Harlow, Essex, p. 305 (1988).
6. T. A. Mansfield, A. M. Hetherington, and C. J. Atkinson, *Annu. Rev. Plant Physiol.*, *41*: 55 (1990).
7. D. L. Rayle and R. E. Cleland, *Plant Physiol.*, *99*: 1271 (1992).
8. E. Schnepf, *Annu. Rev. Plant Physiol.*, *37*: 23 (1986).
9. S. E. Smith and V. Gianinazzi-Pearson, *Annu. Rev. Plant Physiol.*, *39*: 221 (1988).
10. R. T. Kiode and R. P. Schreiner, *Annu. Rev. Plant Physiol.*, *43*: 557 (1992).
11. K. R. Shubert, *Annu. Rev. Plant Physiol.*, *37*: 539 (1986).
12. D. Marmé, *Second Messengers in Plant Growth and Development* (W. F. Boss and D. J. Morré, eds.), Liss, New York, p. 57 (1989).
13. R. T. Leonard and P. K. Hepler, *Calcium in Plant Growth and Development* (R. T. Leonard and P. K. Hepler, eds.), American Society of Plant Physiologists, Rockville, MD, p. 1 (1990).
14. K. J. Einspahr and G. A. Thompson, Jr., *Plant Physiol.*, *93*: 361 (1990).
15. E. Overton, *Naturforsch, Ges. Zurich*, *44*: 88 (1899).
16. E. Gorder and F. Grendel, *J. Exp. Med.*, *41*: 439 (1925).
17. J. F. Danielli and H. A. Davson, *J. Cell Comp. Physiol.*, *5*: 495 (1935).
18. J. D. Robertson, *Cellular Membranes in Development* (M. Locke, ed.), Academic Press, New York, p. 181 (1964).
19. S. J. Singer, *J. Membrane Biol.*, *129*: 3 (1992).
20. J. Kesselmeier and E. Heinz, *Methods Enzymol.*, *148*: 650 (1987).
21. F. Marty, *Biochemistry and Function of Vacuolar Adenosine Triphosphatase in Fungi and Plants* (B. P. Marin, ed.), Springer-Verlag, Berlin, p. 14 (1985).
22. C. Larsson, I. M. Møller, and S. Widell, *The Plant Plasma Membrane—Structure, Function and Molecular Biology* (C. Larsson and I. M. Møller, eds.), Springer-Verlag, Berlin, p. 1 (1990).
23. M.-A. Hartmann and P. Beneveniste, *Methods Enzymol.*, *148*: 632 (1987).
24. J. Lenard and S. J. Singer, *Proc. Natl. Acad. Sci. U.S.A.*, *56*: 1828 (1966).
25. D. F. H. Wallach and P. H. Zahler, *Proc. Natl. Acad. Sci. U.S.A.*, *56*: 1552 (1966).
26. M. K. Jain and D. Zakim, *Biochim. Biophys. Acta*, *906*: 33 (1987).
27. M. L. Jennings, *Annu. Rev. Biochem.*, *58*: 999 (1989).
28. G. Blobel, *Proc. Natl. Acad. Sci. U.S.A.*, *77*: 1496 (1980).
29. J. D. Lear, Z. R. Wasserman, and W. F. Degrado, *Science*, *240*: 1177 (1988).
30. S. Oiki, W. Dauho and M. Montal, *Proc. Natl. Acad. Sci. U.S.A.*, *85*: 2393 (1988).
31. D. J. Morré, *The Plant Plasma Membrane—Structure, Function and Molecular Biology* (C. Larsson and I. M. Møller, eds.), Springer-Verlag, Berlin, p. 76 (1990).
32. W. D. Stein, *Channels, Carriers and Pumps*, Academic Press, San Deigo, CA (1990).
33. C. P. Rochester, P. Kjellbom, and C. Larsson, *Physiol. Plant.*, *71*: 257 (1987).
34. J. A. Traas, *The Plant Plasma Membrane—Structure, Function and Molecular Biology* (C. Larsson and I. M. Møller, eds.), Springer-Verlag, Berlin, p. 269 (1990).
35. A. M. C. Emons, J. Derkson, and M. M. A. Sassen, *Physiol. Plant.*, *84*: 486 (1992).
36. R. Douce, *Mitochondria in Higher Plants: Structure, Function and Biogenesis*, Academic Press, Orlando FL (1985).

37. W. D. Stein, *Transport and Diffusion Across Cell Membranes*, Academic Press, Orlando FL (1986).
38. I. R. Cowan, J. A. Raven, W. Hartung, and G. D. Farquhar, *Aust. J. Plant Physiol.*, *9*: 489 (1982).
39. F. C. Hsu and D. A. Kleier, *Weed Sci.*, *38*: 315 (1990).
40. D. A. Baker and J. L. Hall, *Transport in Plant Cells and Tissues* (D. A. Baker and J. L. Hall, eds.), Loughman Scientific and Technical, Harlow, Essex, p. 1 (1988).
41. D. Sanders and C. L. Slayman, *Plant Membrane Transport: The Current Position* (J. Dainty, M. I. DeMichelis, E. Marré, and F. Rasi-Caldogno, eds.), Elsevier Science Publishers, Amsterdam, p. 3 (1989).
42. H. Marschner, *Mineral Nutrition in Higher Plants*, Academic Press, London (1986).
43. L. V. Kochian and W. J. Lucas, *Plant Physiol.*, *70*: 1723 (1982).
44. L. V. Kochian and W. J. Lucas, *Plant Physiol.*, *73*: 208 (1983).
45. W. Lin, M. R. Schmidt, W. D. Hitz, and R. T. Giaquinta, *Plant Physiol.*, *75*: 936 (1984).
46. P. Nissen, *Annu. Rev. Plant Physiol.*, *25*: 53 (1974).
47. R. T. Leonard and T. K. Hodges, *Plant Physiol.*, *52*: 6 (1973).
48. T. K. Hodges, *Adv. Agron.*, *25*: 163 (1973).
49. M. G. Pitman, *Encyclopedia of Plant Physiology*, Vol. II, Part B (U. Lüttge and M. G. Pitman, eds.), Springer-Verlag, Berlin, p. 57 (1976).
50. D. P. Briskin and J. B. Hanson, *J. Exp. Bot.*, *43*: 269 (1992).
51. H. Sze, J. M. Ward, and S. Lai, *J. Bioenerg. Biomembranes*, *24*: 371 (1992).
52. D. P. Briskin, *Plant Physiol.*, *94*: 397 (1990).
53. D. E. Evans, S. A. Briars, and L. E. Williams, *J. Exp. Bot.*, *42*: 285 (1991).
54. P. A. Rea, Y. Kim, V. Sarafian, R. J. Poole, J. M. Davies, and D. Sanders, *Trends Biochem. Sci.*, *17*: 384 (1992).
55. R. Hedrich and J. I. Schroeder, *Annu. Rev. Plant Physiol.*, *40*: 539 (1989).
56. R. J. Poole, *Solute Transport in Plant Cells and Tissues* (D. A. Baker and J. L. Hall, eds.), Loughman Scientific and Technical, Harlow, Essex, p. 83 (1988).
57. P. M. Mitchell, *J. Biochem.*, *97*: 1 (1985).
58. D. R. Bush, *Photosynth. Res.*, *32*: 155 (1992).
59. A. Tubbe and T. J. Buckhout, *Plant Physiol.*, *99*: 945 (1992).
60. J. L. Ruiz-Cristin and D. P. Briskin, *Arch. Biochem. Biophys.*, *285*: 74 (1991).
61. W. J. Lucas and L. V. Kochian, *Plasma Membrane Oxido-Reductase in Control of Animal and Plant Growth* (F. L. Crane, D. J. Morré and H. Löw, eds.), Plenum Press, New York, p. 219 (1988).
62. Z.-C. Li and D. R. Bush, *Plant Physiol.*, *94*: 268 (1990).
63. E. Blumwald and R. J. Poole, *Plant Physiol.*, *78*: 163 (1985).
64. E. Blumwald and R. J. Poole, *Plant Physiol.*, *80*: 727 (1986).
65. H. P. Getz, *Planta*, *185*: 261 (1991).
66. A. J. Chodera and D. P. Briskin, *Plant Sci.*, *67*: 151 (1990).
67. A. J. Miller and S. J. Smith, *Planta*, *187*: 554 (1992).
68. H. Sze and T. K. Hodges, *Plant Physiol.*, *59*: 641 (1977).
69. D. P. Briskin, *Physiol. Plant.*, *68*: 159 (1986).
70. D. P., Briskin, *The Plant Plasma Membrane—Structure, Function and Molecular Biology* (C. Larsson and I. M. Møller, eds.), Springer-Verlag, Berlin, p. 154 (1990).
71. S. D. O'Neill and R. M. Spanswick, *J. Membrane Biol.*, *79*: 231 (1984).
72. J. L. Giannini, L. H. Gildensoph, and D. P. Briskin, *Arch. Biochem. Biophys.*, *254*: 621 (1987).
73. J. L. Giannini and D. P. Briskin, *Plant Physiol.*, *84*: 613 (1987).
74. D. P. Briskin and I. Reynolds-Niesman, *Plant Physiol.*, *95*: 242 (1990).
75. A. Villalobo, *Folia Microbiol.*, *33*: 407 (1988).
76. R. Serrano, *Annu. Plant Physiol. Plant Mol. Biol.*, *40*: 61 (1989).
77. D. P. Briskin, *Biochem. Biophys. Acta*, *1019*: 95 (1990).
78. S. R. Gallagher and R. T. Leonard, *Plant Physiol.*, *70*: 1335 (1982).
79. S. D. O'Neill and R. M. Spanswick, *Plant Physiol.*, *75*: 586 (1984).
80. D. P. Briskin and R. T. Leonard, *Proc. Natl. Acad. Sci. U.S.A.*, *79*: 6922 (1982).

81. F. Vara and R. Serrano, *J. Biol. Chem.*, *258*: 5334 (1983).
82. D. P. Briskin and R. J. Poole, *Plant Physiol.*, *71*: 507 (1983).
83. D. P. Briskin, *Arch. Biochem. Biophys.*, *248*: 106 (1986).
84. D. P. Briskin, *Plant Physiol.*, *88*: 84 (1988).
85. L. E. Gonzalez de la Vara and G. Medina, *Plant Physiol.*, *94*: 1522 (1990).
86. N. Nelson, *Plant Physiol.*, *86*: 1 (1988).
87. N. Nelson, *Biochim. Biophys. Acta*, *1100*: 109 (1992).
88. R. Serrano, *Biochem. Biophys. Res. Commun.*, *121*: 735 (1984).
89. G. E. Anton and R. M. Spanswick, *Plant Physiol.*, *81*: 1080 (1986).
90. D. P. Briskin, W. R. Thornley, and J. L. Roti-Roti, *Plant Physiol.*, *78*: 642 (1985).
91. D. P. Briskin and I. Reynolds-Niesman, *Plant Physiol.*, *90*: 394 (1989).
92. J. F. Harper, T. K. Surowy, and M. R. Sussman, *Proc. Natl. Acad. Sci. U.S.A.*, *86*: 1234 (1989).
93. J. M. Pardo and R. Serrano, *J. Biol. Chem.*, *264*: 8557 (1989).
94. M. Boutry, B. Michlet, and A. Goffeau, *Biochem. Biophys. Res. Commun.*, *162*: 567 (1989).
95. M. R. Sussman, *Transport and Receptor Proteins of Plant Membranes* (D. T. Cooke and D. T. Clarkson, eds.), Plenum Press, New York, p. 5 (1992).
96. M. G. Palmgren, *Physiol. Plant.*, *83*: 314 (1991).
97. G. E. Schaller and M. R. Sussman, *Planta*, *173*: 509 (1988).
98. A. P. Bidwai, L. Zhang, R. C. Buchmann, and J. Y. Takemoto, *Plant Physiol.*, *83*: 39 (1987).
99. A. P. Bidwai and J. Y. Takemoto, *Proc. Natl. Acad. Sci. U.S.A.*, *84*: 6755 (1987).
100. F. Johansson, M. Sommarin, and C. Larsson, *Plant Cell*, *5*: 321 (1993).
101. D. K. Stone, B. P. Crider, T. C. Südhof, and X.-S. Xie, *J. Bioenerg. Biomembranes*, *21*: 605 (1989).
102. R. R. Walker and R. A. Leigh, *Planta*, *153*: 140 (1981).
103. H. Sze, *Annu. Rev. Plant Physiol.*, *36*: 175 (1985).
104. E. J. Bowman, A. Siebers, and K. Altendorf, *Proc. Natl. Acad. Sci. U.S.A.*, *85*: 7972 (1988).
105. D. L. Schneider, *Biochim. Biophys. Acta*, *895*: 1 (1987).
106. K. A. Churchill and H. Sze, *Plant Physiol.*, *76*: 490 (1984).
107. A. B. Bennett and R. M. Spanswick, *J. Membrane Biol.*, *75*: 21 (1983).
108. S. K. Randall and H. Sze, *J. Biol. Chem.*, *262*: 7135 (1987).
109. J. Guern, Y. Mathieu, A. Kurkdjian, P. Manigault, J. Manigault, B. Gillet, J. Beloeil, and J. Lallemand, *Plant Physiol.*, *89*: 27 (1987).
110. A. B. Bennett and R. M. Spanwick, *Plant Physiol.*, *74*: 545 (1984).
111. A. L. Schmidt and D. P. Briskin, *Arch. Biochem. Biophys.*, *301*: 165 (1993).
112. J. M. Ward, A. Reinders, H.-T. Hsu, and H. Sze, *Plant Physiol.*, *99*: 161 (1992).
113. S. Lai, S. K. Randall, and H. Sze, *J.Biol. Chem.*, *266*: 16731 (1988).
114. K. H. Kaestner, S. K. Randall, and H. Sze, *J. Biol. Chem.*, *263*: 1282 (1988).
115. R. R. Walker and R. A. Leigh, *Planta*, *153*: 150 (1981).
116. P. A. Rea and D. Sanders, *Physiol. Plant.*, *71*: 131 (1987).
117. C. J. Britten, J. C. Turner, and P. A. Rea, *FEBS Lett.*, *256*: 200 (1989).
118. V. Sarafian and R. J. Poole, *Plant Physiol.*, *91*: 34 (1989).
119. C. J. Britten, R. Zhen, E. J. Kim, and P. A. Rea, *J. Biol. Chem.*, *267*: 21850 (1992).
120. V. Sarafian, M. Potier, and R. J. Poole, *Biochem. J.*, *283*: 493 (1992).
121. M. H. Sato, M. Maeshima, Y. Ohsumi, and M. Yoshida, *FEBS Lett.*, *290*: 177 (1991).
122. J. L. Giannini, J. Ruiz-Cristin, and D. P. Briskin, *Plant Physiol.*, *85*: 1137 (1987).
123. T. J. Buckhout, *Planta*, *159*: 84 (1983).
124. T. J. Buchkhout, *Plant Physiol.*, *76*: 962 (1984).
125. R. R. Lew, D. P. Briskin, and R. E. Wyse, *Plant Physiol.*, *82*: 47 (1986).
126. J. L. Giannini, L. H. Gildensoph, I. Reynolds-Niesman, and D. P. Briskin, *Plant Physiol.*, *85*: 1129 (1987).
127. D. R. Bush and H. Sze, *Plant Physiol.*, *80*: 549 (1986).
128. F. Rasi-Caldogno, M. C. Pugliarello, and I. M. DeMichelis, *Plant Physiol.*, *83*: 994 (1987).
129. F. Rasi-Caldogno, M. C. Pugliarello, C. Olivari, and I. M. DeMichelis, *Plant Physiol.*, *90*: 1429 (1989).

130. D. P. Briskin, S. Basu, and I. Ho, *Transport and Receptor Proteins of Plant Membranes* (D. T. Cooke and D. T. Clarkson, eds.), Plenum Press, New York, p. 13 (1992).
131. S. A. Briars and D. E. Evans, *Biochem. Biophys. Res. Commun.*, *159*: 185 (1989).
132. S. A. Briars, F. Kessler, and D. E. Evans, *Planta*, *176*: 283 (1988).
133. P. Dieter and D. Marmé, *FEBS Lett.*, *125*: 245 (1981).
134. P. Askerlund and D. E. Evans, *Plant Physiol.*, *100*: 1670 (1992).
135. L. E. Wimmers, N. N. Ewing, and A. B. Bennett, *Proc. Natl. Acad. Sci. U.S.A.*, *89*: 9205 (1992).
136. E. Perez-Prat, M. L. Narasimhan, M. L. Binzel, M. A. Botella, Z. Chen, V. Valpuesta, R. A. Bressan, and P. M. Hasegawa, *Plant Physiol.*, *100*: 1471 (1992).
137. A. Chanson and L. Taiz, *Plant Physiol.*, *78*: 232 (1985).
138. M. S. Ali and T. Akazawa, *Plant Physiol.*, *81*: 222 (1986).
139. D. Hurley and L. Taiz, *Plant Physiol.*, *89*: 391 (1989).
140. H. Depta, S. E. H. Holstein, D. G. Robinson, M. Lutzelschwab, and W. Michalke, *Planta*, *183*: 434 (1991).
141. D. R. Bush, *Annu. Rev. Plant Physiol. Plant Mol. Biol.*, *44*: 513 (1993).
142. O. Gallet, R. Lemoine, C. Larsson, and S. Delrot, *Biochim. Biophys. Acta*, *978*: 56 (1989).
143. R. Lemoine, S. Delrot, O. Gallet, and C. Larsson, *Biochim. Biophys. Acta*, *778*: 65 (1989).
144. O. Gallet, R. Lemoine, C. Gaillard, C. Larsson, and S. Delrot, *Plant Physiol.*, *98*: 17 (1992).
145. Z.-S. Li, O. Gallet, C. Gaillard, R. Lemoine, and S. Delrot, *FEBS Lett.*, *286*: 117 (1991).
146. Z.-C. Li and D. R. Bush, *Plant Physiol.*, *94*: 268 (1990).
147. N. Sauer, K. Friedländer, and U. Gräml-Wicke, *EMBO J.*, *9*: 3045 (1990).
148. J. W. Riesmeier, L. Willmitzer, and W. B. Frommer, *EMBO J.*, *11*: 4705 (1992).
149. L.-C. Hsu, T.-J. Chiou, L. Chjen, and D. R. Bush, *Proc. Natl. Acad. Sci. U.S.A.*, *90*: 7441 (1993).
150. J. A. Anderson, S. S. Huprikar, L. V. Kochian, W. J. Lucas, and R. F. Gaber, *Proc. Natl. Acad. Sci. U.S.A.*, *89*: 3736 (1992).
151. B. Drøbak, *Plant Physiol.*, *102*: 705 (1993).
152. B. W. Poohvaiah and A. S. N. Reddy, *Crit. Rev. Plant Sci.*, *6*: 47 (1987).
153. J. F. Harper, M. R. Sussman, G. E. Schaller, C. Putnam-Evans, H. Carbonneau, and A. C. Harmon, *Science*, *252*: 951 (1991).
154. M. Rincon and W. F. Boss, *Inositol Metabolism in Plants* (D. J. Morré, W. F. Boss, and F. A. Loewus, eds.), Wiley-Liss, New York, p. 173 (1990).
155. K. S. Schumaker and H. Sze, *J. Biol. Chem.*, *262*: 3944 (1987).
156. B. K. Drøbak, E. F. Allan, J. G. Comerford, K. Roberts, and A. P. Dawson, *Biochem. Biophys. Res. Commun.*, *150*: 899 (1988).
157. K. Hasunuma, K. Furukawa, K. Funadera, M. Kubota, and M. Watanabe, *Photochem. Photobiol.*, *46*: 531 (1987).
158. M. Jacobs, M. P. Thelen, R. W. Farndale, M. C. Astle, and P. H. Rubery, *Biochem. Biophys. Res. Commun.*, *155*: 1478 (1988).
159. M. Dillenschneider, A. Hetherington, A. Graziana, G. Alibert, J. Haiech, and R. Ranjeva, *FEBS Lett.*, *208*: 413 (1986).
160. K. J. Einspahr, T. C. Peeler, and G. A. Thompson, Jr., *Plant Physiol.*, *90*: 1115 (1989).

19
Fruit Development, Maturation, and Ripening

William Grierson
University of Florida, Lake Alfred, Florida

I. INTRODUCTION

A. What Is a Fruit?

The biblical phrase "the precious fruits of the earth" can be taken far more literally than the epistle writer probably imagined. There is very little in agriculture that does not depend on the development of fruits. By definition, a fruit is the end product of a matured ovary. This end product can vary from being a single seed such as a grain of any cereal (e.g., wheat, rice, rye, oats, or barley) to being a fleshy, succulent structure (e.g., peach, pear, or watermelon). All nut crops, including peanuts (or "ground nuts") are technically fruits, as are the products of oil palm, coconuts, rape (canola), flax (linseed), and other plants grown for extraction of edible or industrial oils. Even many root and pasture crops are dependent on fruit setting to provide seed for sowing the next crop. Root and tuber crops grown from vegetative propagules are an obvious exception, but their genetic improvement by plant breeders is dependent on flowering, pollination, and fruit setting to provide seed with which to start improved varieties. It should also be noted that many "vegetables," including tomatoes, peas, beans, cucumbers, squash, peppers (capsicums), eggplant (aubergine), and okra (lady's fingers), are botanically fruits.

B. Scope of This Chapter

For the purposes of this chapter, only those products classified horticulturally as fruits are considered for detailed discussion. In general, these are fleshy products, characteristically high in sugars (the avocado being a notable exception) and although sometimes processed on a very large scale, traditionally eaten raw as dessert. Unlike vegetables, most are perennials grown on trees, vines, or shrubs (strawberries are the fruit of a perennial herbaceous plant). Melons are an exception, being annuals.

Whether annual or perennial, whether classified commercially as a fruit, vegetable, or cereal, it should always be remembered that until the instant of harvesting, a fruit is an integral part of the parent plant, participating in a common physiology and subject to the same ecological influences. As pointed out in Chapter 4, a fruit cannot be considered independent of the growth status of the parent plant or of the environment in which it was grown. A simpleminded quest for a single recommendation as to optimum postharvest conditions for a given type of fruit, regardless of growing district and preharvest climatic conditions, is doomed to failure.

C. Definitions

1. *Fruit* is the product of a matured ovary.
2. *Maturation* is the completion of the development of a fruit to the point at which it is physiologically mature enough to be separated from the parent plant. Typically, this is the point at which its seeds are viable. There is no necessary relationship with market maturity, for which immature fruits may be required (e.g., cucumber, okra) or for which arbitrary legal standards may be set for external color and/or sugar or acid content (e.g., citrus, grapes).
3. *Ripening* and maturation can be synonymous for nonclimacteric fruits (e.g., grapes, strawberries, and citrus) that are edible at the time of picking and have no postharvest ripening cycle. However, they are quite different for climacteric-type fruits: those that are considered unripe until they have entered on a distinctive postharvest respiratory rise in which ethylene is evolved, CO_2 output increases (sometimes as much as tenfold), tissues soften, starch/sugar or acid/sugar changes occur, and typical external color changes may be involved. Tomatoes, apples, pears, avocados, and bananas are typical climacteric-type fruits with distinctive postharvest ripening cycles.
4. *Berry* is used quite differently by plant scientists and by the general public. Botanically, a berry is the product of a single pistil, fleshy throughout, usually indehiscent, and homogeneous in texture [1]. Thus a grape is technically a berry, but a strawberry is not.
5. *Anthesis* is the stage of flowering at which pollination can take place, usually considered to be the initiation of fruit development.
6. *Parthenocarpy* in its narrowest sense is defined as the ability of a plant to develop fruit without sexual fertilization. More broadly, it is the ability to produce fruit without seeds [2].

Readers interested in further details of terminology are referred to two publications: Watada et al. for general terminology relating to developing horticultural crops [3] and Gortner et al. for the biochemical basis for terminology used in maturation and ripening of fruits [4].

II. PREREQUISITES FOR FRUIT FORMATION

As long ago as several hundred years B.C., it was recognized that all fruits came from flowers. The ancient Greeks named one exception, the fig, "the only fruit not preceded by a flower." This was because they did not realize that the fig is an aggregate fruit with many minuscule flowers *inside* the enlarged, fleshy receptacle.

Flowers must be preceded by buds specifically differentiated for flower formation. In deciduous fruits, this starts some 10 or 11 months prior to bloom (i.e., initiation of fruit bud formation for the next year's crop starts almost as soon as the new crop is set). A very recent study of the rate of flower bud development in deciduous fruits indicates that each species follows a sigmoidal growth pattern within a temperature range specific to that species [5]. In citrus fruits, fruit bud differentiation is initiated only a few weeks prior to bloom [6]. For both deciduous and citrus fruits, blossom formation occurs on wood at least 1 year old. ("Fruiting spurs" on apple trees may bear fruit almost every other year for a dozen years or more.) Grapes are in sharp contrast to this pattern. Skilled grape pruners remove almost all woody growth (canes) from the previous year, leaving only a few buds (how many depends on the variety, district, and vigor of the plant). From these few buds grow long canes on which leaf and fruit bud differentiation has to take place rapidly enough to provide for the current crop. A recent paper describes grape flower development in detail [7]. Bud formation in tropical fruits is controlled mainly by water availability and temperature and thus can be less predictable than for deciduous fruits. An extreme example is papaya (*Carica papaya*). Although basically dioecious, under various temperature, moisture, and nutrient stresses, carpels can metamorphosize into stamens, and vice versa [8].

Thus flower bud initiation is a necessary precursor to fruit formation. Particularly after the landmark 1918 paper by Kraus and Kraybill on fruiting in the tomato [9,10], it was believed that flower bud initiation was dependent on the balance between carbohydrates and nitrogenous compounds in developing tissues (the C/N hypothesis). Within the last 50 years, it has been realized that in any plant, flower bud initiation, and hence fruit formation, is controlled by growth regulators (GRs). Development of GRs and the balance between them is, in turn, controlled by environmental forces, notably temperature and light.

Gibberellins were among the first GRs to become available in commercial quantities, thus greatly facilitating research showing that for a very wide range of plants, gibberellins could inhibit flower bud formation and sometimes induce parthenocarpy if applied after flower bud initiation [11–17]. For details of the histology of flower induction in apples, see Buban and Faust [18].

Research on the role of GRs in bud initiation has been facilitated by the finding that tracheal sap is a convenient source of naturally occurring GRs [19]. Abscisic acid (ABA) is now known to be very much involved, not only in flower bud formation, but also in fruit development [20,21]. Growth regulators control messenger RNA (mRNA), which generates necessary enzymes *de novo* for fruit development following anthesis [15,22,23]. Much recent research has involved manipulation of bud differentiation and fruit development using exogenous application of both natural and synthetic GRs [24–26]. But flower bud initiation, and hence the entire cycle of flowering and fruiting, can be controlled solely by intelligent manipulation of temperature and light. A 39-week cycle (repeated at 4-week intervals) has been developed using dark and lighted cold rooms and greenhouse or nursery facilities to provide a continuous supply of three varieties of container-grown apples for year-round harvesting, a remarkable feat [27].

Normally, pollination is necessary for fruit set; however, there are notable exceptions. With the buying public increasingly demanding seedless fruits of various kinds, parthenocarpy has become highly prized for many types of fruits. This is certainly so for citrus fruits, for which pollination had long been deemed unnecessary, even undesirable, as it increases the number of seeds in supposedly "seedless" varieties. That was before the introduction of a number of human-made crosses such as the tangelos (tangerine × grapefruit). Tangelo varieties that were apparently fruitful when grown in small trial plots were almost completely barren when planted in large multihectare blocks. Thus it was found that for some hybrids, such as Orlando tangelo, pollination by some other variety was as necessary as it is for apples and pears [2]. A remarkable example of parthenocarpy is the navel orange, which has a small secondary fruitlet at the stylar end and which is always seedless. Fruit set of navel oranges, which is often uneconomically light in Florida, is sharply affected by ambient temperatures prior to and during fruit set [28]. Since seedless table grapes may fetch more than twice the price of seedy grapes, parthenocarpy is highly valued [29]. In the popular Thompson Seedless variety, fruit set is dependent on GRs involved in pollen tube development, even though the pollen tube does not reach, and hence does not fertilize, the ovule [30].

III. MORPHOLOGICAL CATEGORIES OF FRUITS

Fruits have evolved so many diverse forms that Soule lists 46 different morphological fruit types [1]. Although anatomical and taxonomic considerations cannot be ignored completely, only a few general categories can be considered within this context of fruit physiology. Nondessert fruits are discussed only insofar as is necessary to establish their place in the wide general category of fruits. For a detailed histological treatment of the various tissues that can be involved in fruit development, see Esau [31]. All fruits are the products of matured ovaries. Some, in addition, incorporate other floral parts. This is particularly true for fruits derived from inferior ovaries (epigyny): that is, fruits such as apple and pear, in which the other floral parts (stamens, petals, and sepals) are above the ovary.

A. Achene

An achene is a hard, dry, fully matured simple ovary. Achenes are usually thought of as "seeds" (although some may contain two seeds). A grain of wheat is an achene, each flower within a head of wheat (inflorescence) having matured individually to form an achene. A grain of corn (maize, *Zea mays*) is an achene, corn on the cob being an unusual example of an intact, nondehiscent inflorescence. Achenes are, in general, nutritious and have been utilized as foods since antiquity, not only in the form of our well-known cereal grains, but also as such lesser known species as the sumpweed (*Iva annua*), gathered by native North Americans, and amaranth (*Amaranthus caudatus* and *A. quitensis*), a staple of the pre-Columbian Aztecs, the cultivation of which has persisted in remote Andean valleys and which is currently an interest of "health food" devotees [32].

B. Typical Fruits from Superior Ovaries (Hypogyny)

1. Grape

The grape is the simplest of hypogynous fruits and one that conforms exactly to the botanical definition of a berry. Remnants of floral parts other than the ovary are absent or vestigial and the developed ovary tissue is fleshy, succulent, and homogeneous.

2. Hesperidium

The hesperidium is the highly specialized form of berry specific to citrus fruits. (Etymologically, the term *hesperidium* is a misnomer based on the assumption that the "Golden Apples of the Hesperides" in Greek mythology were oranges. However, citrus, as the etrog, *Citrus medica*, did not reach the Mediterranean area until historical times [33].)

The hesperidium (which is derived entirely from the ovary) has several sharply defined tissues (Figure 1). The usually five-lobed calyx remains attached unless the fruit naturally abscises; then it remains attached to the bearing branch.

The outer layer, or peel, includes the pigmented *flavedo* and the white or colorless *albedo*. The flavedo (Figure 1A, top left) consists of the epicarp proper, hypodermis, and the outer mesocarp. Embedded in it are the so-called "oil glands," containing "essential oils" specific for each citrus species or hybrid. These are principally terpenes (mainly *d*-limonene) and are highly toxic to surrounding tissue if extruded due to rough handling of the fruit. The cells of the single-layered epicarp contain green chloroplasts that metamorphosize into chromoplasts as the fruit degreens. Over the epicarp is the intact cuticle (Figure 1A, lower left), composed largely of cutin, and over it an outer layer of epicuticular wax deposited as easily dislodged platelets. (No citrus fruit is naturally shiny; the shine demanded by retail customers has to be applied as some form of approved wax or resin after washing, an operation that dislodges much

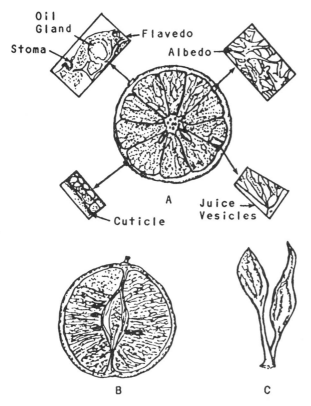

Figure 1 Citrus fruit: (A) transverse section with enlarged views of the flavedo and cuticle on the left and of the albedo and juice vesicles attached to the outer tangential and radial locule walls on the right; (B) longitudinal section showing the lunate locules with seeds attached to the inner tangential wall next to the central axis; (C) separate juice vesicles. (From Ref. 34.)

of the natural, nonshiny wax.) The cuticle is penetrated by numerous stomata, except in a narrow (ca. 3 mm) area around the calyx.

The albedo, or inner mesocarp (Figure 1A, top right), consists of a loose network of parenchymatous cells with large airspaces formed when small, originally spherical albedo cells retained their original points of contact as the fruit expanded. Thickness of the albedo can range from as little as 1 to 2 mm in some limes and tangerine hybrids to 2 cm or more in large shaddocks (pummelos, *Citrus grandis*).

The edible flesh of a mature citrus fruit is divided into segments, each derived from an ovary locule. The number of segments varies widely but is typically between 10 and 15. Each segment is surrounded by a tough endocarp membrane and filled with tightly packed juice sacs or vesicles (Figure 1A, lower right, and C). Each of these thin-walled juice sacs has a minute oil gland in its center and is attached by a fine stalk to vascular bundles in the radial segment walls. Except in parthenocarpic fruit, seeds are within the segments and attached to axial vascular bundles. Despite various varieties being sold as "seedless," few except navel oranges and Persian (Tahiti) limes (*Citrus latifolia*) are truly seedless. Purists prefer the term "sparsely seeded," for which citrus dealers show no enthusiasm at all.

The vascular system is a highly ramified network whereby every cell is connected to, or adjacent to, a cell in contact with a particular sector of the vascular system. In many types of citrus fruits, particularly seedless grapefruit and tangerines, the central "core" bundles separate as the fruit matures, leaving a considerable cavity in the center of the fruit (a complication in specific gravity separation of freeze-damaged fruit). For a more detailed discussion of citrus fruit anatomy, see Soule and Grierson [34].

3. Drupe

Drupes start out as though they were going to be berries but then develop their typical hardened "pit." The resultant fruit is technically described as a "simple fruit with soft exterior, fleshy, usually indehiscent, with heterogeneous texture and the center with a hard, bony, or cartilaginous endocarp enclosing the seed proper" [1].

The most familiar drupe fruits are peach (and its genetic recessive, nectarine), plum, cherry, and apricot, and in the tropics, the mango. In all these fruits, the edible portion is the fleshy mesocarp. Other, less obvious drupe fruits are coffee, in which the fleshy mesocarp (though edible) is discarded. It is an anomalous drupe, having two seeds enclosed in a parchmentlike endocarp, the seeds being the "coffee beans" of commerce. Other drupes grown for their seeds are almond and pistachio. The most atypical of all drupes is the coconut, in which the dry, fibrous epicarp and mesocarp become the husk (the source of coir fiber used in brushes, matting, and rope). The large seed has edible white oily flesh and a liquid endosperm (the "coconut milk").

Drupe crops can be of purely temperate-zone origin with specific winter-chilling requirements (peach, plum, cherry, apricot) or purely tropical (mango, date). Intermediate is the pistachio, which has a brief winter chilling requirement but very limited freeze hardiness [35].

C. Typical Fruits from Inferior Ovaries (Epigyny)

In flowers of epigynous fruits, the other major floral parts, sepals, petals, and stamens, are fused at their bases and located above the ovary. As such fruits develop, nonovarian tissues become intrinsic parts of the fruit. It is often very difficult to discern ovarian from nonovarian tissue.

1. Pome Fruits

All the pome fruits are members of the Rosaceae family: for example, apple, pear, quince, medlar, hawthorn, and the tropical loquat. A pome is defined as a fruit in which the papery or cartilaginous endocarp is embedded in the mesocarp, fused with and completely enveloped by the enlarged fleshy receptacle or the fused base of the sepals; the ripened ovary is only a small part of the total structure [1].

By far the best known and most widely grown pome fruit is the apple (*Malus sylvestris*). Its flower parts are in fives: five sepals, five petals, five stamens, and five carpels making up the deeply embedded ovary (Figure 2). The parenchyma of the fused bases of the calyx, corolla, and stamens constitutes the major portion of the edible tissue of the mature fruit. The nonedible core is largely ovary tissue.

Although parthenocarpy is not unknown in some obscure varieties, fruit development normally starts at pollination. Because most apple varieties are self-infertile, pollen usually has to come from some other variety (cultivar). Fruit development is almost invariably dependent on fertilization and resultant seed formation. (Fortunately, the buying public's prejudice against seedy fruits does not include apples.) The hormonal control of fruit development was first indicated by the common observation that when seeds fail to develop in one or more of the five carpels, the fruit tends to grow lopsided. Most flowers never survive to form fruits. Only about 2 to 4% of the flowers in a normal bloom need to develop to provide as heavy a crop as the trees can bear.

The epidermis of the very young fruit is constantly growing, initially with very active cell

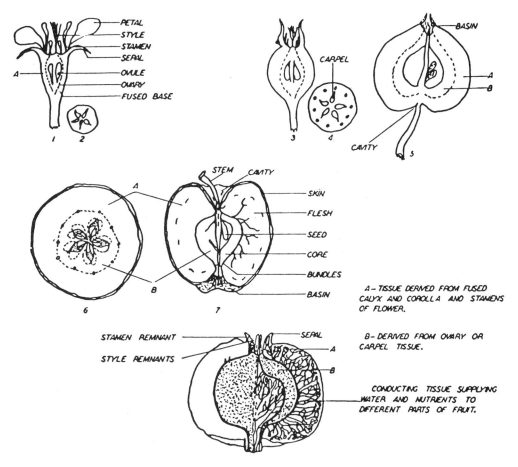

Figure 2 Development of the apple fruit from the flower stage. (From Ref. 36.)

division. After 4 or 5 weeks of development, cell division slows down and then ceases. As the fruit continues to expand, the epidermal cells flatten and elongate. As the fruit matures, these epidermal cells become surrounded by cuticle. The cuticle is covered by a layer of wax which is continuous in varieties with a natural shine but is deposited as irregular platelets in "nonshiny" varieties such as Golden Delicious and Grimes Golden. Today, most apples are artificially waxed, partly to retard shrinkage, but more because of the buying public's fascination with shine (even for fruits that are not naturally shiny). The edible parenchymatous tissue consists of large, thin-walled cells with a surprising volume, estimated at ca. 25%, taken up by airspaces [36].

COMMENT ON THE RELATION BETWEEN FRUIT STRUCTURE AND HANDLING DAMAGE Because most apples have a continuous cuticle, they are very resistant to water damage and have even been stored successfully under water. If fungal spores enter through a break in the cuticle, fungal hyphae tend to spread so slowly that it is common to cut out an infected area and consume the rest of the apple. However, apples and pears have very little resistance to pressure, which can rupture parenchymatous cells. These are rich in oxidases and surrounded by air in the intercellular interstices. The result is an ugly dark brown bruise.

Because of its discontinuous waxy coat and occasional still-dividing epidermal cells [37], a citrus fruit has poor resistance to prolonged submersion in water. The structure of a citrus fruit, with its spongy albedo and radially oriented juice sacs, is very resistant to pressure from smooth surfaces. It does not bruise but is very susceptible to damage from sharp objects. Any rupture of epidermal oil cells releases "peel oil," toxic to adjacent tissue, with resultant ugly lesions ("oleocellosis"). Any fungal spores introduced into the albedo find a perfect culture medium. The spread of fungal mycelium is almost explosive.

2. Banana (Musa sapientum)

An interesting tropical fruit, the banana, is sometimes classified as a berry, which is clearly erroneous since nonovary tissue is involved (be it only as the nonedible skin of the fully mature fruit). Banana flowers are dioecious, the male flowers being borne within conspicuous purple bracts at the end of the long, hanging inflorescence. The female flowers are clustered in groups along the stem of the inflorescence. These groups of flowers develop into clusters of fruits called "hands," in which each individual fruit is referred to as a "finger." The general tendency is that the more hands there are on a bunch, the more fingers there are in each hand.

The female flower is inconspicuous and described as a "tepal," in which the components of the perianth are so similar in size, form, and coloration that sepals cannot be distinguished from petals [1]. The inconspicuous perianth is abscised immediately after the flower opens. Thus only ovary and receptacle remain.

Pollination is necessary for fruit set, but all commercial varieties are, nevertheless, parthenocarpic. Banana breeders thus have a double problem. When, for example, Panama disease was wiping out Gros Michel, the major commercial variety of Central America, they had to cross-fertilize with seedy resistant varieties. With that accomplished, backcrossing was necessary to eliminate the seeds while retaining necessary disease resistance. This was accomplished in a surprisingly short time.

Initially, the peel (which is receptacle tissue) weighs five times as much as the interior pulp. As the fruit grows, the endocarp develops fleshy protusions into the locules forming the edible pulp. At full maturity, the edible pulp typically weighs twice as much as the inedible peel. However, few commercial bananas are allowed to reach full maturity. For long-distance shipment, bananas are picked at stages of development known by such terms as "3/4-full" and "2/3-full," terms describing a somewhat angular cross section. Fortunately, the highly climacteric banana will ripen to good eating quality even when harvested well short of physiological maturity. It is usual commercial practice for bananas to be shipped green and ripened in "ethylene degreening rooms" at destination. In addition to accelerating the natural climacteric ripening process, this ensures uniformity in ripening, a convenience in marketing.

D. Aggregate Fruits

Aggregate fruits are compound fruits produced from many pistils in an inflorescence rather than from a single pistil. Temperate-zone aggregate fruits include strawberry, mulberry, and the various cane and bramble fruits. As mentioned previously, the fig is an aggregate fruit, with its minuscule flowers inside the vaselike receptacle, and so indiscernible without dissection of the fruit. Among tropical fruits, breadfruit, pineapple, and cherimoya are aggregate fruits. Three familiar examples of aggregate fruits are discussed here.

1. Raspberry (Rubus idaeus)

The raspberry is an intrinsically frail fruit, in that unlike its near relative the blackberry, at picking the receptacle remains on the plant. The harvested fruit is thus hollow, formed only from adhering drupelets (miniature drupes), each from a separate floret within an inflorescence.

Improving the inherent structural weakness of the raspberry has become a challenge for research workers [38,39].

2. *Strawberry* (Fragaria virginiana × F. chiloensis)

The strawberry is an accessory fruit, one in which the conspicuous fleshy part is composed of tissues external to the pistil. (The *Annonas*, soursop, sweetsop, and cherimoya are tropical examples of accessory fruits.) The succulent flesh of the strawberry is receptacle tissue. The "seeds" embedded in its exterior surface are achenes and thus true fruits.

3. *Pineapple* (Ananas comosus)

The pineapple is a multiple fruit, one formed from many pistils of an inflorescence. The pineapple fruit develops from separate lavender-colored flowers distributed around the length of the central axis of the inflorescence. The entire flowers become incorporated into the fruit, much of the flesh being formed from the fleshy bracts subtending each flower. Individual varieties are self-infertile; hence pineapples grown in monocultures of a single variety are always seedless. However, in areas such as the Caribbean, where small plots of various varieties are common, it is usual to have pineapples with occasional small black seeds.

IV. PHYSIOLOGICAL DEVELOPMENT

As a general principle, fruit development in terms of weight and volume tends to be sigmoidal. A period of very rapid cell division, but very little increase in fruit size (stage I), is followed by a period of rapid increase in size as small, newly differentiated, dense cells develop vacuoles and assume their roles as specific tissues (stage II). In the final stage, as the fruit reaches physiological maturity, increase in size slows and may even stop, although biochemical changes may continue (stage III). There are about as many variations on this pattern as there are different types of fruit, but the sigmoidal mode is usually discernible. The orange, apple, and apricot are discussed below as typical examples of the development of citrus, pome, and drupe fruits.

A. Orange *(Citrus sinensis)*

The duration of growth and maturation varies sharply with variety. For early varieties such as Hamlin and navels, harvesting commonly starts 6 to 7 months after bloom. For the late Valencia variety, harvesting starts about 12 months after bloom. Harvesting can continue for a "tree storage" period lasting several months, during which late oranges have two crops on the tree at the same time. Herein lies a critical difference between citrus and deciduous fruits. The latter *must* be picked soon after maturation is complete or they will fall from the tree. Citrus fruits have no such sharply defined abscission period, something that is frustrating to would-be developers of mechanical harvesting equipment but an enormous advantage in marketing the crop. Stages of development are shown in Figure 3.

Stage I lasts a month or less, during which cell division is extremely rapid but fruit enlargement is trivial. At this stage the cuticle has not yet developed, making the little fruitlets extremely vulnerable to superficial damage. In growing areas such as Florida, where stage I coincides with the strongest winds of the year, just brushing against an adjacent leaf causes major "windscars" on the mature fruit. This problem is exacerbated in areas such as Brazil and Florida, where rains in the postbloom period facilitate superficial infection of such windscars by waterborne spores of the melanose fungus (*Diaporthe citri*). Although most cell division takes place in this period, some cell division can continue in the peel until maturation, particularly with navel oranges, making such fruit very vulnerable to water damage [37].

Stage II is the period of cell (and hence fruit) enlargement. The fruit expands rapidly, as does CO_2 output per fruit, although CO_2 evolution per unit weight (the usual way of expressing respiration) declines sharply (Figure 3). During this period, the juice sacs are enlarging and developing their distinctive solutes. Such solutes are initially high in organic acids and low in sugars. As the orange matures, sugars increase steadily while acids decline. Legal maturity standards for citrus fruits are usual in major producing areas. In this, every district sets its standards according to what they do best [40]. European citrus districts, South Australia, California, and other districts with Mediterranean-type climates (cool winter nights, bright days, and low rainfall) can rely almost entirely on external standards to sell their oranges. Florida, with its blossom-period winds and humid, subtropical climate, cannot compete on appearance and so relies principally on standards based on the high sugar content of its oranges. These maturity standards are based not only on sugar content but also on the ratio of total soluble solids (TSS, mainly sugars) to acids (titratable as citric acid), with a sliding scale throughout the season [41,42]. At the beginning of the season, Florida oranges must have 8.0% TSS with

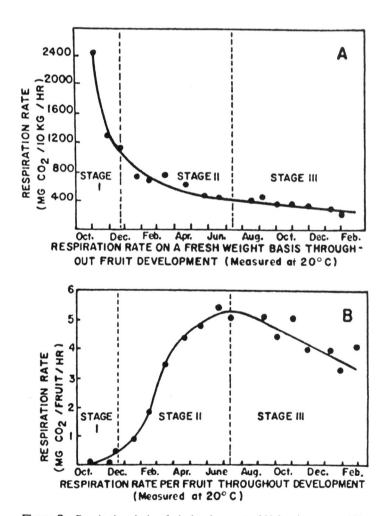

Figure 3   Respiration during fruit development of Valencia orange. (A) expressed as CO_2 evolution per unit fresh weight; (B) expressed as CO_2 evolution per fruit, a form that more clearly defines the stages of fruit development. (From Ref. 34 as adapted from Ref. 66.)

a TSS/acid ratio of 10.5:1. By the end of the season, this ratio may exceed 20:1, but with the proviso that (for fresh fruit sale) acid cannot be below 0.4% lest the oranges taste too insipid.

Regardless of growing district, consistent gradients occur within a citrus fruit, particularly in terms of sugar content. The vascular system extends down the central axis of the fruit, reaching the blossom (stylar, distal) end first, then ramifies back up the carpels to the stem (calyx, proximal) end of the fruit. Apparently as a consequence of this distribution of photosynthates, sugars are higher in the blossom end. A very thorough study reported that the proximal halves of mature California Valencia oranges averaged 7.2 g of sugar per liter of juice as compared to 9.5 g/L for the distal (blossom, stylar) halves, a difference clearly discernible by taste [43]. When sharing a grapefruit, canny citrus people give the stem-end half to their companion, retaining the blossom end half for themselves.

In the jungles of southeast Asia where citrus first evolved, all are still green when mature. The extent to which the expected orange or yellow colors develop depends on the growing area having cold enough nights to stress the fruits [44]. In subtropical areas such as Florida and Brazil, early varieties may mature while still green, necessitating postharvest removal of the green chlorophyll with ethylene [45].

B. Apple

The typical growth curve of any main crop apple variety is only slightly sigmoidal. Very early varieties, such as Early Harvest, Yellow Transparent, and Melba, mature to acceptable eating quality before any deceleration of growth (Figure 4). Apples that mature this early are very

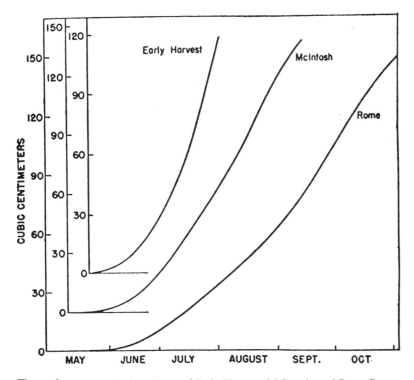

Figure 4 Increase in the volume of Early Harvest, McIntosh, and Rome Beauty apples from full bloom to maturity. (From Ref. 36.)

frail and suitable only for local consumption. The longer it takes an apple variety to reach maturation, the more sigmoidal its growth curve. In general, the later an apple variety matures, the longer its potential marketing life.

Initially, all cells of the apple are alive. Cell division in the epidermis ceases at the end of stage I. Marked elongation and flattening of the epidermal cells occurs throughout stage II, during which period the epidermal cells extrude waxy, cutinous material. In fully mature late-season apples, the epidermal cells are separated, dead or dying, embedded in the continuous cuticle (a heterogeneous polymer of fatty acids overlaid with a layer of wax). The cuticle can continue to develop after harvest. During the stage II growth period, the epidermis is penetrated by stomata that tend to cork over at full maturity. Under the epidermis in some varieties is the periderm, a thin layer of cork cambium. If the epidermis is injured early in stage II growth, as by mechanical abrasion or frost, the periderm develops a protective layer of corky cells: biologically an excellent protection for the fruit but a "grade-lowering defect" for the packer and the consumer.

Parenchyma tissue from the fused bases of the calyx, corolla, stamens, and receptacle constitutes the major part of the edible tissue of the mature fruit. Cell division having ceased at the end of stage I (usually, 3 to 5 weeks after anthesis), the considerable enlargement of the fruit comes from cell enlargement and their partial separation to form a considerable volume of air-filled intercellular spaces. Except for the petals (which abscise and fall after fruit set), all the original parts of the flower persist in the fully developed apple.

C. Apricot *(Prunus armeniaea)*

The growth curve of the apricot, indeed of all fleshy, succulent drupes, is exaggeratedly sigmoidal (Figure 5). Stage II growth is interrupted by "pit hardening," in which the endocarp thickens and lignifies to form the hard, stony "pit" enclosing the seed. During this period, the

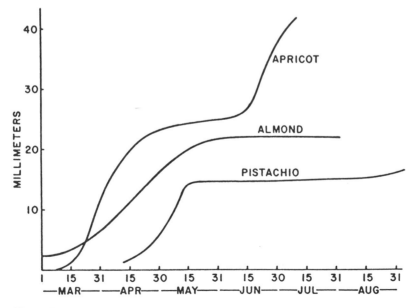

Figure 5   Growth in diameter of fruits of Ne Plus almond, Royal apricot, and Kerman pistachio. (From Ref. 35.)

fruit ceases to increase in size. Biochemical changes continue, but without cell enlargement. Morphological development in the peach (which is essentially similar to apricot) has been reported in considerable detail [46].

The apricot pit is smooth and, at maturity, quite free from the edible mesocarp tissue, being attached only at its proximal end by the persistent vascular system. In other drupes, the pit is seldom so separate, although in "freestone" peaches the deeply incised pit is nearly free from the edible mesocarp. In "clingstone" peaches, the endocarp and mesocarp interfaces adhere.

In the mango (*Mangifera indica*), the ultimate example of a "clingstone drupe," the pit is covered profusely with tough fibrous "hairs" that usually extend into the edible flesh. The date (*Phoenix dactylifera*), the ultimate "oasis crop," is a specialized drupe that develops so much sugar that its cells plasmolyze and ultimately die. When fully mature, all that is still living is the embryo within the stony seed. After harvest, the date is therefore handled as for a confection rather than as a fruit.

Two familiar dessert nuts are the seeds of drupes. The almond (*Prunus communis*) and the pistachio (*Pistacia vera*) are drupes in which the mesocarp fails to develop any further after pit hardening, thus resulting in a growth curve that is definitely not sigmoidal (Figure 5).

V. POSTHARVEST LIFE

It should not be necessary to emphasize that fruits are still alive after harvest. However, a surprising number of people who make their livelihoods growing, packing, shipping, and selling fruit do not realize that they are handling living, breathing creatures, subject to specific diseases and the ravages of senescence. ("Gee, Doc, don't tell me these things are alive. They've been picked!") Moreover, effective postharvest handling is not merely a matter of maintaining the state of fruit quality at time of picking. Properly handled, many fruits improve in eating quality after harvest. Others degenerate rapidly or slowly, depending on their innate physiology and the postharvest conditions to which they are subjected.

A. Climacteric Versus Nonclimacteric Fruits

The first step in proper postharvest handling of a given type of fruit lies in understanding its type of life cycle [47]. The climacteric rise in respiration of fruits such as apple, pear, avocado, mango, and banana represents a rapid depletion of potential postharvest life (Figure 6). For fruits such as pear, banana, and avocado, experiencing the climacteric is essential to the ripening that makes them truly edible. But it should be delayed as much as possible until the consumer is ready to eat that piece of fruit. Very prompt refrigeration is essential for orderly marketing of climacteric-type fruits, to delay or suppress the evolution of endogenous ethylene that initiates the climacteric rise. As the height of the climacteric is reduced, its duration is extended proportionately. Immediate temperature and humidity control is the first line of defense against expensive wastage. Humidity control is important if for no other reason than that a shriveled fruit ceases to be marketable. However, there are other physiological benefits also [48]. Even within a specific variety, response to such storage techniques as controlled atmosphere storage can be sharply influenced by cultural and climatic factors [49]. When the peak of the climacteric rise is past, the fruit becomes senescent. Although adequate reserves of respiratory substrate may be available, cellular organization breaks down, the cell membranes lose their integrity, and the fruit dies of old age [50,51]. Thus the challenge with climacteric-type fruits is to suppress and extend the respiratory rise.

Apples and pears are examples of climacteric-type fruits that have to be harvested within a very brief period but marketed for as long a period as correct storage procedures permit.

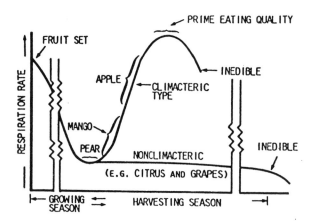

Figure 6   Climacteric and nonclimacteric life cycles for typical fruits. (From Ref. 34.)

Long-storing varieties have ample reserves of respiratory substrate and resilient respiratory systems. Under near-optimum conditions, late varieties such as Winesap can be kept year-round. Some, such as Northern Spy and Winter Banana, improve in eating quality during the first few months of storage.

The avocado (*Persea americana*) is an interesting climacteric-type fruit. Although strongly climacteric, the characteristic respiratory rise will not start until the avocado is picked. For many years research workers were convinced that when their instrumentation improved sufficiently they would be able to identify a preharvest "climacteric inhibitor." Even with modern equipment, it has been impossible to identify any such inhibitor [52].

Most varieties of pears (*Pyrus communis*) do not ripen to acceptable eating quality on the tree. Once picked, pears have to be *either* ripened for immediate use (preferably at 20 to 25 °C) *or* held in cold storage at only a degree or two above their freezing point. Pears, particularly the popular Bartlett variety, will neither ripen nor store at intermediate temperatures, particularly in the range 8 to 12°C. Instead, they become rubbery in texture and virtually inedible.

This is necessarily an abbreviated and simplified account of the complex physiology of climacteric-type fruits. The extraordinary development of nonchemical analytical equipment has stimulated much recent postharvest research. Some surprising results are being encountered, such as a newly developed thornless blackberry being strongly climacteric [53]. Those interested in further reading are referred to a 1985 symposium [54], particularly the paper by McGlasson [55].

Handling of nonclimacteric fruits is very much simpler. There are no significant physiological changes involved in separation from the tree and no postharvest ripening cycle. Those signs along the Florida highways saying "TREE-RIPENED CITRUS," although misleading, are legally defensible, *all* citrus fruits being "tree ripened." With no climacteric rise to suppress, nonclimacteric fruits such as citrus of various types, grapes, and certain vegetables that are botanically fruits do not benefit nearly as much from prompt refrigeration as do climacteric-type fruits. Indeed, for fruits susceptible to chilling injury, delayed storage may be beneficial by enabling the fruit to adapt to lower storage temperatures [44]. Sooner or later, of course, any fruit can be expected to abscise if left on the tree long enough. Modern research shows this to be a surprisingly complicated biochemical and histological process [56].

B. When to Harvest

This discussion is out of chronological order in terms of the life of the fruit because it is necessary to understand something of postharvest fruit physiology before dealing with optimum picking dates for various types of fruits.

1. Citrus Fruits

It is fairly simple to set legal maturity criteria for nonclimacteric fruits such as citrus and grapes. These undergo no considerable physiological change at harvest nor do they abruptly abscise and fall. Maturity standards, either legal or voluntary, can be set in terms of sugar content, sugar/acid ratio, and juice yield. Moreover, citrus fruits can be "tree stored." Early tangerine varieties can be picked over a period of several weeks, at the end of which period they start to dry out rapidly. Orange varieties, particularly the late, main crop Valencia variety, can be picked over a period of 2 to 3 months, sometimes more. Grapefruit from a single bloom can be harvested over a period of 6 months or more. (As this is being written, the same Florida grapefruit that might have been picked in October 1992 are still being harvested during the first week of June 1993.) This is a great convenience in marketing, provided that the shipper does not try to extend marketing by storing grapefruit that has already used up its storage potential during prolonged tree storage [57].

2. Apples (Malus sylvestris)

Deciding on a harvesting date is very much more difficult for climacteric-type, temperate-zone deciduous fruits for which only a narrow window of opportunity is available. "It is exceedingly important that apples be harvested at the right time. The exact degree of maturity at which a given variety should be picked depends in large part on what disposition is to be made of the fruit. . . . If apples are picked too soon and then stored for any length of time they are subject to storage troubles such as bitter pit and scald. . . . Almost every measure or index of maturity has to be defined for not only a given variety but for a given location, season, and soil type (36)." That advice was published 43 years ago, and despite considerable research, not much has really changed since then. In the search for a reliable criterion (or combination of criteria) as a guide to optimum picking date for apples, research workers have investigated days from full bloom, ground color, pull test (ease of separation), pressure test (with various modifications of the original 1925 Magness–Taylor pressure tester [58]), soluble solids, iodine-starch pattern, seed color, and corking of lenticels. No criterion has proved consistent across varieties, growing districts, cultural methods, and seasonal variations in climate. Rootstocks can have a significant effect on maturity criteria [59], as can use of spur-type scion selections. But it is nice to note that organic cultural methods are reported as not significantly affecting maturity criteria [60]. Harvesting criteria for each particular apple variety in each district still has to be based largely on local experience and judgment. A user-friendly computer program has been developed to help growers and packers [61].

A further complication is that "stop-drop sprays" have been used for many years to extend the possible harvesting season for apples [62]. Such prolonging of the harvesting period can be expected to reduce potential storage life. This is particularly true for the highly colored strains that usually have been selected from chance sports regardless of other fruit quality criteria. (This writer has grown weary of attending meetings at which nursery owners and produce merchandisers proclaim that their aim is to "Give the lady what she wants," a policy that all too often sacrifices eating quality for appearance.) Now it appears that the selection of the culturally profitable spur-type strains may also sometimes be at the expense of keeping quality [63].

3. Pears (Pyrus communis)

The situation for pear harvesting is no more promising. Over 50 years ago, this writer was a graduate student participating in a massive 5-year project involving five pear orchards throughout Canada's Niagara Peninsula. A major objective was to establish a reliable maturity standard for harvesting Bartlett pears, particularly for research in a then very new and experimental controlled atmosphere storage. (This method was then called "gas storage," later renamed "controlled atmosphere" by Bob Smock of Cornell University.) As well as pressure test, starch-iodine pattern, and so on, this program included measuring respiration immediately after picking. Although variation among seasons and orchards excluded all other criteria, one remained consistent. The best quality and longest storage life were always from the picking at the nadir of fruit respiration on the tree [64]. Since that can only be determined retroactively, it cannot be used as an indication of when to pick for maximum quality. Recent developments such as growing pears with apple interstocks and on clonal apple roots [65] further complicate the prospect of finding generally applicable criteria to determine optimum picking time for pears. Localized growing areas, particularly in irrigated districts, may use some standard (pressure test is most common), but it seems unlikely that statewide legal maturity standards will ever be established such as have long been enforced for citrus fruits [41–42].

VI. CONCLUSIONS

There is very little in agriculture that one way or another is not dependent on successful fruit development. Among those who make their livelihoods growing and marketing dessert fruits, there are many who could profit from improved understanding of the complex biology of these gracious additions to our diet.

REFERENCES

1. J. Soule, *Glossary for Horticultural Crops*, Wiley, New York (1985).
2. A. H. Krezdorn and F. A. Robinson, *Proc. Fla. State Hortic. Soc.*, *71*: 86 (1958).
3. A. E. Watada, R. C. Herner, A. A. Kader, R. J. Romani, and G. L. Staby, *HortScience*, *19*(1): 20 (1984).
4. W. A. Gortner, G. G. Dull, and B. H. Krauss, *HortScience*, *2*(4): 141 (1967).
5. S. D. Seeley, *HortScience*, *27*(12): 1263 (1992).
6. T. L. Davenport, *Hortic. Rev.*, *12*: 349 (1990).
7. J. M. Gerrath, *Hortic. Rev.*, *13*: 315 (1991).
8. T. D. Arkle, Jr., and N. Y. Nakasone, *HortScience*, *19*(6): 832 (1984).
9. E. J. Kraus and H. R. Kraybill, *Oreg. Agric. Exp. Stn. Bull. 149* (1918).
10. J. S. Cameron and F. G. Dennis, *HortScience*, *21*(5): 1099 (1986).
11. S. P. Monselise and A. M. Halevy, *Proc. Am. Soc. Hortic. Sci.*, *84*: 141 (1964).
12. J. C. Crane, P. E. Primer, and R. C. Campbell, *Proc. Am. Soc. Hortic. Sci.*, *75*: 129 (1960).
13. J. Hull, Jr., and L. N. Lewis, *Proc. Am. Soc. Hortic. Sci.*, *74*: 93 (1959).
14. W. H. Griggs and I. T. Iwakiri, *Proc. Am. Soc. Hortic. Sci.*, *77*: 73 (1961).
15. J. Van Overbeek, *Science*, *152*: 721 (1966).
16. R. Marcelle and C. Sironval, *Nature*, *197*: 405 (1963).
17. J. C. Crane, *Proc. Am. Soc. Hortic. Sci.*, *83*: 240 (1963).
18. T. Buban and M. Faust, *Hortic. Rev.*, *4*: 174 (1982).
19. T. Saidha, E. E. Goldschmidt, and S. P. Monselise, *HortScience 18*(2): 231 (1983).
20. Wen-Shaw Chen, *HortScience*, *25*(3): 314 (1990).
21. R. Rudnicki and M. J. Bukovac, *HortScience*, *19*(5): 655 (1984).
22. D. Grierson, A. Slater, J. Speirs, and G. A. Tucker, *Planta*, *163*: 263 (1985).
23. A. Callahan, P. Morgens, and E. Walton, *HortScience*, *24*(2): 356 (1989).

24. A. Erez, *HortScience*, *22*(6): 1240 (1987).
25. T. J. Gianfagna, R. Marini, and S. Rachmiel, *HortScience*, *21*(1): 69 (1986).
26. S. M. Southwick and F. S. Davies, *J. Am. Soc. Hortic. Sci.*, *107*(3): 395 (1982).
27. M. E. Saltveit, Jr., *HortScience*, *18*(6): 919 (1983).
28. F. S. Davies, *Hortic. Rev.*, *8*: 129 (1986).
29. G. Acuff, *Fruit Grow.*, *113*(2): 34 (1993).
30. C. A. Ledbetter and D. W. Ramming, *Hortic. Rev.*, *11*: 159 (1989).
31. K. Esau, *Plant Anatomy*, Wiley, New York, Chap. 19 (1965).
32. C. B. Heiser, Jr., *Of Plants and People*, University of Oklahoma Press, Norman, Okla., pp. 107–109 and 163–172 (1985).
33. W. Grierson, *HortScience*, *5*(5): 1 (1970).
34. J. Soule and W. Grierson, "Anatomy and Physiology," Chap. 1 in *Fresh Citrus Fruits* (W. F. Wardowski, S. Nagy, and W. Grierson, eds.), AVI, Westport, Conn. (1986).
35. J. C. Crane and B. T. Iwakiri, *Hortic. Rev.*, *3*: 376 (1981).
36. R. M. Smock and A. M. Neubert, *Apples and Apple Products*, Interscience, New York (1950).
37. F. M. Scott and K. C. Baker, *Bot. Gaz.*, *108*: 459 (1947).
38. J. Robbins and T. M. Sjulin, *HortScience*, *24*(5): 776 (1989).
39. J. Robbins and P. P. Moore, *HortScience*, *25*(6): 679 (1990).
40. W. Grierson and S. V. Ting, *Proc. Int. Soc. Citriculture*, pp. 21–27 (1978).
41. W. Wardowski, J. Soule, J. Whigham, and W. Grierson, *Fla. Ext. Spec. Publ. 99* (1991).
42. State of Florida, *The Florida Citrus Code of 1949 (as amended)*, Florida Statutes, Chap. 601 (1949 *et seq.*).
43. A. R. C. Haas and L. J. Klotz, *Hilgardia*, *9*(3): 181 (1935).
44. W. Grierson, *Beneficial Aspects of Stress on Plants*, in *Handbook of Plant and Crop Stress* (M. Pessarakli, ed.), Marcel Dekker, New York, pp. 645–657 (1993).
45. W. Grierson, E. Cohen, and H. Kitagawa, *Degreening*, Chap. 10 in *Fresh Citrus Fruits* (W. F. Wardowski, S. Nagy, and W. Grierson, eds.), AVI, Westport, Conn. (1986).
46. J. Gage and G. Stutte, *HortScience*, *26*(5): 459 (1991).
47. J. B. Biale and R. E. Young, in *Recent Advances in the Biochemistry of Fruits and Vegetables* (J. Friend and R. E. Young, eds.), Academic Press, London, pp. 1–39 (1981).
48. W. Grierson and W. F. Wardowski, *HortScience*, *13*(5): 570 (1978).
49. R. C. Sharples and D. S. Johnson, *HortScience*, *22*(5): 763 (1987).
50. F. R. Harker and I. C. Hallett, *HortScience*, *27*(12): 1291 (1992).
51. D. Grierson, *HortScience*, *22*(5): 859 (1987).
52. G. Zauberman, Y. Fuchs, and M. Ackerman, *HortScience*, *23*(3): 588 (1988).
53. C. S. Walsh, J. Popenoe, and T. Solomos, *HortScience*, *18*(3): 482 (1983).
54. D. Blanpied, S. V. Yang, M. Reid, W. B. McGlasson, A. A. Kader, and M. Sherman, *HortScience*, *20*(1): 39 (1985).
55. W. B. McGlasson, *HortScience*, *20*(1): 51 (1985).
56. L. A. Morrison and B. B. Webster, *Hortic. Rev.*, *1*: 172 (1979).
57. W. Grierson and T. T. Hatton, *Proc. Int. Soc. Citriculture*, *1*: 207 (1977).
58. J. R. Magness and G. F. Taylor, *U.S. Dept. Agric. Circ. 350* (1925).
59. G. R. Brown and D. Wolfe, *HortScience*, *27*(1): 76 (1992).
60. J. R. DeEll and R. K. Prange, *HortScience*, *27*(10): 1096 (1992).
61. C. G. Embree, B. W. MacLean, and R. J. O'Regan, *HortScience*, *26*(12): 1560 (1991).
62. R. P. Marini, R. E. Byers, and D. L. Sowers, *HortScience*, *24*(6): 957 (1989).
63. M. Meheriuk, *HortScience*, *24*(6): 978 (1989).
64. W. R. F. Grierson-Jackson, *The Storage and Ripening of Bartlett Pears*, M.Sc. Agric. thesis, University of Toronto (1940).
65. M. N. Westwood, P. B. Lombard, and H. O. Bjornstad, *HortScience*, *24*(5): 765 (1989).
66. J. M. Bain, *Aust. J. Bot.*, *6*(1): 1 (1958).

20
Dormancy: Manifestations and Causes

Frank G. Dennis, Jr.

Michigan State University, East Lansing, Michigan

I. IMPORTANCE OF DORMANCY

During their life cycles, plants are exposed to periods of stress caused by low or high temperatures, drought, or other environmental factors. In the course of evolution, complex defense mechanisms have developed for protection against such stresses. One such mechanism is dormancy. Simply defined, *dormancy* is the inability of an otherwise viable seed, whole plant, or meristem (a bud, apex, etc.) to grow.

Many plants adapted to the tropics do not become dormant; shoot growth occurs whenever environmental conditions permit. However, growth often occurs in flushes, and certain branches may be growing while others are not. In the dry topics, rainy seasons alternate with dry ones; here plants are adapted to growing when water is available, but growth slows or ceases during the dry season. Where cold and warm seasons alternate, as in the temperate zones, continuous growth is similarly impossible. Plants stop growing in the late summer or autumn, then resume growth again in the spring. In both the temperate and the polar regions another adaptation has occurred—plants develop resistance to low temperatures, or "cold hardiness," to permit survival at temperatures as low as –40°C or below (see Chapter 4). Perennial plants may be *deciduous* or *evergreen*; in the former the leaves abscise before winter begins, in the latter the leaves are functional throughout the year.

Tropical annuals will grow in any climatic zone where the growing season is long enough to allow them to mature. Thus green beans and marigolds can be cultivated from the equator to the arctic circle. In contrast, woody perennials will not survive outdoors if grown in an area where winters are too cold. Peach trees adapted to the temperate zone will grow poorly, or not at all, in the tropics for lack of "chilling" (see below), whereas mangos will not survive the low winter temperatures characteristic of the temperate zone.

Seed physiology may reflect the environmental conditions in the area of origin of the species. The seeds of plants native to the humid tropics need no dormancy provided that conditions are favorable for germination year-round. In contrast, seeds of plants adapted to the temperate zone often exhibit some degree of dormancy. If seeds shed at the end of the growing season were to germinate immediately, they would not survive the winter. Some species have circumvented this problem by having an abbreviated period of fruit development, permitting the shedding of seeds in early summer (silver maple, dandelion). In others, termed *winter annuals*, seeds germinate in late summer/early fall, and the seedlings develop sufficient cold hardiness to survive the winter and produce seed early the following year. Such seeds are dormant when shed but become capable of germination in the fall (see below).

Even when climatic factors do not dictate a need for seed dormancy, the characteristic provides a safeguard for survival. If all seeds germinated immediately, cataclysmic events such as fires and late freezes could destroy entire species, at least in local areas. Differing levels of dormancy in a seed population permit germination over a period of several years or even longer, depending on seed longevity.

II. TYPES OF DORMANCY

Numerous types of dormancy exist. The many types of dormancy exhibited by plant organs have created problems in terminology and definition. This problem was summarized for seeds by Simpson [1]: "A precise definition of dormancy cannot be used in the general sense to apply to all seeds, but can only be given for each individual seed considered in the context of a precisely defined set of environmental conditions." Nevertheless, Lang et al. [2,3] and Lang [4] have attempted to classify the many types of dormancy into three main categories, based on the controlling factor(s): *ecodormancy*, when growth is prevented by environmental conditions, such as low or high temperature; *paradormancy*, when growth is prevented by conditions outside the meristem, but within the plant; and *endodormancy*, when growth is prevented by conditions within the meristem itself. Examples of these types of dormancy are the failure of buds of trees to expand in the late winter, when low temperatures prevent growth (ecodormancy), their failure to grow in early winter, even when held in a warm greenhouse, because they have not been exposed to sufficient "chilling hours" (see below) to permit growth (endodormancy), and the failure of lateral buds to develop in an herbaceous or woody plant when the terminal bud is growing rapidly (paradormancy). In the buds of perennials, dormancy progresses gradually from paradormancy, also called apical dominance, through endodormancy to ecodormancy as the seasons progress from summer to fall to winter and spring.

These definitions are more applicable to whole plants or shoots than they are to seeds, and seed scientists have been less receptive to their use [5]. Is a dry bean seed, which exhibits no dormancy, ecodormant just because it will not grow without water? Does paradormancy exist in a seed? Does a single type of dormancy prevent growth, or are control mechanisms more complex? As we will see, dormancy is indeed a complex phenomenon in many systems.

I have spoken of dormancy in seeds and whole plants, but dormancy can occur in other structures as well. Bulbs, tubers, and corms—all organs that permit plants to survive unfavorable environmental conditions—also exhibit dormancy. This dormancy can be likened to bud dormancy, for all three structures contain buds, and bud development is the primary indication of the ending of their dormant period. In some respects the structures represent intermediates between whole plants and seeds in that they are more compact than the former but less compact than the seed, which has in addition a seed coat surrounding the embryo and closely associated parts. Most of the remainder of this chapter deals with seed and bud dormancy. Given the many aspects of dormancy, I will not address apical dominance in detail. Several recent reviews [6,7] provide information on this topic. Khan [8,9], Bewley and Black [10,11], and Bradbeer [12] provide thorough coverage of seed dormancy; Saure [13], Powell [14,15], and Martin [16] have reviewed many aspects of bud dormancy, and Dennis [17] and Lang [18] offer additional information on dormancy in general.

III.　SEED DORMANCY

A.　Induction of Dormancy

Some seeds do not become dormant until fully mature. The percentage germination of barley seeds increases with maturation up to a certain point, then declines (Table 1). Germination is further reduced when mature seeds are held at room temperature for 1 week, but it is stimulated by a brief exposure of moist seeds to low temperature [19]. Breeders sometimes take advantage of this by harvesting fruits before they reach maturity, when seeds or embryos can germinate without special treatment. Considerable research has focused on the physiological basis for the inability of immature seeds to germinate. This is discussed below, and Kermode [20] provides an analysis of the problem.

B.　Types of Seed Dormancy

Early investigators recognized that many factors could be responsible for the failure of seeds to germinate. One obvious cause of such failure is a nonviable embryo. Death of the embryo can occur during seed development (abortion) or after shedding of the mature seed. Some seeds (silver maple, citrus) are very short-lived; if germination does not occur within a few weeks, the seed does not survive. Other seeds, including many nuts, as well as avocado and cacao, lose viability rapidly when dried; if stored, a high relative humidity should be maintained. Information on methods of evaluating and prolonging seed viability are available elsewhere (e.g., Bewley and Black [10]) and will not be discussed here.

　　By definition, a seed that is dormant has the potential to germinate (is *viable*), but requires exposure to certain treatments or environmental conditions before germination can occur (Table 2). Some fruits contain inhibitors that prevent seed germination. Seeds of tomato and cucumber, for example, will not germinate within the fruit; the pulp must be removed and the seeds washed before germination can occur. In other species (e.g., peach, cherry), the presence of a hard pit (endocarp or inner ovary wall) may limit germination. Although such seeds can germinate following the breaking of endodormancy by chilling (see below), germination is improved by endocarp removal. Neither of these conditions represents true seed dormancy, as control is external to the seed, but they are often discussed in relation to seed dormancy. Some of the conditions that break seed dormancy are given in Table 3.

　　Like the endocarp, the seed coat itself can prevent germination in some species, especially

Table 1 Effect of Stage of Development and Cold Treatment on Germination of 'Cape × Coast' Barley

Stage	Germination (%)
Milk stage	5
Yellow-ripe	60
Mature	36
Mature + stored 1 week	1
Mature + stratified for 2 days	64

Source: Ref. 19.

Table 2 Types of Seed Dormancy, Conditions That Break Dormancy, and Specific Examples

Cause of dormancy	Conditions that break dormancy	Species
A. Control outside the seed		
1. Inhibitors in the fruit	Seed removal, washing	Tomato, cucumber
2. Hard endocarp	Acid treatment, endocarp removal	Stone fruits
B. Control by seed coat		
1. Coat impermeable to H_2O	Acid or mechanical scarification, fire	Some legumes
2. Coat impermeable to O_2	Seed coat removal?	?
C. Morphologically immature embryo	Warm-moist storage Cool-moist storage	Ginkgo, coconut Cowparsnip
D. Physiologically immature embryo		
1. "Shallow" dormancy	Light, alternating temperature, dry storage	Lettuce, celery, oats
2. "Deep" dormancy	Cool, moist storage	Apple, peach
a. Epicotyl dormancy	Cool, moist storage	Tree peony
b. Double dormancy	Cool, moist storage	Trillium
C + D. Hard seed coat plus deep dormancy[a]	Scarification, followed by cool, moist stratification	Redbud

[a]Some authors use the term *double dormancy* for this phenomenon also.

Table 3 Optimum Conditions for Seed Germination in Selected Species

		Conditions during germination				
			Temperature			
Pretreatment	Light	Low	High	Alternating	Species	Ref.
None			+		Bean, tomato	—
	+		+		Birch (*B. pubescens*)	10
		+			Lettuce	10
				+	Broadleafed dock	21
	+			+	*Lythrum salicaria*, tobacco	22
			+(25°C)		*Pinus lambertiana*	23
Dry storage					Wild oats	24
					Rice	25
Scarification					Black locust	26
Chilling					Apple, *Pinus lambertiana*	27, 23
	+				*Pinus strobus*	28
	+	+			*Delphinium ambiguum*	29
Scarification + chilling					Redbud	30

legumes such as alfalfa, locust, and redbud. The structure of the seed coat (testa) prevents the entry of water and its absorption by the embryo (imbibition); thus the embryo cannot germinate. The seed coat must be weakened, either naturally by abrasion, or by exposure to fire or to HCl during passage through the gut of an animal, or artificially by *scarification*, before imbibition can occur. Ground fires damage hard seed coats, thereby permitting germination of seeds that might otherwise remain dormant [31]. Scarification can be either mechanical, by rotating seeds with gravel or filing the seed, or chemical, by brief exposure to concentrated H_2SO_4. "Heat shock" by immersing seeds briefly in boiling water can be more effective than mechanical scarification in some species. For example, Bell et al. [31] reported that germination of seeds of *Acacia divergens* averaged 11, 28, and 90% for no treatment, mechanical scarification, and boiling in water for 30 s, respectively. The coats of some seeds are impermeable to oxygen. In this case, scarification allows oxygen to penetrate to the embryo. Tran and Cavanagh [32] reviewed the structural aspects of seed dormancy, emphasizing seed coat impermeability and methods of increasing it. Microscopic examination of seeds indicated [33] that treatment with boiling water or fire did not soften the seed coat but affected the structure of the "lens" (strophiole) near the hilum, thereby allowing entry of water. In some species the seed coat, although permitting entry of water and oxygen, is a mechanical barrier to germination; on its removal the embryo germinates readily. The seed coat may also contain chemicals that inhibit germination.

Many factors can affect germination. Because of the many interactions possible, Karssen [34] cautions that "an absolute requirement for any stimulatory factor hardly occurs." Therefore, one must be cautious in discussing any one factor in isolation. Nevertheless, several factors, light and temperature in particular, have pronounced effects.

C. Temperature and Seed Dormancy

Dormancy is often temperature dependent. In some cultivars of lettuce and celery, for example, germination occurs readily between 10 and 20°C but declines to nil as temperature increases to 30°C (Figure 1A). In contrast, seeds of birch germinate better in darkness at high than at low temperature (Figure 1B), but exposure to light can markedly affect response. Other

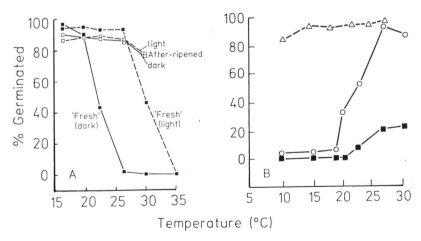

Figure 1   Effect of temperature and light on germination of seeds of (A) lettuce and (B) *Betula pubescens*. 'Grand Rapids' lettuce seeds were tested immediately after harvest ("fresh") or after storage at about 18°C for 18 months ("after-ripened"). Birch seeds were tested in darkness (■), under a 20-h photoperiod (△), or were exposed to red light for 15 min each day (O). (From Ref. 10.)

seeds germinate best when temperature is alternated on a daily cycle (Table 3). For example, when seeds of signal grass [*Brachiaria humidicola* (Rendle) Schweickerdt] are germinated at constant temperatures ranging from 13 to 38°C, germination does not exceed 2%, whereas daily alternation between 13 and 32°C results in 60% germination (Figure 2). Baskin and Baskin [36] reported that freshly harvested seeds of curled dock (*Rumex crispus* L.) remained "nondormant" for 2 years when buried 7 cm deep in moist soil. However, the seeds germinated in the light at alternating temperatures. Few seeds (<1%) germinated while buried. Therefore, the seeds would probably have remained dormant had they been held in darkness at constant temperature.

Seeds of certain species require prolonged exposure to relatively high temperatures before germination can occur. Chickweed (*Stellaria media* L.) and other "winter annuals" remain vegetative in the winter, then flower and produce seeds in the early summer. Such seeds remain dormant until fall, then germinate and repeat the cycle. Experiments have demonstrated that the periods at warm temperatures break dormancy, provided that the seeds are subsequently exposed to appropriate conditions, especially alternating temperatures and light [37]; temperatures below 20°C are ineffective in breaking dormancy regardless of subsequent treatment.

Exposure of such seeds to low soil temperatures in the autumn reintroduces dormancy (see Section III.H), so that they once again become incapable of germination. A seasonal pattern thus develops, with periods of high germinability in autumn alternating with periods of low germinability in the summer. The behavior of such seeds contrasts with that of seeds of summer annuals, such as *Polygonum persicaria* [38], in which chilling is essential for breaking secondary dormancy (see below) and which germinate readily in the late winter and spring but poorly in the summer and fall (Figure 3). Chilling temperatures are required for breaking dormancy in other seeds (see Section III.E).

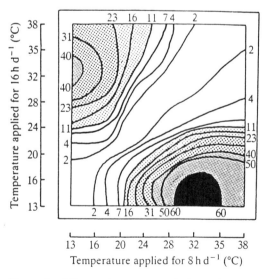

Figure 2 Effects of constant and alternating temperatures on germination of seeds of signal grass (*Brachiaria humidicola*). Seeds were held at indicated temperatures for 40 days. Percentage germination is indicated at points where lines intersect perimeter of square, and is proportional to density of stippling. (From Ref. 35.)

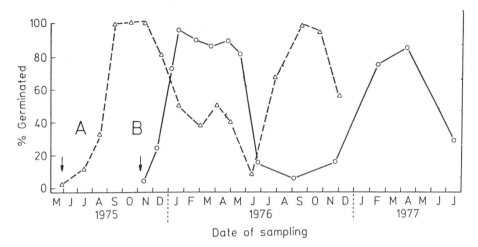

Figure 3 Germination of *Veronica hederofolia* (A) and *Polygonum persicaria* (B) seeds at alternating temperatures following burial in the field for varying periods of time. Arrows indicate dates of burial. *Veronica* seeds were held at 4/10°C for 16/8 h per day in darkness; *Polygonum* seeds were held at 12/22°C for 12/12 h per day and were exposed to light during 12 h at 22°C. (Adapted from Ref. 10, based on data of Karssen [38] and Roberts and Lockett [39].)

D. Light and Seed Dormancy

Seed response to light has been studied intensively in 'Grand Rapids' lettuce. Seeds of this and a number of other cultivars of lettuce and celery germinate readily in the light at 25°C, but fail to germinate in the dark (Figure 1A). A brief exposure of moist seeds to white or red light (660 nm) induces subsequent germination in darkness. The time of exposure required varies with species (Table 4). A brief exposure is effective only at high temperatures in birch, whereas a long exposure time is effective at all temperatures from 10 to 25°C (Figure 1B). However, if the brief red light treatment is followed by a similar brief exposure to far-red light (730 nm), the effect of the red light treatment is nullified (Table 5). Alternating red with far-red light leads to germination or dormancy, depending on the wavelength of last exposure. This is a classic case of a phytochrome-controlled response. Cone and Kendrick [41] provide a thorough review of the role of phytochrome in seed germination. Certain chemicals, especially gibberellic acid, can substitute for red light treatment (see below). In seeds of some species, shade from a plant canopy can reduce germination, relative to seeds held in darkness, by reducing the ratio of red to far-red light [42].

E. Shallow Versus Deep Dormancy

Seeds that will germinate in response to environmental cues (light, alternating temperatures) are considered to have a *shallow* dormancy; those that require prolonged exposure to certain conditions (e.g., moist chilling) are considered to have a *deep* dormancy. Certain seeds will not germinate immediately after harvest, but do so after several weeks or months of dry storage ("after-ripening") at room temperature (Table 6). This characteristic provides a safeguard against premature germination. In genotypes that do not possess this characteristic, germination can even occur on the plant, provided that moisture is abundant or rain occurs. This is an example

Table 4 Time of Illumination Required to Break Dormancy
in Seeds of Selected Species

Time required	Species
Seconds or minutes	'Grand Rapids' lettuce (*Lactuca sativa*)
Several hours	*Lythrum salicaria*
Days	*Kalenchoë blossfeldiana*
Long photoperiods	Begonia (*Begonia evansiana*)
Short photoperiods	Hemlock (*Tsuga canadensis*)

Source: Adapted from Ref. 10.

of *vivipary* (Latin *vivus* = alive, plus *parere* = to give birth). The length of the dormant period in rice seeds is shortened as storage temperature is raised from 27°C to 57°C [43]. Plotting the log of mean dormancy period (*y*) versus storage temperature (*x*) gives a straight line with negative slope (Figure 4). The depth of dormancy declines even at very low temperatures (−75°C) in seeds of orchard grass (*Dactylis glomerata*), although the rate of change is extremely slow [44].

If seed coat removal does not allow germination under favorable conditions, control obviously lies within the embryo (exalbuminous seeds) and/or endosperm (albuminous seeds). Albuminous seeds are composed primarily of endosperm; the embryo is relatively small. Warm, moist conditions for a period of 2 to 4 months following seed dispersal are usually required for coconut and ginkgo embryos to enlarge to the point where they are capable of germination. Some species [e.g., cowparsnip (*Heracleum sphondylium* L.)] require chilling for embryo development [45]; embryos develop very slowly at 15°C.

In exalbuminous seeds, the embryo is fully developed at maturity. However, many such embryos will not germinate, or germinate only sluggishly, when the seed coat is removed. Exposure to moisture and low temperatures (0 to 10°C) for periods of 1 to 20 weeks (cool, moist stratification) is often required to permit germination. Little or no growth of the embryo occurs during this time; the treatment alters the embryo's metabolism without affecting its morphology.

Table 5 Reversible Effects of Brief
Exposures to Red (R = 580 to 680 nm)
and Far-Red (FR > 700 nm) Radiation on
Germination of Lettuce Seed, cv. 'Grand
Rapids', in Darkness

Sequence	Germination (%)
Darkness	8.5
R (640–680 nm)	98
R-FR	54
R-FR-R	100
R-FR-R-FR	43
R-FR-R-FR-R	99
R-FR-R-FR-R-FR	54
R-FR-R-FR-R-FR-R	98

Source: Ref. 40.

Table 6 Time at Room Temperature for Dry After-Ripening of Seeds of Selected Species

Time required (months)	Species	Alternative method
1	Brome grass (*Bromus secalinus*)	Chilling
2–3	Rice (*Oryza sativa*)	—
12–18	Lettuce (*Lactuca sativa*)	Light, chilling
60	Curled dock (*Rumex crispus* L.)	Light, chilling, alternating temperature

Source: Adapted from Ref. 10.

F. Epicotyl Dormancy

Some seeds [e.g., tree peony (*Paeonia suffruticosa* Haw.)] germinate readily without special treatment, but the epicotyl (shoot) will not elongate unless chilled [46]. Chilling prior to germination is ineffective.

G. Double Dormancy

More than one mechanism may prevent the germination of a seed. Certain legumes [e.g., redbud (*Cercis canadensis*)], not only have hard seed coats but their embryos must be chilled before germination can occur (Table 3). Scarification, followed by moist chilling, breaks their dormancy. In other seeds (e.g., *Trillium erectum*) the radicle and the epicotyl both require chilling, but the periods at low temperature must be sequential. The first period permits radicle protrusion, the second shoot emergence [47].

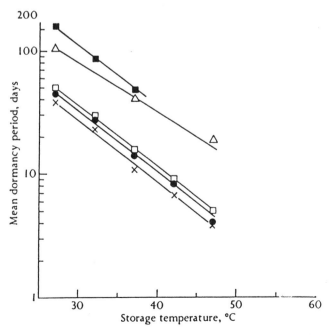

Figure 4 Effect of storage temperature on mean dormancy period in rice. Each line represents a different cultivar. (Adapted from Ref. 43.)

H. Thermodormancy and Secondary Dormancy

All of the types of dormancy described above are examples of *primary dormancy*, in which germination is prevented by conditions within the seed at the time it matures on the plant. *Thermodormancy* can be induced by exposure of seeds that are capable of germination at low temperatures (10 to 15°C) to high temperatures (25 to 30°C). This can occur in lettuce, for example, when soil temperatures are very high. *Secondary dormancy* is induced when a seed that is not dormant when shed, or whose dormancy has been partially broken, is exposed to unfavorable conditions, such as high temperature or drying. In seeds that are chilled for less than the required time, for example, premature exposure to high temperature can eliminate the effects of prior chilling.

IV. BUD DORMANCY

Following bud break in the spring, shoot growth is relatively slow at the beginning of the season, accelerates with time, then slows and eventually stops. This pattern tends to occur even at constant temperature. As noted above, growth tends to be cyclical. Even in the humid tropics flushes of growth occur in a more-or-less random fashion; one shoot on a tree may be growing rapidly while growth of another is negligible or nil. In contrast, growth of perennials in the temperate zone is synchronized. Growth ceases in mid- to late summer and the plants pass through a dormant period lasting for several months.

Fuchigami et al. [48] have described this pattern of growth as a sine wave (degree growth stage model), with 0° representing the end of ecodormancy/beginning of active growth; 90°, the end of active growth (maturity induction point = beginning of paradormancy); 180°, "vegetative maturity" (beginning of endodormancy); 270°, the time of deepest endodormancy; and 315°, the end of endodormancy/beginning of ecodormancy (Figure 5). Note that phase transition is gradual rather than abrupt; endodormancy does not end one day and ecodormancy

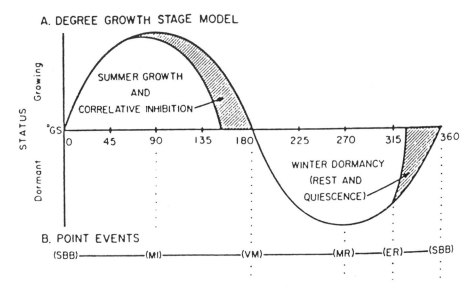

Figure 5 Degree growth state model representing stages in the annual cycle of growth in woody plants. Five sequential growth stages [spring budbreak (SBB), maturity induction point (MI), vegetative maturity (VM) (= onset of endodormancy), maximum endodormancy (MR), and end of endodormancy (ER)] occur at 0, 90, 180, 270, and 315°C, respectively. (From Ref. 48.)

begin the next; rather, there is a gradual transition from one phase to the next. During the early part of the summer, removal of the shoot apex and/or defoliation relieves apical dominance and permits growth of the lateral buds. This is true not only in woody plants but in many herbaceous ones as well. Horticulturists remove the apical portion ("pinch") chrysanthemums and petunias to force branching and thereby create more attractive plants. Arboriculturists use the same practice to stimulate the formation of lateral branches. At this time, the axillary buds are *paradormant* (see above)—they are prevented from growing by the presence of the apex rather than by conditions within the buds themselves. As the summer progresses, the ability of the buds to grow following apex removal declines; paradormancy is gradually becoming endodormancy as control shifts from the apex to the buds themselves. By the end of the season, the buds no longer respond to apex removal; endodormancy is now fully established.

Many woody perennials (e.g., birch) exhibit a marked response to photoperiod, growing rapidly under long photoperiods, slowly or not at all under short photoperiods. This response is truly photoperiodic rather than being a function of total time of exposure to light per se and is an example of ecodormancy. When plants are grown under short days but the long night is interrupted by a brief period of light, they continue their growth. Under natural conditions, the effects of long days are often masked by other environmental limitations, such as water supply or competition among growing points. Thus mature trees of birch stop growth in midsummer, even though daylength is near its maximum.

Chilling temperatures appear to be required for buds to become fully endodormant. In some areas of the tropics and subtropics where temperatures never fall below 20°C, the buds of peaches, grapes, and apples can be forced to grow by defoliation soon after harvest. This permits production of two or more crops per year. The longer the interval between harvest and defoliation, the poorer the response. Trees that are not defoliated may eventually become endodormant; in the absence of chilling, they cease growth entirely and eventually die.

Endodormancy is normally broken by exposure to chilling temperatures. Optimum temperatures vary with species but generally range from 0 to 10°C; temperatures below 0°C have little or no effect. Considerable research has been done to determine the chilling requirements of fruit tree species and cultivars, and several models have been developed to predict when these requirements have been satisfied. For example, according to the Utah model [49], the number of chill units required for 'Elberta' peach and 'Delicious' apple are 800 and 1234, respectively [50]. A *chill unit* is defined as 1 h of exposure to a temperature of 6°C; higher and lower temperatures between 0 and 13°C are less efficient, and temperatures above 13°C are inhibitory; thus adjustments must be made in calculation (Figure 6). This model, developed in the north temperate zone, may not apply in regions where diurnal temperature fluctuations are greater. Israeli scientists have therefore developed a "dynamic" model in which temperatures alternating between about 6 and 13–14°C are considered to have a greater effect than continuous cold in breaking dormancy [52]. Temperatures above 15°C are inhibitory unless the exposure time is less than a critical length. This model was more effective than the Utah model in predicting end of rest when used in Israel [52]. Species and cultivars vary greatly in the number of chilling hours required. For example, the chilling requirements of peach cultivars grown in Florida should not exceed 300 h, whereas those grown in the northernmost parts of the United States may require 800 h or more.

Bud dormancy is not confined to woody plants. Many herbaceous perennials must be chilled before growth can resume in the spring. Ornamental bulbs such as tulips and daffodils are planted in the fall. Cold soil temperatures provide the chilling required to allow normal stem elongation the following spring. If such bulbs are planted indoors, the flower stalks are much shorter and the flowers themselves may abort. Florists meet the demand for these flowers out of season by artificially chilling the bulbs, then forcing them in a warm greenhouse. Note that

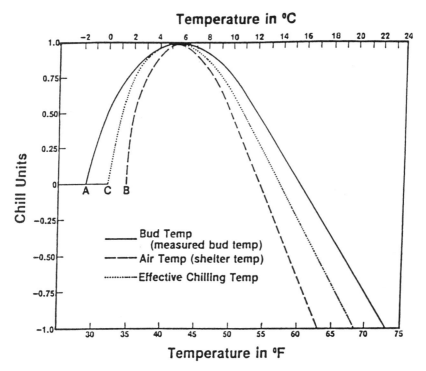

Figure 6 Curve used in estimating chill unit accumulation, based on the Utah model, for the breaking of bud dormancy in deciduous tree fruits. Effective chilling temperature is the mean of the two temperatures measured. Positive values are assigned to temperatures between –2 and 13°C, negative values to higher temperatures. (From Ref. 51.)

this period of cold temperature stimulates elongation of preexisting inflorescences and therefore differs from *vernalization*, in which chilling stimulates the *initiation* of flowers. In some species, however, including Dutch iris (*Iris* spp.) and Easter lily (*Lilium longiflorum*), vernalization indeed occurs. Although the rate of sprout development in onion bulbs is greater at 15°C than at higher or lower temperatures [53], Abdalla and Mann [54] established that the time required for sprouting was independent of storage temperature prior to transfer to 15°C. Thus onion differs from tulip in not requiring chilling for floral stalk elongation.

Similarly, some tubers (e.g., Jerusalem artichoke) must be chilled before buds can grow normally. This, of course, is not the case with crops, such as the potato, that originated in the tropics. Although potato has no chilling requirement, the tubers are dormant at harvest. Dry storage at room temperature for several weeks permits bud development; this parallels the response of seeds of several grains to "after-ripening" (see above).

V.   METHODS FOR BREAKING OR PROLONGING DORMANCY

Dormancy or lack thereof can be troublesome to the plant grower. Waiting 6 to 10 weeks or more for seeds to be after-ripened or buds to be chilled may not be inconvenient in areas where cold temperatures prevent winter production, but can reduce profitability in areas where crops can be grown year-round. In the latter areas, multiple cropping is practiced, with two or more crops being harvested each year. Thus yields will be maximized if no dormant periods intervene.

As noted above, in some areas of the tropics or subtropics, peach, apple, and grape can be multiple cropped, although a brief dormant period intervenes between foliations. The leaves must be removed to stimulate bud break, and chemicals, such as sodium chlorate, copper sulfate, or urea, are often applied to injure the leaves and induce premature abscission. In areas where multiple cropping is impossible, but chilling inadequate to completely relieve dormancy, other chemicals, such as combinations of dinitro-*O*-cresol and oils, are used to hasten bud break and concentrate the bloom period. Hydrogen cyanamide (H_2NCN), which releases HCN within the tissues, is a relatively new compound that has similar effects and has been extensively tested for this purpose [55,56].

In arid regions bud dormancy of some species can be broken by withholding water for several weeks, then irrigating. Asparagus growers in California and Peru can produce crops year-round using this method. Irrigation is also used in combination with rest-breaking chemicals and/or defoliation of deciduous fruit trees in tropical regions [57].

In areas where chilling is adequate, but spring freezes often damage flowers and/or fruits, delaying bloom could provide protection. Evaporative cooling by misting with water can delay bloom; delays of 3 weeks or more are possible in arid climates [58,59]. However, side effects, such as poor fruit set, have limited commercial application. This method has also been tested in warm climates for cooling buds during the winter [60], thereby hastening the breaking of dormancy; again, commercial application has been limited.

Methods of weakening the integuments of seeds with hard seed coats to allow water to penetrate have been discussed above, as well as the effects of light and temperature on seeds with "shallow" dormancy. Several growth regulators, including both gibberellins (GAs) and cytokinins, promote germination in dormant or partially dormant seeds. GA is effective in stimulating germination in seeds with a shallow dormancy. Light-sensitive lettuce seeds, for example, will germinate in darkness when supplied with GA. Cytokinin, although generally effective in stimulating dark germination, can overcome the inhibitory effects of high temperatures. Abscisic acid (ABA) blocks germination in many seeds, regardless of environmental conditions. Khan [61] tested the effects of all three hormones and their combinations on the germination of light-sensitive lettuce seeds. The action of GA was blocked by ABA, but cytokinin counteracted the effect of ABA, thereby permitting germination when all three hormones were applied. From these data, Khan [62] proposed that the roles of GA, ABA, and cytokinin were primary, preventive, and permissive, respectively; GA is the primary stimulus, with cytokinin being essential only when ABA is present. Khan and others [63,64] have confirmed and extended these observations by using inhibitors of GA synthesis to block germination, and demonstrating that in some cases, cytokinin and/or ethylene is required, in addition to GA, to overcome the inhibitory effects of stress caused by water deficit, salinity, and other conditions.

GA will also stimulate germination in some cold-requiring seeds, although some chilling is usually required before maximum response is obtained. Cytokinins are usually less effective. Both GA and cytokinins can hasten release from dormancy in buds of woody plants, as well as overcoming apical dominance during the early growing season. A combination of $GA_{4/7}$ and benzyladenine, for example, is currently available commercially to stimulate growth of lateral buds of conifers used for Christmas trees, thereby providing a more pleasing form.

Ethylene promotes germination in some weed species [e.g., redroot (*Amaranthus retro-flexus*) and lamb's quarters (*Chenopodium album*)], but many species are not responsive [65]. Gibberellins and cytokinins have more general effects. A few cases are known in which ethylene breaks bud dormancy, but again, response is species dependent.

Several chemicals are effective in *prolonging* bud or seed dormancy. Potato tubers are regularly fumigated with 1-methyl-3-chlorophenylcarbamate (CIPC) to delay their sprouting during storage. This compound is currently under review by the Environmental Protection Agency (EPA), and scientists are testing naturally occurring compounds as potential substitutes. Andean natives store potatoes in pits together with leaves of muña (plants of the genera *Minthostachys* and *Satureja*) to delay sprouting and reduce both weight loss and insect injury [66]. Trials with volatile components of readily available essential oils demonstrated that 1,8-cineole, found in eucalyptus oil, has promise in inhibiting both sprouting and fungal growth [67]. Application of maleic hydrazide to the foliage of onion plants several weeks before harvest inhibits sprouting of the stored bulbs [68]. The naturally occurring plant growth inhibitor ABA inhibits seed germination in many species [10], although its cost prohibits commercial use. It is less effective on buds, perhaps because of limited penetration and/or rapid metabolism.

Inhibitors of respiration, or more specifically, of cytochrome oxidase, can break dormancy in some seeds, including rice [69,70], barley [71], and lettuce [72], as well as in isolated apple embryos [73]. Apple embryos also respond to anaerobiosis; holding them in nitrogen for 2 weeks or longer permits subsequent germination in air [74]. Other reports indicate that high oxygen tension relieves dormancy in several grains [71,75–77]. The similar effects of these conditions that restrict versus promote respiration suggest that different mechanisms control dormancy at different times, and dictate that caution be used in assigning causal effects to various external factors that influence dormancy. (See Roberts and Smith [78] for a hypothesis to explain these effects.)

VI. PHYSIOLOGICAL BASIS OF DORMANCY

Despite much effort by scientists, the mechanisms that control dormancy in plants remain a mystery. However, numerous theories have been proposed to account for the phenomenon. All physiological processes are ultimately controlled by genes, and progress is being made in identifying genes associated with dormancy. Seeds of *Arabidopsis thaliana* require dry storage to break dormancy, but mutants have been isolated that produce nondormant seeds [79,80]. The ability of such seeds to germinate has been associated with single-gene differences in their ability to synthesize ABA or GA (see below). In maize, genes have been identified that are responsible for preventing premature germination (vivipary) [81,82]. Again, these genes appear to regulate the synthesis of, or sensitivity to, ABA [83–86]. Skriver and Mundy [87] and Thomas [88] have reviewed the effects of these and related genes during embryo development. Single-gene control of dormancy has also been demonstrated in hazel (*Corylus avellana*) [89] and in peach (*Prunus persica*) [90], although no data are yet available as to the mechanisms involved. More comprehensive information on genetic and molecular approaches to dormancy may be found in Lang [18] and King [91].

Although control of dormancy ultimately lies within the genome, such control must be exerted via physiological mechanisms. The many theories advanced to explain dormancy can be grouped into three general categories: nutritional/metabolic deficiencies, blocks to membrane permeability, and excesses or deficiencies of hormones. Briefly stated, these theories propose that the failure of a seed or bud to grow results from (a) deficiency of a nutrient(s) or of an enzyme(s) able to metabolize such a nutrient, (b) the inability of nutrients to reach shoot and/or root apices within the dormant organ, or (c) an excess of a growth inhibitor(s), a deficiency of a growth promoter(s), or an improper balance between the two within the meristem and/or adjacent tissues. In general, more attention has been devoted to hormone studies than to the other two areas of research. Seeds are more convenient for studying dormancy than are buds, for they are small, self-contained, and thus more easily manipulated.

A. Metabolic Aspects of Dormancy

As Bewley and Black [10] emphasize "Dormancy cannot be equated with overall metabolic inactivity. . . ." Respiration rates of hydrated, dormant seeds of lettuce and cocklebur differ little from those of nondormant seeds prior to germination, and activity of hydrolytic enzymes is unlikely to be crucial, for little mobilization of reserves occurs prior to radicle emergence [10]. Nevertheless, many studies have compared the metabolism of dormant versus nondormant seeds and several investigators have proposed that dormant tissues are deficient in specific enzymes required for metabolism of carbohydrates, fats, and/or proteins.

1. Nutrient Supply

Stokes [92] differentiates between two types of seed dormancy, with embryo dormancy ("true dormancy") being responsible for the first, and lack of nutrients for the second (nonresting embryo). In the former, interruption of chilling by exposure to high temperature can negate the effect of previous chilling by inducing secondary dormancy, and the effects of two or more periods of chilling are less than additive. In the latter, the effects of chilling are additive and irreversible; interruption by high temperature does not negate the effects of prior exposure to low temperature.

The response of seeds of the second type is easier to explain, superficially, at least. The embryo is very small and grows at the expense of the surrounding seed tissues (endosperm and/or nucellus). Chilling stimulates the activity of enzymes that hydrolyze stored reserves, which the embryo cannot otherwise utilize, to compounds that can be used for growth. Thus in seeds of cowparsnip (*Heracleum sphondylium*), embryos of seeds held at 15°C elongate for approximately 6 weeks, then stop growing when approximately half their full size [45]. Although the initial rate of growth is slower in seeds held at 2°C, elongation of the embryo continues logarithmically for 9 weeks. Parallel changes occur in the endosperm, but in reverse (i.e., the endosperm of seeds held at 2°C is consumed by the embryo, while that in seeds held at 15°C is not). If lack of suitable nutrients were responsible for the failure of embryos to develop at 15°C, one would expect that growth of excised embryos in vitro at 15°C could be stimulated by supplying appropriate nutrients. Stokes [93] observed that arginine and glycine concentrations in the endosperm were higher in seeds held at 2°C than in those held at 20°C. When embryos cultured in vitro at 20°C were supplied with glucose plus various sources of nitrogen, arginine and glycine were the most effective amino acids in supporting growth, although KNO_3 was the best source of nitrogen. From these and other data, Stokes [93] concluded that exposure to 2°C stimulated embryo growth by increasing the quantities of arginine and glycine available to the embryo.

A similar situation occurs in both black ash (*Fraxinus nigra*) [94] and European ash (*F. excelsior*) [95,96], except that chilling is not essential for embryo enlargement but is required for germination once embryos have reached full size. Stokes [92] provides other examples of seeds with similar requirements. Axes from dormant hazel embryos will grow in vitro when supplied with inorganic salts and sucrose [97], suggesting that failure of the intact embryo to germinate is due to inability to mobilize nutrients from the cotyledons [98,99]. Application of GA_3 both breaks embryo dormancy and permits mobilization of reserves, suggesting that gibberellin biosynthesis following chilling has a similar effect (see below).

2. Protein Metabolism

A group of proteins termed "late-embryogenesis-abundant" (*Lea*) proteins accumulates as seeds mature and become dehydrated (see Ref. 87). These appear to bind water, thereby protecting macromolecules such as nucleic acids (?) from dehydration and resultant denaturation. *Lea* proteins disappear during germination.

Several facts, summarized by Quatrano [100], suggest that such proteins play a role in dormancy: (a) embryos of viviparous mutants do not synthesize these proteins if cultured on a medium containing ABA; (b) dehydration of immature embryos induces the production of the proteins, possibly by stimulating the synthesis of ABA; and (c) treating mature seeds with ABA prevents both germination and the loss of *Lea* proteins.

Most studies of *Lea* proteins have involved species whose seeds are either nondormant or have a shallow dormancy, and no studies are known involving species with deeply dormant seeds. Therefore, the connection between such proteins and dormancy remains tenuous. ABA blocks germination while inducing or maintaining the synthesis of *Lea* proteins, but these two responses may be unrelated.

Protein metabolism has also been implicated as a factor in the breaking of dormancy. As noted above, holding *Heracleum sphondylium* seeds at 2°C permits the hydrolysis of reserve proteins and their transfer to the embryo, whereas holding them at 20°C does not [45]. In apple embryos, however, hydrolysis of reserve proteins occurs at both 5 and 20°C [101]. Furthermore, no proteolysis is observed in seeds held in the fruit at 0°C, although this treatment also breaks embryo dormancy. Similarly, Chen and Varner [102] reported that dormant and nondormant seeds of wild oats (*Avena fatua* L.) synthesize protein at similar rates.

Lewak et al. [103] suggest that an insufficient supply of amino acids may prevent germination in dormant apple seeds. Protease activity increases with chilling, reaching a maximum after 7 weeks, then declines to the level observed in nonchilled seeds. The authors suggest that germination is dependent on a supply of amino acids released by hydrolysis of proteins. However, they present no data on the effects of amino acids on germination of dormant embryos.

Recent work (see below) has emphasized the effects of dormancy-breaking treatments on the concentrations of specific proteins or polypeptides. The rationale for much of this work is that regardless of what substances control induction or breaking of dormancy, enzymes (proteins) must be synthesized before such compounds can be produced. Therefore, changes in protein content should precede changes in other compounds, be they carbohydrates or hormones. Protein analysis involves electrophoretic separation of extracted proteins, together with the use of radiolabeled amino acids as markers for newly synthesized polypeptides. Although no significant changes were observed in total soluble protein content of pear [104] or apple embryos [105] during chilling, Eichholtz et al. [105] observed an increase in the concentrations of four peptides in the embryonic axes of apple embryos held at 5°C. No changes were evident in the cotyledons at 5°C or in either axes or cotyledons at 20°C. The authors suggested that protein reserves might be mobilized to the axis during the breaking of dormancy.

Despite much research in this field, the picture remains confusing. Some workers have suggested that proteins found in dormant, but not in nondormant, seeds inhibit germination [106–108]. Mahhou and Dennis [109] reported reduced levels of large proteins (36 and 41 kDa) in the cotyledons of peach seeds stratified at 5°C, even when the embryonic axis was excised. These changes did not occur at 20°C. In some tissues, chilling increases the content of certain proteins (e.g., Ref. 110); in others, some proteins increase during chilling while others decrease [111–113]. Ried and Walker-Simmons [114] have presented evidence for heat-stable proteins in embryonic axes of dormant wheat seeds that are induced by treatment with ABA. Much higher concentrations of ABA are required to produce similar levels of proteins in nondormant embryos, suggesting that sensitivity to ABA may play a role in dormancy.

3. Synthesis of Nucleotides

The limited ability of dormant tissues to convert adenosine to nonadenylic nucleotides (NTP = sum of triphosphates of guanidine, cytosine, and uridine) has been suggested as a possible cause of dormancy. Correlations between the ability to convert adenosine to NTP and the dormant state have been reported in Jerusalem artichoke (*Helianthus tuberosum* L.) tubers [115], in apple embryos [116], and in buds or subapical tissues of ash [117], willow, and hazel [118].

B. Permeability Changes

Several investigators have proposed that changes in membrane permeability are responsible for dormancy. To test this hypothesis, tissues are incubated with a weak acid [5,5-dimethyl-2,4-oxazolidinedione (DMO)]; only the undissociated form can pass through the cell membrane. Use of radioactive DMO permits determination of the ratio of the concentration of DMO within the cell (C_i) to the concentration in the intercellular spaces (C_e). Relative membrane permeability parallels the C_i/C_e ratio. Using this method, Gendraud and Lefleuriel [119] observed a higher C_i/C_e ratio in dormant than in nondormant tubers of Jerusalem artichoke. This implies less movement of nutrients to the meristematic tissues of dormant tubers. In similar studies, Ben Ismail [120] compared C_i/C_e ratios in bud versus shoot tissues of apple during the dormant period. Higher ratios occurred in shoots than in buds during the fall and early winter, suggesting limited movement of solutes from shoots to buds. Thereafter, the ratio in the buds rose to levels higher than those observed in the shoots. Although the results parallel the expected response of intact trees or isolated shoots, bud development in single-node cuttings exposed to laboratory conditions was reduced only in samples collected in November.

C. Role of Hormones

The role of hormones in seed dormancy is supported primarily by the effects of applied hormones in both inhibiting the germination of nondormant seeds (ABA) and stimulating the germination of dormant seeds (cytokinins, GAs). However, effective concentrations are often much higher than those found in the seeds themselves, and the response is seldom as great as one might expect. Although treatment with GA is effective in breaking dormancy in lettuce seeds, germination of peach seeds can be maximized only after some chilling has occurred [121]. Even then, the symptoms of insufficient chilling (abnormal leaves, etc.) are not eliminated. Furthermore, despite early reports to the contrary, few good correlations have been established between content of endogenous hormones and dormancy status.

Several hypotheses have been proposed regarding the role of hormones in seed dormancy. Germination is prevented by:

1. High concentrations of growth inhibitors, (e.g., ABA)
2. Inhibitory concentrations of auxin [indole-3-acetic acid (IAA)]
3. Insufficient concentrations of growth promoters (GA, cytokinins)
4. Both (1) and (3)

Modifications of these hypotheses propose that:

5. Promoters are synthesized in seeds requiring chilling only following their return to warm temperatures [122].
6. High levels of promoters are required only temporarily at the beginning of the "trigger" phase that ends dormancy [123].
7. ABA blocks the action of GAs; if both are present, cytokinin must also be present to permit GA to act [61,62].

1. Auxin

Nikolaeva [124] determined the content of presumed IAA (wheat coleoptile segment and mustard seed germination assays) in seeds and/or embryos of several tree species before, during, and after cold stratification. Activity (promotion of coleoptile section growth, inhibition of germination) declined as stratification was prolonged. Inhibitor activity in nondormant seeds was approximately half of that observed in dormant seeds. Nondormant seeds treated with the naturally occurring auxin IAA produced seedlings with symptoms similar to those of seedlings from insufficiently chilled seeds. From these and other data she concluded that high levels of IAA prevented germination of nonchilled seeds, and that chilling reduced the IAA concentration to the levels found in seeds that did not require chilling. Subsequent investigators have found little support for the role of auxin in dormancy. Most recent research on hormones has focused on GAs and ABA.

2. Gibberellins

Amen [123] proposed that seed dormancy could be divided into four phases. During the *induction* phase, levels of growth promoters decline and/or the seed coat becomes impermeable to oxygen; therefore, the seed becomes dormant. During the ensuing *maintenance* phase, germination is prevented by endogenous inhibitors. In the *trigger* phase, a factor that elicits germination, but whose continued presence is not essential (the trigger, e.g., light) induces the production of a *germination agent*, whose continued presence is required for germination. In the final phase (*germination*), the germination agent [growth promoter(s)] provides the stimulus for radicle protrusion.

Much of the evidence for this scheme is based on the effects of exogenous growth regulators on germination; only a few studies have supported the hypothesis in terms of actual increases in seed hormone content following action by "triggers," including chilling, and light. In one such study, Williams et al. [122] could detect little change in GA content of hazel seeds during moist chilling at 5°C. However, levels rose rapidly once dormancy had been broken, provided that the seeds were returned to 20°C.

The gibberellin (GA$_4$) content of apple seeds rises during chilling but is no higher in fully chilled seed than in nonchilled seed [125]. This could, of course, be interpreted as supporting a "trigger" role for GA. Similar roles for both GA and cytokinin have been suggested in maple seeds [126].

3. Abscisic Acid

Considerable effort has been directed toward elucidating the role of ABA in controlling dormancy in seeds. The ABA content of immature seeds of several species, including wheat [127] and rapeseed [128], rises to a maximum, then falls as the seeds mature and dry out. Although the concentration of ABA in the mature seed is low, desiccation reduces water content, thereby preventing germination.

The effects of ABA in preventing the germination of immature embryos in vitro, plus the evidence for the role of ABA in vivipary, noted above, strongly imply that ABA is one of the factors preventing embryo germination. Seeds of the species investigated in these studies (e.g., maize, rapeseed) are nondormant or have only a shallow dormancy at harvest; similar relationships may not apply in seeds that exhibit deep dormancy.

In ash (*Fraxinus*) seeds, ABA content is low in *F. americana* relative to that in *F. ornus* [129]. Seeds of the former are nondormant, whereas the latter require moist chilling to break their dormancy. This dormancy again is correlated with ABA content. While the ABA content of seeds of three species of *Rosa* is negatively correlated with their germinability [130], the ABA content of seeds of several species of pear bears no relationship to depth of dormancy [131], nor does the ABA content of immature or mature seeds of *Avena fatua* (dormant) differ from that of seeds of *A. sativa* (nondormant) [132]. Differences in sensitivity to ABA could, of course, explain some of these discrepancies but have seldom been tested experimentally. Early results indicated that the levels of ABA or ABA-like inhibitors fell during moist chilling of ash [129] and several other species, including apple [133]. However, subsequent investigations indicated that ABA content either did not decline during low-temperature stratification [134] or that the decline was not temperature dependent [135]. The concentration of ABA declines during soaking of lettuce seed, regardless of their germination capacity [136].

In many of these studies the entire seed was extracted. Karssen et al. [79] proposed that the GA and/or ABA content of the embryo may be more important than that of the whole seed. Using selected mutants of *Arabidopsis thaliana*, they demonstrated that embryos recessive for ABA production were nondormant even when the seed coat contained high levels of ABA. Later work with GA-deficient mutants led Karssen et al. [80] to propose that GA content is the critical factor in germination. The seeds of genotypes that cannot synthesize GAs remain dormant regardless of their ABA content.

Inhibitor content of buds has also been quantified in relation to dormancy. Again initial results were promising; the inhibitor content of buds of several species, as measured by bioassay, appeared to rise when plants were transferred from long to short photoperiods [137,138]. This work led to the identification of ABA by Ohkuma et al. [139] and Cornforth et al. [140]. As analytical instruments have become more sensitive and experiments more critical, however, the negative correlation between ABA content and growth response has not been confirmed [141,142]. In fact, one laboratory reported [143,144] that rapidly growing apices contained more ABA than did subapical tissues.

Coleman and King [145] reported a positive correlation between ABA content of tubers of 10 potato cultivars following 2 months of storage at 10°C and the time to 50% sprouting at 20°C. However, ABA content of three other cultivars actually *increased* during storage at three temperatures (2, 10, and 20°C), yet dormancy was broken in all cases, often when ABA content was near maximal.

D. New Approaches to the Understanding of Dormancy

Relatively little is known about how genes control seed and bud dormancy, but research in molecular biology is beginning to open the "black box." Studies of apical dominance, for example, are under way using transgenic plants that differ in the relative amounts of IAA and cytokinin synthesized. Plants with high IAA/cytokinin ratios exhibit strong apical dominance, and vice versa [146,147], suggesting that these hormones may indeed be responsible for this phenomenon. Genes for hormone synthesis in plants that exhibit seed and/or bud dormancy have been identified and can now be cloned. Once these can be inserted in the same or other species, rapid progress may be expected in elucidating the roles of such compounds in controlling dormancy.

VII. SUMMARY

Dormancy serves a protective function in permitting plant survival under extremes of temperature, water deficit, and other environmental stresses, and species differ in their manifestations of dormancy. Several types of dormancy are known, with control sometimes residing within the dormant organ, sometimes outside the organ. As would be expected, the conditions required to break dormancy differ with the type of dormancy exhibited, and vary from changes in light intensity or photoperiod to exposure to low or alternating temperatures. Many theories have been proposed to explain the physiological basis of dormancy, but none has proven valid in accounting for all the facts known. New approaches, especially molecular biology, hopefully will provide new information in this important field.

REFERENCES

1. G. M. Simpson, in *Dormancy and Developmental Arrest* (M. E. Clutter, ed.), Academic Press, New York, p. 167 (1978).
2. G. A. Lang, J. D. Early, N. G. Aroyave, R. L. Darnell, G. C. Martin, and G. W. Stutte, *HortScience, 20*: 809 (1985).
3. G. A. Lang, J. D. Early, G. C. Martin, and R. L. Darnell, *HortScience, 22*: 371 (1987).
4. G. A. Lang, *HortScience, 22*: 817 (1987).
5. O. Juntilla, *HortScience, 23*: 805 (1988).
6. G. C. Martin, *HortScience, 22*: 824 (1987).
7. M. G. Cline, *Bot. Rev., 57*: 318 (1991).
8. A. A. Khan, ed., *The Physiology and Biochemistry of Seed Dormancy and Germination*, North-Holland, New York (1977).
9. A. A. Khan, ed., *The Physiology and Biochemistry of Seed Development, Dormancy, and Germination*, Elsevier Biomedical Press, New York (1982).
10. J. D. Bewley and M. Black, *Physiology and Biochemistry of Seeds in Relation to Germination*, Vol. 2, *Viability, Dormancy and Environmental Control*, Springer-Verlag, New York (1982).
11. J. D. Bewley and M. Black, *Seeds: Physiology of Development and Germination*, Plenum Press, New York (1985).
12. J. W. Bradbeer, *Seed Dormancy and Germination*. Blackie & Son, London (1988).
13. M. C. Saure, *Hortic. Rev., 7*: 239 (1985).
14. L. E. Powell, in *Plant Hormones and Their Role in Plant Growth and Development* (P. J. Davies, ed.), Martinus Nijhoff, Boston, p. 539 (1987).
15. L. E. Powell, *HortScience, 22*: 845 (1989).
16. G. C. Martin, *Plant Physiology: A Treatise*, Vol. X, *Growth and Development* (F. C. Steward and R. G. S. Bidwell, eds.), Academic Press, New York, p. 183 (1991).
17. F. G. Dennis, Jr., *HortScience* (in press) (1994).
18. G. A. Lang, *HortScience* (in press) (1994).
19. B. Moormann, *Kuehn-Arch.*, 56: 41 (1942) (cited by Stokes [79, p. 783]).
20. A. R. Kermode, *Crit. Rev. Plant Sci., 9*: 155 (1990).
21. S. Totterdell and E. H. Roberts, *Plant Cell Environ., 3*: 3 (1980).
22. E. H. Toole, V. K. Toole, H. A. Borthwick, and S. B. Hendricks, *Plant Physiol., 30*: 473 (1955).
23. E. C. Stone, *Forest Sci., 3*: 357 (1957).
24. J. M. Naylor and G. M. Simpson, *Can. J. Bot., 39*: 281 (1961).
25. E. H. Roberts, *J. Exp. Bot., 13*: 75 (1962).
26. A. G. Chapman, *J. Forestry, 34*: 66 (1936).
27. L. C. Luckwill, *J. Hortic. Sci., 27*: 53 (1952).
28. V. K. Toole, E. H. Toole, H. A. Borthwick, and A. G. Snow, *Plant Physiol., 37*: 228 (1962).
29. B. S. Ezumah, Ph.D. thesis, University of London (1980) (cited by Bewley and Black [10]).
30. M. A. Dirr and C. W. Heuser, Jr., *The Reference Manual of Woody Plant Propagation*, Varsity Press, Athens, Ga. (1987).
31. D. T. Bell, J. A. Plummer, and S. K. Taylor, *Bot. Rev., 59*: 24 (1993).
32. V. N. Tran and A. K. Cavanagh, in *Seed Physiology*, Vol. 2, *Germination and Reserve Mobilization* (D. R. Murray, ed.), Academic Press, New York, p. 1 (1984).
33. V. N. Tran and A. K. Cavanagh, *Aust. J. Bot., 28*: 39 (1980).
34. C. M. Karssen, in *The Physiology and Biochemistry of Seed Development, Dormancy and Germination* (A. A. Khan, ed.), Elsevier Biomedical Press, New York, p. 243 (1982).
35. C. O. Goedert and E. H. Roberts, *Plant Cell Environ., 9*: 521 (1986).
36. J. M. Baskin and C. C. Baskin, *Weed Sci., 33*: 340 (1985).
37. J. M. Baskin and C. C. Baskin, *New Phytol., 77*: 619 (1976).
38. C. M. Karssen, *Isr. J. Bot., 29*: 65 (1980/81).
39. A. Roberts and P. M. Lockett, *Weed Res., 18*: 41 (1978).
40. H. A. Borthwick, S. B. Hendricks, M. W. Parker, E. H. Toole, and V. K. Toole, *Proc. Natl. Acad. Sci. USA, 38*: 662 (1952).

41. J. W. Cone and R. E. Kendrick, in *Photomorphogenesis in Plants* (R. E. Kendrick and G. H. M. Kronenberg, eds.), Martinus Nijhoff/Dr. W. Junk, Dordrecht, The Netherlands, p. 443 (1986).
42. J. Silvertown, *New Phytol.*, *85*: 109 (1980).
43. E. H. Roberts, *J. Exp. Bot.*, *16*: 341 (1965).
44. R. J. Probert, R. D. Smith, and P. Birch, *New Phytol.*, *101*: 521 (1985).
45. P. Stokes, *Ann. Bot.*, *16*: 441 (1952).
46. L. V. Barton, *Contrib. Boyce Thompson Inst.*, *5*: 451 (1933).
47. L. V. Barton, *Contrib. Boyce Thompson Inst.*, *13*: 259 (1944).
48. L. H. Fuchigami, C. J. Weiser, K. Kobayashi, R. Timmis, and L. V. Gusta, in *Plant Cold Hardiness and Freezing Stress* (P. H. Li and A. Sakai, eds.), Academic Press, New York, Vol. 2, p. 93 (1982).
49. E. A. Richardson, S. D. Seeley, and D. R. Walker, *HortScience*, *9*: 331 (1974).
50. G. L. Ashcroft, E. A. Richardson, and S. D. Seeley. *HortScience*, *12*: 347 (1977).
51. J. L. Anderson, E. A. Richardson, and C. D. Kesner, *Acta Hortic.*, *184*: 71 (1986).
52. A. Erez, S. Fishman, Z. Gat, and G. A. Couvillon, *Acta Hortic.*, *232*: 76 (1988).
53. S. Komochi, in *Onions and Allied Crops* (H. D. Rabinowitch and J. L. Brewster, eds.), CRC Press, Boca Raton, Fla., Vol. 1, p. 89 (1990).
54. A. A. Abdallah and L. K. Mann, *Hilgardia*, *35*: 85 (1963).
55. Y. Shulman, G. Nir, L. Fanberstein, and S. Lavee, *Sci. Hortic.*, *19*: 97 (1983).
56. A. P. George and R. J. Nissen, *Acta Hortic.*, *279*: 427 (1990).
57. K. Bederski, *Acta Hortic.*, *199*: 33 (1987).
58. J. F. Alfaro, R. E. Griffin, G. R. Hanson, J. Keller, J. L. Anderson, G. I. Ashcroft, and E. A. Richardson, *Trans. Am. Soc. Agric. Eng.*, *17*: 1025 (1974).
59. J. L. Anderson, G. L. Ashcroft, E. A. Richardson, J. F. Alfaro, R. E. Griffin, G. R. Hanson, and J. Keller, *J. Am. Soc. Hortic. Sci.*, *100*: 229 (1975).
60. P. R. Gilreath and D. W. Buchanan, *J. Am. Soc. Hortic. Sci.*, *104*: 536 (1981).
61. A. A. Khan, *Science*, *171*: 853 (1971).
62. A. A. Khan, *Bot. Rev.*, *41*: 391 (1975).
63. A. A. Khan and C. Andreoli, in *4th International Workshop on Seeds* (D. Côme and F. Corbineau, eds.), ASFIS, Paris, France, *2*: 625 (1993).
64. A. A. Khan, H. Xuelin, Z. Guangwen, and J. Prusinski, in *Advances in the Science and Technology of Seeds* (F. Jiarui and A. A. Khan, eds.), Science Press, New York, p. 313 (1992).
65. R. B. Taylorson, *Weed Sci.*, *27*: 7 (1979).
66. T. J. Aliaga and W. Feldheim, *Ernahrung*, *9*: 254 (1985).
67. S. F. Vaughn and F. G. Spencer, *Am. Potato J.*, *68*: 821 (1991).
68. S. H. Wittwer and R. C. Sharma, *Science*, *112*: 597 (1950).
69. E. H. Roberts, *Physiol. Plant.*, *16*: 732 (1964).
70. E. H. Roberts, *Physiol. Plant.*, *16*: 745 (1964).
71. W. Major and E. H. Roberts, *J. Exp. Bot.*, *19*: 77 (1968).
72. R. B. Taylorson and S. B. Hendricks, *Plant Physiol.*, *52*: 23 (1973).
73. C. Perino, E. Simond-Cote, and D. Côme, *C.R. Acad. Sci.*, *299* (Ser. 3): 249 (1984).
74. T. Tissaoui and D. Côme, *Planta*, *111*: 315 (1973).
75. E. H. Roberts, *J. Exp. Bot.*, *12*: 430 (1961).
76. V. Durham and P. S. Wellington, *Ann. Bot.*, *25*: 197 (1961).
77. M. Black, *Can. J. Bot.*, *37*: 393 (1959).
78. E. H. Roberts and R. D. Smith, in *The Physiology and Biochemistry of Seed Dormancy and Germination* (A. A. Khan, ed.), North-Holland, New York, p. 385 (1977).
79. C. M. Karssen, D. L. C. Brinkhorst-van der Swan, A. E. Breekland, and M. Koornneef, *Planta*, *157*: 158 (1983).
80. C. M. Karssen, S. Zagorski, J. Kepczynski, and S. P. C. Groot, *Ann. Bot.*, *63*: 71 (1989).
81. D. S. Robertson, *J. Hered.*, *66*: 67 (1975).
82. D. R. McCarty, C. B. Carson, P. S. Stinard, and D. S. Robertson, *Plant Cell*, *1*: 523 (1989).
83. S. J. Neill, R. Horgan, and A. D. Parry, *Planta*, *169*: 87 (1986).
84. S. J. Neill, R. Horgan, and A. F. Rees, *Planta*, *171*: 358 (1987).

85. C. S. Robichaud, J. Wong, and I. M. Sussex, *Dev. Genet.*, *1*: 325 (1980).
86. C. Robichaud and I. M. Sussex, *J. Plant Physiol.*, *126*: 235 (1986).
87. K. Skriver and J. Mundy, *Plant Cell*, 2: 503 (1990).
88. T. L. Thomas, *Plant Cell*, *5*: 1401 (1993).
89. M. M. Thompson, D. C. Smith, and J. E. Burgess, *Theor. Appl. Genet.*, *70*: 687 (1985).
90. J. Rodriguez-A., W. B. Sherman, R. Scorza, W. R. Okie, and M. Wisniewski, *J. Am. Soc. Hortic. Sci.* (in press) (1994).
91. J. King, *The Genetic Basis of Plant Physiological Processes*, Oxford University Press, New York (1991).
92. P. Stokes, *Encycl. Plant Physiol.*, *15*(2): 746 (1965).
93. P. Stokes, *J. Exp. Bot.*, *11*: 222 (1953).
94. G. P. Steinbauer, *Plant Physiol.*, *12*: 813 (1937).
95. G. Lakon, *Naturwiss. Z. Land- Fortswirtsch.*, *9*: 285 (1911).
96. T. A. Villiers and P. F. Wareing, *J. Exp. Bot.*, *15*: 359 (1964).
97. B. C. Jarvis, B. Frankland, and J. H. Cherry, *Planta*, *83*: 257 (1968).
98. J. W. Bradbeer and B. Coleman, *New Phytol.*, *66*: 5 (1967).
99. J. W. Bradbeer and N. J. Pinfield, *New Phytol.*, *66*: 515 (1967).
100. R. S. Quatrano, in *Plant Hormones and Their Role in Plant Growth and Development* (P. J. Davies, ed.), Martinus Nijhoff, Boston, Chap. E13, p. 494 (1987).
101. M. Bouvier-Durand, A. Dawidowicz-Grzegorsewska, C. Thevenot and D. Côme, *Can. J. Bot.*, *62*: 2308 (1983).
102. S. S. C. Chen and J. E. Varner, *Plant Physiol.*, *46*: 108 (1970).
103. S. Lewak, A. Rychter, and B. Zarska-Maciejewska, *Physiol. Veg.*, *13*: 13 (1975).
104. R. Alscher-Herman and A. A. Khan, *Physiol. Plant.*, *48*: 285 (1980).
105. D. A. Eichholtz, H. A. Robitaille, and K. M. Herrmann, *Plant Physiol.*, *72*: 750 (1983).
106. T. L. Noland and J. B. Murphy, *J. Plant Physiol.*, *124*: 1 (1986).
107. B. A. Hance and J. M. Bevington, *Plant Physiol.*, *96*(Suppl.): 63 (Abstract 414) (1991).
108. B. Li and M. E. Foley, *Plant Physiol.*, *96*(Suppl.): 63 (Abstract 413) (1991).
109. A. Mahhou and F. G. Dennis, Jr., *J. Am. Soc. Hortic. Sci. 119*: 131 (1994).
110. G. A. Lang and J. Tao, *HortScience*, *26*: 733 (Abstract 376) (1991).
111. A. S. Callaway, Ph.D. thesis, University of Georgia, Athens, Ga. (1991).
112. C. H. Lin, L. Y. Lee, and M.-J. Tseng, *Plant Physiol.*, *96*(Suppl.): 404 (Abstract 405) (1991).
113. L. Di Nola, C. F. Mischke, and R. B. Taylorson, *Plant Physiol.*, *92*: 427 (1990).
114. J. L. Ried and M. K. Walker-Simmons, *Plant Physiol.*, *93*: 663 (1990).
115. M. Gendraud, *Physiol. Veg.*, *15*: 121 (1977).
116. F. Thomas, C. Thévenot, and M. Gendraud, *C.R. Acad. Sci.*, *300* (Ser. 3): 409 (1985).
117. S. Lavarenne, M. Champciaux, P. Barnola, and M. Gendraud, *Physiol. Veg.*, *20*: 371 (1982).
118. M. Champciaux, Diplôme d'études approfondies, Clermont-Ferrand, France (1979).
119. M. Gendraud and J. Lafleuriel, *Physiol. Veg.*, *21*: 1125 (1983).
120. M. C. Ben Ismail, Ph.D. thesis, Faculté des Sciences Agronomiques de Gembloux, Belgium (1989).
121. C. W. Donoho, Jr., and D. R. Walker, *Science*, *126*: 1178 (1957).
122. P. M. Williams, J. W. Bradbeer, P. Gaskin, and J. MacMillan, *Planta*, *117*: 101 (1974).
123. R. D. Amen, *Bot. Rev*, *34*: 1 (1968).
124. M. G. Nikolaeva, *Physiology of Deep Dormancy in Seeds* (translated from the Russian by Z. Shapiro), Israel Program for Scientific Translations, Jerusalem (1969).
125. I. Sinska and S. Lewak, *Physiol. Veg.*, *8*: 661 (1970).
126. D. P. Webb, J. van Staden, and P. F. Wareing, *J. Exp. Bot.*, *24*: 741 (1973).
127. R. W. King, S. O. Salminen, R. D. Hill, and T. J. V. Higgins, *Planta*, *146*: 249 (1979).
128. R. R. Finkelstein, K. M. Tenbarge, J. E. Shumway, and M. L. Crouch, *Plant Physiol.*, *78*: 630 (1985).
129. E. Sondheimer, D. S. Tzou, and E. C. Galson, *Plant Physiol.*, *43*: 1443 (1968).
130. G. A. D. Jackson, *Soc. Chem. Ind. (London) Monogr.*, *31*: 127 (1968).
131. F. G. Dennis, Jr., G. C. Martin, P. Gaskin, and J. Macmillan, *J. Am. Soc. Hortic. Sci.*, *103*: 314 (1978).

132. A. M. M. Berrie, D. Buller, R. Don, and W. Parker, *Plant Physiol.*, *63*: 758 (1979).
133. R. Rudnicki, *Planta*, *86*: 63 (1969).
134. J. A. Ozga and F. G. Dennis, Jr., *HortScience*, *26*: 175 (1991).
135. O. Balboa-Zavala and F. G. Dennis, Jr., *J. Am. Soc. Hortic. Sci.*, *102*: 633 (1977).
136. J. W. Braun and A. A. Khan, *Plant Physiol.*, *56*: 731 (1975).
137. I. D. J. Phillips and P. F. Wareing, *J. Exp. Bot.*, *10*: 504 (1959).
138. M. Kawase, *Proc. Am. Soc. Hortic. Sci.*, *78*: 532 (1961).
139. K. Ohkuma, F. T. Addicott, O. E. Smith, and W. E. Thiessen, *Tetrahedron Lett.*, *29*: 2529 (1965).
140. J. W. Cornforth, B. V. Milborrow, G. Ryback, and P. F. Wareing, *Nature*, *205*: 1269 (1965).
141. J. R. Lenton, V. M. Perry, and P. F. Saunders, *Planta*, *106*: 13 (1972).
142. L. E. Powell, *HortScience*, *11*: 498 (1976).
143. L. E. Powell and S. D. Seeley, *HortScience*, *5*: 327 (1970).
144. S. D. Seeley, Ph.D. thesis, Cornell University, Ithaca, N.Y. (1971).
145. W. K. Coleman and R. R. King, *Am. Potato J.*, *61*: 437 (1984).
146. J. I. Medford, R. Horgan, Z. El-Sawi, and H. J. Klee, *Plant Cell*, *1*: 403 (1989).
147. A. C. Smigocki, *Plant Mol. Biol.*, *16*: 105 (1991).

21
Senescence in Plants and Crops

Lola Peñarrubia and Joaquín Moreno

Universitat de València, Burjassot, Valencia, Spain

I. INTRODUCTION

Senescence has been defined as endogenously controlled deteriorative changes which are natural causes of death in cells, tissues, organs, or organisms [1]. The differences with the term *aging* are well established, being that aging comprises all those degenerative changes that occur in time without reference to death as a consequence. Aging takes place during the entire lifespan of an organism, whereas senescence is considered the final developmental phase, which culminates in death [1–3]. Senescence in plants is a very pervasive phenomenon that has been encountered in all plants and at all stages of the life cycle, related with developmental as well as adaptive functions. It shows a variety of patterns, ranging from the death of specific cells to the most extreme case of decline of the entire plant. Senescence in annual plants is part of a process that maintains the equilibrium between the population in the existing environment with the potential for rapid readjustment to new conditions. The loss of leaves in decidous trees during winter, or the reversible leaf senescence in perennial trees during the periods of nitrogen mobilization, are examples of the adaptive role of senescence.

Senescence is a natural developmental process that may be considered as terminal differentiation because it usually takes place at the end of the life cycle of an organ or organism. However, different kinds of stress can induce senescence at any stage of the plant cycle [4]. In many cases the fundamental metabolic aspects of stress-induced senescence are almost identical to those of natural senescence. The reason for this similarity may be that both processes are a result of triggering the same adaptive mechanisms that are constitutively present in plants.

Even if senescence is essentially a degenerative process, it is far from being a chaotic breakdown. On the contrary, senescence occurs as an orderly loss of functions and structures, comprising an array of biochemical and physiological processes whose ultimate goal is the removal of nutrients from the decaying tissues. This sequence of events, known as the *senescence syndrome*, includes the turnover of macromolecules and lipids and the transport of mobilized nutrients out of the senescing structures toward other parts of the plant, be these either growing organs, such as fruits or young leaves, or specialized storage tissues, such as the bark of deciduos trees. In this regard it is not paradoxical that senescence promotes the rise of both degradative and protective enzymatic activities because the ordered dismantling needed for optimal exploitation of nutrients requires both specific degradation and protection against uncontrolled agents, an unavoidable by-product of breakdown.

Fruit ripening is another physiological process that is usually assimilated to senescence because it shares with truly senescent processes several metabolic features, especially the dismantling of chloroplastic components and structures. However, fruit ripening also has many particular metabolic characteristics, and its final goal as a physiological process is different from that of senescence, being the development of physicochemical and organoleptic properties that facilitate the spreading of seeds. This teleological difference is reflected in the fact that fruits usually continue to act as a sink of nutrients during ripening, whereas other senescing organs behave as a source.

Despite the inherent diversity of senescence, three broad phases or stages may be distinguished in a typical senescent process. First, there is a phase of selective degradation of certain molecules, whose lysis does not cause a major impairment of the physiological function of the senescent structure. Therefore, the mobilized molecules may be thought of as nutrient storage materials, and this stage may be termed *storage mobilization*. In some cases senescence may be reversed during this phase by suitable changes in the environmental conditions. The second stage is characterized by the extension and generalization of breakdown to components that are central in maintaining the physiological function, which is consequently lost. Somewhere along this phase, which might be called *generalized breakdown*, the senescent process becomes irreversible and the cells are definitively targeted to death. Finally, once the senescent structure has been emptied of profitable nutrients, there is a third stage of abscission (i.e., shedding of the senescent part from the rest of the plant) and death. Abscission, a biochemically and physiologically complex process, is studied in detail in Chapter 23 and will not be discussed further here. Although exceptions or overlapping of stages may be found in many particular senescent processes, the three-phase scheme above may serve as a developmental outline that emphasizes the strategy of senescence.

Set off and control of senescence are intricate and controversial subjects, involving several metabolic and hormonal signals. Ethylene is the main hormone implicated in both senescence and stress responses [5]. However, this and other signals relevant to senescence are dealt with in Chapter 22 and we refer to them only occasionally. Nevertheless, in the last instance, control of senescence may be expected to be exerted by regulation of gene expression. This aspect is largely unknown at present. As happens in other research fields, studies on plant senescence lag behind the knowledge of aging in animal cells, where the molecular determinants of senescence at the gene expression level are beginning to be understood. Programmed death of noninjured cells (termed apoptosis) has been shown to be regulated in animals by a specific array of "death" genes (for a review, see Ref. 6). It is plausible that a similar, if not homologous genetic organization will be found in plants.

Senescence of crops plants is of special interest because it encompasses phenomena of economic importance that occur both in the field and during storage and handling of plant products of commercial value. Moreover, the still incipient but constantly growing knowledge on the genetic control of senescent processes has already allowed the manipulation of several features of senescence using recombinant DNA technology for improving the quality of the crops. Tomatoes that are bruise resistant [7,8] or that do not overripen [9,10] are among the first offspring of this approach, which will probably revolutionize agricultural practice in the near future.

The aim of this chapter is to provide an outline of the natural patterns, physiological and molecular characteristics of senescence, and experimental approaches to its study. It is not intended to be exhaustive but rather, representative of the current trends in the field, and it is devoted especially to crop plants. The reader interested in more detailed information is referred to the excellent book by Noodén and Leopold [11].

II. PATTERNS OF SENESCENCE IN THE LIFE CYCLE OF PLANTS

Patterns of plant senescence may be illustrated by two extreme behaviors. There are plants, such as trees, in which survival of the individual over a long period (including several reproductive phases) is the most important commitment. These plants (termed *polycarpic*) usually undergo periodic senescence, limited to older organs combined with the growth of new organs. Other plants (e.g., the annuals) sustain only one reproductive phase and die with the development of fruits. The latter species (called *monocarpic*), in which individuals are subordinated to survival of the population, develop whole plant senescence directed to mobilization of nutrients to the growing fruits. Between these extremes, there is a broad spectrum of life cycles displaying intermediate senescence strategies. Nevertheless, the major food crops are typically monocarpic, storing a high proportion of the plant biomass in the harvestable fraction (fruits or seeds).

Leopold [1] distinguishes four possible patterns of senescence in different plants:

1. Senescence of the whole plant at the end of the reproductive phase, which is typical of monocarpic plants.
2. Senescence of the aerial parts of the plant maintaining the underground structures. Bulbs belong to this group.
3. Senescence of the leaves only, the stems and roots remaining alive. This is the case with deciduous plants.
4. Progressive senescence of the leaves and other organs along the stem beginning at the base. Annual plants usually senesce in this way.

Another way to categorize senescence patterns is on the basis of the structural level (i.e., cellular, tissue, organ, or organism level) at which they act, as Noodén [3] has proposed. Senescence at any or all of these levels may occur at different stages of the plant life cycle.

A. Cellular Patterns

In some cases, one cell or a small number of cells may be selected to senesce. Cellular senescence could serve either to clear away the cell or to nurture the neighboring cells with the breakdown products. Whereas death is clearly the consequence of senescence in some specialized cells, in many other cases cells remain alive and the loss of some structures prepares the cells for new functions. The most spectacular cases take place in ovaries after flower pollination, during the first stages of embryo development. These processes involve senescence of specific cells in an orderly sequence of events [3,12].

B. Tissue Patterns

It is a common fact that a group of cells that form a tissue are targeted to degenerate and die. At the tissue level, senescence may have a nurturing function, like the death of cells surrounding the pollen or the embryo sac, but it can also serve to acquire a new function. For example, the aerenchyma is formed when the cells of the root cortex die, and the xylem tracheids are also a result of cell degeneration [3]. Nitrogen fixation in legumes is performed by the endosymbiotic bacterium *Rhizobium*, which binds to the root hairs, producing a nodule, a highly organized hyperplastic, hypertrophic tissue mass derived from root cortical cells. Individual infected cells are able to fix nitrogen in a transient step followed by senescence. Senescent nodule cells are characterized by the formation of large vacuoles that contain autophagocited cytoplasm and degraded bacteroids, with concomitant reductions in nitrogenase activity, nodule respiration, and growth [13,14]. Root nodule senescence is one aspect of nitrogen fixation that is relevant for the prospective improvement of the host–bacteroid interaction.

C. Organ Patterns

Senescence may happen in almost any organ of a plant, such as leaves, roots, flowers, branches, shoots, unfertilized ovaries, and developing fruits.

1. Leaves

Leaf development can be divided into three different periods. The first is leaf expansion, which occurs with import of organic compounds. When the leaf is mature and the photosynthetic capacity is fully developed, it becomes a source of organic material. Senescence takes place as the last step of leaf development, producing a massive mobilization of leaf components toward other plant parts [15].

Leaves may follow different patterns of senescence. For example, in trees the oldest leaves may decline when the new leaves are growing as part of a progressive senescence; or all the leaves may senesce together seasonally. In the first case, the breakdown products serve directly as a nitrogen source for the new organs [16], while in the second case, the nutrients are stored in the branches waiting for the next growing period [17].

2. Flowers

The flower is usually the organ with the shortest longevity [18]. Flower parts such as calyx, perianth, androecium, gynoecium, and peduncle are interrelated but differ from each other in both structure and physiology. Senescence of flowers is a representative example of a kind where several components decay tightly enchained to growth and development of other structures.

Pollination is a central event in flower development. Some parts of flowers, such as the perianth, senesce after pollination, while others, such as ovaries, develop. Changes originated by pollination, collectively termed the *pollination syndrome*, include a number of developmental processes, such as perianth pigmentation changes, ovary maturation, and ovule differentiation, which are crucial to ensure fertilization and embryogenesis [19]. Ethylene may be involved in the pollination syndrome since pollen in flowers has been found to carry large amounts of the ethylene precursor 1-aminocyclopropane-1-carboxylic acid (ACC), responsible for ethylene synthesis after pollination [20,21]. The signal brought forth by pollination may be a requisite for perianth senescence. The most spectacular cases are orchids, where unpollinated flowers can stay fresh for 6 months waiting for a specific insect to be fecundated. However, in different species factors other than pollination could cause flower senescence, possibly regulated by an endogenous clock [22].

Ovaries senesce naturally if not stimulated either by pollination or by hormonal induction of parthenocarpic fruits. Ovary senescence can be induced by emasculating flowers before pollination. Furthermore, senescence of emasculated pea ovaries may be prevented by application of gibberelins [23].

3. Fruits

The most spectacular modifications during fruit ripening probably occur at the cell wall, where important changes in structure and composition take place [24]. Fruit softening is a consequence of the induction of specific cell wall hydrolases [25]. Besides, the ripening-associated color changes of fruits are a result of the transition from chloroplast to chromoplasts which are rich in red or yellow carotenoid pigments [26]. In most fruits there is also a decrease in acidity during ripening, as well as an increase in sweetness and changes in aroma produced by volatile odorant compounds [27].

4. Tubers

The decay process that takes place in potato tubers is considered aging. Potatoes are vegetatively propagated by cutting seed tubers that in some varieties remain viable up to 3 years from harvest. However, from about 7 to 24 months a gradual loss of sprout vigor is caused by aging. During this process, some biochemical changes, such as peroxidative damage of membrane lipids caused by an increase in free radicals, are similar to senescence [28].

D. Whole Plant Senescence

Most studies in whole plant senescence, a typical feature of monocarpic plants, have been carried out in economically important crops. Whole plant senescence is characterized by a general mobilization of nutrients that are transported to the fruits, especially to the seeds, which act as a strong sink. Simultaneous senescence of all plant parts is probably a result of interorgan signaling. The molecular nature of the hormonal effectors that are responsible for whole plant senescence is controversial and is discussed in another chapter.

Even if senescence and death are internally programmed, elimination of flowers and fruits may delay senescence of the whole plant. This fact has been interpreted in diverse ways, being the most accepted idea that hormonal factors produced by those organs affect the levels of signal responsible for senescence in the rest of the plant [2]. There is also evidence of roots playing a role as a source of growth substances during senescence of fruiting plants.

III. EXPERIMENTAL SYSTEMS TO STUDY SENESCENCE

The importance of having appropriate experimental systems and controls deserves special attention. Systems to study senescence have to be well defined and as close as possible to the natural environmental conditions of the plants under study. In this sense, intact plants are the best experimental systems and should be used whenever possible. Usually, plant experiments are carried out under optimal greenhouse conditions. However, stresses due to the field suboptimal supplies of water and mineral nutrients, and to extreme temperatures, salt, ultraviolet light, and wounding, may heavily modify the natural senescence behavior and should be taken into account.

Light conditions are also crucial. For example, chlorophyll breakdown is strongly retarded by continuous illumination when compared with leaves kept in the dark. Even light quality (in relation to red/far red components) is also relevant, due to the participation of phytochrome in light-mediated responses during senescence [29,30]. However, different photosynthetic characteristics respond independently to light throughout senescence in rice leaves [31].

Laboratory studies require the use of simple systems, such as excised parts of a plant, that facilitate the scrutiny of the metabolic aspects and hormonal regulation of senescence under controlled conditions. However, it is always necessary to relate these data to the events in intact organs. Due to the correlative nature of many plant processes, the detachment of plant structures often produces important physiological and biochemical changes, especially in the detached organ but also in the rest of the plant. If the use of a detached system is unavoidable, precautions should be taken in analyzing the results. As information grows, there is more and more evidence of interrelation among all plant parts, probably due to hormonal interference [32].

A. Excised Plant Parts

1. Leaves

Leaves are the most utilized organs for studying senescence, and a great variety of conditions have been tested. The experimental systems that use wounded parts of leaves, such as leaf discs, should be avoided because wound-induced ethylene interferes with natural causes of senescence. Detached leaves are also a common experimental system, but some considerations should be taken into account. For example, excision of leaves combined with darkness may accelerate leaf senescence or may delay it [33]. Becker and Apel [34] have shown that there are differences in gene expression between natural and detached barley leaves incubated in darkness. They conclude that only a minor part of the mRNA changes in detached leaves is related to leaf senescence, whereas stress-related transcripts appear to predominate. On the other hand, the hormonal responses and the redistribution of nutrients are different in detached leaves [35]. Explants have also been used to study leaf senescence. A typical leaf explant consists of terminal segments of vegetative branches bearing several mature leaves placed in vials with water [36,37]. In general, as the experimental system grows in complexity, the results are closer to those of the intact system.

2. Flowers

A common method is the use of flower explants, consisting of the complete flower plus a few centimeters of the peduncle. Petal senescence has been by far the most studied process because of its economical importance. The petal tissue is an excellent model system to study senescence in the absence of the strong influence that chloroplasts exert on the process [18]. Japanese morning glory (*Ipomoea tricolor*), carnation, daylily, and orchids are probably the most studied flowers. In general, flowers follow a strict and predictable sequence of events during senescence [18].

Unpollinated pea ovaries obtained by emasculation 2 days before anthesis have proven to be an excellent system to study senescence. The absence of pollination allows the start of senescence in a natural way in an intact structure. After a few days, senescence takes place in emasculated ovaries, leading to abscission and death. Treatment with gibberelins induces development of the partenocarpic fruit, which is similar to the fruits obtained by pollination except for the absence of seeds, providing an adequate control [38,39]. In this case, senescence seems to occur as a consequence of the absence of developmental signals.

3. Fruits

Fruit ripening represents a special type of adaptative role of senescence in plants. The origin of cells that constitute the edible parts of fruits is different, accounting for the wide range of patterns in ripening. A distinction between climacteric and nonclimacteric fruit should be kept in mind when considering the behavior of fruits during ripening. Climacteric fruits, such as apple, avocado, banana, and tomato, are characterized by a rise in the respiration rate and the production of ethylene at the onset of fruit ripening. In these fruits ethylene is responsible for most of the ripening-related changes [40,41]. To study this process under controlled conditions, fruit detachment from the vine is commonly performed followed by transport and storage of fruits in laboratory chambers. However, detachment of climateric fruits to study ripening has been proven recently to show certain differences in transgenic plants with modified levels of ethylene compared to the behavior of the fruits in the vine [42,43]. In nonclimacteric fruits, such as orange, lemon, cherry, and grapes, where ethylene production is very low and no rise in respiration rate is found, the ripening patterns are quite different and the role of ethylene has been questioned.

Being adapted to survive under unfavorable conditions, seeds probably hold the record for long-term cellular longevity, the oldest seeds germinating after 600 years [1]. The dehydration process happening in seeds would definitively damage most plant tissues. However, in seeds this process contributes to a quiescence period until appropriate germination conditions arrive. Even if the metabolic activity is minimal in dry seeds, the accumulation of events that produce cell death also takes place, producing seed nonviability. Storage conditions such as low temperature and humidity help to increase seed longevity. Senescence may be also studied in imbibed seeds. In this case, anaerobic conditions are necessary to increase seed viability [1].

B. Surgical Experiments

The experiments to test the influence of some parts of a plant on the remaining parts utilize a common procedure consisting of the surgical elimination of these parts. In the study of cotyledon senescence, one of the cotyledons, the roots or the bud, are excised [44]. This is also the case of detopped plants, where the apical meristem, and subsequent axillary growth, has been removed [45]. Surgical experiments are used to understand the influence of certain parts believed to produce growth regulators, such as apex, phloem, young developing fruits, and ovules. Usually, the replacement of the excised organ with an agar block containing the putative hormonal effector, or another type of exogenous application, completes the experiment [44]. The conclusions that can be extracted from this class of work are limited. Exogenous application of any biological material is subjected to strong limitations, such as uptake, sequestration, transport, and metabolism of the active material, and the difficulty of quantitating the amount of it within the target tissue [46].

C. Mutants

The use of mutants to study developmental processes in plants is an excellent tool to understand them at the molecular level [46]. The mutants may be spontaneous or induced by agents altering the DNA. In addition, the use of antisense technology (see below) may be considered a specific form of mutagenesis where the target gene is previously known. Once a mutant affecting senescence has been isolated, the goal is usually to identify the mutated gene and to characterize it at the molecular level in order to evaluate its functional role in the process. In this regard, recent technology that uses map-based cloning, large-scale cDNA sequencing, and tagging systems in plants genetically engineered with transposons [47] or T-DNA from *Agrobacterium tumefaciens* [48,49] facilitates the rescue of the mutated gene (for a review, see Ref. 50).

Several mutations altering senescence have been described. A mutation in a nuclear gene of *Festuca pratensis* results in a drastic reduction of chlorophyll loss without other senescence characteristics, such as decrease of protein and RNA content, rise of proteolytic activities, or degradation of plastid structure, being altered [51–53]. The mutated gene controls the thylakoid membrane disassembly in senescent leaves, impairing the degradation of thylakoid pigments, protein, and lipids [54]. Mutations in nuclear and organular genes altering chlorophyll loss or gas exchange during monocarpic senescence have also been described in soybean [55]. Recent work on plant mutants has frequently been done on the weed *Arabidopsis thaliana*. This plant is specially well suited for the characterization of mutants due to its reduced genome, small size, abundant progeny, and short life cycle [56], which facilitates laboratory culture and screening. Characterization of senescence-related genes in *Arabidopsis* will lead to the isolation (and eventual manipulation) of homologous genes from crop plants. A mutation that affects the light regulation of seedling development but also interferes with the onset of leaf senescence has been described [57]. In addition, several ethylene-insensitive mutants of *Arabidopsis* have been reported [58,59]. The CTR1 gene, responsible for one of these mutations, has been cloned and shown to encode a serine/threonine protein kinase homologous to the Raf protein kinase family [60]. Further cloning of other mutated genes will surely help to elucidate the ethylene transduction pathway and the mechanisms of senescence.

Recently, mutants (acd1) that seem unable to control the rate and extent of cell death when exposed to different senescence-inducing agents have been described [61]. These mutants exhibit accelerated cell death with rapid spreading of necrotic lesions in response to virulent and avirulent pathogens but also during aging of aseptic plants. Since these lesions are characteristic of the hypersensitive defense response to pathogens, analysis of the mutants may provide an understanding at the molecular level of this response and its relationship to natural senescence.

D. Transgenic Plants

The considerable progress achieved in recent years in plant transformation techniques makes the use of transgenic plants frequent in plant laboratories. Nowadays, genetically modified crops cover a wide area of plant research. Plant transformation mediated by *Agrobacterium tumefaciens* [62,63] allows the stable expression of genes in the whole plant, or in specific tissues if the gene is under the control of a selected promoter sequence. On the other hand, plant transformation with mobile genetic elements (transposons) produces gene expression at random stages of development or plant parts, depending on transposon jump [47,64,65].

Especially important has been the sense and antisense approach, which eliminates the expression of a gene by introducing in the plant a copy of the original gene in the same or reverse orientation (transinactivation) [46,66–68]. The use of transgenic plants where hormone metabolism has been altered, thereby modifying hormonal levels endogenously, have shed light on the role of certain hormones, eliminating the problems derived from exogenous applications. The most spectacular case is control over fruit ripening in tomatoes, where different approaches have been taken to reduce ethylene levels by antisense technology. The inhibition of two key enzymes in ethylene biosynthesis, ACC oxidase [9] and ACC synthase [10], reduces ethylene levels and delays fruit ripening [9,10]. Leaf senescence is also retarded in the ACC oxidase antisense plants [43].

Another approach is the use of genes from microbial organisms that are known to interact with plants modifying hormonal levels. Several of those genes have been expressed in plants [69]. Specially interesting have been the assessment of the role of cytokinins in delaying senescence in leaves [70] or the decrease of ethylene levels in tomato fruits by expressing a microbial gene that deaminates the ethylene precursor ACC [71]. Moreover, crossing of different transgenic plants with altered hormonal levels may help to gain further insight into the functional role of hormones [72].

IV. ULTRASTRUCTURAL, PHYSIOLOGICAL, AND BIOCHEMICAL CHANGES DURING SENESCENCE

A. Ultrastructural Changes

Characteristic changes occurring during plant senescence are called collectively the *senescence syndrome*. Senescence-related changes differ depending on tissue and species, but they share common features at the ultrastructural level. In green organs, chloroplasts are the organella where the first symptoms of senescence are visible. Following an ordered sequence of events, the chloroplast dismantling begins with swelling, unstacking, and degradation of thylakoids (first, those of the lamellae, and then, the grana), appearance of lipids droplets and plastoglobuli, and finally, fragmentation of the envelope [73]. In some cases, chloroplasts have been observed to fuse with vacuoles at the late stages of senescence [74]. The number and size of chloroplasts is reduced during senescence and the rate of oxygen evolution decreases approximately in parallel to the chloroplast content [75]. Loss of starch is also characteristic of senescence and may result in deformation of the cells. This may explain the distortion of endocarp and mesocarp cells observed in senescent ovaries of pea [74].

Some extraplastidic membranes, such as those of the endoplasmic reticulum, also undergo early degradation, the smooth and rough fractions being degraded non simultaneously depending on the species [15]. Changes in the properties of lipid phase have been observed in senescing membranes using wide-angle x-ray diffraction and freeze-fracture electron microscopy. Regions of the lipid bilayer switch from liquid-crystalline to gel phase, rendering leaky membranes. Freeze-fracture electron microscopy shows the gel-phase domains as intramembranous particle-free regions that increase in size and number as senescence progresses [76].

In contrast, there are other structures, such as mitochondria, that remain intact until later stages, when some swelling or distortion of cristae become apparent. The plasmalemma integrity is also maintained until the final stages. Cells become progressively more vacuolate with age, and changes in the permeability of the tonoplast membrane, surrounding the vacuole, could allow the transference of cytoplasmic material into the vacuole, favoring its degradation. Autophagic processes in which organelles become engulfed in vacuolelike structures have been observed [74]. In some cases, tonoplast rupture may cause the lysis of cells at very late phases [15,77]. With differences depending on the cell type, these changes proceed until the whole cell is dismanteled.

B. Physiological Changes

1. Leaf Conductance and CO_2 Assimilation

Senescence produces closure of stomata leading to a decline in transpiration. It has been suggested that stomata aperture may control the rate of leaf senescence [78]. The main entrance for CO_2 are stomata, and insufficient CO_2 supply could be the cause of the decreased photosynthetic assimilation observed during senescence. However, experimental measurement of CO_2 concentration in the substomatal cavity [79] suggests that CO_2 does not limit photosynthetic assimilation. Hence stomatal closure may be more a consequence than a cause of lowered photosynthetic activity, according to the optimal variation hypothesis, which proposes that stomatal conductance adapts to the photosynthetic capacity of the leaf [80]. It is remarkable that stomatal guard cells remain functional until very late stages of senescence, far beyond other leaf cells. This may be a result of their lack of symplastic connection with the surrounding cells [81].

2. Respiration

The fruit was the first organ where the rate of respiration during senescence was described. In detached apple fruits, the respiration rate decreases gradually until a sudden burst, termed *climacteric*, followed by another decline in respiratory activity is observed. Fruits are divided into two categories, climacteric and nonclimacteric, depending on whether their respiration show a sudden peak or not. In detached leaves and cut flowers, a climacteric-like rise in respiration has been also observed during senescence in different species, although there are others where no increase have been detected [82]. The only common metabolic feature in climacteric fruits is their ability to produce and respond to ethylene. It appears that the rise in respiration is a consequence of ethylene action and not of senescence as such. The main reason for this conclusion is that inhibition of both the biosynthesis and action of ethylene eliminates the rise in respiration without preventing eventual senescence, while treatment with ethylene enhances respiration of nonripening tomato mutants but does not include the typical changes of ripening. It appears that ethylene enhances plant respiration by activating a preexistent enzymatic potential [82].

Respiratory pathways of senescent plants include glycolysis, pentose pathway, TCA (calcitonin) cycle, and the electron transport pathway, where some changes have been described [82] but also an alternative oxidase pathway which is enhanced during senescence [83]. It has been suggested that the alternative pathway is activated when the cytochrome pathway is saturated or limited, allowing the TCA cycle to function using up excess carbohydrates [84]. During aging of potato tuber slices the alternative oxidase is composed of an integral membrane protein synthetized *de novo* [85]. In addition, changes in the photorespiratory enzymes and glutamate synthases have been described during tomato fruit ripening [86].

In cut carnations, petal senescence is associated with both an increase in ethylene production and sensitivity [87]. Interorgan translocation of both ethylene and ACC has been described in emasculated *Cymbidium* flowers (20). However, the daylily flower does not respond to ethylene, although other senescence-related changes occur, such as the climacteric rise in respiration [88].

C. Biochemical Changes

1. Photosynthetic Pigments

Color changes are important criteria for a visual evaluation of the advance of senescence, specially in fruits [26]. Breakdown of chlorophylls may be one of the earliest symptoms of senescence. However, chlorophyll decline is strongly retarded by continuous illumination in a process regulated by phytochrome [30]. The chlorophyll *a/b* ratio has been shown usually to decline with the advance of senescence [89,90], probably as a result of the nonsynchronous dismantling of lamellae and grana thylakoids and the asymmetrical distribution of photosystems between them.

Carotenoids are lost at a much lower rate than chlorophylls [91]. This difference in degradation rate accounts for most of the color changes associated with leaf senescence, and may reflect the persistence of photoprotective role of carotenoids until later phases of the process. In ripening fruits, senescence is sometimes associated with *de novo* synthesis of both carotenoids and anthocyanins [26].

The pathway of chlorophyll degradation may be distinct for different species or even organs. The activity of chlorophyllase, a thylakoidal enzyme that hydrolyzes the phytyl ester group, has been shown to correlate with chlorophyll degradation in maturing citrus fruits [92]. Chlorophyll oxidase, a complex enzymatic system that renders chlorophyll *a*1 as a first step, may also be involved. The fact that chlorophyllase is hindered by its membrane localization, and that chlorophyll oxidase activity is dependent on free fatty acids, liberated by lipid hydrolysis, may provide a link between thylakoidal membrane and chlorophyll degradation [92,93]. Alternatively, the peroxidase–hydrogen peroxide pathway, which opens the porphyrin ring, can also be involved and has been shown to be the main catabolic pathway in detached spinach leaves [94]. In addition, direct photodamage also represents a contribution, even if minor, to chlorophyll breakdown [73].

The existence of a *Festuca pratensis* mutant that does not exhibit chlorophyll loss during senescence supports the conclusion that the decline in photosynthetic capacity is not a result of the loss of chlorophyll. Studies on the proteins that are abnormally retained in the mutant indicate that all of them possess an associated tetrapyrrole prostetic group (heme or chlorophyll) [53]. Since it has been proposed that the degradation of porphyrins and their associated apoproteins is correlated [95], a lession in the heme/chlorophyll catabolic pathway may be responsible for the phenotype of the mutant [53].

2. Nucleic Acids

The appearance of the nucleus is maintained throughout senescence without major changes, although a decrease in nuclear DNA has been described at the final stages of senescence in soybean cotyledons (about 20%) [96], tobacco, and peanuts leaves [97]. It has been shown that repeated sequences are selectively degraded while coding regions of nuclear DNA remain largely intact [98,99].

Senescence is a process of overall decline in metabolism, including RNA and protein synthesis. Isolated chloroplasts from barley leaves show a decrease in the mRNA and protein synthesis that is not a consequence of DNA rearrangement, gain, loss, or methylation [100]. This decrease supports the view of senescence as a process of inhibition of metabolism and information flow. However, a nonspecific decline in RNA synthesis does not cause senescence. Furthermore, selective activation of RNA synthesis is necessary for the progress of senescence.

The quantitative decline in RNA is explained mainly due to the decrease in ribosomal RNA (rRNA), which is the most abundant cellular RNA in both the chloroplast and cytosol. This decrease correlates with the decline in protein synthesis. Variations in relative amounts of two phenylalanyl transfer RNA (tRNA) have been also detected during senescence [101]. However, other tRNA levels do not change or even increase in senescent tissues, although the variations could be due to differential rates of degradation [102]. tRNA synthase activities are greatly reduced during senescence, probably limiting the translational capacity of senescing chloroplasts (102). A general increase in RNase activities has been described during senescence that could account for the generalized loss of RNA [103]. Nevertheless, qualitative changes in messengers RNA (mRNA) are probably more relevant in the senescence process (see below).

3. Intracellular Proteins

In terms of total protein content, leaf senescence is characterized by a progressive loss of proteins [98]. This loss may be attenuated if additional nitrogen is supplied to the plant [104] or sink organs are removed. In general, the demand of mineral nutrients by growing structures has been described as a regulatory factor in leaf senescence, except for the phosphorus nutrition, which does not show any regulatory control on the process [105].

The patterns of protein loss are characteristic and independent of the cause of senescence. A wide range of specific proteins are degraded at the same time that others remain intact. In green organs, chloroplast proteins are principal targets of degradation during early phases of senescence. The loss of chlorophyll correlates with degradation of chlorophyll-carrying thylakoidal proteins, whose lysis is strongly retarded by continuous illumination. However, stromal proteins rapidly disappear under the same conditions, indicating that breakdown of membrane and soluble proteins is differently regulated by light [30].

The most abundant soluble protein in chloroplasts, ribulose 1,5-bisphosphate carboxylase/oxygenase (Rubisco), represents more than 50% of the chloroplast nitrogen and about 25% of that of the whole cell [106]. The Rubisco is known to be degraded extensively and selectively at early stages of senescence in many plants [16,17,107–110]. The specific proteolysis of this enzyme makes up to 85% of the soluble protein lost in senescing barley leaves [111] and more than 90% of the nitrogen mobilized from leaves before abscission in apple trees [17]. Recent experiments with transgenic plants, where the level of Rubisco has been lowered using the antisense technology, have firmly established the natural excess of Rubisco over the amount needed for performing its catalytic function, and the correlation of the amount of enzyme with the nitrogen status of the plants [112]. Evidence of this "luxury" excess, together with the spectacular contribution of this enzyme to nutrient mobilization during senescence, support the concept of Rubisco as a nitrogen storage protein [112,113].

Protein turnover is a common feature in every living organism; therefore, the loss observed during senescence has to be considered as an imbalance between the rates of protein synthesis and degradation [114]. Probably both reduced synthesis and enhanced proteolysis are responsible for protein loss observed during senescence. In this regard, synthesis of all thylakoidal proteins is known to be severly curtailed in senescing bean leaves except for the D-1 protein of photosystem II [115]. On the other hand, increased protein breakdown may result from different mechanisms: *de novo* synthesis of proteolytic enzymes, activation of preexisting proteases, decompartmentalization of proteases and their substrates, or susceptibilization of the protein substrates to degradation.

References on increased proteolytic activities during senescence are abundant (for reviews, see Refs. 2 and 116); however, they report more frequently enhanced levels of preexistent proteases than the appearance of new activities specific to senescence [117]. Exceptionally, a thiol protease which is differentially expressed in senescing ovaries has been described [118]. The scarcity of reports on proteolytic activities specifically associated with senescent processes may be due to the exceptional nature of the phenomenon, or alternatively, it may reflect the difficulty of detecting these activities due to the interference of the abundant vacuolar proteases, which are ubiquitous in plant cell extracts. The implication of the vacuolar activities in plant senescence seems doubtful for proteins of the organella, at least throughout the first stages of senescence, when the compartmental integrity has been firmly established. During the final phase, the fusion of organella with vacuoles [74] surely results in those activities gaining access to the remaining substrates and producing the ultimate breakdown of the organellar structure.

In other species, no correlation between senescence and increased proteolytic activities has been found [116]. In these cases, the loss of protein during senescence may be due to natural turnover after arrest of synthesis, or to modification of the proteins that may label them for proteolysis. In this regard, the case of the chloroplastic CO_2-fixing enzyme, the Rubisco, appears to be paradigmatic. In some species, such as corn and wheat, the enzyme seems to be degraded as a result of natural turnover [2,119]. However, in most species the turnover rate of the Rubisco is negligible under nonsenescent conditions [120,121], but changes dramatically with the onset of senescence. It has been shown that the susceptibility of the Rubisco to proteases increases markedly through oxidation of sulfhydryl groups belonging to critical cysteine residues of the enzyme [122]. This suggests that the Rubisco degradation may be induced by the oxidative conditions developed inside the functionally impaired chloroplast during senescence [122,123]. Evidence for in vivo oxidation of Rubisco has been found in different organisms under stress-induced senescence [124–127]. Moreover, a chloroplastic proteolytic activity that is activated by oxidative conditions has been described [128]. This suggest that alteration of the redox state of the chloroplast may provide a general mechanism for triggering a selective protein degradation during senescence or other processes that arrest chloroplast function.

Even if usually associated to protein degradation, senescence is known also to cause a rise in some enzymatic activities other than hydrolases. For example, transition of lead peroxisomes to glyoxisomes is a well-characterized phenomenon associated with senescence, Accordingly, enzymatic activities of markers of the glyoxilate cycle increase dramatically during darkness-induced senescence of spinach leaves [129] and pumpkin cotyledons [130]. Cytosolic glutamine synthetase also increases about fourfold during senescence of rice leaves [131]. This rise surely facilitates the mobilization of nitrogen by enhancing the synthesis of the major transported amino acid (glutamine) [132]. In addition, different isozymes may be specifically synthesized during senescence. This is the case for threonine dehydratase, an enzyme that probably plays a role in metabolism leading to nitrogen remobilization in senescing tomato leaves [133].

4. Membrane

One of the most characteristic changes during senescence is the progressive loss of membrane integrity. Unfortunately, the purification of senescent membranes is a difficult task, due to the changes in density, surface charge, and the loss of marker enzymes [134]. The major classes of lipids in plasma membrane and tonoplast are phospholipids, sterols, and ceramide monohexosides [135]. Common changes during senescence include a decrease in the total phospholipid and protein content, an increase in neutral lipids, and generalized oxidation [135]. Sterols also decline with physiological aging [136]. As a consequence of all the above, the physicochemical properties of the membranes, such as lipid fluidity, phase transition temperature, and nonbilayer lipid structure, are progressively altered during senescence [137–139]. The bilayer destabilization leads to a generalized failure of membrane functions, including loss of selective permeability and intracellular compartmentation, as well as membrane-associated enzymes and receptors [140].

The main enzymes implicated in lipid degradation are phospholipase D, phosphatidate phosphatase, and lipolytic acyl hydrolase. Most of the enzymes implicated in this process possess both membranous and cytosolic forms, which are differently regulated. The sequential action of these enzymes produces polyunsaturated fatty acids (PUFA) which are the substrates for the lipoxygenase [141]. Lipoxygenase is a dioxygenase that catalyzes the oxidation of PUFA to fatty acid hydroperoxide, which is a precursor of volatile compounds that provide the typical flavor that characterizes wounded tissues [135]. Lipoxygenase activity has been described to increase during senescence in different plant organs [142,143]. The degree of partitioning of lipoxygenase between cytosol and membranes seems to be an important factor in the peroxidative damage [11] since this enzyme seems to favor the oxidative injury of membranes by superoxide radicals.

The fact that some products of lipid peroxidation could serve as Ca^{2+} ionophores, together with structural changes in the lipid phase, render senescent membranes leaky to Ca^{2+}. Ca^{2+} is stored in compartments such as apoplast and vacuole by the action of ATPases that maintain its cytoplasmic concentration below micromolar levels under steady-state conditions. However, cytosolic Ca^{2+} increases during senescence as a consequence of a decrease in the efficiency of ATPases, along with Ca^{2+} leakage from the storage compartments. In its turn, the increase in cytosolic Ca^{2+} triggers a profusion of secondary effects. Among them, Ca^{2+} may influence directly phospholipase D and phosphatidic acid phosphatase and indirectly influence lipoxygenase [144]. In addition, calcium and spermine has been shown to cause a decrease in membrane fluidity of tomato microsomes and increase phospholipase D activity by a mechanism attributable to the biophysical effect of the cations on the membranes [145].

The relationship between the increase in free radicals and senescence is well established. The breakdown of membranes, nucleic acids, polysaccharides, and proteins is affected by free radicals [146]. Moreover, free radicals influence the ethylene production pathway [147]. Free radicals derived from oxygen such as superoxide, hydroxyl, peroxyl, and alkoxy radicals are known as reactive oxygen intermediates (ROIs). There is a correlation between changes in bulk lipid fluidity of microsomal membranes and changes in ROI production [11]. ROIs are generated as by-products of some enzymatic reactions, but plants specifically produce these radicals as a consequence of the photosynthesis [141,148]. Production of free radicals is enhanced during late stages of senescence, due to impairment of the electron flow between the two photosystems, which limits the availability of photosynthetic power [149]. Under normal conditions, oxygen-detoxifying enzymes such as superoxide dismutases, catalases, peroxidases, and glutathione reductases are present in plant cells to prevent damage by these toxic species [150–153]. Superoxide is eliminated by the superoxide dismutase, which produces hydrogen peroxide. Catalases convert hydrogen peroxide into water and oxygen, whereas peroxidases reduce hydrogen peroxide to water and oxidize a variety of substrates. Ascorbate peroxidase is a hydrogen peroxide–scavenging enzyme that is specific to higher plants and algae. This enzyme protects chloroplasts and other cell constituents from damage by hydrogen peroxide and derived hydroxyl radicals. Ascorbate peroxidase, glutathione reductase, and dehydroascorbate reductase remove hydrogen peroxide through a pathway termed *photoscavenging* [151]. Nevertheless, in certain cases some of these activities decrease with the progression of senescence in parallel with an increase in ROIs [28,154]. In other instances, the soluble activity of protective enzymes decreases but there is an increase of wall-bound activity. This is the case for peroxidase during senescence of stigmas and styles in *Citrus* [155]. The localization of the ROI metabolism in specific compartments such as peroxisomes could also serve to protect the cell under normal conditions [148]. It has been stated that the number of peroxisomes increase with oxidative stress [156] while superoxide radicals have been localized in glyoxisomes, a special kind of peroxisomes [157]. However, membrane deterioration during senescence and the loss of compartimentation could contribute to an extension of the ROI effects.

5. Cell Wall Components

Cell wall is not a static structure but a virtual extension of the cytoplasm. It contains cell surface markers [158], components for signaling and communication through plasmodesmata [159], and it is responsible for the adaptation to osmotic stress [160] and the production of defense molecules against insects [161] as well as fungal and bacterial pathogens [162]. Although there is a wide range of variability and complexity in the cell wall from different tissues, in dicotyledon plants it is composed roughly of 30% cellulose, 30% hemicellulose, 35% pectin, and 5% protein [163]. Different structural models of primary cell walls in flowering plants that are consistent with their physical properties during growth have been proposed [164]. Models of the cell wall have in common cellulose microfibrils embedded in a matrix of noncellulosic polysaccharides and protein. Cellulose self-associates to form the microfibrils that form a complex with hemicellulose, providing mechanical strength to plant cell walls.

Most of the studies on cell walls in relation to senescence have been done during abscission and fruit ripening. Fruit softening is the consequence of changes in the cell wall structure. A number of enzymes that are able to degrade the cell wall have been shown to increase its activity during fruit ripening such as polygalacturonase, pectin methyl esterase, Cx-cellulase, and others (reviewed in Ref. 25). The rise of cellulase activity correlates with the softening of fruits and the decline of break strength in the abscission zone. However, cellulase acts together with other cell wall degradative enzymes [25]. In tomato, the function of polygalacturonase during fruit ripening has been studied in antisense transgenic plants [7,8]. It has been concluded that reduction in the levels of polygalacturonase activity affects the depolymerization of chelator-soluble polyuronides. Tomato fruits obtained from polygalacturonase antisense plants display a measurable improvement in storage life, solids, and the consistency and viscosity of the juice [165].

V. CHANGES IN GENE EXPRESSION

Senescence is a genetically controlled process of disorganization that has to be regulated by a set of genes acting in concert. Studies on the products of mRNA translated in vitro have shown that certain messages disappear while others appear during senescence of wheat and oat leaves [166]. Thus senescence seems to begin with turning off and on of specific genes. The expression of genes can be regulated at several levels, the most studied being transcriptional activation (e.g., Refs. 167 and 168). It is not our aim in this section to include an exhaustive list of references related with transcriptionally activated genes during senescence; therefore, only some illustrative examples are discussed.

Leaf, cotyledon, and fruit are the organs where senescence at the molecular level has been most intensively studied, but the variety of systems and conditions utilized has led to a diversification of results. With regard to the consideration of senescence as a degradative process, the expression of mRNAs coding for proteases (e.g., Ref. 169) as well as proteinase inhibitors [14,170] have been described. The rise of mRNAs corresponding to hydrolytic enzymes that degrade the cell wall during fruit ripening has been reviewed by Fischer and Bennett [25].

Metabolic changes, including the aparison of new senescence-related pathways, usually are also transcriptionally regulated. This is the case for enzymes that participate in the glyoxilate cycle [171] or enzymes for the pathways of pigment synthesis in certain fruits [172].

Special attention is due to a group of genes that are activated as a general plant protection mechanism during senescence and under various kinds of stress. Most of these genes share the fact that they are induced by ethylene and by pathogen attack [173]. Some of them have an unknown function (e.g., Refs. 42 and 174); others have a clear protective role, such as superoxide dismutase, which expressed in transgenic plants reduces cellular damage under oxidative conditions [152].

At present, molecular senescence studies in plants are at the stage of looking for DNA domains at the promoter region shared by several senescence-related genes in order to find the transcription factor(s) able to induce the complete array of genes for protective proteins. This kind of regulator has been isolated and characterized in microorganisms and for animals in the case of oxidative stress [175–177]. As soon as these regulators are cloned in plants, it will be possible to control—and perhaps to improve—the general plant resistance mechanisms.

Some of the most important changes during senescence take place in the chloroplast. In this regard, a chloroplast DNA region that includes four genes homologous to mitochondrial NADH dehydrogenase is known to be activated during senescence, apparently at the level of translation. Accordingly, senescent chloroplasts show high ferricyanide reducing activity [178]. Some of the most abundant proteins in the chloroplast, such as Rubisco and chlorophyll *a/b* binding proteins, are composed of different classes of subunits synthetized in both the nuclear and the chloroplastic genome [179]. A coordination between both genomes has to exist during senescence. Nuclear control of cell senescence has been postulated based on several facts. First, there are mutations in the nuclear genome that alter the senescence syndrome. On the other hand, senescing cells with preserved chloroplasts have been obtained after enucleation. Moreover, selective inhibitors of nuclear RNA synthesis inhibit senescence-related processes. By contrast, specific inhibitors of organelle RNA polymerases do not inhibit senescence. Something similar happens with the inhibitors of protein synthesis; the cycloheximide (80S cytoplasmic ribosome inhibitor) blocks a variety of senescence-related changes, and the chloramphenicol (70S organella ribosome inhibitor) generally does not retard senescence. It has been suggested that control of the chloroplast RNA polymerase (encoded in the nucleus) could be the key step in this coordination [2].

VI. SENESCENCE MEASUREMENTS

To follow senescence it is necessary to find the appropriate parameters to measure the evolution of the process. No single measurement is definitive, although in some cases certain parameters may be adequate for particular tissues. However, single measurements should be checked against other parameters, whenever possible.

The loss of chlorophyll is one of the most obvious changes during senescence of green organs, although the existence of mutants where senescence proceeds without chlorophyll loss indicates that this change may not be crucial to the process [4,53]. Precautions should be taken during the extraction because some protocols (e.g., the acetone) may lead to its degradation [180]. Radiolabeling of chlorophyll has allowed a more sensitive measurement of its disappearance and identification of the degradation products [181].

Other conspicuous features of senescence are the decline in photosynthetic capacity [73] and the lowering of protein content. Since the CO_2-fixing enzyme, Rubisco, is a preferential target of proteases during senescence, its inactivation and degradation are widely used parameters to measure senescence in photosynthetically active organs [37]. Total protein levels may also be measured. To avoid the interferences inherent in some methods, nitrogen can be determined in digests [182] or through dye binding to protein adsorbed on washed paper disks [183]. Leakage of the cell membranes may be a good parameter, but it occurs late in the senescence process, as is the case for water loss and wilting. Leaf abscission in species that shed their leaves can be used as a supplementary measurement.

Finally, in the case where a hormone is directly implicated in senescence, measurement of the hormonal levels can be a good approach to the detection of early symptoms of senescence. This is the case for tomato fruits, where a burst of ethylene precedes the onset of ripening [5]. Probably a combination of parameters may be the best solution for accurate senescence measurement. For example, Pastori and Trippi [184] have utilized chlorophyll loss, lipid peroxidation, and electrolyte leakage from the cell to study senescence in maize. It might be expected that growing knowledge of the regulation of gene expression during senescence will allow the selection of specific messages as markers of senescence stages. The analysis of mRNAs using a set of marker genes may in the near future be the definitive solution to precise tracking of the senescence process.

VII. SUMMARY AND CONCLUDING REMARKS

Senescence appears as an ordered dismantling of structures and components from plant parts whose functional contribution has turned unnecessary and which are therefore directed to abscission and death. Aside from functional advantages that may derive in special cases from senescence of certain structures, the principal goal of senescence is to recover nutrients from the decaying tissues, withdrawing them to the surviving parts before abscission. As such, senescence is essentially a physiological strategy of nutritional economy.

The most general characteristics of senescence include breakdown of selected macromolecules as well as specialized complex metabolites (e.g., chlorophyll), progressive deterioration, and loss of functions of membranes, and at the final stage, degeneration of cell internal structure. These events are brought forth by changes in expression of specific genes and involve multiple intermediate signals that allow precise temporal and spatial patterns.

Senescence regulatory processes at the gene level are still poorly understood, but our knowledge of this aspect is expected to improve in the forthcoming years. This will probably uncover the underlying basic mechanisms, clarify the bounds between senescence and related processes such as stress responses and fruit ripening, and extend the possibilities of genetic engineering of the senescence features of crops for nutritional and commercial benefit.

ACKNOWLEDGMENT

This work was supported by grants PB 92-0018-C02-02 and PB 92-0821 from DGICYT.

REFERENCES

1. A. C. Leopold, in *Senescence in Plants* (K. V. Thimann, ed.), CRC Press, Boca Raton, Fla., p. 1 (1980).
2. R. Sexton and H. W. Woolhouse, in *Advanced Plant Physiology* (M. B. Wilkins, ed.), Longman Scientific & Technical Publication, Wiley, New York, p. 469 (1984).
3. L. D. Noodén, in *Senescence and Aging in Plants* (L. D. Noodén and A. C. Leopold, eds.), Academic Press, London, p. 2 (1988).
4. H. Thomas and J. L. Stoddart, *Annu. Rev. Plant Physiol.*, *31*:83 (1980).
5. A. K. Mattoo and J. C. Suttle, eds., *The Plant Hormone Ethylene*, CRC Press, Boca Raton, Fla. (1991).
6. L. D. Tomei and O. C. Frederick, eds., *Apoptosis: The Molecular Basis of Cell Death*, Cold Spring Harbor Laboratory Press, Cold Spring Harbor, N.Y. (1991).
7. R. E. Sheehy, M. K. Kramer, and W. R. Hiatt, *Proc. Natl. Acad. Sci. USA*, *85*:8005 (1988).
8. C. J. S. Smith, C. F. Watson, J. Ray, C. R. Bird, P. C. Morris, W. Schuch, and D. Grierson, *Nature*, *33*:724 (1988).
9. A. J. Hamilton, G. W. Lycett, and D. Grierson, *Nature*, *346*:284 (1990).
10. P. W. Oeller, L. Min-Wong, L. P. Taylor, D. A. Pike, and A. Theologis, *Science*, *254*:437 (1991).
11. L. D. Noodén and A. C. Leopold, eds., *Senescence and Aging in Plants*, Academic Press London (1988).
12. T. A. Steeves and I. M. Sussex, eds., *Patterns in Plant Development*. 2nd ed., Cambridge University Press, Cambridge (1989).
13. M. A. Egli, S. M. Griffith, S. S. Miller, M. P. Anderson, and C. P. Vance, *Plant Physiol.*, *91*:898 (1989).
14. J. F. Manen, P. Simon, J. C. Van Slooten, M. Osteras, S. Frutiger, and G. J. Hughes, *Plant Cell*, *3*:259 (1991).
15. J. L. Stoddart and H. Thomas, *Nucleic Acids and Proteins in Plants I* (Encyclopedia of Plant Physiology, New Series, Vol. 14A, D. Boulter and B. Parthier, eds.), Springer-Verlag, Berlin, p. 592 (1982).
16. J. Moreno and J. L. García-Martinez, *Physiol. Plant.*, *61*:429 (1984).
17. S. Kang and J. S. Titus, *Physiol. Plant.*, *50*:285 (1980).
18. S. Mayak and A. H. Halevy, in *Senescence and Aging in Plants* (L. D. Noodén and A. C. Leopold, eds.), Academic Press, London, p. 131 (1980).
19. G. N. Drews and R. B. Goldberg, *Trends Genet.*, *5*:256 (1989).
20. E. J. Woltering, *Plant Physiol.*, *92*:837 (1990).
21. A. Singh, K. B. Evensen, and T. H. Kao, *Plant Physiol.*, *99*:38 (1992).
22. H. Kende and B. Baumgartner, *Planta*, *116*:279 (1974).
23. J. Carbonell and J. L. García-Martínez, *Planta*, *147*:444 (1980).
24. C. J. Brady, *Annu. Rev. Plant Physiol.*, *38*:155 (1987).
25. R. L. Fischer and A. B. Bennett, *Annu. Rev. Plant Physiol. Plant Mol. Biol.*, *42*:675 (1991).
26. E. E. Goldschmidt, in *Senescence in Plants* (K. V. Thimann, ed.), CRC Press, Boca Raton, Fla., p. 207 (1980).
27. M. J. C. Rhodes, in *Senescence in Plants* (K. V. Thimann, ed.), CRC Press, Boca Raton, Fla., p. 157 (1980).
28. G. N. M. Kumar and N. R. Knowles, *Plant Physiol.*, *102*:115 (1993).
29. Y. N. Behera and B. Biswal, *Environ. Exp. Bot.*, *30*:181 (1990).
30. K. Okada, Y. Inoue, K. Satoh, and S. Katoh, *Plant Cell Physiol.*, *33*:1183 (1992).
31. J. Hidema, A, Makino, T. Mae, and K. Ojima, *Plant Physiol.*, *97*:1287 (1991).
32. L. D. Noodén, in *Senescence and Aging in Plants* (L. D. Noodén and A. C. Leopold, eds.), Academic Press, London, p. 391 (1988).
33. H. Thomas, *J. Plant Physiol.*, *136*:45 (1990).
34. W. Becker and K. Apel, *Planta*, *189*:74 (1993).
35. L. D. Noodén and A. C. Leopold, in *Phytohormones and Related Compounds: A Comprehensive*

Treatise (D. S. Letham, P. B. Goodwin, and T. J. V. Higgings, eds.), Elsevier/North-Holland, Amsterdam, Vol. 2, p. 329 (1978).

36. J. Riov, *Plant Physiol.*, *53*:312 (1974).
37. L. Peñarrubia, J. Moreno, and J. L. García-Martínez, *Physiol. Plant.*, *73*:1 (1988).
38. J. L. García-Martínez and J. Carbonell, *Planta*, *147*:451 (1980).
39. J. Carbonell and J. L. García-Martínez, *Planta*, *164*:534 (1985).
40. D. DellaPenna and J. J. Giovannoni, in *Developmental Regulation of Plant Gene Expression*, Plant Biotechnology Series, Vol. 2. (D. Grierson, ed.), Blackie & Son, Glasgow, p. 182 (1991).
41. J. Gray, S. Picton, J. Shabbeer, W. Schuch, and D. Grierson, *Plant Mol. Biol.*, *19*:69 (1992).
42. L. Peñarrubia, M. Aguilar, L. Margossian, and R. L. Fischer, *Plant Cell*, *4*:681 (1992).
43. S. Picton, S. L. Barton, M. Bouzayen, A. J. Hamilton, and D. Grierson, *Plant J.*, *3*:469 (1993).
44. L. M. Behera and N. K. Choudhury, *J. Plant Physiol.*, *137*:53 (1990).
45. S. J. Crafts-Brandner, *Physiol. Plant.*, *82*:299 (1991).
46. H. Klee and M. Estelle, *Annu. Rev. Plant Physiol. Plant Mol. Biol.*, *42*:529 (1991).
47. H.-P. Döring and P. Starlinger, *Annu. Rev. Genet.*, *20*:175 (1986).
48. K. A. Feldmann, *Plant J.*, *1*:71 (1991).
49. C. Koncz, K. Németh, G. P. Rédei, and J. Schell, *Plant Mol. Biol.*, *20*:963 (1992).
50. S. Gibson and C. Somerville, *Trends Biotechnol.*, *11*:306 (1993).
51. H. Thomas, *Planta*, *154*:212 (1982).
52. H. Thomas, *Planta*, *154*:219 (1982).
53. T. G. E. Davies, H. Thomas, B. J. Thomas, and L. J. Rogers, *Plant Physiol.*, *93*:588 (1990).
54. H. Thomas, *Theor. Appl. Genet.*, *73*:551 (1987).
55. J. J. Guiamet, E. Schwartz, E. Pichersky, and L. D. Noodén, *Plant Physiol.*, *96*:227 (1991).
56. E. M. Meyerowitz, *Annu. Rev. Genet.*, *21*:93 (1987).
57. J. Chory, P. Nagpal, and C. A. Peto, *Plant Cell*, *3*:445 (1991).
58. A. B. Bleecker, M. A. Estelle, C. Somerville, and H. Kende, *Science*, *241*:1086 (1988).
59. P. Guzman and J. R. Ecker, *Plant Cell*, *2*:513 (1990).
60. J. J. Kieber, M. Rothenberg, G. Roman, K. A. Feldmann, and J. R. Ecker, *Cell*, *72*:427 (1993).
61. J. T. Greenberg and F. M. Ausubel, *Plant J.*, *4*:327 (1993).
62. P. J. J. Hooykaas and R. A. Schilperoort, *Plant Mol. Biol.*, *19*:15 (1992).
63. P. C. Zambryski, *Annu. Rev. Plant Physiol. Plant Mol. Biol.*, *43*:465 (1992).
64. A. M. Bhatt and C. Dean, *Curr. Opin. Biotechnol.*, *3*:152 (1992).
65. A. Gierl and H. Saedler, *Plant Mol. Biol.*, *19*:39 (1992).
66. O. Melton, ed., *Antisense RNA and DNA*, Cold Spring Harbor Laboratory, Cold Spring Harbor, N.Y., p. 149 (1988).
67. J. N. M. Mol, A. R. van der Krol, A. J. van Tunen, R. van Blokland, P. de Lange, and A. R. Stuitje, *FEBS Lett.*, *268*:427 (1990).
68. J. M. Kooter and J. N. M. Mol, *Curr. Opin. Biotechnol.*, *4*:166 (1993).
69. A. Spena, J. J. Estruch, and J. Schell, *Curr. Opin. Biotechnol.*, *3*:159 (1992).
70. C. M. Smart, S. R. Scofield, M. W. Bevan, and T. A. Dyer, *Plant Cell*, *3*:647 (1991).
71. H. J. Klee, M. B. Hayford, K. A. Kretzmer, G. F. Barry, and G. M. Kishore, *Plant Cell*, *3*:1187 (1991).
72. C. P. Romano, M. L. Cooper, and H. J. Klee, *Plant Cell*, *5*:181 (1993).
73. S. Gepstein, in *Senescence and Aging in Plants* (L. D. Noodén and A. C. Leopold, eds.), Academic Press, London, p. 85 (1988).
74. Y. Vercher and J. Carbonell, *Physiol. Plant.*, *81*:518 (1991).
75. M. Kura-Hotta, H. Hashimoto, K. Satoh, and S. Katoh, *Plant Cell Physiol.*, *31*:33 (1990).
76. K. A. Platt-Aloia and W. W. Thomson, *Planta*, *163*:360 (1985).
77. Y. Vercher, A. Molowny, and J. Carbonell, *Physiol. Plant.*, *71*:302 (1987).
78. K. V. Thimann and S. O. Satler, *Proc. Natl. Acad. Sci. USA*, *76*:2295 (1979).
79. V. A. Wittenbach, *Plant Physiol.*, *73*:121 (1983).
80. I. R. Cowan and G. D. Farquhar in *Integration of Activity in the Higher Plant* (D. H. Jenning, ed.), Cambridge University Press, London, p. 471 (1977).
81. E. Zeiger and A. Schwartz, *Science*, *218*:680 (1982).

82. T. Solomos, in *Senescence and Aging in Plants* (L. D. Noodén and A. C. Leopold, eds.), Academic Press, London, p. 112 (1988).
83. A. C. Liden and H. E. Akerlund, *Physiol. Plant.*, *87*:134 (1993).
84. G. G. Laties, *Annu. Rev. Plant Physiol.*, *33*:519 (1982).
85. C. Hiser and L. McIntosh, *Plant Physiol.*, *93*:312 (1990).
86. F. Gallardo, F. R. Canton, A. Garciagutierrez, and F. M. Canovas, *Plant Physiol. Biochem.*, *31*:189 (1993).
87. E. J. Woltering, D. Somhorst, and C. A. Debeer, *J. Plant Physiol.*, *141*:329 (1993).
88. R. L. Bieleski and M. S. Reid, *Plant Physiol.*, *98*:1042 (1992).
89. P. Siffel, J. Kutik, and N. N. Lebedev, *Photosynthetica*, *25*:395 (1991).
90. J. Hidema, A. Makino, Y. Kurita, T. Mae, and K. Ojima, *Plant Cell Physiol.*, *33*:1209 (1992).
91. A. J. Young, R. Wellings, and G. Britton, *J. Plant Physiol.*, *137*:701 (1990).
92. D. Amir-Shapira, E. E. Goldschmidt, and A. Altman, *Proc. Natl. Acad. Sci. USA*, *84*:1901 (1987).
93. B. Luthy, E. Martinoia, P. Matile, and H. Thomas, *Z. Pflanzenphysiol.*, *113*:423 (1984).
94. N. Yamauchi and A. E. Watada, *J. Am. Soc. Hortic. Sci.*, *116*:58 (1991).
95. V. J. Dwarki, V. N. K. Francis, G. J. Bhat, and G. Padmanaban, *J. Biol. Chem.*, *262*:16958 (1987).
96. D. Y. Chang, J. P. Miksche, and S. S. Dhillon, *Physiol. Plant.*, *64*:409 (1985).
97. J. B. Harris, V. G. Schaefer, S. S. Dhillon, and J. P. Miksche, *Plant Cell Physiol.*, *23*:1267 (1982).
98. C. J. Brady, in *Senescence and Aging in Plants* (L. D. Noodén and A. C. Leopold, eds.), Academic Press, London, p. 147 (1988).
99. F. B. Abeles and L. J. Dunn, *Plant Sci.*, *72*:13 (1990).
100. R. Tomas, A. Vera, M. Martin, and B. Sabater, *Plant Sci.*, *85*:71 (1992).
101. H. Pfitzinger, L. Marechal-Drouard, D. T. N. Pillay, J. H. Weil, and P. Guillemaut, *Plant Mol. Biol.*, *14*:969 (1990).
102. C. Jayabaskaran, M. Kuntz, P. Guillemaut, and J. H. Weil, *Plant Physiol.*, *92*:136 (1990).
103. A. Blank and T. A. McKeon, *Plant Physiol.*, *97*:1409 (1991).
104. P. Millard and C. M. Thomson, *J. Exp. Bot.*, *40*:1285 (1989).
105. S. J. Crafts-Brandner, *Plant Physiol.*, *98*:1128 (1992).
106. J. R. Evans and J. R. Seemann, in *Photosynthesis*, (W. S. Briggs, ed.), Alan R. Liss, New York, p. 183 (1989).
107. L. W. Peterson and R. C. Huffaker, *Plant Physiol.*, *55*:1009 (1975).
108. V. A. Wittenbach, *Plant Physiol.*, *62*:604 (1978).
109. R. Shurtz-Swirski and S. Gepstein, *Plant Physiol.*, *78*:121 (1985).
110. S. J. Crafts-Brandner, M. E. Salvucci, and D. E. Egli, *Photosynth. Res.*, *23*:223 (1990).
111. J. W. Friedrich and R. C. Huffaker, *Plant Physiol.*, *65*:1103 (1980).
112. W. P. Quick, K. Fichtner, R. Wendler, E.-D. Schulze, S. R. Rodermel, L. Bogorad, and M. Stitt, *Planta*, *188*:522 (1992).
113. R. C. Huffaker and B. L. Miller, in *Photosynthetic Carbon Assimilation* (H. W. Siegelman, ed.), Plenum Press, New York, p. 139 (1978).
114. H. W. Woolhouse, in *Molecular Biology of Plant Development* (H. Smith and D. Grierson, eds.), Blackwell, Oxford, p. 256 (1982).
115. M. J. Droillard, N. J. Bate, S. J. Rothstein, and J. E. Thompson, *Plant Physiol.*, *99*:589 (1992).
116. R. C. Huffaker, *New Phytol.*, *116*:199 (1990).
117. U. Feller, in *Plant Proteolytic Enzymes*, (M. J. Dalling, ed.), CRC Press, Boca Raton, Fla., Vol. 2, p. 49 (1986).
118. M. Cercós and J. Carbonell, *Physiol. Plant.*, *88*:275 (1993).
119. C.-Z. Jiang, S. R. Rodermel, and R. M. Shibles, *Plant Physiol.*, *101*:105 (1993).
120. L. W. Peterson, G. E. Kleinkopf, and R. C. Huffaker, *Plant Physiol.*, *51*:1042 (1973).
121. T. Mae, A. Makino, and K. Ohira, *Plant Cell Physiol.*, *24*:1079 (1983).
122. L. Peñarrubia and J. Moreno, *Arch. Biochem. Biophys.*, *281*:319 (1990).
123. C. García-Ferris and J. Moreno, *Photosynth. Res.*, *35*:55 (1993).
124. N. P. A. Huner, J. V. Carter, and F. Wold, *Z. Pflanzenphysiol.*, *106*:69 (1982).

125. R. B. Ferreira and D. D. Davies, *Planta*, *179*:448 (1989).

126. R. A. Mehta, T. W. Fawcett, D. Porath, and A. K. Mattoo, *J. Biol. Chem.*, *267*:2810 (1992).

127. C. García-Ferris and J. Moreno, *Planta 193*:208 (1994).

128. T. Kubawara and Y. Hashimoto, *Plant Cell Physiol.*, *31*:581 (1990).

129. R. Landolt and P. Matile, *Plant Sci.*, *72*:159 (1990).

130. L. De Bellis and M. Nishimura. *Plant Cell Physiol.*, *32*:555 (1991).

131. K. Kamachi, T. Yamaya, T. Hayakawa, T. Mae, and K. Ojima, *Plant Physiol.*, *98*:1323 (1992).

132. K. Kamachi, T. Yamaya, T. Mae, and K. Ojima, *Plant Physiol.*, *96*:411 (1991).

133. I. Szamosi, D. L. Shaner, and B. K. Singh, *Plant Physiol.*, *101*:999 (1993).

134. S. Yoshida, M. Uemura, T. Niki, A. Sakai, and L. V. Gusta, *Plant Physiol.*, *72*:105 (1983).

135. G. Paliyath and M. J. Droillard, *Plant Physiol. Biochem.*, *30*:789 (1992).

136. M. M. Olsson and C. Liljenberg, *Phytochemistry*, *29*:765 (1990).

137. G. Paliyath and J. E. Thompson, *New Phytol.*, *114*:555 (1990).

138. K. Lohner, *Chem. Phys. Lipids*, *57*:341 (1991).

139. C. L. Duxbury, R. L. Legge, G. Paliyath, R. F. Barber, and J. E. Thompson, *Phytochemistry*, *30*:63 (1991).

140. A. Carruthers and D. L. Melchior, *Trends Biol. Sci.*, *11*:331 (1986).

141. E. J. Pell and K. L. Steffen, *Active Oxygen/Oxidative Stress and Plant Metabolism*, Current Topics in Plant Physiology, An American Society of Plant Physiologists Series, Vol. 6, ASPP, Rockville, Md. (1991).

142. K. P. Pauls and J. E. Thompson, *Plant Physiol.*, *75*:1152 (1984).

143. M. A. Rouet-Mayer, J. M. Bureau, and C. Laurière, *Plant Physiol.*, *98*:971 (1992).

144. F. Cheour, J. Arul, J. Makhlouf, and C. Willemot, *Plant Physiol.*, *100*:1656 (1992).

145. D. J. McCormac, J. F. Todd, G. Paliyath, and J. E. Thompson, *Plant Physiol. Biochem.*, *31*:1 (1993).

146. B. Halliwell and J. M. C. Gutteridge, *Free Radicals in Biology and Medicine*, 2nd ed., Oxford University Press, New York (1989).

147. A. Kacperska and M. Kubacka-Zebalska, *Physiol. Plant.*, *77*:231 (1989).

148. E. F. Elstner, *Annu. Rev. Plant Physiol.*, *33*:73 (1982).

149. G. I. Jenkins and H. W. Woolhouse, *J. Exp. Bot.*, *32*:989 (1981).

150. H. Greppin, C. Penel, and T. Gaspar, eds., *Molecular and Physiological Aspects of Plant Peroxidases*, University of Geneva, Geneva, p. 468 (1986).

151. I. K. Smith, T. L. Vierheller, and C. A. Thorne, *Physiol. Plant.*, *77*:449 (1989).

152. C. Bowler, M. Van Montagu, and D. Inzé, *Annu. Rev. Plant Physiol. Plant Mol. Biol.*, *43*:83 (1992).

153. L. Guan and J. G. Scandalios, *Plant J.*, *3*:527 (1993).

154. M. Kar, P. Streb, B. Hertwig, and J. Feierabend, *J. Plant Physiol.*, *141*:538 (1993).

155. F. R. Tadeo and E. Primo-Millo, *Plant Sci.*, *68*:47 (1990).

156. J. M. Palma, M. Garrido, M. I. Rodriguez-García, and L. A. Del Rio, *Arch. Biochem. Biophys.*, *287*:68 (1991).

157. L. M. Sandalio, V. H. Fernandez, F. L. Ruperez, and L. A. Del Rio, *Plant Physiol.*, *87*:1 (1988).

158. J. P. Knox, *J. Cell Sci.*, *96*:557 (1990).

159. A. W. Robards and W. J. Lucas, *Annu. Rev. Plant Physiol. Plant Mol. Biol.*, *41*:369 (1990).

160. J.-K. Zhu, J. Shi, U. Singh, S. E. Wyatt, R. A. Bressan, P. M. Hasegawa, and N. C. Carpita, *Plant J.*, *3*:637 (1993).

161. C. A. Ryan, *Annu. Rev. Phytopathol.*, *28*:425 (1990).

162. D. J. Bowles, *Annu. Rev. Biochem.*, *59*:873 (1990).

163. S. C. Fry, *The Growing Plant Cell Wall: Chemical and Metabolic Analysis*, Wiley, New York (1988).

164. N. C. Carpita and D. H. Gibeaut, *Plant J.*, *3*:1 (1993).

165. M. Kramer, R. A. Sanders, R. E. Sheehy, M. Melis, M. Kuehn, and W. R. Hiatt, in *Horticultural Biotechnology* (A. B. Bennett and S. D. O'Neill, eds.), Alan R. Liss, New York, p. 347 (1990).

166. N. S. A. Malik, *Physiol. Plant.*, *70*:438 (1987).

167. J. E. Lincoln, S. Cordes, E. Read, and R. L. Fischer, *Proc. Natl. Acad. Sci. USA*, *84*:2793 (1987).

168. K. A. Lawton, K. G., Raghothama, P. B. Golsbrough, and W. R. Woodson, *Plant Physiol.*, *93*:1370 (1990).
169. A. Granell, N. Harris, A. G. Pisabarro, and J. Carbonell, *Plant J.*, *2*:907 (1992).
170. L. M. Margossian, A. D. Federman, J. J. Giovannoni, and R. L. Fischer, *Proc. Natl. Acad. Sci. USA*, *85*:8012 (1988).
171. I. A. Graham, C. J. Leaver, and S. M. Smith, *Plant Cell*, *4*:349 (1992).
172. C. R. Bird, J. A. Ray, J. D. Fletcher, J. M. Boniwell, A. S. Bird, C. Teuliers, I. Blain, P. M. Bramley, and W. Schuch, *Bio/Technology*, *9*:635 (1991).
173. J. R. Ecker and R. W. Davis, *Proc. Natl. Acad. Sci. USA*, *84*:5202 (1987).
174. K. G. Raghothama, K. A. Lawton, P. B. Goldsbrough, and W. R. Woodson, *Plant Mol. Biol.*, *17*:61 (1991).
175. G. Storz, L. A. Tartaglia, and B. N. Ames, *Science*, *248*:189 (1990).
176. B. Demple, *Annu. Rev. Genet.*, *25*:315 (1991).
177. R. Schreck, P. Rieber, and P. A. Baeuerle, *EMBO J.*, *10*:2247 (1991).
178. A. Vera, R. Thomas, M. Martin, and B. Sabater, *Plant Sci.*, *72*:63 (1990).
179. T. Akazawa, in *Encyclopedia of Plant Physiology* (M. Gibbs and E. Latzdo, eds.), Springer-Verlag, Berlin, Vol. 6, p. 410 (1979).
180. Y. Okatan, G. M. Kahanak, and L. D. Noodén, *Physiol. Plant.*, *52*:330 (1981).
181. C. Peisker, H. Thomas, F. Keller, and P. Matile, *J. Plant Physiol.*, *136*:544 (1990).
182. B. D. Derman, D. C. Rupp, and L. D. Noodén, *Am. J. Bot.*, *65*:205 (1978).
183. H. W. J. Van den Broek, L. D. Noodén, J. S. Sevall, and J. Bonner, *Biochemistry*, *12*:229 (1973).
184. G. M. Pastori and V. S. Trippi, *Physiol. Plant.*, *87*:227 (1993).

22

Physiological Mechanisms of Plant Senescence

Larry J. Grabau
University of Kentucky, Lexington, Kentucky

I. INTRODUCTION

A. Working Definition of Senescence

Senescence processes were reviewed thoroughly in 1988 by numerous authors under the editorship of Noodén and Leopold [1]. Prior to that time, major reviews were done under Thimann's [2] direction in 1980; by Thomas and Stoddart [3], also in 1980; and by Noodén and Leopold [4] in 1978. According to Thimann [5], primary indicators of senescence include chlorophyll loss, proteolysis, changes in nucleic acid levels, altered respiration, and anatomically visible changes (at the organelle level). While much of Noodén's own early work (citations below) focused on chlorophyll degradation (owing to the ease with which that characteristic can be assessed visually at the organ level), it is clear that in at least some species, chlorophyll degradation may be delayed without a substantial effect on the other processes described by Thimann [2]. Thus our working definition of senescence must not be limited to the obvious (chlorophyll degradation) and is therefore stated as follows: "the collective processes resulting in metabolic and anatomic disassembly of photosynthetic plant organs, with the ultimate purpose of providing nutrients needed to support the growth of seeds for the ensuing generation." One of the most important potential agricultural applications of improved understanding of monocarpic leaf senescence is the following: If we can break the senescence code, we may be able to force leaves to remain green and photosynthetically functional longer and thereby increase crop yield. Numerous researchers have tested this stay-green character in various crops; their work is summarized below. However, we must recognize that such an effort is more complicated than modifying the action of a single gene; in fact, it appears likely that several genes control different parts of the senescence program (see below).

B. A Beneficial, Active, Highly Coordinated Process

Sinclair and de Wit [6] in 1975 called senescence a "self-destructive" process in which plant leaves sacrificed their carbon and nitrogen for the benefit of developing seeds. Although it is not possible to deny that the depletion of these nutrients from leaves is a central feature of senescence, it is unfortunate that the term they used to describe this process carried the connotation of disorderly destruction. Indeed, the 1959 article of Leopold et al. [7] had already marked a new appreciation for the complex process of senescence. By 1961, Leopold [8] had described senescence as an orderly, constructive process. Despite Sinclair and de Wit's [6] widely acclaimed work, it is now probably more appropriate to term this process *autodisassembly* or perhaps, *coordinated materials recovery*. Noodén and Leopold [1] called the process an "orderly retreat, at least during the early stages of senescence." In fact, it could be referred to as a remarkably generous system, in which the maternal parent lays down its life in an effort to enhance the chance of reproductive success by her offspring. Not only must those offspring grow, they must reproduce successfully to ensure continued species survival. In the case of senescing monocarpic plants, senescence is much more orderly and purposeful than Sinclair and de Wit's [6] term connotes.

C. Chapter Objective

It is the objective of this chapter, through a review of the available literature, to develop a "snapshot" view of our current collective understanding of the physiological basis of plant senescence. It should be noted that in this chapter we avoid the subject of abscission (treated extensively in Chapter 23 of this volume); in addition, its focus will be primarily on annuals, since the senescence processes of perennial plants could be under quite different controls. We also do not attempt to deal with senescence induced by pathogens, water deficits, or nutrient deficiencies. In short, the focus here is on "normal" plant senescence. It is worth noting that "progressive" plant senescence (as described by Secor et al. [9]) *will* be considered, as will the monocarpic senescence patterns studied so extensively by Noodén and his co-workers (see citations below). Ness and Woolhouse [10] were interested in whether both of these senescence processes could be triggered by a similar signaling agent. If a realistic conceptual model could eventually be developed that would account for both progressive and monocarpic senescence, we would have indeed begun to understand these remarkable processes more fully.

II. PHYSIOLOGIC PROCESS: WHAT HAPPENS WHEN?

A. Early Beginnings?

Some would argue that monocarpic senescence of annuals starts as soon as they are triggered to flower. For example, Burke et al. [11] found that male-sterile soybean [*Glycine max* (L.) Merrill] plants senesced at the same time as their male-fertile sibs even though their floral buds were removed regularly. Burke et al. [11] concluded that the senescence signal would have to have been sent prior to flowering. If their concept holds, phytochrome would potentially be implicated as a very early trigger in the senescence process. Could senescence, in fact, be preprogrammed to occur, with the initial "buttons" pushed at the time of initiation of a plant's reproductive program? Thus our imposed environmental conditions and surgical treatments may merely serve to forestall the inevitable: total plant senescence. A recent article by van Doorn and van Lieburg [12] found that leaf yellowing of *Alstroemeria pelegrina* L. was under phytochrome control, mediated by gibberellin synthesis. Steinitz et al. [13] found phytochrome to be involved in statice (*Limonium sinuatum* L.) stalk senescence. Krizek et al. [14] found that cocklebur (*Xanthium pensylvanicum* Wallr.) leaf senescence was not affected by either flower or fruit removal, as long as the appropriate photoperiod was imposed. This result could also indicate phytochrome involvement.

Extending such thinking to another level, one could perhaps theorize that the original senescence trigger came at the time the seed was induced to germinate by the appropriate combination of warmth, moisture, and perhaps, light. Even a step farther back, it could be argued that the seed was triggered to begin senescence eventually at the point of its initial fertilization, when it received the genetic code for a later senescence process. Such an argument starts to take on the shape of the ancient Russian figure-within-a-figure toys, and is, of course, not entirely satisfying.

Although one can suggest that preflowering light conditions, as monitored by phytochrome, could be responsible for the eventual operation of the package of events we describe as senescence, it is clear that the early post-anthesis activity of many plants does *not* coincide with our working definition of senescence. For example, ribulose biphosphate carboxylase (Rubisco) synthesis often proceeds rapidly, and photosynthetic rates generally do not start to decline, until seedfill begins (see below). Thus if preflowering light quality is involved, it would have to be in an indirect fashion, and almost certainly be mediated by other post-anthesis signals.

B. Comparative Timing of Key Senescence Indicators

Waters et al. [15] concluded that wheat (*Triticum aestivum* L.) senescence is a highly coordinated process. We will examine this coordination at two different levels: anatomic and physiologic.

1. Anatomic Sequence

Chloroplasts were the first organelles to undergo visible degradation in cucumbers (*Cucumis sativus* L.) [16]. Atkin and Sahai Srivastava [17] noted that barley (*Hordeum vulgare* L.) leaf senescence started with chloroplast decay. In brussel sprouts [*Brassica oleracea* (L.) *gemmifera*], chloroplasts were the first organelle to senesce [18]. What leaves had differential organelle degeneration rates [19]. In birch (*Betula verrucosa* Ehrb.), the first anatomical sign of senescence was chloroplast breakdown [20]. Taken together, the studies above firmly establish that chloroplasts are preferentially targeted for senescence. Choe and Thimann [21] have developed a comprehensive view of the sequence of chloroplast and chlorophyll degradation in oats (*Avena sativa* L.). This scheme involves breakdown of chloroplast membranes, attack of the chlorophyll–protein complex by protease(s), and enzymatic degradation of the chlorophyll molecule proper.

The senescence process is apparently initiated outside the chloroplast [21,22]. Working with the aquatic plant *Anacharis canadensis* Planch., Makovetzki and Goldschmidt [23] concluded that the synthesis of a cytoplasmic protein was essential for chloroplast senescence to occur. Kasamo [24] suggested that end-stage senescence of tobacco (*Nicotiana* spp.) was accelerated by mixing the contents of epidermal and mesophyll cells. Clearly, though, Kasamo's [24] proposed process does not appear to account for the initiation of senesence, since most senescence parameters decline before tonoplast degeneration can be observed, and plasmalemma degradation occurs later still.

In summary, chloroplasts are the primary anatomical site of visible organelle disassembly, and further, it appears that this disassembly is under nuclear control.

2. Physiologic Sequence

On a physiologic level, chlorophyll degradation has received much of our collective research attention. For example, Park et al. [25], following chlorophyll catabolites of 12 widely differing species, found that the phytol moiety persists, while N is released. Their data could be interpreted as evidence that the senescence package is designed to scavenge N for developing seeds or for export to establish better positioned leaves. In sunflower (*Helianthus annuus* L.), Purohit [26] and Purohit and Chandra [27] have concluded that chlorophyllase synthesis is the first key step in leaf senescence.

However, most published research seems to make it clear that chlorophyll breakdown is not the first step of senescence, and other senescence processes may continue even if chlorophyll breakdown is somehow delayed. Thomas [28] and Thomas and Stoddart [3,29] found that protein loss of mutant meadow fescue (*Festuca pratensis* Huds.) genotypes was no different than protein loss of normal fescue lines, even though chlorophyll degradation rates of the mutants were much lower. Mondal et al. [30] found depodded soybean plants to show all aspects of senescence, except chlorophyll degradation.

Since oat leaves export amino acids during senescence [31], and since photosynthetic activity is declining, protein degradation must indeed be occurring. Jeschke and Pate [32] observed that both nitrogen movement and photosynthetic decline of castor bean (*Ricinus communis* L.) leaves preceded chlorophyll loss. Thimann and Satler [33] have shown proteolysis of oat leaf proteins to precede chlorophyll breakdown. It has been previously noted that synthesis of Rubisco small and large subunits is closely coordinated in wheat [34]; Speirs and Brady [35] likewise observed a coordinated depression of Rubisco subunit synthesis after full wheat leaf expansion. Remarkably enough, small and large subunit *degradation* of barley is also closely coordinated [36], even though proteolytic activity appears to begin outside chloroplasts. Further, Peterson and Huffaker [36] concluded that Rubisco was being preferentially dismantled compared to other chloroplastic proteins during leaf senescence. With respect to another enzyme, Thomas [37] found that chloroplastic alanine amino transferase (AAT) in *Lolium temulentum* L. was preferentially inactivated in comparison to cytosolic AAT. This finding points to a very specific proteolytic activity that precedes chlorophyll/chloroplast senescence.

Decapitated sunflower went through senescence, although at a slower than normal rate [26]. Defruiting delayed senescence of subtending leaves in rice (*Oryza sativa* L.) [38]. In contrast to the results above, Woodward and Rawson [39] found that depodding soybean did not affect either net photosynthesis or transpiration rates. Apparently, results of defruiting studies should be generalized cautiously.

Secor et al. [9] found that soybean leaves experiencing either monocarpic or progressive senescence showed similar sequences of metabolic changes, concluding that the two processes were under similar control programs. It is noteworthy that monocarpic senescence involves the simultaneous decline of several leaves even though those leaves are not of identical ages [40]. Thus while advancing leaf age could presumably be partially responsible for progressive senescence, it clearly cannot explain monocarpic senescence. Regarding this sequence of events, Crafts-Brandner [41] felt that further study of plant senescence should focus on the degradation of the photosynthetic apparatus itself rather than on nutrient depletion from source leaves. In conclusion, the current research data provide substantial support for the idea that the initial steps of senescence are proteolytic ones. Chlorophyll degradation seems to be a secondary outcome (see below).

III. POTENTIAL SENESCENCE TRIGGERS

A. Nutrient Depletion

Molisch [42] was perhaps the first proponent of nutrient depletion as a cause of senescence. In short, leaves senesce as seeds develop on annual plants because the seeds demand nutrients for their own growth, taking priority over leaf maintenance. Thus, either leaves lose nutrients directly by depletion to the seeds, or do so indirectly by failing to compete with seeds for nutrients coming up from the plant's root system [40,43,44]. Noodén [40,43,44] has effectively dismissed this idea, using a variety of experimental approaches. For example, when petioles of soybean explants were steam-treated to kill the phloem (but not the xylem), thus effectively blocking nutrient depletion from the leaves, the leaves still lost their green color [45,46]. Recently, Crafts-Brandner [41] concluded that in soybean, leaf phosphorus remobilization was not an important part of leaf senescence. While earlier work by Grabau et al. [47] with hydroponically grown soybean had shown that inadequate seedfill-stage P nutrition advanced leaf senescence, Crafts-Brandner [41,48] maintained that this result was related to the artificial methods employed: namely, plants were grown in media containing adequate amounts of P, then moved into solution cultures containing inadequate P levels. Thus, plant metabolism had developed based on adequate P supplies, and when P nutrition was sharply reduced, leaf P remobilization became an important factor.

It is perhaps inappropriate to castigate Molisch [42] for reaching a flawed conclusion in 1938, since his hypothesis did indeed fit the data that he was able to generate. It may, in fact, be more appropriate to call our collective attention to recognizing that he saw senescence as a common process for many diverse plants. In addition, nutrient loss is, at the minimum, a symptom of leaf senescence, even though convincing, broad-based evidence that nutrient loss is the key factor initiating senescence has failed to accumulate. While Seth and Wareing [49] concluded that senescence was not merely due to nutrient depletion, Derman et al. [50] found N to be the primary leaf nutrient redistributed out of senescing soybean leaves. Neumann and Noodén [51] ascribed to soybean leaves control of the export of inorganic P. While leaf nutrient loss is clearly an important outcome of senescence, Okatan et al. [52] concluded that it is not the triggering event. For example, in their study, chlorophyll and protein degradation of soybean were already quite far along before visible leaf yellowing could be observed.

It is important to note here that Choudhuri and co-workers [53–56] are convinced that rice senescence is caused by leaf nutrient depletion. He and his co-workers ruled out ABA, cytokinins, and a seed-sent leaf "death signal." In the work of Mondal and Choudhuri [54,55], modest differences in senescence rates of rice leaves could be accounted for by differences in leaf P remobilization rates. While nutrient depletion (and/or nutrient diversion) still retains some supporters, it appears to be more appropriate to describe nutrient depletion from leaves of most plants as an outcome of senescence rather than as any sort of causal trigger.

B. Plant Growth Regulators

The most extensive and most convincing data regarding a role for plant growth regulators (PGRs) in senescence is for cytokinins [4]. Shaw et al. [57] found kinetin applications delayed leaf senescence of wheat. Similarly, applied benzyladenine (BA) delayed leaf senescence of beans (*Phaseolus vulgaris* L.) [58]. Of course, results from studies following endogenous cytokinin levels may be more enlightening. Ambler et al. [59] found that nonsenescent sorghum [*Sorghum bicolor* (L.) Moench] types had higher endogenous cytokinin levels than did senescent types. Under partial leaf treatment with surface-applied streptomycin, cytokinin levels were inversely related to senescence of different portions of the same (*Streptocarpus molweniensis* subsp. *molweniensis* Hilliard) leaf [60]. When Noodén et al. [61] followed endogenous cytokinin levels of soybean, they found a correlation with senescence progress but could not establish an initiating role for cytokinin.

Some think that cytokinins may inhibit general protease synthesis, thus protecting both Rubisco and the chlorophyll–protein complex from proteolysis. Martin and Thimann [62] concluded that synthesis of a proteolytic enzyme was the first step of oat senescence; in addition, they felt that applied cytokinins may have delayed senescence by suppressing synthesis of said enzyme. In her study of cocklebur, Osborne [63] found kinetin to increase synthesis of both RNA and protein. Perhaps a decline in import of root-produced cytokinins, owing to reduced root activity, releases a latent senescence program in plant leaves [64]. Hsia and Kao [65] concluded that soybean roots delayed full expression of a general senescence package by supplying leaves with cytokinins. Kinetin has also been assigned responsibility for tightening the linkage between oat respiration and photophosphorylation [31]. Similarly, Kao [66] has shown that while cytokinins modulate rice leaf senescence activity, they do not appear to be responsible for triggering its onset. Lindoo and Noodén [67] showed that applications of exogenous cytokinins delayed, but did not stop, soybean senescence. They interpreted this result to mean that cytokinin is involved in the process of senescence, but cannot be implicated as the controlling agent in its onset.

Evidence for a senescence role of other PGRs is less extensive. A role for abscisic acid (ABA) may be more plausible than a role for ethylene, giberellic acid (GA), or indoleacetic acid (IAA); however, with both radishes (*Raphanus sativus* L.) and beans, a consistent relationship between leaf senescence and ABA levels could not be established [68,69]. Radley [70] saw an ABA role in the enhancement of late-stage senescence of wheat and proposed that ABA may act to stimulate greater expression of genes coding for chlorophyll-degrading enzymes. Some feel that ABA may play a role in senescence as induced by declining stomatal conductance; see below for details and citations. Roberts and Osborne [71] felt that ethylene biosynthesis is regulated by the release of a membrane-bound cofactor and that ethylene could be involved in mediating senescence. Beevers [72] and Beevers and Guernsey [73] have found GA to delay leaf senescence; indeed, it has been possible to use broadleaved dock (*Rumex obtusifolius* L.) to assay solution GA levels by monitoring its leaf senescence [74]. A potential IAA role in senescence has been described as an impact on nutrient transport through its elevated concentration in sinklike areas [75].

In conclusion, although many noncited studies have also been done on the role of PGRs as senescence triggers, no convincing support for such a role for ABA, GA, ethylene, or IAA has accumulated. Based on currently available results with cytokinin studies, it appears that cytokinin(s) may play a secondary role in leaf senescence. Perhaps this PGR functions to inhibit synthesis of proteolytic enzymes. Strong support for cytokinin levels as the key senescence trigger has yet to be assembled.

C. Stomatal Control

A key paper proposing a role for stomatal control of the initiation of leaf senescence was that of Gepstein and Thimann [76]. They postulated that stomatal closure of oats caused ABA buildup and thereby triggered the rest of the senescence process. Corroborating evidence has been provided by Colquhoun and Hillman [68,69], who suggested that ABA might play a role in stomata-mediated control of senescence. Thimann and Satler [33,77] postulated that stomatal closure could, in fact, be the *primary* trigger responsible for initiating oat senescence. Gwathmey and coworkers [78–80] felt that the genetically delayed leaf senescence of some cowpea [*Vigna unguiculata* (L.) Walp.] cultivars might explain their enhanced drought tolerance; this enhanced drought tolerance could conceivably have been related to improved stomatal control. Hall and coworkers [81–83] found deflorated pepper (*Capsicum annuum* L.) plants to senesce *before* fruited plants. Hall and Brady [82] concluded that declining leaf conductance provided a better explanation for senescence than did the possibilities of leaf nutrient depletion or a fruit-originated death signal. In the case of Allison and Weinmann [84] with corn (*Zea mays* L.), ear removal was found to enhance leaf senescence. Although this result fits poorly with either a seed-originated senescence signal or leaf nutrient depletion, perhaps the elevated leaf starch and sugar levels they detected in earless corn plants might have resulted in reduced stomatal conductance by a sort of feedback inhibition.

In contrast, Adedipe et al. [85] found BA to maintain both clorophyll levels and photosynthetic rates of bean without affecting stomatal resistance. Haber et al. [86] found light to delay wheat leaf chlorophyll senescence. If, indeed, stomatal closure triggers senescence, the lack of light in lower crop canopies could account for the progressive senescence of lower, shaded soybean leaves in studies like that of Secor et al. [9]. Low light levels indeed appeared to be the key trigger for senescence of lower canopy cucumber leaves [87]. Similarly, Goldthwaite and Laetsch [88] found that light retarded senescence of bean leaves. As bean leaves aged, their stomatal resistance increased [89]. More important, that increase in stomatal resistance *preceded* chlorophyll disappearance [89]. This set of evidence could be appropriately described as "circumstantial," since a clear cause-and-effect relationship between declining stomatal conductance and the onset of leaf senescence has not been established.

A similar physiologic model may explain the response of plants to dry soil or soils that present roots with mechanical impedance [90]. Numerous reports have identified ABA as controlling stomatal conductance under dry soil conditions [91–94]. Neales et al. [93] found that the decline in stomatal conductance preceded any other change in plant water relations. Similarly, Robertson et al. [95] showed that initial drought effects could not be explained by a turgor loss or a lack of available photosynthate. Thus the evidence for declining stomatal conductance as plants' first response to dry or mechanically impeding soil conditions is reasonably good, showing that stomates may serve as critical "front-line" controls of physiologic responses to environmental conditions.

Although some aspects of the stomatal control of leaf senescence are indeed attractive (particularly that this concept could explain both progressive and monocarpic senescence), some problems remain. For example, light-starved lower canopy leaves would seem to fit this model well, since increased stomatal resistance would be the logical result of reduced light availability. However, upper canopy leaves often senesce at maturity, obviously without a lack of available sunlight. Hence if reduced stomatal conductance is triggering a monocarpic leaf senescence program, stomates must be responding to a signal different from the low-light signal they appear to receive in progressive senescence of lower canopy leaves. Interestingly, Malik and Berrie [96] concluded that pea (*Pisum sativum* L.) leaves did *not* induce whole plant senescence, further minimizing the possibility that stomates trigger monocarpic senescence.

D. Genetic Control

As shown several decades ago by Yoshida [22], chloroplasts in enucleate *Elodea* cells maintained their integrity much longer than did chloroplasts in similar but nucleated cells. Thus Yoshida [22] concluded that nuclear DNA was required to drive chloroplast disassembly. In the more recent work of Thomas and Stoddart [29], meadow fescue mutants that remained green longer in the season had delayed chlorophyll breakdown, but their photosynthetic activity declined in a normal fashion. Thomas and Stoddart [29] interpreted this diversion from a "normal" leaf senescence pattern as evidence for multiple genetic controls of different parts of the overall senescence program. Bean leaf chlorophyll persisted in a mutant type studied by Ronning et al. [97]; however, other senescence parameters declined. They felt that a persistent protein was perhaps protecting chlorophyll from degradation. Working with several different soybean mutant types, Guiamét et al. [98] concluded that several different soybean genes play distinct roles in senescence. Abu-Shakra et al. [99] followed a range of senescence parameters for their delayed leaf senescence soybean line and observed a coordinated decline in all measured senescence variables. Woolhouse [100] felt that the gene products responsible for senescent activities (e.g., preferential protein breakdown) were present prior to the onset of senescence but remained quiescent until the appropriate cytosolic trigger was "pulled." Tetley and Thimann [101] determined that oat senescence required protein synthesis as a precondition to breakdown of both Rubisco and chlorophyll. Similarly, in wheat, Laurière et al. [102] measured appreciable protein synthesis during senescence, and in meadow fescue Thomas [103] found protein synthesis inhibitors to reduce the loss of both chlorophyll and Rubisco. Blank and McKeon [104,105] concluded that the three late-developing wheat leaf RNAses they detected were each responsible for developing specific, different, proteolytic activities. Clearly, broad evidence has established a genetic role in leaf senescence: synthesis of proteolytic enzymes.

The evidence for a genetic role in senescence is quite convincing. It seems that many species of plants are genetically programmed to senesce eventually. However, senescence is clearly influenced by environmental parameters: for example, by light levels [86–88]. Thus the genetic machinery appears to be itself triggered into action by other influences: for example, by declining cytokinin levels [59,60,64–67], or by increasing ABA levels [76]. Our quest for a single physiologic trigger is hampered not only by species differences (compare cocklebur [14], soybean [40], and rice [30]) but also by the highly coordinated nature of senescence itself. Perhaps a closer look at the data available for a single species (soybean) will help to clarify a likely senescence trigger (at least for that single species).

IV. SOYBEAN: MODELING SENESCENCE FOR A SINGLE SPECIES

A. A Pair of Cautions

Perhaps we can make more definitive statements as we narrow our focus from plant senescence in general to monocarpic senescence of soybean in particular. Monocarpic senescence of soybean has been studied extensively; Noodén [40] wrote a specific review on this subject area in 1984. Before reviewing this area, allow me to suggest two cautions: (a) a disproportionate amount of the available data comes from the work of Noodén and his co-workers, and (b) a substantial portion of the data available comes from soybean explant studies. Although these two points certainly do not invalidate (or perhaps even diminish) the data, a good deal of interpretive caution ought to be exercised. As noted by Thimann [5] regarding the floating leaf disk system used not only in his own laboratory but also around the world to study leaf senescence patterns, systems that consider isolated or wounded plant portions may yield information with uncertain parallels to whole plant senescence processes. With the above flags of caution prominently erected, consider the following attempt to synthesize the research results available as of late 1993.

B. Sifting Soybean Senescence Studies

1. Nutrient Depletion

Sinclair and de Wit [6] suggested that soybean leaf senescence was a result of N depletion of leaves. Derman et al. [50] followed N, K, Ca, Mg, Mn, and Fe distribution in senescing soybean leaves. They found only N to undergo significant remobilization from leaves; further, N depletion began before visible leaf yellowing. As pointed out earlier, Grabau et al. [47] observed that improved P nutrition delayed leaf senescence and increased yield by improving pod retention and increasing seed size. However, Crafts-Brandner [41] found P nutrition to have no effect on leaf N concentration, CO_2 exchange rate, Rubisco activity, or chlorophyll concentration. He concluded that neither P nutrition nor P remobilization from leaves regulated leaf senescence. In a related paper from the same set of experiments, Crafts-Brandner [48] found extensive N remobilization to occur for all P nutrition levels, and concluded that seed development is independent of net P remobilization from leaves. In a surprising paper, Neumann and Noodén [51] found that sink load size did not affect leaf senescence. For example, ^{32}P

export from leaves was largely unaffected by the number of pods allowed to remain on the plants. The reason that this paper was surprising to this author is that it appears to contradict Noodén's [43] concept of leaf senescence control by seed-sent signals.

In rice, Choudhuri and his co-workers [53–56] are convinced that mineral depletion precipitates senescence. However, in the case of soybean, support for this nutrient depletion idea is not strong. Although under some unusual conditions (e.g., Grabau et al. [47]) data have been developed that appear to show a nutrient depletion relationship, only N seems to be a possible candidate for nutrient-depletion control of monocarpic senescence. Noodén [43] and Mauk and Noodén [106] have further established the concept that leaf senescence is not caused by mineral depletion. In addition to the preponderance of data against nutrient depletion as a trigger of leaf senescence, a logical problem with this theory arises. Since soybean senescence, like that of other plants (e.g., meadow fescue) [3,28,29,37] appears to be under genetic control [11,99], how could simple N depletion (or N diversion) be responsible for initiating gene action toward senescence? To put this another way, if protein synthesis must occur for leaf senescence to begin, it would seem strange indeed that increased N exports (or declining N imports) could trigger that necessary protein synthesis. In short, proponents of the Sinclair and de Wit [6] self-destruction hypothesis must argue that N depletion somehow triggers production of proteolytic enzymes. Clearly, that argument would seem to be missing the logic to make it believable, since enzyme synthesis obviously is dependent on N availability.

Noodén's 1984 review [40] of monocarpic leaf senescence in soybean provided an excellent background on this subject. At that time he saw both hormonal levels and nutrient fluxes as involved in a nonspecific way in the control of monocarpic leaf senescence. He speculated that our ability to increase soybean yield dramatically may be impeded by an apparent inverse relationship between seed load and leaf senescence rates [40]. Thus if we can somehow convince soybean plants to produce more seeds, they may merely respond by senescing sooner, resulting in no substantial net yield increase.

2. Plant Growth Regulators

In their classic 1959 work, Leopold et al. [7] found soybean plants with pods removed to show delayed leaf senescence (as monitored by leaf color) relative to intact plants. Further, as pods were allowed to remain on the plants longer, the delayed senescence response weakened. Leopold et al. [7] reasonably concluded that reproductive structures were signaling leaves to senesce, perhaps via transport of an hormonal messenger. Lindoo and Noodén followed up on this concept with a series of papers [67,107,108], finding the onset of leaf chlorosis to coincide with maximal seed growth rates [108], early seed removal to delay leaf senescence of localized plant leaves [107], and a decline in cytokinin levels to correlate closely with chlorophyll loss [67]. However, these declining cytokinin levels could not be ascribed a controlling role in senescence onset [67]; perhaps cytokinins merely modulate the general senescence program rather than triggering it. In contrast, Mondal et al. [30] found that while leaves of desinked soybean plants retained their chlorophyll, their photosynthetic decline closely paralleled that of intact plants. Thus it appears that any prospective hormonal signal sent by developing seeds may affect only the chlorophyll degradation aspect of leaf senescence, perhaps by impeding breakdown of the protein protecting chlorophyll.

Noodén and Murray [45] and Wood et al. [46] used short-term steam treatments to kill the phloem in soybean petioles and concluded that pods cause leaf senescence (as monitored by chlorophyll levels) by sending a xylem-mobile signal to the leaves. Further, Wood et al. [46] found pods to induce this chlorophyll breakdown independent of net mineral redistribution. Noodén and Noodén [109] tested a series of morphactins for their ability to inhibit auxin transport of soybean, and implicated auxin as a potential regulator of pod development. In the studies of Noodén et al. [110], the combination of α-napthaleneacetic acid (an IAA analog) and BA delayed both chlorophyll and N loss from leaves. Unfortunately, this test confounded potential auxin and cytokinin effects. Based on other studies (see above), it appears more likely that the effects Noodén et al. [110] observed were related to cytokinin than to auxin. Noodén et al. [111] provided indirect evidence for a hormonal messenger sent by seeds, in that some of their plant treatments resulted in seed development without monocarpic senescence. Further, Noodén et al. [61] concluded that soybean pods may act to depress xylem sap cytokinin levels, thus allowing monocarpic senescence to commence. Zhang et al. [112] followed cytokinin metabolism, concluding that molecular alterations of BA could prevent its inactivation within the plant, perhaps enhancing its senescence-retarding effect.

Soybean explants, cultured in media containing various cytokinins, showed delayed chlorophyll loss compared to explants cultured in a nutrient medium devoid of cytokinins [64]. Garrison et al. [64] inferred that root cytokinin production may play an important role in leaf senescence (i.e., if root cytokinin production falls, chlorophyll degradation will soon begin). Hsia and Kao [65] found leaves separated from their root systems to show more rapid chlorophyll loss than did those of intact plants. Their application of BA at the appropriate concentration successfully replaced the senescence-delaying effect of soybean lateral root systems. In the work of Hsia and Kao [65], exogenous GA and ABA each accelerated chlorophyll loss; results with IAA were not as clear-cut.

An extensive body of literature has been developed on the potential role of PGRs in monocarpic senescence, primarily through the efforts of Noodén and his co-workers. It is perhaps presumptuous of this author to summarize Noodén's primary life work; he has already done so ably many times in the past [4,40,43,44,113]. However, it appears impossible to ascribe a causative role for any of the known PGRs in monocarpic leaf senescence of soybean. The best evidence for a modulating role for a PGR appears to be for the cytokinins, perhaps by protecting leaf chlorophyll indirectly from degradation without inhibiting the normal photosynthetic decline of soybean [30].

3. Stomatal Control

Most of the information available on potential stomatal control of senescence onset relates to oats [33,77]. Photosynthetic rates of soybean were closely correlated to stomatal conductance [39]. Thus the comparative photosynthesis and transpiration data of Woodward and Rawson [39] provided little support for stomatal control as the initiating mechanism in leaf senescence.

4. Genetic Control

Abu-Shakra et al. [99] found delayed leaf senescence gene(s) that caused genotypes not only to maintain their chlorophyll levels and Rubisco activities, but also to evidence extended nitrogen fixation activity. Burke et al. [11] found genetically male-sterile soybean plants to show a leaf senescence pattern (as monitored by chlorophyll levels and chloroplast degradation) virtually identical to that of male-fertile plants. They concluded that monocarpic senescence was initiated at or before flowering, and ascribed only a secondary level of regulatory control to any potential signal sent from developing seeds. Guiamét et al. [98] found "stay-green" mutations to exhibit preferentially delayed proteolysis of the chlorophyll a/b binding polypeptides in the light-harvesting complex of photosystem II. This excellent work provided a basis for better understanding of at least part of the leaf senescence process. For example, chlorophyll appears to be protected by a protein; it seems that this mutant line fails to produce a proteolytic enzyme which specifically attacks that chlorophyll-protecting protein. It logically follows then that the other aspects of leaf senescence are controlled by the synthesis of specific proteolytic enzymes. In the work of Okatan et al. [52], leaf protein, leaf N, and leaf starch all began to decline before chlorophyll did, implying that chlorophyll degradation is under a separate control system.

Secor et al. [9] defined senescence onset as the time when photosynthesis begins an irreversible decline and found declines in soluble protein, chlorophyll, and Rubisco activity to coincide for both progressively and monocarpically senescing soybean leaves. When Jiang et al. [114] recently compared progressively and monocarpically senescing soybean leaves, which showed somewhat different patterns of CO_2 exchange rate and Rubisco activity and content in their studies, they found that the primary regulator of photosynthetic activity was simply the Rubisco content of the leaves. Rubisco activation ratio was relatively stable. Interestingly, mRNAs for the small and large Rubisco subunits fell in a coordinated fashion, showing that declines in Rubisco content, at least early in leaf senescence, appear to be due to slower mRNA transcription [114]. Finally, Jiang et al. [114] concluded that progressive and monocarpic leaf senescence are controlled by the same mechanisms.

Obviously, soybean genes exist that provide broad-based control of leaf senescence processes [11,99]. Further, chlorophyll and Rubisco levels both appear to be controlled genetically [98,114]. It is also likely that the various aspects of soybean leaf senescence are under different specific genetic control systems, even though overall controls apparently also exist.

C. Identifying the Trigger in Soybean

Nutrient depletion cannot be supported as a *cause* of leaf senescence; on the other hand, it may very well be the designed *outcome* of monocarpic leaf senescence. Similarly, evidence for a PGR role in triggering senescence appears unconvincing, even though it seems quite likely that root-produced cytokinins at least modulate the disappearance of chlorophyll. Support for stomatal control of monocarpic senescence of soybean does not seem to exist, perhaps because relatively little seems to have been done in this area, possibly because it is difficult to see how a parameter with such dramatic diurnally and environmentally induced variation could trigger a significant whole plant response.

There is no doubt that monocarpic leaf senescence of soybean is under at least two layers of genetic control; however, it is less obvious what triggers those genes into action. Most probably, a complex interaction of physiological and environmental "reads" triggers the genes to act; perhaps both cytokinins and stomatal control play peripheral roles in sensing the conditions under which monocarpic leaf senescence is appropriate. Further efforts ought to follow the leads of Guiamét et al. [98] and Jiang et al. [114] into the molecular basis of gene action in the apparently complex proteolytic mechanisms resulting in the two key disassembly processes: Rubisco decline and chlorophyll degradation.

V. CONCLUSIONS

A. A Highly Coordinated, Therefore, Highly Valuable Process

"There is a time to be born, and a time to die." This phrase from the book of Ecclesiastes in the Bible acknowledges the cyclic nature of life. With regard to plants, it is becoming clear that while death is inevitable, the process of senescence is anything but a random breakdown of integrated control. Instead, senescence is a highly coordinated process, involving assembly of new proteins whose specific function is to preferentially dismantle previously existing proteins. Rather than seeing senescence as a negative, destructive process, perhaps it would be more useful if we treat it as an inevitable, valuable step in the life cycle of the plant species we cultivate. If that is our approach, we may be more likely to unlock the genetic codes that collectively trigger this critical plant process.

B. Gene Action Is Critical; But What Triggers Genes to Act?

This process is undoubtedly under genetic control. However, it is overly simplistic to expect that a single gene triggered by a singular metabolite (or plant growth regulator) could be responsible for initiation of the entire senescence program package. In soybean, for example, at least two layers of genetic control appear to exist. It is possible that our search for a triggering metabolite is hampered by a general failure to recognize that these plants are *designed* to senesce; thus we must simply try to ascertain how they conclude that it is the right "time to die."

REFERENCES

1. L. D. Noodén and A. C. Leopold, eds., *Senescence and Aging in Plants*, Academic Press, New York (1988).
2. K. V. Thimann, ed., *Senescence in Plants*, CRC Press, Boca Raton, Fla. (1980).
3. H. Thomas and J. L. Stoddart, *Annu. Rev. Plant Physiol.*, *31*: 83 (1980).
4. L. D. Noodén and A. C. Leopold, in *Phytochromes and the Endogenous Regulation of Senescence and Abscission* (D. S. Letham, P. B. Goodwin, and T. J. V. Higgins, eds.), Elsevier/North-Holland Biomedical Press, Amsterdam, p. 329 (1978).
5. K. V. Thimann, in *The Senescence of Leaves* (K. V. Thimann, ed.), CRC Press, Boca Raton, Fla., p. 85 (1980).
6. T. R. Sinclair and C. T. de Wit, *Science*, *189*: 565 (1975).
7. A. C. Leopold, E. Niedergang-Kamien, and J. Janick, *Plant Physiol.*, *34*: 570 (1959).
8. A. C. Leopold, *Science*, *134*: 1727 (1961).
9. J. Secor, R. Shibles, and C. R. Stewart, *Can. J. Bot.*, *62*: 806 (1984).
10. P. J. Ness and H. W. Woolhouse, *J. Exp. Bot.*, *31*: 235 (1980).
11. J. J. Burke, W. Kalt-Torres, J. R. Swafford, J. W. Burton, and R. F. Wilson, *Plant Physiol.*, *75*: 1058 (1984).
12. W. G. van Doorn and M. J. van Lieburg, *Physiol. Plant.*, *89*: 182 (1993).
13. B. Steinitz, A. Cohen, and B. Leshem, *Z. Pflanzenphysiol. Bd.*, *100*: 343 (1980).
14. D. T. Krizek, W. J. McIlrath, and B. S. Vergara, *Science*, *151*: 95 (1966).
15. S. P. Waters, M. B. Peoples, R. J. Simpson, and M. J. Dalling, *Planta*, *148*: 422 (1980).
16. R. D. Butler, *J. Exp. Bot.*, *18*: 535 (1967).
17. R. K. Atkin and B. I. Sahai Srivastava, *Physiol. Plant.*, *23*: 304 (1970).
18. D. T. Dennis, M. Stubbs, and T. P. Coultate, *Can. J. Bot.*, *45*: 1019 (1967).
19. M. Shaw and M. S. Manocha, *Can. J. Bot.*, *43*: 747 (1965).
20. J. D. Dodge, *Ann. Bot.*, *34*: 817 (1970).
21. H. T. Choe and K. V. Thimann, *Plant Physiol*, *55*: 828 (1975).
22. Y. Yoshida, *Protoplasma*, *54*: 476 (1961).
23. S. Makovetzki and E. E. Goldschmidt, *Plant Cell Physiol.*, *17*: 859 (1976).
24. K. Kasamo, *Plant Cell Physiol.*, *17*: 1297 (1976).
25. Y. Park, M. M. Morris, and G. Mackinney, *J. Agric. Food Chem.*, *21*: 279 (1973).
26. S. S. Purohit, *Photosynthetica*, *16*: 542 (1982).
27. S. S. Purohit and K. Chandra, *Photosynthetica*, *17*: 223 (1983).
28. H. Thomas, *Planta*, *137*: 53 (1977).
29. H. Thomas and J. L. Stoddart, *Plant Physiol.*, *56*: 438 (1975).
30. M. H. Mondal, W. A. Brun, and M. L. Brenner, *Plant Physiol.*, *61*: 394 (1978).
31. R. M. Tetley and K. V. Thimann, *Plant Physiol.*, *54*: 294 (1974).
32. W. D. Jeschke and J. S. Pate, *J. Exp. Bot.*, *43*: 393 (1992).
33. K. V. Thimann and S. O. Satler, *Proc. Natl. Acad. Sci. USA*, *76*: 2770 (1979).
34. C. J. Brady, *Aust. J. Plant Physiol.*, *8*: 591 (1981).

35. J. Speirs and C. J. Brady, *Aust. J. Plant Physiol.*, *8*: 603 (1981).
36. L. W. Peterson and R. C. Huffaker, *Plant Physiol.*, *55*: 1009 (1975).
37. H. Thomas, *Z. Pflanzenphysiol. Bd.*, *74*: 208 (1975).
38. A. K. Biswas and A. K. Ghosh, *J. Agron. Crop Sci.*, *162*: 342 (1989).
39. R. G. Woodward and H. M. Rawson, *Aust. J. Plant Physiol.*, *3*: 257 (1976).
40. L. D. Noodén, *Physiol. Plant.*, *62*: 273 (1984).
41. S. J. Crafts-Brandner, *Plant Physiol.*, *98*: 1128 (1992).
42. H. Molisch, in *The Longevity of Plants* (E. H. Fulling, transl.), Science Press, Lancaster, Pa. (1938).
43. L. D. Noodén, in *Whole Plant Senescence* (L. D. Noodén and A. C. Leopold, eds.), Academic Press, New York, p. 391 (1988).
44. L. D. Noodén, in *Senescence in the Whole Plant* (K. V. Thimann, ed.), CRC Press, Boca Raton, Fla. p. 219 (1980).
45. L. D. Noodén and B. J. Murray, *Plant Physiol.*, *69*: 754 (1982).
46. L. J. Wood, B. J. Murray, Y. Okatan, and L. D. Noodén, *Am. J. Bot.*, *73*: 1377 (1986).
47. L. J. Grabau, D. G. Blevins, and H. C. Minor, *Plant Physiol.*, *82*: 1008 (1986).
48. S. J. Crafts-Brandner, *Crop Sci.*, *32*: 420 (1992).
49. A. K. Seth and P. F. Wareing, *J. Exp. Bot.*, *18*: 65 (1967).
50. B. D. Derman, D. C. Rupp, and L. D. Noodén, *Am. J. Bot.*, *65*: 205 (1978).
51. P. M. Neumann and L. D. Noodén, *Physiol. Plant.*, *60*: 166 (1984).
52. Y. Okatan, G. M. Kahanak, and L. D. Noodén, *Physiol. Plant.*, *52*: 330 (1981).
53. A. K. Biswas and M. A. Choudhuri, *Plant Physiol.*, *65*: 340 (1980).
54. W. A. Mondal and M. A. Choudhuri, *Physiol. Plant.*, *65*: 221 (1985).
55. W. A. Mondal and M. A. Choudhuri, *Physiol. Plant.*, *61*: 287 (1984).
56. S. Ray and M. A. Choudhuri, *Plant Physiol.*, *68*: 1345 (1981).
57. M. Shaw, P. K. Bhattacharya, and W. A. Quick, *Can. J. Bot.*, *43*: 739 (1965).
58. R. A. Fletcher, *Planta*, *89*: 1 (1969).
59. J. R. Ambler, P. W. Morgan, and W. R. Jordan, *Crop Sci.*, *32*: 411 (1992).
60. J. Van Staden, *J. Exp. Bot.*, *24*: 667 (1973).
61. L. D. Noodén, S. Singh, and D. S. Letham, *Plant Physiol.*, *93*: 33 (1990).
62. C. Martin and K. V. Thimann, *Plant Physiol.*, *49*: 64 (1972).
63. D. J. Osborne, *Plant Physiol.*, *37*: 595 (1962).
64. F. R. Garrison, A. M. Brinker, and L. D. Noodén, *Plant Cell Physiol.*, *25*: 213 (1984).
65. C. P. Hsia and C. H. Kao, *Physiol. Plant.*, *43*: 385 (1978).
66. C. H. Kao, *Plant Cell Physiol.*, *21*: 1255 (1980).
67. S. J. Lindoo and L. D. Noodén, *Plant Cell Physiol.*, *19*: 997 (1978).
68. A. J. Colquhoun and J. R. Hillman, *Z. Pflanzenphysiol. Bd.*, *76*: 326 (1975).
69. A. J. Colquhoun and J. R. Hillman, *Planta*, *105*: 213 (1972).
70. M. Radley, *J. Exp. Bot.*, *27*: 1009 (1976).
71. J. A. Roberts and D. J. Osborne, *J. Exp. Bot.*, *32*: 875 (1981).
72. L. Beevers, *Plant Physiol.*, *41*: 1074 (1966).
73. L. Beevers and F. S. Guernsey, *Nature*, *214*: 941 (1967).
74. P. Whyte and L. C. Luckwill, *Nature*, *210*: 1360 (1966).
75. A. Booth, J. Moorby, C. R. Davies, H. Jones, and P. F. Wareing, *Nature*, *194*: 204 (1962).
76. S. Gepstein and K. V. Thimann, *Proc. Natl. Acad. Sci. USA*, *77*: 2050 (1980).
77. K. V. Thimann and S. O. Satler, *Proc. Natl. Acad. Sci. USA*, *76*: 2295 (1979).
78. C. O. Gwathmey and A. E. Hall, *Crop Sci.*, *32*: 773 (1992).
79. C. O. Gwathmey, A. E. Hall, and M. A. Madore, *Crop Sci.*, *32*: 765 (1992).
80. C. O. Gwathmey, A. E. Hall, and M. A. Madore, *Crop Sci.*, *32*: 1003 (1992).
81. A. J. Hall, *Aust. J. Plant Physiol.*, *4*: 623 (1977).
82. A. J. Hall and C. J. Brady, *Aust. J. Plant Physiol.*, *4*: 771 (1977).
83. A. J. Hall and F. L. Milthorpe, *Aust. J. Plant Physiol.*, *5*: 1 (1978).
84. J. C. S. Allison and H. Weinmann, *Plant Physiol.*, *46*: 435 (1970).
85. N. O. Adedipe, L. A. Hunt, and R. A. Fletcher, *Physiol. Plant.*, *25*: 151 (1971).

86. A. H. Haber, P. J. Thompson, P. L. Walne, and L. L. Triplett, *Plant Physiol.*, *44*: 1619 (1969).
87. J. M. Hopkinson, *J. Exp. Bot.*, *17*: 762 (1966).
88. J. J. Goldthwaite and W. M. Laetsch, *Plant Physiol.*, *42*: 1757 (1967).
89. S. D. Davis, C. H. M. van Bavel, and K. J. McCree, *Crop Sci.*, *17*: 640 (1977).
90. J. B. Passioura, *Aust. J. Soil Res.*, *29*: 717 (1991).
91. K. J. Bradford, *Plant Physiol.*, *72*: 251 (1983).
92. M. Ludewig, K. Dorffling, and H. Seifert, *Planta*, *175*: 325 (1988).
93. T. T. Neales, A. Masia, J. Zhang, and W. J. Davies, *J. Exp. Bot.*, *40*: 1113 (1989).
94. J. Zhang, U. Schurr, and W. J. Davies, *J. Exp. Bot.*, *38*: 1174 (1987).
95. J. M. Robertson, K. T. Hubick, E. C. Yeung, and D. M. Reid, *J. Exp. Bot.*, *41*: 325 (1990).
96. N. S. A. Malik and A. M. M. Berrie, *Planta*, *124*: 169 (1975).
97. C. M. Ronning, J. C. Bouwkamp, and T. Solomos, *J. Exp. Bot.*, *42*: 235 (1991).
98. J. J. Guiamét, E. Schwartz, E. Pichersky, and L. D. Noodén, *Plant Physiol.*, *96*: 227 (1991).
99. S. S. Abu-Shakra, D. A. Phillips, and R. C. Huffaker, *Science*, *199*: 973 (1978).
100. H. W. Woolhouse, *BioScience*, *28*: 25 (1978).
101. R. M. Tetley and K. V. Thimann, *Plant Physiol.*, *56*: 140 (1975).
102. C. Laurière, A. Skakoun, and J. Daussant, *Physiol. Veg.*, *13*: 467 (1975).
103. H. Thomas, *Plant Sci. Lett.*, *6*: 369 (1976).
104. A. Blank and T. A. McKeon, *Plant Physiol.*, *97*: 1402 (1991).
105. A. Blank and T. A. McKeon, *Plant Physiol.*, *97*: 1409 (1991).
106. C. S. Mauk and L. D. Noodén, *J. Exp. Bot.*, *43*: 1429 (1992).
107. S. J. Lindoo and L. D. Noodén, *Plant Physiol.*, *59*: 1136 (1977).
108. S. J. Lindoo and L. D. Noodén, *Bot. Gaz.*, *137*: 218 (1976).
109. L. D. Noodén and S. M. Noodén, *Plant Physiol.*, *78*: 263 (1985).
110. L. D. Noodén, G. M. Kahanak, and Y. Okatan, *Science*, *206*: 841 (1979).
111. L. D. Noodén, D. C. Rupp, and B. D. Derman, *Nature*, *271*: 354 (1978).
112. R. Zhang, D. S. Letham, O. C. Wong, L. D. Noodén, and C. W. Parker, *Plant Physiol.*, *83*: 334 (1987).
113. L. D. Noodén, in *Postlude and Prospects*, (L. D. Noodén, and A. C. Leopold, eds.), Academic Press, New York, p. 499 (1988).
114. C.-Z. Jiang, S. R. Rodermel, and R. M. Shibles, *Plant Physiol.*, *101*: 105 (1993).

23
Abscission

Roy Sexton
Stirling University, Stirling, Scotland

I. GENERAL FEATURES OF ABSCISSION

A. Definitions

During its life a plant will shed many of its organs, such as leaves, fruit, petals, buds, bud scales, and bark. Two distinct processes contribute to this loss:

1. *General attrition* is responsible for the detachment of dead or dying tissues such as bark, old branches, and roots. In these cases large mechanical forces such as the wind or differential growth of the stem, rupture an inherently weak region of tissue, usually producing an irregular tear.

2. *Abscission* is involved in the loss of leaves, fruit, flowers, and floral parts. It is an active metabolic process resulting in the weakening of anything from 1 to 20 rows of cells in genetically determined abscission zones. Cell wall breakdown is an important element in the loss of structural integrity. Unlike attrition, abscission is under very precise internal hormonal control.

This chapter is concerned with the mechanism, regulation, and agricultural importance of abscission.

B. Abscission Zones

The shedding of leaves, flowers, fruit, and so on, occurs as a result of the weakening of abscission zones (AZs). These bisect the base of leaf petioles, leaflets, petals, styles, flower buds, axillary buds, young fruit, and the nodes of very young stems. Abscission zones are not inherently weak and a 0.5-kg weight can be hung on a bean leaf AZ without rupturing it. After abscission has been induced, the same AZ will weaken so that it will break at the slightest touch. This progressive loss of structural integrity occurs over 72 h after an initial 18-h lag [1]. Although leaf and fruit abscission usually takes up to 3 days, the process can be extremely rapid and some petals [2] and flower buds [3,4] are shed 1 to 4 h after the inductive treatment.

C. Weakening Process

The scanning micrograph in Figure 1 shows a fracturing abscission zone at the base of a bean leaflet. The discrete nature of the fracture line reveals that loss of structural integrity is restricted to just one or two rows of cells [1]. If the scar faces are examined after fracture, separation seems to have occurred along the line of the middle lamella, leaving the intact rounded cells covering the surface (Figure 2).

Light-microscope and TEM observations implicate cell wall breakdown as a major factor in the separation of these intact cells [3] (Figure 3). Degradation is particularly prominent in the central region of the wall and involves not only the middle lamella but also the adjacent areas of the primary wall [5]. The fracture bisects all the tissues in a stem or petiole, and studies

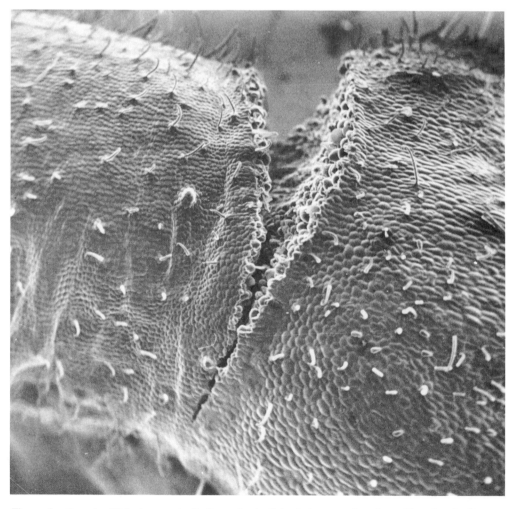

Figure 1 Scanning EM micrograph of a fracturing leaf abscission zone from bean. Note that the fracture is confined to only one or two rows of cells. The tissue on the left is senescing and its diameter contracting (note folds), while that on the right is still turgid and is enlarging. This differential growth causes stresses at the abscission zone interface which help separate the cells in the separation layer and rupture the stele. (From Ref. 1.)

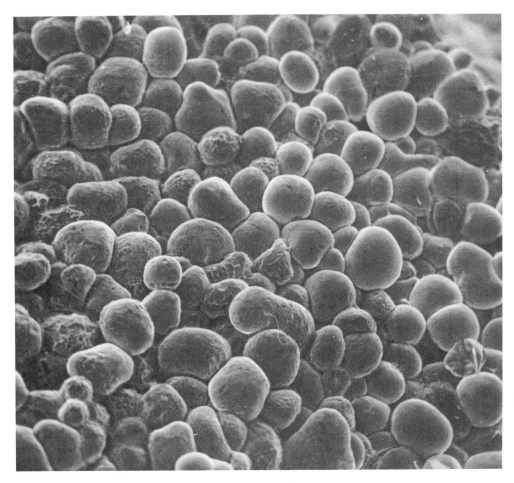

Figure 2 Scanning EM micrograph of part of a bean abscission zone fracture surface. Note that the cells on the exposed scar are round and turgid, having separated as a result of breakdown of the central areas of the wall. (From Ref. 1.)

have shown that wall degradation occurs in all the different living cell classes along the fracture line, including the epidermis and phloem [6].

D. Mechanical Forces and Separation

Although the walls of living cells in the abscission zone (AZ) are enzymically degraded, mechanical forces are necessary both to facilitate cell separation and to rupture the xylem [3]. External forces such as the wind and gravity may be involved, although they are usually not sufficient by themselves. Weisner in 1871 [7] showed that if all the living tissues in a petiole are severed, leaves will often remain attached by the xylem for long periods, despite these external agencies.

Many mechanisms have evolved to generate the forces required to cause complete separation [3]. In bean, the growth of cells on the stem side of the abscission zone, coupled with shrinkage on the distal side, has been implicated in producing stresses at the AZ interface that facilitate rupture [8] (Figure 1). Another common system involves the rounding up and osmotic expansion of the separating cells stretching and breaking the xylem [9] (Figure 3). The squirting cucumber

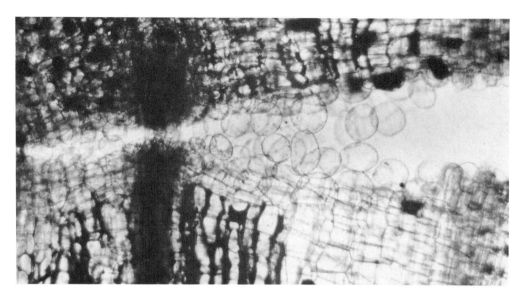

Figure 3 Longitudinal fresh section through an abscising leaf abscission zone. The separation layer that runs through the center of the micrograph is full of round, separated, turgid cells. The expansion of these cells in the petiole cortex results in stretching and subsequent rupture of the xylem vessels in the dark vascular trace. (From Ref. 9.)

provides another rather bizarre example where internal hydrostatic pressures rupture the abscission zone at the base of the fruit, allowing it to shoot away like a water-propelled rocket [10].

E.   Why Is Weakening Restricted to the Separation Layer?

The positions of abscission zones are a genetically determined characteristic of a given species. For instance, the blackberry (*Rubus fruticosus*) is shed by an abscission zone across the base of the fruit, leaving the white receptacle or plug within the berry. In the closely related raspberry (*R. idaeus*), each of the 70 or so drupelets in the berry has an AZ at its base, so when the berry is detached it leaves the receptacle attached to the plant. In hybrids like the loganberry, the blackberry position is dominant [11]. Plant breeders have also produced varieties that lack the normal AZs, such as the lupin cultivar, which cannot abscise its leaves [12].

F.   Are There Specialized Abscission Cells?

Sections through AZs show that their anatomy is not very different from adjacent regions of the petiole or pedicel which they bisect [3]. They often have subtle characteristics that allow the general region to be distinguished [3]. The cells are frequently smaller than those in adjacent tissues, and this close packing can make AZs rather darkly pigmented when viewed externally. The stele usually divides into separate bundles before it enters the zone. Abscission zones lack lignification; sclerenchymatous fibers are often replaced by collenchyma.

  These features are all thought to have evolved to facilitate rupture. It is assumed that lignin is reduced because it makes walls less susceptible to enzymic attack [3]. The close-packed angular AZ cells expand as the walls are degraded and they round up (Figure 3) [9]. Their enlargement has been implicated in producing the forces that rupture the xylem [9]. The branching of the xylem can be explained because under asymmetrical loading a number of thin, separated strands are more readily ruptured than is one large central bundle.

  The cell separation process does not usually involve the entire AZ but only a narrow layer of cells across it. The cells actually involved are known as the *separation layer*. One might expect that the separation layer would contain a distinct specialized class of *abscission cells* that degrade their walls as a result of the abscission signal. However, the cells that will be involved in separation cannot be picked out from their neighbors by simple microscopic examination [3].

One of the enigmas of abscission is why these separation layer cells degrade their walls, when apparently identical cells on either side do not. It has been suggested that separation layer cells are biochemically distinct without there being any structural manifestations of the difference [3,12]. An alternative hypothesis envisages that potentially, all cells can degrade their walls but that the abscission triggering signal is restricted to just a few rows [3]. Recently, McManus and Osborne [13] have used immunological methods to demonstrate specific proteins in the AZ prior to separation. This observation gives credibility to the hypothesis that there are discrete abscission cells.

Abscission zones can be formed very early in the development of some organs; for instance, the minute leaves and flowers inside dormant buds can already have responsive AZs. Why AZs develop at specific positions is not understood. The smaller cells found in them are created by more persistent cell division in the region [14]. Interspecific chimeras between two species of tomato with different AZ positions have been used to study the differentiation of the AZ. It seems that the position of the AZ is dictated by the genetic status of the inner cell lineages, the outer cells responding to signals produced by them [15].

G. Loss of Abscission Zone Responsiveness

Abscission zones are not necessarily active throughout the entire life of the organ. An unfertilized orange flower can be shed by activating the AZ in the pedicel, but after fruit development starts, this abscission zone becomes inactive and will not respond to identical inductive stimuli [16] (also Figure 8). In peach flowers there are three AZs [17], each of which is active at different phases of flower and fruit development.

Two explanations have been put forward to account for the loss of AZ responsiveness. The first envisages that there is a loss of some vital component of the response machinery, such as a hormone receptor. The second proposes that the response still occurs but that the cell walls of the separation layer are modified by substances such as lignin or suberin which make them resistant to attack by wall hydrolases. In orange pedicels, evidence seems to support the latter hypothesis [18].

H. Adventitious Abscission Zones

Although virtually all abscission takes place at precisely predictable sites, there are a few interesting cases where it occurs rather randomly. An example of this type of *adventitious* abscission is the *shot hole effect*, where diseased or damaged areas of the leaf blades of *Prunus* species are abscised, leaving holes [19] (Figure 4). The leaves appear as though a shotgun has been fired through them—hence the name. Adventitious AZs also form in internodes of stems of *Impatiens* [20] and mulberry [21], where fracture can occur at variable positions along the internode.

At first sight the ability to induce abscission at random positions seems to contradict the hypothesis that there are discrete preprogrammed abscission cells at genetically defined positions. Since cell division precedes fracture in adventitious zones, it is possible to argue that these new cells differentiate into an AZ, which is then induced to abscise. This need for differentiation of an AZ before abscission can take place might explain why adventitious stem abscission in *Impatiens* takes so much longer (5 to 14 days) [20] than normal leaf abscission (22 to 36 h) at preformed zones [5]. In a very interesting series of experiments, Warren Wilson

Figure 4 Adventitious abscission zones on the leaf of ornamental cherry. Three areas of the leaf have been wounded by heating, and abscission zones have formed in the living tissue around them. The cell separation, which will result in excision of the damaged area, is evident as a white line. (From R. Sexton, unpublished.)

et al. [20] have shown that the position of adventitious AZs in *Impatiens* can be modified in a predictable manner by manipulating auxin gradients in the tissue.

I. Protection of the Fracture Surface

After fracture has occurred, cells in what remains of the AZ on the plant divide and suberize to form the scar that protects the wound [3,22]. The broken xylem vessels become blocked with gums or tyloses and the phloem sieve plates are callosed over to prevent pathogen entry [3]. The antimicrobial enzymes chitinase and B1–3 glucanase are produced in bean abscission zones [23] and are likely to be one part of an arsenal of compounds produced in the fracture surface to prevent infection. These protective mechanisms must be very efficient, as relatively few diseases originate at abscission scars.

II. REGULATION OF ABSCISSION

A. Inductive Stimuli

Normally, the induction of abscission appears to be an integral part of the senescence program accompanying the yellowing of leaves and ripening of fruit. In most abscission systems the process can be accelerated and will take place prematurely in the absence of senescence. For instance, pollination can dramatically accelerate petal abscission [24,25]. In cyclamen, all pollinated flowers shed their corollas in 5 days, whereas unpollinated flowers retained theirs even after 23 days [24]. Accelerated floral abscission is thought to have evolved to prevent wasted visits of scarce pollinators to fertilized flowers.

Leaf loss in temperate species accompanies senescence, which in turn is induced by environmental factors such as photoperiod changes, low temperatures, and drought. Factors that affect the leaf blade adversely can cause premature shedding. These include frost damage, drought [26], bacterial or fungal attack [27,28], damage by herbivores [29], mineral deficiencies, toxins, excessive shading, darkness [30], and competition with younger leaves. Leaf fall is not invariably linked to lamina senescence, and water-stressed ivy plants will shed leaves with the same chlorophyll content as those still attached to normal healthy plants.

Fruit appear to be abscised at several distinct stages of development [31]. Immature fruit can be shed in large numbers in what appears to be a natural thinning process. Sometimes, this seems immensely wasteful, and in species as diverse as oak and avocado less than 10% of potential fruit mature [32,33]. Some young fruit are shed because seed development is defective, although in avocado many embryos in abscised fruit remain viable [34]. In fruit, water deficits [35], mineral deficiencies, pathogen [36] and herbivore attack, and frost damage can also be factors that precipitate premature abscission.

In many cases the reasons why such huge numbers of young fruit are lost is not clear, although competition between developing fruit is certainly an important factor. The chances of a fruit being shed can be reduced dramatically by removing other fruit from the same plant. In soybean, the fate of fruit that will be shed (50 to 80%) is probably determined before fertilization takes place, on the day that the flower reaches anthesis. In those flowers that will be lost, there is a failure for *sink intensity* to increase, so they do not accumulate photoassimilate from source leaves [37].

During the growth phase of a fruit, abscission seems to be inhibited by the presence of the developing seeds. Mature fruit are usually induced to abscise as one of the terminal events in the ripening program. In some species, abscission is not complete but serves just to loosen the fruit so that birds and small animals can detach them. The manipulation of fruit abscission is very important in developing mechanical harvesting.

Flower bud loss is a serious problem in some crops and decorative plants. In lupins, the development of young fruit at the base of the flower spike seems to induce the loss of buds at the apex [38] (Figure 5). Removing some flower buds usually decreases the likelihood of abscission in those that remain. Disease, water stress, waterlogging, mechanical shaking, and frost damage are also reported to enhance bud abscission.

It is not widely recognized that the cessation of stem growth in several tree species involves abscission of the growing apex. In *Tilia* [39] and *Salix* [40] photoperiod seems to be a primary determinant of when abscission occurs, but position of the branch in the canopy, its orientation, conditions for root growth, competition from other apices, and other climatic factors provide modifying influences [21]. There is very little literature concerned with the inductive conditions that lead to the loss of bud scales, stigmas, anthers, and sepals.

Figure 5 Influence of maturing pods on floral abscission. Lupin flower spike buds open from the base upward (right). If the basal flowers are fertilized, the apical buds abscise, leaving bare stem (center). If the basal flowers are removed, the apical buds remain and develop into pods (left). (From R. Sexton, unpublished.)

B. Experimental Induction of Abscission

Experiments early in the twentieth century showed that removal of the leaf blade resulted in rapid abscission of the subtending petiole (Figure 6). This was a conveniently reproducible system to study abscission, and seedlings of bean, cotton, and *Coleus* were commonly used. The need for faster synchronized abscission led to the *explant* technique. Here, the abscission region was removed by cuts 1 to 2 cm on either side of the zone. The isolated piece of tissue was kept in a sealed container often over 2% agar until it abscised 2 to 3 days later [41]. This explant system has become extremely popular since it provides a lot of material for biochemical studies and regulators can readily be applied directly to the abscising region (Figure 6).

C. Early Experiments with Regulation

After the demonstration that removal of the leaf blade would cause the abscission of the remaining leaf stalk, it was proposed that reduction in photosynthate supply from the leaf caused the AZ cells to collapse and fracture to occur. Kuster in 1916 [42] discovered that a tiny fragment of blade left attached to the petiole was enough to prevent abscission. Since it seemed unlikely that this small area of leaf was providing sufficient nutrients, it was suggested that the healthy blade produced a hormonal factor that prevented abscission [42].

The discovery that auxin was produced by young leaf blades led Laibach and his student Mai [43] to investigate if Kuster's inhibitor was auxin. They demonstrated that pollen rich in auxin applied to the cut end of a debladed petiole delayed abscission. La Rue [44] repeated the experiment with synthetic idoleacetic acid (IAA) (Figure 6). A few years later, workers using

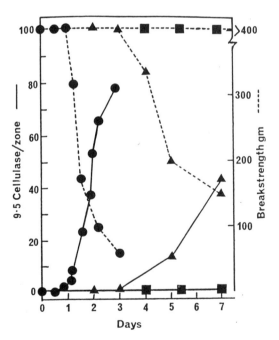

Figure 6 Effects of auxin and ethylene on abscission. The structural integrity (break strength) of bean leaf abscission zones was measured at various times after deblading (triangles). After a 72-h lag, the force necessary to rupture the zone slowly decreases. Adding ethylene speeds up this process by reducing the length of the lag and increasing the rate of weakening (circles). Adding IAA to the end of the petiole inhibits abscission completely (squares). The corresponding levels of 9.5 cellulase in the absission zones are also shown. (From Ref. 1.)

IAA to induce parthenocarpic development of fruit noticed that the treatment also delayed fruit abscission [45]. As a result, a general hypothesis was put forward that abscission resulted from a reduction of the amounts of auxin in the abscission zone caused by a reduced auxin supply from the senescing distal organ (reviewed in Ref. 45).

At the beginning of the twentieth century it had been found that traces of the gas used for illumination would cause the abscission of leaves, petals, and fruits. The active component, identified as ethylene (ethene), promoted abscission at very low concentrations of 1 to 8 ppm [46] (Figure 6). Some 30 years later it was shown that ethylene was synthesized by most plants [47,48], and Milbrath et al. [49] demonstrated that ethylene produced by apples would defoliate roses. However, auxin was in vogue at the time and it was to be 20 years before the role of ethylene as a natural regulator of abscission was taken seriously.

In 1955, Osborne [50] showed that diffusates from senescent petioles contained a soluble factor that accelerated explant abscission. Soon afterward, Van Stevenick [51] demonstrated that developing pods at the base of a lupin inflorescence stimulated abscission of the flowers above them (Figure 5). He succeeded in extracting an abscission stimulator from the young pods [51]. A third group headed by Addicott [52] identified a growth inhibitor that was present in young cotton fruits approaching abscission. The substance was purified [52], shown to accelerate abscission, and characterized as abscisic acid (ABA). Subsequently, lupin pod extracts were also shown to contain ABA [53]. After a number of correlations were reported between increasing ABA levels and abscission [54,55], ABA became accepted as a third potential regulator of the process.

D. IAA and the Control of Abscission

The demonstration that auxin would inhibit leaf absission (Figure 6) was followed by several attempts to measure its levels in naturally abscising systems. It was found that both the extractable and diffusible auxin levels dropped rapidly as the leaves yellowed and abscission approached [45,56,57]. Similar correlations were observed between low auxin levels and fruit abscission [58].

A simple model emerged which suggested that if the auxin levels in the AZ remained above a critical level, abscission was inhibited [58]. Factors that promoted abscission, such as aging, frost damage, and water stress, were thought to lower the levels of free auxin in the zone. It emerged that the rate of auxin transport from the distal organ was a major influence on IAA levels in the AZ [59], although rates of synthesis and degradation are also implicated.

Modifications of the simple auxin concentration theory were necessary when it became clear that the levels of auxin on the stem or proximal side of the AZ also influenced abscission. Jacobs [60] demonstrated that the presence of young auxin-producing leaves on the stem seemed to accelerate loss of debladed petioles. Removal of the apical bud and young leaves delayed abscission below it, the influence of the apical bud being restored if replaced by a supply of IAA [60,61].

As a result of Jacobs's observations, a number of groups showed that if auxin was applied to the proximal side of an AZ, it accelerated abscission [62], while if applied to the distal side, it delayed weakening. This led Addicott et al. [63] to propose the *gradient theory*, where the direction of the auxin gradient across the AZ was important, not the absolute concentration. Auxin approaching the zone from the distal direction inhibited abscission, whereas that moving from the stem accelerated the process [45].

The gradient theory was subsequently challenged since very high levels of auxin applied to the stem side would often inhibit abscission [64]. In 1964, Abeles and Rubinstein [65] reported that auxin applications would promote the synthesis of ethylene, a potent accelerator of abscission. As a consequence it is possible to attribute the accelerating effect of proximal auxin to increased ethylene production coupled to a failure of proximal IAA to reach the AZ before abscission was under way. Abeles [66] argued that proximally applied auxin would move to the zone much more slowly than would distal applications, since its movement by diffusion would be opposed by basipetal auxin transport. Higher auxin concentrations applied proximally would diffuse to the zone more rapidly, accounting for the inhibition sometimes observed. Recently, Morris [67] has shown that the ethylene synthesis inhibitor, aminooxyacetic acid (AOA), will inhibit the accelerating effect of proximal auxin additions, adding weight to Abeles's explanation. Morris [67] also proposes that auxin applied proximally induces the synthesis of the ethylene precursor, aminocyclopropane carboxylic acid (ACC), which in turn diffuses to the zone and promotes ethylene synthesis.

Although ethylene production by proximal auxin applications offers an explanation of the acceleration of abscission, it does not account entirely for the speeding effect of the apex in Jacobs's [60] experiments, and as a result, the gradient theory still has its advocates.

E. Stage 1 and Stage 2 Responses

In 1963 Rubinstein and Leopold [68] discovered that if auxin was added distally more than 12 h after deblading leaves, it accelerated abscission rather than preventing it. They put forth the view that explants went through two stages after excision. In stage 1, auxin additions would inhibit abscission and prolong the stage. If auxin levels fell, stage 2 was entered, when auxin accelerated weakening.

The finding that auxin induced ethylene formation was to provide an explanation of the opposing effects of auxin in stages 1 and 2 [65]. It was proposed that in stage 1 the AZ cells were sensitive to auxin but insensitive to ethylene, and that in stage 2 the reverse applied. If auxin was added in stage 1, it prolonged the length of the ethylene-insensitive condition. If the auxin levels fell below a critical threshold, stage 2 was entered, when auxin no longer had any effect and the additional ethylene accelerated weakening. Some support for this hypothesis has come from experiments that reduced the accelerating effect of auxin in stage 2 by removing ethylene or inhibiting its production [66].

It is widely assumed that a loss of responsiveness to auxin in stage 2 is due to a loss of auxin receptors. Jaffe and Goren [69] have reported that during stage 2 there is not only a reduction in IAA's ability to retard abscission but also in its power to evoke an H^+ ion efflux. They speculate that these two diverse processes may become ineffective because a common component of the response machinery (such as a receptor) is lost. Another possible explanation is that auxin is simply not reaching the separation zone cells if additions are delayed because of a decline in auxin transport. When, after abscission, auxin is added to the cells of the fracture surface, it still inhibits production of the wall-degrading enzymes, suggesting that at least one system is still auxin sensitive [70,71].

The role of auxin in the regulation of abscission has been eclipsed in recent years by work on ethylene. However, it should be remembered that IAA additions will completely prevent any effect of ethylene for extended periods of time and that IAA should therefore be included in any model of abscission control.

F. Ethylene Accelerates Abscission

The ethylene-induced acceleration of abscission is probably the most consistently demonstrated of all plant growth regulator responses (Figure 6). It has been shown to induce shedding of a wide variety of organs and in a huge range of plant species (see the lists in Ref. 72). The threshold concentrations necessary to induce the response are between 0.1 and 5 $\mu L/L$ [72]. Other unsaturated hydrocarbons, such as propene, acetylene, and butene, will act as ethylene substitutes, but they are much less effective [73]. Some analogs, such as 2,5-norbornadiene, are competitive inhibitors [74,75].

There can be dramatic changes in the responsiveness of AZs to ethylene. We have already seen how distal auxin additions to fruit and leaf AZs makes them insensitive to ethylene. There are also well-documented changes in natural sensitivity. For instance, Halevy et al. [24] showed that unfertilized cyclamen flowers will not shed their corollas in ethylene, whereas fertilized flowers will. Similarly, styles of orange would not abscise in the presence of ethylene until fertilization had occurred [76].

It is assumed that the presence of both the gas and its receptors are required for ethylene action to occur.

$$C_2H_4 + receptor \rightarrow ethylene\text{-}receptor\ complex \rightarrow abscission$$

Factors that increase sensitivity, such as water deficit, aging, and ethylene itself, are thought to increase the levels of the receptor, while auxin reduces it [72].

The ethylene receptor has not been isolated, but diazocyclopentadiene (DACP), a potent inhibitor of ethylene action, might prove an effective tool for identifying it [77]. In the light, DACP forms an extremely reactive carbene which is thought to bind covalently to the receptor.

G.   Is Ethylene a Natural Regulator of Abscission?

The early observations that ethylene accelerated abscission were treated as a curious artifact, and even after it was shown that plants naturally evolve ethylene, this gas was not widely envisaged as a natural regulator. Part of the problem was that young leaves seemed to produce more ethylene than older ones [78]. After the demonstration that auxin inhibited the accelerating effects of ethylene, Barlow [79] proposed that it was the auxin/ethylene balance in the tissue that was important. In young leaves there was sufficient auxin to inhibit ethylene's abscission accelerating effect, while in old leaves there was not.

In 1962, a review by Burg [80] argued that the concentrations of ethylene in plants were such that they could easily control abscission. The first claims that the gas was a natural regulator of abscission were based on demonstrations that increased ethylene production rates were correlated with abscission. Such parallels were shown in a wide variety of leaves, fruit, and flowers, although a few authors found no simple correlation (reviewed in Ref. 72).

Of course, correlations do not constitute proof of involvement. To implicate ethylene firmly in abscission, it is necessary to show that endogenous ethylene concentrations increase above the threshold concentrations of 0.5 to 1.0 μL/L that are necessary to cause accelerated abscission if added exogenously [72]. Because AZs are very tiny, it is difficult to extract enough gas to make these measurements. A strong correlation exists between ethylene production rates and internal concentrations. Using this relationship, it was estimated that ethylene production rates of 3 to 5 μL of ethylene per kilogram per hour were necessary to trigger abscission [83], and these were subsequently shown to be exceeded in many abscising systems [72,82]. There have been some direct measurements of the gas concentration in AZs which showed levels above the threshold [83,84]. Raspberry fruit are unusually well suited to these measurements, having 70 to 100 abscission zones enclosed within the fruit. Ethylene levels around these zones showed that concentrations were less than the threshold level of 0.5 μL/L in green fruit, but exceeded it in ripening, abscising fruit [85] (Figure 7).

A clever alternative approach was adopted by Jackson et al. [86]. They measured the rate of ethylene production in senescent bean leaves just prior to abscission and then applied (2-chlorethyl)phosphonic acid (CEPA) to younger petioles to generate similar amounts. This treatment caused abscission.

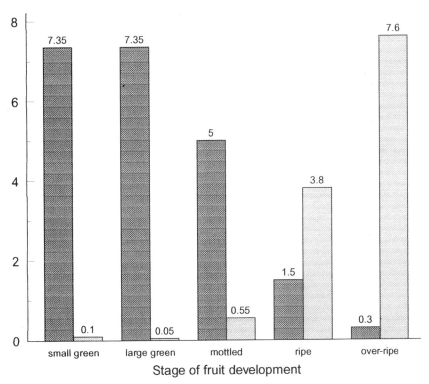

Figure 7 Correlation between the internal ethylene concentrations in raspberry fruit and the onset of abscission. The concentration of ethylene (in μL/L) (pale columns) around the abscission zones within fruit at various stages of ripening are shown. When green fruit progress to the mottled and ripe stages, the concentrations exceed the 0.25 μL/L threshold necessary to induce abscission experimentally in green fruit. The fruit removal force (Newtons) required to break the abscission zones is also plotted (dark columns). Note that it starts to decline in fruit that are mottled or riper, where ethylene levels exceed 0.25 μL/L. (From Ref. 85.)

Reducing the levels of internal ethylene in AZs has also been used to establish ethylene's role. Early experiments employed potassium permanganate or mercuric perchlorate to absorb the gas, and there are several reports of delayed abscission as a result [87]. A more effective approach has been to use hypobaric or low pressures, which increase diffusive loss [72]. Aminoethoxyvinyl glycine (AVG), an inhibitor of ethylene biosynthesis, has been shown to slow natural abscission [89–91]. In the last few years, transgenic tomato plants have been produced that synthesize very little ethylene, but unfortunately, their abscission behavior was not recorded [92].

Inhibitors of ethylene action such as silver ions, which inhibit ethylene responses such as fruit ripening and floral senescence, are also very effective at preventing abscission [93–96]. The mechanism of the Ag^+ effect is not understood, although interaction with the ethylene receptor is assumed. Norbornadiene is a competitive inhibitor and also interferes with natural abscission [74,75]. Recently, Sisler [77] has shown that diazocyclopentadiene inhibits ethylene responses, probably by binding irreversibly to the receptor in the light. This substance inhibits sweet pea abscission (Figure 8).

As a result of this wealth of data, there is widespread acceptance that ethylene is involved in natural abscission. Whether it acts directly or indirectly is a more contentious question. There

Figure 8 Inhibition of sweetpea floral abscission by the ethylene antagonist DACP [77]. Sweetpea buds exposed to DACP for 16 h (far left) are just beginning to abscise after 144 h in a vase; the untreated controls (second left) shed their buds completely 48 h earlier. Opened flowers (far right) had abscised all their petals by 144 h while those exposed to DACP (middle right) were still firmly attached. Note that the unopened buds are shed at an abscission zone at the base of the pedicel, in contrast to older flowers, where this zone is inactive and the petals alone are lost, leaving the remainder of the flower attached. (From R. Sexton, unpublished.)

is recent evidence that the movement of the ethylene precursor ACC in the xylem from water-stressed roots to the leaves may be an important mechanism for inducing abscission [97].

H. Is Ethylene Having a Direct or an Indirect Effect?

It has been shown that ethylene lowers auxin concentrations in the abscission zone by reducing synthesis [98] and transport [99] of the hormone and increasing loss by conjugation and breakdown [100]. As a result, it was proposed that ethylene speeds abscission indirectly by lowering auxin levels [72].

In a series of simple but elegant experiments, Beyer [101] showed that ethylene was actually involved in abscission both directly and indirectly. A system was developed whereby the leaf blade and abscission zone could be exposed to ethylene independently. If the zone or leaf blade were treated alone, abscission did not occur; however, if both were treated, the lamina was shed. Beyer showed that treatment of the leaf blade with ethylene reduced auxin transport down the petiole by over 80%. The effect could be mimicked by auxin transport inhibitors and could be reversed if the supply of auxin from the blade was augmented. This suggested that ethylene served indirectly to reduce the levels of auxin in the AZ by interfering with auxin movement from the blade. Abscission would not occur, however, if auxin transport alone were impeded:

ethylene also had to be present at the zone, suggesting a second direct role in abscission induction.

I. Does Ethylene Induce Abscission or Merely Accelerate It?

The question of whether ethylene induces abscission or just acts as an accelerator has not been resolved. If ethylene is the inducer, removing it or interfering with ethylene action should not just slow the process down, but should stop it entirely. In the majority of experiments of this type, either abscission does eventually occur or observations are not continued long enough to distinguish between stopping and slowing.

J. Abscisic Acid

After Addicott's group had isolated ABA from young cotton fruit, a number of correlations were reported in which increases in ABA seemed to be associated with abscission [54,55,57,102–104]. In contrast, there are a series of papers in which no simple relationship was shown to exist between endogenous ABA and leaf [105,106], fruit, or flower bud [107–109] abscission. However, correlations cannot prove or disprove involvement, and it could be argued that increases in tissue sensitivity to the hormone could induce abscission without a change in ABA concentration. Another general problem encountered when measuring hormone levels is that the value obtained represents an overall mean concentration for the piece of material and ignores important local variations. For instance, a very thin (2 mm) slice made to remove the AZ from a petiole will at best contain only 10% abscission zone cells, and the hormone concentrations in these could be very different from those in the adjacent, contaminating senescing tissue.

In a review of ABA action, Milborrow [110] expressed surprise that among the hundreds of plants sprayed with ABA, more did not show an abscission response. By way of a reply, Addicott [55] lists a considerable number of cases where a response does occur. Abscisic acid is much more effective when applied to explants, and some scientists believe that it acts by promoting ethylene formation. For instance, Sagee et al. [111] have reported that ABA is ineffective in the presence of the ethylene synthesis inhibitor AVG. The current consensus seems to be that ABA is not directly involved in abscission. However, there is some evidence that it may be important in cereal seed shedding, which is one of the few ethylene-insensitive systems [112].

K. Other Potential Regulators

Both gibberellic acid and cytokinins will influence abscission, although they are thought to be less important than the other plant hormones [55,113]. Cytokinins can delay abscission, probably by indirectly delaying senescence [114]. Gibberellic acid will accelerate abscission [115,116], there being some debate as to whether or not the effect is mediated by ethylene [66,117]. Long-chain unsaturated fatty acids such as linolenic acid also enhance abscission [118]. Experiments on bean abscission zones showed that the accelerating effect of the C_{18} unsaturated fatty acids was mediated by the production of fatty acid hydroperoxides and that ethylene was not involved. It is not clear whether these compounds are involved in the regulation of natural abscission, but they do accumulate in some senescent tissues.

L. Regulation of Abscission: Summary

As far as one can judge from the literature, the consensus is that the regulation of abscission directly involves the concentrations of IAA and ethylene and the sensitivity of AZ tissue to them. Other factors that influence the process do so through these agencies.

Figure 9 illustrates a *balance model* based on the relative AZ concentrations of auxin on the right and ethylene on the left. If the left-hand side of the balance goes down, weakening starts, whereas if the right end is down, the process is inhibited. The concentrations of ethylene and auxin in the AZ are influenced by a variety of factors, some of which are shown at the top of the diagram. The balance can also be affected by the position of the fulcrum, which can move from the center toward either end. Moving this to the right represents a decrease in the sensitivity to auxin and an increase in sensitivity to ethylene, and vice versa. Changing sensitivity probably involves the amount of receptors, and some of the factors that change it are illustrated.

III. CELL BIOLOGY OF ABSCISSION

A. Mechanism of Abscission: Early Theories

Early botanists believed that abscission was due to the formation of a corky layer on the stem side of the abscission zone (AZ) which cut the supply of sap to the separation layer and caused

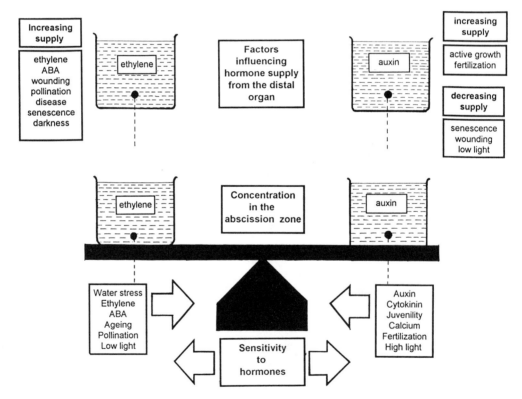

Figure 9 Modern representation of the ethylene auxin balance theory. The relative concentrations of auxin and ethylene in the abscission zone are represented by the weight of beakers on either side of a balance. If the left-hand side goes down, the induction of cell separation occurs and abscission takes place. The weight of the containers is influenced predominately by supply from the distal organ, such as a leaf or fruit. Some factors that influence this supply are shown. The balance can also be tipped in either direction by moving the fulcrum from side to side. This represents changes in the tissue's sensitivity to either hormone. Moving the fulcrum to the right increases sensitivity to ethylene, and some factors that do this, such as water stress and aging, are shown. Moving the fulcrum to the left increases sensitivity to auxin and decreases ethylene responsiveness. Adding auxin itself or cytokinin will cause this to occur. (From R. Sexton, unpublished.)

the cells in it to collapse. The anatomist Inman in 1848 [119] opposed this idea, suggesting that the process was a *vital* one in which the cells of the separation layer remained *plump, fresh, and apparently living*. A few years later, Von Mohl [120] demonstrated that abscission would take place without the formation of a layer of periderm, and as a result, it became widely accepted that the process involved living cells [120].

Two theories emerged to account for the phenomenon. The *turgor theory* proposed that the solute concentration in the separation zone cells increased as a result of starch degradation. The increased turgor pressure generated in the cells caused them to round up, tearing the wall along the line of the middle lamella. Kendall in 1918 [121] claimed to disprove this hypothesis when he showed that cell separation did not always begin at the cell corners and that some separating cells did not round up at all.

One of the first anatomical changes observed to occur after the induction of abscission was increased rates of cell division in the region of the separation layer [3]. It was assumed that this was an important part of the weakening process until Gawadi and Avery [122] showed that abscission would occur in the absence of cell division, as is frequently observed if ethylene is used to accelerate the process.

Beginning in the 1920s, scientists assumed that the newly discovered wall degrading enzymes were involved in cell separation, although some researchers believed that wall acidification was also implicated. Separation zone cells were reported to have very active respiration [117] and protein synthesis [3,123]. These observations fueled speculation that the synthesis and secretion of wall-degrading enzymes was all-important. The turgor mechanism retained some advocates, since it seemed to be the only way of accounting for the abscission of some petals. This took place so rapidly (<1 h) that it was difficult to believe that protein synthesis, secretion, and wall breakdown could all occur in such a short interval [124].

The finding by Horton and Osborne [125] that cellulase (endo-β-1,4-glucan 4-glucan hydrolase) increased in weakening AZs, coupled with the demonstration that both protein and RNA synthesis inhibitors prevented abscission [3,126], led to the current widespread belief that abscission involved the induction of wall-degrading enzymes. It has also been shown recently that protein synthesis inhibitors will stop rapid abscission of petals, removing one of the last objections to the involvement of wall hydrolases [2].

B. Nature of Cell Wall Breakdown

The evidence that wall breakdown is involved in abscission is almost entirely anatomical [3,5]. There are very few biochemical analyses of abscission zone cell walls, although Morre [127] reported an 11% loss of the wall material during weakening and Taylor et al. [128] reported a depolymerization of pectins.

Low-power observations of the fracture surfaces usually show that they are covered in intact rounded cells [6] (Figures 2 and 3). When washed from the surface the cells still have their permeability barriers intact and can be plamolyzed [5]. They are not protoplasts, but retain part of the cell wall, which is still resilient enough to prevent them bursting when turgid (Figure 10). Burst cells sometimes seen over limited areas of the fracture surface are probably ruptured by the mechanical forces that facilitate separation in many abscission systems.

Electron microscope observations show that breakdown of the wall is not restricted to the middle lamella but involves adjacent areas of the primary wall [5] (Figure 10). Both of these swell during AZ weakening, leaving a layer of undigested wall around the protoplast. The swollen areas of wall still contain intact cellulose microfibrils, suggesting that the wall matrix and middle lamella are attacked [3,5,129]. Both x-ray microprobe analysis and autoradiography have shown that Ca^{2+} is lost from the wall during cell separation [130]. It is not clear if Ca^{2+} is lost as a consequence of wall hydrolysis or whether its active removal contributes to wall weakening [3,5].

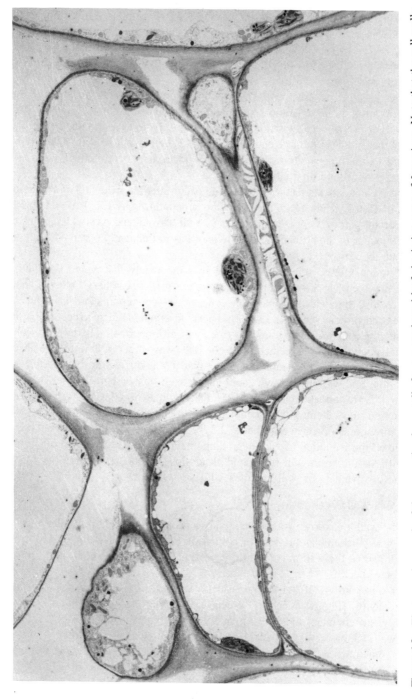

Figure 10 Electron micrograph of the separation layer cells from a weakened leaf abscission zone of *Impatiens*. Note that the cell walls have degraded, allowing the cells to separate. The inner layer of the wall remains intact and the cytoplasm is apparently normal. (From Ref. 5.)

The separation layer bisects the petiole or pedicel and therefore crosses many different tissues. In a study of *Impatiens* leaf abscission, cell wall breakdown was recorded around cells of the epidermis, collenchyma, cortex, xylem parenchyma, phloem seive tubes, and transfer cells [6]. Recent *in situ* hybridization studies [131] indicate that cellulase mRNA is induced in a variety of different cell classes in the separation layer. Comparisons of AZs in leaves and fruit suggest that there may be differences in the nature and extent of wall breakdown [129].

C. Protein Synthesis Is a Prerequisite for Abscission

Anatomists often observed that protein accumulated in separating abscission zone cells [3] and EM observations of the cytoplasm showed it to be rich in organelles, particularly Golgi and RER [70,132]. Studies of the incorporation of labeled amino acids and nucleotides demonstrated very active synthesis of proteins and RNA in the AZ [123, 126, 133]. Some of the proteins synthesized during abscission are thought to play an essential role in the process, since it has been widely demonstrated that both transcriptional and translational protein synthesis inhibitors block abscission [3].

Both qualitative and quantitative changes in the protein profiles of separating AZ cells have been reported [2,134,135]. Complementary abscission-related changes in mRNA populations have also been observed [137,138]. In addition, Poovaiah et al. [136] demonstrated that the pattern of protein phosphorylation is altered in abscising zones.

D. Cell Wall Hydrolases and Their Control

After the initial observations that cellulase increased in separation layers, there was some confusion because it was not appreciated that more than one form of cellulase was found in AZs. Lewis and Varner [139] concluded that a cellulase isoenzyme with an alkaline isoelectric point (9.5 cellulase) was specifically involved in bean leaf abscission. It was formed *de novo* in AZs and its activity seemed correlated with weakening [139] (Figure 6). Antibodies raised against 9.5 cellulase were used to discriminate between it and the other isoenzymes involved in normal growth [140]. The 9.5 cellulase was localized in the separation layer and adjacent stele of bean AZs [140,141], and its production was accelerated by ethylene and inhibited by IAA [142] (Figure 6). This increase in specific cellulase isoforms has now been reported in many different abscission systems, including leaves [140], flowers [4], and fruit [16,143].

Despite its familiar name, 9.5 cellulase will not attack crystalline cellulose. It is assayed by its ability to break down soluble carboxymethylcellulose and is really a Cx-cellulase [144]. Its natural substrate is not known, but is likely to be a β-1,4-glucan in the wall matrix [145]. By itself, bean 9.5 cellulase will not cause cells to separate, but it acts synergistically with pectinase [146].

On the basis of the anatomical observations, one might expect polygalacturonases (PG) to be involved in breakdown of the middle lamella. Recently, PG increases have been reported in *Impatiens*, *Sambucus*, tomato, orange, and peach [5,143,147]. Like cellulase, there are abscission specific isoforms of PG. Transgenic plants have been used to show that the PG associated with ripening in tomato is not the same as that involved in abscission [148]. Both exo- and endo-cleaving PG increase, although the endocleaving enzyme is probably more important.

The cellulases from bean and *Sambucus* abscission zones have been cloned [138,149,150]. The bean cDNA has been sequenced and has 64% identically matched nucleotides to the cellulase from avocado fruit [150]. The partial sequence of the *Sambucus* cDNA shows that it is very similar [138]. They share consensus sequences with a series of E2-type cellulases from microbial and other plant sources [144]. Bonghi et al. [143] have made use of this similarity by employing avocado fruit cellulase cDNA as a heterologous probe. It hybridized to 1.8- and a 2.2-kb mRNAs, which accumulated in ethylene-treated peach abscission zones.

Tucker et al. [149] have studied the expression of bean 9.5 cellulase. In situ hybridization showed that the cellulase mRNA was confined to the separation layer and the adjacent stele [151] (Figure 11). Northern blot analysis indicated that cellulase mRNA was virtually absent from uninduced abscission zones, but increased as they weakened in ethylene [149] (Figure 12). This increase was dependent, at least in the short term, on the presence of ethylene. Indole acetic acid suppressed the increase even in an ethylene atmosphere. Removal of ethylene after cellulase mRNA had started to accumulate, and inhibition of any endogenous ethylene with norbornadiene, caused the cellulase mRNA levels to decline to very low levels [149]. Indole acetic acid administered to the fracture surfaces after abscission had occurred inhibited further accumulation of cellulase mRNA in the fracture surface cells even in the presence of ethylene [71]. This suggests that expression of 9.5 cellulase is under the joint control of both IAA and ethylene. The cellulase gene complete with its upstream sequences has now been cloned [71].

IV. AGRICULTURAL MANIPULATION OF ABSCISSION

A. Methods of Manipulating Abscission

There are a great number of crops for which the control of abscission is desirable. For example, too little natural thinning of young fruit can result in large numbers of small unmarketable fruits, while too much abscission results in uneconomical yields.

Techniques are slowly emerging which allow the manipulation of abscission. They can be categorized as follows:

1. *Understanding the physiological basis of abscission.* Perhaps the best way of controlling abscission is to understand the physiological basis of its induction. For instance,

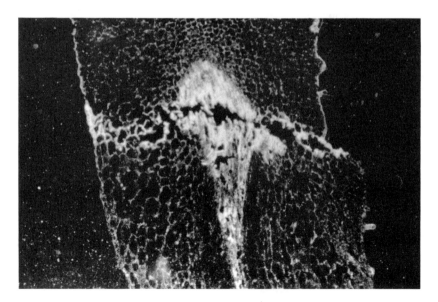

Figure 11 Dark-field micrograph of a thin longitudinal section through an abscising abscission zone of bean. The section was hybridized to a ^{35}S-labeled cellulase cDNA probe. The hybridization signal is seen as bright light reflecting silver grains in the separation layer and in the central vascular traces. (From Ref. 131.)

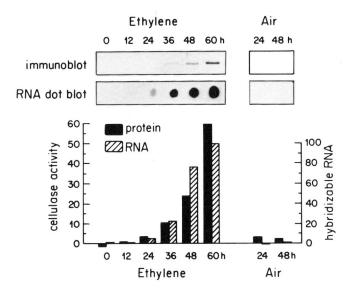

Figure 12 Time course of accumulation of cellulase protein and mRNA. The cellulase activity in bean abscission zones kept in ethylene and air has been plotted against time. The amount of cellulase cDNA probe binding to RNA from the same preparations is also shown. The photographs are of a Western immunoblot of abscission zone proteins probed with cellulase antibodies and RNA dot blots probed with ^{32}P-labeled cellulase cDNA. (From Ref. 149.)

many ornamental plants suffer from bud drop and flower shattering during transit and marketing [152]. This can be reduced by avoiding the inductive conditions that lead to abscission: high temperatures [153], low light intensities [154], ethylene pollution [152], fertilization of flowers [24,25], and mechanical perturbation [155]. In many cases the physiological basis of agriculturally important abscission is not yet fully understood, the extent of young fruit thinning being an important example.

2. *Genetics of abscission behavior.* Geneticists have been able to breed varieties with different abscission characteristics. For instance, among the many raspberry cultivars, there is wide variation in the extent to which fruit abscission has progressed at the time the fruit are ready for harvest [156]. Cultivars that do not drop ripe berries when the bushes are shaken are suitable for hand picking, while more easily detached varieties are selected for mechanical harvesting.

3. *Recombinant DNA technologies and the manipulation of abscission.* Being able to manipulate abscission using transgenic plant technologies is a very real prospect. Oeller et al. [92] have already produced nonripening transgenic tomato plants in which the synthesis of ethylene is blocked. This was achieved by expressing antisense RNA for ACC synthase. One would predict that this strategy should produce slow or nonabscising plants that would abscise to order if treated with ethylene-generating sprays. Similarly, ethylene production has been reduced in plants producing antisense ethylene-forming enzyme RNA [157]. An alternative approach is to overexpress in plants the ACC deaminase gene from bacteria, destroying the ethylene precursor ACC as it is formed [158,159].

It might prove possible to manipulate abscission by producing more IAA in the distal tissues of transgenic plants. The Ti plasmid IAA synthesis genes have already been cloned and there are leaf- and fruit-specific promoters to drive them. An alternative strategy for preventing abscission could be to produce transgenic plants expressing antisense RNA for the wall-degrading enzymes cellulase and polygalacturonase. This approach has been used to reduce softening in fruit [160,161].

4. *Accelerating abscission by increasing ethylene production.* Ethylene-releasing sprays such as Ethephon [(2-chloroethyl)phosphonic acid] [162,163] and Etacelasil [164], which release ethylene spontaneously, are very effective at causing abscission. Their use is not always straightforward, since they can have unfortunate side effects. For instance, ethephon used to promote fruit abscission often induces undesirable leaf fall. Attempts have been made to increase ethylene production using the plant's natural substrate ACC [164], although the method is not widely used.

Ethylene production can be increased by wounding the plant. The protein synthesis inhibitor cycloheximide has been employed to damage and induce consequent abscission of oranges prior to mechanical harvesting. Initial trials with ethylene-releasing sprays were unsuccessful, since they damaged the trees by defoliating them, so a method of localizing ethylene production in the fruit was sought. It was found that cycloheximide damaged the peel of the fruit, causing wound ethylene formation, which in turn induced abscission [165]. The abscission of cotton leaves prior to boll harvest is also achieved by damaging the leaves and inducing ethylene formation [166,167]. The mechanism of action of some thinning agents, such as insecticides carbaryl and oxamyl, which are used to thin apples [168,169], is not fully understood.

5. *Inhibiting abscission by reducing ethylene production and sensitivity.* Abscission can be inhibited by reducing natural ethylene production or interfering with ethylene action. A reduction of natural fruit thinning has been reported using AVG and AOA to inhibit ethylene formation [89,90,170,171]. Silver thiosulfate [152] has been widely employed to prevent abscission of ornamental flowers to such an extent that it is a pollution hazard in some horticultural areas. The new ethylene antagonist DACP [77] also seems extremely effective (Figure 8), but unfortunately, it is an explosive gas, which may limit its field use.

6. *Auxin and control of abscission in the field.* Auxin sprays have been employed to prevent abscission. Indoleacetic acid is not used since it is rapidly degraded in the plant and synthetic auxin analogs such as 1-napththaleneacetic acid (NAA), picloram, and 2,4D are preferred. Preharvest drop of apples has been treated in this way [172]. Rather perversely, NAA sprays are used to thin young apple fruit [169]. In this case, the effect seems to be caused indirectly by interfering with sugar translocation from sprayed leaves to the developing fruit [168,173].

B. Flower Shattering

The abscission or shattering of whole buds or floral parts is a major problem in a number of ornamental plants [152,174]. These include *Zygocactus* [175], *Fuschia, Calceolaria,* snapdragon, sweet peas [176], lilies [154,177], geraniums [155], *Pelargonium* [153], cyclamen [24], *Impatiens, Bourgainvillea,* delphiniums, and foxgloves [25]. There are many other delightful flowers that are not marketed commercially because of these difficulties. The problems usually occur during transit and retailing, when low light intensities, water stress, high temperatures, and ethylene buildup contribute to the problem. The concentration of ethylene in mixed cool stores may reach high levels, particularly if ethylene-generating fruit are enclosed with the flowers. Motor exhausts (0.25% ethylene) have been shown to raise ethylene levels to inductive concentrations in auction halls [152] and the trucks used for transport [179,180].

Silver thiosulfate (0.5 to 2 mM) has been extremely effective at reducing this loss, either when sprayed directly on the plants or *pulsed* through the transpiration stream of cut flowers [152,174]. There are rumors that its use may be banned because of pollution problems. Auxin analogs have been used with flowers [167,178], but petals seem rather unresponsive.

C. Fruit Thinning and Harvesting

Some plants produce vastly more flowers than will mature into fruit. This is particularly true of many fruit trees, such as orange [31], apple, apricot, mango [33], avocado [34,181], and cherry, where a natural thinning process occurs. Only 0.2% of Washington Navel orange flowers develop into fruit [31]. Floral and young fruit abscission is also a particular problem in leguminous crops such as soybean [37,182], field bean (*Vicia*) [166], lupins (38), French bean (*Phaseolus*) [183], and cowpeas [184]. In other crops, such as pistachio, alternate bearing is a difficulty, where a heavy crop of fruit appears to cause excessive abscission of the subsequent year's buds [108]. Flowers and fruit are lost in a succession of abscission episodes [31]. These are classified as follows:

1. *Bud drop* that occurs before the flowers reach anthesis
2. *Flower or young fruit drop* immediately after anthesis
3. *Enlarging fruit or June drop*
4. *Mature fruit or preharvest drop*

The loss of flower buds can be considerable. In oranges, it can be up to 33% and is attributed to nutritional causes such as zinc deficiency [31]. In apricots, a failure to fulfill chilling requirements in warm winter areas such as South Africa, Turkey, and Israel results in almost complete loss of flowers. This can be overcome by spraying the trees with GA to break dormancy or by growing varieties with a short chilling requirement [31].

Loss of opened flowers can usually be attributed to a failure of flower development or failure of fertilization. In Shamouti orange, 77% of the flowers that were shed had abnormalities, often with aborted pistils [31]. Benzyladenine has been used to prevent flower and young fruit drop [182,185].

The June drop of enlarging fruit can be very significant in citrus crops, being as high as 21% in lemons, 45% in Shamouti oranges, and 60% in clementines [31]. The main causes [31] of this drop are:

1. *Abnormalities* in or lack of fertilization and zygote abortion or degeneration.
2. *Competition* for photosynthates and mineral nutrients between fruit and vegetative apices; fruit with fewer seeds on weaker branches tend to be shed first.
3. *Water stress* can be a major cause of abscission [102] in arid areas, since fruit desiccate first, having a higher (less negative) water potential than leaves. Hail and wind damage are also climatic factors.
4. *Invasion of fruit by fungal pathogens* or by herbivorous larvae can cause abscission. *Runoff* of blackcurrants, which results in the premature abscission of apparently healthy fruitlets, can reduce yield by 50% [90]. Recent evidence suggests that symptomless infection with *Botrytis* causes elevated ethylene production, which, in turn, induces abscission [75].

Aminoethoxyvinyl glycine (AVG), the ethylene synthesis inhibitor, has been used to prevent fruit thinning in apples [170,171]. When natural thinning is not vigorous enough and too many fruit are set, sprays are used to reduce the crop and get fewer bigger fruit. Apples have been thinned with ethephon [167,186], carbaryl [167], oxamyl [169], and NAA [173]. Ethephon has also been used with pecans [187], peaches [188], prunes [189], and pears [190].

During maturation of the crop, abscission of the ripe fruit occurs. This preharvest drop is undesirable as far as hand picking is concerned. It is common practice to prevent it by using NAA sprays in pears and apples and with 2,4D [167,172] and NAA in grapefruit and oranges [31]. Trials with NAA have also been conducted to try to reduce the loss of grapes from bunches before and after harvest [191]. Spraying with AVG 1 month before harvest delayed the preharvest drop of apples [89].

Harvesting can be carried out using machines that beat, shake, or blow fruit from plants. The greater the force needed to separate the fruit, the greater the damage to the fruit and the plant. Ethylene-releasing sprays have been used in trials to accelerate and synchronize abscission prior to harvest of grapes [192], oranges [165], apples [172], olives [193], raspberries [194], and blackberries. Ethephon promotes both abscission and reddening in peppers and its feasibility is being assessed as an aid to once-over harvesting [195]. Although these methods have been successful as an aid to mechanical harvesting [196], they can produce unfortunate side effects, such as the shedding of leaves [196], shoot dieback, gummosis [165,196], and excessive fruit drop prior to harvest [192]. The problem of leaf loss has been overcome successfully by the use of calcium acetate sprays [197,198]. Ethephon can give inconsistent results [164] because ethylene production is very dependent on ambient temperature [196] and the pH of the cell sap [163,168]. 2-Chloroethylmethylbisphenylmethoxysilane (CGA) [196] and Etacelasil (164) [2-chlorotris(2-methoxyethoxy)silane] may prove more reliable ethylene-releasing agents.

It is important to realize that using ethylene-generating sprays will accelerate abscission only if natural ethylene production is subsaturating. For this reason, adding more ethylene to ripening raspberries is counterproductive since it has no effect on abscission of ripe fruit, but instead, causes immature green fruit to redden and abscise [85].

D. Leaf Loss

Leaf fall is induced in several crops where the foliage interferes with the mechanical harvesting of fruit. In cotton both desiccants and ethylene-releasing sprays are employed prior to boll harvest [166,167]. Defoliation of nursery trees is also practiced prior to shipping [152,164,174].

Leaf fall is a problem in some display plants, such as *Ficus benjamina* [199]. *Radermachera* [200], *Philodendron* [201], and potted roses. It is induced by low light intensities, water stress [199], and ethylene pollution during retailing. These difficulties are usually solved using silver thiosulfate [200], but the loss of holly leaves is treated using the auxin analog NAA [202].

E. Other Uses

The ability to control abscission has been put to a number of rather unusual uses. During the storage of lemons the short stem (button) left attached to the fruit abscises and allows entry of the fungus *Alternaria* via the scar. This problem is overcome by adding the isopropyl ester of 2,4D [203]. The reverse problem is encountered with bananas, where the failure of perianth abscission makes some varieties of fruit less attractive [204]. Ethephon sprays have been used to reduce mistletoe infestations of Black Spruce by causing the abscission of 90 to 100% of the mistletoe shoots [205]. Twig abscission has been induced in white oak in an attempt to improve timber quality [206].

V. SUMMARY

Abscission occurs at genetically determined abscission zones, where the induction of wall-degrading enzymes such as cellulase and polygalacturonase weaken a restricted band of cells called the separation layer. It seems likely that the separation layer cells are specifically preprogrammed to respond to the inductive stimuli. The generation of mechanical forces to facilitate separation of the loosened separation zone cells and rupture the xylem are an important component of the process. After abscission the wound is protected from microbial attack by the formation of chitinase and β-1,3-glucanase, division and suberization of the surface cells, and blockage of the vascular traces.

The control of abscission seems to depend on the relative concentrations of IAA and ethylene. Auxin is an inhibitor of abscission, and ethylene accelerates and synchronizes the process. It is not clear if one or another of these regulators or the relative concentrations of both are responsible for the induction of abscission. Other hormones, such as ABA, probably have indirect effects via ethylene and IAA.

The manipulation of abscission is important in agriculture. Methods are being developed to control the thinning of fruit, the shattering of flowers, and mature fruit drop as an aid to harvesting.

REFERENCES

1. R. Sexton, M. L. Tucker, E. del Campillo, and L. N. Lewis, in *Cell Separation in Plants* (D. J. Osborne and M. B. Jackson, eds.), NATO ASI Series H35, Springer-Verlag, Berlin, p. 69 (1989).
2. K. B. Eversen, D. G. Clark, and A. Singh, in *Cellular and Molecular Aspects of the Plant Hormone Ethylene* (J. C. Peche et al., eds.), Kluwer, Boston, p. 278 (1993).
3. R. Sexton and J. A. Roberts, *Annu. Rev. Plant Physiol.*, *33*: 133 (1982).
4. J. A. Roberts, C. B. Schindler, and G. A. Tucker, *Planta*, *160*: 164 (1983).
5. R. Sexton, J. N. Burdon, J. S. G. Reid, M. L. Durbin, and L. N. Lewis, in *Structure, Function and Biosynthesis of Plant Cell Walls* (W. M. Dugger and S. Bartnicki-Garcia, eds.), *Proceedings of the 7th Annual Symposium on Botany*, University of California at Riverside, Waverly Press, Baltimore, p. 195 (1984).
6. R. Sexton, *Planta*, *128*: 49 (1976).
7. J. Weisner, 1871, quoted by V. Facey, *New Phytol.*, *49*: *103* (1950).
8. M. Wright and D. J. Osborne, *Planta*, *120*: 163 (1974).
9. R. Sexton and A. J. Redshaw, *Ann. Bot.*, *48*: 745 (1981).
10. M. B. Jackson and D. J. Osborne, *Can. J. Bot.*, *50*: 1465 (1972).
11. K. A. D. Mackenzie, *Ann. Bot.*, *62*: 249 (1988).
12. D. J. Osborne, in *The Plant Hormone Ethylene* (A. K. Matoo and J. C. Suttle, eds.), CRC Press, Boca Raton, Fla., p. 193 (1991).
13. M. T. McManus and D. J. Osborne, *Physiol. Plant.*, *79*: 471 (1990).
14. M. B. Halliday and E. Wangermann, *New Phytol.*, *71*: 649 (1972).
15. E. J. Szymkowiak and I. M. Sussex, in *Cell Separation in Plants* (D. J. Osborne and M. B. Jackson, eds.), NATO ASI Series H35, Springer-Verlag, Berlin, p. 363 (1989).
16. J. Greenberg, R. Goren, and J. Riov, *Physiol. Plant.*, *34*: 1 (1975).
17. N. Rascio, G. Casadoro, A. Ramina, and A. Masia, *Planta*, *164*: 1 (1985).
18. M. Huberman, R. Goren, and E. Zamski, *Physiol. Plant.*, *59*: 445 (1983).
19. G. Samuel, *Ann. Bot.*, *41*: 375 (1927).
20. J. Warren Wilson, P. M. Warren Wilson, and E. S. Walker, *Ann. Bot.*, *62*: 235 (1988).
21. T. Suzuki, *Physiol. Plant.*, *82*: 483 (1991).
22. A. R. Biggs and J. Northover, *Can. J. Bot.*, *63*: 1547 (1985).
23. E. del Campillo and L. N. Lewis, *Plant Physiol.*, *98*: 955 (1992).
24. A. H. Halevy, C. S. Whitehead, and A. M. Kofranek, *Plant Physiol.*, *75*: 1090 (1984).
25. A. D. Stead and K. G. Moore, *Planta*, *157*: 15 (1983).
26. W. R. Jordan, P. W. Morgan, and T. L. Davenport, *Plant Physiol.*, *50*: 756 (1972).
27. D. L. Ketring and H. A. Melouk, *Plant Physiol.*, *69*: 789 (1982).
28. L. S. Jankiewicz, *Bull. Soc. Bot. Fr.*, *127*, *Actual. Bot.*, *1*: 165 (1980).
29. R. M. Hendrickson and R. J. Dysart, *J. Econ. Entomol.*, *76*: 1075. (1983).
30. L. E. Cacker, S. Y. Zhao, and D. R. Decoteau, *J. Exp. Bot.*, *38*: 883. (1987).
31. N. Kaska, in *Cell Separation in Plants* (D. J. Osborne and M. B. Jackson, eds.), NATO ASI Series H35, Springer-Verlag, Berlin, p. 309 (1989).
32. P. P. Feret, R. E. Kreh, S. A. Merkle, and R. G. Oderwald, *Bot. Gaz.*, *143*: 216 (1982).

33. R. Nunez-Elisea and T. L. Davenport, *Plant Physiol.*, *82*: 991 (1986).
34. K. G. M. Skene and M. Barlass, *Ann. Bot.*, *52*: 667 (1983).
35. G. Guinn, *Crop Sci.*, *22*: 580 (1982).
36. B. Williamson, R. J. McNicol, and K. Young, *Grower*, Oct.: 21 (1989).
37. W. A. Brun and K. J. Betts, *Plant Physiol.*, *75*: 187 (1984).
38. R. F. N. Van Stevenick, *J. Exp. Bot.*, *8*: 373 (1957).
39. C. D. Pigott, *New Phytol.*, *97*: 575 (1984).
40. O. Junttila, *Physiol. Plant.*, *38*: 278 (1976).
41. F. T. Addicott, R. S. Lynch, G. A. Livingston, and J. K. Hunter, *Plant Physiol.*, *24*: 537 (1949).
42. E. Kuster, *Dtsch. Bot. Ges.*, *34*: 184 (1916).
43. G. Mai, *Jahr. Wiss. Bot.*, *79*: 681 (1934).
44. C. D. La Rue, *Proc. Natl. Acad. Sci. USA*, *22*: 254 (1936).
45. F. T. Addicott, *Biol. Rev.*, *45*: 485 (1970).
46. S. Doubt, *Bot. Gaz.*, *63*: 209 (1917).
47. P. W. Zimmerman, W. Crocker, and A. E. Hitchcock, *Proc. Natl. Soc. Hortic. Sci.*, *27*: 53 (1930).
48. W. Crocker, *Growth of Plants*, Reinhold, New York (1948).
49. J. A. Milbrath, E. Hansen, and H. Hartman, *Science*, *91*: 100 (1940).
50. D. J. Osborne, *Nature*, *176*: 1161 (1955).
51. R. F. M. Van Stevenick, *Nature*, *183*: 1246 (1959).
52. K. Ohkuma, J. L. Lyon, F. T. Addicott, and O. E. Smith, *Science*, *142*: 1592 (1963).
53. J. W. Cornforth, B. V. Milborrow, G. Ryback, K. Rothwell, and R. L. Wain, *Nature*, *211*: 742 (1966).
54. L. A. Davis and F. T. Addicott, *Plant Physiol.*, *49*: 644 (1972).
55. F. T. Addicott, *Abscission*, University of California Press, Berkeley, Calif. (1982).
56. J. A. Roberts and D. J. Osborne, *J. Exp. Bot.*, *32*: 875 (1981).
57. M. Elkinawy, *Physiol. Plant.*, *62*: 593 (1984).
58. H. R. Carns, *Annu. Rev. Plant Physiol.*, *17*: 295 (1966).
59. T. L. Davenport, P. W. Morgan, and W. R. Jordan, *Plant Physiol.*, *65*: 1023 (1980).
60. W. P. Jacobs, *Annu. Rev. Plant Physiol.*, *13*: 403 (1962).
61. W. P. Jacobs, *Plant Hormones and Plant Development*, Cambridge University Press, Cambridge (1979).
62. F. T. Addicott and R. S. Lynch, *Science*, *114*: 688 (1951).
63. F. T. Addicott, R. S. Lynch, and H. R. Carns, *Science*, *121*: 644 (1955).
64. B. K. Guar and A. C. Leopold, *Plant Physiol.*, *30*: 487 (1955).
65. F. B. Abeles and B. Rubinstein, *Plant Physiol.*, *39*: 963 (1964).
66. F. B. Abeles, *Physiol. Plant.*, *20*: 442 (1967).
67. D. A. Morris, *J. Exp. Bot.*, *44*:261:807 (1993).
68. B. Rubinstein and A. C. Leopold, *Plant Physiol.*, *38*: 262 (1963).
69. M. T. Jaffe and R. Goren, *Bot. Gaz.*, *140*: 378 (1979).
70. D. J. Osborne, M. T. McManus, and J. Webb, in *Ethylene and Plant Development* (J. A. Roberts and G. A. Tucker, eds.), Butterworth, London, p. 197 (1985).
71. M. J. Tucker, G. L. Matters, S. M. Koehler, E. C. Kemmerer, S. L. Baird, and R. Sexton, in *Cellular and Molecular Aspects of the Plant Hormone Ethylene* (J. C. Peche et al., eds.), Kluwer, Boston, p. 265 (1993).
72. R. Sexton, L. N. Lewis, A. J. Trewavas, and P. Kelly, in *Ethylene and Plant Development* (J. A. Roberts and G. A. Tucker, eds.), Butterworth, London, p. 173 (1985).
73. F. B. Abeles and H. E. Gahagan, *Plant Physiol.*, *43*: 1255 (1968).
74. E. C. Sisler, R. Goren, and M. Huberman, *Physiol. Plant.*, *63*: 114 (1985).
75. R. J. McNicol, B. Williamson, and K. Young, *Acta Hortic.*, *262*: 209 (1989).
76. E. E. Goldschmidt and B. Leshem, *Am. J. Bot.*, *58*: 14 (1971).
77. E. C. Sisler, S. M. Blankenship, J. C. Fearn, and R. Haynes, in *Cellular and Molecular Aspects of the Plant Hormone Ethylene* (J. C. Peche et al., eds.). Kluwer, Boston, p. 182 (1993).
78. W. C. Hall, *Bot. Gaz.*, *113*: 310 (1952).
79. H. W. B. Barlow, *Rep. 13th Int. Hortic. Congr.*, *1*: 145 (1952).

80. S. P. Burg, *Annu. Rev. Plant Physiol.*, *13*: 265 (1962).
81. S. P. Burg, *Plant Physiol.*, *43*: 1503 (1968).
82. P. W. Morgan, C.-J. He, and M. C. Drew, *Plant Physiol.*, *100*: 1587 (1992).
83. S. Ben-Yehoshua and B. Aloni, *Bot. Gaz.*, *135*: 41 (1974).
84. P. W. Morgan and J. I. Durham, *Plant Physiol.*, *66*: 88 (1980).
85. J. N. Burdon and R. Sexton, *Ann. Bot.*, *66*: 111 (1990).
86. M. B. Jackson, C. Hartley, and D. J. Osborne, *New Phytol.*, *72*: 1251 (1973).
87. M. B. Jackson and D. J. Osborne, *Nature*, *225*: 1019 (1970).
88. R. Sexton, J. N. Burdon, and J. M. Bowmer, in *Cellular and Molecular Aspects of the Plant Hormone Ethylene* (J. C. Peche et al., eds.), Kluwer, Boston, p. 317 (1993).
89. F. Bangerth, *J. Am. Soc. Hortic. Sci.*, *103*: 401 (1978).
90. M. W. Williams, *HortScience*, *15*: 76 (1980).
91. T. L. Davenport and M. M. Manners, *J. Exp. Bot.*, *33*: 815 (1982).
92. P. W. Oeller, L. M. Wong, L. P. Taylor, D. A. Pike, and A. Theologis, *Science*, *254*: 437 (1991).
93. E. M. Beyer, *Plant Physiol.*, *58*: 268 (1976).
94. A. C. Cameron and M. S. Reid, *Sci. Hortic.*, *19*: 373 (1983).
95. M. S. Reid, *Acta Hortic.*, *167*: 57 (1985).
96. D. L. Ketring and H. A. Melouk, *Plant Physiol.*, *69*: 789 (1982).
97. D. Tudela and E. Primo-Millo, *Plant Physiol.*, *100*: 131 (1992).
98. L. C. Ernest and J. G. Valdovinos, *Plant Physiol.*, *48*: 402 (1971).
99. E. M. Beyer, *Plant Physiol.*, *52*: 1 (1973).
100. J. Riov and R. Goren, *Plant Cell Environ.*, *2*: 83 (1979).
101. E. M. Beyer, *Plant Physiol.*, *55*: 322 (1975).
102. G. Guinn, *Crop Sci.*, *22*: 580 (1982).
103. G. Guinn and D. L. Brummett, *Plant Physiol.*, *86*: 28 (1988).
104. G. Guinn, *Plant Physiol.*, *69*: 349 (1982).
105. T. L. Davenport, W. R. Jordan, and P. W. Morgan, *Plant Physiol.*, *59*: 1165 (1977).
106. J. C. Peterson, J. N. Sacalis, and D. J. Durkin, *J. Am. Soc. Hortic. Sci.*, *105*: 793 (1980).
107. A. Ramina and A. Masia, *J. Am. Soc. Hortic. Sci.*, *105*: 465 (1980).
108. F. Takeda and J. C. Crane, *J. Am. Soc. Hortic. Sci.*, *105*: 573 (1980).
109. N. G. Porter, *Physiol. Plant.*, *40*: 50 (1977).
110. B. V. Milborrow, *Annu. Rev. Plant Physiol.*, *25*: 259 (1974).
111. O. Sagee, R. Goren, and J. Riov, *Plant Physiol.*, *66*: 750 (1980).
112. J. A. Sargent, D. J. Osborne, and S. M. Dunford, *J. Exp. Bot.*, *35*: 1663 (1984).
113. R. Sexton and H. W. Woolhouse, in *Advanced Plant Physiology* (M. B. Wilkins, ed.), Longman, Harlow, Essex, England, p. 469 (1984).
114. A. Kuang, C. M. Peterson, and R. R. Dute, *J. Exp. Bot.*, *43*: 1611 (1992).
115. S. K. Chatterjee and A. C. Leopold, *Plant Physiol.*, *39*: 334 (1964).
116. P. W. Morgan, *Planta*, *129*: 275 (1976).
117. M. C. Marynick, *Plant Physiol.*, *59*: 484 (1977).
118. J. Ueda, Y. Morita, and J. Kato, *Plant Cell Physiol.*, *32*: 983 (1991).
119. T. Inman, *Proc. Lit. Philos. Soc. Liverpool*, *36*: 89 (1848).
120. H. von Mohl, *Bot. Z.*, *18*: 273 (1860).
121. J. N. Kendall, *Univ. Calif. Publ. Bot.*, *5*: 347 (1918).
122. A. G. Gawadi and G. S. Avery, *Am. J. Bot.*, *37*: 172 (1950).
123. F. B. Abeles and R. E. Holm, *Plant Physiol.*, *41*: 1337 (1966).
124. R. Sexton, W. A. Struthers, and L. N. Lewis, *Protoplasma*, *116*: 179 (1983).
125. R. F. Horton and D. J. Osborne, *Nature*, *214*: 1086 (1967).
126. F. B. Abeles, G. R. Leather, L. E. Forrence, and L. E. Craker, *HortScience*, *6*: 371 (1971).
127. D. J. Morre, *Plant Physiol.*, *43*: 1545 (1968).
128. J. E. Taylor, S. T. J. Webb, S. A. Coupe, G. A. Tucker, and J. A. Roberts, *J. Exp. Bot.*, *44*: 93 (1993).
129. C. Zanchin, G. Bonghi, G. Casadoro, A. Ramina, and N. Rascio, *New Phytol.*, *123*: 555 (1993).
130. R. Stosser, H. P. Rasmussen, and M. J. Bukovac, *Planta*, *86*: 151 (1969).

131. M. L. Tucker, S. L. Baird, and R. Sexton, *Planta*, *186*: 52 (1991).
132. R. Sexton, G. G. C. Jamieson, and M. H. I. L. Allan, *Protoplasma*, *99*: 55 (1977).
133. F. B. Abeles and R. E. Holm, *Ann. N.Y. Acad. Sci.*, *144*: 367 (1967).
134. A. S. N. Reddy, M. Friedmann, and B. W. Poovaiah, *Plant Cell Physiol.*, *29*: 179 (1988).
135. E. del Campillo and L. N. Lewis, *Plant Physiol.*, *98*: 955 (1992).
136. B. W. Poovaiah, M. Friedmann, A. S. N. Reddy, and J. K. Rhee, *Physiol. Plant.*, *73*: 354 (1988).
137. P. Kelly, A. J. Trewavas, L. N. Lewis, M. L. Durbin, and R. Sexton, *Plant Cell Environ.*, *10*: 11 (1987).
138. J. A. Roberts, J. E. Taylor, S. A. Coupe, N. Harris, and S. T. J. Webb, in *Cellular and Molecular Aspects of the Plant Hormone Ethylene* (J. C. Peche et al., eds.), Kluwer, Boston, p. 272 (1993).
139. L. N. Lewis and J. E. Varner, *Plant Physiol.*, *46*: 194 (1970).
140. R. Sexton, M. L. Durbin, L. N. Lewis, and W. W. Thompson, *Nature*, *283*: 873 (1980).
141. E. del Campillo, P. D. Reid, R. Sexton, and L. N. Lewis, *Plant Cell*, 2: 245 (1990).
142. M. L. Durbin, R. Sexton, and L. N. Lewis, *Plant Cell Environ.*, *4*: 67 (1981).
143. C. Bonghi, N. Rascio, A. Ramina, and G. Casadoro, *Plant Mol. Biol.*, *20*: 839 (1992).
144. C. C. Lashbrook and A. B. Bennett, in *Cellular and Molecular Aspects of the Plant Hormone Ethylene* (J. C. Peche et al., eds.), Kluwer, Boston, p. 123 (1993).
145. R. Hatfield and D. J. Nevins, *Plant Cell Physiol.*, *27*: 541 (1986).
146. L. N. Lewis, A. E. Linkins, S. O'Sullivan, and P. D. Reid, *Proceedings of the 8th International Conference on Plant Growth Substances*, Hirokawa Publishing Co., Tokyo, p. 708 (1974).
147. J. E. Taylor, S. T. J. Webb, S. A. Coupe, G. A. Tucker, and J. A. Roberts, *J. Exp. Bot.*, *44*: 93, 253 (1993).
148. J. E. Taylor, G. A. Tucker, Y. Lasslett, C. J. S. Smith, C. M. Arnold, C. F. Watson, W. Schuch, D. Grierson, and J. A. Roberts, *Planta*, *183*: 133 (1990).
149. M. L. Tucker, R. Sexton, E. del Campillo, and L. N. Lewis, *Plant Physiol.*, *88*: 1257 (1988).
150. M. L. Tucker and S. B. Milligan, *Plant Physiol.*, *95*: 928 (1991).
151. M. L. Tucker, S. L. Baird, and R. Sexton, *Planta*, *186*: 52 (1991).
152. M. S. Reid, *Acta Hortic.*, *167*: 57 (1985).
153. A. S. Cameron and M. S. Reid, *Sci. Hortic.*, *19*: 373 (1983).
154. A. J. B. Durieux, G. A. Kamerbeek, and U. Van Meeteren, *Sci. Hortic.*, *18*: 287 (1983).
155. A. M. Armitage, R. Heins, S. Dean, and W. Carlson, *J. Am. Soc. Hortic. Sci.*, *105*: 562 (1980).
156. J. N. Burdon and R. Sexton, *Sci. Hortic.*, *43*: 95 (1990).
157. J. E. Gray, S. Picton, R. Fray, A. J. Hamilton, H. Smith, S. Barton, and D. Grierson, in *Cellular and Molecular Aspects of the Plant Hormone Ethylene* (J. C. Peche et al., eds.), Kluwer, Boston, p. 82 (1993).
158. R. E. Sheehy, V. Ursin, S. Vanderpan, and W. R. Hiatt, *Cellular and Molecular Aspects of the Plant Hormone Ethylene* (J. C. Peche et al., eds.), Kluwer, Boston, p. 106 (1993).
159. H. J. Klee, M. B. Hayford, K. A. Kretzmer, G. F. Barry, and G. M. Kishore, *Plant Cell*, *3*: 1187 (1991).
160. C. J. S. Smith, C. F. Watson, J. Ray, C. R. Bird, P. C. Morris, W. Schuch, and D. Grierson, *Nature*, *334*: 724 (1988).
161. R. E. Sheehy, M. Kramer, and W. R. Hiatt, *Proc. Natl. Acad. Sci. USA*, *85*: 8805 (1988).
162. L. J. Edgerton and W. J. Greenhalgh, *J. Am. Soc. Hortic. Sci.*, *94*: 11 (1969).
163. S. Lavee and G. C. Martin, *J. Am. Soc. Hortic. Sci.*, *106*: 14 (1981).
164. K. Lurssen and J. Konze, in *Ethylene and Plant Development* (J. A. Roberts and G. A. Tucker, eds.), Butterworth, London, p. 363 (1985).
165. W. C. Cooper and W. H. Henry, *J. Agric. Food Chem.*, *19*: 559 (1971).
166. D. J. Osborne, *Outlook Agric.*, *13*: 97 (1984).
167. F. T. Addicott, in *Herbicides: Physiology, Biochemistry and Ecology.* (L. J. Audus, ed.), Academic Press, London, Vol. 1, p. 191 (1976).
168. J. N. Knight, *J. Hortic. Sci.*, *58*: 371 (1983).
169. R. H. Meyer, *HortScience*, *17*: 658 (1982).
170. D. W. Greene, *J. Am. Soc. Hortic. Sci.*, *108*: 415 (1983).
171. R. D. Child and R. R. Williams, *J. Hortic Sci.*, *58*: 365 (1983).

172. L. J. Edgerton, *HortScience*, *6*: 378 (1971).

173. G. W. Schneider, *J. Am. Soc. Hortic. Sci.*, *103*: 455 (1978).

174. M. S. Reid, *HortScience*, *20*: 45 (1985).

175. A. C. Cameron and M. S. Ried, *HortScience*, *16*: 761 (1981).

176. Y. Mor, M. S. Reid, and A. M. Kofranek, *J. Am. Soc. Hortic. Sci.*, *109*: 866 (1984).

177. U. van Meeteren and M. de Proft, *Physiol. Plant.*, *56*: 236 (1982).

178. Y. Mor, A. H. Halevy, A. M. Kofranek, and M. S. Ried, *J. Am. Soc. Hortic. Sci.*, *109*: 494 (1984).

179. F. B. Abeles, in *Ethylene and Plant Development* (J. A. Roberts and G. A. Tucker, eds.), Butterworth, London, p. 287 (1985).

180. S. P. Schouten, in *Ethylene and Plant Development* (J. A. Roberts and G. A. Tucker, eds.), Butterworth, London, p. 353 (1985).

181. I. Adato and S. Gazit, *J. Exp. Bot.*, *28*: 636 (1977).

182. C. M. Peterson, J. C. Williams, and A. Kuang, *Bot. Gaz.*, *151*: 322 (1990).

183. M. S. Zehni, D. G. Morgan and F. A. Saad, *Nature*, *227*: 628 (1970).

184. O. O. Ojehomon, *J. Exp. Bot.*, *23*: 751 (1972).

185. A. Kuang, C. M. Peterson, and R. R. Dute, *J. Exp. Bot.*, *43*: 1611 (1992).

186. K. M. Jones, T. B. Koen, and R. J. Meredith, *J. Hort. Sci.*, *58*: 381 (1983).

187. B. W. Wood, *HortScience*, *18*: 53 (1983).

188. G. C. Martin and C. Nishijima, *J. Am. Soc. Hortic. Sci.*, *97*: 561 (1972).

189. G. S. Sibbett and G. C. Martin, *HortScience*, *17*: 665 (1982).

190. J. N. Knight, *J. Hortic. Sci.*, *57*: 61 (1982).

191. F. Ergenoglu, in *Cell Separation in Plants* (D. J. Osborne and M. B. Jackson, eds.), NATO ASI Series H35, Springer-Verlag, Berlin, p. 323 (1989).

192. P. R. Hedberg and P. B. Goodwin, *Am. J. Enol. Vitic.*, *31*: 109 (1980).

193. G. C. Martin, in *Cell Separation in Plants* (D. J. Osborne and M. B. Jackson, eds.), NATO ASI Series H35, Springer-Verlag, Berlin, p. 331 (1989).

194. J. N. Knight, *Acta Hortic.*, *60*: 99 (1976).

195. K. M. Batal and D. M. Granberry, *HortScience*, *17*: 944 (1982).

196. W. C. Olien and M. J. Bukovac, *J. Am. Soc. Hortic. Sci.*, *107*: 1085 (1982).

197. G. C. Martin, R. C. Cambell, and R. M. Carlson, *J. Am. Soc. Hortic. Sci.*, *105*: 34 (1980).

198. S. Iwahori and J. T. Oohata, *Sci. Hortic.*, *12*: 265 (1980).

199. J. C. Peterson, J. N. Sacalis, and D. J. Durkin, *J. Am. Soc. Hortic. Sci.*, *105*: 788 (1980).

200. Y.-T. Wang and J. R. Dunlap, *HortScience*, *25*: 233 (1990).

201. F. J. Marousky and B. K. Harbaugh, *J. Am. Soc. Hortic. Sci.*, *104*: 876 (1979).

202. J. A. Milbrath and H. Hartman, *Oreg. Agric. Stn. Bull.*, *413* (1942).

203. J. W. Einset, J. L. Lyon, and P. Johnson, *J. Am. Soc. Hortic. Sci.*, *106*: 531 (1981).

204. Y. Israeli and A. Blumenfeld, *HortScience*, *15*: 187 (1980).

205. W. H. Livingston and M. L. Brenner, *Plant Dis.*, *67*: 909 (1983).

206. W. R. Chaney and A. C. Leopold, *Can. J. Forest Res.*, *2*: 492 (1972).

24

Plant Growth Hormones: Growth Promotors and Inhibitors

Syed Shamshad Mehdi Naqvi

Atomic Energy Agricultural Research Center, Tando Jam, Pakistan

I. INTRODUCTION

Since the dawn of agriculture, one of principal aims of human beings has been the control and promotion of plant growth to satisfy human needs. These two important aspects of people's work with plants, in the struggle to increase production, are by no means synonymous. Humans soon realized that lush green growth does not always produce the best crop in the form of fruit and seeds and hence were forced to evolve such well-known cultural methods as pruning, balance manuring, and mineral fertilizers to regulate the nature and luxuriance of plant growth. This control of the plant development pattern—this adjustment of balance between root and shoot and between leaf and flowering—was until about seven decades ago mediated by appropriate and very large empirical combinations of what may be called dietary and surgery. More than 60 years ago proof was given of what Julius Sachs [1] in 1880 postulated: that endogenous substances regulate the growth of various plant organs. In 1926, Went [2], in Holland, provided convincing evidence of a diffusible substance (auxin) from oat (*Avena sativa*) seedlings that promoted growth of these seedlings. At the same time, Kurosawa [3], in Japan, discovered another substance (gibberellin) from cell-free fungus (*Gibberella fujikuroi*) filterate which promoted growth of rice (*Oryza sativa*) seedlings. But it was not until 1955 that Skoog and his associates [4] discovered kinetin in an autoclaved sample of herring sperm DNA, which was active in what Wiesener in 1892 called cell division factor [5].

In both scientific and popular literature, these chemicals in plant and crop physiology have come to be known as plant growth regulators. Starting, as do many great scientific advances, in a few unobtrusive laboratories as purely academic observations, the volume of research on these substances by both plant physiologists and chemists has swelled to enormous proportions.

The naturally occurring (endogenous) growth substances are commonly known as plant hormones, while the synthetic ones are called growth regulators. A *plant hormone* (synonym: phytohormone) is an organic compound synthesized in one part of a plant and translocated to another part, where in very low concentrations it causes a physiological response. Plant hormones are identified as promotors (auxin, gibberellin, and cytokinin), inhibitors (abscisic acid, xanthoxin, and violaxanthin), ethylene, and other hypothetical growth substances (florigen, death hormone, etc.). They usually exist in plants and crops at a concentration lower than 1 μM; above this they are generally considered supraoptimal [6].

The mechanism(s) by which hormones trigger a response is still far from well understood. Specific receptors have been proposed, but no proof for their function in mediating hormone action has been given. There is, however, considerable evidence that gene expression is controlled by these hormones, but how it is done biochemically is largely unknown [7]. These hormones are found in all actively growing plant parts; young leaves and apical buds are particularly high in auxin, whereas young roots are high in gibberellins and cytokinins. Fruits and seeds are generally rich in all growth hormones. Therefore, these hormones are ubiquitous in plants and crops and are generally not species specific.

In general, a deficiency of hormone must be created experimentally (as by removing young leaves or using a hormone-deficient mutant) to show that adding a hormone has an effect. In this respect, the Mitscherlich law of diminishing returns [8] can be modified as follows: the increase in plant response produced by a unit increment of a deficient (limiting) hormone is proportional to the decrement of that hormone from the maximum.

To deal in depth and breadth with the entire field within our space limitations here is not possible. Therefore, we aim in this chapter to discuss advances in our understanding that are relevant to plant and crop management.

II. AUXINS

Auxin is a Greek word derived from *auxein*, which means "to increase." It is a generic term for chemicals that typically stimulate cell elongation, but auxins also influence a wide range of growth and development response. The existence of growth-regulating chemicals that control plant growth, and the interrelations between their parts, was the outcome of experiments on root and shoot responses to external stimuli. Ciesielski, working with roots, and Charles and Francis Darwin, working with shoots, observed that in both organs the tip controlled the growth rate of the immediate growing axes as well as the regions located some distance away [9]. The Darwins performed simple experiments on the photoresponse of canary grass (*Phalaris canariensis*) and oat coleoptiles. When the tip was unilaterally illuminated, a strong positive curvature (growth toward light) along the growing axes resulted. If the tip was shielded by an opaque cap and only the lower part was exposed unilaterally, curvature generally did not result. From these and other empirical experiments, the Darwins concluded that "when seedlings are freely exposed to a lateral light, some influence is transmitted from the upper to the lower part, causing the latter to bend." Boysen-Jensen [10], working with *Avena* coleoptiles, concluded that "the transmission of the irritation is of a material nature produced by concentration changes in the coleoptile tip." Paál [11] corroborated his findings by demonstrating that "the stem tip is the seat of the growth regulating center. In it, a substance (or mixture) is formed and internally secreted, and this substance equally distributes on all sides, moving downwards through the living tissue." But it remained for Went [2] to make the definitive discovery of auxin and to determine it quantitatively by *Avena* curvature bioassay (Figure 1). The chemical isolation and characterization was, however, done by Kögl et al. [12]. The details of the development of the concept and discovery of indoleacetic acid (IAA), as outlined above, is described in two classical works [10,13].

It was not until 1946 that a good chemical identification of IAA was made in a higher plant [14]. IAA has come to be recognized as perhaps the only true auxin of plants and crops.

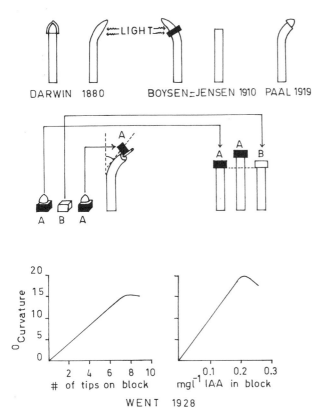

Figure 1 Diagrammatic representation of the major experiments leading to the discovery and quantification, by bioassay, of the auxin.

This auxin has also been isolated from culture filterates of bacteria, fungi, and yeast plasmolysate, but its role in these organisms is less clear.

Besides IAA, plants contain three other compounds which are structurally similar and elicit many of the same responses as that of IAA: 4-chloroindoleacetic acid (ClIAA), phenylacetic acid (PAA), and indolebutyric acid (IBA). However, their physiological significance and transport properties remain obscure at present. Engvild [15] has advanced the idea of death hormones and suggested ClIAA as one of them. Four additional compounds—indoleacetaldehyde (IAALD), indoleacetamide (IAM), indoleacetonitrile (IAN), and indole ethanol—are also found in a range of plants, but they are readily converted to IAA in vivo. The enzymes aldehyde dehydrogenase and indoleacetamide hydrolase, which catalyze the conversion of IAALD and IAM, respectively, to IAA, are active in plant tissues in which workers have detected IAALD or IAM. Similarly, IAN, found in the Cruciferae and Gramineae families, is also accompanied by the enzyme nitrilase, which is involved in the conversion of IAN to IAA. These similar circumstances, in a range of plants, indicate that IAA is the true active free auxin in plants. Furthermore, free auxin forms are probably the most immediately utilizable by plants in their growth processes.

A. Chemical Nature

In 1882, Nencki and Sieber discovered that indole-3-acetic acid was a constituent of human urine, which in 1934 was confirmed by Kögl and his co-workers, with the additional information that it was active in promoting the growth of some plant tissues or organs [16]. Within a year, it was also isolated from yeast plasmolysate and from the culture filterates of *Rhizopus suinus* [10,13]. However, its first isolation from a crop plant [i.e., from immature maize (*Zea mays*) kernels] was made by Haagen-Smit et al. [14]. It is now commonly accepted that IAA is perhaps the only endogenous auxin in plants and crops. Interestingly, chemists were aware of IAA long before plant scientists became aware of it. IAA was first synthesized by a German chemist in 1904 [17], but it was not suspected to have biological activity.

There are many purely synthetic compounds which mimic physiological actions similar to that of IAA. They are chemically diverse but can be classified in five major categories: indole acids, naphthalene acids, chlorophenoxy acids, benzoic acid, and picolinic acid derivatives. Two compounds belonging to the first group, indolebutyric acid and indolepropionic acid, are not exclusively synthetic; they have also been reported to be present in some plant species. The well-known naphthaleneacetic acid and β-naphthoxyacetic acid belong to the second group. The best known among the chlorophenoxy acid group are 2,4-dichlorophenoxyacetic acid (2,4-D), 2,4,5-trichlorophenoxyacetic acid (2,4,5-T), and 2-methyl-4-chlorophenoxyacetic acid (MPCA), which are known to be very powerful defoliants and herbicides when used in higher concentrations. In the benzoic group, the common synthetic auxins are the 2,3,6- and 2,4,6-trichlorobenzoic acids and dicamba, which is a powerful herbicide and is effective in some species of deep-rooted perennials, which are not readily killed by 2,4-D. Among the picolinic acid series, the best known is picloram (4-amino-3,5,6-trichloropicolinic acid), which is known to be the most powerful selective herbicide. However, the trend in recent years is to use natural plant hormones to regulate crop growth for greater production. This is because they act in low concentration and are fully degraded, therefore do not pose environmental and/or ecological threats.

B. Metabolism

The hormone IAA has been studied for more than six decades, yet it remains unclear how it is synthesized or degraded in plants. Because of the structural similarities the amino acid tryptophan is commonly considered to be a precursor to IAA. To date, biosynthetic pathways from L-tryptophan by way of tryptamine [18], indole-3-pyruvate and indole-3-acetaldehyde [19], indole-3-aldoxine and indole-3-acetonitrile [20], and indole-3-acetamide [21] have been proposed. Nevertheless, some workers have reported that D-tryptophan may also be an effective precursor for IAA biosynthesis [7,21]. However, in *Lemna gibba*, D-tryptophan was not converted to IAA and also the rate of conversion from L-tryptophan was far lower than expected for a direct precursor [22]. However, it remains unclear which pathway does function in plants. Wright et al. [23], using the tryptophan auxotroph maize mutant *orange pericarp*, have questioned the idea that tryptophan is a precursor of auxin and suggest a nontryptophan pathway as a primary route of IAA biosynthesis. It is therefore possible that plants and crops use more than one route for in vivo IAA biosynthesis.

It is reasonable to assume that plants have mechanisms to regulate the levels of auxin to maintain balanced growth. This is done by controlling the rate of synthesis as well as by degradation or by forming conjugates (bound). The enzyme IAA-oxidase with its several isoenzymes, which usually have the characteristics of peroxidases, is known to catalyze the reaction. Two pathways of degradation are known in many plants. The first involves oxidation by O_2, leading to the loss of carboxyl group as CO_2 and usually 3-methyleneoxyindole as a principal product. In the second pathway the carboxyl group of IAA remains intact, but carbon at the second position of the heterocyclic ring is oxidized to oxindole-3-acetic acid. In some species, however, carbons 2 and 3 are oxidized to form dioxindole-3-acetic acid [24]. Lee and

Starratt [25] have shown that soybean (*Glycine max*) callus and hypocotyl tissues were capable of oxidizing [^{14}C]IAA via the carboxylative pathway to indole-3-methanol glucoside as a major product. However, details of these degradative pathways are still unclear. Synthetic auxins and IAA conjugates are also not destroyed by these enzymes.

In auxin conjugates the carboxyl group is covalently combined with other molecules in the cell to form derivatives that do not allow for easy extraction. Many IAA conjugates are known, including the indole-3-acetylaspartic acid (IAAsp), indole-3-acetylglutamic acid (IAAGlu), and the esters IAA-*myo*-inositol (IAIns) and indole-3-acetylglucose (IAGlu). These conjugates, along with the free IAA, have also been found in IAA overproducing transgenic and wild-type tobacco (*Nicotiana tabacum*) plants [26]. These conjugates are not active per se but by hydrolysis release free IAA.

C. Transport

The integrity of the complicated structure of plants depends, to a great extent, on regulations that coordinate the various parts of the whole plant. Since production centers and action sites are often located at different places in the plant body, auxin transport takes place. Ever since Went first demonstrated the basipolar movement of auxin and its quantitative description by van der Weij [13], physiologists have been interested in studying various parameters of its transport as a way to understand the general phenomena of polarity in plant development. With the availability of high-specific-activity ^{14}C-labeled auxin coupled with liquid scintillation spectrometry (>90% counting efficiency), it became possible to perform more complex experiments and to obtain more reliable data from a single segment than was feasible previously [27,28]. In dicots and monocots alike, auxin moves predominantly in the basipolar direction [28]. However, in the young vegetative *Coleus blumei* internode, the auxin applied moves with a 3:1 ratio in the basipetal-to-acropetal direction, but this changed to 1.3:1.0 when the plants flowered [27,29]. Using paper chromatography, evidence was obtained, for the first time, that the auxin collected apically was the auxin applied at the basal end [27–29]. In excised root segments the polar transport was in the acropetal direction, but it appears that in the apical segments of intact roots (with caps), auxin moves basipetally [30]. However, in both organs, polarity was maintained regardless of the tissue orientation.

The other characteristics of auxin transport are that it is metabolically dependent and moves basipolarly with a velocity generally ranging from 10 to 20 mm/h, depending on the plant species tested. In short-term (4 h or so) experiments, it moved unchanged with similar velocity in intact and isolated segments. It could also move laterally following tropic stimulation [27,31,32].

In intact plants, auxin moves in two distinctly different systems. From the apex it moves toward the roots, and movement from the young leaves and meristematic regions of shoots resembles polar transport. This transport requires living cells and is interferred with by inhibitors such as triiodobenzoic acid (TIBA) and abscisic acid (ABA), anoxia, and low temperature. Auxin was not usually translocated through the epidermis, cortex, pith, or vascular bundles, but instead through parenchyma cells in contact with vascular bundles [33]. Cell-to-cell auxin transport across the tissues is now considered by some investigators to be chemiosmotically coupled to the electrochemical potential of auxin and proton [7]. The second mode of transport is along with the assimilates exported from the leaves. This movement lacks polarity and the auxin can move in any direction with a velocity of 100 to 240 mm/h, depending on the location of the metabolic sink and the water status of plants. This indicates that auxin supplied to or from the mature leaves enters the sieve tubes and is transported rapidly with assimilates. The physiological importance of this system for delivering endogenous auxin over long distances has not been investigated.

D. Biological Activity

Biological activities of the applied auxin are so diverse that compiling a complete list is quite difficult. A number of responses at the molecular, cellular, organ, and whole plant levels have been described which are known to be influenced by the exogenous application of IAA. But to what extent these are under the control of endogenous IAA has not been established unequivocally. However, there are a few examples, such as control of the elongation of stamen filament of *Gaillardia grandiflora* and the photoinhibition of mesocotyl growth, which correlate well with the endogenous IAA levels [32].

Auxin response is related to concentration, which is normally extremely low. In plants, free IAA is on the order of 10^{-8} g/kg fresh weight. The endogenous level of auxin is important in determining the course of development [34] (Figure 2). Changing concentration can convert root meristem to shoot meristem, and vice versa [35]. A high concentration is inhibitory, while a low concentration is stimulatory, and both are important. Commonly, the highest concentration of auxin is found in the meristematic regions.

1. Bioassay

Only a bioassay can detect the physiological activity of a substance at hormonal concentrations, and thus development of a quantitative bioassay provided the beginning for all hormonal work [13]. A number of bioassays have been developed to measure the activity of auxins. Well known among them are the (a) *Avena* coleoptile curvature test, (b) *Avena* coleoptile section test, (c) split pea (*Pisum sativum*) stem curvature test, and (d) cress root inhibition test. Modern instruments of separation and quantitation, such as high-performance liquid chromatography (HPLC) and gas chromatography coupled with mass spectrometry (GC/MS), are commonly

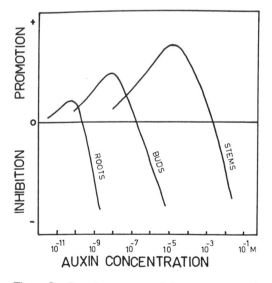

Figure 2   Growth responses of plant organs to various concentrations of auxin. (From Ref. 34.)

used. Another extremely sensitive detection method is immunoassay (a type of bioassay), and commercial kits are available for determining picogram quantities of plant hormones [36].

2. Tropisms

In nature, the orientation of shoot and root is of crucial importance to seedlings developing from seeds oriented at all angles in the soil. For survival, shoots therefore need to be oriented toward the light so that photosynthesis can begin before the stored food reserves are depleted, and roots must be oriented toward the gravitational vector, to obtain water and ions and to secure anchorage and mechanical support. Shoots are thus considered to be positively phototropic, and roots, negatively phototropic. On the other hand, shoots are negatively gravitropic and roots positively gravitropic. These two responses are of great ecological importance and also have relevance to plant and crop productivity.

Tropisms (from the Greek word *trope*, "turn") have been divided into three phases: perception, transduction, and response. To explain the transduction and response phases, caused by photo or gravity stimulation, the Cholodny–Went theory states: "Growth curvatures, whether induced by internal or by external factors, are due to an unequal distribution of auxin between the two sides of the curving organ. In the tropisms induced by light or gravity the unequal auxin distribution is brought about by a transverse polarization of the cells, which results in lateral transport of the auxin."

Although the validity of the Cholodny–Went theory has been questioned without an alternative explanation [37], others have come to its defense [38,39]. Recently, Li et al. [40], using auxin-responsive mRNAs called small auxin up RNAs (SAURs) as a molecular probe, have supported the idea of asymmetric distribution of auxin at the sites of action during tropistic response. However, Jaffe et al. [41], using the pea mutant 'Ageotropum', found that roots appear to be neither negatively phototropic nor positively gravitropic but grow in the direction of increasing soil moisture.

a. PHOTOTROPISM. When the exposure of light falling on plant organs becomes differential, a curvature develops, so that reorientation takes place in such a way that the organ is evenly illuminated. This response of the plant or its organ(s), where the plane of curvature is determined by the spatial relationship of the organ and the light stimulus, is known as *phototropism*.

The Cholodny–Went theory suggests that unilateral illumination causes auxin to move laterally to the darkened side, causing the organ to curve toward light. Bioassay and ^{14}C-labeled auxin studies using etiolated maize coleoptile tips have shown that there was no difference in auxin yield between evenly illuminated segments (first or second positive range) and their dark control segments. But unilateral illumination caused asymmetry in auxin yield between the two halves of the tip segments [42,43]. These observations demonstrated the consequence rather than the cause of the lateral asymmetry in auxin transport. In such a cause-and-effect relationship, it is important to differentiate between the two. Using [^{14}C]IAA and maize coleoptiles, Naqvi and Gordon [44], studying the [^{14}C]IAA transport kinetics of etiolated maize coleoptiles, provided the first evidence that bilateral illumination, in the first positive range, caused retardation of basipetal auxin transport intensity (capacity) without affecting the velocity. The lateral asymmetry observed was thus a consequence of the resultant concentration gradient. Based on their observations and other evidence, Naqvi and Engvild [31] proposed that "photolysis of a carotenoid (violaxanthin) produces compounds (similar to or identical with ABA) that inhibit the basipetal auxin transport. Unilateral stimulation produces an asymmetry of inhibition and, hence, the curvature." It is now known that in maize coleoptiles, ABA is the most dominant hormone after IAA [45]. Therefore, any change in its concentration would influence auxin transport. Further support for the lateral transport theory came from the observation that in bean (*Phaseolus vulgaris*), auxin transport inhibitor, DPX-1840 inhibited basipetal as well as lateral transport, resulting in growth retardation and the loss of phototropic response [46]. These results suggest that basipetal and lateral transports are essential for photo-stimulated differential growth.

b. GRAVITROPISM. Frank [47] was perhaps the first to establish that curvatures due to gravitational stimulus were directly connected with growth and described this phenomenon by the term *geotropism*. The precise gravity-perceiving mechanism is still unknown. Since the root cap (terminal 500μ) appears to be the seat of perception, it was thought for a long time that amyloplasts (organelles filled with starch grains) were, in fact, the gravity-perceiving mechanism. However, the use of mutant lacking amyloplasts in the cap, yet exhibiting nearly normal gravitropic response, has ruled out this mechanism [48]. Whatever may be the mechanism of perception, the ultimate response is asymmetrical growth, resulting in a positive (root) or negative (shoot) curvature.

Strong evidence suggests that auxin controls the gravitropic response and the Cholodny–Went theory has been invoked to explain the growth asymmetry which results in the shoot bending away from, and the root toward, the gravity vector. It is well established that when shoot segments are placed horizontally, more auxin is recovered from the lower than from the upper half. Using bioassay and ^{14}C-labeled auxin, differences between the upper and lower halves of the maize coleoptiles were demonstrated, but the activity on the upper halves did not differ from that of the vertical halves [38]. However, Naqvi and Gordon [49], working on ^{14}C-labeled auxin transport kinetics, demonstrated that horizontal reorientation of maize coleoptiles reduced the transport intensity (capacity) of the upper half without affecting the velocity. Thus a lateral auxin concentration gradient enhanced movement from the upper toward the lower half. This lateral movement has also been demonstrated in the maize mutant amylomaiz [50]. McClure and Guilfoyle [51], using a molecular biology approach, have shown a clear correlation between auxin-controlled gene expression and the gravitropic response of soybean hypocotyls. Molecular genetic studies on the phenotype of auxin-resistant mutants have further substantiated that auxin played an important role in root gravitropism [52].

3. Apical Dominance

The integrity of the complicated form of higher plants depends to a great extent on regulations that integrate the various component parts. This gives a characteristic form or shape that is repeatable in time and space. The stems assume characteristic geometry due largely to the extent of biochemical influence exerted by the apex on the development of lateral (axillary) bud meristems. This phenomenon of apical dominance (growth correlation or compensatory growth, i.e., preventing or slowing of lateral bud growth by the apex) is of major importance in integration of the plant body, and parallel examples are found in mosses and ferns [53]. Awareness of this role of the apex has undoubtedly influenced pruning practices in horticulture and crop production. Interested readers are referred to excellent reviews on the subject [54,55].

In a pioneering work using *Vicia faba*, Thimann and Skoog [56] demonstrated that auxin diffused in agar blocks from the excised shoot apices, *Rhizopus* filterate, or human urine can partially inhibit lateral bud growth. Later, Leopold [57] showed that when the shoot apex of 'Wintex' barley (*Hordeum vulgare*) was destroyed, tillering was profuse unless the apex destroyed was replaced by an auxin. These observations indicated that auxin from the apex exerts an influence on *Vicia* lateral bud growth as well as on barley tillering. Studies on inhibiting auxin emanating from an apical bud, using inhibitors such as TIBA or morphactin, have shown that the lateral bud growth was effectively enhanced. These studies thus provide evidence in support of a direct role, but others question it and assign an indirect role to auxin in controlling this phenomenon [54]. In intact tobacco, petunia, and *Arabidopsis thaliana* plants, use of transgene-mediated auxin and/or ethylene deficiencies, along with mutants insensitive to auxin or ethylene, supports the idea that apical dominance is the result of the auxin/cytokinin ratio rather than auxin-induced ethylene production [58].

4. Root Formation

The most apparent auxin control of cell division is the formation of roots. Early evidence indicating the presence of active buds on cuttings to promote root development below led to the identification of auxin as the root-forming hormone [13]. The ease with which roots can form on cuttings varies enormously; shoot cuttings from some plants produce roots simply if their basal cut end is left in water, whereas other species do so only rarely. Root formation shows polarity and is always formed at the morphological basal end, even if the cuttings are inverted upside down. Since auxin moves basipolarly, it was logical to believe that root formation at the basal end is a consequence of the movement of auxin to the lower tissues. Removal of rich sources of auxin (i.e., buds and young leaves) reduces the number of lateral roots formed. This capacity is restored, however, if auxin is substituted for these organs. Tissue culture studies have provided further support by showing that a higher auxin/cytokinin ratio induces root formation and that changing it converts shoot meristem to root meristem, and vice versa [35].

To preserve genetic homogeneity, vegetative propagation (cloning) is important in horticulture, floriculture, forestry, and in the conventional breeding and/or biotechnology of higher plants. Therefore, basal treatment with indolebutyric acid (IBA) and synthetic auxin naphthaleneacetic acid (NAA) is commonly used to induce adventitious root formation in hard-to-root species.

III. GIBBERELLINS

The unique property of gibberellins (GAs)—that of increasing the growth of plants by greatly elongating the cells—was discovered by Kurosawa [3]. Studying the symptoms of the rice disease "bakanaebyo" ("foolish seedling disease"), Takahashi [59] observed that the causal pathogen was a soilborne fungus, *G. fujikuroi*, the sexual or perfect stage of *Fusarium moniliforme*, which caused infected seedlings to grow abnormally taller and to fall over due to their spindly stem structure. He observed further that when a pure culture filtrate was sprayed onto rice seedlings, it produced the same abnormal growth. This suggested that the abnormal growth of the infected seedlings was caused by a soluble substance(s) produced by the fungus. Other Japanese biologists showed that the excessive growth was not confined to rice but that the filtrate could induce it in many other species. According to Takahashi et al. [59,60], in 1938 Yabuta and Sumiki isolated two crystalline active substances from the culture filtrates and called them gibberellin A and B.

Western scientists became interested in gibberellin research in early 1950 and succeeded in isolating an active principle from *G. fujikuroi*. The growth-promoting activity of this compound was similar to that of the GAs isolated by Japanese investigators, but the chemical nature was clearly different. Therefore, it was named gibberellic acid (GA_3) [59]. The concentration of GAs is usually highest in immature seeds, reaching up to 18 mg/kg fresh weight in *Phaseolus* species [61]. However, it decreases rapidly as the seeds mature. In general, roots contain higher amounts of GAs than the shoots, and vegetative tissues contain a comparatively low level of GAs, depending on the types of tissues and their stages of development.

A. Chemical Nature

Eighty-four gibberellins are currently (1993) listed, of which 25 are from fungi, 73 from higher plants, and 14 are common to both [7,59]. Among these, 68 are free and 16 are known to occur in conjugated form [60]. All gibberellins are acidic diterpenoids having an *ent*-gibberellane carbon skeleton and are designated GA_1, GA_2, GA_3, . . ., GA_{84}. They differ from one another mainly in the numbers and positions of substituent groups on the ring system and in the degree of saturation in the A ring. Free GAs are divided into two groups: those possessing an *ent*-gibberellane skeleton (20 carbons) or *ent*-20 nongibberellane (19 carbons) mono-, di-, or tricarboxylic acids. The terms *C-20* and *C-19* denote compounds that have retained and lost, respectively, carbon atom 20, and generally, C-19 GAs are more active than C-20 GAs. They are grouped in either four- or five-ring systems. The fifth ring is the lactone ring attached to ring A, which is not present in the *ent*-gibberellane. The carboxyl group at C-7 seems to be essential for biological activity. They also seem to be rather stable in plants and are readily interconverted to form glycosides by conjugating with sugars.

B. Metabolism

Gibberellins are diterpene, belonging to a large group of naturally occurring compounds in plants known as terpenoids. All terpenoids are basically built up from isoprene units, which are five-carbon (5C) compounds.

$$\begin{array}{c} C \\ \diagdown \\ \diagup \\ C \end{array} \!\!\!> C\text{–}C\text{–}C$$

The linking of two units yields a monoterpene (C-10), of three a sesquiterpene (C-15), of four a diterpene (C-20).

Our knowledge regarding the biosynthesis of GA in plants and crops stems from feeding ^{14}C-labeled acetic acid and mevalonate to *G. fujikuroi* through the culture medium. It was observed that ^{14}C from these two compounds was incorporated into gibberellic acid (GA_3) [59]. Graebe et al. [62], using a cell-free system [the endosperm nucellus of wild cucumber (*Marah macrocarpus*; formerly *Echinocystis macrocarpa* Greene)], reported incorporation of [^{14}C]mevalonic acid into kaurene, kaurenol, and geranylgeraniol. The pathway commonly accepted is 3-acetyl-CoA → mevalonic acid → isopentenyl pyrophosphate (a five-carbon terpenoid) → geranylgeranyl pyrophosphate (a 20-carbon compound) → copalyl pyrophosphate → kaurene → kaurenol → kaurenal → kaurenoic acid → GA_{12}-aldehyde. The GA_{12}-aldehyde is a branch point to the formation of various GAs. Pathways to various GAs differ mainly in the position and sequence of hydroxylation, and more than one pathway can operate in the same plant. The details of the pathways are covered comprehensively in several excellent works [59,60].

The metabolism of GAs in plant tissue is not well understood, and very meager information exists regarding its eventual fate. There is evidence that considerable interconversion of gibberellins (i.e., one GA can be converted to another GA) takes place in the plant. Immature seeds from "summer"-grown *Pisum sativum* were fed with GA_9, which was metabolized to GA_{51} and dihydro-GA_{31} and its conjugate. But in "winter"-grown seeds, the metabolites were GA_{20} and GA_{51}. Another metabolite, gibberellethione, was isolated from immature seeds of *Pharbitis nil* [60]. Degradation of commonly used GA_3 appears to be slow. However, during the active growth phase, most of the gibberellins are metabolized to inactive forms by hydroxylation or by conjugation with glucose to form glucosides.

C. Transport

Gibberellins are known to be synthesized in all young, actively growing organs, vegetative or reproductive, including immature and mature seeds. Understanding their transport within the plant pertains primarily to work with excised coleoptile, stem, or petiole segments in a donor–tissue–receiver system. Transport has generally been observed to be nonpolar, but occasionally, basipolar movement has been reported [28] with a velocity up to 1 mm/h. However, information regarding endogenous movement is rather indirect. It has been noted to occur in the phloem by the same mechanism and in a pattern similar to that with which other assimilates move. Gibberellins have been isolated from phloem sieve tube saps as well as from the xylem stream. Experimental evidence using ^{14}C-labeled GA shows an interchange between phloem and xylem [17]. This suggests that GA is transported both symplastically and apoplastically. Its phloem transport rate was similar to that of other assimilates. In analogy with the source-to-sink movement of assimilates in phloem, perhaps the polar movement observed was to a growth center rather than to the morphological base.

D. Biological Activity

The physiological properties of these highly active compounds are wide ranging, but extensive growth and *de novo* enzyme synthesis is the most significant. The GAs act synergistically with other hormones in what might be called a system approach. The best known response is the stimulation of internode growth of dwarf maize, pea, and bush bean, which after treatment with GA attains the normal height. In some cases, but not all, dwarfism does in fact seem to be correlated with endogenous GA deficiency. The most detailed analysis, at the molecular level, has been done with dwarf mutant of maize, known as dwarf-5 (d_5) [59,60]. The height of the dwarf-5 mutant is about one-fifth that of its parent, due to a single gene mutation causing a deficiency of GA. When treated with GA, the mutant attains the height of normal maize. The action of many GAs is similar to that of IAA, including cell elongation, promotion of cambial activity, induction of parthenocarpy, and stimulation of nucleic acid and protein synthesis. The GAs vary greatly in their biological activity, and GA_3 and GA_7 are considered to have the widest range. In ferns, algae, and fungi, GAs have also been shown to influence growth and development.

The content of GAs varies depending on the types of tissues and their stages of growth. Tissues other than seeds usually contain very low amounts (e.g., 0.3 µg/kg in young bamboo shoots) [60]. Roots are considered to be the richest source of GAs, and most GAs, as such or in bound form, are supplied to the shoot.

1. Bioassay

A number of bioassays, such as elongation of dwarf maize, pea, and rice seedlings, and of lettuce (*Lactuca sativa*) hypocotyl, *Avena* leaf segments, and chlorophyll retention in *Rumex* leaf disks [63,64], have been used to determine gibberellin activity in plant extracts. These basic responses have generally been used in various modifications with different plant materials. These assays are convenient, specific, easy to use, and detect nanogram levels of gibberellins. Nishijama and Katsura [65] have improved the dwarf rice bioassay to detect picogram quantities of gibberellins. Among the commonly used bioassays, the barley endosperm assay [66] has generally been preferred. In this bioassay, sterile de-embryoed barley (cv. Himalaya) half-seed (endosperm) is incubated in GA_3 solutions for 24 to 48 h. The presence of GA_3 stimulates α-amylase synthesis in the aleurone layers (two to four layers of live but undividing cells), which breaks up the starch and builds up the reducing sugars. The amount of reducing sugar, analyzed colorimetrically, is dependent on the GA_3 concentration in the test solution.

2. Growth Promotion

The isolation of gibberellins from plants, combined with their physiological responses to applied GAs, suggests that they do play a role in the regulation of various phases of their development.

At the same time, if a process fails to respond to a certain GA, it could not be used as evidence that GA is not required. It may be that a different GA is required to elicit the response. In *Silene*, GA_3 fails while GA_7 induces flowering under noninductive conditions, suggesting the involvement of gibberellin in this process. Several species of conifers show little or no elongation to GA_3 treatment, but they do respond to a mixture of GA_4 and GA_7 [67].

During the vegetative growth phase, mitotic activity in subapical meristem is regulated by gibberellins. A reduction in its level causes a severe imbalance between internode and leaf growth, resulting in a form of growth called a *rosette*, first noted in *Hyoscymus niger* and later in many other plants. In plants such as cabbage (*Brassica oleracea capitata*), leaf development is profuse and internode growth is retarded during the vegetative phase. But before the start of reproductive growth, a marked elongation of the internode, called *bolting*, takes place. When treated with GA_3 during their rosette phase, such plants bolt and flower, whereas nontreated plants remain rosetted. There is evidence that endogenous gibberellin levels are higher in the bolted plants than in the rosetted plants. In addition, higher concentrations have been found in the bolted long-day *Rudbeckia speciosa* and cold-requiring *Chrysanthimum morifolium* cv. Shuokan than in their nonbolted forms [63]. Thus it appears that the influence of gibberellin in such a response includes the stimulation of cell division as well as cell elongation.

For many crop species there are genetic dwarf mutants that are deficient in gibberellin. Dwarfs of rice, maize, and peas phenotypically attained the height of normal varieties when treated with gibberellin. These mutants have been used successfully for gibberellin bioassay and in breeding programs for increasing crop productivity. Dwarf rice responded to as little as 4 pg of GA_3 per plant [64]. Five different gibberellin-synthesis mutants are known which are underproducing dwarf mutants. Each mutant has a mutation on a different gene, and each gene controls a different enzyme needed for gibberellin synthesis. The work of MacMillan and Phinney [68] suggests that only GA_1 controls elongation in maize, and all five dwarf mutants lack the enzyme(s) that can convert other gibberellins to GA_1. Other evidence also indicates that GA_1 is the main gibberellin needed by dwarf rice, rape (*Brassica napus*), peas, sweetpeas, tomato (*Lycopersicon esculentum*), and some wheat (*Triticum aestivum*) cultivars for stem elongation. Mutants are not only lacking GA_1, but GA_1-overproducing mutants with abnormally long internodes have been reported in *Brassica rapa* (syn. *campestris*) [69].

3. Dormancy of Buds and Seeds

Buds and seeds of many plants show the ability to retain viability while having limited metabolic activity and no observable growth during an unfavorable season. This physiological condition is commonly known as dormancy, and plants that grow in regions with a pronounced season usually adopt this strategy in late summer or early fall. Buds in dormant conditions are relatively more cold and drought tolerant than are actively growing buds. Similarly, seeds of many noncrop plant species remain dormant when they mature and will not germinate even if favorable conditions are provided. Dormancy of buds and seeds must be broken at a time when conditions are suitable for their growth and germination, respectively, during the spring. Long-day or brief red-light exposures have been found to break seasonal dormancy in many species. Gibberellins have also been found effective in overcoming both kinds of dormancy in buds as well as in seeds. Treatment with gibberellins has been observed to substitute effectively for long-day, low-temperature, or red-light exposure requirements. Due to ease in handling, much more is known about seed dormancy, but it is likely that much of this information may also be applicable to buds.

4. Mobilization of Nutrients

Endosperm is the nutrient storage organ of the seed for the developing embryo. Soon after the axis becomes active and root and shoot develop, the nutrient reserve (minerals, fats, starch, and proteins) is mobilized to support the juvenile seedling growth. This supply continues until the root develops the ability to absorb nutrient ions from the soil and the shoot system begins the photosynthetic process. Large molecules of proteins, fats, and starch have problems getting translocated; therefore, they need to be metabolized to smaller molecules such as amino acids or amides and sugars, which can readily be translocated from source to sink. Gibberellins are known to play a key role in the hydrolysis of starchy endosperm of cereal seeds. In 1960, Yomo in Japan and Paleg in Australia independently observed that GA_3 stimulated the degradation of de-embryoed barley endosperm [60]. A few years later it was demonstrated that barley embryo axis produces GAs, and it has further been shown by GC/MS as well as by immunochemical methods that grains also contain a large number of GAs [70]. The type of gibberellin is species specific, but GA_1 and GA_3 are important in barley. The involvement of a living aleurone layer in the degradation of starchy endosperm in grasses, including barley, was recognized more than a century ago, but the experimental evidence was not provided until much later.

When isolated aleurone layers are incubated with GA_3, a large number of enzymes (i.e., α-amylase, protease, ribonuclease, esterase, β-1,3-glucanase, acid phosphatase, glucosidase, peroxidase, etc.) were secreted in the incubation media [70]. Later studies confirmed that GA_3 was required for the *de novo* synthesis of α-amylase, β-1,3-glucanase, protease, and ribonuclease. Aleurone layers of barley, wheat, and wild oat respond to GA_3 by synthesizing hydrolytic enzymes, but most maize cultivars and some cultivated oat cultivars do not respond identically. Thus there is considerable genetic variability regarding the gibberellin response in cereal seeds. However, the role of gibberellins in the mobilization of food is not as clear in dicots and gymnosperms. The food reserves in these two classes of plants could be starch or fat, where added GAs may or may not influence degradation.

IV. CYTOKININS

The discovery of cytokinins was an outgrowth of tissue culture research by Skoog and associates. The isolation and identification of kinetin (6-furfurylaminopurine) from aged or autoclaved herring sperm DNA, and its promotion of cytokinesis (cell division) at concentrations as low as 1 μg/L, greatly stimulated research in the field of plant growth and development. Although kinetin does not occur naturally, its discovery greatly supported the concept of the existence of a cell division factor, postulated by Wiesner in 1892.

Haberlandt is generally credited as the pioneer in providing experimental evidence for the hypothetical cell division factor. In 1913 he demonstrated that phloem diffusates could cause cell division in potato (*Solanum tuberosum*) parenchyma cells. Later, in 1921, he reported that cell division induced by wounding was prevented if the cut surface was washed, and that leaf juice spread over the washed cut surface would restore it [71]. In the early 1940s, Van Overbeek [71a] observed that coconut milk could sustain the growth of isolated *Datura* embryo. According to Koshizima and Iwamura [72], subsequent work by Stewart and his associates established that coconut milk markedly stimulated the growth of carrot (*Daucus carota*) root explants by cell division. They did succeed in isolating the growth-inducing factor, but it was a mixture rather than a single compound that was effective in their carrot bioassay system. According to Koshizima and Iwamura [72], Skoog and associates, in the late 1940s, using aseptically isolated slabs of nondividing mature stem piths of tobacco plants (var. Wisconsin No. 38), observed cell division in the presence of vascular tissues. However, the first endogenous cytokinin was isolated from maize kernels and was named zeatin (Z) [6-(4-hydroxy-3-methyl-2-*trans*-butanylamino) purine]. Germinating seeds, roots, sap streams, developing fruits, and tumor tissues are rich in cytokinins [73]. There are 25 free cytokinins reported from higher plants [72] and some are active in causing maximum tobacco callus growth at concentrations as low as 0.004 μM.

A. Chemical Nature

All the known endogenous cytokinins are substituted purines attached to the N^6 position of the adenine ring. These N^6-substituted adenines can be classified into two groups according to their carbon skeleton of N^6 substituents: as N^6-isoprenoid and N^6-benzyladenine analogs.

The major group of cytokinins are N^6-isoprenoid adenine analogs and can be divided into three groups. The first subgroup consists of zeatin and its derivatives—ribosides, glucosides, and nucleotides—whose N^6-isoprenoid side chain is either 4-hydroxy-3-methyl-2-*trans*-butenylaminopurine or its *cis* isomer. The second subgroup of N^6-isoprenoid analogs consists of dihydrozeatin (diH)Z and its ribosides and glucosides. The third subgroup includes 6-(3-methyl-2-butenylamino)purine and N^6-(Δ^2-isopentenyl)adenine (2iP) and its ring substitution products.

The second and minor group of cytokinins (i.e., N^6-benzyladenine analogs) was first synthesized as 6-benzylaminopurine (BAP) with high biological activity. Later it was found to exist in a number of plant species.

Certain nonpurine compounds, such as 8-azakinetin, benzimidazole, N,N'-diphenylurea, and 2-benzthiozolyloxyacetic acid, have also been reported to have cytokininlike activity. Of these, three are synthetic and only N,N'-diphenylurea occurs naturally in plants. It has been suggested by some workers that these so-called "urea cytokinins" may also be considered true cytokinins, but most hormone physiologists do not agree [72,73]. These compounds may not be active as such, but they may serve as precursors or inducers for the commonly accepted cytokinins.

B. Metabolism

Miura and Miller [74] have suggested that all plant cells are capable of synthesizing cytokinins provided that the mechanisms to do so are "switched on." However, this does not mean that cytokinins are biosynthesized in the entire plant. Recent evidence suggests that actively dividing regions of plants are the sites of cytokinin biosynthesis. Since the root system possesses the most actively dividing regions, these regions are considered to be the major sites of cytokinin production.

Compared with other aspects of cytokinin physiology, little is known about their biosynthesis, which is comparatively quite complicated. The circumstances of its discovery and its effect on cell division and protein synthesis have somehow closely associated free cytokinins with RNA and DNA. The production in plants can be accounted for either by the turnover of cytokinin-containing tRNA, by *de novo* biosynthesis, or by both mechanisms. It has also been reported to be present in rRNA [75].

The major cytokinin-active base in tRNA, [9R]iP, is formed by the condensation of adenine with an appropriate donor of the N^6 substituent during post-transcriptional processing. The Δ^2-isopentenyl pyrophosphate (IPP) is the immediate precursor (donor) of the Δ^2-isopentenyl side chain of N^6-(Δ^2-isopentenyl)adenosine in tRNA. A cell-free enzyme system, isopentenyl AMP synthase, has been isolated from cultured autotrophic tobacco tissue which forms cytokinin from adenosine monophosphate (AMP) and IPP as substrate [76]. The Δ^2-isopentenyl pyrophosphate is a product of mevalonic acid (MVA) (an important precursor of carotenoids, abscisic acid, gibberellins, sterols, and other isoprenoid compounds) via Δ^3-IPP.

Besides the formation of free cytokinins from tRNA, there is strong evidence that they are also formed by *de novo* biosynthesis. Beutelmann [77] supplied labeled adenine to moss callus cells and obtained labeled cytokinin which co-chromatographed with 2iP, but no labeled cytokinin was detectable from tRNA. Similar results were obtained from the cytokinin-autotroph tobacco callus tissues, *Vinca rosea* crown gall tissues, and synchronously dividing tobacco callus cells [78].

The amount of cytokinin present in the tissue is regulated by conversion to a diversity of metabolites by the following reactions: (a) trans-hydroxylation of the terminal methyl group on the side chain, (b) side-chain reduction, (c) isoprenoid side-chain cleavage, (d) O-glucosylation, (e) N-glucosylation, (f) ring substitution by alanine moiety, and (g) base–ribonucleoside–ribonucleotide interconversion. These types of reactions have been observed in a number of plant species as well as in crown gall tissues. These reactions have also been obtained from tissues exogenously supplied with the hormone. Three enzymes, cytokinin oxidase (MW 88,000), cytokinin 7-glucosyltransferase (MW 46,500), and (9-cytokinin)alanine synthase (MW 64,500), have been purified and characterized [78]. The free base, nucleotide, and nucleoside forms of cytokinins appear to be easily interconverted in plant tissues. Incorporation of labeled cytokinin bases into ribosides (ribonucleoside) and ribotides (riboside 5'-phosphates) have been observed in a number of plant species. Five enzyme systems, purified from wheat germ, may be responsible for this interconversion. These are (a) adenosine phosphorylase, (b) adenosine kinase, (c) adenine phosphoribosyltransferase, (d) (5'-ribonucleotide phosphohydrolase) 5'-nucleotidase, and (e) adenosine ribohydrolase (adenosine nucleosidase) [72]. The ribosides are considered to be the translocation form, while ribotides are associated with uptake and transport across the cell membrane forms of cytokinins. Thus enzymatic regulation of bases, ribosides, and ribotides plays an important role in keeping adequate levels of free and active forms of cytokinins in plants and crops.

C. Transport

It is paradoxical that plant parts which are meristematic or which otherwise have growth potential (young leaves, buds and internodes, and developing fruits and seeds) are known to be the primary center of production as well as the main sink for the endogenous cytokinins or its metabolites.

The detection of cytokinin, from the xylem exudate and phloem sap of a large number of plants and crops, clearly indicates that nonliving as well as living tissues are involved in translocation. The polarity of cytokinin movement is acropetal in the xylem, whereas it moves bidirectionally in the phloem. Thus in phloem, cytokinins not only move from organ to organ in the aerial portion, but also from shoot to root, and vice versa [79]. Thus the velocity of its movement is the same as that of other assimilates. However, under in vitro conditions (donor–tissue–receiver system), 6-benzylaminopurine (BAP) is very poorly translocated [80].

D. Biological Response

Cytokinins are known to evoke a diversity of responses when applied exogenously to whole plants, plant tissues, or plant organs. Like other hormones, cytokinins have been used as a tool to investigate their role as endogenous controllers of plant growth and development. Like other hormones, cytokinins influence a multitude of morphological and physiological processes, among them seed germination, cell division and cell elongation, promotion of cotyledonary and leaf growth, control of apical dominance, delayed senescence, and morphology of cultured tissues.

As pointed out earlier, deficiency of a hormone must exist either experimentally or genetically to show that adding hormone has an effect. Since cytokinins occur in all meristematic as well as in potential growing tissues and organs, it is not possible to create experimental deficiency. Genetically engineered cytokinin-overproducing tobacco and *Arabidopsis* plants have been used to study the phenomena of apical dominance [81]. But in the absence of comparison with cytokinin-deficient mutants, the conclusion is equivocal. Thus clear evidence is yet lacking, which shows that specific physiological processes in plants and crops are under the control of endogenous cytokinins.

1. Bioassay

The physicochemical methods for isolation, purification, and identification of endogenous cytokinins have improved rapidly, but it appears that bioassays will always be an integral part of the identification process. Therefore, a number of bioassays have been developed and used by various investigators to test the bioactivities of endogenous as well as synthetic cytokinins. They include (a) lettuce seed germination, (b) radish (*Raphanus sativus*) leaf disk expansion, (c) *Xanthium* leaf disk (chlorophyll preservation), (d) soybean callus, (e) carrot phloem, (f) tobacco pith callus, (g) cucumber (*Cucumis sativus*) cotyledon greening, (h) radish cotyledon expansion, (i) *Amaranthus* betacyanin, (j) barley leaf senescence, and (k) oat leaf senescence [82,83].

2. Germination

Seeds that require preexposure to light for germination are called photodormant. Red-light exposure stimulates, and far-red exposure inhibits, germination of a number of species, including lettuce (cv. Grand Rapids) seeds. But the seeds of most crops do not require light because of natural selection against such a requirement. Cytokinin-imbibed seeds germinate better in dark than do unimbibed lettuce seeds. Similarly, cytokinin together with gibberellin effectively breaks the photodormancy of celery (*Apium graveolens*) seeds, but it was not as effective alone [84]. This indicates that red-light exposure may cause enhancement in the hormone level either by biosynthesis or by release from a bound form. However, such information is lacking because the level of hormones in the radicle or hypocotyl cells, responsible for germination, has not yet been determined.

3. Organ Development

The totipotency of plant cells was demonstrated in a classic paper by Skoog and Miller [85] showing that a balance between cytokinin and auxin controlled bud and root formation in tobacco pith explant. At high concentrations of both hormones, cells often grow amorphously without differentiation. But a high cytokinin/auxin ratio causes induction of shoots, while a high auxin/cytokinin ratio enhances root formation [35].

Lateral development of axillary buds is inhibited by the presence of apical bud. This was shown by excision of the apical bud to remove correlative inhibition. But in the presence of apical bud, soaking the entire shoot in cytokinin enhanced lateral bud growth to a large extent [86]. Treatment with hadacidin, an inhibitor of purine synthesis, inhibited axillary bud growth following decapitation [87]. Since adenine treatment could not reverse this inhibition, the inhibitor may not be specific. In the absence of cytokinin-deficient mutants in higher plants, there is only indirect evidence that an endogenous cytokinin level regulated the development of axillary buds. Medford et al. [81], using genetically engineered cytokinin-overproducing tobacco and *Arabidopsis* plants, observed that the most significant morphological change of high cytokinin levels was that it caused extensive growth of the axillary buds (Figure 3). Thus there is strong evidence that cytokinin and auxin balance is important in organ differentiation and its further development.

4. Delayed Senescence

In plants and crops, the process of senescence is encountered at all stages of their life cycle. When a functional mature leaf is excised from the main body of a plant, it switches on to its

A   B

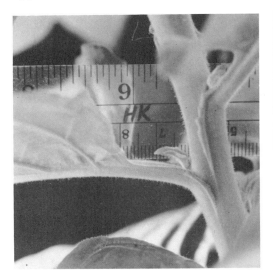 

Figure 3   Axillary bud growth (twelfth node from apex) in (A) wild-type and (B) transgenic cytokinin-overproducing tobacco plants. (From Ref. 81; with permission from copyright owner.)

death program. Progressive degradation of RNA, proteins, lipids, and chloroplast leading to the loss of chlorophylls start when the leaf dies. Once started, the degradation of cell constituents continues even if the cut end is dipped in mineral salts solution. This process of senescence (i.e., breakdown of cell constituents and yellowing of leaf), leading to ultimate death, is accelerated further if the leaf is kept in darkness. In many plant species, adventitious roots are formed at the cut end of the petiole, which decelerates the degradative process of the metabolites in the leaf blades. Since the supply of mineral salts did not influence the degradative process, roots being the major source of cytokinin supply [79], this hormone may have been responsible for delaying the process. But different species show a diversity of response to cytokinins, auxins, or gibberellins, in terms of loss of chlorophyll and protein, in experimental systems using detached leaves or leaf disks [88]. However, two lines of evidence suggest that cytokinins may play an important role in delaying senescence. The cytokinin contents of rooted leaf blades rise substantially and cytokinins can partially replace these roots [7]. Thimann [89] observed that when cut leaves of many species, including oats, were floated in cytokinin solution in the dark, the light requirement for delaying senescence was effectively replaced. He suggested that treatment with cytokinins maintained the integrity of the cell membrane. Studies have also shown that cytokinin, auxin, and/or ABA influence stomatal movement [6] and that its closure accelerates senescence [89].

Cytokinins are also known to act as a sink for the transport of solutes from older to younger part(s) of a plant. Leopold and Kawase [90] demonstrated this very clearly. They painted the primary (oldest) leaves of a bean plant with benzyladenine at 4-day intervals. These leaves started senescing as soon as the trifoliate leaves above expanded, and thus died off first. However, the treated leaves, in their experiment, lived longer than the untreated first trifoliate leaves because cytokinins do not readily move except in the xylem stream. Many variations in the experimental design and test species have indicated that they act as a sink for solutes in the potential or actively growing parts of the plant [7].

V. ETHYLENE

The effect of ethylene on plant growth was noted as early as 1858 by the behavior of plants exposed to illuminating gas [91]. Nevertheless, the Russian scientist Neljubow [92] is credited with having identified the active growth-regulating component of the illuminating gas as ethylene. In the presence of ethylene, etiolated pea plants exhibit inhibition of elongation, an increase in diameter, and horizontal growth of shoots. In the literature, these three responses are known as the *triple response* and are still sometimes used to identify and measure ethylene response. However, Cousins [93] was the first to observe that gases released from oranges caused premature ripening of banana. But it was not until 1934 that Gane [94] provided evidence that ethylene was produced autocatalytically by ripening fruits.

A. Chemical Nature

Ethylene is the simplest organic compound.

$$
\begin{array}{ccc}
H & & H \\
\diagdown & & \diagup \\
& C{=}{=}{=}C & \\
\diagup & & \diagdown \\
H & & H
\end{array}
$$

Its structural simplicity and the fact that it is gaseous in nature makes it a unique plant hormone. It is a symmetric molecule having one double bond; the biological activity seems to relate to its unsaturated bond, which is attached to a terminal carbon atom.

B. Metabolism

The task of unraveling the biosynthesis of ethylene was not an easy one. Feeding experiments using various radioactive materials with ethylene-producing plant tissues was unsuccessful in identification of the pathway. However, based on model nonenzymatic system, Lieberman and Mapson [95] proposed methionine as the precursor. Subsequently, Lieberman et al. [96] demonstrated the in vivo conversion of $[^{14}C]$methionine to $[^{14}C]$ethylene in apple (*Malus domestica*) tissues. Studies with $[^{14}C]$methionine have shown that the C-1 atom is converted to CO_2, C-2 to formic acid, and C-3 and C-4 to ethylene and the sulfur atom is retained in the tissue. Since the ethylene production system is extremely labile and is completely lost by tissue disruption, the characterization has been made at the living tissue level (Figure 4) [97]. In these studies, climactaric fruit slices or plugs and auxin-treated stem segments of etiolated pea and mungbean (*Phaseolus aureus*) seedlings have been used extensively [98].

Earlier studies on the metabolism of ethylene, conducted with improper precautions, had led to the conclusion that the compound was metabolically inert. However, Beyer [99], employing proper precautionary measures, convincingly demonstrated that ethylene was metabolized by plants, and the metabolic products of the dark-grown aseptic pea seedlings were identified as CO_2 and ethylene oxide. In addition to these two gaseous metabolites, ethylene was metabolized to a number of nonvolatile soluble products, including free ethylene glycol and its glucose conjugate, plus oxalate and a number of unidentified products [98]. However, metabolism to volatile products and metabolism of nonvolatile products are independent of each other. Pea, tomato, cotton, carnation, and morning glory tissues have also been observed to metabolize ethylene [98]. Studies based on nonbiological systems suggest that copper (Cu^+) is involved in ethylene oxidation.

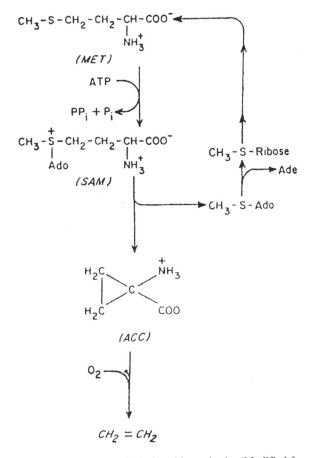

Figure 4 Pathway of ethylene biosynthesis. (Modified from Ref. 97.)

C. Biological Response

Ethylene elicits biologically spectacular responses at very low concentrations. As a plant hormone, it is unique in its structural simplicity and in being gaseous in nature. Whether the term *hormone* should be applied to ethylene, in that its *translocation* in the gas phase seems nonspecific, is a subject of debate. But being moderately water soluble, it moves rapidly between tissues, with minimum hindrance, in either the gaseous or the liquid phase. Therefore, there can be no doubt that it is a natural mobile growth regulator. Unlike other hormones, it is not transported directionally but accomplishes its integrative function by diffusing rapidly through the tissues.

At physiological concentrations, ethylene inhibits stem and root extension growth, but there are instances where it increases the growth rate in *Callitriche platycarpa* stem, in *Helianthus* petiole, and in rice stems and roots [100]. The myriad of plant responses and functions, including seed germination, cell division, epicotyl curvature, seedling growth, flowering, fruit ripening, response to stress, and senescence, are known to be influenced by ethylene. The diversity of the processes, in a wide variety of plants, makes it difficult to assign the hormone a definitive role.

Its production is regulated by a number of developmental and environmental factors. Ethylene production is induced at germination, ripening of fruits, and senescence (abscission) by auxins and by wounding and other chemical stress. It is also produced autocatalytically in the climactaric fruits. Many of the effects of IAA, such as apical dominance and stomatal movement, are attributed to IAA-induced ethylene production [58,101]. However, recent evidence, with transgene-mediated auxin/or ethylene deficiencies and mutants insensitive to either of the hormones, have ruled out the notion that auxin-induced ethylene was involved in the inhibitory influence that the apical bud exerts on the growth of lateral buds [58].

1. Emergence and Seedling Growth

A seed is considered to have germinated when its radicle emerges through the outer covering. Cell division and elongation start at about this time, and by most seed scientists it is the completion of germination. But from the crop physiologists' point of view it is extended to include the processes that ensure seedling establishment. Several fungal species and some bacteria are known to produce ethylene, including those growing in the soil [98]. The ethylene released by these soilborne microorganisms are suspected to influence seed germination, retard soilborne diseases, and regulate seedling growth. Thus these stages can be separated in two phases: radicle protrusion (phase I) and the subsequent events relating to seedling growth, dependent on seed reserves (phase II).

Two types of germination have been recorded in plants and crops: *epigean*, whereby the cotyledons emerge above ground (*Phaseolus vulgaris*; garden bean), and *hypogean* (*Pisum sativum*; garden pea), where the cotyledon remains below the soil surface. In the former case, the cotyledons emerge above ground with the growing tip, due to the elongation of hypocotyl forming a hook or arch. With the emergence of hypocotyl hook and exposure to light, symmetrical growth takes place and the hook straightens up. In the second case, the plumule is arched or recurved near the apex, to protect shoot tip, and when it is pushed through the soil, exposure to light causes the epicotyl to straighten up. As it turned out, localized production of ethylene, at a rate of about 6 μL/kg per hour (etiolated pea seedlings), was found responsible for the formation and maintenance of the hook [17]. As the etiolated seedling emerges from the soil or exposed to white light, a transient decrease in ethylene is observed with concomitant straightening of the shoot [17]. A similar explanation applies to seedlings that show *epigean* germination (garden bean) and develop hypocotyl hook. It has further been observed that green tissues of seedlings are not as sensitive to ethylene as are etiolated tissues. Ethylene responses of emerging dicot seedlings have a survival value for the crop. Shortly after germination the hook is formed, in response to endogenous ethylene, which helps the cotyledons or young leaves to emerge safely out of the normal soil. Under compacted soils the hook and the primary root become unusually thick (i.e., they grow in diameter). This response is elicited by the organs, perhaps due to enhanced ethylene synthesis by imposed mechanical stress.

However, plants growing with their roots and stems submerged respond to ethylene by eliciting enhanced growth. This occurs due to accumulation of ethylene because of slower diffusion of the gas out of the tissue and through the water. Among the species are *Regnellidium diphyllum* (water fern), *Nymphoides peltata*, *Rananculus sceleratus*, and *Callitriche platycarpa* (star wort), which experience submergence at least part of the time during their growth. During submergence the stems elongate rapidly to keep leaves and upper stem parts buoyant. Submergence causes ethylene accumulation, which causes stems or petioles to grow rapidly [102,103]. Deepwater rice exhibits similar phenomena, and internode lengths of up to 0.6 m have been recorded and the plant completes its life cycle in several meters of water [102]. These contrasting responses support the notion that depending on the condition, similar cells respond differently to the same hormone [104].

2. Stress Ethylene

A low level of ethylene is produced by plant species, but when tissues are injured by a variety of stresses (wounding, pathogens, mechanical, chemical, temperature extremes, etc.), ethylene production increases severalfold. This enhanced production, frequently referred to as wound- or stress-induced ethylene, injures not only the tissues but also the site of ethylene production [105]. The intensity of enhancement in stress ethylene has been related to gamma radiation dosage, and it has also been suggested as a rapid assay method for bacterial toxins [106]. The enhancement in its production has also been observed to correlate with the number and size of foliar lesions induced by tobacco mosaic virus (106). Visual injury was suggested as the most sensitive and rapid technique to evaluate the plant response to acute pollution abuses [107]. Although visual evaluation is rapid, it is also subjective. Measurement of pollution-induced ethylene surge before any visual symptom appears points to its sensitivity and superiority for physicochemical determinations. However, the ethylene production surge is not long-lasting; rather, it is a short-lived phenomenon. Therefore, it has been suggested by Craker [108] that it acts as a trigger mechanism that initiates the biochemical change(s) expressing the response.

Auxin at supraoptimal concentrations (10^{-5} to $10^{-3}M$) acts as a natural factor in enhancing ethylene production, and depending on species and severity of stress, its biosynthesis starts after a lag of 30 to 60 min, can continue up to 48 h, and then declines to normal level. The mechanism of IAA-induced ethylene production has been studied in pea and mungbean seedlings. The detailed studies have shown that IAA stimulates ethylene production by enhancing conversion of SAM (*S*-adenosylmethionine) to ACC (1-aminocyclopropane-1-carboxylic acid) through its effect on the enzyme ACC synthase [109]. Thus at supraoptimal concentrations, auxin per se does not cause growth inhibition, but it is the induction of enhanced ethylene production that inhibits growth [91].

A number of developmental processes have been listed by Abeles [110], where auxin-induced ethylene synthesis is considered to mediate auxin action. Therefore, it seems reasonable to assume that like other stresses, auxin acts similarly in accelerating ethylene production.

3. Senescence

It can be considered that programmed changes in the metabolic processes may ultimately lead to the death of a tissue, organ, or the whole plant. In nature, we experience three categories of senescence: sequential, where the oldest leaves senesce first; synchronous, where all the leaves senesce simultaneously (as in deciduous trees); and senescence of the whole plant after the completion of seed production (as in monocarpic crops). Simon [111] believes it is likely that the various types of senescence may result from different control mechanisms in the leaves. Studies have commonly been conducted with detached leaves, where experimental conditions may influence the result. Thus it is very difficult to relate results from a single experimental system to the system operating in situ. Two major biochemical events, extensive proteolysis and chlorophyll loss, have been observed consistently at the beginning of the process. Leaf senescence may be induced or accelerated by a number of environmental factors, including competition for space, light, and nutrients; pollution; biotic or abiotic stresses; or it may be genetically programmed. At the cellular level it does seem to be controlled by endogenous growth regulators. It is now known that in most cases, auxin, cytokinin, ethylene, and ABA play a role in the regulation of senescence in plants and cytokinins, and auxin can delay senescence in a number of plant species. Thus, according to their actions, they have been classified as senescence promotors and retardants [112]. Ethylene plays an important role in accelerating leaf, petals, and fruit senescence, and auxin and cytokinins act as retardants. It is now well accepted that a balance between auxin and ethylene is a crucial factor in the retention or nonretention of leaves and/or fruits. Premature fruit drop is common in a number of important fruit trees, such as apple and mango (*Mangifera indica*). In mango, fruit drop occurs at all stages of its development but is extensive (>90%) during the first 2 to 4 weeks after fertilization [113]. This stage coincides with the maximum ethylene production by the fruitlet pericarp [114]. Therefore, treatments with Ag^+, Co^{2+}, or synthetic auxin NAA (naphthaleneacetic acid), which regulates ethylene balance, have been observed to enhance the number of harvested fruits significantly [113,115]. In some recent studies, salicylic acid (an inhibitor of ethylene biosynthesis) has also been observed to reduce mango fruit drop (S. S. M. Naqvi, unpublished results).

VI. ABSCISIC ACID

The pioneering work, pointing to the possibility that plant growth and development is regulated by levels of both promotor (auxin) and inhibitor, is generally credited to Hemberg. Using *Avena* bioassay, he observed that potato peels contained high levels of growth inhibitors [116]. In the same year, he demonstrated further the presence of a similar inhibitor that could be correlated with the levels and degree of ash (*Fraxinus excelsior*) bud dormancy [117].

Employing paper chromatography to analyze plant growth substances from plant extracts, Bennet-Clark et al. [118] observed growth inhibitory activity at R_f 0.6 to 0.7. This was later shown to be present in a number of plant species and the levels responded to the changes in environmental conditions. As pointed out by Hemberg [116,117], these results supplemented the physiological importance of the growth inhibitor, which was named "inhibitor β" [119,120].

Independent investigations of two different physiological phenomena in two different laboratories across the Atlantic Ocean led to the identification and discovery of abscisic acid (ABA) as the causal agent. Wareing and his associates at Wales worked for two decades or more on the seasonal changes in bud dormancy in woody plants, particularly of sycamore (*Acer pseudoplatanus*), and identified a compound that was named *dormin* [121]. Concurrently, a team led by Carns and Addicott in California, working on natural control of abscission in cotton, identified two compounds, which they named abscisin I and abscisin II [122,123]. However, by 1965 these two independent but diverse paths converged on the discovery that ABA was the hormone involved in both phenomena [124]. Like other hormones, abscisic acid is also ubiquitous among vascular plants and has been found to occur in some mosses, algae, and fungi.

A. Chemical Nature

The naturally occurring enantiomorph is (S)-ABA, which is a sesquiterpenoid (a 15-carbon compound) and by its biogenesis is related to monoterpenes, diterpenes (gibberellins), carotenoids, and triterpenes. Endogenous (S)-ABA is optically active, having one center of asymmetry at C-1', while synthetic ABA is racemic and composed of equal amounts of (S)- and (R)-enantiomers. The synthetic (R)-ABA accounts for 50% of the racemic mixtures of ABA and has biological activity equal to that of the natural (S)-ABA (Figure 5) in most cases, except in stomatal closure, where it is inactive. Since the catabolism of the (S)- and (R)-enantiomers is different, it is necessary to identify which compound is being used (R, S, or RS). In such a situation, care must be taken to use only natural (S)-ABA for metabolic studies.

B. Metabolism

The typical sesquiterpene nature of ABA indicates that its endogenous synthesis is through mevalonic acid (MVA) as a precursor. Two pathways for its biosynthesis have been suggested. First is via farnesyl pyrophosphate, from which GAs are also derived. Through this pathway,

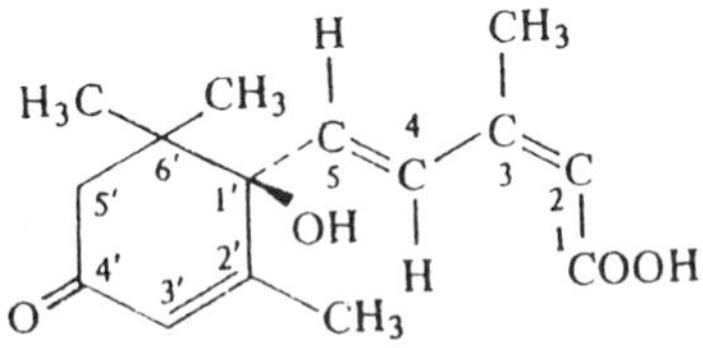

Figure 5 Structure of natural abscisic acid.

MVA is converted to mevalonate 5-phosphate → mevalonate-5-pyrophosphate → Δ^3-isopentenyl pyrophosphate (IPP). This compound is either directly converted to geranyl pyrophosphate or through 3,3-dimethylallyl pyrophosphate (DMAPP) to geranyl pyrophosphate → farnesyl pyrophosphate and finally, to ABA. However, use of radioactive MVA has yielded low amounts of ABA in only a few systems [125]. The second pathway is known to occur through the degradation of certain (40-carbon) carotenoids. Although this pathway is indirect, it seems to produce major amounts of ABA via ABA-aldehyde in perhaps all plants [126,127]. Recently, Zeevaart et al. [128], using various tissues incubated in an atmosphere containing $^{18}O_2$, have demonstrated that xanthophylls rather than farnesyl pyrophosphate are the precursors of ABA. In the xanthophyll cycle, 9'-*cis*-neoxanthin is converted to xanthoxin → ABA-aldehyde, which is finally oxidized to ABA.

Abscisic acid is catabolized to more polar compounds by conjugation, oxidation, hydroxylation, or isomerization. However, it seems that each plant species has its own system to regulate its free ABA level. This regulation is further dependent on the kind of organ tested as well as the physiological state. This regulation of ABA may operate through conjugation with sugar(s) to form glucoside or glycosyl ester or an acylated form. It can also be inactivated by oxidation to more polar free acids such as phaseic and dihydrophaseic acids. Both of these metabolites possess low or no growth-regulating activities and are derived via 6'-hydroxymethyl ABA.

C. Transport

Abscisic acid is known to be translocated through the xylem and phloem to the actively growing regions (apical buds, root tips) and also in the parenchyma outside the vascular tissues. Thus ABA is not translocated polarly; instead, it moves bidirectionally short as well as long distances [129].

D. Biological Activity

It is reasonable to assume that if growth and developmental processes proceeded uncontrolled, the result could be a distinct disadvantage. Plants have localized areas of cell division, the meristem, which is usually located at the tips of the growing organ(s). There are obvious structural limitations to an unlimited elongation of stem internodes, leaves, or in fact nearly any plant organ. Thus the programmed plasticity, exhibited by growth and developmental processes, is certainly advantageous for plant survival, as shown by their ability to become dormant or otherwise restrain their growth or reproductive activities to match the alterations in their external environment. Plants do this by producing ABA to adjust their shoot and root growth accordingly. Like other hormones, abscisic acid also has multiple physiological effects in influencing plant growth and development. The concentrations of ABA varies widely, from 3 to 5 μg/kg in aquatic plants to 10 μg/kg in avocado (*Persea americana*) fruit mesocarp, and in the leaves of temperate crop plants it is usually between 50 and 500 μg/kg [130].

1. Bioassay

The main impact of bioassays has been to study the growth-regulating properties of endogenous substances and aid in their isolation and identification in a pure form by existing physicochemical methods. Despite the superiority of the physicochemical methodologies, the sheer simplicity of the bioassay is likely to continue indefinitely to be used as an analytical tool. Recently, an enzyme-amplified immunoassay, with a sensitivity of 0.05 to 2.5 pg of ABA, has been developed to measure the ABA content of mesophyll and/or guard cells [131]. The majority of the bioassays have exploited the growth-inhibiting properties of the ABA, and such diverse materials as *Lemna*, oat first internode or wheat coleoptile, lettuce and cucumber hypocotyls, and rice seedlings were used. Stomatal closure response of *Commelina communis* and barley and inhibition of hydrolases in barley aleurone layers have also been used effectively to detect the presence or absence of ABA [132, 133]. Depending on the test employed, these bioassays have been able to detect ABA levels ranging from 10^{-6} to 10^{-11} M.

2. Growth

Abscisic acid was the first inhibitory hormone known to be involved in the regulation of growth along with growth promotors. At a concentration of 10^{-7} to 10^{-5} M it inhibits the growth of wheat coleoptiles, barley shoots, bean axes, and the second leaf sheaths of rice seedlings [134]. It is generally accepted that abscisic acid inhibits shoot growth, but its effect on root growth is contradictory and needs satisfactory resolution. Exogenous ABA has been observed to promote as well as inhibit root growth [135,136], and endogenous ABA has been shown to elicit similar responses. Under water stress conditions, the endogenous ABA level increases manyfold, which in turn has been implicated to affect shoot and root growth differentially (i.e., reducing shoot while maintaining root growth) [137].

Mulkey et al. [135] observed a triphasic response to ABA treatment: a period of promotion lasting 12 h, followed by a similar period of inhibition (12 h) and gradual recovery to about 80% of the normal growth rate after 24 h. Robertson et al. [138] dried sunflower roots to enhance the endogenous ABA level, or treated them with exogenous ABA and observed a similar triphasic response. The initial transitory increase in length was related to the initial rise in water potential in the root apices, followed by an inhibition and then a partial recovery in root elongation. On the other hand, Saab et al. [137], using fluridone (an inhibitor of carotenoid biosynthesis) and a mutant deficient in carotenoid biosynthesis (*vp5*) to reduce the endogenous ABA level in maize seedlings, concluded that ABA played a direct role in the inhibition of shoot growth and in the maintainance of root elongation. However, Creelman et al. [139] concluded that under water-deficit conditions, at all the internal ABA concentrations tested, root growth was inhibited less than hypocotyl growth. But Plaut and Carmi [140] attributed the root response to the hydrotropic nature of the organ, which induces it to reach to the wet soil rather than to any other factor(s).

Abscisic acid was also projected as causing differential growth in gravitropic responses of roots. But later work using norflurazon or fluridone (inhibitors of carotenoid biosynthesis) and viviparous maize mutant *vp*-9 (lacking ABA biosynthesis) showed that a drastic reduction in endogenous ABA level did not alter root gravitropic response [141,142].

3. Dormancy

Dormancy can be considered as the ability to retain viability while having minimal metabolic activity and no visible growth. Plants and crops have evolved this strategy as a mechanism of survival to cope with the pronounced seasonal changes unfavorable for their normal growth and development. Walton [143] concluded that "a role for ABA on the induction of bud and seed dormancy has been neither unequivocally demonstrated nor disproven." This is still valid with regard to bud dormancy in woody species. A seasonal change in the ABA content of leaves, stem apices, and xylem sap of *Salix viminalis* [144] and in the buds and stems of *Acer saccharum* [145] was observed. But these workers concluded that ABA did not play a role in the photoperiodic control of bud dormancy. These works have received further support by the recent evidence that cessation of seedling growth in *Salix* spp. is not regulated through the effect of daylength on ABA levels [146,147]. It is, however, possible that short-day conditions may have altered tissue sensitivity to ABA [146]. These studies indicate further that in the control of bud dormancy, factor(s) other than ABA are possibly involved.

The results in the case of seeds are, however, different. Several studies indicate that ABA treatment prevents vivipary (precocious germination of the developing embryo) in immature seeds. In vitro studies have shown that a high percentage of germination was obtained when the ABA content of immature soybean embryo was less than 4 μg/g fresh weight [148]. Similar studies with cultured immature embryos of wheat [149], soybean [148,150,151], cotton [152], rapeseed [153], and maize [154] have shown that exogenous ABA not only prevented precocious germination but often caused embryo growth and storage protein accumulation. However, with the maturity of embryos, the endogenous level of ABA and the sensitivity to exogenous hormone declined. Convincing evidence for the control of seed dormancy by ABA has been provided by a reciprocal cross between wild-type and ABA-deficient *Arabidopsis* mutants as well as with the treatments of wild-type young maize kernels with fluridone. The reciprocal crosses indicated that maternal ABA had a minor role [155], while treatment with fluridone induced precocious germination [156].

4. Stomatal Control

The discovery that ABA plays a leading role in the regulation of stomatal movement generated the interest of many workers [157,158]. Abscisic acid–deficient mutants are known in tomato, potato, pea, and *Arabidopsis* [125], which reverts phenotypically to the wild types when treated with ABA. The response is quite rapid, and after exogenous ABA application to the cut leaf bases, it takes 3 to 9 min to close stomata in maize, sugar beet (*Beta vulgaris*), and *Rumex obtusifolia* [17]. The magnitude of stomatal response to ABA is, however, dependent on the concentration of K^+ in the incubation media [159]. It has been estimated that when the stomates are closed, K^+ concentration of the epidermal cells ranges from 250 to 450 mol/m^3, but when K^+ concentration falls to about 100 mol/m^3, it opens. Harris and Outlaw [160] have measured ABA levels in isolated guard cells using an enzyme-amplified immunoassay and observed that water stress caused at least a 20-fold increase (up to 8 fg per cell). This may suggest that ABA causes stomates to close by inhibiting an energy-dependent (ATP/cAMP) proton pump in the guard cell plasma membrane. Thus ABA exerts two major biochemical effects. One is its effect on altering plasma membranes, which by shutting off the proton pump stops influx of K^+, causing K^+ and water to leak out. This reduces guard cell turgor, causing the stomates to close. However, the evidence of ABA role in stomatal regulation discussed above is not unequivocal and there is evidence which suggests the involvement of other factor(s), including hormone(s) and/or modification in tissue sensitivity [6,161].

VII. CONCLUSIONS

Currently, five classes of hormone—auxin, gibberellin, cytokinin, ethylene, and abscisic acid—are known to be ubiquitous in higher plants and crops. Some of them have also been found to be produced by bacteria, fungi, bryophytes, and pteridophytes. They influence a myriad of plant functions and responses, and presumably any one process is influenced by the balance of the existing complement of hormones. Hormone physiologists generally classify auxin, gibberellin, and cytokinin as growth promotors and ethylene and abscisic acid as growth inhibitors. Although plant and crop hormones regulate a wide range of growth and developmental processes, their diversity makes it difficult to assign a definitive role to them from observations on plant responses. They influence each other's level and thus play important roles in a network of feedback control mechanisms modulating normal growth and development and thus preventing odd overgrowths. At times, each of them can act as a promotor or inhibitor, or vice versa, in this network of feedback control mechanisms. Therefore, the categorization seems rather conjectural.

New research in molecular biology and biotechnology/genetic engineering has opened the door to exciting approaches. Mutants are available that are either synthesis or response mutants, and genetically engineered plants, which overproduce auxin, cytokinin, and/or ethylene, have also been developed. Using a molecular biological approach, investigators have provided support to the classical Cholodny–Went theory of differential growth elicited by tropistic responses. Similarly, transgenic plants have been used to support the theory that the auxin/cytokinin ratio, not the auxin-stimulated ethylene, controls the process of correlative growth (apical dominance, compensatory growth). Thus recent studies using advanced technologies have supported theories advanced many years ago with the response techniques developed earlier. It is therefore encouraging that use of such mutants and/or transgenic plants are generating information on the biochemical and cellular processes that modulate plant growth and development [162].

REFERENCES

1. J. Sachs, *Arb. Bot. Inst. (Wurzburg)*, 2: 452 (1880).
2. F. W. Went, *Recl. Trav. Bot. Neerl.*, 25: 1 (1928).
3. E. Kurosawa, *J. Nat. Hist. Soc. (Formosa)*, 16: 213 (1926).
4. C. O. Miller, F. Skoog, M. H. von Saltza, and F. M. Strong, *J. Am. Chem. Soc.*, 77: 1392 (1955).
5. J. Wiesner, *Die Elementarstrucktur und das Waschstum der labenden Substanz*, Holder, Vienna (1892).
6. S. S. M. Naqvi, in *Handbook of Plant and Crop Stress* (M. Pessarakli, ed.), Marcel Dekker, New York, p. 383 (1994).
7. F. B. Salisbury and C. W. Ross, *Plant Physiology*, Wadsworth, Belmont, Calif. (1992).
8. F. P. Gardner, R. B. Pearce, and R. L. Mitchell, *Physiology of Crop Plants*, Iowa State University Press, Ames, Iowa (1985).
9. J. Heslop-Harrison, in *Plant Growth Substances 1979* (F. Skoog, ed.), Springer-Verlag, Berlin, p. 3 (1980).
10. P. Boysen-Jensen, in *Growth Hormones in Plants* (translated and revised by G. S. Avery, Jr., and P. R. Burkholder, eds.), McGraw-Hill, New York (1936).
11. A. Paál, *Jahr. Wiss. Bot.*, 58: 406 (1919).
12. F. Kögl, A. J. Haagen-Smit, and H. Erxleben, *Zeitschr. Physiol. Chem.*, 288: 90 (1934).
13. F. W. Went and K. V. Thimann, *Phytohormones*, Macmillan, New York (1937).
14. A. J. Haagen-Smit, W. B. Dandliker, S. H. Wittwer, and A. E. Murneek, *Am. J. Bot.*, 33: 118 (1946).
15. K. C. Engvild, *Physiol. Plant.*, 77: 282 (1989).
16. R. L. Wain and C. H. Fawcett, *Plant Physiology: A Treatise*, Vol. 5A (F. C. Steward, ed.), Academic Press, New York, p. 231 (1969).
17. T. C. Moore, *Biochemistry and Physiology of Plant Hormones*, Springer-Verlag, New York (1979).
18. R. H. Phelps and L. Sequeira, *Plant Physiol.*, 42: 1161 (1967).
19. R. A. Gibson, E. A. Schneider, and F. Wightman, *J. Exp. Bot.*, 23: 381 (1972).
20. J. Ludwig-Muller and W. Hilgenberg, *Physiol. Plant.*, 74: 240 (1988).
21. M. Kawaguchi, S. Fujioka, A. Sakurai, Y. T. Yamaki, and K. Syono, *Plant Cell Physiol.*, 34: 121 (1993).

22. B. G. Baldi, B. R. Maher, J. P. Slovin, and J. D. Cohen, *Plant Physiol.*, *95*: 1203 (1991).
23. A. D. Wright, M. B. Sampson, M. G. Neuffer, L. Michalczuk, J. P. Slovin, and J. D. Cohen, *Science*, *254*: 988 (1991).
24. D. M. Reinecke and R. S. Bandurski, in *Plant Hormones and Their Role in Plant Growth and Development* (P. J. Davies, ed.), Martinus Nijhoff, Boston, p. 24 (1987).
25. T. T. Lee and A. N. Starratt, *Physiol. Plant.*, *84*: 209 (1992).
26. F. Sitbon, A. Ostin, S. Sundberg, O. Olsson, and G. Sandberg, *Plant Physiol.*, *101*: 313 (1993).
27. S. M. Naqvi, Ph.D. thesis, Princeton University, Princeton, N.J. (1963).
28. W. P. Jacobs, *Plant Hormones and Plant Development*, Cambridge University Press, Cambridge (1979).
29. S. M. Naqvi and S. A. Gordon, *Plant Physiol.*, *40*: 116 (1965).
30. M. L. Evans, *Plant Physiol.*, *95*: 1 (1991).
31. S. M. Naqvi and K. C. Engvild, *Physiol. Plant.*, *30*: 283 (1974).
32. R. S. Bandurski and H. M. Nonhebel, in *Advanced Plant Physiology* (M. B. Wilkins, ed.), Pitman, London, p. 1 (1984).
33. R. Aloni, in *Plant Hormones and Their Role in Plant Growth and Development* (P. J. Davies, ed.), Martinus Nijhoff, Boston, p. 363 (1987).
34. A. C. Leopold, *Auxins and Plant Growth*, University of California Press, Berkeley, Calif. (1960).
35. V. J. Philip and J. Padikkala, *J. Plant Physiol.*, *135*: 233 (1989).
36. V. C. Pence and L. Caruso, in *Plant Hormones and Their Role in Plant Growth and Development* (P. J. Davies, ed.), Martinus Nijhoff, Boston, p. 240 (1987).
37. R. D. Firn and J. Digby, *Annu. Rev. Plant Physiol.*, *31*: 131 (1980).
38. B. G. Pickard, *Annu. Rev. Plant Physiol.*, *36*: 55 (1985).
39. I. R. MacDonald and W. Hart, *Plant Physiol.*, *84*: 568 (1987).
40. Y. Li, G. Hagen, and T. J. Guilfoyle, *Plant Cell*, *3*: 1167 (1991).
41. M. J. Jaffe, H. Takahashi, and R. L. Biro, *Science*, *230*: 445 (1985).
42. W. R. Briggs, *Plant Physiol.*, *38*: 237 (1963).
43. B. G. Pickard and K. V. Thimann, *Plant Physiol.*, *39*: 341 (1964).
44. S. M. Naqvi and S. A. Gordon, *Plant Physiol.*, *42*: 138 (1967).
45. E. W. Weiler, R. S. Jourdan, and W. Conrad, *Planta*, *153*: 561 (1981).
46. S. Marumo, in *Chemistry of Plant Hormones* (N. Takahashi, ed.), CRC Press, Boca Raton, Fla., p. 9 (1986).
47. A. B. Frank, *Beitrage zur Pflanzenphysiologie. I. Ueber die durch Schwerkeft Verusachte Bewegung von Pflanzentheilen*, W. Engelmann, Leipzig (1868).
48. M. L. Evans, R. Moore, and K.-H. Hasenstein, *Sci. Am.*, *255*(6): 100 (1986).
49. S. M. Naqvi and S. A. Gordon, *Plant Physiol.*, *41*: 1113 (1966).
50. R. Hertel, R. K. de la Fuente, and A. C. Leopold, *Planta*, *88*: 204 (1969).
51. B. A. McClure and T. Guilfoyle, *Science*, *243*: 91 (1989).
52. H. Klee and M. Estelle, *Annu. Rev. Plant Physiol. Plant Mol. Biol.*, *42*: 529 (1991).
53. K. V. Thimann, *Annu. Rev. Plant Physiol.*, *14*: 1 (1963).
54. G. C. Martin, *HortScience*, *22*: 824 (1987).
55. I. A. Tamas, in *Plant Hormones and Their Role in Plant Growth and Development* (P. J. Davies, ed.), Martinus Nijhoff, Boston, p. 393 (1987).
56. K. V. Thimann and F. Skoog, *Proc. R. Soc.* (*London*), *B114*: 317 (1934).
57. A. C. Leopold, *Am. J. Bot.*, *36*: 437 (1949).
58. C. P. Romano, M. L. Cooper, and H. J. Klee, *Plant Cell*, *5*: 181 (1993).
59. N. Takahashi, B. O. Phinney, and J. MacMillan, *Gibberellins*, Springer-Verlag, Berlin (1990).
60. N. Takahashi, I. Yamaguchi, and H. Yamane, in *Chemistry of Plant Hormones* (N. Takahashi, ed.), CRC Press, Boca Raton, Fla., p. 57 (1986).
61. R. C. Durley, J. MacMillan, D. M. Reid, and B. H. Most, *Phytochemistry*, *10*: 1891 (1971).
62. J. E. Graebe, D. J. Dennis, C. D. Upper, and C. A. West, *J. Biol. Chem.*, *240*: 1847 (1965).
63. R. M. Devlin and F. H. Witham, *Plant Physiology*, PWS Publishers, Boston (1983).
64. Y. Murakami, *Jpn. Agric. Res. Quart.*, *5*(2): 5 (1970).
65. T. Nishijima and N. Katsura, *Plant Cell Physiol.*, *30*: 623 (1989).

66. R. L. Jones and J. E. Varner, *Planta*, *72*: 155 (1967).
67. R. P. Pharis, L. T. Evans, R. W. King, and L. N. Mander, in *Plant Production: From Floral Induction to Pollination* (E. Lord and G. Bernier, eds.), American Society of Plant Physiologists Symposium Series, Vol. 1, ASPP, Rockville, MD., p. 29 (1989).
68. J. MacMillan and B. O. Phinney, in *Physiology of Cell Expansion During Plant Growth* (D. J. Cosgrove and D. P. Knievel, eds.), American Society of Plant Physiologists, Rockville, Md., p. 156 (1987).
69. S. B. Rood, P. H. Williams, D. Pearce, N. Murofushi, L. N. Mander, and R. P. Pharis, *Plant Physiol.*, *93*: 1168 (1990).
70. R. L. Jones and J. MacMillan, in *Advanced Plant Physiology* (M. B. Wilkins, ed.), Pitman, London, p. 21 (1984).
71. G. Haberlandt, *Beitr. Allg. Bot.*, *2*: 1 (1921).
71a. J. Van Overbeek, *Science*, *152*: 721 (1966).
72. K. Koshizima and H. Iwamura, in *Chemistry of Plant Hormones* (N. Takahashi, ed.), CRC Press, Boca Raton, Fla., p. 153 (1986).
73. D. S. Letham, *Phytochemistry*, *5*: 269 (1966).
74. G. A. Miura and C. O. Miller, *Plant Physiol.*, *44*: 1035 (1969).
75. B. J. Taller, N. Murai, and F. Skoog, *Plant Physiol.*, *83*: 755 (1987).
76. C. M. Chen and D. K. Melitz, *FEBS Lett.*, *107*: 15 (1979).
77. P. Beutelmann, *Planta*, *112*: 181 (1973).
78. D. S. Letham and L. M. S. Palni, *34*: 163 (1983).
79. J. Van Staden and J. E. Davey, *Plant Cell Environ.*, *2*: 93 (1979).
80. P. E. Pilet, in *Biochemistry and Physiology of Plant Growth Substances* (F. Wightman and G. Setterfield, eds.), Runge Press, Ottawa, Ontario, Canada, p. 993 (1968).
81. J. I. Medford, R. Horgan, Z. El-Sawi, and H. J. Klee, *Plant Cell*, *1*: 403 (1989).
82. D. S. Letham, *Annu. Rev. Plant Physiol.*, *18*: 343 (1967).
83. R. Horgan, in *Advanced Plant Physiology* (M. B. Wilkins, ed.), Pitman, London, p. 53 (1984).
84. T. H. Thomas, *J. Plant Growth Regul.*, *8*: 255 (1989).
85. F. Skoog and C. O. Miller, *Symp. Soc. Exp. Biol.*, *11*: 118 (1957).
86. M. Wickson and K. V. Thimann, *Physiol. Plant.*, *11*: 62 (1958).
87. P. K. W. Lee, B. Kessler, and K. V. Thimann, *Physiol. Plant.*, *31*: 11 (1974).
88. R. Sexton and H. W. Woolhouse, in *Advanced Plant Physiology* (M. B. Wilkins, ed.), Pitman, London, p. 469 (1984).
89. K. V. Thimann, in *Plant Senescence: Its Biochemistry and Physiology* (W. W. Thomson, E. A. Nothnagel, and R. C. Huffaker, eds.), American Society of Plant Physiologists, Rockville, Md., p. 1 (1987).
90. A. C. Leopold and M. Kawase, *Am. J. Bot.*, *51*: 294 (1964).
91. S. P. Burg, *Proc. Natl. Acad. Sci. USA*, *70*: 591 (1973).
92. D. N. Neljubow, *Beih. Bot. Centralbl.*, *10*: 128 (1901).
93. H. H. Cousins, "Agricultural Experiments: Citrus," *Jamaica Department of Agriculture Annual Report*, p. 7 (1910).
94. R. Gane, *Nature*, *134*: 1008 (1934).
95. M. Lieberman, and L. W. Mapson, *Nature*, *204*: 343 (1964).
96. M. Lieberman, A. Kunishi, L. W. Mapson, and D. A. Wardale, *Plant Physiol.*, *41*: 376 (1966).
97. S. F. Yang, *HortScience*, *20*: 41 (1985).
98. H. Imaseki, in *Chemistry of Plant Hormones* (N. Takahashi, ed.), CRC Press, Boca Raton, Fla., p. 249 (1986).
99. E. M. Beyer, in *Ethylene and Plant Development* (J. A. Roberts and G. A. Tucker, eds.), Butterworth, London, p. 125 (1985).
100. M. Zeroni and M. A. Hall, in *Encyclopedia of Plant Physiology* (*NS*): *Hormonal Regulation of Development I*, Vol. 9 (J. MacMillan, ed.), Springer-Verlag, Berlin, p. 511 (1980).
101. L. K. Levitt, D. B. Stein, and B. Rubinstein, *Plant Physiol.*, *85*: 318 (1987).
102. M. B. Jackson, in *Ethylene and Plant Development* (J. A. Roberts and G. A. Tucker, eds.), Butterworth, London, p. 241 (1985).

103. I. Ridge, in *Ethylene and Plant Development* (J. A. Roberts and G. A. Tucker, eds.), Butterworth, London, p. 229 (1985).

104. D. J. Osborne, M. T. McManus, and J. Webb, in *Ethylene and Plant Development* (J. A. Roberts and G. A. Tucker, eds.), Butterworth, London, p. 197 (1985).

105. F. B. Abeles, *Ethylene in Plant Biology*, Academic Press, New York (1973).

106. D. T. Tingey, C. Standley, and R. W. Field, *Atmos. Environ.*, *10*: 969 (1976).

107. J. A. Dunning and W. W. Heck, *Environ. Sci. Technol.*, *7*: 824 (1973).

108. L. E. Craker, *Environ. Pollut.*, *1*: 299 (1971).

109. H. Yoshii and H. Imaseki, *Plant Cell Physiol.*, *22*: 369 (1981).

110. F. B. Abeles, in *Ethylene and Plant Development* (J. A. Roberts and G. A. Tucker, eds.), Butterworth, London, p. 1 (1985).

111. E. W. Simons, *Symp. Soc. Exp. Biol.*, *21*: 215 (1967).

112. L. D. Noodén, in *Senescence in Plants* (K. V. Thimann, ed.), CRC Press, Boca Raton, Fla., p. 219 (1980).

113. S. S. M. Naqvi, S. M. Alam, and S. Mumtaz, *Aust. J. Exp. Agric.*, *30*: 433 (1990).

114. R. Nunez-Elisea and T. L. Davenport, *Plant Physiol.*, *82*: 991 (1986).

115. S. S. M. Naqvi, S. M. Alam, and S. Mumtaz, *Pak. J. Bot.*, *24*: 197 (1992).

116. T. Hemberg, *Physiol. Plant.*, *2*: 24 (1949).

117. T. Hemberg, *Physiol. Plant.*, *2*: 37 (1949).

118. T. A. Bennet-Clark, M. S. Tambiah, and N. P. Kefford, *Nature*, *169*: 452 (1952).

119. T. A. Bennet-Clark and N. P. Kefford, *Nature*, *171*: 645 (1953).

120. N. P. Kefford, *J. Exp. Bot.*, *6*: 245 (1955).

121. P. F. Wareing, C. F. Eagles, and P. M. Robinson, in *Régulateurs naturels de la croissance végétale* (J. P. Nitsch, ed.), CNRS, Paris, Vol. 9, p. 377 (1964).

122. W. C. Liu and H. R. Carns, *Science*, *134*: 384 (1961).

123. K. Ohkuma, J. L. Lyon, F. T. Addicott, and O. E. Smith, *Science*, *142*: 1592 (1963).

124. F. T. Addicott, J. L. Lyon, K. Ohkuma, W. E. Thiessen, H. R. Carns, O. E. Smith, J. W. Cornforth, B. V. Milborrow, G. Ryback, and P. F. Wareing, *Science*, *159*: 1493 (1968).

125. J. A. D. Zeevaart and R. A. Creelman, *Annu. Rev. Plant Physiol. Plant Mol. Biol.*, *39*: 439 (1988).

126. C. D. Rock and J. A. D. Zeevaart, *Plant Physiol.*, *93*: 915 (1990).

127. R. K. Sindhu, D. H. Griffin, and D. C. Walton, *Plant Physiol.*, *93*: 689 (1990).

128. J. A. D. Zeevaart, C. D. Rock, F. Fantauzzo, T. G. Heath, and D. A. Gage, in *Abscisic Acid: Physiology and Biochemistry* (W. J. Davies and H. G. Jones, eds.), BIOS Scientific Publishers, Oxford, p. 39 (1991).

129. O. Wolf, W. D. Jeschke, and W. Hartung, *J. Exp. Bot.*, *41*: 593, (1990).

130. B. V. Milborrow, in *Advanced Plant Physiology* (M. B. Wilkins, ed.), Pitman, London, p. 76 (1984).

131. M. J. Harris, W. H. Outlaw, R. Mertens, and E. W. Weiler, *Proc. Natl. Acad. Sci. USA*, *85*: 2584 (1988).

132. K. Dörffling and D. Tietz, in *Abscisic Acid* (F. T. Addicott, ed.), Praeger, New York, p. 23 (1983).

133. C. H. Lin, Y. L. Lin, and Y. J. Chow, *J. Plant Growth Regul.*, *7*: 161 (1988).

134. N. Hirai, in *Chemistry of Plant Hormones* (N. Takahashi, ed.), CRC Press, Boca Raton, Fla., p. 201 (1986).

135. T. J. Mulkey, M. L. Evans, and K. M. Kuzmanoff, *Planta*, *157*: 150 (1983).

136. P. E. Pilet and P. W. Barlow, *Plant Growth Regul.*, *6*: 217 (1987).

137. I. N. Saab, R. E. Sharp, J. Pritchard, and G. S. Voetberg, *Plant Physiol.*, *93*: 1329 (1990).

138. J. M. Robertson, K. T. Hubick, E. C. Yeung, and D. M. Reid, *J. Exp. Bot.*, *41*: 325 (1990).

139. R. A. Creelman, H. S. Mason, R. J. Benson, J. S. Boyer, and J. E. Mullet, *Plant Physiol.*, *92*: 205 (1990).

140. Z. Plaut and A. Carmi, *Plant Physiol.* (Abstr.)., *99*: 160 (1992).

141. R. Moore and K. Dickey, *J. Exp. Bot.*, *36*: 1793 (1985).

142. L. J. Feldman and P. S. Sun, *Physiol. Plant.*, *67*: 472 (1986).

143. D. C. Walton, *Annu. Rev. Plant Physiol.*, *31*: 453 (1980).

144.   R. Alvim, S. Thomas, and P. F. Saunders, *Plant Physiol.*, *62*: 779 (1978).

145.   E. B. Dumbroff, D. B. Cohen, and D. P. Webb, *Physiol. Plant.*, *45*: 211 (1979).

146.   R. S. Barros and S. J. Neill, *Planta*, *168*: 530 (1986).

147.   L. G. Johansen, P.-C. Oden, and O. Juntilla, *Physiol. Plant.*, *66*: 409 (1986).

148.   R. C. Ackerson, *J. Exp. Bot.*, *35*: 414 (1984).

149.   B. A. Triplett and R. S. Quatrano, *Dev. Biol.*, *91*: 491 (1982).

150.   R. C. Ackerson, *J. Exp. Bot.*, *35*: 403 (1984).

151.   A. J. Eisenberg and J. P. Mascarenhas, *Planta*, *166*: 505 (1985).

152.   D. L. Hendrix and J. W. Radin, *J. Plant Physiol.*, *117*: 211 (1984).

153.   R. R. Finkelstein, K. M. Tenbarge, J. E. Shumway, and M. L. Crouch, *Plant Physiol.*, *78*: 630 (1985).

154.   S. J. Neill, R. Horgan, and A. F. Rees, *Planta*, *171*: 358 (1987).

155.   C. M. Karseen, D. L. C. Brinkhorst-van der Swan, A. E. Breekland, and M. Koornneef, *Planta*, *157*: 158 (1983).

156.   F. Fong, J. D. Smith, and D. E. Koehler, *Plant Physiol.*, *73*: 899 (1983).

157.   S. T. C. Wright and R. W. P. Hiron, in *Plant Growth Substances 1970* (D. J. Carr, ed.), Springer-Verlag, Berlin, p. 291 (1972).

158.   K. Raschke, in *Stomatal Function* (G. Zeiger, D. Farquhar, and I. R. Cowan, eds.), Stanford University Press, Stanford, Calif., p. 253 (1987).

159.   P. J. Snaith and T. A. Mansfield, *Plant Cell Environ.*, *5*: 309 (1982).

160.   M. J. Harris and W. H. Outlaw, *Physiol. Plant.*, *78*: 495 (1990).

161.   E. V. Kearns and S. M. Assmann, *Plant Physiol.*, *102*: 711 (1993).

162.   J. D. Hamill, *Aust. J. Plant Physiol.*, *20*: 405 (1993).

25
Physiological Adaptation of Plants to Environmental Stresses

G. N. Amzallag and H. R. Lerner
The Hebrew University of Jerusalem, Jerusalem, Israel

I. INTRODUCTION

The majority of studies on the physiological response of plants to stress are based on the assumption that the only possible way that a plant can survive under stressing conditions is to express preexisting genetic information that counteracts the effect of the stress. In this perspective there is no difference between resistance (also called preadaptation), accommodation, acclimation, and adaptation; in all cases the stress is considered as a trigger of the expression of a preexisting defense program. This outlook considerably limits the field of investigation as well as the interpretation of results. Since the response is considered to be preprogrammed, it is also assumed to be practically immediate; it is presumed that time is not a crucial factor in the elaboration of the response. The kinetics of growth is therefore usually short, a few days to 2 weeks, and to avoid secondary effects, proteins and mRNAs are extracted soon after exposure to stress. The aim of this chapter is to show that in addition to the preexisting defense program, there is another type of response to stress, which we call *adaptation*. Adaptation is a long-term response during which the plant adjusts its physiology to the environmental conditions in an environment-oriented manner.

II. ADAPTATION OF PLANTS TO STRESS

Following exposure to a sublethal level of an environmental stress, many plants have been shown to accommodate to the stress conditions. In the examples presented in Table 1, exposure to mild stress for a period of a few days up to 3 weeks resulted in a marked increase in tolerance. Amzallag et al. [1] showed that in the case of *Sorghum* the adapted plant can develop and produce seeds under stress conditions that are lethal for the unadapted plant. All responses induced by the environment are not similar. As reported by Apriyanto and Potter [20], Gottstein and Kuc [21], and McIntyre et al. [16], in the induction of systemic resistance to pathogens, the presence of a specific pathogen induces the expression of defense against numerous other

Table 1 Induction of Acclimation to stress by Pretreatment with Sublethal Stress Conditions[a]

Stress	Induction time days	Window	Effect	Remark	Refs.
			Sorghum bicolor		
NaCl	21	Yes	Tolerance ↗	Medium must be stable, cultivar dependent	1,2
			Mesembryanthemum crystallinum		
NaCl	20	No	C3 to C4 metabolic shift	Induction threshold of 100 m*M* NaCl	3
			Hedisarum carnosum		
NaCl	n.r.	n.r.	Tolerance ↗		4
			Zea mais		
Waterlogging	4	n.r.	Aerenchyma formation	Effector is ethylene	5
			Helianthus annus		
Waterlogging	n.r.	n.r.	Aerenchyma formation	Induction of cellulase	6
			Holcus lanatus		
Cadmium	10	n.r.	Root growth ↗		7
			Betula verrucosa		
Low nitrogen	20	n.r.	Adaptation to low nitrogen	Leaves become yellow; after adaptation leaves become green	8
			Lolium perenne		
Low nitrogen	n.r.	n.r.	Plants adapt their uptake to the nitrate concentration present in the medium	For adapted plants RGR is independent of nitrate concentration	9
			Alocasia macrorrhiza		
Light intensity	15–20	Yes	Anatomical and physiological adaptation		10
			Fragaria virginiana		
Light intensity	7–13	Yes	Anatomical and physiological adaptation		11

Table 1 (*Continued*)

Stress	Induction time days	Window	Effect	Remark	Refs.
			Many species (review article)		
Cold hardening	3–30	n.r.	Acclimation	Photoperiod sensitive	12
			Medicago sativa		
Cold hardening	n.r.	n.r.	Acclimation	Acclimation is cultivar dependent	13
			Cucumis sativus		
Pathogen	6	Yes	Locally induced resistance within 3 days; systemic resistance within 6 days	Signal for systemic resistance comes from contaminated leaf	14
			Nicotiana tabacum		
TMV and pathogens	12	Yes	Induced systemic resistance	Nonspecific resistance	15,16
			Sinapis alba		
Drought	12	n.r.	Wide lateral root initiation	For other plants see Refs. 18 and 19	17

[a]Induction time, time required for the induction of acclimation; window, existence of a developmental window; ↗ indicates an increase; n.r., not reported.

pathogens. It seems that in this case the plant not only reacts against the presence of a particular pathogen, but as suggested by Ye et al. [15], responds to general metabolic perturbations induced by pathogens. The accommodation is oriented toward improvement of the metabolism of the plant rather than being a response against a specific pathogen. This example shows that the response of the plant to an environmental perturbation is not always specifically oriented by the environment.

We shall use the term *adaptation* for responses induced by the environment which are oriented by the perturbation. Examples can be found in the development of tolerance toward a specific factor by the presence of that factor at sublethal concentrations; see, for example, the adaptation of *Sorghum* to salinity or adaptation to cadmium (Table 1). When there is a lack of precise information concerning the response of the plant following the induction phase, or when the response is not specific, it is not possible to show that the plant has adapted specifically to the environmental conditions. We shall in such cases use the term *acclimation* to describe the accommodation of the plant to the environment. We use the term *acclimation* in a less precise

Table 2 Adaptation of Cultured Plant Cells

Stress	Effect[a]	Remark	Refs.
	Nicotiana		
NaCl	Tolerance ↗	Cells adapted to 500 nM NaCl	22, 23
	Nicotiana tobacum		
NaCl	Tolerance ↗	Cells adapted to 600 mM NaCl	24
	Citrus sinensis		
NaCl	Tolerance ↗	Cells adapted to 200 mM NaCl	25
	Medicago sativa		
NaCl	Tolerance ↗	Cells adapted to 170 mM NaCl; adaptation correlated with multiple chromosomal aberrations	26
	Medicago sativa		
L-Phosphinothricin (L-PPT) inhibitor of glutamine synthase (GS)	Tolerance to L-PPT ↗ 100-fold	Adaptation correlated with amplification of the GS gene	27
	Daucus carota		
Methotrexate (Mtx) inhibitor of dihydrofolate reductase (DHFR)	Tolerance to Mtx ↗ 30-fold	Adaptation correlated with a threefold ↗ in DHFR activity and a 15-fold ↘ in Mtx uptake	28
	Solanum commersonii		
Cold	Tolerance to freezing temperature ↗	Adaptation induced by cold or by ABA treatment	29
	Bromus inermis		
ABA	Tolerance to freezing temperature ↗ as long as ABA is present	Addition of CK or GA counteracts effect of ABA	30, 31
	Many species (review article)		
In vitro *culture*	Cells lose the requirement for a hormone (habituation)	Habituation is an epigenetic change	32

[a] ↗ indicates an increase, ↘ indicates a decrease.

way than adaptation. Acclimation includes not yet demonstrated responses oriented specifically by the environment, as well as changes induced by stresses but not oriented by it.

Adaptation of cultured plant cells from different species to several stress conditions (Table 2) shows that adaptation is a general response of plant cells. Adaptation can be induced rapidly if the stress is applied progressively. For example, in our laboratory, a wild-type *Nicotiana* cell suspension was adapted within six generations (two transfers) to 35 mM NaCl. By repeating this treatment 10 times, cells were adapted to 350 mM NaCl within 60 generations (unpublished results). At each increase in salinity the bulk of the cells follow the adaptation process, indicating that this is a property of the majority of the cells. When the wild-type cells were exposed directly to 350 mM NaCl, however, they died. For this reason we use the term *adaptation*, not *selection* (used by some authors). The possibility of selection of highly tolerant cells, preexisting in the plant or appearing in the culture by rare random mutation, cannot explain the rapid response of the cells.

Environmentally oriented changes in genome expression have been reported to occur in *Escherichia coli* by Shapiro [33], Cairns et al. [34], and Hall [35]. Under nutritional stress, advantageous and nonrandom mutations occur. The cells change their genotype; they become able to grow in a medium that was previously incompatible for growth. Hall [36] has obtained similar results with a eukaryote, yeast. These environment-oriented changes in genome expression in microorganisms show that adaptation is a general characteristic of organisms.

Regeneration of whole plants from cultured cells reveals chromosomal aberrations [37] and changes in genome expression called somaclomal variations. When cells are adapted to salinity, regenerated plants are less vigorous than plants regenerated from control cells. As reported by McCoy [26], plants regenerated from NaCl-adapted cells show a very high frequency of multiple chromosomal aberrations. It seems that imposing culture conditions is a stress for the cell which results in changes in DNA organization (so-called somaclonal variation). Exposure of cultured cells to NaCl enhances the stress level and increases the frequency of aberrations.

Since adaptation is obtained so readily with cultured cells, why is it so difficult to demonstrate it on whole plants? For adaptation to occur, the cell, organ, or organism has to be in the proper physiological state; it must be "competent" to undergo the change. We have shown that in *Sorghum* the plant is competent to undergo adaptation to salinity only during a short time period, between days 5 and 10, following germination [2]. While this "developmental window" is open, the adaptive response of the plant is a function of the intensity of the stress [38] and of the stability of the environment [2]. Hamza [4] showed that the response to NaCl of *Hedysarum carnosum* was a function of the mode of salinization. Ingestad and Lund [8] and Clement et al. [9] reported a requirement for constant nitrate concentration for adaptation to nitrate. As reported by Jurik et al. [11] and Sims and Pearcy [10] for adaptation to light intensity, anatomical and physiological adaptation is possible only for leaves that have not reached an advanced stage of development when exposed to the environmental change. In the development of systemic resistance, Ye et al. [15] showed that leaves which expand after exposure of the plant to the inducer are better protected than leaves that were fully expanded before induction.

In *Sorghum*, when the developmental window is closed the plant does not adapt; it responds to salinity by "preexisting resistance" mechanisms (in the literature sometimes called "pre-adaptation"). Adaptation can be observed only within narrow experimental conditions: (a) the developmental window must be open; (b) environmental conditions must remain constant until the process of adaptation is completed; (c) good growth conditions are essential: plants must be healthy and be growing vigorously, which requires high light intensity, not too low relative humidity, and appropriate temperature; and (d) prolonged growth studies are required to recognize the phenomena. These requirements considerably limit the probability of observing adaptation following exposure of plants to stress. Moreover, as shown by Baker et al. [39], the capacity for adaptation seems to vary considerably among species. It also varies between genotypes within the same species, as shown by Amzallag et al. [2], Durrant [40], and Rikin et al. [13]. The ease by which plant cells adapt to salinity indicates that the developmental window is open at least once each growth cycle (or perhaps once each cell cycle) and that adaptation is a widespread property of plants. Similarly to cultured plant cells, adaptation enables whole plants to accommodate to new environmental conditions. It results in rapid creation of new genotypes. The data presented by Al-Hiyaly et al. [41] show that wild plants can accommodate to zinc-contaminated soils within 17 to 30 years. The authors assume that the mechanism of accommodation is through selection of variants having the appropriate gene (or genes). They state that "there has clearly been rapid evolution of zinc tolerance" and seem to be surprized that evolution can act with such rapidity. If the mechanism of accommodation is, as is our opinion, adaptation (a rapid, environmentally triggered change in genome structure and expression), it should not take more than several generations for a new genotype to develop and establish itself in the contaminated environment. The authors [41] also report that tolerant individuals of all species are not always found in all contaminated areas even when the species is present in the surrounding vegetation. Our interpretation is that not all genotypes have the capacity to adapt. Although adaptation at the whole plant level is in many ways similar to adaptation of cultured cells, there are, nevertheless, differences, since in the whole plant the cells must respond in an integrated fashion. Each cell and each organ must play its appropriate role at the end of the adaptation process. The response of each cell must be integrated in the new physiological mode of the plant. This implies that the expression of adaptation is different for cells having different functions. Hormones play a central role in the regulation of expression at the level of the whole plant.

III. ROLE OF HORMONAL BALANCE IN THE ADAPTATION OF PLANTS TO STRESS

One of the characteristics of integrated whole plant physiology is the discrepancy between the small number of hormones and the vast number of functions they perform. This probably results from control by the "hormonal balance" of the integrated response of the plant (not individual hormones having specific functions). At the same time, as discussed by Osborne [42], Trewavas [43], and Guern [44], changes in competence modulate the response of the plant to the hormonal balance; it results in an increased number of functions that can be controlled. Environmental factors perturb the hormonal balance (Table 3) as well as modifying the competence of the plant to respond to the hormonal balance. In this way it guides the response of the plant to changes occurring in the environment (Table 4). Too often, studies on the effect of a single exogenous hormone, or of a change in concentration of one endogenous hormone, are interpreted as a direct effect of that hormone. This may often result in contradictory reports. Both the balance of all the hormones as well as the physiological state of the plant (competence to respond) must be taken into consideration in the interpretation of results.

   Several authors have reported data showing that hormones originating in the root and transferred to the shoot via the xylem controlled physiological activities in the shoot. Under stressed soil conditions, there is a modification of hormones transferred to the shoot, the root acting as an organ sensing the soil environment. Wareing et al. [62] showed that cytokinin (CK) and gibberellin (GA) of root origin affected photosynthesis. In a review article, Vaadia [63] reported that osmotic and saline stresses resulted in decreased CK in xylem exudate. Lachno and Baker [64] showed that abscisic acid (ABA) of root origin increased under osmotic and saline stresses. Gowing et al. [65] showed that water stress of part of the root results in a signal of root origin that controls shoot growth in the absence of decreased shoot water status. Davies

Table 3 Effect of Environmental Stress on the Hormonal Balance of the Plant[a]

Stress	Effect	Remark	Ref.
	Barley, cotton, *Atriplex spongiosa*		
NaCl	ABA ↗	In barley and cotton exposure to 75 mM NaCl causes ABA ↗; in Atriplex 75 mM does not affect ABA; 150 mM is required for ABA ↗	45
	Helianthus annus		
NaCl, drought	CK ↘	CK was measured in xylem exudate, suggesting that it plays a role in root-to-shoot communication	46
	Nicotiana rustica		
Mineral nutrients	Transfer from 0.5 × HS to water causes ABA ↗		47
	Tomato		
Mineral nutrients	8.5 × increase in nutrients causes transient ABA ↗ and permanent CK ↘	*cis*- and *trans*-Zeatin ↘, zeatin riboside ↗	48
	Many species (review literature)		
Mineral nutrients	Deviation from optimal concentration of NO_3^-, NH_4^+, P_i, or K^+ causes CK ↘, GA ↘, IAA ↘		Table 1 in Ref. 49
	Lemna gibba		
Mineral nutrients	Changes in nitrogen cause changes in CK		50
	Many species (review article)		
Drought	ABA ↗	ABA transported from root to shoot	51

Table 3 (*Continued*)

Stress	Effect	Remark	Ref.
		Zea mays	
Root temperature	ABA max. between 13 and 18°C; CK and GA max. at 28°C	ABA, CK, and GA measured in xylem exudate, suggesting that they play a role in root-to-shoot communication	52
		Wheat	
Cold	Max. shoot GA in plants grown at 20 or 6°C; min. shoot GA at 2°C	Plants first grown at 2°C, then transferred to 20°C, had intermediate shoot GA concentration	53

[a] ↗ indicates an increase, ↘ indicates a decrease

and Zhang [51] reviewed the data on drought, indicating increased ABA transport to the shoot. Jackson [66] reported that in tomato root flooding results in increase ACC (1-aminocyclopropane-1-carboxylic acid), which is transported to the leaves, where it is transformed in ethylene. These data, together with those found in Table 3, show that under soil stress conditions there is an increase in ABA and ACC, a decrease in CK and changes in the GAs transported from root to shoot.

Exposure to exogenous ABA in the absence of stress can sometimes induce acclimation. Hwang and VanToai [59] reported that ABA treatment enhances tolerance to anoxia in corn plants. Lee et al. [29] showed that ABA induced tolerance to freezing in cultured potato cells; however, it is not identical to that induced by cold treatment. The ABA-induced tolerance shows a decline within a few days although the ABA treatment is continued, while cold treatment (4°C) induces freezing tolerance throughout the 11-day treatment. Moreover, the changes in translatable mRNAs are not exactly the same for the ABA and cold treatments. The reports of Reaney et al. [30] and of Ishikawa et al. [31] show that ABA induces cold tolerance in cultured bromegrass cells. However, upon transferring the cells to ABA-free medium, they lose a large part of their tolerance. Addition of a mixture of 100 μM of GA_4, GA_7, and GA_9 during the treatment counteracts the effect of ABA. Cytokinin also counteracts the effect of ABA, while GA_3 has no effect. A whole plant study on *Arabidopsis thaliana* by Lang et al. [67] shows that new polypeptides are induced by cold and by ABA treatment. Some polypeptides are common to both treatments, but not all. It seems that cold acclimation of ABA is not identical to that induced by cold itself.

Another indication of the "nonspecificity" of the response to stress produced by ABA is *cross-adaptation*, by which a given stress confers increased resistance toward other, apparently unrelated stresses. For example, as shown by Rikin et al. [68], the increase in ABA induced by an environmental stress such as drought or salinity increases tolerance to cold. Or, as shown by Radin and co-workers [69–71], nitrogen or P_i deficiency results in a faster response to water stress (stomatal closure occurs at a lower stress because of less negative leaf water potential). Boussiba et al. [72] reported that stresses which result in increased ABA concentration, such as drought, mineral deprivation, salinity, or boron toxicity, induced enhanced tolerance toward

Table 4 Effect of Exogenous Hormones on the Adaptation of Whole Plants, Organs, or Cultured Plant Cells to Stress[a]

Stress	Hormone	Effect	Refs.
		Sorghum bicolor	
NaCl	ABA	Accelerates adaptation	54
		Sorghum bicolor	
NaCl	CK	Inhibits adaptation	49
	GA	Perturbs adaptation	
		Nicotiana tobacum cells	
NaCl	ABA	Accelerates adaptation	55
		Medicago sativa	
Cold	ABA	ABA and SD ↗ acclimation	13
	GA	GA and LD ↘ acclimation	
		Bromus inermis cells	
Cold	ABA	Enhances cold hardening	30
	CK	Prevents cold hardening	
	GA	Prevents cold hardening	
		Bromus inermis cells	
Cold	ABA	Enhances cold hardening	31
		Solanum commersonii	
Cold	ABA	Enhances cold hardening	29
		Tobacco detached leaves, wheat and barley plants	
Cold	CK	Enhances cold hardening	56
		Pisum sativum	
Cold	CK	Enhances cold hardening, during October-to-April period	57
		Acer negundo	
Cold	GA	Inhibits cold hardening	58

Table 4 (*Continued*)

Stress	Hormone	Effect	Refs.
		Medicago sativa	
Cold	ABA	Enhances cold hardening	13
	GA	Inhibits cold hardening	
		Zea mays	
Anoxia	ABA	Enhances tolerance	59
		Zea mays	
Anoxia	Ethylene	Enhances adaptation to anoxia by promoting aeren-chyma formation	5,60
		Phaseolus vulgaris	
Pathogens	ABA	Enhances resistance	61

[a] ↗ indicates an increase, ↘ indicates a decrease; SD, short-day photoperiod; LD, long-day photoperiod.

low temperature and reduced root oxygen. One hypothesis suggests that ABA induces the expression of "protector" genes; these early stress proteins are thought to be unspecific protectors of cellular metabolism. Other explanations are that ABA modifies the water status of the plant, for example, by decreasing water losses due to stomatal closure, by making cell sap water potential more negative due to accumulation of solutes such as proline [73], or by increasing water uptake by increasing water permeability [74]. Guy and co-workers [75] showed that exposure of spinach to cold stress or to water stress resulted in increase of the same proteins, including proteins of 85 and 160 kDa. Sequencing of the 85-kDa protein indicates that it is similar to ABA-induced proteins and to LEA proteins (late abundant embryogenesis). The authors interpret their results as indicating common mechanisms in response to the two stresses since in both cases there is a decrease in water potential.

While the role of ABA in adaptation is still not understood, it does seem that ABA plays several important roles. With initiation of the stress, the increase in ABA helps the plant to face the change in the environment through stomatal closure, osmotic adjustment, and expression of stress proteins. However, ABA not only helps the plant to survive the immediate changes but plays a role in the enhancement of tolerance of the plant (adaptation). In *Sorghum*, at the time of the first exposure to salinity (which induces adaptation while the developmental window is open), exogenous ABA enhances growth and accelerates the process of adaptation. If the plant is further stressed, by increasing salinity, exogenous ABA no longer enhances growth but is inhibitory [54]. This unique effect of exogenous ABA on a plant undergoing adaptation (during the first exposure to salinity) indicates that ABA plays an active role in the process of adaptation. The reestablishment of a hormonal balance plays a key role in the response of plants to stress. Resistance (preadaptation) to stress seems to be linked to the capacity of the plant to maintain the hormonal balance close to that of the prestress state. In contrast, adaptation to stress involves the establishment of a new hormonal balance necessary for integrated development of the plant, which probably includes changes in competence. The new hormonal balance is not the same as the previous one. This is illustrated by the report of Reid et al. [53], which shows that wheat acclimated to cold at 2°C has a lower GA concentration than that of control plants (maintained at 20°C). When the cold-acclimated plants are brought back to 20°C, their GA levels do not return to those of controls.

The readjustment of the hormonal equilibrium to the new environmental conditions probably plays a central role in the response of the plant to stress. The preadaptive response (resistance) often seems to depend on maintenance of the preexisting hormonal balance in spite of the disturbance (stress). This is particularly evident in responses to pathogens, which seems to depend on maintenance of a high level of CK in the presence of the parasite [76]. The decreased growth observed following stress seems in many cases to result from a decrease in CK transported from root to shoot rather than from a direct effect of the stress, as indicated by the studies of Horgan and Wareing [77], Kuiper et al. [78], and Thorteinsson et al. [50]. This effect on shoot growth is similar to the control of photosynthesis by CK and GA demonstrated by Wareing and co-workers [62].

IV. ROLE OF HORMONES IN MECHANISMS OF ADAPTATION

A. Fundamental Mechanisms Responsible for Sensitivity and for Tolerance to Stress

Many authors have tried to elucidate the fundamental modifications that allow a plant to survive stress conditions. It was thought that comparison between varieties, or ecotypes, having different levels of tolerance would indicate which mechanisms bestow tolerance. In the case of salinity, for example, workers have been trying to characterize the gene, or genes, responsible for salt tolerance. Several mechanisms, such as turgor maintenance (required for cell elongation), reduced Na^+ and Cl^- uptake and accumulation in the leaves (to prevent inhibition of photosynthesis), compartmentation of Na^+ and Cl^- in the vacuole (where it is assumed that it will not disturb metabolism), and accumulation of compatible solutes in the cytoplasm, were suggested as factors responsible for tolerance [79]. However, as pointed out by Cheeseman [80] and by Munns [81] and as is becoming more evident with time, these hypotheses have not improved our understanding of the mechanisms responsible for the inhibitory effects of salinity. The publication of McCoy [82] showing that there is no correlation between tolerance at the whole plant level and at the cultured cell level for alfalfa genotypes very much weakens the concept that mechanisms at the cellular level are responsible for salt tolerance at the whole plant level. Our demonstration that in salt-adapted *Sorghum* there is no correlation between growth and shoot Na^+ concentration weakens the concept that salt tolerance simply depends on reduced Na^+ accumulation in the shoot [1] (Amzallag et al., unpublished results). The comparison between polypeptides expressed by the wild-type cells in the absence of salt and those expressed by salt-adapted cells in the presence of salt for different species by Ben-Hayyim et al. [25] shows that the change in tolerance does not result from the expression of a limited number of proteins (genes). In each species the change in genome expression is different, indicating that in plants tolerance to salinity does not depend on the expression of a specific set of genes. Furthermore, additional lines of wheat having enhanced salt tolerance were obtained by crosses between salt-sensitive Chinese spring wheat and halophytic *Lophopyrum elongatum*. Studies by Dvorak et al. [83] show that the correlation between salt tolerance and the presence of individual *L. elongatum* chromosomes indicate that a number of loci on different chromosomes, each having a relative small effect and able to act independently, were responsible for the enhancement of tolerance. These data suggest that induction of salt tolerance, during adaptation, results from reorganization of genome expression.

Colonization by several species of soils contaminated by heavy metals demonstrates the capacity of plants for adaptation to such toxic conditions. For example, the report of Wu and Lin [84] shows the adaptation of the legume *Lotus pushianus* and its symbiont *Rhizobium loti* to copper mine waste. *Rhizobium* isolated from adapted plants demonstrate greater tolerance to copper than that of unadapted *Rhizobium*, as well as decreased copper uptake. Adapted *Lotus pushianus* also shows enhanced tolerance but accumulates as much copper as do control (unadapted) plants. It is evident that the bacteria adapt to copper by a mechanism that is very different from that of the plant. As pointed out by Baker [85] in a review on heavy metal tolerance in plants, the proposed mechanisms do not give a clear picture of the nature of the adaptation process.

When some of the changes brought about by acclimation can be measured biochemically, it is easier to follow the process. For example, in cold hardening Wang et al. [86] showed changes in membrane lipid composition in squash, while Sarhan and Chevrier [87] showed increases in RNases and in the RNA polymerases I/RNA polymerase II ratio in winter wheat. Besford [88] demonstrated modification in Calvin cycle enzyme activities in tomato, and Davies et al. [89] showed changes in electron transport and thylakoid ATPase activity during acclimation to irradiance. These changes occur within a few days. However, these changes represent only a fraction of the changes occurring during acclimation.

The development of necrotic lesions in response to infection by pathogens leads to the induction of systemic acquired resistance of the whole plant. Salicylic acid is thought to be the molecule that is transported from the infected leaf to the rest of the plant, where it induces general resistance to pathogens. Ward et al. [90] showed that exogenous salicylic acid induces resistance as well as the expression of several mRNAs. However, as pointed out by Ye et al. [15], salicylic acid does not induce all the response triggered by pathogen infection.

From the data in the literature, for the response to a stress to be oriented (to result in adaptation), the stressing factor itself must be present. The presence of hormones such as ABA or ethylene or of salicylic acid trigger only general, predetermined responses. It seems that the environment is directly involved in achieving a precise response. Since the precise, oriented response to a stress does not have to depend on the expression of a well-defined group of genes whose expression is triggered by identifiable factors, it seems that the environment itself modifies, in an oriented way, the expression of the genome.

B. Repetitive DNA and Genome Expression

Repetitive DNA, which is not translated, may represent more than 80% of the total DNA present in the plant cell. Except for some repetitive rDNA, the rest of this repetitive DNA is not transcribed. For a long time no function was ascribed to it, and it was thought to be "parasite," "junk," or "selfish" DNA. However, the important quantitative changes occurring in repetitive DNA in plant cells has attracted new interest to these supposed useless sequences. In her review articles, Bassi [91,92] describes the changes, underrepresentation or overrepresentation, occurring in the quantity of repetitive DNA during development (differentiation). It seems that modifications of genome expression, such as in the transition from juvenile to adult phase or vegetative to flowering, are generally also modifications in repetitive DNA. The awareness of a relation between repetitive DNA and genome expression resulted from the demonstration that repetitive DNA played a role in DNA tertiary structure, which modulates gene expression as suggested by Nagl [93]. There is a certain level of "plasticity" to genome conformation, resulting in a modulation of its expression at certain stages of plant development (stages when repetitive DNA undergoes modifications). This resembles the developmental window for adaptation in *Sorghum* demonstrated by Amzallag et al. [2]; the plant shows plasticity, capacity for adaptation, at a very precise stage of development. Another mechanism resulting in modulation of genome expression, described by Levy and Walbot [94], is movement of transposable elements; this also occurs during particular stages of plant development, while a developmental window is open.

C. Environmental Effects on Repetitive DNA

The comparative studies of the genomes in species of the genus *Lathyrus* by Narayan and Durrant [95] and of the genus *Allium* by Evans et al. [96] indicate that the main difference resides in the organization of their repetitive DNA. These authors suggest that changes in repetitive DNA during evolution seem to have played a role in the development of new species. Bennett [97] showed that there is a positive correlation between the quantity of DNA per nucleus and the latitude at which the plant grows: the higher the latitude, the higher the amount of DNA. This increase in DNA reflects mainly an increase in repetitive DNA. If we suppose that the environment becomes less foreseeable as we move from the equator to the higher latitudes, these data would suggest that repetitive DNA plays a role in increasing the capacity for adaptability of the plant. A study by Flavell et al. [96] on *Triticum dicoccoides* in Israel shows that rDNA diversities are correlated with climatic variables. Several other reports in the literature show that stresses affect DNA either by causing a change in amount per cell or by modifying it (see Table 5). Bassi [92] suggests that repetitive DNA sequences may act as mediators between the environment and gene expression. They might be affected by information from the environment resulting in DNA modification and hence modulate the expression of specific genes.

Adaptation of cultured cells to environmental factors is related to changes in genome expression (Table 2) and perhaps also to modifications in repetitive DNA. Indeed, Landsmann and Uhrig [107] reported that changes in repetitive rDNA accompanied somaclonal variations in plants regenerated from cultured potato cells. These data suggest that the stress imposed on the cells during accommodation to culture conditions induced the DNA modification.

D. Effect of Hormones on Repetitive DNA

Although we do not have information concerning all the hormones, and the information we do have is only partial, several reports do show that exogenous phytohormones induce changes in repetitive DNA (see Table 6). Although not always determined, it is probable that in most cases it is replication of repetitive DNA that is affected by the hormone. The lack of information concerning ABA, which has been shown to play an important role in acclimation, is rather unfortunate. However, from the data available we can presume that the effect of stress on CK, GA, and indoleacetic acid (IAA) (and probably ABA and ethylene) modulates the level of amplification of specific families of repetitive DNA. This change in the noncoding DNA results in structural changes in the vicinity of genes, thereby modifying DNA conformation, which results in changes in gene expression. Adaptation of a plant to a stress could proceed according to the following sequence of events: under the constraints imposed by the environmental stress the hormonal balance of the plant is modified; this induces specific changes in repetitive DNA which modulate the expression of specific genes.

V. SUMMARY AND CONCLUSION

From the variety of examples found in the literature it is evident that adaptation is a widespread property of plants. It has, however, not been studied seriously by most workers in the field of plant stress. In fact, most authors use the expressions *accommodation*, *acclimation*, and *adaptation* as synonyms, and therefore are not able to differentiate between adaptation and

Table 5 Effect of Environmental Stress on Amount of DNA per Cell or DNA Modification[a]

Stress	Adaptation	Type of affected repetitive DNA	Ref.
		Linum	
Nutrients	n.r.	rDNA content ↗ or ↘	99
		Triticum aestivum	
Cold	Yes	rDNA content ↗	100
		Cruciferae species	
Cold	Yes	rDNA content ↗	101
		Zea mays	
UV	n.r.	Transposable element movement ↗	102
		Hordeum vulgare	
Aluminum	n.r.	DNA content ↗	103
		Lobularia maritima	
NaCl	Yes	DNA content per cell ↗	104
		Hordeum vulgare	
NaCl	n.r.	Repetitive DNA ↘	105
		Solanum tuberosum	
Cutting	Yes	DNA ↗	106

[a] ↗ indicates an increase, ↘ indicates a decrease; n.r., not reported.

resistance (preadaptation). It is, however, possible in certain cases to differentiate between the preadaptive response and adaptation when sufficiently detailed information is available.

When a plant responds to a stress through its resistance (preadaptive response) it expresses a preexisting program that enables it to survive the stress while maintaining (more or less) its original developmental program. Resistance reflects the capacity of the plant to express, under stress, its original developmental program. In general, expression of a preexisting program occurs relatively rapidly, within 48 h of exposure to stress. The reaction is not specific to the particular stress. There is usually an increase in ABA in the organs exposed to the stress. Growth decreases and rapidly stabilizes at a new level. The decrease in growth is proportional to stress intensity and inversely proportional to the tolerance (preadaptive capacity) of the plant.

Since most of the studies in the literature describing the response of plants to stress relate to the preadaptive response, we have focused our attention on the less known response,

Table 6 Effect of Exogenous Hormones on DNA Content per Cell[a]

Type of plant material	Hormone	Effect	Ref.
		Cucumis melo	
Callus	Auxin	Change in amplification of satellite DNA	108
		Nicotiana tabacum	
Excised stem segment	Auxin	On defoliated flowering segment, IAA inhibits DNA synthesis; on defoliated, vegetative segment, IAA enhances DNA synthesis	109
		Cymbidium (Orchid hybrid)	
In vitro protocorm pieces	Auxin	Amplification of AT-rich satelite DNA	110
	CK	No effect	
	GA	Amplification of GC-rich satelite DNA	
		Cucumis sativus	
Hypocotyl section	Auxin	DNA ↗	111
	GA	DNA ↗	
		Phaseolus vulgaris	
Chloroplast	CK	Chloroplast DNA ↗	112
		Glycine max	
Cultured cell suspension	CK	Replication of certain repetitive sequences ↗	113
		Daucus carota	
Whole Plant	GA	In root, changes in amounts of intermediate-repeated DNA	114
		Lens culinaris	
Epicotyl	GA	Amount of DNA ↗ during epicotyl elongation (in absence of cell division)	115

[a] ↗ indicates an increase.

adaptation. During adaptation the plant establishes a new developmental program as a function of the precise stress conditions. At the beginning of the process of adaptation there is considerable decrease in growth, which is usually more important than in preadaptation (G. N. Amzallag et al., unpublished results). Once the plant is adapted, the growth rate increases and can reach levels similar to the mean relative growth rate ($\overline{RGR}$) of control plants of the same age [1]. The increase in $\overline{RGR}$ at the end of the process of adaptation indicates that the environment is no longer stressing for the plant [1,8,116]. The adaptive response includes modifications of the hormonal balance, metabolic processes, and expression of the genome.

In contrast with the preadaptive response, the adaptive response is not completely preprogrammed in the genome. The plant adapts to the precise environmental conditions provided that these conditions remain stable during the period of adaptation. Development of adaptation requires more time than the preadaptive response. It takes between 5 and 20 days to develop. However, once adapted the plant is no longer in a stressed environment. Based on the data reported in this review, the following schema summarizes our hypothesis of the various stages in the process of adaptation.

Hypothetical Schema of the Sequence of Events Occurring During Adaptation

Stress applied while developmental window is opened
↓
Modification of hormonal balance (ABA ↗, CK ↘, GA modified)
↓
Short-term response: synthesis of stress proteins
Long-term response: modification of repetitive DNA
↓
Changes in repetitive DNA affect DNA conformation
↓
Changes in genome expression
↓
Improvement of the physiology of the plant, decrease in ABA, increase in CK (or of sensitivity toward CK), restoration of GA (or change in sensitivity toward GA)
↓
Adaptation; the environment is no longer stressing

While the adaptive response is not programmed, the capacity to respond to stress by adaptation is genetically controlled. Not all species are able to adapt. Within a species that does adapt, not all genotypes (cultivars) are capable of adaptation. Moreover, within a species not all the genotypes that adapt do so to the same extent. We can suppose that in natural populations there are a variety of responses toward stress between preadaptation and adaptation. There is no relation between the capacity of a plant to respond to stress by preadaptation when the developmental window is closed (its resistance) and the capacity of the plant to respond by adaptation when the window is open. The capacity to respond by adaptation is an independent genetic parameter. Geneticists and plant breeders have not taken into consideration the capacity of plants to adapt. Even for genotypes that adapt well, triggering of the process of adaptation is under strict control. There is (at least for the Poaceae) a very limited period during which the plant is competent to respond to stress by adaptation. This phenomenon seems to be similar to developmental windows in animal embryology. Moreover, even during the period of competence, triggering of adaptation requires a certain threshold of stress, and the stress must remain at a stable level during the process for adaptation to occur. The requirement for a constant stress level may result from the fact that the adaptation response is a function of a particular change in the environment and that its induction requires time. These requirements are probably one of the reasons why adaptation has not been described previously. Nevertheless, this phenomenon probably plays an important role in the development of ecotypes adapted to new environments and in the evolution of species. Since reorganization of repetitive DNA, which modifies genome expression, is an important part of the adaptation process, it seems possible that these changes in DNA may be transmitted to offspring. Indeed, environmentally induced rapid changes of the phenotype due to imbalance of mineral nutrients, which affect DNA organization and which are heritable, have been reported by Durrant [40,117] and Cullis [99] for flax and by Hill [118] for tobacco. The modification is maintained even when the offspring are grown in balanced mineral medium. Modifications of the genome are in the repetitive DNA. Moreover, Highkins [119], using an experimental approach similar to that of Durrant, showed that mineral imbalance increases diversity in the offspring in pea. The modification induced in flax by Durrant [40] does not occur with all varieties. Furthermore, it does not seem to occur in all species; Hill [118] reported that he had not been able to obtain the effect with *Arabidopsis*. It is interesting to note that *Arabidopsis* is extremely poor in repetitive DNA. In a study on pea, Moss and Mullett [120] showed that optimum temperature for germination is modified by maintaining temperature for five consecutive generations at a set value. A change in optimum germination temperature does not occur with all genotypes tested.

At present, we do not understand the mechanism by which an organism is able to modify the expression of its genome as a function of the environment. It seems, however, to be a property of many organisms. Shapiro [33], Hall [35], and Cairns [34] have shown that in bacteria certain mutations are guided by the environment. Hall [36] has also demonstrated this for yeast. An understanding of this property is particularly interesting. It would open the possibility of developing new varieties of crop plants by first selecting adaptable cultivars, then adapting them and selecting those variants that are better adapted to a particular environment. We could in this way develop new adapted varieties.

In the past, adaptation has been confused with preadaptation (resistance). Stress was considered to trigger the expression of a preexisting genetic program. The lack of references to the phenomenon of adaptation in journals as well as in monographs specialized in plant response to stress [121,122] shows the extend to which it has remained neglected even though evidence has been accumulating for some time. Unfortunately, the work on flax by Durrant [40,117] has not been taken seriously, even though it was backed by classical genetic studies and detailed information at the molecular level published by Cullis [99]. The reviews by Bassi [91,92] show that many hundreds of publications demonstrate that changes in repetitive DNA occur during plant development and as a function of environmental stress. These works did not receive the attention they deserved, mainly because of preconceived dogmas originating in Darwinian concepts, the idea that genetic changes can result only from random mutations and that environment cannot modify, adaptively, genetic information.

ACKNOWLEDGMENT

This work was supported by the U.S.–Israel Binational Agricultural Research and Development (BARD) Foundation Project US-1869-90R.

REFERENCES

1. G. N. Amzallag, H. R. Lerner, and A. Poljakoff-Mayber, *J. Exp. Bot.*, *41*: 29 (1990).
2. G. N. Amzallag, H. Seligmann, and H. R. Lerner, *J. Exp. Bot.*, *44*: 645 (1993).
3. K. Winter, *Ber. Dtsch. Bot. Ges.*, *86*: 467 (1974).

4. M. Hamza, *Bull. Soc. Bot. Fr. Actual. Bot.*, *45*: (3/4) 51 (1978).
5. H. Konings, *Physiol. Plant.*, *54*: 119 (1982).
6. M. Kawase, *Am. J. Bot.*, *66*: 183 (1979).
7. H. Brown and M. H. Martin, *New Phytol.*, *89*: 621 (1981).
8. T. Ingestad and A. B. Lund, *Physiol. Plant.*, *45*: 137 (1979).
9. C. R. Clement, M. J. Hopper, and L. H. P. Jones, *J. Exp. Bot.*, *29*: 453 (1978).
10. D. A. Sims and R. W. Pearcy, *Am. J. Bot.*, *79*: 449 (1992).
11. T. W. Jurik, J. F. Chabot, and B. F. Chabot, *Plant Physiol.*, *63*: 542 (1979).
12. D. Graham and B. D. Patterson, *Annu. Rev. Plant Physiol.*, *33*: 347 (1982).
13. A. Rikin, M. Waldman, A. E. Richmond, and A. Dvorat, *J. Exp. Bot.*, *26*: 175 (1975).
14. R. A. Dean and J. Kuc, *Phytopathology*, *76*: 966 (1986).
15. X. S. Ye, S. Q. Pan, and J. Kuc, *Physiol. Mol. Plant Pathol.*, *35*: 161 (1989).
16. J. L. McIntyre, J. A. Dodds, and J. D. Hare, *Phytopathology*, *71*: 297 (1981).
17. G. Sabatier and N. Vartanian, *Physiol. Plant.*, *59*: 501 (1983).
18. N. Vartanian, *C.R. Acad. Sci. Paris*, *274*: 1497 (1972).
19. S. Da, C. Hubac, and N. Vartanian, *Can. J. Bot.*, *55*: 1236 (1977).
20. D. Apriyanto and D. A. Potter, *Oecologia*, *85*: 25 (1990).
21. H. D. Gottstein and J. A. Kuc, *Phytopathology*, *79*: 176 (1989).
22. A. A. Watad, L. Reinhold, and H. R. Lerner, *Plant Physiol.*, *73*: 624 (1983).
23. A. A. Watad, H. R. Lerner, and L. Reinhold, *Physiol. Veg.*, *23*: 887 (1985).
24. M. L. Binzell, P. M. Hasegawa, A. K. Handa, and R. A. Bressan, *Plant Physiol.*, *79*: 118 (1985).
25. G. Ben-Hayyim, Y. Vaadia, and B. G. Williams, *Physiol. Plant.*, *77*: 332 (1989).
26. T. J. McCoy, *Plant Cell Rep.*, *6*: 417 (1987).
27. G. Donn, E. Tischer, J. A. Smith, and H. M. Goodman, *J. Mol. Appl. Genet.*, *2*: 621 (1984).
28. R. Cella, D. Albani, M. G. Biasini, D. Carbonera, and B. Parisi, *J. Exp. Bot.*, *35*: 1390 (1984).
29. S. P. Lee, B. Zhu, T. H. H. Chen, and P. H. Li, *Physiol. Plant.*, *84*: 41 (1992).
30. M. J. T. Reaney, L. V. Gusta, S. R. Abrams, and A. J. Robertson, *Can. J. Bot.*, *67*: 3640 (1989).
31. M. Ishikawa, A. J. Robertson, and L. V. Gusta, *Plant Cell Physiol.*, *31*: 51 (1990).
32. F. Meins, *Annu. Rev. Genet.*, *23*: 395 (1989).
33. J. A. Shapiro, *Mol. Gen. Genet.*, *194*: 79 (1984).
34. J. Cairns, J. Overbaugh, and S. Miller, *Nature*, *335*: 142 (1988).
35. B. G. Hall, *Genetics*, *126*: 5 (1990).
36. B. G. Hall, *Proc. Natl. Acad. Sci. USA*, *89*: 4300 (1992).
37. M. W. Bayliss, *Int. Rev. Cytol. Suppl.*, *11A*: 113 (1980).
38. H. Seligmann, G. M. Amzallag, and H. R. Lerner, *Aust. J. Plant Physiol.*, *20*: 243 (1993).
39. A. J. M. Baker, C. J. Grant, M. H. Martin, S. C. Shaw, and J. Whitebrook, *New Phytol. 102*: 575 (1986).
40. A. Durrant, *Philos. Trans. R. Soc. London*, *B292*: 467 (1981).
41. S. A. K. Al-Hiyaly, T. McNeilly, and A. D. Bradshaw, *New Phytol.*, *114*: 183 (1990).
42. D. J. Osborne, in *Hormone Action in Plant Development*, (G. V. Hoad, J. R. Lenton, M. B. Jackson, and R. K. Atkin, eds.), Butterworth, London, pp. 265–274 (1987).
43. A. Trewavas, *Plant Cell Environ.*, *14*: 1 (1991).
44. J. Guern, *Ann. Bot.*, *60* (Suppl.), *4*: 75 (1987).
45. Z. Kefu, R. Munns, and R. W. King, *Aust. J. Plant Physiol.*, *18*: 17 (1991).
46. C. Itai, A. E. Richmond, and Y. Vaadia, *Isr. J. Bot.*, *17*: 187 (1968).
47. Y. Mizrahi and A. E. Richmond, *Plant Physiol.*, *50*: 667 (1972).
48. M. A. Walker and E. B. Dumbroff, *Z. Pflanzenphysiol.*, *101*: 461 (1981).
49. G. N. Amzallag, H. R. Lerner, and A. Poljakoff-Mayber, *J. Exp. Bot.*, *43*: 81 (1992).
50. B. Thorteinsson and L. Eliasson, *Plant Growth Regul.*, *9*: 171 (1990).
51. W. J. Davies and J. Zhang, *Annu. Rev. Plant Physiol. Plant Mol. Biol.*, *42*: 55 (1991).
52. R. T. Atkin, G. E. Barton, and D. K. Robisson, *J. Exp. Bot.*, *79*: 475 (1973).
53. D. M. Reid, R. P. Pharis, and D. W. A. Roberts, *Physiol. Plant.*, *30*: 53 (1974).
54. G. N. Amzallag, H. R. Lerner, and A. Poljakoff-Mayber, *J. Exp. Bot.*, *41*: 2934 (1990b).

55. P. C. LaRosa, P. M. Hasegawa, D. Rhodes, J. M. Clither, A. A. Watad, and R. A. Bressan, *Plant Physiol.*, *85*: 174 (1987).
56. Z. Dror, A. Rikin, and A. E. Richmond, *Isr. J. Bot.*, *25*: 96 (1976).
57. S. Kuraishi, T. Tezuka, T. Ushijuma, and T. Tazaki, *Plant Cell Physiol.*, *7*: 705 (1966).
58. R. M. Irving and F. O. Lanphear, *Plant Physiol.*, *43*: 9 (1968).
59. S. Y. Hwang and T. T. VanToai, *Plant Physiol.*, *97*: 593 (1991).
60. M. C. Drew, M. B. Jackson, S. C. Giffard, and R. Campbell, *Planta*, *153*: 217 (1981).
61. R. M. Dunn, P. Hedden, and J. A. Bailey, *Physiol. Mol. Plant Pathol.*, *36*: 3393 (1990).
62. P. F. Wareing, M. M. Khalifa, and K. J. Treharne, *Nature*, *220*: 453 (1968).
63. Y. Vaadia, *Philos. Trans. R. Soc. London, B273*: 513 (1976).
64. D. R. Lachno and D. A. Baker, *Physiol. Plant.*, *68*: 215 (1986).
65. D. J. Gowing, W. J. Davies, and H. G. Jones, *J. Exp. Bot.*, *41*: 1535 (1990).
66. M. B. Jackson, in *Hormone Action in Plant Development*, (G. V. Hoad, J. R. Lenton, M. B. Jackson, and R. K. Atkin, eds.), Butterworth, London, pp. 189–199 (1987).
67. V. Lang, P. Heino, and E. T. Palva, *Theor. Appl. Genet.*, *77*: 729 (1989).
68. A. Rikin, A. Blumenfeld, and A. E. Richmond, *Bot. Gaz.*, *137*: 307 (1976).
69. J. W. Radin, *Plant Physiol.*, *67*: 115 (1981).
70. J. W. Radin, L. L. Parker, and G. Guinn, *Plant Physiol.*, *70*: 1066 (1982).
71. J. W. Radin, *Plant Physiol.*, *76*: 392 (1984).
72. S. Boussiba, A. Rikin, and A. E. Richmond, *Plant Physiol.*, *56*: 337 (1975).
73. P. Pesci, *Physiol. Plant.*, *86*: 209 (1992).
74. Z. Glinka and L. Reinhold, *Plant Physiol.*, *48*: 103 (1971).
75. C. Guy, D. Haskell, L. Neven, P. Klein, and C. Smelser, *Planta*, *188*: 265 (1992).
76. J. D. Johnson, *Plant Growth Regul.*, *6*: 193 (1987).
77. J. M. Horgan and P. F. Wareing, *J. Exp. Bot.*, *31*: 525 (1980).
78. D. Kuiper, P. J. C. Kuiper, H. Lambers, J. Schmit, and M. Staal, *Physiol. Plant*, *75*: 511 (1989).
79. H. Greenway and R. Munns, *Annu. Rev. Plant Physiol.*, *31*: 149 (1980).
80. J. M. Cheeseman, *Plant Physiol.*, *87*: 547 (1988).
81. R. Munns, *Aust. J. Plant Physiol.*, *15*: 717 (1988).
82. T. J. McCoy, *Plant Cell Rep. 6*: 31 (1987).
83. J. Dvorak, M. Edge, and K. Ross, *Proc. Natl. Acad. Sci. USA*, *85*: 3805 (1988).
84. L. Wu and S. L. Lin, *New Phytol*, *116*: 531 (1990).
85. A. J. M. Baker, *New Phytol.*, *106* (Suppl.): 93 (1987).
86. C. Y. Wang, G. F. Kramer, B. D. Whitaker, and W. R. Lusby, *J. Plant Physiol.*, *140*: 229 (1992).
87. F. Sarhan and N. Chevrier, *Plant Physiol.*, *78*: 250 (1985).
88. R. T. Besford, *J. Exp. Bot.*, *37*: 200 (1986).
89. E. C. Davies, W. S. Chow, J. M. Le Fay, and B. R. Jordan, *J. Exp. Bot.*, *37*: 211 (1986).
90. E. R. Ward, S. J. Uknes, S. C. Williams, S. S. Dincher, D. L. Wiedehold, D. C. Alexander, P. Ahl-Goy, J. P. Metraux, and J. A. Ryals, *Plant Cell*, *3*: 1085 (1991).
91. P. Bassi, *Biol. Rev.*, *65*: 185 (1990).
92. P. Bassi, *Biol. Zentralbl.*, *110*: 1 (1991).
93. W. Nagl, *Biol. Zentralbl.*, *102*: 257 (1983).
94. A. A. Levy and V. Walbot, *Science*, *248*: 1543 (1990).
95. R. K. J. Narayan and A. Durrant, *Genetica*, *61*: 47 (1983).
96. I. J. Evans, A. M. James, and S. R. Barnes, *J. Mol. Evol.*, *170*: 803 (1983).
97. M. D. Bennett, *Environ. Exp. Bot.*, *16*: 93 (1976).
98. R. B. Flavell, M. O'Dell, P. Sharp, E. Nevo, and A. Beiles, *Mol. Biol. Evol.*, *3*: 547 (1986).
99. C. A. Cullis, *Adv. Genet.*, *28*: 73 (1990).
100. E. Paldi and M. Devay, *Plant Sci. Lett.*, *30*: 61 (1983).
101. A. Laroche, X. M. Geng, and J. Singh, *Plant Cell Environ.*, *15*: 439 (1992).
102. V. Walbot, *Mol. Gen. Genet.*, *234*: 353 (1992).
103. M. Sampson, D. Clarkson, and D. D. Davies, *Science*, *148*: 1476 (1965).
104. I. Capesius and S. Loeben, *Z. Pflanzenphysiol.*, *110*: 259 (1983).
105. L. E. Murry, M. L. Christianson, S. H. Alfinito, and S. J. Garger, *Am. J. Bot.*, *74*: 1779 (1987).

106. A. Watanabe and H. Imesaki, *Plant Physiol.*, *51*: 772 (1973).
107. J. Landsmann and H. Uhrig, *Theor. Appl. Genet.*, *71*: 500 (1985).
108. J. Grisvard and A. Tuffet-Anghileri, *Nucleic Acids Res.*, *8*: 2843 (1980).
109. W. L. Wardell and F. Skoog, *Plant Physiol.*, *52*: 215 (1973).
110. W. Nagl and W. Rucker, *Nucleic Acids Res.*, *3*: 2033 (1976).
111. Y. Degani, D. Atsmon, and A. H. Halevy, *Nature*, *228*: 554 (1970).
112. I. Kinoshita and I. Tsuji, *Plant Physiol.*, *76*: 575 (1984).
113. M. Caboche and K. G. Lark, *Proc. Natl. Acad. Sci. USA*, *78*: 1731 (1981).
114. A. Schafer and K. H. Neumann, *Planta*, *143*: 1 (1978).
115. J. Nitsan and A. Lang, *Plant Physiol.*, *41*: 965 (1966).
116. K. Wignararajah, *Environ. Exp. Bot.*, *30*: 141 (1990).
117. A. Durrant, *Heredity*, *17*: 27 (1962).
118. J. Hill, *Nature*, *207*: 732 (1965).
119. H. R. Highkins, *Nature*, *182*: 1460 (1958).
120. G. I. Moss and J. H. Mullett, *J. Exp. Bot.*, *33*: 1147 (1982).
121. J. Levitt, *Response of Plants to Environmental Stresses*, Academic Press, New York (1980).
122. M. G. Hale and D. M. Orcutt, *The Physiology of Plants Under Stress*, Wiley, New York (1987).

26
Adaptive Components of Salt Tolerance

James W. O'Leary

University of Arizona, Tucson, Arizona

I. INTRODUCTION

Plants have evolved two very different strategies in adapting to high levels of sodium salts in their environments. One strategy is to exclude the salts from the interior of the leaf cells, and the other includes the salts within the leaf cells but sequesters most of them in the vacuoles of those cells. In both cases, the end result is to maintain the cytoplasmic sodium concentration relatively low. This is accomplished in the former case by either preventing entry of the ions into the plant at the root surface or preventing them from being transported in the xylem from the roots to the leaves. In the latter case, entry and transport to the leaves is not prevented nor severely restricted, and the problem is handled primarily at the tonoplast level of the leaf cells themselves. The latter strategy seems to have been more effective when adapting to the most extreme saline habitats, but the former seems to have been manipulated more successfully during directed selection by plant breeders. Even though the exclusion process is not perfect in most plants, and some might argue that there is no sharp line separating the two categories, for simplicity's sake in the following discussion, these two broad categories of plants will be referred to as *excluders* and *includers*, respectively.

Although both strategies are effective, there are some important differences between the two types of plants. Those that exclude salts from the leaf cells are able to tolerate high levels of those salts in the root environment, but at the expense of reduced growth. That is, as cultivars or ecotypes within a species are developed with increasing ability to exclude the sodium salts and thereby survive at increasingly higher concentrations of those salts in their environment, growth is reduced to well below what it is in the absence of those salts, even within the range of the relatively low salt concentrations characteristic of irrigated agriculture. On the other hand, the plants that not only allow the salts to reach the leaves but contain them at relatively high concentrations typically show increased growth with increasing level of external salinity within this range of salinities. It seems to me that this difference provides a unique opportunity to address the question of what constitutes salt tolerance, a philosophy developed further later in this discussion.

The excluders include virtually all crop plants and most, if not all, monocotyledonous halophytes, plus many dicotyledonous halophytes. The includers are limited to a relatively small number of dicotyledonous halophytes. It is somewhat surprising that the domesticated plants whose salt tolerance has been increased most successfully by selection and breeding are the monocotyledonous species, all of which are excluders to some degree, but the plants that have been selected by nature to tolerate the most saline habitats are the includers, not the excluders. Why has increased ability to tolerate high salt concentration in leaf cells not been a target for plant breeders? The answer to that question, and my feelings about the prospects of using that approach successfully to increase salt tolerance in crop plants will also be addressed in the following discussion.

The typical approach to studying salt tolerance is to compare plants (both sensitive and tolerant plants) subjected to excess salinity with plants not subjected to salinity, looking for responses to the added salt. Examples of such responses are production of unique proteins or large amounts of presumed compatible osmotic solutes, such as proline and glycinebetaine. The difficulty with such an approach is that it is difficult to distinguish those responses that are truly adaptive from those that are reflections of metabolic lesions.

For example, even though there has been a substantial amount of research over the years devoted to comparison of plant responses to growth-inhibiting salinity and nonsaline conditions, and the production and accumulation of putative compatible osmotic solutes such as proline and glycinebetaine have been investigated almost exhaustively in numerous plant species [1–4], the role of those solutes in salt tolerance has not yet been clearly demonstrated. The pathways, and control points therein, of synthesis and degradation of such solutes have been studied in great detail, yet there is now growing concern whether production of those solutes for osmotic adjustment is of any adaptive or other beneficial value [5,6]. That is, they may be produced in large quantities as a result of disruptions in metabolism (metabolic lesions) in response to stress, or they may simply accumulate as a result of a lower utilization of photosynthate in stressed plants.

It seems appropriate to suggest that it is time to take a fresh look at the salt tolerance question and consider some new approaches. For example, rather than continuing to focus on those plants that are not especially salt tolerant and trying to decide how to make them more tolerant, it may be more productive to devote more effort to trying to find out what makes the highly salt tolerant plants so tolerant. The plants to which I refer are halophytes, and one advantage they provide is the opportunity to compare growth at suboptimal salinity with optimal salinity, an approach that is not possible with present crop plants or other glycophytes. The optimum salinity for growth in crop plants and other glycophytes is zero, with decreased growth as salinity increases beyond a few mol/m^3. In contrast, in many (but not all) halophytes, the optimum salinity for growth has shifted to 50 to 200 mol/m^3, with decreased growth occurring at both higher and lower salinities. Thus if one compares the responses of such plants to less than optimum salinity with responses in plants grown at optimum salinity, it might be possible to distinguish those responses that are truly adaptive from those that are the result of lesions or other types of damage. The hypothesis is that adaptation in these halophytes involves some processes having optimum performance at a salinity level well above zero, while at lower salinities these processes do not function as well. A process that does not function as well in a plant growing at 50 mol/m^3 as it does in a plant growing at 200 mol/m^3 certainly is not being altered by excess salinity, and a priori would seem to be involved in the better growth of that plant at the higher salinity. The challenge, then, is to identify those processes.

Unfortunately, most research so far has focused on the effects of excess salinity rather than inadequate salinity, but there are a few examples in which those studies did include the suboptimal salinity levels as well. A quick survey of some of those studies may give us a hint about where the attention should be concentrated in studies involving comparison of suboptimal and optimal salinity levels.

II. PLANT RESPONSES TO SALINITY

A. Effect of Salinity Level on Growth

That some halophytes grow better at an appreciable salinity level than they do in fresh water was acknowledged by Chapman [7] and Waisel [8] in their comprehensive reviews of halophytes, although they implied that this represented a minority of halophytes. Barbour's [9] thorough review of the older literature specifically relating growth to salinity, however, revealed that such observations were fairly numerous, and the classic review by Flowers et al. [10] listed several species that showed greater growth at salinity levels equivalent to 50 to 200 mM NaCl than in nonsaline conditions. In fact, the general feeling now that this is such a common response among halophytes is reflected in the latest review by Flowers et al. [11], in which they emphasized a few species that do not show such a response rather than listing those that do.

In general, monocotyledonous halophytes do not have growth optima at substantial salinity levels (i.e., they show a steady decline in growth with any increase in salinity). There are a few reports in the literature (e.g., Refs 12 and 13) indicating that some monocots do have greater growth at salinity levels greater than 50 mM NaCl, but the overwhelming body of experimental evidence supports the generalization that monocotyledonous halophytes do not require substantial salinity levels for optimum growth [11,14].

Thus this review and discussion are concerned primarily with dicotyledonous halophytes. There are some problems with interpreting results of previous studies, however. In most cases, the intent was not to determine optimum salinity levels for growth, so the intervals between imposed salinity treatments often were large. Some halophytes have been reported to respond to extremely small amounts of Na, with growth increases two- or threefold in response to 1 mM NaCl [15,16]. So if the lowest treatment level is 50 or 100 mM NaCl, for example, the response to this relatively high level cannot be distinguished from the response due to satisfying the "need" for the trace amount of Na. However, unless specific measures are taken to exclude Na, it is usually present in most nutrient solutions at such low levels due to contamination. This should be verified by analysis of the base nutrient solution used, and if there is no Na present, NaCl should be added to give 1 to 2 mM Na in the control solution. Further complicating interpretation is the failure, in many cases, to correct for the weight of salt in the tissue. Halophytes characteristically accumulate substantial quantities of salt in their shoots, easily 30 to 50% of the total dry weight [17], so much of the difference in dry weight between plants grown in nonsaline conditions versus those grown in some substantial salinity can be due to the increased salt content in the latter. Nevertheless, when care is taken to account for the weight of salt in the tissue, and other factors, it is clear that there are many halophytes for which growth is maximum at salinity levels on the order of 50 to 200 mM NaCl or equivalent (ca. 3000 to 12,000 ppm). As part of an intensive halophyte domestication program [18], we screened 150 diverse species, in several families, and 57 (38%) of them had greater growth at 170 mM NaCl (ca. 10,000 ppm) than they did on nonsaline nutrient solution. We termed those Euhalophytes, and the others Miohalophytes. A detailed report of 10 of each type, representing 19 genera and 10 families, is given in Glenn and O'Leary [19].

B.   Cause of the Growth Reduction at Low Salinity

The emphasis in the reviews by Barbour [9] and Flowers et al. [10,11] was on growth *stimulation* between 1–2 m*M* and 50–200 m*M* NaCl. Even in the latest review by Rozema [20], the difference is still viewed as a stimulation effect. This is not surprising because that focus is emphasized in virtually all graphical comparisons of growth at various salinities. Growth usually is plotted as a percentage, with growth at zero salinity equal to 100%. When this is done, the extreme halophytes always show growth greater that 100% over the salinity range between 1–2 or zero NaCl and 50–200 m*M* NaCl. However, when actual growth (either as rate or final biomass) is plotted, it is clear that what has happened is a shift of the response curve, similar to what occurs in plants adapted to extremes of other environmental parameters, such as light or temperature. It is difficult, if not impossible, to generate a pair of curves comparing adapted and nonadapted plants for salinity as has been done for light and temperature [21]. However, a comparison of relative growth rates (RGRs) for 10 euhalophytes and 10 miohalophytes (as defined above) [19] showed that the average of the maximum RGR for the miohalophytes was 0.43 g/g per week, and it occurred at zero salinity. The maximum RGR for the euhalophytes occurred at 180 m*M* salinity, and it was 0.42 g/g per week, almost exactly equal to the maximum RGR for the miohalophytes. The average RGR for the euhalophytes at zero salinity was 0.33 g/g per week. If these data were presented by comparing them on a percentage basis, setting growth at zero salinity equal to 100%, the miohalophytes would show a steady decline in RGR with increasing salinity, but the euhalophytes would show a "stimulation" effect, having growth at 180 m*M* salinity equal to 127% of their RGR at zero salinity. What actually happened is that the euhalophytes still had about the same maximum RGR as the miohalophytes, but they were able to achieve it at a substantial salinity level (180 m*M*), while the miohalophytes had a much reduced RGR at that salinity level. The "price" paid by the euhalophytes, however, is loss of the ability to maintain that same RGR at lower salinity. That is, the entire response curve has shifted.

Thus, it should not be surprising that there is not much information available about the cause for the growth *reduction* at low salinity. Munns et al. [14] did acknowledge that maybe we should think in terms of a growth reduction at zero NaCl rather than a growth stimulation at 50 to 200 m*M*. Their suggestion for the cause of the difference in growth is increased water deficit at the lower salinity. They feel that "the improvement in growth above 1 m*M* NaCl is most likely related to improved water relations of the leaves, due to accumulation of Cl and Na." This feeling that less growth of halophytes at suboptimal levels of salinity is due to lack of sufficient solutes for generating turgor or even possibly to reduced root hydraulic conductivity is widely shared [10,20,22].

C.   Water Relations at Low Salinity

As mentioned above, there is a paucity of data on water relation parameters that allow one to compare the water status of plants at suboptimal versus optimal salinity. As one might expect, the osmotic potential in leaves typically declines with increasing salinity of the growth medium (e.g., Refs. 23 and 24), although Matoh et al. [25] found there to be no difference in osmotic potential of *Phragmites* at 0 and 100 m*M* NaCl. The calculated turgor pressure in *Atriplex* fell to almost 0 by 9:00 A.M. in control plants, but in the salinized plants it stayed positive all day long, albeit dropping slightly at midday [23]. However, when Clipson et al. [26] measured turgor directly with a pressure probe in *Suaeda*, they found it be about the same at all salinities. Downton [27] measured osmotic and water potentials in *Avicennia* and calculated turgor pressure. In all salt treatments (10 to 100% seawater), turgor was about 0.8 MPa, but at zero salinity, it was only about 0.2 MPa. Growth was less at zero than at all salinities. In young seedlings of *Salicornia* grown at low light in the lab, Stumpf et al. [28] found the turgor to be almost zero (0.02 MPa) when grown on nonsaline conditions, but at salinities of 170 and 340 m*M*, the turgor pressures were 0.53 and 0.99 MPa, respectively. On the other hand, Weeks [29] found that in greenhouse-grown *Salicornia* the osmotic and water potentials both paralleled the decline in salinity, with the result being almost no difference in calculated turgor pressure across the entire range from 17 to 1020 m*M* salinity. The turgor pressure was close to 1.0 MPa at all treatment levels, but the growth at the lowest salinity level was very poor, and the growth was high and no different between 170 and 1020 m*M* salinity. Some of the discrepancies are undoubtedly due to methodology problems. It is difficult enough to obtain reasonably accurate measures for water and osmotic potentials in any plants, so that the calculated turgor pressures can be accepted with reasonable confidence, but with halophytes it is even more problematic, especially succulent halophytes. Nevertheless, based at least partly on the direct measure of turgor with a pressure probe in *Suaeda* by Clipson et al. [26] mentioned above, Munns [30] recently acknowledged that inadequate turgor probably is not a likely cause for the lower growth of halophytes at suboptimal salinity.

Hydraulic conductivity of roots typically falls with salinity, in both halophytes [23,31] and nonhalophytes [32–34]. Thus it would not be expected to find that the hydraulic conductivity of halophytes at suboptimal salinity is less than at optimal salinity. However, there really have not been enough measurements of conductivity at the appropriate salinity levels to allow for any conclusive statements at this point. Nevertheless, Munns et al. [14] speculated that the relatively low values for root hydraulic conductivity in halophytes, in general, coupled with the low root/shoot ratio in halophytes, in general, could account for the low turgor, if it in fact occurs, in the plants at suboptimal salinity. The difficulty with such a scenario is that at least in the few observations cited above, the root hydraulic conductivity in halophytes is usually higher in the plants at suboptimal salinity, and also, the root/shoot ratio is usually higher in those plants [35,36]. Thus the plants at optimal salinity would be more likely to be at a disadvantage in this context. It is clear that there is a need for measurement of all of these parameters in the same plants at optimal and suboptimal salinity levels.

D. Growth Component Analysis

Most of the difference in dry weight production between optimal and suboptimal salinities is accounted for by difference in shoot growth (e.g., Refs. 35 and 36), but there are some reports of root growth being affected as well [27,37]. In addition to leaf size being reduced at suboptimal salinity, the leaf number can be less [38]. However, Longstreth and Strain [39] found no differences in leaf area or specific leaf weight in *Spartina* at 10 versus 0.5 ppt salinity. The difference in total leaf area between the two salinities is thought to be a major cause of the growth difference between the two by Munns et al. [14]. Osmond et al. [21] feel the same way. They analyzed the data from the studies of Gale et al. [40] and Kaplan and Gale [23] and concluded that even though photosynthetic rate per unit leaf area was less at 72 m*M* NaCl than at zero NaCl, the leaf area was greater enough that the total photosynthetic capacity per plant was increased. Plotting of photosynthetic capacity per plant as a function of salinity over the range from 0 to 360 m*M* NaCl yielded a curve that closely paralleled the growth response curve over that same salinity range. It should be noted, however, that Clipson [41] found that leaf area in *Suaeda* decreased with all additions of salinity. Nevertheless, this may indicate the importance of photosynthate partitioning in determining the growth response. It may be that an important difference between the plants at the two salinity levels is how much photosynthate is reinvested in new photosynthetic surface. Unfortunately, there aren't many data available that bear on that question. A thorough growth component analysis, similar to that done by

Aslam et al. [42] in which they compared growth at optimal with growth at supraoptimal salinities, is needed.

E. Photosynthesis at Low Salinity

Since growth depends on substrate availability as well as sufficient turgor, it is reasonable to question the effect of the less optimum salinity on photosynthesis. The effect of salinity on photosynthesis in halophytes has been investigated, and despite the fact that the focus usually was on comparing optimum versus excess salinity, in some cases data were obtained for a range of salinity levels that enable one to compare photosynthesis rates at less than optimum salinities with those at optimum levels. In *Salicornia*, arguably the most salt-tolerant C_3 vascular plant, photosynthesis was higher at -3.2 MPa osmotic potential [43] or 342 mM salinity [35] when measurements were made at several salinity levels. However, Kuramoto and Brest [44] found that photosynthesis decreased at all levels of salinity, and Pearcy and Ustin [45] found no differences from 0 to 450 mM. Kuramoto and Brest [44] found the same response with *Batis maritima*, *Spartina foliosa*, and *Distichlis spicata*. Kemp and Cunningham [46] also found a steady decrease in photosynthesis with increasing salinity in *Distichlis spicata*. Longstreth and Strain [39] found no difference in photosynthesis at different salinity levels in *Spartina alterniflora*. In *Atriplex nummularia*, photosynthesis was higher at leaf water potentials of -1.5 to -2.0 MPa than at either higher or lower water potentials [47], and in *Sporobolus airoides* photosynthesis was higher at 1.0 MPa than at zero salinity [48]. In *Lepochloa fusca*, grown in the absence of NaCl or with NaCl at 250 mM, photosynthesis was higher in the presence of added NaCl than when it was absent at 32 or 39° C, but the reverse was true when the temperature was 19°C [49]. The data of Hajibagheri et al. [50] showed that photosynthesis in *Suaeda* was lower at 170 mM than at either 340 or 680 mM salinity.

In those few cases where photosynthesis was found to be lower at the lower salinity, there was insufficient information to determine whether the lower photosynthetic rates at the lower salinites (or higher water potentials) were due to stomatal or nonstomatal effects. In fact, even in the cases where investigators have demonstrated reduced photosynthetic rates at excessive salinity levels in halophytes, the picture is unclear. Some have attributed the reduced photosynthesis to reduced leaf conductance [51–54], while others have concluded that photosynthesis was reduced independently of changes in stomatal conductance [45,55,56]. The results of Schwartz and Gale [57] in which *Atriplex halimus* growing at 170 mM NaCl had a much greater increase in growth in response to increasing CO_2 than that of plants growing in nonsaline conditions is often cited as support for the view that the stomatal effect predominates at the higher salinity. On the other hand, Demming and Winter [58] found that even isolated chloroplasts from *Mesembryanthemum crystallinum* growing at different salinites showed reduced CO_2 fixation with increasing salinity. They found that electron transport was much less sensitive than CO_2 fixation, suggesting a direct effect of salinity on biochemical processes. Pearcy and Ustin [45] found that salinity did not affect the initial slope of the CO_2 response curve, but it did affect the CO_2-saturated photosynthetic capacity in *Spartina*. However, *Salicornia* photosynthetic capacity seemed to be relatively independent of salinity.

Furthermore, in some cases it has been concluded that growth was reduced by some factors other than photosynthesis, and the net effect was due to reduced photosynthetic surface rather than reduced photosynthetic rate per unit leaf surface [59–61]. Flowers [62] concluded that the general interpretation of the available evidence is that photosynthetic rates per unit leaf area are decreased or little affected by increases in salinity, and the rates per unit of chlorophyll appear to be either unaffected or to *increase*. He also indicated that the results of Kemp and

Cunningham [46] with *Distichlis* showed that the reduced photosynthesis at increasing salinity is due to changes in stomatal frequency.

Some of the differences may be due to real differences among species, but some are also the result of differences in experimental conditions and differences in the manner in which the data are expressed [41,56]. It depends on whether the photosynthetic rate is expressed on a leaf area, leaf weight, or chlorophyll basis. Depending on which is used, the photosynthetic rate may be shown to be lower or higher [61].

F. Summary

It should be clear from this brief survey that it currently is no easier to explain the cause for the reduced growth at suboptimal salinity that it is to explain the reduced growth at supraoptimal salinity. In the former case, however, the reason is largely due to the fact that not nearly as much research has been directed at the question as in the latter case. Nevertheless, based on the limited data base available, it does not seem likely that the reduced growth at suboptimal salinity is due to insufficient turgor. Neither does it seem that there is insufficient production of substrates for growth. In fact, with the few exceptions noted above, the evidence seems primarily to indicate that the rate of photosynthesis at suboptimal salinity is higher than at optimal salinity. If that is the case, and yet the plants are significantly smaller than those growing at optimal salinity, the obvious question that comes to mind is: Where is all the carbon going? Part of the problem may be that all of the photosynthetic rate measurements are instantaneous values, whereas the growth data are integrated values. The total carbon fixed per day, as well as the total carbon lost to respiration during the ensuing night period, needs to be determined under those salinity levels. A total carbon balance needs to be determined, in other words.

Partitioning of the photosynthate may be more important than the total amount produced. As described above, there is very little information available that bears on that point, particularly as concerns the plants growing at suboptimal salinity. There is almost no information available on hormone metabolism at the various salinity levels. Since the reduced growth at suboptimal salinity seems to be due to an apparent overall stunting of the plant, the problem may be one of growth regulation. There is a lack of information on this topic. Much needs to be done yet. That is clear.

Despite it being no easier to explain the cause of reduced growth at supoptimal salinity than it is to explain the reduced growth at supraoptimal salinity, it is clear that there are physiological differences between the plants growing at those salinities. It is likely that the causes for the growth reduction in each case are different. Thus increased attention to the response of highly salt tolerant halophytes to suboptimal salinity is warranted.

III. CONCLUSIONS AND RECOMMENDATIONS

The decreasing availability of fresh water for agriculture, coupled with the increasing demand for plant-based agricultural commodities, makes the eventual use of increasingly saline water in agriculture a certainty. There are abundant reserves of saline water within reasonable pumping distance from the surface available in many areas of the world, especially in areas where successful crop production depends on irrigation. The limitation to eventual use of that water to irrigate crops is the availability of sufficiently salt tolerant crops. The long-term survival of agriculture in such areas is dependent on development of such crops. Even though progress has been made in increasing salt tolerance of some crops over the years, the pace of continued improvement and the ultimate maximum tolerance that can be achieved through conventional breeding approaches may not be sufficient to fulfill this need. Thus other approaches that offer promise are required. The ability now to transfer genetic information between widely different types of plants makes the possibility of moving traits associated with increased tolerance of environmental stresses from alien genotypes into crop plants highly likely, *if* the required traits can be identified. Study of the physiology of highly salt tolerant "wild" plants (halophytes) is a necessary but not sufficient step. For the reasons described above, a completely different approach to studying salt tolerance in those plants is desirable, as an *additional* approach to this important problem. This paradigm shift in our conceptual approach to analysis of salt tolerance has the strong potential for opening up a new line of research that could be highly productive. It could provide the means whereby the emerging molecular genetic techniques can be applied to what heretofore has been viewed as one of the most intractable problems at the whole plant level—stress resistance.

REFERENCES

1. D. Aspinall and L. G. Paleg, in *The Physiology and Biochemistry of Drought Resistance in Plants* (L. G. Paleg and D. Aspinall, eds.), Academic Press, New York, pp. 205–241 (1981).
2. C. R. Stewart, in *The Physiology and Biochemistry of Drought Resistance in Plants* (L. G. Paleg and D. Aspinall, eds.), Academic Press, New York, pp. 243–259 (1981).
3. R. G. Wyn Jones and R. Storey, in *The Physiology and Biochemistry of Drought Resistance in Plants* (L. G. Paleg and D. Aspinall, eds.), Academic Press, New York, pp. 171–204 (1981).
4. N. C. Turner, *Aust. J. Plant Physiol.*, *13*: 175 (1986).
5. R. G. Wyn Jones and J. Gorham, in *Physiological Plant Ecology*, Vol. III, *Responses to the Chemical and Biological Environment* (O. L. Lange, P. S. Nobel, C. B. Osmond, and H. Ziegler, eds.), Springer-Verlag, Berlin, pp. 35–58 (1983).
6. R. Munns, *Aust. J. Plant Physiol.*, *15*: 717 (1988).
7. V. J. Chapman, *Salt Marshes and Salt Deserts of the World*, Interscience, New York, (1960).
8. Y. Waisel, *Biology of Halophytes*, Academic Press, New York (1972).
9. M. Barbour, *Am. Midl. Nat.*, *84*: 105 (1970).
10. T. J. Flowers, P. F. Troke, and A. R. Yeo, *Annu. Rev. Plant Physiol.*, *28*: (1977).
11. T. J. Flowers, M. A. Hajibaheri, and N. J. W. Clipson, *Quart. Rev. Biol.*, *61*: 313 (1986).
12. D. A. Adams *Ecology*, *44*: 445 (1963).
13. A. J. Macke and I. Ungar, *Can. J. Bot.*, *49*: 515 (1971).
14. R. Munns, H. Greenway, and G. O. Kirst, in *Physiological Plant Ecology III: Responses to the Chemical and Biological Environment* (O. L. Lange, P. S. Nobel, C. B. Osmond, and H. Ziegler, eds.), Springer-Verlag, Berlin, pp. 59–135 (1983).
15. R. F. Black, *Aust. J. Biol. Sci.*, *13*: 249 (1960).
16. M. C. Williams, *Plant Physiol.*, *35*: 500 (1960).
17. J. W. O'Leary, in *Arid Lands Today and Tomorrow* (E. Whitehead, C. Hutchinson, B. Timmermann, and R. Varady, eds.), Westview Press, Boulder, Colo., pp. 773–790 (1988).
18. J. W. O'Leary, in *Salinity Tolerance in Plants: Strategies for Crop Improvement* (R. C. Staples and G. H. Toenniessen, eds.), Wiley, New York, pp. 285–300 (1984)
19. E. P. Glenn and J. W. O'Leary, *Plant Cell Environ.*, *7*: 253 (1984).
20. J. Rozema, *Aquat. Bot.*, *39*: 17 (1991).
21. C. B. Osmond, O. Bjorkman, and D. J. Anderson, *Physiological Processes in Plant Ecology: Toward a Synthesis with Atriplex*, Springer-Verlag, Berlin (1980).
22. D. H. Jennings, *Biol. Rev.*, *51*: 453 (1976).
23. A. Kaplan and J. Gale, *Aust. J. Biol. Ser.*, *25*: 895 (1972).
24. J. A. Bolanos and D. J. Longstreth, *Plant Physiol.*, *75*: 281 (1984).
25. T. Matoh, N. Matsushita, and E. Takahashi *Physiol. Plant*, *72*: 8 (1988).
26. N. J. W. Clipson, A. D. Tomos, T. J. Flowers, and R. G. Wyn Jones, *Planta*, *165*: 392 (1985).
27. W. J. S. Downton, *Aust. J. Plant Physiol.*, *9*: 519 (1982).
28. D. K. Stumpf, J. T. Prisco, J. R. Weeks, V. A. Lindley, and J. W. O'Leary, *J. Exp. Bot.*, *37*: 160 (1986).
29. J. R. Weeks, Ph.D. dissertation, University of Arizona, Tucson, Ariz. (1986).

30. R. Munns, *Plant Cell Environ.*, *16*: 15 (1993).
31. R. S. Ownbey and B. E. Mahall. *Physiol. Plant.*, *57*: 189 (1983).
32. J. W. O'Leary, *Isr. J. Bot.*, *18*: (1969).
33. J. W. O'Leary, in *Structure and Function of Primary Root Tissues* (J. Kolek, ed.), Veda, Bratislava, Czechoslovakia, pp. 309–314 (1974).
34. R. J. Joly, *Plant Physiol.*, *91*: 1261 (1989).
35. F. S. Abdulrahman and G. J. Williams III, *Oecologia*, *48*: 346 (1981).
36. G. Naidoo and R. Rughunanan, *J. Exp. Bot.*, *41*: 497 (1990).
37. K. C. Blits and J. L. Gallagher, *Plant Cell Environ.*, *13*: 419 (1990).
38. T. F. Neales and P. J. Sharkey, *Aust. J. Plant Physiol.*, *8*: 165 (1981).
39. D. J. Longstreth and B. R. Strain, *Oecologia*, *31*: 191 (1977).
40. J. Gale, R. Naaman, and A. Poljakoff-Mayber, *Aust. J. Biol. Sci.*, *23*: 947 (1970).
41. N. J. W. Clipson, D. Phil. thesis, University of Sussex (1984); cited in T. J. Flowers, *Plant Soil*, *89*: 41 (1985).
42. Z. Aslam, W. D. Jeschke, E. G. Barrett-Lennard, T. L., Setter, E. Watkin, and H. Greenway, *Plant, Cell Environ.*, *9*: 571 (1986).
43. B. L. Tiku, *Physiol. Plant*, *37*: 23 (1976).
44. R. T. Kuramoto and D. E. Brest, *Bot. Gaz.*, *140*: 295 (1979).
45. R. W. Pearcy and S. L. Ustin, *Oecologia*, *62*: 68 (1984).
46. P. R. Kemp and G. L. Cunningham, *Am. J. Bot.*, *68*: 507 (1981).
47. A. T. Pham Thi, C. Pimentel, and J. Vieria da Silva, *Photosynthetica*, *16*: 334 (1982).
48. H. M. El-Sharkawi and B. E. Michel, *Photosynthetica*, *9*: 277 (1975).
49. J. Gorham, *Plant Cell Environ.*, *10*: 191 (1987).
50. M. A. Hajibagheri, D. M. R. Harvey, and T. J. Flowers, *Plant Sci. Lett.*, *34*: 353 (1984).
51. G. D. Farquhar, M. C. Ball, S. von Caemmerer, and Z. Roksandic, *Oecologia*, *52*: 121 (1982).
52. R. D. Guy, D. M. Reid, and H. R. Krouse, *Can. J. Bot.*, *64*: 2693 (1986).
53. R. D. Guy, D. M. Reid, and H. R. Krouse, *Can. J. Bot.*, *64*: 2700 (1986).
54. L. B. Flanagan and R. L. Jeffries, *Plant Cell Environ.*, *11*: 239 (1988).
55. M. C. Ball and G. D. Farquhar, *Plant Physiol.*, *74*: 1 (1984).
56. D. J. Longstreth, J. A. Bolanos, and J. E. Smith, *Plant Physiol.*, *75*: 1044 (1984).
57. M. Schwarz and J. Gale, *J. Exp. Bot.*, *35*: 193 (1984).
58. B. Demmig and K. Winter, *Planta*, *159*: 66 (1983).
59. T. M. DeJong, *Oecologia*, *36*: 59 (1978).
60. J. Gale and A. Poljakoff-Mayber, *Aust. J. Biol. Sci.*, *23*: 937 (1970).
61. K. Winter, in *Ecological Processes in Coastal Environments* (R. L. Jeffries and A. J. Davy, eds.), Blackwell, Oxford, (1979).
62. T. J. Flowers, *Plant Soil*, *89*: 41 (1985).

27

Photosynthesis in Plant/Crops Under Water and Salt Stress

Zvi Plaut

Agricultural Research Organization, Institute of Soils and Water, Bet Dagan, Israel

I. INTRODUCTION

A. What Is Stress?

Under optimal growing conditions productivity of crop plants may largely exceed the productivity of these crops under actual growing conditions [1]. A great variation in productivity can be found due to three groups of factors: (a) large differences in the potential productivity of varieties or cultivars; (b) diversity in human skills, background, and knowledge; and (c) exposure to environmental stresses. Our present interest is mainly in the third group of factors, and it should be noted that one of the major functions of agricultural technology is to eliminate such stresses.

Stress conditions is the term adopted by biologists, ecologists, and agronomists to describe environmental factors potentially unfavorable to living organisms. It differs from mechanical stress in two main ways. Biological stress is measured in units of energy, not in units of force, and second, it always has a connotation of possible injury [2]. Since the biological stress is not necessarily a force, the strain, which is the response of the organism to stress, is also not necessarily a change in dimensions. This strain is either chemical, such as changes in metabolism, or physical, such as change in position of organs or of cells (stomates).

B. Photosynthesis and Stress

Photosynthesis is the source of organic carbon and energy for plants and is needed for their growth, production of biomass, and bearing of yield. Knowing the control of photosynthesis by environmental stress factors is thus of importance to understand the relationship between productivity and the environment.

Photosynthesis begins with the absorption of light energy and utilizing the energy resulting in a charge separation into photochemical reaction centers within microseconds. These reactions are linked to membrane-bound electron transport reactions, and those are linked to an oxidation-reduction process catalyzed by enzymes. These lead to the oxidation of water to O_2

and ultimately, the reduction of CO_2 to carbohydrates. All these reactions occur within the specialized structure of the chloroplast. The oxidation of H_2O, the capture and transport of electrons up to a secondary reductant $NADP^+$, and the coupled ATP synthesis are known as *light reactions*. The reductant and the ATP are then utilized for the conversion of CO_2 to carbohydrate in a process directly independent of light and known as the *dark reaction*. Light is, however, of importance in control processes of the dark reaction.

The different environmental stresses may thus influence three major groups of reactions: (a) input or utilization of light energy, (b) activity of enzymes that link the photochemical energy ultimately to CO_2 fixation, and (c) the concentration and availability of CO_2 at the site of its fixation. The scope of this chapter is to relate the activity of these groups of reactions and the entire process of photosynthesis to the various environmental stresses.

II. WATER OR DROUGHT STRESS

A. The Concept and Its Significance

Water stress occurs as a result of water deficit or drought. *Drought* is a meteorological term and can be defined as a period without significant rainfall. *Water stress* is usually the result of water withholding by human beings: namely, eliminating or retricting irrigation water. However, the terms *water stress* and *drought* are often used interchangeably. The ultimate result of this stress is evaporative dehydration due to water loss at a rate that exceeds the rate of uptake. The restricted rate of water uptake over its loss may also be due to the osmotic potential of the growing media (see Section III.A), not due to limited water availability, and is thus known as *osmotic stress*.

Water stress is commonly assessed by the change in water potential measured in the environment or within the plant. Water potential (ψ_w) is always negative and the accepted units are pascal (Pa), but in practice MPa (megapascal; $= 10^6$ Pa) is used, as the Pa units are too small. The unit used in the past was the bar (1 bar $= 10^5$ Pa). Values of plant water potential documented as decreasing photosynthesis are in the range -1.0 to -3.0 MPa [3,4]. The decrease in plant ψ_w, which is due primarily to transpirational water losses, may bring about loss of cell turgor pressure (ψ_p). In many cases turgor will still be maintained above zero, as a result of reversible osmotic and elastic cell dehydrations. The decrease in plant ψ_w associated with changes in osmotic potential (ψ_s) and/or cell turgor may be found on a daily basis during hours of peak irradiance and seasonally prior to water application or rainfall.

Short periods of water stress during middays are usually periods of high irradiance, high temperature, and a reduced ambient water vapor pressure. The response of photosynthesis to leaf water status can therefore not be discussed independently from a possible response to water vapor differences between leaf and air [5,6].

B. Stomatal Conductance

Most plant leaves are poor capacitors for water and the difference in water vapor pressure between leaf and atmosphere results in severe damage to leaves unless controlled by stomatal conductance. Stomatal closure, which decreases their conductivity, is indeed one of the earliest and most sensitive responses during plant exposure to water stress [7]. In fact, stomatal conductance was found to be reduced, for instance, at -0.6 MPa, net photosynthesis only at -1.2 MPa [2].

1. Carbon Assimilation Control by Stomatal Conductance

The simplest explanation for the inhibition of CO_2 assimilation (A) under water stress is the closure of stomata or at least a decrease in stomatal conductance (g_s). This results in a decrease in the interecellular CO_2 partial pressure (C_i), which is the direct cause for the decrease in photosynthesis [8,9]. This was suggested as early as 1923 [10] and since by many investigators up to very recently [11,12,13]. Levitt [2] has defined plant species that reduce their stomatal conductance at an early stage of drought development as "drought avoiders of the saving type," since desiccation is postponed. In such genotypes the decrease in stomatal conductance exceeds the decrease in photosynthesis, indicating an increase in water use efficiency (WUE) [14]. In fact, this can be seen from the equation relating C_i to the ratio of A and g_s:

$$C_i = C_a - \frac{1.6A}{g_s} \qquad (1)$$

in which C_a is the ambient CO_2 partial pressure and 1.6 is the ratio of CO_2 and water vapor diffusion rates. It can be seen that at a given C_a a decrease in C_i, which is an outcome of the change in g_s, will cause a decrease in A but is proportionally smaller than that in g_s. The plot of A against g_s for corn plants gave a convex curve rather than a linear relationship when plants were rapidly exposed to water stress [15]. A linear relationship between A and g_s was, however, obtained when water stress was applied more gradually [15,16]. The relative changes in A versus g_s under water stress may differ among species [17]. We have presented a plot of A versus g_s calculated from these data (Figure 1). This was curvilinear for lupins and the slope was much smaller during the morning when water stress was minimal than during midday when stress increased. Similar results were obtained for eucalyptus and sunflowers. In grapevine A and g_s changed in parallel and a linear relation was thus obtained. Differences between stomatal conductance and CO_2 assimilation leading to differences in water use efficiency were even documented for different genotypes of the same cultivar [18]. Although water use efficiency seems to increase slightly under water stress in all genotypes when g_s was decreasing, the increase was most remarkable in one cultivar, IN-15 (Figure 2). The different responses of stomatal conductance and carbon assimilation to water stress was thus recommended as a screening tool for drought-tolerant genotypes.

2. Stomatal Patchiness

Doubts have been raised during the last decade concerning the validity of using proportinal changes in A and g_s. The calculation of C_i [equation (1)] is based on uniform CO_2 and water vapor exchange over the entire leaf area. Terashima et al. [19] found heterogeneous ("patchy") stomatal closure when ABA was exogeneously applied to attached or detached sunflower leaves and to detached *Vicia faba* leaves. Patchiness was also reported by Downton et al. [20] as a response to exogenous hormone application and by Sharkey and Seemann in bean plants exposed to water stress [21]. On the basis of autoradiograms of $^{14}CO_2$-fed leaves, they proposed that portions of the leaves had stomata that were completely closed while others were open, and thus incorporated ^{14}C. These studies have led to the conclusion that the $A:C_i$ relationship deduced from gas exchange measurements may be invalid since this relationship is based on uniform response over the entire leaf.

The generality of heterogeneous stomatal response over a leaf area when exposed to water or salt stress has been questioned. It is possible that this phenomenon is related to the species subjected to stress, to the treatment or type of stress, or to the rate that stress is being imposed and possibly also to its intensity. While a patchy CO_2 assimilation pattern was found using autoradiograms of bean leaves, no patchiness was evident in spinach and wheat leaves when exposed to water stress [22]. No patchiness in CO_2 assimilation pattern was found even in bean leaves when ψ_w declined at a low rate (approximately 1 MPa in 17 days) as compared to the patchiness found when stress was imposed at a high rate (4 days). Bunce [23] deduced on the basis of preliminary studies on three different species that heterogeneous stomatal closure influenced the $A:C_i$ relationship.

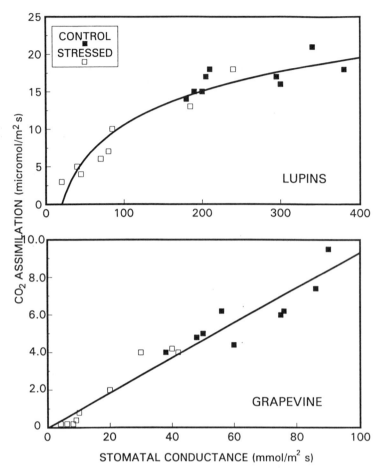

Figure 1   Changes in *A* (CO_2 assimilation rates) versus changes in g_s (stomatal conductance) calculated for stressed and nonstressed lupin and grapevine plants. (Based on recalculated data from Ref. 17.)

In heterobaric cotton leaves that were water stressed under field conditions, a homogeneous distribution of ^{14}C-labeled photosynthates was evident, even at a leaf ψ_w of -1.34 MPa [24]. Patchiness could, however, be induced only by uprooting plants and allowing the shoot to dehydrate within a few minutes. Similarly, no patchiness was detected in salt-stressed cotton when salinity was gradually increased [25]. This conclusion was based on both autoradiograms and statistical analysis of ^{14}C distributed over the leaf (Table 1). The standard deviation (SD) between locations on the same leaf averaged for different leaves is very similar to the SD between leaves averaged for the same location. The ANOVA *F* values for leaves and locations are identical when calculated for treatments or replications. In four woody angiosperms a patchy stomatal closure was found under mild water stress but not under more severe water stress or under unstressed conditions [26]. In conclusion, it seems that spatial heterogeneities in photosynthesis do not occur frequently under natural drought conditions.

3. Factors Involved in Stomatal Closure

The classical concept associates stomatal closure under stress to leaf water status [4]. Moreover, a threshold level of leaf ψ_w or relative water content (RWC) was indicated above which stomatal

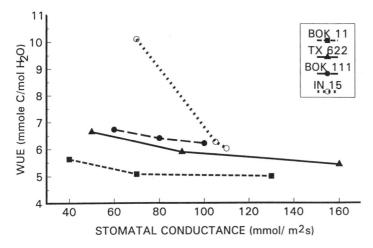

Figure 2 Changes in WUE (water use efficiency) versus changes in g_s (stomatal conductance) calculated for four genotypes of sorghum subjected to different water stresses. (Based on recalculated data from Ref. 18.)

Table 1 Rates of CO_2 Fixation in Leaves of Salinity-Acclimated and Nonacclimated Cotton Plants and the Distribution of These Rates Between 10 Different Locations Within a Leaf and Between Six Different Leaves[a]

Treatment	Mean $^{14}CO_2$ fixation rate (μmol/m/s)	Average SD (μmol/m/s)	
		Between locations	Between leaves
Control	28.3 E	2.99	2.89
Acclimated			
−0.3 MPa	26.1 D	2.76	2.67
−0.6 MPa	23.4 C	2.43	2.34
Nonacclimated			
−0.3 MPa	21.7 B	2.24	2.20
−0.6 MPa	14.8 A	1.61	1.54

F values	Treatment	Replication
Leaves	228	0.18
Locations	229	0.17
Individual measurements	222	0.09

Source: Ref. 25.

[a] Leaf discs (area = 1 cm^2) were removed from the leaf immediately after exposure of the leaf to $^{14}CO_2$ for 30 s, and the ^{14}C incorporated was extracted and analyzed. Numbers followed by different letters differ significantly at $p = 0.05$ (according to Duncan's multiple range test).

conductance remains constant. The overall threshold values of ψ_w presented for different species were in the range -0.7 to -1.6 MPa and the threshold values for RWC were 80 to 85% [7]. Once the threshold water status for stomatal closure is reached, their conductance is believed to decrease sharply to $1/20$ or $1/30$ of unstressed plants caused by a drop of 0.5 MPa or even less. The conductance may continue to decrease with further increase in water deficit up to a complete stomatal closure.

Another factor ascribed as responsible for stomatal closure is water vapor pressure deficit (VPD) between the leaf and the atmosphere (see review in Ref. 9). This is probably the main factor for midday stomatal closure, although leaf ψ_w and temperature effects cannot be excluded. The effect of VPD on stomatal closure was reported for both C3 and C4 species but varied among species [27–29]. Stomatal closure may be caused by water stress in the epidermal tissue and guard cells due to excessive loss of water by transpiration. Closure of stomata under such conditions could also be a result of a decrease in bulk leaf ψ_w. It was, however, indicated that stomata close at high VPD without changes in bulk leaf water potential [30] and that it was due to a direct loss of water from epidermal and guard cells [31]. Another possibility to be considered is that high VPD directly affects carbon assimilation or photochemical processes, leading to a rise in C_i which would, in turn, decrease g_s (see Sections II.C and II.D). In a recent extensive study by Dai et al. [32] the effect of high VPD on the decrease in photosynthesis and increase in photorespiration under different environmental conditions could be totally accounted for by stomatal closure and decrease in C_i. Although the determinations of C_i by gas-exchange measurements did not account for the entire VPD effect on photosynthesis, the conclusion was obtained from measurements of quantum yield and quenching of chlorophyll fluorescence.

Another factor involved in stomatal closure is based on the accumulated evidence showing a direct relationship between stomatal sensitivity to water deficit and ABA content in leaves. Earlier work has shown, for instance, that wheat genotypes which are accumulators of ABA have a higher WUE than that of low ABA accumulators. This was ascribed to lower g_s in high-ABA genotypes [33]. Evidence for nonhydraulic root-to-shoot communication under a water shortage which caused stomata to close without changes in leaf ψ_w or ψ_p had accumulated during the last decade. It was suggested that increased synthesis of ABA which moved in the transpiration stream to the shoot was responsible for decreased stomatal conductance in sunflower and maize [34,35], but not explicitly in two genotypes of beans [36].

It was recently shown that stomatal response to the concentration of ABA in xylem sap was not independent of environmental conditions. In the laboratory, the g_s reduction was only by 60%, even for an increase of ABA of more than 1000 μmol/m^3 [34]. The sensitivity was less than 20% when leaf ψ_w was maintained at high values in a pressurized root system [37]. In contrast, a decrease of more than 90% of g_s was observed under field conditions for a lower xylem-ABA concentration [38]. A revised model for the control of g_s by xylem ABA transported from the roots was thus recently developed [38,39]. According to this model, the chemical signal (ABA) and the hydraulic signal (leaf ψ_w) must be integrated to control g_s. The chemical signal produced by the roots depends on their ψ_w but also on water flux through the soil–plant–atmosphere continuum. This flux may dilute the concentration of ABA in the xylem sap. Stomatal sensitivity to the chemical signal is increased when leaf ψ_w drops. In conclusion, the operation of the chemical signal is based on root and on leaf ψ_w. This explains that during the morning when leaf ψ_w is usually high, the response to ABA is much lower than in the afternoon, when leaf ψ_w is low.

C. Effect of Stress on Thylakoid-Bound Reactions

The light-driven reactions, which were found to respond to water stress, include (a) chlorophyll fluorescence, which is an in vivo probe for PSII photochemical activity; (b) electron transport; and (c) photophosphorylation.

1. Chlorophyll Fluorescence Response to Water Stress

When a dark-adapted leaf is exposed to light, the chlorophyll fluroescence exhibits a characteristic series of changes. Minimum fluorescence (F_0) is induced by preillumination with either a far-red light (700 to 720 μm) or a weakened modulated light. F_0 is considered to represent emission by excited antenna chlorophyll a molecules occurring before the excitions have migrated to the reaction centers. Following illumination with actinic light fluorescence increases rapidly to a peak P (consists of a rise to an inflection point I, a slight dip D, and a rise to P). Fluorescence between F_0 and P is termed variable fluorescence (F_v). This reflects the reduction of the primary electron acceptor of PSII-Q, which in the oxidized state quenches fluorescence. Quenching of F_v depends on the reoxidation of Q, as carbon assimilation becomes active and/or another mechanism of energy dissipation is operative. Changes in F_0 and F_v reflect effects on the primary photochemical events, and quenching of F_v also includes the effect on the entire photosynthetic process.

Changes in chlorophyll fluorescence were shown mostly in leaves subjected to rapid dehydration [40]. In many studies the response of chlorophyll fluorescence to water stress was found only at relatively extreme water losses [41–43]. In oak leaves, for instance, no change in chlorophyll fluorescence was detected before leaf water deficit reached values of 30%. At more severe dehydration nonphotochemical quenching of chlorophyll fluorescence was increased while photochemical quenching was not changed, indicating that thermal deexcitation of PSII caused reduced photochemical activity.

The use of chlorophyll fluorescence to detect differences among genotypes in drought resistance was recently suggested [40,44–46]. F_v increased from well-watered to water-stressed maize plants at different sites and genetic variability, but the correlation between F_v and grain yield was poor [45]. It thus appears that F_v may be of some use as a selection criterion for drought tolerance in wheat, but only in combination with additional factors.

In potatoes an effect of drought on F_v quenching was found and a significant difference between genotypes for $F_v/(F_0 + F_v)$. Genotype $\times$ treatment interactions were, however, not significant for all variables of fluorescence that were examined [44]. In beans as well, the response of F_v as determined from chlorophyll fluorescence of different genotypes could not be related to differences in their leaf dehydration [47].

In tobacco cultivars the activities of the photosystems declined during plant water stress development as determined by chlorophyll fluorescence [46]. PSII activity was found to be more sensitive than PSI activity, mainly in drought-sensitive cultivars. The ratio I/P was found to be the most sensitive parameter to distinguish between sensitive and tolerant cultivars.

2. Thylakoid-Mediated Electron Transport

The response of electron transport and photophosphorylation to water or osmotic stress were studied in isolated chloroplasts, protoplasts, or intact cells. Electron transport from water to methylviologen was found to be insensitive when isolated chloroplasts were dehydrated in hypertonic sorbital solutions [48,49]. Little uncoupling occurred under these conditions and it was concluded that membrane integrity and photophosphorylative activity were also hardly impaired in dehydrated chloroplasts and protoplasts.

When thylakoids were isolated from intact plants exposed to severe water stress and to natural light intensities the rate of open-chain electron transport was reduced [50]. Thylakoids membranes isolated from sunflower leaves at ψ_w above -0.7 MPa exhibited PSII electron transfer which was nearly sufficient to account for maximum CO_2 fixation rate; it was reduced by 50% when leaf ψ_w was about -2.0 MPa [51]. It should, however, be noted that CO_2 fixation rate was reduced by 90% under those conditions, suggesting that additional or alternative processes were involved. Other investigators found that electron transport was unchanged in thylakoids isolated from water-stressed bean leaves [52,53]. In conclusion, it seems that thylakoid-mediated reactions are not very sensitive to water stress and that activity may be affected only under extreme conditions of water stress and high irradiance.

3. Photophosphorylation and Energy Conservation

The possible causes for inhibited phosphorylation can be categorized into three general groups: an effect on electron transport resulting in lack of driving force; uncoupling, that is, the possibility that the thylakoid membrane will maintain an energized state, accompanied by decaying proton accumulation; and possible inhibition of actual formation of the covalent bond of ATP by the membrane-bound ATPase enzyme complex.

Electron transport was just shown to be relatively insensitive to water stress levels, which reduces CO_2 fixation. Rapid wilting of detached leaves under dim room light to approximately 50% of leaf relative water content (RWC) did not impair thylakoid energization [54]. This finding is in agreement with the absence of uncoupling in osmotically dehydrated chloroplasts [48]. On the other hand, ATPase isolates from water-stressed spinach and sunflower leaves were markedly decreased compared with well-watered plants [55]. It was suggested that conformational changes of ATPase occurred so that the enzyme was unable to bind nucleotides and synthesize ATP. On the other hand, Ortiz-Lopez et al. [56] and Meyer and Kouchkovsky [57] concluded that water stress decreased ATPase activation, probably by affecting the enzyme in its oxidized state. The later steps of the process, like ATP synthesis by the oxidized enzyme or hydrolysis by the reduced enzyme, seem not to be affected by stress.

D. Dark Reactions of Photosynthesis

The dark reactions of photosynthesis dominate the fixation of CO_2 and the synthesis of carbohydrates. The CO_2 reduction consumes nearly 95% of the energy conserved in ATP and the reductant NADP. A small portion is used for nitrate and sulfate reduction. Enzyme activity determines the rate of CO_2 fixation but also the balance between processes such as starch synthesis in the chloroplast, export of reduced carbon compounds out of the chloroplast, and photorespiration. Since these reactions are conducted primarily in the stroma, the stromal environment is an important factor controlling enzyme activity and is responsible for both the rate of CO_2 fixation and the preferred pathways. Evaluation of the stromal dark reactions under water stress have thus to be accounted for by direct responses of various key enzymes.

1. Carboxylation in Intact Leaves

The most common procedure for determining the response of carboxylation reactions to water stress and eliminating stomatal responses is the plot of A–C_i curves for intact leaves at different values of ψ_w. The validity of this line of research, which became questionable during the last decade due to possible stomatal patchiness, can probably be reconsidered as an appropriate tool in light of more and more recent information (see Section II.B.2). The A–C_i curve consists of a linear portion in which A increases linearly as C_i increases, namely, when CO_2 is rate limiting at the reaction site due to low Rubisco (RuBP carboxylase) activity or limited affinity to CO_2 [15]. This is followed by a curvilinear part and a plateau where A is limited by the regeneration of RuBP. A difference in the slope of the linear portion would suggest inhibition of CO_2 binding at the reaction site under different leaf ψ_s. A decrease in V_{max} would suggest a shortage in RuBP supply due either to a shortage of ATP or a limitation of other steps in the Calvin cycle. Slopes of A versus C_i at the linear portion which were calculated for a wide range of species mostly from recent data show that those were decreased significantly at decreased leaf ψ_w (Table 2). The R^2 are mostly at a 0.95 level or higher.

Table 2 Slopes of A (μmol/m^2/s) Against C_i (μL/L) for the Linear Portion of the A–C_i Curve, R^2 Values for the Linear Regression Coefficient, and V_{max} Values (μmol/m/s) of Different Species and at Different Leaf ψ_w Values[a]

Plant species	Leaf ψ_w	Slope	R^2	V_{max}	Ref.
Acer saccharum	NS	0.0427	0.95	NP	26
	−2.36	0.0070	0.96	NP	
Quercus stellata	NS	0.0748	0.97	NP	
	−2.64	0.0160	0.99	NP	
Quercus alba	NS	0.0445	0.99	NP	
	−2.55	0.0087	0.96	NP	
Juglans nigra	NS	0.0155	0.99	NP	
	−1.84	0.0057	0.98	NP	
Sunflower					
Hybrid SH	−0.6	0.198	0.99	45.0	70
	−2.6	0.023	0.98	3.5	
Hybrid	−0.6	0.103	0.92	35.0	
sungro	−2.0	0.024	0.77	5.0	
Bean	−0.2	0.094	0.97	43.0	22
(cv. Top Crop)	−1.0	0.045	0.97	24.0	
Cotton	−0.04	0.136	0.95	55.0	25
(cv. Acala SJ1)	−0.60	0.085	0.91	NP	
	−0.90	0.053	0.89	NP	
Barley	−0.5	0.121	0.99	42.0	107
	−0.9	0.113	0.99	40.0	
	−1.2	0.112	0.98	37.0	
	−1.4	0.107	0.99	32.0	
	−1.7	0.103	0.98	23.0	
	−1.9	0.087	0.98	16.0	

[a] Values were calculated from the original data given by the references listed. NS, unstressed controls but ψ_w not presented; NP, V_{max} was not presented.

2. Rubisco Activity and Concentration

Changes of Rubisco from leaves of different plants which were gradually stressed were demonstrated by several investigators [58–61]. These changes include a decrease in both activity and enzyme concentration, due to a regulating mechanism. Since the inhibited Rubisco activity under stress was the same at all CO_2 concentrations up to saturation, it was suggested that the responsibility of the enzyme is independent of stomatal conductance [58]. Another carboxylating enzyme, PEP-carboxylase, which is a cytosolic enzyme, was also found to be inhibited by stress. In contrast, Raschke and Fischer [62] and Robinson et al. [63] found no difference in Rubisco activity isolated from stressed and nonstressed plants.

The carboxylation activity probably does not respond directly to stress but to the microenvironment within the chloroplast stroma. This might at least partly resolve the conflicting findings in the literature. Reduced leaf ψ_w may result in a volume decrease of chloroplast or protoplasts [64] in order to achieve equilibrium of stromal ψ_w with external ψ_s. This dehydration may result in biophysical changes inhibiting enzyme activity. The extent of protoplast volume reduction in leaves by low ψ_w was correlated with their photosynthetic activity in wheat cultivars of different drought sensitivity [65]. Correlation between the extent of protoplast volume reduction and photosynthetic activity in vitro at low ψ_w values was also shown by Kaiser [66] for various species.

Few studies on in situ protoplast volume changes and photosynthesis of water-stressed leaves are available. CO_2 assimilation of cells or chloroplasts isolated from plants acclimated to water or salt stress were not as inhibited in high osmotic strength media as in those isolated from unstressed plants [67,68]. These studies indicate that the different chloroplast activity may be a result of different potential control of their volume in situ. Such volume changes in chloroplast and/or proptoplast in situ are probably controlled by transport of ions and compatible molecules of low molecular weight. Kaiser has shown that under water stress, total solute concentrations differ considerably in leaves of different species and in subcellular compartments [54].

Another aspect to be considered is that of changes in ion composition in chloroplasts of stressed plants rather than changes in concentration. It was shown that among stroma solutes the divalent anions sulfate and phosphate were potentially inhibitory at concentrations existing in dehydrated chloroplasts [54]. In fact, Rubisco was found to be inhibited in vitro at lower concentrations of these anions than those found in the dehydrated leaves. If this indeed is the dominating factor controlling photosynthesis under water stress, one would expect a more severe inhibition, in stressed chloroplasts, which is probably avoided by Rubisco concentration being higher than the K_m value determined in vitro [54]. Stromal acidification has also been reported to occur during leaf desiccation and might also contribute to enzyme inhibition [69].

Another factor responsible for inhibited nonstomatal photosynthetic activity under stress is the decrease in RuBP concentration. In a drought-sensitive hybrid of sunflowers, the decreased photosynthesis was found to be associated with a sharp decline in RuBP, while in a nonsensitive hybrid the decline was much less [70]. While these two hybrids were very different in RuBP concentration and photosynthesis in unstressed leaves, they were similar under stress, indicating that the major difference between them was potential concentration and CO_2 fixation under unstressed conditions.

It was suggested that at conditions of severe water stress, reactive oxygen intermediates may be generated as H_2O_2 and may play a role in the inhibition of photosynthesis [71]. It seems, however, that such injurious effects may be alleviated by enzymatic and nonenzymatic reactions scavenging oxygen free radicals. In fact, CO_2 fixation under drought was found at much higher leaf ψ_w values than the occurrence of reactive O_2 intermediates [72].

E. Photoinhibition and Water Stress

When water stress limits photosynthesis, excess excitation energy originating from light energy is absorbed by the light-harvesting system. As this energy cannot be utilized in photosynthetic processes, it may cause harm to the photosynthetic system known as *photoinhibition* [73,74]. The major site of photoinhibition is located in PSII, and its sensitivity has been related to loss of variable chlorophyll fluorescence. Leaves are mostly susceptible to photoinhibition under specific environmental conditions, which are in essence bright light and low temperature. Prolonged exposure to bright light may also lead to chlorophyll photooxidation, an oxygen-dependent chlorophyll bleaching [75]. Several mechanisms were proposed and can be considered as protecting mechanisms.

Osmond et al. [15] claimed that internal recycling of CO_2 plays a major role in such protection of leaf tissue under stress. The carboxylation and oxygenation cycles are linked by the activity of RuBP carboxylase–oxygenase in C3 plants. A CO_2 compensation point in air of approximately 50 to 70 $\mu L/L$ CO_2 is maintained at atmospheric CO_2 concentration. When stomata close in the light, the integration of the two reactions may maintain the CO_2 compensation point in illuminated leaves by an internal recycling of CO_2. Whereas illumination of C3 leaves at full sunlight under low O_2 which prevented photorespiration resulted in rapid photoinhibition [76], a decay of photoinhibition was obtained under ambient O_2 concentration [15]. The implication of these findings is that at the compensation point, carbon recycling in C3 plants is adequate to permit dissipation of photochemical energy, but only under gradual stress application. In C4 plants only bundle sheath cells contain all the enzymes of the RuBP carboxylation and oxygenation cycles, and oxygenation is conducted at a lower rate. Some protection against photoinhibition may thus still be provided, as in C3 plants. The mesophyll cells of C4 plants lack the RuBP carboxylase–oxygenase cycles and CO_2 recycling is minimal. In fact, in *Zea mays* plants subjected rapidly to stress, a substantial loss of chlorophyll from light-harvesting chlorophyll–protein complex of mesophyll chloroplast was found [77]. Under mild stress and its gradual development, some protection against photoinhibition may be obtained in C4 plants by recycling of metabolites between mesophyll and bundle sheath cells [15].

A morphological device for protection against inhibition under water stress was suggested primarily for desert plants. Some desiccation-tolerant species have evolved the ability to endure extreme water losses and resume metabolic activity upon rehydration. Stem curling while plants dehydrate was shown to serve as a protection-avoiding device. In *Salaginella lepidophylla* the microphyll bearing stem, for instance, was found to curl very tightly and to form a sphere during dehydration [75]. Excessive light energy can in some species be avoided by leaf movement, so that the apex of the leaf is oriented parallel to the sun and thus less light is being absorbed [78].

Finally, nonradiative dissipation of light energy may be an efficient mechanism for protection [79]. It was suggested that the xantophyll in the light-harvesting complex undergo a reversible light-induced interconversion. Violaxanthin is convered to zeaxanthin by a deepoxidase, which is a water-soluble enzyme. The back reaction is catalyzed by oxygenase-consuming NADPH and oxygen. Zeaxanthin differs from violaxanthin in that it readily loses light energy as heat. This might be an important protecting mechanism, although it needs further clarification [79].

F. CAM Plants Photosynthesis and Water Stress

CAM plants exhibit a special strategy that allows the performance of photosynthesis even under severe drought based on stomatal opening in the dark when water and radiation stresses are minimal. These plants, which are mostly succulents, accumulate malic acid during the dark period by fixation of external and respiratory CO_2 under relatively low stress. When stress is severe, only respiratory CO_2 is refixed. The acid is then decarboxylated in the light and the CO_2 is refixed by Calvin cycle enzymes while stomata are closed. When the malic acid is fully decarboxylated, internal cycling of CO_2 via photorespiration may occur as in C3 plants and protect the system against photoinhibition.

A number of succulents have been found to have the ability to switch from C3 to CAM metabolism [80,81] and thereby reduce losses of water by transpiration. This switch is characterized by increased affinity of PEP-carboxylase (PEPC) to the substrate PEP, reduced sensitivity to inhibition by malate, and increased activation by glucose-6-phosphate [82]. In *Sedum telephiun* other factors in addition to water stress, such as light-intensity photoperiod and age, were found to change the activity of PEPC [83,84]. Light intensity and water stress were additive in inducing activity of PEPC and accumulating titratable acid overnight.

Mesembryanthemum crystallimum, which is normally a succulent C3 plant, was also found to close stomata during the light period, accumulating malic acid and deacidifying it in the light as CAM plants [85]. *Crassula sieberiana* is another example of a succulent in which CAM metabolism could be induced by subjecting plants to abrupt droughts [86], and PEPC activity was reversible under progressively increased drought.

Although the C3–CAM switch appears to be a quite universal phenomenon, the trigger for the changes observed is not known. Moreover, a good correlation could not be found between PEPC synthesis parameters and leaf water status [86].

III. SALT STRESS

A. Osmotic and Specific Ion Effects

Salt stress refers to an excess of salts or ions in the soil or growing media. The stress is measured in energy units as water stress, in concentration, or in electrical conductivity. Most salt stresses in nature are due to Na salts, particularly NaCl. The term *halophytes* is used to specify that plants can grow in the presence of high concentrations of Na salts. Plants that cannot grow under such conditions are known as *glycophytes* [2]. Most crop plants are glycophytes.

There is an inseparable relation between water and salt stress. This is due to the lowered ψ_s value of the water when salt content rises, which is also the result of drought. A plant that is exposed to high salt concentration in the growing medium is therefore subjected to conditions which are similar to dehydration. Because of this similarity, the salt stress of plants can be measured in units of ψ_w, similar to water stress. The phenomena of salt and water stress are found in the same regions, as the source of salinity is due primarily to the high rates of evaporation in those regions.

In addition to their osmotic effect, salts exert specific ion effects. This might be on the external cell membrane, but occurs mostly after penetration into the protoplast. The specific effect can be a toxic effect of Cl, but is generally of Na and sometimes of both. Specific effects can also result in nutritional imbalances or deficiencies, due to competition of salt ions and nutrients. The specific effects are in many cases dependent on the composition of the salt and are reduced when the content of Ca salts is relatively high.

The uptake of salts will counteract the osmotic effect, leading to osmotic regulation. Therefore, if the specific ion effects are small, they may lead to a release of the osmotic stress, but probably only after a lag period.

In practice, no real distinction is made between stress originated from water shortage or salinity, so that the osmotic potential caused by salts and the matric potential caused by water shortage are added together. These are then identified as total water potential in the growing media [87]. This probably also refers to the response of photosynthesis to salinity when specific ion effects are negligible. Specific ion effects should, however, be taken into consideration, as these may lead to different responses of photosynthesis to salinity than to water stress, to an aggrevation of the osmotic effect, or to its weakening.

Similar to the effect of water stress, salinity is known to inhibit CO_2 assimilation [88–93]. This inhibition was attributed by many investigators to a reduced stomatal conductance [90–92]. Other investigators attributed it to biochemical or photochmeical activities within the chloroplasts, and we shall refer to these two aspects.

B. CO_2 Assimilation Under Salt Stress

A decrease in stomatal conductance of plants grown under salinity can be interpreted as a combination of osmotic and specific Na effects. Recovery of CO_2 assimilation was found when cowpea plants were transferred to nonsaline conditions, indicating the accountability of an osmotic effect. On the other hand, a linear relationship between Na content and CO_2 assimilation indicates the involvement of a specific effect [91]. More and more information can be withdrawn from recent studies which emphasize the importance of an ion effect on photosynthesis. A decrease in C_i under salinity was markedly more than under an identical ψ_w value obtained by water stress [94]. WUE was clearly enhanced, at least under low and intermediate salt levels, much more than under similar water stress [94,95]. Leaf turgor was found not to decrease under salinization and leaf ψ_s became lower than external ψ_s, but despite the osmotic adjustment, CO_2 assimilation was decreased significantly [93]. Acclimation of CO_2 assimilation to salinity was much better than to similar levels of external ψ_w [25]. Nonphotochemical fluorescence quenching and zeaxanthin formation were not evident under salinity, in contrast to water stress [95]. The response of carbon assimilation was found to be inhibited very differently by different salt compositions at identical concentrations [96,97].

It may thus be deduced that the immediate effect of salinity on carbon assimilation is osmotic, while the more specific effect takes over with time after ion accumulation. This explains the decrease in slope of the linear portion of the A–C_i curve with time of pepper plants under 100 to 150 mmol NaCl but not under 50 mmol [93]. It also explains the CO_2 assimilation rate of salt acclimated versus nonacclimated plants [25] and the severe decline of CO_2 fixation by a saline-tolerant plant such as sugar beets, but only during the first day of exposure to salinity [98].

C. Photochemical and Biochemical Responses to Salinity

The response of photosynthetic photochemical and biochemical processes to salinity is expected to be different from their responses to water stress, if, indeed, leaf ion content controls photosynthesis under salinity. In isolated chloroplasts of *Suaeda maritima*, the Hill reaction was not altered by salinity, while CO_2 fixation was increased [99]. Alia et al. [100] showed enhanced PSII-mediated photochemical activity in thylakoids of *Brassica juncece* seedlings grown in 200 mM NaCl compared with control plants. They elaborated this finding by showing that the higher PSII activity induced by salinity was at pH values in the range 6.0 to 9.0. In more basic pH activity, photochemical activity was inhibited, but the recovery when thylakoids were resuspended at pH 7.5 was much higher in thylakoids from salinity-raised plants. Relative electron transport rate at low light compared to saturating intensity was higher in thylakoids from salinity-raised plants compared to those from control plants, suggesting a stimulated quantum yield. It is therefore interesting that in barley, which is known as one of the most salt-tolerant glycophytes, PSII activity was not changed by salinity in most leaves [98]. Moreover, some apparently healthy and growing leaves exhibited even sudden decreases in PSII photochemistry. This might be a result of the very high salt levels in the case of barley and accumulation of high Na concentrations after a more extended period.

The alteration in the photochemical activity of salinity-exposed leaves is possibly related to their ability to dissipate excess energy. Development of excessive energy that is not consumed by carboxylation or oxygenation activities deepoxidates violaxanthin to zeaxanthin and is involved in nonphotochemical quenching of chlorophyll fluorescence (see Section B.5). This might be the case under severe water stress and under extreme and probably extended salinity [101]. It was, however, not the case in other studies on photochemical responses to salinity [95,100,102].

Limited information is available for salinity effects on photosynthesis dark reactions. An increase in the rate of photorespiration and in oxidase activity accompanied by an alteration in synthesis and distribution of photoassimilation was found under conditions of NaCl and Na_2SO_4 salinity [103]. The efficiency of Rubisco was found to be decreased in bean plants grown at

100 mmol NaCl [104]. The RuBP pool size was also reduced by an effect on RuBP regeneration at a similar salt level. As far as a change in Rubisco is concerned, Miteva et al. [105] showed that even in salt-tolerant barley, Rubisco content was 20% of the control at 100 mmol NaCl. The polypeptide profile of soluble leaf protein (out of which Rubisco is a weighty component) was changed significantly by this NaCl concentration. A shift in carbon metabolism from C3 to CAM was documented to be induced by salinity in *Mesembryanthemum crystallinum* similar to the induction by water stress [106].

IV. CONCLUDING REMARKS

A major effort in the study of the response of photosynthesis to stress was to identify a trigger or a specific site in the complex chain of reactions that is of special sensitivity. This is of importance from a theoretical point of view to elucidate the sequence of events taking place during exposure of plants to water or salt stress. A better insight into the mechanism of strain development as a result of the applied external stress on plants may thus be gained. It is certainly of importance for breeding, selection, and genetical studies aimed to obtain crop genotypes of higher salinity and water stress tolerance. It may serve practical purposes in the determination of critical growth stages of a crop when it is subjected to severe damage by water and salt stress. It will assist in quantifying the potential damage by stress, which is important for an overall analysis of the effects of water shortage and salinity. The difficulties in detecting a key reaction or a compound that controls the response of photosynthesis to stress was outlined in this chapter. There is evidence for control by stomatal versus nonstomatal reactions, by CO_2 depletion versus hormonal signals, by light versus dark reactions, by photochemical versus biochemical processes, by recycling of compounds, or by changes in enzyme activities. We believe that the solution to this problem is very difficult for two groups of reasons:

1. Partial processes of photosynthesis are very tightly linked and a minimal change in one may lead almost instantaneously to a chain of reactions of the entire system. As long as our devices are not adequate and sensitive enough to detect a very slight response, it would be difficult to assess the sequence of events.

2. Water stress, as well as salinity, cannot be separated from other environmental stresses, such as radiation, temperature, and nutrition. The interaction thus becomes very complicated and difficult to analyze. Moreover, differences among species and genotypes, stages of plant development, and previous events may add to this complication, so the enigma seems difficult to solve.

REFERENCES

1. A. San Pietro, F. A. Greer, and T. J. Army, *Harvesting the Sun: Photosynthesis in Plant Life*, Academic Press, New York (1967).
2. J. Levitt, *Responses of Plants to Environmental Stresses, Water Radiation, Salt and Other Stresses*, Vol. 2. Academic Press, New York, pp. 3, 25 (1980).
3. J. S. Boyer, *Philos. Trans. R. Soc. London, B273*: 501 (1976).
4. K. J. Bradford and T. C. Hsiao, in *Physiological Plant Ecology II: Water Relations and Carbon Assimilation*, in *Encyclopedia of Plant Physiology*, Vol. 12B (O. L. Lange, P. S. Nobel, C. B. Osmond and H. Ziegler, eds.), Springer-Verlag, Berlin, p. 246 (1982).
5. J. A. Bunce, *Oecologia (Berlin), 71*: 117 (1986).
6. E. D. Schulze and A. E. Hall, in *Physiological Plant Ecology II: Water Relations and Carbon Assimilation*, in *Encyclopedia of Plant Physiology*, Vol. 12B (O. L. Lange, P. S. Nobel, C. B. Osmond, and H. Ziegler, eds.), Springer-Verlag, Berlin, p. 181 (1982).
7. T. C. Hsiao, *Annu. Rev. Plant Physiol., 24*: 519 (1973).

8. G. P. Farquhar and T. D. Sharkey, *Annu. Rev. Plant Physiol.*, *33*: 317 (1982).
9. E. D. Schulze, *Annu. Rev. Plant Physiol.*, *37*: 247 (1986).
10. W. S. Iljin, *Flora (Jene)*, *116*: 360 (1923).
11. Y. Castonguay and A. H. Markhart III, *Crop Sci.*, *32*: 980 (1992).
12. B. Martin and N. A. Ruiz-Torres, *Plant Physiol.*, *100*: 733 (1992).
13. D. S. Ellsworth and P. B. Reich, *Tree Physiol.*, *10*: 1 (1992).
14. C. B. Osmond, M. P. Austin, J. A. Berry, W. D. Billings, J. U. S. Boyer, J. W. H. Dacey, P. S. Nobel, S. D. Smith, and W. E. Winner, *Bioscience*, *37*: 338 (1987).
15. C. B. Osmond, K. Winter, and S. B. Powles, in *Stress Physiology* (N. C. Turner and P. J. Kramer, eds.), Academic Press, New York (1979).
16. S. C. Wong, I. R. Cowan, and G. D. Farquhar, *Plant Physiol.*, *78*: 830 (1985).
17. W. P. Quick, M. M. Chaves, R. Wendler, M. David, M. L. Rodrigues, J. A. Posseharinho, J. S. Pereira, M. D. Adcock, R. C. Leegood, and M. Stitt, *Plant Cell Environ.*, *15*: 25 (1992).
18. S. H. Al-Hamdani, J. M. Murphy, and G. W. Todd, *Can. J. Plant Sci.*, *71*: 689 (1991).
19. I. Terashima, S. D. Wong, C. B. Osmond, and C. D. Farquhar, *Plant Cell Physiol.*, *29*: 385 (1988).
20. W. J. S. Downton, B. R. Loveys, and W. J. R. Grant, *New Phytol.*, *108*: 263 (1988).
21. T. D. Sharkey and J. R. Seemann, *Plant Physiol.*, *89*: 1060 (1989).
22. D. Gunasekera and G. A. Berkowitch, *Plant Physiol.*, *98*: 660 (1992).
23. J. A. Bunce, *Plant Physiol.*, *86*: 5 (1988).
24. R. R. Wise, A. Ortiz-Lopez, and D. R. Ort, *Plant Physiol.*, *100* (1992).
25. Z. Plaut and E. Federman, *Plant Physiol.*, *97*: 515 (1991).
26. Ni Bing-Rui and S. G. Pallardy, *Plant Physiol.*, *99*: 1502 (1992).
27. D. A. Grantz, *Plant Cell Environ.*, *13*: 667 (1990).
28. F. Loreto and T. D. Sharkey, *Tree Physiol.*, *6*: 409 (1990).
29. P. J. Aphalo and P. C. Jarvis, *Plant Cell Environ.*, *14*: 127 (1991).
30. M. A. El-Sbarkawy and J. H. Cock, in *Biological Control of Photosynthesis* (R. Marcelle, M. Clijsters, and M. V. Poucke, eds.), Martinus Nijhoff, Boston, p. 187 (1986).
31. K. A. Mott and D. F. Parkhurst, *Plant Cell Environ.*, *14*: 509 (1991).
32. Z. Dai, G. D. Edwards, and M. S. B. Ku, *Plant Physiol.*, *99*: 1426 (1992).
33. P. Innes, R. D. Blackwell, and S. A. Quarrie, *J. Agric. Sci. Cambridge*, *102*: 341 (1984).
34. J. Zhang and W. J. Davies, *Plant Cell Environ.*, *12*: 73 (1989).
35. J. Zhang and W. J. Davies, *Plant Cell Environ.*, *13*: 277 (1990).
36. C. L. Trejo and W. J. Davies, *J. Exp. Bot.*, *42*: 1507 (1991).
37. U. Schurr, T. Gollan, and E. D. Schulze, *Plant Cell Environ.*, *15*: 561 (1992).
38. F. Tardieu, J. Zhang, and W. J. Davies, *Plant Cell Environ.*, *15*: 185 (1992).
39. F. Tardieu and W. J. Davies, *Plant Cell Environ.*, *16*: 341 (1993).
40. M. Havaux and R. Lannoye, *J. Agric. Sci. Cambridge*, *104*: 501 (1985).
41. D. Epron and E. Dreyer, *Ann. Sci. For.*, *47*: 435 (1990).
42. D. Epron and E. Dreyer, *Tree Physiol.*, *10*: 273 (1992).
43. T. Graan and J. S. Boyer, *Planta*, *181*: 378 (1990).
44. R. H. Jefferies, *Potato Res.*, *35*: 25 (1992).
45. A. Selmani and C. E. Wassom, *Field Crops Res.*, *31*: 173 (1993).
46. R. van Rensburg and G. H. Kruger, *J. Plant Physiol.*, *141*: 357 (1993).
47. Y. Castonguay and A. H. Markhart III, *Crop Sci.*, *31*: 1605 (1991).
48. W. M. Kaiser, G. Prachuab, G. Kaiser, S. G. Wildmann, and U. Heber, *Planta*, *153*: 416 (1981).
49. T. D. Sharkey and M. R. Badger, *Planta*, *156*: 199 (1982).
50. O. Bjorkman and S. B. Powles, *Planta*, *161*: 490 (1984).
51. D. R. Ort and J. S. Boyer, in *Changes in Eukaryotic Gene Expression in Response to Environmental Stresses* (B. Atkinson and D. Walker, eds.), Academic Press, New York, p. 279 (1985).
52. S. von Caemmerer and G. D. Farquhar, *Planta*, *160*: 320 (1984).
53. Z. Plaut, *Plant Physiol.*, *48*: 591 (1971).
54. W. M. Kaiser, *Physiol. Plant.*, *71*: 142 (1987).
55. H. M. Younis, J. S. Boyer, and Govindjee, *Biochim. Biophys. Acta*, *548*: 328 (1979).

56. A. Ortiz-Lopez, D. R. Ort, and J. S. Boyer, *Plant Physiol.*, *96*: 1018 (1991).

57. S. Meyer and Y. de Kouchkovsky, *FEBS Lett.*, *303*: 233 (1992).

58. W. M. Kaiser, *J. Exp. Bot.*, *35*: 1 (1984).

59. J. C. V. Vu, L. H. Allen, Jr., and G. Bowes, *Plant Physiol.*, *83*: 573 (1987).

60. J. C. V. Vu and G. Yelenosky, *Plant Physiol.*, *88*: 375 (1988).

61. M. Speer, J. E. Schmidt, and W. M. Kaiser, in *Plant Membranes: Structure, Assembly and Function* (J. L. Harwood and T. J. Walton, eds.), The Biochemical Society, London, p. 209 (1988).

62. K. Raschke and E. Fischer, in *Progress in Photosynthesis Research* (J. Biggin, ed.), Martinus Nijhoff, Dordrecht, The Netherlands, Vol. IV, p. 5 (1987).

63. S. P. Robinson, W. J. Grant, and B. R. Loveys, *Aust. J. Plant Physiol.*, *15*: 495 (1988).

64. A. Sen Gupta and G. A. Berkowitz, *Plant Physiol.*, *88*: 200 (1988).

65. M. Santakumari and G. A. Berkowitz, *Plant Physiol.*, *92*: 733 (1990).

66. W. M. Kaiser, *Planta*, *154*: 538 (1982).

67. Z. Plaut, M. Grieve, and E. Federman, *Plant Physiol.*, *91*: 493 (1989).

68. G. A. Berkowitz, *Plant Cell Rep.*, *6*: 208 (1987).

69. G. A. Berkowitz and M. Gibbs, *Plant Physiol.*, *72*: 1100 (1973).

70. C. Gimenez, V. G. Mitchell, and D. W. Lawlor, *Plant Physiol.*, *98*: 516 (1992).

71. S. R. Chowdhury and M. A. Choudhuri, *Physiol. Plant.*, *65*: 503 (1985).

72. J. J. Irigoyen, D. W. Emerich, and M. Sanchez-Diaz, *Physiol. Plant.*, *84*: 67 (1992).

73. S. B. Powles, *Annu. Rev. Plant Physiol.*, *35*: 15 (1984).

74. C. Schafer and O. Bjorkman, *Planta*, *178*: 367 (1989).

75. J. G. Lebkuecher and W. G. Eickmeier, *Can. J. Bot.*, *70*: 205 (1992).

76. G. Cornic, *Can. J. Bot.*, *56*: 2128 (1978).

77. R. S. Alberte and J. P. Thornber, *Plant Physiol.*, *59*: 351 (1977).

78. M. M. Ludlow and O. Bjorkman, *Planta*, *161*: 505 (1984).

79. B. Demmig-Adams and W. W. Adams III, *Annu. Rev. Plant Physiol. Plant Mol. Biol.*, *43*: 599 (1992).

80. H. S. J. Lee and H. Griffiths, *J. Exp. Bot.*, *38*: 834 (1987).

81. H. Greenway, K. Winter, and U. Luttge, *J. Exp. Bot.*, *29*: 547 (1978).

82. D. L. Sipes and I. P. Ting, *Plant Physiol.*, *91*: 1051 (1989).

83. S. H. Cheng and G. E. Edwards, *Plant Cell Environ.*, *14*: 271 (1991).

84. A. M. Borland and H. Griffiths, *J. Exp. Bot.*, *43*: 353 (1992).

85. K. Winter and U. Luttge, in *Water and Plant Life: Problems and Modern Approaches* (O. L. Lange, L. Koppen, and E.-D. Schulze, eds.), Springer-Verlag, Berlin, p. 323 (1976).

86. Y. Brulfert, S. Güclü, and M. Kluge, *Plant Physiol.*, *138*: 685 (1991).

87. P. W. Rundel and W. M. Jarrell, in *Plant Physiological Ecology* (A. W. Percy, J. Ehleringer, H. A. Mooney, and P. W. Rundel, eds.), Chapman & Hall, New York, p. 29 (1989).

88. H. Greenway and R. Munns, *Annu. Rev. Plant Physiol.*, *31*: 149 (1980).

89. A. R. Yeo, S. J. M. Capron, and T. J. Flowers, *J. Exp. Bot.*, *36*: 1240 (1985).

90. L. B. Flanagan and R. L. Jefferies, *Planta*, *178*: 377 (1989).

91. Z. Plaut, C. M. Grieve, and E. V. F. Maas, *Physiol. Plant.*, *79*: 31 (1990).

92. E. Brugnoli and M. Lauteri, *Plant Physiol.*, *95*: 628 (1991).

93. P. C. Bethke and H. C. Drew, *Plant Physiol.*, *99*: 219 (1992).

94. Z. Plaut, in *Structural and Functional Responses to Environmental Stresses* (K. H. Kreeb, H. Richter, and T. M. Hinckley, eds.), SPB Academic Publishing, The Hague, The Netherlands, p. 155 (1989).

95. E. Brugnoli and I. Bjorkman, *Planta*, *187*: 335 (1992).

96. J. Banuls and E. Primo-Millo, *Physiol. Plant.*, *86*: 115 (1992).

97. R. R. Walker, D. H. Blackmore, and Sun Qing, *Aust. J. Plant Physiol.*, *20*: 173 (1993).

98. B. Heuer, *Isr. J. Bot.*, *40*: 275 (1991).

99. H. Genard, J. L. Saos, J. P. Billard, A. Tremoliers, and J. Bocaud, *Plant Physiol. Biochem.*, *29*: 421 (1991).

100. Alia, Saradhi Pardha Saradhi, P., *Biochem. Physiol. Pflanz.*, *188*: 1 (1992).

101. F. Morales, A. Abadia, J. Gomex-Aparizi, and J. Abadia, *Physiol. Plant.*, *86*: 419 (1992).

102. Alia, Saradhi Pardha Saradhi, P. and P. Mohanty, *Physiol. Plant.*, *86*: 189 (1992).

103. S. Yoshi and D. Nimbalkar, *Photosynthetica*, *18*: 128 (1981).

104. J. R. Seemann and T. D. Sharkey, *Plant Physiol.*, *82*: 555 (1986).

105. T. S. Miteva, N. Zh. Shelev, and L. P. Papova, *J. Plant Physiol.*, *140*: 46 (1992).

106. M. Pipenbrock and J. M. Schmitt, *Plant Physiol.*, *97*: 998 (1991).

107. D. W. Lawlor, in *The Effects of Stress on Photosynthesis* (R. Marcelle, H. Clijsters, and M. van Pousse, eds.), Dr. W. Junk, The Hague, The Netherlands, p. 35 (1983).

28
Physiological Mechanisms of Nitrogen Absorption and Assimilation in Plants Under Stressful Conditions

R. S. Dubey
Banaras Hindu University, Varanasi, India

Mohammad Pessarakli
The University of Arizona, Tucson, Arizona

I. INTRODUCTION

Nitrogen is one of the most essential elements for plant growth and development. It is a constituent of many biomolecules, such as proteins, nucleic acids, amino acids, coenzymes, vitamins, and pigments. Due to its high requirement by plants and its complete absence in the bedrock, it has a special place in plant nutrition. Nitrogen supply in the soil is often the most important factor limiting plant growth and yield. In the soil, N availability is due to the application of N fertilizers, biological action of N_2-fixing organisms, or natural fertilization. With the advent of modern agricultural practices, inorganic N fertilizers have become the major input to the soil. In our quest to achieve sustainable food production, to meet the increasing food requirements for a global population, excessive use of various forms of N fertilizers is likely in the near future.

Both ammonium (NH_4^+) and nitrate (NO_3^-) are available forms of N that can be absorbed by plants. However, NO_3^- is the predominant form of N available to the most cultivated plants grown under normal field conditions. Availability of nitrogenous nutrients, especially NO_3^-, is considered as rate limiting for plant growth and crop production. Application of NO_3^- in the soil medium induces NO_3^+ uptake and its assimilation to ammonia by assimilatory enzymes.

Plants are often exposed to various kinds of harsh environmental conditions which adversely affect their growth and metabolism. Adverse environmental conditions, such as soil salinity, drought, heat, cold, and excessive heavy metal content in the soil, create considerable stress in growing plants and severely affect N absorption by the roots and its assimilation in the plant. To adapt to changing environments, higher plants show well-defined metabolic alterations toward nutrient availability in the environment [1]. For instance, NO_3^- in the soil induces the system in its uptake, assimilation, transport and so on [1]. The biochemical events leading to the uptake of NO_3^- by plants are not well defined [2]. However, the process of NO_3^- reduction involving the enzyme nitrate reductase (NR) and NO_2^- reductase (NIR) has been studied extensively in the diverse plant species, and these enzymes have been well characterized with regard to their physicochemical properties and subcellular localizations [3,4]. Environmental stresses influence N nutrition in plants by inhibiting N uptake [5–21] as well as its assimilation [3,12,16,18–20,23–47]. In this chapter, various N sources, their mode of absorption and assimilation by plants, and the effects of different conditions on the uptake process are discussed.

II. NITROGEN SOURCES, ABSORPTION, AND ASSIMILATION

A. Sources of Nitrogen

Different forms of N which are absorbed by plants from the soil are NO_3^-, NH_4^+, and organic compounds such as amino acids and urea. The two major forms of soil N are NO_3^- and NH_4^+. Of these two forms, the NO_3^- is more abundant, and under normal conditions most of the N absorbed by plant roots from the soil is in NO_3^- form, which is reduced further, to NH_4^+ form, in the plant tissue. Frota and Tucker [15], Saad [16], and Pessarakli et al. [30] found that beans (*Phaseolus vulgaris* L.), C3 plants, under either normal or stress conditions absorbed more of the NO_3^- than the NH_4^+ form of N. Sometimes the NH_4^+ form of N is abundant in the soil, due to biological N_2 fixation by symbiotic association of N_2^- fixing organisms, free-living soil bacteria, and blue-green algae. Although N fertilizers in ammoniacal forms are widely used in agricultural fields, ammonium in the soil is readily oxidized to NO_3^- by nitrifying bacteria present in the soil. Certain plant species, such as those inhabiting acidic soils and all C4 plants (i.e., grasses), show a preference for the NH_4^+ form of N. Because of the deficiency of NO_3^- in acid soils, and due to the specific physiological, metabolic, and photosynthetic pathways of C4 plants, such plants prefer NH_4^+ over NO_3^-. Plants that have low intrinsic NO_3^- reductase activity also absorb NH_4^+ preferably over NO_3^-. Since ammonia is toxic to the plants and interferes with various metabolic processes inside the cell, after absorption, NH_4^+ ions are rapidly assimilated into amino acids, amides, and so on, in the roots. Plant species that have an efficient ammonia detoxifying system grow well on the NH_4^+ form of N [48]. These plants can detoxify NH_3 by forming NH_4^+ salts of organic acids. The majority of the plant species grow better when N is supplied as a mixture of both the NO_3^- and NH_4^+ forms in the soil. Certain plants can absorb either NH_4^+ or NO_3^- ions, depending on the pH of the nutrient medium.

It has been suggested by certain groups of investigators that nutrient media containing both the NO_3^- and NH_4^+ forms of N in a proper combination are more suitable for the growth of the cells as well as the plants compared to the either form alone [49]. In many vegetable crops, NH_4^+ is taken up in preference to NO_3^- when its concentration is above 10% of the total N in the nutrient solution [50]. Especially at low root temperatures, NH_4^+ is regarded as a safe source of N [50]. Genotypic diversity occurs in plants for N use efficiency. Kafkafi [50], when evaluating 12 genotypes of pearl millet (*Pennisetum glaucum*) plants, observed that genotypes varied greatly for uptake, translocation, and N use efficiency. Genotypes with lower N use efficiency and translocation indices for N showed lower grain yields.

In plant roots, the initial product of NH_4^+ assimilation is glutamine, whether NH_4^+ or NO_3^- is absorbed by the roots. Other products of assimilation are asparagine, citrulline, amino acid, allantoin, and certain other soluble nitrogenous compounds [51]. The assimilation products are then translocated to various organs of the plants through xylem and phloem vessels. In some plants (i.e., tomatoes), NO_3^- after absorption through the roots is reduced to NH_4^+ in the root itself, while in others (i.e., grasses) it may be transported as NO_3^- to different organs.

B. Absorption and Assimilation of Nitrogen

Nitrate (NO_3^-) is the predominant form of N available in the soil, regardless of the NO_3^- or NH_4^+ forms of fertilizers used. The availability of NO_3^- in the soil is considered rate limiting for plant growth [1]. Systems related with NO_3^- uptake, its intracellular transport and translocation are directly affected by the soil NO_3^- level [2]. In response to environmental NO_3^- root tissues are proliferated and a general increase in root growth and metabolism occurs [1]. The uptake of NO_3^- by plant roots is an active process involving a high-affinity NO_3^- transport system where NO_3^- acts as a signal for these events [52]. The uptake system is inducible with NO_3^- and can be blocked with inhibitors of RNA and protein synthesis [2] as well as amino acid–modifying reagents [53]. This suggests that plasma membrane proteins are involved in the transport process. According to Redinbaugh and Campbell [1], when plant roots sense exogenous NO_3^-, the primary response involves the transcription of genes encoding NO_3^- transport proteins followed by the synthesis of these proteins. A model proposed by these investigators suggests that a constitutive NO_3^- sensor protein system is the first component that detects and senses NO_3^- in the environment. The binding of the NO_3^- with a sensor induces certain regulatory proteins, which, in turn, initiate the transcription of primary response genes by RNA polymerase II. The resulting transcripts are further translated into proteins such as NO_3^- transporters, NO_3^- translocators, and NO_3^- assimilatory enzymes. Biochemical events leading to secondary responses such as root-proliferation-enhanced respiration in response to environmental NO_3^- are not yet clearly understood [1]. Since environmental NO_3^- is the only inducer for the synthesis of NR, NIR, and NO_3^- transport proteins, it appears that NO_3^- induces these processes at the plasma membrane level before entering the cell [2]. However, in plants, the specific receptor for NO_3^- has not yet been identified.

Certain investigators suggest that the prime enzyme of NO_3^- reduction, nitrate reductase (NR), plays a significant role in the uptake of nitrate [54]. A plasma membrane–bound NR has been detected in the roots of barley (*Hordeum vulgare* L.) seedlings, and it is observed that NO_3^- transport is inhibited by anti-NR IgG fragments [55]. Uptake of NO_3^- by plant roots is also dependent on the NO_3^- reduction process or reduced NO_3^- products in the shoot [56]. At a high intracellular concentration of NO_3^- or in the presence of NH_4^+ in the growth medium, NO_3^- uptake tends to decline. Production of malate in the shoot also influences NO_3^- uptake. To neutralize the alkaline conditions due to NO_3^- reduction in shoots, malate is produced. This is further transported to roots with K^+ as a counterion and, in turn, due to oxidation, bicarbonate ions are formed in roots. These are exchanged for NO_3^- of the external environment.

Nitrate uptake and its reduction activities in the plant tissues are coordinately regulated. In plant tissues, NO_3^- induces increase in NR activity. The activity of NR increases in root cells in response to exogenous NO_3^-. Besides the NO_3^- reduction process, which is the primary response of NO_3^- uptake, plants also possess the systems for translocation of NO_3^- within and between the cells [54]. After uptake, NO_3^- may be translocated to the vacuole in the cells, where it can be accumulated and serve as a NO_3^- reserve (57). The intracellular NO_3^- translocation process possibly requires a tonoplast NO_3^- translocator, which is different from a membrane NO_3^- transporter [1]. The distinct NO_3^- translocators present at the symplasm/xylem interface control the translocation of NO_3^- from root to xylem and then to various organs of the plant [1]. Although the translocator proteins appear to be different from transport proteins and are encoded by different genes, all three processes—NO_3^- transport, its translocation, and its reduction—are coordinately regulated.

Although both forms of N (NO_3^- and NH_4^+) are taken up by plants, only NH_4^+ is incorporated into organic molecules in the plant tissues by an enzymatic process. The primary step in the reduction process involves the reduction of NO_3^- to NO_2^- catalyzed by the enzyme NR. Ammonium, either directly absorbed by plant roots or as a result of NO_3^- reduction, is further assimilated and incorporated into the amide amino group of glutamine by the action of glutamine synthetase and subsequently into glutamic acid by glutamate synthase. These two enzymes are responsible for the assimilation of most of the NH_4^+ derived from NO_3^- reduction under normal growth conditions. An alternative route of NH_4^+ assimilation into glutamate involves the reductive amination of α-ketoglutarate catalyzed by a mitochondrial enzyme glutamate dehydrogenase. Other amino acids, such as alanine and aspartic acid, are further synthesized from glutamic acid by transamination reactions.

III. NITROGEN ABSORPTION AND ASSIMILATION UNDER DIFFERENT STRESSES

Crops growing in adverse environmental conditions of salinity, drought, high or low temperature, low light, and heavy metal–containing soils suffer severe losses in yield. Harsh environmental situations interfere with normal growth and metabolism of plants, and plants respond to these stresses by different types of physiological and biochemical adjustments. Like various physiological processes, N uptake, as well as its translocation and assimilation, are severely affected due to various types of stresses [5–47,58–60]. Since the availability of nitrogenous nutrients in the soil, and their uptake and assimilation, are directly related to each other as well as to the growth and yield of the crops, considerable effort has been made by various groups of investigators to study the possible implications of various stress conditions on N nutrition in plants [5–47,57]. In the following sections, the influence of diverse environmental stresses on the overall process of N uptake is summarized.

A. Salinity

Soil salinity is one of the major environmental stresses affecting crop productivity. The effect of salinity on plants may vary depending on the developmental stage of the plant as well as the types and concentration of salts. The responses of salinity on uptake differ in different plant species and also depend on the type and extent of salinity. In the majority of the plant species studied, salinization in the soil affects N uptake, whereas in the halophytes and in many salt-tolerant crop species, no significant effect of NaCl on NO_3^- uptake is observed [61]. Barley (*Hordeum vulgare* L.) plants growing under saline conditions show reduced growth [61,62] as well as decreased N uptake [61]. In young barley seedlings, salinity severely inhibited NO_3^- uptake, whereas little effect was observed on NO_3^- reduction [58]. In wheat (*Triticum aestivum* L.) plants, reduction in growth was even more than that in barley at higher NaCl salinity levels [62], and the uptake of N decreased with increasing salinity [61]. However, by increasing the N supply to the soil the effect of salinity was alleviated [61]. Khalil et al. [63] found similar results for cotton (*Gossypium hirsutum* L.) and corn (*Zea mays* L.) plants. Soltani et al. [59] observed that when barley seedlings were grown in the presence of 200 mM NaCl, the growth of the seedlings decreased with a concomitant reduction in the uptake and translocation of N compared to nonsalinized seedlings.

When seedlings of maize genotype differing in drought resistance were grown under –0.84 MPa NaCl salinity, the supply of reduced N for the synthesis of amino acids and proteins in the tissues was reduced [60]. The effect was more pronounced in drought-resistant genotypes, where salinity reduced the activity of the metabolic pathway, supplying reduced N accompanied by a corresponding reduction in the relative growth rate of the seedlings. Reduced growth in terms of dry matter production and a decrease in the absorption of N have been reported by several investigators for various plant species with different degrees of salt tolerance [7–11,14–17,20,21,26,27,29–32,36].

Under salinization, reduced uptake of N by crops appears to be due to more intake of Na^+ as well as Cl^- by the roots. Increased levels of Na^+ in the plant tissues cause nutrient imbalance and displace Ca^{2+} from exchange sites on the membranes and cell walls [61]. Chloride present in more than 100 mM concentration in the saline medium inhibited NO_3^- uptake possibly due to increased accumulation of Cl^- in the roots [55]. Smith [64] observed that NO_3^- uptake in barley was dependent on the internal concentrations of Cl^- rather than the external. Reduced uptake of N could lead to N deficiency in plants, and thus could become a limiting factor for growth of plants under saline conditions [55]. Since salinity leads to N deficiency, fertilization of plants growing under saline environments with increasing doses of nitrogenous fertilizers has proven to be beneficial. It minimizes salt-induced damage and apparently provides salt tolerance [61]. However, in certain crop species, such as corn, rice (*Oriza sativa* L.), wheat, and spinach (*Spinacia oleracea* L.), a decrease in salt tolerance has been observed following excess application of nitrogenous fertilizers [61].

The presence of Ca^{2+} in the medium increases NO_3^- uptake under saline conditions. Ward et al. [55] observed that NaCl decreased NO_3^- uptake in barley seedlings, whereas the uptake rate increased with increasing level of Ca^{2+} in the saline medium, between 1.0 and 3.0 mM. These investigators observed a 31 to 35% greater uptake of NO_3^- by increasing Ca^{2+} in the medium compared to salt-free (control) medium. Manganese and Mg^{2+} also enhanced NO_3^- uptake under saline conditions, but Ca^{2+} was more effective than either of the two ions [55]. The presence of Ca^{2+} in the saline medium possibly decreased Na^+ as well as Cl^- uptake and also reduced membrane disruption in saline solutions, leading to increased NO_3^- uptake [55]. Calcium plays a significant role in maintaining the integrity of the root membranes; thus its deprivation, under salinization, decreases ion transport and NO_3^- uptake by disrupting the NO_3^- transporter which is located in the plasmalemma of roots [55]. Under saline conditions, Ca^{2+} has been shown to increase the activity of the NO_3^- transporter [55].

In halophytes, salinity either induces uptake and accumulation of NO_3^- or has no effect on these processes [65]. Halophytes have the capacity to accumulate inorganic ions such as Na^+, K^+, Cl^-, and NO_3^- much in excess compared to nonhalophytes, under saline conditions. *Atriplex*, *Salicornia*, and *Suaeda maritima* plants show a higher uptake of Na^+, Cl^-, SO_4^{2-}, and NO_3^- in saline environments than in nonsaline conditions [65]. According to Flowers et al. [65], even in conditions of low salinity, the levels of K^+, NO_3^-, and SO_4^{2-} were much higher in halophytes than in other plants. These investigators believe that the high uptake of NO_3^- by halophytes under salinization is related to the intrinsic property of these plants to adapt, grow, and show normal metabolic functions at high ion concentrations.

The prime enzyme of NO_3^- assimilation, NR, which catalyzes the conversion of NO_3^- to NO_2^-, is shown to play a major role in the uptake of NO_3^- by serving as NO_3^- transporter [66]. This enzyme has been studied extensively by various groups of investigators for its behavior in different plant species, under salinization. Evidence indicates that the uptake and utilization of applied NO_3^- is largely dependent on its assimilation inside the plant tissues [67]. The effects of salinity on NR activity are varied and depend on the type of salinity as well as the plant species. Nitrate reductase is a highly sensitive enzyme to various types of environmental stresses, including salinity [61]. In many salt-sensitive plant species, NR activity decreased under NaCl salinity [68–70]. Plaut [68] observed decreased activity of NR in cell-free extracts as well as intact tissues of wheat (*Triticum aestivum* L.) seedlings when NaCl was applied to the nutrient medium and the enzyme was assayed after a 24-h exposure to salinity stress. Similarly, while studying the effects of salinity on N metabolism on wheat plants, Abdul-Kadir and Paulsen [69] observed decreased NR activity under salinization. In pea (*Pisum sativum* L.) seedlings, an isoosmotic concentration of NaCl suppressed NR activity and caused the accumulation of NO_3^- in the plant tissue, and in wheat seedlings, an isoosmotic salinity level decreased NR activity without a significant accumulation of NO_3^- in the tissues [61]. In young barley plants [58] and rice (*Oriza sativa* L.) seedlings [71], decreased NR activity has been observed under salinization. In pasture plants, NaCl salinity reduced growth, with a concomitant decrease in NR activity [72].

Lal and Bhardwaj [70] observed that after 15 days of salinization of field pea (*Pisum sativum* L.) with NaCl/CaCl$_2$ (1:1) there appeared a significant suppression in NR activity accompanied by a decrease in total N as well as protein N and an increase in NO$_3$-N and NH$_4$-N. According to these investigators, NaCl as well as CaCl$_2$ salinity impaired NO$_3^-$ assimilation in pea plants, leading to accumulation of NO$_3^-$ and NH$_4^+$ in the tissues. While conducting stress studies in lentil (*Lens esculenta* Moench), Tewari and Singh [73] observed that by increasing exchangeable sodium percentage (ESP) in the cell, there appeared a continuous decrease in NR as well as NIR activities in plants up to 60 days after sowing. Genotypes of rice plants differing in salt tolerance show varying behaviors of NR as well as NIR under salinity stress [24]. While studying the mode of N assimilation under salinization in the seedlings of two sets of rice cultivars differing in salt tolerance, Katiyar and Dubey [24] observed a decrease in NR activity in seedlings of salt-sensitive cultivars. When desalted enzyme extracts from nonsalinized rice seedlings were assayed for NR activity in the presence of 1 *M* NaCl in the assay medium, a strong suppression of enzyme activity was observed. Other investigators also noticed similar suppression of NR activity in salt-sensitive genotypes of rice seedlings when grown in saline medium [23,74].

Several possible explanations have been suggested for the decreased NR activity in salt-sensitive plants under saline stress [68,75]. The plausible reason appears to be the inhibition of enzyme induction under salinization [73]. As NR is a substrate-inducible enzyme, under saline conditions, NO$_3^-$ uptake by the plants is reduced. This causes a limited NO$_3^-$ availability in the plant tissues and, thereby, NR induction is suppressed, which results in decreased NR activity [75].

Several investigators have emphasized that under salinity stress, enhanced translocation of NO$_3^-$ and assimilates takes place from roots to shoots and from flag leaves to developing grains [58,76]. In certain plant species, an increase in NR activity has been observed due to salinity [23,24,61,77]. *Salicornia europeaca* plants and corn (*Zea mays* L.) seedlings showed increased NR activity when grown in salinized medium [61]. While conducting experiments on *Cajanus cajan* plants, Joshi [77] observed that NaCl salinity stimulated NR activity in the leaves of plants, whereas Na$_2$SO$_4$ salinity inhibited the enzyme activity. In *Cajanus* plants, a gradual increase in NR activity was observed, with the increase in NaCl salinity of the soil in the range 2.5 to 10.0 dS/m [77]. In the seedlings of *Phaseolus aconitifolius*, Sankhla and Huber [78] observed an increased in vivo NR activity under salinization.

Rice plants differing in salt tolerance showed varying behaviors of NR activity [23,24,74]. Salt-sensitive genotypes of rice plants showed a decrease in NR activity under salinization, whereas an increase in NR level was observed due to salinity in salt-tolerant genotypes [23,24]. Katiyar and Dubey [24] observed a marked increase in in vivo NR activity in roots as well as in the shoots of salt-tolerant rice cultivars, CSR-1 and CSR-3, with a salinity level up to 14 dS/m NaCl compared to nonsalinized plants. The higher NR level in seedlings of salt-tolerant cultivars suggests that salinity possibly promotes the synthesis/induction of the enzyme in seedlings of such cultivars. Salt-tolerant crop cultivars thus appear to have better adaptability to saline stress by exhibiting efficient NO$_3^-$ reduction under salinization.

The GS/GOGAT pathway, which is the route of ammonium assimilation in plants under normal conditions, is adversely affected under salinization [79,80]. Miranda-Ham and Loyola-Vargas [79] observed that when *Canavalia ensiformis* plants were subjected to NaCl salinity stress the activity of GS decreased markedly in the roots as well as in the shoots. In the roots of salt-sensitive pea plants, decreased GS activity was observed under salinization [61]. While studying the behavior of GOGAT in situ in two sets of rice cultivars differing in salt tolerance,

Katiyar [80] observed that a NaCl salinity level of 14 dS/m was inhibitory to the enzyme. In tolerant genotypes of crop plants, GOGAT is comparatively more tolerant to NaCl than in sensitive genotypes [80]. Decreased activities of GS and GOGAT under salinization suggests a possible impairment of N assimilation/amino acid biosynthesis by this pathway due to salinity.

The GDH enzyme plays an important role in ammonium assimilation under stress conditions by detoxifying ammonia, which tends to accumulate under such conditions. In many crop species examined under salinization, the activity of GDH increases [61,80–82]. However, in certain cases it remains as stable as the controls [79] or is decreased due to NaCl salinity [61,70]. In an obligate halophyte *Suaeda maritima*, Boucaud and Billard [81] observed an increase in GDH activity with 25 mM NaCl. Similarly, in peanut leaves a salinity-induced increase in GDH activity was observed by Rao et al. [82]. Sharma and Garg [83], while studying the amination and transamination events in wheat plants, observed that plants grown under 8 and 16 dS/m NaCl showed an increase in the activity of GDH in leaves as well as in roots. Seedlings of rice cultivars differing in salt tolerance when raised under increasing levels of NaCl salinity showed a marked increase in GDH activity both in vivo and in vitro compared to controls [80]. The effects were greater in the sensitive than in the tolerant cultivars. Increased activity of GDH under salinization suggests a possible role of this enzyme in ammonium assimilation under saline conditions [84]. It appears that saline conditions favor an increased accumulation of ammonium and related compounds. Thus the GS/GOGAT pathway of ammonium assimilation is impaired, and under such conditions, the increased level of GDH imparts adaptive value to plants by detoxifying and assimilating more ammonium (84). Rice (*Oryza sativa* L.) crop cultivars appear to have better adaptability to saline stress by exhibiting efficient NO_3^- reduction under salinization.

The effects of soil salinity on the N content of plant species are varied and dependent on the species, the organs studied, and the type of salinity. Lal and Bhardwaj [70] observed a decrease in total-N as well as protein-N content of 15-day-old *Pisum sativum* seedlings salinized with a mixture of NaCl and $CaCl_2$ with 4 and 8 dS/m salinity. However, an increase in soluble forms of N (NO_3^- and NH_4^+) was observed due to salinity. A similar decrease in the content of pea seedlings growing under isoosmotic levels (–0.1 to –0.5 MPa) of NaCl, $CaCl_2$, and Na_2SO_4 salts was observed by Singh et al. [85]. These investigators observed a decrease in N content with an increase in salt stress and found that the effect of salt stress was more harmful to N content than was water stress.

While investigating the effects of salinization on nodulation and N fixation in pea plants, Siddiqui et al. [86] reported a decrease in nodule N, total plant N, and a significant reduction in the N_2-fixing efficiency of the nodules with an increase in the level of salinity. In *Vigna radiata* plants, a salinity level of 8 dS/m was lethal to the plant growth and nodulation. Increasing the salinity level in such plants from 0 to 4 and 6 dS/m decreased the nodulation and the N content of roots, stem, and leaves [87].

In certain cases, increased N contents due to salinity have been observed in various plant species by several investigators [7–20,27–32,34,36,63,69–77]. Pessarakli and Tucker [11] found that N contents of cotton shoots and roots increased under NaCl salinity up to –0.8 MPa osmotic potential of the nutrient solution. At the low level of salinity (–0.4 MPa osmotic potential), plants contained significantly higher total N [11] and crude protein [32] than did controls (nonsalinized plants). Khalil et al. [63] reported a similar increase in the total-N concentration of cotton and corn under salt-stress conditions. In *Cajanus cajan*, a protein-rich leguminous crop, salinity treatment of 10 dS/m NaCl caused about a 43% increase in the total-N and protein content of the leaves of 3-month-old plants over that of controls, whereas similar salinity treatment with Na_2SO_4 caused a decrease in the N and protein content compared to controls [77]. Joshi [77] suggested that the two salts NaCl and Na_2SO_4 show a specific ion effect on the N metabolism of *Cajanus cajan*. When grown under greenhouse conditions and irrigated with water containing 44, 88, and 132 mM NaCl, *Phaseolus vulgaris* plants showed an increase in the total-N content of the leaves with an increase in salinity [88]. The addition of 4 or 8 mM $CaCl_2$ or $CaSO_4$ in the NaCl treatment medium further increased the leaf N content in such plants, indicating that the addition of Ca^{2+} helps in maintaining the selective permeability of the membranes [88].

B. Water Stress

Water availability is one of the most limiting environmental factors affecting crop productivity. In the semiarid tropics, the occurrence of drought or water deficit in the soil is common; in temperate and tropical regions, crop plants undergo seasonal periods of water stress, especially during the summer. The plant responses to water stress depend on the severity and duration of the stress and the growth stage of the plant [89]. A low water potential in the soil as well as inside the plant inhibits the growth of the plants, reduces developmental activities of its cells and tissues, decreases the uptake of essential nutrient elements, and causes a variety of morphological and biochemical modifications. Plants growing in water-stressed environments show a reduced N uptake from the soil and decreased activities of N assimilatory enzymes [90–92].

When water potential inside the plant declines below a threshold value, stomata closure takes place, which causes a reduction in transpiration and a reduction of water transport through the plant. This, in turn, affects the roots directly, so that the roots are unable to accumulate or absorb NO_3^- as effectively as when transpiration is normal [93]. At low water potential, the ability of roots to supply NO_3^- to the transpiration stream decreases, leading to a decrease in NO_3^- concentration of the xylem sap [93]. Under nonstressed conditions, in freely transpiring plant, a continuous movement of NO_3^- from the roots to the leaves (NO_3^- flux) is maintained. This NO_3^- flux decreases during water stress.

It was suggested by Viets [94] that under water stress conditions, roots are unable to take up much nutrient from the soil due to a lack of root activity, slow rates of ion diffusion, and water movement. While examining the uptake of various nutrients by wheat varieties, Rao and Ramamoorthy [95] observed a 39% drop in N uptake by six improved varieties of wheat when moisture stress was imposed at different stages of plant growth. According to these investigators, the uptake of N was affected with applied stress due primarily to the restricted movement of water under such conditions.

According to Shaner and Boyer [93], water stress caused a decrease in leaf NO_3^- content as well as a decrease in NO_3^- flux from the root to the leaves. Upon rewatering the water-stressed plants, NO_3^- flux increased but not the leaf NO_3^- content. When water-stressed plants were fertilized with more NO_3^-, NO_3^- flux increased and plant performance as well as grain yield were improved [91]. Kathju et al. [91] observed that when wheat plants were grown under low (N_0P_0)- and high ($N_{80}P_{80}$)-fertility conditions and water stress was imposed at various stages of a plant's life cycle, increasing intensities of stress adversely affected leaf metabolism and plant performance. However, the performance of plants was better under high-fertility conditions at all stages, under different intensities of water stress. Similar observations by other investigators also indicate that NO_3^- application can partly alleviate water stress–associated damage in plants [96,97]. Lahiri [96] demonstrated that N application to the soil reduced the adverse effect of drought on dry matter and grain yield of pearl millet. When fertilized with N, Sorghum (*Sorghum halepense* L.) plants recovered faster after relief from water stress [97]. Although fertilized plants experienced water stress severely, they recovered from stress more quickly than do unfertilized ones. Such observations have far-reaching consequences in the sense that in dryland agriculture, where water is a limiting factor, fertilizer application can be considered for drought mitigation management [97].

Numerous studies have been undertaken by various groups of investigators to examine the behaviors of NO_3^- assimilatory enzymes in plants under water-stress conditions. Among these studies, NR has received maximum attention. Shaner and Boyer [93] found that NR activity was sensitive to the water potential of the plant and decreased with the lowering of water potential. Even under mild water-stress conditions, NR activity declined rapidly compared to other N assimilatory enzymes [48].

In various crop species examined, NR activity has often been shown to decline due to water stress [68,95,98,99]. In field-grown wheat plants, imposition of water stress caused a gradual decline in NR activity in leaves [91]. Kathju et al. [91] observed that in wheat plants increasing the intensity of water stress progressively for 3 to 9 days reduced NR activity. These investigators also reported that under both low and high NP fertility conditions, water stress reduced NR activity at different growth stages of plants. However, activity was always greater in highly fertilized plants than in low-fertilized treatments. Slow decline in NR activity due to water stress is possibly attributed to partially maintained NO_3^- flux inside the plant despite increased stomatal resistance and decreased transpiration rate under stress conditions.

In maize plants desiccation leads to a steady decrease in NR activity with a concomitant decrease in leaf water potential, leaf NO_3^- content, and the NO_3^- flux [98,99]. When rewatered, water-stressed maize plants recovered partially and showed an increase in NR activity and NO_3^- flux [93]. In maize callus tissue, a decrease in relative humidity caused a gradual decrease in NR activity [97]. While examining NR activity in different organs of two chickpea (*Cicer arietinum* L.) varieties in relation to soil moisture stress, Wasnik et al. [98] observed a significant reduction in leaf NR activity due to moisture stress. In cacao (*Theobroma cacao* L.) plants, which experience a periodic drought from January to May in the coastal regions of India, water stress induced by withholding irrigation for 7 days caused a substantial decrease in NR activity in the seedlings [100].

Several explanations have been put forth for decreased NR activity in plant parts subjected to water stress [48,93,99]. The most plausible explanation, suggested by Morilla et al. [99], indicates that a reduction in NR activity in *Zea mays* plants subjected to water stress is due to a decline in the rate of synthesis of NR protein rather than its increased rate of degradation or a direct effect of the water potential on enzyme activity. According to these investigators, desiccation of plants leads to a decrease in leaf water potential. This, in turn, decreases NO_3^- flux and causes slow delivery of NO_3^- to the transpiration stream. Thus movement of NO_3^- to the induction site is prevented, resulting in decreased NR activity. These investigators believe that decreased NR activity in water-stressed plants is due primarily to a decrease in NO_3^- flux, not to a decrease in the water potential or NO_3^- content of leaves [93]. A similar interpretation suggested by Singh and Sawhney [48] indicates that the decline in NR activity during water stress is due to a lowered capacity of tissues to synthesize NR protein due to degradation of polyribosomes to monoribosomes. More conclusive evidence is still required to ascertain whether decreased NR activity in water-stressed plants is due to a decreased rate of enzyme synthesis or an increased rate of enzyme degradation.

Certain measures have been suggested by different groups of investigators to overcome partially the effects of drought stress. By increasing soil fertility, especially with nitrogenous fertilizers, the adverse effects of drought can be alleviated substantially [96]. More N fertilizer application to water-stressed plants improved NO_3^- uptake and increased NR activity. Such plants showed better performances and grain yield than those of low-fertilized plants [91]. Similarly, in plants such as sesame (*Sesamum indicum* L.), moderate water stress, when imparted at the early vegetative stage, helped partially to overcome the adverse effect of subsequent severe stress [101]. Such prestressed plants maintained a high water status, showed higher activities of NR, and had better plant performance. Foliar application of chemicals such as chlormequat, cycoel, and ABA also increased relative water content in wheat [102] and cacao [100] plants, and such plants showed increased NR activity.

In water-stressed plants, activities of enzymes of ammonium assimilation remain high, as seen from little or no accumulation of NH_4^+ in the leaves of such plants [103]. However, the pathway of ammonium assimilation under stress conditions depends on the plant species, growth stage, and the plant organs studied. It has been shown that water stress lowers the activity of GOGAT in the root nodules of alfalfa (*Medicago sativa* L.) and cicer plants [104,105]. In these plants, GOGAT is more sensitive than GS to water stress. Koundal and Chopra [105] reported a decline in NADH-GOGAT and GS activities in nodules of chickpea plants exposed to water stress (with a greater percentage decline in the activity of GOGAT than of GS) compared to the nodules of unstressed plants. The rewatering of such plants caused an increase in GOGAT and GS activities, but NR activity remained comparable to controls [105]. These observations indicate that in alfalfa and chickpea nodules, ammonium is possibly assimilated by the GDH pathway under water stress. At the flowering stage in *Brassica* and in the shoots of *Poterium*, increased GDH activity has been observed under water stress [84].

C.   Light

Light remarkably influences N uptake and its assimilation. Decreased light intensity reduces the uptake of NO_3^-, causes NO_3^- accumulation in the tissues, and decreases the rate of its reduction by lowering the activities of NO_3^- reductase enzymes [48,76]. Plants grown under low light intensity show decreased NO_3^- uptake up to the extent that the NO_3^- uptake per gram of fresh weight production is also decreased [76]. During the day, roots show higher rates of NO_3^- uptake than they show during the night [106]. Blom-Zandstra et al. [76] observed that when grown under light with decreasing intensity, lettuce genotypes (*Lactuca sativa* L.) differing in NO_3^- accumulation showed decreased NO_3^- uptake with a concomitant decrease in growth. In such plants, NO_3^- uptake per plant decreased proportionally even more than fresh-weight production, due to a decline in light intensity.

It has been shown that the uptake of NO_3^- by roots is dependent on the continued flux of soluble carbohydrates from the shoot [107]. During the day, due to the metabolic activity of the roots as well as a greater demand of carbohydrates from the shoot pool, translocation of carbohydrates from shoot to the root is greater, which parallels with higher uptake of NO_3^- by roots under daylight conditions [108]. Increased rates of NO_3^- uptake by roots are observed due to diurnal variations associated with changes in day/night or seasonal conditions [109]. With only hours interruption of the dark period, using light of low intensity by incandescent lamp, NO_3^- uptake increased twofold in soybean (*Glycine max* L.) plants compared to the day period [108]. Raper et al. [108] suggested that light-induced increase in NO_3^- uptake by plant roots is phytochrome mediated. This, in turn, alters the permeability of plasma membranes and enhances starch degradation by increasing the activity of starch-degrading enzymes. This leads to an increase in the availability of soluble carbohydrates for translocation from the shoots to the roots.

In plant cells, the bulk of NO_3^- is stored in vacuoles in the form of a "storage pool" [110]. This represents a metabolically inactive pool of NO_3^- and is not available for the induction of cytosolic NR; however, it plays a significant role as an osmoticum along with organic acids and sugars which are located in the vacuoles [76]. Metabolically, the "active pool" of NO_3^- is present in the cytosol [49]. It is believed that light affects the movement of NO_3^- from storage to the metabolic pool [85]. Nitrate taken up in the dark accumulates largely in the vacuoles and when such dark-kept plants are illuminated, the proportion of NO_3^- in the metabolic pool increases [48]. In the light, NO_3^- taken up by plants enters the metabolic pool, where it is available for NR induction. Thus the processes of NO_3^- uptake and NR induction are interrelated and both are dependent on light. It was suggested by Aslam et al. [111] that the transfer of NO_3^- from storage to the metabolic pool is mediated by phytochrome. Light thus regulates the availability of NO_3^- in the metabolic pool.

Plants grown at low light intensities accumulate NO_3^- largely in vacuoles, where the NO_3^- serves as an osmoticum [76]. Accumulation of NO_3^- is inversely related to the accumulation of organic compounds and in this way accumulating NO_3^- may compensate for the shortage of photosynthates as a result of a decreased rate of photosynthesis under shade conditions [76]. Plants growing under insufficient light conditions thus show a twofold demand for N, one for the metabolic pool, which after reduction can be used for protein synthesis, and the other for the storage pool, which acts as an osmoticum [76]. The distribution of N between organic-N and nitrate-N changes in plants grown in the shade. Decreasing light intensity decreases the organic-N level in the vacuoles and increases the nitrate-N level. Lettuce genotypes differing in the extent of NO_3^- accumulation, when grown under shade conditions, show an increased NO_3^- concentration in the cell sap in both sets of cultivars accompanied by a decreased concentration of organic N [76].

Light exerts a marked stimulatory effect on the reduction of NO_3^- by regulating both the synthesis and the functioning of NR. Leaves of shade-grown plants show a very low level of NR activity, but when such plants are transferred to light, NR activity increases severalfold [48]. Similar to NR, in photosynthetic tissues light plays a significant role in regulating the activity of NiR [48].

Several postulations have been put forth as regulatory mechanisms for light-mediated enhancement of NR activity. Based on the inhibitor studies and the labeling experiments, it has been suggested that light promotes *de novo* synthesis of both NR and NiR. Illumination of leaves leads to increased protein synthesis, indicating that light enhances the production of NR in leaves [49]. Certain investigators suggest that light-mediated enhancement of NR activity is due to the enhanced uptake of NO_3^- by plants in the light [112]. Light enhances the movement of NO_3^- from the storage pool to the metabolic pool [111], where NO_3^- becomes available for the induction of NR activity. Sharma and Sopory [113] observed that in maize seedlings NR activity increased by more than 300% when treated with red light and kinetin. These investigators suggested that the light-induced increase in NR activity is mediated via phytochrome. Phytochrome action does not appear to be mediated via hormones; however, there appears to be an overlap in signal transduction chains of phytochrome and plant hormones [113]. According to Sawhney and Naik [114], some early events of photosynthesis, like the Hill reaction, cause redox changes in green tissues and create favorable intracellular conditions for the synthesis of NR. These findings indicate that light influences the level of NR and NiR in plants, but a unified mechanism leading to the mode of action of light in regulating the activities of these enzymes is still awaited.

D. Temperature

Like all other physiological and biochemical processes, nitrogen uptake, metabolism, and assimilation are strongly related to temperature. Optimum absorption and assimilation occur under normal temperature. Any deviation from the normal temperature range adversely affects N absorption, metabolism, and assimilation.

1. High Temperature

High temperatures affect seed germination, plant growth, and yield attributes, and induce many metabolic alterations in crops. Different plant species have different optimum temperature ranges for better growth and yield. Even a slight increase in soil temperature affects growth and nutrient uptake in plants. A rise in temperature beyond optimum growth temperature impairs the rate of uptake as well as the assimilation of nutrients by exerting a profound influence on the activities of assimilatory enzymes. In many plant species, the effects of day/night temperatures on the uptake of various nutrient elements have been studied extensively [115,116].

The C4 plants grown under conditions of high temperature and high humidity show enhanced efficiency in N use compared to C3 plants [116]. C4 plants such as corn (*Zea mays* L.) and sorghum (*Sorghum bicolor* L.) and C3 plants such as barley, rice, wheat, and oats (*Avena sativa* L.) were grown either for 7 days at 20 or 28°C, or for 3 weeks at 26°C. A greater accumulation of NO_3^- was observed in C3 than in C4 plants in all three conditions tested [116]. However, N supplied as NO_3^- was more efficiently assimilated into protein in C4 than in C3 plants (116). Lowering the temperature in both sets of plants from 28°C to 20°C caused accumulation of NO_3^- as well as a lower protein/nitrate ratio [116]. Greater efficiency of C4 cereals for NO_3^- uptake and assimilation compared to C3 plants at all temperature levels tested appears to be due to highly organized cellular structure and spatial organization of N assimilatory enzymes in C4 plants [116].

In young corn seedlings, the day–night temperature of 30/30°C is regarded as optimum for NO_3^- uptake [115]. While examining the effect of three temperature treatments (30/20°C, 30/30°C, and 35/35°C day/night temperatures on NO_3^- uptake in corn seedlings), Polisetty and Hageman [115] observed that the amount of NO_3^- taken up during the night was about four- and threefold greater for 30/30°C over 30/20°C and 35/35°C, respectively, whereas during the light period, NO_3^- uptake increased by 1.5- and 1.3-fold, respectively. This suggests that optimum NO_3^- uptake by corn seedlings occurs at 30/30°C and that either increase or decrease in the temperature leads to a decrease in NO_3^- uptake. The possible reasons for decreased NO_3^- uptake above optimum growth temperatures appear to be due to the impairment of the NO_3^- uptake process, as well as the inhibition in root and shoot development occurring due to the increase in temperature [115].

Partitioning of N in different parts of the plants is affected by an increase in temperature. In rice, the absolute N content per kernel was comparatively stable during the temperature range from 24/19°C to 33/28°C, whereas beyond this temperature a decline in N content of the kernels was observed [117]. The varieties of rice differ in sensitivity toward higher temperature. Japonica varieties of rice were more sensitive than indica types during kernel development [118]. The highest concentration of N in terms of percent of dry weight is recorded in rice kernels in the temperature range from 33/28°C to 39/34°C [117]. Decrease in N content of shoots in soybean plants has been reported by Hafeez et al. [119] due to an increase in temperature from 30°C to 48°C compared to the control plants growing at 30°C.

The NR is sensitive to higher temperatures. Temperatures above a certain optimum affect the level of NR in plants as well as inhibit its activity. The magnitude of inactivation or lowering of NR level by higher temperature varies according to the species [22]. Chandra and Pareck [120] observed that in sorghum plants an increase in temperature of 6 to 11°C caused more than a 60% reduction in NR activity at the vegetative stage and a 30% reduction at the anthesis stage. Corn seedlings maintained at 15 to 20°C showed six times more NR activity than at 25 to 30°C [48]. In barley seedlings, induction of NR did not take place when seedlings were maintained at 41°C. When transferred to 43°C, seedlings growing at 24°C lost 70% of their NR activity [48]. Whether high temperature causes inactivation, decreased synthesis, or increased degradation of NR remains to be investigated.

Among the enzymes of ammonium assimilation, GS appears to be sensitive to higher temperatures, whereas GDH is comparatively heat stable (50 to 70°C) in many plant species [83]. Stability of GDH at higher temperatures appears to be of adaptational significance for plants growing at elevated temperatures, as such plants may possibly assimilate NH_4^+ by GDH pathway instead of the normal GOGAT/GS pathway.

2. Low Temperature

Low-temperature treatment of plants below optimum growth temperature reduced N uptake [57, 120,121], decreased N partitioning in the young shoots [122], induced remobilization of N from older leaves to younger ones [122], and adversely affected the process of N assimilation [123].

In *Lolium multiflorum* and *L. perenne* grasses, a decrease in the rate of NO_3^- uptake was observed due to the short-term exposure of the roots to low-temperature treatment by decreasing the temperature from 25°C to 15°C [57]. A similar decrease in NO_3^- uptake by *Cicer arietinum* plants was observed at 16°C soil temperature compared to 22°C [120]. Macduff and Jackson [121] observed that when the root temperature of barley plants was lowered by 3°C (maintaining a common day/night air temperature of 25/15°C), NO_3^- uptake by the roots decreased with a concomitant decrease in the total-N content of the plant. In barley plants, at all different temperatures tested, NH_4^+ uptake was more than NO_3^- uptake [121]. At low root temperatures, NH_4^+ is regarded as a safe source of N, whereas it appears to be harmful at higher temperatures [50]. Decreased uptake of NO_3^- with lowering temperature indicates that NO_3^- uptake is sensitive to temperature.

Low root temperatures drastically affect the partitioning of N within the whole plant [122]. Walsh and Layzell [122] reported that when 35-day-old soybean plants were exposed to a 15°C temperature for 4 days, N partitioning in the young shoots decreased 52 to 61% compared to control plants grown at 25°C. In treated plants, mature leaves maintained a nitrogen level similar to that of the controls. In another experiment, Rufty et al. [106] observed similar N partitioning pattern in soybean plants when roots were treated with low temperature. Besides reduced N uptake as well as disproportionate partitioning of N, low-temperature treatment of roots causes remobilization of N from the older leaves to the young shoots. Walsh and Layzell [122] observed about 22% remobilization of N from mature leaves of soybean plants by day 11 of 15°C temperature treatment compared to N present in leaves on day 4 of treatment. It appears that the remobilized N from older leaves supports growth of the new shoots under low-temperature stress conditions. Increased remobilization of N to the new shoots and proportionally less N partitioning indicates that cold-tolerant cultivars possess increased partitioning of N in the shoots. This also suggests that tolerance to low temperature can be increased by increasing the N supply to young shoots [122].

Low-temperature treatment of roots decreased the rate of NO_3^- flux to the leaves and, in turn, decreased NR activity [49]. When grown at 20°C for 7 days, barley and maize seedlings show a drastic reduction in NR activity compared to seedlings grown at 28°C [116]. Since NR is a substrate-inducible enzyme, the level of NO_3^- in the active pool has a major role in regulating leaf NR activity. Decreased NO_3^- uptake by the roots as a result of the chilling treatment would ultimately lead to a decrease in NR activity. However, contrary to this in certain cases, an increased rate of ion uptake and root NR activity has been observed at a low temperature [123]. Vogel and Dawson [123] reported that when 2-week-old black alder (*Alnus glutinosa*) seedlings were exposed to chilling temperatures of −1 to 4°C for 2 h during the night, immediately after chilling in vivo, NR activities of the roots and shoots increased significantly compared to activities in prechilled plants. The reason for the apparent increase in NR activity following chilling appears to be due to the increased activity of constitutive NR enzyme, which is reported to be present in many N_2-fixing plants.

E. Metal Toxicity

Heavy metals such as Cd^{2+}, Zn^{2+}, Cu^{2+}, Pb^{2+}, and Al^{3+} are major environmental pollutants which spread to the soil by sewage sludge, waste disposal practices, or via airborne pollution.

These elements cause plant growth to deteriorate, reduced NO_3^- uptake by plants, and have direct inhibitory effects on the N assimilatory enzymes. In corn seedlings, a direct adverse effect of cadmium on NO_3^- uptake was reported by Volk and Jackson [124]. Industrial areas in many countries which are polluted with the heavy metals show a reduced N concentration in leaves of plants growing in such areas. While investigating the effects of pollutants in two industrialized belts of Sweden, Pahlsson [125] observed reduced N content in the leaves of polluted trees compared to those growing in nonpolluted areas. This investigator [125] suggested that the elevated level of heavy metals in the soil has a direct deteriorative effect on growth of the finer roots and root hairs of the plants, contributing to reduced N uptake from the soil. Nitrogen deficiency occurs in plants growing under a high level of heavy metals in the soil, which also results in disturbed carbohydrate metabolism, total carbohydrate level, and especially, starch and sucrose increase in the leaves of such plants [125].

Indiscriminate use of acid-forming nitrogenous fertilizers causes acidity in the soil. In acid soils below pH 5.0, Al^{3+} toxicity is a major problem. In such soils the NH_4^+ form of N predominates and NO_3^- availability is limited. Uptake of many essential nutrients, including NO_3^-, is reduced due to NH_4^+ and Al^{3+} toxicity in the soil. It has been suggested that plants differ in their sensitivity to Al^{3+} and that Al^{3+}-tolerant plants are characterized by the efficient use of NO_3^- in the presence of NH_4^+. Such plants have the capacity to increase the pH of their growth medium [126]. When genotypes of sorghum plants, differing in Al^{3+} tolerance, were grown with different NO_3^-/NH_4^+ ratios (39:1, 9:1, and 3:1) with O or 300 μM Al^{3+} in the medium, Al^{3+}-sensitive cultivar ICA-Natiama showed a greater reduction in NO_3^- and NH_4^+ uptake than did the Al^{3+}-tolerant cultivar SC-283 when the plants were grown with Al^{3+}, whereas when the plants were grown without Al^{3+}, the sensitive cultivar showed greater NH_4^+ uptake than the tolerant one [127]. This shows that due to Al^{3+} toxicity, uptake of NO_3^- and NH_4^+ is reduced. It is suggested that the difference in NO_3^- and NH_4^+ uptake by plants is associated with changes in solution pH. As long as NH_4^+ is in solution, pH decreases, and it increases when NH_4^+ is depleted from the solution [127].

Among the N assimilatory enzymes, NR is most sensitive to heavy metal toxicity. Cadmium ion (Cd^{2+}), Cu^+, and Pb^{2+} drastically inhibit NR activity. Elemental cadmium (Cd^{2+}) has a strong affinity for $-SH$ groups and thus it inhibits the activity of many enzymes, including NR. Muthuchelian et al. [128] reported that when elioated leaf segments of *Vigna sinensis* were treated with Cd^{2+}, up to 10 μM stimulation in NR activity was observed. Beyond this level Cd^{2+} strongly suppressed NR activity and there was complete inhibition in activity with 1 mM Cd^{2+}. In similar experiments, Cu^{2+} caused 91% inhibition in NR activity. Purified NR preparation from barley seedlings was inhibited up to 80 to 100% with 1 mM Cu^{2+}, Zn^{2+}, and Co^{2+} [129]. In general, the inhibitory effect of Cd^{2+} and Cu^{2+} appears to be due to interference with the sulfhydryl sites of the enzymes [128].

In germinating pea (*Pisum sativum*) seeds, Pb^{2+} retarded the utilization of N reserves from cotyledons and decreased the activities of the N assimilatory enzymes NR, GS, and GDH, whereas NiR remained relatively insensitive [130]. Mittal and Sawhney [130] reported about a 50% depression in NR activity of pea seeds 5 days after germination with a medium containing 1.0 mM Pb^{2+}. Decreased activities of NR, GDH, and aminotransferase in germinating pea seeds disturb the respiratory activity, due to restricted generation of organic acids from amino acids. This would otherwise facilitate operation of the TCA cycle even under the partially anaerobic conditions existing during germination of seeds.

IV. ACCUMULATION OF NITROGENOUS COMPOUNDS IN STRESSED PLANTS

Plants subjected to environmental stresses accumulate a number of soluble nitrogenous compounds. These compounds accumulate in a high concentration and have specific roles in plants under stress conditions. Several investigators have observed the accumulations of these compounds in a variety of plant species [7–18,26–32,34–36,38–44,46,47,55,58–60,63,70,77,85,88,91). Several review articles have also been published dealing with the effects of various stresses on the accumulation and metabolism of these compounds (20,25,33,34,131–133).

The soluble nitrogenous compounds that accumulate most widely in stressed plants are the amino acids proline, arginine, glycine, serine, alanine, and leucine; the quaternary ammonium compounds glycine betaine, β-alanine betaine, stachydrine, trigonelline, homostachydrine; the amides glutamine and asparagine; the imino acids pipecolic acid and 5-hydroxypipecolic acid; the diamines putrescine, *N*-carbamyl putrescine, and agmatine; and the polyamines spermine and spermidine. For the complete list of these compounds, readers are referred to the recently published review article by Rabe [33]. When subjected to stress, plant species show an accumulation of these compounds, depending on the type of stress, extent of stress, and types of plant species. In most of the stresses, the amino acid proline and the quaternary ammonium compound glycine betaine accumulate, which are regarded as components of a stress tolerance mechanism. These compounds also contribute to osmotic balance in the cytoplasm when electrolytes are lower than vacuoles in cytoplasm and impart a protective role on enzymes in the cytoplasm in the presence of a high level of electrolytes [133].

A. Amino Acids

Plants subjected to most stressful environments show an increased level of total free amino acids [34,131,134]. Proline seems to be the amino acid accumulated in the largest amounts in response to salinity [131,134], drought [135], temperature stress [136], mineral deficiency [132], pathogenesis, and anoxia [137]. In most of the plants studied, salinity and water stresses caused substantial increases in the proline levels of the plant tissues. Proline, together with the other soluble nitrogenous compounds, serves as an osmoregulator in plants. A proline level up to 600 mM did not inhibit enzyme activities [133]. Higher plants differ markedly in their capacity to accumulate proline. When grown in NaCl-free environments, proline-accumulating species contain low level of proline, but the level increases in the presence of salinity [131]. In salt-stressed plants, proline accumulation results from increased synthesis and decreased utilization [131]. As a result of water stress, free proline accumulated appreciably in leaves and other tissues. The functional role of proline accumulation appears to be as a cytoplasmic osmoticum to lower cell water potential, provide hydration to biopolymer, and serve as an energy and N source under adverse environmental conditions [138].

In addition to proline, other amino acids that accumulate under salt and water stress are arginine, glycine, serine, alanine, leucine, and valine. Salt-stressed rice plants accumulated arginine, alanine, leucine, and valine in addition to proline [134]. Under salinity stress conditions, the level of these amino acids is higher in salt-tolerant plants than in salt-sensitive species [134]. Water-stressed plants accumulate proline, alanine, arginine, and phenylalanine, which have a distinct correlation with the stress tolerance mechanism [139]. When grown under low-temperature conditions, crop plants such as barley and radish accumulated a substantial amount of proline [131,136]. Other amino acids, such as serine, glycine, and alanine, also accumulated appreciably in several plant species grown under low-temperature conditions [131]. Barley and radish plants exposed to high temperatures accumulated substantial amounts of proline [136]. In lemon and orange leaves, infection by *Phytophthora* spp. or anaerobiosis conditions caused increased levels of proline, arginine, and total free amino acids [137]. Most of the mineral deficiencies cause an increase in the level of free amino acids. The types of amino acids accumulated depend on the nature of the mineral deficiency. Copper deficiency caused a substantial accumulation of proline and serine in citrus plants [140], whereas an iron deficiency resulted in an accumulation of arginine, lysine, histidine, and serine in citrus and macadamia plants [141]. The basic amino acid arginine has been shown to accumulate under a variety of stress conditions, such as Mg, K, S, Ca, Fe, Mn, and Zn deficiencies, osmotic stress, acid stress, excess ammonium in the growth medium, and infection by the pathogens [34].

B.   Quaternary Ammonium Compounds

Among quaternary ammonium compounds, glycine betaine (trimethylammonio-2-acetic acid) accumulates most widely in stressed plants. It is the most predominant nitrogenous compound accumulating under salinity stress [142]. Together with proline, glycine betaine serves as a compatible cytoplasmic solute and has an important role as an osmoregulator in salinity stress. Another compound, β-alanine betaine (trimethyl ammonio-3-propanoic acid), also accumulates in laboratory-grown salinized plants and in plants growing in saline habitats. Glycine betaine accumulated in many species, whereas β-alanine betaine was restricted to halophytes of *Plumbaginaceae* [131]. Plants differ in their capacity to accumulate betaine. Certain other quaternary ammonium compounds, such as stachydrine, homostachydrine, and trigonelline, were accumulated in alfalfa plants in response to water stress, salinity, and abscisic acid treatments, respectively [143].

C.   Amides and Imino Acids

Glutamine and asparagine, together with amino acids and certain imino acids such as pipecolic acid and 5-hydroxypipecolic acid, accumulate in plants subjected to saline stress; however, their levels are much lower than those of proline or betaine. In plants such as *Agrostis stolonifera*, asparagine accumulation was greater than proline under a high salinity level [131]. Accumulation of amides may sometimes exceed the proline in certain species during water stress [131]. Mineral deficiencies of K^+, P, Mg^{2+}, S, Cl^-, and Zn^{2+} caused accumulation of asparagine and glutamine along with the basic amino acid arginine [34]. Deficiencies of many micronutrients cause an accumulation of asparagine. In certain cases (i.e., Zn^{2+} deficiency) asparagine accumulation may occur up to 50-fold [131]. The imino acid pipecolic acid accumulated appreciably in Mg^{2+}-, Cl^--, K^+-, and Fe^{2+}-deficient plants [34,131]. Saline conditions also favor the accumulation of pipecolic acid. Hydroxypipecolic acid accumulated in *Limonium* plants subjected to salt stress [131].

D.   Nonprotein Amino Acids

Among nonprotein amino acids, citrulline, ornithine, and γ-aminobutyric acid accumulate in plants under certain stresses. Deficiency of mineral nutrient elements such as K^+ in *Sesamum* and P in *Citrus* caused the accumulation of citrulline and ornithine [34]. Barley plants subjected to water stress accumulated appreciable levels of ornithine in the leaves [131]. In conditions of anaerobiosis, the most striking response was the accumulation of γ-aminobutyric acid [144]. Following anaerobiosis, γ-aminobutyric acid accumulated rapidly in leaves, due to increased decarboxylation of glutamate and decreased transamination of 4-aminobutyrate [144]. Accumulation of γ-aminobutyric acid has also been observed in copper-deficient citrus plants [140] and in the leaves of tomato plants infected with tobacco mosaic virus [145].

E.   DI and Polyamines

The accumulation of diamine putrescine and stimulation of its biosynthetic enzyme arginine decarboxylase have been reported in several forms of environmental stresses: for example, nutrient deficiencies such as K^+ and Mg^{2+} [146] and Ca^{2+} [65], salinity stress [65,147], drought [135], ammonium toxicity [148], SO_2 fumigation [149], and acid stress [150]. It has been suggested that putrescine accumulation under stress conditions has adaptive significance, as it serves as an organic ion and can compensate partly for K^+ in K^+-deficient plants [131]. Many plants accumulated the polyamines spermine, spermidine, and agmatine under a deficiency of K^+ [146], P [151], Mg^{2+} [65], S, Ca^{2+}, and Mn^{2+} [151]; salinity stress [147], acid stress [142]; and ammonium toxicity [148]. Concentrations of these amines are very low in nonstressed plants, but stress conditions induce a severalfold increase in their level. Accumulation of the type of polyamines depends on the type of stress as well as the plant species. In detached oat leaves, osmotic treatment induced a rise in the level of putrescine and stimulation of arginine decarboxylase activity, whereas other species showed an increase in the levels of spermidine and spermine and a decline of putrescine as well as its biosynthetic enzymes [150]. Smith and Sinclair [150] suggested that changes in the level of putrescine under stress conditions might be important in regulating the ionic environment within the cell.

It appears from the preceding discussion that various environmental stresses induce the accumulation of soluble nitrogenous compounds. The extent and nature of the compounds accumulated depend on the type of stress and the plant species. The levels of accumulation of some of these compounds (i.e., amino acids, betaine) are associated with stress sensitivity or tolerance of the plant species. For instance, salt-tolerant plants possess in-built higher levels of amino acid proline [131], glycine betaine [142], and β-alanine betaine [131]. The accumulation of these compounds is greater in tolerant plants than in sensitive ones. The sensitive species have a low level of these compounds in the nonstressed plants and show less accumulation under salinization [131]. The functions of these nitrogenous compounds are diverse. Amino acids and betaine accumulating under salt and water stresses serve as osmoregulators, protect biomolecules, decrease the water potential of the cytoplasm, and improve moisture uptake. The accumulation of amides and the amino acid arginine appears to have a potential role in detoxifying ammonia, which attains elevated levels during mineral deficiencies, water stress, low-temperature stress, and so on. Rabe [34] has advocated that most of the nitrogenous compounds that accumulate during the environmental stresses serve to detoxify the cell of ammonia.

V. CONCLUDING REMARKS

Nitrogen is one of the most essential elements in plant nutrition; however, its availability is limited under harsh environmental conditions of salinity, water deficit, extremes of temperature, metal toxicities, and others. These stresses reduce NO_3^- uptake, metabolism, and protein synthesis considerably, and drastically affect crop yields. The process of NO_3^- uptake, its translocation inside and between the cells, and its reduction are coordinately regulated. The pronounced effects of most of the stresses include decreased NO_3^- uptake and inhibition in the activity of the key enzyme of nitrate assimilation, NR. The NR is NO_3^- inducible and its activity is subject to regulation by a variety of environmental parameters which are readily influenced under stresses. The levels of NO_3^- and NR inside the tissues are directly related to plant health and yield. Genotypes of plants differing in stress tolerance show different behaviors of NR and other N-assimilatory enzymes. For instance, a salt-tolerant variety shows an increased NO_3^- uptake and a high level of NR, which is further stimulated under salinization, whereas a salt-sensitive variety shows a decreased NO_3^- uptake and a decreased activity of NR under salinization. This suggests that stress tolerance and sensitivity are complex phenomena that depend on the genetic and biochemical makeup of the species.

Over the last decades, despite extensive studies, our knowledge regarding the biochemical mechanisms underlying the uptake of NO_3^- by plants, the process of its assimilation, and the regulation of enzymes of NO_3^- assimilation is still incomplete. Little information is available regarding molecular events of NO_3^- uptake, NO_3^- sensor protein system, signal transduction of environmental NO_3^-, NO_3^- induction regulatory proteins, primary responsive genes which are transcribed and translated as a result of NO_3^- induction, and other factors. In addition, the nature of NO_3^- transporters, NO_3^- translocaters, events involving overall induction of NR by NO_3^-, regulation of NR and other enzymes of NO_3^-, and NH_4^+ assimilation under various environmental conditions (e.g., light, water, temperature, salinity stresses) need to be examined in detail.

Although all the adverse environmental conditions so far discussed reduce NO_3^- uptake and inhibit NR activity, the precise biochemical mechanisms involved in these events remain to be investigated. More extensive investigations are required to unveil the role of light and other factors in the regulation of NR level in plants. Environmental stresses adversely affect the behavior of enzymes of NO_3^- and NH_4^+ assimilation. In certain cases, such as salinity and water stresses, suppression of the GS/GOGAT pathway and a sustained level of induction of the GDH pathway of ammonium assimilation is observed. Uptake and metabolism of N is so triggered under stressed environments that specific soluble nitrogenous compounds accumulate which provide adaptive value to the plants. Further information is required regarding molecular structures and catalytic properties of enzymes involved in NO_3^- and NH_4^+ assimilation. The precise effects of various environmental conditions both in vivo and in vitro on the behavior of these enzymes, the sequences of events leading to an accumulation of nitrogenous compounds, and the functional roles of these compounds in stressed plants need to be studied in greater detail.

REFERENCES

1. M. G. Redinbaugh and W. H. Campbell, *Physiol. Plant.*, *82*: 640 (1991).
2. C. M. Larsson and B. Ingemarsson, in *Molecular and Genetic Aspects of Nitrate Assimilation* (J. R. Kinghorn and J. L. Wray, eds.), Oxford Science Publishers, New York, p. 4 (1989).
3. W. H. Campbell, *Physiol. Plant.*, *74*: 214 (1988).
4. J. L. Wray, in *Molecular and Genetic Aspects of Nitrate Assimilation* (J. A. Kinghorn and J. L. Wray, eds.), Oxford Science Publications, New York, p. 244 (1989).
5. S. M. Alam, in *Handbook of Plant and Crop Stress* (M. Pessarakli, ed.), Marcel Dekker, New York, pp. 227–246 (1994).
6. S. R. Grattan and C. M. Grieve, in *Handbook of Plant and Crop Stress* (M. Pessarakli, ed.), Marcel Dekker, New York, pp. 203–226 (1994).
7. M. Pessarakli, *Crop Sci.*, *31*(6): 1633 (1991).
8. M. Pessarakli, J. T. Huber, and T. C. Tucker, *J. Plant Nutr.*, *12*(3): 279 (1989).
9. M. Pessarakli and T. C. Tucker, *Soil Sci. Soc. Am. J.*, *52*(6): 1673 (1988).
10. M. Pessarakli and T. C. Tucker, *Soil Sci. Soc. Am. J.*, *52*(3): 698 (1988).
11. M. Pessarakli and T. C. Tucker, *Soil Sci. Soc. Am. J.*, *49*(1): 149 (1985).
12. G. Palfi, *Plant Soil*, *22*: 127 (1965).
13. V. Hernando, L. Jimeno, and C. Cadahia, *An. Edafol. Agrobiol.*, *26*: 1147 (1967).
14. T. S. Mahajan and K. R. Sonar, *J. Maharashtra Agric. Univ.*, *5*: 110 (1980).
15. J. N. E. Frota and T. C. Tucker, *Soil Sci. Soc. Am. J.*, *42*: 753 (1978).
16. R. Saad, Ph.D. dissertation, University of Arizona; University Microfilms, Ann Arbor, Mich.; *Diss. Abstr. B*, *40*: 4057 (1979).
17. M. Pessarakli, Ph.D. dissertation, University of Arizona; University Microfilms, Ann Arbor, Mich.; *Diss. Abstr. B*, *42*: 286 (1981).
18. G. W. Langdale and J. R. Thomas, *Agron. J.*, *63*: 708 (1971).
19. A. A. Luque and F. T. Bingham, *Plant Soil*, *63*: 227 (1981).
20. M. Pessarakli, in *Handbook of Plant and Crop Stress* (M. Pessarakli, ed.), Marcel Dekker, New York, pp. 415–430 (1994).

21. C. Torres and F. T. Bingham, *Proc. Soil Sci. Soc. Am.*, *37*(5): 711 (1973).
22. H. S. Srivastava, *Phytochemistry*, *19*: 725 (1980).
23. U. K. Pandey and R. D. L. Srivastava, *Indian J. Plant Physiol.*, *32*: 175 (1989).
24. S. Katiyar and R. S. Dubey, *J. Agron. Crop Sci.*, *169*: 289 (1992).
25. R. S. Dubey, in *Handbook of Plant and Crop Stress* (M. Pessarakli, ed.), Marcel Dekker, New York, pp. 277–299 (1994).
26. R. S. Dubey and M. Rani, *J. Agron. Crop Sci.*, *162*(2): 97 (1989).
27. A. S. Gupta (A. Sen-Gupta), *Diss. Abstr. Int. B Sci. Eng.*, *47*(12): I 4728B (1987).
28. R. Krishnamurthy and K. A. Bhagwat, *Indian J. Exp. Biol.*, *27*(12): 1064 (1989).
29. M. Pessarakli and J. T. Huber, *J. Plant Nutr.*, *14*(13): 283 (1991).
30. M. Pessarakli, J. T. Huber, and T. C. Tucker, *J. Plant Nutr.*, *12*(11): 1361 (1989).
31. M. Pessarakli, J. T. Huber, and T. C. Tucker, *J. Plant Nutr.*, *12*(10): 1105 (1989).
32. M. Pessarakli and T. C. Tucker, *J. Plant Nutr.*, *8*(11): 1025 (1985).
33. E. Rabe, *Handbook of Plant and Crop Stress* (M. Pessarakli, ed.), Marcel Dekker, New York, pp. 261–276 (1994).
34. E. Rabe, *J. Hortic. Sci.*, *65*(3): 231 (1990).
35. S. Ramagopal, *Plant Cell Rep.*, *5*(6): 430 (1986).
36. J. N. E. Frota and T. C. Tucker, *Soil Sci. Soc. Am. J.*, *42*: 743 (1978).
37. A. Golan-Goldhirsh, B. Hankamer, and S. H. Lips, *Plant Sci.* (*Limerick*), *69*(1): 27 (1990).
38. S. M. Abdul-Kadir and G. M. Paulsen, *J. Plant Nutr.*, *5*: 1141 (1982).
39. G. V. Udovenko, V. N. Sinel'nikova, and G. V. Khazova, *Dokl. Akad. Nauk. USSR*, *192*(6): 1395 (1970).
40. G. V. Udovenko, V. N. Sinel'nikova, and G. V. Khazova, *Agrokhimiya*, *3*: 23 (1971).
41. H. M. Helal and K. Mengel, *Plant Soil*, *51*: 457 (1979).
42. I. Kahane and A. Poljakoff-Mayber, *Plant Physiol.*, *43*: 1115 (1968).
43. R. Krishnamurthy, M. Anbazhagan, and K. A. Bhagwat, *Indian J. Plant Physiol.*, *30*(2): 183 (1987).
44. J. Wieneke and R. Fritz, *Acta Univ. Agric. Brno Fac. Agron.*, *33*(3): 653 (1985).
45. C. A. Morilla, J. S. Boyer, and R. H. Hageman, *Plant Physiol.*, *51*: 817 (1973).
46. J. T. Prisco and J. W. O'Leary, *Rev. Brazil Biol.*, *30*: 317 (1970).
47. N. M. Iraki, R. A. Bressan, and N. C. Carpita, *Plant Physiol.*, *91*(1): 54 (1989).
48. P. Singh and S. K. Sawhney, in *Recent Advances in Plant Biochemistry* (S. L. Mehta, M. L. Lodha, and P. V. Sane, eds.), ICAR Publications, New Delhi, India, p. 141 (1989).
49. L. Beevers and R. H. Hageman, in *The Biochemistry of Plants*, Vol. 5 (B. J. Miflin, ed.), Academic Press, New York, p. 115 (1980).
50. U. Kafkafi, *J. Plant Nutr.*, *13*: 1291 (1990).
51. C. A. Atkins, *Proceedings of the International Congress of Plant Physiology*, New Delhi, India, pp. 1022–1026 (1990).
52. A. D. M. Glass, M. Y. Siddiqui, T. J. Ruth, and T. W. Rufty, *Plant Physiol.*, *93*: 1585 (1990).
53. M. Ni and L. Beevers, *J. Exp. Bot.*, *41*: 987 (1990).
54. W. A. Jackson, W. L. Pan, R. H. Moll, and E. J. Kamprath, in *Biochemical Basis of Plant Breeding*, Vol. 2 (C. Neyra, ed.), CRC Press, Boca Raton, Fla., p. 73 (1986).
55. M. R. Ward, M. Aslam, and R. C. Huffaker, *Plant Physiol.*, *80*: 520 (1986).
56. A. Benzioni, Y. Vaadia, and S. H. Lips, *Physiol. Plant.*, *24*: 288 (1971).
57. D. T. Clarkson, in *Fundamental, Ecological and Agricultural Aspects of Nitrogen Metabolism in Higher Plants* (H. Lambers, J. J. Neeteson, and I. Stulen, eds.), Martinus Nijhoff, Dordrecht, The Netherlands p. 3 (1988).
58. M. Aslam, R. C. Huffakar, and D. W. Rains, *Plant Physiol.*, *76*: 321 (1984).
59. A. Soltani, M. Hajji, and C. Grignon, *Agronomie*, *10*: 857 (1990).
60. Y. I. Mladenova, "Influence of Salt Stress on Primary Metabolism of *Zea mays* L. seedlings of Model Genotypes," *Proceedings of the 3rd International Symposium on Genetic Aspects of Plant Mineral Nutrition*, Braunschweig, Germany; *Plant Soil*, *123*: 217 (1990).
61. S. K. Sharma and I. C. Gupta, *Saline Environment and Plant Growth*, Agro Botanical Publishers, India, p. 92 (1986).
62. M. Pessarakli, T. C. Tucker, and K. Nakabayashi, *J. Plant Nutr.*, *14*(4): 331 (1991).

63. M. A. Khalil, A. Fathi, and M. M. Elgabaly, *Soil Sci. Soc. Am. Proc.*, *31*: 683 (1967).
64. T. A. Smith, *Phytochemistry*, *12*: 2093 (1973).
65. T. J. Flowers, P. F. Troke, and A. R. Yeo, *Annu. Rev. Plant Physiol.*, *28*: 89 (1977).
66. R. G. Butz and W. A. Jackson, *Phytochemistry*, *16*: 409 (1977).
67. T. V. R. Nair, S. R. Chatterjee, and Y. P. Abrol, *Plant Physiol. Biochem.*, *10(s)*: 176 (1983).
68. Z. Plaut, *Physiol. Plant*, *30*: 212 (1974).
69. S. M. Abdul-Kadir and G. M. Paulsen, *J. Plant Nutr.*, *5*: 1141 (1982).
70. R. K. Lal and S. N. Bhardwaj, *Indian J. Plant Physiol.*, *30*: 165 (1987).
71. R. S. Dubey, S. Katiyar, and R. Mittal, in *Proceedings of the International Conference of Plant Physiology*, Banaras Hindu University, Varanasi, India, pp. 189–194 (1991).
72. G. R. Smith and K. R. Middleton, *New Physiol.*, *84*: 613 (1980).
73. T. N. Tewari and B. E. Singh, *Plant Soil*, *136*: 225 (1991).
74. R. Krishnamurthy, M. Anbazhgan, and K. A. Bhagwat, *Curr. Sci.*, *56*: 489 (1987).
75. M. Lacuesta, B. Gonzalez-More, C. Gonzalez-Marue, and A. Munoz-Rueda, *J. Plant Physiol.*, *136*: 410 (1990).
76. M. Blom-Zandstra, J. E. M. Lampe, and F. H. M. Ammerlaan, *Physiol. Plant.*, *74*: 147 (1988).
77. S. Joshi, *Indian J. Plant Physiol.*, *30*: 223 (1987).
78. N. Sankhla and W. Huber, *Z. Pflanzenphysiol.*, *76*: 467 (1975).
79. M. de L. Miranda-Ham and V. M. Loyola-Vargas, *Plant Cell Physiol.*, *29*: 747 (1988).
80. S. Katiyar, Ph.D. dissertation, Banaras Hindu University, Varanasi, India (1990).
81. J. Boucaud and J. B. Billard, *Physiol. Plant.*, *44*: 31 (1978).
82. G. G. Rao, J. K. Ramiah, and G. R. Rao, *Indian J. Exp. Biol.*, *19*: 771 (1981).
83. S. K. Sharma and O. P. Garg, *Indian J. Plant Physiol.*, *38*: 407 (1985).
84. H. S. Srivastava and R. P. Singh, *Phytochemistry*, *21*: 997 (1982).
85. M. Singh, B. B. Singh, and P. C. Ram, *Biol. Plant.*, *32*: 232 (1990).
86. S. Siddiqui, S. Kumar, and H. R. Sharma, *Indian J. Plant Physiol.*, *28*: 369 (1985).
87. N. Garg, I. S. Dua, S. K. Sharma, and O. P. Garg, *Res. Bull. Pubjab Univ. (India)*, *39*: 187 (1988).
88. M. Khavan-Kharazian, W. F. Campbell, J. J. Jurinak, and L.M. Dudley, *Arid Soil Res. Rehabil.*, *5*: 97 (1991).
89. J. Levitt, *Plant Responses to Environmental Stress*, Vol. 2, Academic Press, New York (1980).
90. J. A. Morgan, *Plant Physiol.*, *76*: 112 (1984).
91. S. Kathju, S. P. Vyas, B. K. Garg, and A. N. Lahiri, "Fertility Induced Improvement in Performance and Metabolism of Wheat under Different Intensities of Water Stress" *Proceedings of the International Congress of Plant Physiology*, *'88*, New Delhi, India, pp. 854–858 (1990).
92. S. K. Sinha and D. J. Nicholas, in *Physiology and Biochemistry of Drought Resistance in Plants* (L. G. Paleg and D. Aspinall, eds.), Academic Press, New York, p. 145 (1981).
93. D. L. Shaner and J. S. Boyer, *Plant Physiol.*, *58*: 505 (1976).
94. F. G. J. Viets, *Irrigation of Agriculture Lands* (R. M. Hagan et al., eds.), *Agronomy*, Vol. 11, American Society Agronomy, Madison, Wis., p. 458 (1967).
95. A. C. S. Rao and B. Ramamoorthy, *Indian J. Plant Physiol.*, *23*: 269 (1980).
96. A. N. Lahiri, in *Adaptation of Plants to Water and Higher Temperature Stress* (N. C. Turner and P. J. Kramer, eds.), Wiley, New York, p. 341 (1980).
97. D. G. Rao and V. Balasubramanian, *Indian J. Plant Physiol.*, *29*: 61 (1986).
98. K. G. Wasnik, P. B. Varade, and A. K. Bagga, *Indian J. Plant Physiol.*, *31*: 324 (1988).
99. G. A. Morilla, J. S. Boyer, and R. N. Hageman, *Plant Physiol.*, *51*: 817 (1973).
100. D. Balasimha, *Plant Physiol. Biochem.*, *10*: 69 (1983).
101. S. P. Vyas, B. K. Garg, S. Kathju, and A. N. Lahiri, in *Proceedings of the International Congress of Plant Physiology*, New Delhi, India, pp. 880–884 (1990).
102. R. Sairam, P. S. Deshmukh, and D. S. Shukla, *Anal. Plant Physiol.*, *3*: 98 (1989).
103. A. D. Hanson and W. D. Hitz, *Annu. Rev. Plant Physiol.*, *33*: 163 (1982).
104. R. G. Groat and C. P. Vance, *Plant Physiol.*, *67*: 1198 (1981).
105. R. K. Koundal and R. K. Chopra, *Biochem. Physiol. Pflanzen.*, *163*: 69 (1989).
106. T. W. Rufty, C. D. Raper, and W. A. Jackson, *New Phytol.*, *88*: 607 (1981).
107. C. D. Raper, Jr., D. L. Osmond, M. Wann, and W. W. Weeks, *Bot. Gaz.*, *139*: 289 (1978).

108. C. D. Raper, Jr., J. K. Vessey, L. T. Henry, and S. Chaillou, *Plant Physiol. Biochem.*, *29*: 205 (1991).
109. C. R. Clement, M. J. Hooper, L. H. P. Jones, and E. L. Leafe, *J. Exp. Bot.*, *29*: 1173 (1978).
110. R. C. Granstedt and R. C. Huffaker, *Plant Physiol.*, *70*: 410 (1982).
111. M. Aslam, A. Oaks, and R. C. Huffaker, *Plant Physiol.*, *58*: 588 (1976).
112. L. Beevers and R. H. Hageman, *Photophysiology* (A. G. Giese, ed.), Vol. VIII, Academic Press, New York, p. 85 (1972).
113. A. K. Sharma and S. K. Sopory, *Plant Physiol. Biochem.*, *15*: 107 (1988).
114. S. K. Sawhney and M. S. Naik, *Biochem. J.*, *130*: 475 (1972).
115. R. Polisetty and R. H. Hageman, *Indian J. Plant Physiol.*, *32*: 359 (1989).
116. A. Oaks, X. He, and M. Zoumadakis, "Nitrogen Use Efficiency in C_3 and C_4 Cereals," *Proceedings of the International Congress of Plant Physiology '88*, New Delhi, India, pp. 1038–1045 (1990).
117. T. Tashiro and I. F. Wardlaw, *Aust. J. Plant Physiol.*, *18*: 259 (1991).
118. S. Yoshida and T. Hara, *Soil Sci. Plant Nutr.*, *23*: 93 (1977).
119. F. Y. Hafeez, S. Asad, and K. A. Malik, *Environ. Exp. Bot.*, *31*: 285 (1991).
120. R. Chandra and R. P. Pareck, *Legume Res.*, *13*: 95 (1990).
121. J. H. Macduff and S. B. Jackson, *J. Exp. Bot.*, *41*: 237 (1991).
122. K. B. Walsh and D. B. Layzell, *Plant Physiol.*, *0*: 249 (1986).
123. C. S. Vogel and J. S. Dawson, *Physiol. Plant.*, *85*: 551 (1991).
124. R. J. Volk and W. A. Jackson, *Environ. Health Perspect.*, *4*: 103 (1973).
125. A. M. Balsberg Pahlsson, *Can. J. Bot.*, *67*: 2106 (1989).
126. C. D. Foy, R. L. Chaney, and M. C. White, *Annu. Rev. Plant Physiol.*, *29*: 511 (1978).
127. L. Galvez and R. B. Clark, *Plant Soil*, *134*: 179 (1991).
128. K. Muthuchelian, S. N. V. Rani, and K. Paliwal, *Indian J. Plant Physiol.*, *31*: 169 (1988).
129. Y. Oji, T. Hamano, Y. Ryema, Y. Mike, N. Wakichi, and S. Okamoto, *J. Plant Physiol.*, *119*: 247 (1985).
130. S. Mittal and S. K. Sawhney, *Plant Physiol. Biochem.*, *17*: 75 (1990).
131. G. R. Stewart and F. Larher, in *The Biochemistry of Plants* (B. J. Miflin, ed.), Academic Press, New York, p. 609 (1980).
132. G. R. Stewart, in *Physiology and Biochemistry of Drought Resistance in Plants* (L. G. Paleg and D. Aspinall, eds.), Academic Press, New York, p. 243 (1981).
133. H. Greenway and R. Munns, *Annu. Rev. Plant Physiol.*, *31*: 149 (1980).
134. R. S. Dubey and M. Rani, *J. Agron. Crop Sci.*, *162*: 97 (1989).
135. N. N. Barnett and A. W. Naylor, *Plant Physiol.*, *41*: 1222 (1966).
136. T. M. Chu, D. Aspinall, and L. G. Paleg, *Aust. J. Plant Physiol.*, *1*: 87 (1974).
137. C. A. Labaanuskas, L. H. Stolzy, and M. F. Handy, *J. Am. Soc. Hortic. Sci.*, *99*: 497 (1974).
138. D. Aspinall and L. G. Paleg, in *Physiology and Biochemistry of Drought Resistance in Plants* (L. G. Paleg and D. Aspinall, eds.), Academic Press, New York, p. 205 (1981).
139. V. K. Rai, G. Singh, P. S. Thakur, and S. Banyal, *Plant Physiol. Biochem.*, *10*: 161 (1983).
140. I. Stewart, *Proc. Am. Soc. Hortic. Sci.*, *81*: 244 (1962).
141. I. N. Gilfillan and W. W. Jones, *Proc. Am. Soc. Hortic. Sci.*, *93*: 210 (1968).
142. R. Storey and R. G. Wyn-Jones, *Plant Sci. Lett.*, *4*: 161 (1975).
143. G. Parameshwara, L. G. Paleg, D. Aspinall, and G. P. Jones, in *Proceedings of the International Congress of Plant Physiology*, New Delhi, India, pp. 1014–1021 (1990).
144. J. G. Streeter and J. F. Thompson, *Plant Physiol.*, *49*: 572 (1972).
145. P. Cooper and I. W. Selman, *Ann. Bot.*, *38*: 625 (1974).
146. L. C. Basso and T. C. Smith, *Phytochemistry*, *13*: 875 (1974).
147. S. Katiyar and R. S. Dubey, *Trop. Sci.*, *30*: 229 (1990).
148. H. Klein, A. Priebe, and H. J. Jager, *Z. Pflanzenschutz*, *83*: 555 (1979).
149. A. Priebe, H. Klein, and H. J. Jager, *J. Exp. Bot.*, *29*: 1045 (1978).
150. T. A. Smith and C. Sinclair, *Ann. Bot.*, *31*: 103 (1967).
151. C. Sinclair, *Nature*, *213*: 214 (1967).

29
Protein Synthetic Responses to Environmental Stresses

Timothy S. Artlip
Cornell University, Ithaca, New York

Edward A. Funkhouser
Texas A&M University, College Station, Texas

I. INTRODUCTION

The responses of plants and crops to environmental stresses are varied but generally involve some alteration in protein synthesis. Protein-based responses include overall changes in protein synthesis or, of particular interest, changes in the level of specific proteins. These changes depend on the nature, duration, and severity of the stress and are manifested by increases or decreases in an already existing pool of protein(s) or the *de novo* appearance of protein(s). The primary focus of this chapter will be on proteins that are induced or increase in abundance in response to an environmental stress. While the physiology behind decreases of specific proteins is of interest, most research has centered on inducible proteins on the assumption that they confer enhanced survival to the organism.

Two broad areas of stress are considered: biotic (pathogen/herbivore) and abiotic (physical environment). Each confronts the plant with a particular set of challenges. In general, biotic stresses engender an active defensive response and may cause modification of the plant's developmental plan. Abiotic stresses also call on the plant to respond with some sort of defense response, but chronic exposure typically modifies its developmental pattern in some manner. As will be seen, biotic stresses engender similar responses, frequently in common with abiotic ones. As such, the biotic stresses are considered first.

It must be noted that due to the rapid, continuing progress in the field, a comprehensive treatise on all the relevant literature is beyond the scope of this review. It is hoped that the researchers whose work is not mentioned will not take offense, and that the reader will seek out the appropriate references for further information.

II. BIOTIC STRESS

Among the first symptoms that an incompatible interaction between a plant and a pathogen generates is the hypersensitive response, which is a localized, rapid necrosis of the tissue at the infection site. It is not known whether this response is a symptom or a cause of the changes in metabolism (e.g., protein synthesis). However, Jakobek and Lindgren [1] have shown that the hypersensitive response appears to result from a parallel signaling pathway aside from the induction of various plant defense genes.

Changes in protein synthesis in response to fungal, bacterial, and viral pathogens are similar; that is, many protein products or biosynthetic pathways are induced in common between the pathogens. Several proteins and enzymes that accumulate appear to lignify the cell wall, thus limiting the spread of pathogens. Another general response is the production of nonproteinaceous compounds, the phytoalexins, which appear to act as antibiotics. The compounds are structurally complex, with chemical derivation paralleling the species in which they appear. Specific protein products and biosynthetic pathways may also be mutually induced by a particular pathogen. The incompatible reaction between a plant and pathogen may induce protein products (including enzymes), referred to as pathogenesis-related (PR) proteins. In addition, polypeptides known as thionins and proteinase inhibitors may be induced. Each of these defensive factors is considered below.

A. Cell Wall Modifications

The cell-wall-modifying proteins encompass two groups, proteins that alter the cell wall matrix, and enzymes involved in the deposition of phenolic materials (lignification). The cell-wall-matrix proteins are typically hydroxyproline-rich (extensins) or glycine-rich. They appear to function by providing a framework for the cross-linking of carbohydrate (pectin, cellulose) or polyphenolic (lignin, suberin) moieties [2,3]. These proteins are inducible by ethylene (wounding), fungal elicitors, or viral infection [3]. The cell-wall-modifying enzymes are, in general, peroxidases, which are thought to catalyze the suberization or lignification of cell walls [3]. They are, of course, active in the normal synthesis of cell walls, but are also inducible by fungal elicitors or pathogenic attack [2] and may act in concert with enzymes involved with the biosynthesis of phenolic compounds [3]. The thickening of the cell walls then serves to wall off the pathogen and acts as a deterrent to further invasion.

Phenolic compounds utilized in the modification of cell walls have the same biosynthetic origin as that of the isoflavonoid phytoalexins: the phenylpropanoid pathway. The phenylpropanoid pathway is, in turn, a branch of the shikimic acid pathway, responsible for the synthesis of aromatic amino acids [4]. The precursor to the phenylpropanoid pathway, phenylalanine, is converted to 4-coumaroyl-CoA by the actions of phenylalanine-ammonia lyase (PAL), cinnamate-4-hydroxylase (C4H), and 4-coumaroyl-CoA ligase (4CL) [4]. These enzymes, present during normal development, increase dramatically and rapidly upon exposure to pathogens. This has been shown to occur at the level of both transcription and translation [3]. 4-Coumaroyl-CoA serves as a branch point for the synthesis of lignin as well as of (iso)flavonoids. Hydroxylation of 4-coumaroyl-CoA produces caffeic acid, which is successively modified to make ferulic and sinapic acids and coniferyl alcohol [4]. Peroxidase-catalyzed polymerization of the alcohols corresponding to ferulic and sinapic acids, as well as coniferyl alcohol, gives rise to lignin [3,5]. Cinnamyl alcohol dehydrogenase is integral to this process and has been reported to be induced by fungal elicitors in several systems [6–9] and by ozone in spruce [10]. Inducibility by ozone suggests that the enzyme superoxide dismutase (SOD) may also be involved in this process. One of the products of SOD is H_2O_2, a substrate for peroxidases. Fungal attack or elicitor application also induces SOD in several plant species or cultivars that display the hypersensitive response. Bowler et al. [11, and references therein] hypothesize that SOD activation may aid in the strengthening of cell walls, or possibly kill pathogens directly through H_2O_2. However, evidence does exist that H_2O_2 involved in lignification arose from other sources (reviewed in Ref. 12).

B. Phytoalexins

Phytoalexins have been isolated from a number of plant families and appear to be characteristic for a particular family. For example, the Leguminosae use (iso)flavonoids predominantly, the Solanaceae emphasize sesquiterpenes, and the Umbelliferae utilize coumarin derivates primarily [3]. Phytoalexins are toxic not only to pathogens, but to the plants themselves, and thus may contribute to the necrosis associated with the hypersensitive response [13].

1. (Iso)flavonoid-Derived Phytoalexins

Chalcone synthase (CHS) is the start of the (iso)flavonoid branch of the phenylpropanoid pathway. The enzyme is active in the normal growth, development, and metabolism of plants, and is ultimately reponsible for many of the plant pigments (e.g., anthocyanins) [4]. In Leguminosae, where it has been studied extensively, it is highly inducible by pathogen attack [14]. This is accomplished by the differential regulation of several isozymes, some of which are constitutive while others are specific to pathogenic attack [15]. Although evidence for the specific mechanisms by which isoflavonoid phytoalexins achieve their toxicity is generally lacking, accumulated evidence suggests that they cause dysfunctions in the plasma membrane or tonoplast [16].

2. Coumarin-Derived Phytoalexins

Coumarins are also derivatives of the phenylpropanoid pathway, but the branch mechanisms remain unresolved [9]. Hahlbrock and Scheel [9] suggest that glucosides or glucose esters are key intermediates, providing a means of safely sequestering potentially self-toxic compounds until they are needed. As with the (iso)flavonoid-synthesizing enzymes, coumarin pathway enzymes are also stimulated by elicitor treatment [9].

3. Terpenoid-Derived Phytoalexins

The other major family of phytoalexins derive from the terpenoid biosynthetic pathway. This pathway also operates during normal development and metabolism, producing such compounds as ABA, giberrellins, chlorophyll, carotenoids, and phytosterols [17]. Terpenoids arise via mevalonic acid, with hydroxymethyl glutarate reductase (HMGR) as the putative key regulatory enzyme for the entire pathway. The activity of several critical enzymes in the central pathway, including HMGR, increases as a result of pathogen attack, as evidenced by increases in transcription and translation that accompany the increased activity [17]. Data also exist concerning the *de novo* synthesis of several enzymes specific to terpenoid phytoalexin biosynthesis. For example, farnesyl pyrophosphate transferase and casbene synthase have been shown to be synthesized *de novo* in response to elicitors, leading to the formation of diterpenoid phytoalexins [18,19].

C. Pathogenesis-Related Proteins

Pathogenesis-related proteins are quite diverse, falling into several groups or families of proteins, enumerated as 1, 2a, 2b, 3, 4, 5a, and 5b [20]. The functions of groups 1, 4, and 5b are currently unknown, but it is assumed that they do act in the defense of the plant in some manner. Group 5a has been given a putative role in pathogen resistance, while groups 2 (a and b) and 3 have been identified as β-1,3 glucanases and chitinases, respectively.

1. Thaumatin/Osmotin Protein Family

The group 5a PR proteins are structurally related to the sweet-tasting protein named thaumatin, originally isolated from *Thaumatococcus daniellii* Benth. [21]. Thaumatin, and other members of the group, have sequence homology with the maize trypsin/α-amylase inhibitor family of proteins [22]. Those inhibitors have reactive sites against both serine proteinases (see below) and α-amylases, with varying specificity and activity against those enzymes in insects, animals, and pathogens [23]. They are considered a portion of the defense response of plants. However, the 5a group has been shown to have another defensive capability as well. A well-characterized member of this family of proteins is osmotin, a 26-kDa salinity-inducible protein initially described isolated from cultured tobacco cells [24]. In whole tobacco plants, the expression of this protein is complex. Ethylene and salt induce the accumulation of osmotin in a tissue-specific manner [25]. In contrast, osmotin mRNA, but not protein, is inducible by ABA, wounding, and tobacco mosaic virus [25]. Vigers et al. [26] showed that osmotin and the serologically related proteins zeamatin (from maize) and PR-5 (from tobacco) had antifungal activity, causing the rapid bursting of hyphal tips. This is probably due to a membrane-permeating ability that osmotin, along with a related class of proteins termed permatins, has been shown to have [27]. The evolution of this family or proteins has yet to be determined, but should prove interesting in light of osmotins multiple gene induction.

2. Glucanases/Chitinases

Perhaps the best characterized of the PR proteins are the chitinases and glucanases. Chitinases hydrolyze β-1,4-acetylglucosamine linkages of chitin polymers, which is a primary constituent of fungal walls [28]. While plant secondary cell walls reportedly contain chitin, those linkages resist chitinase, possibly the result of glycolipid modification of the linkages [29]. Chitinases are also induced by pathogens and their elicitors, as well as by abiotic stresses such as heavy metals and salt [30].

Glucanases act by hydrolyzing β-1,3-glucan residues. This type of carbohydrate dominates fungal cell walls along with chitin [31]. Glucanase expression, as determined by mRNA and protein analysis, increases greatly upon exposure of plant tissue to pathogens or fungal elicitors [2]. As with the chitinases, several isoforms of glucanase are known to exist, with their genes demonstrating differential regulation [32–34]. Nonpathogen-induced expression of glucanases have been reported in roots, stems, and flowers [2], with certain isoforms specific to the normal development of the plant [34]. Glucanases are apparently induced in concert with chitinases in a number of species [2]. The combination of the two enzymes has been shown to have direct antifungal properties in vitro [35].

D. Non-PR Proteins

1. Proteinase Inhibitors

While proteinase inhibitors are generally considered to be polypeptides rather than proteins, they are also classified as inducible defense molecules and are worthy of mention. Proteinase inhibitors act against the proteolytic activity of microbe-secreted proteases as well as the proteinases found in the digestive tract of animals, particularly insects. Consequently, animals are not deterred in the short term from consuming plant tissue. However, prolonged exposure to the inhibitors will contribute to starvation of the animal [23].

There are several classes of proteinases, based on mechanism of action [36]. They are named after the active residue or cofactor responsible for the proteolytic cleavage (i.e., serine, cysteine, aspartic, and metalloproteinases). Not surprisingly, there are inhibitors for each of these proteinase classes, and all apparently work by a similar mechanism, competitive inhibition [23].

The predominance of serine proteinases over other proteinases in pathogens and phytophagous animals [36] is reflected by an equal abundance of inhibitors in plants [23]. Cysteine proteinases are also widespread, occurring in bacteria, eukaryotic microorganisms, plants, and animals [36]. In contrast to the serine proteinase inhibitors, the cysteine proteinase inhibitors of plants are less well characterized [23]. Even less is known about the aspartic and metalloproteinases inhibitors. For all inhibitor classes, pathogen attack, fungal elicitors, or wounding appear to be sufficient for their induction [23].

2. Thionins

A second class of polypeptides that have a putative role in plant defense are the thionins, shown to exist in several families of plants [37]. Thionins were originally described as toxic factors in many cereal seeds [38]. However, they have recently been shown also to be synthesized in leaves, with the highest abundance occurring in the epidermis and being localized to the cell wall [38]. Since the cell wall of the epidermis is often the primary site of pathogenic attack, thionins could act as a first line of defense. Thionin expression is regulated both developmentally and by pathogen attack, with pathogenic attack causing enhanced transcription and translation of thionin mRNA [38]. Nonpathogen-induced expression of thionins is limited to the period of growth preceding emergence from the soil, with high mRNA and protein levels seen. Thionin toxicity is hypothesized to arise from an amphipathic structure, which may cause an increase in membrane permeability [37]. It is also believed that self-toxicity to the plant cell is minimized by the presence of a proteolytically cleaved precursor, which may shield the cell's membrane system from the active domains of the thionins [38].

E. Signal Transduction

1. Elicitors

An aspect of alterations in protein synthesis that has been alluded to but not described is that of the signal transduction pathways responsible for the expression of the various pathogen-induced proteins. It is known that specific molecules (i.e., elicitors) are important in these processes. Fragments of fungal cell walls can cause the synthesis of phytoalexins, as can fragments of plant cell walls hydrolyzed by pathogens [39]. It is likely that the plasma membrane contains receptors for these elicitors and that a signal transduction pathway(s) ultimately stimulates synthesis of the appropriate proteins [40].

2. Oxygen Radicals

The enzyme superoxide dismutase (SOD) may also be involved in signal transduction. As noted above, pathogen attack or fungal elicitors induce SOD in several plant species or cultivars that display the hypersensitive response. Bowler et al. [11, and references therein] hypothesize that SOD activation may possibly generate H_2O_2 as a secondary messenger. Ozone induction of the mRNAs for several PR proteins [41] provide evidence of this possibility.

3. Signal Molecules

JASMONIC ACID. Three putative plant-generated signal molecules have recently received much attention. Farmer et al. [42] have shown that proteinase inhibitor genes can be induced by methyl jasmonate or jasmonic acid, with or without wounding. They suggest that wounding may cause the release of the jasmonate precursor linolenic acid. Linolenic acid is converted to jasmonic acid via a series of enzymes, including lipoxygenase [43]. Lipoxygenase has also been shown to be induced by pathogenic attack in *Arabidopsis* [44,45]. Jasmonic acid could then act in some manner to activate proteinase inhibitor expression. Indeed, jasmonic acid has been reported to induce a variety of PR proteins [43].

SYSTEMIN. The second molecule is an 18-amino acid peptide known as systemin, which has been shown to be released in tomato leaves in response to pathogenic attack [46,47]. It arises from a 200-amino acid precursor [47], and acts, in turn, to induce the synthesis of proteinase inhibitors. In contrast to many of the other signal molecules, which act locally, systemin appears to act as a long-range signal.

SALICYLIC ACID. The third molecule is salicylic acid, itself derived from the phenylpropanoid pathway and widespread in plant species [48]. It is apparently inducible by pathogen attack or fungal elicitors and has been shown to be transportable [48]. Exogenous application of salicylic acid induces the synthesis of many of the PR proteins, while endogenous levels are also correlated with PR-protein induction [48]. However, it should be noted that exogenous application of saliyclic acid prevents the induction of the proteinase inhibitor II gene in tomato and potato leaves, apparently by blocking the synthesis of jasmonic acid [49]. This contradictory report indicates that additional work is necessary to unravel the signal transduction pathway(s) with result from pathogenesis.

It is also known that abiotic stresses can induce the expression of some of the defense-related genes. In particular, the enzymes responsible for phenylpropanoid-derived phytoalexins have been shown to be induced by ultraviolet light and heavy metals. As mentioned above, osmotin is induced by salt stress. This suggests that many of the genes involved in pathogen defense may have evolved for a particular purpose, and in the course of evolution, been converted for other purposes, but retain the sequences necessary to respond to the original stimulus.

III. ABIOTIC STRESSES

In contrast to biotic stresses, abiotic challenges are more diverse, engendering a wider array of inducible proteins. In some cases, similar proteins are induced by different stresses. As noted above, biotic and abiotic stresses may induce similar or identical proteins as well. Frequently considered abiotic factors include light, temperature, nutrients, salinity, and air and water pollutants. It is well known that an overabundance or deficit of any of these factors can greatly reduce actual growth and reproduction [50].

A. Heat Stress

Of the abiotic stresses, perhaps the most universal response is to that of elevated temperature. Indeed, many of the protein synthetic responses are found throughout the eukaryotic kingdom, as well as in prokaryotes.

1. Decreased Translation

Quantitatively, a decline in overall protein synthesis occurs, a result of the translational repression of most mRNAs [51]. Many explanations have been suggested, including general instability of the polysomes at high temperature [52], loosely bound translational factors, changes in the cytoskeleton, and inhibition at the initiation or elongation steps [51]. Some of the translationally repressed mRNAs are sequestered during the heat shock and are expressed after the stress is relieved [51].

2. Heat-Shock Proteins/Molecular Chaperones

Qualitatively, several classes of proteins with different molecular masses are rapidly (20 min to 3 h) and preferentially translated [51,53]. These proteins have been termed *heat-shock proteins*

(HSPs) and referred to by their molecular masses: low (LMW, 15 to 30 kDa), which are seen only in plants, and the HSP60, HSP70, HSP90, and HSP110 classes. The appearance of these proteins has been positively correlated with enhanced thermotolerance, as well as some measure of cross protection to other environmental stresses [51,54]. The HSPs are typically seen when tissue temperature exceeds 32 to 33° C [55], regardless of whether the temperature rise is rapid or slow [56], or under field conditions [56,57]. In plants, 27- and 70-kDa proteins produced in response to elevated temperature also appear in response to a variety of environmental stresses [58,59]. It has been speculated that these commonly produced proteins constitute a form of general-purpose stress tolerance. Conversely, other proteins appear which seem to be unique for particular environmental stresses.

The intensive study of heat shock proteins has led to a more general biological concept, that of molecular chaperones. Molecular chaperones are considered to act by assisting the self-assembly of nascent polypeptides into their correctly folded tertiary structures [60,61]. HSPs are believed to function in a similar fashion, preventing the formation or aggregation of nonfunctional proteins resulting from heat denaturation [62]. Homologs of the large MW HSPs have been reported in the cytoplasm, mitochondria, and chloroplasts. The HSP60 class appears to be restricted to mitochondria and chloroplasts, despite its nuclear origin [55]. In contrast, the function of the LMW HSPs is currently unknown, even thought they show a distribution similar to the higher MW HSPs [55].

Evidence supports the hypothesis that control of the synthesis of HSPs exists both transcriptionally and translationally [63]. A model of the transcriptional regulation for HSP70 was presented by Morimoto [64]. HSP70 is thought to bind to its own transcriptional factor, thus blocking transcription. Elevated temperature then provides a pool of aberrant proteins, which competes with the transcription factor for HSP70. The factor then trimerizes, allowing binding to the appropriate promoter element, followed by phosphorylation. The increased levels of HSP70 eventually lead to the binding of the trimer by HSP70, release from the promoter element, and monomerization of the factor. In concert with this, HSP mRNA is stabilized at elevated temperatures and is efficiently translated, unlike most other mRNAs [63].

3. Other Inducible Proteins

It must be noted that other proteins or mRNAs also increase in abundance during elevated temperature but are not typically considered HSPs. They include several glycolytic enzymes [63], protein kinases [65], and ubiquitin [63,66,67]. Vierling [55] suggests that the glycolytic enzymes and protein kinases may be involved with metabolic readjustment. The activation or deactivation of regulatory proteins and enzymes by phosphorylation could be especially important. Ubiquitin is involved in protein degradation, and its enhanced expression is probably required to remove aberrant proteins, resulting from damage to translational machinery, or thermally denatured proteins.

B. Changes Due to Water Deficit

1. Physiology

Resistance to water deficit is manifested in four general ways: timing to avoid water deficit, morphological adaptations, physiological adaptations, and metabolic alterations. The first three are complex processes and are poorly understood. The last means of resistance has met with the most success in understanding. This area includes synthesis of proteins involved in osmotic adjustment, considered in Section III.C.

Quantitative and qualitative changes in the synthesis of proteins have been reported to occur in plants in response to water deficit. Reductions in polyribosome stability have been reported

[68–71], as well as changes in transcription [72]. Many of these proteins also appear in response to ABA application, supporting the role of ABA as a mediator in stress responses. Isolation of the genes responsible for these proteins has been accomplished in a number of species by cDNA cloning techniques. The proteins have been placed in several gene families based on sequence homology and similar spatial and temporal expression.

2. LEA/RAB/Dehydrin Protein Families

Three families of ABA responsive genes and gene products have been described: LEAs (late embryogenesis abundant), RABs (responsive to ABA), and dehydrins. The gene families were initially isolated from maturing seeds of cotton (LEA [73]), rice embryos (RAB [74]), and barley aleurone (dehydrins [75]). Many of the mRNAs for these could also be detected in ABA-treated or ABA-stressed vegetative tissues. The gene families appear in a number of agronomically important crops [72,76,77]. It is likely that these gene families occur in noncrop species as well but have not been described at this time.

It is apparent that there is a degree of overlap among these gene families. Dure et al. [76], Skriver and Mundy [72], and Dure [78] compared the gene sequences of ABA or water-deficit-induced genes and noted three major homology groups. Conserved amino acid domains within each group have been postulated to function in a protective role during desiccation. These domains, and the cellular localization of the gene products, are suggested to enhance protein stability and membrane integrity. The actual mechanisms by which this may be accomplished are currently the subject of much debate. However, it has been suggested that protein stability may be mediated in a fashion analogous to that observed with heat-shock proteins [72]. In contrast, Dure [78] speculates that ion sequestration, to counter the highly ionic cellular environment resulting from water removal, may be their function.

It has been shown that many of these genes contain conserved sequences within the promotor (5′ regulatory region), upstream of the ABA responsive genes [72]. These are postulated to be abscisic acid–response elements (ABREs), which would bind transcriptional factors unique to the ABA signal transduction pathway [72]. Mundy et al. [79] have shown the existence of nuclear proteins that bind specifically to several putative ABREs. Guiltinan et al. [80] have further identified a protein, which recognizes an ABRE, as a member of the class of transcriptional factors known as leucine zippers.

Additional evidence for ABA-mediated gene expression during water deficit was provided by Bray [81]. Her elegant approach employed a normal tomato cultivar, Ailsa Craig, and an ABA biosynthesis mutant, *flacca*, that fails to generate additional ABA during water deficit. She compared the proteins produced under water deficit to proteins synthesized in response to ABA in both genotypes. The protein pattern synthesized by Ailsa Craig under water deficit did not correspond to the proteins seen in *flacca*, but after ABA application the pattern from *flacca* was restored to that of Ailsa Craig. This was true for both in vivo and in vitro translated proteins. Later work [82] has demonstrated that some of the ABA- and water-deficit-induced proteins belong to the LEA and RAB families.

3. Vegetative Storage Proteins

There is also evidence that the signal transduction of water deficit may not be synonymous with ABA. Bozarth et al. [83] observed that upon transfer of soybean seedlings to dry rooting media, 28- and 31-kDa proteins appeared in the cell wall fraction of cellular extracts. Mason et al. [84] determined that these proteins were identical to vegetative storage proteins (VSPs). Mason and Mullet [85] found that VSP mRNAs are weakly induced by ABA in a tissue-dependent manner, but are strongly induced by jasmonic acid (see above).

C. Salinity Stress

1. Physiology

Salinity stress is related to water deficit in that a reduction in water status occurs. In addition, the presence of excess ions appears to be detrimental to many plant processes. Thus plants subjected to salt stress appear to face two stresses at the same time. Plants vary in their ability to survive salt stress, with tolerant plants generally either sequestering ions in the vacuole or synthesizing osmotically active compounds (i.e., osmoregulation). Nontolerant plants typically attempt to exclude excess ions via active transport.

2. Multiple Protein Responses

Like water deficit, salt stress results in a general decrease in protein synthesis (e.g., Ref. 86), which could be correlated with a loss of polysomes in vitro [87]. In turn, many proteins and mRNAs have been reported to increase or be synthesized *de novo* in response to salt stress. Not surprisingly, some of the proteins or mRNAs are also inducible by water deficit or ABA. For example, several members of the lea family of mRNAs show dual inducibility; Em [88], lea B19.1 [89], and rab 21 (now referred to as rab 16) [74,79].

It is possible that alterations in water status by water deficit and salt are not equivalent in their signal transduction pathways. Espelund et al. [89] also reported that lea genes related to B19.1 have differential expression by other forms of osmotic stress, as well as ABA. Bostock and Quatrano [88] showed that while Em is salt inducible, the induction appears to be a synergistic reaction with ABA. In addition, Vernon et al. [90] demonstrated that several salt-inducible mRNAs from the facultative halophyte *Mesambryanthemum crystallinum* respond differentially to various treatments that lower the water status of the plants. Together, these reports suggest that alternative signal transduction pathways probably exist, with the potential for some commonly mediated steps. It must be noted, however, that changes in expression at the transcriptional level do not necessarily equate to changes at the translational level. LaRosa et al. [25] reported that in tobacco, osmotin mRNA is strongly induced by NaCl, water deficit, wounding, ABA, ethylene, and tobacco mosaic virus. In contrast, osmotin protein levels are weakly stimulated by all but NaCl and water deficit.

3. Osmotic Adjustment

Another aspect shared by salt stress and water deficit is osmotic adjustment, wherein organic or inorganic osmotically active solutes are accumulated. This accumulation produces lower solute potential, which allows a plant cell to maintain a higher water content than otherwise. Many different organic molecules have been described as accumulating during water deficit or salt stress, including quaternary amines, polyols, and sugars [91], as well as inorganic K^+ and Cl^- ions [92]. Proline and glycine betaine, two quaternary amines, have received the most attention. In addition to balancing the osmotic potential between the ions sequestered in the vacuole, they have been shown in vitro to maintain the stability of various enzymes incubated with salt solutions [93]. This ability has earned these cytoplasmic compounds term *compatible solutes*.

PROLINE. Delauncey and Verma [94] summarized information indicating that proline accumulation is active, rather than passive, as a result of inhibited metabolism. Further, Hu et al. [95] showed that mRNA for the enzyme Δ^1-pyrroline-5-carboxylate synthetase is increased in salt-stressed roots, and Delauncey and Verma [94] suggest that it may also be a rate-limiting step for proline synthesis. Delauncey and Verma [94] also indicated that Δ^1-pyrroline-5-carboxylate reductase activity and mRNA also increase with salt stress, but there is doubt as to whether it is involved in the NaCl-dependent regulation of proline synthesis.

GLYCINE BETAINE. Glycine betaine originates from choline rather than via amino acid biosynthesis, and is also found in many plant species. McCue and Hanson [96] have shown that betaine aldehyde dehydrogenase (BADH), the last step in the synthesis of glycine betaine, is salt inducible at both the protein and mRNA levels in sugar beets and spinach. ABA can also stimulate BADH protein and mRNA synthesis, but at levels lower than via salt stress [97]. A biochemical messenger other than ABA was suggested by this work, providing some evidence for multiple signal transduction pathways.

4. Ca^{2+}-ATPase

Na^+ has been implicated in some of the difficulties faced by plants during salt stress. It reportedly displaces Ca^{2+} from membranes, possibly reducing membrane stability [98]. However, it is more likely the primary injuries are from displaced Ca^{2+} increasing the cytoplasmic Ca^{2+}, which could cause a disruption of the many signal transduction pathways requiring moderated levels of the ion [99]. Wimmers et al. [100] have shown that the mRNA for a Ca^{2+}-ATPase is increased in abundance in response to elevated NaCl concentrations in tomato. They suggest that this ATPase may act to maintain proper levels of Ca^{2+}, thus mitigating the effects of Na^+.

D. Cold Stress

1. *Physiology*

Cold stress may be considered as a composite of two separate stresses: chilling (generally, temperatures from 4 to 15°C) and freezing. Chilling stress has in general been attributed to effects at the plasma membrane, manifested by electrolyte leakage from tissues (e.g., Minchin and Simon [101]). Williams [102] summarized data that indicated that leakage could be due to phase transitions caused by the presence of minor lipid components of the membrane, or alternatively, failure to seal critical intrinsic membrane proteins into the cell membrane by nonbilayer forming lipids. Another explanation, that of lipid peroxidation (e.g., during photoinhibition), does not appear to be a cause of leakage. Hodgson and Raison [103] demonstrated that neither superoxide dismutase activity (see Section III.F) nor lipid peroxidation appears to increase at moderate photon flux levels.

Freezing stress appears to be the result of two components. The first is intracellular, in which ice crystals can pierce the plasma membrane (immediately lethal). The second is extracellular, in which the low water potential of ice in the intercellular spaces and cell wall can remove water from the cell (i.e., desiccation) [104,105]. Plants that achieve freezing tolerance apparently mitigate the first stress component through biochemical means, and the second stress component through physical means [106].

Tolerance to chilling is apparently a prerequisite for tolerance to freezing. Tolerance is an inducible response, dependent on daylength and temperature [107], and is accompanied by an increase in the ABA content of the cells (e.g., Wright [108]). Numerous proteins or their mRNAs are also induced by cold temperatures, and evidence exists that some of these proteins are necessary for cold tolerance. For example, Mohapatra et al. [109] compared alfalfa cultivars and cold-inducible gene products. They found a high correlation coefficient (0.993) between the LT_{50} (the temperature that is lethal to 50% of the treated plants) and the relative amounts of a particular cold-induced mRNA.

2. *Comparisons with Heat Shock*

Unlike heat shock, general protein synthesis does not appear to cease [105 and references therein]. In addition, there appears to be little conserved in terms of the types of synthesized proteins [105]. It is interesting to note that some of the heat-shock proteins, or their mRNAs, are also cold inducible [110–113]. Jaenicke [114] indicated that the stability of proteins is limited by both high *and* low temperatures. As such, the presence of heat-shock proteins may not be that unusual. However, conclusive proof of a role for heat shock proteins during cold stress is currently lacking.

In two excellent reviews, Guy [105] and Thomashow [115] summarized numerous reports of cold-inducible proteins, including enzymes, and the inducibility of these proteins by other stimuli. In conjunction with the desiccating effects of extracellular freezing, it is perhaps not surprising that some of the cold-induced proteins or their mRNAs are also inducible by ABA, water deficit, or osmotic stress. For example, Hahn and Walbot [116] noted that the rab 16A gene (also a member of the lea family) is inducible by all three treatments.

3. Cryoprotective Proteins

Since those reviews were published, putative cryoprotective genes and proteins have been reported, which theoretically could prevent intracellular ice crystal formation. *Arabidopsis thaliana* has been shown to transcribe at least two genes, *kin1* and *2*, that show sequence homology to the antifreeze (thermal hysteresis) proteins of certain cold water fish [117,118]. These proteins act by lowering the freezing point of water, apparently by providing an alternative, and limiting, ice crystal nucleation seed [119]. Antifreeze proteins are apparently widespread throughout the plant kingdom, as in vitro recrystallization assays have demonstrated [120,121]. However, the proteins or polypeptides responsible for the activity have not been isolated.

Thomashow [115] coined the gene name *cor* (cold-regulated) to describe numerous cold-inducible proteins from *Arabidopsis thaliana*. The COR proteins are a heterogeneous group, ranging in molecular mass and relatedness to other protein families. This reflects the apparently unconserved nature of the proteins synthesized in various plant species in response to cold stress. For example, Gilmour et al. [122] reported that *cor47* has sequence homology to some of the *lea* genes and is also ABA and water-deficit inducible. Gilmour et al. [122] also reported that the sequence of the *cor6.6* bears a striking similarity to the *kin1* gene described above. Lin and Thomashow [123] showed that the COR15 protein can extend the activity of the cold-labile enzyme lactate dehydrogenase in an in vitro chilling assay, suggesting that it may act as a compatible solute.

E. Oxygen Deprivation

1. Physiology

The phrase *oxygen deprivation* is a very general term for an area of study that has had a considerable variation of terminology and should be explained briefly before continuing. *Anaerobic* means O_2-free, *anoxia* refers to O_2 levels so low that ATP production by oxidative phosphorylation is essentially nil, and *hypoxia* defines O_2 levels that limit ATP production by mitochondria [124]. The actual levels of O_2 that correspond to these states are highly dependent on the tissue utilized and the physiological process under investigation. In general, O_2 levels from 2 to 10% (compared to the normal atmospheric concentration of 21%) results in a hypoxic state. Low-O_2 environments are associated with excess water in the soil and the relatively low diffusibility of O_2 in water compared to air. Proper soil aeration is prevented, which leads subsequently to consumption of available O_2 by aerobic organisms.

The effects of O_2 deprivation on protein synthesis are similar to those encountered during heat shock. In maize, there is a decrease in normal aerobic protein synthesis, associated with a loss of polysomes [125–127]. This is followed by the concommittant synthesis of approximately 20 proteins [125] under transcriptional and posttranscriptional regulation [127,128]. There may be one protein commonly expressed by both stresses [126].

2. Anaerobic Polypeptides

The preferentially synthesized proteins may be divided into two temporally regulated groups. Members of the first group are translated primarily during the first 5 h of anoxia and are referred to as transition polypeptides (TPs) [125]. These proteins are stable, lasting long after their synthesis declines [126]. The second group, the anaerobic polypeptides (ANPs), begin to appear after approximately 90 min of anoxia, with synthesis continuing for several days, until cell death [125].

Rather than acting to ameliorate protein denaturation, as in heat shock, most of these proteins are apparently involved in maintaining the ATP levels of the cells. In particular, several of the ANPs have been identified as glycolytic or fermentative enzymes. They include sucrose synthase, phosphoglucoisomerase, aldolase, alcohol dehydrogenase (ADH), and pyruvate decarboxylase (PDC) (summarized by Drew [129]).

Of these enzymes, ADH is the best characterized. Recently, Andrews et al. [130] examined *Adh* gene expression and enzyme activity in several tissues of maize, under different O_2 concentrations. They show that *Adh* gene expression is maximal with anoxia or extreme hypoxia (i.e., 0 to 4% O_2) in both root tips and axes. However, due to the limited viability of root tips, their period of transcription was not as long as the root axes. In addition, *Adh* transcripts did not always parallel ADH activity. The authors conclude that hypoxia is apparently crucial to increased ADH induction and activity. They suggest that delay between *Adh* induction and enhanced activity provides a mechanism for survival during the anoxic state that would follow hypoxia. Indeed, Drew [129] indicates that in maize, tolerance to anoxia can be improved by exposure to hypoxic conditions. The regulatory pathway by which *Adh* is induced is not currently known; however, common sequence elements have been found in the *Adh1* and aldolase gene promotor regions [128].

3. Ethylene Synthesis

Another important enzyme that increases during O_2 deprivation is 1-aminocarboxylate-1-cyclo-propane synthase (ACC synthase). This enzyme catalyzes the rate-limiting step in the synthesis of ethylene, which increases dramatically in response to hypoxia [131]. One of ethylene's actions is to stimulate the formation of aerenchyma in the stem (and possibly roots), thus providing more O_2 to deprived tissues. Zarembinski and Theologis [132] report that in rice, several ACC synthase genes are induced by anoxia, and that differential expression occurs under different hormonal and environmental signals. They suggest that the multiplicity of responses is consistent with the various aspects of ethylene in development and responses to other environmental stimuli (e.g., wounding).

4. Oxygen Radicals

A third enzyme worth noting is superoxide dismutase (SOD). The action of SOD in a low-O_2 environment may seem counterintuitive; however, the generation of superoxide radicals, leading to lipid peroxidation, has been implicated in postanoxic tissue damage [133]. In anoxia-intolerant *Iris germanica* rhizomes, lipid peroxidation was widespread compared to anoxia-tolerant *Iris pseudoacorus* rhizomes. Monk et al. [134] demonstrated that SOD activity rises after extended anoxia in *Iris pseudoacorus* rhizomes but not in *Iris germanica* rhizomes or *Glyceria maxima*. This suggests that one of the strategies of tolerance to anoxia may be to synthesize proteins that anticipate the return of an aerobic environment [129].

A final note concerns the difference in anoxia tolerance between maize and soybean. Soybean is less tolerant than maize to flooding [135]. Hwang and VanToai [136] reported that the activities of proteinases with alkaline and neutral pH optima increased in anoxic maize roots. In contrast, neutral pH optima proteinase activity declined in anoxic soybean roots.

Russell et al. [137] reported that soybean roots synthesize fewer ANPs than are synthesized by maize. Both Sachs [126] and Hwang and VanToai [136] speculate that a possible reason for the difference in anoxia tolerance is the number or types of ANP's synthesized.

F. Air Pollution

Air pollution is still emerging as an area of research for alterations of protein synthesis. Ozone (O_3) and sulfur dioxide (SO_2) are usually considered to be the primary culprits in damage due to air pollution. However, these molecules eventually generate toxic oxygen species via the reaction $H_2O_2 + O_2^- \Rightarrow OH^- + O2 + OH\cdot$.

Changes in both protein (e.g., Ref. 138) and mRNA (e.g., Refs. 41 and 139) are known to occur, generally resulting in the action of antioxidants or detoxifying enzymes to minimize the damage to cellular membranes or macromolecules. Of the enzymes, superoxide dismutase (SOD) is the best characterized, catalyzing the reaction $2O_2^- + 2H^+ \Rightarrow H_2O_2 + O_2$. Bowler et al. [11] summarized information indicating that SOD protection in response to O_3 stress is often contradictory and extremely dependent on the conditions and plant species under consideration. Indeed, Badiani et al. [140] report that with *Phaseolus vulgaris* L., fluctuations in the level of antioxidants and detoxifying enzymes occur during the day. Previously, SOD in *P. vulgaris* was reported to be unaffected by ozone treatment [141]. In contrast, there appears to be much better evidence for a role in SO_2 protection.

Of the antioxidant molecules, glutathione (GSH) has received the most attention. Glutathione is a polypeptide of the sequence γ-glutamyl-cysteinyl-glycine, and acts to maintain the redox state of cysteine groups in proteins via its cysteinyl side chain. Reduction of protein disulfide bonds results in the formation of oxidized glutathione (GSSG). Glutathione, its synthetic enzymes, and glutathione reductase are known to be induced by O_3 [10]. As with SOD, however, the actual value of enhanced glutathione levels is equivocal, again being species and condition dependent.

G. Heavy Metal

Heavy metal stress is frequently encountered by plants in areas of industrial pollution or as a result of mining activity. In animals, the primary means of heavy metal sequestration is by the metallothionein family of proteins [142]. These proteins are Cys rich, relying on those residues to chelate metal ions. While evidence of such proteins exist in plants, the primary defense against heavy metal toxicity relies on polypeptide equivalents, the phytochelatins. These polypeptides are inducible by heavy metals and are found throughout the plant kingdom [143]. Like metallothioneins, phytochelatins are Cys rich and appear to act in a similar fashion. Recent evidence suggests that phytochelatins and the bound metal ligands are localized in the central vacuole, thus removing the metal ions from more sensitive areas of the cell [143].

Phytochelatins are related to glutathione, having a primary structure of (γ-Glu-Cys)$_n$-Gly or (γ-Glu-Cys)$_n$-β-Ala, where $n = 2$ to 11 (142). These peptide bonds are not known to be synthesized on ribosomes, but phytochelatins appear to share some of the biosynthetic enzymes of glutathione [143]. Evidence for a metal-inducible phytochelatin synthase exists, catalyzing the transfer of γ-glutamylcysteine to glutathione, thus generating the (γ-Glu-Cys)$_n$ portion of the molecule [144].

H. UV Radiation

1. UV-Absorbing Compounds

The responses of plants to UV radiation is of increasing concern due to the depletion of UV-absorbing ozone in the upper atmosphere. Research on the changes in protein synthesis due to UV has centered on the transcription and translation of enzymes involved in the flavonoid and anthocyanin pathways. Aside from their roles in plant defense against pathogens and as pigments, flavonoids also absorb UV radiation. Chalcone synthase (CHS), and to some extent, phenylammonia lyase (PAL) are the best studied. Wingender [145] examined the promotor region of the *chs* gene from parsley and determined that two elements exist for the induction of CHS by UV, in addition to an element for elicitor induction. Subsequent work indicates that UV photoreceptors are responsible for the initial perception, while additional photoreceptors are required for anthocyanin or flavonoid synthesis in parsley [146].

2. DNA Repair

A second area of interest are DNA repair mechanisms. As is well known, UV radiation induces various lesions in DNA. The best studied are cyclobutane-type pyrimidine dimers, which have been the only type of DNA lesions reported in plants [146]. The dimers can be repaired via photoreactivation (photolyase), excision repair, or recombinatorial repair [147,148]. The latter type of repair has not been reported in plants [146], and very little research has apparently been reported for excision repair systems in plants. Photoreactivation has been reported in several species (e.g., wild carrot [149], tobacco [150], pinto bean [151], and maize pollen [152]). Pang and Hays [153] have reported on the presence of a photolyase activity in *Arabidopsis thaliana*. They indicate that the putative photolyase has a requirement for visible light, with an optimum of between 375 and 400 nm, which is similar to maize pollen photolyase. Pang and Hays [153] further suggest that *Arabidopsis* may actually have two photolyases, one similar to that found in *E. coli* and one similar to that reported in pinto bean.

IV. SUMMARY

As has been shown, the protein synthetic responses of plants to environmental stresses is diverse, in many cases yielding a specific set of proteins that presumably aid the plant during the stress. While the function of some of these proteins is understood, many more await a complete understanding. The untangling of the signal transduction pathways will prove to be quite interesting, illuminating the evolution of many of these proteins and why some of them are induced by multiple stresses. The nonequivalence of transcription and translation of some of the proteins provides a caution not only to physiological interpretations of gene induction, but also toward efforts to manipulate these proteins in other plant systems via recombinant DNA technology.

REFERENCES

1. J. L. Jakobek and P. B. Lindgren, *Plant Cell, 5*: 49 (1993).
2. J. F. Bol, H. J. M. Linthorst, and B. J. C. Cornelissen, *Annu. Rev. Phytopathol., 28*: 113 (1990).
3. R. A. Dixon and M. J. Harrison, *Adv. Genet., 28*: 165 (1990).
4. T. W. Goodwin and E. I. Mercer, *Introduction to Plant Biochemistry*, Pergamon Press, Elmsford, N.Y. (1983).
5. R. B. Herbert, *The Biosynthesis of Secondary Metabolites*, Chapman & Hall, New York (1981).
6. C. Grand, F. Sarni, and C. J. Lamb, *Eur. J. Biochem., 169*: 73 (1987).
7. M. H. Walter, J. Grima-Pettenati, C. Grand, A. M. Boudet, and C. J. Lamb, *Plant Mol. Biol., 15*: 525 (1990).
8. M. M. Campbell and B. E. Ellis, *Planta, 186*: 409 (1992).
9. K. Hahlbrock and D. Scheel, *Annu. Rev. Plant Physiol. Plant Mol. Biol., 40*: 347 (1989).
10. I. K. Smith, A. Polle, and H. Rennenberg, in *Stress Responses in Plants*: *Adaptation and*

Acclimation Mechanisms (R. G. Alscher and J. R. Cumming, eds.), Wiley-Liss, New York, p. 201 (1990).

11. C. Bowler, M. Van Montagu, and D. Inze, *Annu. Rev. Plant Physiol. Plant Mol. Biol.*, *43*: 83 (1992).

12. A. R. Cross and O. T. G. Jones, *Biochim. Biophys. Acta*, *1057*: 281 (1991).

13. A. A. Bell, *Annu. Rev. Plant Physiol.*, *32*: 21 (1981).

14. T. B. Ryder, C. L. Cramer, J. N. Bell, M. P. Robbins, R. A. Dixon, and C. J. Lamb, *Proc. Natl. Acad. Sci. USA*, *81*: 5724 (1984).

15. T. B. Ryder, S. A. Hedrick, J. N. Bell, X. Liang, S. D. Clouse, and C. J. Lamb, *Mol. Gen. Genet.*, *210*: 219 (1987).

16. D. A. Smith and S. W. Banks, *Phytochemistry*, *25*: 979 (1986).

17. D. R. Threlfall and I. M. Whitehead, in *Ecological Chemistry and Biochemistry of Plant Terpenoids* (J. B. Harborne and F. A. Tomas-Barberan, eds.), Clarendon Press, Oxford, p. 159 (1991).

18. M. W. Dudley, M. T. Dueber, and C. A. West, *Plant Physiol.*, *81*: 335 (1986).

19. M. W. Dudley, T. R. Green, and C. A. West, *Plant Physiol.*, *81*: 343 (1986).

20. B. Fritig, S. Kauffmann, B. Dumas, P. Geoffroy, and M. Kopp, in *Plant Resistance to Viruses* (D. Evered and S. Harnett, eds.), Wiley, Chichester, West Sussex, England, 92 (1987).

21. H. Van Der Wel and K. Loeve, *Eur. J. Biochem.*, *31*: 221 (1972).

22. M. Richardson, S. Valdes-Rodriguez, and A. Blanco-Labra, *Nature*, *327*: 432 (1987).

23. C. A. Ryan, *Annu. Rev. Phytopathol.*, *28*: 425 (1990).

24. N. K. Singh, C. A. Bracker, P. M. Hasegawa, A. K. Handa, S. Buckel, M. A. Hermodson, E. Pfankoch, F. E. Regnier, and R. A. Bressan, *Plant Physiol.*, *85*: 529 (1987).

25. P. C. LaRosa, Z. Chen, D. E. Nelson, N. K. Singh, P. M. Hasegawa, and R. A. Bressan, *Plant Physiol.*, *100*: 409 (1992).

26. A. J. Vigers, W. K. Roberts, and C. P. Selitrennikoff, *Mol. Plant Microbe Interact.*, *4*: 315 (1991).

27. A. J. Vigers, S. Wiedemann, W. K. Roberts, M. Legrand, C. P. Selitrennikoff, and B. Fritig, *Plant Sci.*, *83*: 155 (1992).

28. S. Bartnicki-Garcia, *Annu. Rev. Microbiol.*, *22*: 87 (1968).

29. N. Benhamou and A. Asselin, *Biol. Cell*, *67*: 341 (1989).

30. D. B. Collinge, K. M. Kragh, J. D. Mikkelsen, K. K. Nielsen, U. Rasmussen, and K. Vad, *Plant J.*, *3*: 31 (1993).

31. J. H. Burnett, in *Fungal Walls and Hyphal Growth* (J. H. Burnett and A. P. J. Trinci, eds.), Cambridge University Press, New York, p. 1 (1979).

32. D. A. Ward and D. W. Lawlor, *J. Exp. Bot.*, *41*: 309 (1990).

33. M. D. van de Rhee, R. Lemmers, and J. F. Bol, *Plant Mol. Biol.*, *21*: 451 (1993).

34. F. Cote, J. R. Cutt, A. Asselin, and D. F. Klessig, *Mol. Plant Microbe Interact.*, *4*: 173 (1991).

35. F. Mauch, B. Mauch-Mani, and T. Boller, *Plant Physiol.*, *88*: 936 (1988).

36. A. J. Barrett, in *Proteinase Inhibitors* (A. J. Barrett and G. Salvesen, eds.), Elsevier, Amsterdam, p. 3 (1986).

37. H. Bohlmann and K. Apel, *Annu. Rev. Plant Physiol. Plant Mol. Biol.*, *42*: 227 (1991).

38. K. Apel, H. Bohlmann, and U. Reimann-Philipp, *Physiol. Plant*, *80*: 315 (1990).

39. A. G. Darvill and P. Albersheim, *Annu. Rev. Plant Physiol.*, *35*: 243 (1984).

40. C. A. Ryan and E. E. Farmer, *Annu. Rev. Plant Physiol. Plant Mol. Biol.*, *42*: 651 (1991).

41. D. Ernst, M. Schraudner, C. Langebartels, and H. Sandermann, Jr., *Plant Mol. Biol.*, *20*: 673 (1992).

42. E. E. Farmer, R. R. Johnson, and C. A. Ryan, *Plant Physiol.*, *98*: 995 (1992).

43. G. Sembdner and B. Parthier, *Annu. Rev. Plant Physiol. Plant Mol. Biol.*, *44*: 569 (1993).

44. M. A. Melan, X. Dong, M. Endara, K. R. Davis, F. M. Ausubel, and T. K. Peterman, *Plant Physiol. Suppl.*, *99*: 37 (1992).

45. M. A. Melan, X. Dong, M. E. Endara, K. R. Davis, F. M. Ausubel, and T. K. Peterman, *Plant Physiol.*, *101*: 441 (1993).

46. G. Pearce, D. Strydom, S. Johnson, and C. A. Ryan, *Science*, *253*: 895 (1991).

47. B. McGurl, G. Pearce, M. Orozco-Cardenas, and C. A. Ryan, *Science*, *255*: 1570 (1992).

48. I. Raskin, *Annu. Rev. Plant Physiol. Plant Mol. Biol.*, *43*: 439 (1992).

49. H. Pena-Cortes, T. Albrecht, S. Prat, E. W. Weiler, and L. Willmitzer, *Planta*, *191*: 123 (1993).
50. J. S. Boyer, *Science*, *218*: 443 (1982).
51. S. Lindquist, *Annu. Rev. Biochem.*, *55*: 1151 (1986).
52. V. A. Bernstam, *Annu. Rev. Plant Physiol.*, *29*: 25 (1978).
53. F. Schoffl, G. Baumann, E. Raschke, and M. Bevan, *Philos. Trans. R. Soc. London B Biol. Sci.*, *314*: 453 (1986).
54. M. M. Sachs and T.-H. D. Ho, in *Plants Under Stress* (H. G. Jones, T. J. Flowers, and M. B. Jones, eds.), Cambridge University Press, New York, p. 157 (1989).
55. E. Vierling, *Annu. Rev. Plant Physiol. Plant Mol. Biol.*, *42*: 579 (1991).
56. J. A. Kimpel and J. L. Key, *Plant Physiol.*, *79*: 672 (1985).
57. J. J. Burke, J. L. Hatfield, R. R. Klein, and J. E. Mullet, *Plant Physiol.*, *78*: 394 (1985).
58. E. Czarnecka, L. Edelman, F. Schoffl, and J. L. Key, *Plant Mol. Biol.*, *3*: 45 (1984).
59. J. J. Heikkila, J. E. T. Papp, G. A. Schultz, and J. D. Bewley, *Plant Physiol.*, *76*: 270 (1984).
60. R. J. Ellis and S. M. van der Vies, *Annu. Rev. Biochem.*, *60*: 321 (1991).
61. A. A. Gatenby, *Plant Mol. Biol.*, *19*: 677 (1992).
62. H. Pellham, *Science*, *332*: 776 (1988).
63. S. Lindquist and E. A. Craig, *Annu. Rev. Genet.*, *22*: 631 (1988).
64. R. I. Morimoto, *Science*, *259*: 1409 (1993).
65. S. Moisyadi and H. M. Harrington, *Plant Physiol. Suppl.*, *93*: 88 (1990).
66. A. H. Christensen and P. H. Quail, *Plant Mol. Biol.*, *12*: 619 (1989).
67. T. J. Burke, J. Callis, and R. D. Vierstra, *Mol. Gen. Genet.*, *213*: 435 (1988).
68. R. S. Dhindsa and R. E. Cleland, *Plant Physiol.*, *55*: 778 (1975).
69. R. S. Dhindsa and R. E. Cleland, *Plant Physiol.*, *55*: 782 (1975).
70. P. R. Rhodes and K. Matsuda, *Plant Physiol.*, *58*: 631 (1976).
71. H. S. Mason and K. Matsuda, *Physiol. Plant.*, *64*: 95 (1985).
72. K. Skriver and J. Mundy, *Plant Cell*, *2*: 503 (1990).
73. J. Gomez, D. Sanchez-Matinez, V. Stiefel, J. Rigau, P. Puigdomenech, and M. Pages, *Nature*, *334*: 262 (1988).
74. J. Mundy and N.-H. Chua, *EMBO J.*, *7*: 2279 (1988).
75. T. J. Close, A. A. Kortt, and P. M. Chandler, *Plant Mol. Biol.*, *13*: 95 (1989).
76. L. Dure III, M. Crouch, J. Harada, T.-H. D. Ho, J. Mundy, R. Quatrano, T. Thomas, and Z. R. Sung, *Plant Mol. Biol.*, *12*: 475 (1989).
77. T. J. Close and P. M. Chandler, *Aust. J. Plant Physiol.*, *17*: 333 (1990).
78. L. Dure III, *Plant J.*, *3*: 363 (1993).
79. J. Mundy, K. Yamaguchi-Shinozaki, and N.-H. Chua, *Proc. Natl. Acad. Sci. USA*, *87*: 1406 (1990).
80. M. J. Guiltinan, Jr., W. R. Marcotte, and R. S. Quatrano, *Science*, *250*: 267 (1990).
81. E. Bray, *Plant Physiol.*, *88*: 1210 (1988).
82. A. Cohen and E. A. Bray, *Planta*, *182*: 27 (1990).
83. C. S. Bozarth, J. E. Mullet, and J. S. Boyer, *Plant Physiol.*, *85*: 261 (1987).
84. H. S. Mason, F. D. Guerrero, J. S. Boyer, and J. E. Mullet, *Plant Mol. Biol.*, *11*: 845 (1988).
85. H. S. Mason and J. E. Mullet, *Plant Cell*, *2*: 569 (1990).
86. S. Ramagopal and J. B. Carr, *Plant Cell Environ.*, *14*: 47 (1991).
87. C. J. Brady, T. S. Gibson, E. W. R. Barlow, J. Speirs, and R. G. Wyn Jones, *Plant Cell Environ.*, *7*: 571 (1984).
88. R. M. Bostock and R. S. Quatrano, *Plant Physiol.*, *98*: 1356 (1992).
89. M. Espelund, S. Saeboe-Larssen, D. W. Hughes, G. A. Galau, F. Larsen, and K. S. Jakobsen, *Plant J.*, *2*: 241 (1992).
90. D. M. Vernon, J. A. Ostrem, and H. J. Bohnert, *Plant Cell Environ.*, *16*: 437 (1993).
91. R. G. Wyn Jones, *Chem. Br.*, *21*: 454 (1985).
92. J. M. Morgan, *Annu. Rev. Plant Physiol.*, *35*: 299 (1984).
93. A. Pollard and R. G. Wyn Jones, *Planta*, *144*: 291 (1979).
94. A. J. Delauncey and D. P. S. Verma, *Plant J.*, *4*: 215 (1993).
95. C.-A. Hu, A. J. Delauncey, and D. P. S. Verma, *Proc. Natl. Acad. Sci. USA*, *89*: 9354 (1992).

96. K. F. McCue and A. D. Hanson, *Plant Mol. Biol.*, *18*: 1 (1992).
97. K. F. McCue and A. D. Hanson, *Aust. J. Plant Physiol.*, *19*: 555 (1992).
98. A. C. Leopold and R. P. Willing, in *Salinity Tolerance in Plants: Strategies for Crop Improvement* (R. C. Staples and G. H. Toenniessen, eds.), Wiley-Interscience, New York, p. 67 (1984).
99. A. Lauchli, in *Calcium in Plant Growth and Development* (R. T. Leonard and P. K. Hepler, eds.), American Society of Plant Physiologists, Rockville, Md., p. 26 (1990).
100. L. E. Wimmers, N. N. Ewing, and A. B. Bennett, *Proc. Natl. Acad. Sci. USA*, *89*: 9205 (1992).
101. A. Minchin and E. W. Simon, *J. Exp. Bot.*, *24*: 1231 (1973).
102. W. P. Williams, *Philos. Trans. R. Soc. London B Biol. Sci.*, *326*: 555 (1990).
103. R. A. J. Hodgson and J. K. Raison, *Planta*, *185*: 215 (1991).
104. P. L. Steponkus, *Ann. Rev. Plant Physiol.*, *35*: 543 (1984).
105. C. L. Guy, *Annu. Rev. Plant Physiol. Plant Mol. Biol.*, *41*: 187 (1990).
106. W. D. Miller, *South. Lumberman*, *188*: 32 (1954).
107. C. J. Weiser, *Science*, *169*: 1269 (1970).
108. S. T. C. Wright, *J. Exp. Bot.*, *26*: 161 (1975).
109. S. S. Mohapatra, L. Wolfraim, R. J. Poole, and R. S. Dhindsa, *Plant Physiol.*, *89*: 375 (1989).
110. R. K. Yacoob and W. G. Filion, *Biochem. Cell Biol.*, *65*: 112 (1986).
111. R. K. Yacoob and W. G. Filion, *Can. J. Genet. Cytol.*, *28*: 1125 (1986).
112. V. V. Kuznetsov, J. A. Kimpel, G. Goekjian, and J. L. Key, *Sov. Plant Physiol.*, *34*: 685 (1987).
113. M. Cabane, P. Calvet, P. Vincens, and A. M. Boudet, *Planta*, *190*: 346 (1993).
114. R. Jaenicke, *Philos. Trans. R. Soc. London B Biol. Sci.*, *326*: 535 (1990).
115. M. F. Thomashow, *Adv. Genet.*, *28*: 99 (1990).
116. M. Hahn and V. Walbot, *Plant Physiol.*, *91*: 930 (1989).
117. S. Kurkela and M. Franck, *Plant Mol. Biol.*, *15*: 137 (1990).
118. S. Kurkela and M. Borg-Franck, *Plant Mol. Biol.*, *19*: 689 (1992).
119. P. L. Davies and C. L. Hew, *FASEB J.*, *4*: 2460 (1990).
120. M. E. Urrutia, J. G. Duman, and C. A. Knight, *Biochim. Biophys. Acta*, *1121*: 199 (1992).
121. J. G. Duman and T. M. Olsen, *Cryobiology*, *30*: 322 (1993).
122. S. J. Gilmour, N. N. Artus, and M. F. Thomashow, *Plant Mol. Biol.*, *18*: 13 (1992).
123. C. Lin and M. F. Thomashow, *Biochem. Biophys. Res. Commun.*, *183*: 1103 (1992).
124. A. Pradet and J. L. Bomsel, in *Plant Life in Anaerobic Environments* (D. D. Hook and R. M. M. Crawford, eds.), Ann Arbor Science, Ann Arbor, Mich., p. 89 (1978).
125. M. M. Sachs, M. Freeling, and R. Okimoto, *Cell*, *20*: 761 (1980).
126. M. M. Sachs, in *Plant Life Under Oxygen Deprivation: Ecology, Physiology and Biochemistry* (M. B. Jackson, D. D. Davies, and H. Lambers, eds.), SPB Academic Publishing, The Hague, The Netherlands, p. 129 (1990).
127. J. Bailey-Serres and M. Freeling, *Plant Physiol.*, *94*: 1237 (1990).
128. E. S. Dennis, J. C. Walker, D. J. Llewellyn, J. G. Ellis, K. Singh, J. G. Tokuhisa, D. R. Wolstenhome, and W. J. Peacock, in *NATO Advanced Study Institute on Plant Molecular Biology (1987: Carlsberg Laboratory)* (D. von Wettstein and N.-H. Chua, eds.), Plenum Press, New York, p. 407 (1987).
129. M. C. Drew, *Soil Sci.*, *154*: 259 (1990).
130. D. L. Andrews, B. G. Cobb, J. R. Johnson, and M. C. Drew, *Plant Physiol.*, *101*: 407 (1993).
131. M. C. Drew, M. B. Jackson, and S. Giffard, *Planta*, *147*: 83 (1979).
132. T. I. Zarembinski and A. Theologis, *Mol. Biol. Cell*, *4*: 363 (1993).
133. M. I. S. Hunter, A. M. Hetherington, and R. M. M. Crawford, *Phytochemistry*, *22*: 1145 (1983).
134. L. S. Monk, K. V. Fagerstedt, and R. M. M. Crawford, *Plant Physiol.*, *85*: 1016 (1987).
135. F. T. Turner, J. W. Sij, G. N. McCauley, and C. C. Chen, *Crop Sci.*, *23*: 40 (1983).
136. S. Y. Hwang and T. T. VanToai, *Plant Soil*, *126*: 127 (1990).
137. D. A. Russell, D. M. L. Wong, and M. M. Sachs, *Plant Physiol.*, *92*: 401 (1990).
138. P. Guillemaut, F. Weber-Lotfi, D. Blache, M. Prost, B. Rether, and A. Dietrich, *Physiol. Plant.*, *85*: 215 (1992).
139. S. Karpinski, G. Wingsle, B. Karpinska, and J.-E. Hallgren, *Physiol. Plant.*, *85*: 689 (1992).
140. M. Badiani, G. Schenone, A. R. Paolacci, and I. Fumagalli, *Plant Cell Physiol.*, *34*: 271 (1993).

141. C. P. Chanway and V. C. Runeckles, *Can. J. Bot.*, *62*: 236 (1984).

142. W. E. Rauser, *Annu. Rev. Biochem.*, *59*: 61 (1990).

143. J. C. Steffins, *Annu. Rev. Plant Physiol. Plant Mol. Biol.*, *41*: 553 (1990).

144. E. Grill, S. Loffler, E. L. Winnaccer, and M. H. Zenk, *Proc. Natl. Acad. Sci. USA*, *86*: 6838 (1989).

145. R. Wingender, H. Rohrig, C. Horicke, and J. Schell, *Plant Cell*, *2*: 1019 (1990).

146. A. E. Stapleton, *Plant Cell*, *4*: 1353 (1992).

147. K. C. Smith, in *The Science of Photobiology* (K. C. Smith, ed.) Plenum Press, New York, p. 111 (1989).

148. A. Kornberg and T. A. Baker, *DNA Replication*, W. H. Freeman, New York, p. 771 (1989).

149. G. P. Howland, *Nature*, *254*: 160 (1975).

150. J. E. Trosko and V. H. Mansour, *Radiat. Res.*, *36*: 333 (1968).

151. N. Saito and H. Werbin, *Radiat. Bot.*, *9*: 421 (1969).

152. M. Ikenaga, S. Kondo, and T. Fujii, *Photochem. Photobiol.*, *19*: 109 (1974).

153. Q. Pang and J. B. Hays, *Plant Physiol.*, *95*: 536 (1991).

30
Plant/Crop Hormones Under Stressful Conditions

Syed Shamshad Mehdi Naqvi
Atomic Energy Agricultural Research Center, Tando Jam, Pakistan

I. INTRODUCTION

Plants growing in nature and crops in an agricultural system seldom experience ideal environmental conditions to express their genetic potential for reproduction. Any environmental perturbation unbalances the supply of various resources and limits the overall productivity of food, fiber, fuel, drugs, forest products, and so on. Thus plants and crops growing under suboptimal biotic or abiotic environments are "stressed." Stress can therefore be considered as any change in the environment that evokes response in the form of adjustment of both the rate at which plants grow and the rate and pattern of its development until productivity becomes uneconomical and ultimately ceases.

At the present state of our understanding, optimum productivity is mainly dependent on a favorable balance between water and carbon. Any environmental disturbance, regardless of its nature, that disturbs this balance affects it unfavorably. A common pattern is observed in plants exposed to any stressful environmental condition. For example, as the salinity stress increases it approaches a *threshold* above which the productivity starts declining until it becomes toxic (Figure 1) [1]. Some plants, particularly halophytes and mangroves, have evolved strategies to exclude salt efficiently from the tissue [2]. However, in several plants salt is able to affect the ionic balance, osmotic potentials, and several other processes [3]. Our crop species belong predominantly to the group known as glycophytes, and their growth and biological yield are invariably affected by suboptimal environmental conditions. There is, however, flexibility in these adjustments which appears to be governed by Shelford's *law of tolerance* [4]. According to these, as long as increase in the factor enhances response, it is considered to be *deficient*, and when no further response is elicited, it is *optimum*. When the response starts declining it is *inhibitory* or at the *toxic* level. It is thus obvious that plants are stressed at the levels of both *deficiency* and *inhibitory/toxicity*. But plants or crops seldom experience variation in a single environmental factor, and covariation of and interactions between them is a norm rather than

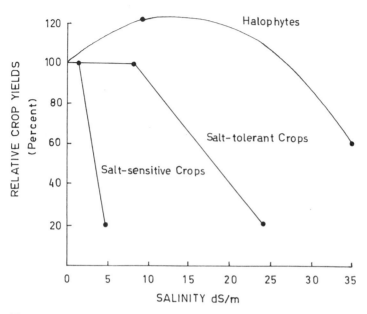

Figure 1 Growth response of plants and crops to salinity stress. (From Ref. 1.)

an exception. Thus it becomes complicated to determine these levels exactly; therefore, we have to compromise to get the optimum productivity under our natural or agricultural system.

Certain common suites of traits have been noted that characterize plants from suboptimal environments. These include slow growth, low photosynthetic rate, and low capacity for nutrient uptake [5]. The earlier the response, the more likely that it has to do with the primary response of stress itself rather than indirect responses to one or more of the early events. It has been observed that under sublethal stress(es) the earliest response of glycophytes are that (a) stomates close to maintain a water balance, and (b) leaf growth is slowed down, and ultimately shoot, root, and reproductive growth is affected. These effects are obvious even in the short term, before shoot solutes would have build up to the high levels of changes in the water balance of the tissues are detectable [6,7].

Shoots or roots are the organs that are the receptors of the above- or underground stressful environmental stimulus. They send signals, as the case may be, and a response is elicited that affects the whole plant body. Thus there are three steps in such a cause-and-effect relationship phenomenon: *perception*, *transduction*, and *response*. Under stressful environment (anaerobiosis, nutrients, osmotic/water, temperature extremes, etc.), marked and often rapid changes in the hormonal level are commonly observed. Since a given stress induces resistance to unrelated stress(es) [5,8] and plant hormones also influence stress resistance, such changes in their level(s) may enable the plant to adjust its growth despite stressful environments [9]. Nearly everything a plant does, from seed germination to programmed senescence, is influenced by plant hormones. Plants or crops respond by adjusting their hormonal balance according to the changing environment, frequently producing less auxin and cytokinin and more abscisic acid [10]. These observations lend support to the idea that independent of osmotic/ionic influences, plant hormones act as a chemical signal in root-to-shoot communication, or vice versa. Recent evidence, however, suggests that as soon as the root perceives drying, it sends a signal, which is not an enhancement in the ABA level, and the stomata close [6]. Since levels of other hormones were not determined, their role in eliciting stomatal closure cannot be ruled out.

Indeed, in a large number of instances it appears that the levels of hormones in a tissue *relative* to one another are a more important consideration than are their absolute concentrations. At present we know a great deal about the response of plants and crops to a stressful environment and have some knowledge of the mode of transduction, but information on the perception mechanism is lacking. But what are these hormones, and what sets them apart from other naturally occurring metabolites that support plant growth and development? Chapter 24 in this volume and a chapter in an earlier volume [10] cover various aspects of the subject. Plant hormones (synonym; *phytohormone*) are *organic compounds, synthesized in one part of a plant and translocated to another part, which in very low concentrations cause a physiological response*. Three characteristics are defined that separate them from other nutrients and metabolites: they (a) are produced endogenously, (b) are transported to a place of action other than the site of production, and (c) act in low concentrations. Generally, they are divided into promoters (auxins, gibberellins, and cytokinins), inhibitors (abscisic acid, xanthoxin, and violaxanthin), ethylene, and other hypothetical growth substances much as "florigen," "death hormone," and so on.

From the outset, plant growth hormones have been heavily involved with "action at a distance." Their ability to move within the plant body has been a paramount consideration. Their transport characteristics are unlike those of most other substances. Water moves up from the root hairs to the transpiring surface along a gradient of (negative) hydrostatic pressure; sugars move from the leaf chlorenchyma down to the cambium or roots along a gradient of both (positive) hydrostatic pressure and sucrose concentration. Ions can be accumulated against concentration gradients, but these are not growth substances. Auxin does not appear to move along an auxin gradient but what is really a morphological gradient: polarly and predominantly from apex to base in leaves and shoots, and from apex to short distance behind the apex in roots in most of the species. As a rule, only slight polarity is shown by gibberellins, which in most concentrations travel freely in both directions in plants. Cytokinins may be carried in small amounts in the transpiration stream or in bleeding sap, but they mostly appear to remain close to the site of their formation. Abscisic acid also seems to move with only slight directionality.

Plant hormones are not only involved in cell division and/or cell differentiation, but there is also a wealth of information about many other processes, including induced gene expression and biochemical changes (Table 1). These changes point to the nature of the control exercised by hormones at the subcellular or molecular level. Unlike most animal hormones, which may have relatively specific types of physiological regulatory functions, plant hormones have considerable interplay between the various groups in the overall regulatory process. For instance, cytokinin interacts with auxin to produce callus and regenerate shoot and/or root; it opposes auxin in lateral bud development (apical dominance, compensatory growth); it resembles auxin in inhibiting root elongation; does strongly what auxin does weakly in promoting protein synthesis; and acts in the same way as auxin to cause cell division. Similarly, gibberellin acts like auxin in promoting elongation of etiolated stems and formation of parthenocarpic fruit (although it generally delays fruit set); reacts with auxin in producing elongation of isolated green stems; acts far more powerfully than auxin in elongation of intact stems; and does what auxin cannot do in causing flowering of long-day plants on noninductive photoperiods and the elongation of monocotyledonous leave sheaths. Yet it acts in the opposite direction to auxin regarding root formation by leaves and stem cuttings. It was suggested that auxin exerted its influence on lateral bud growth via enhanced ethylene synthesis, but recent evidence using transgenic-overproducing auxin, cytokinin, or ethylene tobacco (*Nicotiana tabaccum*) and *Arabidopsis thaliana* plants has shown that the auxin/cytokinin ratio controlled lateral bud growth and not ethylene [11,12]. In this chapter we cover the hormonal relations elicited under moderate stresses and not at the levels that cause injury/toxicity.

Table 1 Some Physiological Processes Affected by Known Classes of Hormones[a]

	Auxin	Gibberellin	Cytokinin	Abscisic acid	Ethylene
Dormancy		+	+	+	
Germination	+	+	+	+	+
Cell division	+	+	+		+
Cell expansion	+	+	+	+	+
Cell permeability	+		+	+	+
Cell differentiation	+		+		+
Transpiration		+	+	+	
Stomata	+		+	+	
Senescence	+	+	+	+	+
Protein metabolism	+	+	+	+	+
Nucleic acid metabolism	+	+	+	+	
Gene expression	+	+	+	+	+

Source: Ref. 10.
[a]Absence of a plus sign indicates a lack of information in the literature.

II. HORMONES

The adaptive responses elicited by plants experiencing stress is that cellular growth (irreversible cell enlargement) is reduced and stomata close. These responses take place before any measurable change in the turgor potential or total water potential is detected [6,7]. It is well known that both these responses are under hormonal control. Five class of hormones—auxins, gibberellins, cytokinins, abscisic acid, and ethylene—are commonly accepted to influence the rate at which plants grow and the rate and pattern of their development.

A. Auxins

Auxins is a generic name for compounds that typically stimulate cell elongation, but they also influence a wide range of other growth and developmental processes [10,13]. It has been observed that a reduction of −0.1 MPa in external water potential may result in a perceptible decrease in cellular growth and thus in shoot and root growth [14,15]. Considering its importance in keeping normal system functioning, it is surprising that very little work has been done to elucidate its role in stress responses of plants.

The available information shows that under salinity stress, the free or diffusible auxin indoleacetic acid (IAA) level decreases (Table 2) in maize (*Zea mays*) seedlings [16]. It also decreases with decreasing soil water potential in sunflower (*Helianthus annuus*) and *Anastatica hierochuntica* [17]. Guinn and Brummett [18,19] also observed a decrease in free auxin

Table 2 Effect of Salinity on Diffusible Auxin Recovered from Maize Coleoptile Tips

Treatment	Mean curvature (deg) + SE
100 µg/L IAA	24.3 + 0.6
Control	18.4 + 0.3
1% Salinity	12.6 + 0.4

Source: Ref. 30.

recovery, from the cotton (*Gossypium hirsutum*) boll abscission zone, under water-stress conditions. They explained the reduction as due to enhancement of auxin conjugation. Fluctuations in auxin concentration in field-grown *Sorghum bicolor* was observed by Kannangara et al. [20]. They concluded that the change did not bear a significant relation to diurnal changes in ABA or in the leaf water potential. However, an increase in the auxin level, with decreasing leaf water potential, has been reported in squash (*Cucurbita maxima*) hypocotyls [21]. Mechanical stress, imposed by shaking [22] or in the form of compacted soil [23,24], reduced or enhanced, respectively, the auxin levels in maize. Wounding of tobacco leaves also caused a reduction in endogenous IAA level [25].

The changes reported in the free auxin levels may be caused either by an overpowering pressure of stress on metabolism or on the transport kinetics or simultaneously on both parameters. In pea (*Pisum sativum*), decreasing water potential increased the in vitro IAA-oxidase activity, which was correlated with a reduction in the endogenous auxin level [26,27]. But in wheat (*Triticum aestivum*), a decrease in the enzyme activity has been observed [28]. Oat (*Avena sativa*) seeds, imbibed for 48 h in various Na$^+$ salt solutions, also recorded a decrease in the in vitro IAA-oxidase activity in all the treatments. Exposure to longer time periods, however, recorded a sharp increase in the enzyme activity, which was proportional to an increase in the salts concentration [29]. However, Naqvi et al. [30] observed no effect on in vivo IAA-oxidase activity in salinity-stressed maize coleoptiles. These observations lend support to the idea that the organization of the cytoplasm is such that enzymes in vivo do not respond to salts as they do in vitro [31].

Since the reduction in free auxin level was not due to its breakdown, Naqvi [32] studied the absorption and transport properties of [^{14}C]IAA in maize seedlings raised under NaCl or Na$_2$SO$_4$ salinity. Under these conditions no effect of the stress was observed on these parameters. Detailed studies on the kinetics of [^{14}C]IAA transport, in maize coleoptile segments, showed that a salinity of 0.5 or 1.0% did not materially affect either the velocity or the intensity (capacity) (Table 3), even though the stress reduced the seedling growth [30].

Transport studies using excised internode segments of water-stressed pea plants showed that stress reduced the transport capacity but did not affect the velocity of [^{14}C]IAA movement [33]. Further studies with cotton cotyledonary petioles also showed a reduction in auxin transport capacity from 30% to 15% when the stress was enhanced from –8 bar to –12 bar [34]. Studying the effect of aging, Davenport et al. [35] excised petiole segments from the upper, middle, and lower canopy of mature water-stressed cotton plants and obtained similar results regardless of the age of the petioles. However, a close analysis of their data show no effect up to –20 bar, which does not support their conclusion. Sheldrake [36] raised *Avena* seedlings for 4 to 6 days and stressed (0.5 *M* sorbitol) isolated mesocotyl segments for 2 h. In these stressed segments, he observed an enhancement of auxin transport. However, he could not explain the results in the absence of an effect on either absorption or the velocity of auxin transport. In a

Table 3 Salt Stress Effect on Various Transport Parameters Estimated by Linear Regression

Transport parameter	Control	0.5%	1.0%
Velocity (mm/h)	16.5	15.9	16.7
Intensity (cpm/h)	1134	1198	1243
Density (cpm/mm)	70.5	71.8	70.8

Source: Modified from Ref. 30.

gravity-compensated experiment, Dedolph et al. [37] observed that horizontal rotation did not affect either endogenous auxin production or absorption and transport of exogenously applied [^{14}C]IAA.

Treatment with IAA has been observed to reduce the salinity inhibition of wheat coleoptile segment elongation [38,39]. Khan [40] extended the work further and showed that salinity influenced the elastic as well as the plastic extensibility of the cell, and treatment with IAA affected these parameters favorably.

To counter the inhibitory effect of stress on germination, seedling growth, and the yield of crops, a number of workers have reported success with auxin pretreatment. Under salinity stress, IAA soaking of wheat seeds prior to sowing affected germination favorably and improved rooting as well as its length [41–44]. Besides improving these parameters, IAA treatment enhanced nutrient uptake in wheat [45], and in maize it also significantly enhanced dry matter and the grain yield [46]. In salinity-stressed wheat crop, spraying weekly with IAA (5 mg/L) solution, begining 4 weeks after sowing until maturity, significantly enhanced the shoot dry weight (50%) and the grain yield by 31% [42,47]. When water stress coincided with head formation, flowering, or the milk stage of the wheat crop, spraying with IAA or 2,4-D (200 μg per plant) also had a significant beneficial effect. A wheat crop experiencing water stress for a shorter period also exhibited an increased yield [48].

Salama et al. [49] sprayed pot-grown tomato (*Lycopersicon esculentum*) plants (8 weeks old) with 50 mg/L IAA and observed improvement in growth under salinity stress. But in the same experiment, the rocket plant (*Eruca sativa*) did not respond to the treatment. Treatment with IAA has also been shown to improve significantly the growth of seawater (*Caloglossa lepreuii*, *Bostrychia binderi*, and *Rhizoclonium implexum*) as well as freshwater (*Pithophora*; species unidentified) algae under salinity stress [50].

B.   Gibberellins

Up to now the presence of 84 characterized gibberellins (i.e., GA_1, GA_2, GA_3, . . ., GA_{84}) have been established from fungi, plants, and crops. But all the gibberellins have biological activity on plants and crops. Besides other actions, the primary action of gibberellins is on stem elongation, which is a consequence of both increased cell multiplication and increased cell size.

Under desiccating stress, excised lettuce (*Lactuca sativa*) leaves recorded a rapid decline in gibberellin-like activity which was proportional to the intensity and duration of the stress [51]. The reduction was also closely related to the enhancement in leaf water saturation deficit and a concomitant rise in ABA level. After 6 h of desiccation, the gibberellin-like activity was barely detectable but returned to the control level after 4 h of recovery from stress. Aharoni and Richmond [52] showed further that a 10% reduction in the relative water content (RWC) of detached lettuce leaves did indeed enhance the decline in tissue gibberellin. Reduction in gibberellin level under water stress conditions can be deduced indirectly from the studies on gladiolus (*Gladiolus psittacinus*) flower bud growth [53]. Mechanical stress imparted to bean (*Phaseolus vulgaris*) seedlings resulted in decreased stem growth and a decline in gibberellin concentration [54]. But recent data using HPLC and dwarf rice (*Oryza sativa*) bioassay have shown that drought-stressed but otherwise aeroponically grown sunflower plants showed neither a change in total gibberellin level nor a change in their distribution [55]. These studies also showed that the gibberellin content of shoots was lower than that of the roots, supporting the contention that roots are the major source of gibberellins in plants and crops.

Transport experiments through cotton petiole segments employing ^{14}C-labeled gibberellin in a donor–tissue–receiver system did not show the influence of either water or anaerobic stress on the transport capacity of the labeled material [56]. These studies also did not show any change from the 1 mm/h transport velocity.

Studies using GA$_3$ to improve seed germination and seedling growth under stress conditions have shown that the treatment enhanced these parameters in lettuce [57], flax (*Linum usitatissimum*), sesame (*Sesamum indicum*), and onion (*Allium cepa*) [58] and in the polymorphic seeds and seedlings of *Atriplex triangularis* [59]. In wheat seeds the stress reduction of α-amylase activity can be enhanced effectively by gibberellin [60,61]. Under salinity stress, germination of barley (*Hordeum vulgare*) (cv. Natans Schubl) seeds was enhanced by GA$_3$ treatment, which also increased the radical and coleoptile lengths [62].

Treatments with GA$_3$-enhanced coleoptile lengths of the stressed wheat seedlings [61] and that of bean seedling growth at lower but not at higher levels [63]. In field- or pot-grown cotton, spray application of GA$_3$ has been reported to have beneficial effects on dry matter yield and enhancement of stress tolerance [64,65]. Recently, Rao and Ram [53], using spikes of gladiolus, observed that the bud opening was sensitive to stress and required GA$_3$, indicating that the stress interfered with the gibberellin supply system.

Improvement in wheat crop yield and nutrient uptake, under stress conditions, has been observed with GA$_3$ treatment [45]. Starck and Kozinska [66] observed that gibberellin-treated bean plants absorbed more P and Ca and less Na and partially reestablished the monovalent/divalent ratio, which increased in the apical parts and leaves due to salinity stress.

C. Cytokinin

Currently, 25 free cytokinins are reported from plants and crops, among which some are active in causing maximum tobacco callus growth, at concentrations as low as 4 n*M* [67]. Which of the cytokinin(s) should be measured under stressful conditions may have been a problem. They have been classified by substituted base into three groups: zeatin (Z), dihydrozeatin [(diH)Z], and N^6-(Δ^2-isopentenyl)adenine (2iP). McGaw [68] has concluded that of the hormonally active cytokinins, the nucleotides are probably associated with uptake, the "active" form comprising the ribosides and the bases. Letham and Palni [69] observed that ribosides were the major forms transported through the xylem and phloem. But *O*-glucosides and nucleotides were observed to be the major constituents in bean xylem exudates [70].

Actively dividing regions of plants are known to synthesize cytokinins [71,72]. Since there are more meristematic regions in the root system, the major portion of the cytokinins are produced there and transported to the shoot [72,73]. These root-originated cytokinins, along with the locally synthesized cytokinins, influence the control of both development and senescence of the whole plant [74]. Despite their established role in plant and crop development, information as to how endogenous cytokinins are affected under stressful conditions is meager.

Vascular exudates and/or leaves of stressed plants exhibit reduced cytokinin activity, and the response is known to be rapid [7,10,55,72,73]. Tomato plants exposed to salinity stress for 8 days showed reduced cytokinin activity which was correlated with reduced growth. When the stress was relieved, the cytokinin regained normal levels in the next 4 days (Figure 2) [75]. Neuman et al. [76] observed that xylem sap did contain less cytokinin under root hypoxia, but zeatin riboside (ZR), dihydrozeatin riboside (DHZR), and their equivalents were not reduced in the leaves of bean and poplar (*Populus trichocarpa* × *P. deltoides*). But Bano et al. [7] have shown that rice seedling roots, experiencing 30 h of drying, exhibited a reduced level of cytokinins, isopentenyladenine (2iP) + isopentenyladenosine (2iPA), and zeatin (t-Z) + zeatin riboside (t-ZR) in xylem sap, which started increasing after rewatering.

The explanation for the detection of reduced cytokinin activity is either enhanced degradation or a reduction in its biosynthesis [76]. However, critical evaluation of the evidence

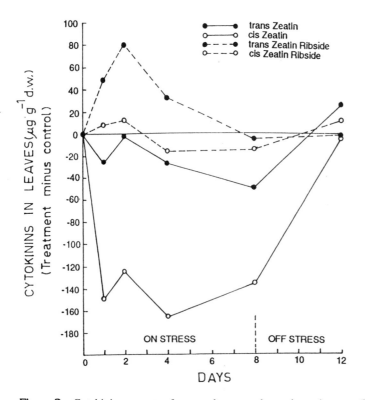

Figure 2 Cytokinin contents of tomato leaves under prolonged stress. (From Ref. 75.)

reveals that they are largely inferential and not tested directly. Enzymes such as cytokinin oxidase, cytokinin-7-glucosyl transferase, and β-(9-cytokinin)alanine synthase, which affect the metabolism of cytokinins have been purified and characterized [69]. But we have no data on the activity of these enzymes in a stressful environment. Transport studies using cotton leaf petiole segments from plants exposed to water or anaerobic stress have revealed that there was no effect of the stress on the kinetin transport capacity. It was observed further that the labeled material moved nonpolarly with a velocity of 1 mm/h [56]. But Hubick et al. [55] observed reduced transport of root-to-shoot cytokinin under stress conditions. These workers also reported no material difference as to the cytokinin metabolism under stressful conditions. Indirect evidence on the flux of cytokinin movement in the xylem sap also indicates that stress does not enhance cytokinin metabolism; rather, it reduces its transport [76].

In lettuce, treatment with kinetin largely overcame the salinity stress effect and enhanced the fresh weight of early seedlings at the higher salinity levels [77,78]. However, Hegarty and Ross [79] observed that although kinetin treatment effectively enhanced the germination of lettuce and red clover (*Trifolium incarnatum*) seeds under water stress conditions, it inhibited root growth. In tomato and pea, kinetin treatment enhanced seed germination and seedling growth under salinity stress [80,81]. Naqvi et al. [82] observed that kinetin partially countered the inhibitory effect of salinity on wheat seedling growth. It now seems well documented that treatment with kinetin enhances seed germination under stressful conditions [83].

Spray applications of cytokinin caused more rapid delay of senescence of detached leaves in stressed plants that could otherwise senesce than in nonstressed plants [84]. It was further shown that as regards kinetin and benzyladenine (BA), only BA enhanced the fresh and dry weights of the salinity-stressed bean plants. Partial or complete reversal of the inhibitory effect of stress on CO_2 fixation, leucine incorporation, chlorophyll content, and stomatal conductance in *Nicotiana rustica* leaf discs has been reported [85]. Prevention of leaf necrosis induced by stress has also been achieved with kinetin treatment [86].

D. Ethylene

Plants and crops normally produce ethylene, which is the simplest organic compound and is biologically active in trace amounts. Due to the diversity of its action and being a gaseous hormone, it is difficult to assign a definitive role to this hormone in growth responses. However, the classical *triple response*, characterized by growth retardation, increase in diameter, and horizontal growth of shoots, is still sometimes used by workers to identify and measure ethylene.

When subjected to stressful conditions, ethylene production is enhanced. Since it is a gaseous hormone, its emissions can increase from 2 to 50 times or more, depending on the sensitivity of the tissue and the severity of the stress. However, this surge in ethylene is short-lived, peaking quickly and returning to the normal levels within 24 hours or less [87]. Tissues injured by a variety of stresses (wounding, mechanical, chemicals, temperature extremes, pathogens, etc.) are the site of such enhanced ethylene production. This induced ethylene is apparently produced by the same biosynthetic pathway that produces the normal hormonal ethylene. However, Kacperska and Kubacka-Zebalska [88] suggested two prerequisites for such stress-induced in vivo ethylene biosynthesis: promotion of ACC (1-aminocyclopropane-1-carboxylic acid) synthesis and activation of a free-radical generating system. The latter system was needed for the nonenzymatic conversion of ACC to ethylene and depended on the activation of the membrane-associated lipoxygenase, caused by stress-induced alterations in membrane properties.

It is interesting to note that whenever detached plant organs were stressed, an enhanced production of ethylene was observed [89,90]. This enhancement was more than 30-fold at a water loss of 9% of the initial fresh weight of detached wheat leaves in 4 h, after which it rapidly declined at or even higher stress [89]. On the other hand, when intact plants of sunflower, bean, cotton, and miniature rose (*Rosa hybrida*) were stressed moderately, no enhancement of ethylene production was shown [55,91].

Ethylene can promote cell elongation, but generally it is known to inhibit elongation growth [92,93]. Cells of different sizes and shapes can be produced by influencing the internal levels of ethylene or auxin [94]. The inhibitory effect of stress has been either reduced or prevented in many cultivars of lettuce [95,96]. In rice, ethylene production has been suggested as a marker for screening stress-tolerant lines because it correlated well with seedling growth [97].

E. Abscisic Acid

Regulation and maintainance of normal growth and development of plants and crops is dependent on the balance between the level of hormones and the changing sensitivity of the tissues throughout the life cycle. Abscisic acid was identified originally as a dormancy-inducing and abscission-accelerating hormone, but soon it was realized that it also had many other physiological functions [10,13]. Since the discovery that endogenous abscisic acid (ABA) level increases in response to stress and that ABA inhibits transpirational water loss [98–100], later workers generally called it a "stress hormone." Increasing ABA concentration leads to many physiological changes, and Quarrie's [101] remark that "a plant that cannot make abscisic acid (ABA) is in serious trouble" signifies its importance. Abscisic acid is a typical sesquiterpene, consisting of three isoprene units, and is known to be synthesized through carotenoid pathways with xanthophylls being immediate precursors [102].

Under stressful environments, ABA not only accumulates in higher plants and crops but also in algae, fungi, and bryophytes [103–105]. Some excellent reviews describing the role of ABA in plants under stressful environments have been published [10,106,107].

Abscisic acid is found in all the organs and tissues tested; therefore, it is easily available for a physiological role. Its role as a stress hormone was suggested primarily by Wright and Hiron [99]. They observed that when excised wheat leaves were wilted, ABA content increased 40-fold during the first 0.5 h. Later, excised leaves of over 70 species of plants and crops were surveyed and all recorded, an increase in ABA content following a period of wilting [108]. Zabadal [109] exposed two species of *Ambrosia* to a desiccating atmosphere (62,400 lux, 28°C, relative humidity 24%) and observed that ABA started accumulating when leaf water potential declined to –10 to –12 bar. Sharp and Davies [110] have also shown that a slight decrease in isolated root-tip turgor potential triggers a buildup of ABA. These results suggest that there is a critical threshold water potential value that switches on the ABA accumulating system.

In short-term experiments, isolated segments stressed for various lengths of time generally showed a rapid buildup of ABA (Figure 3). It has been estimated that detached leaves of a rice line can accumulate ABA at a rate of 4 nmol/g fresh weight per hour [107]. Analysis of xylem sap from 6- 7-week-old intact rice seedlings has further substantiated the observation that the ABA level is enhanced when the roots were stressed for 30 h [7]. Although the workers concluded that bound ABA increased more than free ABA, their data showed a 13-fold increase in free ABA as against a sevenfold increase in bound ABA (see Table 1). But in long-term studies, tobacco plants stressed for 10 to 12 days showed a major peak at the fourth day and thereafter declined rapidly, reaching a control level at days 6 through 12 [111]. Similarly, using tomato plants, Walker and Dumbroff [75] observed that even though osmotic potential and water content continued to decline over 8 days of stress, the accumulation of ABA declined rapidly after 2 days, reaching a normal level in the next 6 days (Figure 4). Recently, Robinson and Barritt [112] have also observed that by the third day of a 21-day stress period, the ABA level in apple (*Malus domestica*) seedlings was almost the same as that of the unstressed control. The conclusion reached was that the continued shoot growth could not be correlated with the turgor pressure or the ABA levels of mature leaves. In young bean seedlings, Trejo and Davies [6] have also observed that an increase in ABA level started 6 days after the imposition of stress and reached a twofold level in the next 6 days. But the stomates close much earlier, on the fourth day of the stress. A number of workers have demonstrated that the accumulation of ABA lags behind stomatal closure under a stressful environment [6,113] and upon release of stress, leaf ABA declines ahead of stomatal opening [113,114]. Thus there is enough evidence to conclude that stress-enhanced ABA may not be correlated with growth or the stomatal response and is also not sustained if the plants are stressed for an extended period.

Bellandi and Dörffling [115] observed that ^{14}C-labeled ABA moved in the phloem of turgid pea seedlings, and Zeevaart [116] showed further that ABA and its metabolites could be recovered from phloem exudates of stressed *Ricinus communis* leaves. Also, in white lupin (*Lupinus albus*), Hoad [117] has observed further that ABA is actively translocated out of mature leaves in the phloem and transported to stem apices, fruits, and seeds. There is substantial evidence that ABA produced in roots is rapidly translocated over long distances in plants via phloem and xylem [10,118]. Abscisic acid moves bidirectionally and not polarly in the plants and is also transported rapidly from cell to cell. Thus the presence of ABA in a tissue may not be indicative of its formation there.

Abscisic acid affects shoot and root growth differentially, which may be an advantage for survival in a stressful environment. Inhibition of shoot growth is unequivocal, but promotion of root growth is contradictory and needs satisfactory resolution. It causes stunted shoot growth by influencing cell wall extensibility, leading to a decline in cell elongation [6,21]. Abscisic

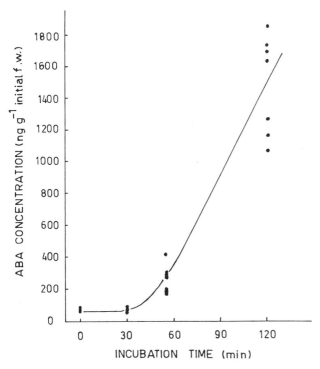

Figure 3 Accumulation of abscisic acid (ABA) in detached leaves of rice under partial dehydration. (From Ref. 107.)

acid also inhibited the growth of wheat coleoptiles, barley shoots, bean axes, and second leaf sheaths of rice seedlings at concentrations between 10^{-7} and 10^{-5} M [119]. In maize coleoptiles, ABA, the next dominant hormone after IAA, reduced elongation rate and inhibited auxin-induced growth by affecting cell wall loosening [120,121]. But in the case of mesocotyl, ABA plays a protective role in the growth inhibition caused by supramaximal auxin concentrations [122]. Abscisic acid can promote [123,124] or even have little effect [125] on root growth. But Saab et al. [126], using fluridone treatment (an inhibitor of carotenoid biosynthesis) and *vp*5 mutant (deficient in carotenoid biosynthesis) of maize, showed that ABA played a direct role in both the maintenance of primary root growth and the inhibition of shoot growth at a low water potential.

Why such a difference exists between shoot and root cells with regard to ABA response has not been investigated. It is significant that in a tissue culture system, these two divergent organs have the capacity to regenerate the entire plant under suitable conditions. It has also been shown that ABA alters gene expression by stimulating the synthesis of new proteins, which seems to be associated with enhanced stress tolerance [127,128].

III. HORMONE BALANCE

Plant growth and development can be profoundly influenced by the relative levels of hormones rather than by their absolute concentrations. From the preceding discussion it appears that plants and crops that experience moderately stressful conditions do modify their hormonal

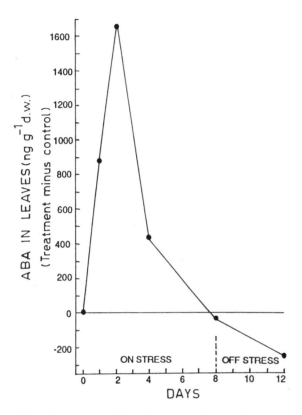

Figure 4 Free abscisic acid content of tomato leaves under prolonged stress. (From Ref. 75.)

levels. Producing more abscisic acid and less cytokinin and auxin, gibberellin and ethylene are usually unaffected. Enhancement in ABA and ethylene is generally transient, but reductions in cytokinins and auxin persist until the stress is relieved. Since auxin, cytokinin, and abscisic acid are known to influence a number of vital growth and developmental processes, including maintainance of cell membrane integrity, induction of hydrolases, cell growth, stomatal behavior, and programmed senescence, they have attracted more attention. Many scientists are now working on abscisic acid physiology under stressful conditions, and the information gleaned by each scientist is beginning to merge, like pieces of a jigsaw puzzle, into a larger, more detailed picture. But information on cytokinin and auxin lags far behind. Hormones not only influence various processes but regulate each other's levels. Auxin and cytokinin stimulate endogenous ethylene production [12,90] whereas ABA inhibits it [129], and cytokinin is known to inhibit ABA biosynthesis in *Cercospora rosicola* [130] and to act in regulating ABA and IAA levels in maize [131]. From such an interlinked control system it is difficult to assume that enhancement in one hormone can be looked on as a mechanism imparting stress tolerance. Tissue sensitivity may also be an important factor, which could respond immediately to a change in the relative levels of the hormone(s). Therefore, the more we have tried to learn about hormonal relations under stressful conditions, the more it is realized that there are a number of missing pieces of information that are needed to solve the puzzle. Advances in molecular biology techniques, together with the use of hormone-deficient and/or overproducing mutants, may further elucidate the unresolved issues regarding the role of hormone(s) in stress responses.

IV. CONCLUSIONS

Reasonable evidence is now available, indicating that plants and crops experiencing a stressful environment respond by closing their stomata and reducing the expansive growth. Since plants are sessile, they cope with the hostile environment by modifying their hormonal balance to regulate these two processes.

Presently, five classes of plant hormones are known to influence a plethora of plant functions which ultimately regulate their growth and development. Unlike animal hormones, they do not act independently but interact to exert their influence on growth regulation. Information is almost nonexistent regarding the perception of stressful environment, but we do know that transduction is elicited by changes in the hormonal level. In general, levels of auxin and cytokinin are reduced and ABA is enhanced. There is strong evidence suggesting that the stomatal movement is regulated by interaction among auxin, cytokinin, and ABA. The tissues' sensitivity to these hormones, under a stressful environment, may also be an important factor.

Alleviation of adverse effects of stress have also been achieved, by exogenous hormonal treatments, in a number of plant and crop species. The accumulation of ABA, in a stressful environment, has captured the imagination of many investigators, but the reduction in auxin and cytokinin levels has not been investigated adequately. Unless a balanced approach is taken, it will be a futile exercise to understand the transduction mechanism that elicits the primary response in a stressful environment.

REFERENCES

1. *Saline Agriculture: Salt Tolerant Plants for Developing Countries*, Report of a Panel of the Board on Science and Technology for International Development, Office of International Affairs, National Research Council, Washington, D.C. (1990).
2. Y. Waisel, *Biology of Halophytes*, Academic Press, London (1972).
3. D. Pasternak, *Annu. Rev. Phytopathol.*, *25*: 271 (1987).
4. F. B. Salisbury and C. W. Ross, *Plant Physiology*, Wadsworth, Inc., Belmont, California (1992).
5. F. S. Chapin III, *BioScience*, *41*: 29 (1991).
6. C. L. Trejo and W. J. Davies, *J. Exp. Bot.*, *42*: 1507 (1991).
7. A. Bano, K. Dörffling, D. Bettin, and H. Hahn, *Aust. J. Plant Physiol.*, *20*: 109 (1993).
8. S. Kaku, *Plant Cell Physiol.*, *34*: 535 (1993).
9. A. Boussiba, A. Raikin, and A. E. Richmond, *Plant Physiol.*, *56*: 337 (1975).
10. S. S. M. Naqvi, in *Handbook of Plant and Crop Stress* (M. Pessarakli, ed.), Marcel Dekker, Inc., New York. p. 383 (1994).
11. J. I. Medford, R. Horgan, Z. El-Sawi, and H. J. Klee, *Plant Cell*, *1*: 403 (1989).
12. C. P. Romano, M. L. Cooper, and H. J. Klee, *Plant Cell*, *5*: 181 (1993).
13. A. C. Leopold, in *Hormone Action in Plant Development: A Critical Appraisal* (G. V. Hoad, J. R. Lenton, M. B. Jackson, and R. K. Atkin, eds.), Butterworth, London, p. 3 (1987).
14. P. M. Neumann, E. Van Volkenburgh, and R. E. Cleland, *Plant Physiol.*, *88*: 223 (1988).
15. N. Sakurai and S. Kuraishi, *Plant Cell Physiol.*, *29*: 1337 (1988).
16. S. M. Naqvi and R. Ansari, *Experientia*, *30*: 350 (1970).
17. W. Hartung and J. Witt, *Flora (Jena) Abt.*, *B157*: 603 (1968).
18. G. Guinn and D. L. Brummett, *Plant Physiol.*, *83*: 199 (1987).
19. G. Guinn and D. L. Brummett, *Plant Physiol.*, *86*: 28 (1988).
20. T. Kannangara, R. C. Durley, and G. M. Simpson, *Z. Pflanzenphysiol.*, *106*: 55 (1982).
21. N. Sakurai, M. Akiyama, and S. Kuraishi, *Plant Cell Physiol.*, *26*: 15 (1985).
22. P. L. Neel and R. W. Harris, *Science*, *175*: 718 (1972).
23. D. R. Lachno, R. S. Harrison-Murray, and L. J. Audus, *J. Exp. Bot.*, *136*: 943 (1982).
24. D. R. Lachno, in *Growth Regulators in Root Development* (M. B. Jackson, and A. D. Stead, eds.), British Plant Growth Regulation Group, Monograph 10, Bristol, England, p. 37 (1983).

25. R. W. Thornburg and X. Y. Li, *Plant Physiol.*, *96*: 802 (1991).
26. B. Darbyshire, *Plant Physiol.*, *47*: 65 (1971).
27. B. Darbyshire, *Physiol. Plant.*, *25*: 8 (1971).
28. V. M. Mills and G. W. Todd, *Plant Physiol.*, *51*: 1145 (1973).
29. A. K. Shukla and B. D. Baijal, *Indian J. Plant Physiol.*, *20*: 157 (1977).
30. S. S. M. Naqvi, R. Ansari, and M. Hanif, in *Prospects for Biosaline Research* (R. Ahmad and A. San Pietro, eds.), Department of Botany, University of Karachi, Karachi, Pakistan, p. 307 (1986).
31. H. Greenway and R. Munns, *Annu. Rev. Plant Physiol.*, *31*: 149 (1980).
32. S. M. Naqvi, *Experientia*, *28*: 1246 (1972).
33. H. Kaldewey, U. Ginke, and G. Wawczyniak, *Ber. Dtsch. Bot. Ges.*, *81*: 563 (1974).
34. T. L. Davenport, P. W. Morgan, and W. R. Jordan, *Plant Physiol.*, *59*: 554 (1977).
35. T. L. Davenport, P. W. Morgan, and W. R. Jordan, *Plant Physiol.*, *65*: 1023 (1980).
36. A. R. Sheldrake, *Planta*, *145*: 113 (1979).
37. R. R. Dedolph, S. M. Naqvi, and S. A. Gordon, *Plant Physiol.*, *41*: 897 (1966).
38. S. S. M. Naqvi, K. Samo, and S. A. Ala, in *Environmental Stress and Plant Growth* (S. S. M. Naqvi and R. Ansari, eds.), Atomic Energy Agricultural Research Center, Tando Jam, Pakistan, p. 19 (1986).
39. K. B. Samo, M.Phil. thesis, University of Sindh, Jamshoro, Pakistan (1987).
40. M. A. Khan, M.Phil. thesis, University of Sindh, Jamshoro, Pakistan (1990).
41. M. N. Sarin, *Proc. Natl. Acad. Sci. (India) Sec. B*, *XXXI*: 287 (1961).
42. M. N. Sarin, *Agra Univ. J. Res. (Sci.)*, *XI*: 187 (1962).
43. B. L. Darra, S. P. Seth, H. Singh, and R. S. Mendiratta, *Agron. J.*, *65*: 292 (1973).
44. R. B. Chippa and P. Lal, *J. Indian Soc. Soil Sci.*, *26*: 390 (1978).
45. A. S. Balki and V. W. Padole, *J. Indian Soc. Soil Sci.*, *30*: 361 (1982).
46. B. L. Darra and S. N. Saxena, *Plant Soil*, *38*: 657 (1973).
47. M. N. Sarin and I. M. Rao, *Agra Univ. J. Res. (Science)*, *X*: 7 (1961).
48. T. G. Kudrev, *Wiss. Z. Univ. Rostock*, *16*: 4 (1967).
49. P. M. Salama, S. E. A. Khodary, and M. M. Heikal, *Phyton (Austria)*, *21*: 177 (1981).
50. J. Nowak, B. Sonaike, and G. W. Lawson, *Environ. Pollut.*, *51*: 213 (1988).
51. N. Aharoni, A. Blumenfeld, and A. E. Richmond, *Plant Physiol.*, *59*: 1167 (1977).
52. N. Aharoni and A. E. Richmond, *Plant Physiol.*, *62*: 224 (1978).
53. I. R. Rao and H. Y. M. Ram, *J. Plant Physiol.*, *122*: 181 (1986).
54. H. Suge, *Plant Cell Physiol.*, *19*: 1557 (1978).
55. K. T. Hubick, J. S. Taylor, and D. M. Reid, *Plant Growth Regul.*, *4*: 139 (1986).
56. T. L. Davenport, W. R. Jordan, and P. W. Morgan, *Plant Physiol.*, *63*: 152 (1979).
57. A. A. Khan, in *The Physiology and Biochemistry of Seed Dormancy and Germination* (A. A. Khan, ed.), Elsevier/North-Holland Biomedical Press, Amsterdam, p. 283 (1977).
58. M. M. Heikal, A. M. Ahmed, and M. A. Shadad, *Biol. Plant.*, *24*: 124 (1982).
59. M. A. Khan and I. A. Ungar, *Physiol. Plant.*, *63*: 109 (1985).
60. R. Ansari, S. M. Naqvi, and A. R. Azmi, *Pak. J. Bot.*, *9*: 163 (1977).
61. K. Kabar and S. Baltepe, *Doga Turk Biyol. Derg.*, *11*(3): 108 (1987).
62. M. N. Sarin and A. Narayanan, *Physiol. Plant.*, *21*: 1201 (1968).
63. R. H. Nieman and L. Bernstein, *Am. J. Bot.*, *46*: 667 (1959).
64. D. Agakishiev, *Fiziol. Rast.*, *11*: 201 (1964).
65. A. A. Ibrahim, *Ann. Agric. Sci. Moshotohor*, *21*: 519 (1984).
66. Z. Starck and M. Kozinska, *Acta Soc. Bot. Poloniae*, *49*: 111 (1989).
67. K. Koshimizu and H. Iwamura, in *Chemistry of Plant Hormones* (N. Takahashi, ed.), CRC Press, Boca Raton, Fla., p. 153 (1986).
68. B. A. McGaw, in *Plant Hormones and Their Role in Plant Growth and Development* (P. J. Davies, ed.), Martinus Nijhoff, Dordrecht, The Netherlands, p. 76 (1987).
69. D. S. Letham and L. M. S. Palni, *Annu. Rev. Plant Physiol.*, *34*: 163 (1983).
70. M. V. Palmar and O. C. Wong, *Plant Physiol.*, *79*: 269 (1985).
71. C. M. Chen, J. R. Ertl, S. M. Leisner, and C.-C. Cheng, *Plant Physiol.*, *78*: 510 (1985).

72. L. D. Incoll and P. C. Jewer, in *Cytokinins: Plant Hormone in Search of a Role* (R. Horgan and B. Jeffcoat, eds.), British Plant Growth Regulator Group, Monograph 14, Bristol, England, p. 85 (1987).

73. P. G. Blackman and W. J. Davies, *J. Exp. Bot.*, *36*: 39 (1985).

74. P. F. Wareing, R. Horgan, I. E. Hanson, and W. Davis, in *Plant Growth Regulation* (P. E. Pilet, ed.), Springer-Verlag, Berlin, p. 147 (1977).

75. M. A. Walker and E. B. Dumbroff, *Z. Pflanzenphysiol.*, *101*: 661 (1981).

76. D. S. Neuman, S. B. Rood, and B. A. Smit, *J. Exp. Bot.*, *41*: 1325 (1990).

77. O. A. Odegbaro and O. E. Smith, *J. Am. Soc. Hortic. Sci.*, *94*: 167 (1969).

78. M. R. Kaufmann and K. J. Ross, *Am. J. Bot.*, *57*: 413 (1970).

79. T. W. Hegarty and H. A. Ross, *Isr. J. Bot.*, *29*: 83 (1980/81).

80. S. Bozcuk, *Ann. Bot.*, *48*: 81 (1981).

81. R. C. Setia and S. Narang, *Flora (Jena)*, *177*: 369 (1985).

82. S. S. M. Naqvi, R. Ansari, and A. N. Khanzada, *Plant Sci. Lett.*, *26*: 279 (1982).

83. K. Kabar and S. Kudrat, *J. Plant Physiol.*, *128*: 179 (1987).

84. J. W. O'Leary and J. T. Prisco, *Adv. Front. Plant Sci.*, *25*: 129 (1970).

85. K. Katz, K. Dehan, and C. Itai, *Plant Physiol.*, *62*: 836 (1978).

86. A. Benzioni, Y. Mizrahi, and A. E. Richmond, *New Phytol.*, *73*: 315 (1974).

87. D. T. Tingey, *HortScience*, *15*: 630 (1980).

88. A. Kacperska and M. Kubacka-Zebalska, *Physiol. Plant.*, *77*: 231 (1989).

89. A. Apelbaum and S. F. Yang, *Plant Physiol.*, *68*: 594 (1981).

90. S. T. C. Wright, *Planta*, *148*: 181 (1980).

91. P. W. Morgan, C. J. He, J. A. De Greef, and M. P. De Proft, *Plant Physiol.*, *94*: 1616 (1990).

92. M. Lieberman, *Annu. Rev. Plant Physiol.*, *30*: 533 (1979).

93. W. Eisinger, *Annu. Rev. Plant Physiol.*, *34*: 225 (1983).

94. D. G. Osborne, in *Perspective in Experimental Biology* (N. Sunderland, ed.), Pergamon Press, Oxford, Vol. 2, p. 89 (1976).

95. F. B. Abeles, *Plant Physiol.*, *81*: 780 (1986).

96. J. Prusinski and A. A. Khan, *J. Am. Soc. Hortic. Sci.*, *115*: 294 (1990).

97. A. A. Khan, M. Akbar, and D. V. Seshu, *Crop Sci.*, *27*: 1242 (1987).

98. C. H. A. Little and D. C. Eidt, *Nature*, *220*: 498 (1968).

99. S. T. C. Wright and R. W. P. Hiron, *Nature*, *224*: 719 (1969).

100. C. J. Mittelheuser and R. F. M. Van Steveninck, *Nature*, *221*: 281 (1969).

101. S. A. Quarrie, *Biol. Vestn.*, *39* (1/2): 67 (1991).

102. J. A. D. Zeevaart, C. D. Rock, F. Fantauzzo, T. G. Heath, and D. A. Gage, in *Abscisic Acid: Physiology and Biochemistry* (W. J. Davies and H. G. Jones, eds.), BIOS Scientific Pubishers, Oxford, p. 39 (1991).

103. W. Hartung, E. W. Weiler, and O. H. Volk, *Bryology*, *90*: 393 (1987).

104. R. Hirsch, W. Hartung, and H. Gimmler, *Bot. Acta*, *102*: 326 (1989).

105. O. Werner, R. M. R. Espin, M. Bopp, and R. Atzorn, *Planta*, *186*: 99 (1991).

106. W. J. Davies and T. A. Mansfield, in *Abscisic Acid* (F. T. Addicott, ed.), Praeger, New York, p. 237 (1983).

107. S. A. Quarrie, in *Environmental Stress in Plants*, NATO ASI Series, Vol. G 19 (J. H. Cherry, ed.), Springer-Verlag, Berlin, p. 27 (1989).

108. S. T. C. Wright, in *Phytohormones and Related Compounds: A Comprehensive Treatise* (D.S. Letham, P. B. Goodwin, and T. J. V. Higgins, eds.), Elsevier/North-Holland Biomedical Press, Amsterdam, p. 495 (1978).

109. T. J. Zabadal, *Plant Physiol.*, *53*: 125 (1974).

110. R. E. Sharp and W. J. Davies, in *Plants Under Stress* (H. G. Jones, T. J. Flowers, and M. B. Jones, eds.), Cambridge University Press, Cambridge, p. 71 (1989).

111. Y. Mizrahi, A. Blumenfeld, and A. E. Richmond, *Plant Cell Physiol.*, *13*: 15 (1972).

112. T. L. Robinson and B. H. Barritt, *J. Am. Soc. Hortic. Sci.*, *115*: 991 (1991).

113. M. F. Beardsell and D. Cohen, *Plant Physiol.*, *56*: 207 (1975).

114. M. M. Ludlow, T. T. Ng, and C. W. Ford, *Aust. J. Plant Physiol.*, *7*: 299 (1980).

115. P. M. Bellandi and K. Dörffling, *Plant Physiol.*, *32*: 365 (1974).
116. J. A. D. Zeevaart, *Plant Physiol.*, *59*: 788 (1977).
117. G. V. Hoad, *Planta*, *142*: 287 (1978).
118. O. Wolf, W. D. Jeschke, and W. Hartung, *J. Exp. Bot.*, *41*: 593 (1990).
119. N. Hirai, in *Chemistry of Plant Hormones* (N. Takahashi, ed.), CRC Press, Boca Raton, Fla., p. 201 (1986).
120. E. W. Weiler, P. S. Jordan, and W. Conrad, *Planta*, *153*: 561 (1981).
121. U. Kutschera and P. Schopfer, *Planta*, *169*: 437 (1986).
122. B. V. Milborrow, in *Advanced Plant Physiology* (M. B. Wilkins, ed.), Pitman, London, p. 440 (1984).
123. T. J. Mulkey, M. L. Evans, and K. M. Kuzmanoff, *Planta*, *157*: 150 (1983).
124. J. M. Robertson, K. T. Hubick, E. C. Yeung, and D. M. Reid, *J. Exp. Bot.*, *41*: 325 (1990).
125. R. L. Creelman, H. S. Mason, R. J. Bensen, J. S. Boyer, and J. E. Mullet, *Plant Physiol.*, *92*: 205 (1990).
126. I. N. Saab, R. E. Sharp, J. Pritchard, and G. S. Voetberg, *Plant Physiol.*, *93*: 1329 (1990).
127. N. K. Singh, P. C. LaRosa, A. K. Handa, P. M. Hasegawa, and R. A. Bressan, *Proc. Natl. Acad. Sci. USA*, *84*: 739 (1987).
128. Z. Xin and H. Li, *Plant Physiol.*, *101*: 277 (1993).
129. T. Zhi-Yi and K. V. Thimann, *Physiol. Plant.*, *75*: 13 (1989).
130. S. M. Norman, R. D. Bennett, V. P. Maier, and S. M. Poling, *Plant Sci. Lett.*, *28*: 255 (1982).
131. M. Bourquin and P. E. Pilet, *Physiol. Plant.*, *80*: 342 (1990).

31

Salinity-Induced Changes in Rice: Effects on Plant–Insect Interactions

Muhammad Salim and Muhammad Akbar
National Agricultural Research Centre, Islamabad, Pakistan

I. INTRODUCTION

Coastal or inland salinity is one of the major rice production constraints in southern and southeastern Asia [1,2]. The importance of coastal saline soils as prospective rice land lies in their widespread occurrence and existence in humid tropical deltas, river basins, and estuaries in physical environments conducive to rice growth [3]. In coastal saline soils, salinity, acidity, P and Zn deficiencies, and B and Fe toxicities limit high yields of rice. In inland saline soils, constraints on rice production are salinity and deficiencies of N, P, K, and Zn [4].

Saline soil is one whose electrical conductivity in the saturation extract (EC) exceeds 4 dS/m at 25°C. It contains sufficient salts in the root zone to impair the growth of crop plants. The dominant ion species in salt-affected soils are Na, Ca, Mg, Cl, SO_4, and HCO_3. Of these, NaCl is predominant. Most rice cultivars are adversely affected at an EC value of 8 to 10 dS/m.

Various morphological, physiological, and biochemical changes occur in rice plants grown under saline conditions. There is thus an urgent need to understand the mechanism of salinity-induced charges and their effects on plant–insect interactions. This knowledge will help scientists to develop effective procedures for rapidly screening rice genotypes for salt tolerance, and rice varieties with stable resistance against biotic and abiotic factors.

II. MORPHOLOGICAL EFFECTS

Salinity affects the growth of rice in varying degrees at different growth stages [5–7]. Rice is generally tolerant during germination, becomes very sensitive at early seedling stage, gains tolerance during vegetative stage, again becomes sensitive during pollination and fertilization, and then becomes comparatively more tolerant at maturity [8–10]. Salinity delays germination but does not significantly reduce the final germination percentage [11,12]. Even at EC 25 to 30 dS/m, 100% germination of rice was recorded [13]. However, high salt concentrations (2 to 4%) reduced final germination percentate [14,15]. Salinity decreased seedling height, root length, and emergence of new roots at EC 5 to 6 dS/m [9,12]. It suppressed leaf elongation and the formation of new leaves [15,16]. Root development is affected adversely by salinity [12,17]. Plant height and number of tillers per plant are reduced by salinity [18,19]. Salinity severely affected panicle initiation, spikelet formation, fertilization of florets and germination of pollen grains, and subsequently increased the number of sterile florets [19]. Moreover, salinity reduced panicle length, number of primary branches per panicle, number of spikelets per panicle, seed setting percentage, and panicle weight and thereby reduced grain yield [13,19]. It also affected the quality of rice adversely [20].

III. PHYSIOLOGICAL AND BIOCHEMICAL EFFECTS

The mechanisms by which salinity affects the growth and yield of rice are not very well understood. However, it is found that under saline conditions, water stress, ion imbalance, or a combination of these factors affect growth and yield. Among various factors that contribute to sensitivity to salinity is the inability of the plant to regulate the amount of Na and Cl that reaches the leaves [21].

Responses to salinity are displayed both at cellular and tissue levels. At the cellular level, the very first response is to exchange intracellular Na and H with extracellular K. The second mechanism is exchange with vacuole. Exchange between vacuole and cytoplasm is similar to that between extracellular space and cytoplasm for Na, K, and H. In addition, NO_3, HPO_4, Cl, and organic anions are also transported to the vacuole [22]. Cellular as well as whole tissue responses together determine the type of response exhibited in the case of salinity stress (e.g., exclusion of salt into mesophyll vacuole, tolerance to salt within tissue, and mobilization of K to younger leaves and deposition of Na in older leaves, etc.) [23].

The ion species present in excess in saline soils have specific effects on plants [24–27]. Varietal differences in relative growth rate (RGR) in salinized conditions were dependent on differences in shoot Na content. The shoot Na content was affected by Na selectivity in root and by leaf area ratio (LAR). The contribution of LAR was almost as important as that of root selectivity against Na uptake under a higher salinization condition, where root selectivity against Na may be decreased due to reduced root activity [28]. Varieties with high NAR (net assimilation rate) in the control showed a faster rate of decrease in RGR. The LAR showed a continuous decrease with increasing shoot Na concentration. A lower RGR under salinization conditions can therefore be attributed to both NAR and LAR, the latter due primarily to decreased leaf area development [28]. Dwarf IR varieties usually have a high LAR, whereas salt-tolerant varieties such as Pokkali and NonaBokra have a low LAR. Dwarf varieties cannot become salt tolerant unless a lower LAR is incorporated. Salinization with EC 8 dS/m within 6 to 12 days reduced the RGR of Pokkali and its variant to 4% and those of susceptible cultivars to 35%. The salinity level at EC 15 dS/m reduced the RGR of Pokkali and variants to 34%, whereas those of susceptible cultivars were reduced to 80% [29].

The changes following exposure of plants to increased salinity level are not confined to physiological adjustments. Biochemical changes such as synthesis of osmoregulators [30], changes in total proteins, nucleic acids [31], and activities of a number of enzymes [32] are also well documented.

Zongli et al. [33] reported that the content of malonyldialdehyde (MDA) was highly correlated with plasmalemma permeability. The MDA accumulation induced by higher activity of peroxidase occurred before the appearance of salt injury symptoms in younger leaves. The accumulation of MDA was greater in sensitive genotypes than in tolerant genotypes. However, the increase in peroxidase activity was not proportional to the increase in plasmalemma permeability and MDA accumulation.

In another study, Zongli et al. [34] found that the quantity of nucleic acid was reduced with prolonged salinity treatment. The RNA degradation was faster than DNA breakdown. The activity of RNAase in sensitive variety was much higher than that of the tolerant variety. There was positive correlation between the amount of soluble protein and RNA and negative correlation between the activity of RNAase and the quantity of RNA. The RNA/DNA ratio was higher in the salt-tolerant variety. Salinity caused a greater reduction in the embryonic RNA content of the susceptible cultivar (Ratna) than in the salt-tolerant cultivar CSR1 [35]. The RNA reduction under saline conditions was caused not only by less synthesis but also by faster degradation due to the enzymatic action of RNAase [33].

According to Zongli et al. [33], the activities of both superoxide dismutase (SOD) and catalase decreased markedly with an increase in NaCl concentration in the medium and the prolonged duration of salinity treatments. A very significant correlation between enzyme activities and MDA was obtained. Varietal differences in response to different levels of NaCl were shown by changes in enzyme activities and MDA content. The activities of SOD and catalase in the tolerant variety were affected slightly at 0.2% NaCl. At higher NaCl concentrations (0.5 and 0.8%), the inhibition of activities of two protective enzymes and the accompanying increase of MDA in the tolerant variety were less affected than activities in the sensitive variety.

Recent studies by Naqvi [36] have shown changes in *de novo* syntesis of proteins. Induction of two new polypeptides (27 and 25.5 kDa) and heavy accumulation in 70-kDa regions in roots of Nona Bokra in response to salt has been observed in roots. Such proteins may serve as a tool for molecular screening for salt tolerance. The phenomenon of salt-induced polypeptides has far-reaching consequences, as the study of their gene and upstream regulatory sequences may reveal more genes under such control and help in finding homologies with other plants [36].

Because ethylene is involved in growth responses under stress, its potential as an indicator of salt tolerance has been studied in rice [37]. In this study [37] the capacity of salt-tolerant rice varieties Nona Bokra and Pokkali to produce ACC (1-aminocyclopropane-1-carboxylic acid)-dependent ethylene was a cultivar trait and correlated well ($r = 0.91$) with salt tolerance at the seedling stage. The results of this study [37] indicated that ethylene can serve as a biochemical marker for mass screening of large breeding populations of rice for salinity tolerance at the seedling establishment phase.

A. Photosynthetic Effects

Cellular growth is most sensitive to decreased water potential. Both roots and shoots stop growing and cell wall synthesis is inhibited. The salt accumulation is correlated with decrease in protochlorophyll formation, and thus reduced photosynthetic activity and ultrastructural and metabolic damages occur [38]. Zongli et al. [34] reported that a reduction in net photosynthetic rate caused by salinity in the susceptible variety was due to a higher stomatal resistance and low relative water content. Ota and Yasue [39] found that photosynthetic function and chlorophyll content decreased in proportion to increased salt concentration. According to these investigators [39], a decrease in the size of stomata, which indicates a low carbon dioxide concentration in the leaves containing more Nacl, resulted in a lower photosynthetic rate. Kang and Titus [40] observed that NaCl and KCl enhanced the degradation of chlorophylls and proteins in detached rice leaves in a concentration dependent manner. These salts increased the activity of ribulose 1,5-biphosphate carboxylase (Rubisco)-degrading endoproteinases. The amount of ninhydrin-positive compounds measured from HCl-hydrolyzates of trichloroacetic acid–soluble supernatent, but decreased the activity of hemoglobin- and Rubisco-degrading exoproteinases (the amount of ninhydrin-positive compounds measured directly from trichloroacetic acid–soluble supernatent) [40].

B. Osmotic Effects

Salt accumulation in the soil lowers soil water potential, consequently lowering the water potential gradient from the soil to the plant cell. This decreases water availability and turgor pressure [41–43]. Plants tolerant to salinity adjust osmotically (osmotic adjustment) to a saline root medium to maintain the osmotic uptake of water [44–46]. Osmotic adjustment might be induced by salt uptake into the cells or organic solute accumulation in plant cells [47].

Rice growth decreased with increase in osmotic pressure [48] as water uptake decreased [49,50], Na and Cl ions in the leaves and stems increased [39,48,51,52], and the absorption of K and Ca ions by rice plants decreased [48].

Osmotic adjustment in rice is associated with a high electrolyte content in the roots and a low content in shoots. The shoots probably attain osmotic balance by generating harmless soluble organic substance such as proline [53]. When salinity is increased suddenly, water uptake by the plant may be impaired temporarily, due to the low osmotic potential of the soil solution. However, the plant is able to reduce the osmotic potential of the cells to avoid dehydration and death [54]. Yoshida [54] found that the water potential in the first leaf of the salt-tolerant variety was more negative than that of sensitive ones, indicating better adaptation and osmoregulation, which could be a major factor in maintaining stomatal opening and allowing more stomatal CO_2 diffusion and achieving a higher rate of photosynthesis. This may serve as a criterion for selecting salt-tolerant rice varieties [33].

C. Specific Ion Toxicity

The ion species present in excess in saline soils have specific effects on plants [24,26,27]. Salinity damage in rice is attributable to excessive ion entry or to water deficit imposed by the colligative properties of dissolved salts in the environment [55–57]. Water deficit imposed internally due to high inorganic salt concentrations in the extracellular spaces of the leaves [58] is classified as a consequence of excessive ion entry [59,60]. Excessive salt injury may be mitigated by the extrusion of ions from the roots or retention in the cortical vacuoles in preference to long-distance transport. If fluxed to the xylem, salts may be reabsorbed anywhere along with water pathway and possibly retranslocated back to the medium, or saline ions may be compartmentized on tissue [59,60].

According to Ota and Yasue [39], salt injury was more severe at a high temperature (30.7°c) and low humidity (63.5%), because of increased transpiration and uptake of water and salt by the rice plants. Several investigators [50–52] have reported that most of the Cl up take by plants is localized in leaf sheaths, leaf blades, and stems. The roots contain the least Cl. Ion concentration in the shoot can be a useful indicator of salt stress during vegetative growth if NaCl is the major salt. Salt tolerance in rice can be improved either by reducing the concentration of Na in the shoot or by increasing K uptake into the shoot or both [61]. Aslam and Qureshi [62] found that salt-tolerant genotype of rice had a higher K/Na ratio under NaCl treatment and therefore had a smaller amount of Na ions in the leaves than did the sensitive genotype. The concentration of Na ions in the salt-tolerant rice cultivar may be due to low Na transport from root to shoot or its exclusion [33,62]. The lower Na content and high K/Na ratio may play an important role in imparting tolerance to salinity in rice plants. Sodium accumulation in the shoot could lead to a decrease in superoxide dismutase and catalase activities, thus lowering the protective role against the active oxygen toxicity. Presumably, the salt tolerance in rice varieties relates to the inhibitory nature of the enzyme activities responsible for the protective and scavenging system used to detoxify the active oxygen [33].

D. Nutrient Relations

Nutrient imbalances caused by salinity may adversely affect mineral nutrition and growth in rice plants. Some of the important changes in the nutritional status of rice plants and their effects on different parameters of rice plants are discussed next.

1. Nitrogen

Nitrogen in rice plants may increase or decrease under saline conditions. A decrease in the concentration of N was found with increasing salinity levels [63–65]. In contrast, N increased with increasing salinity [66–68]. Rashid [69] found that the N content in the shoot, straw, and grain increased with increasing salinity. The increase of N was more pronounced in the shoot than in the grain and straw. Salim [70] also found an increase in N content in rice cultivars due to salinity (Table 1).

2. Phosphorus

The effect of salinity on the P content of rice is not consistent. Increasing soil salinity may increase the concentration of P in the shoot, grain, and straw [65,69]. However, in other studies [64,67] salinity decreased the P concentration in rice plants.

3. Potassium

Salt tolerance in crop plants is generally correlated with maintenance of a high K/Na ratio in the cytoplasm and a high Na/K ratio in the vacuole, exclusion of Na at the plasma membrane, mobilization of Na to older leaves, and transport of K to younger leaves [71]. The accumulation of K in rice plants decreased with increased salinity [10,65,67,70,72–77]. Salinity reduced K concentration in the shoot, straw, and grain [78].

Appreciable differences among the varieties with respect to K accumulation in straw have been reported. The IR42 cultivar of rice accumulated the highest amount of K, a fairly constant concentration within the salinity range EC 2.5 to 10 dS/m. This indicates a greater ability of the IR42 cultivar to absorb sufficient K from saline media. Salt-sensitive varieties, the IR28 and IR34, accumulated much higher amounts of Na than did the IR42 cultivar [65].

4. Sodium

Salinity increased the concentration of Na^+ in the rice plant (Table 2) [67,69,73,74,77,79–81]. Increase in Na content in the shoot of rice plants due to salinity is important and affects various physiological and biochemical processes in rice plant. Akita and Cabuslay [28] found

Table 1 Effect of Salinity (EC 12 dS/m) on N, P, and K Content of Rice Plants at Vegetative Stage

Cultivar	N (%)		P (%)		K (%)	
	Control	Salinity	Control	Salinity	Control	Salinity
TNI	1.25	1.64	0.37	0.33	2.10	2.01
IR2035-117-3	0.93	1.45	0.32	0.41	2.08	1.34
Nona Bokra	0.91	1.19	0.28	0.30	2.60	1.80
Pokkali	1.05	1.42	0.34	0.33	2.96	2.20

Source: Ref. 70.

Table 2 Mean Na Content in the Stem +
Leaf Sheath of Rice Cultivars Differing in Salt
Tolerance at Different Levels of Salinity

| | | Salinity level (ds/m) | |
Group	Control	EC 4	EC 8
Susceptible	0.32	2.32	12.25
Medium	0.18	2.00	9.09
Tolerant	0.19	1.71	8.89

Source: Ref. 81.

that varietal differences in relative growth rate in salinized conditions were dependent primarily on differences in shoot Na content. The shoot Na content was affected by Na selectivity in root and by leaf are ratio [28]. The relative growth rate decreased sharply as shoot Na concentration increased, and this pattern of decrease in relative growth rate differed with variety. The relative growth rate of varieties with a lower relative growth rate in the control increased or remained at similar levels (EC 8 dS/m) as those in the control. However, varieties with higher relative growth rate in the control showed a continuous decrease in relative growth rate with increased shoot Na content. Varieties with higher growth rate in the control were affected severely by salinity. Cultivars with high salt tolerance (Nona Bokra, Pokkali) had less shoot Na concentration at EC 15 dS/m [28]. Other studies [65,79,82] also showed that the tolerant varieties accumulate less Na in their tissue than do those of sensitive varieties. Aslam and Qureshi [62] found that the rate of Na ion transport from roots to shoots in salt-sensitive rice varieties was four times higher than that of salt-tolerant varieties. The lower uptake rate of Na may be attributed to high absorption selectivity against Na ions in the roots. The low concentration of Na ions in tolerant cultivars may be due to salt exclusion or by less absorption. The uptake of Na is dependent partly on passive absorption via the lipid bilayer of the plasma membrane [59,60]. Sodium uptake is also affected by plant leaf area [28]. Restriction of the entry of Na ion is a major factor in salt tolerance. Other factors include differences in tissue tolerance to absorb Na ions [83] and synchronization of Na ion compartmentation by the leaf cells, with a high rate of ion transport to the shoot [56]. Flowers and Yeo [84] reported that individual variability in Na uptake and survival of rice seedlings was negatively correlated.

The relationship between quantity of salt leaving the root and the resulting concentration in the shoot is determined by the growth rate of the shoot in relation to the net transport of ion out of the root [85]. Pokkali has a substantially lower Na content, which contributes to its salinity tolerance. Initially, the net Na transport by Pokkali and IR22 was similar. Later, Pokkali had a higher rate of transport of ions out of the root. It was reported that [59,60] the lower shoot Na content in Pokkali was not due to better control of Na transport by its roots but was directly attributable to the dilution effect of its rapid vegetative growth.

According to Yeo and Flowers [60], the Na ions uptake was greater in roots of IR28 than that of Nona Bokra at various salinity and humidity levels. At a relative humidity (RH) of 60% and EC 8 dS/m, in Nona Bokra the maximum uptake of Na was in sheath and culm, while IR28 blade contained the maximum Na. Sodium concentration in sheaths of Nona Bokra was more or equal with IR28 at control and EC 4 dS/m, but at EC 8 dS/m, it was more in IR28. Sodium uptake in younger blades of IR28 was greater than that in Nona Bokra, but in sheaths it was greater in Nona Bokra. Rashid [78] stated that Nona Bokra has some mechanism to keep more Na in sheaths than blades, and hence blades of Nona Bokra are less affected than IR28 by toxic effects of Na.

5. Calcium

Calcium is required in the external medium to maintain the selectivity and integrity of the cell membrane [86,87]. Calcium is also required to counteract the adverse effects of toxic ions in the external solution of plants grown [87]. Calcium is also needed for selective transport of ions such as K across membranes [87]. Poor root growth in saline environment may also be associated with Ca deficiency [87,88]. A high Na/Ca ratio in the external solution may impair the selectivity of root membrane and result in passive accumulation of Na in the root and shoot [89]. Poor absorption of Ca due to salinity has been reported by various investigators [65,74,82]. The salinity of the root environment interferes with the absorption and translocation of Ca by plants [90].

An increase in the Ca/Na ratio through addition of Ca to the saline root environment (40 mM NaCl), slightly increased the growth and K uptake in rice [91]. According to the 1985 annual report of the International Rice Research Institute [92], the addition of Ca improved the survival time of Nona Bokra but not that of IR28.

At a given salinity level, Ca increased while Na decreased in the shoot with decreasing Ca/Na ratio in the growth medium [93]. This supports the role of Ca in reducing the permeability of root cell membrane to Na, resulting in decreased Na uptake by rice. In contrast, Yeo and Flowers [94] observed no effect of Ca/Na ratios on the uptake of Na. This may be due to a lower salinity level and shorter period of salinization (7 days) used in their studies.

6. K/Na

In the nutrition of plants in saline media, the relative quantity of K and Na in plants is more important than their absolute quantity. A low Na/K ratio is important for stomatal movement, photosynthesis, and control of undue evaporation [95].

Several reports [10,18,65,75] indicated that the K content of rice straw decreased with increasing Na content, suggesting an antagonism between Na and K (Table 3). Salt-tolerant varieties accumulated less Na and maintained higher K/Na ratios in their tissues than did sensitive varieties [65,75,82]. Ponnamperuma [88] reported that the ability to accumulate K in the shoot correlated well with salt tolerance in rice.

The higher K/Na ratio in salt-tolerant genotypes indicates its greater capacity to adapt to salinity and suggests that excessive accumulation of Na is the major cause of injury to rice under salt treatment. Salt-tolerant genotypes had a higher K/Na ratio under NaCl treatment and therefore had a smaller amount of Na ions in the leaves [33,34]. Salt tolerance in rice can be improved either by reducing the concentration of Na in the shoot or increasing the K uptake into the shoot, or both [61]. A lower Na content and a higher K/Na ratio may play important

Table 3 K/Na Ratios in Rice Straw of Three Rice Varieties as Affected by Five Salt Levels.

Salt level, EC (ds/m)	IR34	IR42	IR9884-54-3
2.0	4.3	4.6	4.5
2.5	2.7	3.8	2.8
5.0	1.6	4.9	2.1
7.5	1.0	3.2	1.5
10.5	1.0	2.5	1.1

Source: Ref. 65.

roles in imparting tolerance in rice [33]. Moreover, salt tolerance in plants might be due to very efficient selectivity toward K during absorption [96].

7. Chloride

Several studies [15,66,69,77–80] have shown that salinity increased the chloride content in rice (Table 4). The accumulation of Cl paralleled that of Na [94]. The NaCl and KCl enhanced the degradation of chlorophyll and proteins in detached rice leaves in a concentration-dependent manner. Kang and Titus [40] found that NaCl at 50 mM increased proteolysis by 21% over the control. According to Muhammad et al. [93], decrease in Cl concentration in the shoot was observed with a decrease in the Na/Ca ratio.

IV. PLANT–INSECT INTERACTIONS UNDER SALINE ENVIRONMENTS

Plants serve not only as food but also as a microhabitat, with shelter and protection for phytophagous insects. Ecology, biology, and behavior of insect pests depend largely on the physical and chemical makeup of host plants. Chemical factors of plants are very important in plant–insect interactions. Insects use chemical factors (allelochemicals) as signals for several purposes: (a) to attract adult insects to suitable host plants, (b) to aid the adult in locating opposite partners, and (c) to help larvae in choosing an appropriate host plant via pheromones left by the female in the eggs [97]. Even minor changes in the physical and chemical attributes of plants can affect their suitability as hosts [98,99].

Interactions between insects and plants are complex phenomenon and are the result of a series of interactions between them. Such interactions influence the ultimate degree of establishment of insect populations on plants [100]. The factors that determine insect establishment on plants can be categorized into two groups: (a) plant factors influencing insect responses, and (b) insect responses to plants. Plant factors are both morphological (shape, color, toughness, hairiness, etc.) and biochemical. The insect responses include orientation, feeding, metabolic utilization of ingested food, growth, and development, adult longevity, fecundity, and oviposition. Of these, orientation, feeding, and oviposition are behavioral responses of the insect, while the others are physiological responses. The relationships between phytophagous insects and their plant hosts are subtle and intimate and are a consequence of prolonged coevolutionary interactions.

Stresses reduce the size of energy and nutrient budget of plants or require their expenditure to accommodate the stress and may result in a lower commitment for defense. The level of defensive substances in plants may change in response to stress [101]. However, effects of plant stress on the production of defensive substances (allelochemicals) in plants are complex [102].

Table 4 Effect of Different Levels of Salinity on the Cl Contents (%) of Rice Straw

Salt level, EC (ds/m)	IR34	IR42	IR9884-54-3
2.0	0.4	0.2	0.3
2.5	0.5	0.3	0.4
5.0	0.8	0.4	0.6
7.5	0.8	0.7	0.7
10.0	1.0	0.7	0.9

Source: Ref. 65.

Salinity is one of the most important stresses which induces various changes in rice plant and as a result can alter significantly host–insect interactions. However, the research work relating to the effects of salinity on the host–insect interactions is yet at infancy stage. Salim [70] and Salim et al. [103] are pioneers in studies on the effects of salinity on rice plant–insect interactions. They investigated the salinity effects on various parameters of whitebacked planthopper, *Sogatella furcifera* (Horvath), by growing rice plants in culture solution at EC 10 and 12 dS/m. The research work was carried out in a phytotron at the International Rice Research Institute Philippines (IRRI) at 29/21°C day/night temperatures and 70% RH. A brief description of their work follows.

A. Growth and Development

Growth duration of *S. furcifera* was longer and significantly fewer nymphs became adults on insect-resistant IR2035-117-3 and salt-tolerant Pokkali and Nona Bokra rice cultivars than on susceptible to insect Taichung Native 1 plants (Table 5). Nymphal growth and development of the insect increased significantly on all cultivars grown at EC 12 dS/m but not at EC 10 dS/m, except on insect-resistant and salinity-sensitive cultivar IR2035-117-3 plants. The growth index of insects on all test varieties increased significantly due to salinity stress. The index, calculated

Table 5 Growth and Development of *S. furcifera* Nymphs on Salinity-Stressed Plants of Four Rice Cultivars[a]

Cultivar	EC 10	Control	Difference	EC 12	Control	Difference
			Nymphs becoming adults (%)			
Taichung Native 1	95.0a	92.5a	2.5NS	97.5a	92.5a	5.0*
IR2035-117-3	42.5	30.0a	12.5*	42.5c	32.5c	10.0*
Nona Bokra	82.5b	77.5b	5.0NS	80.0b	70.0b	10.0*
Pokkali	80.0b	75.0b	5.0NS	85.0b	75.0b	10.0*
			Developmental period (days)			
Taichung Native 1	14.6c	14.8c	–0.2NS	14.8c	15.3c	–0.5*
IR2035-117-3	18.2a	18.9a	–0.7**	18.4a	19.1a	–0.7**
Nona Bokra	15.6b	15.7b	–0.1NS	15.7b	16.3b	–0.6*
Pokkali	15.6b	15.7b	–0.1NS	15.9b	16.3b	–0.4*
			Growth index[b]			
Taichung Native 1	6.49a	6.24a	0.25NS	6.58a	6.06a	0.52*
IR2035-117-3	2.34c	1.60c	0.74**	2.32c	1.71c	0.61**
Nona Bokra	5.28b	4.93b	0.35NS	5.09b	4.31b	0.78**
Pokkali	5.14b	4.79b	0.35NS	5.35b	4.5ab	0.76**

Source: Ref. 103.
[a]Analysis is based on values transformed to arcsin [SQR(X)] for percent of first-instar nymphs becoming adults.
[b]Growth index = percent nymphs becoming adults divided by mean developmental period.
NS,*,**, nonsignificant and significant at the 0.05 and 0.01 probability levels, respectively, by the least significant difference test. Values followed by the same letter in each column are not significantly different at the 0.05 probability level by Duncan's multiple range test.

by dividing percent nymphs becoming adults by the average nymphal developmental period (growth period) in days, indicates the comparative suitability of insect development on a particular host. A higher insect growth index indicates that the host plant is more suitable for insect. The insect growth index on all test rice cultivars increased significantly due to salinity stress.

B. Adult Longevity and Fecundity

Adult longevity was determined as the period from the date the insect nymph molted into an adult until its death. The total number of nymphs that emerged and the number of unhatched eggs represented the fecundity of the insect. Salinity stress at EC 10 dS/m did not significantly increase the adult longevity of *S. furcifera* on test cultivars except on IR2035-117-3 (Table 6). However, the increase in adult longevity of both males and females on rice plants of all test cultivars grown at a high level of salinity (EC 12 dS/m) was significant. Insect fecundity increased significantly on IR2035-117-3 and Nona Bokra, but the effect was not significant on Taichung Native 1 and Pokkali at EC 10 ds/m (Table 7). However, at EC 12 dS/m, fecundity increased significantly on all test cultivars. It can be concluded that an increase in the level of salinity prolonged the adult life of the insect and increased fecundity.

C. Intake and Assimilation of Food

The amount of intake and assimilation of food was calculated as follows [104]:

food intake = food assimilated + excreta

$$\text{food assimilated} = W_1 \times \frac{C_1 - C_2}{C_1} + (W_2 - W_1)$$

where W_1 is the initial weight of the female, W_2 the final weight of the female, C_1 the initial weight of the control female, and C_2 the final weight of the control female.

Table 6 Longevity of *S. furcifera* Adults on Salinity-Stressed Plants of Four Rice Cultivars[a]

Cultivar	EC 10	Control	Difference	EC 12	Control	Difference
			Male longevity (days)			
Taichung Native 1	16.4ab	15.8ab	0.6NS	16.4a	13.8a	2.6**
IR2035-117-3	9.4c	5.5c	3.9**	10.0b	6.3b	3.7**
Nona Bokra	16.7a	16.3a	0.4NS	15.2a	13.0a	2.2**
Pokkali	15.5b	14.9b	0.6NS	15.3a	13.3a	2.0**
			Female longevity (days)			
Taichung Native 1	22.7a	21.0a	1.7*	22.9a	19.8a	3.0**
IR2035-117-3	10.4c	6.9c	1.5*	11.4c	6.9c	4.5**
Nona Bokra	19.5b	18.7b	0.8NS	19.3b	16.8b	2.5**
Pokkali	19.5b	18.7b	0.8NS	19.3b	16.8b	2.7**

Source: Ref. 103.
[a]NS,*,**, nonsignificant and significant at the 0.05 and 0.01 probability levels, respectively, by the least significant difference test. Values followed by the same letter in each column are not significantly different at the 0.05 probability level by Duncan's multiple range test.

Table 7 Fecundity of *S. furcifera* Females on
Salinity-Stressed Plants of Four Rice Cultivars[a]

Cultivar	Fecundity[b]		
	Salinity	Control	Difference
EC 10 dS/m			
Taichung Native 1	1089a	1058a	31NS
IR2035-117-3	192c	118c	74*
Nona Bokra	971b	911b	60*
Pokkali	1042ab	1005a	35NS
EC 12 dS/m			
Taichung Native 1	1068a	994a	74*
IR2035-117-3	199c	138c	61*
Nona Bokra	916b	835b	81*
Pokkali	1012a	940a	72*

Source: Ref. 103.
[a]NS,*, nonsignificant and significant at 0.05 probability level,
respectively, by the least significant difference test. Values followed
by the same letter in each column are not significantly different at
the 0.05 probability level by Duncan's multiple range test.
[b]Number of eggs laid by 10 females.

Salinity stress at EC 10 dS/m did not increase intake and assimilation of food by *S. furcifera* on test rice cultivars (Table 8). Food intake increased significantly on rice plants grown at EC 12 dS/m, except on IR2035-117-3 plants. Assimilation of food by insects did not increase on salt-tolerant Pokkali plants even at EC 12 dS/m. However, food assimilation of the insect was significantly higher on plants of other test rice cultivars grown at EC 12 dS/m than on plants of the same cultivars grown in standard culture solution without salinity.

D. Population Buildup

Population of *S. furcifera* is considered to be an important criterion for assessing the suitability of rice cultivars to the insect because it represents the combined effects of growth and development, adult longevity and fecundity, intake and assimilation of food and orientational and settling responses. Population of *S. furcifera* significantly increased on IR 2035-117-3 plants grown at EC 10 dS/m compared with the plants grown in standard culture solution without salinity (Table 9). At high level of salinity (EC 12 dS/m) population of the insect increased significantly on all test rice cultivars.

V. CAUSES OF CHANGES IN PLANT–INSECT INTERACTIONS

Salinity stress at EC 12 dS/m increased suitability (susceptibility) of rice plants for food intake, growth, adult longevity, fecundity, and population increase of *S. fucifera* [70,103]. Salinity-induced changes in rice plants to reduce the level of resistance to insect may be attributed to (a) increase in N content, (b) decrease in K content, and (c) reduction in the production of allelochemicals in rice plants (Table 10). A brief description of each factor follows.

Table 8 Intake and Assimilation of Food by *S. furcifera* Females on Salinity-Stressed Plants of four Rice Cultivars[a]

Cultivar	EC 10	Control	Difference	EC 12	Control	Difference
		Food intake (mg/female/24 h)				
Taichung Native 1	14.1a	12.5a	1.6NS	16.5a	13.4a	3.1**
IR2035-117-3	5.3c	3.9c	1.4NS	4.4c	3.2c	1.2NS
Nona Bokra	9.8b	9.3b	0.5NS	10.5b	8.2b	2.3*
Pokkali	9.1b	9.0b	0.1NS	11.7b	8.9b	2.8*
		Food assimilated (mg/female/24 h)				
Taichung Native 1	0.62a	0.59a	0.03NS	0.56a	0.51a	0.05*
IR2035-117-3	0.18c	0.14c	0.04NS	0.17c	0.12c	0.05*
Nona Bokra	0.51b	0.50b	0.01NS	0.39b	0.34b	0.05*
Pokkali	0.54b	0.52b	0.02NS	0.38b	0.35b	0.03NS

Source: Ref. 103.

[a]NS,*,**, nonsignificant and significant at the 0.05 and 0.01 probability level, respectively, by the least significant difference test. Values followed by the same letter in each column are not significantly different at the 0.05 probability level by Duncan's multiple range test.

A. Increase in N Content

Nitrogen plays an important role in plant–insect interactions. It is considered a limiting factor for many herbivores, including insects which, as a consequence of selection pressure, have evolved behavioral, physiological, morphological, and other adaptations to utilize fully the available N in their host plants [105,106]. It has been reported that N enhances the suitability of plants to insect pests [98,105,107–113].

A positive correlation had been observed between fecundity of leafhoppers and the level of N in the host plants [107,108]. Fecundity of the brown planthopper, *Nilaparvata lugens*, increased with increase in the application of N [110–112]. The insect digestion capacity

Table 9 Population of *S. furcifera* Females on Salinity-Stressed Plants of Four Rice Cultivars for 40 Days After Infestation[a]

Cultivar	Number of insects[b]					
	EC 10	Control	Difference	EC 12	Control	Difference
Taichung Native 1	491a	472a	19NS	463a	435a	28*
IR2035-117-3	54c	29c	25*	60c	27c	33**
Nona Bokra	255b	232b	23NS	282b	255b	27*
Pokkali	297b	282b	15NS	281b	252b	29**

Source: Ref. 103.

[a]NS,*,**, nonsignificant and significant at the 0.05 and 0.01 probability level, respectively, by the least significant difference test. Values followed by the same letter in each column are not significantly different at the 0.05 probability level by Duncan's multiple range test.

[b]Total number of nymphs and adults recovered from five pairs of newly emerged males and females released on caged plants.

Table 10 Effect of Salinity Stress on Production of Allelochemicals by Four Rice Cultivars

Cultivar	EC 10 (mg/kg leaf sheath)	Control	Decrease over control (%)	EC 12 (mg/kg leaf sheath)	Control	Decrease over control (%)
Taichung Native 1	35.5	40.0	11.25	30.75	35.70	18.00
IR2035-117-3	52.5	62.5	16.00	45.75	63.75	28.24
Nona Bokra	42.0	45.5	7.69	35.70	45.00	20.67
Pokkali	43.0	47.5	9.47	37.25	45.75	18.58

Source: Ref. 103.

increased at a higher level of N in the food [113]. Intake and assimilation of food, growth and development, adult longevity and fecundity, and population of *S. furcifera* increased with increase of N content in rice plants [98].

The peaks of populations of insects on *Holcus mollis* corresponded well with the peaks of N availability in the plants [105]. At least 115 different studies indicate that insect growth or fecundity or population buildup of insects was higher with increase in plant N [113]. An increase in N due to salinity seems to be an important factor in inducing susceptibility or reducing the level of resistance of rice plants to insect pests.

B. Decrease in K Content

Salinity has been found to decrease the K content of rice plants [67,68,70,73,74,103]. Potassium is needed by plants for the synthesis of sugars, protein, and starch. Deficiencies of K in soil or nutrient solution increased the N content in plant tissues, which enhanced insect populations [70,98,114–116]. Potassium deficiency increased soluble carbohydrates, reducing sugars and amino acids, and impaired synthesis of starch, glycogen, and proteins [117].

Increase in K content in rice plants decreased the intake and assimilation of food, growth, and population buildup of *S. fucifera* [70,98]. Higher levels of K in rice plants also suppressed growth and population buildup of *N. lugens* and *Nephotettix* spp. [118–120]. Based on the results of various investigations, it can be concluded that a reduction in K content in rice plants due to salinity stress profoundly influences rice plant–insect interactions.

C. Reduction in Allelochemicals

Plants have developed many defenses against insect pests, but most of them can be classified as physical or chemical. Morphological modifications are of minor importance from a defense perspective. The best defense a plant has against herbivores, including insects, is allelochemicals. These include kairomones, allomones, and synomones. *Kairomones* are chemical substances that benefit the receiver, *allomones* benefit the emitter, and a *synomone* is a substance produced or acquired by an organism such that when it contacts another species it evokes a behavioral or physiological reaction from the receiver that is adaptively favorable to both emitter and receiver [121]. Many of these substances are deposited outside the living cells or within spaces between the cells, or in the glandular hairs found on the surface of many plants.

Various interactions between allelochemicals and nutrients may affect the suitability of insect food. Thus not only the presence of nutrients but also their "bioavailability" may be significant [121]. Allelochemicals also influence the availability of N to herbivores by forming nonutilizable chemical complexes with proteins [122,123] and through effects on the herbivore's nutritional physiology [124,125]. The allelochemicals may be considered ecological chemical requirements of plants since they serve to tie the insects to their hosts and protect plants from other insects, pathogens, and general herbivores that have not broken the chemical defenses of these plants [121].

Various studies have shown the importance of allelochemicals as controlling agents of insects and other pests [70,123,124,126–132]. These have also been demonstrated to play an important role in the control of insect pests of rice [98,99,103,133–135]. Although the allelochemicals have not been identified in rice plants, a large group of low-molecular-weight compounds, such as essential oils, terpenoids, alcohols, aldehydes, fatty acids, esters, and waxes, have been obtained [136]. Being insoluble in water, most of the compounds are not likely to be translocated in the vascular tissue of plants [130]. They strongly influence the chemical environment of plants and hence play an important role in determining the susceptibility or resistance of plants to insects.

Even the most susceptible plants are defended physically and/or chemically against insect attack [137]. However, the susceptible plants have smaller quantities of allelochemicals than do resistant plants of the same species [70]. This means that low levels of allelochemicals in a plant indicate low level of resistance. Salinity stress reduced the quantity of allelochemicals in both insects-resistant and insect-susceptible rice cultivars (Table 10) [103]. Reduction in the quantity of allelochemicals in rice plants due to salinity stress may be a major factor in reducing the level of resistance or inducing susceptibility to insect.

VI. CONCLUSIONS

The effects of salinity on the growth of rice plants are very complicated. It is not feasible to select a single parameter that could isolate tolerant varieties from others. Varietal differences regarding tolerance to salinity do exist in rice. Breeding efforts for salt tolerance are being pursued, but progress is slow because of the complexity of the problem. For evolving rice salt-tolerant varieties, it is imperative to screen and evaluate germplasm at different levels of salinity under a wide range of temperature and relative humidity regimes, taking various parameters into consideration. Salinity stress induced susceptibility in rice to the whitebacked planthopper, a major pest of rice in southern and southeastern Asia. However, the mechanism of this susceptibility is not yet fully understood. It is therefore suggested that further studies be carried out to determine the effects of salinity-induced morphological and physiological/bio-chemical changes in rice on biological parameters of various insect pests. Such information would be helpful in evolving rice varieties with stable resistance against insect pests and other biotic and abiotic factors and to achieve sustainability in production. Such effects of salinity must be considered for breeding insect- and disease-resistant varieties for salinity-prone areas.

REFERENCES

1. M. Akbar, and F. N. Ponnamperuma, "Saline soil of South and Southeast Asia as Potential Rice Lands," *Proceedings of Rice Research Strategies for the Future*, International Rice Research Institute, Manila, The Philippines, pp. 265–281 (1982).
2. G. E. Boje-Klein, "Problem Soils as Potential Areas for Adverse Soils-Tolerant Rice Varieties in South and Southeast Asia," *IRRI Research Paper Series No. 119* (1986).
3. F. N. Ponnamperuma and A. K. Bandyopadhya, "Soil Salinity as a Constraint on Food Production in the Humid Tropics," *Proceedings of Priority for Alleviating Soil-Related Constraints to Food Production in the Tropics*, International Rice Research Institute, Manila, The Philippines, pp. 203–216 (1980).
4. M. Akbar, "Varietal Improvement for Salt Tolerance in Rice," *Proceedings of Improvement and Management of Winter Cereals Under Temperature, Drought and Salinity Stresses* (E. Acevedo,

E. Fereres, C. Gimenez, and J. P. Srivastava, eds.), Institute Nacional de Investigaciónes Agrarias, Madrid, pp. 337–350 (1991).

5. E. V. Maas and G. J. Hoffman, "Crop Salt Tolerance: Evaluation of Existing Data," *Proceedings of Managing Saline Water for Irrigation*, International Salinity Conference, Aug. 16–20, 1976, Texas Technical University, Lubbock, Texas, pp. 187–198 (1977).
6. R. U. Castro and S. R. Sabado, *Grains J.*, *11*: 43 (1978).
7. G. M. Panaullah and F. N. Pannamperuma, "Effect of Salt Stress Applied at Four Development Stages on the Growth and Yield of Four Rices," *Proceedings of 11th Annual Scientific Meeting of the Crop Science Society of the Philippines*, Babay, Leyte, The Philippines, Apr. 27–29 (1980).
8. G. A. Pearson and A. D. Ayers, *USDA Res. Serv. Prod. Res. Rep.*, *43*: 1 (1960).
9. G. A. Pearson, A. D. Ayers, and D. L. Eberkard, *Soil Sci.*, *102*: 151 (1966).
10. *Annual Report for 1966*. International Rice Research Institute, Manila, The Philippines (1967).
11. M. T. Kaddah, *Soil Sci.*, *96*: 105 (1963).
12. M. Akbar and T. Yabuno, *Jpn. J. Breed.*, *24*(4): 176 (1974).
13. G. A. Pearson, *Int. Rice Commun. Newsl.*, *10*: 1 (1961).
14. S. Iwaki, *Mem. Ehima Univ. (Sec. 6 Agric.)*, *2*: 10,156 (1956).
15. M. Akbar, *J. Agric. Res.*, *13*(1): 341 (1975).
16. T. Tagawa and N. Ishizaka, *Mem. Fac. Agric. Bakkaido Univ.*, *5*: 77 (1964).
17. K. T. Ota, T. Yasue, and M. Nakagawa, *Res. Bull. Fac. Agric. Gifu Univ.* (1958).
18. *Annual Report for 1967*, International Rice Research Institute, Manila, The Philippines (1968).
19. M. Akbar, T. Yabuno, and S. Nakao, *Jpn. J. Breed.*, *22*: 277 (1972).
20. J. J. Siscar Lee, B. O. Juliano, R. H. Qureshi, and M. Akbar, *Plant Foods Hum. Nutr.*, *40*: 31 (1990).
21. T. J. Flowers, F. M. Salama, and A. R. Yeo, *Plant Cell Environ.*, *11*: 453 (1988).
22. J. Gorham, R. G. W. Jones, and E. McDonnell, *Plant Soil*, *89*: 15 (1985).
23. A. R. Yeo, S. A. Flowers, and T. J. Flowers, *Theor. Appl. Genet.*, *79*: 377 (1990).
24. J. V. Lagerwerff, *Plant Soil*, *31*: 77 (1969).
25. P. A. Layae and E. Pestein, *Science*, *166*: 395 (1969).
26. A. T. Ayoub and H. M. Ishac, *J. Agric. Sci.*, *82*: 339 (1974).
27. L. Bernstein, *Am. J. Bot.*, *50*: 360 (1975).
28. S. Akita and G. S. Cabuslay, "Physiological Basis of Differential Response to Salinity in Rice Cultivars," *Proceedings of the 3rd International Symposium on Genetic Aspects of Plant Mineral Nutrition*, June 19–24, Braunschweign, Germany, pp. 19–24 (1988).
29. G. S. Cabuslay and S. Akita, *J. Crop Sci.*, *14* (1): 26 (1989).
30. D. M. Veron and H. J. Bohhnert, *EMBO J.*, *11*(6): 2077 (1992).
31. L. Prakash, M. Dutt, and G. Prathapasenan, *Aust. J. Plant Physiol.*, *15*: 769 (1988).
32. R. S. Dabey and M. Rani, *Aust. J. Plant Physiol.*, *7*: 215 (1991).
33. W. Zongli, L. Jaiankum, and W. Zhixia, *Jiangsu, J. Agric. Sci. (China)*, 6(2): 106 (1990).
34. W. Zongli, D. Quujie, L. Xiaozhong, W. Zhixia, and K. L. Jain, *Philipp. J. Crop Sci.*, *14*(Suppl. 1): 526 (1989).
35. R. S. Dubey, *Plant Physiol. Biochem.*, *12*(1): 9 (1985).
36. S. M. S. Naqvi, Ph.D. Thesis, Middle East Technical University, Ankara, Turkey (1993).
37. A. A. Khan, M. Akbar, and D. V. Seshu, *Crop Sci.*, *27*: 1242 (1987).
38. A. R. Yeo, K. S. Lee, P. Izard, P. J. Boursier, and T. J. Flowers, *J. Exp. Bot.*, *42*(240): 881 (1991).
39. K. Ota and T. Yasue, *Res. Bull. Fac. Agric. Gifu Univ.*, *16*: 1 (1962).
40. S. Kang and J. S. Titus, *Plant Physiol.*, *91*: 1232 (1989).
41. H. B. Peterson, "Some Effects on Plants of Sodium from Saline and Sodic Soils," *Proceedings of Symposium on Salinity Problems in Arid Zones*, UNESCO, Paris (1961).
42. A. Meiri and L. Poljakoff-Mayber, *Soil Sci.*, *109*: 26 (1970).
43. H. Lessani and H. Marschner, *Aust. J. Plant Physiol.*, *5*: 27 (1978).
44. L. Bernstein, *Am. J. Bot.*, *48*: 909 (1961).
45. L. Bernstein, *Am. J. Bot.*, *50*: 360 (1963).
46. B. E. Janes, *Soil Sci.*, *101*: 180 (1966).

47. E. V. Mass and R. H. Neiman, in *Crop Tolerance to Suboptimal Land Condition*, SA Spec. Publ. 32, Madison, Wis., pp. 277–299 (1978).
48. N. Shimose, *J. Soil Fert. Jpn. 34*: 107 (1963).
49. S. Iwaki, K. Ota, and T. Ogo, *Crop Sci. Soc. Jpn. Proc., 22*: 13 (1953).
50. T. Tagawa and N. Ishizaka, *Proc. Crop Sci. Soc. Jpn., 31*: 249 (1963).
51. N. Shimose, *J. Sci. Soil Tokyo, 29*: 158 (1958).
52. T. Tagawa and N. Ishizaka, *Proc. Crop Sci. Soc. Jpn., 31*: 337 (1963).
53. *Annual Report for 1977*, International Rice Research Institute, Manila, The Philippines (1978).
54. S. Yoshida, *Fundamentals of Rice Crop Science*, International Rice Research Institute, Manila, The Philippines (1981).
55. T. J. Flowers, P. F. Troke, and A. R. Yeo, *Annu. Rev. Plant Physiol., 28*: 89 (1977).
56. H. Greenway and R. H. Munns, *Annu. Rev. Plant Physiol., 31*: 149 (1980).
57. H. U. Neue, "Adverse Soil Tolerance of Rice: Mechanisms and Screening Techniques," *Proceedings of Rice Production on Acid Soils of the Tropics* (P. Dgturck and F. N. Ponnam Peruma, eds.), Institute of Fundamental Studies, Kendy, Sri Lanka, pp. 243–250 (1991).
58. J. J. Oertli, *Agrochemica, 12*: 461 (1968).
59. A. R. Yeo and T. J. Flowers, in *Salinity Tolerance in Plants: Strategies for Crop Improvement* (R. C. Staples and G. H. Toenniessen, eds.), Wiley, New York, pp. 151–170, (1984).
60. A. R. Yeo and T. J. Flowers, *Plant Physiol., 75*: 298 (1984).
61. D. P. Heenan, L. G. Lewin, and D. W. McCaffery, *Aust. J. Exp. Agric., 28*: 343 (1988).
62. M. Aslam and R. H. Qureshi, *Int. Rice Res. Newsl., 14*(3): 25 (1989).
63. C. Torres and F. T. Bingham, *Soil Sci. Soc. Am. Proc., 37*: 711 (1973).
64. H. Lam and E. O. McLean, *Commun. Soil Sci. Plant Anal., 10*: 969 (1979).
65. G. M. Panaullah, M.Sc. thesis, University of the Philippines at Los Banos, The Philippines (1980).
66. N. Shimose, *J. Sci. Soil Manure, 27*: 193 (1956).
67. K. S. Murty and C. Narasinga Rao, *Oryza, 2*: 87 (1965).
68. K. S. Murty and C. Narasinga Rao, *Oryza, 4*: 42 (1967).
69. M. Rashid, Ph.D. thesis, University of the Philippines at Los Banos, The Philippines (1983).
70. M. Salim, Ph.D. thesis, University of the Philippines at Los Banos, The Philippines (1988).
71. M. G. Pitman, in *Salinity Tolerance in Plants: Strategies for Crop Improvement* (R. G. Staples and G. H. Toenniessen, eds.), Wiley, New York, pp. 93–123 (1984).
72. D. W. Rains, *Annu. Rev. Plant Physiol., 23*: 367 (1972).
73. S. A. Korkor and R. M. Abdel-Aal, *Agric. Res. Rev. (Cairo), 52*(5): 73 (1974).
74. P. C. Paricha, C. J. Patra, and P. Sahoo, *J. Indian Soc. Soil, 23*: 344 (1975).
75. K. Giriraj, A. S. P. Murty, and K. V. Janardhan, *SABRAO J., 8*: 47 (1976).
76. V. Balasubramanian and S. Rao, *Riso, 26*: 291 (1977).
77. C. D. John, V. Limpinuntana, and H. Greenway, *J. Exp. Bot., 28*(102): 133 (1977).
78. M. Rashid, *Effect of Relative Humidity and Temperature on Salt Tolerance of Rice*, terminal report, Oct. 1988 to Oct. 1990, Post Doctoral Scientist, Soil and Water Sciences Division, International Rice Research Institute, Manila, The Philippines (1990).
79. M. T. Kaddah, W. F. Lehman, B. D. Meek, and F. E. Robinson, *Agron. J., 67*: 436 (1975).
80. Y. Osotsapar and B. T. Mercado, *Philipp. J. Sci., 107*(1/2):51 (1978).
81. M. Ibrahim Aly El-Naggar, Ph.D. thesis, Soil Department, Faculty of Agriculture, Cairo University, Egypt (1990).
82. B. A. Hedge and A. V. Joshi, *Plant Soil, 41*: 423 (1974).
83. A. R. Yeo and T. J. Flowers, *Physiol. Plant., 59*: 189 (1983).
84. T. J. Flowers and A. R. Yeo, *Phytology 88*: 363 (1981).
85. M. T. Pitman and W. J. Cram, *Regulation of Iron content in Whole Plants*, Soc. Exp. Biol. Symp. 31, Cambridge University Press, Cambridge, pp. 391–424 (1977).
86. N. K. Fageria, *Plant Soil, 70*: 309 (1983).
87. R. G. Wyn Jones and O. R. Lunt, *Bot. Rev., 33*: 407 (1967).
88. F. N. Ponnamperuma, in *Salinity Tolerance in Plants: Strategies for Crop Improvement* (R. C. Staples and G. H. Toenniessen, eds.), Wiley, New York, pp. 255–271 (1984).
89. A. Kramer, A. Lauchi, A. R. Yeo, and J. Gullasch, *Ann. Bot., 41*: 1031 (1977).

90. P. J. C. Kuiper, in *Salinity Tolerance in Plants: Strategies for Crop Improvement* (R. C. Staples and G. H. Toenniessen, eds.), Wiley, New York, pp. 77–91 (1984).

91. T. Kawasaki and M. Moritsugu, *Ber. Ohara Inst. Landwirtsch. Biol. Okayama Univ.*, *17*(2): 73 (1978).

92. *Annual report for 1984*, International Rice Research Institute, Manila, The Philippines (1985).

93. S. Muhammed, M. Akbar, H. U. Nue, *Plant Soil*, *101*: 57 (1987).

94. A. R. Yeo and T. J. Flowers, *New Phytol.*, *99*: 81 (1985).

95. H. Beringer, "The Role of Potassium in Crop Production," *Proceedings of International Seminar on the Role of Potassium in Crop Production*, Pretoria, South Africa, pp. 25–32 (1980).

96. D. Kramer, in *Salinity Tolerance in Plants: Strategies for Crop Improvement* (R. C. Staples and G. H. Toenniessen, eds.), Wiley, New York, pp. 3–15 (1984).

97. P. J. Edwards and S. D. Wratten, *Ecology of Insect-plant Interactions*, The Institute of Biology, Studies in Biology No. 121, Edward Arnold, London (1980).

98. M. Salim and R. C. Saxena, *Crop Sci.*, *31*: 797 (1991).

99. M. Salim and R. C. Saxena, *Crop Sci.*, *31*: 1620 (1991).

100. K. N. Saxena, *Entomol. Exp. Appl.*, *12*: 751 (1969).

101. D. F. Rhoades, in *Variable Plants and Herbivores in Natural and Managed Systems* (R. F. Denno and M. S. McClure, eds.), Academic Press, New York, pp. 155–220 (1983).

102. D. F. Rhoades, in *Herbivores: Their Interaction with Secondary Plant Metabolites* (G. A. Rosenthal and D. H. Janzen, eds.), Academic Press, New York, pp. 3–54 (1979).

103. M. Salim, R. C. Saxena, and M. Akbar, *Crop Sci.*, *30*(3): 634 (1990).

104. R. C. Saxena and M. D. Pathak, "Factors Affecting Resistance of Rice Varieties to Brown Planthopper, *Nilaparvata lugens*," in *8th Annual Conference of the Pest Control Council of the Philippines*, May 18–20, Baclod City, The Philippines (1977).

105. S. McNeill and T. R. E. Southwood, in *Biochemical Aspects of Plant and Animal Coevolution* (J. B. Harborne, ed.), Academic Press, London, pp. 77–98 (1978).

106. W. J. Mattson, *Ann. Ecol.*, *11*: 119 (1980).

107. A. D. Hinckley, *Bull. Entomol. Res.*, *24*: 467 (1963).

108. R. G. Fennah, in *Pests of Sugarcane* (J. R. Williams, ed.), Elsevier, Amsterdam, pp. 367–389 (1969).

109. A. Regupathy and A. Subramanian, *Oryza*, *9*: 81 (1972).

110. M. B. Kalode, "Recent Changes in Relative Pest Status of Rice Insects as Influenced by Cultural, Ecological and Genetic Factors," *International Rice Research Conference*, International Rice Research Institute, Manila, The Philippines (1974).

111. M. D. Pathak, in *Insects, Science and Society* (D. Pimentel, ed.), Academic Press, New York, pp. 121–148 (1975).

112. V. A. Dyck, B. C. Misra, S. Alam, C. N. Chen, C. Y. Hsieh, and R. S. Rejesus, "Ecology of the Brown Planthopper in the Tropics," *Proceedings of Brown Planthopper: Threat to Rice Production in Asia*, International Rice Research Institute, Manila, The Philippines, pp. 61–98 (1979).

113. J. M. Scriber, in *Nitrogen in Crop Production*, American Society Agronomy, Madison, Wis., pp. 441–460 (1984).

114. H. F. Van Emden, *Entomol. Exp. Appl.*, *9*: 444 (1966).

115. J. R. Metcalfe, *Bull Entomol. Res.*, *60*: 309 (1970).

116. M. L. Sharma, *Ann. Soc. Entomol. Que.*, *15*: 88 (1970).

117. P. Baskaran, P. Narayanasamy, and A. Pari, "The Role of Potassium on Incidence of Insect Pests Among Crops Plants with Particular Reference to Rice," *Proceedings of Potassium in Crop Resistance to Insect Pests*, RRII Research Review Series 3, Potash Research Institute of India, Gurgaon, Haryana, India, pp. 63–68 (1982).

118. R. Subramanian and M. Balasubramanian, *Madras Agric. J.*, *63*: 561 (1976).

119. C. Vaithilingam, M. Balasubramanian, and R. Subramanian, *Madras Agric. J.*, *63* (8/10): 571 (1976).

120. P. J. Ittyavirah, K. P. Vasudevan Nair, and M. J. Thomas, *Agric. Res. J. Kerala*, *17*(1): 118 (1979).

121. K. S. Hagen, R. H. Dadd, and J. Reese, in *Ecological Entomology* (C. B. Huffaker and R. L. Rabb, eds.), Wiley, New York, pp. 79–112 (1984).

122. R. G. Cates and D. F. Rhoades, *Biochem. Syst. Ecol.*, *5*: 1985 (1977).

123. T. Swain, *Annu. Rev. Plant Physiol.*, *28*: 479 (1977).

124. S. D. Beck and J. C. Reese, in *Biochemical Interactions Between Plants and Insects* (J. Wallace and R. Maxwell, eds.), *Rec. Adv. Phytochem.*, *10*: 41 (1976).

125. J. C. Reese, *Entomol. Exp. Appl.*, *24*: 425 (1978).

126. V. G. Dethier, *Evolution*, *8*: 330 (1954).

127. G. Fraenkel, *Science*, *129*: 1466 (1959).

128. P. R. Ehrlich and R. H. Raven, *Evolution*, *18*: 586 (1964).

129. L. M. Schoonhoven, *Symp. R. Entomol. Soc. London*, *6*: 87 (1972).

130. D. McKey, in *Herbivores: Their Interactions with Secondary Plant Metabolites*, G. A. Rosenthal and D. H. Janzen, eds., Academic Press, New York, pp. 56–133 (1979).

131. P. Feeny, *Rec. Adv. Phytochem.*, *10*: 1 (1976).

132. D. F. Rhoades and R. G. Cates, *Rec. Adv. Phytochem.*, *10*: 168 (1976).

133. M. D. Pathak and R. C. Saxena, in *Breeding Plants Resistant to Insects* (F. G. Maxwell and P. R. Jennings, eds.), Wiley, New York, pp. 421–455 (1980).

134. Z. R. Khan and R. C. Sexena, *J. Econ. Entomol.*, *78*: 562 (1985).

135. R. C. Saxena and S. H. Okech, *J. Econ. Entomol.*, *11*: 1601 (1985).

136. R. C. Saxena, in *Natural Resistance of Plants to Pests: Roles of Allelochemicals* (M. B. Green and O. Hedin, eds.), American Chemical Society, Washington, D. C., pp. 142–149 (1986).

137. J. C. Reese, in *Plant Resistance to Insects* (P. A. Hedin, ed.), ACS Symposium Series 208, American Chemical Society, Washington, D.C., pp. 231–243 (1983).

32
Physiological Responses of Cotton (*Gossypium hirsutum* L.) to Salt Stress

Mohammad Pessarakli

The University of Arizona, Tucson, Arizona

I. INTRODUCTION

No plant species or animals are immune from stress. Any species at least once during its life cycle is subjected to stress. Nutrient uptake and utilization as well as water absorption by plants are adversely affected under stressful conditions. Plant growth and metabolism are usually impaired under such conditions, resulting in decreased crop yields.

Among the essential nutrient elements, nitrogen is one of the most widely limiting elements for crop production, and when plants are subjected to stress, N uptake will probably be affected more severely than any other mineral nutrient.

In recent years, several studies indicated that decreases in plant growth and crop yields under stress conditions have been associated with impairment in nutrient and water uptake, abnormal metabolism, and inhibition of plant protein synthesis [1–42]. In these studies, salt and/or water stress impaired growth and the incorporation of nutrients (i.e., N) into the protein, and increased accumulation of inorganic N in plants. Reduction in nutrient uptake and utilization by plants was reported by several investigators in earlier studies [43–49]. Uptake of N and P by plants was inhibited under high concentrations of NaCl and Na_2SO_4 in the root medium, and the excess amount of absorbed Na^+ depressed NH_4^+ absorption in these studies. The absorption and metabolism of ammonium (NH_4^+) and nitrate (NO_3^-) in red kidney beans (*Phaseolus vulgaris* L.) were reduced significantly under salt or water stress [47–49]. In all of the studies noted above, reduction in root permeability and consequent decrease in water and nutrient uptake under high electrolyte concentrations was stated as the cause of this abnormality in water and nutrient absorption and metabolism. Nevertheless, low levels of salts in the presence of N, P, and K stimulated growth and increased the yield of cotton, *Gossypium hirsutum* L. [28,29,50–53]. With further increase in salinity, dry matter yield decreased, but it increased with the addition of N at each salinity level. Moreover, plants continued to accumulate N under saline conditions despite the reduction in yield and dry matter production.

Soil salinity did not inhibit N absorption by bermudagrass (*Cynodon dactylon* L.), a high-salt-tolerant plant [54], and stress had little or no effect on the rate of NO_3^- uptake by barley (*Horedeum vulgare* L.), another high-salt-tolerant crop, except at the highest osmotic pressure and lowest osmotic potential value (–0.54 MPa) of the rhizosphere [9,55]. Also, NaCl in the culture solution did not influence NO_3^- uptake by tomatoes (*Lycopersicon esculentum* Mill.), a medium-salt-tolerant plant [45].

Abdul-Kadir and Paulsen [56] reported that the soluble protein and free amino acid content of wheat (*Triticum aestivum* L.) plants were not affected consistently by $MgSO_4$, $MgCl_2$, and NaCl. Udovenko et al. [57,58] found that under salt stress the non-protein-N fraction increased in beans, peas (*Lathyrus hirsutus* L.), barley, and wheat, whereas the protein-N fraction changed irregularly. These investigators concluded that the response of N metabolism to salt stress is similar in plants with varying salt tolerance. Increases in soluble-N fractions and free amino acid levels and decreases in protein-N content of cotton plants under medium (–0.8 MPa osmotic potential) and high (–1.2 MPa osmotic potential) levels of salinity was reported by Pessarakli and Tucker [29]. However, these investigators found that the low level of salinity (–0.4 MPa osmotic potential) slightly enhanced the dry matter production and protein content of the plants. On the other hand, this level of salinity (–0.4 MPa osmotic potential) and lower level (–0.25 MPa osmotic potential) in the culture solution substantially decreased the protein content of red kidney beans [48,49], green beans [23,24,59], and alfalfa, *Medicago sativa* L. [22]. Impaired N metabolism and decreased protein content of a number of plants under stress conditions have also been reported by several other investigators [60–66]. Recently, Rabe [30] and Dubey [6] have reviewed altered N metabolism and protein synthesis, respectively, in plants under stressful conditions. These authors have reported that N metabolism and protein synthesis in plant species were severely affected under stress.

Water stress induced by Carbowax also caused marked reduction in protein synthesis by plants [48,49]. Although these studies were conducted with red kidney beans, a salt-sensitive plant, salt (NaCl) stress resulted in appreciably greater reduction in ^{15}N incorporation in the protein fraction than water stress created by the Carbowax treatment. These results indicated inhibition of N utilization caused by an ionic effect in addition to the osmotic effect of either NaCl or Carbowax.

Although investigations of the effects of salt and/or water stress on nutrient (i.e., N) absorption, utilization, metabolism, and protein synthesis indicated primarily a reduction in the absorption rate of N and decreases in the protein content of plants, a few controversial results make generalization difficult. These and other inconsistent results that demonstrated either an increase or no effect on the nutrient (i.e., N) absorption and metabolism can probably be explained as resulting from a dilution effect. Frota and Tucker [47,48], Saad [49], Pessarakli and Tucker [26–29], and Pessarakli [21,51,59], suggested that plant growth was affected more than the nutrient uptake and metabolism, and as a result the relative concentration of N was higher for the stressed plants.

While the mechanisms by which salinity stress or drought adversely affect plant growth are still controversial, it is generally agreed that impairment of N absorption and metabolism is a critical factor. For a detailed review of the adverse effects of stress on plants and crops, readers are referred to the most comprehensive source, the *Handbook of Plant and Crop Stress* [67].

If it could be determined at what particular stage of growth high salinity affects plant growth and metabolism most negatively, the mechanisms by which these adverse effects occur might be identified and the detrimental effects prevented. In this regard, in addition to this report, several investigators have already attempted to study the effects of stress at various stages of plant growth [28,29,51,53,68–89].

The purpose of this investigation was to determine the physiological effects of salt stress on growth in terms of dry matter production, nitrogen ($^{15}NH_4^+$) absorption and metabolism, protein synthesis, and water absorption by cotton plants at two stages of growth.

II. RESULTS AND DISCUSSION

A. Dry Matter Production of Cotton Plants

At both the vegetative and reproductive stages of growth, salt stress (particularly at medium and higher NaCl levels) drastically reduced dry matter production (Table 1). Kurth et al. [90] have also observed the adverse effects of both NaCl and $CaCl_2$ salinity on cotton growth in terms of cell enlargement and cell production. In the present report, the dry matter production of the stressed plants was highly negatively correlated with increasing levels of salinity at both stages of growth (r of -0.98 to -0.96). Reduction in plant growth at higher levels of salinity has also been reported by other investigators for other salt-tolerant plants, such as barley [9,91,92], mangrove, *Avicennia marina* (17), and other halophytes (i.e., *Suaeda maritima* L.) [93]. In stress physiology, several other investigators found that the growth of various plant species substantially decreased under stressful conditions [2,3,5,7,10,11,13,14,16–27,34–41,46–51,53,59,67–89,94–116]. The present study showed that the shoot dry weight was reduced more than root dry weight by increasing salinity. This is supported by the findings of several other investigators [21–27,47–49,51,59,67,91,92], and is consistent with the common knowledge in plant physiology that plant roots under stress conditions grow more and penetrate deeper in the soil in search of water and nutrients. Other studies also indicated a substantial reduction in shoot growth under stress conditions. For example, sodium chloride stress severely decreased the shoot growth of rice, *Oryza sativa* L., cultivar GR-30 [117], and *Lactuca sativa* plants [118]. In the present study, the effect of salinity was more pronounced at the vegetative growth than at the reproductive growth stage. Other studies have also indicated that plants at earlier stages of growth were more sensitive to stress than at later stages of growth [16,51,53,68–89,110,119–123]. Abnormal plant growth was also observed in experiments using sufficient amounts of other types of salts than sodium chloride [46,86,90,93,124,125], as well

Table 1 Dry Matter Production of Cotton Plants Subjected to NaCl Salinity During Vegetative and Reproductive Stages of Growth

Growth stage	Treatment, osmotic potential (MPa)	Plant dry weight/pot (two plants) (g)		
		Shoots	Roots	Total
Vegetative	Control	5.42	0.93	6.35
	−0.4	3.79	0.97	4.76
	−0.8	2.71	0.77	3.48
	−1.2	1.71	0.39	2.10
	LSD (0.05)[a]	1.43	0.20	1.58
Reproductive	Control	20.13	3.90	24.03
	−0.4	16.72	3.98	20.70
	−0.8	12.11	3.52	15.63
	−1.2	7.60	2.42	10.02
	LSD (0.05)[a]	4.22	0.83	4.60

Source: Ref. 28.

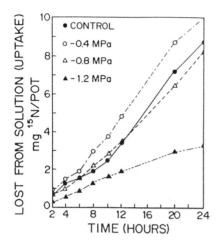

Figure 1 Solution loss of ^{15}N (uptake) by cotton plants under various NaCl salinity conditions during the vegetative stage of growth. (From Ref. 28.)

as under drought stress conditions [36,70,72,87,88,126,127]. This is an indication of the adverse effects of stress on plant growth, regardless of the source of the salt or the type of stress.

B. Nitrogen Absorption by Cotton Plants

1. Nitrogen (^{15}N) Absorption and Concentration in Plant Tissues

The mean values of $^{15}NH_4^+$ absorption by cotton plants for 24-h uptake time under normal Hoagland solution (control) and salt (NaCl) stress conditions obtained by analyzing solution samples indicated that low and medium levels of salinity did not significantly decrease the rate of ^{15}N absorption (Figures 1 and 2). In fact, absorption was increased slightly at the vegetative stage with a low salinity level (–0.4 MPa osmotic potential). Similar amounts of NaCl drastically reduced the uptake rate of ^{15}N in red kidney beans [47,49], green beans [21,23,24,59], alfalfa

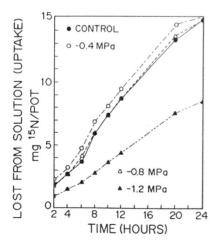

Figure 2 Solution loss of ^{15}N (uptake) by cotton plants under various NaCl salinity conditions at the beginning of the reproductive stage of growth. (From Ref. 28.)

[22], and eggplant, *Solanum melongena* L. (26). Such differences reflect variations in the salt tolerance of these different plant types. The high level of salinity (–1.2 MPa) appears to have caused a substantial reduction in the N absorption rate by cotton plants. The effect of high salinity level on ^{15}N uptake was more pronounced at the vegetative stage than at the reproductive stage of growth. The values for ^{15}N uptake obtained by total N analysis of the plant materials (Tables 2 and 3) indicated essentially the same pattern as the solution loss data (Figures 1 and 2). Total amounts of ^{15}N recovered in plants generally accounted for 95 to 99% of the apparent solution loss.

The concentration of ^{15}N in roots was higher than in shoots for all treatments, at both stages of growth (Table 4). Adsorption of ammonium ions onto the root surfaces or infusion of ions into apparent free space within roots, as suggested by Pessarakli and Tucker [26–29] and Pessarakli [21,51,59], could be a possible reason for the higher ^{15}N concentrations in cotton roots. At both stages of growth, the ^{15}N concentration of shoots significantly increased for salinized plants (–0.8 MPa) compared with controls. This can be explained as a dilution effect, as suggested by Frota and Tucker [47], Saad [49], Pessarakli and Tucker [26–29], and Pessarakli [21,59], as being due to a greater reduction in plant growth than ^{15}N absorption under stress conditions. The relative translocation of ^{15}N from roots to shoots was not affected appreciably by salt concentration.

2. Total N Uptake by Plants

Total N uptake decreased as the culture medium became more saline (Table 5). The reduction in N uptake values at –0.8 MPa osmotic potential was to approximately 50 and 70% of the control values at the vegetative and the reproductive stages of growth, respectively. At the higher salinity level (–1.2 MPa), the reduction in N uptake was proportionately greater at the

Table 2 Distribution of [15]N Absorbed as Ammonium in Cotton Shoots and Roots Under Different NaCl Salinity Levels During Vegetative Stage of Growth

Plant parts	Treatment, osmotic potential (MPa)	15N uptake/pot (two plants), (mg) for uptake time (h):		
		6	12	24
Shoots	Control	0.98	1.61	4.97
	–0.4	0.82	1.71	4.78
	–0.8	1.02	1.63	4.36
	–1.2	0.36	0.88	1.81
	LSD (0.05)[a]	0.43	0.49	0.45
Roots	Control	0.73	1.29	3.22
	–0.4	1.39	1.84	4.20
	–0.8	0.67	1.27	3.23
	–1.2	0.34	0.57	1.24
	LSD (0.05)[a]	0.48	0.93	0.77
Total	Control	1.71	2.90	8.19
	–0.4	2.21	3.55	8.98
	–0.8	1.69	2.90	7.59
	–1.2	0.70	1.45	3.05
	LSD (0.05)[a]	0.32	0.95	0.98

Source: Ref. 28.
[a]The least significant difference among the means at the 0.05 probability level.

Table 3 Distribution of [15]N Absorbed as Ammonium in Cotton Shoots and Roots Under Different NaCl Salinity Levels at the Beginning of the Reproductive Stage of Growth

Plant parts	Treatment, osmotic potential (MPa)	15N uptake/pot (two plants), (mg) for uptake time (h)		
		6	12	24
Shoots	Control	1.77	4.11	8.57
	−0.4	1.91	3.78	8.64
	−0.8	1.49	3.55	8.45
	−1.2	1.15	2.18	4.67
	LSD (0.05)[a]	0.43	0.64	1.67
Roots	Control	1.32	2.44	5.49
	−0.4	1.62	3.07	5.68
	−0.8	1.58	2.85	5.65
	−1.2	1.01	1.81	3.39
	LSD (0.05)[a]	0.39	0.65	1.80
Total	Control	3.09	6.54	13.99
	−0.4	3.53	6.85	14.32
	−0.8	3.07	6.40	14.10
	−1.2	2.16	3.99	8.06
	LSD (0.05)[a]	0.21	0.60	1/71

Source: Ref. 28.
[a]The least significant difference among the means at the 0.05 probability level.

Table 4 Nitrogen (^{15}N) Concentration of Cotton Plants During Vegetative and Reproductive Stages of Growth as Influenced by NaCl Salinity

Growth stage	Treatment, osmotic potential (MPa)	[15]N concentration (mg ^{15}N/kg dry wt)	
		Shoots	Roots
Vegetative	Control	917	3462
	−0.4	1261	4330
	−0.8	1609	4195
	−1.2	1059	3180
	LSD (0.05)[a]	384	586
Reproductive	Control	422	1408
	−0.4	517	1427
	−0.8	698	1605
	−1.2	615	1401
	LSD (0.05)[a]	109	245

Source: Ref. 28.
[a]The least significant difference among the means at the 0.05 probability level.

Table 5 Total N Uptake of Cotton Plants During Vegetative and Reproductive Stages of Growth as Influenced by NaCl Salinity

Growth stage	Treatment, osmotic potential (MPa)	Total N/pot (two plants) (mg)		
		Shoots	Roots	Total
Vegetative	Control	185.1	20.5	205.6
	−0.4	114.8	23.0	137.8
	−0.8	89.8	17.8	107.6
	−1.2	43.1	7.2	50.3
	LSD (0.05)[a]	15.9	5.3	17.7
Reproductive	Control	371.2	63.6	434.8
	−0.4	333.8	66.7	400.6
	−0.8	245.9	62.0	307.9
	−1.2	156.5	41.6	198.1
	LSD (0.05)[a]	96.1	15.0	105.6

Source: Ref. 28.
[a]The least significant difference among the means at the 0.05 probability level.

vegetative stage than at the reproductive stage of growth. Reduced N uptake at the vegetative stage was largely due to the reduction in dry matter production, except at the high salinity level where N concentration was reduced appreciable (Tables 1 and 5). At the reproductive stage, N concentration was essentially the same at all salinity levels, indicating that plants had adjusted somewhat to salinity and its effect on N uptake. The dry weights at the high salinity level still were less than 50% of the controls, as were values for the total N uptake. Although generally these observations are similar to the previous discussion of ^{15}N data, some small deviations are apparent. Concentration data for ^{15}N (Table 4) indicate that short-term ^{15}N concentrations increased in the shoots at the −0.8 MPa salinity level.

C. Nitrogen Metabolism and Assimilation in Cotton Plants

1. Protein-N Content of Plants

After 24-h exposure to ^{15}NH$_4^+$, the protein-^{15}N content of plants treated with a high level of NaCl (−1.2 MPa osmotic potential) was significantly less than in either the control, low, or medium NaCl treatments (Table 6). The depressing effects of salt on protein content of cotton plants under high levels of NaCl could be attributed to the decreased amino-N incorporation into protein reported for red kidney beans [48,49], rice [7,44,65], and other plants [31]. The decrease in polyribosome levels reported for corn, *Zea mays* L. [128], and for barley and pea shoots [129] is probably another reason for the decrease in protein synthesis in cotton shoots. The low level of NaCl (−0.4 MPa osmotic potential of the nutrient solution) significantly increased the protein content of cotton shoots at the vegetative stage of growth. The same level (−0.4 MPa) of salt stress substantially decreased protein synthesis in a number of plants with lower degrees of salt tolerance [22–24,48,49,59,62]. Impaired N metabolism, with the consequence of reduced protein content of several other plant species with various degrees of salt tolerance under stress conditions, has been reported by several investigators [6,7,9,15,16, 30,31,33,37,42,44,54,56–58,60,61,63–66,128,130].

Water stress is also known to impair N metabolism and reduce protein synthesis in plants [6,30,31,33,36,48,49,62,63,129]. In addition to the osmotic effect of salt, the specific ion

Table 6 Concentration of ^{15}N Fractions and Protein-^{15}N/Nonprotein-^{15}N Ratio of Cotton Shoots Influenced by NaCl Stress for Two Stages of Growth After 24-h Uptake

Osmotic potential (MPa)	Protein-N (mg)	Total soluble N (mg)	Ammonium plus amide-N (mg)	Free amino acid-N (mg)	Protein-^{15}N/ nonprotein-^{15}N Ratio
Vegetative					
Control	489.6	318.6	6.97	129.0	1.54
–0.4	560.9	463.9	9.37	189.1	1.21
–0.8	544.3	741.3	21.96	287.1	0.73
–1.2	230.4	523.2	13.67	273.1	0.44
LSD (0.05)[a]	68.3	120.4	2.10	51.9	0.24
Reproductive					
Control	229.9	145.5	4.87	60.6	1.58
–0.4	232.8	204.8	5.31	65.9	1.14
–0.8	217.3	363.6	19,71	96.4	0.60
–1.2	98.8	393.6	9.64	108.1	0.25
LSD (0.05)[a]	46.6	76.2	1.83	11.6	0.47

Source: Ref. 29.
[a]The least significant difference among the means at the 0.05 probability level.

effect of Na^+ and/or Cl^- has certainly contributed appreciably to the inhibition of ^{15}N incorporation into protein. However, this study was not designed to distinguish specifically between osmotic and ionic effects.

The rates of ^{15}N incorporation into protein as measured by the concentration of ^{15}N in the protein fraction at 6, 12, and 24 h of exposure to $^{15}NH_4^+$ appear to have been influenced only by the high level of NaCl at both growth stages (Table 7). The rate of incorporation was reduced at the –1.2 MPa salt level by factors of 2.5 and 2.8 at the vegetative and reproductive stages of growth, respectively. The rate of ^{15}N incorporation into protein at the vegetative stage was greater by approximately 2.5 times the rate at the reproductive stage, based on the ^{15}N concentration. The total ^{15}N in the protein fraction was greater at the reproductive than at the vegetative stage of growth, apparently due to the much higher amount of shoot dry weight (Table 1). Furthermore, the total protein-^{15}N decreased with increased salt level at both stages of growth. This reflects the combined effect of salt on shoot growth and ^{15}N incorporation into protein.

2. Total Soluble-N Content of Plants

Soluble N compounds that should be in an ethanol extract of plant tissues include NO_3^-, NH_4^+, amides, amino acids, amine, amino sugars, peptide, alkaloids, nucleotide, chlorophyll, and even some fats. Since only the NH_4^+ form of N was used in this investigation, ^{15}N from $^{15}NH_4^+$ exposure should not be found in the NO_3^- form in the plant tissues in this study.

Total soluble-^{15}N concentration of the plant tissues increased with NaCl concentration at –0.8 MPa osmotic potential after 24 h of exposure to $^{15}NH_4^+$ at the vegetative growth stage, then declined at –1.2 MPa (Table 6). At the reproductive growth stage, total soluble ^{15}N increased in a similar manner, but did not decline at the highest salinity level. Thus the decrease in protein-^{15}N at the –1.2 MPa salinity level did not result from a shortage of soluble-^{15}N compounds. The rates of ^{15}N incorporation into the total soluble-^{15}N fraction as indicated by the slope in the regression of ^{15}N tissue concentration versus time (Table 7) followed the same pattern as described above for the 24-h uptake time. Accumulation of soluble-N compounds in

Table 7 Slope (b) and Intercept (a) of the Regression Lines for Concentration of ^{15}N Fraction in Cotton shoots Influenced by NaCl Salinity Versus 6, 12, and 24 h of Exposure Time for Two Stages of Growth[a]

^{15}N Fraction	Treatment, osmotic potential (MPa)	Vegetative		Reproductive	
		b	a	b	a
Protein	Control	24.22	−99.08	10.38	24.65
	−0.4	26.44	−81.19	10.43	23.18
	−0.8	24.71	−61.10	9.36	−11.58
	−1.2	9.89	−6.80	3.61	+11.56
Soluble	Control	13.20	−13.70	5.14	17.53
	−0.4	21.28	−58.50	0.06	−1.79
	−0.8	30.64	−19.25	16.36	−28.95
	−1.2	20.88	+30.25	16.17	+2.36
Ammonium polus amide	Control	0.32	− 0.53	0.19	−0.06
	−0.4	0.40	−0.16	0.23	−0.17
	−0.8	0.93	−0.49	0.80	+0.58
	−1.2	0.51	+1.71	0.30	+2.52
Free amino	Control	6.03	−22.79	2.65	−1.05
	−0.4	9.63	−48.44	2.73	+1.13
	−0.8	13.88	−58.63	4.41	−9.63
	−1.2	14.24	−76.61	5.12	−15.17

Source: Ref. 29.
[a]Correlation coefficient, *r*, values lie between 0.92 and 1.00.

plants under stress conditions has also been reported by several other investigators for various plant species [9,15,19,20–27,30,31,34,35,37–41,44,46–51,54–60,65,66,116].

The amounts of soluble ^{15}N reflect concentrations in both the tissue and dry matter production (Table 1). Significantly less total soluble ^{15}N was found after 24 h with −1.2 MPa salinity at the vegetative growth stage. With the −0.8 MPa salinity, a larger amount of total soluble ^{15}N resulted than with other treatments. At the reproductive stage, the amounts of total soluble ^{15}N were equal at the high and at low levels of salinity, with the intermediate salinity levels resulting in higher quantities of total soluble ^{15}N in the plant parts.

Although the rate of ^{15}NH$_4^+$ absorption was severely curtailed by high salinity at both growth stages [28], growth was not restricted by decreased total soluble-N concentration. However, impairment of soluble-N utilization at high salinity was reflected in a severe decrease in protein concentration of plants. It is not clear whether or not this lower protein concentration was a cause of reduced growth. Growth was reduced at the lower salinity levels without a reduction in protein concentration.

The ratio of protein-N to nonprotein-N (soluble N) is further evidence for the decrease in protein-N content of the NaCl-treated plants. A substantial decrease in protein-^{15}N/nonprotein-^{15}N ratio was observed for plants subjected to a high level of NaCl (−1.2 MPa osmotic potential) compared with controls at both stages of growth (Table 6). At both stages of growth, values for the −0.8 MPa osmotic potential of the NaCl-treated plants were significantly lower than the controls.

3. Ammonium Plus Amide-N Content of Plants

At both stages of growth, significantly higher concentrations of ammonium plus amide-^{15}N accumulated in the shoots of plants subjected to NaCl stress compared with controls (Table 6). The concentration of ammonium plus amide-^{15}N increased with increasing salinity to a maximum at –0.8 MPa osmotic potential. Since the absorption rate of ^{15}NH$_4^+$ did not change appreciably at these salinity levels, this increased accumulation of ammonium plus amide-N must have resulted from a reduced rate of utilization; however, reduced growth is another possible consideration. The concentrations of ammonium plus amide-N at –1.2 MPa stress were lower than at –0.8 MPa osmotic potential. These values reflect markedly reduced absorption rates at –1.2 MPa stress [28]. The rate of ^{15}N utilization decreased also, allowing a higher ^{15}N concentration than commensurate with absorption rate. Slopes for regressions of ammonium plus amide-^{15}N and time of uptake for each salinity level (Table 7) indicate a pattern for the rate of accumulation similar to the concentrations indicated for 24-h exposure (Table 6).

4. Free Amino-N Content of Plants

Free amino acids would be expected to constitute the major portion of the total ethanol soluble-N compounds from plant tissues. In this study the amino-N and ammonium plus amide-N accounted for 30 to 55% of the total soluble-N. The ninhydrin release method for free amino-N determination was used in this investigation. This method, however, can result in poor recoveries of a number of amino acids [131]. In Kennedy's [131] investigation, recoveries varied from 2 to 60% for 12 amino acids, with complete recovery of 14 others. In the present study, the apparent low recovery of amino-^{15}N from cotton tissues by the ninhydrin release method is consistent with the results of Kennedy [131] when all aspects of methodology are considered. Even with the low recovery, however, the relative effects of NaCl salinity on amino acid formation and utilization should be valid.

After a 24-h exposure to ^{15}NH$_4^+$, a higher concentration of amino-^{15}N accumulated in the NaCl-stressed plants than in the controls at both growth stages (Table 6). This increased concentration was sufficient to equal or exceed the reduction in dry weight, except at the highest NaCl level (–1.2 MPa osmotic potential), when dry weight was reduced most drastically. Yet at the higher NaCl salinity level, the amino-^{15}N concentration either remained constant or increased slightly as protein concentration values declined to less than half the values for all other treatments. Slopes of the regression lines of the amino acid and exposure time (Table 7) reflect the rate of amino acid accumulation. Thus the incorporation of amino acids into protein was impaired by a high level of NaCl. The level of salinity that was required for interference with protein formation in cotton, a relatively high salt-tolerant plant, was much higher than was reported for green beans [23,24,59], red kidney beans [48,49,62], soybeans (*Glycine max* L.) [66], peas [61], alfalfa [22], corn [33,128], rice [7,15,44,65], and wheat [56], which all have lower degrees of salt tolerance than that of cotton.

D. Total Water Uptake by Plants

At both stages of growth (except for the –0.4 MPa osmotic potential during the reproductive stage), salt-stressed plants absorbed significantly less water than the controls (Table 8). Plants at –0.4 MPa stress did not exhibit a statistically significant difference for water uptake during the reproductive stage of growth compared with controls. Reduction in water absorption by plants due to salinity stress has been reported by many investigators [21–27,39,47,49,51,59,92, 132–135]. These investigators generally agreed that the root permeability of plants (expressed as hydraulic conductivity of the root system) was decreased significantly under salt stress. This

Table 8 Influence of NaCl Salinity on Water Absorption by Cotton
Plants During the 24-h 15N Uptake Period for the Vegetative and
Reproductive Stages of Growth

Growth stage	Treatment, osmotic potential (MPa)	Water uptake/pot (two plants) (mL) for uptake time (h)		
		6	12	24
Vegetative	Control	125.1	160.0	202.5
	−0.4	87.5	122.5	172.5
	−0.8	57.5	120.5	135.0
	−1.2	40.0	70.0	87.5
	LSD (0.05)[a]	46.0	23.4	28.2
Reproductive	Control	165.0	275.0	490.0
	−0.4	147.5	245.0	430.0
	−0.8	130.0	167.5	215.0
	−1.2	75.0	107.5	145.0
	LSD (0.05)[a]	78.7	102.8	205.5

Source: Ref. 28.
[a]The least significant difference among the means at the 0.05 probability level.

is an explanation for the reduction in water absorption rate and may contribute to a similar reduction in nutrient uptake under salinity conditions.

III. SUMMARY AND CONCLUSIONS

Cotton plants grown in normal (control) and NaCl-treated Hoagland solutions were studied at two stages of growth (vegetative and reproductive). Plant growth in terms of dry matter production was measured. Nitrogen absorption (total N and ^{15}N) and water uptake were determined. Plant parts (shoots and roots) were analyzed separately for N content and distribution of ^{15}N in NH_4^+ plus amide-N, free amino-N, total soluble N, and protein-N after the plants were provided $^{15}NH_4NO_3$ in nutrient solutions for 6, 12, and 24 h.

Dry matter production of the cotton plants was significantly reduced by decreasing the osmotic potential (increasing salinity) of the nutrient solution. The low and medium levels of salinity did not exhibit a significant effect on the ^{15}N absorption rate, but the high salt levels caused a substantial reduction in the ^{15}N uptake rate. The ^{15}N concentration of the roots was higher than that of the shoots, particularly under stress conditions. The ^{15}N percentage in plants increased with increasing salinity levels. The concentration of ^{15}N in plants in terms of the ratio of plant total ^{15}N content to dry matter produced (mg ^{15}N/kg dry matter) was significantly higher for moderately stressed than for control plants. This indicates that plants continued to accumulate ^{15}N under salt stress conditions despite the reduction in dry matter production. Total water absorbed by plants decreased linearly with increasing salinity. This reduction was even more appreciable than the reduction in ^{15}N absorption rate. The effect of salinity was more pronounced at the vegetative than at the reproductive stage of growth.

The metabolism of ^{15}N in salinized cotton plants was adversely affected under medium and high levels of NaCl at both the vegetative and reproductive stages of growth. Significant accumulations of all soluble-^{15}N fractions occurred when plants were subjected to medium and high levels of NaCl compared with controls. The −0.4 MPa osmotic potential of the culture solution enhanced protein synthesis at the vegetative growth stage. Only the −1.2 MPa osmotic potential decreased the protein-^{15}N content of plants significantly compared with controls and any other level of NaCl. Protein synthesis was impaired by a large excess of NaCl in the nutrient solution which inhibited NH_4^+ metabolism.

Consequently, under salt stress conditions of sufficient magnitude, plant growth, N absorption and metabolism, protein synthesis, and water absorption will be altered. This will result in the failure of plants to fully utilize nutrients and water. Salinity levels in excess of those causing drastic interference with plant growth, nutrient (i.e., N) absorption and metabolism, and water uptake in salt-sensitive plants such as beans do not interfere appreciably with these factors in cotton, a plant that is relatively high in salt tolerance. This indicates a link between salt tolerance, growth, nutrient (i.e., N) absorption and metabolism, and water uptake. Although the contribution of osmotic and specific ion effects cannot be distinguished from this study, it is probable that both were involved.

REFERENCES

1. S. M. Alam, in *Handbook of Plant and Crop Stress* (M. Pessarakli, ed.), Marcel Dekker, New York, pp. 227–246 (1993).
2. S. M. Alam, *Pak. J. Sci. Ind. Res.*, *33*(7): 292 (1990).
3. S. M. Alam, S. S. M. Naqvi, and A. R. Azmi, *Pak. J. Sci. Ind. Res.*, *32*(2): 110 (1989).
4. M. Aslam and R. H. Qureshi, *Int. Rice Res. Newsl.*, *14*(3): 25 (1989).
5. V. N. Bhivare and J. D. Nimbalkar, *Plant Soil*, *80*(1): 91 (1984).
6. R. S. Dubey, in *Handbook of Plant and Crop Stress* (M. Pessarakli, ed.), Marcel Dekker, New York, pp. 277–299 (1993).
7. R. S. Dubey and M. Rani, *J. Agron. Crop Sci.*, *162*(2): 97 (1989).
8. S. R. Grattan and C. M. Grieve, in *Handbook of Plant and Crop Stress* (M. Pessarakli, ed.), Marcel Dekker, New York, pp. 203–226 (1993).
9. A. S. Gupta (A. Sen-Gupta), *Diss. Abstr. Int. B Sci. Eng.*, *47*(12): I4728B (1987).
10. F. Y. Hafeez, Z. Aslam, and K. A. Malik, *Plant Soil*, *106*(1): 3 (1988).
11. A. Hamdy, *Proceedings of the 15th ICID European Regional Conference*, Dubrovnik, Yugoslavia; *Int. Comm. Irrig. Drain.*, *1988*(2): 144 (1988).
12. P. B. Kavi-Kishor, *Plant Cell Environ.*, *12*(6): 629 (1989).
13. A. H. Khan and M. Y. Ashraf, *Acta Physiol. Plant.*, *10*(3): 257 (1988).
14. R. Krishnamurthy, M. Anbazhagan, and K. W. Bhagwat, *Oryza*, *24*(1): 66 (1987).
15. R. Krishnamurthy and K. A. Bhagwat, *Indian J. Exp. Biol.*, *27*(12): 1064 (1989).
16. N. Mehta and S. Bharti, *Indian J. Plant Physiol.*, *26*(3): 322 (1983).
17. G. Naidoo, *New Phytol.*, *107*(2): 317 (1987).
18. A. S. Nigwekar and P. D. Chavan, *Acta Soc. Bot. Polon.*, *56*(1): 93 (1987).
19. T. A. Omran, *Alexandria J. Agric. Res.*, *31*(2): 449 (1986).
20. R. Pandey and P. S. Ganapathy, *J. Exp. Bot.*, *35*(157): 1194 (1984).
21. M. Pessarakli, *Crop Sci.*, *31*(6): 1633 (1991).
22. M. Pessarakli and J. T. Huber, *J. Plant Nutr.*, *14*(3): 283 (1991).
23. M. Pessarakli, J. T. Huber, and T. C. Tucker, *J. Plant Nutr.*, *12*(11): 1361 (1989).
24. M. Pessarakli, J. T. Huber, and T. C. Tucker, *J. Plant Nutr.*, *12*(10): 1105 (1989).
25. M. Pessarakli, J. T. Huber, and T. C. Tucker, *J. Plant Nutr.*, *12*(3): 279 (1989).
26. M. Pessarakli and T. C. Tucker, *Soil Sci. Soc. Am. J.*, *52*(6): 1673 (1988).
27. M. Pessarakli and T. C. Tucker, *Soil Sci. Soc. Am. J.*, *52*(3): 698 (1988).
28. M. Pessarakli and T. C. Tucker, *Soil Sci. Soc. Am. J.*, *49*(1): 149 (1985).
29. M. Pessarakli and T. C. Tucker, *J. Plant Nutr.*, *8*(11): 1025 (1985).
30. E. Rabe, in *Handbook of Plant and Crop Stress* (M. Pessarakli, ed.), Marcel Dekker, New York, pp. 261–276 (1993).
31. E. Rabe, *J. Hortic. Sci.*, *65*(3): 231 (1990).
32. R. K. Rabie and K. Kumazawa, *Soil Sci. Plant Nutr.*, *34*(3): 385 (1988).

33. S. Ramagopal, *Plant Cell Rep.*, *5*(6): 430 (1986).

34. M. Salim, *J. Agron. Crop Sci.*, *66*(3): 204 (1991).

35. M. Salim, *J. Agron. Crop Sci.*, *162*(1): 35 (1989).

36. J. Shalhevet and T. C. Hsiao, *Irrig. Sci.*, *7*(4): 249 (1986).

37. M. C. Shannon, J. W. Gronwald, and M. Tal, *J. Am. Soc. Hortic. Sci.*, *112*(3): 416 (1987).

38. S. K. Sharma, *Indian J. Plant Physiol.*, *32*(3): 200 (1989).

39. S. M. Shukr-Almashhadany and S. M. S. Almashhadany, *Diss. Abstr. Int. B Sci. Eng.*, *47*(2): 442B (1986).

40. G. Singh, H. S. Gill, I. P. Abrol, and S. S. Cheema, *Field Crops Res.*, *26*(1): 45 (1991).

41. A. Sinha, S. R. Gupta, and R. S. Rana, *Plant Soil*, *95*(3): 411 (1986).

42. S. Subbanaidu-Ramagopal (S. Ramagopal), *J. Plant Physiol.*, *132*(2): 245 (1988).

43. H. E. Dregne, *N. M. Agric. Exp. Stn. Res. Rep. 94* (1964).

44. G. Palfi, *Plant Soil*, *22*: 127 (1965).

45. V. Hernando, L. Jimeno, and C. Cadahia, *An. Edafol. Agrobiol.*, *26*: 1147 (1967).

46. T. S. Mahajan and K. R. Sonar, *J. Maharashtra Agric. Univ.*, *5*: 110 (1980).

47. J. N. E. Frota and T. C. Tucker, *Soil Sci. Soc. Am. J.*, *42*: 753 (1978).

48. J. N. E. Frota and T. C. Tucker, *Soil Sci. Soc. Am. J.*, *42*: 743 (1978).

49. R. Saad, Ph.D. dissertation, University of Arizona; University Microfilms, Ann Arbor, Mich., *Diss. Abstr. B*, *40*: 4057 (1979).

50. M. A. Khalil, A. Fathi, and M. M. Elgabaly, *Soil Sci. Soc. Am. Proc.*, *31*: 683 (1967).

51. M. Pessarakli, Ph.D. dissertation, University of Arizona; University Microfilms, Ann Arbor, Mich., *Diss. Abstr. B*, *42*: 286 (1981).

52. A. Golan-Goldhirsh, B. Hankamer, and S. H. Lips, *Plant Sci. (Limerick)*, *69*(1): 27 (1990).

53. G. L. Maliwal and K. V. Paliwal, *Agric. Sci. Dig. (India)*, *4*(3): 147 (1984).

54. G. W. Langdale and J. R. Thomas, *Agron. J.*, *63*: 708 (1971).

55. A. A. Luque and F. T. Bingham, *Plant Soil*, *63*: 227 (1981).

56. S. M. Abdul-Kadir and G. M. Paulsen, *J. Plant Nutr.*, *5*: 1141 (1982).

57. G. V. Udovenko, V. N. Sinel'nikova, and G. V. Khazova, *Dokl. Akad. Nauk USSR*, *192*(6): 1395 (1970).

58. G. V. Udovenko, V. N. Sinel'nikova, and G. V. Khazova, *Agrokhimiya*, *3*: 23 (1971).

59. M. Pessarakli, ed., *Handbook of Plant and Crop Stress* Marcel Dekker, New York, pp. 415–430 (1993).

60. H. M. Helal and K. Mengel, *Plant Soil*, *51*: 457 (1979).

61. I. Kahane and A. Poljakoff-Mayber, *Plant Physiol.*, *43*: 1115 (1968).

62. J. T. Prisco and J. W. O'Leary, *Rev. Brazil Biol.*, *30*: 317 (1970).

63. N. M. Iraki, R. A. Bressan, and N. C. Carpita, *Plant Physiol.*, *91*(1): 54 (1989).

64. S. Katiyar and R. S. Dubey, *J. Agron. Crop Sci.*, *165*(1): 19 (1990).

65. R. Krishnamurthy, M. Anbazhagan, and K. A. Bhagwat, *Indian J. Plant Physiol.*, *30*(2): 183 (1987).

66. J. Wieneke and R. Fritz, *Acta Univ. Agric. Brno A Fac. Agron.*, *33*(3): 653 (1985).

67. M. Pessarakli, ed., *Handbook of Plant and Crop Stress*, Marcel Dekker, New York (1993).

68. M. Ashraf and E. Rasul, *Plant Soil*, *110*(1): 63 (1988).

69. A. R. Azmi and S. M. Alam, *Acta Physiol. Plant.*, *12*(3): 215 (1990).

70. A. R. Bal and N. C. Chattopadhyay, *Biol. Plant.*, *27*(1): 65 (1985).

71. A. R. Bal, Y. C. Joshi, and A. Qadar, *Curr. Agric.*, *10*(1–2): 65 (1986).

72. L. K. Chugh, M. S. Kuhad, and I. S. Sheoran, *Anal. Biol.*, *4*(1–2): 20 (1988).

73. P. S. Curtis and A. Lauchli, *Crop Sci.*, *25*(6): 944 (1985).

74. K. S. Datta and J. Dayal, *Indian J. Plant Physiol.*, *31*(4): 357 (1988).

75. E. B. Dumbroff and A. W. Cooper, *Bot. Gaz.*, *135*(3): 219 (1974).

76. L. E. Francois, E. V. Maas, T. J. Donovan, and V. L. Young, *Agron. J.*, *78*(6): 1053 (1986).

77. L. E. Francois, T. J. Donovan, K. Lorenz, and E. V. Maas, *Agron. J.*, *81*(5): 707 (1989).

78. D. P. Heenan, L. G. Lewin, and D. W. McCaffery, *Aust. J. Exp. Agric.*, *28*(3): 343 (1988).

79. E. V. Maas and J. A. Poss, *Irrig. Sci.*, *10*(4): 313 (1989).

80. E. V. Maas and J. A. Poss, *Irrig. Sci.*, *10*(1): 29 (1989).

81. G. L. Maliwal and K. V. Paliwal, *Legume Res.*, *5*(1): 23 (1982).
82. A. Mozafar and J. R. Goodin, *Plant Soil*, *96*(3): 303 (1986).
83. A. Nukaya, M. Masui, and A. Ishida, *J. Jpn. Soc. Hortic. Sci.*, *53*(2): 168 (1984).
84. O. A. A. Osman, A. Lauchli, and A. El-Beltagy, *Proceedings of the First Conference of the Agricultural Development Research*, Cairo, Dec. 19–21, 1987, Vol. II, *Agronomy, Horticulture, Soil Science, and Rural Sociology*, Ain Shams University, Cairo, pp. 126–145 (1987).
85. L. Prakash and G. Prathapasenan, *Aust. J. Plant Physiol.*, *15*(6): 761 (1988).
86. H. Y. Ryu, H. C. Choi, C. H. Cho, and S. T. Lee, *Res. Rep. Rural Dev. Adm. (Suweon)*, *30*(3 Rice): 1–15 (1988).
87. U. Schmidhalter and J. J. Oertli, *Plant Soil*, *132*(2): 243 (1991).
88. A. R. Sepaskhah, *Can. J. Plant Sci.*, *57*(3): 925 (1977).
89. M. C. Shannon, G. W. Bohn, and J. D. McCreight, *HortScience*, *19*(6, Sec. 1): 828 (1984).
90. E. Kurth, G. R. Cramer, A. Lauchli, and E. Epstein, *Plant Physiol.*, *82*(4): 1102 (1986).
91. S. Al-Khafaf, A. Adnan, and N. M. Al-Asadi, *Agric. Water Manage.*, *18*(1): 63 (1990).
92. M. Pessarakli, T. C. Tucker, and K. Nakabayashi, *J. Plant Nutr.*, *14*(4): 331 (1991).
93. N. J. W. Clipson, *J. Exp. Bot.*, *38*(197): 1996 (1987).
94. V. Balasubramanian and S. K. Sinha, *Physiol. Plant.*, *36*(2): 197 (1976).
95. V. A. Bastianpillai, C. Stark, and J. Unger, *Beitr. Trop. Landwirtsch. Veterinaermed.*, *20*(4): 359 (1982).
96. S. Bouraima, D. Lavergne, and M. L. Champigny, *Agronomie*, *6*(7): 675 (1986).
97. G. S. Cabuslay, L. C. Blanco, and S. Akita, *Jpn. J. Crop Sci.*, *60*(2): 271 (1991).
98. G. F. Craig, C. A. Atkins, and D. T. Bell, *Plant Soil*, *133*(2): 253 (1991).
99. P. S. Curtis, H. L. Zhong, A. Lauchli, and R. W. Pearcy, *Am. J. Bot.*, *75*(9): 1293 (1988).
100. K. S. Datta, J. Dayal, and C. L. Goswami, *Anal. Biol.*, *3*(1): 47 (1987).
101. W. J. S. Downton, *Aust. J. Plant Physiol.*, *9*: 519 (1982).
102. J. Gorham, E. McDonnell, E. Budrewicz, and R. G. Wyn-Jones, *J. Exp. Bot.*, *36*(168): 1021 (1985).
103. S. Joshi and J. D. Nimbalkar, *Plant Soil*, *74*(2): 291 (1983).
104. S. Kannan and S. Ramani, *J. Plant Nutr.*, *11*(4): 435 (1988).
105. D. K. Kishore, R. M. Pandey, and R. Ranjit-Singh, *Prog. Hortic.*, *17*(4): 289 (1985).
106. O. A. M. Lewis, E. O. Leidi, and S. H. Lips, *New Phytol.*, *111*(2): 155 (1989).
107. A. F. Radi, M. M. Heikal, A. M. Abdel-Rahman, and B. A. A. El-Deep, *Rev. Roum. Biol.*, *33*(1): 27 (1988).
108. R. Rai and S. V. Prasad, *Soil Biol. Biochem.*, *15*(2): 217 (1983).
109. P. Reddell, R. C. Foster, and G. D. Bowen, *New Phytol.*, *102*(3): 397 (1986).
110. B. A. Roundry, J. A. Young, and R. A. Evans, *Agric. Ecosyst. Environ.*, *25*(2–3): 245 (1989).
111. A. M. Shaheen and M. M. El-Sayed, *Minufiya J. Agric. Res.*, *8*: 363 (1984).
112. M. G. T. Shone and J. Gale, *J. Exp. Bot.*, *34*(146): 1117 (1983).
113. C. Stark, *Beitr. Trop. Landwirtsch. Veterinaermed.*, *23*(1): 33 (1985).
114. E. L. Taleisnik, *Physiol. Plant.*, *71*(2): 213 (1987).
115. C. Torres and F. T. Bingham, *Proc. Soil Sci. Soc. Am.*, *37*(5): 711 (1973).
116. E. Zid and M. Boukhris, *Oecologia Plant.*, *12*(4): 351 (1977).
117. L. Prakash and G. Prathapasenan, *Biochem. Physiol. Pflanz.*, *184*(1–2): 69 (1989).
118. D. Lazof, N. Bernstein, and A. Lauchli, *Bot. Gaz.*, *152*(1): 72 (1991).
119. P. S. Curtis, *Diss. Abstr. Int. B Sci. Eng.*, *47*(2): 476B (1986).
120. L. E. Francois, *J. Am. Soc. Hortic. Sci.*, *112*(3): 432 (1987).
121. C. R. Hampson and G. M. Simpson, *Can. J. Bot.*, *68*(3): 529 (1990).
122. K. S. Gill, *Plant Physiol. Biochem. (India)*, *14*(1): 82 (1987).
123. N. H. Karim and M. Z. Haque, *Bangladesh J. Agric.*, *11*(4): 73 (1986).
124. E. Aceves-N, L. H. Stolzy, and G. R. Mehuys, *Plant Soil*, *42*(3): 619 (1975).
125. E. H. Hansen and D. N. Munns, *Plant Soil*, *107*(1): 95 (1988).
126. T. J. Keck, R. J. Wagenet, W. F. Campbell, and R. E. Knighton, *Soil Sci. Soc. Am. J.*, *48*(6): 1310 (1984).
127. W. J. Zimmerman, *Anal. Bot.*, *56*(5): 689 (1985).

128. C. A. Morilla, J. S. Boyer, and R. H. Hageman, *Plant Physiol.*, *51*: 817 (1973).

129. P. R. Rhodes and K. Matsuda, *Plant Physiol.*, *58*: 635 (1976).

130. T. S. Gibson, *Plant Soil*, *111*(1): 25 (1988).

131. I. R. Kennedy, *Anal. Biochem.*, *11*: 105 (1965).

132. J. W. O'Leary, in *Structure and Function of Primary Root Tissues* (J. Kolek, ed.), Veda, Publishing House of the Slovak Academy of Science, Bratislava, Czechoslovakia pp. 309–314 (1974).

133. V. R. Babu and V. Ramesh-Babu, *Seed Res.*, *13*(1): 129 (1985).

134. M. Salim, *J. Agron. Crop Sci.*, *166*(4): 285 (1991).

135. R. Tipirdamaz and H. Cakirlar, *Doga Biyol. Serisi*, *14*(2): 124 (1990).

33
Metal Tolerance Aspects of Plant Cell Wall and Vacuole

Jian Wang and V. P. Evangelou
University of Kentucky, Lexington, Kentucky

I. INTRODUCTION

A. Occurrences of Heavy Metal Toxicity in Plants

Heavy metal toxicity in plants has been reported for almost a century. However, early reports focused more on expression of metal toxicity than on mechanistic causes [1]. Only during the past few decades have plant and soil scientists begun to study the causes of heavy metal toxicity in plants from the perspectives of soil properties and plant reactions to metal accumulation [2]. Metal-enriched soils that cause metal toxicity effects in plants result either from natural processes such as geochemical anomalies or human activities such as exploitation of mineral resources and/or disposal of industrial waste [3]. Metal toxicity effects have been observed to vary widely among plant species and genotypes within a specific species. Interestingly, metal toxicity effects are even different among plant species that accumulate equal quantities of a particular heavy metal. The latter indicates that although plants may accumulate similar amounts of heavy metals, they react to or tolerate heavy metal accumulations differently.

Frequently, heavy metals and/or metalloids causing toxicity in plants are biologically nonessential. Such metals and/or metalloids include aluminum (Al), cadmium (Cd), arsenic (As), uranium (U), lead (Pb), thallium (Tl), chromium (Cr), mercury (Hg), silver (Ag), and gold (Au). However, it is also common that biologically essential metals such as manganese (Mn), iron (Fe), zinc (Zn), copper (Cu), and molybdenum (Mo) as well as heavy metals such as nickel (Ni), cobalt (Co), and vanadium (V) (which are biologically essential only to a limited number of species) can be toxic when they are accumulated in plants at relatively high concentrations.

Occurrence of heavy metal toxicity in plants, particularly in agricultural and economic crops, presents a challenge for plant scientists concerned with yield and quality in crop production. On the other hand, plant heavy metal accumulation potentials may also provide us with the opportunity to remediate heavy metal-contaminated soils biologically through growing metal-accumulating species. Metal-accumulating plants and metal-tolerant crop cultivars may be produced either by conventional breeding or by genetic engineering techniques. In any case, success in either of the two goals of research noted above requires understanding the mechanisms of heavy metal accumulation, toxicity, and tolerance in plants.

In this chapter we focus on the roles of the cell wall and vacuole in plant heavy metal tolerance and discuss possible mechanisms involved. Before these topics are approached, other aspects of heavy metal tolerance will be discussed to provide an overview of plant heavy metal tolerance.

B. Ecological Aspects of Heavy Metal Tolerance

The observation that different plant species are found at heavy metal–contaminated and noncontaminated soil sites has attracted much attention to the evolutionary aspect of plant heavy metal tolerance for the past 30 years [1]. The evolution of plants in tolerating heavy metals is similar to the evolution of plants in tolerating other environmental stress [4]. However, it appears that adaptation of plants to naturally occurring environmental stress may take place over a long period of time, whereas evolution of plants in terms of tolerating anthropogenic heavy metal–contaminated habitats may occur over a very short period, less than a few years [5,6]. In addition, evolution of plant heavy metal tolerance can be very localized. For instance, replicated evolution of Zn tolerance in the populations of *Agrostis capillaris* growing beneath electricity pylons in Great Britain were restricted to an area as small as 10 m^2 [7].

Evolution of plant heavy metal tolerance may be governed by the availability of genes conferring the appropriate adaptation. In the example given above concerning Zn tolerance, although some degree of tolerance was found in five different plant species growing beneath electricity pylons, only one of those plant species showed a level of tolerance equal to that found on zinc mines [4]. Apparently, there are some limits in the occurrence of Zn tolerance. As indicated by Bradshaw [8], the restriction in supply of genetic variability may directly limit the evolutionary process and extent of adaptation.

The terms *sensitivity* and *resistance* have often been used to describe consequences and reactions of plants exposed to heavy metals in growth media [1,9]. *Sensitivity* refers to the effects of heavy metal stress, which results in injury or death of a plant; *resistance* is the reaction of a plant to heavy metal stress to allow survival and its reproduction. According to Baker [1], plant resistance to heavy metal stress can be achieved by either of two strategies: (a) avoidance, by which a plant is protected externally from the influence of the stress, and (b) tolerance, by which a plant survives the effects of internal stress. Tolerance, therefore, is conferred by the possession of specific physiological mechanisms which collectively enable the plant to function normally even in the presence of high concentrations of potentially toxic elements. Therefore, a genetic basis for tolerance is apparently implied, in that mechanisms of plant tolerance to heavy metals are heritable attributes of tolerant mutants and genotypes. It should be noted that the distinction between avoidance and tolerance is, to some extent, arbitrary. As indicated by Verkleij and Schat [2], tolerance mechanisms may also be considered as avoidance mechanisms operating at the subcellular level.

Plant populations growing on normal soils rarely contain tolerant individuals. Studies indicate that the advantage enjoyed by tolerant individuals growing under toxic soil conditions is generally associated with disadvantages when the same tolerant individuals are growing under normal soil conditions [4]. In other words, there is a cost involved in the evolutionary development of heavy metal tolerance in plants. For instance, Lolkema et al. [10] reported that a Cu-tolerant population of *Silene cucubalus* exhibited a much lower biomass production at low nontoxic levels of Cu exposure than did a Cu-sensitive population exposed to the same conditions. Therefore, this cost of tolerance suggests that in a normal population, growing on normal soils, any individual showing tolerance is likely to be at a selective disadvantage.

While population differences between tolerant and sensitive species suggest the evolutionary process of plant heavy metal tolerance, research indicates that a constitutional tolerance to heavy metal stress may also exist [1]. Evidence indicates that plant species differ substantially in their sensitivity to heavy metals. Some plant species exhibit a high threshold of resistance; others exhibit an extremely low threshold of resistance. Comparative studies by McNaughton et al. [11] and Taylor and Crowder [12] on a population of *Typha latifolia* from sites contaminated severely with Zn and Cu/Ni smelter, respectively, and control populations from uncontaminated soils provided no evidence for population differentiation. These reports along with others [1,13,14] suggest the occurrence of plant heavy metal tolerance without the evolution of tolerant plant races. This concept of constitutional metal tolerance indicates that metal tolerance in plants is not necessarily only "plastic," as suggested in some early reports [15].

Although in most cases plant metal tolerance appears to be determined genetically, recent laboratory studies employing suspension culture techniques have also shown inducible tolerance to heavy metals in plants, due primarily to the production of metal-complexing compounds such as organic acids [16] and phytochelatins [17,18] after plant exposure to heavy metals. For instance, Cd tolerance of suspension cultured cells was increased by prior exposure of cultured cells to Cd [19,20]. These studies at least suggest that heavy metal tolerance is controlled partially by metal-induced changes in plant physiological traits. However, one should keep in mind that heavy metal tolerance in intact plants acquired through evolution may not be comparable to metal tolerance in plant cell suspension cultures [15].

Heavy metal tolerance in plants is largely metal specific and involves primarily metals occurring at toxic conditions [15]. Multiple-metal resistance is often associated with occurrence of high levels of those metals in soils [21]. However, there are also cases of cross-resistance or co-tolerance, where resistance to one or more metals confers resistance to another metal present at a low and nontoxic level [22]. In addition, low-level tolerance in plants to metals other than the primary contaminants has also been observed [1,15,23]. Clearly, plant heavy metal tolerance is a complex phenomenon.

C. Compartmentation Theory of Heavy Metal Tolerance

Various mechanisms have been proposed to explain heavy metal tolerance in plants, based on evidence obtained from whole plant, tissue, cellular, and subcellular observations. These mechanisms include metal accumulations in trichomes, increased exudation of metal-chelating substances, cell wall metal binding, restriction of metal transport from the root to the shoot, alteration of membrane structure and permeability, alteration of cellular metabolism, production of intracellular metal-sequestering compounds, active pumping of metal ions into vacuoles, and so on [1,2,24–26]. Evidence also suggests that plant metal tolerance may be achieved by the simultaneous operation of more than one mechanism [1].

A broad theory that may best summarize most of the heavy metal tolerance mechanisms in plants is compartmentation theory. It is based on the ability of plants to store excess metals in organs or subcellular compartments where few or no sensitive metabolic activities take place. At the plant organ level, it has often been observed that some plants are able to translocate excess metals into old leaves just before shedding [2]. Excess metal accumulation (e.g., Mn) is also observed in metabolically inactive plant trichomes [25]. In addition, a number of plant species, particularly monocotyledonous species, are able to maintain relatively constant, low heavy metal concentration in their leaves over a broad range of external metal concentrations [3,27,28].

A significant fraction of the heavy metals stored in plant roots is found to be located in the apparent free space, probably associated with the pectin and protein fractions of the cell wall [2,29]. In fact, it has been suggested that the positive correlation observed between metal resistance and root/shoot metal concentration ratios may result from changes in cell wall metal-binding capacity in resistant plants [30–32]. However, restriction of root-to-shoot metal transport in some plants does not mean that metal-sensitive metabolic processes would take place primarily in the shoot. Tolerant plants also accumulate heavy metals in the shoot [33,34]. In many cases, tolerant species or genotypes accumulate more heavy metals in the shoot than in the root, depending on the types of heavy metals to which plants are exposed [35,36].

More recent evidence at the subcellular level obtained from advance techniques such as microprobe analysis, compartmental flux analysis, and vacuole/extravacuole distribution analysis reveals that in tolerant plant species or genotypes, excess heavy metals are stored in metabolically less active vacuoles [37–41]. Heavy metals (depending on their complexing potentials with organic/inorganic ligands and availability of the latter) may exist in different forms inside vacuoles [42,43].

While current literature points out the vacuole's role in heavy metal resistance, especially in terms of internal tolerance, recent evidence also indicates that certain properties of cell walls, particularly surface characteristics of root cell walls [44–47] may influence metal availability for uptake. They may do so by affecting the electrochemical environment of plasma membranes and the specific function of cell wall enzymes, which, in turn, may influence plant heavy metal tolerance [2,44,48]. In the following sections, characteristics of cell walls and vacuoles and their roles in heavy metal uptake, accumulation, and tolerance are discussed.

II. PLANT CELL WALL AND METAL TOLERANCE

A. Cell Wall Structure and Characteristics

A plant's cell wall is generally considered to provide rigidity and protection to the cell without preventing diffusion of water and ions from the external environment to the plasma membrane. The latter acts as a selective barrier to transfer metal ions or other chemical species into the cell. However, the cell wall is no longer viewed just as a rigid structure but as a dynamic, pliable portion of the living cell [36,44]. The cell wall is known to form a single continuous extracellular matrix throughout the body of plant; this continuity has implications for cell signaling and communication considering that there are many enzymes at the cell wall surface participating in various metabolic reactions [44,49].

According to Taiz and Zeiger [50], a typical primary cell wall of a dicotyledonous plant consists of 25 to 30% cellulose, 15 to 25% hemicellulose, 35% pectin, and 5 to 10% glycoprotein or extensin on a dry weight basis. The precise molecular composition and structure of the cell wall depend on type of cell, tissue, and plant species. The primary wall consists of two phases, a crystalline microfibrillar phase embedded in a noncrystalline amorphous matrix phase [50]. The microfibrillar phase is composed of cellulose looped with glycoprotein (extensin) in a "warp and weft" structure [51]. Glycoprotein molecules are mostly in a helix conformation (although they do contain some nonhelical sections, which could allow the glycoproteins to bend) and are cross-linked by intermolecular isodityrosine groups [52]. The high tensile strength of the microfibrils (which are cellulose chains held together by hydrogen bonds of their sugar OH groups) reinforces the wall by introducing shear interactions between the microfibrils and the matrix [50]. The matrix materials of primary walls are mainly pectins and hemicelluloses.

Overall, the cell wall structure is viewed as a meshwork with various sizes of tortuous pores provided by the interstices of bundles of cellulose microfibrils which are deposited at angles to one another [36,49,52]. Pores of plant cell walls are less than 10 nm in diameter, with a rare maximum of 20 nm [49,53]. These pores provide a pathway for water and solutes, leading to the plasma membrane. Apparently, cell wall pores are the major constituent of "free space" or apoplasm [36], a concept used to describe initial absorption of ions by plants [54].

Pectins and glycoproteins are considered to be the major cell wall components, generating surface charges through proton association/dissociation of their functional groups [36,55]. Recent findings also suggest that pectins as well as glycoproteins may have a role in regulating the porosity of the overall cell wall meshwork [49,56].

Like other wall polysaccharides, pectin is heterogeneous with respect to chemical structure and molecular weight. It is a polymer of D-galacturonic acid and composed primarily of D-galactopyranosyluronic acid units joined in α-D (1→4) glycosidic linkages [57]. Free (surface exposed) pectin carboxyls are highly hydrated and become negatively charged as pH is increased. This negative charge can be counterbalanced by cations such as metals. Pectins in plant cell walls are often esterified to various degrees. Since esterification decreases the number of free carboxyls, the process results in less metal ion retention within cell walls. Glycoprotein is another wall component that generates charges due to its residue functional groups of $-COOH$, $-SH$, $-(CH_2)_6 \cdot OH$, and $-NH_2$. Therefore, glycoprotein is also expected to be important in metal binding, although its presence in the cell wall generally is smaller in quantities than those of pectin [55].

According to Haynes [36], 70 to 90% of the cation-exchange capacity (CEC) of a typical plant root may result from carboxyls of root cell wall pectins. The remaining 10 to 30% of plant root CEC may be attributed to functional groups of cell wall glycoproteins. Anion-exchange capacity (AEC) is believed to result from free amino groups of glycoproteins [36,58].

B. Cell Wall Surface Chemistry

The overall surface of cell walls includes both internal and external surfaces. The external surface is farthest from the plasma membrane; the internal surface may be considered the innermost wall surface next to the plasma membrane plus the surface of wall pores. It is expected that the interaction of metal ions with both surfaces affects the plant extracellular environment, certain functions of the cell plasma membrane, and cellular metabolism.

As discussed in Section II.A, cell walls carry mainly negative charges generated from proton dissociation of free pectic carboxyls and other functional groups of glycoproteins. Charge density depends on the quantity of functional group per unit surface of cell well mass and is affected by carboxyl esterification and external medium pH. The negatively charged walls can attract an adjacent layer of mobile cations, and an electrical double layer is thus formed [36].

In general, affinity between cations and a given plant cell wall is expected to follow the strength order $M^{3+} > M^{2+} > M^+$ (e.g., $Al^{3+} > Mg^{2+} > Na^+$) [36]. In other words, cations with higher positive charges are more likely to be preferred by the negatively charged cell walls than cations of lower charge. However, an exception of this is H^+, which exhibits a similar affinity to M^{3+} [36]. For cations with similar valence, preference is given to those with least hydration (e.g., $K^+ > Na^+ > Li^+$). It should be noted that although polyvalent cations exhibit stronger affinity for the cell well surface, they are expected to be less well adsorbed in molar quantity to cell walls because their greater neutralization effectiveness of cell wall negative charges than monovalent cations [36].

The general expectations for affinities between cations and plant cell walls desribed above do not necessarily occur in the same order for all plant species. Depending on the specific plant species or genotype, interactions between cations and cell walls may vary appreciably. For instance, the cell wall surface of *Brassica oleracea* shows less affinity for Ca^{2+} than for K^+ [59], but the cell wall surface of *Regnellidium disphyllum* and *Nymphoides peltate* exhibitis much stronger affinity for C_a^{2+} than for K^+ [46]. These observations clearly demonstrate that cell walls of different plant species possess different metal-ion selectivities.

Differences in metal-ion selectivity of plant cell wall is determined by differences in the nature of the complexes formed at cell wall surface. It is conceivable that characteristics of both metal ions and plant cell walls affect metal–wall complex formation. Two broad categories of surface complexes can be distinguished on structural grounds: if no molecule of bathing solvent (e.g., H_2O) is interposed between the surface functional group and the metal it binds, the complex formed is called an inner-sphere complex; if at least one solvent molecule is interposed between the functional group and the bound metal, the complex formed is called an outer-sphere complex [60]. In general, outer-sphere surface complexes involve only electrostatic bonding mechanisms, whereas inner-sphere surface complexes involve either ionic or covalent bonding or some combination of the two; outer-sphere complexes are less stable than inner-sphere complexes. Van Cutsem and Gillet [61,62] reported that whereas almost all Zn^{2+} and Ca^{2+} retained by *Nitella flexilis* L. cell walls may simply form outer-sphere complexes with wall carboxyls, only 60% of all Cu^{2+} retained by the wall formed similar complexes and the remaining 40% formed inner-sphere complexes by chelation with oxygen and nitrogen of pectic carboxylate and glycoprotein functional groups of cell walls, respectively. Copper bound to cell walls of unripe apple fruit cortex was also observed to lose its hydration shell and to form inner-sphere complexes with two carboxyls on adjacent wall polymer chains [63]. In general, metal ions that are able to form wall surface inner-sphere complexes are more selective by the wall than those that are only able to form wall surface outer-sphere complexes. Therefore, Cu exhibits stronger affinity for the cell wall than Zn or Ca [61,62].

By employing Fourier transform-infrared spectroscopy and potentiometry, Wang et al. [47] investigated surface–metal ion interactions of purified cell walls of two tobacco (*Nicotiana tabacum* L.) genotypes that exhibit different sensitivity to Mn toxicity. They found that the strength of binding in terms of covalent character between various metals with the cell wall surface was in the order Cu > Ca > Mn > Mg > Na and that for a given metal ion the bonding strength of KY 14 (Mn-sensitive genotype) is greater than that of T.I. 1112 (Mn-tolerant genotype) (Figure 1). In addition, KY 14 exhibited a higher percentage ratio of $COO^-/COOH$ and thus a lower pK_a (negative logarithm of acidity constant) value of cell wall surface carboxyls than T.I. 1112 [47]. Since CEC is inversely correlated to pK_a [64], the lower pK_a Mn-sensitive KY 14 possesses as much as 18 $cmol_c/kg$ more capacity of metal binding than the higher pK_a Mn-tolerant T.I. 1112 (Figure 2). It should be noted, however, that no difference in pK_a between other metal-tolerant and metal-nontolerant clones have also been observed for other plant species [65].

External factors such as medium ionic strength and pH also affect cell wall metal-binding capacity. For example, when external solution ionic strength is at 0.01 M, maximum Cu binding capacity (MCBC) of isolated red clover root cell walls is 3.8 and 2.4 times higher than that of Italian ryegrass at pH 4.5 and 5.5, respectively [66]. However, as external ionic strength was increased to 0.1 M, the MCBC of red clover root cell walls became 4.4 times higher at pH 4.5 and 1.9 times higher at pH 5.5 than that of Italian ryegrass [66]. These observations clearly indicate that the chemical interaction between the cation and the plant cell wall surface is determined by both a given plant's specific endogenous cell wall characteristics and external environmental conditions.

Two antagonistic effects may result from the physicochemical interaction between metals and cell walls with respect to plant metal-ion uptake and accumulation. A stronger cell wall binding with a metal increases the overall fraction of the metal in the pool of cations present in the pores of wall, or apoplast. This is expected to increase the potential for the metal ion to be taken up by the plant. For example, stronger cell wall Cu binding has been found to increase

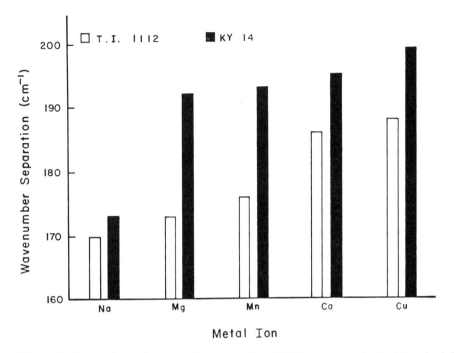

Figure 1 Comparison of wavenumber separations (FT-IR spectroscopically determined difference between antisymmetric and symmetric COO^- stretching) due to different metal-ion saturations for T. I. 1112 and KY 14 root cell walls. The greater the wavenumber separation, the stronger the bonding between metal ion and cell wall in terms of covalent character. (From Ref. 47.)

Cu uptake in *Silene cucubalus* and other plant species [2]. On the other hand, stronger metal–cell wall binding is also expected to decrease metal ion activity within the wall or apoplast, which would reduce the metal's contribution to the ionic fluxes [67]. For instance, Fe accumulated in the cell wall free space of barley roots has been found to contribute little to the Fe in the xylem flow [68]. Apparently, the outcome of metal-wall interaction with respect to metal availability, uptake rate, and accumulation is, at least to some extent, affected by the balance between these two apparent antagonist effects. The magnitude of either effect may depend on cell wall charge density, metal selectivity, pore size, enzyme activity, and other factors.

C. Cell Wall Influence on Metal Uptake and Tolerance

For water and solutes (ions and neutral molecules) to move from the soil medium to the top part of the plant, they must first be transported radially through various root cell types, then enter the xylem and move up to the leaves via the transpiration stream. The radial transport path in the plant root is shown schematically in Figure 3. According to Haynes [36], there are two pathways for solute movement in a plant; one is through the symplast and the other is through the apoplast. After solutes diffuse into the cell walls of the epidermal cells, active uptake may occur at the plasmalemma of these cells. Solutes may then be transported across the cortex, endodermis, and pericycle in the symplasm (through plasmadesmatal connections between cells).

Solutes may also move passively into the free space of the cortical cell walls (apoplast) and then be transported across the cortical and endodermal cell plasmalemma and enter the

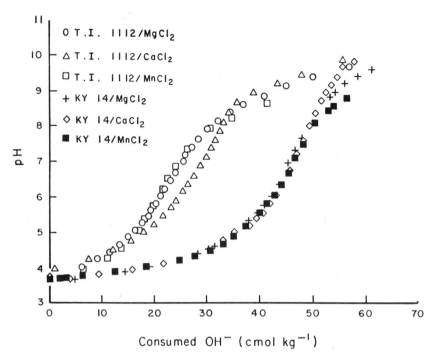

Figure 2 Comparisons of titration curves obtained in 20 mN of MgCl$_2$, MnCl$_2$, and CaCl$_2$ for T.I. 1112 and KY 14 root cell walls. The quantity difference in consumed OH^{-1} between the two cell walls at equivalent points at pH 6.5 to 7.0 represents capacity difference in the most easily accessible cell wall sites and is about 18 cmol$_c$/kg. (From Ref. 47.)

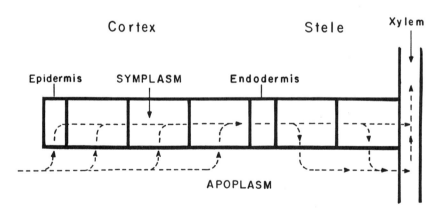

Figure 3 Schematical illustration of radial transport of solutes in the plant root. (From Ref. 36.)

cortical symplasm. Since hydrophobic bands of suberin deposited in the radial walls of the endodermal cells (Casparian bands) essentially restrict apoplastic movement from the free space of the cortex to the free space of the stele, movement of solutes and water across the endodermis into the stele must proceed in the symplasm [36]. Irrespective of where solutes enter the symplasm, they must first pass through the free space of cell walls. It appears logical that any chemical species that exhibits limited interaction with the symplasm and cell walls of apoplast should easily be taken up by plant roots and transported to shoots.

As discussed in Section II.B, polyvalent cations are more likely to interact with cell walls than monovalent cations, due to their stronger electrostatic attraction to cell wall negative charge sites. Most heavy metals are divalent or trivalent cations and therefore are expected to undergo adsorption/exchange reactions with wall surfaces before they move into the symplast. In fact, accumulation of heavy metals in plant roots seems to be a widespread phenomenon. In general, Mn, Cu, and Zn appear to accumulate slightly (one to three times) more in the root than in the shoot; heavy metal ions such as Pb, Ag, Hg, Cd, Al, and Fe accumulate much more in the root than in the shoot [32]. Such a widespread metal-ion accumulation phenomenon apparently suggests some form of localization or compartmentation of metal ions in plant roots. Cell wall localization is found to be responsible for root extracellular accumulation of many heavy metals, including Hg, Pb, Cd, Cu, Zn, and Al in various plants [32,36].

In addition, various lines of evidence indicate that heavy metals enter plant roots passively. A number of studies reported that Pb uptake by some plants is not influenced by metabolic inhibitors and low temperature, suggesting that the process does not require energy expenditure [69,70]. Jarvis et al. [71] indicated that uptake of Cd, at least in the short term, is entirely a physicochemical process. Apparently, the cell wall's charge and pore characteristics and the heavy metal's polyvalence are expected to influence greatly passive physiochemical absorption of these heavy metal ions by plant roots.

Because root accumulation of heavy metals is a widespread phenomenon, restriction of heavy metal transport from root to shoot has been thought of as a mechanism of plant tolerance to heavy metal toxicity. Accumulation of heavy metals in roots due to extracellular cell wall compartmentation along with intracellular vacuolar compartmentation (see Section III) of plant root cells may be a strategy employed by plants to, at least in part, accomplish this process. Cell walls involved in localization of heavy metal ions are expected to include both cell walls in radial movement and vertical movement. In the latter case it has been reported that xylem cell walls of tomato can retain twice as much free Cd as that present in the xylem [72].

The direct involvement of the plant cell wall in heavy metal resistance was proposed by Turner and Marshall [30], who reported a linear correlation between the Zn-binding capacity of isolated root cell walls and Zn tolerance index of various populations of *Agrosis tenuis* Sibth (Figure 4). Based on these results, they suggested that cell wall binding of heavy metals could be a major heavy metal tolerance mechanism in higher plants. Using ^{65}Zn, Peterson [29] also found that a pectate extract of cell walls of a Zn-tolerant clone of *A. stolonifera* contained five to six times more Zn than a similar extract of cell walls from a sensitive clone of the same species. Metal binding of cell wall as a possible heavy metal tolerance mechanism has been proposed for various plants in resisting Zn, Al, Mn, Pb, and possibly other metal toxicities [2,24,35].

The results of Turner and Marshall's work, as described in Figure 4, imply that heavy metal tolerance is mainly determined by the cell wall metal-binding capacity of metal-tolerant species. It should be noted, however, that although metal binding to cell walls may be important in resisting heavy metal toxicity in certain plants, it is unlikely that such a mechanism is the only major tolerance mechanism in higher plants, as suggested by Turner and Marshall [30]. As discussed previously, major cation-binding functional groups of cell walls are carboxyls of

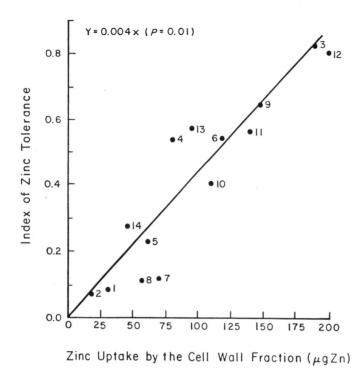

Figure 4   Relationship between zinc accumulation by the cell wall and the index of zinc tolerance of 16 clones of *Agrostis tenuis*. The tolerance index defined as the mean length of the longest root in distilled water. The numbers against points on the graph are reference numbers to the clones. (From Ref. 30.)

pectins and glycoproteins. If the carboxyls of cell walls are the only major functional group responsible for plant Zn tolerance of Zn-tolerant clones, Zn-tolerant clones should also exhibit Cu tolerance, since Cu is known to form stronger complexes with cell wall carboxyls than Zn [62]. However, it is well known that Zn-tolerant species are not necessarily also Cu-tolerant [65].

Wainwright and Woolhouse [73,74] later reported that Zn resistance differences between Zn-tolerant and Zn-sensitive plants persist even after Zn^{2+} saturation of the root cell wall exchange sites. This result clearly implies that heavy metal tolerance could not be solely dependent on the cell wall metal-binding capacity of the tolerant plant as suggested by Turner and Marshall [30]. Additional mechanisms must exist to resist further metal toxicity. There are, however, two possibilities in Wainwright and Woolhouse's studies, which may still support the dominant role of cell wall metal binding in plant tolerance. One possibility is that under Wainwright and Woolhouse's experimental conditions, co-ion exclusion may take place and still keep more Zn from being taken up by Zn-tolerant plants [75], if the Zn-tolerant clone's cell wall charge density was high and the size of cell wall pores was small. The other possibility is that under Wainwright and Woolhouse's experimental conditions, the tolerant clones may continue to grow and provide new sites for binding additional Zn [32]. Whether or not these two processes actually took place is not currently known. Nonetheless, other studies showed that for some plants, Cu and Al tolerance may or may not correlate with root cation-exchange capacity (CEC) [2,76]. Recently, Wang et al. [47] reported that a Mn-sensitive tobacco genotype exhibited a higher Mn-binding strength in its root cell walls than that of a Mn-tolerant genotype (Figure 1).

As discussed in Section II.B, metal binding by cell walls may result in two opposite effects with respect to metal uptake: (a) reduction of free metal ion activity [67] near and around the plasmalemma environment, which might be expected to limit uptake; and (b) increase of total metal ion concentration present near and around the plasmalemma environment, which under certain conditions might increase uptake. As to which effect would predominate depends on cell wall charge density, metal-ion selectivity, pore size, and enzyme activity. Since plant metal resistance is mainly dependent on the ability to keep free metal ions from interfering with metabolic reactions, the potential effect of the cell wall in reducing metal ion activity through binding can thus contribute to the plant's metal resistance. This effect may be particularly dominant in plants that could retain large quantities of heavy metals in their cell walls. For instance, *Anemone nemorosa* was found to retain 63 times more Pb in its roots than in its shoots, and metal binding to its cell wall was considered a decisive mechanism for such plants to resist Pb toxicity [2,32]. In this case the effect of reducing Pb^{2+} activity through cell wall binding is apparently the action the plant takes for reducing the toxicity of this biologically nonessential element. On the other hand, Mn-tolerant tobacco genotype T.I. 1112 with relatively low-cell-wall Mn-binding strength and capacity may employ other Mn-toxicity resistance mechanisms [47].

The plant cell wall contains many dynamic enzyme activities. Wainwright and Woolhouse [73] reported that wall-located acid phosphatases of Cu-tolerant *Agrostis tenuis* exhibited a greater tolerance to Cu than those of nontolerant plants of same species. Cox and Thurman [77] also found that wall-located acid phosphatases of Zn-tolerant *Anthoxanthum odoratum* clones exhibited higher Zn tolerance than those of nontolerant clones of the same species. It appears that such a phenomenon is specific to cell wall enzymes, since most cytoplasmic enzymes of tolerant plants are as metal sensitive as those of nontolerant plants [78]. Verkleij and Schat [2] indicated that it is conceivable that heavy metal resistance may be affected by enzyme adaptation in the extracellular compartment (cell wall) if other protective mechanisms may not be available or sufficient. Nonetheless, no apparent difference was observed for Zn tolerance of cell wall–located acid phosphatases between Zn-tolerant and Zn-nontolerant clones of *Agrostis tenuis* [74]. This suggests that enzyme difference in metal tolerance may not be a widespread phenomenon and it is unique only for certain metal-tolerant clones of certain plant species.

It has frequently been suggested that Ca plays an important role in plant heavy metal tolerance. Baker [79] indicated that due to Ca interaction with wall pectate, Ca may play a role in Zn tolerance. It was found that in long-term culture Ca stimulates more Zn accumulation in tolerant clones of *Silene maritima* than in nontolerant clones. It was also reported that higher Ca uptake by Al-tolerant plants took place in cell walls and other parts of plants and that Al-tolerant cultivars were more likely to produce high Ca-containing exudates after Al exposure than were Al-nontolerant cultivars [35]. Wang et al. [47] reported that a Mn-tolerant tobacco genotype did show a greater Ca preference over Mn by its root cell walls than that of a Mn-sensitive genotype. It is well known that Ca plays an important role in switching on and off cellular normal processes through a messenger function [80]. Therefore, Ca preference by cell wall binding may be one of the traits that tolerant plants possess to resist heavy metal toxicity.

In Table 1, some aspects of the cell wall contributing to plant metal tolerance, as discussed in this section, are summarized. It should be kept in mind that although these aspects of cell walls in plant heavy metal resistance are important to various tolerant plant species resisting various metal toxicities, they cannot explain every case of metal resistances. Other mechanisms, particularly those emphasizing internal tolerance, may also coexist with mechanisms of cell wall metal resistance. Under some conditions, where plants accumulate an appreciable amount of heavy metals inside cells, internal mechanisms may even dominate.

Table 1 Metal Tolerance Aspects of Plant Cell Walls

Possible resisting function	Toxicity metals	Plant species	Ref.
Binding to wall	Zn	Agrosis tenuis	30
	Al	Pea	35
	Pb	Anemone nemorosa	32
Accommodating tolerant enzymnes	Cu	Agrosis tenuis	73
Selecting Ca preferably	Zn	Silene maritima	79
	Al	Dade snapbean	35
	Mn	Tobacco T.I. 1112	47

III. PLANT CELL VACUOLE AND METAL TOLERANCE

A. Vacuole Characteristics and Functions

Number and size of vacuoles per cell vary, depending on type of plant tissues. Whereas young meristematic cells have a number of small vacuoles, mature mesophyll cells have one large central vacuole, which can encompass more than 80 to 90% of the cell volume. In a herbaceous vascular plant, vacuoles occupy at least half of the plant's volume and wet weight [81].

While the primary function of vacuole appears to be associated with statics of plant form and structure, plant vacuoles also have many dynamic aspects [81]. Vacuoles change form and structure during growth and differentiation. Proteins and other macromolecules can accumulate in vacuoles and may later be degraded. Ions and metabolites are not simply deposited in vacuoles, but rather, remain in continual interplay and exchange with cytoplasmic constituents [82]. Therefore, vacuoles form an internal environment for the cytoplasm, essential to its homeostasis [81].

The vacuole is a multifunctional organelle that is unique to higher plants. Major functions of vacuoles include lytic, turgor regulation and space filling, and storage; and they are considered important for normal cell expansion and growth [82]. Among the vacuole's major functions, the storage function is probably the most important for substance balance and health maintenance in the plant. Isolated vacuoles contain inorganic ions such as Na^+, K^+, Ca^{2+}, Cl^-, SO_4^{2-}, NO_3^-, PO_4^{3-} and so on, in concentrations of up to 300 mM or even 800 mM in the case of halophytes [83–85]. It is assumed that a steady exchange of the stored ions takes place between the cytosol and the vacuole, since many of these ions play important roles in cellular processes such as signal transduction (Ca^{2+}), building of proton gradients, or maintenance of electron potentials at the plasmalemma [82]. Vacuoles are also rich in sugars, organic acids, and amino acids, with concentrations sometimes reaching as high as 200 mM (extremes > 1000 mM) [82,83].

One of the important aspects of the vacuole's storage function is its role in signal, defense, and detoxification. Many secondary metabolites stored in vacuoles are found to play an important role in signaling and defending microorganisms, viruses, and/or herbivores [82]. In addition, heavy metals have also been found to be accumulated and sequestered in plant vacuoles [41]. Increasing evidence suggests that heavy metals in vacuoles may exist in different forms, depending on type of ligands present in the vacuole [42,43].

B. Vacuole Metal Detoxification Mechanisms

Physical evidence that heavy metals are recovered from vacuoles of plants after exposure to toxic levels has brought attention to the possible role of the vacuole in plant heavy metal tolerance. A decade ago, Brookes et al. [37], using compartmental flux analysis, found that Zn-resistant clones of *Deschampisa* were able to pump Zn actively into the vacuole of root cells, whereas Zn-sensitive clones seemed to have a much lower capacity to do so. Later, Memon et al. [86] and Memon and Yatazawa [87], using x-ray microprobe analysis, demonstrated that Mn was accumulated primarily in leaf cell vacuoles of *Acanthopanax sciadophylloides* and a tea plant. In vivo ^{31}p NMR studies by Pfeffer et al. [88] also showed that excess Mn^{2+} was entrapped in corn root tissues after exposure to toxic levels of Mn. Furthermore, electron microscopy studies have revealed the existence of Cu in vacuoles of both tolerant and nontolerant ecotypes after exposure to Cu [89], and scanning transmission electron microscopy (STEM) have shown that Cd is localized in a granular form in root cell vacuoles of *Agrostis gigantea* after Cd exposure [39]. Most recently, Vögeli-Lange and Wagner [41] have localized all of the leaf protoplast Cd in the vacuole of tobacco plants. Salt and Wagner [90] also observed transfer of free Cd^{2+} into the vacuole via a tonoplast Cd/H antiporter in oat roots.

The evidence described above clearly indicates that vacuolar compartmentation may play an important role in plant internal metal tolerance. Since the vacuole exhibits essentially little or no metabolic activity, it seems logical for plants to adapt vacuolar metal compartmentation as a metal tolerance mechanism. A number of heavy metal detoxification mechanisms proposed, based on metal-complexing ligands, are associated with vacuolar compartmentation. Among those mechanisms, production of phytochelatins (or metal-binding peptides) and of organic acids has received extensive attention in the literature.

Specific metal-binding peptides were discovered in plants following the discovery of metal-binding proteins, so-called metallothioneins (MTs), in animals and fungi. The MTs can selectively bind Hg, Cu, Zn, Co, Ni, and so on, and play an important role in intracellular regulation of heavy metals [18]. In the early 1980s, proteins similar to the animal MTs were isolated from various plants and suspension culture cells of plants. Later, however, it was found that mammalian-MT-like plant proteins are actually polypeptides, now referred to commonly as phytochelatins (PCs) [17,91]. Various evidence suggests that the general structural formula for phytochelatins can be characterized as $(\gamma\text{-Glu-Cys})_n\text{-Xa}$, where $n = 2$ to 11, depending on the source, and the carboxy-terminal Xa is either Gly or β-Ala [18]. Although phytochelatins contain the Cys-X-Cys sequence, which is like that of mammalian MTs, there are major differences between the two. Phytochelatins are not translational products and are a mixture of small peptides that form metal-binding complexes, whereas mammalian MTs are translational products and are composed of a single molecule that binds metals in two clusters [18].

Phytochelatins are induced in plants after exposure to a number of heavy metals, including Cd, Cu, and Zn [91]. The strongest metal-binding phytochelatin reported so far is that of Cd-binding peptides (CdBPs) produced in intact plants or suspension cultured cells after exposure to toxic levels of Cd [18,26]. Plant cell suspension culture studies indicate that Cd-tolerant cells induce production of more CdBPs than Cd-sensitive cells [92], and Cd-tolerant cells bind more than 80% of the cellular Cd as CdBP-Cd [18]. Early and rapid formation of CdBP-Cd has generally been found to correlate highly with metal tolerance of cultured cells [18]. Comparison of clones of Cd-sensitive and Cd-tolerant populations of *Silene vulgaris* plants has also showed that more phytochelatins are produced in tolerant clones than in sensitive ones, and the PCs in the tolerant clones bind twice as much Cd as those in the sensitive clones [93]. Clearly, the evidence above suggests that Cd detoxification or tolerance in plants may be, at least partially, achieved by CdBP induction and CdBP-Cd complex formation after metal exposure. Further evidence also suggests that CdS solids may be formed and be coated by CdBP-Cd complexes after exposure of cultured suspension cells to high levels of Cd [18]. The co-occurrence of CdS and CdBP-Cd has been thought to play a role in stabilizing CdBP-Cd complexes [18,94]. However, the mechanism of phytochelatins described above seems to have less importance in plant Cu or Zn tolerance. Few occurrences of Cu-PCs complexes have been reported, and the Zn-PCs complexes observed so far in both suspension cultured cells and intact plants exhibit low stability [15,18,26].

Recent compartmental analysis of tobacco cells has demonstrated that all of the Cd and CdBPs could be recovered from vacuoles of tobacco leaves [41]. The occurrence of virtually all the CdBP and Cd in vacuoles of tobacco leaf protoplast after Cd exposure led Vögeli-Lange and Wagner [41] to propose that phytochelatins such as CdBPs may be involved in heavy metal transport across the tonoplast for vacuolar compartmentation, although Cd and CdBPs might be transported separately and then form CdBP-Cd in the vacuole sap. Since it is conceivable that sequestration of heavy metals occurs in metabolically less active compartments, such as vacuoles, for effective detoxification the observation by Vögeli-Lange and Wagner [41] above suggests that simple cytoplasmic metal chelation by phytochelatins is unlikely even though phytochelatin synthesis takes place extravacuolarly. It is not known, however, if phytochelatin is also accumulated in vacuoles in other plant species. Further studies on determining subcellular localization of phytochelatins are needed.

While production of phytochelatins may be a mechanism that plants employ in internal detoxification of heavy metals, it is unlikely that such a mechanism is the only mechanism in heavy metal tolerance, even in the case of Cd tolerance by CdBPs, the best evidence available. This is because induction of PCs is associated with exposure to high external metal concentration [26]. For example, CdBPs are only found to bind Cd in intact tobacco plants or cultured suspension cells containing Cd levels that vary from a couple of hundreds to as much as 1400 times that of the Cd found in tobacco plants growing in soils [18]. Apparently, at relatively low Cd exposure, plants employ mechanisms other than phytochelatin production to tolerate the metal internally.

Complexing of heavy metals with organic acids is another internal metal tolerance mechanism associated with vacuolar compartmentation. As discussed in Section III.A, vacuoles are often rich in low-molecular-weight organic acids. Certain organic acids are also generally known to form stable complexes with various heavy metal ions [95,96]. Therefore, it seems logical that organic acids accumulated in vacuoles may be involved in metal detoxification.

In the late 1970s, even before much of the evidence for vacuolar compartmentation of heavy metals became available, Mathys [97] proposed a model to explain tolerance of certain plant clones to potentially phytotoxic concentrations of Zn. In this model, Mathys [97] proposed that Zn (after entering the cytoplasm) forms complexes with malate, which are then transferred to the vacuole, where the Zn is exchanged with oxalate and sequestered as stable Zn-oxalate. After metal exchange between malate and oxalate, the malate would be recycled in the cytosol to complex more Zn (Figure 5). Mathys [97] indicated that malate would serve as a shuttle to compartmentalize Zn into plant vacuoles. Ernst [98] also suggested that the mechanism of Zn tolerance was based on Zn chelation by malate, since he found that Zn tolerance was correlated with high levels of malate in some Zn-tolerant plants.

There has been little direct evidence to support Ernst and Mathys' model for the role of malate in heavy metal tolerance. Studies on the comparison of clones of Zn-tolerant and Zn-nontolerant *Deschampsia caespitosa* show that both citrate and malate are increased more in tolerant clones than in nontolerant ones after Zn exposure, but the increase of citrate is much greater than that of malate. This observation [16] was considered to be not entirely consistent with the hypothesis of Ernst and Mathys, even though it was later suggested that the citrate may be compartmentalized in vacuoles, whereas malate was not [99]. *Silence vulgaris* was also found to induce more citrate

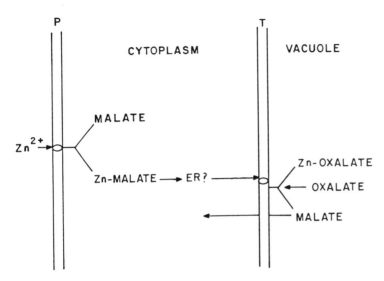

Figure 5 Model of a mechanism of zinc tolerance in herbaceous plants. The model is based on measurements made on *Silene cucubalus* and incorporates contemporary ideas on ion uptake. *P*, plasmalemma; *T*, tonoplast; *ER*, endoplasmatic reticulum. (From Ref. 97.)

and malate synthesis in its tolerant clones than in its nontolerant ones after Zn exposure [100]. In the case of Ni tolerance, a positively charged Ni-citrate complex was isolated and identified as the major form of Ni complexes from Ni-accumulating plants [101,102], whereas in the case of Mn tolerance, Mn-oxalate complexes were the only low-molecular-weight organic acid complexes detected in Mn-accumulating *Acanthopanax sciadophylloides* [87].

Studies with suspension-cultured tobacco cells showed that both Zn- and Cu-tolerant cells induce about 14 and 7 times more citrate and malate, respectively, than Zn- and Cu-nontolerant cells; 90% of cellular Zn and 84% of cellular Cu were complexed with low-molecular-weight organic acids (primarily citrate and malate) [103]. Furthermore, Al tolerance has also been observed to correlate with increased levels of citrate in selected Al-resistant carrot cells [104]. Although there has been little direct experimental evidence in support of the hypothesis of Ernst and Mathys, organic acid levels often seem to correlate with heavy metal tolerance, and therefore they may be involved in plant internal metal detoxification.

Another mechanism of heavy metal tolerance based on metal-sequestering ligands (probably also associated with vacuolar metal compartmentation) is that heavy metals may form metal–phytate complexes and deposit in globular bodies of plant subcellular compartments, particularly vacuoles. This mechanism has been proposed only recently, and the evidence is limited to Zn-tolerant *Deschampisa casepitosa* and aquatic *Lemna minor* L. [40,94]. Van Steveninck et al. [94] observed an alternative switch of CdS/Cd-PCs formation and Zn-phytate formation in *Lemna minor* L. after the plant was exposed to toxic levels of Cd and Zn, respectively. This observation suggested that the metal-phytate mechanism may be Zn-tolerant specific and the CdS/Cd-PCs mechanism may be Cd-tolerant specific. However, little is known about the physiology and biochemistry of the Zn-phytate formation.

C. Computer Simulation of Vacuole Chemistry

The potential involvement of various ligands in complexing/chelating heavy metals in vacuoles as discussed in Section III.B is well accepted by many researchers. However, the conditions under which a particular ligand would be significant in sequestering heavy metals in a plant vacuole is not clearly understood. For example, Cd-binding peptides (CdBPs) have been determined to play a role in plant Cd detoxification; however, CdBP production can be induced only after a moderate-to-high Cd exposure. Therefore, the role of CdBPs in Cd tolerance at low Cd exposure (encountered in most agricultural conditions) may be questioned [26,42,105]. It is suggested that although various internal heavy metal–tolerant mechanisms based on metal-sequestering ligands in plant vacuoles have been proposed, the relative importance of each of the ligands at a specific in vivo condition is unknown.

Wang et al. [42] pointed out that currently used experimental methods of determining compartmentation of ions, including heavy metals, estimate only the vacuole contents and do not accurately identify the chemical species of ion complexes present in vivo in vacuoles. Wang et al. [42], however, evaluated the significance of vacuolar metal-sequestering ligands through computer-simulated metal speciation by using published vacuolar contents and thermodynamic stability constants of complexes as inputs. In their work [42,43], metal-ion species distribution in the plant vacuole under various pH conditions was calculated using the GEOCHEM-PC [106,107]. The underlying assumption in these simulations was that the plant vacuole obeys thermodynamic equilibrium. Such an assumption was considered reasonable by the authors since chemical reactions were expected to be rather fast relative to the movement of ligands in and out of the vacuole.

Figure 6 shows computer-predicted Cd species distribution in vacuoles of suspension-cultured tobacco cells (exposed to 45 μM Cd after 4 h) at pH 5 (a commonly found vacuolar sap pH) as a function of CdBPs varying in concentration from 10^{-7} to $10^{-3.5}$ M. The induction of CdBPs has been known to be dependent on duration and concentration of Cd exposure [18,26]. Clearly, if concentrations of all other ligands in the vacuole system remain constant, the amount of Cd complexed by peptides depends on the amount of peptide present (Figure 6). Vacuolar

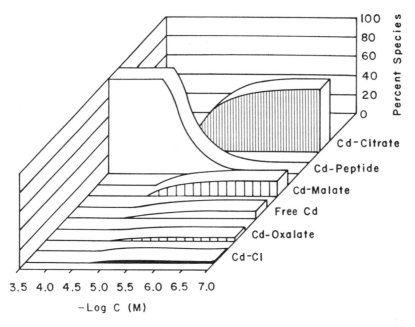

Figure 6   Simulated vacuolar distribution of Cd species with respect to concentration of Cd-binding peptide in the vacuole at pH 5 for cultured tobacco cells exposed to 45 μM Cd for 4 h. C (M) is the molar concentration of Cd occurring as peptide-bound Cd. (From Ref. 42.)

concentration of CdBP ligands need to be only $10^{-5.1}$ M (in thiol content) in order to complex 30% of the Cd recovered in vacuoles, based on the Cd-peptide formation constant of 10^{19} [108]. To complex all of the Cd present in the vacuole would require concentrations of about $10^{-4.65}$ M of Cd-binding peptide. At low Cd-binding peptide concentrations, organic acids, mainly citrate, become dominant in complexing Cd. Simulation of vacuolar speciation of Zn species as a function of pH for cultured tobacco cells (exposed to 300 μM of zinc) indicated that citrate sequestered most Zn (71–99%) over pH 4 to 7 (Figure 7), although only 7% of the citrate present in the vacuole was predicted to participate in the complexation process [43]. Computer simulations by Wang et al. [42,43] indicated that citrate had the highest potential for complexing Cd and Zn relative to malate and oxalate between vacuolar pH 4.0 and 7.0.

Malate is the most abundant vacuolar organic acid present in the tobacco cultured cells (75% of total organic acids) [109], but speciation simulations [42,43] have shown that malate becomes important in sequestering Cd only at vacuolar pH < 4.5. Furthermore, malate appears to have little potential for sequestrating Zn over the entire expected vacuolar pH range (Figure 7). The latter suggests that it is unlikely that malate plays any significant role in complexing Zn even in the cytoplasm, as proposed by Mathys [97], unless other stabilizing factors are involved. In addition, it appears that citrate rather than malate would more likely serve as a shuttle for transporting heavy metals into vacuole in Mathys's model, as well as in sequestering these metals in the vacuole. The central role of malate in plant vacuoles may be for charge maintenance.

Inorganic sulfide, phosphate, and chloride may also form either soluble complexes or insoluble solids with metal ions under certain vacuolar conditions [42,43]. While formation of soluble Cd-Cl complexes is expected due to the relatively high concentration of Cl⁻ ($\approx$0.01 M) encountered in tobacco vacuoles, precipitation of CdS and Cd-phosphate is due to the high formation constants of the mineral precipitates. The predicted CdS deposition [42] seems to be consistent with the co-occurrence of CdS and CdBPs reported for *Lemna minor* L. [18,94].

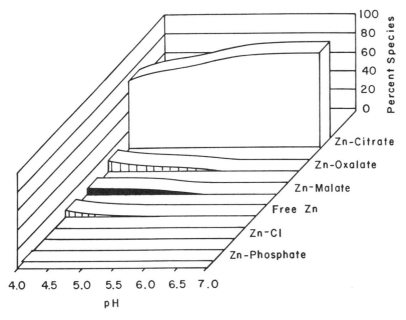

Figure 7 Simulated vacuolar distribution of Zn species as a function of pH for cultured tobacco cells exposed at 300 μM. (From Ref. 43.)

Wagner and Krotz [105] have argued that endogenous organic acids may be sufficient to complex Cd and Zn under conditions of low-metal-level exposure and absence of phytochelatins. This argument seems to be supported by the computer simulations of vacuolar chemistry. In general, the overall evaluation of vacuolar metal-sequestering ligands through computer simulations by Wang et al. [42,43] clearly suggests that under low levels of heavy metal exposure (as generally occurring in agricultural soils), production of metal-complexing organic acids, mainly citrate, may be a major plant mechanism to internally tolerate heavy metal toxicity. Under high levels of heavy metal exposure (as occur in a high-metal-polluted or high-metal experimental environment), production of phytochelatins may function as a predominant mechanism to internally detoxify heavy metals. At a very high level of metal exposure (probably growth inhibitory), precipitation of inorganic metal sulfides in vacuoles and perhaps in other compartments may be an additional mechanism to internal resistance of heavy metal toxicity.

Additional evaluation of vacuolar ligands are needed for various plant species growing under various environmental conditions. Although malic acid is often the most abundant organic acid in plants and citric acid is usually also present in substantial amounts, additional organic acids, such as isocitric, aconitic, succinic, oxalic, and malonic, may be abundant in some plants [110]. Whether or not these organic acids profoundly affect vacuolar metal complexation patterns is not known. In addition, the conditions for significant involvement of other metal-complexing ligands, such as Zn-phytate in plant metal tolerance, should also be evaluated in an integrated system.

The overall potential major detoxification mechanisms based on vacuolar compartmentation and ligand complexation as discussed above and in Section III.B are presented in Figure 8. In general, while vacuolar compartmentation keeps heavy metals away from metal-sensitive metabolic centers in the cytoplasm, sequestering ligands seem to safeguard them from readily moving by reducing their chemical activity. For an effective and efficient internal metal tolerance, both roles are important, although in reality plants may not have evolved to do so. While the mechanism of low-molecular-weight endogenous or induced organic acids, particularly citrate, may be employed by plants as a strategy to detoxify and low-level exposure of heavy metals, additional mechanisms of producing large-molecule and more metal-specific

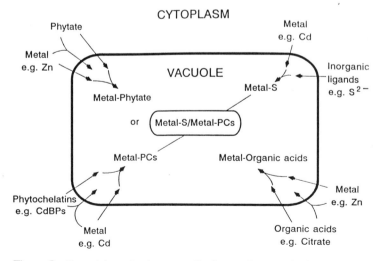

Figure 8 Potential mechanisms contributing to plant metal tolerance role of vacuoles; "or" indicates the possibility of Metal-PCs co-occurrence with Metal-S.

compounds such as phytochelatins and phytate may be employed by plants to fight high-level exposure to heavy metals. Inorganic ligands such as S^{2-} produced from metabolic process may also contribute to metal detoxification by forming insoluble metal sulfides. The latter may also be involved in stabilizing metal-phytochelatins.

IV. SUMMARY AND CONCLUSIONS

Heavy metal toxicity and tolerance are widespread phenomena in the planta kingdom. While adaptation of plants to naturally occurring environmental stress (e.g., low/high temperatures, drought) may take place over a long period of time, evolution of plants in terms of resisting heavy metal toxicity may occur in a very short time. Plant heavy metal tolerance appears to be determined genetically, although inducible heavy metal tolerance in laboratory suspension cell cultures has also been observed. One of the major important phenomena in heavy metal tolerance is the compartmentation of heavy metals in metabolically less active compartments of plants. At the subcellular level, these compartments are mainly extracellular cell walls and intracellular vacuoles.

The role of the cell wall in plant metal tolerance may be primarily through metal binding to reduce metal-ion activity for uptake. Other aspects of the cell wall's metal tolerance role may include accommodation of metal-tolerant enzymes and preferential selectivity. The plant cell wall's metal-binding capacity is determined by its porous meshwork structure and its surface charges, owing to functional groups carried by the cell wall pectins and glycoproteins (extensin). Cell wall surface chemical property differences with respect to metal-ion selectivity between various tolerant and nontolerant species/genotypes may be attributed to endogenous plant differences in the quantity, composition, and spatial structural arrangements of cell wall components, particularly pectins and glycoproteins.

The vacuole's role in heavy metal detoxification is heavy metal storage in the form of soluble complexes and/or insoluble precipitates. Ligands participating in complexing heavy metals in vacuoles may be organic acids (mainly citrate), phytochelatins, and phytates; all are produced extravacuolarly and may be transported with or without metals into the vacuole. Ligands participating in forming insoluble metal precipitates are mainly sulfides.

Vacuolar compartmentation appears to play an important role in keeping heavy metals away from the cytoplasm, where important metabolic functions are performed. Once metal ions are inside the vacuole, formation of organometallic complexes and inorganic metal solids may play the role of safeguarding the metal ions by lowering their activities. Both formation of soluble complexes and solids are important for internal metal tolerance. While induced and/or endogenous organic acids (particularly citrate) appear to play an important role in internally tolerating heavy metals at low to moderate levels of metal exposure, induced phytochelatins, phytates, and inorganic sulfide may play an important role in internal heavy metal tolerance at moderate to very high metal exposure.

In the current literature it seems that the emphasis is given to internal tolerance mechanisms [111], particularly vacuolar metal compartmentation and sequestration by phytochelatins, even though Cd-binding peptides appear to be the only good example available at the present time. It is evolutionarily conceivable that plants possess both external and internal mechanisms (e.g., extracellular cell wall compartmentation for external metal avoidance and intracellular vacuole compartmentation for internal metal tolerance). It would seem to be an unwise evolutionary selection for higher plants to possess only internal tolerance mechanisms. Execution of the internal tolerance mechanisms is energy expensive because the process allows heavy metals (particularly those nonessential nutrient elements) to cross the plasma membrane, then produces and transports necessary complexing (big) compounds, and finally, complexes the heavy metals in the vacuole.

It may be reasonable to assume that plants resist heavy metals first with a non-energy-consuming mechanism of extracellular compartmentation (cell wall binding) to a level dictated by the magnitude of the cell wall's negative charges, and second, inactivate those heavy metals passing the plasmalemma by internal mechanisms such as production of sequestering compounds and vacuolar metal compartmentation. This assumption does not in any way imply that the cell wall is synthesized solely to resist heavy metals. Rather, inherent characteristic regarding metal binding in cell walls is considered as one of the possible mechanisms in plant metal resistance. Further studies are needed to elucidate the relation between extracellular and intracellular metal resistance mechanisms hypothesized above.

Different mechanisms have been proposed for the same plant species tolerating the same heavy metal [2,24,26]. Such an inconsistency may suggest that plant metal tolerance is not simple execution of a single mechanism. Instead, expression of metal tolerance may be an integration of all possible mechanisms that a plant may possess and operate at a plant-specific, growth stage–specific, and external medium environment–specific condition. This conclusion seems to be supported by computer evaluation of vacuolar ligand-based mechanisms. It is also conceivable that the role of cell wall compartmentation may be important in tolerant plants which accumulate much higher quantities of heavy metals in roots than in shoots, with the majority of the metals binding to the cell walls. On the other hand, the role of vacuolar compartmentation may dominate in tolerant plants which accumulate much higher levels of heavy metals in shoots than in roots, with the majority of the metals residing in the vacuoles.

REFERENCES

1. A. J. M. Baker, *New Phytol. (Suppl.)*, *106*: 93 (1987).
2. J. A. C. Verkleij and H. Schat, in *Evolutionary Aspects of Heavy Metal Tolerance in Plants* (J. Shaw, ed.), CRC Press, Boca Raton, Fla., pp. 179–193 (1990).
3. J. Antonovics, A. D. Bradshaw, and R. G. Turner, *Adv. Ecol. Res.*, 7: 1 (1971).
4. A. D. Bradshaw and T. McNeilly, in *Ecological Responses to Environmental Stresses* (J. Rozema and J. A. C. Verkleij, eds.), Kluwer Academic Publishers, Dordrecht, The Netherlands, pp. 2–5 (1991).
5. L. Wu, A. D. Bradshaw, and D. A. Thurman, *Heredity (London)*, *34*: 165 (1975).
6. W. H. O. Ernst, J. A. C. Verkleij, and R. Vooijs, *Environ. Monit. Assess.*, *3*: 297 (1982).
7. S. A. Al-Hiyaly, T. McNeilly, and A. D. Bradshaw, *New Phytol.*, *110*: 571 (1988).
8. A. D. Bradshaw, in *Origins and Development of Adaptation* (D. Evered and G. M. Collins, eds.), CIBA Foundation Symposium 102, Pitman, London, pp. 4–19 (1984).
9. J. Levitt, *Response of Plants to Environmental Stress*, 2nd ed., Vol. 2, Academic Press, New York (1980).
10. P. C. Lolkema, M. H. Donker, A. J. Schouten, and W. H. O. Ernst, *Planta*, *162*: 174 (1984).
11. S. J. McNaughton, T. C. Folsom, T. Lee, F. Park, C. Price, D. Roeder, J. Schmitz, and C. Stockwell, *Ecology*, *55*: 1163 (1974).
12. G. J. Taylor and A. A. Crowder, *Can. J. Bot.*, *62*: 1304 (1984).
13. R. D. Reeves and A. J. M. Baker, *New Phytol.*, *98*: 191 (1984).
14. P. L. Fiedler, *Am. J. Bot.*, *72*: 1712 (1985).
15. A. J. C. Verkleij, P. C. Lolkema, A. L. DeNeeling, and H. Harmens, in *Ecological Responses to Environmental Stresses* (J. Rozema and J. A. C. Verkleij, eds.), Kluwer Academic Publishers, Dordrecht, The Netherlands, pp. 8–19 (1991).
16. A. D. Thurman and A. J. Rankin, *New Phytol.*, *19*: 629 (1982).
17. E. Grill, E.-L. Winnacker, and M. H. Zenk, *Science*, *230*: 674 (1985).
18. W. E. Rauser, *Annu. Rev. Biochem.*, *59*: 61 (1990).

19. P. J. Jackson, E. J. Roth, P. R. McClure, and C. M. Navanjo, *Plant Physiol.*, *75*: 915 (1984).
20. P. J. Jackson, C. J. Undefer, J. A. Doolen, K. Watt, and N. J. Robinson, *Proc. Natl. Acad. Sci. USA*, *84*: 6619 (1987).
21. R. P. G. Gregory and A. D. Bradshaw, *New Phytol.*, *64*: 131 (1965).
22. R. M. Cox and T. C. Hutchinson, *New Phytol.*, *84*: 631 (1980).
23. L. Symeonidis, T. McNeilly, and A. D. Bradshaw, *New Phytol.*, *101*: 309 (1985).
24. H. W. Woolhouse, in *Encyclopedia of Plant Physiology* (O. L. Lange, P. S. Nobel, C. B. Osmond, and H. Ziegler, eds.), New Series, Vol. 12C, Springer-Verlag, Berlin, pp. 245–300 (1983).
25. F. P. C. Blamey, D. C. Joyce, D. G. Edwards, and G. J. Asher, *Plant Soil*, *81*: 171 (1986).
26. G. J. Wagner, *Adv. Agron.* (in press).
27. A. J. M. Baker, *New Phytol.*, *80*: 635 (1978).
28. A. J. M. Baker, *J. Plant Nutr.*, *3*: 643 (1981).
29. P. J. Peterson, *J. Exp. Bot.*, *20*: 863 (1969).
30. R. G. Turner and C. Marshall, *New Phytol.*, *71*: 671 (1972).
31. P. J. Peterson, in *Metals and Micronutrients* (D. A. Robb and W. S. Pierpoint, eds.), Academic Press, New York, pp. 51–69 (1981).
32. A. H. Fitter and R. K. M. Hay, *Environmental Physiology of Plants*, 2nd ed., Academic Press, New York (1987).
33. A. Reilly and C. Reilly, *New Phytol.*, *72*: 1041 (1973).
34. D. A. Baker, in *Metals and Micronutrients* (D. A. Robb and W. S. Pierpoint, eds.), Academic Press, New York, pp. 3–19 (1981).
35. C. D. Foy, R. L. Chaney, and M. C. White, *Annu. Rev. Plant Physiol.*, *29*: 511 (1978).
36. R. J. Haynes, *Bot. Rev.*, *46*: 75 (1980).
37. A. Brookes, J. C. Collins, and D. A. Thurman, *J. Plant Nutr.*, *3*: 695 (1981).
38. E. Heuillet, A. Moreau, S. Halpern, N. Jeanne, and S. Puiseux-Dao, *Biol. Cell*, *58*: 79 (1986).
39. W. E. Rauser and C. A. Ackerley, *Can. J. Bot.*, *65*: 643 (1987).
40. R. F. M. Van Steveninck, M. E. Van Steveninck, D. R. Fernando, W. J. Horst, and H. Marschner, *J. Plant Physiol.*, *131*: 247 (1987).
41. R. Vögeli-Lange and G. J. Wagner, *Plant Physiol.*, *92*: 1086 (1990).
42. J. Wang, B. P. Evangelou, M. T. Nielsen, and G. J. Wagner, *Plant Physiol.*, *97*: 1154 (1991).
43. J. Wang, B. P. Evangelou, M. T. Nielsen, and G. J. Wagner, *Plant Physiol.*, *99*: 621 (1992).
44. J. Richard, *J. Theor. Biol.*, *128*: 253 (1987).
45. D. L. Allan and W. M. Jarrell, *Plant Physiol.*, *89*: 823 (1989).
46. P. O'Shea, J. Walters, I. Ridge, M. Wainright, and A. P. J. Trinci, *Plant Cell Environ.*, *13*: 447 (1990).
47. J. Wang, B. P. Evangelou, and M. T. Nielsen, *Plant Physiol.*, *100*: 496 (1992).
48. L. Taiz, *Annu. Rev. Plant Physiol.*, *35*: 585 (1984).
49. M. C. McCann and K. Roberts, in *The Cytoskeletal Basis of Plant Growth and Form*, Academic Press, London, pp. 109–129 (1991).
50. L. Taiz and E. Zeiger, *Plant Physiology*, Benjamin/Cummings, Redwood City, Calif. (1991).
51. D. T. A. Lampton, in *Cellulose: Structure, Modification and Hydrolysis* (R. A. Young and R. M. Rowell, eds.), Wiley, New York (1986).
52. L. G. Wilson and J. C. Fry, *Plant Cell Environ.*, *9*: 239 (1986).
53. N. Carpita, D. Sabularse, D. Montezinos, and D. P. Delmer, *Science*, *205*: 1144 (1979).
54. G. E. Briggs, *New Phytol.*, *56*: 305 (1957).
55. S. C. Fry, *The Growing Plant Cell Wall: Chemical and Metabolic Analysis*, Longman Scientific and Technical, Harlow, Essex, England (1988).
56. B. Keller, *Plant Physiol.*, *101*: 1127 (1993).
57. J. BeMiller, in *Chemistry and Function of Pectins* (M. L. Fishman and J. J. Jen, eds.), ACS Symposium Series 310, American Chemical Society, Washington, D.C., pp. 2–12 (1986).
58. A. Lauchli, in *Encyclopedia of Plant Physiology, Transport in Plants*, Vol. II, Part B, Tissues and Organs (U. Luttge and A. M. G. Pitman, eds.), Springer-Verlag, Berlin, pp. 3–34 (1976).
59. D. S. Bush and J. G. McColl *Plant Physiol.*, *85*: 247 (1987).
60. G. Sposito, *The Surface Chemistry of Soils*, Oxford University Press, New York (1984).

61. P. Van Cutsem and C. Gillet, *Plant Soil*, *62*: 367 (1981).
62. P. Van Cutsem and C. Gillet, *J. Exp. Bot.*, *33*: 847 (1982).
63. P. L. Irwin, M. D. Sevilla, J. J. Shieh, and C. L. Stoudt, in *Chemistry of Function of Pectins* (M. L. Fishman and J. J. Jen, eds.), ACS Symposium Series 310, American Chemical Society, Washington, D.C., pp. 175–188 (1986).
64. P. R. Bloom, in *Chemistry in the Soil Environment*, ASA Special Publication 40, American Society of Agronomy, Madison, Wis., pp. 129–150 (1981).
65. D. A. Thurman, in *Effect of Heavy Metal Pollution on Plants* (N. W. Lepp, ed.), Vol. 2, Applied Science Publishers, Barking, Essex, England, pp. 239–249 (1981).
66. K. Iwasaki, K. Sakurai, and E. Takahashi, *Soil Sci. Plant Nutr.*, *36*: 431 (1990).
67. V. P. Evangelou and G. J. Wagner, *J. Exp. Bot.*, *38*: 1637 (1987).
68. D. T. Clarkson and J. Sanderson, *Plant Physiol.*, *61*: 731–736 (1978).
69. A. Goren and H. Wanner, *Ber. Schweiz. Bot. Ges.*, *80*: 334 (1971).
70. J. H. Arvik and R. L. Zimdahl, *J. Environ. Qual.*, *3*: 374 (1974).
71. S. G. Javis, L. H. P. Jones, and M. J. Hopper, *Plant Soil*, *44*: 179 (1976).
72. H. Th. Wolterbeek, P. Bode, and M. De Bruin, *Plant Cell Environ.*, *10*: 297 (1987).
73. S. J. Wainwright and H. W. Woolhouse, in *The Ecology of Resource Degradation and Renewal* (M. J. Chadwick and G. T. Goodman, eds.), Blackwell Scientific Publications, Oxford, pp. 231–257 (1975).
74. S. J. Wainwright and H. W. Woolhouse, *J. Exp. Bot.*, *29*: 525 (1978).
75. D. A. Thurman and J. C. Collins, "Metal Tolerance Mechanisms in Higher Plants: A Review," *Proceedings of the International Conference on Heavy Metals in the Environment*, Heidelberg, CEP Consultants, Edinburgh, pp. 298–304 (1983).
76. G. J. Taylor, *Commun. Soil Sci. Plant Anal.*, *19*: 1179 (1988).
77. R. M. Cox and D. A. Thurman, *New Phytol.*, *80*: 17 (1978).
78. W. Mathys, *Physiol. Plant.*, *33*: 161 (1975).
79. A. J. M. Baker, *New Phytol.*, *81*: 321 (1978).
80. P. K. Hepler and R. O. Wayne, *Annu. Rev. Plant Physiol.*, *36*: 397 (1985).
81. T. Boller, and A. Wiemken, *Annu. Rev. Plant Physiol.*, *37*: 137 (1986).
82. M. Wink, *J. Exp. Bot.*, *44*: 231 (1993).
83. C. A. Ryan and M. Walker-Simmons, *Methods Enzymol.*, *96*: 580 (1983).
84. J. A. Smith, in *Plant Vacuoles, Their Importance in Solute Compartmentation in Plant Cells and Their Applications in Plant Biotechnology* (B. Marin, ed.), NATO ASI Series, Plenum Press, New York, pp. 79–87 (1987).
85. T. Matoh, J. J. Watanabe, and E. Takahashi, *Plant Physiol.*, *84*: 173 (1987).
86. A. R. Memon, M. Chino, K. Hara, and M. Yatazawa, *Soil Sci. Plant Nutr.*, *27*: 317 (1981).
87. A. R. Memon and M. Yatazawa, *J. Plant Nutr.*, *7*: 961 (1984).
88. P. E. Pfeffer, S. Tu, W. V. Gerasimowicz, and J. R. Cavanaugh, *Plant Physiol.*, *80*: 77 (1986).
89. M. Mullins, K. Hardwick, and D. A. Thurman, "Heavy Metal Location by Analytical Electron Microscopy in Conventionally Fixed and Freeze-Substituted Roots of Metal Tolerant and Nontolerant Ecotypes," *Proceedings of the International Conference on Heavy Metals in the Environment*, CEP Consultants, Edingburgh, pp. 43–46 (1985).
90. D. Salt and G. J. Wagner, *J. Biol. Chem.*, *268*: 12297 (1993).
91. E. Grill, E.-L. Winnacker, and M. H. Zenk, *Proc. Natl. Acad. Sci. USA*, *84*: 439 (1987).
92. H. Obata and M. Umebayashi, *Soil Sci. Plant Nutr.*, *35*: 479 (1989).
93. J. A. P. Verkleij, P. Koevoets, J. V. Riet, R. Bank, Y. Nijdam, and W. H. O. Ernst, *Plant Cell Environ.*, *13*: 913 (1990).
94. R. F. M. Van Steveninck, M. E. Van Steveninck, D. R. Fernando, L. B. Edwards, and A. J. Wells, *C. R. Acad. Sci. Paris*, *310* (Ser. III):671 (1990).
95. A. E. Martell and R. M. Smith, *Critical Stability Constants*, Vol. 3, Plenum Press, New York (1977).
96. A. E. Martell and R. M. Smith, *Critical Stability Constants*, Vol. 5, Plenum Press, New York (1982).
97. W. Mathys, *Physiol. Plant.*, *40*: 130 (1977).

98. W. H. O. Ernst, in *Effects of Air Pollutants on Plants* (T. A. Mansfield, ed.), Cambridge University Press, Cambridge, pp. 67–99 (1976).
99. D. L. Godbold, W. J. Horst, J. C. Collins, D. A. Thurman, and H. Marschner, *J. Plant Physiol.*, *116*: 59 (1984).
100. H. Harmens, J. A. C. Verkleij, P. Koevoets, and W. H. O. Ernst, in *Heavy Metals in the Environment* (J. P. Vernet, ed.), Vol. 2, CEP Consultants, Edingburgh, pp. 178–181 (1987).
101. J. Lee, R. D. Reeves, R. R. Brooks, and T. Jaffré, *Phytochemistry*, *16*: 1503 (1977).
102. J. Lee, R. D. Reeves, R. R. Brooks, and T. Jaffré, *Phytochemistry*, *17*: 1033 (1978).
103. I. Kishinami, and J. M. Widholm, *Plant Cell Physiol.*, *28*: 203 (1987).
104. K. Ojima, H. Abe, and K. Ohira, *Plant Cell Physiol.*, *25*: 855 (1984).
105. G. J. Wagner and R. M. Krotz, in *Metal Ion Homeostasis: Molecular Biology and Chemistry*, Alan R. Liss, New York, pp. 325–336 (1989).
106. G. Sposito and S. V. Mattigod, *GEOCHEM: A Computer Program for Calculation of Chemical Equilibria in Soil Solutions and Other Natural Water Systems*, The Kearney Foundation of Soil Science, University of California, Riverside, Calif. (1979).
107. D. R. Parker, W. A. Novell, and R. L. Chaney, in *Chemical Equilibrium and Reaction Models* (R. H. Leoppert, ed.), Soil Science Society of America Special Publication, American Society of Agronomy, Madison, Wis. (in press).
108. N. R. Reese and G. J. Wagner, *Biochem. J.*, *241*: 641 (1987).
109. R. M. Krotz, B. P. Evangelou, and G. J. Wagner, *Plant Physiol.*, *91*: 780 (1989).
110. R. B. Clark, *Crop Sci.*, *9*: 341 (1969).
111. A. B. Tomsett and D. A. Thurman, *Plant Cell Environ.*, *11*: 383 (1988).

34
Calcium as a Messenger in Stress Signal Transduction

A. S. N. Reddy
Colorado State University, Fort Collins, Colorado

I. INTRODUCTION

Plant growth in the natural environment is often adversely affected by a number of factors. These include environmental factors such as low temperature, heat, drought, and high salinity, and biological factors such as pathogens (bacteria, viruses, and fungi). Abiotic and biotic factors that limit growth and development of plants and eventually, productivity are considered stress factors. Crop losses due to those various biotic and abiotic stresses amount to billions of dollars annually. It has been estimated that stress factors (biotic and abiotic) depress the yield of agronomically important crops in the United States by 78%, of which about 70% is due to unfavorable environmental conditions [1]. Plant scientists have been studying the effects of various stresses on plants to better understand the mode of action of stress signals. It is hoped that the knowledge derived from increased understanding of plant responses to biotic and abiotic stresses would eventually help in developing new plant varieties that are resistant to these stress factors. Recent advances in molecular and cellular biology are offering a variety of new approaches to investigate plant responses to stresses.

In recent years there has been increasing interest in understanding the biochemical and molecular basis of stress tolerance and in the identification of genes involved in stress tolerance. Several genes that are involved in biotic stress tolerance have been identified and used to obtain transgenic plants with enhanced resistance to these stresses [2,3]. Because of the complex nature of plant responses to abiotic stresses, very little is known about the biochemical and molecular mechanisms that contribute to resistance. Recent studies suggest that the regulation of expression of specific genes as well as changes in the levels of certain osmolytes is an integral part of plant adaptation to stressful environmental conditions [4–9]. Effects of different stresses on various physiological processes and gene expression are reviewed in other chapters in this book, hence are not covered here. The mechanisms by which stress signals induce changes in gene expression and affect biochemical pathways are poorly understood.

Because of their inability to move, plants have developed mechanisms to sense and respond to the stress signals so that they can adapt or develop resistance to environmental variables. Different plant species differ in their ability to adapt to environmental variables. The mechanisms by which plant cells perceive and transduce stress signals are not well understood. In animal cells, messengers such as cyclic nucleotides and calcium play a vital role in signal transduction pathways. The role of calcium ions as one of the key signaling molecules in animal systems has been known for a long time. In plants also, calcium has been implicated, for decades, in regulating various physiological processes during growth and development. However, research during the last 15 years strongly indicates that calcium plays an important messenger role in transducing a variety of hormonal and environmental signals. Several comprehensive reviews on various aspects of the calcium messenger system in plants have appeared [10–19]. In this chapter, I focus primarily on the role of calcium as a messenger in stress signal transduction. A number of criteria have been used to consider a chemical or an ion as a messenger in signal transduction. These include (a) changes in the concentration of putative messenger in response to a signal prior to a response, (b) the presence of receptors to sense changes in the level of messenger, (c) induction of a signal-induced response by changing the levels of putative messenger in the absence of a primary signal, and (d) blocking of a signal-induced response by blocking the changes in the level of putative messenger. Evidence obtained in recent years indicates that calcium satisfies all these criteria to consider it as a messenger molecule in transducing stress signals.

II. CHANGES IN CYTOSOLIC CALCIUM IN RESPONSE TO STRESS SIGNALS

Despite initial technical problems in measuring cytosolic calcium in plant cells, tremendous progress has been made in this area in recent years [12,19]. Various methods using calcium-binding fluorescent dyes, calcium-selective electrodes, and aequorin, a calcium-binding photoprotein, have been used to measure signal-induced changes in cytosolic calcium [20–22]. Using these methods, several laboratories have demonstrated that hormonal, environmental, and stress signals elevate cytosolic calcium. Calcium measurement studies indicate that the concentration of calcium in the cytoplasm of plant cells, as in animal cells, is maintained in the micromolar range (0.1 to 1 μM). However, calcium concentration in the cell wall and in organelles is in the millimolar range [12]. Despite a large electrochemical gradient for calcium entry into the cytoplasm, cytosolic calcium concentration is maintained low (0.1 to 1 μM). Maintenance of low cytosolic calcium levels requires active pumping of calcium to the apoplast or organelles. Using different approaches, a number of signals, including stress signals, have been shown to elevate cytosolic calcium. Table 1 shows various signals that have been shown to change the level of cytosolic calcium. Here I describe only stress-induced changes in cytosolic calcium levels. Elevation of cytosolic calcium in response to other signals has been discussed elsewhere [19].

There are several reports indicating that cold and salt stress affect calcium homeostasis in plants [23,31–33]. Increasing evidence obtained during the last two years suggest that abiotic stress signals rapidly elevate the level of cytosolic calcium. Treatment of corn protoplasts with sodium chloride elevated cytosolic calcium from about 1.1 μM to 1.8 μM [23]. Knight et al. [22] used transgenic plants expressing apoaequorin to reconstitute aequorin and measure changes in cytosolic calcium in response to various signals. These studies with transgenic plants have shown that signals such as cold, touch, and wind, which are known to influence plant growth and development markedly [5,34,35], elevate the levels of cytosolic calcium in seedlings [22,24,36]. In vivo imaging of cold-induced changes in cytosolic calcium indicate that cotyledons and roots of a seedling are highly responsive, whereas hypocotyls are relatively insensitive to cold shock [36]. Furthermore, it has been shown that cold shock causes wavelike

Table 1 Environmental, Hormonal, and Stress Signals That Induce Changes in Cytosolic Calcium Level in Higher Plants

Signal	Mode of free calcium measurement	Effect on cytosolic calcium[a]	Refs.
Cold	Aequorin[b]	↑	22
Salt stress	Calcium-binding fluorescent dye	↑	23
Wind	Aequorin[b]	↑	24
Gravity	Calcium-binding fluorescent dye	↑	25
Light	Calcium-binding fluorescent dye	↑	25
Touch	Aequorin[b]	↑	22
Fungal elicitors	Aequorin[b]	↑	22
Abscisic acid	Calcium-binding fluorescent dye	↑ ↓	26–29
Auxin	Calcium-binding fluorescent dye	↑	30
Gibberlic acid	Calcium-binding fluorescent dye	↑	21

[a] ↑ denotes an increase in cytosolic calcium; ↓ denotes a decrease in cytosolic calcium.
[b] Aequorin was constitutively expressed in transgenic seedlings.

calcium increases in the cells of cotyledons. Fungal elicitors that induce defense response in plants also elevate cytosolic calcium [22]. However, the magnitude and kinetics of calcium transients induced by touch, cold, and fungal elicitors are found to be different [22,24,36]. Whether all the cells or specific cell types in a seedling respond to these stress signals is not known at present.

The levels of ethylene and abscisic acid (ABA) are known to change as the plants cope with stress factors [37,38]. Using calcium-binding fluorescent dyes, it has been demonstrated that ABA, gibberlic acid, and auxin increase cytosolic calcium [21,26,27,39,40]. However, transgenic plants with reconstituted aequorin did not show changes in cytosolic calcium in response to hormones and heat shock [22]. This could be due to various factors, such as sensitivity of the method, localized nature of the response due to restricted target cells, and stability of aequorin at high temperature.

It is clear from various studies that stress and hormonal signals cause transient increases in cytosolic calcium. The mechanisms by which increased cytosolic calcium could affect biochemical, and eventually, physiological processes are shown in Figure 1. Various components of calcium-mediated signal transduction pathways are discussed later in this chapter. Influx of calcium from extracellular calcium as well as release of calcium from internal stores seems to contribute to signal-induced changes in cytosolic calcium. Depending on the cell type or the signal, either one or both processes could be involved in raising calcium level in the cytosol. Monitoring of signal-induced changes in the presence of calcium channel blockers of plasma membrane or organelles indicate that different signals use distinct calcium stores in elevating cytosolic calcium. For instance, cold-induced calcium increase is inhibited by plasma membrane channel blockers but is not affected by organellar channel blockers. However, wind-induced calcium increase is blocked by organellar calcium channel blockers, whereas plasma membrane channel blockers did not have any effect [24]. These studies indicate that

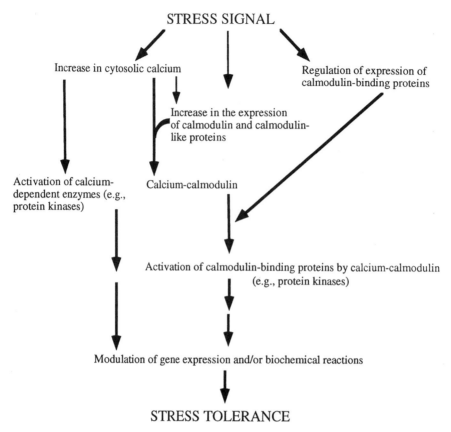

Figure 1 Schematic illustration of the proposed events involving calcium, calmodulin, and calmodulin-binding proteins in stress tolerance/resistance. The model is based on several recent reports [19,22,23,36,41–50].

the extracellular calcium contributes to cold-induced elevation of cytosolic calcium, and internal calcium stores contribute to wind-induced increase in cytosolic calcium. The ABA-induced changes in cytosolic calcium have been attributed to both calcium release from internal stores and calcium influx [27,40]. However, the mechanism(s) by which stress and mechanical signals such as cold, salt, and wind cause changes in cytosolic calcium are not clear.

III. MODE OF CALCIUM ACTION IN PLANTS: CALCIUM-MODULATED PROTEINS

Changes in free calcium concentration in the cytoplasm are believed to regulate various cellular processes at the biochemical and molecular levels, eventually leading to a physiological response. Transmission of the calcium signal to the metabolic machinery is accomplished through intracellular calcium receptors or calcium-binding proteins. All the calcium-binding proteins, except annexins, contain a 29-residue helix–loop–helix structure called an EF hand that binds to calcium with high affinity [51,52]. However, different calcium-binding proteins differ in their affinity to calcium, with K_d values ranging from 10^{-5} to 10^{-9} M. Binding of calcium to these calcium receptors results in a conformational change in the receptor which enables them to interact with other proteins and alter their function. A number (over 150) of calcium-binding proteins have been identified and characterized in animals [15,52]. Of these, only a few are present in all eukaryotic cells and are believed to be involved in mediating calcium action, whereas the majority of them (e.g., troponin C and paravalbumin) are found in specific tissues and play restricted roles. In plants, so far, only two calcium-binding proteins—calmodulin, and a calcium-dependent, calmodulin-independent protein kinase (also called CDPK, discussed in the Section III.B)—have been well characterized [12,15,17,19]. In addition, recent studies indicate the presence of calmodulinlike proteins and other calcium-binding proteins in plants [41,53,54]. However, the function of these proteins in calcium signaling pathways is not known. Calmodulinlike proteins show limited homology with calmodulin and occasionally are larger than the calmodulin [53]. Hence it is likely that these proteins are functionally distinct and are involved in controlling various cellular functions.

A. Calmodulin

Calmodulin is found in all eukaryotic organisms and is a primary mediator of calcium action in animal cells. In plants, calmodulin and another calcium-modulated protein called calcium-dependent protein kinase are believed to sense the changes in cytosolic calcium. Calmodulin was first discovered in animals as an activator of cyclic nucleotide phosphodiesterase [55], and subsequently, it has been isolated and characterized from all eukaryotic organisms. The discovery of calmodulin in plants led plant scientists to propose a messenger role for calcium in plants. Calmodulin has been isolated and characterized from many different plants [12,15]. Gene structure and expression of calmodulins from a number of plants have been analyzed [19]. Calmodulin is a low-molecular-weight protein of 148 amino acids that is highly conserved between plants and animals. Calmodulins from all eukaryotes, except from budding yeast [56], have four helix–loop–helix motifs (also known as EF-hand domains) that bind to four calcium ions with high affinity. Plant calmodulin is structurally and functionally very similar to animal calmodulin [15]. Calmodulin has no enzyme activity, but when bound to calcium it can modulate the activity and function of many enzymes and certain structural proteins. The crystal structure of calmodulin has revealed that it has two globular domains, each with a pair of EF hands, connected by a central helix [57]. The binding of calcium to calmodulin results in a conformational change in such a way that the hydrophobic pockets are exposed in each globular end, which can then interact with and regulate the activity of target enzymes. The hydrophobic pockets that are exposed upon calcium binding are believed to be involved in its interaction with hydrophobic regions along the amphipathic helix of calmodulin-target proteins [58,59].

1. Effect of Stress Signals on the Expression of Calmodulin

In plant cells, calmodulin and calmodulinlike proteins are highly responsive to physical and hormonal signals such as touch, wounding, light, and auxin [41,60]. The development of strawberry fruit is dependent on auxin from the achenes. Application of auxin to deachened (auxin-deprived) fruits induced calmodulin mRNA [60]. In *Arabidopsis*, the expression of calmodulin and calmodulinlike proteins is rapidly induced by several stimuli, such as touch, wind, and wounding [41]. Signal-induced changes in calmodulin and calmodulinlike proteins may play a significant role in cell growth and physiology. Small changes in calmodulin levels have been shown to affect drastically the progression of cell cycle in animal cells [61] (see Chapter 6). Touch and wind signals have been shown to cause rapid and transient increases in cytosolic calcium which occur prior to observed changes in gene expression [22,24,41,42]. Hence, it is likely that signal-induced changes in cytosolic calcium level could be involved in the expression of calmodulin and calmodulinlike proteins. In several plant systems there are multiple calmodulin genes which code for proteins that are identical or contain a few conservative changes [18,62–64]. Hence, they may not differ much in their function or affinity toward calcium. However, there seems to be differential regulation of multiple calmodulin genes [62,63], indicating the presence of different regulatory elements in their promoters.

2. Calmodulin-Binding Proteins

Calmodulin itself has no enzymatic activity. It controls various cellular activities by modulating the activity or function of a number of proteins. Hence, the role of calmodulin in a given cell or a tissue is determined by its target proteins. Calmodulin is multifunctional because of its ability to interact with and control the activity of many target proteins (also called calmodulin-binding proteins). In animal systems, the activity of more than 20 enzymes has been shown to be regulated by calmodulin in a calcium-dependent manner [59,65,66]. These include several protein kinases, a protein phosphatase (also known as calcineurin), a plasma membrane calcium ATPase, adenyl cyclases, cyclic $3',5'$-nucleotide phosphodiesterase, inositol trisphosphate kinase, nitric oxide synthease, and some structural proteins. Identification and characterization of the calmodulin target protein in animal cells has helped in understanding the mechanisms by which calcium/calmodulin regulate various biochemical and molecular events leading to a physiological response. The amino acid sequences of the calmodulin-binding domain in different calmodulin target proteins is not conserved. However, calmodulin-binding motifs from different calmodulin-binding proteins form amphipathic α-helices [58]. When arranged in a helical wheel, the amino acids in the calmodulin-binding domain form an amphipathic helix with several positive residues on one side and a number of hydrophobic residues on the other side. In most cases, binding of calmodulin to its target proteins requires calcium. However, calmodulin binds to a couple of proteins (e.g., neuromodulin) in the absence of calcium [67]. Furthermore, a calmodulin with no calcium-binding activity has been shown to rescue a mutation in the calmodulin genes in budding yeast [68], suggesting that calmodulin may perform some functions in the absence of calcium. Gel overlay studies with plant proteins indicate the presence of a number of calmodulin-binding proteins in plants [69,70]. A few enzymes that are activated by calmodulin have been identified in plants. These include NAD kinase, calcium ATPase, and nuclear NTPases [17]. In fact, calmodulin was discovered in plants as an activator of NAD kinase [71,72]. A new approach of isolating calmodulin-binding proteins by screening expression libraries with ^{35}S-labeled calmodulin should greatly aid in isolating and characterizing cDNAs for calmodulin-binding proteins (Figure 2). Several cDNAs have been isolated with this approach [43,50,73], and one of the isolated clones is found to have significant similarity with *E. coli* glutamate decarboxylase [73]. The expression of some of the calmodulin-binding proteins is regulated by heat and wind [43,50,73].

B. Calcium-Regulated Protein Kinases

Signals either directly or through messengers such as calcium regulate the activity of protein kinases and protein phosphatases, which, in turn, regulate the phosphorylation status of many proteins in the cell. The enzyme activity or biological properties of many proteins is strongly influenced by phosphorylation status. It has been well established that calcium-regulated protein phosphorylation plays a pivotal role in signal transduction in animal cells. In animal cells, protein kinase C, a calcium- and phospholipid-dependent protein kinase [75], and several calcium/calmodulin-dependent protein kinases [66,76] mediate the calcium effects on protein phosphorylation. Calcium-regulated protein phosphorylation is believed to be involved in signal amplification, in obtaining sustained responses, and in producing diverse responses to transient raises in cytosolic calcium [66,76]. Demonstration of a central role for calcium-regulated protein kinases in animal cells in a calcium signaling system led plant scientists to investigate the presence of calcium-regulated protein kinases. Current evidence indicates that there are two

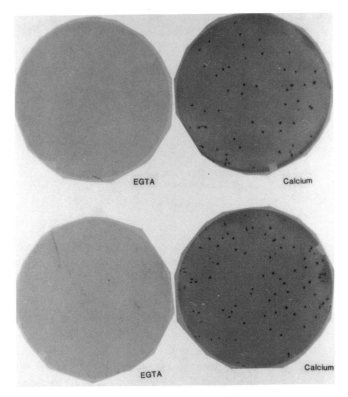

EGTA   Calcium

EGTA   Calcium

Figure 2 Calcium-dependent binding of [^{35}S]calmodulin to a positive control ICM1 (top) and a calmodulin-binding protein cDNA clone isolated from corn root tip (bottom) expression library. A purified bacteriophage-containing cDNA insert for corn root calmodulin-binding protein and a positive control ICM1 that encodes a calmodulin-binding protein from mouse [74] were plated with *E. coli* Y1090 and incubated at 42°C for 3 h, and the fusion protein was induced by applying the IPTG-soaked nitrocellulose filters. Duplicate filters were made for each clone. One filter was incubated with binding buffer containing [^{35}S]calmodulin and 1 m*M* calcium chloride (right) for 12 h and the duplicate filter was incubated for the same time in the binding buffer containing [^{35}S]calmodulin and 5 m*M* EGTA (left). Filters were washed three times, dried, and exposed to film. (From Reddy, Fromm, and Poovaiah, unpublished.)

types of calcium-regulated protein kinases: a calcium-dependent and calmodulin-independent protein kinase (CDPK) and a calcium/calmodulin-dependent protein kinase (CaM kinase). CDPK is unique to plants and has been well characterized compared to CaM kinase [17].

Calcium-dependent and calmodulin-independent protein kinase (CDPK) is found in many plants. This protein kinase has been purified to homogeneity from soybean [77] and partially purified from a number of other plant systems [17]. Calcium binds directly to the CDPK and stimulates the kinase activity by about 100-fold, whereas calmodulin did not have any significant effect on the kinase activity [78]. Using primers that correspond to amino acid sequences obtained from proteolytic fragments of a purified CDPK, a cDNA that codes for CDPK has been isolated [45]. The predicted amino acid sequence of soybean CDPK has revealed a unique structural organization that is different from all the known protein kinases, indicating that it represents a new family of protein kinases that are unique to plants. The deduced primary structure of the CDPK contained a protein kinase catalytic domain followed by a calmodulinlike region with four calcium-binding motifs (see Figure 3). The kinase domain of CDPK showed

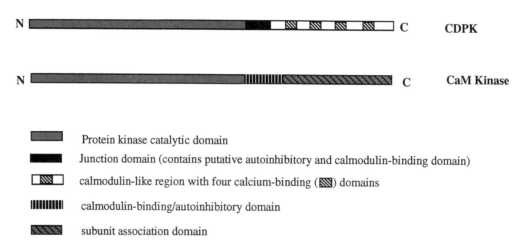

Figure 3 Diagrammatic representation of catalytic and regulatory domains of calcium-dependent and calmodulin-independent protein kinase (CDPK) and calcium/calmodulin-dependent protein kinase (CaM kinase) [45,46,66,81,82]. The amino and carboxyl termini are denoted by N and C, respectively.

significant homology with mammalian calcium/calmodulin-dependent protein kinase II (CaM KII) catalytic domain. The region that joins the kinase domain to the calmodulinlike region corresponds to the autoinhibitory/calmodulin-binding region of CaM KII. The regulatory role of this junction region is currently not known. Complementary DNAs (cDNAs) that code for CDPK have been isolated from other systems also [46,47]. The activity of an *Arabidopsis* CDPK that is expressed in *E. coli* is stimulated by calcium [46,47]. Immunological and cloning studies as well as Southern analysis of soybean and *Arabidopsis* genomic DNA suggest that there are several isoforms of CDPK in plants [17]. Recent studies indicate that, besides calcium, lipids are involved in the regulation of CDPK activity [46,79]. Studies with antibodies to CDPK and activity assays indicate that the CDPK is widely distributed in plants. The molecular weight of purified CDPKs from different plant systems ranges from 35,000 to 90,000 [17]. A wide range in the size and differences in their substrate specificity suggest that there could be multiple isoforms. It is also possible that some of the small enzymes are derived from larger ones by proteolytic cleavage, which has been shown to occur in oat [79].

In vitro and in vivo protein phosphorylation studies have demonstrated calcium-regulated protein phosphorylation in a number of plant systems. Based on the inhibition of protein phosphorylation by calmodulin antagonists and the modest stimulation of protein phosphorylation by calmodulin, it was concluded that plants contain calcium/calmodulin-dependent protein kinases [12]. However, the observations that calmodulin antagonists inhibit CDPK activity and have nonspecific effects raised doubts about the presence of calcium/calmodulin-dependent protein kinases [17]. Studies with antibodies raised to a synthetic peptide corresponding to a mammalian calcium/calmodulin-dependent protein kinase suggest the presence of a calmodulin-dependent protein kinase in plants [80]. The amino acid stretch that was used in raising the antibodies is unique to CaM KII and is not present in CDPK or other protein kinases. A cDNA that shows significant similarity to mammalian CaM KII has been isolated from plants by screening an expression library with radiolabeled calmodulin [81,82]. These studies indicate the presence of a CaM kinase in plants, although it is not clear how widely these kinases are distributed. The structure and organization of regulatory domains of CDPK and CaM KII are shown in Figure 3.

There is some evidence to indicate that calcium-regulated protein phosphorylation is involved in host–pathogen interaction. Treatment of parsley suspension cultures with fungal elicitors results in rapid and transient phosphorylation of specific proteins. These fungal elicitor–induced changes in phosphorylation are shown to be dependent on the presence of calcium in the medium [83]. Fungal elicitor–induced protein phosphorylation, phytoalexin production, and mRNAs for phenylalanine ammonia-lyase and 4-coumarate:CoA ligase mRNAs are greatly reduced in calcium-deprived cells [83]. Furthermore, readdition of calcium to the cultures restored the inhibitory effects of calcium deprival, suggesting calcium involvement in fungal elicitor–induced responses.

IV. CALCIUM AND GENE EXPRESSION

It is becoming increasingly clear that the regulation of expression of specific genes is involved in the ability of plants to adapt or develop resistance to abiotic and biotic stress factors, such as cold, heat, salinity, and pathogens [5–9]. The role of calcium in regulating gene expression, at both the transcriptional and translational levels, has been well documented in animal cells [84–91]. Much of the gene regulation by calcium is accomplished by calcium-regulated protein kinases. Transcriptional regulation of expression of specific genes by calcium is caused by the regulation of phosphorylation of transacting factors by calcium/calmodulin-dependent protein kinase II (CaM KII) [87] or calcium/phospholipid-dependent protein kinase (protein kinase C) [91]. Calcium influences translation at both the initiation and elongation steps. One of the elongation factors (eEF-2, eukaryotic elongation factor-2) is a substrate for a calcium/calmodulin-dependent protein kinase III (CaM KIII), and phosphorylation of eEF-2 makes it inactive, resulting in the inhibition of translation [92–94]. Depletion of calcium results in a decrease in the rate of initiation of protein synthesis [95–97]. Inhibition of protein synthesis initiation by calcium depletion is correlated with dephosphorylation of a ribosome-associated protein (26 kDa) [98].

Although there is a great deal of information on the involvement of calcium in regulating various physiological processes [11,12], very little is known about the role of calcium in regulating gene expression in plants. Manipulation of cytosolic calcium by various means is shown to affect the expression of specific genes in plants. Lam et al. [99] presented some evidence indicating that the induction of expression of chlorophyll *a/b*-binding (cab) protein is mediated by calcium and calmodulin. Ten percent of light-induced cab mRNA could be induced in the dark by increasing the intracellular level of calcium by ionomycin [99]. Microinjection studies with tomato phytochrome mutant (*aurea*) clearly show that calcium is involved in the expression of specific genes [100]. In *aurea* mutant, light-regulated genes are not expressed. Injection of a plasmid construct containing cab promoter, a light-responsive promoter, fused to GUS reporter (cab-GUS) into cells of *aurea* mutant, showed no expression of the GUS gene. However, coinjection of cab-GUS with calcium or calcium-activated calmodulin resulted in the expression of the GUS gene. These results suggest the involvement of calcium and calmodulin in regulating the cab promoter activity [100]. Partial development of chloroplasts in *aurea* mutant, which requires the expression of several genes, could be obtained by microinjection of calcium into these cells. These results indicate that calcium regulates the expression of other genes involved in chloroplast development. In *Arabidopsis*, calmodulin and calmodulin-related genes (TCH1, TCH2, TCH3, and TCH4) are strongly induced by various mechanical stimuli such as touch and wounding [41]. It has been demonstrated that increased external calcium or heat shock rapidly induce the expression of touch-induced calmodulin-related genes (TCH2, TCH3, and TCH4), whereas TCH1 gene, which codes for calmodulin, is not significantly induced [42]. Heat shock, in the presence of EGTA, a calcium chelator, did not show induction of TCH genes. This EGTA effect is reversed by calcium replenishment. Based on these results, it was suggested that heat shock elevates cytosolic calcium, which, in turn, regulates the expression of TCH2, 3, and 4 genes. Heat shock is known to increase cytosolic calcium in animal cells [101–103]. However, increase in cytosolic calcium in plant cells in response to heat shock has not yet been demonstrated. The calcium effect on touch-induced calmodulin-related genes is specific, since magnesium, another divalent ion, did not have any effect. Furthermore, increased calcium did not affect the expression of a heat-shock-induced gene. In plants, calmodulin is coded by a multigene family [19]. Using transgenic tobacco plants containing a promoter of one of the rice calmodulin genes (CaM-2) fused to a reporter gene, we found that calcium depletion by EGTA and ionomycin decreased the expression of the reporter gene driven by calmodulin promoter [80]. Furthermore, verapamil, a calcium channel blocker, reduced the activity of calmodulin promoter [80]. Similar treatments did not have a significant effect on CaMV 35S promoter, suggesting that the cytosolic calcium level has a specific effect on CaM-2 promoter activity. However, it is not clear whether all the calmodulin genes in a plant are affected by calcium. In the case of *Arabidopsis*, the expression of one of the calmodulin genes (TCH1) is not affected significantly by an increase in external calcium. There are multiple calmodulin genes in *Arabidopsis*, and it is not known whether any of these genes are affected by cytosolic calcium changes [62–64].

Recent studies show that calcium is involved in stress-induced (both biotic and abiotic) gene expression. Plant–pathogen interaction, chemical elicitors, and a number of other stress factors have been shown to stimulate ethylene production in plants [37,104]. Ethylene is involved in the expression of some of the pathogenesis-related proteins, including a chitinase. Depletion of calcium by a calcium chelator blocked ethylene-induced chitinase synthesis, whereas artificial elevation of cytosolic calcium with calcium ionophore (ionomycin) or an inhibitor of microsomal Ca^{2+} ATPase induced chitinase synthesis in the absence of ethylene [48]. These results indicate that the induction of a chitinase gene by ethylene is mediated by calcium. More recently, it has been shown that ethylene induced the phosphorylation of specific proteins, although it is not known whether calcium is involved in this ethylene-regulated protein phosphorylation [49]. Using inhibitors of protein kinases and protein phosphatase, it was concluded that protein phosphorylation is one of the intermediate events involved in ethylene signal transduction. Fungal elicitors that are known to induce pathogenesis-related proteins have been shown to elevate cytosolic calcium [22].

In some plants, freezing tolerance can be developed by exposing them to nonfreezing low temperature [5]. This process, which is known as cold acclimation or cold hardening, is associated with changes in gene expression [5,105–107]. Although the function of the cold-regulated genes is unknown, expression of some of these genes is positively correlated with the ability of plants to develop freezing tolerance [106,107]. Earlier studies have shown elevation of cytosolic calcium in response to cold shock [22,44]. In alfalfa, low-temperature-induced freezing tolerance is completely abolished by calcium channel blockers—lanthanum and verapamil—and partially by EGTA [108]. Furthermore, a calmodulin inhibitor (W7), but not its inactive analog (W5), inhibited the development of cold acclimation–induced freezing tolerance. These results suggest that increase in cytosolic calcium is necessary for developing cold-induced freezing tolerance and that calmodulin is involved in freezing tolerance. Accumulation of cold-induced mRNAs in alfalfa was partially blocked by lanthanum, a calcium channel blocker, and calmodulin inhibitor (W7) completely blocked the expression of cold-regulated genes [108]. Low-temperature-induced changes in protein phosphorylation are affected by lanthanum and W7. However, the effects of these antagonists on phosphorylation are more severe and are not restricted to cold-induced changes.

As described above, there are several reports implicating calcium in regulating the expression of specific genes, including its own receptors, in plant cells. However, the mode of calcium in regulating gene expression is not yet understood. Calcium-regulated protein phosphorylation is likely to be involved in calcium-regulated gene expression, although there is no evidence currently to support this hypothesis. The availability of calcium-responsive genes and promoters should help in identifying the regulatory elements and elucidating the mechanism(s) by which calcium regulates the expression of genes. It is not yet known to what extent the changes in cytosolic calcium are reflected in changes in free calcium concentration in the nucleus. Recent studies show that there is a calcium gradient between the nucleus and cytoplasm, indicating the presence of regulatory mechanisms that control calcium movement into and out of the nucleus [109–111]. ATP stimulates calcium uptake into nuclei, and studies implicate calmodulin involvement in this uptake process [110]. Currently, nothing is known regarding nuclear calcium concentration and the influence of increased cytosolic calcium on the level of nuclear calcium in plants.

V. CONCLUSIONS

The reports described above indicate a messenger role for calcium in transducing a variety of stress signals. Stress-induced rapid and transient changes in cytosolic calcium have been documented in several cases. At least two proteins (calmodulin and CDPK) that sense calcium changes in the cytoplasm have been well characterized. These proteins are found in all the plants that are tested, suggesting their broad role in mediating calcium action. In addition, several proteins that bind to calcium have been detected in plants, although the identity and function of most of these proteins are not known. Stress signals have been shown to modulate various components involved in the calcium signaling pathway. It has been demonstrated that some of the responses induced by stress signals can be blocked by preventing changes in calcium levels in the cells or by blocking the activity of components involved in the signaling pathway. Furthermore, at least in some instances, stress responses at the molecular level could be induced by increasing cytosolic calcium in the absence of a stress signal. Taken together, these results indicate an important messenger role for calcium in stress signal transduction. The fact that diverse physiological processes in plants are regulated by calcium indicates a complex mechanism by which this ion could regulate an array of processes.

ACKNOWLEDGMENTS

I thank Dr. Farida Safadi for critical reading of the manuscript and Nadine Kuehl for typing the references section. Research in my laboratory is supported by a grant from Agricultural Experiment Station (Project 702), Plant Biotechnology Laboratory, Colorado Biotechnology Research Institute, Colorado RNA Center, and a Biomedical Research Support Grant.

REFERENCES

1. J. S. Boyer, *Science*, *218*: 443 (1982).
2. R. Fraley, *Bio/Technology*, *10*: 40 (1992).
3. B. J. C. Cornelissen and L. S. Melchers, *Plant Physiol.*, *101*: 709 (1993).
4. J. A. Hellebust, *Annu. Rev. Plant Physiol.*, *27*: 485 (1976).
5. C. L. Guy, *Annu. Rev. Plant Physiol. Plant Mol. Biol.*, *41*: 187 (1990).
6. C. Lin, W. W. Guo, E. Everson, and M. F. Thomashow, *Plant Physiol.*, *94*: 1078 (1990).
7. S. Ramagopal, *Proc. Natl. Acad. Sci. USA*, *84*: 94 (1987).
8. E. Vierling, *Annu. Rev. Plant Physiol. Plant Mol. Biol.*, *42*: 579 (1991).
9. J. P. Carr and D. F. Klessing, in *Genetic Engineering: Principles and Methods* (J. K. E. Setlow, ed.), Plenum Press, New York (1989).

10. S. J. Roux and R. D. Slocum, in *Calcium and Cell Function* (W. Y. Cheung, ed.), Academic Press, New York (1982).
11. P. K. Hepler and R. O. Wayne, *Annu. Rev. Plant Physiol.*, *36*: 397 (1985).
12. B. W. Poovaiah and A. S. N. Reddy, *CRC Crit. Rev. Plant Sci.*, *6*: 47 (1987).
13. B. W. Poovaiah, J. J. McFadden, and A. S. N. Reddy, *Physiol. Plant.*, *71*: 401 (1987).
14. B. W. Poovaiah, G. M. Glenn, and A. S. N. Reddy, *Hortic. Rev.*, *10*: 107 (1988).
15. D. M. Roberts, T. J. Lukas, and D. M. Watterson, *CRC Crit. Rev. Plant Sci.*, *4*: 311 (1986).
16. A. J. Trewavas and S. Gilroy, *Trends Genet.*, *7*: 356 (1991).
17. D. M. Roberts and A. Harmon, *Annu. Rev. Plant Physiol. Plant Mol. Biol.*, *43*: 375 (1992).
18. B. W. Poovaiah, A. S. N. Reddy, G. An, Y. J. Choi, and Z. Q. Wang, in *Progress in Plant Growth Regulation* (C. M. Karssen, L. C. VanLoon, and D. Vreugdenhil, eds.), Kluwer Academic Publishers, Dordrecht, The Netherlands, (1992).
19. B. W. Poovaiah and A. S. N. Reddy, *CRC Crit. Rev. Plant Sci.*, *12*: 185 (1993).
20. D. S. Bush and R. L. Jones, *Plant Physiol.*, *93*: 841 (1990).
21. S. Gilroy and R. L. Jones, *Proc. Natl. Acad. Sci. USA*, *89*: 3591 (1992).
22. M. R. Knight, A. K. Campbell, S. M. Smith, and A. J. Trewavas, *Nature*, *352*: 524 (1991).
23. J. Lynch, V. W. Polito, and A. Läuchli, *Plant Physiol.*, *90*: 1271 (1989).
24. M. R. Knight, S. M. Smith, and A. J. Trewavas, *Proc. Natl. Acad. Sci. USA*, *89*: 4967 (1992).
25. C. A. Gehring, D. A. Williams, S. H. Cody, and R. W. Parish, *Nature*, *345*: 528 (1990).
26. M. R. McAinsh, C. Brownlee, and A. M. Hetherington, *Nature*, *343*: 186 (1990).
27. M. R. McAinsh, C. Brownlee, and A. M. Hetherington, *Plant Cell*, *4*: 1113 (1992).
28. S. Gilroy, M. D. Fricker, N. D. Read, and A. J. Trewavas, *Plant Cell*, *3*: 333 (1991).
29. M. Wang, B. Van Duijn, and A. W. Schram, *FEBS Lett.*, *278*: 69 (1991).
30. C. A. Gehring, H. R. Irving, and R. W. Parish, *Proc. Natl. Acad. Sci. USA*, *87*: 9645 (1990).
31. P. V. Minorsky, *Plant Cell Environ.*, *8*: 75 (1985).
32. J. Lynch and A. Läuchli, *Plant Physiol.*, *87*: 351 (1988).
33. E. Perez-prat, M. L. Narasimhan, M. L. Binzel, M. A. Botella, Z. Chen, V. Valpuesta, R. A. Bressan, and P. M. Hasegawa, *Plant Physiol.*, *100*: 1471 (1992).
34. J. Grace, *Plant Responses to Wind*, Academic Press, London (1977).
35. M. J. Jaffe, *Planta*, *114*: 143 (1973).
36. M. R. Knight, N. D. Read, A. K. Campbell, and A. J. Trewawas, *J. Cell Biol.*, *121*: 83 (1993).
37. S. Y. Wang, C. Y. Wang, and A. R. Welburn, *Stress Responses in Plants: Adaptation and Acclimation Mechanisms*, Wiley-Liss, New York (1990).
38. P. W. Morgan, *Stress Responses in Plants: Adaptation and Acclimation Mechanisms*, Wiley-Liss, New York (1990).
39. S. Gilroy, M. D. Fricker, N. D. Read, and A. J. Trewavas, *Plant Cell*, *3*: 333 (1991).
40. J. I. Schroeder and S. Hagiwara, *Proc. Natl. Acad. Sci. USA*, *87*: 9305 (1990).
41. J. Braam and R. W. Davis, *Cell*, *60*: 357 (1990).
42. J. Braam, *Proc. Natl. Acad. Sci. USA*, *89*: 3213 (1992).
43. Y. T. Lu and M. H. Harrington, *Plant Physiol. (Suppl).* *96*: 397 (1991).
44. M. R. Knight, A. K. Campbell, S. M. Smith, and A. J. Trewavas, *FEBS Lett.*, *282*: 405 (1991).
45. J. F. Harper, M. R. Sussman, G. E. Schaller, C. Putnam-Evans, H. Charbonneau, and A. C. Harmon, *Science*, *252*: 951 (1991).
46. J. F. Harper, B. M. Binder, and M. R. Sussman, *Biochemistry*, *32*: 3282 (1993).
47. K.-L. Suen and J. H. Choi, *Plant Mol. Biol.*, *17*: 581 (1991).
48. V. Raz and R. Fluhr, *Plant Cell*, *4*: 1123 (1992).
49. V. Raz and R. Fluhr, *Plant Cell*, *5*: 523 (1993).
50. A. S. N. Reddy, D. Takezawa, H. Fromm, and B. W. Poovaiah, *Plant Sci.* *94*: 109–117.
51. T. N. Davis, *Cell*, *71*: 557 (1992).
52. N. D. Moncrief, R. H. Krestinger, and M. Goodman, *J. Mol. Evol.*, *30*: 522 (1990).
53. V. Ling and R. E. Zielinski, *Plant Mol. Biol.*, *22*: 207 (1993).
54. D. Bartling, H. Butler, and E. W. Weiler, *Plant Physiol.*, *102*: 1059 (1993).
55. W. Y. Cheung, *Science*, *207*: 19 (1980).
56. T. N. Davis, M. S. Urdea, F. R. Masiarz, and J. Thorner, *Cell*, *47*: 423 (1986).

57. Y. S. Babu, C. E. Bugg, and W. J. Cook, *J. Mol. Biol.*, *204*: 191 (1988).
58. K. T. O'Neil and W. F. DeGrado, *Trends Biochem. Sci.*, *15*: 59 (1990).
59. A. R. Means, M. F. A. VanBerkum, I. Bagchi, K. P. Lu, and C. D. Rasmussen, *Pharm. Ther.*, *50*: 255 (1991).
60. P. K. Jena, A. S. N. Reddy, and B. W. Poovaiah, *Proc. Natl. Acad. Sci. USA*, *86*: 3644 (1989).
61. K. P. Lu and A. R. Means, *Endocr. Rev.*, *14*: 40 (1993).
62. V. Ling, I. Perea, and R. E. Zielinski, *Plant Physiol.*, *96*: 1196 (1991).
63. I. Y. Perea, E. Zielinski, *Plant Mol. Biol.*, *19*: 649 (1992).
64. M. C. Gawienowski, D. Szymanski, I. Y. Perera, and R. E. Zielinski, *Plant Mol. Biol.*, *22*: 215 (1993).
65. C. B. Klee, *Neurochem. Res.*, *16*: 1059 (1991).
66. R. J. Colbran and T. R. Soderling, *Curr. Top. Cell Regul.*, *31*: 181 (1990).
67. K. A. Alexander, B. T. Wakim, G. S. Doyle, K. A. Walsh, and D. Storm, *J. Biol. Chem.*, *263*: 7544 (1988).
68. J. R. Geiser, D. van Tuinen, S. E. Brockerhoff, M. M. Neff, and T. N. Davis, *Cell*, *65*: 949 (1991).
69. V. Ling and S. M. Assmann, *Plant Physiol.*, *100*: 970 (1992).
70. S.-H. Oh, H.-Y. Steiner, D. K. Dougall, and D. M. Roberts, *Arch. Biochem. Biophys.*, *297*: 28 (1992).
71. S. Muto and S. Miyachi, *Plant Physiol.*, *59*: 55 (1977).
72. J. M. Anderson, H. Charbonneau, H. P. Jones, R. O. McCann, and M. J. Cormier, *Biochemistry*, *19*: 3113 (1980).
73. G. Baum, Y. Chen, T. Arazi, H. Takatsuji, and H. Fromm, *Plant Physiol. (Suppl.) Abstr. 69*, *102*: 14 (1993).
74. J. M. Sikela and W. E. Hahn, *Proc. Natl. Acad. Sci. USA*, *84*: 3038 (1987).
75. Y. Nishizuka, *Nature*, *334*: 661 (1988).
76. A. C. Nairn, H. C. Hemmings, and P. Greengard, *Annu. Rev. Biochem.*, *54*: 931 (1985).
77. C. Putnam-Evans, A. C. Harmon, and M. J. Cormier, *Biochemistry*, *29*: 2488 (1990).
78. A. C. Harmon, C. Putnam-Evans, and M. J. Cormier, *Plant Physiol.*, *83*: 830 (1987).
79. G. E. Schaller, A. C. Harmon, and M. R. Sussman, *Biochemistry*, *31*: 1721 (1992).
80. A. S. N. Reddy, Z. Q. Wang, Y. J. Choi, G. An, A. J. Czernik, and B. W. Poovaiah, "Calmodulin Gene Expression and Calcium/Calmodulin-Dependent Protein Kinase II in Plants," *Proceedings of Cold Spring Harbor Symposium on Plant Signal Transduction*, Cold Spring Harbor, N.Y. (1991).
81. B. Watillon, R. Kettmann, P. Boxus, and A. Burny, *Plant Sci.*, *61*: 227 (1992).
82. B. Watillon, R. Kettmann, P. Boxus, and A. Burny, *Plant Physiol.*, *101*: 1381 (1993).
83. A. Dietrich, J. E. Mayers, and K. Hahlbrock, *J. Biol. Chem.*, *265*: 6360 (1990).
84. E. J. Resendez, J. W. Attenello, A. Grafsky, C. S. Chang, and A. S. Lee, *Mol. Cell Biol.*, *5*: 1212 (1985).
85. C. Stratowa and W. J. Rutter, *Proc. Natl. Acad. Sci. USA*, *83*: 4292 (1986).
86. B. A. White and C. Bancroft, *Calcium and Cell Function*, Academic Press, New York, Vol. VII (1987).
87. P. K. Dash, K. A. Karl, M. A. Colicos, R. Prywes, and E. R. Kandel, *Proc. Natl. Acad. Sci. USA*, *88*: 5061 (1991).
88. M. S. Kapiloff, J. M. Mathis, C. A. Nelson, C. R. Lin, and M. G. Rosenfeld, *Proc. Natl. Acad. Sci. USA*, *88*: 3710 (1991).
89. B. M. Rayson, *J. Biol. Chem.*, *266*: 21335 (1991).
90. M. Sheng, M. A. Thompson, and M. E. Greenberg, *Science*, *252*: 1427 (1991).
91. M. Wegner, Z. Cao, and M. G. Rosenfeld, *Science*, *256*: 370 (1992).
92. A. G. Ryazanov, E. A. Shestakova, and P. G. Natapov, *Nature*, *334*: 170 (1988).
93. R. Z. Ryazanov and A. S. Spirin, *New Biol.*, *2*: 843 (1990).
94. O. Nygard, A. Nilsson, U. Carlberg, L. Nilsson, and R. Amons, *J. Biol. Chem.*, *266*: 16425 (1991).
95. C. O. Brostrom, K. V. Chin, W. L. Wong, C. Cade, and M. A. Brostrom, *J. Biol. Chem.*, *264*: 1644 (1989).

96. M. A. Brostrom, X. Lin, C. Cade, D. Gmitter, and C. O. Brostrom, *J. Biol. Chem.*, *264*: 1638 (1989).

97. K.-V. Chin, C. Cade, C. O. Brostrom, E. M. Galuska, and M. A. Brostrom, *J. Biol. Chem.*, *262*: 16509 (1987).

98. E. H. Fawell, I. J. Boyer, M. A. Brostrom, and C. O. Brostrom, *J. Biol. Chem.*, *264*: 1650 (1989).

99. E. Lam, M. Benedyk, and N.-H. Chua, *Mol. Cell. Biol.*, *9*: 4819 (1989).

100. G. Neuhaus, C. Bowler, R. Kern, and N.-H. Chua, *Cell*, *73*: 937 (1993).

101. M. A. Stevenson, S. K. Calderwood, and G. M. Hahn, *Biochem. Biophys. Res. Commun.*, *137*: 826 (1986).

102. I. A. S. Drummond, S. A. McClure, M. Poenie, R. Y. Tsein, and R. A. Steinhardt, *Mol. Cell. Biol.*, *6*: 1767 (1986).

103. S. K. Calderwood, M. A. Stevenson, and G. M. Hahn, *Radiat. Res.*, *113*: 414 (1988).

104. S. F. Yang, *HortScience*, *29*: 41 (1985).

105. C. Lin and M. F. Thomashow, *Plant Physiol.*, *99*: 519 (1992).

106. S. S. Mohapatra, L. Wolfraim, R. J. Poole, and R. S. Dhindsa, *Plant Physiol.*, *89*: 375 (1989).

107. A. F. Monroy, Y. Castonguay, S. Laberge, F. Sarhan, L. P. Vezina, and R. S. Dhindsa, *Plant Physiol.*, *102*: 873 (1993).

108. A. F. Monroy, F. Sarhan, and R. S. Dhindsa, *Plant Physiol.*, *102*: 1227 (1993).

109. D. A. Williams, P. L. Becker, and F. S. Fay, *Science*, *235*: 1644 (1988).

110. P. Nicotera, D. J. McConkey, D. P. Jones, and S. Orrenius, *Proc. Natl. Acad. Sci. USA*, *86*: 453 (1989).

111. B. D. Birch, D. L. Eng, and J. D. Kocsis, *Proc. Natl. Acad. Sci. USA*, *89*: 7978 (1992).

35

Regulation of Gene Expression During Abiotic Stresses and the Role of the Plant Hormone Abscisic Acid

Elizabeth A. Bray

University of California, Riverside, California

I. INTRODUCTION

The plant response to the environment is a complex set of processes. Changes in the environment result in plant responses at many different levels: morphological, physiological, cellular, and metabolic. The type of response depends on the source of the stress, the duration and severity of the stress, the genotype of the stressed plant, the stage of development, and the organ and cell type in question. To fully understand plant stress, the mechanism by which these responses are regulated and the function of the responses must be characterized and understood. To further understand plant stress, researchers have turned to the study of molecular responses. In this chapter, stresses discussed are water deficit, salt stress, and low-temperature stress. A common link between these stresses is loss of cellular water.

As a result of the complex nature of the responses, it is difficult to determine the role of the various responses with respect to resistance of the plant to the loss of water. Responses may be adaptive, contributing to the ability of plant to withstand stress, may not be involved in adaptation, or may be a result of injury. Among the many responses to changes in the environment are alterations in the pattern of gene expression. Changes in gene expression are an important part of the plant response to the environment. Although some responses, possibly short-term metabolic and physiological responses, may not require changes in gene expression, the majority of responses to the environment are predicted to require alterations in gene expression.

As there are a complex of responses to the environment, the mechanisms that control the plant response to the environment are also expected to be complex. It is important to understand the cues from the environment that are detected as stress and to understand the signaling mechanism(s) within the plant at the whole plant and cellular levels. One plant signal that is prominent in plant stress studies is the plant hormone abscisic acid (ABA). The concentration of ABA increases in the plant during stress [1]. The best studied of the changes in ABA concentration is in response to drought stress. ABA levels increase in response to lowered water potential, and it is postulated that the loss of turgor is the trigger that induces ABA biosynthesis [2]. ABA levels have also been observed to rise in response to salt stress and low temperature. The signal ABA is common to all of the stresses discussed in this chapter, although there are differences in the pattern and magnitude of ABA accumulation in response to the different stresses. These similarities have led to the suggestion that some of the responses to the various stresses are similar and play similar roles in the ability of the plant to withstand periods of water deficit imposed by different environmental stresses.

In recent years, a number of genes have been identified that are induced when plants or plant parts are subjected to stresses resulting in cellular water deficit. Yet the function or role that most of these gene products play is still elusive. Some of the changes in gene expression may be adaptive, having functions that promote plant survival during water deficit, but this cannot be assumed. The function of many genes can be predicted based on the amino acid sequence of the gene product deduced. Clues to function may also be obtained from expression characteristics such as timing of expression, and organ, tissue, cellular, and subcellular location. However, the significance of gene expression with respect to stress tolerance cannot always be predicted.

At this time, most researchers have concentrated on identifying genes that respond to stress, beginning to determine the function, and studying their regulation. In this chapter, genes that are regulated by water deficit will be categorized based on functional predictions. Utilization of the techniques of plant molecular biology are providing new insights into plant responses to the environment with respect to the function of these changes and the regulation of these responses, and the commonalty of the response among different stresses and among different species.

II. PREDICTED FUNCTIONS OF STRESS- AND ABA-INDUCED GENE PRODUCTS

Many different genes, encompassing many classes of gene products, are induced by abiotic stresses. To fully understand the significance of these changes in gene expression, the function of the stress-induced gene products and the mechanism by which these genes are regulated must be understood. Many of these stress-induced genes are responsive to ABA application. In terms of function, these genes can be divided into different classes based on their DNA sequence, expression characteristics, and/or predicted functions. Unfortunately, at this writing, there are few examples in which the in vivo function of the gene product has been demonstrated. In many cases, functions have been predicted based on deduced amino acid sequence, but no biochemical or physiological data has been obtained to prove the function in vivo. Therefore, several of the gene classifications are based on predicted functions, and as the true functions are determined these categories may need to be corrected. The classifications used for abiotic-stress-induced genes are hydrophilic gene products (Figure 1), enzymes, those with other predicted functions, and those for which a function has not been predicted.

A. Hydrophilic Proteins Predicted to Have a Protective Function

A number of genes induced during periods of water deficit have been identified that encode proteins which are overwhelmingly hydrophilic and are soluble upon boiling, and are therefore expected to be located in the cytosol. These characteristics have led to the prediction that these gene products are involved in protecting cellular structures and components from dehydration associated with water deficit, salinity, and low-temperature stress. Many of these genes were first shown to be expressed during seed desiccation, the period of seed development following maturation, and are referred to as late embryogenesis abundant (*lea*) genes [3,4]. Dure et al. [4] established three groups of *lea* genes based on the publication of homologous genes found

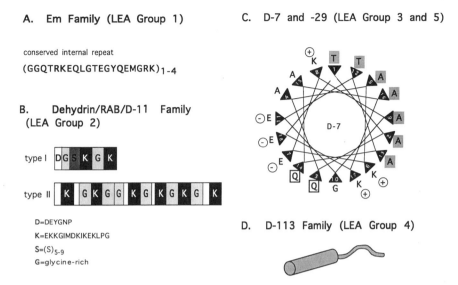

A. Em Family (LEA Group 1)

conserved internal repeat

(GGQTRKEQLGTEGYQEMGRK)₁₋₄

B.   Dehydrin/RAB/D-11   Family
(LEA Group 2)

type I

type II

D=DEYGNP
K=EKKGIMDKIKEKLPG
S=(S)₅₋₉
G=glycine-rich

C. D-7 and -29 (LEA Group 3 and 5)

D-7

D. D-113 Family (LEA Group 4)

Figure 1 Unique aspects of several of the overwhelmingly hydrophilic proteins that accumulate during periods of water deficit. (A) Conserved element that may be repeated in the Em family of proteins [3–7]. (B) Conserved amino acid motifs found in the dehydrin/RAB/D-11 family of hydrophilic proteins. Two types have been noted, those with the polyserine region (S) and those without [3,8–10,19,24]. (C) Conserved 11-amino acid repeat found in D-7 which may form an amphipathic α-helix. At position 12 in the wheel, the amino acid sequence is repeated. Positive and negative amino acids are noted with plus and minus signs. Amino acids with an amide group are boxed. Apolar amino acids are shown in the shaded box. Note that one face is apolar and the other has a pattern on negatively charged, amide-containing, positively charged amino acids [3,25]. A similar pattern of amino acids with these characteristics is repeated in D-29, although the exact amino acids are not conserved [3,4,25,26]. (D) The D-113 family of proteins has a conserved structure with an α-helix in the amino-terminal domain followed by a random coil region at the carboxy terminus [19,26].

in seeds of other species. This classification is useful because of the preponderance of different names for genes within these classes. However, it must be acknowledged that the name *lea* for genes expressed during stress in vegetative organs may not be appropriate because a number of these stress-induced genes are not expressed during seed development. It was suggested by Dure et al. [4] that the name WSP (water stress proteins) be applied to proteins of this class for that reason. However, this naming system has not been adopted, possibly because the name WSP also has problems; this name does not acknowledge that these proteins may be induced by other stresses, such as low-temperature or salt stress, and it does nothing to add to our understanding of the function of these genes. The genes discussed in this section on hydrophilic gene products include genes that have been identified after periods of stress in vegetative organs and share significant DNA sequence homology with *lea* genes from cotton. In addition, several new genes have been identified that are stress induced, overwhelmingly hydrophilic, yet have not been found in desiccating cotton seeds.

1. Em Family (Group 1)

This family of proteins is hydrophilic with no notable structural domains predicted by the amino acid sequence [5]. Cysteine and tryptophan are not found in the proteins of this group. The entire deduced protein sequence is well conserved among all the genes that have been identified in monocots and dicots. An internal repeat of a hydrophilic amino acid motif GGQTRKEQLGEEGYREMGHK was found in the barley genes and may be repeated up to four times [6] (Figure 1). A similar 20-amino acid motif is duplicated in the cotton gene *leaA2*, corresponding to the cDNA D-32 [7], although most of the genes identified thus far contain only one of these motifs, including the gene first published, D-19 [3]. It has been suggested that group 1 proteins function in a water-binding capacity, creating a protective aqueous environment [6].

2. Dehydrin/RAB/D-11 Family (Group 2)

Group 2 proteins have been called dehydrin [8], RAB (Responsive to ABA) [9], and D-11 [3]. These proteins are also overwhelmingly hydrophilic. There is a characteristic lysine-rich region with the consensus amino acid sequence EKKGIMDKIKEKLPG, which is repeated at least two times, once at the carboxy terminus and once internally [10]. These genes are expressed in seeds and in response to water deficit, salt stress, and low-temperature treatments in vegetative tissues [11,12]. These genes are also ABA regulated. The mRNA has been found in all stressed organs that have been investigated. In tomato, maize, and *Arabidopsis*, family members have been shown to be regulated by elevated levels of endogenous ABA during periods of dehydration in leaves [13–15]. In *Craterostigma plantagineum*, the protein DSP16 is localized in the cytoplasm, as determined by immunocytolocalization [16]. Close et al. [10] have observed that dehydrins are localized in the cytoplasm and the nucleus of aleurone layers, but primarily in the cytoplasm of root and shoot cells.

This class of genes has been identified more frequently than any other stress-induced gene in plants. Members of this family of proteins have been found in grasses and in dicots, and there is also immunological evidence that these proteins accumulate in *Anabeana* [17,18]. Tables of these genes are presented in Close et al. [10] and Dure [19]. It has recently been recognized that this class of proteins can be divided into at least two types. In type I, the internal lysine-rich signature motif is adjacent to a polyserine region. The polyserine region is a site for phosphorylation [20,21]. The lysine-rich signature motif is also found at the carboxy terminus. There is another conserved motif, DEYGNP, that is found one or two times near the amino terminus. However, gene products with the consensus lysine-rich repeat have been identified that do not have the polyserine region, referred to as type II in Figure 1. These genes are made up of the consensus lysine-rich repeat alternating with glycine-rich repeats. The glycine-rich repeats do not have a consensus sequence; they are characterized only by the abundance of glycine residues. These genes have been identified in wheat [22,23] and in alfalfa [24]. Interestingly, the alfalfa genes are induced by low-temperature treatments but not by drought stress or by ABA application [24]. In wheat, the gene, which encodes a 39-kDa protein, is induced preferentially by low-temperature treatments.

3. Groups 3 and 5

Group 3 proteins, represented by D-7 from cotton, and group 5 proteins, represented by D-29 from cotton, contain repeated tracts of 11 amino acids [4]. Although the 11-mer repeat has diverged among different species, a functional consensus was derived based on the polarity, charge, or methylation of the amino acid (Figure 1). The periodicity of 11 indicates that there may be an amphiphilic α-helix formed by polar and apolar amino acids aligned on different faces of the α-helix [3,4,25]. The D-7 protein is found uniformly in the mature cotton embryo, calculated to be from 200 to 300 μM in the cytoplasm immediately prior to desiccation [26]. Interestingly, in *Craterostigma plantagineum*, a protein with a similar 11-mer repeat was found to be localized in the chloroplast [16]. The abundance of these proteins has been used to rule out several possible functions—these proteins are not expected to function as enzymes, structural (architectural) proteins, regulatory proteins, or as ion or water transport proteins [26]. The prediction has been made that these proteins function in the sequestration of ions during cellular dehydration [25].

4. Group 4

It has been discovered that another gene that is expressed in drying cotton seeds accumulates to high levels [26]. This protein, D-113, has a homolog in tomato, LE25 [11,27], whose mRNA is expressed in vegetative tissues in response to drought stress, and elevated levels of endogenous ABA are required for its expression [14]. Members of the D-113 family are biased toward alanine residues and contain the random-coil-promoting residues glycine and threonine. It is predicted that the amino-terminal portion of the molecule is made up of an α-helix (Figure 1). The remainder of the molecule has no predicted structure [19]. It has been proposed that D-113 behaves as a surrogate water film in the desiccated state, stabilizing the intracellular surface of seeds [26]. It is uncertain if this would also be a role for D-113-like proteins in vegetative tissues of plants such as tomato because these tissues are not capable of surviving desiccation.

5. LTI78, 65/RD29 Family

An additional family of genes, which have thus far only been identified in *Arabidopsis*, are also induced by water-deficit-based stresses and ABA [28–30]. There are two genes in *Arabidopsis* that are adjacent to each other in the genome. The deduced amino acid sequences result in 77.9- and 64.5-kDa proteins that are overwhelmingly hydrophilic [30]. These genes were identified by three different laboratories and have been named *lti78* and *lti65* [28], *rd29A* and *rd29B* [29,30], and *cor78* [31]. Thus far, no specific predictions about the function of these proteins have been made.

6. Glycine-Rich Proteins

A family of genes in alfalfa has been characterized which are regulated by ABA, low temperature, and drought. The gene products are predicted to be hydrophilic and are characterized by glycine-rich repeats [31–33]. There are no specific predictions for their role other than a general role in protection from cellular dehydration.

7. KIN1 and KIN2/COR6.6

Another protein family was identified during low-temperature stress in *Arabidopsis* and is encoded by at least two genes in *Arabidopsis* [34–36]. These proteins are of low molecular weight (6.6 kDa), are rich in alanine, glycine, and lysine, and are largely hydrophilic. These genes are also induced by ABA treatments [35]. The *kin2* gene is expressed strongly in response to drought stress and salinity [36]. A low degree of similarity with fish antifreeze protein was noted [35]; however, there is no functional evidence for the role of these proteins in a stress response.

A variety of genes that encode hydrophilic proteins are expressed in response to abiotic stresses. Many of the protein families have different structures, leading to predictions that these proteins are involved in ion sequestration, water binding, and other protective roles in the cytoplasm. At this time, biochemical studies are needed to confirm these predictions.

B. Genes Encoding Enzymes

The activity or the amount of many enzymes has been shown to be altered during drought stress [37]. In many cases the activity of an enzyme has been studied during stress but the gene has not been characterized. In this chapter, only enzymes in which the gene has been cloned and the mRNA corresponding to that gene has been shown to be elevated in response to drought are discussed. In the coming years, the number of enzymes that are cloned and shown to be regulated during stresses that impose water deficits is sure to grow. Although an enzymatic function can be predicted from the deduced amino acid sequence of a cloned gene, the role during stress may not be obvious or ensured by the presence of a transcript during stress. Until additional experiments are done, it cannot be determined if the activity of these enzymes promotes stress adaptation. Enzymes induced by water stress and involved in osmolyte accumulation, protein degradation, CAM, ion transport, and signal transduction are described.

1. Osmolyte Accumulation

During periods of water deficit, compatible solutes accumulate in the cell, resulting in a lower cellular osmotic potential. If the cellular water potential is more negative than that of the cell's environment, water will be taken up by the cell. This process, called osmotic adjustment, may occur in the field after a long-term drought stress [38]. Compatible solutes, or osmolytes, include inorganic ions; organic ions; soluble carbohydrates, including polyols; amino acids, particularly proline; and quaternary ammonium compounds such as betaines [37]. Several genes have been identified which code for enzymes that may be involved in the accumulation of osmolytes during osmotic stress; these include enzymes in the biosynthetic pathway of proline [39–41], glycine betaine [42], and polyols [43,44]. The plant biosynthetic pathway for proline contains two genes; one is a bifunctional enzyme, Δ^1-pyrroline-5-carboxylate synthetase, which has both γ-glutamyl kinase and glutamic-γ-semialdehyde dehydrogenase activities [40], and the other is Δ^1-pyrroline-5-carboxylate reductase (P5CR) [39–41]. Both of these genes are induced by salt stress, indicating that they may play a role in osmotic adjustment. In pea, P5CR is induced in roots, but not in shoots, of pea seedlings in response to salt stress [41]. The gene coding for betaine aldehyde dehydrogenase, the last step in glycine betaine synthesis, has been isolated and is induced by salt stress [42]. A gene encoding *myo*-inositol *O*-methyl transferase (*imt1*) was isolated from the facultative CAM plant *Mesembryanthemum crystallinum* and is induced by salt stress [44], drought, and low temperature [45]. This enzyme is involved in synthesis of the polyol, pinitol. Tobacco plants transformed with *imt1* driven by the 35S promoter accumulated ononitol, an osmolyte not synthesized in wild-type tobacco. Although stress studies were not reported, these results indicate that the enzyme encoded by *imt1* is a step in the biosynthetic pathway of pinitol [46]. Aldose reductase, an enzyme involved in sorbitol synthesis, has been identified in barley seeds, although this gene is not expressed in dehydrated barley leaves [43]. An NADPH-dependent aldose reductase is induced by ABA in bromegrass suspension cells with increased freezing tolerance [47]. In response to a number of stresses, enzymes involved in osmolyte accumulation are induced. These enzymes are probably involved in the accumulation of osmolytes, which promotes uptake of water into the cells through osmotic adjustment.

Vegetative storage protein genes, *vsp*, are induced by environmental stresses such as water deficit and wounding. The VSPs have been shown to accumulate preferentially in the vacuoles of paraveinal mesophyll cells of soybean. It has recently been shown that two of the proteins, VSPα and VSPβ, are acid phosphatases with the highest substrate specificity for tetrapolyphosphates [48]. Staswick et al. [49] reported that although the VSPs have acid phosphatase activity, they are not the major acid phosphatase in leaves. DeWald et al. [48] propose that VSPs are involved in amino acid uptake and temporary sequestration of amino acids in paraveinal mesophyll cells. Therefore, these proteins may play a role in osmotic adjustment during stress.

2. Proteases

Proteases have also been shown to be induced by abiotic stresses. A thiol protease is induced in pea by water deficit [50]. Two additional proteinases, cysteine proteinases, were identified in drought-stressed *Arabidopsis* [51]. These genes have the catalytic sites typical of cysteine proteinases and have amino-terminal signal peptides. They are not induced by ABA, or temperature stress, but are strongly induced by salt and drought stress [51]. These genes are similar to a cysteine protease induced by low temperature in tomato [52]. Their function during drought stress is not confirmed, but they may be involved in the degradation of polypeptides denatured during stress, processing of precursor proteins to the mature form, or degradation of vacuolar proteins, after which the amino acids may be used in synthesis of stress-induced proteins or osmotic adjustment [50,51].

3. Induction of CAM

In plants that have the capacity for Crassulacean acid metabolism (CAM), a switch from C3 metabolism to CAM may occur during periods of stress or in response to development. In *Mesembryanthemum crystallinum*, there is a 10- to 20-fold increase in phosphoenolpyruvate carboxylase (PEPCase) activity in response to salt stress [53]. PEPCase is the enzyme responsible for the primary fixation of CO_2 into oxaloacetate, which is subsequently converted to malate during CAM. In *M. crystallinum*, PEPCase is represented by two genes, one of which is salt-stress induced. However, induction does not occur in salt stressed suspension cultures of *M. crystallinum* [54]. CAM coupled with stomatal conductance provides a means for the improved stress tolerance that may occur in specialized plants.

4. Plasma-Membrane H^+-ATPase

During salt stress or water deficit, the concentration of ions in the cytoplasm must be controlled. The import of ions into the cytoplasm must be limited to the capacity to compartmentalize Na^+ and Cl^- into the vacuole. Active transport of ions is driven by an H^+ electrochemical gradient that is generated by the plasma-membrane H^+-ATPase [55]. Carriers or channels facilitate the active or passive transport of these ions, with active transport driven by the H^+-ATPase. During development of water deficit in soybean seedlings, there is an increase in H^+-ATPase mRNA levels in the roots only. This correlates with the ability of growth maintenance in this organ; in shoots in which growth is not maintained during water stress, there is not an increase in H^+-ATPase mRNA levels [56]. This may indicate that growth during stress is dependent on increased activity of ATPase and is associated with ion transport. In the halophyte, *Atriplex nummularia*, it has been demonstrated that the plasma-membrane H^+-ATPase is regulated by NaCl [57]. The ATPase activity may have an important role in stress tolerance and should be studied further.

5. Protein Kinase

For plants to respond to the environment, mechanisms must have evolved that signal changes from the environment at the cellular level. The pathway of information transfer from the environment to the cell, resulting in alterations in gene expression, is the signal transduction pathway. Many different component pathways may be required to achieve alterations in gene expression. In one type of signal transduction pathway, a protein kinase controls activation or deactivation of proteins by phosphorylation. A cDNA, PKABA1, corresponding to a protein kinase that is induced by ABA, has been isolated [58]. The deduced amino acid sequence has 12 catalytic subdomains found in serine/threonine protein kinases that are thought to be critical for phosphorylation. Two mRNAs hybridize to PKABA1 in dehydrated seedlings, and the accumulation of these mRNAs corresponds to an increase in ABA concentration. The rate of accumulation is similar to the accumulation of other drought- and ABA-induced genes; therefore, it is proposed that this kinase is involved in the phosphorylation of other ABA-induced gene products [58]. Although there are many other possibilities for the action of this gene product, the function cannot be determined until its substrate specificity is characterized.

C. Genes That Have Other Predicted Cellular Functions

For a plant to survive periods of stress, many developmental, physiological, and metabolic functions may need to be altered. A number of unique changes in gene expression have been identified, using gene cloning techniques. Many of these would not have been identified using other types of studies. Genes have been identified that are involved in antifungal activity, protection from freeze–thaw inactivation, transport of water, RNA binding, and gene regulation.

1. Antifungal Proteins

In addition to the direct effects of water stress, environmental stress may contribute to the susceptibility of the plant to pathogens. Genes encoding osmotin and nonspecific lipid transfer proteins have antifungal activity and are induced in response to water-deficit-based stresses.

A protein was first discovered that was prominent in tobacco cells adapting to high concentrations of salt. The accumulation of the protein is correlated with osmotic adaptation and therefore was named osmotin [59,60]. This protein has a signal sequence and is localized in vacuole inclusion bodies. It is a basic homolog of family 5 pathogenesis-related proteins. This protein family, including osmotin, has been shown to have antifungal activity [60–62]. Members of this family have been shown to permeabilize the fungal plasma membrane. Transgenic tobacco plants overproducing osmotin have been shown to be more tolerant of fungal attack than are control plants [60]. It is proposed that there is a specific interaction between osmotin and the membrane, possibly with a portion of the molecule interacting with the membrane, with the rest forming an ion or water channel that permeabilizes the fungal membrane [60]. It is not certain if this is the only role of osmotin or if it also plays a direct role in salt tolerance.

Proteins that have the capacity to transfer lipids from liposomes to mitochondria in vitro have been studied in plants. A class of these proteins transfers a number of different classes of lipids and have been called nonspecific lipid transfer proteins (nsLTPs) [63]. Three genes that are homologous to nsLTP genes have been found to be induced by stress in aerial plant parts: two from tomato [64,65] and one from barley [66]. However, the role for these genes during stress has not been determined. Since it was determined that nsLTPs are expressed in the epidermis of the shoot, it has been suggested that they play a role in cuticle formation [67]. It has now been demonstrated that an nsLTP-like protein isolated from radish seeds has antifungal activity; it inhibited fungal hyphae growth but did not affect spore germination [68]. Therefore, the role of nsLTPs during stress may be in the protection of the shoot from fungal attack. However, further studies are required to elucidate the function of nsLTP-like proteins during stress.

2. Protection from Freeze–Thaw Inactivation

Another type of protection is the protection of enzymes from freeze–thaw inactivation. The cor15 gene is induced by cold acclimation in *Arabidopsis* [69]. The polypeptide that is encoded by this gene is targeted to the chloroplast [70]. It is 100 times more effective than BSA at protecting lactate dehydrogenase from freeze–thaw inactivation [71]. This preliminary evidence indicates that this gene encodes a protein with a potential protective role in freeze-induced dehydration.

3. Protein Degradation

Cellular function may also be protected during stress by preventing protein degradation or degrading proteins that are no longer functional. Heat-shock proteins that are induced by water deficit [72,73] may be involved in the refolding of proteins to regain their function, or the prevention of protein aggregation [74] during stress. Protease inhibitors induced by water deficit may protect against proteases released after cellular disruption and membrane disorganization as a result of stress. The gene *Bnd22*, induced after prolonged dehydration stress in *Brassica napus*, has some characteristics of Künitz trypsin inhibitor, although it does not have all the signature amino acids. It is also expressed in a manner that is different from other Künitz trypsin inhibitors; it is not expressed in seeds, it is only expressed in the aerial parts of the plant after a prolonged drought stress [75]. Further studies are needed to determine if this protein is a protease inhibitor that plays a role during water deficit. Ubiquitin, which was also shown to be induced by water deficit [72], may target proteins for degradation that cannot regain function after water deficit.

4. Major Intrinsic Proteins

A class of proteins, major intrinsic proteins, have been identified which may form transmembrane channels and be involved in the transport of ions, other metabolites, or water across membranes. These proteins have six membrane-spanning domains that are postulated to form a channel. Drought-induced examples of this class of proteins have been isolated from pea [52] and *Arabidopsis* [76–79]. These genes are most closely related to each other, but are also similar to *nod26*, tonoplast intrinsic protein (TIP) from bean, bovine major intrinsic proteins (MIP), and glycerol facilitator protein [77]. *Arabidopsis* has at least four genes in this family, each with different expression characteristics. Two members of this family are induced by decreased turgor, TMPα and TMPβ [78,79]. It is predicted that the expression pattern reflects functional specialization of the various family members [77]. During drought stress, these channels may facilitate transport of ions, metabolites, or even water. Maurel et al. [80] demonstrated that γ-TIP is a water-channel protein when expressed in *Xenopus* oocytes. Although γ-TIP is not expressed during water stress, these results present some interesting possibilities for the drought-induced members of this class of proteins. It must be determined if the water-stress-induced proteins are also water-channel proteins, and in which membranes these proteins are located.

5. RNA-Binding Protein

A gene from maize, MA16, with many characteristics of the *leas*, contains a glycine-rich repeat and is developmentally regulated in seeds and induced in dehydrated vegetative tissues [81]. Unlike the LEAs, this protein contains a consensus sequence, RGFGFVTF, which is conserved in RNA-binding proteins. Ludevid et al. [82] used in vitro ribohomopolymer binding assays to confirm that MA16 has RNA-binding activity. In these studies, MA16 preferentially binds poly(G). There are many different functions for ribonucleoproteins, including regulation of alternative splicing, pre-mRNA processing, and mRNA translation and stability. At this time the stress response cannot be predicted because the RNAs that are associated with MA16 in vivo have not been determined. But it is possible that this protein has an important regulatory or protective function during plant stress.

6. Gene Regulation

In addition to signaling changes in the environment, the signal transduction pathway must include the induction or activation of transcription factors and activators that are required for transcription induction. A transcription factor that is involved in the induction of some of the

ABA-induced genes has been identified and isolated [83]. This protein EmBP-1 is a bZIP-type transcription factor and recognizes DNA sequences containing the DNA sequence CACGTGGC. A gene, *Alfin-1*, has been isolated from salt-tolerant alfalfa cells that has zinc finger motifs and may be a sequence specific-DNA-binding protein [84]. An H1 histone-like protein [85] and a nonhistone chromosomal protein [86] have also been shown to be induced by drought stress. The roles for these proteins have not been elucidated.

D. Genes for Which a Function Has Not Been Predicted

In addition to the genes which have a function that can be predicted from the DNA sequence and amino acid sequence, there are also many stress-induced genes for which no function has been predicted. A few of these types of genes are discussed below.

1. LEA Group 6

With the identification of *rab28* [87], a new *lea* group can be formed. Two representatives of group 6 have been identified, one from cotton, D-34 [3], and the other from maize, *rab28* [87]. The predicted gene product is different from the other LEA proteins, which are overwhelmingly hydrophilic; it has a balanced hydrophobicity plot. RAB28 is predicted to have four to six α-helices and a globular structure. Functional predictions have not been made.

2. RD22

A gene, *rd22*, induced by water stress, salinity, and ABA in *Arabidopsis* [88], was determined to be similar to an unidentified seed protein (USP) from *Vicia faba* [89]. RD22 is induced early during seed development but not during the late stages of embryogenesis, like USP. The homology between RD22 and USP is found in the carboxy-terminal portion of the protein [88]. RD22 has an amino-terminal hydrophobic region with five repeated sequences in the amino terminus. The consensus is TnVnVGnGGVnnnnnnKGK, which is predicted to contain a β sheet–turn–β sheet structure [88]. The significance of this structure or the function of this gene product is not known.

3. Germinlike Proteins

A root-specific protein that is induced in barley by salt stress was found to be similar to germin [90]. Germin is a protein of unidentified function that accumulates during the onset of growth following germination in wheat [91]. Germinlike proteins of barley accumulated only in roots after salt stress. The proteins are soluble upon boiling and are found mainly in the soluble fraction using cellular fractionation studies [90]. A similar protein was found in *Mesembryan-themum crystallinum* roots; however, it was found to decrease in response to salt stress. It is proposed that germinlike proteins are sensors of water status, and the expression of these genes may be involved in the control of growth, depending on plant water status [92].

4. Early Light-Induced Proteins

In desiccated *Craterostigma plantagineum* leaves, a gene, *dsp*-22, that is targeted to the chloroplast is induced [93]. The deduced amino acid sequence is similar to a protein called early light-induced protein (ELIP) from pea and barley [94,95]. The protein consists of alternating regions of hydrophobic and hydrophilic domains. There is a putative transit peptide at the amino terminus. Light is essential for the accumulation of this protein. ABA induction does not occur in the dark. DSP-22 may be involved in photoprotection of the photosystems or in maintaining assembled photosynthetic structures essential for resuming active photosynthesis following rehydration [93].

This description of genes that are induced by water deficit elicited by different types of abiotic stresses serves to indicate that there are indeed many changes in genes expression in response to changes in the environment. There is a complex of molecular processes that are altered by the environment. The function of the majority of the genes that have been characterized fall under the broad category of protection of cellular function. Protection is predicted to come from hydrophilic proteins in the cytoplasm, osmolyte accumulation, degradation of denatured proteins, and protection from pathogens. Genes involved in the regulation of other genes that are induced by stress have also been identified. These responses occur in many different plants, species that are tolerant and those that are not, and in response to different stresses that cause water deficit.

III. METHODS TO EVALUATE THE ADAPTIVE ROLES OF STRESS-INDUCED GENES

The expression of specific genes during stress implies that the genes are involved in stress tolerance. However, this may not be a valid assumption. It is possible that gene induction, in addition to promoting stress tolerance, is a result of an injury or a coincidence because of similar signal transduction pathways initiated by different stresses. Therefore, techniques must be developed to evaluate the adaptive significance of the expression of these genes. In many cases, to begin this evaluation it must first be determined if the gene products in fact accumulate during stress. In many of the studies that have been completed thus far, only the accumulation of transcripts has been studied. This does not always ensure that the protein will accumulate. Using osmotin as an example, it was shown that osmotin mRNA is not always translated even though it accumulates [96]. Once it has been established that the protein accumulates, further studies can be completed at the biochemical, genetic, and molecular levels. Several molecular techniques may be exploited to study the role of specific genes.

A. Under- and Overexpression of Specific Genes

One method to investigate the function of drought-induced gene is to over- and underexpress the genes in transgenic plants. The rationale behind this strategy is that the altered expression of these genes will alter stress tolerance if these genes play an essential role in tolerance. Three *lea*-like genes from *Craterostigma plantagineum* were overexpressed in tobacco driven by the 35S promoter from cauliflower mosaic virus (CaMV) [97]. The plants were tested for physiological traits that might indicate a difference in stress tolerance. An ion leakage test after PEG stress of leaflets was used to test for drought tolerance. However, there were no differences observed between the transgenic plants and the wild type. It is possible that the parameters measured were not useful for detecting differences in drought tolerance, that these proteins alone are not sufficient by themselves and need other proteins and/or osmolytes [97], or that these proteins do not function in drought tolerance.

The antisense strategy, in which antisense RNAs accumulate in transgenic plants in order to eliminate a specific protein, might also be used to evaluate gene function. This technique can be used to construct specific single gene mutants. At this time, antisense plants for stress-induced genes have not been evaluated and reported. However, the same problems as encountered above may also occur. There are multiple responses to stress, and the elimination of a single aspect of the stress response may not have a significant effect on plant stress tolerance.

B. Introduction of Foreign Genes

When a function is understood that promotes adaptation to stress, genes that have been identified previously may be exploited to improve plant stress tolerance. Thus far, there is one example of this. A gene isolated from *E. coli*, *mtlD*, which encodes mannitol-1-phosphate dehydrogenase, is involved in mannitol catabolism and leads to the production of fructose-6-phosphate. It was hypothesized that if this gene were expressed in plants, the enzyme may catalyze the reverse reaction, resulting in the synthesis of mannitol-1-phosphate, which would be a substrate for general phosphatases, resulting in the synthesis and accumulation of mannitol [98]. Transgenic tobacco plants, with *mtlD* driven by the 35S CaMV or NOS promoter, resulted in the accumulation of mannitol in young leaves and roots [98]. After 30 days of exposure to salinity, transgenic plants producing mannitol had a greater shoot height and root length than those of control plants [99]. Therefore, this is the first example of a transgenic plant, altered with a microbial gene, which has a greater resistance to plant osmotic stress. Further trials are required to determine the agricultural applicability of these plants. Other genes might also be exploited, especially for other strategies of osmotic adjustment. The use of these strategies in combination may improve stress tolerance of transgenic plants as well as our understanding of the importance of water-deficit responses to plant adaptation to the environment.

IV. ABA INDUCES SPECIFIC GENES DURING WATER DEFICIT

The concentration of ABA is altered by changes in the environment. Large increases in ABA concentration have been documented in response to drought stress, with lesser increases in ABA concentration occurring in response to salt and low-temperature stress. These changes occur at the cellular level, but the ABA that is transported throughout the plant at the whole plant level is also changed.

The ABA biosynthetic pathway occurs through the carotenoid biosynthetic pathway. The compound 9'-*cis*-neoxanthin is cleaved to result in the postcleavage intermediate to ABA, xanthoxin. Xanthoxin is oxidized to ABA-aldehyde, which is converted to ABA by ABA-aldehyde oxidase [100]. The step in the ABA biosynthetic pathway that is regulated by stress is likely to be the cleavage step, although this has not been proven because this gene has not been isolated and/or an assay has not been developed for that enzymatic step. The rate of ABA biosynthesis during stress is limited by the production of xanthoxin, not the conversion of xanthoxin to ABA [100].

The trigger that is recognized by the cell to induce ABA biosynthesis is not understood. ABA accumulation is correlated with a reduction in turgor to near zero [2]. Therefore, it is thought that the cellular mechanism for turgor perception is linked to the ABA biosynthetic pathway through a signal transduction pathway. Inhibition of transcription and translation prevent ABA accumulation in response to stress [10], indicating that these processes are required for the cell to recognize stress, or the ABA biosynthetic enzymes must be synthesized for ABA to accumulate. More effort is needed to understand the mechanism of stress-induced ABA accumulation.

Although it is not certain how ABA biosynthesis is controlled, it has been demonstrated that ABA is part of the signaling mechanism during stress that induces specific genes. This has been demonstrated using mutants that are deficient in ABA biosynthesis [14,15,87]. For most studies it is convenient to apply ABA and determine if the application of ABA causes an accumulation of specific transcripts. In the cases where it does, those genes are found to be ABA responsive, indicating that gene induction may occur in response to ABA. However, these studies do not prove that ABA is an endogenous signal used during specific stresses to induce particular genes. Application studies only indicate that ABA is one of the signals that the gene is capable of responding to. Inhibitors of carotenoid biosynthesis result in a decreased level of ABA [100], and responses of the plant that are reduced by inhibitor application have been used to analyze the role of ABA. But as with the use of other inhibitors, effects that are not caused directly by the reduction in ABA concentration may also occur, because carotenoid concentration is also reduced after application of these inhibitors. Mutants that are deficient in ABA biosynthesis and cannot accumulate ABA during stress can be used to identify genes that require elevated levels of endogenous ABA for expression. Mutants in tomato, maize, and *Arabidopsis*, which have specific blocks in the ABA biosynthetic pathway, have been used for this purpose. In each of these cases, several genes have been identified which require ABA for expression [14,15,87].

A. Identification of Genes That Require ABA for Expression

Mutants that are blocked in the ABA biosynthetic pathway have proven to be useful in the identification of genes that require elevated levels of ABA for expression. mRNAs for specific water-deficit-induced genes do not accumulate in response to a water deficit in ABA-deficient mutants as they do in the wild type. For example, the ABA-deficient mutant of tomato, *flacca*, is blocked in the last step of the ABA biosynthetic pathway [102]. This mutant does not accumulate as much ABA during periods of water deficit as does the wild type. After stress, the mutant accumulates 6% of the ABA that the wild type accumulates [103]. There are fewer proteins accumulating in *flacca* leaves than in the wild type during drought stress. Application of ABA to *flacca* restores the accumulation of this set of ABA-induced proteins [103]. The accumulation of three mRNAs was shown to be dependent on the accumulation of ABA during stress [14]. These mRNAs accumulated only in the drought-stressed wild type and were not detected in the drought-stressed ABA-deficient mutant. Mutants in maize and *Arabidopsis* have been used similarly to identify additional ABA-requiring genes (Table 1).

B. Genes That Are Responsive to ABA but Do Not Require ABA

The ABA-deficient mutants have been used to define an additional set of genes, those that are responsive to ABA but do not require ABA for expression [30,104,105]. These genes are induced by ABA application, but unlike the ABA-requiring genes, they are induced by low-temperature and water-deficit treatments in the ABA-deficint mutants of *Arabidopsis*. Therefore, it has been concluded that these genes do not require elevated levels of endogenous ABA for expression, but are capable of responding to ABA and may be called ABA-responsive genes. These results indicate that there are two pathways that can be followed to induce these

Table 1 Genes That Have Been Demonstrated to Be Regulated by Elevated Levels of Endogenous ABA Resulting from Environmental Stress Using ABA-Deficient Mutants of Maize, *Arabidopsis*, and Tomato

Genotype	Gene designation	Gene family	Stress induction[a]	Refs.
Maize	*rab*17	*dhn/rab*/group 2	D	13
	*rab*28	D-34/group 6	D	87
Arabidopsis	*rab*18	*dhn/rab*/group 2	D, L	15
	*lti*65	*rd29/cor78*	D, L	28, 29, 104
Tomato	*le*4	*dhn/rab*/group 2	D, S, L	11,14
	*le*16	nsLTP	D, S, L	14, 64
	*le*20	H1-histone	D, S, L	85
	*le*25	D-113/group 4	D, S, L	11, 14

[a] D, water deficit; L, low-temperature stress; S, salinity.

genes, but it is unknown if the pathways converge or if there are two entirely separate pathways. ABA applications have also been used to show that there are a number of water-deficit-induced genes that do not respond to ABA application [52,76]. These genes may be induced directly by the drought stress, or they may be controlled by other signaling mechanisms operating during water deficit.

C. DNA Elements That Confer ABA Responsiveness

The conditions under which a gene is induced is controlled by the DNA elements acting within each gene. Therefore, to understand the mechanism of regulation of a specific gene, the DNA elements that confer responsiveness and the factors that recognize those elements must be identified and characterized. Studies have been initiated on genes that are regulated by ABA to understand the basis of ABA-regulated gene expression during stress and seed development. A region of *Em*, a member of the group 1 *lea* family from wheat, was identified which confers ABA inducibility upon a minimal 35S CaMV promoter [106,107]. A chimeric gene was constructed with a 646-bp segment of *Em* and the reporter gene β-glucuronidase (GUS). When this gene was introduced into rice protoplasts, GUS activity was increased 15- to 30-fold after ABA application [106]. Further delineation of the 5′-flanking DNA of *Em* identified a 50-bp DNA sequence that is sufficient for ABA induction of GUS activity [107]. A DNA element, CACGTGGC, was conserved among other ABA-induced genes, includ- ing *rab16* [108]. However, this element is also related to G-box motifs, which are involved in light-induced gene expression [109]. This conserved element, which has been referred to as an ABA-responsive element (ABRE), has been shown to bind nuclear proteins [83,108]. A leucine zipper protein, EmBP-1, was identified and a cDNA cloned whose gene product binds the ABRE of *Em* in vitro [83]. This protein, EmBP-1, contains a leucine-zipper DNA binding motif, thought to be responsible for dimer formation, adjacent to a basic domain which is a cluster of positively charged amino acids responsible for sequence-specific recognition. The combination of the basic domain and the leucine zipper domain has been termed the bZIP domain. The basic domain of EmBP-1 is similar to that found in other transcription factors that bind the DNA sequence element T/CACGTGGC, including TAF-1 from tobacco, which binds the ABRE conserved in *rab16* genes [110], and GBF, which binds the G-box found in *rbcS* genes of tomato, *Arabidopsis* and pea [109,111] (Figure 2). It is similar to other transcription factors that bind a similar DNA sequence element, TCCACGTAGA [112–114]. An element has been identified that confers ABA inducibility in a transient assay system, and a factor has been identified that can bind this DNA ele- ment. However, the identification of several similar transcription factors that bind similar DNA elements indicates that additional aspects of ABA inducibility are yet to be under- stood.

Additional specificity of the DNA elements may be derived from nucleotides surrounding the core DNA sequence. Williams et al. [115] characterized the sequences flanking the G-box

```
EmBP-1    MDERELKRERRKQSNRESARRSRLRKQQ        5'-GACACGTGGC-3'
TAF-1     QNERELKREKRKQSNRESARRSRLRKQA        5'-GCCACGTGGC-3'
02        KMPTEERVRKRKESNRESARRSRYRKAA        5'-TCCACGTAGA-3'
```

Figure 2   Basic domain motif of three bZIP proteins: EmBP-1 [83], TAF-1 [110], and O2 [112]. Identical amino acids are shown in bold face type. The DNA-binding site for each of the bZIP is shown to the right of the amino acid sequence.

or ABRE core CACGTG to determine how those DNA sequences affected binding. Based on the flanking sequence, the G-box elements have been divided into two different classes to which bind two distinct classes of G-box binding proteins. Further characterizations were made by using ACGT as the core and defining three different types of boxes by the nucleotide surrounding the core. EmBP-1, TAF-1, and HBP-1a were found to have the greatest affinity for the CACGTGC sequence [116].

TAF-1 from tobacco [110] and EmBP-1 [83] from wheat both bind DNA elements with the core sequence CACGTGGC and have the same DNA-binding basic motif. However, neither motif I, TACGTGGC, [108] nor the hex tetramer, GGTGACGTGGC, can confer ABA responsiveness on a GUS reporter gene in transgenic tobacco [117]. But a hex mutant, GGACGCGTGGC, with greatly reduced affinity for TAF-1 can confer ABA responsiveness on a GUS reporter gene in transgenic tobacco [117]. These results indicate that there are multiple factors that can bind similar DNA motifs and that the exact DNA sequence determines which factors will recognize it. In addition, although factors have similar amino acid sequences within the DNA-binding domain, this information cannot be used to predict the exact DNA elements these factors will bind in vivo. Other regions of the transcription factor besides the DNA binding domain must also be important for determining DNA-binding specificity.

Since the studies on *Em*, other ABA-regulated genes have also been investigated. Another *lea* gene, *rab28*, is known to be regulated by endogenous ABA in maize [87]. This gene is expressed in vegetative organs during periods of water deficit, and it has been demonstrated that an ABRE is involved in the regulation of *rab28* by ABA and water stress [118]. Chimeric genes, with a portion of the *rab28* 5′-flanking DNA containing the ABRE and GUS, transfected into rice protoplasts were ABA responsive. An in vitro dimethyl sulfate footprinting experiment identified guanine residues within the ABRE that are involved in binding nuclear proteins [118]. Interestingly, when electrophoretic mobility shift assays were completed with proteins isolated from seeds or from drought-stressed leaves, complexes of two different sizes were found. Both complexes were shown to bind the ABRE [118]. Therefore, it is proposed that the same DNA sequence element or ABRE is involved in regulation of the expression of *rab28* in the seed during development and in the leaf during stress, but the transcription factors and activators that are involved in these two different types of regulation are not identical. Similarly, it has been shown that factors that bind the G-box are complexes made up of at least two different proteins [119]. A complex array of events is required for ABA-regulated expression to occur in response to stress and/or developmental cues.

In transgenic tobacco plants it has been shown that expression of drought-regulated genes are properly regulated in seeds, but not in response to stress [120,121]. When –482 to +184 of *rab16B* from rice was translationally fused to GUS, expression was limited to developing seeds and was not induced by ABA or water stress in vegetative tissues. Expression of a related gene from rice, *rab16A*, was not detected in seeds or vegetative tissues [120]. These results may indicate that elements recognized in rice for ABA-regulated expression cannot be used in tobacco. However, another family member derived from maize, *rab17*, is correctly expressed in transgenic tobacco plants when –1330 to +29 is included in the reporter gene fusion constructs [122]. This gene also contains an ABRE, and only when that sequence is present in deletion constructs is the reporter gene responsive to ABA [122].

Promoter deletion analyses for several additional ABA-regulated genes are in progress [121]. In the resurrection plant, *Craterostigma plantagineum*, DNA elements required for regulation of the gene CDeT27-45, which is similar in amino acid sequence to the cotton gene *lea14*, were characterized using promoter deletion analyses. Using a transient expression system and *C. plantagineum* protoplasts, a region between –282 and –197 of the promoter was demonstrated to be required for ABA-regulated expression. Similar studies were completed on transgenic tobacco plants carrying this gene and it was found that there was not ABA-induced expression in tobacco leaves, although the genes were expressed during development in seeds and anthers. In the region required for ABA induction of CDet27-45 expression, there are no ABRE-like elements. Therefore, although the ABRE is found in many genes that are expressed during seed development and in response to ABA application, it is not found in all genes that are in the ABA-requiring category. As another example, *le16*, a gene expressed in wild-type tomato but not in the ABA-deficient mutant, does not contain a consensus ABRE [64]. Therefore, it is expected that there are multiple DNA elements involved in ABA-regulated expression. In addition, genes that are expressed during drought are also expressed during specific developmental stages and in specific cell types. Therefore, additional elements are required to control tissue- and organ-specific expression and other specific aspects of the expression pattern during water deficit.

D. Recognition of ABA at the Cellular Level

Although the responsiveness of the gene is controlled by the DNA elements within the gene, the pathways that lead to transcription factor binding are also important to understand. For the ABA-requiring and ABA-responsive genes, there must be an ABA recognition event followed by the activation of a pathway that leads to gene induction. Cellular conditions that are required for any of these events to occur are not understood. The response of the cell to ABA may be altered by the physiological state of the cell. For example, sensitivity to ABA is altered by the osmotic potential of the cells; there is increased sensitivity to ABA with increased osmotic stress. In some cases, osmoticum can completely replace exogenous ABA. For *Em* mRNA accumulation in rice cell cultures, increasing concentrations of NaCl increased the accumulation of *Em* mRNA in response to suboptimal concentrations of ABA [123]. Therefore, the cells' response to ABA may be altered by the water potential or water content of the cell. However, another possibility should be considered. Because the cells are induced to accumulate ABA in response to osmotic stress, it becomes difficult to determine if the newly synthesized ABA is contributing to the induction of genes. The ABA that is synthesized in the cell may not be located in the same compartment within the cell as ABA that is applied to the cell [124]. Therefore, the plant may be more sensitive to endogenous ABA than to applied ABA. It is known that high levels of ABA must be applied to elicit a response similar to that stimulated by endogenous ABA concentrations.

In addition to understanding the mechanism of gene induction at the gene level, it must also be understood how the cell recognizes ABA and what signal transduction pathway is taken to gene induction. It is important to understand the aspects of the ABA molecule that are required for gene induction (Figure 3). The strategy has been taken to use ABA analogs to identify parts of the ABA molecule that are required for gene induction. Walker-Simmons et al. [125] compared optically pure ABA analogs in the induction of *rab*, *Em*, and *lea* group 3. The induction of *rab* and *lea* group 3 was similar with similar analogs; however, *Em* induction differed. These results support the conclusion that there is more than one mechanism for ABA regulation of gene expression. In the induction of *rab16* and *basi* in barley aleurone protoplasts, methylation of the carboxyl group had the least effect on the level of gene expression [126]. Removal of the carboxyl group, the 1'-hydroxyl, and the 4'-carbonyl had the greatest reduction in gene expression [126]. The ABA molecule is still recognized for gene induction if the 1'-hydroxyl is removed (Figure 3).

In another attempt to determine what is required for ABA regulation of gene expression, protein synthesis has been inhibited by the application of cycloheximide to determine if proteins must be synthesized for ABA action. Interestingly, the requirement for protein synthesis in the

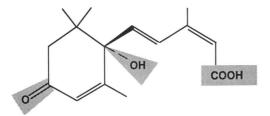

Figure 3 Structure of (+)-abscisic acid. Modifications made to the molecule to study the molecular structure required for gene regulation are shown shaded. (From Ref. 126.)

response to ABA is dependent on the gene studied. Protein synthesis is required for ABA induction of *rd22* but not for *rd29* [88]. It was also found that ABA induction did not require protein synthesis for *rab16* [9]. Therefore, there are at least two different pathways of gene induction in response to ABA.

V. FUTURE DIRECTIONS AND POSSIBLE AGRICULTURAL BENEFITS

Much progress has been made in the identification, isolation, and characterization of genes induced by different abiotic stresses. Studies on the isolation of genes and the regulation of specific genes have indicated that there are many similarities between stresses that result in cellular dehydration. Many of the genes that are induced by these stresses are also induced by ABA application, and several of these have been shown to require elevated levels of ABA for expression. Many of the genes induced during stress are predicted to play a protective role through direct protection of cellular contents or by altering the cellular water content.

The major challenge of the future is to obtain biochemical and genetic evidence that these gene products function in stress tolerance, improving the adaptability of plants to the environment. If adaptive gene products are characterized, these may have promise for use in the development of crop plants with increased stress tolerance. The use of different protective traits in combination, such as enhanced osmotic adjustment and overproduction of a hydrophilic gene product, may improve the chances of developing transgenic crop plants with an agricultural benefit.

REFERENCES

1. J. A. D. Zeevaart and R. A. Creelman, *Annu. Rev. Plant Physiol. Plant Mol. Biol.*, *39*: 439 (1988).
2. M. Pierce and K. Raschke, *Planta*, *148*: 174 (1980).
3. J. C. Baker, C. Steele, and L. Dure III, *Plant Mol. Biol.*, *11*: 277 (1988).
4. L. Dure III, M. Crouch, J. Harada, T.-H. D. Ho, J. Mundy, R. Quatrano, T. Thomas, and Z. R. Sung, *Plant Mol. Biol.*, *12*: 475 (1989).
5. J. C. Litts, G. W. Colwell, R. L. Chakerian, and R. S. Quatrano, *Nucl. Acids Res.*, *15*: 3607 (1987).
6. M. Espelund, S. Saebøe-Larssen, D. W. Hughes, G. A. Galau, F. Larssen, and K. S. Jakobsen, *Plant J.*, *2*: 241 (1992).
7. G. A. Galau, H. Y.-C. Wang, and D. W. Hughes, *Plant Physiol.*, *99*: 783 (1992).
8. T. J. Close, A. A. Kortt, and P. M. Chandler, *Plant Mol. Biol.*, *13*: 95 (1989).
9. J. Mundy, and N.-H. Chua, *EMBO J.*, *7*: 2279 (1988).
10. T. J. Close, R. D. Fenton, A. Yang, R. Asghar, D. A. DeMason, D. E. Crone, N. C. Meyer, and F. Moonan, in *Plant Responses to Cellular Dehydration During Environmental Stress* (T. J.

Close and E. A. Bray, eds.), Current Topics in Plant Physiology, Vol. 10, American Society of Plant Physiologists, Rockville, Md., p. 104. (1993).

11. A. Cohen, Á. L. Plant, M. S. Moses, and E. A. Bray, *Plant Physiol.*, *97*: 1367 (1991).
12. J. A. Godoy, J. M. Pardo, and J. A. Pintor-Toro, *Plant Mol. Biol.*, *15*: 695 (1990).
13. M. Pla, A. Goday, J. Vilardell, J. Gómez, and M. Pagès, *Plant Mol. Biol.*, *13*: 385 (1989).
14. A. Cohen and E. A. Bray, *Planta*, *182*: 27 (1990).
15. V. Lång and E. T. Palva, *Plant Mol. Biol.*, *20*: 951 (1992).
16. K. Schneider, B. Wells, E. Schmelzer, F. Salamini, and D. Bartels, *Planta*, *189*: 120 (1993).
17. T. J. Close and P. J. Lammers, *Plant Physiol.*, *101*: 773 (1993).
18. J. Curry, and M. K. Walker-Simmons, in *Plant Responses to Cellular Dehydration During Environmental Stress* (T. J. Close and E. A. Bray, eds.), Current Topics in Plant Physiology, Vol. 10, American Society of Plant Physiologists, Rockville, Md., p. 128. (1993).
19. L. Dure III, in *Plant Responses to Cellular Dehydration During Environmental Stress* (T. J. Close and E. A. Bray, eds.), Current Topics in Plant Physiology, Vol. 10, American Society of Plant Physiologists, Rockville, Md., p. 91. (1993).
20. J. Vilardell, A. Goday, M. A. Freire, M. Torrent, M. C. Martínez, J. M. Torné, M. Pagès, *Plant Mol. Biol.*, *14*: 423 (1990).
21. M. Plana, E. Itarte, R. Eritja, A. Goday, M. Pagès, and M. C. Martínez, *J. Biol. Chem.*, *266*: 22510 (1991).
22. M. Houde, J. Danyluk, J.-F. Laliberté, E. Rassart, R. S. Dhindsa, and F. Sarhan, *Plant Physiol.*, *99*: 1381 (1992).
23. W. Guo, R. W. Ward, and M. F. Thomashow, *Plant Physiol.*, *100*: 915 (1993).
24. L. A. Wolfraim, R. Langis, H. Tyson, and R. S. Dhindsa, *Plant Physiol.*, *101*: 1275 (1993).
25. L. Dure III, *Plant J.*, *3*: 363 (1993).
26. J. K. Roberts, N. A. DeSimone, W. L. Lingle, and L. Dure III, *Plant Cell*, *5*: 769 (1993).
27. A. Cohen and E. A. Bray, *Plant Mol. Biol.*, *18*: 411 (1992).
28. K. Nordin, T. Vahala, and E. T. Palva, *Plant Mol. Biol.*, *21*: 641 (1993).
29. K. Yamaguchi-Shinozaki and K. Shinozaki, *Plant Physiol.*, *101*: 1119 (1993).
30. K. Yamaguchi-Shinozaki and K. Shinozaki, *Mol. Gen. Genet.*, *236*: 331 (1993).
31. M. Luo, L. Lin, R. D. Hill, and S. S. Mohapatra, *Plant Mol. Biol.*, *17*: 1267 (1991).
32. M. Luo, J.-H. Liu, S. Mohapatra, R. D. Hill, and S. S. Mohapatra, *J. Biol. Chem.*, *267*: 15367 (1992).
33. S. Laberge, Y. Castonguay, and L.-P. Vezina, *Plant Physiol.*, *101*: 1411 (1993).
34. S. J. Gilmour, N. N. Artus, and M. J. Thomashow, *Plant Mol. Biol.*, *18*: 13 (1992).
35. S. Kurkela and M. Franck, *Plant Mol. Biol.*, *15*: 137 (1990).
36. S. Kurkela and M. Borg-Franck, *Plant Mol. Biol.*, *19*: 689 (1992).
37. A. D. Hanson and W. D. Hitz, *Annu. Rev. Plant Physiol.*, *33*: 163 (1982).
38. T. C. Hsiao, *Annu. Rev. Plant Physiol.*, *24*: 519 (1973).
39. A. J. Delauney and D. P. S. Verma, *Mol. Gen. Genet.*, *221*: 299 (1990).
40. C.-A. A. Hu, A. J. Delauney, and D. P. S. Verma, *Proc. Natl. Acad. Sci. USA*, *89*: 9354 (1992).
41. C. L. Williamson and R. D. Slocum, *Plant Physiol.*, *100*: 1464 (1992).
42. E. A. Weretilnyk and A. D. Hanson, *Proc. Natl. Acad. Sci. USA*, *87*: 2745 (1990).
43. D. Bartels, K. Engelhardt, R. Roncarati, K. Schneider, M. Rotter, and F. Salamini, *EMBO J.*, *10*: 1037 (1991).
44. D. M. Vernon and H. Bohnert, *EMBO J.*, *11*: 2079 (1992).
45. D. M. Vernon, J. A. Ostrem, and H. J. Bohnert, *Plant Cell Environ.*, *16*:437 (1993).
46. D. M. Vernon, M. C. Tarczynski, R. G. Jensen, and H. J. Bohnert, *Plant J.*, *4*: 199 (1993).
47. S. P. Lee and T. H.-H. Chen, *Plant Physiol.*, *101*: 1089 (1993).
48. D. B. DeWald, H. S. Mason, and J. E. Mullet, *J. Biol. Chem.*, *267*: 15958 (1992).
49. P. E. Staswick, C. Papa, and J.-F. Huang, *Plant Physiol.*, *102S*: 27 (1993).
50. F. D. Guerrero, J. T. Jones, and J. E. Mullet, *Plant Mol. Biol.*, *15*: 11 (1990).
51. M. Koizumi, K. Yamaguchi-Shinozaki, H. Tsuiji, and K. Shinozaki, *Gene*, *129*: 175 (1993).
52. M. A. Schaffer and R. L. Fischer, *Plant Physiol.*, *87*: 431 (1988).

53. J. C. Cushman, G. Meyer, C. B. Michalowski, J. M. Schmitt, and H. J. Bohnert, *Plant Cell, 1*: 715 (1989).

54. J. C. Thomas, R. L. De Armond, and H. J. Bohnert, *Plant Physiol., 98*: 626 (1992).

55. H. Sze, *Annu. Rev. Plant Physiol., 36*: 175 (1985).

56. T. K. Surowy and J. S. Boyer, *Plant Mol. Biol., 16*: 251 (1991).

57. X. Niu, J.-K. Zhu, M. L. Narasimhan, R. A. Bressan, and P. M. Hasegawa, *Planta, 190*: 433 (1993).

58. R. J. Anderberg and M. K. Walker-Simmons, *Proc. Natl. Acad. Sci. USA, 89*: 10183 (1992).

59. N. K. Singh, A. K. Handa, P. M. Hasegawa, and R. A. Bressan, *Plant Physiol., 79*: 126 (1985).

60. A. K. Kononowicz, K. G. Ragothama, A. M. Casas, M. Reuveni, A.-E. Watad, D. Liu, R. Bressan, and P. M. Hasegawa, in *Plant Responses to Cellular Dehydration During Environmental Stress* (T. J. Close and E. A. Bray, eds.), Current Topics in Plant Physiology, Vol. 10, American Society of Plant Physiologists, Rockville, Md., p. 144 (1993).

61. W. K. Roberts and C. P. Selitrennikoff, *J. Gen. Microbiol., 136*: 1771 (1990).

62. A. J. Vigers, S. Weidemann, W. K. Roberts, M. Legrand, C. P. Selitrennikoff, and B. Fritig, *Plant Sci., 83*: 155 (1992).

63. M. Yamada, *Plant Cell Phyiol., 33*: 1 (1992).

64. Á. L. Plant, A. Cohen, M. S. Moses, and E. A. Bray, *Plant Physiol., 97*: 900 (1991).

65. S. Torres-Schumann, J. A. Godoy, and J. A. Pintor-Toro, *Plant Mol. Biol., 18*: 749 (1992).

66. M. A. Hughes, M. A. Dunn, R. S. Pearce, A. J. White, and L. Zhang, *Plant Cell Environ., 15*: 861 (1992).

67. P. Sterk, H. Booij, G. A. Schellekens, A. Van Kammen, and S. C. De Vries, *Plant Cell, 3*: 907 (1991).

68. F. R. Terras, I. J. Goderis, F. Van Leuven, J. Vanderleyden, B. P. A. Cammue, and W. F. Broekaert, *Plant Physiol., 100*: 1055 (1992).

69. R. K. Hajela, D. P. Horvath, S. J. Gilmour, and M. F. Thomashow, *Plant Physiol., 93*: 1246 (1990).

70. C. Lin and M. F. Thomashow, *Plant Physiol., 99*: 519 (1992).

71. M. F. Thomashow, in *Plant Responses to Cellular Dehydration During Environmental Stress* (T. J. Close and E. A. Bray, eds.), Current Topics in Plant Physiology, Vol. 10, American Society of Plant Physiologists, Rockville, Md., p. 137. (1993).

72. C. Borkird, B. Claes, A. Caplan, C. Simoens, and M. Van Montagu, *J. Plant Physiol., 138*: 591 (1991).

73. C. Almoguera and J. Jordano, *Plant Mol. Biol., 19*: 781 (1992).

74. E. Vierling, *Annu. Rev. Plant Physiol. Plant Mol. Biol., 42*: 579 (1991).

75. W. L. Downing, F. Mauxion, M.-O. Fauvarque, R.-P. Reviron, D. de Vienne, N. Vartanian, and J. Giraudat, *Plant J., 2*: 685 (1992).

76. K. Yamaguchi-Shinozaki, M. Koizumi, S. Urao, and K. Shinozaki, *Plant Cell Physiol., 33*: 217 (1992).

77. H. Höfte, L. Hubbard, J. Reizer, D. Ludevid, E. M. Herman, and M. J. Chrispeels, *Plant Physiol., 99*: 561 (1992).

78. D. Bar-Zvi and T. Shagan, *Plant Physiol., 101*: 1397 (1993).

79. T. Shagan, D. Meraro, and D. Bar-Zvi, *Plant Physiol., 102*: 689 (1993).

80. C. Maurel, J. Reizer, J. I. Schroeder, and M. J. Chrispeels, *EMBO J., 12*: 2241 (1993).

81. J. Gómez, D. Sanchez-Martínez, V. Stiefel, J. Rigau, P. Puigdomènech, and M. Pagès, *Nature, 334*: 262 (1988).

82. M. D. Ludevid, M. A. Freire, J. Gómez, C. G. Burd, F. Alberico, E. Giralt, G. Dreyfuss, and Pagès. *Plant J., 2*: 999 (1992).

83. M. J. Guiltinan, W. R. Marcotte, Jr., and R. S. Quatrano, *Science, 250*: 267 (1990).

84. I. Winicov, *Plant Physiol., 102*: 681 (1993).

85. E. A. Bray, M. S. Moses, R. Imai, A. Cohen, and Á. L. Plant, in *Plant Responses to Cellular Dehydration During Environmental Stress* (T. J. Close and E. A. Bray, eds.), Current Topics in Plant Physiology, Vol. 10, American Society of Plant Physiologists, Rockville, Md., p. 167. (1993).

86. N. D. Iusem, D. M. Bartholomew, W. D. Hitz, and P. A. Scolnik, *Plant Physiol.*, *102*: 1353 (1993).
87. M. Pla, J. Gómez, A. Goday, and M. Pagès, *Mol. Gen. Genet.*, *230*: 394 (1991).
88. K. Yamaguchi-Shinozaki and K. Shinozaki, *Mol. Gen. Genet.*, *238*: 17 (1993).
89. R. Bassuner, H. Bäumlein, A. Huth, R. Jung, U. Wobus, T. A. Rapoport, G. Saalbach, and K. MÜntz, *Plant Mol. Biol.*, *11*: 321 (1988).
90. W. J. Hurkman, H. P. Tao, and C. K. Tanaka, *Plant Physiol.*, *97*: 366 (1991).
91. E. Dratewka-Kos, S. Rahman, Z. F. Grzelczak, T. D. Kennedy, R. K. Murray, and B. G. Lane, *J. Biol. Chem.*, *264*: 4896 (1989).
92. C. B. Michalowski and H. J. Bohnert, *Plant Physiol.*, *100*: 537 (1992).
93. D. Bartels, C. Hanke, K. Schneider, D. Michel, and F. Salamini, *EMBO J.*, *8*: 2771 (1992).
94. W. Kolanus, C. Scharnhorst, U. Kühne, and F. Herzfeld, *Mol. Gen. Genet.*, *209*: 234 (1987).
95. B. Grimm, E. Kruse, and K. Kloppstech, *Plant Mol. Biol.*, *13*: 583 (1989).
96. P. C. LaRosa, Z. Chen, D. E. Nelson, N. K. Singh, P. M. Hasegawa, and R. A. Bressan, *Plant Cell*, *4*: 513 (1992).
97. G. Iturriaga, K. Schneider, F. Salamini, and D. Bartels, *Plant Mol. Biol.*, *20*: 555 (1992).
98. M. C. Tarczynski, R. G. Jensen, and H. J. Bohnert, *Proc. Natl. Acad. Sci. USA*, *89*: 2600 (1992).
99. M. C. Tarczynski, R. G. Jensen, and H. J. Bohnert, *Science*, *259*: 508 (1993).
100. A. D. Parry, *Methods Plant Biochem.*, *9*: 381 (1993).
101. F. D. Guerrero and J. E. Mullet, *Plant Physiol.*, *80*: 588 (1986).
102. I. B. Taylor, in *Abscisic Acid: Physiology and Biochemistry* (W. J. Davies and H. G. Jones, eds.), Environmental Plant Biology Series, Bios Scientific Publishers, Oxford, p. 23 (1991).
103. E. A. Bray, *Plant Physiol.*, *88*: 1210 (1988).
104. S. J. Gilmour and M. F. Thomashow, *Plant Mol. Biol.*, *16*: 1233 (1991).
105. K. Nordin, P. Heino, and E. T. Palva, *Plant Mol. Biol.*, *16*: 1061 (1991).
106. W. R. Marcotte, Jr., C. C. Bayley, and R. S. Quatrano, *Nature*, *335*: 454 (1988).
107. W. R. Marcotte, Jr., S. H. Russell, and R. S. Quatrano, *Plant Cell*, *1*: 969 (1989).
108. J. Mundy, K. Yamaguchi-Shinozaki, and N.-H. Chua, *Proc. Natl. Acad. Sci. USA*, *87*: 1406 (1990).
109. G. Giuliano, E. Pichersky, V. S. Malik, M. P. Timko, P. A. Skolnik, and A. R. Cashmore, *Proc. Natl. Acad. Sci. USA*, *85*: 7089 (1988).
110. K. Oeda, J. Salinas, and N.-H. Chua, *EMBO J.*, *10*: 1793 (1991).
111. U. Schindler, A. E. Menkens, H. Beckmann, J. R. Ecker, and A. R. Cashmore, *EMBO J.*, *11*: 1261 (1992).
112. R. J. Schmidt, F. A. Burr, M. J. Aukerman, and B. Burr, *Proc. Natl. Acad. Sci. USA*, *87*: 46 (1990).
113. R. J. Schmidt, M. Ketuadat, M. J. Aukerman, and G. Hoschek, *Plant Cell*, *4*: 689 (1992).
114. M. J. Varagona, R. J. Schmidt, and N. V. Raikhel, *Plant Cell*, *4*: 1213 (1993).
115. M. E. Williams, R. Foster, and N.-H. Chua, *Plant Cell*, *4*: 485 (1992).
116. T. Izawa, R. Foster, and N.-H. Chua, *J. Mol. Biol.*, *230*: 1131 (1993).
117. E. Lam and N.-H. Chua, *J. Biol. Chem.*, *266*: 17131 (1991).
118. M. Pla, J. Vilardell, M. J. Guiltinan, W. R. Marcotte, Jr., M. F. Niogret, R. S. Quatrano, and M. Pagès, *Plant Mol. Biol.*, *21*: 259 (1993).
119. N. C. de Vetten, G. Lu, and R. J. Ferl, *Plant Cell*, *4*: 1295 (1992).
120. K. Yamaguchi-Shinozaki, M. Mino, J. Mundy, and N.-H. Chua, *Plant Mol. Biol.*, *15*: 905 (1990).
121. D. Michel, F. Salamini, D. Bartels, P. Dale, M. Bagga, and A. Szalay, *Plant J.*, *4*: 29 (1993).
122. J. Vilardell, J. Mundy, B. Stilling, B. Leroux, M. Pla, G. Freyssinet, and M. Pagès, *Plant Mol. Biol.*, *17*: 985 (1991).
123. R. M. Bostock and R. S. Quatrano, *Plant Physiol.*, *98*: 1356 (1992).
124. E. A. Bray and J. A. D. Zeevaart, *Plant Physiol.*, *80*: 105 (1986).
125. M. K. Walker-Simmons, R. J. Anderberg, P. A. Rose, and S. R. Abrams, *Plant Physiol.*, *99*: 501 (1992).
126. R. M. Van der Meulen, F. Heidekamp, B. Jastorff, R. Horgan, and M. Wang, *J. Plant Growth Regul.*, *12*: 13 (1993).

36

Physiological Responses of Plants and Crops to Ozone Stress

D. P. Ormrod and Beverley A. Hale
University of Guelph, Guelph, Ontario, Canada

I. INTRODUCTION

Ozone (O_3) is a strong oxidant that may adversely affect all living things. It is the most widespread phytotoxic air pollutant in the lower troposphere [1]. Ozone is the most abundant of the photochemical oxidants; ultraviolet radiation from sunlight is required for its formation in ambient air [2]. The outdoor concentration of ozone has a diurnal pattern, typically increasing on sunny days with warm temperatures which foster the photochemical synthesis reactions involving oxygen, nitrogen oxides (NO_x), volatile organic compounds (VOCs), and peroxides. The VOCs arise from human activities and natural sources. Nitrogen oxides are primarily from the burning of fossil fuels [2]. Ozone and its precursors are readily transported from urban areas to rural locations, where the ozone can cause adverse effects on the growth of crops, forest trees, and other natural vegetation [3] through its effects on the physiology and biochemistry of plants.

Ozone is extremely phytotoxic. It oxidizes plant surfaces and tissues and affects many important physiological processes [4]. It is particularly injurious to membranes, and notably inhibits photosynthesis and biomass accumulation, suppresses phloem loading, and reduces carbon allocation to the roots. These actions have resulted in ozone being rated as the pollutant of major concern in crop production and in forest tree growth. Because of its ubiquitous nature, numerous researchers have studied and modeled vegetation response, particularly in the past two decades [5]. In addition, tropospheric ozone has implications for climate change; it contributes to global warming and intercepts ultraviolet-B radiation.

II. BIOCHEMISTRY

A. Cellular Responses

To exert their oxidative effects, ozone or its reactive breakdown products must reach sensitive sites within plant tissues [3]. Stomates are the primary route of uptake into the interior of the leaf.

The strong oxidizing properties of ozone govern its impact on cell constituents. It can react with a large number of biological components and metabolites present in or on plant cells. It oxidizes sulfhydryl and fatty acid double bonds and other cellular scavenging compounds, and increases membrane permeability [6,7]. Free sulfhydryl groups on enzymes are highly susceptible to such oxidation.

There is uncertainty as to whether the effects of exposure are attributable to molecular ozone itself, which reactions are most likely to occur, and where the reactions occur in the pathway of ozone from outside the leaf to the interior of the leaf [3]. Many of the anatomical effects could be due to disturbed osmotic relationships within the cell. The rapidity of observed changes does suggest that ozone penetrates beyond the plasmalemma into the cell [6].

The effect of ozone in plant cells is dependent on the localization of its reactions with cellular constituents [3]. Glutathione and ascorbate are among the primary reactants and scavengers of ozone because of their susceptibility to oxidation, mobility, high cellular concentrations, and dispersion throughout the cell.

Active decomposition products, including superoxide and peroxyl and hydroxyl radicals, are produced by the dissolution of ozone in water. Substrates that interact with ozone to produce hydrogen peroxide can lead to generation of the highly reactive hydroxyl radicals [3]. Production of the superoxide anion and hydroxyl radicals is thought to form the basis for the initial effects of ozone in the cell [8].

Ozone may also react at the leaf surface and within the apoplastisc space outside the plasmalemma. The surface of the leaf includes the cuticle, and in many species, trichomes, glands, and wax deposits. Although these components are less reactive with ozone than are many cellular constituents that have unsaturated bonds, oxidation can occur, changing the properties of the leaf surface [3]. Large amounts of ozone can be sorbed by the leaf surface, adding to the apparent leaf diffusive conductance. The intercellular airspace surfaces and constituents may also be affected by ozone; ascorbate may be a particularly important reactive cell wall constituent outside the plasmalemma.

B.   Photosynthesis and Respiration

Ozone disrupts membrane-bound photosynthetic systems [6,7], resulting in decreases in carbon fixation in photosynthesis as tropospheric ozone increases [9]. Ozone has great potential for disruption of chloroplast metabolism. The enzymes of the reductive pentose cycle are particularly vulnerable [3]. Ozone decreases the activity and quantity of ribulose-1,5-bisphosphate carboxylase/oxygenase (Rubisco), a key photosynthetic enzyme [10]. Among the components of the functioning chloroplast that are sensitive to ozone are ascorbate, ATP, nitrate reductase, DNA, and lipids. The most widely observed change induced by ozone is chlorophyll destruction. Metabolic processes in photosynthesis may also be affected in nonspecific ways, through effects on integrity of the chloroplast envelope and thylakoid membranes that disrupt the chemiosmotic balance [3].

The sensitivity of photosynthesis to air pollutants varies widely among species and with environmental conditions [4]. Greater inhibition of photosynthesis does occur when exposure is for longer periods, but there is evidence for acclimation to pollutant exposure [4]. Photosynthesis effects may play a major role in stomatal behavior in ozone- and water-stressed plants; a reduction in rate of photosynthesis may lower the requirement for carbon dioxide, resulting in stomatal closure [4].

Respiratory processes are also sensitive to ozone. As net photosynthesis declines, dark respiration usually increases [11]. This stimulation of dark respiration by ozone has been reported in shoots in both the presence and the absence of visible injury [12]. The effects on dark respiration are the result of ozone effects on mitochondria and glycolysis enzymes, together with the effects related to the key role of the respiratory process in providing energy for maintenance and repair [3]. In contrast to dark respiration, light respiration is inhibited only by higher concentrations of ozone [4].

C. Assimilate Distribution

Ozone induces alterations in the partitioning of biomass among plant parts that can be related largely to the translocation of carbohydrates. Consistent with impairment of photosynthesis and increased dark respiration, foliar sugars and polysaccharide levels are lowered by ozone exposure of leaves [13]. At the same time, the allocation patterns of the carbohydrate change, even at ozone concentrations at which photosynthesis inhibition is minimal. The roots become the least powerful sink in O_3-stressed plants [14,15]. Developing fruits may receive less assimilate than leaves. Possible causes of the changed allocation include ozone-induced malfunctioning of the phloem loading and translocation processes, reduction in photosynthetic carbon fixation, and greater demand for assimilate at the source [16]. Reduced translocation of carbohydrates to the roots results in a smaller root mass, which may become important if there is substantial soil water stress.

The ozone-induced reduction in photosynthesis and the preferential distribution of reduced levels of photosynthate to developing leaves rather than to roots or older leaves has been demonstrated by carbon labeling [17,18]. These changes in allocation to the roots are reflected in increased shoot/root ratios that favor leaf development and maintenance of growth in the face of ozone stress [19,20]. These changes, which favor leaf development over root growth, can compensate for a decline in net assimilation rate up to a certain point [4].

In a detailed study with alfalfa, ozone's greatest effect was in reducing root growth, but it also reduced the number of leaves per plant and leaf area per plant, although the size of individual leaves was increased [21]. In this case photosynthate was being retained in older leaves and not being used to develop new leaves. Increased ozone increases specific leaf weight in sensitive plants [9], suggesting that affected leaves are thicker and/or more dense than those grown in a low-ozone environment.

D. Visible Injury, Pigmentation, and Senescence

The symptoms of visible injury take many forms and usually include changes in pigmentation. Typically, the visible ozone injury symptoms vary with species and may include chlorotic or necrotic spotting, flecking or blotching, and browning or reddening. The chlorosis may appear as an upper leaf surface stipple and the necrosis may be bifacial. Leaf visible injury often takes two forms. One form is the demonstration of distinctive ozone symptoms on leaves as noted above, and the other form premature senescence and leaf abscission [22]. More subtle changes in pigmentation, not visible to the naked eye, can be revealed by tissue analysis or surface reflectance techniques. In general, plants with visible injury symptoms have greater decreases in growth than those with invisible or latent injury [22]. Unfortunately, visible injury, when used as a simple screen of ozone sensitivity, does not relate well to differences in growth and yield among cultivars [5]. Plant responses resulting in visible changes in foliar characteristics are considered to be secondary processes of ozone toxicity that appear after the cellular biochemical and physiological changes that affect critical metabolic functions have established the plant's sensitivity to ozone [3].

Accelerated leaf senescence is a widespread response to ozone [23–25]. This premature senescence is associated with ozone-induced peroxidation of the lipids of plant membranes [26]. The ozone treatment enhances lipoxygenases, enzymes involved in membrane deterioration during plant aging, suggesting that lipoxygenases are involved in the precocious aging induced by ozone [27]. Impaired photosynthetic activity is closely related to this accelerated senescence, which includes decreases in chlorophyll content and decreased overall growth [28].

III. GROWTH AND DEVELOPMENT

A. Vegetative and Reproductive Phases

Vegetative growth is impaired through decreases in the leaf area of sensitive plants [9] as well as by reductions in dry matter accumulation associated with decreased net photosynthesis. Growth analysis techniques have been employed to facilitate the study of ozone effects on growth. Ozone stress has been shown to reduce total plant dry weight; leaf area; leaf area duration; relative growth rate of leaves, stems, and roots; and net assimilation rate [29]. With increased ambient ozone, there is decreased floral development, delayed fruit setting, and fewer and smaller fruit [9].

B. Yield

Yield decreases with increased tropospheric ozone have been demonstrated for many crop plants [5,9]. For forage crops, this may be a reflection of impacts of ozone on total aboveground dry matter production; for other crops the yield represents photosynthate and nutrient allocation to particular economic yield components, such as fruit, seeds, or roots. Photosynthesis and assimilate distribution have a close relationship to yield [30]. Ozone effects on photosynthesis that directly impinge on yield have been summarized as follows: short-term exposures to low ozone levels may reduce photosynthesis, but plants may recover quickly; higher concentrations may cause severe injury that permanently reduces photosynthesis; and long-term exposures reduce photosynthesis by reducing chlorophyll, accelerating senescence, and impairing cellular activities [30].

IV. WATER AND NUTRIENT RELATIONS

A. Plant Water Relations, Stomata, Transpiration, and Humidity

Stomata play a major role in regulating plant responses to ozone [31]. They control the entry of both carbon dioxide and ozone into the substomatal cavity, around which lies the mesophyll tissue. Stomatal action is controlled by CO_2 concentration in the substomatal cavity, leaf water status, epidermal cell characteristics, ion fluxes, and water [32].

Stomatal conductance is the principal rate-determining step in regulating ozone entry into leaves. In general, those species with higher stomatal conductances are more sensitive to ozone [4]. Environmental factors that are conducive to high rates of stomatal conductance, including high relative humidity and optimal soil water, increase ozone uptake and increase plant susceptibility [33]. As well as interactions with water stress, carbon dioxide levels, temperature, leaf age, and the presence of other pollutants are known to affect stomates. Ambient humidity during ozone exposure has major effects on stomatal response; high humidities may reduce the magnitude or speed of stomatal closure in response to ozone or reverse the direction of an ozone response [4].

The effects of water stress and ozone together on plant processes are complex and depend on both biological and physicochemical factors [34]. With increased soil water stress, plants become less sensitive to ozone but sensitive to the water stress itself [9]. Water stress inhibits photosynthesis either by direct effects on hydration of mesophyll tissues or by closing stomata and reducing carbon dioxide uptake. Decreased sensitivity to ozone in water-stressed plants may be attributable to lower stomatal conductance reducing ozone uptake, a more rapid closing response upon exposure to ozone, or some change(s) in the internal biochemistry of the cell [4]. This modification of stomatal conductance has implications for the amount of ozone actually received by tissues. It is considered that water-stressed plants are less sensitive to ozone because of stomatal closure rather than because of changed tissue sensitivity [4]. Exposure of sensitive plants to increased ozone results in coincident and long-term reductions in stomatal conductance and transpiration [9,35–37]. In general, stomatal opening occurs at low concentrations of ozone, below the threshold for effects on photosynthesis, and closure occurs at injurious concentrations, often following photosynthesis inhibition [4]. Cultivars may differ in stomatal response; closure occurred in an insensitive bean cultivar but not in a sensitive one [38].

The transpiration rate decreases with stomatal closure, so ozone-induced changes in stomatal diffusive resistance will affect transpiration as well as the exchange of other gases. Stomatal closure may conserve water as well as reduce ozone flux [3].

B. Mineral Nutrition and Tissue Nutrient Composition

The uptake, distribution, and assimilation of nutrients in response to ozone has not received much study. Some information is available on mineral nutrition effects on ozone response. In general, increased mineral nutrient deficiency results in greater susceptibility to ozone [9]. Ozone may suppress infection of roots by an ectomycorrhizal fungus [39,40] and suppress sporulation [41], thus indirectly interfering with root nutrition.

V. OTHER ENVIRONMENTAL INTERACTIONS

A. Radiation

Ozone responses of plants are influenced by ambient radiant energy in terms of intensity, quality, and duration. The morphology and physiology of plants differ between leaves grown in high and low light intensities. Thus the light intensity under which a plant has been grown may be expected to affect plant sensitivity to ozone [4]. Plants are generally more sensitive to ozone when grown under low light intensities [42,43]. At the time of exposure, plants may be most sensitive at both low and high light intensities and least sensitive at intermediate levels [44]. The explanation is that the energy available for repair processes is limited at low light intensities, while at high light intensities, the light-induced production of injurious reactive oxygen species is implicated, in addition to the injurious free radicals produced by ozone.

The interaction of ozone and potentially injurious ultraviolet radiation (UV-B) is complex. Ambient concentrations of ozone exhibit temporal and spatial variability that will result in increased variability in UV-B because ozone attenuates UV-B [1]. Ozone thus has some protective action by decreasing the UV-B impinging on plants on those days with high ambient ozone levels. Studies of the effects of alternating UV-B and ozone exposure of plants are needed to determine if the physiological processes in ozone-exposed plants have carryover effects on UV-B response on low-ozone days.

B. Temperature

Studies of temperature responses have been limited, precluding generalizations. Many observations of temperature effects have been confounded with humidity differences. When the effects of humidity are removed in such cases, little or no temperature effect can be demonstrated [45]. The direct effects of temperature are often mediated through stomate action. Low temperatures were found to decrease ozone sensitivity by decreasing leaf stomatal conductance and promoting stomatal closure [4].

C. Carbon Dioxide

Ambient carbon dioxide concentrations can influence air pollutant response through effects on photosynthesis or on stomatal opening. At elevated carbon dioxide concentrations, there may be a degree of stomatal closure that reduces the uptake of pollutant. Also, the stomatal response to carbon dioxide may reduce or eliminate closing responses due to ozone [4].

VI. PLANT SENSITIVITY

There is much variability in sensitivity to increased tropospheric ozone among species and among cultivars, strains, or ecotypes within species [9]. The variation in sensitivity among plants treated alike is clearly due to genetics factors, but variability from location to location and day to day is also a response to differences in environmental conditions both prior to and during exposure interacting with the genetic characteristics of the plant [4].

Factors that control maximum rates of stomatal conductance may largely determine relative sensitivity to ozone in many species [36]. Sensitive and insensitive strains of the same species may differ in the amount of carbohydrate allocation to the roots with ozone exposure [17]. Visible injury evaluation may demonstrate wide differences in cultivar sensitivity to ozone, but it may be difficult to relate this visible injury to differences in plant growth and yield [5,46]. Sensitivity may also be viewed from the perspective of the history of ozone exposures to which the plant has been subjected. Acclimation to ozone exposure has been demonstrated; this will change sensitivity ratings [4,47].

VII. EXPERIMENTAL METHODOLOGY

Several field exposure systems for ozone have been devised: free air exposure systems and open-top chambers. In free air exposure systems, significant spatial variability in pollutant concentrations within the study plots will occur during windy, turbulent conditions. Similarly, with open-top chambers, the influence of a single variable such as ozone is often optimized while altering the influence of other variables encountered in the ambient environment, because of a chamber effect and the experimental methodology used [1]. Indoor controlled-environment facilities often serve to establish physiological responses to ozone without the climatic variation that exists in outdoor ambient conditions, permitting more precise determination of cause–effect relationships but with less certainty that these relationships exist under outdoor conditions.

VIII. SUMMARY AND CONCLUSIONS

Ozone is an important pollutant gas in the lower troposphere. A strong oxidizing agent, it is formed by photochemical reactions of nitrogen oxides and volatile organic compounds on warm, sunny days. Ozone enters plant tissue primarily through the stomates. The first step toward control of ozone injury thus depends on the stomatal conductance. After stomatal entry a plant's sensitivity to ozone is established by cellular biochemical and physiological changes that affect critical metabolic functions. Leaf tissues are injured, plant processes are impaired, and growth and yield suffer. The primary cellular effects of ozone are the creation of reactive free radicals that must be scavenged if injury is to be limited. Membrane-based photosynthetic systems are disrupted and chlorophyll is destroyed, while respiration rates increase with the onset of repair processes. Carbon allocation patterns change; leaf development is favored over root and fruit development.

Ozone injury is enhanced by high ambient humidity and decreased by soil water stress through their effects on stomatal conductance. Radiation levels, temperatures, and carbon dioxide levels all influence ozone sensitivity. However, the greatest sources of variation in ozone sensitivity are associated with the plants themselves, their genetic makeup, stage of development, and history of ozone exposures.

Current interests and concerns focus on the establishment of science-based air quality regulations for the control of ozone precursors; the interactions of ozone with other emerging stressors, such as acid rain, UV-B radiation, and "greenhouse" warming and climate change; and the genetic alteration of sensitive plants through the application of molecular biology and biotechnology. The earlier emphasis on ozone effects in agriculture and forestry has broadened to include entire ecosystems. Studies of cellular effects now include those antioxidant enzymes and compounds that are common responses to a wide range of environmental stresses. The importance and concentrations of tropospheric ozone in the environment are not expected to decrease in the foreseeable future. In this context the study of ozone perturbation of plant functions is expected to remain at the leading edge of scientific discovery for many years to come.

REFERENCES

1. S. V. Krupa and R. N. Kickert, *Vegetatio, 104–105*: 223 (1993).
2. U.S. Environmental Protection Agency, *Ozone and Its Effects and Control*, OPA 3/9 (undated).
3. V. C. Runeckles and B. E. Chevone, in *Surface Level Ozone Exposures and Their Effects on Vegetation* (A. S. Lefohn, ed.), Lewis Publishers, Chelsea, Mich., p. 189 (1992).
4. N. M. Darrall, *Plant Cell Environ.*, *12*: 1 (1989).
5. W. W. Heck, D. T. Tingey, and O. C. Taylor, eds., *Assessment of Crop Loss from Air Pollutants*, Elsevier Applied Science, London (1988).
6. R. Guderian, D. T. Tingey, and R. Rabe, in *Air Pollution by Photochemical Oxidants Formation: Transport, Control and Effects on Plants* (R. Guderian, ed.), Springer-Verlag, Berlin, p. 127 (1985).
7. J. B. Mudd, in *Effects of Gaseous Air Pollution in Agriculture and Horticulture* (M. H. Unsworth and D. P. Ormrod, eds.), Butterworth Scientific, London, p. 189 (1982).
8. R. G. Alscher and J. M. Amthor, in *Air Pollution and Plant Metabolism* (S. Schulte-Hostede, N. M. Darrall, L. W. Blink, and A. R. Wellburn, eds.), Elsevier Applied Science, London, p. 94 (1988).
9. S. V. Krupa and R. N. Kickert, *Environ. Pollut.*, *61*: 263 (1989).
10. L. G. Landry and E. J. Pell, *Plant Physiol.*, *101*: 1355 (1993).
11. R. L. Barnes, *Environ. Pollut.*, *3*: 133 (1972).
12. E. J. Pell and E. Brennan, *Plant Physiol.*, *51*: 378 (1973).
13. J. E. Miller, R. P. Patterson, W. A. Pursley, A. S. Heagle, and W. W. Heck, *Environ. Exp. Bot.*, *29*: 477 (1989).
14. R. J. Oshima, J. P. Bennett, and P. K. Braegelmann, *J. Am. Soc. Hortic. Sci.*, *103*: 348 (1978).
15. K. F. Jensen, *Environ. Pollut. (Ser. A)*, *26*: 147 (1981).
16. D. R. Cooley and W. J. Manning, *Environ. Pollut.*, *47*: 95 (1987).
17. S. B. McLaughlin and R. B. McConathy, *Plant Physiol.*, *73*: 630 (1983).
18. K. Okano, O. Ito, G. Takeba, A. Shimizu, and T. Totsuka, *New Phytol.*, *97*: 155 (1984).
19. W. E. Hogsett, M. Plocher, V. Wildman, D. T. Tingey, and J. P. Bennett, *Can. J. Bot.*, *63*: 2369 (1985).
20. A. H. Chappelka and B. I. Chevone, *Can. J. For. Res.*, *16*: 786 (1986).
21. D. R. Cooley and W. J. Manning, *Environ. Pollut.*, *49*: 19 (1988).
22. D. F. Karnosky, Z. E. Gagnon, D. D. Reed, and J. A. Witter, *Can. J. For. Res.*, *22*: 1785 (1992).
23. P. B. Reich, *Plant Physiol.*, *73*: 291 (1983).

24. R. D. Noble and K. F. Jensen, *Am. J. Bot.*, *67*: 1005 (1980).
25. P. B. Reich and J. P. Lassoie, *Plant Cell Environ.*, *7*: 661 (1984).
26. K. P. Pauls and J. E. Thompson, *Physiol. Plant.*, *53*: 255 (1981).
27. M. Maccarrone, G. A. Veldink, and J. F. G. Vliegenthart, *FEBS Lett.*, *309*: 225 (1992).
28. J. E. Miller, *Recent Adv. Phytochem.*, *21*: 55 (1987).
29. A. G. Endress and C. Grunwald, *Agric. Ecosyst. Environ.*, *13*: 9 (1985).
30. J. E. Miller, in *Assessment of Crop Loss from Air Pollutants* (W. W. Heck, O. C. Taylor, and D. T. Tingey, eds.), Elsevier Applied Science, London, p. 287 (1988).
31. T. A. Mansfield, *Curr. Adv. Plant Sci.*, *2*: 11 (1973).
32. T. A. Mansfield and P. H. Freer-Smith, in *Gaseous Air Pollutants and Plant Metabolism* (M. J. Koziol and R. F. Whatley, eds.), Butterworth, London, p. 131 (1984).
33. S. Rich and N. C. Turner, *J. Air Pollut. Control Assoc.*, *77*: 718 (1972).
34. B. E. Chevone, J. R. Seiler, J. Melkonian, and R. G. Amundson, *Stress Responses in Plants: Adaptation and Acclimation Mechanisms* (R. G. Alscher and J. R. Cumming, eds.), Wiley-Liss, New York, p. 311 (1990).
35. A. C. Hill and N. Littlefield, *Environ. Sci. Technol.*, *3*: 52 (1969).
36. P. B. Reich and R. G. Amundson, *Science*, *230*: 566 (1985).
37. P. I. Coyne and G. E. Bingham, *For. Sci.*, *28*: 257 (1982).
38. L. K. Butler and T. W. Tibbitts, *J. Am. Soc. Hortic. Sci.*, *104*: 211 (1979).
39. P. M. McCool, J. A. Menge, and O. C. Taylor, *J. Am. Soc. Hortic. Sci.*, *107*: 839 (1982).
40. P. M. McCool and J. A. Menge, *New Phytol.*, *94*: 241 (1983).
41. P. F. Brower and A. S. Heagle, *Phytopathology*, *73*: 1035 (1983).
42. W. M. Dugger, O. C. Taylor, C. R. Thompson, and E. Cardiff, *J. Air Pollut. Control Assoc.*, *13*: 423 (1963).
43. W. W. Heck and J. A. Dunning, *J. Air Pollut. Control Assoc.*, *17*: 112 (1967).
44. R. W. Carlson, *Environ. Pollut.*, *18*: 159 (1979).
45. A. G. Todd, D. P. Ormrod, B. A. Hale, and S. N. Goodyear, *Biotronics*, *20*: 43 (1991).
46. P. J. Temple, *Environ. Exp. Bot.*, *30*: 283 (1990).
47. L. Walmsley, M. R. Ashmore, and J. N. B. Bell, *Environ. Pollut.*, *23*: 165 (1980).

37
Physiological Responses of Plants and Crops to Ultraviolet-B Radiation Stress

D. P. Ormrod and Beverley A. Hale
University of Guelph, Guelph, Ontario, Canada

I. INTRODUCTION

The depletion of stratospheric ozone has resulted in increased ultraviolet (UV)-B penetration to the earth's surface. The amount of increase is dependent on latitude, with the greatest increases in arctic and antarctic regions. Many questions have been raised about the possible consequences of enhanced UV-B radiation in crop production and in natural ecosystems. The ultraviolet radiation that is present in sunlight is divided into three classes: UV-A, UV-B, and UV-C. The UV-A, with wavelengths from 320 to 390 nm, is not attenuated by ozone and thus is not affected by depletion of the stratospheric ozone layer. Plants are not as sensitive to UV-A as they are to UV-B [1]. The UV-C, with wavelengths shorter than 280 nm, does not reach ground level and this is not expected to change. It is the UV-B radiation, with wavelengths from 280 to 320 nm, that has received most attention because UV-B is absorbed by ozone; the daily fluence at the earth's surface increases as stratospheric ozone decreases.

Many different plant responses to supplemental UV-B radiation have been observed, mostly injurious but sometimes beneficial. The injury to plant processes can be classified into two categories: injury to DNA, which can cause heritable mutations, and injury to physiological functions [2]. In this chapter we are concerned largely with the latter.

The response of various plant processes to UV-B is wavelength dependent, so that action spectra are used to weight the measured UV-B dose relative to its effects on particular processes [3]. Plant sensitivity to UV-B varies widely among species and processes. Some of this differential sensitivity may be due to adaptive UV-B protective mechanisms, including increases in UV-B absorptive substances, increases in UV-B reflective properties, and increases in cuticle and leaf thickness with increasing incident UV-B [4].

II. BIOCHEMISTRY

A. Cellular Responses

The effects of UV-B that ultimately result in changed plant growth and productivity are initially felt at the cellular level, where both general and specific and direct and indirect effects are found. The direct effects of UV-B can include membrane changes and protein denaturation. These result in the impairment of enzyme action, translocation, photosynthesis, and other essential processes. The mechanistic details of many of these responses are as yet unclear. As well as disruptions of physiological functions, injury to genetic molecules can be expected. Induction of DNA injury and the formation of pyrimidine dimers are probable consequences of increased UV-B radiation. In a study of two soybean cultivars of differing UV-B sensitivity in the field, there was a decrease in the integrity of DNA, as measured by strand breaks, in both cultivars [5]. DNA from the more sensitive cultivar had a greater pyrimidine dimer concentration. The sensitivity of plant enzymes is illustrated by a study of rose cell plasma membrane ATPase [6]. The action spectrum of this response peaks at 290 nm. The inactivation of this ATPase probably results from singlet oxygen-mediated destruction of tryptophan residues in the enzyme's protein.

The stimulation of the synthesis of protecting substances by UV-B radiation takes place at the cellular level (see also Section II.D). Individual cells in many species are capable of encoding and synthesizing a range of UV-B-absorbing compounds [7]. Much information is still needed on tissue and developmental differences in UV-B radiation responses and how tissue responses are integrated to increase plant tolerance of UV-B.

B. Photosynthesis and Respiration

Photosynthesis is sensitive to increased UV-B radiation. Many studies have demonstrated reductions in photosynthesis in both C3 and C4 plants [1]. Direct injury to the photosynthetic apparatus has been studied extensively. Photosystem II is sensitive to UV-B, as indicated by the increase in variable chlorophyll fluorescence observed after exposure to UV-B radiation [2]. The action spectrum for photosynthesis is broad, suggesting multiple sites of action rather than a specific target molecule. The photosynthetic capacity of rice cultivars was measured in terms of oxygen evolution and found to be sensitive to UV-B [8]. Both rice and pea showed marked decreases in the ratios of variable to maximum chlorophyll fluorescence yield and in the quantum yield of photosynthetic oxygen evolution in response to UV-B radiation. The greater declines occurred in pea leaves [9].

Indirect injury effects can be demonstrated by reductions in photosynthesis on a per plant basis that may be due to overall growth reduction of leaves and photosynthesizing surfaces [7]. Another indirect reason for reductions in photosynthesis might be stomatal closure by elevated UV-B irradiance.

The role of level of photosynthetically active radiation is an important one in determining injury to photosynthesis. Growth chamber studies, using low levels of white light which limit photosynthesis in most plants, often demonstrate more photosynthesis injury than would be found in the field or greenhouse, where higher levels of white light are available [7].

Respiration effects have not had the amount of study accorded to photosynthesis. In jack pine, UV-B treatment decreases the dark respiration rate and light compensation point [10]. In this species, UV-B induces increases in ribulose-1,5-bisphosphate carboxylase/oxygenase efficiency and ribulose-1,5-bisphosphate regeneration.

C. Plant Composition

Few studies of UV-B effects on the chemical composition of plants have been undertaken. The plant's composition has often been only part of a broader study of UV-B impact. For example, cucumber seedlings were used to study surface lipid changes with UV-B treatment [11]. The distribution of the main wax compounds, the alkanes and alcohols, was shifted toward compounds with shorter chain lengths. This effect was due to direct UV-B effects on wax biosynthesis, not simply to photo-oxidation. In a UV-B and water stress interaction study, the protein content of radish cotyledons was unaffected by UV-B or water stress. However, the protein content of cucumber cotyledons was increased by exposures to enhanced UV-B radiation and water stress compared with well-watered plants exposed to control levels of UV-B [12].

D. Pigmentation and UV-B-Absorbing Substances

Increased UV-B radiation has effects on visible plant pigments. On the one hand, both artificially and naturally supplied enhanced UV-B radiation decrease the chlorophyll and carotenoid content of plant tissue. Chlorophyll destruction is a function of the amount of UV-B radiation, but only in UV-B-sensitive plants. Carotenoids are similarly affected in sensitive species. The basis for chlorophyll and carotenoid changes may alter the interpretation. For example, UV-B may decrease chlorophyll content on a per plant basis, but it increases chlorophyll content when expressed on a leaf area basis.

On the other hand, a group of phenylalanine-derived secondary products (such as flavonoids) are induced by UV-B exposure in many species; among the many functions of flavonoids in plants, screening against UV radiation has long been recognized as important [13]. Activation of the phenylpropanoid and interrelated biosynthetic pathways leading to increased synthesis of flavonoids and related compounds appears to be a general response to UV-B radiation. Flavonoid induction has been demonstrated in many species [2]. The flavonoids and other phenylpropanoids generally accumulate in the epidermis, where these UV-B-absorbing compounds can help prevent UV-B radiation from reaching inner cell layers and photosynthesizing tissues. The epidermis blocks 95 to 99% of incoming UV-B radiation. The UV-B-absorbing compounds are mainly phenylpropanoids, such as cinnamoyl esters, flavones, flavonols, and anthocyanins esterified with cinnamic acids [7]. Over 30% of the UV-screening pigments in the *Brassica napus* leaf are found in the adaxial epidermal layer [14].

Anthocyanins accumulate under enhanced UV-B in many species but have only weak UV-B absorption and thus provide little or no protection as UV-B screens. In addition to phenylpropanoids, other products of the shikimic acid pathway and terpenoids accumulate under increased UV-B. Such compounds may be toxic to other organisms, including insects, so changes in their concentration might affect the plant's resistance to pests or change herbivory habits for insect populations. The increased flavonoid contents might also provide the basis for UV-B effects on food quality [7].

The accumulation of UV-B-absorbing compounds as an adaptive response has been clearly shown to be specifically UV induced and dependent on wavelength and radiation intensity. The increase in flavonoid concentration is due to higher activity of the enzyme phenylalanine ammonia lyase, which catalyzes the first step in phenylalanine-derived secondary aromatic compound biosynthesis [4] and/or higher rates of biosynthesis of this enzyme. The activation of this response has been the subject of considerable enquiry and may involve phytochrome, a blue light receptor and a UV-B receptor [2]. Visible light which is absorbed by phytochrome and a blue light photoreceptor control this response to UV-B irradiation in many species [15]. In other species, a UV-absorbing receptor controls the response, with phytochrome and a blue light photoreceptor playing supporting roles. Another avenue to flavonoid synthesis has been demonstrated in bean leaves [16]. Formation of the isoflavonoid coumestrol was mediated via UV-induced pyrimidine dimer formation in the plant's DNA. In rye, PAL activity increased within minutes of UV-B treatment by decreasing the PAL inhibitor, transcinnamic acid, through its isomerization to the cis form [7]. This trans–cis shift has implications in rapid responses to UV-B and changing wavelength sensitivity of epidermal cells because cis isomerization shifts absorption to shorter wavelengths. In parsley, individual epidermal cells accumulating flavonoids have been shown to contain the sequence of mRNA encoding the enzymes of flavonoid synthesis, the enzymes themselves, and the end products [7]. Plant responses to UV-B radiation at the level of gene expression have been included in a review of stress responses [17].

Arabidopsis thaliana mutants with defects in the synthesis of these UV-B-absorbing compounds have been tested for UV-B sensitivity [4]. The transparent testa mutants, which have reduced flavonoids and either normal or reduced sinapate esters, were found to be sensitive to UV-B. The synthesis of these UV-absorbing compounds may form a basis for cultivar sensitivity to UV-B. In a field test of two soybean cultivars, enhanced UV-B caused changes in concentrations of UV-absorbing compounds in both cultivars, one increasing and the other decreasing [3]. The time course of UV-B treatment can also have important implications. For example, young cucumber cotyledons increased their flavonoid content, but after longer UV-B treatment, the deleterious effects of UV-B overcame the biosynthesis of flavonoids [12].

III. GROWTH AND DEVELOPMENT

A. Photomorphogenesis

Photomorphogenesis is a radiation-induced change in plant form. It has been demonstrated that enhancing UV-B can alter the growth form of several species independently of phytochrome without reducing shoot dry weight. For example, UV-B exposure causes positive geotropism similar to that induced by blue light [15]. Alfalfa hypocotyls maintained in red light (so that phytochrome was not activated) and exposed to UV-B demonstrated phototropism [18] (see also Section III.B). In dense canopies, the photomorphogenic effects of UV-B alter leaf placement and thereby influence competition for light.

B. Growth Regulation

Growth reductions induced by UV-B may be associated with changes in cell division and/or cell elongation [7]. Interaction of UV-B with indoleacetic acid (IAA) would be expected and has been demonstrated in the hypocotyls of sunflower seedlings. Hypocotyl epidermal cell elongation was inhibited by UV-B radiation as a result of the photooxidation of IAA [2]. The inhibiting effects of this photooxidation of IAA have been demonstrated both in vivo and in vitro. The photo product 3-methyleneoxindole inhibits hypocotyl growth when applied exogenously [7].

As well as the primary UV-B events involving hormonal inactivation, growth changes could be controlled by phytochrome, which photoconverts in response to UV-B and/or by changes in other UV-B photoreceptors [2].

C. Leaf and Plant Growth and Development

Enhanced UV-B radiation has many effects on leaves. Epidermal deformation and cuticular wax deposition and erosion have been seen after UV-B treatment [2]. A number of studies have demonstrated reductions in leaf area under enhanced UV-B [1]. For example, the leaf area of cucumber responds to the UV-B fluence [7]. In many plants the reduced leaf area is associated with increased leaf thickness and increases in specific leaf weights [1]. In jack pine, stomatal density is increased by high UV-B radiation [10].

The development pattern of the plant may also be affected. A decrease in seedling height has been demonstrated in a number of studies [1]. Plant species and cultivars vary in height growth sensitivity. Hypocotyl and stem lengths of cucumber are decreased by increasing UV-B [7]. Reductions in plant height and leaf blade length have been found in wheat and wild oats grown in monoculture [19]. These reductions were not associated with reduced photosynthesis. Reduced leaf blade and internode lengths and increased leaf and axillary shoot production have been found in several species, but they were not associated with reductions in total shoot dry matter production [20]. In jack pine, high UV-B decreases seedling dry weight and shifts biomass partitioning in favor of leaf production [10].

Morphological changes may be mainly responsible for competitive balance shifts between plants in ecosystems subjected to enhanced UV-B (see also Section III.F). In a study of wheat and wild oat grown as a competing crop/weed pair, wild oat showed greater reductions in stem elongation and leaf blade length under enhanced UV-B than did wheat [3]. Photosynthesis was unchanged in both species and was not responsible for the competitive change. It may be difficult to distinguish a UV-B response that is developmental from one that results from nonspecific injury, such as injury to photosynthesis or photooxidation of proteins [15]. Growth inhibition alone may not be evidence for nonspecific injury, since UV-B photoexcitation of a blue light receptor and of phytochrome also decrease growth rate.

D. Reproduction

Enhanced UV-B radiation may inhibit flowering in some species but stimulate flowering in others [1]. In general, flowering is decreased by UV-B [7]. Inhibition of photoperiodic flower induction which is dependent on UV-B fluence has been observed in a long-day plant.

Decreases in the percentage of pollen germination have also been observed [21]. Poor pollen viability may decrease reproductive potential, although the overall impact of UV-B on the reproduction potential is unclear. The ovules may well be sufficiently well protected from UV-B radiation, because they are well hidden in the ovaries. Pollen enclosed in the anthers would also be well protected against UV-B radiation, since the anther walls will filter out much of the UV-B. The pollen wall itself also contains UV-absorbing compounds which lend protection prior to pollen germination. Potentially the most sensitive of reproductive processes are germination of the pollen and growth of the pollen tube on the stigma, which may expose the generative nuclei to UV-B, especially if there is a long time curve for germination and stigma penetration. The results of in vitro experiments indicate that even moderate UV-B levels inhibit pollen germination in some species [7].

E. Biomass and Yield

Increased UV-B radiation can result in decreases in biomass or total dry matter production and marketable yield in some cultivars of some species [1,2]. Such decreases are often, but not necessarily always, connected with UV-B-dependent reductions in growth parameters. For example, field studies with six soybean cultivars showed reductions in relative growth rate, height growth, and net assimilation rate [7], while in grasses such as wheat and wild oat, reductions in plant height or leaf area were not always correlated with total biomass reductions. In these species, leaf area is displayed vertically rather than horizontally as is the case in dicotyledonous plants [7]. While global plant growth is clearly at risk, many species have not yet been studied under field conditions. It appears that crop maturation rates are not affected by UV-B radiation.

There are large cultivar differences in yield effects. In a comparison of two soybean cultivars in the field, one cultivar had less biomass in major plant organs than the other, with the greatest response in seed pods and stems [5]. In a test of 16 rice cultivars, about one-third had a significant decrease in total biomass with UV-B radiation [8]. For these sensitive cultivars, leaf area and tiller number were also significantly reduced. Little is known of the reasons for the large variation in UV-B yield and biomass responses among cultivars. Studies of the heritability and genetic bases for UV tolerance and sensitivity are needed.

Studies on the effects of UV-B radiation on biomass and yield have provided conflicting information on species and cultivar responses, depending on whether the research was performed in a growth chamber, a greenhouse, or in ambient field plots [7]. Studies of biomass accumulation and plant yield of crop plants and other high-light species should be conducted outdoors. Apparently, the injury from UV-B radiation is magnified when plants are given levels of visible light that are below saturation for photosynthesis. The UV-B protective mechanism, known generally as photorepair, is thought to be not fully developed in low levels of visible light. This problem is complicated by differences in light sources, temperature, daylength, nutrition, and the presence of other stresses.

F. Competitive Ability

Interactions among species in mixed populations may be sensitive to UV-B as the plants compete for the resources needed for growth, even when a particular species did not have growth reductions when exposed in a pure stand to enhanced UV-B [1]. Shifts in competitive balance between species involving differential growth responses have been demonstrated for some species pairs [7]. Those competition studies suggest that there is a potential risk of changes in ecosystem composition in enhanced-UV-B regimes. The nature of the species mixture is expected to have some effect on competitive balance resulting from differential UV-B effects on leaf placement [20]. More changes are expected when monocots are involved in mixtures, rather than mixtures of dicots only.

IV. WATER RELATIONS

A. Stomatal Responses

Treatment with UV-B can affect stomatal conductance, altering the rate of water loss by transpiration and the uptake rate of CO_2 for photosynthesis [1]. Stomatal closure by enhanced UV-B and increased leaf diffusive resistance has been demonstrated with the action spectrum peaking below wavelengths of 290 nm [7]. Stomatal closure is thought to be caused by a loss of turgor pressure with ion leakage from the guard cells. Transpiration has been demonstrated to be reduced in some UV-B-sensitive seedlings [7]. The time course for stomatal closure is rapid even at low UV-B levels. Stomatal opening is slowed by higher UV-B levels.

B. Water Stress

With drought stress, plants become less sensitive to UV-B as the applied water stress increases [1]. Several experiments have served to elucidate some of the water stress/UV-B interactions. Well-watered soybean plants grown in the field under enhanced UV-B conditions had reductions in growth, dry weight, and net photosynthesis compared with ambient UV-B, while no UV-B effect could be detected in water-stressed plants [22]. Photosynthesis recovery after water stress was greater and more rapid in UV-B-treated soybeans and associated with UV-B effects on stomatal conductance rather than with internal water relations. Water-stressed cucumber plants lost their capacity to close stomata at midday with increasing UV-B [12]. Radish seedlings were less sensitive to UV-B under water stress than cucumber. Radish had higher leaf flavonoid contents, possibly protecting seedlings by absorbing UV-B in the leaf epidermis (see also Section II.D).

V. OTHER ENVIRONMENTAL INTERACTIONS

A. Visible Light

The level of visible or white light (400 to 700 nm) to which experimental plants are exposed has been found to have a very great effect on UV-B injury. Growth chamber experiments have demonstrated that UV-B injury is greater with low levels of photosynthetic photon flux (PPF) (less than 200 μmol/m^2 per second) than with high levels [7]. High levels of white light as well as UV-A radiation with blue light mediate photorepair mechanisms and ameliorate the UV-B injury. The relationship of PPF to UV-B effects is further complicated by the fact that a source of UV-B, whether natural or simulated, can exhibit not only different total output energies but also varying spectral composition within the range 280 to 320 nm [1]. Growth chamber studies have been particularly criticized because greater negative effects on the plant in response to UV-B exposure have been found in growth chambers than when a similar exposure takes place under field conditions. Perhaps the most studied interaction of UV-B with another environmental variable has been with concurrent visible light; this is likely to continue as exposure facilities are fitted with high PPF irradiance sources to permit realistic simulation of field conditions on a year-round basis. Studies of photorepair processes that mitigate against otherwise accumulatory UV-B injury in the plant tissue are needed in the various UV-B exposure facilities.

B. Nutritional Status

The few studies that have focused on possible interactions of UV-B with plant nutrient dynamics indicate that some plants become less sensitive to UV-B with mineral stress, while others become more sensitive [1]. For example, the sensitivity of soybean to UV-B is dependent on the phosphorus status [23]. Deficient plants are less sensitive to UV-B than are plants at optimum P levels, due at least in part to the accumulation of flavonoids and to leaf thickening in P-deficient plants. On the basis of current information, it is difficult to know the applicability of such small-scale research projects on mineral deficiency to natural field conditions [7].

C. Temperature

Little consideration has been given to temperature effects on UV-B response, even though global climate change is expected to invoke increases in both UV-B and temperature [1,7]. It will be desirable to know whether the generally deleterious UV-B effects will be altered by changes in temperature. However, it should be noted that temperature changes are likely to affect plant growth both directly and through altered water supply, in addition to any effects on UV-B action.

D. Carbon Dioxide

With global climate change, increases in CO_2 concentration will be superimposed on increases in UV-B radiation [3]. Increased CO_2 generally stimulates plant growth, while increased UV-B generally decreases plant growth. Higher ambient CO_2 may allow more repair processes to proceed in some plants, so that sensitivity to increased UV-B may be offset to some extent [3]. In keeping with this premise a research report has indicated that in jack pine, high UV-B inhibited photosynthesis at 350 but not at 750 μmol/mol CO_2 [10]. In another study, however, the stimulating effects of elevated CO_2 on rice plant photosynthesis were eliminated or reduced by increased UV-B radiation [24].

E. Air Pollutants

Little consideration has been given to the interaction of enhanced UV-B with tropospheric gaseous air pollutants [1]. Of increasing interest to the welfare of plants are the pollutants ozone, sulfur dioxide, and nitrogen oxides associated with urbanization and industrial development and already known to have damaging effects on crops, forests, and natural ecosystems [7]. The negative effects of these air pollutants can be exacerbated by increasing UV-B radiation. For example, the combination of UV-B and ozone reduced pollen tube growth more than did either stress alone [25].

Tropospheric ozone in particular may play a role in ameliorating UV-B effects on vegetation by exerting a protective role in the troposphere by absorbing UV-B. When ambient tropospheric ozone is high, theoretical considerations suggest that UV-B at plant surfaces will be low. In fact, similar but opposite temporal and spatial variability is expected in the surface levels of both ozone and UV-B [3]. This phenomenon could take place downwind of northern midlatitude cities and metropolitan areas [1]. Experimental protocols for both outdoor and indoor studies of UV-B and ozone interaction should consider this model of alternating exposures.

VI. PLANT SENSITIVITY

There are wide differences in sensitivity to UV-B among species and among cultivars within species [1]. Evidence for negative sensitive responses, positive stimulatory responses, and tolerance has been obtained in numerous studies. Such evidence may be on the basis of a variety of physiological and morphological responses or on yield and biomass accumulation. There are no well-established criteria for recognition of sensitivity of a species or a cultivar other than by exposing plants to enhanced UV-B. In such direct exposure studies, differences in sensitivity for the same species among studies have been attributed to differences among cultivars, growth stages, light sources and amounts of UV-B, and environmental conditions during exposure [1]. In particular, controlled environments, growth chambers, greenhouses, and the field often provide conflicting information on plant sensitivity.

Adaptation to UV-B radiation may occur and will affect sensitivity ratings. Photorepair, accumulation of UV-B-absorbing pigments, cuticular changes, and morphological changes are potential adaptive mechanisms that will function as protective mechanisms that confer insensitivity [7].

Sensitivity to UV-B is known to change as a plant goes through its life cycle. The role of the cuticle in protecting plants from UV-B may explain growth stage effects on sensitivity. Greatest sensitivity is expected in early stages of growth, with progressive decreases in response as the cuticle thickens [3]. The epicuticular wax on *Brassica napus* leaves is more densely arrayed on the adaxial leaf surface of UV-B-treated plants [14]. Growth studies with soybean in a greenhouse indicated that the most UV-B-effective period occurs between the vegetative and reproductive phases. At the end of the vegetative phase even intermediate levels of UV-B reduced plant height, leaf area, and total dry weight [7]. The cuticular properties of greenhouse-grown plants are different from those of similar plants grown under field conditions.

Greenhouse-grown plants have a much thinner cuticle and may demonstrate greater sensitivity to UV-B [3].

Differential accumulation of UV-B-screening pigments may account for wide differences in sensitivity (see also Section II.D). For example, cucumber cotyledons are more sensitive than radish, possibly due to lower concentrations of protective flavonoid pigments in cucumber. Flavonoid pigment in radish was increased three- to four-fold by UV-B compared with control plants, especially in older leaves [1]. On the other hand, a study of rice cultivar sensitivity to UV-B indicated that the highest concentration of UV-B-absorbing substances was present in the leaves of the most sensitive cultivar [9].

VII. EXPERIMENTAL METHODOLOGY

The design, installation, and operation of experimental systems for physically simulating UV-B radiation is challenging whether under the controlled environment conditions of growth chambers or greenhouses or in the field, where ambient natural UV-B radiation is to be altered. The principle in all experiments to determine the effects of UV-B on plants involves the use of UV sources (lamps) coupled with different types of filters to exclude UV wavelengths not desired in particular treatments [1]. The intensity of UV-B is varied by changing the distance between lamps and plants or by an electronic method that varies lamp output. The total ultraviolet radiation is generally confined to UV-A and UV-B by the use of cellulose acetate filters which do not transmit UV-C wavelengths. Control plants are normally irradiated with UV lamps that are shielded with a plastic film such as Mylar which transmits only UV-A wavelengths. Growth chamber studies usually combine lamps with high output in the UV-B range fitted with selected-wavelength cutoff filters of plastic films with lamps supplying sufficient photosynthetically active radiation [1]. Similar systems are used in greenhouses, without necessarily providing supplemental photosynthetic radiation. For field exposure, UV lamps and filters are used to supplement natural UV-B radiation. Many studies have increased UV-B levels to simulate conditions that would exist with a defined reduction (typically 10 to 20%) in the stratospheric ozone layer [2].

The sensitivity of biological processes to UV-B is wavelength dependent. Accordingly, mathematical response functions called action spectra are used as weighting factors to adjust the measured UV-B radiation and provide quantitative UV-B dosages that relate to the biological response. Action spectra indicate the degree of response at each wavelength and may be used to identify the active chromophore or UV-B-absorbing compound(s). Where many factors are involved in a response to UV-B, such as in plant growth, a complex action spectrum can be used to estimate the effects of a given UV-B dose [2]. Measurement of the lag time between UV-B irradiation and the appearance of a response can also provide information about the mechanism of the response.

VIII. SUMMARY AND CONCLUSIONS

Ultraviolet-B (UV-B) radiation effects are of increasing interest in plant physiology as questions are raised about the impact of enhanced UV-B in sunlight resulting from stratospheric ozone depletion. The UV-B wavelength band of the radiant energy spectrum of sunlight is capable of much injury to tissues and processes. Many plants are protected from UV-B by leaf components that reflect and absorb the radiation before it can reach the metabolizing tissues. In many plants the synthesis of additional protective substances is triggered by UV-B radiation. Radiation that penetrates this protective zone impinges on the structural systems and metabolic processes in the tissue, impairing protein function, enzyme action, photosynthesis, and other activities, upsetting the equilibria and normal functions of cells. As a result, changes in carbon fixation, pigmentation, and composition can be detected which ultimately lead to growth retardation and changes in plant development. Biomass and yield are reduced in sensitive species. Interactions with other environmental factors are readily demonstrated; water status, visible-light conditions, nutrition, temperature, carbon dioxide, and air pollutants can all influence UV-B response.

Remarkable differences in sensitivity to UV-B exist among species and among cultivars within species. Variation in UV-B-absorbing substances, cuticular characteristics, and cellular tolerance of UV-B radiation all contribute to a rich pool of research findings and opportunities that point to great potential for genetic manipulation of UV-B sensitivity. Great advances in research methodology have been made in recent years. The need to measure action spectra for UV-B-sensitive processes and to provide adequate visible radiation for photorepair has set rigorous demands on researchers for acceptable instrumentation and facilities.

Much research remains to be accomplished [7]. Field validations of yield and ecosystem effects are needed to complement controlled environment exposure work. A particular need is for research that integrates UV-B effects with those of climate change, especially CO_2 and temperature increases, and with the impact of increasing tropospheric ozone. Information is needed on whether UV-B radiation will offset some of the potential benefits and exacerbate the detrimental effects of climate change. Coupled with the need for information on effects is the need for verifying mechanistic models of the responses at the cellular and molecular level. This will assist geneticists and biochemists in their search for functional sources of insensitivity that can be incorporated into UV-B sensitive plants.

REFERENCES

1. S. V. Krupa and R. N. Kickert, *Environ. Pollut.*, *61*: 263 (1989).
2. A. E. Stapleton, *Plant Cell*, *4*: 1353 (1992).
3. S. V. Krupa and R. N. Kickert, *Vegetatio*, *104–105*: 223 (1993).
4. J. Li, T.-M. Ou-Lee, R. Raba, R. G. Amundson, and R. L. Last, *Plant Cell*, *5*: 171 (1993).
5. S. J. D'Surney, T. J. Tschaplinski, N. T. Edwards, and L. R. Shugart, *Environ. Exp. Bot.*, *33*: 347 (1993).
6. T. M. Murphy, *Physiol. Plant.*, *58*: 381 (1983).
7. M. Tevini and A. H. Teramura, *Photochem. Photobiol.*, *50*: 489 (1989).
8. A. H. Teramura, L. H. Ziska, and A. E. Sztein, *Physiol. Plant.*, *83*: 373 (1991).
9. J. He, L.-K. Huang, W. S. Chow, M. I. Whitecross, and J. M. Anderson, *Aust. J. Plant Physiol.*, *20*: 129 (1993).
10. J. D. Stewart and J. Hoddinott, *Physiol. Plant.*, *88*: 493 (1993).
11. M. Tevini and D. Steinmuller, *J. Plant. Physiol.*, *131*: 111 (1987).
12. M. Tevini, W. Iwanzik, and A. H. Teramura, *Z. Pflanzenphysiol.*, *110*: 459 (1983).
13. H. A. Stafford, *Flavonoid Metabolism*, CRC Press, Boca Raton, Fla., p. 239 (1990).
14. Y.-P. Cen and J. F. Bornman, *Physiol. Plant.*, *87*: 249 (1993).
15. P. A. Ensminger, *Physiol. Plant.*, *88*: 501 (1993).
16. C. J. Beggs, A. Stolzer-Jehle, and E. Wellmann, *Plant Physiol.*, *79*: 630 (1985).
17. M. M. Sachs and T.-H. D. Ho, *Annu. Rev. Plant Physiol.*, *37*: 363 (1986).
18. T. I. Baskin and M. Iino, *Photochem. Photobiol.*, *46*: 127 (1987).
19. P. W. Barnes, P. W. Jordan, W. G. Gold, S. D. Flint, and M. M. Caldwell, *Funct. Ecol.*, *2*: 319 (1988).
20. P. W. Barnes, S. D. Flint, and M. M. Caldwell, *Am. J. Bot.*, *77*: 1354 (1990).
21. S. D. Flint and M. M. Caldwell, *Ecology*, *65*: 792 (1984).
22. N. S. Murali and A. H. Teramura, *Photochem. Photobiol.*, *44*: 215 (1986).
23. N. S. Murali and A. H. Teramura, *Physiol. Plant.*, *63*: 413 (1985).
24. L. H. Ziska and A. H. Teramura, *Plant Physiol.*, *99*: 473 (1992).
25. W. A. Feder and R. Shrier, *Environ. Exp. Bot.*, *30*: 451 (1990).

38
Developmental Genetics in Lower Plants

John C. Wallace

Bucknell University, Lewisburg, Pennsylvania

I. INTRODUCTION

A chapter on lower plant developmental genetics may seem out of place in a handbook devoted largely to crop plants, but so much about the basic processes of plant development remains unknown that study of any organism that can aid in the discovery of such processes is worthy of inclusion. The apparent simplicity of lower plants and ease of genetic analysis in the haploid state can make lower plants a model system to rival even *Arabidopsis thaliana* for some aspects of flowering plant development. In the age of genetically engineered plants, when knowledge of fundamentals will be exploited in ways that are not yet dreamed of, such an inclusion is certainly appropriate.

This review focuses on the two lower plants (a volvox and a moss) for which there exists substantial developmental genetic data (i.e., a variety of mutants whose normal ontogeny is altered have been isolated and characterized). A brief section on a third plant (a fern) is also included.

Recent results have shown that the simplest of the three, *Volvox carteri*, is perhaps as closely related to animals as to plants [1], and one of the most fundamental questions that one hopes to answer by studying *Volvox*, that of the origin of the germ–soma dichotomy, is not normally even relevant to plants. But it remains classically defined as a plant, has fascinated biologists for a very long time (e.g., Ref. 2), and recent work on it has produced some exciting results.

The moss *Physcomitrella patens* is much more obviously a model for crop plants, and a healthy array of developmental mutants from it have been isolated [3]. The ease with which it can be transformed [4,5] and the transformants grown to maturity suggests that many developmentally interesting genes from *Physcomitrella* are soon likely to be characterized at the molecular level.

The most complex of the three, the fern *Ceratopteris richardii*, has only recently emerged as a model organism for plant developmental genetics, but as it has been declared "a model plant for the 90s" [6], we shall probably be hearing a great deal more about it in the future.

II. VOLVOX

A. Volvocales: A Variety of Developmental Potentials

Although probably only distantly related to other plants [1], the order Volvocales contains organisms that display an interesting variety of developmental potentials. These range from single-celled free-living organisms (such as those in the genus *Chlamydomonas*), through colonial organisms consisting of a single cell type (genera *Gonium*, *Pandorina*, *Eudorina*), to organisms in which different cells specialize for separate somatic and reproductive functions (genus *Volvox*). Thus within this order are organisms ideal for study of the genetic changes involved in the evolution of multicellularity and of cell specialization. Since the volvox are the most studied in terms of developmental genetics, in this chapter I concentrate on that group and mention others only for purposes of comparison. Several reviews [7–9] give more detailed accounts of volvox development than are included here.

B. Asexual Life Cycle and Mutants

The adult of the most studied volvox, *V. carteri*, comprises a spheroid of about 2000 terminally differentiated somatic cells that contains, within an extracellular glycoprotein matrix, about 16 developing embryos. Reproduction can be either asexual or sexual. In the asexual cycle (Figure 1), large reproductive cells (the gonidia) housed within the spheroid undergo six cycles of mitosis to produce small embryonic spheres of 32 morphologically indistinguishable cells. At the next division the cells in the anterior half of each embryo divide asymmetrically to produce 16 larger cells that will give rise to the next generation of gonidia, and 16 smaller cells that will, together with the remainder of the organism, form the somatic cells. The former divide asymmetrically twice more, releasing more small somatic cell precursors, and the latter undergo several more rounds of cell division, until about 2000 cells are formed, including the 16 larger gonidia. Intitially, the gonidia are located on the outside of the sphere and the cilia of the somatic cells point inward, but in the process known as inversion the adult form, consisting of internal gonidia and external, ciliated somatic cells, is produced. After a period of expansion and cytodifferentiation, the somatic cells of the parental spheroid self-destruct, and the now juvenile volvox are dispersed and live independently.

The haploid nature of all metabolically active cells in volvox is a great aid to mutant isolation, and many have been described that affect the asexual cycle described above [10–12]. Their properties suggest several intriguing possibilities for how the initial split between somatic and germ cell lines is controlled. Mutants at the *pcd* (premature cessation of division) [10] and the several *mul* (multiple gonidia) [11] loci both display altered cleavage patterns and result in a greater than normal number of gonidia being formed. This led to the hypothesis that gonidial determination is a direct consequence of the larger cell size resulting from the asymmetrical cell division [10]. The competing theory [13] is that a cytoplasmic determinant that is partitioned into the larger cells is responsible for initiating reproductive cell development—a phenomenon similar to the pole plasm found to determine germ cell development in *Drosophila* [14]. A recent summation of several lines of evidence suggests that the former hypothesis is correct [12]. For example, if heat shock is used to alter the cleavage of gonidial cells after only two asymmetric divisions, making the third cleavage symmetric, the larger-than-normal offspring, which would normally become somatic, differentiate as extra gonidia. The issue is still clouded,

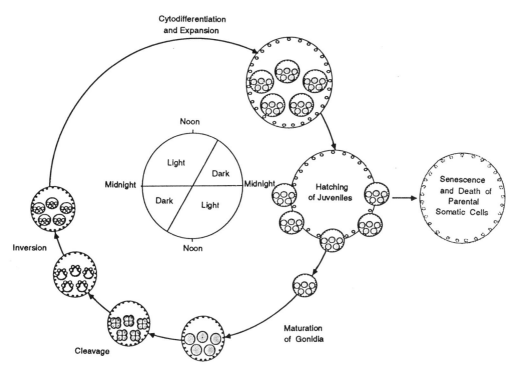

Figure 1 Asexual life cycle of *Volvox carteri*, as synchronized by a light–dark cycle. Mature gonidia (asexual reproductive cells) undergo a rapid series of cleavage divisions, certain of which are asymmetric. The larger cells resulting from these unequal divisions will become the gonidia of the next generation, while the small cells will become part of the somatic cell population. At the end of cleavage, all cells that will be present in the adult are present in undifferentiated form, but the embryo is inside out with respect to the adult configuration. The adult orientation is achieved through a process called inversion. Following inversion, both the parental spheroid and the juveniles contained within it expand by deposition of extracellular matrix. Midway through expansion, the juveniles hatch and swim away, leaving a "hulk" of parental somatic cells that will undergo programmed cell death. The juvenile spheroids continue to expand while their gonidia mature, preparing for a new round of embryogenesis. Under the synchronizing influence of the light–dark cycle, one asexual life cycle is completed every 48 h; cleavage (which takes about 7 h) begins near the end of a light period and inversion (which takes less than an hour) occurs in the dark period. (From Ref. 8.)

however, by the observation that different *Volvox* species appear to utilize different mechanisms to direct this differentiation (reviewed in Ref. 9).

In addition to the *pcd* and *mul* loci mentioned above, several other mutant phenotypes have been isolated that alter the normal developmental sequence. In the Lag (late gonidia) series of mutants, the asymmetric cleavages that lead to gonidia occur as usual, but the resultant large cells at first differentiate somatically, producing cilia. Only later do they redifferentiate and follow the germ cell pathway. A similar phenotype is seen in the RegA mutants. In this type, gonidial cell development is normal but the somatic cells first differentiate normally, then redifferentiate to form small gonidia. Clearly, the *lag* and *regA* genes are critically important to regulation of the germ–soma dichotomy. Finally, mutants of a third gene, *gls* (gonidialess), undergo no asymmetric cleavages and all cells develop somatically; clearly, this mutant cannot reproduce and to be maintained must be carried in a RegA⁻ background.

An especially noteworthy feature of the *regA* locus is its hypermutability [15]. It was found that RegA mutants appeared at an exceptionally high frequency after treatment with agents that interfere with DNA recombination or repair functions. The timing of treatment was also critical and the hypermutability appeared only at two times: a few hours before cleavage began in the gonidial cells, and then shortly after the asymmetric cleavage.

Based on the nature of the mutant phenotypes described above and on the unusual properties of the *regA* locus, a model for how these genes control volvox development has been proposed by David Kirk and colleagues (Figure 2). In the model the *gls* gene is directly responsible for the asymmetric cleavages; the *mul* genes specify the exact times and places for these divisions to occur. In the larger cells the *lag* genes become active, and their products keep those cells from undergoing somatic differentiation. In the smaller cells the *regA* gene product prevents development as a reproductive cell, and in the absence of *lag* gene products the pathway of somatic development is followed. The complementary nature of the *lag* and *regA* genes is particularly apparent here: the former prevent expression of somatic cell genes, and the latter prevents expression of germ cell genes. To ensure that the *regA* gene cannot be expressed in gonidia, it is proposed to be inactivated by an actual physical rearrangement of the gene itself. The

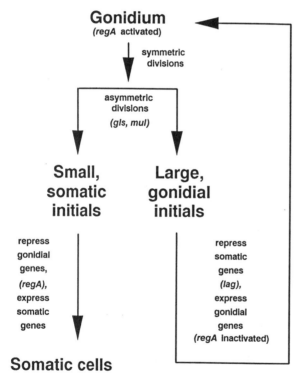

Figure 2 Diagrammatic representations of the asexual life cycle of *Volvox carteri*. Vegetative and reproductive functions are divided between somatic cells and gonidia, respectively. After a number of symmetric divisions have occurred, the *gls* gene acts (at times and places specified by the various *mul* genes) to cause a set of asymmetric cleavage divisions that generate large/small sister cell pairs. In the small somatic initials, the *regA* gene is expressed and acts to suppress expression of genes required for gonidial differentiation. These cells thus become terminally differentiated as somatic cells. In the large gonidial initials, meanwhile, the *lag* genes are expressed and the *regA* gene is inactivated. This leads to repression of the genes required for differentiation of somatic cell features and expression of the genes required for gonidial differentiation. The *regA* gene becomes reactivated in mature gonidia some time prior to the first cleavage. (Modified from Ref. 9.)

hypermutability of the locus is suggested to be a consequence of these DNA rearrangements: the gene is inactivated in pregonidial cells so that they can express the genes for reproductive function, then reactivated in mature gonidia prior to the first cleavage as one-celled embryos [15].

Recently, several somatic- and gonidia-specific cDNA clones have been obtained [16], and analysis of their expression in normal and mutant volvox is proceeding apace [17,18]. Surely this direct analysis of gene expression, and of course the cloning and analysis of the genes responsible for the mutant phenotypes described above, will eventually shed light on how the germ–soma dichotomy is achieved. Even though not directly paralleled in plants, this age-old conundrum is surely of interest to all biologists.

C. Sexual Life Cycle and Mutants

A good description of the sexual cycle in volvox remained elusive until W. H. Darden finally determined the conditions for its sexual propagation in culture [19]. The cycle is summarized in Figure 3. During asexual development males and females are morphologically indistinguishable from one another. After exposure to the exceedingly potent sexual inducer (active at concentrations lower than 10^{-16} M), however, both sexes undergo one more round of asexual development and then initiate gamete formation. For female development, the gonidia divide symmetrically up to the 64-cell stage, and then up to 48 of the cells divide asymmetrically and the larger daughter cells form eggs. If fertilization does not occur, the eggs have the capability of developing into gonidia and continuing development in an asexual manner. After induction, a male gonidium divides symmetrically up to the 256-cell stage, and then an unequal cleavage of all the cells produces 256 sperm initials along with somatic cells; the former undergo six or seven further divisions to produce packets containing 64 or 128 mature sperm. Soon after they form, the sperm packets are released into the surrounding medium and attach to the somatic cells of female spheroids. They subsequently make a hole in the spheroid wall and the individual sperm are released into the interior, where they fertilize the eggs. The resulting zygote develops into a cold- and drought-resistant dormant zygospore that, in culture, becomes activated only when fresh medium is added. In the meiotic division that follows activation only one new germling is produced; the other meiotic products are polar bodies.

Similarly to the asexual cycle, many mutants in sexual development have been isolated

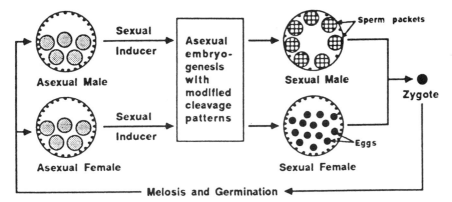

Figure 3 Sexual cycle of *Volvox carteri*. Asexual males and females (which are morphologically indistinguishable from one another) respond to the sexual inducer by undergoing another round of asexual embryogenesis in which the patterns of asymmetric division are modified and in which the germ cells that are formed develop not as gonidia, but as sperm packets or eggs. Sperm–egg fusion results in formation of a dormant, resistant zygote. When dormancy is broken by washing with fresh medium, each zygote undergoes meiosis to form one viable germling and three polar bodies. (From Ref. 8.)

and characterized [11,20]. Somewhat surprisingly, many of these mutants mix and match the asexual, male, and female patterns of cell division with the production of gonidia, sperm packets, and eggs: gonidia can result from cleavages in either the male or female pattern, and sperm or eggs can result from the pattern normally seen in the opposite sex. Thus it is clear that there is no tight linkage between the specific cleavage pattern and the specific type of cellular differentiation that ensues. Explanation and understanding of the controlling factors in these processes will, as above, have to await cloning of the relevant genes.

III. MOSSES

Bryophytes clearly are much more closely related to higher plants than are the Volvocales, both biochemically and functionally. Mosses respond to many of the same growth regulators as higher plants (reviewed in Ref. 21)—in fact, one of the first direct demonstrations of the effect of a cytokinin on plant development was its stimulation of bud development in a moss [22]. They have a multicellular, differentiated sporophyte, and there is no separation, early in development, of discrete germ line and somatic tissue. Thus the study of developmental and physiological processes in mosses is potentially quite relevant to flowering plants.

As with volvox, genetic analysis of bryophytes, especially the generation of mutants, is greatly aided by the fact that the dominant stage of the life cycle is haploid. Unlike the situation for volvox, however, most differentiated cells of mosses, like those of higher plants, are totipotent, remaining capable of dedifferentiating and thence giving rise to an entire new plant. Thus mutations that produce a sterile phenotype can still be maintained and studied. They can even be further mutagenized, since techniques of somatic mutagenesis and regeneration from protoplasts have been developed [23]. Complementation assays and dominance relationships can also be studied in sterile mutants by the production, via protoplast fusion, of somatic hybrids that behave as diploids [24].

A. Basic Life Cycle and Developmental Program

Most of the genetic analysis that has been performed on bryophytes has been in the moss *Physcomitrella patens*, so much of what follows will be based on studies of that organism (recently reviewed in Ref. 3). *Physcomitrella* is a relatively simple moss, but its basic developmental pathways appear to be quite similar to those of all mosses [25,26].

Although mature mosses certainly contain far more than two cell types, and are thus considerably more complex than *volvox*, in the early stages of the life cycle moss development is rather simple. Following germination of a haploid spore, the initial growth pattern consists of a two-dimensional filamentous network of cells called a protonema (Figure 4). In most mosses, including *Physcomitrella*, the initial protonemal cell type is the chloronema. These are chloroplast-rich cells about 115 μm in length that divide about every 20 h. Growth of protonemal tissue is solely at the apices, and after three or four cell divisions an apical chloronemal cell begins to differentiate into a second protonemal cell type, the caulonema. Caulonemal cells contain fewer chloroplasts, are about 160 μm in length, and divide every 8 h. Thus after several days of growth, caulonemal filaments and their derivatives dominate the culture. Essentially all subapical caulonemata will divide to produce side branches, and it is these side branches that can give rise to the next stage of development. Most (ca. 90%) become new chloronemal filaments, a few give rise to new caulonemata, but about 3% begin more complex two- and three-dimensional growth and become buds. Exogenously added cytokinin can cause essentially all side-branch initials to become buds. A bud develops into a gametophore, which consists primarily of the small, leaflike structure that is the most conspicuous part of a moss gametophyte.

Figure 4 Caulonemal filaments from a 3-week-old culture of *P. patens*, which has grown under standard conditions. The main filament axes are composed of caulonemal cells. Almost every subapical caulonemal cell has divided to produce one or more side branches. Most of these have developed into filaments of secondary chloronema. A few have produced buds, which may be seen at various stages of development. Bar 1 mm. (From Ref. 3.)

A young *Physcomitrella* gametophore comprises only a few cell types: the leaf, which is only one cell thick and of course contains no vascular tissue; its supporting stem; and the rhizoid, a filament resembling a caulonemal cell that extends from the base. Other mosses may contain more complex structures and even primitive conducting cells. Figure 5 shows a cell lineage chart of the basic early developmental pattern for *P. patens*.

When gametophores have developed sufficiently and been exposed to the right environmental conditions (cool temperatures for *Physcomitrella*), they are induced to make antheridia and archegonia, the organs that produce sperm and eggs, respectively. After the presence of water allows the motile sperm to effect fertilization, the diploid sporophyte grows out of the archegonium that housed the egg. The sporophyte remains largely dependent on the gametophyte for most of its nutritional requirements. It differentiates a sporangium, or capsule, in which spore mother cells undergo meiosis and generate spores. Moss sporophytes appear to be rather complex morphologically, and very little is known of the factors influencing their development. In *Physcomitrella* the sporophyte is all but invisible; only the mature spore capsule can easily be seen, sitting among the gametophores.

B. Hormone Responses and Mutants

As was the case with volvox, many mutants have been isolated that affect the moss developmental pathway. Since the moss can, unlike volvox, be propagated from single, nonreproductive cells, mutants that block the normal reproductive pathway can be maintained. Cal⁻ mutants, for example, are unable to undergo the first differentiation from chloronema to caulonema. For some isolates with this phenotype, the defect can be overcome by the addition of auxins or cytokinins to the medium, showing that these substances are involved in the switch. In these cases the lesions are most likely in genes involved in the biosynthesis of these compounds. Other Cal⁻ isolates, however, are not responsive to the addition of hormones and thus are likely to be more directly related to the developmental response itself. Similarly, Gam⁻ mutants are unable to form buds, and some, but not all isolates of this mutant can be countered by the exogenous addition of cytokinins. The existence of separate cytokinin-sensitive Cal⁻ and Gam⁻ mutants suggests that the cytokinin response threshold is different for the two types of differentiations: in Gam⁻ mutants of this type some cytokinin must be present for caulonemal cells to form, but it apparently is not enough to trigger the development of buds. A similar two-tiered response to cytokinin has been found for induction of caulonemal branching and of bud formation in the moss *Funaria hygrometrica* [27].

Finally, a third type of hormone-response mutant is given the name Ove. These mutants overproduce buds as a direct consequence of supernormal levels of cytokinin. Somewhat surprisingly, mutations in at least three separate loci give this phenotype, indicating that cytokinin concentrations must be controlled very carefully by the moss [28]. Light also has important effects on many of the developmental switches outlined above (Figure 5 and Ref. 29), but the functions of genes involved in this regulation have not yet been identified.

C. Tropisms and Mutants

Tropisms in a moss protonema occur in the apical cells only (i.e. they involve a change in the direction of growth of a single cell) (Figure 6). This makes the moss particularly attractive for the study of the intracellular signaling events involved in tropisms, since the effects of treatments or manipulations that affect an individual cell can be monitored directly using time-lapse video microscopy.

As to phototropism, chloronemata, caulonemata, and gametophores, all show phototropic and polarotropic responses, and the different cell types vary in their reactions to different

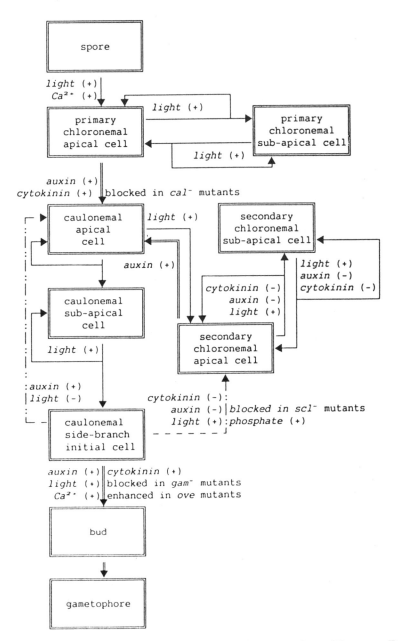

Figure 5 Cell lineages in the development of the gametophyte of *P. patens*. Transitions between stages which are connected by a broken arrow do not require cell division. The branched arrows with single lines represent developmental steps involving a cell division; the arrows indicate the two products of the division. Stages connected by a double-lined arrow require more than one cell division. The (+) sign beside a signal indicates that it is required for or enhances the frequency of the transition; the (–) sign, that it decreases the frequency of the transition. (From Ref. 3.)

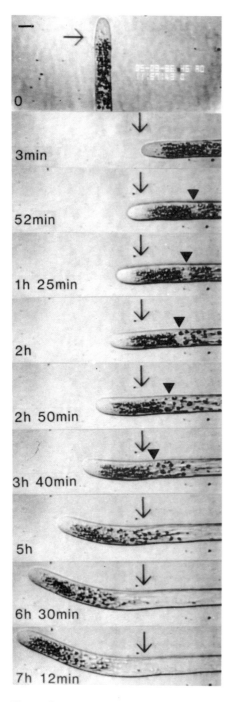

Figure 6   Video images showing the response to 90° reorientation of an apical protonemal cell grown in infrared light. Times after reorientation are shown. The large arrow provides a reference point for the measurement of growth rate and was repositioned immediately after reorientation. The arrowhead denotes the position of the nucleus. (From Ref. 33.)

wavelengths and light intensities [29,30]. Phytochrome is clearly involved in most of these responses. In *Physcomitrella*, mutants in at least three loci have been isolated that have lost phototropism in gametophores (Ptr mutants). Caulonemal cells are similarly affected in these mutants, but the chloronemal response is more complex (reviewed in Ref. 31). These mutations do not alter the light effects of the normal developmental pathways, so they are apparently not involved in light perception per se. The variation in the tropisms of the different cell types in both wild-type and mutant strains makes it clear that no simple controls of phototropism operate in mosses, and more work is needed to determine the mechanisms, both genetic and biochemical, involved.

Caulonemal cells and gametophores also show a negative gravitropic response, but this can be observed only in darkness or infrared light since the phototropic response will override it [32,33]. As is true of the phototropic response, it is only the apical cell of a caulonemal filament that responds to gravity (Figure 6). Several mutants with altered or reduced caulonemal gravitropism (Gtr mutants) have been isolated [32]. Some of these show no gravitropism whatsoever, and surprisingly, one shows a *positive* gravitropism, the opposite of normal. Unlike the phototropic mutants, however, gametophores in the Gtr mutants isolated thus far display a normal gravitropic response.

In contrast to the proposed situation for flowering plants [34], it would appear that the gravitropic response in *Physcomitrella* is not mediated by auxins, since mutants defective in auxin synthesis still show a normal gravitropic response [35].

IV. FERNS

Fern development, especially of the gametophyte, has been a favorite object of study for many years (the subject is reviewed thoroughly in Ref. 36). When compared to *Volvox* and *Physcomitrella*, however, genetic analysis of fern development is very new. Recent studies with the fern *Ceratopteris richardii*, however, show great promise for an understanding of some fundamental features of its developmental genetics [6,37]. Like mosses, ferns have a free-living, independent gametophyte stage that greatly aids mutant isolation and analysis. Like *Physcomitrella*, *Ceratopteris* is self-fertile, so homozygous sporophytes can easily be obtained, and its life cycle can be completed in about 3 months. Also like mosses, if a *Ceratopteris* mutant proves to be sterile or otherwise unable to complete the normal life cycle, it can be maintained indefinitely as either gametophyte or sporophyte via vegetative propagation. The highly developed vascular system and the independence of the sporophyte, however, clearly make ferns substantially more similar to higher plants than are mosses; as a model system, perhaps, they represent the best of both worlds.

Fern gametophyte development, like that of mosses, begins with spore germination producing a protonema. Protonemal development in ferns, however, is one-dimensional and very limited: after only a few cell divisions have resulted in a linear filament, the apical cell begins two-dimensional divisions and becomes the meristem of the flat, heart-shaped prothallus. The sexual archegonia and antheridia eventually develop on the prothallus. After fertilization is effected, the sporophyte rapidly outgrows the gametophyte and assumes an independent existence.

As mentioned above, the genetic study of ferns is still in its infancy, so although others were mentioned briefly in Ref. 6, as of this writing, description of only one developmentally altered mutant has been published [38]. In this Dkg (dark germinating) mutant the normal light requirement for spore germination has been lost; in fact, exposure to light actually inhibits germination. Remarkably, it appears that phytochrome mediates both the light requirement in wild-type spores and the light inhibition in Dkg spores [39]; somehow in the mutant the effects of phytochrome are reversed. Clearly, elucidation of the molecular nature of the Dkg mutant, as well as discovery of other interesting mutants in *Ceratopteris*, are eagerly awaited by the plant developmental biology community.

V. OUTLOOK

In the explosion of interest and excitement in plant biology that has come with the advent of the new biotechnology, nonflowering plants perhaps have been left somewhat by the wayside. No doubt this is at least partly because they have less obvious commercial utility than do crop plants and their relatives. However, as pointed out above, the ease of genetic analysis due to haploidy and rapid generation time, coupled with the apparent simplicity of their development, make lower plants excellent model systems both for some practical aspects of crop plant development and for the study of very fundamental biological processes. In volvox the most basic questions about how one cell can produce daughters of two very different natures is poised to be answered; from mosses such as *Physcomitrella* we shall probably soon learn fundamental truths about hormone signal transduction and tropisms; and study of *Ceratopteris* will aid considerably in an understanding of pattern formation and the development of complex, integrated organ systems. One can only hope that work with these and related systems gets the attention and support that it deserves.

REFERENCES

1. H. Rausch, N. Larsen, and R. Schmitt, *Mol. Evol.*, *29*: 255 (1989).
2. J. H. Powers, *Trans. Am. Microsc. Soc.*, *28*: 141 (1908).
3. D. A. Cove, in *Development, the Molecular Genetic Approach* (V. E. A. Russo, S. Brody, D. Cove, and S. Ottolonghi, eds.), Springer-Verlag, Berlin, p. 179 (1992).
4. D. Schaefer, J.-P. Zryd, C. D. Knight, and D. J. Cove, *Mol. Gen. Genet.*, *226*: 418 (1991).
5. W. Sawahel, S. Onde, C. Knight, and D. Cove, *Plant Mol. Biol. Rep.*, *10*: 315 (1992).
6. R. Chasan, *Plant Cell*, *4*: 113 (1992).
7. D. L. Kirk and J. F. Harper, *Int. Rev. Cytol.*, *99*: 229 (1986).
8. D. L. Kirk, *Trends Genet.*, *4*: 32 (1988).
9. R. Schmitt, S. Fabry, and D. L. Kirk, *Int. Rev. Cytol.*, *139*: 189 (1992).
10. M. L. Pall, in *Developmental Biology: Pattern Formation, Genetic Regulation* (D. McMahon and C. F. Fox, eds.), Vol, 2, Benjamin/Cummings, Menlo Park, Calif., p. 148 (1975).
11. R. J. Huskey, B. E. Griffin, P. O. Cecil, and A. M. Callahan, *Genetics*, *91*: 229 (1979).
12. D. L. Kirk, M. R. Kaufman, R. M. Keeling, and D. A. Stamer, *Development (Suppl.)*, *1*: 67 (1991).
13. G. Kochert, in *The Developmental Biology of Reproduction* (C. I. Markert and J. Papaconstantinou, eds.), Academic Press, New York, p. 55 (1975).
14. S. F. Gilbert, *Developmental Biology*, Sinauer Associates, Sunderland, Mass., p. 276 (1991).
15. D. L. Kirk, G. J. Baran, J. F. Harper, R. J. Huskey, K. S. Huson, and N. Zagris, *Cell*, *48*: 11 (1987).
16. L.-W. Tam and D. L. Kirk, *Dev. Biol.*, *145*: 51 (1991).
17. L.-W. Tam, K. A. Stamen, and D. L. Kirk, *Dev. Biol.*, *145*: 67 (1991).
18. L.-W. Tam and D. L. Kirk, *Development*, *112*: 571 (1991).
19. W. H. Darden, *J. Protozool.*, *13*: 239 (1966).
20. A. M. Callahan and R. J. Huskey, *Dev. Biol.*, *80*: 419 (1980).
21. D. J. Cove and N. W. Ashton, in *The Experimental Biology of Bryophytes* (A. F. Dyer and J. G. Duckett, eds.), Academic Press, London, p. 177. (1984).
22. B. S. Gorton and R. E Eakin, *Bot. Gaz.*, *119*: 31 (1957).
23. P. J. Boyd, N. H. Grimsley, and D. J. Cove, *Mol. Gen. Genet.*, *211*: 545 (1988).
24. N. H. Grimsley, N. W. Ashton, and D. J. Cove, *Mol. Gen. Genet.*, *154*: 97 (1977).

25. M. Lal, in *The Experimental Biology of Bryophytes* (A. F. Dyer and J. G. Duckett, eds.), Academic Press, London, p. 97 (1984).
26. B. Knoop, in *The Experimental Biology of Bryophytes* (A. F. Dyer and J. G. Duckett, eds.), Academic Press, London, p. 143 (1984).
27. M. Bopp and H. J. Jacob, *Planta*, *169*: 462 (1986).
28. D. R. Featherstone, D. J. Cove, and N. W. Ashton, *Mol. Gen. Genet.*, *222*: 217 (1990).
29. E. Hartmann and G. I. Jenkins, in *The Experimental biology of Bryophytes* (A. F. Dyer and J. G. Duckett, eds.), Academic Press, London, p. 203 (1984).
30. D. J. Cove, A. Schild, N. W. Ashton, and E. Hartmann, *Photochem. Photobiol.*, *27*: 249 (1978).
31. D. J. Cove and C. D. Knight, in *Developmental Mutants in Higher Plants* (H. Thomas and D. Grierson, eds.), Cambridge University Press, Cambridge, p. 181 (1987).
32. G. I. Jenkins, G. R. M. Courtice, and D. J. Cove, *Plant Cell Environ.*, *9*: 637 (1986).
33. C. D. Knight and D. J. Cove, *Plant Cell Environ.*, *14*: 995 (1991).
34. Y. Li, G. Hagen, and T J. Guilfoyle, *Plant Cell*, *3*: 1167 (1991).
35. J. R. M. Courtice, unpublished data cited in Ref. 26.
36. V. Raghavan, *Developmental Biology of Fern Gametophytes*, Cambridge University Press, Cambridge (1989).
37. L. G. Hickok, T. R. Warne, and M. K. Slocum, *Am. J. Bot.*, *74*: 1304 (1987).
38. R. J. Scott and L. G. Hickok, *Can. J. Bot.*, *69*: 2616 (1991).
39. T. J. Cooke, L. G. Hickok, W. J. Van Der Woude, J. A. Banks, and R. J. Scott, *Photochem. Photobiol.*, *57*: 1032 (1993).

39

Transpiration Efficiency: Avenues for Genetic Improvement

G. V. Subbarao, Chris Johansen, and R. C. Nageswara Rao

International Crops Research Institute for the Semi-Arid Tropics (ICRISAT), Patancheru, Andhra Pradesh, India

Graeme C. Wright

Queensland Department of Primary Industries, Kingaroy, Queensland, Australia

I. INTRODUCTION

In view of the increasing demand for water for nonagricultural uses (such as for urban and industrial uses), and to rationally redeploy available water resources for more areas of crop production, it is important to optimize the use of water for crop production. Agricultural research has a major responsibility to develop and use techniques and practices that will result in more effective use of water in farming systems. This involves improvement of *water use efficiency* (WUE), defined here as aerial dry matter production of a crop per unit of evapotranspiration (ET). *Transpiration efficiency* (TE) is a component of WUE, being aerial dry matter production per unit of water transpired by the crop. The difference between WUE and TE is important, as suppression of soil evaporation and transpiration by weeds can improve WUE without improving TE, which is a direct measure of the crop species performance. Plant attributes (such as canopy structure, rate of canopy development, etc.) and management means (such as manipulating plant population, optimizing planting dates, fertilizer management, etc.) can modify soil evaporative losses (E_s) relative to transpiration (T), and can therefore affect WUE to a greater extent than can TE.

Generally, any means (either genetic or management) that promotes early canopy development and radiation interception will reduce E_s and increase T (as evaporational losses would be negligible once the canopy closes), often with little or no increase in total ET [1,2]. For example, in Syria, erect chickpea lines intercept less solar radiation, thus permitting greater evaporative water losses during early growth, and consequently, they had a lower WUE value than that of chickpea lines with a prostrate habit [3]. Similarly, leafless pea had a lower WUE value than that of either semileafless or conventionally leafed types [4]. Leafless pea intercepts less radiation than semileafless or conventionally leafed pea, and therefore the crop suffers greater E_s losses. Fertilizer application can increase WUE [5], as it promotes greater leaf area development and reduces E_s relative to T. In many legumes, a basal dose of nitrogen and phosphorus promotes early growth rate and thus minimizes E_s [2]. Other management options, such as improving water delivery systems, nutrient management approaches, and improved cultural practices, could enhance WUE by minimizing E_s.

Also, vapor pressure deficits (vpd) during the growing season play a major role in determining the WUE. When other factors are nonlimiting, the cost of producing dry matter (in terms of water) would be much higher under high vpd (i.e., results in low WUE) compared to low vpd (i.e., results in high WUE) conditions. For instance, in Mediterranean environments, the seasonal WUE varies from 8.5 g/kg (grams of dry matter produced per kilogram of water evaporated or transpired) in midwinter to only 2.5 g/kg in midsummer [6]. Thus management (by early planting, optimizing the plant population and fertility requirements, etc.) and genetic means (such as early vigor, rapid canopy development, cold tolerance, and tolerance to diseases such as *Ascochyta*) that would permit full canopy development, and rapid dry matter accumulation during periods when the vpd is low, would maximize WUE for the growing season. Early planting (i.e., winter planting) in Mediterranean climates usually allows rapid canopy development and dry matter production when the vpd is low and thus results in higher WUE of both dry matter production and grain yield [2,7].

However, once options for minimizing E_s relative to T are exhausted, further improvements in WUE are possible for a given crop only by genetically improving the TE value of that crop. Under water-limited environments, yield is a function of T, TE, and harvest index (HI) [8]. Increased production may result from increased TE if other components (i.e., T and HI) are independent [9] and not affected. By reducing T or by allowing more efficient use of transpirational water in photosynthesis, available soil moisture could be better rationed during the cropping period, which should increase productivity [8].

Plants lose water as they fix carbon dioxide (CO_2) from the air. The loss is inevitable because it is necessary for CO_2 to dissolve in water in order to become available for photosynthesis [10]. This would lead to evaporation as the wet cell surface inside the leaf is exposed to the atmosphere. CO_2 diffuses down a concentration gradient to the leaf interior and water diffuses outward along a decreasing humidity gradient [10]. The lower the external humidity, the higher will be the evaporation, when all the other factors are constant. This two-way diffusion of CO_2 and water forms the basis of improving TE [10]. Cultivars with improved TE are those with inherent characteristics that will allow increased production of dry matter per unit of water transpired [11]. This chapter focuses on exploring the opportunities for genetic improvement of the various morphological, physiological, and biochemical factors that determine TE in C3 crop plants and assesses the scope for exploiting this trait in plant breeding programs.

II. FACTORS AFFECTING TE

Transpiration efficiency is a function of both environmental and plant attributes related to resistances to CO_2 fixation by leaves. Under some circumstances, the environment can have a significant influence on TE. Variation in humidity and temperature can influence TE [12]. TE is governed by three factors: (a) the vpd between air and leaf, (b) the CO_2 gradient from the air to the leaf, and (c) the diffusion resistances for both CO_2 and water [13]. The first factor is mainly abiotic, although the surface temperature of the leaf will actually respond to the atmosphere (e.g., radiation and vpd). The last two factors are largely plant-controlled factors. Also, incident irradiance has an important effect on TE [14]. There is an optimum irradiance for maximum efficiency of water use which is usually less than the irradiance incident on a leaf [15] (see Section II.C for further discussion on this aspect).

A variety of morphological, anatomical, physiological, phenological, and biochemical processes enable crop plants to regulate and ration water for production of dry matter and yield in a given agroecological production system. These are discussed below.

A. Stomatal Behavior

Stomata may exert relatively greater control on water loss than that exerted by CO_2 uptake. This is because the rate of biochemical reactions involved in CO_2 assimilation (A) influences removal of CO_2 from cell solutions and thereby affects CO_2 gradients [16]. This is in addition to resistances faced by CO_2 in its transport, with stomatal resistance perhaps being a smaller component of the total resistance for CO_2 than for water [16]. Stomatal aperture plays a key role in maintaining the balance between taking up CO_2 and losing water [17]. Stomatal movements are the most rapid means by which plants can adjust to changes in the environment [17]. In particular, stomata respond directly to ambient humidity [18], thereby strongly influencing plant TE.

For C3 crop plants, optimization of TE normally requires midday stomatal closure [12]. Such behavior has been observed frequently and is at least partly attributable to the effect of water deficit [19] or is a direct stomatal response to vpd [20]. If diurnal variation in a natural environment were regular and predictable, optimization would require only an appropriate circadian rhythm for stomatal movement [17]. However, this is usually not the case, and therefore optimization requires that the plant respond directly to the changing environment [17]. This demands that stomata respond to changes in external environmental conditions, which in turn influences rates of T and A. Thus stomata should be capable of controlling gas exchange by a feedforward process, making it possible for T to decrease when environmental changes tend to enhance the rate of T (e.g., under high vpd), or for intercellular partial pressure of CO_2 (P_i) to increase when environmental changes would tend to enhance A [21].

Reduced stomatal aperture increases TE because the rate of A is reduced proportionately less than that of T [22–24]. This often happens when plants are subjected to moderate levels of water stress. Factors such as osmotic adjustment (OA) can significantly influence stomatal aperture and thus determine TE under moisture stress. For example, the critical leaf water potential for stomatal closure varies with the level of OA [25,26]. Crop plants show genetic variation for stomatal characteristics such as stomatal density, aperture size, opening patterns, and sensitivity to changes in internal plant water status and soil water status [27–30]. This, in turn, affects their ability to regulate and optimize water use [31,32]. The existence of genetic variation in stomatal characteristics suggests that it may be possible to develop cultivars that utilize water more efficiently, thus contributing to their adaptation to moisture-limiting environments [33,34].

B. Canopy Structure

The aerodynamic resistance of a crop can play a role in determining the relative importance of stomatal conductance (g_s) to TE. If the canopy resistance to heat and water vapor diffusion is large, an increase in g_s would tend to cool and humidify the air in the boundary layer, thus lowering the leaf-air vpd; TE would then increase [35,36]. Thus cultivars with greater g_s could assimilate more at the same level of TE [21,37]. Under field conditions, the boundary layer that forms over crop canopies could cause gas exchange to be less dependent on g_s, and is thus one of the important factors affecting TE [38].

A plant with high TE may be able to decrease the aerodynamic conductance of its canopy boundary layer through greater rigidity of the canopy, while maintaining a high g_s value [39]. This provides it with ready access to CO_2 within the canopy, which is not depleted compared to the bulk atmosphere, while retaining water vapor within the canopy. Boundary layer resistance is a function of the thickness of the unstirred air boundary layer adjacent to the leaf, which in turn is determined by the leaf size [40]. Smaller leaves have a thinner unstirred boundary layer [40]. Thus boundary layer resistance at the canopy level depends on canopy architecture, which is determined by leaf size, leaf arrangement, growth habit (i.e., prostrate versus erect), and height of the canopy. With a low canopy conductance, leaf water equilibrates with an adjacent airspace of higher humidity than the bulk atmosphere [39]. However, such a canopy structure may create sufficiently high levels of humidity within the canopy to be conducive to fungal disease development, thus negating the positive effects of higher TE on biomass production or yield. For instance, in chickpea the closed canopy types, which have greater WUE than that of open canopy types [3], also provide a conducive microenvironment for the development of *Botrytis* and *Ascochyta* blight diseases [41]. Thus the positive effects of such closed canopies on improving the TE of a crop and its production would depend on the availability of sources of resistance to such diseases, which could be incorporated into cultivars forming closed canopies if they lack disease resistance.

C. Leaf Movements and Surface Reflectance

Incident radiation is completely absorbed by the canopy once 100% ground cover is achieved and the incident energy is partitioned between T and A [10]. The proportional allocation differs between species and climates and from year to year [42]. The optimum irradiance for maximum TE is usually less than the irradiance incident upon a leaf oriented normal to the sun's rays [15,43,44]. This is primarily because T normally shows a positive relationship (linear or curvilinear) with increasing irradiance (due to rising leaf temperature and falling stomatal resistance), while A shows a downward curvilinearity with increased irradiance [6]. Leaf movements and surface reflectance provide a means of optimizing this radiation load on the leaf for the maximization of TE. This can be particularly advantageous in water-deficit environments, to dissipate the energy as latent heat, to minimize heat damage, and to optimize TE and radiation use efficiency (RUE) [45–48]. The main advantage of leaf movements is that they would allow maximum exposure of leaf area to direct radiation when evaporative demand is low and thus improve TE. Almost all crop plants show some degree of leaf movement in response to radiation, soil, and plant water status. However, the degree of leaf movement, and the threshold soil and plant water status that triggers these movements, varies among and within crop species, which could contribute partially to their growth performance in water-limited environments [31,49–51].

Leaf pubescence and surface reflectance can provide additional means of controlling leaf temperature and water balance, apart from stomatal control and leaf movement [52–54]. In near isogenic lines of soybean it was shown that lines with pubescent leaves had significantly lower T than either normal or glabrous isolines [52,55]. Leaf pubescence in *Encelia farinosa* reduced absorbance of irradiance as much as 56% compared with the nonpubescent plant *E. californica* [56]. This reduced absorbance can result in lower leaf temperatures and lower T [57]. However, leaf hairs can reflect radiation, which may reduce A. Nevertheless, it appears that in climates with high irradiance and temperatures, beneficial effects of reduced leaf temperature would more than counterbalance the effect of decreased light on A [58]. Other morphological features, such as cuticle thickness and wax deposits on the leaf surface, can to some extent control evaporational losses from the leaf surface [59–62]. There is genetic variability in a number of crop species for leaf surface wax levels and cuticle thickness [60–62].

D. Specific Leaf Area

Variation in TE in crop plants can result from changes in water vapor flux through stomata or by changes in photosynthetic capacity [28,63]. In wheat, variation in TE is caused by stomatal mechanisms [28,64], while in groundnut it appears to be caused by variation in photosynthetic capacity [63,65]. Genotypic variation in photosynthetic capacity on a unit leaf area basis has been observed in many crops [66,67], and a significant negative correlation has been shown between photosynthetic capacity and specific leaf area [68]. This evidence suggests indirectly that the basis of variation in TE through specific leaf area (i.e., leaf thickness) may result from differences in photosynthetic capacity on a unit leaf area basis (see Section V.B for more discussion of this).

E. Root Systems

Root distribution, density, and resistance can influence water use in space and time. Thus WUE can be affected by the rate of growth and spread of roots, particularly during early stages of crop growth. Under receding residual moisture situations, profligate water use during early crop growth might lead to water-deficit conditions during reproductive growth stages. In such circumstances, induction of a large resistance within the plant to the flow of water through selection for smaller metaxylem vessel diameters in the seminal roots should change the pattern of water use for different growth phases [69]. Thus the same amount of water can be transpired to produce more grain yield. Selection for increased root resistance has been shown to be amenable to genetic manipulation in cereals [70,71]. Differences in root radial resistance to water flux have been suggested to occur among groundnut genotypes [72].

III. ASSESSMENT OF GENOTYPIC DIFFERENCES IN TE

Measurement of T in the field is quite complex [73]. Even the field measurement of ET is difficult in many situations where drainage from the root zone, water uptake from saturated zones, and runon and runoff from the area are difficult to measure both temporally and spatially. Transpiration is usually estimated from evapotranspiration measurements such as by (a) subtraction of an estimate of soil evaporation (E_s), which is often a seasonal constant, from the measured seasonal ET [745]; (b) daily water balance simulation using empirical functions to calculate T separately from daily calculations of ET, using measured plant parameters such as leaf area index (LAI) or ground cover [75,76]; or (c) measuring E_s and subtracting it from measurements of ET [77]. All of these measurement techniques, however, result in indirect estimates of T. Direct long-term estimates of TE require accurate measurements of the water used. Rates of water movement through plants can be measured using heat-pulse velocity techniques [78], but difficulties in volume calibrations have limited the accurate estimation of transpiration flux. However, recent improvements in heat-pulse instrumentation have reduced the calibration problems [79,80]. Technical problems related to data collection limit the number of plants that can be measured using this technique. This limits its use in genetic improvement programs where large numbers of plants and genotypes need to be characterized. Pot experiments can give reliable estimations of TE, as they allow accurate measurement of T and dry matter production, including roots. However, these experiments are extremely laborious and are not realistically applicable to screening germ plasm or to genetic studies associated with cultivar improvement [81].

Assessment of genetic variation in TE has often been made based on instantaneous measurements of CO_2 fixation and T from single leaves [82]. However, both of these processes vary markedly during the day and according to leaf and plant age. Thus these instantaneous measurements do not integrate performance throughout the life of the plant. Also, these instantaneous measurements of TE cannot assess the impact of morphological or physiological adaptations to drought that may influence season-long TE and plant performance under water-limited conditions [83,84]. Further, these measurements have large coefficients of variation and are thus usually not suitable for screening and selection studies [85]. It is therefore apparent that breeding for improved TE has been constrained by difficulties in measuring TE on a large number of plants under field conditions [86]. Selection criteria and methods are therefore needed that are efficient and can be used at least indirectly to select genotypes with high TE from large populations in the field.

IV. CARBON ISOTOPE DISCRIMINATION AND ITS RELATION TO TE

A. Theoretical Background

Carbon occurs naturally as two stable isotopes, ^{12}C and ^{13}C. Most carbon is ^{12}C (98.9%), with 1.1% being ^{13}C. As the ^{12}C isotope is lighter than ^{13}C, $^{12}CO_2$ diffuses faster than $^{13}CO_2$. Ribulose 1,5-biphosphate carboxylase (Rubisco) fixes the lighter isotope faster, thus discriminating against the heavier isotope ^{13}C [87]; these two effects cause the $^{13}C/^{12}C$ ratio to be lower in plants than in the ambient atmosphere. The link between TE and $^{13}C/^{12}C$ discrimination (Δ) is *via* the gas-exchange characteristics of the leaves [88]. Since the isotopes are stable, the information inherent in the ratio of abundance of carbon isotope ($^{13}C/^{12}C$) is invariant [88]. The extent of discrimination against the naturally occurring stable isotope ^{13}C during photosynthetic CO_2 fixation in C_3 plants is determined largely by the ratio of intercellular to atmospheric partial pressure (P_i/P_a) of CO_2 [81,88]. As Rubisco actively discriminates against $^{13}CO_2$ [35], $^{13}CO_2$ is concentrated relative to $^{12}CO_2$ in the intercellular spaces as P_i decreases. This concentrating effect results in Rubisco fixing an increased proportion of ^{13}C relative to ^{12}C, and Δ decreases. This is reflected in the carbon isotope ratio of C3 plants, which shows a ^{13}C value of around −25‰ [37]. Therefore, Δ normally correlates positively with P_i/P_a in C3 plants and not in C4 plants (Figure 1), where Rubisco plays a relatively minor role in overall CO_2 fixation. Thus according to theory, in C3 plants a lower ^{13}C discrimination is associated with a higher TE. Variation exists among C3 crop species in their photosynthetic rates (A). This leads to variation in P_i/P_a and is reflected in ^{13}C discrimination values ranging from −22 to −40‰, depending on the crop species [89]. For C4 crops, which have a higher TE than that of C3 crops, ^{13}C discrimination values range from −9 to −19‰; however, these lower values are due mainly to the alternative pathways of CO_2 fixation in C4 crops, such as PEP carboxylase, which does not discriminate between C_{13} and C_{12} [89].

The carbon isotope ratio ($\delta^{13}C$) can be calculated by comparing the ^{13}C to ^{12}C composition of a sample (R_{sample}) relative to the Pee–Dee–Belemnite (PDB) standard (R_{PDB}).

$$\delta^{13}C_{\text{sample}} = \left(\frac{R_{\text{sample}}}{R\text{PDB}} - 1\right) \times 1000 \tag{1}$$

These $\delta^{13}C$ values can be used to calculate isotope discrimination (Δ), as described by Farquhar and Richards [28] and Hubick et al. [63]:

$$\Delta = \frac{\delta^{13}C_{\text{air}} - \delta^{13}C_{\text{sample}}}{1 + \delta^{13}C_{\text{sample}} 1000} \tag{2}$$

The absolute isotopic composition of a sample is not easy to measure directly; the mass spectrometer measures the deviation of the isotopic composition of the material from the standard.

$$\delta P = \frac{R_p - R_s}{R_s} = \frac{R_p}{R_s} - 1 \tag{3}$$

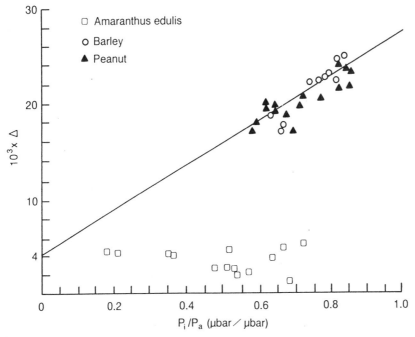

Figure 1 Carbon isotope discrimination, Δ, versus the ratio of intercellular and ambient partial pressures of CO_2 (P_i/P_a), when both are measured simultaneously in a gas-exchange system. Peanut and barley are C3 species and *A. edulis* is a C4 species. (From Ref. 37.)

$$\delta a = \frac{R_a - R_s}{R_s} = \frac{R_a}{R_s} - 1 \tag{4}$$

where δP is the carbon isotope composition of the plant sample, δa the carbon isotope composition of air, R_s the molar abundance ratio of $^{13}C/^{12}C$ of the standard, and R_p and R_a the molar abundance ratio of $^{13}C/^{12}C$ of the plant sample and air, respectively.

The reference material in determinations of carbon isotope ratios has traditionally been in CO_2 generated from a fossil PDB. The carbon isotope composition (δ) is standardized against PDB; atmospheric CO_2 has a value of $-8\permil$ relative to PDB [90].

The carbon isotopic technique can also be used to quantify internal CO_2 levels of leaves on a long-term basis. Internal CO_2 levels (C_i) represent a balance between A and T. The existence of variation in C_i confirms the existence of genotypic differences in TE. Carbon isotope discrimination and TE are related through independent relationships with P_i/P_a [9]. This depends to different extents on the way in which plants coordinate leaf conductance to water vapor with the capacity for photosynthetic CO_2 uptake. Variation in coordination of leaf g_s and A can give rise to variation in P_i/P_a [9]. This, in turn, results in variation in TE and carbon isotope discrimination. It has been stated that if plant breeding is to affect detectable changes in TE of dry matter production, $(1-P_i/P_a)$ needs to be modified substantially [91]. In theory, greater TE will be associated with low Δ if the leaf-to-air vpd remains constant [9].

Farquhar et al. [88] have suggested that Δ can be expressed based on gas exchange, as follows:

$$\Delta = a\frac{P_a - P_i}{P_a} + b\frac{P_i}{P_a} = a + (b - a)\frac{P_i}{P_a} \tag{5}$$

where a is the fractionation occurring due to diffusion in air, which is about -4.4% [92], b the net fractionation caused by carboxylation, which is about -27% [28], and P_a and P_i the ambient and intercellular partial pressures of CO_2, respectively.

The significance of b in equation (5) is that when g_s is small in relation to CO_2 fixation, P_i is small and Δ tends toward a (-4.4%); when conductance is comparatively large, P_i approaches P_a, and Δ approaches b (-27 to -30%; i.e., becomes more negative) [88]. Thus ^{13}C discrimination measurements should be useful in studying the genetic control of g_s in relation to A. Measurements of Δ in C3 crops may contribute to selection for TE. Theory [88] and supporting empirical evidence have shown that differences in intrinsic TE were associated with Δ in a range of crops [9,28,63,65,84,93,94].

The instantaneous ratio of CO_2 assimilation rate of a leaf (A) to its T is given approximately by

$$\frac{A}{T} = \frac{P_a - P_i}{1.6v} \tag{6}$$

where v is the difference in partial pressure of water vapor between the intercellular spaces and the surrounding air. The factor 1.6 is the ratio of the diffusivity of water vapor and CO_2 in air [35].

Farquhar et al. [35] suggested that equation (6) may be rewritten as

$$\frac{A}{T} = \frac{P_a(1 - P_i/P_a)}{1.6v} \tag{7}$$

Equation (7) emphasizes that a small value of P_i/P_a would result in an increase in TE for a constant vpd. Selecting for lower P_i/P_a thus should equate with selecting for greater TE [35]. Therefore, carbon isotope composition ($^{13}C/^{12}C$) of C3 plant tissues provides a long-term integrated measure of photosynthetic capacity [95].

To account for losses of carbon and water due to metabolic and physical processes, Farquhar et al. [35] modified equation (7) to describe the molar ratio, W, of carbon gain by a plant-to-water loss:

$$W = \frac{P_a(1 - P_i/P_a)(1 - \phi c)}{1.6v(1 + \phi w)} \tag{8}$$

where ϕc is the proportion of carbon lost due to respiration, and ϕw is the proportion of water lost other than through stomata (i.e., cuticular transpiration, etc.).

The presence of vpd (v) in equation (8) suggests that TE is affected by environment as well as by physiological responses of the plant [37]. Thus v can vary because of alterations in canopy interception and absorption of radiation via changing leaf angle and surface reflection properties (see Section II.C for more details) and increases or decreases in their coupling to ambient temperature by decreasing or increasing leaf size respectively.

Equation (8) also explains that TE is likely than Δ to be more affected by processes independent of those resulting in variation in P_i/P_a [9]. For example, genetic differences in respiratory losses of carbon, and nonstomatal water losses such as cuticular transpiration, may affect TE independent of P_i/P_a [9]. Thus equations (8) and (5) can be combined to show that Δ is largely dependent on P_i and vpd. Plants with higher TE values will therefore show less negative ^{13}C values or lower Δ values, giving a negative correlation between TE and Δ [35]. This theoretical relationship between Δ and TE in plants with a C3 photosynthetic pathway has been confirmed for several crops in pot [9,28,63,65,81,96–98] and field experiments [72,94,99] (Figure 2).

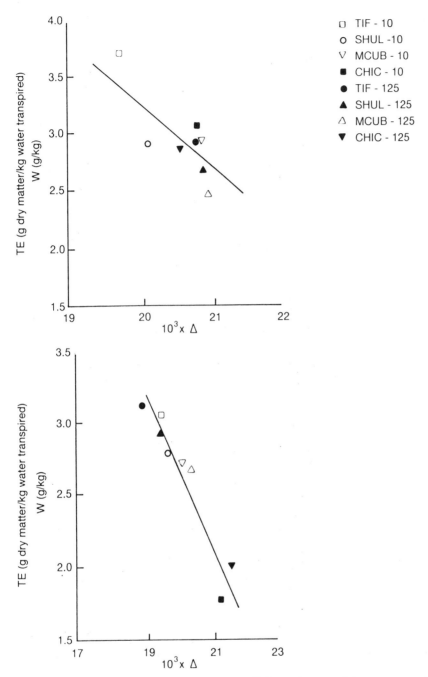

Figure 2 Relationship between transpiration efficiency (grams of dry matter per kilogram of water transpired) and carbon isotope discrimination (Δ) under well-watered and moisture-deficient conditions for a range of peanut cultivars grown under field conditions. (From Ref. 72.)

B.   Water Deficit and TE

The degree of stomatal closure induced by water stress depends on the level of stress and the ability of the crop to meet evapotranspirational demands [100]. Direct measurement of TE using whole plant carbon and water balances have shown that moderate drought can cause an increase in TE of up to 100%, while extreme drought could substantially decrease TE [101]. A common response to water stress is a simultaneous decrease in A and T and an increase in leaf temperature [102]. If T decreases faster than A, then P_i will decrease [23,103]. This response results in water savings to the plant and a subsequent increase in TE. As Rubisco discriminates against $^{13}CO_2$, the proportion of $^{13}CO_2$ to $^{12}CO_2$ also increases within the leaf. Thus $^{13}CO_2$ discrimination decreases as stress becomes more pronounced [104]. In long-term observations in both growth chamber and field conditions, plants under water deficit had lower P_i, as indicated by ^{13}C discrimination analysis [28,105–108]. Several studies with a number of crop species have shown that moderate water stress leads to an increase in TE as indicated by the level of ^{13}C discrimination (Δ) [86,96,109,110]. Water stress resulted in about 2‰ lower Δ than that in well-irrigated plants of chickpea [105]. Similarly, for cowpea (*Vigna unguiculata*), it was shown that leaves sampled from field-grown plants in a dry environment had about a 1.5‰ lower Δ than that of plants from irrigated conditions [111].

Under severe water deficit, TE is reported to decrease [101]. This is because leaves become less efficient with respect to water and CO_2 exchange; water can still be lost through the cuticle, but CO_2 entry through stomata is severely restricted, causing reduced TE [17]. In groundnut, the relation between Δ and TE can break down under severe drought conditions, which could be related to increased respiratory losses of carbon [72]. A similar response has been reported for sunflower [98]. Respiratory losses of carbon can be as much as 40% under severe drought conditions [101].

C.   Influence of Crop Canopy on Δ and TE

The negative relationship between Δ and TE might hold with individual plants in pots [63], or for small plots in the field [65,72] or field-grown crops [94], but might become inconsistent when results are extended to a large area, depending on the crop and microclimate [35]. First, the microclimate in field canopies is usually different than that of isolated plants in pots. This could lead to potential differences in stomatal control of T as influenced by environmental factors, and thus to a breakdown in the relationship between TE and Δ. This emphasizes the problem in the field, where the aerodynamic resistance of the crop has to be taken into account if the canopy and leaf boundary layer resistances to energy flux are very large [37,72]. Because of this it is possible that under high atmospheric evaporative demands, plants can have a high g_s, and thus a high Δ, but also high TE, due to complete closure of the canopy [112]. However, this is less likely to occur when crops have small LAIs, as would be the case under conditions where stress occurs early in the cropping season, because under these conditions the crop is more closely coupled to the atmosphere [38,112]. However, if the source of variation in Δ is the capacity for photosynthesis, the effects of boundary layers are unimportant [112], as seems to be the case for groundnut [9,72]. Therefore, at the crop level, identification of the causes underlying differences in Δ may become important.

Second, the nonstomatal loss of water (i.e., cuticular transpiration, soil evaporation) ($\emptyset w$) could vary with leaf area development and the level of wax deposition on the cuticle, and thus is not an independent fixed proportion of transpiration. This could influence Δ, as $\emptyset w$ is an important component of WUE [equation (8)]. Also, since vpd is an important component of equation (8), any fluctuation in vpd during the growing season and the growth rate of a given variety during the growing season could influence TE. For example, those genotypes that grow faster when vpd is small because of their adaptation to low temperatures could show a greater TE for the same Δ.

V. SCOPE FOR GENETIC IMPROVEMENT OF TE IN C3 CROP PLANTS

A. Relation Between Transpiration and Photosynthesis

Since the stomatal diffusion pathway is the same for both water vapor and CO_2 exchange, water is inevitably lost when stomata open and CO_2 is absorbed. Stomatal conductance is believed to adjust according to the assimilatory capacity of the mesophyll tissue [113]. That is, other factors being similar (i.e., nonlimiting), stomata open to the extent required to provide CO_2 at rates sufficient to meet the CO_2 fixation requirements of the metabolic pathway [114]. Close coupling between A and T is expected since CO_2 and H_2O move simultaneously through the stomata [115]. The diffusive conductance of the stomatal opening imposes a major control on the rates of both processes, although C_i concentration and the external water vapor concentration determine the magnitude of the respective gradients [115]. However, changes in g_s may not necessarily affect T and A similarly [23].

There is a strong correlation between A and g_s over a wide variety of plant species and under a diversity of environmental conditions [114,116]. This implies some level of regulation between CO_2 demand by chloroplasts and CO_2 supply, via stomatal control. Generally, leaf conductance and photosynthesis are correlated at low conductance levels but are uncoupled at high conductance levels [117]. If there is no deviation from the slope of photosynthesis versus conductance relationships, and if the intercept is zero (as is assumed initially), the P_i values of all crop plants should be constant, depending only on the photosynthetic pathway [83]. Although many studies have shown a significant tendency for photosynthesis and conductance to be correlated [114], many of these data sets exhibit some deviation from a linear relationship or nonzero intercept [118,119].

Genotypic variation in TE can result from variation in g_s but with the genotypes having the same level of photosynthetic capacity [56]. The slopes of the regression line of g_{max} (stomatal conductance maximum) versus A_{max} vary substantially among C3 plants [56,120]. For high evaporative environments, it has been shown that genotypic differences in P_i, based on long-term gas-exchange studies, as well as on ^{13}C discrimination analysis offer the possibility of genetically modifying TE [56]. However, for low evaporative environments, it appears that A is highly dependent on leaf g_s suggesting little possibility of improvement of TE [56].

B. Mechanisms by Which Genotypes Differ in TE

Any factor that influences genetic variation in either g_s or A in a disproportionate manner would influence Δ and thus TE [37]. If variation in A was the only cause of variation in P_i, increasing photosynthetic capacity should lower P_i/P_a and therefore lower Δ. In this situation, TE would increase and the relationship between Δ and plant biomass should be negative [121]. In groundnut, differences in A are reported to be largely responsible for TE variation, as dry matter production is negatively correlated with Δ in pots [9,72] and at the canopy level [65,94]. Significant variation in A per unit leaf area have been reported in groundnut genotypes and there is also heterosis for this trait [67,122–124]. Similarly, in cowpea, genotypic means for TE were positively correlated with A but only weakly correlated with g_s, indicating that genotypic differences in TE were due primarily to differences in A [110].

A strong positive correlation has been observed between Δ and specific leaf area (SLA) among groundnut genotypes [99]. This is consistent with the foregoing hypothesis that high TE genotypes have higher A. Indeed, the genotypes with thicker leaves (low SLA) had significantly higher leaf nitrogen contents, again indicative of higher photosynthetic capacity. The significant application of these observations is that breeders could use the inexpensively measured SLA, in lieu of Δ, to screen for high TE among groundnut genotypes within specific environments [72].

However, if g_s is the main source of variation in P_i/P_a, greater g_s should increase P_i/P_a, and therefore increase Δ. In adequately irrigated coffee, higher TE values of some of the genotypes tested was associated with reduced stomatal aperture rather than increased A at a given g_s [97]. This suggests that high TE may restrict yield when water supply is not limiting. Thus in this case, as in wheat, selection for higher Δ could lead to increased biomass production but with decreased TE [125]. For example, in crested wheatgrass, greater TE in low Δ clones resulted from a proportionately greater decline in g_s than in A [104]. Similar results were reported for chickpea [105]. However, variation in P_i/P_a among wheat genotypes is approximately equal to variation in leaf g_s and in A [64,126–128]. In wheat, it was reported that g_s covaried with A, with the change in g_s being relatively greater [128]. This means that there could be a positive correlation between A and P_i/P_a. The effect of this on growth may be compounded if genotypes with large P_i/P_a partition more carbon into shoots [129].

Cultivar differences in Δ may also result indirectly from genetic variation in root characteristics affecting the level of water stress experienced by the canopy [96,130]. Differences in root growth affect the degree of dehydration postponement, and this could prolong gas-exchange activity and the maintenance of relatively high P_i and thus Δ [130].

C. Genetic Variation and Genetics of TE and Δ

Genetic variation in TE and Δ has been reported in wheat [64,121,125,131], barley [93], tomato [84], sunflower [98], chickpea [105], groundnut [63,65,72], cowpea [86], and coffee [97]. In wheat, variation in Δ among genotypes is typically in the range 2×10^{-3} [131]. This is equivalent to a variation in TE of 59% [131]. In groundnut, genotypic variation in TE is estimated as about 65% [63]. Based on extreme cases of genotypes which differ in TE, it was reported that cowpea genotypes such as vita 7 and 8049 had nearly 67% higher TE values than those of other genotypes tested [109]. Also, earliness is generally associated with low TE in cowpea; however, significant genotypic differences were noted within any given maturity group, suggesting that these two traits are not necessarily linked [109]. Similarly, tall landrace genotypes of wheat, which are also late maturing, had higher TE values than did the modern dwarf and semidwarf genotypes [121]. However, among Australian wheats, low values of Δ and thus high TE have been found to be strongly associated with the WW15 genetic background, which was introduced into Australia from CIMMYT as a major source of the dwarfing gene in Australian wheats.

The utility of a trait for selection in plant breeding programs is strongly enhanced by the consistency of genotypic ranking across environments [110]. Based on studies with wheat, cowpea, crested wheat grass, groundnut, and beans, it was found that genotypic ranking for Δ across environments is consistent [36,37,99,107,109,110,125,131,132]. For crops such as groundnut, it was shown that genotypic ranking for Δ was maintained during ontogeny [72] (Figure 3). However, in crops such as wheat, genotypic ranking could change between the early vegetative stage and the heading and grain filling stages [131]. This could be due to a number of factors, including hormonal imbalance, causing loss of stomatal control on water loss after heading. Also, the plant material used for Δ analysis could determine the level of heritability [131]. It was shown in a number of crops that the Δ value of leaf material is a better indicator of differences in TE than that of grains [9,36,63,107,109,121]. One of the main reasons could be genotypic differences in the ability to translocate preanthesis-stored carbohydrate reserves for grain filling [133].

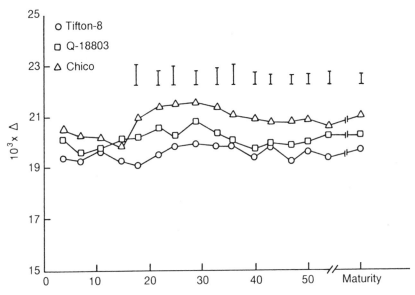

Figure 3 Change in carbon isotope discrimination in leaves and error variation versus time for well-watered groundnut cultivars of Tifton-8, Q 18803, and Chico grown in a greenhouse. (From Ref. 72.)

The effectiveness of indirect selection for TE using Δ will depend partly on the magnitude of the heritabilities for TE and Δ and the genotypic correlation between these characters [134]. Broad-sense heritability, which is the proportion of total phenotypic variance that is attributable to genotypic differences [131], is a measure of the repeatability of the expression of those genotypic differences [131]. In many crops, heritabilities for Δ are above 80% [9,37, 107,109,121,131].

D. Advantages of Using Δ for TE Evaluations

Breeding for improved TE has been limited by the lack of screening tools for identifying desirable genotypes under field conditions [110]. The ^{13}C discrimination technique makes it possible to survey a large number of plants with a simple, albeit expensive analysis of the leaf tissue [10]. As Δ provides an integrated estimate of TE, it has been suggested that measurement of Δ may better differentiate among genotypes than most instantaneous physiological assays [121]. Genotypic ranking based on Δ is much more consistent than that based on gas-exchange measurements [110], and thus should be easier to select for in breeding programs. Also, as Δ remains reasonably constant throughout crop ontogeny, selection could be made during crop development [72].

Further, Δ is faster and easier to measure than total growth relative to total water use [28]. It is readily determined on field-grown plants because it does not require the plant to be sheltered from rain, or that any other special experimental treatment be maintained. Measurements can be made on small plant samples collected at maturity with minimal problems of storage and handling. The material can be either leaf, stem, or grain. Leaves and stems are easier to grind, and use of vegetative material has the potential advantage that selection can be made early in the crop growth cycle and thus could assist in improving selection efficiency and reducing the time and maintenance costs [28,131].

E. Limitations of Using Δ to Select for TE

Carbon partitioning and Δ would not be expected to be stable across all environments and with changes in plant hormonal balance. For example, cytokinins and ABA can affect both leaf gas exchange and carbon allocation [104]. Also, there are some problems of assessment of TE through carbon isotope estimations: (a) it is a "ratio" and not correlated directly with yield or productivity; (b) the small sample size may introduce subsampling errors and careful grinding is required; and (c) the technique requires considerable capital investment in equipment and technical expertise [104].

Also, there are a number of potential sources of nongenetic variability in the measurement of Δ. Some can be readily overcome by technical or sampling precautions, as they are associated with the composition of plant dry matter [135] and the size and storage of the dry matter sample used in the measurement [37]. Other sources of variation in Δ among plant organs result from temporal variation in the growth environment. Increased salinity [136,137], decreased soil water availability [28,65,106], soil compaction [129], and a decrease in vpd [138] could all result in lower values of Δ.

Genotypic variation for Δ measured under field conditions could be complicated by inherent differences in root growth [130]. This would affect the degree of dehydration postponement that could allow prolonged maintenance of relatively large g_s, thus decreasing TE but increasing growth and yield. Positive correlations between root length density and Δ have been reported in crops such as beans [130,139] (Figure 4), and thus selection for low Δ (high TE) may lead to selection of genotypes with poor root attributes, such as shallow rooting and low root densities. Bean genotypes that had a deeper root system had high Δ values compared to the shallow-rooted genotypes [130]. Thus leaf physiology (as measured by Δ) is not independent of root activity, and it seems that there is a close correlation between gas exchange under water-deficit environments and root attributes [130]. One way to overcome this problem of differences in root attributes is to evaluate germ plasm lines under irrigated conditions, where differences in root growth do not affect the leaf gas-exchange characteristics, and thus Δ. In many crop species, variation in P_i/P_a and Δ has been reported among genotypes under irrigated conditions, indicating the existence of genetic variation in the "baseline C_i" that is expressed under nonstress conditions [130].

In crops such as groundnut, there is a moderately positive correlation ($r = 0.55$) between Δ and HI, and thus selecting for low Δ (high TE) could lead to selection of genotypes with low partitioning [9,65,94]. This indicates that selection for high TE and HI, and thus yield potential, could be difficult because of this negative association. However, the possibility of combining high HI and high TE requires further research [9,94]. This highlights the need for physiologists and breeders to be aware of the potential for negative associations between traits such as TE, partitioning of biomass, and root water uptake attributes of roots.

As several factors can alter plant dry weight independently of Δ, there may not always be a direct association between Δ and productivity [35]. However in many crops, the general trend in relationship between Δ and dry matter productivity is negative; that is, higher productivity under optimum conditions (e.g., irrigated) is associated with lower Δ [132]. Thus in crops where there is a positive association between Δ and dry matter production, it may be that high TE and potential for dry matter productivity are incompatible. For crops such as wheat, barley, and beans, where differences in TE are due mainly to differences in g_s, there appears to be a positive correlation between Δ and dry matter production [125]. This indicates that selection for low Δ could lead to selection of genotypes with low dry matter accumulation capability and thus potential productivity. It was suggested that selection for low Δ will improve adaptation to drought [28], whereas selection for high Δ should improve yield potential [125]. However, it should still be possible to identify genotypes that do not comply with this general relationship.

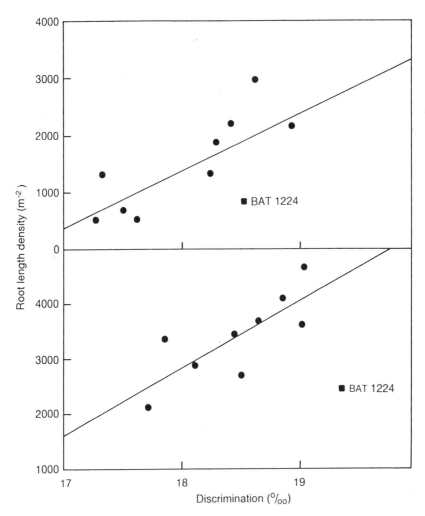

Figure 4 Relations between leaf carbon isotope discrimination and root length density for rain-fed bean genotypes at two locations: Palmira (upper graph) and Quilichao (lower graph). (From Ref. 130.)

For example, in barley, although there is generally a negative relationship between TE and dry matter accumulation among the genotypes tested, certain genotypes deviate from this relationship (Figure 5) [140].

For crops such as groundnut, and in cool-season grasses, where photosynthetic rates are the main source of variation in TE, selection for low Δ should lead to genotypes with high dry matter production capabilities [9,36,65,132]. Thus it is interesting to note that the usefulness of Δ in selection for high TE could vary depending on the crop species and the target environment; in one case it could lead to improving productivity, and in other cases it could be detrimental to productivity.

F. Role of TE in Improving Drought Resistance of Crops

Crop plants have evolved a variety of strategies to cope with water-deficit conditions [141]. The seasonal progression of temperature, the distribution and intensity of rainfall, and the

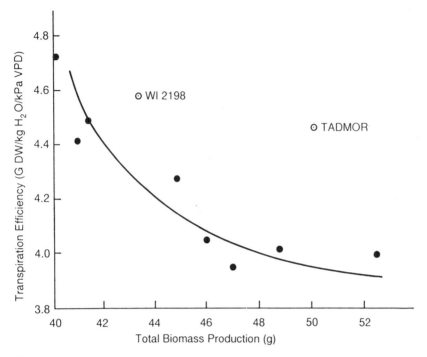

Figure 5 Transpiration efficiency and total biomass production in barley genotypes grown in a greenhouse. (From Ref. 140.)

availability of soil moisture will largely determine the plant attributes that need to be altered beneficially to improve the efficiency of water use [142]. Transpiration efficiency is one of the components involved in adaptation to drought by potentially extending the period of soil moisture availability, and is thus expected to contribute to improving adaptation to drought-prone environments. This is particularly so if the crop is raised on finite amounts of stored moisture. A drought-resistant groundnut genotype (drought resistance defined here as relative total dry matter production under drought conditions), Tifton-8, was found to be very efficient in its water use compared to a sensitive *A. villosa* [143]. Chico, a short-season groundnut variety, had a lower TE value than those of long-season groundnut varieties [96], which are also found to be more drought resistant than the short-season varieties. In wheat, barley, cowpea, and groundnut, TE is positively correlated with days to heading, which indicates that selection for early maturity might result in decreased TE values [63,94,109,121,125,144]. However, in groundnut, there is still considerable variation in TE/Δ within similar maturity groups, indicating that the variation in TE could be located in any given maturity group [63,94]. Thus simultaneous selection for TE and phenological characteristics should be practiced to improve TE within an optimum maturity group. Tall landrace wheat genotypes had greater total dry matter and TE, but were later in maturity than the modern dwarf and semidwarf genotypes [121].

In many cropping systems where irrigation water is not readily available, yield stability can be affected by intermittent droughts [9]. Ideally, maximum growth with the water available is a goal. One possibility for improving productivity in low-rainfall and drought-prone areas is to select and breed plants that require less water for growth without losing their yield potential (i.e., to improve their TE value). However, there is a distinction between TE and drought resistance as a whole, and it needs to be recognized that the development and use of drought-resistant plants can lead to the effective use of limited soil water that would otherwise be unavailable. In effect, WUE would be increased for the entire land area even if the drought-resistant crops grown actually transpire more water per unit of dry matter than do nonresistant crops.

In rain-fed environments, TE alone may not play a key role in determining the level of drought resistance of a given cultivar. The negative correlations between reduced Δ, biomass, yield, and LAI indicate that greatest growth under rain-fed conditions would occur in cultivars best able to postpone desiccation and maintain relatively large stomatal conductance (i.e., mostly to deal with the efficiency with which the water is extracted rather than utilized), thus showing less reduction in C_i than that occurs in irrigated treatments [130]. However, high levels of TE and efficient root systems (deep root system, uniform root length distribution through the soil profile, efficient water uptake from low soil water potentials, etc.) are independent attributes of a plant; therefore, they need not be incompatible. Thus one could improve the TE of a given variety through breeding even if it is found to have a more efficient root system but a low TE. In groundnut, some of the genotypes that have deep rooting attributes and are more efficient in water uptake also had higher levels of TE than those of genotypes poor in both attributes [145].

Assuming that the traits contributing to drought resistance are independent attributes, it would be necessary to develop ideotypes to suit the requirements of specific target production environments [146,147]. Then genetic improvement would depend largely on the local variety that needs to be improved, which can be guided by using the ideotype as a basis for the evaluation of traits that need to be incorporated [146]. Thus genetic improvement for better adaptation to moisture-deficit environments could be focused on a few selected traits rather than considering adaptation as a single component of improvement. This would assist in quantifying progress and devising appropriate strategies for further improvement, apart from being able to use genetic stocks developed during the process in related breeding programs in other production environments.

VI. FUTURE OUTLOOK

Large sums of money have been spent to develop irrigated cropping systems throughout the world, but relatively little attention has been paid to research on improving WUE, let alone genetically improving the TE values of crop species [148]. Although differences among and within crop species in their TE values (thus in their total water requirements to produce a given amount of yield) were demonstrated 80 years ago [149], very little progress has been made since in initiating breeding programs specifically targeted at improving TE values in any crop species. This is due mainly to the lack of appropriate means of characterizing and quantifying genotypic variation in TE and the inability to handle the large number of samples required in a breeding program. The recent finding that TE is negatively related to ^{13}C discrimination (Δ) has led to a renewed interest in TE as a potentially exploitable trait, and thus Δ has been proposed as a selection criterion for improving TE in plant breeding programs [28]. It has now been shown that genetic variation in TE exists for many crop species under both well-watered and moisture-deficit environments. The high levels of heritability for Δ have further strengthened the argument that Δ is amenable to genetic improvement. This opens the way for developing crop varieties that require less water to produce the same amount of yield according to their present potential. This also provides scope for much more rational deployment of irrigation water.

However, ^{13}C discrimination analysis of plant samples requires mass-spectrometer facilities, and it is beyond the ability of many breeding programs to acquire and maintain such expensive and sensitive equipment. This is particularly so in developing countries, which are located mostly in semiarid regions, where improving crop TE could play a crucial role in improving and stabilizing crop production. Thus this would presently be the limiting factor for the use of this technology in breeding programs focused specifically toward genetic improvement of TE. Nevertheless, it could still be handled by having centralized facilities in selected institutes where analyses could be done. Also, once the equipment is installed and maintained, the actual analysis costs may be within the capability of many breeding programs. Correlated traits such as specific leaf area, which has been shown to be related with Δ, could thus be used as a surrogate to ^{13}C discrimination analysis. Measuring specific leaf area could be relatively inexpensive and requires no special equipment. However, it needs to be proved that selection programs based on specific leaf area could lead to genetic enhancement of TE and its heritability needs to be established clearly before proposing this as surrogate to Δ in a selection program. There are indications in groundnut that it could be used effectively as an alternative to Δ in selecting for TE [99], but this needs to be proved convincingly. It will therefore be interesting to observe the extent to which this new tool (i.e., Δ) is put into use in developing varieties that are better adapted to moisture-deficit environments without a loss in yield potential.

ACKNOWLEDGMENT

We wish to acknowledge editorial assistance from the ICRISAT Editorial Committee in improving the structure and presentation of this manuscript.

REFERENCES

1. P. J. M. Cooper, J. D. H. Keatinge, and G. Hughes, *Field Crops Res.*, *7*: 299 (1983).
2. P. J. M. Cooper, G. S. Campbell, M. C. Heath, and P. D. Hebblethwaite, in *World Crops: Cool Season Food Legumes* (R. J. Summerfield, ed.), Kluwer Academic Publishers, London, pp. 813–829 (1988).
3. G. Hughes, J. D. H. Keatinge, P. J. M. Cooper, and N. F. Dee, *J. Agric. Sci.*, *(Camb.)*, *108*: 419 (1987).
4. M. C. Heath and P. D. Hebblethwaite, *Ann. Appl. Biol.*, *107*: 309 (1985).
5. F. G. Viets, Jr., *Adv. Agron.*, *14*: 223 (1962).
6. R. A. Fischer and N. C. Turner, *Annu. Rev. Plant Physiol.*, *29*: 277 (1978).
7. M. C. Saxena, in *The Chickpea* (M. C. Saxena, and K. B. Singh, eds.), CAB International, Wallingford, Berkshire, England, pp. 207–232 (1987).
8. J. B. Passioura, *J. Aust. Inst. Agric. Sci.*, *43*: 117 (1977).
9. K. T. Hubick, R. Shorter, and G. D. Farquhar, *Aust. J. Plant Physiol.*, *15*: 799 (1988).
10. J. S. Boyer, in *Plant Breeding in the 1990s* (H. T. Stalker, and J. P. Murphy, eds.), CAB International, Wallingford, Berkshire, England, pp. 181–200 (1992).
11. D. K. Barnes, in *Limitations to Efficient Water Use in Crop Plants* (H. M. Taylor, W. R. Jordan, and T. R. Sinclair, eds.), American Society of Agronomy, Madison, Wis., pp. 127–136 (1983).
12. W. J. Davies, in *Plant Physiology: A Treatise*, Vol. IX, *Water and Solutes in Plants* (F. C. Steward, J. F. Sutcliffe, and J. E. Dale, eds.), Academic Press, New York, pp. 49–154 (1986).
13. D. Hillel, in *Irrigation of Agricultural Crops* (B. A. Stewart, and D. R. Nielson, eds.), American Society of Agronomy, Madison, Wis., pp. 5–30 (1990).
14. S. M. A. Faiz and P. E. Weatherley, *New Phytol.*, *81*: 19 (1978).
15. H. G. Jones, *J. Appl. Ecol.*, *13*: 605 (1976).
16. D. N. Moss, J. T. Wooley, and J. F. Stone, *Agric. Meteorol.*, *14*: 311 (1974).
17. W. Wenkert, in *Limitations to Efficient Water Use in Crop Plants* (H. M. Taylor, W. R. Jordan, and T. R. Sinclair, eds.), American Society of Agronomy, Madison, Wis., pp. 137–172 (1983).
18. G. D. Farquhar, *Aust. J. Plant Physiol.*, *5*: 787 (1978).

19. H. Meidner and T. A. Mansfield, *Physiology of Stomata*, McGraw-Hill, London (1968).
20. E. D. Schulze, O. L. Lange, M. Evenari, L. Kappen, and M. Evenari, *Oecologia*, *17*: 159 (1974).
21. I. R. Cowan and G. D. Farquhar, in *Integration of Activity in the Higher Plants* (D. H. Jennings, ed.), Cambridge University Press, Cambridge, pp. 471–505 (1977).
22. K. J. Bradford, T. D. Sharkey, and G. D. Farquhar, *Plant Physiol.*, *72*: 245 (1983).
23. I. R. Cowan and J. H. Troughton, *Planta*, *97*: 323 (1971).
24. J. I. L. Morison, *Plant Cell Environ.*, *8*: 467 (1985).
25. K. W. Brown, W. R. Jordan, and J. C. Thomas, *Physiol. Plant.*, *37*: 1 (1976).
26. E. Fereres, E. Acevedo, D. Henderson, and T. C. Hsiao, *Planta*, *44*: 261 (1978).
27. M. M. Ludlow, in *Adaptations of Plants to Water and High Temperature Stress* (N. C. Turner and P. J. Kramer, eds.), Wiley, New York, pp. 123–138 (1980).
28. G. D. Farquhar and R. A. Richards, *Aust. J. Plant Physiol.*, *11*: 539 (1984).
29. A. H. Markhart, *Plant Physiol.*, *77*: 113 (1985).
30. D. Vignes, A. Djekoun, and C. Planchan, *Can. J. Plant Sci.*, *66*: 247 (1986).
31. R. J. Lawn, *Aust. J. Agric. Res.*, *33*: 481 (1982).
32. R. G. Henzell, K. J. McCree, C. H. M. van Bavel, and K. F. Schertz, *Crop Sci.*, *16*: 660 (1976).
33. L. Riccardi and P. Steduto, *FABIS Newsl.*, *20*: 21 (1988).
34. M. J. Hattendert, D. W. Evans, and R. N. Peaden, *Agron. J.*, *82*: 873 (1990).
35. G. D. Farquhar, J. R. Ehleringer, and K. T. Hubick, *Annu. Rev. Plant Physiol.*, *40*: 503 (1989).
36. J. J. Read, R. C. Johnson, B. F. Carver, and S. A. Quarrie, *Crop Sci.*, *31*: 139 (1991).
37. G. D. Farquhar, K. T. Hubick, A. G. Condon, and R. A. Richards, in *Stable Isotopes in Ecological Research* (J. R. Ehleringer and K. A. Nagy, eds.), Ecological Studies, Vol. 68, Springer-Verlag, New York, pp. 21–40 (1988).
38. P. G. Jarvis and K. G. McNaughton, *Adv. Ecol. Res.*, *15*: 1 (1986).
39. C. D. Walker and R. C. N. Lance, *Aust. J. Plant Physiol.*, *18*: 411 (1991).
40. D. F. Parkhurst and O. L. Loucks, *J. Ecol.*, *60*: 505 (1972).
41. C. Johansen, B. Baldev, J. B. Brouwer, W. Erskine, W. A. Jermyn, Lang Li-Juan, B. A. Malik, A. Ahad Miah, and S. N. Silim, "Biotic and Abiotic Stresses Constraining Productivity of Cool Season Food Legumes in Asia, Africa and Oceania," *Proceedings of the International Food Legume Research Conference II*, Apr. 12–16, 1992, Cairo (F. J. Muehlbauer and W. J. Kaiser, eds.), Kluwer, Dordrecht, The Netherlands (in press).
42. R. J. Hanks, in *Limitations to Efficient Water Use in Crop Production* (H. Taylor, W. R. Jordan, and T. R. Sinclair, eds.), American Society of Agronomy, Madison, Wis. pp. 393–411 (1983).
43. J. F. Beerhuizen and R. O. Slatyer, *Agric. Meteorol.*, *2*: 259 (1965).
44. R. W. Downes, *Aust. J. Biol. Sci.*, *23*: 775 (1970).
45. K. A. Shackel and A. E. Hall, *Aust. J. Plant Physiol.*, *6*: 265 (1979).
46. M. M. Ludlow and O. Bjorkman, *Planta*, *161*: 505 (1984).
47. I. N. Forseth and A. H. Teramura, *Ecology*, *67*: 564 (1986).
48. V. S. Berg and S. Heuchelin, *Crop Sci.*, *30*: 631 (1990).
49. R. C. Muchow, *Field Crops Res.*, *11*: 309 (1985).
50. G. R. Squire, *The Physiology of Tropical Crop Production*, CAB International, Wallingford, Berkshire, England (1990).
51. R. B. Matthews, D. Harris, J. H. Williams, and R. C. Nageswara Rao, *Exp. Agric.*, *24*: 203 (1988).
52. S. R. Ghorashy, J. W. Pendleton, and M. E. Bernard, *Crop Sci.*, *11*: 426 (1971).
53. J. Ehleringer, in *Adaptation of Plants to Water and Temperature Stress* (N. C. Turner and P. J. Kramer, eds.), Wiley-Interscience, New York, pp. 295–308 (1980).
54. M. Ashraf and F. Karim, *Trop. Agric. (Trinidad)*, *68*: 57 (1991).
55. D. D. Baldocchi, S. B. Verma, and N. J. Rosenberg, *Agric. For. Meteorol.*, *34*: 53 (1985).
56. E. D. Schulze and A. E. Hall, in *Physiological Plant Ecology II: Water Relations and Carbon Assimilation* (O. L. Lange, P. S. Nobel, C. B. Osmond, and H. Ziegler, eds.), Springer-Verlag, New York, pp. 181–230 (1982).
57. J. R. Ehleringer, O. Bjorkman, and H. A. Mooney, *Science*, *192*: 376 (1976).
58. J. R. Ehleringer and H. A. Mooney, *Oecologia*, *37*: 183 (1978).

59. J. M. Clarke and R. A. Richards, *Can. J. Plant Sci.*, *68*: 975 (1988).
60. M. C. M. Paje, M. M. Ludlow, and R. J. Lawn, *Aust. J. Agric. Res.*, *39*: 363 (1988).
61. J. M. Clarke, I. Ramagosa, S. Jana, J. P. Srivastava, and T. N. McCaig, *Can. J. Plant Sci.*, *69*: 1075 (1989).
62. Y. Castonguay and A. H. Markhart, *Crop Sci.*, *31*: 1605 (1991).
63. K. T. Hubick, G. D. Farquhar, and R. Shorter, *Aust. J. Plant Physiol.*, *13*: 803 (1986).
64. A. G. Condon, G. D. Farquhar, and R. A. Richards, *Aust. J. Plant Physiol.*, *17*: 9 (1990).
65. G. C. Wright, K. T. Hubick, and G. D. Farquhar, *Aust. J. Plant Physiol.*, *15*: 815 (1988).
66. D. H. Wallace, J. L. Ozbun, and H. M. Munger, *Adv. Agron.*, *24*: 97 (1972).
67. A. S. Bhagsari and R. H. Brown, *Peanut Sci.*, *3*: 1 (1976).
68. G. M. Dornhoff and R. M. Shibles, *Crop Sci.*, *16*: 377 (1976).
69. J. B. Passioura, *Aust. J. Agric. Res.*, *23*: 745 (1972).
70. R. A. Richards and J. B. Passioura, *Crop Sci.*, *21*: 249 (1981).
71. R. A. Richards and J. B. Passioura, *Crop Sci.*, *21*: 253 (1981).
72. G. C. Wright, K. T. Hubick, G. D. Farquhar, and R. C. Nageswara Rao, in *Stable Isotopes and Plant Carbon–Water Relations* (J. E. Ehleringer, A. E. Hall, and G. D. Farquhar, eds.), Academic Press, New York, pp. 247–267 (1993).
73. N. L. Klocke, D. F. Heermann, and H. R. Duke, *Trans. ASAE*, *28*: 183 (1985).
74. R. J. Hanks, H. R. Gardner, and R. L. Florian, *Agron. J.*, *61*: 30 (1969).
75. T. A. Howell, K. R. Davis, R. L. McCormick, H. Yamada, V. T. Walhood, and D. W. Meek, *Irrig. Sci.*, *5*: 195 (1984).
76. R. J. Hanks, in *Advances in Evapotranspiration*, ASAE, St. Joseph, Mich., pp. 431–438 (1985).
77. R. J. Lascane, C. H. M. van Bavel, J. L. Hatfield, and D. R. Upchurch, *Soil Sci. Am. J.*, *51*: 113 (1987).
78. M. E. Bloodworth, J. B. Page, and W. R. Cowley, *Soil Sci. Soc. Am. Proc.*, *19*: 411 (1955).
79. T. Sakuratini, *Agric. Meteorol.*, *40*: 273 (1984).
80. J. M. Baker and C. H. M. van Bavel, *Plant Cell Environ.*, *10*: 779 (1987).
81. J. R. Evans, T. D. Sharkey, J. A. Berry, and G. D. Farquhar, *Aust. J. Plant Physiol.*, *13*: 281 (1986).
82. A. B. Frank, R. E. Barker, and J. D. Berdahl, *Agron J.*, *79*: 541 (1987).
83. T. R. Sinclair, C. B. Tanner, and J. M. Bennett, *Bioscience*, *34*: 36 (1983).
84. B. Martin and Y. R. Thorstenson, *Plant Physiol.*, *88*: 213 (1988).
85. H. G. Jones, in *Stomatal Function* (E. Ziegler, G. D. Farquhar and I. R. Cowan, eds.), Stanford University Press, Stanford, Calif. (1984) (cited in Ref. 24).
86. A. M. Ismail and A. E. Hall, *Crop Sci.*, *32*: 7 (1992).
87. R. D. Guy, M. F. Fogel, J. A. Berry, and T. C. Hoering, in *Progress in Photosynthetic Research III* (J. Biggins, ed.), Martinus Nijhoff, Dordrecht, The Netherlands, pp. 597–600 (1987).
88. G. D. Farquhar, M. H. O'Leary, and J. A. Berry, *Aust. J. Plant Physiol.*, *9*: 121 (1982).
89. J. H. Troughton, in *Photosynthesis II: Photosynthetic Carbon Metabolism and Related Processes* (M. Gibbs and E. Latzko, eds.), Springer-Verlag, Berlin, pp. 140–149 (1979).
90. W. G. Mook, M. Koopmans, A. F. Carter, and C. D. Keeling, *J. Geophys. Res.*, *88*: 10915 (1983).
91. C. B. Tanner and T. R. Sinclair, in *Limitations to Efficient Water Use in Crop Production* (H. Taylor, W. R. Jordan, and T. R. Sinclair, eds.), American Society of Agronomy, Madison, Wis., pp. 1–28 (1983).
92. H. Craig, *J. Geol.*, *62*: 115 (1954).
93. K. T. Hubick and G. D. Farquhar, *Plant Cell Environ.*, *12*: 795 (1989).
94. R. C. Nageswara Rao, J. H. Williams, K. D. R. Wadia, K. T. Hubick, and G. D. Farquhar, *Ann. Appl. Biol.*, *122*: 357 (1993).
95. W. S. F. Schuster, S. L. Philips, D. R. Sandquist, and J. R. Ehleringer, *Am. J. Bot.*, *79*: 216 (1992).
96. G. C. Wright, T. A. Sarwanto, A. Rahmianna, and D. Syarefuddin, in *Peanut Improvement: A Case Study in Indonesia* (G. C. Wright and K. J. Middleton, eds.), ACIAR Proc. No. 40, Canberra, Australia, pp. 74–84 (1993).

97. F. C. Meinzer, G. Goldstein, and D. A. Grantz, *Plant Physiol.*, *92*: 130 (1990).

98. J. M. Virgona, K. T. Hubick, H. M. Rawson, G. D. Farquhar, and R. W. Downes, *Aust. J. Plant Physiol.*, *17*: 207 (1990).

99. R. C. Nageswara Rao and G. C. Wright, *Crop Sci.*, *34*(1) (in press) (1993).

100. R. D. Guy, P. G. Warne, and D. M. Reid, in *Ecological Studies 68: Stable Isotopes in Ecological Research* (P. W. Rundel, J. R. Ehleringer, and K. A. Nagy, eds.), Springer-Verlag, Berlin, pp. 55–75 (1988).

101. K. J. McCree and S. G. Richardson, *Crop Sci.*, *27*: 543 (1987).

102. G. D. Farquhar and T. D. Sharkey, *Annu. Rev. Plant Physiol.*, *33*: 317 (1982).

103. I. R. Cowan, in *Physiological Plant Ecology II: Water Relations and Carbon Assimilation* (O. L. Lange, P. S. Nobel, C. B. Osmond, and H. Ziegler, eds.), Encyclopedia of Plant Physiology, New Series, Vol. 12B, Springer-Verlag, Berlin, pp. 589–613 (1982).

104. D. A. Johnson, K. H. Assay, L. L. Tieszen, J. R. Ehleringer, and P. G. Jefferson, *Crop Sci.*, *30*: 338 (1990).

105. K. Winter, *Z. Pflanzenphysiol.*, *101*: 421 (1981).

106. K. T. Hubick and G. D. Farquhar, *Aust. Cotton Grow.*, *8*: 66 (1987).

107. J. R. Ehleringer, in *Research on Drought Tolerance in Common Bean* (J. W. White, G. Hoogenboom, F. Ibarra, and S. P. Singh, eds.), Working Document 41, CIAT, Cali, Colombia, pp. 165–191 (1988).

108. J. R. Ehleringer and T. A. Cooper, *Oecologia*, *76*: 562 (1988).

109. A. E. Hall, R. G. Mutters, K. T. Hubick, and G. D. Farquhar, *Crop Sci.*, *30*: 300 (1990).

110. A. E. Hall, R. G. Mutters, and G. D. Farquhar, *Crop Sci.*, *32*: 1 (1992).

111. W. R. Kirchhoff, A. E. Hall, and W. W. Thomson, *Crop Sci.*, *29*: 109 (1989).

112. I. R. Cowan, in *Flow and Transport in the Natural Environment: Advances and Applications* (O. T. Denmead, ed.), Springer-Verlag, New York, pp. 160–172 (1988).

113. C. B. Osmond, K. Winter, and H. Ziegler, in *Physiological Plant Ecology II: Water Relations and Carbon Assimilation* (O. L. Lange, P. S. Nobel, C. B. Osmond, and H. Ziegler, eds.), Encyclopedia of Plant Physiology, New Series, Vol. 12B, Springer-Verlag, Berlin, pp. 479–547 (1982).

114. S. C. Wong, I. R. Cowan, and G. D. Farquhar, *Nature*, *282*: 424 (1979).

115. T. A. Howell, in *Irrigation of Agricultural Crops* (B. A. Stewart and D. R. Nielson, eds.), American Society of Agronomy, Madison, Wis., pp. 391–434 (1990).

116. J. Goudriaan and H. H. van Laar, *Photosynthetica*, *12*: 241 (1978).

117. D. R. Krieg, in *Limitations to Efficient Water Use in Crop Production* (H. M. Taylor, W. R. Jordan, and T. R. Sinclair, eds.), American Society of Agronomy, Madison, Wis., pp. 319–330 (1983).

118. C. Ramos and A. E. Hall, *Photosynthetica*, *16*: 343 (1982).

119. M. H. O'Leary, I. Treichel, and M. Rooney, *Plant Physiol.*, *80*: 578 (1986).

120. S. D. Wullschleger, *J. Exp. Bot.*, *44*: 907 (1993).

121. B. Ehdaie, A. E. Hall, G. D. Farquhar, H. T. Nguyen, and J. G. Waines, *Crop Sci.*, *31*: 1282 (1991).

122. J. E. Pallas, Jr., and Y. B. Samish, *Crop Sci.*, *14*: 478 (1974).

123. J. E. Pallas, Jr., *Peanut Sci.*, *9*: 14 (1982).

124. W. D. Branch and J. E. Pallas, Jr., *Peanut Sci.*, *11*: 56 (1984).

125. A. G. Condon, R. A. Richards, and G. D. Farquhar, *Crop Sci.*, *27*: 996 (1987).

126. D. Shimshi and J. Ephrat, *Agron. J.*, *67*: 326 (1975).

127. R. C. Johnson, H. Kebede, D. W. Mornhinweg, B. F. Carver, A. Lanerayburn, and H. T. Nguyen, *Crop Sci.*, *27*: 1046 (1987).

128. R. L. Dunstone, R. M. Gifford, and L. T. Evans, *Aust. J. Biol. Sci.*, *26*: 295 (1973).

129. J. Masle and G. D. Farquhar, *Plant Physiol.*, *86*: 32 (1988).

130. J. W. White, J. A. Castillo, and J. Ehleringer, *Aust. J. Plant Physiol.*, *17*: 189 (1990).

131. A. G. Condon and R. A. Richards, *Aust. J. Agric. Res.*, *43*: 921 (1992).

132. R. C. Johnson and L. M. Bassett, *Crop Sci.*, *31*: 157 (1991).

133. P. C. Pheloung and K. H. M. Siddique, *Aust. J. Plant Physiol.*, *18*: 53 (1991).

134. D. S. Falconer, *Introduction to Quantitative Genetics*, Longman Group, New York (1981).
135. M. H. O'Leary, *Phytochemistry*, *20*: 553 (1981).
136. W. J. S. Downton, W. J. R. Grant, and S. P. Robinson, *Plant Physiol.*, *78*: 85 (1985).
137. R. D. Guy and D. M. Reid, *Plant Cell Environ.*, *9*: 65 (1986).
138. K. Winter, J. A. M. Holtum, G. E. Edwards, and M. H. O'Leary, *J. Exp. Bot.*, *33*: 88 (1982).
139. B. N. Sponchiado, J. W. White, J. A. Castillo, and P. G. James, *Exp. Agric.*, *25*: 249 (1989).
140. E. Acevedo, in *Physiology: Breeding of Winter Cereals for Stressed Mediterranean Environments* (E. Acevedo, A. P. Conesa, P. Monneveux, and J. P. Srivastava, eds.), INRA, Paris (1991).
141. T. C. Hsiao and E. Acevedo, *Agric. Meteorol.*, *14*: 59 (1974).
142. G. H. Heichel, in *Limitations to Efficient Water Use in Crop Production* (H. M. Taylor, W. R. Jordan, and T. R. Sinclair, eds.), American Society of Agronomy, Madison, Wis., pp. 375–380 (1983).
143. T. A. Coffelt, R. O. Hammons, W. D. Branch, P. M. Mozingo, P. M. Phipps, J. C. Smith, R. E. Lynch, C. S. Kvien, D. L. Ketring, D. M. Porter, and A. C. Misen, *Crop Sci.*, *25*: 203 (1985).
144. P. Q. Craufurd and A. B. Austin, *Annual Report of the Agriculture and Food Research Council*, Institute of Plant Science Research and John Innes Institute, John Catt Ltd., England, pp. 12–14 (1987).
145. G. C. Wright, R. C. Nageswara Rao, and H. B. So, "Variation in Root Characteristics and Their Association with Water Uptake and Drought Tolerance in Four Peanut Cultivars," paper presented at the *Australian Agronomy Conference*, Sept. 19–24, The University of Adelaide, Adelaide, South Australia (1993).
146. G. V. Subbarao, C. Johansen, A. E. Slinkard, R. C. Nageswara Rao, N. P. Saxena, and Y. S. Chauhan, *CRC Crit. Rev. Plant Sci.* (in preparation).
147. S. Ceccarelli, E. Acevedo, and S. Grando, *Euphytica*, *56*: 169 (1991).
148. M. N. Christiansen, in *Breeding Plants for Less Favorable Environments* (M. N. Christiansen and C. F. Lewis, eds.), Wiley, New York, pp. 1–11 (1982).
149. L. J. Briggs and H. L. Shantz, *USDA Bur. Plant Ind. Bull.*, *285*: 1 (1913).

40
Physiological Mechanisms Relevant to Genetic Improvement of Salinity Tolerance in Crop Plants

G. V. Subbarao and Chris Johansen
International Crops Research Institute for the Semi-Arid Tropics (ICRISAT), Patancheru, Andhra Pradesh, India

I. INTRODUCTION

Crop species differ widely in their ability to grow and yield under saline conditions. However, almost all crop plants belong to the glycophytic category, except for a few crop species such as sugar beet, which has halophytic ancestors. By ecological definition, halophytes are the native flora of saline habitats [1]. From a crop improvement perspective, the variability for salinity tolerance within a crop species or among its wild relatives is important. It is also important to understand the physiological mechanisms of salinity tolerance operating within a crop species so that suitable breeding strategies can be developed for improving salinity tolerance. There are several reviews covering the general responses of plants to salinity stress and the mechanisms available in halophytes and glycophytes which allow them to cope with saline habitats [2–7]. However, little attempt has been made to integrate information on these physiological aspects into genetic improvement concepts.

Salinity creates stress by reducing the osmotic potential of the rooting medium and increasing ambient concentrations of ions such as Cl, SO_4, CO_3, HCO_3, Na, Ca, and Mg. Being glycophytes, crop species have no appendages, such as salt glands, bladders, or hairs, that excrete salts absorbed in excess from their shoot tissues. The limited compartmentation ability of the shoot demands strict regulation of ionic delivery to the shoot. Physiological mechanisms controlling salt absorption and distribution in crop plants, and the osmotic adjustment that is essential for turgor-driven water uptake, are covered in this chapter. We specifically address the question of how information on these physiological mechanisms could be utilized in genetic improvement programs as an integrated approach to improving salinity tolerance in a given crop.

II. REGULATION OF ION TRANSPORT

Plants regulate their intracellular ionic composition to maintain a suitable ionic environment for the physiological and biochemical processes that proceed within a cell. This internal environment needs to be maintained within acceptable limits if plant growth and function are to proceed in saline environments [8]. Salinity under field conditions is characterized by a mixture of salts. However, Na and Cl are predominant in most situations. Therefore, most studies of salinity effects refer to NaCl salinity as a model system, although the effects of all ions that are in excess in a saline environment on nutrient uptake are recognized. Similarly, due to the importance of K in plant nutrition and because the effects of Na on K uptake have been studied extensively, we refer primarily to this interaction in our discussion of ion uptake mechanisms.

A. Regulation at Root Membranes

The concept of dual mechanisms of ion transport is a useful framework for describing ion uptake [9]. At low concentrations of K in the external solution, below 1 mM, uptake of K, described by a discrete Michaelis–Menten kinetic equation, is thought to operate at the plasmalemma. We shall call this mechanism 1. At K concentrations in the range 1 to 50 mM, mechanism 2 operates. Mechanism 2 is thought to involve diffusive, or at least nonselective ion movement across the plasmalemma with the rate limitation inward from the plasmalemma, probably at the tonoplast [9]. For mechanism 1, there is a high selectivity of the active transport mechanism for K over competing cations such as Na. For mechanism 2, this level of selectivity is not present. Mechanism 1 is not influenced by the concomitant counteranion, but mechanism 2 is. For example, compared with Cl, SO$_4$ severely depresses K absorption at K concentrations in the range of mechanism 2 but not in the range of mechanism 1. This dual phenomenon of ion uptake has been described for various plant and ionic species (see p. 136 of Ref. 9).

Selective ion transport, at least in the range of mechanism 1, depends on metabolic energy derived from adenosine triphosphate (ATP). This allows charge separation across cell membranes, through primary transport of H$^+$, thus creating a localized electrochemical gradient for other ions to traverse the membrane [10–12]. Cations move in the opposite direction to H$^+$ (antiport), while anions are co-transported with it (symport) or move as antiport to OH$^-$ or HCO$_3^-$ [13].

Selectivity between ionic species is governed by the particular binding properties of cell membrane constituents. Little is known about this process, due to limited knowledge of plant membrane structure and function [13–15]. Breakthroughs in this regard will allow for an understanding of the molecular basis of ion transport and the effects of salinity on this process. The entry of Na or other ions in excess in the ambient solution can be controlled by this selective binding. An alternative for regulating K/Na levels inside root cells is by means of an outwardly directed Na pump at the plasmalemma [2,16–18].

In most situations, saline or otherwise, Na movement across the plasmalemma into root cells is thought to be passive down an electrochemical gradient [7]. For example, the membrane leakage of Na accounts for the cytoplasmic Na levels found in rice [19]. Jeschke [6] has proposed a model to explain K/Na exchange at the plasmalemma (Figure 1), the components of which are:

1. A proton pump powered by ATP generates an electrical potential difference and proton gradient across the plasmalemma.
2. The electrical charge of H is compensated by an influx of K at a specific site or channel. This site has a lower affinity for Na.
3. The proton gradient provides energy for extrusion of Na from the cytoplasm by a H/Na antiport; this site is reported to have a lower affinity for K.

There is variation among crop species in their K/Na exchange capability [6,20]. Barley, wheat, and rye showed efficient K/Na exchange compared to sensitive species such as *Allium cepa* and *Helianthus annuus* [20]. The existence of genotypic differences in this trait within a

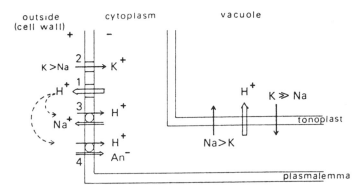

Figure 1 Model of the proton-mediated K/Na exchange system at the plasmalemma and Na/K exchange system at the tonoplast. 1, Proton pump; 2, K uniport (i.e., system 1 of K influx); 3, H-Na antiport; 4, H^+-anion symport. (From Ref. 6)

crop species and its relation to salinity tolerance is not known. Such information is vital to an evaluation of the potential utility of this trait in genetic improvement programs for salinity tolerance. The relation between K/Na selectivity and salt tolerance has been reviewed [2–4,21–23]. Variation in K/Na exchange suggests at least quantitative differences in membrane properties among different crop species [6]. The general response of many crop plants to a moderate increase in external salinity is increased plant K levels and reduced Na concentrations in tolerant relative to nontolerant genotypes [24–29].

For the high-affinity system mediating K influx (Epstein's mechanism 1), a proton pump appears to be present in the plasmalemma of root cortical cells [6]. However, the graded response of Na-efflux to added K suggests quantitative differences between species, and perhaps among genotypes of a crop species, in the number and efficiency of sites mediating the H/Na antiport [6]. The number of sites for the H/Na antiport needs to be quantified and the existence of genotypic variation within a crop species estimated to determine the feasibility of favorable genetic manipulation of this trait.

At K concentrations above 1 mM, in the range of mechanism 2, selectivity is diminished in the face of competition from other ions, such as Na, in the ambient medium. Whether this is due to increased passive movement of all ambient ions across the plasmalemma, down an electrochemical gradient, or less selectivity in an active transport process remains unclear [9]. Eventually, however, if ambient salt concentrations reach high enough levels, membranes would become completely permeable. Information on species or genotypic differences regarding the level at which such physical disruption occurs may also provide a guide to selection for salinity tolerance [30,31].

Most of the kinetic studies referred to above were carried out on tissue previously starved of salts (low salt status). However, as cytoplasmic concentrations of absorbed ions increase, influx rates slow down, indicating a feedback mechanism controlling active influx of ions [13,32]. For example, K concentrations in the cytoplasm of normally growing plants are maintained in the range 90 to 110 mM [21]. Although there is considerable speculation on the nature of such feedback mechanisms [13], their further understanding would also assist in selection of genotypes that better control their ion transport processes at the plasmalemma.

B. Intracellular Compartmentation in Roots

Vacuoles occupy more than 80% of a mature root cell's volume and thus provide a means of osmotic regulation for root tissue [33]. This is achieved by compartmentation of inorganic salts, primarily because these are metabolically inexpensive compared to organic solutes. Salt ions move across membranes more easily than do molecules of large molecular weight. There are considerable metabolic costs in transporting photosynthates from the shoots for use as osmotica in roots [34].

Inorganic ions contribute substantially to osmotic adjustment in root cells of glycophytes under saline conditions. However, the amount of osmotic adjustment varies from one species to another and could be an important factor in determining salinity tolerance. Roots of many glycophytic crop species contain sustantially higher levels of Na and Cl under saline conditions than do shoots [29,35]. In pigeonpea (*Cajanus cajan*) and its wild relatives, the most tolerant genotypes retained higher levels of Na and Cl in the roots, and this was associated with salinity tolerance in this crop [29,36]. Ability to retain Na and Cl in roots breaks down at a given concentration, leading to large-scale translocation of these ions to the shoot, with resultant plant mortality. This critical level varies between pigeonpea genotypes and between pigeonpea and its wild relatives, and is considered a determinant of the level of salinity tolerance [29].

The cytoplasm shows a strong selectivity for K over Na, Mg over Ca, and P over Cl or NO_3 [27,37]. Optimal concentrations for various ions vary in the cytoplasm; thus when ions enter the protoplast above this concentration, they may be actively transported through the tonoplast into the vacuole. However, these ions could be recovered from the vacuole, depending on the metabolic requirements in other plant parts. Retranslocation of K is one example [6].

Vacuoles play an important role in maintaining stable levels of various inorganic ions in the cytoplasm, by acting as a storage reservoir for these ions [33]. Under NaCl salinity, Na and Cl are normally the predominant ions entering the protoplast of root cells. These ions are actively pumped into the vacuole after reaching a threshold concentration in the cytoplasm. This would reduce the flow into the xylem of Na and Cl and of other ions associated with salinity (e.g., Ca, Mg, SO_4, CO_3) and thus restrict their translocation to the shoot.

The general hypothesis is that Na and Cl must be excluded from the cytoplasm. This is based on the sensitivity of enzyme activities to high NaCl levels in vitro [7]. High levels of Na in the cytoplasm are reported to interfere with K metabolism, resulting in ionic toxicity; but it is not known what Na levels are biochemically compatible with other cytoplasm solutes [7]. In corn, cytoplasmic Na concentrations can reach 40 to 70 mM under nonsaline conditions [28], but can rise to 140 mM under 100 mM NaCl external salinity and become toxic to the plant. In roots of the halophyte *Triglochin maritima* exposed to 500 mM NaCl, the Na/K ratio was only 2 in the cytoplasm compared to 15 in the vacuole, although there was approximately 150 mM Na in both compartments [38]. Thus the tolerance of the cytoplasm to Na can vary between species. As long as tissue Na concentration is below the level acceptable for the cytoplasm, more sophisticated compartmentation may not be necessary [7].

There are several factors that could mitigate the adverse effects of excess ions in the cytoplasm. One is the type and quantity of organic solutes that could modify the tolerance level of cytoplasm to monovalent cations such as Na. Another is the existence of isoenzymes for many enzyme systems, which may have different tolerance thresholds in the cytoplasm. In *Zea mays*, although the total acid phosphatase activity was slightly reduced under salinity, certain isoenzymic forms of acid phosphatase increased in different plant parts [39]. Similarly, the relative proportions of malate dehydrogenase isoenzymes were changed during salinity stress in pea seedlings [40].

In sunflower, a plastome mutant line that has higher resistance to salinity than that of its parental line, reportedly produced a unique isoenzyme of peroxidase under saline conditions [41]. This isoenzyme was found to be resistant to NaCl or Na_2SO_4 salinity up to 1.2% and 2.4% respectively, in vitro. Cavalieri and Huang [42] reported that enzymes isolated from roots were distinctly more tolerant to Na than those from the shoots; these results might reflect other differences between shoots and roots and in compartmentation between cytoplasm and vacuole [33]. Another possibility is that certain isoenzymes exist only in certain plant parts; for example, the isoenzyme patterns of shoots could be different from those of roots [43]. Thus the statement which is often made—that "there are no differences in enzyme systems of halophytes and nonhalophytes in their tolerance to monovalent cations in vitro" [3,21,44–46], needs to be reexamined.

Another aspect of the adaptation of higher plants to salinity is compartmentation within the cytoplasm, because the cytosol is particularly sensitive to fluctuating salt levels [35]. For cells involved in salt transport, the rough endoplasmic reticulum (RER) provides a compartment within the cytoplasm in which salt may be sequestered [35]. Substances can be transported symplastically through the RER, via desmotubules. This may also provide a means of ion transfer to vacuoles without disrupting ion concentrations in the cytosol, as RER cisternae may fuse with the tonoplast, releasing their contents into the vacuole [35].

Several hypotheses have been proposed to explain the mode of ion transport from cytoplasm to vacuole through the tonoplast. Pitman and Saddler [16] located an inwardly directed Na pump at the tonoplast that would effectively deplete Na levels in the cytoplasm. Jennings [47] proposed a very similar model for transport of Na from the cytoplasm into the vacuole by means of Na/K exchange. Proton pumps powered by ATP are also thought to play a crucial role in generating the transmembrane electrochemical potential differences required to energize tonoplast ion transport [48,49]. Two types of proton pump are reported to be located in the tonoplast; they are catalyzed by functionally and physiologically distinct phosphohydralases: tp-ATPase and tp-PPase (tonoplast pyrophosphatase) [48].

Exchange of Na and K at the tonoplast can occur only while K remains in the vacuole [6]. Thus distribution of K and Na between vacuole and cytoplasm appears to be crucial for salt tolerance [21,23], and since vacuolar K concentration represents a potential reservoir that could be removed by exchange for Na, the allocation of these ions needs to be regulated. However, the vacuole of root cortical cells is in some respects a dead end; continued selective transport across the root depends on selective transport at the point of entry of salts into the cytoplasm, which depends on the ability of the plasmamembrane to restrict passive influx of sodium and maintain high K/Na selectivity [50]. Thus without control of the quantity of salt that is allowed into the root or that reaches the leaves, intracellular compartmentation either at root cortex or in the shoot would in any case be a very limited option [7]. The vacuole's role may be more in using Na as an osmoticum instead of K and in providing a source of stored K under salinization rather than as part of a selective system of salt transport across the root [50].

C. Regulation of Long-Distance Transport to Shoots

Beyond the plasmalemma, there are several other possible barriers that could minimize transport of excess salts to the shoots. An important one is movement of salts from xylem parenchyma cells into the xylem stream. Evidence favors this process being mediated by active transport [9], with the possibility of further selectivity in ion transport. Xylem parenchyma cells can be differentiated as transfer cells (XPTs) with well-developed wall protuberances adjacent to the bordered pits of xylem vessels in the proximal region of roots and stems. These are reported in *Phaseolus coccineus* [35], *Glycine max* [51], maize [52,53], and squash [54]. These transfer cells accumulate K in the absence of NaCl in the growth medium and Na under saline (NaCl) conditions [51].

A salt-induced formation of wall ingrowths has been reported for xylem parenchyma cells in soybean [51,55] and for the root epidermis cells of *Phaseolus coccineus* [35]. Xylem parenchyma cells and transfer cells are both capable of restricting solutes, particularly Na, by exchange with K from the transpiration stream [31,56]. These XPTs have been reported to accumulate Na selectively from the transpiration stream and then transfer it to the phloem pathway to be extruded by the roots [57]. In *Lycopersicon*, XPTs in the leaf petiole remove Na from the xylem stream before it enters the leaf lamina [58]. It appears that the entire xylem transport pathway has a backup reabsorption system [6].

The cytoplasm of these transfer cells contains cisternae of RER which increase under NaCl or Na_2SO_4 salinity in *Phaseolus coccineus* hypocotyl and epicotyl [35] and in *Zea mays* [56]. RER could permit a large flow of ions through the cytoplasm of xylem parenchyma cells, assuming that ions are localized mainly in the vacuole [3]. The quantitative significance of this reabsorption process from the xylem in regulating Na-ion transport to the shoot is not known.

The ability of XPTs to absorb Na is finite and could be exhausted rapidly under saline conditions [59]. Some lateral redistribution is possible, however, but this may not be sufficient to prevent Na from eventually reaching the shoot [60]. However, XPTs have a limited capability to store Na, and this Na needs to be removed to the lateral tissue for XPT to continue absorbing Na from the transpiration stream. This Na could be loaded into the phloem and translocated to the roots, where it could either be further compartmentalized or extruded. Such Na extrusion has been reported in *H. vulgare* [16,18,61] and *P. vulgaris* [62]. Thus the practical significance of XPT cells in the basal part of the stem may be limited in controlling Na flow into the shoot to a low degree or a short duration of salinity stress [63,64]. The existence of quantitative variation in XPTs among genotypes in relation to differences in salinity tolerance is not known. Such knowledge is necessary to evaluate the usefulness of this trait from a genetic improvement perspective.

D. Apoplastic Salt Accumulation

Oertli [65] predicted that apoplastic salt load could cause water deficit and turgor loss in leaf cells and proposed it as a mechanism of salinity damage. This concept has received renewed interest [4,66–68]. Under saline conditions, Na and Cl can bypass the ion transport control mechanisms discussed earlier, be carried upward in the xylem stream, and be delivered to the apoplasts of leaf cells [69]. If shoot protoplast accumulates these ions beyond levels that are tolerated in the cytoplasm and its compartmentation capacity of the vacuole, disruption of the metabolic functions due to ionic toxicity would result [70]. On the other hand, a failure to do so would lead to ion accumulation in the apoplast, which could reach very high levels in a short time, as the apoplast occupies only 1% of the cell's volume [65,70]. For instance, even if 90% of the NaCl arriving in the xylem (plants grown at 50 mM NaCl external solution) is accumulated in the protoplast, the apoplastic concentrations could reach 500 mM within 7 days [70] and cause cell death, although the average tissue Na and Cl concentrations may not reach 100 mM. Because of the small apoplast volume, such ion concentrations in the apoplast could occur at overall low tissue concentrations and would thus escape detection in standard tissue analysis [70]. Excessive accumulation of salts in the leaf apoplast would cause turgor loss, stomatal closure, and cell dehydration.

Water deficits in a particular leaf, as opposed to the plant as a whole, could be an inevitable consequence of increasing apoplastic salt load [65] and will occur whenever the rate of arrival of NaCl in the xylem is greater than the rate of accumulation of these ions in leaf cells [67]. Thus arguments that plants have adjusted osmotically to external salinity, which are based on comparisons of solute concentrations in tissue water with external salinity, need to be viewed with caution [71]. The success of a crop species in surviving and reproducing under saline conditions depends considerably on its ability to regulate ion delivery into the xylem stream without causing ion toxicity in leaf protoplasts or apoplastic salt buildup [70]. Genotypes that could more effectively transfer NaCl from leaf apoplast into leaf cells would be at an advantage. Although this increases their protoplast salt concentrations due to the relative volumes of protoplast and apoplast, this is considered to be less serious than the consequences of apoplastic salt buildup [19,70].

E. Phloem Retranslocation

When Na or Cl levels in the cytoplasm of mesophyll cells reach a tolerance threshold and their compartmentation capacity becomes saturated, additional Na or Cl ions can immediately be transported by intraveinal recycling so as to prevent apoplastic buildup of Na or Cl or ion toxicity in the cytoplasm [64]. Since there is no barrier between the xylem and the leaf apoplast [72], ions can be actively loaded into phloem vessels [73]. This mechanism may play a significant role in the regulation of Na or Cl ions in the shoot [54,62,74]. Based on cytoplasmic Na concentrations, it has been estimated that nearly 25% of the Na entering the leaf can be retranslocated by the phloem [33]. However, phloem loading and retranslocation of Na or Cl is seen as metabolically expensive. Large quantities of Na or Cl in phloem reflects poor control at the root level in regulating ion flow into the xylem. This was found in studies by Lessani and Marschner [75], where phloem translocation of Na or Cl is greatest in sensitive species such as bean, and least in tolerant species such as barley and sugar beet [27].

Among a range of species, there was a significant correlation between a decrease in dry matter production at 100 mM NaCl in the medium and Na retranslocation from leaves, particularly, efflux from roots (Figure 2) [75]. If incoming ions are excessive to the shoot's compartmentation ability and the phloem translocation capacity, overloading of Na or Cl ions into the phloem parenchyma transfer cells could occur. This would result in destruction of phloem transfer cells [64,76]. Although phloem retranslocation does contribute to regulation of Na or Cl levels in the shoot, it appears to have a limited role in this regard and thus in determining the level of salinity tolerance. Regulation of Na and Cl levels in the shoot lies primarily with the root's ability to regulate Na or Cl flow into the xylem, rather than the shoot's ability to retranslocate to the root [59].

Availability of sufficient K in growing and expanding regions of the shoot and root is crucial to maintenance of K/Na selectivity and subsequent Na compartmentation in the root cortex. In addition to efficient K/Na selectivity at the plasmamembrane, phloem transport of K reserves within the plant plays an important role in salinity tolerance. Potassium is remobilized from mature leaves by removal of vacuolar K through Na/K exchange at the tonoplast of mesophyll cells. This K is then retranslocated to the growing regions of the root, shoot, and expanding leaves, where there is little vacuolar space and the cytoplasm occupies a major portion of the cell. These growing zones require large quantities of K to meet their demands for osmotic adjustment in the rapidly expanding vacuolar space. Leaves develop and expand close to the shoot apex and derive their mineral nutrient supply from the phloem (which is rich in K), particularly since phloem tissue differentiates prior to xylem elements [77]. With increasing leaf age, minerals are imported mainly by the xylem, which is high in Na levels compared to phloem supply. This Na is compartmentalized through Na/K exchange at the tonoplast; thus K is recovered from the vacuole to provide a major source of K for retranslocation [23].

Nearly 20% of K arriving in the shoot through the xylem could be retranslocated to the growing regions of the root, where high K levels are essential [6]. Such K retranslocation has been reported in barley [78–80], tomatoes, and lupins [6]. The ability to remobilize and retranslocate K into the growing region of the root and shoot plays an important role in Na compartmentation in the root cortex and in maintaining a high K/Na ratio in shoot growing

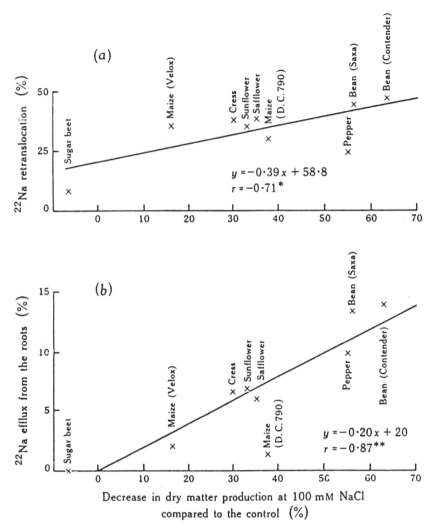

Figure 2 Relationship between decrease in dry matter production and (a) Na retranslocation and (b) efflux of Na from roots, in species differing in their tolerance to salinity. (From Ref. 75)

regions, thus protecting them from Na or Cl toxicity. Most tolerant crop species, such as barley and sugar beet, have a very efficient K recirculation system which is tightly linked to Na regulatory mechanisms. This mechanism may also be important in determining genotypic differences in salinity response.

F.   Role of Transpiration

Shoot ion concentrations are a product of transpiration rate, xylem ion concentrations, and growth rate [81]. Under high evapotranspirational demands, transpiration increases, while K/Na selectivity decreases, resulting in increased Na and Cl uptake [82,83]. Alternatively, a reduction in transpiration can decrease ion (Na and Cl) uptake [6,27,60].

There are a number of hypotheses proposed to explain increased xylem sap Na and Cl levels under high evapotranspiration rates in saline growth media. Enhanced water flow interacts with ion flow across membranes of root cells at more than one site, thus interfering with processes that regulate the balance between ion accumulation in the root cell vacuole and transport to the shoot [6]. Increased water flow due to transpiration promotes passive ion movements where there is no active transport barrier [84].

Water flow can promote the ion flow across the cortex toward a pump that secretes ions into xylem vessels [80]. Ions could either be moved by water along an apoplastic pathway at high concentrations or be coupled to water flow during symplastic passage across the root [85]. High transpiration rates increased Na transport more than K, thus shifting the selectivity toward Na [86,87]. Potassium ions absorbed in roots may be released through Na/K exchange, mainly from vacuoles, for transport to the shoot at times of high evaporative demand [6].

In halophytes, the entry of ions such as Na or Cl into the roots or their release to the xylem sap is tightly regulated at high evapotranspirative demand under saline conditions, thus regulating ion supply to the shoot. Certain morphological features, such as increased width or early development of Casparian strips [88] or formation of a double endodermis [89,90], have been reported to develop under high evaporative demands, thus minimizing the passive influx and bypass flow of Na and Cl ions into the xylem. We are not aware of such anatomical changes reported in any crop species under saline conditions. It may be worthwhile to examine genotypes that show high salinity tolerance for such kinds of adaptive features. Since water use is tightly linked to ion uptake and selectivity, the morphological and physiological traits that increase water use efficiency (WUE) in a given genotype could have a role in determining salinity tolerance [81]. In rice, genotypes that showed higher WUE also have a higher level of salinity tolerance [81].

III. ORGANIC SOLUTE ACCUMULATION

A wide variety of organic solutes have been reported to accumulate in plant tissues during water and salt stress conditions and contribute to osmotic adjustment [91]. The chemical nature of the compatible solutes varies from one taxonomic group to another, but most are derivatives of polyols or nitrogen dipoles [27] (Table 1). Osmotic adjustment by the plant promotes turgor maintenance and is thus associated with adaptation to both high soil salinity and low soil moisture [4,92]. Compatible solutes are an important factor in the osmotic balance of the cytoplasm under salt stress [21], where sodium salts are sequestered to play a complementary osmotic role in the vacuole [3,21,23]. However, this is considered to be a halophytic mode of osmoregulation [93], which is energetically more efficient than overall osmoregulation by organic solutes [5,94], which is a common feature of glycophytes [91].

These organic solutes may comprise common metabolites such as sugars, amino acids such as proline [95–98], and organic acids such as proline betaine [99] and other aliphatic quaternary ammonium compounds [100]. There is evidence that solute accumulation is a regulated process and not merely the result of a discrepancy between the sensitivity of the growth process and photosynthesis to stress [101]. Nevertheless, metabolites such as glucose and sucrose accumulate in tissues whose growth has been inhibited by stress [91].

The type of stress would determine which compounds act as osmotic solutes [102]. In grain sorghum, betaine accumulates only under moderate levels of salt stress, not under water stress [102]. However, in crops such as wheat, barley, and rye, betaine accumulates under water stress as well as salinity stress [102]. Nevertheless, salinity is reported to be the more effective stimulator of betaine accumulation [103]. In barley, more glycine betaine is accumulated under gradual stress, but proline is the predominant solute under sudden stress [104].

Table 1 Types of Compatible Solutes That Could Accumulate Under Salinity Stress in Various Plant Species

Solute	Structure	Distribution
D-Sorbitol		Plantaginaccaec Rosaccae
D-Mannitol		Combretaccae Myrsinaccae Rublaccae
D-Pinitol		Leguminoseae Rhizophoraccae Caryophyllaceae
L-Quebrachitol		Euphorblaceae
Glycine betaine	$(CH_3)_3N^oCH_3COO^-$	Chenopodlaceae Amaranthaccae Asteraceae Solanaccae Gramlneae Avicennisceae
β-Alanine betaine	$(CH_3)_3N^oCH_3CH_3COO^-$	Plumbaginaccae
Proline		Juncaginacceae Asteraccae Gramincae
Proline betaine (stachydrine)		Lablaicae Capparidaceae Leguminoscae
3-Dimethylsulfonio proplonate	$(CH_3)_3S^oCH_3CH_3COO^-$	Asteraccae Gramincae

Source: Ref. 27.

A. Role in Osmoregulation

High concentrations of organic solutes in the cytoplasm could contribute to the osmotic balance when electrolytes are lower in the cytoplasm than in the vacuole [95,105]. These compatible solutes could also act as a nitrogen source [106] or protect membranes against salt inactivation [107,108]. These proposed activities may complement each other within the integrated metabolic and ontogenic pattern of a particular species [109].

Under saline conditions, the large quantity of Na, K, and Cl and other ions that are translocated to the shoot and contribute to the osmotic adjustment are believed to accumulate mainly in the vacuole after reaching threshold levels in the cytoplasm [7]. This concentration of inorganic ions could be considered as a threshold level at which accumulation of such organic solutes as proline, betaine, or other compounds begins in the cytoplasm, thus maintaining the intracellular osmotic balance between cytoplasm and vacuole [102]. For instance, in wheat, proline began accumulating when Na + K exceeded a threshold value of 200 mol/g fresh weight [110]. Also, in grain sorghum, a moderate level of salt stress (0.4 MPa or more) is required to induce a significant betaine concentration [111,112]. Further studies are needed to determine the extent to which this threshold level varies among genotypes of a given species.

The total quaternary ammonium compounds (QACs) in the leaf tissue in wheat species (*Triticum aestivum* and *T. durum*) shows a high positive correlation with salinity treatment [102]. The capacity to accumulate betaine in grasses has been reported to be correlated with basal levels of betaine in unstressed plants [113]. Crops such as oats and rice, which have very low betaine levels under nonsaline conditions, accumulated very little under stress conditions [102].

The relatively small increases in glycine betaine with increasing external salinity, together with the high levels found in many halophytes at very low external salinity, implies that this solute may be redistributed between the vacuole and cytoplasm, depending on tissue electrolyte concentrations [105]. However, in crop plants such as sorghum, it is reported that betaine is relatively nonlabile compared to compounds such as proline [114,115]. A sixfold increase in glycine betaine levels in isolated chloroplasts of spinach under saline conditions was observed, which could account for 36% of the osmotic adjustment in chloroplasts [116].

Proline levels can change quickly in response to abrupt stress, while other organic solutes accumulate more slowly [112]. Thus when stress is applied slowly, less proline accumulates, but the total accumulation of organic solutes remains predictable on the basis of tissue Na and Cl levels [104]. Accumulation of free proline has been correlated with tissue Na concentration in a number of crop species [117–119]. A level of 25 mol proline per gram fresh weight could produce a concentration of 280 mM if confined to cytoplasm, thus making a significant contribution to the cytoplasmic solute potential [3].

Proline concentrations were reported to be directly proportional to Na concentrations [120]; each increase in Na concentration is reported to be balanced by an increase in proline concentration equal to about 4% of the rise in Na [121]. This relationship between steady-state proline concentrations and Na levels indicates its role as a cytoplasmic solute [121]. Proline levels for various grasses (*Sorghum bicolor, Agrostis stolonifera, Cyanodendactyla, Paspalum vaginatum*, etc.) increased in response to Na accumulation [120]. However, overall proline levels and accumulation rates were highly variable among grasses and therefore are not reliable indicators of relative tolerance levels [120].

In pigeonpea, proline levels increased with increasing external salinity in two genotypes differing in their salt tolerance. The highest proline levels are observed at 10 dS/m, where both genotypes died subsequently [29]. Among the wild species related to pigeonpea, there is a steady increase of proline levels with increasing external salinity in only a few species (Figure 3). There

Proline concentration (μg/g fresh Wt.)

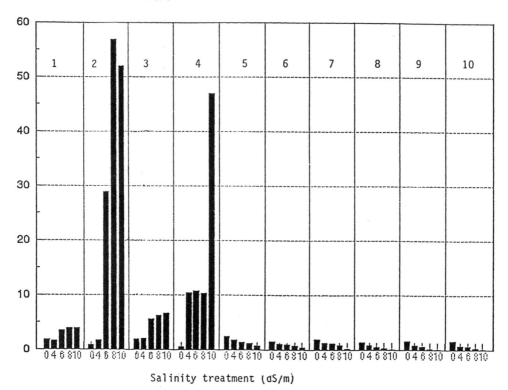

Salinity treatment (dS/m)

Figure 3 Proline accumulation in the wild relatives of pigeonpea (*Atylosia* sp.) at various salinity levels. (Leaf samples were collected at 50 days after sowing.) 1, *A. albicans*; 2, *A. sericea*; 3, *A. acutifolia*; 4, *A. lineata*; 5, *A. cajanifolia*; 6, *A. volubilis*; 7, *A. reticulata*; 8, *A. Grandifolia*; 9, *A. goensis*; 10, *A. lanceolata*. (From Refs. 29 and 157.)

was no clear relationship between salinity tolerance and proline accumulation, as proline accumulated to higher levels in both sensitive and tolerant species [29]. Similarly, some tolerant and sensitive species did not accumulate significant levels of proline at any level of external salinity, thus defying any simple relationship between proline accumulation and salinity tolerance [29].

It is usually assumed that the cytoplasm comprises about 5% of the cell's volume, proline is accumulated in the cytoplasm, and Na is largely sequestered in the vacuole [121]. Under these conditions, proline alone could merely osmotically balance the Na. However, other ions and organic solutes are also likely to be involved, as field salinity is often caused by a mixture of salts. Thus a variety of ions, particularly K, Mg, or Ca, can accumulate in the cytoplasm under those conditions. Given the wide range of organic solutes that can accumulate in different crop species (Table 1) or even among different genotypes within a crop species, which may have a functionally similar role, it would be unrealistic to expect any direct correlation between salinity tolerance and accumulation of any particular organic solute, either qualitatively or quantitatively.

B. Role in Ion Compartmentation

Compatible solutes or cytosolic solutes could play an important role in regulating intracellular ion distribution under salt stress, thus inducing Na accumulation in the vacuole [120]. Externally applied glycine betaine is reported to increase vacuolar Na concentration in barley roots [114]. The salt concentration required for proline accumulation could be the same as is required for salts to be sequestered into the vacuole [112]. The reported threshold of about 200 mol (Na + K/g fresh weight) is only slightly above (Na + K) levels measured in unstressed leaves [112]. In sorghum, proline accumulation seems to be related to total monovalent cation concentration whether either Na or K salts were used in the salinity treatment [111]. An ion pump at the tonoplast could become active at about the same cytoplasmic salt concentration that activates the accumulation of proline or other organic solutes [112].

C. Role in Protecting Enzymes Against Monovalent Cations

Apart from the purpose of osmoregulation, organic solutes can accumulate to protect cell metabolism from the toxic effects of accumulated ions [3,118–120,122]. Pollard and Wyn Jones [123] demonstrated such protection using glycine betaine and, in barley leaves, with the enzyme malate dehydrogenase (decarboxylating). Glycinebetaine has been reported to partially stabilize enzymes and membranes against a range of perturbations [124]. Proline levels up to 600 mM did not inhibit enzyme activity in vitro [125]. In barley, 1000 mM proline did not inhibit dehydrogenase activity [91]. Polyribosomes are stable in vitro in glycine betaine and proline concentrations up to about 1000 mM [126].

Thus the effect of proline and glycine betaine on enzyme systems in the presence of inhibitory ion concentrations may be an expression of a wider role of such compounds in protein stability [91]. Most organic solutes that accumulate under stress conditions are compatible with enzyme activity and continued metabolism [91].

Osmoregulators can not only be compatible with cytoplasmic enzymes, but can either promote or inhibit enzyme activity, depending on the enzyme source [127]. The affinity of phosphoenolpyruvate carboxylase (PEP Case) (extracted from *Cynodon dactylon* and *Sporobolus pungens* grown on saline soil) for PEP was increased by betaine and proline, which resulted in full protection against NaCl inhibition [127]. However, proline did not protect PEP Case against NaCl when it was extracted from *Salsola soda*, although betaine did provide protection [127]. These differences could be due to the existence of isoenzymes.

Although organic-compatible solutes may ameliorate some of the effects of accumulated ions, it seems that ion compartmentation is of greater significance in preserving metabolic activities. In some cases the effects of compatible solutes are apparent only under severe stress and act merely as a survival trait rather than having any beneficial effect on growth during stress [128]. But they may promote growth recovery if these solutes protect enzyme systems against stress-induced degradation, so that they can recommence synthetic function rapidly [91].

D. Metabolic Costs of Organic Solute Accumulation

Despite active accumulation of organic osmotica, there is no evidence of an additional cost, and thus osmotic adjustment exists as an energy-efficient and physiologically effective device for alleviation of drought and salinity stress [129]. However, synthesis of organic molecules such as proline or betaine does put an additional metabolic load on the plant. When sugars are used for osmotic adjustment, they are not available for growth [129]. The accumulation of nonstructural carbon is associated with osmotic adjustment and turgor maintenance [7]. Turner [130] considers that the carbon required for osmotic adjustment would be only a small fraction of that produced by the plant. However, the metabolic cost of storing photosynthate and using it for osmotic adjustment is less than the cost of converting it to new biomass, which the nonstressed plants were better able to do [129]. This explanation was confirmed by the fact that there was a large increase in the respiration rate accompanied by a rapid increase in leaf area when stressed plants were irrigated [129].

From the above it appears that a variety of organic solutes accumulate under salinity or drought stress conditions. Some of these compounds could be the result of passive accumulation (i.e., due to the general reduction in growth processes). Carbon and nitrogen compounds are simply diverted from growth-related activities to produce compounds such as proline, sucrose, or others as a way of storing them. This avoids the formation of toxic compounds, such as ammonia or putricine, from excess nitrogen metabolites. However, there is evidence that solute accumulation is an active process and is very strongly regulated according to immediate plant needs as influenced by external salinity and the plant's ability to regulate ion entry into the transpiration stream. Also, apart from acting as an organic osmoticum in the cytoplasm, these compatible solutes accelerate the compartmentation of Na and Cl into the vacuole, thus playing a significant role in determining the crop species level of salinity tolerance. However, it needs to be realized that organic solute accumulation is only one component in the overall maintenance of a stable internal ionic environment in the cytoplasm, which would ultimately determine the survival and production potential of a crop species grown in a saline environment. Thus the ability to accumulate organic solutes would have a positive functional role only if a genotype has the "genetic know-how" to regulate ion entry, particularly of Na and Cl, into the transpiration stream.

IV. ORGANISM INTEGRATION

Although various processes that play a role in ionic and osmotic regulation at the whole plant level have being discussed separately, the level of salinity tolerance of a given crop species or genotype is the collective expression of a number of processes: influx selectivity, K/Na exchange and Na extrusion, Na compartmentation in the root cortex, Na and Cl regulation at the endodermis, retrieval of Na from the xylem stream by XPT, transpiration efficiency, preventing apoplastic accumulation, phloem retranslocation of Na and Cl, K retranslocation, organic solute accumulation, Na and Cl compartmentation in the leaf, and others. For this reason it is not surprising that no single physiological mechanism/trait shows a clear-cut direct relationship to salinity tolerance. Genotypes may differ in one or many processes that regulate entry of Na or Cl ions into the plant or qualitative or quantitative differences in the organic solutes. These processes interact at the organism level to determine the ultimate level of tolerance.

V. CONCEPTUAL FRAMEWORK FOR INTEGRATING PHYSIOLOGICAL ASPECTS INTO GENETIC IMPROVEMENT PROGRAMS

There is a substantial amount of information on the physiological responses of crop plants to salinity (i.e., mostly NaCl) stress. A major portion of this information deals merely with the effects of excess salts on various metabolic functions of the plants. As Munns et al. [33] point out, most of this information describes only the consequences rather than the causes of reduced growth or injury and is thus of limited use for integration into genetic improvement programs. We believe that there is scope for more directed physiological research that would be more relevant to genetic improvement considerations. Emphasis should be given to understanding the interactions among the many possible processes involved, and thus "organism integration." The two main approaches that we see for achieving this are the "black box" and "physiological ideotype" approaches.

A. Black Box Approach

The black box approach attempts to proceed from established phenotypic differences (i.e., response to salinity) to the underlying differences in physiological mechanisms contributing to higher levels of tolerance [93,131]. Once a source of a higher level of salinity tolerance is identified in the cultivated species or its wild relatives, the next step would be to transfer this tolerance to agronomically acceptable varieties through a conventional breeding approach. Since salinity tolerance is a complex physiological trait, governed by different genes or groups of genes, the problem is how best to transfer this type of trait or ensemble of traits from the donor parent to the recipient. A black box approach is therefore enhanced by an understanding of the specific physiological traits operating in the donor parent by conducting comparative physiological studies between donor and recipient parents. This will facilitate design of the most appropriate genetic improvement procedures. In particular, simple and effective means of screening segregating populations for salinity tolerance are needed, rather than having to rely on the measurement of growth or yield reduction under given levels of salinity. Identification of the predominant physiological trait or traits responsible for the genotypic differences measured is desirable.

In pigeonpea and its related wild species, there appears to be either a curvilinear or a linear relationship between dry matter and tissue Na or Cl levels ($r^2 = 0.76$; $r^2 = 0.70$; $p < 0.001$; Figure 4a and b). However, this relationship is stronger for Na than for Cl. There is a significant positive linear relationship between tissue Na and Cl levels in both shoots and roots ($r^2 = 0.66$; $p < 0.001$; Figure 4e and f). Although the overall relationship between growth reduction and tissue Na or Cl levels appears to be positive, there is considerable variation among various wild species in the level of ionic tolerance within their tissues. This is indicated by the scatter of points. For instance, for a 50% reduction in growth, tissue Cl levels ranged from <1% to about 4%, and for Na it varied from 0.02% to about 1%. For tissue K levels, we did not find any significant relationship ($r^2 = 0.008$; Figure 4c), however, there is a positive relation between K/Na in shoot and shoot growth ($r^2 = 0.73$; $p < 0.001$; Figure 4d). These data points are also very scattered, which indicates a wide range of variation among species for their optimum K/Na requirements at a given level of growth reduction under salinity. This is not surprising given the complexity of physiological mechanisms operating in Na, K, and Cl regulation and the number of mitigating factors that could change the metabolic tolerance of Na and Cl levels in the tissues.

However, in comparing genotypes that differ in their tolerance, especially among the wild relatives of pigeonpea, we have noticed that the ability to retain higher levels of Na and Cl in the roots could be one of the crucial factors in regulating their levels in the shoot. This regulatory ability breaks down at salinity thresholds that vary across species and genotypes [29,36]. Further studies have shown that this regulatory ability is expressed in the F_1 hybrids of crosses between a tolerant wild relative (*Atylosia albicans*) and a sensitive pigeonpea genotype (ICP 3783) (Figure 5) [36]. Thus this trait is heritable. Further studies are required on the segregating F_2 and F_3 generations, including analysis of the ionic constituents, to establish the inheritance pattern of these physiological traits.

B. Physiological Ideotype and Pyramiding Approach

An *ideotype* is defined as "a hypothetical plant described in terms of traits that are thought to enhance genetic yield potential" [132]. Thus a physiological ideotype for salinity tolerance could be defined in terms of the specific physiological traits that are expected to contribute functionally in maintaining ionic and osmotic relations under saline conditions. As expressed

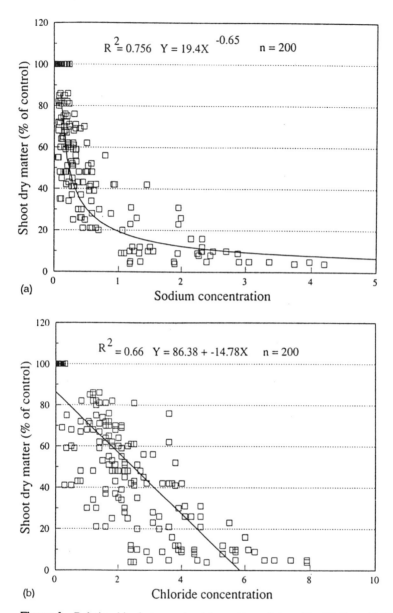

Figure 4   Relationships between shoot dry matter and tissue Na, Cl, K, and K/Na levels (a–d) and between Cl and Na levels in root and shoot (e and f). (Plant samples were collected for growth and chemical analysis at 55 days after sowing; plants were grown at 0-, 4-, 6-, 8-, and 10-dS/m salinity levels.) (From Ref. 157.)

on a relative yield basis, salinity tolerance is the collective expression of a number of physiological traits, as described earlier.

Salinity stress normally varies over time within a crop cycle, from season to season, and from site to site. Different landraces/genotypes/varieties that show a given level of tolerance to salinity are expected to have evolved a variety of mechanisms that contribute to yielding ability under those conditions. For instance, *T. aestivum*, *Secale cereale*, and *Aegilops squarrosa* have an efficient K/Na selectivity character because of the D genome but are less tolerant than crop

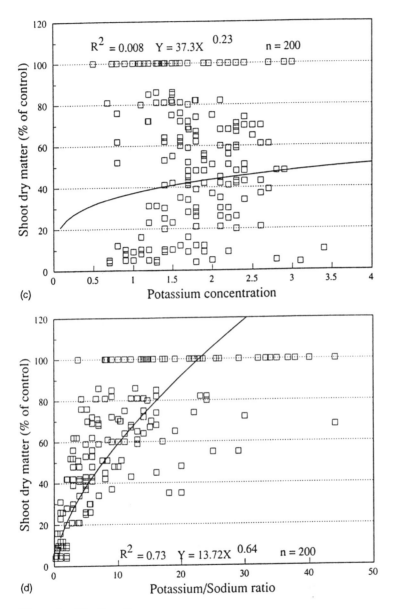

Figure 4 *(Continued)*

species such as *H. vulgare* and *T. durum*, which are less efficient in K/Na selectivity but more efficient in their compartmentation ability [133,134]. Similarly, such differences can be observed among genotypes within a crop species, which is reflected in contradictory reports for various crop species either confirming or disputing a direct correlation between K/Na selectivity and level of salinity tolerance [4].

The underlying philosophy is that although different genotypes may show the same level of tolerance to salinity, they could attain this level of tolerance (i.e., phenotype) through different physiological mechanisms or traits. Lack of sufficient phenotypic variation for salinity tolerance is a serious problem in many crops, and this is particularly so with rice [135]. Even after

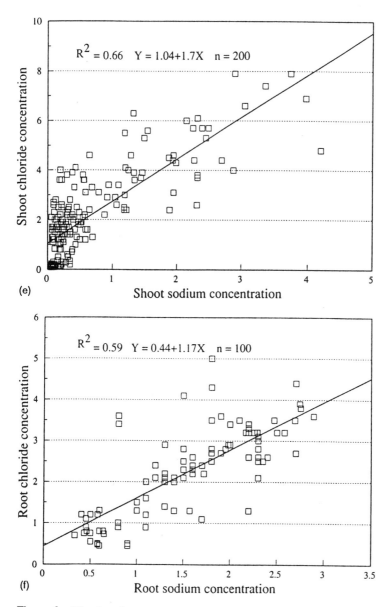

Figure 4   (*Continued*)

screening the entire world collection of rice germplasm, the most tolerant genotypes would still suffer about a 50% yield reduction at 5 dS/m [135–137]. Conceptually, the physiological approach for improving salinity tolerance in crop plants should be to bring together the relevant traits that would complement each other in a pyramidic manner ("building block" approach) by their selective incorporation into a single genotype or variety under improvement (i.e., optimization of several, probably independent, physiological mechanisms into a single variety) [19].

    An analogy can be drawn from disease resistance breeding. In breeding for disease resistance, horizontal resistance (which can be defined as resistance to a number of physiological

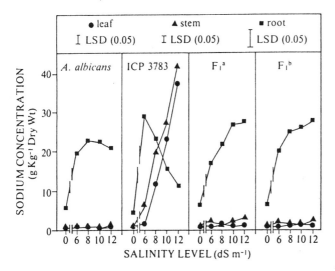

Figure 5 Effect of salinity on tissue Na concentration (g/kg dry weight) of *Atylosia albicans*, *Cajanus cajan* (ICP 3783), and their reciprocal F_1, hybrids (a and b), 75 days after transplanting. Data are means of two replications. (From Ref. 36.)

races of a disease) can be achieved by pyramiding different genes specifically resistant to individual physiological races. This contributes to the stability of a genotype across years in disease-prone environments. The same concept could also be applied to the genetic improvement of salinity tolerance, whereby pyramiding of genes that regulate various specific physiological traits into a single genotype or variety could provide that genotype with the necessary genetic means to respond to various types and levels of salinity stress that it is likely to experience at different locations, sites, and over years. This would contribute to its stability of production as well as widening its adaptability to a greater range of saline environments.

The various steps involved in this kind of approach are:

1. Define the various physiological traits having functional significance in determining the tolerance and productivity of a given crop under saline environments.
2. Establish genetic variability and locate sources of high efficiency for each physiological trait in the germ plasm. Selection should be directed toward the individual components of salinity tolerance on a trait-by-trait basis irrespective of phenotype.
3. Establish the genetic basis for each physiological trait under consideration by studying its inheritance pattern and estimating its heritability, which would determine the feasibility of using that particular trait in a breeding program.
4. Develop restriction fragment length polymorphisms (RFLP) markers if easily identifiable morphological, physiological, or other markers are not readily available for each physiological trait, as this would streamline the selection process of segregating materials in a breeding program.
5. Identify genotypes for each physiological trait which have good combining ability.
6. Incorporate relevant traits into an agronomically acceptable background basis.

Information generated through this exercise could be stored in a database system which would be made available to breeders interested in incorporating salinity tolerance in their breeding programs. This is similar to information databases that are available for morphological traits from the germ plasm evaluation exercises at CGIAR (Consultative Group for International Agricultural Research) centers.

Selection of traits to be introduced into a given genotype/variety under improvement depends on the target environment in which it will be grown, and the specific traits a particular variety may be lacking. For instance, a variety may be very efficient in Na and Cl compartmentation in the root as well as in the shoot, but may be lacking effective Na or Cl regulation at the plasmalemma. There is evidence of genotypic variation within crop species in Na compartmentation in shoots [19,138,139] and tolerance to high internal Na and Cl levels [69]. In this case, only the trait that is lacking needs to be introduced. Similarly, a given variety may be very efficient in ion regulation, but lacks the necessary genetic means to produce organic solutes.

Development of RFLP markers for each of these physiological components of salinity tolerance could play a crucial role in the incorporation of these physiological traits into a genotype/variety under improvement. Salinity tolerance traits are controlled by a number of genes located throughout the chromosome complement [140]. Each gene of a polygenic system may contribute only a small amount to the trait of interest. Clear dominance is not likely to be exhibited and the phenotype (i.e., the specific trait in this case) would have a large component of environmental variance. All these characteristics conspire to make physiological traits very difficult to analyze. Thus conventional Mendelian methods of analysis, which are suitable for traits controlled by a single or a few genes, cannot be applied to analysis of these physiological traits. This is one reason that physiological traits have not been used extensively in the genetic improvement programs for salinity or drought tolerance, although a number of them having functional significance for determining level of tolerance have been identified [6,141].

With the development of RFLP mapping techniques (for a detailed discussion of RFLP techniques, see Tanksley et al. [142]), it is possible to analyze complex polygenic characters, such as physiological traits, as ensembles of single Mendelian factors. Since RFLP markers can be used to follow simultaneously the segregation of all chromosome segments during a cross, the basic idea is to look for correlations between physiological traits and specific chromosome segments marked by RFLPs. If correlations exist, the inference is that the chromosome segment must be involved in the quantitative trait. The difficult part in this procedure is to establish correlations between the trait and specific chromosome segments. The RFLP markers can easily be scored, but the physiological trait must be characterized in a conventional fashion [142]. Once this most difficult process is completed and specific chromosome segments are implicated in the trait, RFLP markers with a positive effect on a quantitative trait can be selected from a population of plants and incorporated into a single genotype. This is possible because of the ability to score for several RFLP markers simultaneously in a single plant in a manner that is free of environmental influence or gene interactions. Carbon isotope (^{13}C) discrimination, which is an indicator of water use efficiency, could be predicted satisfactorily from three RFLPs in tomato [143]. We have not found any other reports implicating RFLP markers for physiological traits contributing to salinity tolerance. The feasibility of using RFLP markers for physiological traits could bridge the gap between plant physiology and breeding, to facilitate integration of these two disciplines and thus expedite development of varieties that are higher yielding and more stable across environments affected by salinity.

VI. FUTURE OUTLOOK

The past 30 years of research (after the report of dual mechanisms of ion transport by Epstein et al. [144]) on physiological aspects of salinity tolerance has contributed substantially to an understanding of the mechanisms by which plants cope with excess salts in their habitat. In recent times, efforts have been initiated to identify genes responsible for specific physiological mechanisms. Location of the K/Na selectivity character on the 7a chromosome of the D genome in wheat is one such example [145–147]. Similarly, Na exclusion capability and K/Na discrimination were enhanced in *T. aestivum* by the incorporation of a *Lophopyrum* genome [148]. The K/Na discriminating locus has been located on the 3E chromosome in *Lophopyrum elongatum* [140]. An association with the higher level of salinity tolerance in *Agropyron junceum* has been located in the 5J chromosome [149]. In rice, at least three groups of genes were found to be involved in the inheritance of Na and Ca levels in the plant; Na and Ca levels in shoots and roots were reported to show additive affects with a high degree of heritability [150].

Similarly, Cl translocation is under genetic control [151,152]. Accumulation of organic solutes such as betaine has been reported to be regulated by a limited number of genes [153–156]. Our research with pigeonpea has shown that the higher levels of salinity tolerance, and the associated physiological mechanisms identified in the wild relative *Atylosia albicans*, could be expressed in the reciprocal crosses of F_1 hybrids of this species with the cultivated species (Figure 5) [36]. Information on the genetic control of specific mechanisms is essential for proper integration of physiological research into breeding programs. Recent developments in biotechnology, particularly with genetic markers such as RFLPs, could accelerate this integration of disciplines. Wild relatives have been inadequately explored for their potential to contribute unique physiological mechanisms of salinity tolerance. We hope future efforts would be directed toward generating information on these areas.

REFERENCES

1. D. H. Jennings, *Biol. Rev.*, *51*: 453 (1976).
2. D. W. Rains, *Annu. Rev. Plant Physiol.*, *23*: 367 (1972).
3. T. J. Flowers, P. F. Troke, and A. R. Yeo, *Annu. Rev. Plant Physiol.*, *28*: 89 (1977).
4. H. Greenway and R. Munns, *Annu. Rev. Plant Physiol.*, *31*: 149 (1980).
5. A. R. Yeo, *Physiol. Plant.*, *58*: 214 (1983).
6. W. D. Jeschke, in *Salinity Tolerance in Plants: Strategies for Crop Improvement* (R.C. Staples and G. H. Toenniessen, eds.), Wiley, New York, pp. 37–66 (1984).
7. J. M. Cheeseman, *Plant Physiol.*, *87*: 547 (1988).
8. H. Boyko, *Salinity and Aridity: New Approaches to Old Problem*, Dr. W. Junk, The Hague, The Netherlands (1966).
9. E. Epstein, *Mineral Nutrition of Plants: Principles and Prospectives*, Wiley, New York (1972).
10. P. Mitchell, *Nature*, *191*: 144 (1961).
11. R. N. Robertson, *Protons, Electrons, Phosphorylation, and Active Transport*, Cambridge University Press, Cambridge, (1968).
12. D. P. Briskin and J. B. Hanson, *J. Exp. Bot.*, *43*: 269 (1992).
13. D. T. Clarkson and C. Grignon, in *Phosphorus Nutrition of Grain Legumes in the Semi-arid Tropics* (C. Johansen, K. K. Lee, and K. L. Sahrawat, eds.), International Crops Research for the Semi-arid Tropics, Patancheru, Andhra Pradesh, India, pp. 49–62 (1991).
14. A. C. Leopold and R. P. Willing, in *Salt Tolerance in Plants: Strategies for Crop Improvement* (R. C. Staples and G. A. Toenniessen, eds.), Wiley, New York, pp. 67–76 (1984).
15. P. J. C. Kuiper, in *Salinity Tolerance in Plants: Strategies for Crop Improvement* (R. C. Staples and G. A. Toenniessen, eds.), Wiley, New York, pp. 77–91 (1984).
16. M. G. Pitman and H. D. W. Saddler, *Proc. Natl. Acad. Sci. USA*, *57*: 44 (1967).
17. D. H. Jennings, *New Phytol.*, *67*: 899 (1968).
18. W. D. Jeschke, in *Ion Transport in Plants* (W. P. Anderson, ed.), Academic Press, New York, pp. 285–290 (1973).
19. A. R. Yeo and T. J. Flowers, *Aust. J. Plant Physiol.*, *13*: 161 (1986).

20. W. D. Jeschke, in *Genetic Aspects of Plant Nutrition* (N. E. L. Bassam, M. Dambroth, and B. C. Loughman, eds.), Kluwer Academic Publishers, Boston, pp. 71–86 (1990).

21. R. G. Wyn Jones, C. J. Brady, and J. Speirs, in *Recent Advances in the Biochemistry of Cereals* (D. L. Laidman and R. G. Wyn Jones, eds.), Academic Press, New York, pp. 63–104 (1979).

22. R. G. Wyn Jones, in *Genetic Engineering of Osmoregulation: Impact on Plant Productivity for Food, Chemicals, and Energy* (D. W. Rains, R. C. Valentine, and A. Hollaender, eds.), Plenum Press, New York, pp. 155–170 (1980).

23. W. D. Jeschke, in *Recent Advances in the Biochemistry of Cereals* (D. L. Laidman and R. G. Wyn Jones, eds.), Academic Press, New York, pp. 37–61 (1979).

24. J. Rozema, E. D. Rozema, A. H. J. Freijsen, and J. J. L. Huber, *Oecologia (Berlin), 34*: 329 (1978).

25. J. Gorham, L. L. Hughes, and R. G. Wyn Jones, *Oecologia (Berlin), 3*: 309 (1980).

26. I. Ahmad, S. J. Wainwright, and G. R. Stewart, *New Phytol., 87*: 615 (1981).

27. J. Gorham, R. G. Wyn Jones, and E. McDonnel, *Plant Soil, 89*: 15 (1985).

28. M. A. Hajibagheri, D. M. R. Harvey, and T. J. Flowers, *New Phytol., 105*: 367 (1987).

29. G. V. Subbarao, C. Johansen, M. K. Jana, and J. V. D. K. Kumar Rao, *J. Plant Physiol., 137*: 64 (1990).

30. A. Scarpa, and J. deGiez, *Biochem. Biophys. Acta, 241*: 789 (1971) (cited in Ref. 6).

31. A. Lauchli, in *Transport and Transport Processes in Plants* (I. F. Wardlaw and J. B. Passioura, eds.), Academic Press, New York, pp. 101–112 (1976).

32. C. Johansen, D. G. Edwards, and J. F. Loneragan, *Plant Physiol., 45*: 601 (1970).

33. R. Munns, H. Greenway, and G. O. Kirst, in *Encyclopedia of Plant Physiology*, New Series, Vol. 12C, *Physiological Plant Ecology*, Vol. III (O. L. Lange, P. S. Nobel, C. B. Osmond, and H. Ziegler, eds.), Springer-Verlag, New York, pp. 59–135 (1983).

34. E. Epstein, in *Genetic Engineering of Osmoregulation: Impact on Plant Productivity for Food, Chemicals, and Energy* (D. W. Rains, R. L. Valentine, and A. Hollaender, eds.), Plenum Press, New York, pp. 7–21 (1980).

35. D. Kramer, A. Lauchli, A. R. Yeo, and J. Gullasch, *Ann. Bot., 41*: 1031 (1977).

36. G. V. Subbarao, C. Johansen, J. V. D. K. Kumar Rao, and M. K. Jana, *Crop Sci., 30*: 785 (1990).

37. R. Behl and W. D. Jeschke, *Physiol. Plant., 53*: 95 (1981).

38. R. L. Jeffries, in *Ion Transport in Plants* (W. P. Anderson, ed.), Academic Press, New York (1973).

39. S. M. Pan and Y. R. Chen, *Bot. Bull. Acad. Sinica, 29*: 33 (1988).

40. R. Weinberg, *J. Biol. Chem., 242*: 3000 (1967).

41. Y. D. Beletskii, T. B. Karnaukhova, and N. I. Shevyakova, *Sov. Plant Physiol., 33*: 890 (1986).

42. A. J. Cavalieri and A. H. C. Huang, *Physiol. Plant., 41*: 79 (1977).

43. S. D. Tanksley, *Isozymes in Plant Genetics and Breeding*, Vol. 1, Elsevier, Amsterdam (1983).

44. J. L. Hall and T. J. Flowers, *Planta, 110*: 361 (1973).

45. T. J. Flowers, *J. Exp. Bot., 23*: 310 (1972).

46. T. J. Flowers, *Phytochemistry, 11*: 1881 (1972).

47. D. H. Jennings, *Ecological Aspects of the Mineral Nutrition of Plants* (I. H. Rorison, ed.), Blackwell Scientific Publications, Oxford, pp. 261–279 (1969).

48. P. A. Rea and D. Sanders, *Physiol. Plant., 71*: 131 (1987).

49. M. G. Palmgren, *Physiol. Plant., 83*: 314 (1991).

50. M. G. Pitman, in *Salinity Tolerance in Plants: Strategies for Crop Improvement* (R. C. Staples and G. H. Toenniessen, eds.), Wiley, New York, pp. 93–123 (1984).

51. A. Lauchli and J. Wieneke, "Salt Relations of Soybean Mutants Differing in Salt Tolerance: Distribution of Ions and Localization by X-ray Microanalysis," *Plant Nutrition: Proceedings of the 8th International Colloquium on Plant Analysis and Fertilizer Problems*, Auckland, New Zealand, Government Printer, Wellington, pp. 275–282 (1978).

52. J. G. Johansen and J. M. Cheeseman, *Plant Physiol., 73*: 153 (1983).

53. J. G. Johansen and J. M. Cheeseman, *Plant Physiol., 73*: 159 (1983).

54. B. J. Cooil, R. K. Dela Fuente, and R. S. Dela Pena, *Plant Physiol.*, *40*: 625 (1965).
55. A. Lauchli and J. Wieneke, *Z. Pflanzenernaehr. Bodenkd.*, *124*: 3 (1979).
56. A. R. Yeo, D. Kramer, A. Lauchli, and J. Gullasch, *J. Exp. Bot.*, *28*: 17 (1977).
57. B. Jacoby, *Ann. Bot.*, *43*: 741 (1979).
58. R. T. Besford, *Plant Soil*, *58*: 399 (1978).
59. R. Munns, D. B. Fisher, and M. L. Tennet, *Aust. J. Plant Physiol.*, *13*: 757 (1986).
60. R. R. Walker, *Aust. J. Plant Physiol.*, *13*: 293 (1986).
61. H. Nassery and D. A. Baker, *Ann. Bot.*, *38*: 141 (1974).
62. H. Marschner and O. Ossenberg-Neuhaus, *Z. Pflanzenernaehr. Bodenk.*, *139*: 129 (1976).
63. U. Luttge, in *Encyclopedia of Plant Physiology*, New Series, Vol. 15 (A. Lauchli and R. L. Bieleski, eds.), Springer-Verlag, Berlin, pp. 181–211 (1983).
64. E. Winter, *Aust. J. Plant Physiol.*, *9*: 239 (1982).
65. J. J. Oertli, *Agrochimica*, *12*: 461 (1968).
66. R. A. Leigh and R. G. Wyn Jones, *New Phytol.*, *97*: 1 (1984).
67. R. Munns and J. B. Passioura, *Aust. J. Plant Physiol.*, *11*: 479 (1984).
68. N. J. W. Clipson, A. D. Tomos, T. J. Flowers, and R. G. Wyn Jones, *Planta, 165*: 392 (1985).
69. A. R. Yeo, S. J. M. Caporn, and T. J. Flowers, *J. Exp. Bot.*, *36*: 1240 (1985).
70. T. J. Flowers and Y. R. Yeo, *Aust. J. Plant Physiol.*, *13*: 75 (1986).
71. T. J. Flowers, M. A. Hajibagheri, and A. R. Yeo, *Plant Cell Environ.*, *14*: 319 (1991).
72. J. S. Pate, in *Transport and Transfer Processes in Plants* (I. F. Wardlaw and J. B. Passioura, eds.), Academic Press, New York, pp. 253–281 (1976).
73. D. R. Geiger, in *Transport and Transfer Processes in Plants* (J. F. Wardlaw and J. B. Passioura, eds.), Academic Press, New York, pp. 167–183 (1976).
74. E. Levi, *Physiol. Plant.*, *21*: 213 (1968).
75. H. Lessani and H. Marschner, *Aust. J. Plant Physiol.*, *5*: 27 (1978).
76. E. Winter, *L. Micron.*, *2*: 519 (1980).
77. J. A. Webb and P. R. Gorham, *Plant Physiol.*, *39*: 663 (1964).
78. H. Greenway and M. G. Pitman, *Aust. J. Biol. Sci.*, *18*: 235 (1965).
79. J. S. Pate, *Encyclopedia of Plant Physiology*, New Series, Vol. 1 (M. H. Zimmerman and J. A. Milburn, eds.), Springer-Verlag, Berlin, pp. 451–473 (1975).
80. M. G. Pitman, in *Ion Transport in Plant Cells and Tissues* (D. A. Baker and J. L. Hall, eds.), North-Holland, Amsterdam, pp. 267–308 (1975).
81. T. J. Flowers, F. M. Salama, and Y. R. Yeo, *Plant Cell Environ.*, *11*: 453 (1988).
82. H. Greenway, *Aust. J. Biol. Sci.*, *18*: 249 (1965).
83. M. G. Pitman, *Aust. J. Biol. Sci.*, *18*: 987 (1965).
84. D. T. F. Bowling and Weatherley, *J. Exp. Bot.*, *16*: 732 (1965).
85. M. G. Pitman, *Annu. Rev. Plant Physiol.*, *28*: 71 (1977).
86. F. Malek and D. A. Baker, *Planta*, *135*: 297 (1977).
87. M. G. Pitman, *Aust. J. Biol. Sci.*, *19*: 257 (1966).
88. A. Poljakoff-Mayber, in *Plants in Saline Environments* (A. Poljakoff-Mayber and J. Gale, eds.), Springer-Verlag, Berlin, pp. 97–117, (1975).
89. R. Stelzer and A. Lauchli, *Z. Pflanzenphysiol.*, *84*: 95 (1977).
90. M. A. Hajibagheri, J. L. Hall, and T. J. Flowers, *New Phytol.*, *94*: 125 (1983).
91. D. Aspinall, *Aust. J. Plant Physiol.*, *13*: 59 (1986).
92. N. C. Turner and M. M. Jones, in *Adaptation of Plants to Water and High Temperature Stress* (N. C. Turner and P. J. Kramer, eds.), Wiley, New York, pp. 87–103 (1980).
93. M. Tal, *Plant Soil*, *89*: 199 (1985).
94. A. R. Yeo, *J. Exp. Bot.*, *32*: 487 (1981).
95. G. R. Stewart and J. A. Lee, *Planta*, *120*: 279 (1974).
96. G. R. Stewart and A. D. Hanson, in *Adaptation of Plants to Water and High Temperature Stress* (N. C. Turner and P. J. Kramer, eds.), Wiley, New York, pp. 173–189 (1980).
97. S. Treichel, *Z. Pflanzenphysiol.*, *76*: 56 (1975).
98. A. D. Hanson and W. D. Hitz, *Annu. Rev. Plant Physiol.*, *33*: 163 (1982).
99. G. Parameshwara, Ph.D. thesis, University of Adelaide, Australia (1984).

100. R. Storey and R. G. Wyn Jones, *Phytochemistry*, *16*: 447 (1977).

101. H. Greenway, R. Munns, and J. Gibbs, *Plant Cell Environ.*, *5*: 405 (1982).

102. C. M. Grieve and E. V. Maas, *Physiol. Plant.*, *61*: 167 (1984).

103. R. G. Wyn Jones and R. Storey, in *Physiology and Biochemistry of Drought Resistance in Plants* (L. G. Paleg and D. Aspinall, eds.), Academic Press, Sydney, Australia, pp. 171–203 (1981).

104. R. G. Wyn Jones and R. Storey, *Aust. J. Plant Physiol.*, *5*: 817 (1978).

105. Z. Kefu, R. Munns, and R. W. King, *Aust. J. Plant Physiol.*, *18*: 17 (1991).

106. J. F. Thomson, C. R. Stewart, and C. J. Morris, *Plant Physiol.* (*Lancaster*), *41*: 1578 (1966).

107. I. Ahmad, F. Larher, A. F. Mann, S. F. McNally, and G. R. Stewart, *New Phytol.*, *91*: 585 (1982).

108. L. G. Paleg, T. J. Douglas, A. vonDaal, and D. B. Kuch, *Aust. J. Plant Physiol.*, *8*: 107 (1981).

109. R. L. Jeffries and T. Rudmik, in *Salinity Tolerance in Crop Plants: Strategies for Crop Improvement* (R. C. Staples and G. H. Toenniessen, eds.), Wiley, New York, pp. 213–227 (1984).

110. R. Weinberg, *Physiol. Plant.*, *73*: 418 (1988).

111. R. Weimberg, H. R. Lerner, and A. Poljakoff-Mayber, *Physiol. Plant.*, *55*: 5 (1982).

112. G. Voetberg and C. R. Stewart, *Plant Physiol.*, *76*: 567 (1984).

113. W. D. Hitz and A. D. Hanson, *Phytochemistry*, *19*: 2371 (1980).

114. N. Ahmad and R. G. Wyn Jones, *Plant Sci. Lett.*, *15*: 231 (1979).

115. A. D. Hanson and C. E. Nelson, *Plant Physiol.*, *62*: 305 (1978).

116. S. P. Robinson and G. P. Jones, *Aust. J. Plant Physiol.*, *13*: 659 (1986).

117. R. Weimberg, H. R. Lerner, and A. Poljakoff-Mayber, *Physiol. Plant.*, *62*: 472 (1984).

118. J. Levitt, *Responses of Plants to Environmental Stresses*, Vol. II *Water, Salt and Other Stresses*, 2nd ed., Academic Press, New York (1980).

119. R. G. Wyn Jones, R. Storey, R. A. Leigh, N. Ahmad, and A. Pollard, in *Regulation of Cell Membrane Activities in Plants* (E. Marre and O. Cieferri, eds.), North-Holland, Amsterdam, pp. 121–135 (1977).

120. W. A. Torello and L. A. Rice, *Plant Soil*, *93*: 241 (1986).

121. T. J. Flowers and A. R. Yeo, in *Environmental Stress in Plants: Biochemical and Physiological Mechanisms* (J. H. Cherry, ed.), NATO ASI Series, Series G, Ecological Sciences, Vol. 19, Springer-Verlag, New York, pp. 101–119 (1989).

122. H. Greenway and A. P. Sims, *Aust. J. Plant Physiol.*, *1*: 15 (1974).

123. A. Pollard and R. J. Wyn Jones, *Planta*, *144*: 291 (1979).

124. R. G. Wyn Jones, *Recent Advances in Phytochemistry*, Vol. 13, *Phytochemical Adaptations to Stress* (B. N. Timmerman, C. Steelink, and F. A. Loewis, eds.), Plenum Press, New York, pp. 55–78 (1984).

125. M. I. Lone, J. S. H. Kuch, R. G. Wyn Jones, and S. W. J. Bright, *J. Exp. Bot.*, *38*: 479 (1987).

126. C. J. Brady, T. S. Gibson, E. W. R. Barlow, J. Speirs, and R. G. Wyn Jones, *Plant Cell Environ.*, *7*: 371 (1984).

127. Y. Manetas, Y. Petropoulore, and G. Karabourniotis, *Plant Cell Environ.*, *9*: 145 (1986).

128. C. Itai and L. G. Paleg, *Plant Sci. Lett.*, *25*: 329 (1982).

129. K. J. McCree, *Aust. J. Plant Physiol.*, *13*: 33 (1986).

130. N. C. Turner, in *Stress Physiology in Crop Plants* (H. Mussell and R. C. Staples, eds.), Wiley, New York, pp. 344–372 (1979).

131. R. A. Fischer, *Plant Soil*, *58*: 249 (1981).

132. D. C. Rasmusson, *Crop Sci.*, *27*: 1140 (1987).

133. J. Gorham, *J. Exp. Bot.*, *41*: 623 (1990).

134. J. Gorham, A. Bristol, E. M. Young, R. G. W. Jones, and G. Kashour, *J. Exp Bot.*, *41*: 1095 (1990).

135. A. R. Yeo, M. E. Yeo, S. A. Flowers, and T. J. Flowers, *Theor. Appl. Genet.*, *79*: 377 (1990).

136. S. Yoshida, *Fundamentals of Rice Crop Science*, International Rice Research Institute, Los Banos, The Philippines (1981).

137. A. R. Yeo, K. S. Lee, P. Izard, P. J. Boursier, and T. J. Flowers, *J. Exp Bot.*, *42*: 881 (1991).

138. T. J. Flowers, E. Dugue, M. A. Hajibagheri, T. P. McGenigle, and A. R. Yeo, *New Phytol.*, *100*: 37 (1985).

139. Y. Saranga, J. Rudich, and D. Zamir, *Acta Hortic.*, *200*: 203 (1987).
140. J. Dvorak, E. Epstein, A. Galvez, P. Gulick, and J. A. Omielan, "Genetic Basis of Plant Tolerance to Soil Toxicity," *Plant Breeding in the 1990s* (H. T. Stalker and J. P. Murphy, eds.), *Proceedings of the Symposium on Plant Breeding in the 1990s*, North Carolina State University, Mar. 1991, CAB International, Wellingford, Berkshire, England, pp. 201–217 (1992).
141. M. M. Ludlow and J. M. Muchow, *Adv. Agron.*, *43*: 107 (1990).
142. S. D. Tanksley, N. D. Young, A. H. Paterson, and M. W. Bonierbale, *Biotech*, *7*: 257 (1989).
143. B. Martin, J. Nienhuis, G. King, and A. Schaeffer, *Science*, *243*: 1725 (1989).
144. E. Epstein, D. W. Rains, and O. E. Elzam, *Proc. Natl. Acad Sci. USA*, *49*: 684 (1963).
145. R. Storey, R. G. Graham, and K. W. Shepherd, *Plant Soil*, *83*: 323 (1985).
146. S. H. Shah, J. Gorham, B. P. Forster, R. G. W. Jones, and R. G. Wyn Jones, *J. Exp. Bot.*, *38*: 254 (1987).
147. J. Gorham, C. Hardy, R. C. W. Jones, L. R. Joppa, C. N. Law, and R. G. Wyn Jones, *Theor. Appl. Genet.*, *74*: 584 (1987).
148. J. Gorham, "Genetics of Sodium Uptake in Wheat," *Proceedings of the 7th International Wheat Genetics Symposium*, pp. 817–821 (1988).
149. D. P. Schachtman, A. J. Bloom, and J. Dvorak, *Plant Cell Environ.*, *12*: 47 (1989).
150. M. Akbar, G. S. Khush, and D. Hilleristambers, in *Rice Genetics*, International Rice Research Institute, Manila, The Philippines, pp. 339–409 (1986).
151. G. H. Abel, *Crop Sci.*, *9*: 697 (1969).
152. A. V. Venables and D. A. Wilkins, *New Phytol.*, *80*: 613 (1978).
153. R. Grumet, T. G. Isleib, and A. D. Hansen, *Crop Sci.*, *25*: 618 (1985).
154. R. Grumet and A. D. Hansen, *Aust J. Plant Physiol.*, *13*: 353 (1986).
155. S. A. Quarrie, *Plant Cell Environ.*, *4*: 147 (1981).
156. S. A. Quarrie, in *Abscisic Acid* (F. T. Addicott, ed.), Praeger, New York, pp. 365–419 (1983).
157. G. V. Subbarao, Ph.D. dissertation, Indian Institute of Technology, Kharagpur, India (1989).

41
Whole-System Research Complements Reductive Research

Donald H. Wallace

Cornell University, Ithaca, New York

Richard W. Zobel

Agricultural Research Service, U.S. Department of Agriculture, Ithaca, New York

I. COMPLEXITY OF BIOLOGICAL SYSTEMS

Trewavas [1] stated: "In biology we deal with the most complex systems known. . . . Complex situations are surely going to require complex explanations if the explanation is to be accurate. Or if we retain simple views should we not honestly admit that they may achieve little when faced with biological complexity." Trewavas additionally stated: "Biological systems include a great density of connections between parts of the system." He recommended more research at the whole system level. Zobel [2,3] indicates that the complexities of biology are compounded by difficulty to measure, document, and interpret the interactions among the parts of the system.

Trewavas [1] recognized that the notion of scientific advance through refutation of hypotheses originated in biochemistry and physics, for which reductive simplicity enables "all or none" experimental approaches. He doubted "all or none" confirmation or rejection of hypotheses based on whole biological systems, because the number of variables and of interactions among parts of the system is almost infinitely large. He suggested that conclusions can be relative but not definitive, and therefore acceptance of biological hypotheses can only be conditional. These first two paragraphs are from the introduction to Wallace et al. [4] entitled "A whole-system reconsideration of paradigms about photoperiod and temperature control of crop yield."

II. OBJECTIVES OF THIS CHAPTER

Briefly summarized, the objectives of this chapter are to:

1. Synthesize current knowledge about the complex biological system that accumulates crop yield.
2. Indicate current insufficiency of whole system research of plant biology.
3. Demonstrate that whole system research complements reductive research.
4. Describe needed changes from current social constraints for whole system research

III.   WHOLE SYSTEM AND REDUCTIVE RESEARCH ARE COMPLEMENTARY

Reductive research studies one component of a system at a time or, infrequently, a few at a time. The reduction is used because it avoids most of the feedforward and feedback effects that result through the great density of connections between parts of the system. Circular feedforward and feedback effects arise from almost any difference in genotype or any change of environment. Reductive isolation of a subsystem from the whole system reduces both genetic and environmentally caused influences through other parts of the system, thereby facilitating "yes or no" refutation of hypotheses.

The alternative to reductive research as a way to affirm or reject hypotheses is use of statistical analyses to quantify the probability that a measured apparent difference reflects a true difference and not sampling error. Proper experimental design and statistical analysis quantifies a main effect on the measured trait by each genotype, and a main effect by each environment, plus a level of contribution by each tested genotype and by each tested environment to each genotype × environment (G × E) interaction effect on the expressed level of the measured trait (see Chapters 42 and 43). Statistical analysis is essential to whole system research because the output from the system is an integration of all consequences from all the genetic and environmental effects in addition to those from the feedforward and feedback effects that result through the great density of connections between parts of the system. To synthesize understanding of a complex biological system, whole system research is required in addition to reductive research.

IV.   SOCIAL ATTITUDES ABOUT WHOLE SYSTEM VERSUS REDUCTIVE RESEARCH

The current social and philosophical attitude [1–6] of plant scientists tends to designate reductive research as "basic" and whole system plant research as "applied." The attitude that applied research benefits science less than reductive research has arisen because, until now, most whole system research is directed toward raising economic profitability. To illustrate, most crop yield trials have the single objective of identifying the cultivars (genotypes), or levels of environmental factors (fertilizer nutrients, water, pesticides, soil types, land preparation practices), or the ecological zones (latitude and elevation, i.e., daylength and/or mean temperature) that give the highest yield. The sole objective is to learn which cultivar (genotype) to recommend for planting by farmers, or what level of fertilizer or other environmental factor(s) to apply, or which of alternative possibilities to use in preparing and cultivating the land. Yield trials conducted with the sole objective of supporting and enhancing the economic goals of agriculture are, without question, purely applied research.

V.   WHOLE SYSTEM RESEARCH CAN BE BOTH BASIC AND APPLIED

Research is basic research if it tests hypotheses of why and how a genotype and/or environmental factor causes a given level of output from a biological system whereas another genotype or environmental level results in a different level of the output. Basic research enlarges knowledge about the biological process(es). Enhanced knowledge about plant biology can be derived from virtually every yield trial, simultaneous with its applied goal. Why and how the various genotypes and environments result in different levels of the system output (crop yield as an example) can be asked and answered for every yield trial [4,7] (see Chapter 44). A specific example of this is detailed in the following paragraphs.

VI.   SOURCE OF THE DENSITY OF CONNECTIONS AMONG PARTS OF THE SYSTEM THAT ACCUMULATES CROP YIELD

Whole system research derived from yield trials has revealed at least 25 traits that are controlled by a single photoperiod gene [8] and the modulations of its activity by daylength and/or by temperature [4,7–10]. This great density of interconnections between parts of the yield system is due to a largely unrecognized pivotal role of the photoperiod gene activity. This many interconnections between parts of the crop yield system have not and cannot be revealed in full by the abundant highly reductive research on the photoperiod response of plants.

Some of the 25 often seemingly independent traits altered by photoperiod gene activity include the days to flowering, node to first flower, rate of growth of flower buds, leaf area, total plant biomass, rate of accumulation of biomass, days to harvest maturity, rate of yield accumulation, cultivar adaptation to the environment, number of nodes and leaves, net accumulated biomass, harvest index, and yield. The level of each of at least 25 traits is altered simultaneously by the feedforward plus feedback effects that arise as daylength and temperature differently but jointly modulate the gene activity of a photoperiod-insensitive genotype, in contrast to their modulation of the activity of a photoperiod-sensitive genotype. How photoperiod gene activity controls such a great density of connections between parts of the crop yield system is described later in this chapter.

VII.   DEVELOPMENTAL STAGES OF A BIOLOGICAL SYSTEM

A.   Primordial Stage of Development

Crop yield arises from thousands of genetic instructions. The building blocks for these instructions are DNA. Specific segments of the strands of DNA constitute a gene. The gene is transcribed to give mRNA, which in turn is translated to give a protein that functions as an enzyme. The enzyme coded by each gene implements the genetic instruction of that gene. There are tens of thousand of genes in a plant [11–13]. Therefore, the block of DNA that codes for a gene is but one of tens of thousands of primordial beginnings within the biological system. Each gene–enzyme-implemented primordial beginning must be integrated with all the others.

Reductive disciplines such as molecular genetics and biotechnology are currently capable of cloning some of the specific genes. However, contributions to plant improvement using this DNA to transform plants is limited by lack of understanding of the role of most genes within the yield system. Applications of biotechnology are limited further by not knowing how that role is altered by its interactions with the roles of the other genes, and by not knowing how each gene's activity is modulated by each of the many environmental factors. There is insufficient understanding about how to select among the tens of thousands of genes.

B.   Basal Stage of Development

All the individual genes are integrated into long chains of DNA which combine with specific proteins to form chromosomes. Only at this basal (chromosomal) stage is a gene capable of being transcribed so that the enzyme activity can implement its genetic instruction and thereby advance development of the system.

Duplication of the paired DNA strands that constitute each chromosome and the attendant meiosis lead to the phenomena of inheritance. Inheritance is altered by segregation of the parental chromosomes, in addition to crossing over of the paired strands of DNA, and by deletions, substitutions, or other mutations of the genetic information. The laws of inheritance patterns are understood relatively well.

C. Meso Stage of Plant Growth and Development

The enzymatic implementations of the tens of thousands of individual genetic instructions, collectively, constitute the meso stage of plant development. These implementations [13] precede, and their integrations result in, a few penultimate outputs as well as the ultimate output from the biological system. During the meso stage of plant development, subsets of the enzymatic activities result in competition between processes, such as photosynthesis in contrast to dark- and photo-respiration, or alternative pathways such as C3 photosynthesis in contrast to C4 photosynthesis, water uptake balanced against its transpiration, mineral uptake and utilization in contrast to mineral deficiency or toxicity, and nitrate uptake and utilization in contrast to nitrogen fixation and its utilization. Other alternatives are response to photoperiod in comparison with insensitivity to photoperiod, large versus small or many versus few stomata, large compared with small leaves and total leaf areas, thick versus thin leaves, vertical in contrast to horizontal orientation of leaves, small in contrast to large mesophyll resistances within leaves, many versus few or close versus distantly spaced vascular bundles.

The competing or environmentally modulated processes or contrasting levels of the traits mentioned in the preceding paragraphs are but a few of many plant traits, including a few specific enzymatic activities, that have been identified for use in attempts to raise the genetic potential for crop yield. Few such attempts have resulted in any degree of success. Lack of effectiveness is illustrated by the conclusion of Gifford [14] that selection for rate of photosynthesis will not lead to higher yield until we understand "feedback regulations arising from environmental limitations in the field." As described by Wallace et al. [4,7–10] (see Chapters 42 to 45), these limitations are the environmental influences on the G × E interaction effect on the yield. Included are additional influences on the yield due to interaction effects among the gene activities.

In earlier publications we indicated that crosses intended to incorporate genes that give a higher level of one of the many traits of the meso stage of development virtually always result in simultaneous segregation of other genes that affect photosynthesis or other processes and morphological traits that affect yield [13,15–18]. We suggested that these simultaneous segregations, combined with the resulting shifts in the interaction effects among the gene activities, are an additional reason why selection for rate of photosynthesis does not result in yield gains. The segregations of many genes for many interacting traits result in a multitude of both positive and negative gene × gene interaction effects and G × E interaction effects on the yield of the segregates. The integrated effect is not known in advance.

D. Ultimate Developmental Stage

1. Ultimate System Output

The ultimate stage of development is when the many processes and traits that constitute the system have all been fully integrated to give the system output. For the crop yield system, the ultimate stage of development is harvest maturity. Harvestable yield is the system output.

2. Penultimate Outputs

There are just three nearly fully integrated penultimate outputs from the yield system. It is the integration of these penultimate outputs that results in the ultimate output (the yield). The penultimate outputs from the crop yield system, described more fully below, are:

1. The total accumulated biomass
2. The partitioning of that biomass between the plant organs that constitute yield and all the other organs of the plant
3. The time used to implement both this net accumulation of the biomass and its partitioning.

Physiological-genetic implementation of all three penultimate outputs attends the advances in plant development toward harvest maturity. We include the penultimate outputs within the ultimate stage of plant development because each attains its final level at the same time as attainment of the final level of the ultimate output from the system (i.e., at harvest maturity).

VIII. ENVIRONMENTAL CONTROLS OVER GENE ACTIVITY

Environmental factors modulate the levels of each of the penultimate outputs. The modulations alter the contribution to the yield by each of the penultimate outputs. These modulations usually alter the ultimate output (i.e., the yield). They may, however, just alter which of the penultimate outputs is predominantly causal of the expressed level of yield. Of the modulating environmental factors, the daylength and temperature are emphasized in this chapter and in Chapters 42 to 45. Daylength and temperature cause a large variation in the expressed crop yield [19–27]. The only other environmental factors that may cause as large or larger modulations of the genetic implementation are deficient versus optimal versus excess availability of moisture in addition to those of the many essential mineral and fertilizer elements [25]. Every effect on the level of system ouput by an environmental factor results from modulation by that factor of one or more to virtually all of the gene activities of the plant.

It is necessary to recognize that the effect of temperature is ubiquitous, because it affects virtually all gene activities. Biochemistry recognizes that each enzyme reacts to a change of temperature with a change of the level of its enzymatic activity. This changes the level of implementation of the genetic instruction. The magnitude of change is quantified as the Q10 response of the enzyme. On average across all enzymes, each 10°C higher temperature doubles the rate of enzymatic activity (i.e., has a Q10 of 2). Considering that tens of thousands of enzyme activities function within the meso developmental stage, this number of separate Q10 effects (responses to temperature) are all integrated in giving the level of system output. To understand the effect of temperature on yield, it must also be recognized that some enzyme actions and physiological processes cause feedforward effects while others result in feedback effects [4,8,19,21–27]. Similarly, some Q10 effects enhance gene actions that alter the system output negatively, while other effects by that same temperature alter that yield positively. Fully understanding the net effect of temperature requires understanding all the Q10 effects across the whole system.

The available evidence indicates that daylength modulates the activity of only a few photoperiod genes, whereas temperature modulates all gene activities, including the activities of the photoperiod gene(s) [4,7,8,18,20] (Chapters 43 to 45). Because daylength and temperature modulate photoperiod gene activity simultaneously while the temperature modulates all other gene activities, there is a strong daylength × temperature interaction effect on the yield.

IX. PENULTIMATE COMPONENTS OF YIELD ACCUMULATION

A. Interconnections Among the Penultimate Components

Basic whole system research applied to ongoing yield trials [4,7,8] (Chapter 44) has revealed that the thousands of primordial, genetically instructed, enzyme-implemented processes and resultant morphological traits of the meso stage of plant development, in addition to all of the environmental modulations of these implementations, are integrated into only three differentiable, penultimate outputs from the yield system. Again, these major physiological-genetic components of the process of accumulating crop yield are:

1. The net accumulation of biomass
2. The partitioning of this net biomass between the reproductive growth (the actual accumulation of the yield) versus any competitive continuation of vegetative growth and its consequent continuation of vegetative development
3. The duration of the plant growth (i.e., the days from planting to harvest maturity), which is the time the system uses (needs) to accumulate the expressed yield

In listing these as the three major physiological-genetic components of the yield system, the word *major* is used to indicate that the penultimate outputs of a system are that smallest number of differentiable physiological-genetic components into which the biological system can be subdivided while remaining functionally inclusive of the system's fully integrated (ultimate) output.

Basic whole system research as applied to yield trials [7] (Chapter 44) and as detailed further below, measures each penultimate output from the yield system of every genotype in the yield trial. The correlations across the genotypes of the changes in level among the three penultimate components and their levels of correlation with the yield indicate the following. Intrinsic interconnection occurs between each pair among the three penultimate components of the yield system [9]. These intrinsic interconnections result from control by the photoperiod gene activity and other maturity genes over the proportion of the net accumulated photosynthate (the physiological genetic component of yield 1) that is partitioned (this partitioning is the physiological genetic component 2) to support the reproductive growth and its consequent reproductive development. The reproductive growth and development stage involves all the processes that result in the actual accumulation of the yield. The basis for the intrinsic interconnections with the time to harvest maturity (the physiological genetic component of yield 3) is described next.

Partitioning a larger proportion of the available photosynthate to the flower buds causes them to grow faster. Their faster growth causes the buds to develop to flowering in fewer days. Postflowering continuation of partitioning the larger proportion of the photosynthate toward growth of the seeds causes them to grow fast also, which reinforces development to harvest maturity in fewer days, attended by higher rates per day of the accumulation of yield. The partitioning of any photosynthate to support growth of the reproductive organs (support of actual accumulation of the yield), competitively, deprives the potential continuation of vegetative growth of that photosynthate. Vice versa, any photosynthate partitioned to support continued vegetative growth (development of more nodes, leaves, branches, and shoot elongation) is, competitively, removed from support of the reproductive growth and yield accumulation.

B. Genetic Control over the Penultimate Components of the Yield System

Virtually all of the tens of thousands of genes of the plant, including the relatively few genes that implement the partitioning of the pool of photosynthate, partially control the net accumulation of biomass (the major physiological genetic component 1 of the crop yield system) [4,7,8]. This genetic control over the partitioning of the photosynthate between the reproductive (yield) organs versus the vegetative organs is the control over the major physiological genetic component 2 and, this control is by a subset of a few photoperiod genes, plus a few other maturity genes that do not respond to the daylength [4,7,8]. One indirect consequence of this control over partitioning, as discussed in the preceding paragraphs, is that these same few maturity genes simultaneously control the physiological genetic component of yield 3 (i.e., control over the partitioning of the available photosynthate by these few genes establishes the time the cultivar needs to accumulate and partition its biomass and develop to harvest maturity). Thus all the genetic instructions from all the thousands of genes are integrated into just these two (1 and 2) major physiological processes. Major physiological genetic component 3, days to harvest maturity, is not a process but a time span. The rate of plant development associated with this time span is simply 1/days to flowering or 1/days to maturity. In other words, control over the rate of development to flowering and maturity by the photoperiod and other maturity genes is an indirect consequence of their genetic control over the partitioning of the photosynthate between the reproductive and vegetative organs.

X. DEFINITION OF BASIC WHOLE SYSTEM RESEARCH

The penultimate outputs of the yield system listed above are:

1. The net accumulation of biomass
2. Partitioning of that net accumulated biomass between reproductive growth and continued vegetative growth
3. The time duration of the system activities that accumulate the expressed yield (i.e., the duration of the plant growth and development) [7]

The partitioning is quantified as the harvest index, that is, the percent of the aerial biomass that is the yield. These three penultimate system outputs are the three "major" physiological genetic components of the process of accumulating crop yield.

We suggest that basic and truly whole system research requires measurement of the ultimate output and at least all of the penultimate outputs from the system under study, and there must be measurements across both multiple genotypes and multiple environments. The measured level of each genotype's expressed level of each penultimate and ultimate system output should be compared to quantify the negative and/or positive correlations of all the genetically directed changes and all the environmentally modulated changes among these system outputs, and to relate these changes to the known (or knowable) differences among the controlling genetic factors and the correlations with the known (or knowable) differences among the environmental factors. With this definition, a physiological study remains reductive rather than whole system research, if comparisons of both multiple genotypes and multiple environments are not incorporated into the research design.

XI. YIELD SYSTEM ANALYSIS COMBINED WITH AMMI ANALYSIS PROVIDES TRULY BASIC WHOLE SYSTEM RESEARCH

A yield system analysis has been developed and is described fully in Chapter 42. Yield system analysis measures the three penultimate outputs plus five ante-penultimate outputs (see Section XII) from the yield system. Yield and these eight other system outputs are all measured for each cultivar (genotype) in the yield trial. Each of the eight is compared against the other seven and against the yield [7] (Chapter 44). By definition, a cultivar yield trial involves multiple genotypes. A single yield trial involves a single growing environment, however. Therefore, two or more yield trials that include the same genotypes are needed to provide the multiple environments required for basic whole system research on crop yield.

Comparison of any one of the penultimate outputs or of the yield of multiple genotypes grown across multiple environments invariably involves G × E interaction effects. Measurement of all the major physiological genetic components (1 to 3) of the process of yield accumulation and of the yield, compared across all the genotypes and across the multiple environments of each yield trial facilitates limited interpretation of the G × E interaction effects on the yield. Thorough quantification and interpretation of the G × E interaction effect on yield is facilitated if the yield system analysis of each yield trial is followed by statistical analysis of all the data from all the yield trials, using the additive main effects and multiplicative interaction effects (AMMI) analysis [29]. AMMI analysis quantifies a contribution by each of the tested genotypes (cultivars) and a contribution by each different environment to the G × E interaction effect on the expressed penultimate output or yield (Chapter 42). The G × E interaction effect on the yield and the G × E interaction effect on each of the eight penultimate outputs can be compared visually and simultaneously. (The next paragraph and Section XVII explain this reference to eight rather than just three penultimate outputs.) This visual comparison is achieved by plotting the contributions to the G × E interaction effect on the penultimate output or yield against the main effect on that trait that is due to each genotype and each environment (Chapters 42 to 45). The visually observable graph presents comparisons of the physiological genetic pathway that gives the yield expressed by each combination of one of the genotypes grown in each one of the environments. The alternative pathways to yield simply consist of which of components 1 to 3 is predominantly responsible for the expressed yield and which is of secondary or tertiary importance. The pathway used depends on both the different genotypes and different environments [4] (Chapters 42 and 43).

Listed below are the three penultimate plus the five ante-penultimate components of the biological process of accumulating crop yield. All eight are quantified by the yield system analysis of a yield trial. In addition to direct measurement of the yield, yield system analysis requires direct measurement of at least the days to harvest maturity (the major physiological-genetic component 3) and the net accumulated aerial biomass (component 1). Additionally, we recommend direct measurement of at least the days to flowering. The other five traits measured by yield system analysis are all calculated. One is the major physiological genetic component 2, the partitioning of the photosynthate; it cannot be measured directly but must be calculated. Each of the other four calculations quantifies an ante-penultimate output of the system.

Another direct measurement that is not included in the yield system analysis presented, the biomass at flowering, could provide further information about the system. Because the time of flowering divides the biomass accumulation during the time span of the vegetative stage of development from that during the later time span of the reproductive stage, measurement of the biomass at flowering facilitates precise quantification of any actual continuation of vegetative growth plus quantification of its competitive effect on the reproductive growth and development and on the yield. Partitioning of the pool of available photosynthate between reproductive and vegetative growth begins just prior to flower bud initiation, but inexpensive quantification of this partitioning is not feasible in yield trials until the postflowering (reproductive) stage of plant development.

A. Yield System Analysis as an Adjunct to Each Yield Trial

1. Direct quantification for each cultivar of four traits:
   a. *Days to flowering*: time used for development to flowering
   b. *Days to harvest maturity*: time used to develop to harvest maturity (major system component 3)
   c. *Aerial biomass*: overall photosynthetic efficiency (major system component 1)
   d. *Yield*: economically important system output
2. Five calculations from the four direct measurements:
   e. *Days of seedfill*: time used for actual yield accumulation
   f. *Yield/day to maturity*: efficiency of yield accumulation (rate of partitioning)
   g. *Yield/day of seedfill*: efficiency of yield accumulation (rate of partitioning)
   h. *Biomass/day of plant growth*: efficiency of photosynthesis
   i. *Harvest index*: endpoint efficiency of partitioning to yield (major system component 2)

XII. PENULTIMATE OUTPUTS FROM THE YIELD SYSTEM

In addition to yield, each of the other eight traits quantified by a yield system analysis is either one of the three penultimate outputs (1c = output 1, 2i = output 2, and 1b = output 3) or is a major subcomponent of at least two of them. Again, the term *major* implies that the trait is one of the smallest number of subcomponents into which the referenced penultimate component can be subdivided while the collective subcomponents (ante-penultimate system outputs) give rise to the penultimate components. Such relationships are illustrated as follows. The days to flowering (1a) and the days of seedfill (2e) constitute the smallest number of time spans (subcomponents) into which the days to harvest maturity (1b = penultimate output 3) can be subdivided while remaining inclusive of the total days to harvest maturity. Similarly, the yield accumulated per day of seedfill (2f) plus the days of seedfill (2e) are the two major subcomponents of the harvest index (2i = penultimate output 2). The days to harvest maturity (1b = penultimate output 3) plus the biomass accumulated per day of plant growth (2h) are the two major subcomponents of the net accumulation of biomass (1c = penultimate output 1).

Measuring yield plus each of the other traits for every one of the genotypes (cultivars) in a yield trial facilitates asking and testing how each of these physiological-genetic components of the process of accumulating yield contributes to each genotype's larger or smaller yield as compared with every other genotype in the yield trial. Doing this across multiple yield trials allows asking how each of the eight traits is modulated by the different environments. Applying AMMI analysis across multiple yield trials elucidates the G × E interaction effect on the yield plus on each of the eight penultimate or ante-penultimate outputs from the system.

XIII. ALTERNATIVE PHYSIOLOGICAL-GENETIC PATHWAYS TO YIELD

Within one yield trial any differences in level among the penultimate and ante-penultimate outputs of the genotypes, and any negative or positive correlations in the changes of levels between these major physiological-genetic components of yield, indicate the differences in the genetic instruction for the physiological pathway to yield of the various genotypes (cultivars). Across several yield trials, any change of a genotype from one pathway to yield toward an alternative pathway quantifies the environmental modulations of the implementation of that cultivar's genetic instruction.

Plant physiologists report that most of the gain in yield of modern cultivars results from partitioning a larger part of the net accumulated biomass toward the yield (reproductive organs) and a competitively smaller proportion toward the vegetative organs. Therefore, the pathways to yield of the old and new cultivars differ primarily in the quantitative proportions of the partitioning of the photosynthate. Usually attending this is a shorter duration of growth which arises because this predominant partitioning to the reproductive organs results in rapid development to harvest maturity, which maximizes the yield if the growing season is short. The alternative extreme of the continuum of the partitioning is predominant partitioning to continuation of the vegetative growth. This is the extreme opposite pathway to yield; its longer duration of the growth required to develop to harvest maturity maximizes the yield if the growing season is long enough. All other alternative pathways are but quantitatively intermediate variations of the relative competition for and receipt of the available photosynthate by the reproductive organs versus the vegetative organs. The physiological pathway toward higher yield that can arise from increasing the rate of biomass accumulation to thereby achieve a larger net accumulation of biomass across the same time span has not resulted from the direct selection for yield that has been practiced in the past.

Wallace et al. [7] indicate a physiological-genetic explanation as to why selection for yield per se usually results in a higher harvest index but seldom results in a larger biomass. They suggest that the net accumulated biomass has a very low heritability because it is controlled by virtually every gene of the plant. On the contrary, they believe the harvest index has a very high heritability because of its control by a small number of photoperiod and other maturity genes (see Section IX.B). Wallace et al. [7] suggest that appending yield system analysis to all yield trials in order to identify the germ plasm for use as parents that has the highest rates of biomass accumulation per day can result in gains of the net accumulated biomass. The selection should be for rate of accumulation of the biomass because at each given location, direct selection for higher biomass will indirectly result in later maturity than that which maximizes the yield for the growing-season duration of that location. AMMI analysis of the results from the yield system analysis will synergistically advance the gains.

XIV. EXAMPLES OF BASIC WHOLE SYSTEM PLANT RESEARCH

Wallace et al. [4–9] (also, Chapters 43 to 45) all describe pivotal effects on yield by photoperiod gene activity and by the modulation of this activity by daylength and temperature. Chapter 45 elucidates yet more complexity, which is due to interaction between photoperiod gene(s) and additional control by the activity of another maturity gene which is not modulated by the daylength. These pivotal influences on yield encompass the long-recognized adaptation to maladaptation due to effects by both the photoperiod and the temperature over whether the time the cultivar requires to develop to harvest maturity is shorter or longer than the growing-season duration made available by the site season. The data demonstrate that the larger the photoperiod gene activity, the larger its negative control [7] over the rate of growth of the reproductive (yield) organs. The negative control results from reducing the competitive access of the reproductive (yield) organs to the pool of photosynthate. Further, the rate of growth of the reproductive organs is causal of the predominant control by the reproductive organs over the rate of development to harvest maturity. The rate of development is 1/days needed to develop to the referenced stage (i.e., the days to flowering or the days to harvest maturity). Negative control over the growth of the reproductive organs results from photoperiod-gene activity because it reduces the proportion of the photosynthate partitioned to the reproductive organs, resulting in competitively larger receipt by the vegetative organs, with consequent greater vegetative growth (increase of the biomass) and vegetative development (more nodes, leaves, and shoot elongation) but lower yield unless the growing season is long enough to compensate for the lower rate of reproductive growth and development and of the consequent accumulation of yield.

We cite Witzenberger et al. [30] and Linnemann [31] as research by others, of which we are aware, that approaches our definition of basic whole system research on the biological system that accumulates crop yield. Witzenberger et al. [30] grew peanut (*Arachis hypogaea*) in yield trials. One planting was in the natural short day of a tropical location. A second planting was a minimal distance away under a nonoverlapping 22-h daylength created by incandescent lighting above the canopy. Their summary follows: "Yield differences between the photoperiods were largely explained by changes in crop growth rate, partitioning, and the length of the effective pod-filling period. Long days resulted in increased crop growth rates, but generally decreased partitioning and the duration of the crop's effective pod-filling phase. However, it was dependent on the genotype as to which of the yield-determining processes had been more influenced by day-length conditions. In some cases, partitioning contributed most to yield differences; in others, the duration of the effective pod-filling phase contributed most." Obviously, Witzenberger et al. [30] arrive at essentially the same conclusions as summarized in this chapter, in Chapter 43, and in Wallace et al. [7].

Linneman [31] grew bambara groundnut (*Vigna subterranea*) under daylengths of 10, 12, 12.5, 13, 14, and 16 h. The progressively longer days progressively delayed the appearance of the first flower and also progressively delayed the subsequent flowers. Photoperiod caused a stronger effect on fruit set than on flowering, with no set under 14 and 16 h. The 14- and 16-h daylengths resulted in the largest total aerial biomass, leaf number and area, but in the smallest percent of the biomass as yield.

For peanut, Flohr et al. [32] conclude that the time of fruit initiation and of the developmental changes are reflected by subsequent partitioning of assimilates to the pods. This reasoning is the inverse of ours. Their reasoning is based on the paradigm that photoperiod gene activity controls the times to initiation of flower buds, to flowering, and to maturity and that partitioning is a subsequent process. This paradigm fails [4] (Chapter 44) to recognize that the rate of growth of the reproductive organs controls the time they require to grow to the referenced stage of development. Our interpretation that the time needed for development is a consequence of the growth rate is a simple shift of the original paradigm. The shift is based on evidence, from whole system research [4,7,8] (Chapter 44), that partitioning is causal rather than a subsequent process. Our interpretation agrees with the effects of partitioning suggested by the following theoretical hypothesis of Trewavas and Allen [5].

XV. SUGGESTED SHIFT OF PARADIGM ABOUT CONTROL OVER PLANT DEVELOPMENT

In the first paragraph of this chapter, Trewavas [1] is cited. He suggests that preparation, comprehension, and acceptance of research proposals and research reports plus their review and acceptance by peers are all interpreted in accordance with the attitudes engendered by training, background, experience, and personality. Not indicated in our introduction is that Trewavas' suggestion for more research at the whole system level arises from recognition by him that the paradigm that the concentration of one or more hormones controls plant development fails to explain most developmental variabilities of plants. In a later paper, Trewavas and Allen [5] state: "The problem with all such embracing views is that they can be (and are) used to explain any phenomenon in development and thus become constrictive of alternative approaches to understanding the process of development itself. The classical view (assumption of major control by hormones) survives largely by ignoring critical or disagreeable data and probably also as much by lacking challenge from an alternative, equally all-embracing hypothesis."

Trewavas and Allen [5] suggest that competition between the growing and metabolically active parts of a plant provides a communication between the competing parts of the plant that supplements the hormonal controls. They state: "There are two features of this competition model which are not widely appreciated. First, what one growing area removes from circulation is not available to another. This represents a form of communication between the two areas which may be termed negative control and would be very difficult to distinguish, except with further analysis, from a positive hormonal model. Second, since the different growing areas (e.g., leaves, meristem, fruits, and stem) require qualitatively different growth materials, reduction of the growth rate (competitive ability) of some by alteration of photoperiod, environment, etc., will cause qualitative changes in what is left to circulate in the vascular system. Such qualitative changes then could act as effective signals for developmental change, although these would be subject to the constraints on accuracy and precision described earlier."

In a 1991 paper [33] Trewavas suggests a systems view of regulation. He proposes that control over plant development is by the system as a whole. The control is holistic because the interconnections among virtually all parts of the biological system result in gene × gene and G × E interaction effects among many of the parts of the system. The system as a whole is greater than the parts of the system. In Chapter 43 we suggest that Trewavas's suggestion to measure the proportional control by these parts is achieved through yield system analysis followed by AMMI analysis.

We have cited Trewavas and Allen [5] verbatim because their hypothesis corresponds with our interpretation that photoperiod gene activity controls the partitioning of photosynthate between the reproductive (yield) and vegetative organs, with consequent indirect control over both the rate and the duration of plant development. Before we became aware of this hypothesis of Trewavas and Allen [5], we published data that supports their hypothesis in at least three ways. First, the activity of a single photoperiod gene as modulated by daylength and temperature was shown to control at least 25 traits of bean simultaneously, including all the traits measured for yield system analysis [8] (Section XI). Second, positive correlations, but more obviously the negative correlations in change of level between pairs among all 25 traits, are fully explainable by competitive sharing of the pool of photosynthate between the reproductive organs versus the vegetative organs. Third, the data demonstrate [8] that photoperiod insensitivity attends an inherent ability of the plant to develop rapidly to flowering in any environment, whereas the activity of the photoperiod sensitive gene causes negative control over this intrinsic ability to develop rapidly to flowering [4,8,9]. All the variations in level of the 25 correlated traits are fully explainable by the effects expected from competitive sharing of the photosynthate. Nevertheless, we (Wallace et al. [8]) and Linnemann [31] also invoked the hormone paradigm. We suggested that the photoperiod gene control over partitioning probably arises at the molecular level, through hormonal control over the relative sink activities of the reproductive versus vegetative organs.

XVI. MORE ABOUT SOCIOLOGY OF REDUCTIVE VERSUS WHOLE SYSTEM RESEARCH

We refer again to the statement by Trewavas [1] that comprehension and writing of research proposals and research reports and their acceptance by peer reviewers are interpreted in accordance with each plant scientist's attitudes as engendered by training, background, experience, and personality. Our experience indicates that the interpretations of most reviewers are based on paradigms derived from reductive research. Reductive research usually is basic research, because it tests hypotheses about how a biological endpoint is achieved. Insufficient basic whole system research exists to demonstrate repeatedly its synergistic complementarity to the abundant reductive research. This prevents ready acceptance of proposals for research on intact, complex, biological systems.

Philosophical consequences of the paucity of whole system research are illustrated by the photoperiod literature. Virtually all basic research on photoperiodism has been done on plant species that have been selected deliberately for a qualitative response to daylength. One or but a few short or long days results in initiation of flower buds. Use of such ecotypes limits the span of genotypes tested. Focus on flower bud initiation and the molecular role of phytochrome reduces the system further. It reduces the number of components from the whole system that are included in the biology that is investigated. The few genotypes are selected, because their "all or none" flowering bypasses the variable quantitative response to photoperiod that is expressed by most crops.

Among the 435 pages of the most often cited text on photoperiodism by Vince-Prue [34], about 10 pages discuss photoperiod control over partitioning to storage organs, including stem tubers, root tubers, and bulbs (leaves). As her conclusion in a later text [35], Vince-Prue states: "Daylength has a number of effects on all aspects of plant reproduction; it can modify all stages of development of the flower and is often an important factor controlling the onset and development of vegetative reproduction." She [35] also states: "The emphasis of interest in the switch from vegetative to reproductive growth, often using model systems which require only a single photoinductive cycle to effect the switch, has tended to obscure the fact that the primordia formed after one or a few inductive cycles frequently do not complete their development into fertile flowers." (This is the negative control over continued reproductive development described above; see Ref. 4.) Readers of text that is 95% about flower bud initiation and phytochrome end their relationships to hormonal concentrations do not automatically associate such statements in but a few pages with control over partitioning and yield accumulation. Vince-Prue's statements agree with our published [4,7,8] conclusions, however. These conclusions are reinforced in Chapters 43 to 45 and by the research on peanuts [30,32] and bambara groundnut [31]. In Ref. 4 we cite evidence that photoperiod controls partitioning and yield of at least 24 different crops.

This chapter focuses on photoperiod control over partitioning of the available photosynthate to the yield of seed crops or of crops for which the fruits or pods constitute the yield. However, the photoperiod control over partitioning to stem tubers, root tubers, bulbs (leaves), and other storage organs described in [34,35] indicates control over the yield of these crops also.

Stated paradoxically, but therefore forcefully, and illustrating the circularity within the yield system, the evidence from whole system research indicates that photoperiod gene activity strongly controls the rate of accumulation of yield (Chapter 44). Through this, the rates of growth of the flower buds and (of more obvious relevance to the yield) of the fruits and seeds are controlled, and these growth rates result indirectly in control by photoperiod over the days to flowering and to maturity. The message most readers of texts that present basic and reductive research on photoperiodism acquire is far less holistic. It is that photoperiod directly controls only the time to initiation of the flower buds. Based on this paradigm, many earlier reviewers rejected our conclusions [8] outright, because they interpreted them as disagreeing with the accepted paradigm. It became evident that the interpretation derivable from our whole system research would never rise above the current paradigm of some journal editors, leaving no recourse except to submit the paper to a different journal. Nevertheless, we emphasize that the reviews and rejections motivated attempts toward more comprehensible writing of the research results and their interpretation.

XVII. CONSEQUENCES OF PREDOMINANCE OF REDUCTIVE RESEARCH

We suggest that to the detriment of advancement in the understanding of plant biology, in comparison with reductive research, basic whole system research is neglected, underfunded, and not encouraged. Illustrating this, a recent three-volume publication [36] is titled *Models in Plant Physiology and Biochemistry*. In that publication, molecular models of 83 physiological processes are presented. None are derived from whole system research. Nevertheless, most are presented as a potential approach toward achieving higher yield. In the present volume the paucity of basic whole system research is illustrated by the ratio of chapters derived via reductive research versus those derived from basic whole system research.

Basic whole system research and basic reductive research are complementary. Together, they synergistically advance plant science. Most paradigms derived from reductive research are not negated by whole system research; they are simply modified and enlarged by the additional interpretations. The paucity to date of basic whole system research prevents expectation by the plant science community of the potential complementation and synergism.

The current sociological and philosophical attitudes arise from a proper conclusion that "basic research" advances understanding of biological science more than does research applied solely toward enhancing economic goals. The current attitudes ignore, however, that the why and how hypotheses that make research basic can be tested simultaneously with and/or in justaposition to the testing of applied hypotheses.

Basic research is often defended as allowing each scientist to pursue her or his personal academic interests, with no application in view. The presumption is that this academic freedom fosters. serendipitous technological advance in unforeseen applications [37,38]. This attitude ignores many advances in plant biology that have arisen through research directed toward the goals of agriculture [39–41]. This goal-directed research has lowered food costs for all the people of technologically developed countries. As the term is currently used, *basic research* is essentially synonymous with *reductive research*. Therefore, the current attitude ignores the synergistic advance in scientific knowledge derivable from combining whole system research with reductive research, and the additional synergistic gain in competitive economic advantages for applications of reductive, whole system, and applied research.

XVIII. SOCIOLOGICAL ADJUSTMENTS TOWARD BASIC WHOLE SYSTEM RESEARCH

Balanced social and philosophical acceptance of basic whole system research requires attitude adjustments by plant scientists. As compared with reductive research, it must become accepted that longer time and more effort will continue to be required to develop a grant proposal, to conduct the research, to test the hypotheses, and to interpret the results. A complex whole system research proposal should not be rejected on the basis that it is not likely to be completed in the two or three years commonly allocated for completion of a reductive research proposal. Basic whole system research requires accepting an array of research agendas, because of the many interconnections and multiple, circularly acting feedforward and feedback effects within the system. Writing research reports will require more time per publication, because the interconnections and circularly acting feedforward and feedback effects must be described and interpreted with appropriate balance and intricate illustration of these interactions and one effect followed by another and another (i.e., tandem effects). The writing will require more peer review to clarify the descriptions and interpretations of the great density of connections between parts of the system.

More constructive suggestions for manuscript improvement are needed for reports of basic whole system research than for reductive research. Outright rejection based on existing paradigms is not constructive. Reviewers must assist authors to achieve the clarity of writing needed for ease of reader comprehension. The reader must recognize and accept inherent complexity and be willing to reflect and ponder in order to achieve his or her comprehension and cognitive integration of the multiple agendas of the system. The multiple agendas of papers that present synthesis of the system necessarily reflect the complexity of the system. Basic whole system research requires team effort and collaboration as much or more than basic reductive research. Describing and interpreting the multiple feedforward and feedback effects within a biological system necessarily returns circularly to processes and phenomena discussed earlier, because circularity occurs within the system. Reviewers should be receptive to shifts of paradigm.

Reductive research has given rise to an attitude that a research report should address only one or a few concepts. This must give way to acceptance of research reports that synthesize the complexity of the system, its multiple biological phenomena, and their interactions. Reviewers accustomed to reductive simplicity should not invariably consider circular return to an already discussed concept as evidence of poor writing and circular reasoning by the author(s). Full understanding of the relationships and interactions among the components of complex biological systems cannot come simply from considering the parts one by one. The system is greater than its parts. Publication of basic whole system research requires acceptance by some journals of manuscripts subdivided in accordance with the components of the system rather than according to a predetermined format for the journal.

XIX SUMMARY AND CONCLUSIONS

Plant scientists agree that crop yield results from most if not all of the thousands of gene (enzyme) activities of the plants. Nevertheless, almost all yield trials that compare genotypes are directed solely toward the applied economic goal of determining which genotype yields the most. Then that cultivar is recommended for planting by farmers. This is essential but purely applied research. It is inefficient use of the yield trial data, however, because every yield trial can advance understanding about the biology of how crop yield is achieved.

All that is needed to fulfill the goals of basic research is to use each yield trial to ask basic questions. Why do some of the genotypes yield more and some less? How does the environment cause another cultivar to yield most for another year? Why are some cultivars adapted to the daylengths and/or temperatures of other environments? We need understanding of these genetic and environmental factors. The biological theory plus inexpensive procedures described in this chapter can result in fullfillment of both applied and basic research objectives from every yield trial.

It is shown that results from whole system research using yield trials are complementary to current basic research, which is almost always reductive research that investigates only one or a few parts of the plant system. Currently, reductive research is favored because it more readily provides "yes or no" answers to research hypotheses, so that it is easier to interpret the research data and to both write and read the research reports. These research steps are easier because the reductive research avoids the circular feedforward and feedback effects of the system. These circular effects are consequences that arise from the other parts of the system and/or from the variation of the environmental factors. Basic whole system research reveals biological understanding that cannot arise from reductive research, thereby complementing basic reductive research. More basic whole system research is essential to rapid advance of understanding the complexities of the yield system as a whole.

REFERENCES

1. A. Trewavas, *Aust. J. Plant Physiol.*, *13*: 447 (1986).
2. R. W. Zobel, in *Genotype-by-Environment Interaction and Plant Breeding* (M. S. Kang, ed.), Louisana State University, Baton Rouge, La., pp. 126–140 (1990).
3. R. W. Zobel, *Adv. Soil Sci.*, *19*: 27 (1992).
4. D. H. Wallace, K. S. Yourstone, and R. W. Zobel, *Theor. Appl. Genet.*, *86*: 17 (1993).
5. A. Trewavas and E. Allan, in *Plant Growth Modeling for Resource Management*, Vol. 2, *Quantifying Plant Processes* (K. Wisiol and J. D. Hesketh, eds.), CRC Press Boca Raton, Fla., pp. 25–45 (1987).
6. D. E. Koshland, Jr., *Science*, *259*: 732, 1379 (1993).
7. D. H. Wallace, J. P. Baudoin, J. Beaver, D. P. Coyne, D. E. Halseth, P. N. Masaya, H. M. Munger, J. R. Myers, M. Silbernagel, K. S. Yourstone, and R. W. Zobel, *Theor. Appl. Genet.*, *86*: 27 (1993).
8. D. H. Wallace, K. S. Yourstone, P. N. Masaya, and R. W. Zobel, *Theor. Appl. Genet.*, *86*: 6 (1993).
9. D. H. Wallace, P. A. Gniffke, P. N. Masaya, and R. W. Zobel, *J. Am. Soc. Hortic. Sci.*, *116*: 534 (1991).
10. D. H. Wallace and G. A. Enriquez, *J. Am. Soc. Hortic. Sci.*, *105*: 583 (1980).

11. R. B. Goldberg, G. Hoschek, and J. C. Kamalay, *Cell, 14*: 123 (1978).
12. G. U. Jurgens, R. A. Mayer, R. A. Torres Rioz, T. Berleth, and S. Misera, *Dev. Suppl., 1*: 27 (1991).
13. D. Rhodes, G. C. Ju, W.-J. Yang, and S. Samaras, *Plant Breed. Rev., 10*: 53 (1992).
14. R. M. Gifford, in *Progress in Photosynthesis* (J. Biggins, ed.), Martinus Nijhoff, Dordrecht, The Netherlands, pp. 377–384 (1987).
15. D. H. Wallace, M. M. Peet, and J. L. Ozbun, in CO_2 *Metabolism and Plant Productivity* (R. H. Burris and C. C. Black, eds.), University Press, Baltimore, pp. 43–58 (1976).
16. D. H. Wallace, "Well-Being of Mankind and Genetics," *Vol. I, Book 2, Proceedings of the 14th International Congress on Genetics*, MIR Publishers, Moscow, pp. 306–317 (1980).
17. D. H. Wallace and R. W. Zobel, in *Handbook of Crop Productivity* (M. Rechcigl, Jr., ed.), CRC Press, Boca Raton, Fla., pp. 137–142 (1982).
18. D. H. Wallace, *Plant Breed. Rev., 3*: 21 (1985).
19. G. R. Squire, *The Physiology of Tropical Crop Production*, Bookcraft, Bath, Somersetshire, England (1990).
20. T. Hodges, *Predicting Crop Phenology*, CRC Press, Boca Raton, Fla. (1991).
21. V. P. Abrol, P. Mohanty, and Govindjee, *Photosynthesis: Photoreactions to Plant Productivity*, Kluwer Academic Publishers, Dordrecht, The Netherlands (1992).
22. D. O. Hall, J. M. Scurlock, H. R. Bolhar, R. C. Leegood, and S. P. Long, eds., *Photosynthesis and Production in a Changing Environment*, Chapman & Hall, New York. (1993).
23. K. Wisiol and J. D. Hesketh, eds., *Plant Growth Modeling for Resource Management*, Vol. 2 *Quantifying Plant Processes*, CRC Press, Boca Raton, Fla. (1987).
24. H. Lambers, M. L. Cambridge, H. Konings, and T. L. Pons, *Causes and Consequences of Variations in Growth Rate and Productivity of Higher Plants*, SPB Academic Publishing, The Hague, The Netherlands (1990).
25. J. R. Porter and D. W. Lawlor, *Plant Growth: Interactions with Nutrition and Environment*, Cambridge University Press, Cambridge (1991).
26. P. M. Gresshoff, *Plant Responses to the Environment*. CRC Press, Boca Raton, Fla. (1992).
27. F. D. Whisler, B. Acock, D. N. Baker, R. E. Fye, H. F. Hodges, R. R. Lambert, H. E. Lemmon, J. M. McKinion, and V. R. Reddy, *Adv. Agron., 40*: 141 (1986).
28. D. P. S. Verma, *Control of Plant Gene Expression*, CRC Press, Boca Raton, Fla. (1992).
29. H. G. Gauch, Jr., *Statistical Analysis of Regional Yield Trials: AMMI Analysis of Factorial Designs*, Elsevier, New York (1993).
30. A. Witzenberger, J. H. Williams, and F. Lenz, *Field Crops Res., 18*: 89 (1988).
31. A. R. Linnemann, *Ann. Bot., 71*: 445 (1993).
32. M.-L. Flohr, J. H. Williams, and F. Lenz, *Exp. Agric., 26*: 397 (1990).
33. A. Trewavas, *Plant Cell Environ., 14*: 1 (1991).
34. D. Vince-Prue, *Photoperiodism in Plants*, McGraw-Hill, New York. (1975).
35. D. Vince-Prue, in *Strategies of Plant Reproduction* (W. J. Meudt, ed.), Allanheld, Osmun, Totowa, N. J., p. 73 (1983).
36. D. W. Newman and K. G. Wilson, eds., *Models in Plant Physiology and Biochemistry*, CRC Press, Boca Raton, Fla. (1983).
37. G. E. Brown, Jr., *Science, 258*: 200 (1992).
38. P. H. Abelson, *Science, 252*: 625 (1991).
39. J. H. Meyer, *Science, 260*: 881 (1993).
40. P. Lyrene, *Science, 259*: 162 (1993).
41. K. A. Dahlberg, *Science, 259*: 163 (1993).

42

AMMI Statistical Model and Interaction Analysis

Richard W. Zobel
Agricultural Research Service, U.S. Department of Agriculture, Ithaca, New York

Donald H. Wallace
Cornell University, Ithaca, New York

I. INTRODUCTION

A. Sources of Interaction

A primary thrust of reductive research is to limit the number of variables that can affect the results of an experiment so that "yes or no" questions can be asked and answered unequivocally. Uncontrolled factors such as environmental variables and unplanned treatment effects cause disturbances in the pattern of response, making "yes/no" answers ambiguous. Disturbances in response pattern are termed *interaction effects*. Whole system research must deal with these interaction effects: interaction between competing physiological processes, interaction between differing genetic systems, and interaction between the environment and these processes and their genetic control.

It is convenient to conceive of the impact of environment on biological processes as modulation of those processes. The modulation can be so drastic as to disrupt the normal processes, such as hail damage, drought, aluminum toxicity, insect feeding, and the like. Normally, interactions and modulations are far less drastic, and can be quantified by measuring the output of the biological process, its time or duration, and the characteristics of the surrounding environmental factors.

Chapters 41 and 43 to 45 deal with interpretation of the genotype × environment (G × E) interaction effects on plant physiological responses. We felt it imperative, therefore, to describe thoroughly the statistical procedure used and to demonstrate how to derive the interpretations. In this chapter we present this procedure in mathematical terms; the need for the procedure is discussed in Chapter 41. The reader may find it useful to refer to this chapter while reading Chapters 43 to 45, where interpretations are presented of the G × E interaction effects on time to flowering and crop yield.

B. Need for a Nonadditive Statistical Model

In terms of classical statistics, average duration, average output, and average environmental values can be calculated and appropriate statistics calculated. However, none of this characterizes the interactions between modulations and the biological processes being modulated. Nor are the interactions between factors of a multifactorial experiment characterized by a purely additive statistical procedure. The focus of this chapter is on a statistical procedure that allows interpretation of experimental results in the face of uncontrolled or poorly controlled environmental variables, and multiple interacting factors.

II. DEVELOPMENT OF AN ADDITIVE/MULTIPLICATIVE MODEL

A. History

Zobel et al. [1], Gauch [2], Gauch and Zobel [3–5], Gauch [6], and Zobel [7] reintroduced a statistical procedure first proposed and investigated in the 1960s [8–11]. With this procedure [called the additive main effects and multiplicative interaction (AMMI) procedure] they demonstrated that interaction could be quantified, and frequently, the underlying causal physiological and environmental patterns can be identified.

A subcase of the AMMI analysis, linear regression, was introduced by Mandel [12] followed by Finlay and Wilkinson [13] to describe a specific interaction pattern which, with some data sets, more accurately models the data than the more general AMMI procedure. The relation between linear regression and AMMI is discussed elsewhere [10,14] and is not included here. This discussion is restricted to the more general AMMI analysis.

B. Format of AMMI analysis

AMMI analysis is a two-factor analysis of variance (ANOVA) where the variance due to factor mean deviations from the grand mean is removed from the data matrix cells along with the grand mean, and the resulting matrix subjected to the matrix algebra procedure called *singular value decomposition* (SVD). This process is followed by representing the results on a two-factor scatter diagram called a *biplot*, a name that derives from representing two different factors on the same plot. In an AMMI biplot, the x-axis represents the factor mean deviations from the grand mean and the y-axis represents their interaction with each other, or alternatively, the x- and the y-axis each represents interaction (more on this later).

III. MATHEMATICAL BASIS FOR AMMI

A. Additive Model

Two-factor data can be represented as a matrix with columns representing the different levels of one factor and rows representing the different levels of the second factor. When an experiment is replicated (the normal case), each replicate can be represented as a separate matrix (Table 1a). The first step in an analysis of a replicated data set is to determine the means of each cell across all replicates (Table 1b). The variance of the replicate cell values from that mean comprises the error variance. Throughout this discussion we assume that the error variance has been removed and that we are dealing with a single matrix of cell means (Table 1b). An additive analysis of the data matrix in Table 1b would have the following mathematical representation:

$$Y_{rc} = \mu + \alpha_r + \beta_c + \rho_{rc} \tag{1}$$

where Y_{rc} is the actual cell value, μ the grand mean (i.e., the mean across all cells of the matrix), α_r the row mean deviations from the grand mean, β_c the column mean deviations from the grand mean, and ρ_{rc} the residual after removing the previous three values from the cell value Y_{rc}. The ANOVA Table from an analysis of the data in Table 1b is shown in Table 2.

Table 1a Artificial Data Set with Five Rows, Four Columns, and Three Replications, Containing Only Additive Variance and Replication Variance

Row	Column			
	1	2	3	4
	Replication 1			
1	1000	2000	3000	4000
2	2000	3000	4000	5000
3	3000	4000	5000	6000
4	4000	5000	6000	7000
5	5000	6000	7000	8000
	Replication 2			
1	1050	2050	3050	4050
2	2050	3050	4050	5050
3	3050	4050	5050	6050
4	4050	5050	6050	7050
5	5050	6050	7050	8050
	Replication 3			
1	950	1950	2950	3950
2	1950	2950	3950	4950
3	2950	3950	4950	5950
4	3950	4950	5950	6950
5	4950	5950	6950	7950

Table 1b Row and Column Cell Means for Table 1a

Row	Column				
	1	2	3	4	Mean
1	1000	2000	3000	4000	2500
2	2000	3000	4000	5000	3500
3	3000	4000	5000	6000	4500
4	4000	5000	6000	7000	5500
5	5000	6000	7000	8000	6500
Mean	3000	4000	5000	6000	4500

B. Singular Value Decomposition

Two-way data tables like Table 1b are matrices, and these matrices can be evaluated using matrix algebra procedures. One matrix algebra procedure that is effective in characterizing a data matrix is singular value decomposition (SVD). A mathematical representation of a complete SVD of a data matrix is

Table 2 ANOVA Table for an Additive Analysis of the Data in Tables 1a and 1b

Source	df	SS	MS	Probability
Total	59	195,100,000.00	3,306,779.66	
TRT	19	195,000,000.00	10,263,157.89	0.000***
Row	4	120,000,000.00	30,000,000.00	0.000***
Column	3	75,000,000.00	25,000,000.00	0.000***
Residual	12	0.00	0.00	1.000
Error	40	100,000.00	2,500.00	
Grand mean = 4500.0				

***, p < 0.001.

$$Y_{rc} = \sum_{i=1}^{N} \lambda_i \gamma_{ir} \delta_{ic} \tag{2}$$

where N is the smaller of the number of rows or the number of columns; λ_i is the singular value (square root of the eigenvalue, which, in turn, is the sums of squares divided by the number of replications) for axis i (if the data is kg/ha, the eigenvalue is in terms of kg/ha); γ_{ir} is the eigenvector for row r in axis i; and δ_{ic} is the eigenvector for column c in axis i.

We will restrict discussion to least-squares procedures. The process of a least-squares SVD can best be visualized by assuming that a data set is a multidimensional cloud of points (number of dimensions equals N). SVD then determines the least-squares line through that cloud of points. This is axis 1. A least-squares line perpendicular to the previous line is then drawn through the cloud and becomes axis 2. This process is continued for N axes, with each succeeding axis being perpendicular to all the previous axes. This partitions all the data into the complete set of axes. This operation removes the largest singular value first and then the second largest, and so on, until all the data have been accounted for. The eigenvectors are the proportion of that singular value which is due to a given row or column: the eigenvectors are scaled such that

$$\sum_{r=1}^{R} \gamma_{ir}^2 = \sum_{c=1}^{C} \delta_{ic}^2 = 1$$

C. Principal Components Analysis

If prior to applying SVD to a data matrix the grand mean is removed, the mathematical representation for *principal components analysis* (PCA) is obtained:

$$Y_{rc} = \mu + \sum_{i=1}^{N} \lambda_i \gamma_{ir} \delta_{ic} \tag{3}$$

Thus PCA is nothing other than the application of SVD to a matrix after the removal of the grand mean from each cell. This removal results in what is commonly called a *covariance matrix*. If as a result of a test of statistical significance, fewer than N axes are statistically different than noise (error variance), the last few axes that are not significant can be left in a residual matrix, giving the following mathematical representation of PCA:

$$Y_{rc} = \mu + \sum_{i=1}^{N} \lambda_i \gamma_{ir} \delta_{ic} + \rho_{rc} \tag{4}$$

where N is the number of significant axes and is less than the smallest of the number of rows or columns and ρ_{rc} is the cell residual after removing N axes from the matrix.

D. Derivation of the Additive Model

1. Mathematical Proof of the Model

Now, if in equation (4), $N = 2$, and, when $i = 1$, all δ_{1c} eigenvectors are constant, and when $i = 2$, all γ_{2r} eigenvectors are constant, the following can be obtained:

$$Y_{rc} = \mu + \lambda_1 \gamma_{1r}\delta_{1c} + \lambda_2 \gamma_{2r}\delta_{2c} + \rho_{rc} = \mu + k_1\gamma_{1r} + k_2\delta_{2c} + \rho_{rc} = \mu + \alpha_r + \beta_c + \rho_{rc} \quad (5)$$

By defining $\alpha_r = k_1\gamma_{1r} = \lambda_1\gamma_{1r}\delta_{1c}$ and $\beta_c = k_2\delta_{2c} = \lambda_2\,\gamma_{2r}\,\delta_{2c}$, the final product of equation (5) becomes identical to equation (1).

2. Visual Proof: Biplots

This is not just a numbers game, Tables 1 to 3 and Figures 1 to 3 provide proof of this identity. Table 1b is a 5 × 4 data set after determining the cell means across replications. Table 2 is the additive ANOVA table for this data set, and Table 3 is the PCA ANOVA table for this data set. This is a strictly additive data set with only two significant PCA axes. Figure 1 is a biplot of axis 1 (*x*-axis) and axis 2 (*y*-axis) eigenvector scores. [These scores are calculated by multiplying the eigenvector value times the square root of the singular value (i.e., $\lambda_1^{0.5}\gamma_{1r}$ and $\lambda_1^{0.5}\delta_{1c}$). Why this relationship is important will be discussed later.] It has been demonstrated that for biplots of PCA analyses [15], the cross-type pattern of the biploted scores in Figure 1 (i.e., the lines for the rows crosses the line for the columns) indicates that the first two axes have extracted additive data.

Figure 2 is the biplot for axis 1 (*y*-axis) and the cell means (*x*-axis), and Figure 3 is the corresponding biplot for axis 2. In Figure 2 the eigenvector scores for columns are constant (Table 4a; note that in Figure 2 the eigenvector scores for columns are below the 0 line for the *y*-axis and parallel to it, i.e., are constant relative to that axis), and in Figure 3 the eigenvector scores for rows are constant (Table 4b; note that in Figure 3 the eigenvector scores for rows are above the 0 line for the *y*-axis and parallel to it). This data set clearly fits the assumptions of equation (5).

3. Biplot Manipulations

It should be noted that the signs for the eigenvectors are chosen such that the first one calculated is positive and the rest are assigned a sign based on their relation to the first. This allows the statistician to multiply all eigenvector scores by -1, resulting in an inversion of the biplot while maintaining exactly the same pattern and relationship between eigenvector scores, but upside

Table 3 ANOVA Table for a PCA Analysis of the Data in Tables 1a and 1b

Source	df	SS	MS	Probability
Total	59	195,100,000.00	3,306,779.66	
TRT	19	195,000,000.00	10,263,157.89	0.000***
PCA 1	5	120,000,000.00	24,000,000.00	0.000***
PCA 2	5	75,000,000.00	15,000,000.00	0.000***
Residual	9	0.00	0.00	1.000
Error	40	,100,000.00	2,500.00	
Grand mean = 4500.0				

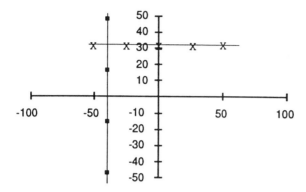

Figure 1   Biplot of PCA row and column scores. The *x*-axis is the PCA 1 scores and the *y*-axis is the PCA 2 scores (X represents individual rows, and ■ individual columns).

down relative to the original. Multiplying the eigenvectors of Figure 2 by −1 would result in a biplot visually similar to Figure 3.

E.   AMMI Model

1.   Real Data

In Table 5 we have a data set (cell means) for a soybean yield trial of seven cultivars grown for 8 years at the same location (these data are extracted from the set used in Zobel et al. [1]). If we apply PCA analysis to this data set we get the ANOVA table in Table 6 and the eigenvector scores of Tables 7a and 7b. Tables 7a and 7b and the respective biplots for axes 1 and 2 (Figures 4 and 5) suggest that the largest portion of the variance is additive and can best be described by an additive analysis.

2.   Multiplicative Terms

The PCA ANOVA table (Table 6), however, shows that there is also a third significant axis. Equation (5) [and, therefore, equation (1)] is based on the assumption that there are only two significant axes. When two axes are additive and there are additional axes, the appropriate

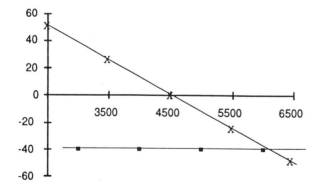

Figure 2   Biplot of PCA row and column scores. The *y*-axis is the PCA 1 scores and the *x*-axis the row and column means (X represents individual rows, and ■ individual columns).

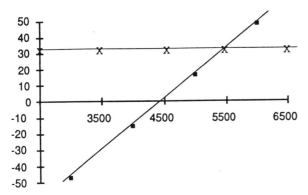

Figure 3 Biplot of PCA row and column scores. The *y*-axis is the PCA 2 scores and the *x*-axis the row and column means (X represents individual rows, and ■ individual columns).

mathematical model is the one in equation (6), which shows the appropriate derivation from equation (4):

$$Y_{rc} = \mu + \sum_{i=1}^{N} \lambda_i \gamma_{ir} \delta_{ic} + \rho_{rc} = \mu + \alpha_r + \beta_c + \lambda_3 \gamma_{3r} \delta_{3c} + \rho_{rc}$$

$$= \mu + \alpha_r + \beta_c + \sum_{i=3}^{N} \lambda_i \gamma_{ir} \delta_{ic} + \rho_{rc} \tag{6}$$

where $N = 3$ in this specific case but can equal the smaller of $r - 1$ or $c - 1$, or the number of significant axes. This is the equation for AMMI analysis.

3. AMMI Model

As mentioned earlier, AMMI analysis consists of first removing the additive terms, then applying SVD to the residual. Equation (6) can therefore be written as

Table 4a PCA Row Scores

Row	PCA 1	PCA 2
1	50.297	31.622
2	25.148	31.622
3	0.000	31.622
4	−25.148	31.622
5	−50.297	31.622

Table 4b PCA Column Scores

Column	PCA 1	PCA 2
1	−39.763	−47.434
2	−39.763	−15.811
3	−39.763	15.811
4	−39.763	47.434

Table 5 Soybean Yield Data for Seven Cultivars Grown at the Same Location over 12 Years

	1977	1979	1982	1983	1984	1985	1986	1987
CHIP	2333.3	2490.8	1884.3	2020.3	3187.5	942.7	2815.0	2443.5
CORS	3106.5	2877.0	2561.8	2057.8	3424.3	1162.3	3644.0	2772.0
EVAN	2725.3	3127.3	2182.5	2585.8	3159.0	1137.5	3164.0	2559.5
HODG	2741.0	3202.8	2440.8	2314.0	3860.5	1426.3	3407.3	2914.3
S200	2843.0	3432.3	2440.3	2195.3	3374.3	1540.0	3188.3	2787.8
WELL	2745.3	2788.3	2238.8	1967.8	3547.5	1115.3	3589.8	2834.8
WILK	2470.5	2621.3	1869.5	2123.8	2717.5	910.0	2605.8	2043.5

Table 6 ANOVA Table for a PCA Analysis of the Data in Tables 5

Source	df	SS	MS	Probability
Total	213	126,352,791.30	593,205.59	
TRT	55	110,427,827.05	2,007,778.67	0.000***
PCA 1	8	94,410,340.91	11,801,292.61	0.000***
PCA 2	8	11,719,216.21	1,464,902.02	0.000***
PCA 3	8	2,235,094.06	279,386.75	0.006**
Residual	31	2,063,175.86	66,554.06	
Error	158	15,924,964.25	100,790.91	

Grand mean = 2547.51042 kg/ha

Table 7a Cultivar PCA Scores for Three Axes

Cultivar	Axis 1	Axis 2	Axis 3
CHIP	24.972	−20.152	6.602
CORS	29.392	10.364	10.689
EVAN	24.825	−1.884	−14.845
HODG	28.184	15.508	−3.886
S200	23.833	11.464	−12.544
WELL	30.665	4.472	13.133
WILK	21.251	−28.331	−4.928

Table 7b Cultivar PCA Scores for Three Axes

Year	Axis 1	Axis 2	Axis 3
1977	6.672	11.235	0.588
1979	14.648	13.294	−21.814
1982	−11.248	17.027	2.993
1983	−14.073	4.224	−12.686
1984	30.447	14.333	1.794
1985	−51.484	18.514	2.390
1986	26.123	16.651	8.504
1987	3.799	16.595	4.485

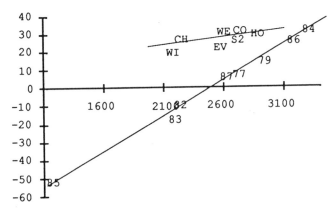

Figure 4 Biplot of PCA axis 1 scores (*y*-axis) and means (*x*-axis) for cultivars (two-character letter codes corresponding to the first two letters of the cultivar name) and years (two-digit number codes corresponding to the last two digits of the year).

$$Y_{rc} = \mu + \alpha_r + \beta_c + \sum_{i=1}^{N} \lambda_i \gamma_{ir} \delta_{ic} + \rho_{rc} \tag{7}$$

where N is equal to $r - 1$, $c - 1$, or the number of significant axes, whichever is less, and ρ_{rc} is the residual from pooling nonsignificant axes. This, then, is the general AMMI model.

The result of applying AMMI analysis to the data of Table 5 is shown in the corresponding ANOVA table (Table 8) and the means and eigenvector score tables (Tables 9a and 9b), in addition to the biplot of the first AMMI axis (Figure 6).

IV. INTERPRETATION OF AMMI ANALYSES

A key to the merit of AMMI analysis is the interpretability of the corresponding biplots. In Figure 6 the ■'s and ×'s of the first three figures have been replaced by two-character (letter and/or number) codes to identify the cultivar and year associated with a given score. It is

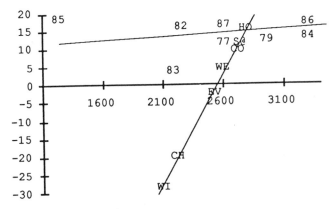

Figure 5 Biplot of PCA axis 2 scores (*y*-axis) and means (*x*-axis) for cultivars (two-character letter codes corresponding to the first two letters of the cultivar name) and years (two-digit number codes corresponding to the last two digits of the year).

Table 8 ANOVA Table for an AMMI Analysis of the Data in Table 5

Source	df	SS	MS	Probability
Total	213	126,352,791.30	593,205.59	
TRT	55	110,427,827.05	2,007,778.67	
cult	6	10,866,030.69	1,811,005.11	0.000***
styr	7	93,184,525.27	13,312,075.03	0.000***
$c \times s$	42	6,377,271.09	151,839.78	
IPCA 1	7	3,660,015.62	522,859.37	0.000***
Residual	35	2,717,255.46	77,635.87	
Error	158	15,924,964.25	100,790.91	

Grand mean = 2547.5 kg/ha

immediately obvious from Figure 6 that 1985 was a remarkably low-yielding year. From Figure 6 it can be determined that Wells and Corsoy had strong positive interactions with the years 1986 and 1984, had relatively little interaction with years 1977 and 1982, and had strong negative interactions with 1979 and 1983. On the other hand, Wilkin and Evans had positive interactions with 1983 and 1979 and negative interactions with 1986 and 1984. This is the visual interpretation, which is substantiated by the following numerical demonstration of those statements.

Table 9a Cultivar Means and Axis Scores

Cultivar	Mean	Axis 1
CHIP	2264.64	−0.076
CORS	2700.69	13.718
EVAN	2580.09	−13.064
HODG	2788.35	4.705
S200	2725.13	−6.818
WELL	2603.41	17.013
WILK	2170.22	−15.478

Grand mean 2547.5 kg/ha $LSD_{0.05}$ = 156.76

Table 9b Year Means and Axis Scores

Year	Mean	Axis 1
1977	2709.25	0.422
1979	2934.22	−11.976
1982	2231.11	2.168
1983	2180.64	−18.388
1984	3324.35	10.979
1985	1176.30	−6.582
1986	3202.00	15.748
1987	2622.17	7.627

Grand mean 2547.5 kg/ha $LSD_{0.05}$ = 167.58

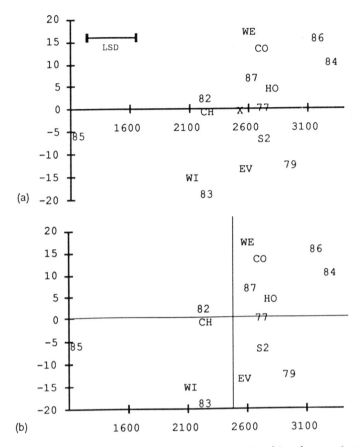

Figure 6 (a, b) Biplots of AMMI axis 1 scores (*y*-axis) and means (*x*-axis) for cultivars (two-character letter codes corresponding to the first two letters of the cultivar name) and years (two-digit number codes corresponding to the last two digits of the year). In (a) the X near the center of the *x*-axis represents the grands mean; in (b) the grand mean is represented by the vertical line (format used by Zobel et al. [1]). (c, d) Alternative formats for the biplots of AMMI axis 1 scores (*y*-axis) and deviations of means from the grand mean (*x*-axis) for cultivars and years [denoted as in (a) and (b)].

A. Fitting a Predictive Model

With reference to equation (7), each of the terms in the equation, except Y_{rc} and ρ_{rc} is presented on the graph in Figure 6. If we rewrite equation (7) to fit this data set, we get

$$Y_{rc} = \mu + \alpha_r + \beta_c + \sum_{i=1}^{1} \lambda_i \gamma_{ir} \delta_{ic} + \rho_{rc} = \mu + \alpha_c + \beta_y + \lambda_1 \gamma_{1c} \delta_{1y} + \rho_{cy} = Y_{cy} \qquad (8)$$

where the *c* for cultivars replaces the *r* for rows and the *y* for years replaces the *c* for columns, and only one axis is represented. Since ρ_{cy} is the nonsignificant variance, if we remove it from the equation, we get an equation for the estimate (or a prediction) of the true value of a given cell (*c,y*):

$$\hat{Y}_{cy} = \mu + \alpha_c + \beta_y + \lambda_1 \gamma_{1c} \delta_{1y} \qquad (9)$$

where $\hat{Y}_{cy}$ is the predicted (estimated) true cell value.

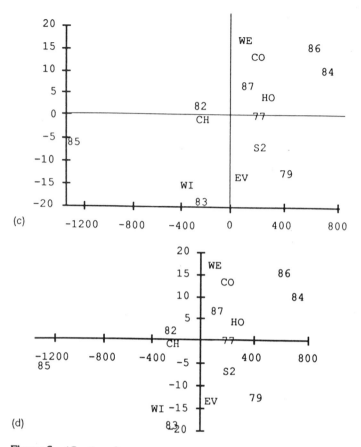

Figure 6 *(Continued)*

B. Predicting True Values: Model Based

Previously, we mentioned that the scores that are derived from an AMMI or PCA analysis include the square root of the singular value (Table 9 and $\lambda_1^{0.5}\gamma_{1r}$ and $\lambda_1^{0.5}\delta_{1c}$). Since the data set in Table 5 is yield (kg/ha), the singular value is in kg/ha and the product of the scores is therefore in kg/ha. By referencing Table 9, we find that the predicted value for Wells in 1986 is $Y_{6,7} = \mu + \alpha_6 + \beta_7 + \lambda_1\gamma_{1,6}\delta_{1,7} = 2548 + (2603 - 2548) + (3202 - 2548) + (17 \times 16) = 3526$ kg/ha (272), while the predicted value for Wells in 1979 is $2548 + (2603 - 2548) + (2934 - 2548) + (17 \times -12) = 2786$ kg/ha (−204). You will note that $\mu + \alpha_6 + \beta_7$ (this is the means-based predicted value) gives the same value as the cultivar mean plus the year mean deviation from the grand mean (i.e., in 1986, Wells = 2603 + 654 = 3257 and = $\mu + \alpha_6 + \beta_7$ = 2548 + 55 + 654 = 3257; and in 1979, Wells = 2603 + 386 = 2989). The difference between Wells in 1979 and Wells in 1986 is 268 kg/ha when the interaction is not taken into account. When the interaction is taken into account, the difference is 268 + 272 + 204, or a total difference of approximately 740 kg/ha.

C. Predicting True Values: Biplot Interpretation

These calculations demonstrate that when the row is on one side of the *y*-axis zero point and the column is on the opposite side, their product is negative. Thus a negative interaction indicates a decrease in the predicted value. If both are on the same side, their product is therefore positive, indicating a positive interaction and an increase in the predicted value. The larger the eigenvector score (also called the interaction score), the greater the shift in predicted value. Whereas the means-based predicted yield for Wells in 1979 (2989) is not significantly different from the means-based predicted yield for Wells in 1986 (3257), $LSD_{0.05} = 443$, the predicted values *are* significantly different when the interaction is taken into account.

D. Interpreting the Physiological/Environmental Bases for Interaction Axes

Often the underlying cause of an interaction can be deduced from knowledge about genetic differences between the cultivars involved and/or the different environments involved. In this example, the interactions represented by the biplot (Figure 6) are due to the interaction of this set of cultivars with the amount of total rainfall during the growing season. The year 1983 had approximately 140 mm of rainfall during the 90 days of the main growing season, while 1979 had 230 mm of rainfall. On the other hand, 1984 and 1986 had 300 and 390 mm of rainfall, respectively. The remaining four years had intermediate amounts of rainfall. It can be postulated from this that at this experimental location, the cultivars Wells and Corsoy interacted positively with wet years and negatively with dry years, and that Wilkin and Evans had the opposite interaction response.

Wilkin and Evans were originally selected out of the same F_8 population and would therefore be expected to respond to their environment in a similar fashion. It should be noted that Wilkin and Evans are maturity group 0 cultivars and Wells and Corsoy are maturity group 2 cultivars. These differences in maturity group result in different developmental timing. Wilkin and Evans flower the first week of July at this location, while Corsoy and Wells flower the last week in July. The years 1983 and 1979 had little rain during the first three weeks of July, while the other years had adequate rain during this critical time of early seed development for Wilkin and Evans. On the biplot it was an easy task to identify that Wilkin and Evans and Corsoy and Wells responded differently, and that 1979 and 1983 and 1984 and 1986 represented different response patterns. With that information it was an easy task to find the environmental values and the pedigree differences, which easily explain the underlying physiological characteristics that result in the yield differences observed.

AMMI analysis with its biplot presentation of results provides insights into experimental results for which it might require years of additional and complex experimentation to develop adequate hypotheses. This ease of interpretation and more rapid advancement of physiological genetic understanding of the quantitative effect by each different genotype and environment facilitates rapid development of hypotheses from normal field experiments. These hypotheses can be defined sufficiently narrowly to be tested using a few well-designed reductive or whole system experiments.

E. Alternative Biplot Forms

Finally, Figure 6 provides four different presentation formats for biplot of axis 1 against the means for the data set above. Any of these formats is correct. We use Figure 6b because we feel it gives a clearer, more easily understood graph. Figure 6d is perhaps the format most often used in the theoretical literature. Figure 6c and d are constructed directly from the elements of equation (9), while the others combine the grand mean and the additive mean deviations before graphing.

V. SUMMARY

The most commonly used statistical model, the analysis of variance (ANOVA), involves only additive effects and does little to describe interaction between factors. The additive main effects and multiplicative interaction (AMMI) statistical model extends the ANOVA analysis by applying principal component analysis (PCA) to the ANOVA interaction effects. It is shown that the mathematical models describing the ANOVA and AMMI analyses are simply subsets of the more general linear algebra procedure singular value decomposition (SVD), with the additive ANOVA additionally being a subset of AMMI.

AMMI analysis partitions the interaction effects into separate axes according to the pattern of interaction between the levels of the data factors. The procedure also calculates the percentage of the variance (score) for each axis which is due to individual levels of each factor. [The term *score* as used in this chapter is synonomous with the term *contribution* (to the G × E interaction effect) used in Chapters 41 and 43 to 45.] These scores are plotted on a biplot (a scatter diagram that contains both factors) to provide visual interpretation of the interaction, or they can be used to calculate the predicted or estimated true value of the data. The methods for these calculations are presented, and the interpretation of the biplot is described.

REFERENCES

1. R. W. Zobel, M. W. Wright, and H. G. Gauch, *Agron. J., 80*: 388 (1988).
2. H. G. Gauch, Analysis of the data matrix in *Theor. Appl. Genet., 76*: 1 (1988).
3. H. G. Gauch and R. W. Zobel, *Theor. Appl. Genet., 77*: 473 (1989).
4. H. G. Gauch and R. W. Zobel, *Theor. Appl. Genet., 79*: 753 (1990).
5. H. G. Gauch and R. W. Zobel, in *Statistics in Agriculture* (G. Milliken and J. R. Schwenke, eds.), Kansas State University Symposium, pp. 205–213 (1990).
6. H. G. Gauch, *Statistical Analysis of Regional Yield Trials: AMMI Analysis of Factorial Designs*, Elsevier, Amsterdam (1992).
7. R. W. Zobel, In *Genotype-by-Environment Interaction and Plant Breeding* (M. Kang, ed.), Louisiana State University, Baton Rouge, La., pp. 126–140 (1990).
8. J. Tukey, *Ann. Math. Stat. 33*: 1 (1962).
9. H. Gollub, *Psychometrika, 33*: 73 (1968).
10. J. Mandel, *J. Res. Natl. Bur. Std., 74B*: 1 (1969).
11. J. Mandel, *Technometrics, 13*: 1 (1971).
12. J. Mandel, *J. Am. Stat. Assoc., 56*: 878 (1961).
13. K. Finlay and G. Wilkinson, *Aust. J. Agric. Res., 14*: 742 (1963).
14. R. Kempton, *J. Agric. Sci., 103*: 123 (1984).
15. D. Bradu and R. Gabriel, *Technometrics, 20*: 47 (1978).

43

Photoperiod × Temperature Interaction Effects on the Days to Flowering of Bean (*Phaseolus vulgaris* L.)

Donald H. Wallace
Cornell University, Ithaca, New York

K. S. Yourstone
Pioneer Seed Company, Urbandale, Iowa

J. P. Baudoin
Phytotechnie des Regions, Gembloux, Belgium

J. Beaver
University of Puerto Rico, Mayaguez, Puerto Rico

D. P. Coyne
University of Nebraska, Lincoln, Nebraska

J. W. White
Centro Internacional de Agricultura Tropical, Cah, Colombia

Richard W. Zobel
Agricultural Research Service, U.S. Department of Agriculture, Ithaca, New York

I. INTRODUCTION

In this chapter we briefly review our earlier research on the control by photoperiod gene activity over the days to flowering (DTF) maturity and yield of bean. Thereafter, previously unpublished data are analyzed to identify the physiological-genetic bases for the variations in DTF expressed across the known range of differences in genetically controlled sensitivity to photoperiod when bean is grown across tropical-to-temperate daylengths (latitudes) and temperatures (elevations).

A. Review of the Literature

1. This Research on Photoperiodism Originated from Attempts to Breed for Higher Yield

The research discussed in this and Chapters 41, 44, and 45 began in 1967, with no intention to study photoperiodism. The red-kidney-type bean cultivar Redkote was crossed with the yelloweye-type cultivar Charlottetown, which flowers and matures earlier than Redkote. The objective of the cross was to combine Charlottetown's higher harvest index with the larger total aerial biomass of Redkote. The harvest index is the percentage of the net accumulated aerial biomass that is yield (i.e., the proportion that has been partitioned to the seed). From that cross, cultivar Redkloud was released in 1973. Redkloud develops to harvest maturity in about 85 days, while its parent Redkote requires 105. The average yield of Redkloud equals that of Redkote, so Redkloud accumulates yield at a rate about 20% higher per day of plant growth. In our breeding program, early flowering and harvest maturity, higher rate of accumulation of yield per day of plant growth, and higher harvest have been inherited as a syndrome through many subsequent crosses and generations of segregation.

Wallace and Enriquez [1] reported in 1980 that the syndrome of earlier flowering, higher rate of accumulation of yield, higher harvest index, lower total aerial biomass, and smaller total leaf area all attend a photoperiod-insensitive genotype. The contrasting syndrome of later flowering and maturity, lower rate of accumulation of yield, lower harvest index, larger net accumulation of aerial biomass, and larger total leaf area always attend the photoperiod-sensitive genotype.

In 1993, Wallace et al. [2] confirmed that the contrasting syndromes of Redkloud and Redkote are pleiotropically controlled by a single photoperiod gene. The pleiotrophy results because the photoperiod gene activity controls the partitioning of the available photosynthate between growth of the reproductive (yield) organs versus competitive growth of the vegetative organs. The contrasting levels of traits of the alternative syndromes occur because any photosynthate received by the reproductive (yield) organs becomes unavailable to support growth of the vegetative organs, and vice versa. The earlier flowering and maturity, higher harvest index, and higher rate of yield accumulation but smaller aerial biomass and smaller leaf area attend the photoperiod-insensitive genotype, because this genotype maximizes the proportion of the photosynthate that is partitioned to the reproductive organs, which results in their rapid growth with consequent need of shorter time for these organs to develop to flowering and harvest maturity.

2. Controlled Environment Demonstration of a U-Shaped Curve of the Days to Flowering

Our 1980 report [1] of control by photoperiod gene activity was based on controlled environments. The report demonstrates that both daylength and temperature strongly modulate activity of the photoperiod-sensitive gene(s), whereas temperature predominantly modulates the activity of the insensitive genotype. With daylength of 9 h, both the photoperiod-insensitive and photoperiod-sensitive genotypes flower earlier as the temperature rises from 21° C through 24, 27, and 30°C (Figure 1). These rises also cause the photoperiod-insensitive genotype (Redkloud) to flower earlier under 12, 14, and 16 h daylength. In contrast, a 12-h daylength causes the photoperiod-sensitive genotype to respond to these same temperatures with a U-shaped curve of its days to flowering (DTF) (Figure 1). The rises from averages of 21 to 24 to 27°C cause progressively smaller DTF, but 30°C delays DTF as compared with 27°C. The simultaneous modulation of DTF of the photoperiod-sensitive genotype by both daylength and temperature, in contrast to strong modulation by temperature alone for the insensitive allele of the same photoperiod gene, demonstrates a complex photoperiod × temperature × genotype interaction effect on the DTF.

Research on bean by Coyne in controlled environments in 1966 [3] and 1978 [4] suggested that the largest photoperiod-caused delay of DTF occurs when the day/night difference in temperature is large. This was supported by our 1980 controlled environment study (compare Figure 1). Further research in controlled environments, completed in 1983 [5], involved the full range of known sensitivities to photoperiod but failed to establish a direct relationship

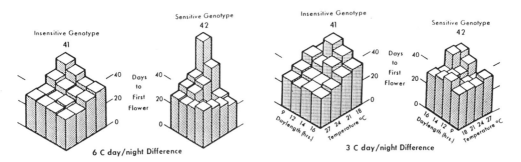

Figure 1 Response surfaces for an early-maturing photoperiod-insensitive bean genotype and a late-maturing photoperiod-sensitive bean genotype grown in the 32 environments of a factorial set of 4 daylengths × 4 night temperatures × 2 day/night differences in temperature. (From Ref. 1.)

between the photoperiod-caused delay of DTF and the day/night difference in temperature [5]. Changing either the day or night temperature also changes both the mean temperature and the diurnal difference; all these aspects of the temperature effect on the DTF are confounded.

The 1983 study [5] demonstrated that most genotypes express a U-shaped response of DTF if a wide enough range of average temperatures is spanned, and verified the synergistic delay of DTF by progressively higher temperature combined with a more delaying (longer) daylength. Further, both the U-shaped response and the synergistic delay of DTF were enlarged by increase of sensitivity to genotype photoperiod. These growth chamber results reinforced interpretation of a large and complex photoperiod × temperature × genotype interaction effect on the DTF.

In 1985 a review was published [6] of the photoperiod and temperature effects on flowering of many crops. That review recognized that photoperiod gene activity causes simultaneous effects on both the reproductive (yield) and vegetative organs, and that the U-shaped response of DTF to temperature occurs in many plant species. Two 1993 papers [2,7] establish that the syndrome of earlier flowering, higher rate of accumulation of yield, higher harvest index, lower total aerial biomass, and smaller total leaf area for the photoperiod-insensitive genotype and the opposite syndrome (lower versus higher levels) for the photoperiod-sensitive genotype are due to control over the alternatives of partitioning the photosynthate to reproductive versus vegetative growth and development. For Redkloud and Redkote the control is by a single photoperiod gene.

3. Conclusions from Field Demonstration of a U-Shaped Curve of the Days to Flowering

Based on the results in a controlled environment for bean reviewed above plus the review for other crops [6], three essentially simultaneous experiments were conducted. Each was directed toward determining if, as for controlled environments, both the U-shaped curve of DTF in response to temperature plus the complex photoperiod × temperature × genotype interaction effect on DTF associated with it do occur across field environments.

Interpretations from two of the three field studies were published in 1991 in a common research report [8]. As illustrated in Figure 2, that report demonstrates the following:

1. A U-shaped response to temperature does occur across field environments, and its maximum expression requires a long daylength and a photoperiod-sensitive genotype.

2. The delay of DTF of photoperiod-sensitive bean caused by long daylength is synergistically enlarged by a higher temperature and/or by a yet-longer photoperiod and/or by greater sensitivity to photoperiod of the genotype.

3. The fewest days to flower along the U-shaped curve of DTF in response to temperature indicates an optimum temperature for flowering. That optimum temperature causes the most rapid rate of development to flowering. This rate is simply 1/DTF.

4. The optimum temperature and its fastest rate of development to flowering (smallest DTF) differ for each genotype.

5. The optimum temperature for flowering tends to be highest for the genotypes that are the most insensitive to photoperiod, and generally, it becomes progressively lower with both increasing sensitivity of the genotype to photoperiod and a more delaying daylength.

6. One of the two physiological causes of the U-shaped response of DTF to temperature is interpreted as being due to modulation by progressively higher temperatures of progressive increases in the activity of the photoperiod gene(s). This interpretation arises because every rise of temperature delays the node to flowering (from Figure 3).

7. From Figures 1 to 3 it is clear that the increase in photoperiod gene activity modulated by a higher temperature is small for a photoperiod-insensitive genotype, intermediate for a moderately sensitive genotype, and large for a highly sensitive genotype.

8. From Figure 3 we infer that simultaneous with its enlargement of DTF through enhancement of the photoperiod gene activity, the second cause of the U-shaped curve of DTF is that the same higher temperature causes a decrease in the DTF by decreasing the number of days needed to develop an additional node on the plant. This means that for a photoperiod-sensitive genotype, a higher temperature simultaneously delays the DTF (i.e., decreases the rate of reproductive development while increasing the rate of vegetative development).

9. From Figure 3 (compare Figures 1 and 2) we infer that the optimum temperature is one at which the effect it causes toward delay of flowering by its modulation of photoperiod gene activity is exactly canceled by the opposite effect that temperature simultaneously causes toward earlier flowering through its modulation of the rate of vegetative development (i.e., through the fewer days needed to develop a node) if the temperature is raised.

10. Expanding on conclusion 9 (from Figure 3, compare Figures 1 and 2), we infer that regardless of whether the growing temperature is below or above the optimum for flowering of the genotype growing in the given daylength, every rise in temperature decreases the rate (1/DTF) of development to flowering by delaying the node to flower (enhancing the photoperiod gene activity), but every rise simultaneously accelerates the rate of vegetative development (1/days per node), thereby increasing the rate of development to flowering.

11. From Figure 3 and conclusions 9 and 10, we infer that if the growing temperature is below the optimum for flowering, a rise in temperature simultaneously causes a relatively large decrease in DTF through its enhancement of the rate of development of the nodes and a comparatively smaller delay of DTF through the photoperiod gene activity. On the contrary, if the temperature is above the optimum for flowering, a rise in temperature causes a comparatively small decrease in DTF through the faster rate of development of the nodes but causes a relatively larger delay of DTF through the photoperiod-gene-activity-caused delay of the node to flower.

4. Vegetative and Reproductive Bases for the U-Shaped Curve of the Days to Flowering

Conclusions 12 to 15 are based on simultaneous modulations by every temperature of both the reproductive rate and the vegetative rate of development. More explicitly, they arise from conclusions 10 and 11, which are interpretations derived from Figures 3 and from Figure 1 and 2.

12. For all levels of genetically controlled photoperiod insensitivity versus photoperiod sensitivity, if the average growing temperature is below the optimum for flowering of that genotype and daylength, the relatively small additional delay of the node to flowering modulated

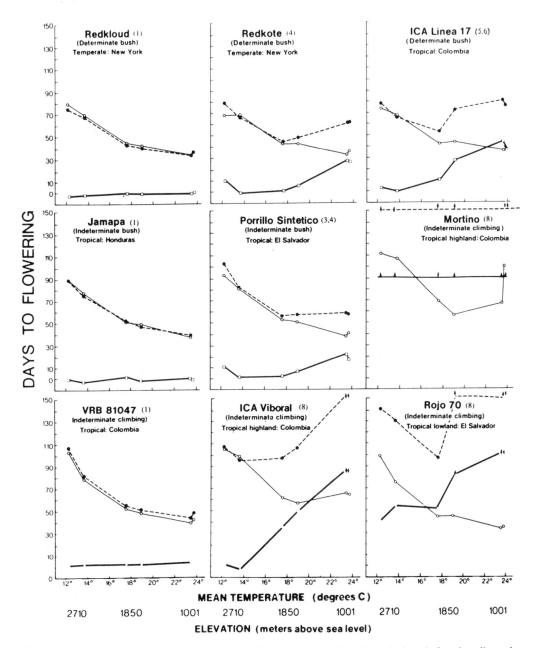

Figure 2 Days to first flower and the delay of flowering caused by long daylength for nine diversely adapted bean genotypes grown under 12.3- and 18.0-h daylengths during two successive growing seasons at mean temperatures near 12, 18, and 24°C. A rating for sensitivity to photoperiod is given in parentheses adjacent to the cultivar name of each genotype; the sensitivity increases progressively from 1 to 8. Days to flower (DTF) under 12.3-h natural daylength(○),DTF under incandescent-lamp-extended 18-h daylength (●), days delay by the long daylength (□). Failure to flower during the observed 100, 150, or 180 days is indicated by ↑. (From Ref. 8.)

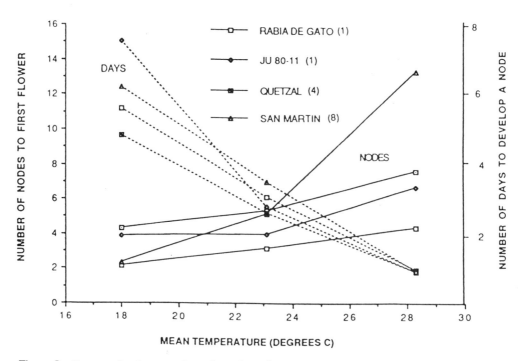

Figure 3 Days to develop a node and number of nodes to flowering of four bean genotypes with indeterminate bush habit grown at mean temperatures of 18.0, 23.1, and 28.3°C.

by a higher temperature which remains below the optimum for flowering will not delay DTF as much as the DTF will be made earlier by that temperature's modulation of a faster rate of development of the nodes.

In contrast, if the average growing temperature is above the optimum for flowering of that genotype when it is grown in the given daylength, the delay of the node to flowering due to amplification of the photoperiod gene activity modulated by the higher temperature will delay DTF more than that DTF is made earlier by that higher temperature's modulation of the need for fewer days to development of each node.

13. From interpretation 12, the rate of development to flowering (1/DTF) is controlled predominantly by the rate of vegetative development (1/days per node) whenever the original growing temperature is, and the higher temperature remains, below the optimum temperature for flowering.

If the original average growing temperature is above, or if a rise brings the temperature above the optimum for flowering, however, the change in rate of development to flowering (1/DTF) will be controlled predominantly by the rate of development toward flowering of the reproductive organs, which is controlled by the photoperiod-gene-caused delay of the node to flowering, which is controlled by the proportion (2) of the photosynthate that the photoperiod gene activity allows to be partitioned to the reproductive organs.

14. If the genotype is photoperiod insensitive, each rise of temperature will enlarge the photoperiod gene activity in proportion to any small photoperiod activity of that genotype. For a moderately photoperiod-sensitive genotype, the same temperature rise will enlarge the photoperiod gene activity an intermediate amount in proportion to its intermediate level of photoperiod gene activity, which will cause an intermediate delay of the DTF (Figure 2). For a more highly photoperiod-sensitive genotype, the same higher temperature will cause a

correspondingly proportionally larger quantitative enlargement of the photoperiod gene activity, and the consequent temperature-caused delay of DTF will be larger (Figure 2).

15. Beginning from any temperature that is below the optimum for flowering, a sufficient rise in temperature will amplify the photoperiod activity of almost all, if not all, insensitive genotypes enough that their photoperiod gene activity will cause a larger delay of the DTF than the change through the earlier DTF through the rate of node development [8]. For the most photoperiod-sensitive genotypes, such a higher temperature will cause a state of never flowering unless the daylength is shortened [9].

B. Objective

The objective of this chapter is further elucidation of the daylength × temperature × genotype interaction effects on bean DTF.

II. MATERIALS AND METHODS

A. Experimental Procedure

Forty-eight bean genotypes (cultivars) were distributed in 1983 and 1984 for an international bean flowering and adaptation nursery. Distribution was by the Centro Internacional de Agricultura Tropical of Palmira, Colombia [10]. At each location the 48 genotypes were grown in a randomized block design with three replications. A replication included two sampling units (hills) spaced 50 cm apart with 1 m between rows. Five seeds were sown per hill. After emergence each hill was thinned to three plants. The DTF recorded for each hill was the number of days from planting to the day that two of its three plants had at least one flower [10]. The DTF analyzed for a replication was the average of its two sampling units.

B. Genotypes

The 48 genotypes were selected to span the known range of sensitivity to photoperiod, plus the known range of growth habits. Cultivar names, their 13 countries of origin, and their growth habit plus an independently determined score for sensitivity to photoperiod are presented in Table 1, along with the statistically quantified effects on DTF by each genotype.

C. Environments

A description of all 15 environments is presented (Table 2), along with the statistically quantified contributions to the largest G × E interaction effect on the DTF caused by each environment. The 15 environments represent six temperate and four tropical locations. There were two plantings at each of three tropical locations and two at each of two temperate locations. Differences in latitude, elevation, and time of year result in a different combination of temperature and daylength for each of the 15 environments.

D. Statistical Analysis

While the physiological genetic conclusions about DTF reviewed above were being derived, Zobel and Gauch [11] were independently developing a statistical procedure to quantify the contribution by each genotype and by each environment to the G × E interaction effect. Their goal was to quantify the G × E interaction effects on the crop yields measured across multiple yield trials. Advanced descriptions of the additive main effects and multiplicative interaction effects (AMMI) analysis which they developed are presented in Chapter 42 and in the book by Gauch [12]. Interpretations of the daylength × temperature × genotype interaction effects on DTF are greatly facilitated by using AMMI analysis of the data.

The AMMI analysis results for the genotypes are presented in Table 1 along with the sensitivities and habits of the genotypes. The results for the environments are shown in Table

Table 1 Variety Name, Country of Origin, Growth Habit, and Score for Sensitivity to Photoperiod, Plus the Attendant Main Effect (Average Days to Flowering Across all 15 Environments) by Each Genotype (Variety) on the Days to Flowering and That Genotype's Contribution to the Largest G × E Interaction Effect on These Days

Variety name	Country of origin	Growth habit[a]	Main effect (days)	Contribution to G × E (days 0.5)	Photoperiod sensitivity score[b]
Rojo 70	El Salvador	IV A	Few flowers at temperate sites		8
Sangretoro	Colombia	IV B	Few flowers at temperate sites		8
ICA Viboral	Colombia	IV B	Few flowers at temperate sites		8
ICA L 32983	Colombia	IV A	Few flowers at temperate sites		8
E 1056	Ecuador	IV A	Few flowers at temperate sites		8
Flor de Mayo	Mexico	IV A	Few flowers at temperate sites		8
Gloriabamba	Peru	IV A	75.7	5.99	8
Ojo de Cabra	Mexico	III B	68.5	4.81	8
IAN 5091	Guatemala	III B	68.3	4.52	7
ICA L 33341	Colombia	II B	69.9	4.15	5
ICA Linea 17	Colombia	I A	56.9	2.57	6
Diacol Andino	Colombia	I A	57.3	2.24	6
Comp. Alajuela	Costa Rica	III B	53.0	1.23	8
Antioquia-8	Colombia	I A	52.7	1.03	8
Canario Divex	Peru	I A	57.4	0.88	3
Diacol Calima	Colombia	I A	48.9	0.87	5
Magdalena-3	Colombia	IV A	60.6	0.65	3
Redkote	United States	I A	45.4	0.57	6
Pompadour 2	Dominican Republic	I A	47.5	0.34	4
N 555	El Salvador	II A	54.9	0.22	4
Pompadour Checa	Dominican Republic	I A	46.0	0.08	7
51051	Costa Rica	II A	51.0	0.03	4
Puebla 152-C	Mexico	III A	52.0	−0.10	4
Puebla 152-N	Mexico	III A	54.3	−0.14	4
Mexico 80	Mexico	II B	46.6	−0.19	4
ICTA Quetzal	Guatemala	III A	49.0	−0.29	4
A 132	Colombia	II	50.4	−0.30	3
Porrillo Sintetico	El Salvador	II B	50.6	−0.37	4
Sanilac	United States	I B	44.4	−0.46	3
Porrillo-1	El Salvador	II A	51.4	−0.69	3
Miss Kelly	Jamaica	I A	40.1	−0.81	6
Brasil-2	Brazil	I B	48.3	−0.93	3
ICA Tui	Colombia	II A	48.7	−0.95	3
Rojo de Seda	El Salvador	III B	46.3	−0.98	3
Zamorano	Honduras	III A	45.2	−1.02	3
VRB 81047	Colombia	IV A	51.5	−1.25	2
Rabia de Gato MG	Guatemala	III A	43.4	−1.58	1
Brunca (BAT 304)	Colombia	III B	43.9	−1.08	3
Rio Tibagi	Brazil	II B	47.9	−1.66	1
Redkloud	United States	I A	36.2	−1.85	1
Jutiapa 80-13	Guatemala	II A	46.1	−1.89	1

Table 1 (*Continued*)

Variety name	Country of origin	Growth habit[a]	Main effect (days)	Contribution to G × E (days 0.5)	Photoperiod sensitivity score[b]
Carioca	Brazil	III B	46.6	−1.91	1
C 63 (S 630B)	Costa Rica	II B	47.9	−1.93	1
Revolcion 79 (BAT 41)	Nicaragua	II B	41.8	−1.93	1
Aroana	Brazil	II B	46.7	−1.94	1
P 326 (PI 310740)	Guatemala	IV A	45.1	−1.94	1
ICA Pijao	Colombia	II A	48.0	−1.95	1
Jamapa	Mexico	II A	44.6	−1.99	1

[a]Growth habit as defined by Centro Internacional Agricultura Topical (CIAT), where:

IA   is determinate with erect branches
IB   is determinate with open branches
IIA is indeterminate bush with no climbing ability
IIB is indeterminate bush with weak climbing ability
IIIA is indeterminate sprawling with no climbing ability
IIIB is indeterminate sprawling with weak to moderate climbing ability
IVA is indeterminate with strong climbing ability
IVB is indeterminate with very strong climbing ability

The numerical representation of vegetativeness of the plant habit as used to correlate it with the contribution to third G × E effect on the days to flowering:
1
2
3
4
5
6
7
8

[b]The scores were independently assigned by CIAT, where:
1 delay less than 4 days by the 18 h compared with 12 h environment at Palmira Colombia
2 delay between 4 and 10 days by the 18 h compared with 12 h environment at Palmira Colombia
3 delay between 11 and 19 days by the 18 h compared with 12 h environment at Palmira Colombia
4 delay between 20 and 39 days by the 18 h compared with 12 h environment at Palmira Colombia
5 delay between 40 and 59 days by the 18 h compared with 12 h environment at Palmira Colombia
6 delay between 60 and 79 days by the 18 h compared with 12 h environment at Palmira Colombias
7 delay between 80 and 99 days by the 18 h compared with 12 h environment at Palmira Colombia
8 delay greater than 100 days by the 18 h compared with 12 h environment at Palmira Colombia

2 along with the daylength and temperatures of the environments. The correlations between the daylengths or the temperatures at each of the 15 environments and the AMMI-quantified environmental effects on DTF are used to assist physiological genetic interpretation of the effects by the environments. Correlation between the AMMI results and the known sensitivity to photoperiod or plant habit of the genotypes are used to derive physiological genetic interpretations for the changes in DTF across the genotypes and across the environments.

III. EXPERIMENTAL RESULTS: AMMI STATISTICAL ANALYSIS OF VARIATION IN DAYS TO FLOWERING

A. Proportions of the Variation Due to Environment, Genotype, and G × E Interaction

Six of the 48 genotypes, all being among the most highly sensitive to photoperiod and having extreme vegetativeness of plant habit (Table 1), failed to flower in all but the cooler tropical environments with natural short daylength (SD). These six genotypes are not included in the AMMI

Table 2 Latitude, Elevation, Mean Temperature, Mean Photoperiod of 15 Environments, Each Environment's Average Days to Flowering Across all 42 Genotypes (Environment's Main Effect on the Days to Flowering of Each Treatment[a]), and Each Environment's Contribution to the Largest G × E Interaction Effect on These Days

Geographical location	Latitude	Elevation (m)	Mean temp. (°C)	Photo-period (h)	Main effect (days)	Contribution to G × E (days$^{0.5}$)
Palmira, Colombia[b]	3.31	1001	23.6	18.0[c]	62.7	7.26
Gembloux, Belgium	50.34	180	18.0	15.7	46.7	3.01
Aurora, NY 1984	42.43	240	21.2	14.7	58.3	2.64
Scottsbluff, Nebraska	41.42	1204	21.5	14.9	55.9	2.39
Mt. Pleasant, NY	42.27	528	20.8	14.8	52.3	2.26
Aurora, NY (1983)	42.43	240	22.5	14.8	53.7	2.02
Lincoln, Nebraska	40.49	1184	26.2	14.8	64.5	0.04
Popayan, Colombia[b]	2.25	1700	15.1	18.0[c]	54.0	0.02
Elora, Ontario	43.40	376	20.4	15.0	53.5	−0.05
Mt. Pleasant, NY	42.27	528	18.7	14.8	53.5	−0.16
Palmira, Colombia	3.31	1001	23.6	12.4	38.4	−3.36
Popayan, Colombia	2.25	1700	15.1	12.4	48.9	−3.86
Puerto Rico	18.30	150	25.4	11.3	35.3	−3.67
Chiclayo, Peru	6.44	37	19.8	12.6	44.9	−4.15
Puerto Rico	18.30	150	23.6	11.9	38.3	−4.40

[a] A treatment is one of the 42 genotypes grown in one of these 15 environments, so there are 630 treatments in this study.

[b] Natural 12.4-h photoperiod was extended to 18 h with incandescent light.

[c] Incandescent lights extended the national 12 h daylength to 18 h.

analysis. The variation in DTF across the 630 treatments (the 15 environments × the 42 genotypes that flowered in all or most of the 15 environments) is about equally due to the main effect across all environments (34.8% of the total variation), the main effect across all genotypes (31.7%), and the total G × E interaction effect (33.5%) (Table 3). All three effects are statistically significant ($p < 0.001$). The statistical main effect of a genotype is its average DTF across all 15 environments. The statistical main effect of an environment is its average DTF across all 42 genotypes.

The largest pattern of a G × E interaction effect on the DTF (IPCA-1 in Table 3) accounts for 59.6% of the total G × E interaction effect. The second largest (IPCA-2, Table 3) accounts for 13.6%, and the third largest (IPCA-3) for 11.3%. The AMMI analysis indicates three additional patterns of a statistically significant G × E interaction effect on DTF (IPCA-4, IPCA-5, and IPCA-6, Table 3). These three smallest patterns account for only 11.1% of the G × E interaction effect on the DTF, which is but 3.7% (11.1% × 33.5%) of the total variation in DTF. Thus only the 1.4% of the total G × E interaction remaining in the residual G × E interaction (from 4.2% × 33.5%) plus the 4.8% of the error term are not allocated by the AMMI analysis to a statistically significant physiological-genetic cause of the total variation in DTF across the 630 combinations of genotype and environment (Table 3). Hereafter, the goal is for the researcher to identify or interpret a physiological genetic basis for each of these patterns of change of the DTF.

B. Main Effects by Environment

The main effect by each environment on the DTF of each treatment (Table 2) is its average DTF measured across all 42 genotypes included in the AMMI analysis. For the four lowland

Table 3 Proportional Sources of the Sums of Squares Plus the Mean Sums of Squares and Its Statistical Significance from an AMMI Analysis of the Days to Flowering of 42 Bean Genotypes Grown in 15 Environments

Source	Degrees of freedom	Sums of squares	Percent of total sums of squares	Percent of treatment sums of squares	Mean sums of squares
Total	1829	398,477			218
TRT	629	379,182	95.20		603***
Genotype	41	120,180		31.70	293***
Environment	14	131,831		34.80	9417***
G × E interaction[a]	574	127,171		33.50	222***
				100	
IPCA 1	41	75,888		59.60	1851***
IPCA 2	41	17,307		13.60	422***
IPCA 3	41	14,397		11.30	351***
IPCA 4	41	7,261		5.70	177***
IPCA 5	41	4,199		3.30	102***
IPCA6	41	2,791		2.10	68***
Residual	328	5,329		4.20	16
				100	16
Error	1200	19,295	4.80%		

[a]IPCA.1 is the largest interaction principal component axis, IPCA2. the next largest, etc. (see Chapter 42).
***$p < 0.001$.

tropical environments with short daylength (SD) near 12 h and mean temperature between 20 and 25°C, these averages range from 35 to 45 days (Table 2, Figure 4a); all are smaller than the average DTFs of 52 to 65 for the eight temperate-zone locations with long daylength (LD) near 15 h. For the highest-elevation tropical location, the average for its natural SD and 15°C is 49 days, which is enlarged to 54 days by incandescent-light extension to LD of 18 h. This average 6-day delay of DTF by 18 h daylength at 15.1°C becomes a 25-day delay at the SD tropical location with 23.6°C mean temperature (Table 2, Figure 4a).

C. Contributions by Short Daylength Versus Long Daylength to the Largest G × E Interaction Effect

The contribution to the largest G × E interaction effect on DTF by each environment (i.e., that environment's IPCA-1 in Table 2) is plotted against the average days to flowering for that environment across all 42 genotypes (Figure 4a). Contributions by the five SD environments range from −4.4 to −3.3 days$^{0.5}$ (Figure 4a). Contributions by four LD environments are near zero (−0.13 to 0.06 day$^{0.5}$), while the other six range from 2.1 to 2.7 days$^{0.5}$. The contribution by the highland tropical location with natural SD and 15°C is changed toward a positive level (from −3.86 to 0.04 day$^{0.5}$) by the incandescent-light extension to an LD of 18 h. Extension to 18 h causes an even larger change toward a more positive contribution by the tropical location with 23.6°C; the change is from −3.34 to 7.32 days$^{0.5}$. The negative contributions by all five SD tropical environments accompany the smaller average DTFs across all the genotypes (the smaller main effects of the environments), while the more positive contributions of the LD temperate environments attend the larger average DTF across all the genotypes.

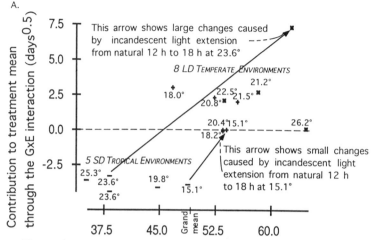

The average days to flowering across all 42 genotypes, which is the main effect by the environment on the days to flowering of each treatment

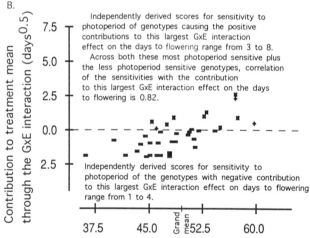

The average days to flowering across all 15 environments, which is the main effect by the genotype on the days to flowering of each treatment.

Figure 4 (A) Contribution to each treatment's days to flowering through each environment's main effect plotted against that environment's contribution to the largest cause of a genotype × environment interaction effect on the days. (B) Contribution to each treatment's days to flowering through each genotype's main effect plotted against that genotype's contribution to the largest cause of a genotype × environment interaction effect on the days. A treatment is one of the 42 genotypes grown in one of these 15 environments. In (A) the symbol – is used for all environments that contribute negatively; + is used for environments with positive contribution to both this largest and the second-largest G × E effect, and × indicates environments that contribute positively to the largest but negatively to the second-largest G × E effect. In (B) the symbol – is used for all genotypes that contribute negatively; + is used for genotypes with positive contribution to both this largest and the second-largest G × E effect, and × indicates genotypes that contribute positively to the largest but negatively to the second-largest G × E effect; scores of 1 to 8 indicate progressively more sensitivity to photoperiod.

D. Daylength × Temperature Interaction Effect Within the Largest G × E Interaction Effect on DTF

Under SD, the trend is for the average DTF of the environments across all the genotypes to decrease with higher temperature (Figure 4a); the correlation is –0.99. On the contrary, under LD the trend is for the average DTF to increase with higher temperature (Figure 4a); this correlation is 0.70 across the eight environments with natural LD, and 0.73 if the two with incandescent-light extension to 18 h daylength are included (Table 4). Correlation between the average DTF and the daylength is 0.75 across all environments, 0.16 across just the LD environments, and 0.81 across the five SD environments.

Correlation between contributions to the largest G × E interaction effect on DTF and temperature is 0.04 across all 15 environments, 0.18 across just the SD environments, and 0.29 across just LD environments (Table 4). Correlation between this contribution and the photoperiod is 0.83 across all 15 environments, 0.43 across the LD environments, and –0.10 across the five SD environments.

The largest of all contributions (7.32 days$^{0.5}$) is associated with the second largest average DTF (62.7 days). It is at the 23.6°C tropical location (Figure 4a), but it results from the incandescent-light extension to 18 h daylength from the natural 12 h daylength (where the contribution is –3.36 days and the average DTF is 38.4 days). At the 15.1°C tropical location this extension to 18 h causes a change from a contribution of –3.86 (in association with an average DTF of 48.9 days) to a near-zero contribution of 0.02 day$^{0.5}$ (in association with an average DTF of only 54 days). Interpreted from Figure 4b is that short versus long daylengths are the major environmental cause of the largest G × E interaction effect on the DTF. The differences in temperature cause the second largest effect on the TF.

E. Contributions by the Genotypes to the Largest Cause of G × E Interaction

1. Genotype for Sensitivity to Photoperiod

The largest negative contributions to the G × E interaction effect on DTF are caused by the most photoperiod insensitive of the 42 genotypes (Figure 4b). With change from most insensitive toward intermediate sensitivity, the contribution shifts toward being less negative (nearer to zero contribution to the G × E interaction effect). Across all the genotypes that contribute negatively, the correlation between their contribution to the largest source of the G × E effect on the DTF and their independently derived scores for sensitivity (Table 1) is 0.88 (Table 5). All the genotypes that contribute positively to the G × E interaction effect on DTF (Figure 4b) are intermediately to highly sensitive to photoperiod, and the correlation between their contribution and sensitivity (Table 1) is 0.50 (Table 5).

All negative contributions to this largest G × E interaction effect by a genotype are accompanied by a relatively small main effect. Similarly, as the known photoperiod sensitivity of the genotype increases to an intermediate level, this causes a nearer-zero contribution, and as the sensitivity is yet higher, it causes a positive to a large positive contribution. The more and more positive contributions are attended by progressively larger averages of the DTF.

Across all 42 genotypes, correlation between the independently derived scores for sensitivity to photoperiod and the contribution by the genotype to the largest G × E interaction effect on DTF is 0.82, and its correlation with the main effect on these days is 0.59 (Table 5). There is a yet higher correlation of 0.89 between the main effect and the contribution to the G × E interaction effect on these DTF.

2. Genotype for Plant Habit

Small correlation (only 0.054, Table 5) between vegetativeness of the plant habit as described in Table 1 and the contribution by the genotype to the largest G × E effect on DTF occurs,

Table 4 Correlations Among the Contributions by 15 Environments to the Three Largest Physiological Genetic Sources of a Genotype × Environment Interaction Effect on Days to Flowering and Correlations of These Contributions with the Average Days to Flowering of the Environments Across 42 Genotypes and with Both the Photoperiod and the Temperature of the Environments

G × E interaction effect or environmental modulator thereof	Environments	Mean days to flowering	Largest G × E effect (IPCA-1)	Next-largest G × E effect (IPCA-2)	Third-largest G × E effect (IPCA-3)	Temperature
IPCA-1	All 15	0.732				
	10 long daylength	0.228				
	5 short daylength	-0.24				
IPCA-2	All 15	0.337	0			
	10 long daylength	-0.862	-0.091			
	5 short daylength	0.919	-0.477			
IPCA-3	All 15	-0.132	0	0		
	10 long daylength	-0.125	-0.115	0.009		
	5 short daylength	-0.368	-0.851	-0.692		
Temperature	All 15	-0.043	0.041	-0.636	-0.079	
	10 long daylength	0.731	0.294	-0.782	0.145	
	5 short daylength	-0.986	0.18	-0.93	-0.079	
Photoperiod	All 15	0.748	0.825	0.074	-0.135	-0.266
	10 long daylength	0.157	0.43	0.036	-0.144	-0.324
	5 short daylength	0.708	-0.096	0.426	0.386	-0.615

Table 5 Correlations Among the Contributions by 42 Genotypes to the Three Largest Physiological Genetic Sources of a Genotype × Environment Interaction Effect on Days to Flowering and Correlations of These Contributions with the Average Days to Flowering of the Genotypes Across 15 Environments and with Both the Photoperiod Sensitivity and Plant Habit of the Genotypes

G × E interaction effect or genotypic modulator thereof	Genotypes	Average days to flowering	Largest G × E effect (IPCA-1)	Next-largest G × E effect (IPCA-2)	Third-largest G × E effect (IPCA-3)	Score for sensitivity to photoperiod
IPCA-1	Across all 42 genotypes	0.89				
	Across positive contributors to largest G × E effect	0.90				
	Across negative contributors to largest G × E effect	0.52				
IPCA-2	All 42 genotypes	0.09	0			
	Across positive contributors to largest G × E effect	0.68	0.646			
	Across negative contributors to largest G × E effect	-0.58	-0.725			
IPCA-3	All 42 genotypes	-0.08	0	0		
	Across positive contributors to largest G × E effect	-0.25	-0.247	-0.247		
	Across negative contributors to largest G × E effect	-0.49	-0.471	-0.841		
Score	All 42 genotypes	0.59	0.081	-0.183	0.016	
	Across positive contributors to largest G × E effect	0.23	0.502	-0.396	-0.193	
	Across negative contributors to largest G × E effect	0.29	0.877	-0.584	-0.360	
Habit	All 42 genotypes	0.30	0.054	0.201	-0.507	0.217
	Across positive contributors to largest G × E effect	0.73	0.574	0.588	-0.519	0.251
	Across negative contributors to largest G × E effect	0.36	0.024	-0.297	-0.49	-0.049

because a full range of sensitivity to photoperiod is associated with several levels of the plant habit (Table 1) [7, 8,12]. Nevertheless, among the 48 genotypes, most with a very vegetative habit (habits IV.A or IV.B, Table 1) are highly sensitive. Elsewhere (7), this is interpreted as a consequence of selection for yield within the 6- to 9-month growing season durations and low mean temperature of the highland tropical sites. In these environments it is suggested that selection for yield results in indirect selection for extreme sensitivity to photoperiod, because a long delay of flowering and maturity are necessary to maximize yield in the long growing seasons. Only the most photoperiod-sensitive genotypes have their photoperiod genes activated by the relatively short but longest daylengths of the tropics in association with the low mean temperatures of the highland elevations (Figure 2).

IV. INTERPRETING MAIN EFFECTS AND G × E INTERACTION PATTERNS

Our previous experience, which is reviewed in the introduction, led us to focus on a temperature effect on the DTF in addition to the effect by the photoperiod. The effect by temperature on the DTF is less well described in the literature than the effect by the photoperiod. Our interpretation of the largest G × E effect on the DTF therefore begins with the expected effects by the temperature.

A. Physiological Bases for Temperature Effects

A concept of biochemistry is that every enzyme activity increases in rate as temperature rises. The increase is measured as a Q_{10} effect (i.e., the "fold" increase in rate with a rise of 10°C). Most enzymes have a Q_{10} near 2.00. All biological processes are directed by genetic instruction, implemented by the consequent enzymatic activity (or several integrated activities); the rate is always modulated by the temperature.

The variations in DTF presented in this chapter show a Q_{10} effect on the photoperiod gene activity, and repeat the long time understanding that the genetic instruction of a photoperiod-sensitive genotype of a short-day plant implements early flowering in SD, whereas LD modulates the activity of that photoperiod-sensitive gene to cause delay of DTF. The data show that daylength is a larger environmental modulator of the DTF, but they also demonstrate that a higher temperature amplifies (intensifies) the genetic instruction that is implemented by both the insensitive and the sensitive photoperiod gene(s). That is (from Figures 1 to 4), with reduction (or prevention) of the photoperiod gene activity by either the genotype or a sufficiently short daylength, a higher temperature which remains below the optimum temperature for flowering amplifies the gene actions (in the absence of photoperiod gene activity) that lead to early flowering, thereby causing a smaller DTF. Similarly, in LD, higher temperature amplifies the photoperiod gene activities that implement the delay of flowering, thereby enlarging DTF. The optimum temperature for flowering is defined in Sections I.A.3 and I.A.4.

B. Daylength × Temperature Interaction Effect

In addition to our focus on simultaneous effects by temperature and daylength on the DTF, daylength × temperature interaction is emphasized here because a much published model for quantifying both photoperiod and temperature effects on DTF considers only a near-linear effect by both. That model is based on reductive research with many plant species (see Chapter 41); Chapter 44 presents additional discussion of the insufficiency of that model. The authors of that model [13,14] explain that the linear response applies only for the range $T_b < T < T_o$, where T is the effective mean temperature, T_b the base temperature (at and below which there is no progress toward flowering), and T_o the optimum temperature (at which DTF is minimum, i.e., the rate of development to flowering is maximum). The model ignores that part of the variation in DTF for which $T_o < T$, because these "supraoptimal temperatures" (the authors' term) cause large daylength × temperature interaction effects on the DTF.

All measures of DTF in this chapter are from environments where bean is a commercial crop. Chapter 44 presents the bean yields for two sites where $T_o < T$; one with a mean of 23.5°C is a commercial bean production area. The other is a nonproduction environment because its 28.3°C mean is too far above the optimum temperature for flowering. In Chapter 44 there is only a temperature × genotype interaction effect on the DTF and yield, because the five mean temperatures (elevations) compared were all planted the same day and have the same daylength. That chapter shows that the T value which is optimal for DTF is also optimal (or near optimal) for (maximizes) the rate of accumulation of yield.

Fully demonstrated in this chapter is that the daylength × temperature interaction effect cannot be elucidated without elucidating the effects by (the gene actions of) the genotype(s). Understanding the gene action is required because this gene action implements the responses to the daylength and to the temperature. The overall effect by temperature is the summed integration of all its Q_{10} effects on the gene (enzymatic) activities that implement the genetic instructions that control the DTF.

For many crops [2,6,15], it has been shown that the activities of but a few "photoperiod genes" are modulated by the daylength. In contrast, temperature modulates the activity of these few photoperiod genes [1,2,7–10,13–15] (Chapters 44 and 45) plus the activities of most, if not all, other genes.

C. Interpreting Genotype for Sensitivity to Photoperiod × Daylength × Temperature Interaction Effects on DTF

1. Multiple Causes of the G × E Interaction Effect on DTF

AMMI analysis divided (Table 3) the total sums of squares for the G × E interaction among six different physiological-genetic statistically significant causes of a G × E interaction effect on DTF. As presented in the interpretation of the largest G × E effect that follows, it is not readily obvious that the sums of squares allocated to this largest, and also to each of the five progressively smaller physiological genetic causes of a G × E effect, is the summation of an individual G × E interaction effect caused by every one of the 630 treatments entered into the AMMI analysis. The 630 treatments arise from growing all 42 genotypes in all 15 environments. Thus with 630 treatments and six statistically significant physiological genetic causes of a G × E interaction effect, the total G × E interaction effect on DTF is the summation of 6 × 630 = 3780 individual G × E interaction effects; all 3780 influence the pattern of the variation in DTF across the 630 treatments.

Each G × E effect by one of the 630 treatments (one of the 42 genotypes grown in one of the 15 environments) for each of the six physiological genetic causes is the product of (the contribution by the genotype to that physiological-genetic cause of a G × E interaction effect) multiplied by (the contribution by the environment to that G × E effect). All levels of contribution (except zero) have either a plus or a minus sign (Figures 4 to 6, Tables 1 and 2). Therefore, most of the 630 individual G × E interaction effects also have a plus or a minus sign.

An individual treatment's G × E interaction effect on DTF can be calculated. It is positive (increases the DTF) whenever both the contribution to this G × E effect by the specific genotype and the contribution by the specific environment are positive or else both are negative (see Chapter 42). A treatment's G × E effect is negative (decreases DTF) whenever the contributions to it by the genotype and environment have opposite sign (i.e., if either is positive while that of the other is negative).

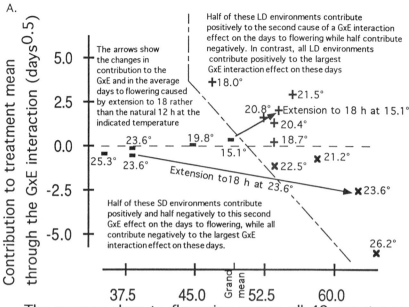

The average days to flowering across all 42 genotypes, which is the main effect by the environment on the days to flowering of each treatment.

Figure 5 (A) Contribution to each treatment's days to flowering through each environment's main effect plotted against that environment's contribution to the second-largest cause of a genotype × environment interaction effect on these days. (B) Contribution to each treatment's days to flowering through each genotype's main effect plotted against that genotype's contribution to the second-largest cause of a genotype × environment interaction effect on the days. A treatment is one of these 42 genotypes grown in one of the 15 environments. Symbols for the environments and genotypes are as described in Figure 4.

2. Patterns of Change for the Largest G × E Interaction Effect

The pattern of change of DTF across the genotypes is that the more sensitive the genotype is to photoperiod, the larger and more positive is its contribution to the largest physiological genetic source of a G × E interaction effect on DTF (Figure 4b). Correlation between this contribution by the genotypes and their independently derived scores for sensitivity to photoperiod is 0.82 (Table 5). This high positive correlation indicates that the less sensitive the genotype is to the photoperiod, the more negative is its contribution to this largest G × E effect. Stated inversely, the less insensitive (more photoperiod sensitive) the genotype, the more positive its contribution.

The pattern of change across the environments follows. All five SD environments contribute negatively to the largest G × E effect, while eight of the 10 LD environments contribute positively (Figure 4a). The two other LD environments have contributions near zero, which are intermediate to the contributions by the other eight LD environments and the contributions by the five SD environments.

3. Calculation of the G × E Interaction Effect on DTF

As described in Section IV.C.2 and discussed mathematically in Chapter 42, the individual G × E interaction effect for just one of the 630 treatments is the product of the contribution to this G × E by the specific genotype (the unit is days$^{0.5}$) multiplied by the contribution by the specific environment (this unit is also days $^{0.5}$). Therefore, a highly photoperiod-sensitive

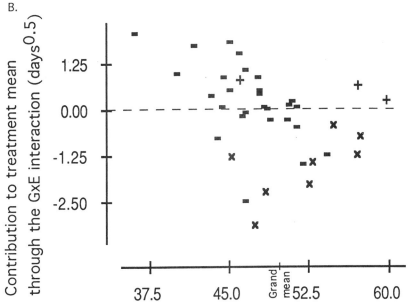

B.

The average days to flowering across all 15 environments, which is the main effect by the genotype on the days to flowering of each treatment.

genotype grown in an LD environment causes a positive G × E interaction effect on the DTF, because the contribution by the genotype (Figure 4b) and by the environment (Figure 4a) are both positive. A positive G × E effect on the DTF causes an enlargement of the DTF expressed in days (see Chapter 42). This calculated individual G × E effect agrees with long-time understanding of the photoperiod response; a photoperiod-sensitive genotype of a short-day species interacts with an LD environment to delay the DTF. On the contrary, an insensitive genotype (Figure 4b) of a short-day species grown in an LD environment (Figure 4a) results in a negative G × E interaction effect (i.e., a decrease of DTF).

For a sensitive genotype in LD its AMMI-predicted expressed DTF is the positive days of its G × E effect plus its relatively large average days plus the large average days of the environment minus (the grand mean days of all the data). For the insensitive genotype in LD, the DTF is the negative days of the G × E effect plus the relatively large average days of the LD environment plus the relatively small average days of the genotype minus (the grand mean days of all the data).

4. Final Interpretation of the Largest G × E Interaction Effect

PHOTOPERIOD-SENSITIVE AND PHOTOPERIOD-INSENSITIVE GENOTYPES CAUSE OPPOSITE G × E INTERACTION EFFECT IF IN THE SAME ENVIRONMENT. One conclusion about the largest physiological-genetic cause of the G × E interaction follows. A photoperiodinsensitive and a photoperiod-sensitive genotype will cause opposite (earlier versus later) G × E interaction effects on DTF whenever both are grown in the same (LD or SD, or low-compared with high-temperature) environment. Clearly, the major difference in gene action that causes this G × E interaction effect is due to the relative insensitivity versus photoperiod sensitivity of the photoperiod gene(s). Differences in daylength interact with the range of

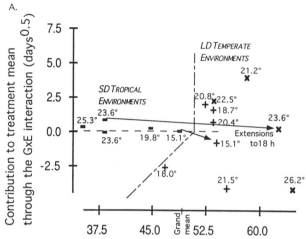

The average days to flowering across all 42 genotypes, which is the main effect by the environment on the days to flowering of each treatment.

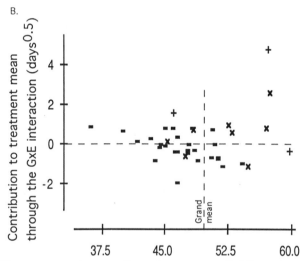

The average days to flowering across all 15 environments, which is the main effect by the genotype on the days to flowering of each treatment.

Figure 6 (A) Contribution to each treatment's days to flowering through each environment's main effect plotted against that environment's contribution to the third-largest cause of a genotype × environment interaction effect on the days. (B) Contribution to each treatment's days to flowering through each genotype's main effect plotted against that genotype's contribution to the third cause of a genotype × environment interaction effect on the days. A treatment is one of these 42 genotypes grown in one of the 15 environments. Symbols for the environments and genotypes are as described in Figure 4.

intensity of the photoperiod gene activity to cause this largest G × E interaction effect on the DTF.

A second conclusion is that a smaller environmental effect is apparent within this daylength-modulated largest G × E effect on the DTF. In LD, the Q_{10} effect by a higher temperature on the low level of photoperiod gene activity expressed by an insensitive genotype causes it to flower earlier, but that higher temperature amplifies the photoperiod gene activity of a sensitive genotype to cause later (delayed) flowering. In SD, a higher temperature causes the insensitive genotype to flower relatively and slightly earlier, and because no or reduced photoperiod gene activity occurs under the shorter daylength, this same higher temperature may also cause the sensitive genotype to flower relatively earlier (rather than later).

A third conclusion is: To elucidate the daylength × temperature interaction effect on DTF, one essential step is to know the variation in the genotype and to understand the effect of this genotypic variation. This is required because it is the gene activity that implements the response to any environmental factor. Therefore, perhaps the most important of all the conclusions in this chapter follows. Understanding the environmentally modulated variation(s) of the gene action(s) is absolutely essential to explaining any G × E interaction effect.

POSITIVE G × E INTERACTION EFFECT ON DTF ENLARGES THE MAIN EFFECT ON DTF WHILE NEGATIVE G × E INTERACTION DECREASES IT. A series of further conclusions about the graph for the largest G × E interaction effect follows. When a photoperiod-sensitive genotype is growing in an LD environment that has a higher temperature than another, the higher temperature both enlarges the contribution to the G × E interaction effect and the average days to flowering of the genotype (Figure 4; Table 1). The simultaneous enlargements are reflected by the contribution of the environments having correlation with their average days of 0.73 (Table 4). This average days is enlarged also, as the contribution to the G × E interaction by the genotype is of larger positive rather than negative magnitude (Figure 4; Tables 1 and 3) (i.e., as the genotype is more sensitive to the photoperiod). The correlation between the contribution by the genotype and the independently derived score of the genotypes for sensitivity to photoperiod is 0.82 across all 15 environments, and the correlation with the main effect is 0.59 (Table 5).

The largest G × E interaction effect causes 20% of the total variation in DTF across the 15 environments × 42 genotypes (630 treatments) (Table 1). This compares with the main effect × the 15 environments, causing 34.8% of the total variation in DTF, and the main effect × the 42 genotypes, causing 31.7%.

On moving from consideration of the largest physiological-genetic cause of a G × E interaction effect to considering the progressively smaller physiological genetic causes, it has to be recognized that the average DTF for each environment and genotype is identical on the AMMI graph for every statistically significant physiological genetic cause of a G × E interaction effect (Figures 4 to 6). These main effects are identical because for an environment this average DTF is the consequence of every effect by that total environment on every genotype. Further, this average incorporates an interaction effect by that environment with every one of the 42 genotypes tested. Similarly, for a genotype its main effect is its average across every environment, which includes the G × E effect of that genotype with every environment tested. Therefore, a preponderance in a trial of sensitive genotypes and/or of LD environments will enlarge all the main effects. Preponderance of insensitive genotypes and/or SD environments will lower the main effects. Consequently, the quantitative number for each average DTF and for each contribution to each specific physiological genetic cause of the total G × E interaction effect on DTF applies specifically and only to the analyzed subset of genotypes grown in only the analyzed subset of environments. This does not limit comparison of these numbers with the contributions and main effects derived from all other subsets of genotypes and environments, however.

For other subsets that include some previously tested genotypes, the previously compared relatively higher-than versus lower-than averages will maintain the same relationships, and the contributions by the genotypes will retain the same more-negative-than versus more-positive-than relationships.

A subset of genotypes is more repeatable than a subset of environments. Fluctuating climatic conditions in the field cause virtual impossibility to repeat an identical subset of field environments. Nevertheless, it remains possible to compare the similar and dissimilar environments.

5. Interpretation of the Second Largest G × E Interaction Effect

INTERPRETATION IS BASED ON PRIOR UNDERSTANDING OF SIMULTANEOUS INCREASES AND DECREASES OF DTF BY EVERY TEMPERATURE. Interpretation of the second largest G × E interaction effect on DTF is based on the simultaneous effect of temperature on both reproductive and vegetative development. These are summarized in Section I.A.4. The concepts are based on previously published [8] data which are presented in Figure 3.

The first concepts are that every higher temperature amplifies any photoperiod gene activity, and this enhanced photoperiod gene activity delays the node to flower, which results in a lower rate (1/DTF) of the development toward flowering (i.e., in more days to flowering). Additional concepts are that every temperature simultaneously modulates the rate of the vegetative development, which is quantifiable (Figure 3) as the average number of days needed to develop each node on the plant. The average rate of this vegetative development is 1/average days per node.

It is concluded (from Figure 3 as reviewed in Section I.A.4) that simultaneous modulations by every temperature of both the rate of vegetative development (the days per node) and the rate of reproductive development (the node to flowering) result in every change of temperature causing a change toward smaller DTF plus a simultaneous opposite change toward larger DTF [8]. Consequently, the expressed change in DTF between two temperatures is the larger minus the smaller of the opposing increases and decreases of the DTF.

Section I.A.4 also presents the concept that the simultaneous increases and decreases of DTF caused by every temperature interact for every genotype to cause a U-shaped curve of its DTF across a properly centered and sufficient range of temperatures. This U-shaped response to temperature has been documented repeatedly [1,5,8,13,14] in environments with constant daylength (Figures 1 and 2). Indicated also is that the smallest DTF along the U-shaped curve occurs at and identifies the optimum temperature for flowering. The optimum temperature is interpreted [8] (Section I.A) as that temperature at which the Q_{10} effects on the days per node (rate of vegetative development) and on the node to flower (rate of reproductive development) cause increases and decreases of the DTF which are equal and opposite, so that they cancel each other (Figure 3). A further interpretation is that the optimum temperature is lowered for a genotype if the daylength is made more delaying, and this optimum is lowest for genotypes that are the most highly sensitive to photoperiod and highest for genotypes that are the most insensitive to the photoperiod (Figure 2).

A still further interpretation is that at all temperatures that are below the optimum for flowering, the larger of the contrasting increases versus decreases of the DTF is due to the change in the days needed to develop a node (i.e., is through the rate of vegetative development). In contrast (Figure 3), at all temperatures that are above the optimum for flowering, the larger of the increase versus decrease of the DTF is through the node to flower [i.e., through the rate of development of the reproductive (yield) organs].

AMMI ANALYSIS SUPPORTS SIMULTANEOUS INCREASES AND DECREASES OF DTF BY EVERY TEMPERATURE. All 15 environments of this study involve a different combination of temperature and daylength. Therefore, the previously observed U-shaped response to temperature of the DTF under the same daylength is not demonstrable for any of the 42 genotypes of this study. However, eight of the nine genotypes for which it is shown (Figure 2) using previously published data are included in this study. Further, the results of AMMI analysis do show (Figure 5) that the second-largest cause of the G × E interaction is interpretable as resulting from simultaneously occurring increases versus decreases of the DTF of every genotype by every temperature. Further, the AMMI results show that delay of the DTF by photoperiod gene activity at high temperature is the predominant cause of late DTF for the more photoperiod-sensitive genotypes, while slowed rate of vegetative growth is the predominant cause of late DTF for the less sensitive genotypes.

All five SD environments cause a near-zero contribution to the second largest G × E effect interaction effect on the DTF (Figure 5a). However, two of the five cause slightly positive contributions, while three cause slightly negative contributions. Similarly, approximately half (six) of 10 LD environments cause positive contributions, while four cause negative contributions. This nearly equal number of positive and negative contributions for both the SD and LD environments indicates that short versus long daylength is not (Figure 5a) the major environmental cause of this second source of G × E interaction as it is for the primary cause of a G × E interaction effect on the DTF (Figure 4a).

TEMPERATURE-CAUSED G × E EFFECTS ON DTF ARE REVERSED FOR THE SD VERSUS LD ENVIRONMENTS. The range of contributions to the second cause of G × E interaction across the LD environments is from –6.06 to 3.60 days$^{0.5}$, a total of 9.66 days$^{0.5}$. This contrasts with the range of only 0.70 day$^{0.5}$ (from –0.08 to 0.32 day$^{0.5}$) across the SD environments (Table 1; compare Figure 5a). Therefore, the second source of G × E interaction occurs mostly across the LD environments, these being the environments capable of promoting photoperiod gene activity. Across the SD environments that prevent photoperiod gene activity, the second source of G × E interaction causes relatively only small effects on the DTF (Figure 5a).

Contributions by the environment correlate closely with the temperature (Table 4), being negative across the SD ($r = –0.93$), LD (–0.78), and all 15 environments (–0.64). Thus across just the SD, just the LD environments, or all the environments, the contribution becomes more negative as the temperature rises. These large negative correlations indicate that temperature is the major environmental cause of the second largest physiological-genetic source of a G × E interaction effect on DTF. The contributions to this G × E by the environments have only a weak correlation with the photoperiod, 0.43 across the five SD environments, 0.04 across the 10 LD environments, and 0.07 across all 15 environments (Table 4).

TEMPERATURE-CAUSED G × E EFFECTS ON DTF ARE OPPOSITE FOR THE LESS VERSUS THE MOST PHOTOPERIOD-SENSITIVE GENOTYPES. The contributions to the second-largest cause of a G × E effect on DTF are negatively correlated with the average DTF of the more insensitive genotypes ($r = –0.58$) (Figure 5b, Table 5), but this correlation is positive ($r = 0.68$) across the more photoperiod-sensitive genotypes and is 0.09 across all 15 genotypes. Negative versus positive correlations for the lower and higher levels of photoperiod sensitivity contrast with always-positive correlations between the contributions and the average DTF of these ranges of photoperiod sensitivity for the largest cause of a G × E interaction effect on DTF. These positive correlations are 0.89 across all the genotypes, 0.52 across the less photoperiod-sensitive genotypes, and 0.90 across just the more photoperiod-sensitive genotypes (Figure 4b, Table 5).

In summary, the average DTF of the more photoperiod-sensitive genotypes becomes larger as the genotype's contribution to the second cause of a G × E effect on the DTF changes toward being more positive (Figure 5b), whereas the average DTF of the photoperiod-insensitive genotype becomes larger as their contribution becomes more negative. Thus a more positive contribution to both the largest and the second-largest G × E interaction effect enlarges the averaged DTF of the more photoperiod-sensitive genotypes, while the average of the less photoperiod-sensitive genotypes is enlarged by positive contribution to the largest G × E effect but is decreased by positive contribution to the second-largest G × E effect. These opposite effects by the same temperature indicate that the genetic basis (gene action) that causes the largest delay of DTF of the less photoperiod-sensitive genotypes differs from the gene action that causes the later DTF of the more photoperiod-sensitive genotypes.

OVERALL INTERPRETATION OF THE SECOND CAUSE OF A G × E INTERACTION EFFECT ON THE DAYS TO FLOWERING. Interpretation of the second source of a G × E effect on DTF requires accounting for the following from its AMMI graph. First, the LD environments with the highest temperatures cause the largest negative contribution to this G × E effect, while the LD environments with the lowest temperatures cause the largest positive contribution (Figure 5a). Second, most of the less photoperiod-sensitive genotypes cause a positive contribution, but their contributions grade toward being zero and then negative as the genotype changes toward intermediate levels of sensitivity to photoperiod; while most of the more photoperiod-sensitive genotypes cause negative contributions to this second-largest source of G × E interaction, but their contributions grade toward being zero and then positive as the sensitivity of the genotype changes toward the intermediate level (Figure 5b).

Multiplying the large positive contributions to the second source of G × E interaction of the LD environments with lowest temperatures (Figure 5a) times the large positive contributions (Figure 5b) of the genotypes that have the lowest sensitivity to photoperiod (from Table 1 and Figure 4b) indicates a large positive G × E interaction effect on the DTF. We infer that this G × E caused delay of the DTF of the less photoperiod-sensitive genotypes occurs when temperatures are low, because more days are required for completion of the vegetative development of each additional node. These insensitive genotypes have minimal potential for delay of the DTF through photoperiod gene activity (compare Figure 3); therefore, even the LD environments with the higher temperatures do not delay the node to flowering and DTF of the most insensitive genotypes beyond the delay due to the slowed rate of the vegetative development of the nodes. Supporting evidence is that for the insensitive cultivars Redkloud and Porillo Sintetico the 15.1°C temperature with incandescent-light extension to 18 h causes the same intermediate DTF that occurs under that same 15.1°C with a natural 12-h daylength (Table 6).

In contrast to the increases in DTF for the less photoperiod-sensitive genotypes through only lowering of the rate of vegetative development by a low temperature, multiplying the large positive contributions of these same low-temperature LD environments times the negative contributions of the genotypes with the higher sensitivities to photoperiod indicates a negative G × E interaction effect on the DTF. We interpret the indicated decrease of the DTF as being due to reduction of the photoperiod gene activity by the Q_{10} effect of the lower temperature. However, the delay of DTF due to this photoperiod gene activity remains larger than the delay of the DTF due to the Q_{10}-caused reduction of the rate of vegetative development (i.e., due to the increase in the days needed to develop a node). Consequently, the expressed DTF of the more photoperiod-sensitive genotypes is larger in almost all LD environments than it is in the SD environment with the lowest temperature (Table 6).

In yet a further contrast, multiplying the large negative contributions of the LD environments with the highest temperatures (Figure 5a) times the also large negative contributions of the

Table 6 Days to Flowering of One Photoperiod-Insensitive and Three Photoperiod-Sensitive Genotypes of *Phaseolus vulgaris* in Five SD and 10 LD Environments[a]

	Daylengths and Temperatures of the Environments														
	Short Daylengths and Temperatures					Long Daylengths and Temperatures									
Symbol for environments	SD5	SD4	SD3	SD2	SD1	LD1	LD2	LD3	LD4	LD5	LD6	LD7	LD8	LD9	LD10
Mean temperature (°C)	15.1	19.8	25.4	23.6	23.6	26.2	23.6	18.0	21.2	21.5	22.5	20.8	18.7	20.4	15.1
Daylength (h)	12.4	14.8	11.3	11.0	12.0	14.8	18.0	15.7	14.7	14.9	14.8	14.7	14.8	15.0	18.0
Genotype[b]	Treatments (combinations of a genotype growing in an environment) and their days to flowering														
Environment Redkloud [1]	SD5 43.7	SD4 34.7	SD3 29.7	SD2 31	SD1 32.7	LD2 32	LD1 32.3	LD3 34	LD4 35.3	LD5 37	LD6 37	LD7 38.7	LD8 41	LD9 41	LD10 42.7
Environment Redkloud [6]	SD5 43.7	SD4 38.7	SD1 33.7	SD3 30.3	SD2 30.7	LD3 35.7	LD10 44.3	LD7 44.5	LD9 45	LD8 45.2	LD5 49.7	LD4 52.8	LD6 52.8	LD1 62.7	LD2 70.7
Environment Linea 17 [6]	SD5 45	SD4 39	SD1 36.3	SD2 35.3	SD3 35	LD3 54.3	LD8 56.2	LD5 58	LD7 59	LD9 61	LD10 63.3	LD6 69.7	LD1 73	LD4 74.7	LD2 95.5
Environment Gloriabamba [8]	SD5 49.7	SD4 42	SD1 37	SD2 35	SD3 35	LD10 72.7	D1 77.5	LD6 79.5	LD8 86	LD9 96.7	LD7 99	LD4 102	LD5 103	LD3 109	LD2 112
Environment Environmental mean	SD5 48.9	SD4 44.9	SD1 38.4	PR4 38.3	PR3 35.3	LD3 46.7	LD7 52.3	LD8 53.5	LD9 53.5	LD6 53.7	LD10 54	LD5 55.9	LD4 58.3	LD2 62.7	LD1 64.5

[a]Earlier flowering in response to a higher temperature in SD but later flowering in LD is shown by all genotypes.
[b]The larger the number in parentheses, the more photoperiod sensitive the genotype (see Table 2).

genotypes with the highest levels of sensitivity to photoperiod (Figure 5b) indicates a large positive G × E interaction effect. We interpret these large G × E-interaction-caused delays of the DTF to result because the most photoperiod-sensitive genotypes most strongly express their potential for a high level of photoperiod gene activity in the environments with both LD and the highest growing temperatures. This photoperiod gene activity delays the node to flowering, and thereby the DTF, more than the high temperatures decrease the DTF through the need for fewer days to develop a node. For some of the highly sensitive genotypes, the delay of DTF by LD, combined with high temperature, is more than 100% larger than the DTF in SD with the lowest temperature (Table 6). The smallest DTF for every genotype is in an SD environment combined with a high temperature (Table 6), which are the optimal daylength temperatures and cause the most rapid rate of development to flowering.

SUMMARY OF EFFECTS ON DTF THROUGH INTERACTION OF THE LARGEST AND SECOND-LARGEST CAUSES OF A G × E INTERACTION EFFECT. The largest G × E interaction effect on the DTF results because the LD environments modulate increasingly larger photoperiod gene activity by insensitive to moderately sensitive to highly photoperiod-sensitive genotypes, whereas modulation by the SD environments allows little, if any, expression of the photoperiod gene activity (Figure 4). From within the AMMI graph (Figure 4) for this largest daylength-modulated photoperiod-gene-directed G × E effect on the DTF, we inferred (Section III.D) that a higher temperature causes a small decrease in the average DTF within the SD environments, accompanied by a comparatively small change in the contribution to the G × E interaction, but that within the LD environments a higher temperature causes a later average DTF. Here we infer that a comparatively small temperature effect within the largest G × E interaction effect is actually the second-largest G × E interaction effect, for which the AMMI graph (Figure 5) indicates that the gene activity is modulated by the temperature.

During our interpretation of the largest G × E interaction effect on the DTF (Section III.D), we searched for and paid special attention to the modulation by the temperature that is observable within the larger modulation of the DTF by the daylength. We did so because our accumulated prior experience (reviewed in Section I) has indicated that there is always a daylength × temperature interaction effect on the DTF. Summarizing the current interpretation, the daylength modulation of the DTF through the largest G × E interaction effect and the temperature modulation of the DTF through the second-largest G × E effect interactively control the DTF.

One genetic basis for the second-largest G × E interaction effect is the same variation in photoperiod gene activity due to the range of sensitivity to photoperiod that is the genetic cause of the largest G × E effect. However, the modulation of the photoperiod gene activity that is one genetic basis for the second-largest G × E interaction effect is by the temperature rather than by the daylength.

There is a second gene action which is a partial cause of the second-largest G × E interaction effect. This simultaneously occurring second gene action is modulated by the same temperature that modulates the photoperiod gene activity; it is the activity of any genes that control the time needed to complete the vegetative development of a node. Enhancement by higher temperature of this gene activity means that fewer days are needed to develop a node, which tends to cause a smaller DTF, as that temperature simultaneously also tends to cause a later DTF through its enhancement of the photoperiod gene activity and the resultant delayed node to flowering. Consequently, the expressed effect on the DTF of a genotype due to a different temperature but the same daylength results from just that part of the larger of the increase or decrease in the DTF that is not canceled by the smaller and opposite decrease or increase (Figure 3, compare Ref. 8). We are not aware of specific genes with demonstrated control over the rate of vegetative development of the nodes that have been identified by others. We did find that lack of photoperiod gene activity slightly accelerates the rate of node development [15], presumably because a larger proportion of the photosynthate is partitioned to the vegetative growth and the vegetative development can therefore occur faster.

A summarized interpretation for the last four paragraphs follows. The largest G × E interaction effect on DTF results because the daylength modulates the photoperiod gene activity and thereby accounts for 59.6% of the environmentally modulated variation in the DTF (of the total G × E interaction effect on the DTF), while the second-largest G × E effect accounts for 13.9% of the modulated variation in DTF and results because temperature strongly modulates the photoperiod gene activity of the most photoperiod-sensitive genotypes, moderately modulates the intermediate activity for the moderately sensitive genotypes, less strongly modulates the smaller photoperiod gene activity of the less sensitive genotypes, and simultaneously modulates the activity of all the genes that control the days needed for a node to develop on the plant.

The node to flowering and the days needed to develop a node were not measured in this study. Therefore, our interpretation of the second cause of a G × E interaction source as being due to simultaneous Q_{10} effects by temperature on the rate of the reproductive development (node to flowering) and on the rate of the vegetative development (days needed to develop a node) is entirely dependent on previously published [8] external data. These data are presented again in Figures 1 to 3 and are reviewed in Section I of this chapter.

The following supports the foregoing interpretations. Table 6 presents 15 different DTFs predicted by the AMMI results for each of the five genotypes when grown in every one of the 15 environments. The predictions are calculated as discussed in Chapter 42. The predictions across the 15 environments show a U-shaped curve of the DTF of every bean genotype.

6. Interpretation of the Third-Largest G × E Interaction Effect

The third-largest cause of a G × E interaction effect is nearly as large (11.3% of the total G × E interaction effect) as the second (13.6%), so interpreting it should have similar merit. The AMMI graphs (Figure 3) for it are presented, but neither our prior experience nor the current data facilitate an adequate interpretation of this G × E interaction effect. However, the correlation data suggest a partial explanation.

The contributions to this third-largest cause of a G × E interaction effect have a correlation of only −0.08 to 0.15 with temperature and only −0.14 to 0.39 with photoperiod (Table 4). The contributions are consistently correlated with only the vegetativeness of the plant habit (Table 6). As defined (Table 1), the vegetativeness is increased by a larger number of nodes on the plant shoots plus larger numbers of axillary branches. The correlation between increasing vegetativeness and the contribution is −0.51 across all 42 genotypes, 0.49 across only the 27 less sensitive genotypes, and 0.52 across the 15 more sensitive genotypes (Table 5). These correlations imply genetic control over the vegetativeness of the plant habit as causal of the third cause of a G × E interaction effect.

The only larger correlation is 0.84 between these contributions and contributions to the second cause of the G × E interaction (Table 6). This high correlation occurs across just the more insensitive genotypes, however. Across the more sensitive genotypes, this correlation is −0.25, and it is zero across all 42 genotypes. The 0.84 correlation suggests a physiological genetic basis for the third G × E effect that is similar to that of the second G × E effect. It is reasonable to think that the vegetativeness of the plant habit, the rate of nodal development, and the node to flowering all have some common or overlapping genetic controls. Additional understanding of the genetic control and of the plant habit is required to interpret more fully this third cause of a G × E interaction effect on the DTF [16,17].

7. Additional Uninterpreted G × E Interaction Effects

No attempt is made to interpret the fourth, fifth, or sixth statistically significant physiological-genetic components of progressively smaller G × E interaction effects on DTF because, respectively, they account for only 1.9, 1.1, and 0.7% of the total variation in the DTF. Also, it is progressively more difficult to interpret the progressively smaller G × E interaction effects.

V. SUMMARY AND CONCLUSIONS

A. Days to Flowering Is Controlled by the Metabolic Network

The joint interpretation of the largest and second-largest G × E interaction effects on DTF requires recognition that there is a continuum of the daylengths of the environments, a continuum of the temperatures of these environments, and a continuum of the genotypic sensitivity to photoperiod. Jointly, the two quantitatively variable environmental factors and single quantitatively variable genetic factor result in a quantitative continuum of the photoperiod gene activity and its effect on delay of the node to flowering (i.e., on the rate of reproductive development). It must be recognized additionally that the same temperature also modulates the rate of the vegetative development (i.e., the days needed to develop a node, which is another part of the joint control over the DTF).

Consideration of the following adds to interpretation of the joint control. The modulations of the photoperiod gene activity by the daylength and temperature control the quantitative proportion of the available photosynthate that is partitioned to the reproductive organs and is thereby removed from support of the vegetative growth (see Chapters 41, 44, and 45). Therefore, the rate of the vegetative development of the nodes is partially controlled by the photoperiod gene activity. We are not aware of other specific genes that control this rate of development, but they must exist.

The AMMI analysis results (Table 3) indicate that the main effect due to differences among the environments accounts for 34.8% of the variation in DTF across the 15 environments and 42 genotypes. The main effect due to differences among the genotypes accounts for 31.7%, and the total G × E interaction effect accounts for 33.5%.

The 33.5% of the total variation in DTF caused by the G × E interaction is all due to modulation of the gene actions by the factors of the environment. Of this environmentally modulated 33.5% of the total variation, 59.6% is due to modulation by daylength of the quantitatively different photoperiod gene activity of the genotypes, and 13.6% is due to modulation of this photoperiod gene activity by the temperature and by simultaneous modulation by that temperature of the rate (days to develop a node) of vegetative development. Differences among the genotypes in plant habit as represented by the number of nodes on the shoots and the number of branches are interpreted as the genetic basis for a further 11.3% of environmental modulation of the DTF. Three additional environmentally modulated and gene-directed G × E effects are identified as statistically significant controls over the DTF, even though they are responsible for only 1.9, 1.1, and 0.7% of the total variation in the DTF. For these we have not identified either the modulating environmental factors or the causal gene activities.

Multiple controls over the DTF constitute what Trewavas [18] calls "a complex interactive network with a surprising density of interconnection but coherence in control." The coherent control is over the rate of development to flowering and the reciprocally related rate of vegetative growth and development. Trewavas states further: "The control of network behaviour is shared to some extent by all the interconnected parts and is thus a holistic character. Therefore, there is no single, metabolic, controlling element." Trewavas refers to this as a "systems view," with the control being by the "metabolic network."

B. Control Strengths Within a Metabolic Network

Trewavas [18] suggests that the systems view calls for quantitative comparison of the proportion of the total control that is due to each of the interconnected parts of the metabolic network. He labels each proportion as the control strength of that part of the system. The AMMI analysis presented herein of the DTF across 15 environments and 42 daylengths provides quantification and comparison of the control strengths. These control strengths are summarized in Section V.A, and all are listed in Table 3.

REFERENCES

1. D. H. Wallace and G. A. Enriquez, *J. Am. Soc. Hortic. Sci.*, *105*: 583 (1980).
2. D. H. Wallace, K. S. Yourstone, P. N. Masaya, and R. W. Zobel, *Theor. Appl. Genet.*, *86*: 6 (1993).
3. D. P. Coyne, *J. Am. Soc. Hortic. Sci.*, *89*: 350 (1966).
4. D. P. Coyne, *J. Am. Soc. Hortic. Sci.*, *103*: 606 (1978).
5. A. F. H. Muhammad, *Ph.D. dissertation*, Cornell University, Ithaca, N.Y. (1983).
6. D. H. Wallace, *Plant Breed. Rev.*, *3*: 21 (1985).
7. D. H. Wallace, J. P. Baudoin, J. Beaver, D. P. Coyne, D. E. Halseth, P. N. Masaya, H. M. Munger, J. R. Myers, M. Silbernagel, K. S. Yourstone, and R. W. Zobel, *Theor. Appl. Genet.*, *86*: 27 (1993).
8. D. H. Wallace, P. A. Gniffke, P. N. Masaya, and R. W. Zobel, *J. Am. Soc. Hortic. Sci.*, *116*: 534 (1991).
9. J. W. White and D. R. Laing, *Field Crops Res.*, *22*: 113 (1989).
10. K. S. Yourstone, Ph.D. thesis, Cornell University, Ithaca, N.Y. (1988).
11. R. W. Zobel, M. J. Wright, and H. Gauch, *Agron. J.*, *80*: 388 (1988).
12. H. G. Gauch, Jr., *Statistical Analysis of Regional Yield Trials: AMMI Analysis of Factorial Designs*, Elsevier, New York (1993).
13. R. D. Summerfield and E. H. Roberts, in *World Crops: Cool Season Legumes* (R. J. Summerfield, ed.), Kluwer Academic Publishers, Norwell, Mass., pp. 911–922 (1988).
14. R. J. Summerfield, S. T. Collinson, R. H. Ellis, E. H. Roberts, F. W. T. Penning de Vries, *Ann. Bot.*, *69*: 101 (1992).
15. K. S. Yourstone and D. H. Wallace, *J. Am. Soc. Hortic. Sci.*, *115*: 824 (1990).
16. J. Kornegay, J. W. White, and O. Ortiz de la Cruz, *Euphytica*, *62*: 171 (1992).
17. J. Kornegay, J. W. White, R. R. Dominguez, G. Tejada, and C. Cajiao, *Crop Sci.* (in press).
18. A. Trewavas, *Plant Cell Environ.*, *14*: 1 (1991).

44

Genotype, Temperature, and Genotype × Temperature Interaction Effects on Yield of Bean (*Phaseolus vulgaris* L.)

Donald H. Wallace
Cornell University, Ithaca, New York

Porfirio N. Masaya
Interamerican Institute for Cooperation for Agriculture, San Jose, Costa Rica

Rafael Rodríguez
Instituto de Ciencia y Tecnologia Agricolas, Guatemala City, Guatemala

Richard W. Zobel
Agricultural Research Service, U.S. Department of Agriculture, Ithaca, New York

I. INTRODUCTION

Days to flowering and maturity are phenological traits, since their durations are influenced by climate. Richards [1] states: "Crop phenology is the most important single factor influencing yield and adaptation." The research on these phenological traits of bean presented in this chapter arose from studies on the physiological genetics of crop yield that began in 1954 (Chapter 43). Progress has been summarized and interpreted periodically since 1972 [2–6]. Recent results elucidate the inherent interconnections between yield and the days to flowering and days to maturity [7–12].

Wallace et al. previously [7] (Chapters 43 and 45) used the additive main effects and multiplicative interaction effects (AMMI) analysis procedure [13] (Chapter 42) to quantify the changes in days to flowering due to the main effect by the daylength, the main effect by the environment (daylength and temperature), and the main effect by the genotype. The advantage of AMMI statistical analysis is that it also quantifies a contribution by each different genotype and by each different environment to the geneotype × environment (G × E) interaction effect [13] (Chapter 42). Based on the pattern of change of these G × E interaction effects across the genotypes and across the daylengths and temperatures, a model [10–12] was developed to explain control by the genotype, by the photoperiod, and by the temperature plus by the genotype × daylength × temperature interaction effect on the days to flowering. For seed crops, the model indicates that the effect on the days to flowering caused by daylength and temperature and by genotype results from a preceding control over the proportion of the available photosynthate that is partitioned to the reproductive organs. This proportion becomes the biomass that is the crop yield. The model indicates that this proportion controls the rate of growth of the flower buds and of the seeds, and thereby controls the rate of accumulation of the yield [11]. That is, the larger the proportion partitioned to the reproductive organs, the faster they grow and the fewer the days they use (need) to develop to flowering and to harvest maturity.

The physiological-genetic model led to a proposed yield system analysis, which is an adjunct procedure applied to yield trials [10]. The model states [6,10–12] (Chapter 41) that there are three major physiological-genetic components of the process of accumulating the yield of a seed crop. Each major component is a penultimate output from the yield system. Integration of the three major components results directly in the yield, which is the ultimate output from the system [6,10] (Chapter 41). The three major components of the overall process of accumulating seed yield are:

1. The net accumulated aerial biomass.
2. The proportion of the aerial biomass that is partitioned to the reproductive organs that constitute the yield.
3. The time the plant cultivar uses (needs) to complete components 1 and 2. That time is the days from planting to harvest maturity.

Partitioning a proportion of the accumulated biomass to the yield begins at or just before flowering, so the duration of the seedfill is the days from flowering to harvest maturity. The alternative partitioning is for the photosynthate to go to support of the growth of more nodes, leaves, axillary shoots, and shoot elongation.

A complete yield system analysis [10] (Chapter 41) measures the following for each plot of a yield trial:

1. Yield
2. Days to flowering
3. Days to harvest maturity (major component 3)
4. Aerial biomass (major component 1)

These four measures facilitate calculation of:

5. Rate per day of the accumulation of the aerial biomass
6. Rate per day to maturity of the accumulation of the yield
7. Rate per day of seed fill of the accumulation of the yield
8. Seedfill duration
9. Harvest index (major component 2)

In this chapter we present a partial yield system analysis. Focus at the time the yields were measured on the effect by the temperature (*T*) on the phenological traits of days to flowering and days to harvest maturity resulted in failure to measure the aerial biomass. The measured yield plus days to flowering and maturity facilitates calculation of the rate of yield accumulation per day to maturity and per day of seedfill plus the calculated rate of development to flowering and to maturity. These are all compared across 10 genotypes grown at five elevations with temperatures that span 15.7 to 28.3°C. The results from the phenological traits led to description in 1993 [10] of a complete yield system analysis. The only additional measurement required is of the aerial biomass accumulated on each yield trial plot.

II. RESEARCH OBJECTIVE

The objective of this chapter is to compare the effects by five temperatures (*T*) on the yield with the effects by these temperatures on the phenological traits of days to flowering and days to maturity. The comparisons strongly support Richard's [1] statement that crop phenology is one of the most important factors influencing yield and adaptation.

III. MATERIALS AND METHODS

Ten black-seeded Guatemalan bean genotypes, with adaptations to moderate to low mean T (moderate to high tropical elevations) and sensitivity or insensitivity to photoperiod, were grown in yield trials in Guatemala. The 10 genotypes (Figure 1) were planted in fields at five elevations (2200, 1786, 895, 478, and 50 m) on the same day of May 1983. Respective mean T across the growing season was 15.7, 18.0, 23.1, 25.4, and 28.3°C. At 14 to 15° north latitude, the daylength was 13 h in May, 13.3 h in late June, and 13.0 h in August. A randomized complete block with four replications was grown at each elevation. Days from planting to flowering, harvest maturity and yield were measured for each genotype in each replicate. The summarized yields are expressed as the kilograms per hectare or the kilograms per hectare per day.

Additive main effects and multiplicative interaction (AMMI) analysis [13] was used previously [7] to compare the main and G × E interaction effects on the days to flowering caused by each of the 10 genotypes and each of the five mean temperatures. Herein the effects on yield and on rate of yield accumulation are compared with the effects on days to flowering, days to maturity, and the seedfill duration. Quantification of the statistical main and interaction effects and treatment estimations were according to the procedures of Zobel et al. [14] and Zobel [15].

For a 1982 and a 1983 planting, both reported previously [7], the immediate objective was to determine if, in the field, bean expresses the U-shaped curve of days to flowering observed in response to controlled T [16]. All the mean temperatures of 1982 were at elevations where beans are grown commercially. A rise from 23.1°C to 25.7° (change of only 2.6°C) caused a slight delay in days to flowering for six of the 10 genotypes (Figure 1) [7]. For four genotypes, continued but smaller change toward earlier flowering occurred. These results suggested that a yet-higher mean growing T would reveal the U-shaped curve in response to T. To evaluate this, the 1983 plantings included an elevation of 50 m with a mean of 28.3°C.

The data reported in this chapter were collected in 1983 for days to flowering and days to harvest maturity at all five levels of T. Yields were measured at 25.4, 23.1, 18.0, and 15.7°C

Numerical identity in text and Figure 2	Name of the cultivar	Symbolic identity in Figure 2	Sensitivity to photoperiod
1	Jutiapa	—□—	1
2	80-13	·······◇········	1
3	78-12	·····○·····	1
4	80-11	----△-----	1
5	Rabia de Gato	---⊞---	1
6	Negro Pacoc	····◆·····	2
7	Pata de Zopa	～～◈～～	4
8	San Martin	--▽--	7
9	Negro Patzicia	---◪----	4
10	Turrialba	——◆——	1

Figure 1 Numerical identities in Figure 2, symbolic identities in Figure 3, and names and sensitivities to photoperiod of 10 Guatemalan cultivars of bean (*Phaseolus vulgaris* L.). The higher the number, the more sensitive the genotype is to the photoperiod.

but were so low at 28.3° for all 10 genotypes that the yield was not measured. It was reported previously [7] that all 10 genotypes did express a U-shaped curve of days to flowering across the five mean T's of this study.

After AMMI analysis of the yields and rates of yield accumulation at the mean growing T of 15.7, 18.0, 23.1, and 25.4°C, two AMMI analyses were made across these four plus 28.3°C. One analysis assumed zero yield at 28.3°C for all genotypes. The second incorporated yields measured in 1984 for three. The other seven genotypes were treated as having missing data, using the AMMI estimation maximization routine of Gauch and Zobel [17].

The nine additional traits for which the AMMI analyses are compared with the AMMI analyses for yield are:

1. Rate of yield accumulation per day to maturity (Table 1B, E, H)
2. Rate of yield accumulation per day of seedfill (Table 1C, F, I)
    plus the phenological traits of
3. Node to flower (Table 1J)
4. Days to flowering (Table 1M)
5. Days to harvest maturity (Table 1K, N)
6. Days of seedfill (Table 1L, O)
7. Growing degree days to flowering (Table 1P)
8. Growing degree days to harvest maturity (Table 1Q)
9. Yield per growing degree day (Table 1R)

IV. MEASURED RESULTS AND THEIR AMMI ANALYSES

A. Measured and Estimated Yields

At 28.3°C in 1983 there was little yield for any genotype, so the focus in 1983 [7] (Chapter 43) being on the days to flowering, the yields were not harvested at this lowest elevation with its highest of the five mean T's. The estimation maximization (EM) algorithm [17] of the AMMI analysis was used to derive after-the-fact estimates of expected yields at 28.3°C. The estimation used yields of genotype 3, 5, and 8 that were measured in 1984 at the 28.3°C location. These were incorporated into the AMMI analysis as if they were 1983 data, and missing data were assumed for the other seven genotypes. The AMMI estimation maximization routine [13,17] uses both the additive main effects and the multiplicative G × E interaction effects (Chapter 42) to estimate the missing data. For another alternative the yields at 28.3°C were entered as zero for all genotypes. The difference in result is that the estimated average yield at 28.3°C is 60 (data not shown) rather than zero (Figure 2A). Both zero and 60 at 28.3°C are markedly below the average yield of 657 at 25.4°C in 1983, this being a lower actually measured yield per hectare than the progressively higher yield at 23.1, 18.0, and 15.7°C (Figure 2D).

One change in the biplot graph for assumed zero yield at 28.3°C (Figure 2A, B, and C) as compared with just the actually measured yields at the four lower T's (Figure 2D, E, and F) is a lower grand mean yield and rate of accumulation of the yield. The second change is a lower average for each genotype. Relatively, the averages and contributions to the G × E interaction provide the same interpretation.

B. Main and G × E Effects on Yield

For all three analyses of yield, the partitioned sums of squares shows the G × E effect on the variation in yield to be about three times larger than the main effect by the genotypes (Table 2; Table 1A, D, and G). In turn, the main effect by T (elevations) is two to five times larger

Table 1 Percentages of the Variation in Days to Flowering Among the Treatments That AMMI Analysis Allocates to Cause by the Genotype, Daylength, and the G × E Interaction, and the Percentages Allocated to the Two Largest Physiological-Genetic Causes (IPCA-1 and IPCA-2) of the Total G × E Interaction Effect[a,b]

Analyzed data at 15.7, 18.0, 23.1, and 25.4°C

A — Yield	B — Yield/day to maturity	C — Yield/day of seedfill
Geno 8.9%***	Geno 10.3%ns	Geno 13.2%**
Env 64.0%***	Env 38.7%***	Env. 43.2%***
G × E 27.1%***	G × E 51.0%***	G × E 43.6%***
IPCA-1 76.8%***	IPCA-1 80.2%***	PCA-1 71.4%***
IPCA-2 3.9%ns	IPCA-2 13.5%ns	IPCA-2 21.2%ns

Analyzed yields assuming zero yield at 28.3°C

D — Yield	E — Yield/day to maturity	F — Yield/day of seedfill
Geno 3.4%**	Geno 3.2%ns	Geno 4.3%**
Env 82.9%***	Env 76.5%***	Env 76.8%***
G × E 13.7%***	G × E 20.3%***	G × E 18.9%***
IPCA-1 72.2%***	IPCA-1 78.2%***	IPCA-1 69.5%***
IPCA-2 5.4%*	IPCA-2 13.9%ns	IPCA-2 21.0%*

Analyzed data incorporated 1984 yields at 28.3°C for three genotypes

G — Estimated yield	H — Est. Yield/day to maturity	I — Est. Yield/day of seedfill
Geno 4.6%**	Geno 4.1%*	Geno 18.9%***
Env 81.6%***	Env 75.2%***	Env 73.6%***
G × E 13.8%***	G × E 20.7%***	G × E 19.1%***
IPCA-1 74.6%***	IPCA-1 79.0%***	IPCA-1 70.9%***
IPCA-2 20.4%ns	IPCA-2 14.2%ns	IPCA-2 21.5%*

Table 1 (*Continued*)

Analysis of phenological data for five mean temperatures

J	K	L
Node to flowering	Days to harvest maturity	Seedfill duration
Geno 21.7%**	Geno 3.2%*	Geno 0.1%**
Env 60.2%***	Env 95.3%***	Env 96.3%***
G × E 18.2%***	G × E 1.4%***	G × E 2.8%***
IPCA-1 40.0%*	IPCA-1 63.9%***	IPCA-1 44.8%***
IPCA-2 37.0%*	IPCA-2 18.7%***	IPCA-2 30.8%***

Analysis of phenological data for four mean temperatures

M	N	O
Days to flowering	Days to harvest maturity	Seedfill duration
Geno 10.6%**	Geno 3.9%*	Geno 1.4%**
Env 79.9%***	Env 95.2%***	Env 96.3%***
G × E 9.8%***	G × E 1.0%***	G × E 2.3%***
IPCA-1 93.8%***	IPCA-1 68.8%***	IPCA-1 71.4%***
IPCA-2 0.7%***	IPCA-2 24.6%***	IPCA-2 28.3%***

Analysis of phenological growing degree days

P	Q	R
GDDs to flowering	GDDs to harvest maturity	GDDs of seedfill
Geno 14.6%**	Geno 18.2%*	Geno 6.2%**
Env 62.6%***	Env 75.8%***	Env 83.3%***
G × E 22.6%***	G × E 6.0%***	G × E 10.5%***
IPCA-1 77.4%***	IPCA-1 7.7%***	IPCA-1 61.9%***
IPCA-2 19.5%***	IPCA-2 27.0%***	IPCA-2 32.5%***

[a]The percentages for genotype, environment, and the G × E interaction add up to 100%, except for rounding error. On the contrary, the percentages for IPCA-1 and IPCA-2 add up to only that proportion of the percentage of the variation due to the G × E interaction effect of which its two largest sources are jointly causal. The difference between this total proportion and 100% is the residual G × E, shown in Table 2.

[b]*, **, ***: significant at 0.05, 0.01, and 0.001 levels, respectively; ns, not significant.

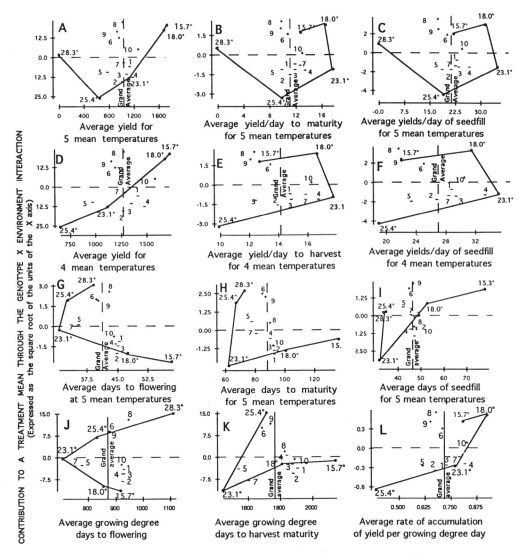

Figure 2   Contribution to a treatment mean of the main effect by each genotype and each temperature plotted against the contribution to this mean of the interaction effect by that same genotype or temperature. (See Figure 1 for the identity and photoperiod sensitivity of genotypes 1 to 10.)

than the G × E effect on the yield. Differences among the 10 genotypes account for 8.9% of the variation in yield among the 40 treatments (10 genotypes × 4 environments) across 15.7, 18.0, 23.1, and 25.4°C, but for only 3.4% across these four mean *T*'s and assumed zero yield for all 10 genotypes at 28.3°C, and for only 4.6% when the yield in 1984 at 28.3°C for three of the 10 genotypes is incorporated into the analysis. The respective three analyses (Table 1A, D, and G) show the mean *T* (environment) to cause 64.0, 82.9, and 81.6% of the variation in yield. The G × E interaction effect causes 27.1, 13.7, and 13.8% of the variation in yield.

All of the positive or negative G × E interaction effects compared below arise from multiplying the negative or positive contribution to the G × E interaction by the referenced genotype times the positive or negative contribution to that G × E interaction effect by the

Table 2 ANOVA tables from AMMI analyses of yield at four or five mean temperatures

Source[a]	Degrees of freedom	Sums of squares	Mean sums of squares	Probability
		Yields measured at 15.7, 18.0, 23.1, and 24.5°C		
Total	159	66,608,869	418,923	
Treatment	39	48,784,792	1,250,892	0.0000000***
Genotype	9	4,345,745	482,860	0.0014388**
Environment	3	31,235,720	10,411,906	0.0000000***
G × E interaction	27	13,203,325	489,012	0.0000041***
IPCA 1	11	10,146,407	922,400	0.0000001***
Residual	16	15,667,163	979,197	0.2168602
Error	120	17,824,076	148,533	
		Yields measured at 15.7, 18.0, 23.1, and 24.5°C and assumed zero at 28.3°C		
Total	199	120,537,135	605,714	
Treatment	49	102,713,059	2,096,185	0.0000000***
Genotype	9	3,476,596	386,288	0.0012291**
Environment	4	85,163,987	21,290,997	0.0000000***
G × E interaction	36	14,072,476	390,288	0.0000002***
IPCA 1	12	10,161,192	846,766	0.0000000***
IPCA 2	10	2,701,722	270,172	0.0165736***
Residual	10	1,968,861	89,494	0.7446532
Error	22	17,824,077	118,827	
		Yields measured at 15.7, 18.0, 23.1, and 24.5°C and assumed zero at 28.3°C		
Total	178	116,246,387	653,070	
Treatment	49	98,422,269	2,008,618	0.0000000***
Genotype	9	4,511,647	501,294	0.0004604**
Environment	4	80,301,895	20,075,474	0.0000000***
G × E interaction	36	13,608,727	378,020	0.0000182***
IPCA 1	12	10,147,102	845,591	0.0000000***
IPCA 2	10	2,773,105	277,311	0.0375439***
Residual	14	688,519	49,179	0.9841667
Error	129	17,824,118	138,171	

[a]The sources are offset to indicate that each value for genotypes, environments, and G × E is a subset of the treatments, and IPCA-1, IPCA-2, and the residual are each a subset of the total G × E interaction.

referenced environment. All of these are shown on the figure cited. Theoretical bases for these calculations of the G × E interaction effect are discussed in Chapter 42. Discussed in Chapter 43 is that the fact that the single point showing the contribution of each of the 10 genotypes plus the single point showing the contribution of each of the five (or four) environments (i.e., the 14 or 15 points of each of the biplots in Figure 2) facilitate more rapid comparison of the G × E effects than direct comparison of the calculated 50 (or else 40) G × E effects. Each calculable G × E interaction effect is attributable to one treatment. A treatment is but one of the 10 genotypes grown at but one environment. Each G × E effect is a calculable number (magnitude) with either a positive or a negative sign. If the reader has difficulty following the interpretations of the G × E effects below, Chapters 42 and 43 should be consulted before continuing to read this chapter.

The biplot for yield (Figure 2A and D) indicates a near-zero contribution to the G × E interaction effect on the yield by genotypes 1, 7, and 10. That is, multiplying their near-zero contribution times the contribution of any T indicates a near-zero G × E interaction effect on the yield. The near-zero G × E interaction effect indicates that the yield of these three genotypes does not change markedly from one environment (T) to another as the yield does for the genotypes with either a large positive or a large negative contribution. Thus, multiplying the large positive contribution to the G × E interaction of genotypes 6, 8, and 9 times the large negative contribution of 23.1° and 25.4°C indicates a large negative G × E interaction effect on the yield of these treatments; this means that these genotypes interact with 23.1 and 25.4°C to strongly depress their yield below their average yield across all the environments (T's of 15.7, 18.0, 23.1, and 25.4°C). On the contrary, the large positive contribution to the G × E interaction of genotype 6, 8, or 9 times the positive contribution of 18.0 and 15.7°C results in a positive G × E interaction effect (i.e., the yields at these T's are above the genotype's average yield across all the temperatures as shown in Figure 3D). Genotypes 2, 3, 4, and 5 interact in an inverse manner; their negative contribution causes their yields to be above their average at 23.1 and 25.4°C but below their average at the 18.0 and 15.7°C environments.

The yields are predicted by adding the genotype's main effect (its average across all the environments) to the environment's main effect, followed by adding the G × E interaction effect and subtracting the grand mean as explained further in Chapters 42 and 43. The predictions across 15.7, 18.0, 23.1, 25.4, and 28.3°C indicate the highest yield per hectare to be at 15.7° for all 10 genotypes. Also, all genotypes yield progressively less at each higher temperature. These predicted yields closely approximate the measured yields (Figure 3D).

C. G × E Interaction Is the Predominant Control over the Rate of Accumulation of Yield

The biplots for rate of accumulation of yield per day to harvest (Figure 2B) and per day of seedfill (Figure 2C) suggest that if a T lower than 15.7°C had been included, that environment's average rate across all genotypes and its contribution to the G × E interaction effect would tend toward the near-zero level observed at 28.3°C and toward the near-zero yield (Figure 2A). The highest rate of yield accumulation averaged across all 10 genotypes occurs between 23.1 and 18.0°C (Figure 2B, C, E, and F). This indicates the optimum temperature (T_o), that T which gives the highest rate of accumulation of yield, to be about 22°C, which agrees with the actually measured rates (Figure 3E and H). Above and below this T_o the progressive rises of T cause reverse effects on the average rate of accumulation of yield. Also reversed above versus below this T_o is a positive versus negative contribution by T to the G × E interaction effect on the rate of accumulation of yield (Figure 2B, C, E, and F).

Across 15.7, 18.0, 23.1, and 25.4°C the G × E interaction is causal of 51.0% of the variation in rate of accumulation of yield per day to maturity (Table 1A; compare with Figures 2E and 3E), and for 43.6% of the variation in rate of yield accumulation per day of seedfill (Table 1C; compare with Figures 2F and 3C). Thus the G × E interaction causes about half of all the variation in rate of yield accumulation among the treatments. A smaller 38.7 and 43.2% of the variation in rate is due to T and only 10.3 and 13.2% is due to genotype.

G × E causes about half as large an effect on yield per se (27.1% of the total variance, Table 1A) as on the rate at which the yield accumulates (51%, Table 1B). For yield, the main effect by T causes 64% of the variance in comparison with only 38 to 43% for the rate of accumulation. Differences among the genotypes cause about the same proportion of the variance

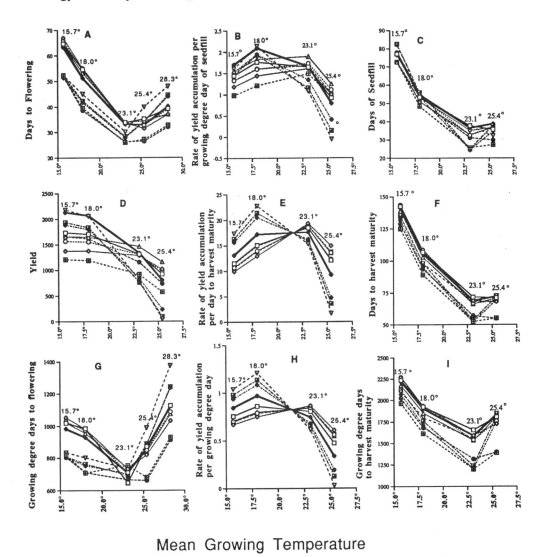

Mean Growing Temperature

Figure 3 Comparison of the days to flowering, days of seedfill, and days to maturity of 10 genotypes at five (or four) mean growing temperatures with their attendant growing degree days, rate of accumulation of yield, and yield. (See Figure 1 for definition of symbols and lines.)

for yield (8.9%), yield per day to maturity (10.3%), and yield per day of seedfill (13.2%) (Table 1A, B, E, C, and F). That the G × E causes 50% of the variation in rate of accumulation of yield and 27% of the variation in yield, but smaller proportions of all of the phenological traits, suggests that the G × E effect on all of them occurs through the rate of yield accumulation (the rate of growth of the reproductive organs). It suggests that the effect on the days to flowering and days to maturity is indirect rather than direct.

D. Temperature Predominantly Controls Phenology

The experimental design of this study eliminated any effect of daylength, because the plantings at all the elevations were on the same day and at the same latitude. With this design, the phenological time durations are largely controlled by the main effect of T; the differences in T account for 79.9% of the variation in days to flower among all the treatments (Table 1M), 95.2% of the days to harvest maturity (Table 1K and N), 96.3% for days of seedfill (Table 1L and O), and 60% of the node to flowering (Table 1J). The G × E interaction effect on days to flowering and days to maturity is always significant, even though it accounts for only 1 to 18% of the variation. Indicated in Chapter 43 is that the fact that daylength can account for a larger proportion of the variation in days to flowering than the T accounts for.

E. Photoperiod-Sensitive and Photoperiod-Insensitive Genotypes Interact Differently with Temperature

The three photoperiod-sensitive genotypes (6, 8, and 9) all have higher daily rates of accumulation of yield at 15.7 and 18.0°C than do any of the seven insensitive genotypes, but all three have lower rates at 23.1 and 25.4°C (Figure 3E). The reversal in rank occurs at about 22°C (Figure 3E; compare Figure 2B, C, E, and F). Additionally, both between the sensitive and insensitive groups and even among the genotypes within each, comparison of most pairs among the 10 genotypes reveals change from higher-than or lower-than to the reverse at above versus below about 22°C.

The average rate of accumulation of yield across all 10 genotypes increases progressively from 28.3°C to 25.4 to 23.1°C (Figure 2B, C, D, and F; compare Figure 3B and E). The increases reverse to become decreases of the average rate as the T decreases from about 22°C to 18.0 and 15.7°C (Figure 2B, C, E, and F; compare Figure 3B and E). Also at above versus below about 22°C, the contribution by T to the G × E interaction reverses in sign. Temperatures of 15.7 and 18.0°C contribute positively to the G × E interaction effect, while 23.1 and 25.4°C contribute negatively (Figure 2B, C, E, and F).

The actually measured rates of the photoperiod-sensitive genotypes 6, 8, and 9 show that their highest rates of accumulation of yield occur at 18.0°C, with smaller rates at both the lower T of 15.7°C and the higher 23.1°C (Figure 3E). Further, their lowest measured rates are at 25.4°C, and their near-zero yield at 28.3°C indicates yet lower rates. The AMMI analyses also indicate this (Figure 2B, C, E, and F); the sensitive genotypes 6, 8, and 9 interact positively with 15.7 and 18.0°C, which raises their rate of yield accumulation above each genotype's average across all of the T's. On the contrary, these sensitive genotypes interact negatively with 23.1 and 25.4°C, indicating that these higher T's lower their rates below the average rate of the genotype across all the T's.

As compared with the positive contribution to the G × E interaction effect on the rate of accumulation of yield by each of the sensitive-genotypes 6, 8, and 9, the photoperiod-insensitive genotypes 1, 2, 3, 4, 5, and 7 always contribute negatively to this G × E interaction effect (Figure 2B, C, E, and F). Consequently, in T's immediately higher than the optimum of about 22°C (at 23.1 and 25.4°C) the G × E interaction effect causes the rate of yield accumulation to be above the average rate of each of these photoperiod-insensitive genotypes, while the G × E interaction at temperatures below about 22°C (18.0 and 15.7°C) lowers the rate below the sensitive genotype's average across all T's (Figure 2A, compare with Figure 2B and C).

The insensitive genotype 10 has an actual rate of accumulation of yield that at all T's remains intermediate to rates of the sensitive group 6, 8, and 9 compared against rates of the other six insensitive genotypes 1, 2, 3, 4, 5, and 7 (Figure 3D). Additionally, at above versus below about 22°C the rate for genotype 10 reverses from being higher or else lower in comparison to both the insensitive and sensitive groups. This always-intermediate rate of genotype 10 is indicated on the biplot by its near-zero contribution to the G × E interaction effect on both the rate of yield accumulation (Figure 2B, C, E, and F) and the yield (Figure 2A and D), plus on most of the phenological traits (Figure 2G to K). The near-zero contribution by genotype 10 to the G × E interaction effect on the rate of accumulation of yield (Figure 2B, C, E, F, and L) and on the yield (Figure 2A and D) enables genotype 10 to have yield as high as the average of the insensitive genotypes at 25.4 and 23.1°C, but as high or higher than the average sensitive genotype at 18.0 and 15.7°C. Genotypes 7 and 1 cause a likewise near-zero interaction effect on the rate (Figure 2B, C, E, and F) and the yield (Figure 2A and D); like genotype 10, they have stability of yield across the temperatures, but their average yields are lower (Figure 3B).

F. Phenology as Growing Degree Days Rather than Days

The days needed to develop to flowering, to maturity, and to completion of seedfill are phenological traits, because each is influenced by climate. In this study, climatic differences are largely differences in T, because all elevations had the same daylength and none had serious stress due to too little or too much rainfall or nutrients.

When growing degree days (GDDs) are used in lieu of calendar days, additional interpretations emerge. The major difference in interpretation is that nearly 100% more days are needed to develop to flowering (Figure 3A versus G) and to harvest maturity at 15.7 and 18.0°C than at 23.1°C (Figure 3F versus H), but only about 50% more GDDs are needed. Averaged across all 10 genotypes, for both days to flowering (Figure 2G) and GDDs (Figure 2J), development to flowering is most rapid at an optimum near 23.1°C.

The graph of actually measured rates of accumulation of yield per growing degree day (Figure 3H) is nearly indistinguishable from that for rate per day (Figure 3E). However, comparing them above and below the near-22°C at which all higher-than versus lower-than rankings reverse, comparisons between most pairs of genotypes reveals the following. Below about 22°C at 18.0 and 15.7°C, the rate of accumulation of yield per growing degree day is relatively higher than the rate per day. AMMI analysis confirms this; the average rate per growing degree day is 20 and 11% higher at 15 and 18.0°C, respectively, than at 23.1°C (Figure 2L), but the average rate per day is about 6 and 40% lower (Figure 2B and E). The GDDs directly integrate the Q_{10} effects by temperature on the actions of photoperiod and all other genes that affect maturity. On the practical side, the farmer sees doubling of the time from sowing to harvest.

GDD phenology more effectively separates the genotypes among differential responses to T than does phenology expressed as days. Five groups among the 10 genotypes are differentiated by AMMI analysis of the GDDs:

1. Genotypes 6 and 9 cause large positive contribution to the G × E interaction for the GDD to both flowering (Figure 2J) and maturity (Figure 2K), while maintaining a moderate average GDD for both.
2. Genotype 8 causes the highest average GDD to flowering and maturity combined with the largest of all positive contributions to the interaction with T, but causes the sixth ranking average and virtually no interaction with T for the GDD to maturity. Genotype 8 was known in advance of this study to be the most photoperiod sensitive of the 10 genotypes. Its genetic control over days to flowering has been shown to differ from that of genotype 5 by a single photoperiod-sensitive gene [20] (Chapter 45).
3. Genotypes 5 and 7 cause the lowest average GDD to flowering and maturity combined with low contribution to the interaction, but both are like genotype 8 in that their contribution is more negative for GDD to maturity than for the days to maturity.
4. Genotypes 10 and 1 cause moderate average days to flowering and little interaction.

5. Genotypes 2, 3, and 4 cause moderate average GDD but large negative interaction. The largest contributions to the G × E interaction caused by the photoperiod-insensitive genotypes 2, 3, and 4 are negative in comparison with the largest contributions caused by genotypes 6, 8, and 9 being positive.

The genotypes are separable among a larger number of different responses to temperature when the phenology is expressed as GDD rather than days. This is mostly because the AMMI of GDDs spreads the averages of the genotypes. A wider spread occurs for flowering (Figure 2G versus J), maturity (Figure 2H versus K), and rate of yield accumulation (Figure 2E versus L). Also, some slight to large shifts in the ranked averages of the genotypes for GDDs to flowering compared with days suggest additional differences in the genotypes, while the averages of the T's do not change in rank.

Another basis for more separable differences among the responses of the genotypes to the GDDs is change of the contribution to the G × E interaction by some genotypes. For instance, the AMMI analysis of both the GDDs to maturity (Figure 2K) and days to maturity (Figure 2H) show a near-zero contribution to the G × E by the highly photoperiod-sensitive genotype 8, in contrast to large positive contribution by the sensitive genotypes 6 and 9. This contrasts with both the days (Figure 2G) and the GDDs to flowering (Figure 2J) having a large contribution by genotypes 6, 8, and 9, the largest being by genotype 8. Genotypes 5 and 7 remain together in the AMMI result for both days and GDDs to both flowering and maturity. For both genotypes, however, the GDDs (compare Figure 2H and K) contribute slightly more negatively to the G × E interaction effect than does the near-zero contribution for the days to flowering. Thus the photoperiod-insensitive genotypes 5 and 7 respond like sensitive genotype 8, in that they contribute more negatively to the G × E interaction effect on GGDs than days to maturity. Relatively, the photoperiod-insensitive genotypes 1, 2, 3, 4, and 10 all contribute less negatively to the GDDs to maturity, while the sensitive genotypes 6 and 9 cause the largest positive contribution for both the days and growing GDDs to maturity.

That the most highly photoperiod-sensitive genotype 8 (Figure 1) causes the largest of all positive contributions to the G × E (G × T) interaction effect on both the days (Figure 2G) and GGDs (Figure 2J) to flowering repeats prior results for days to flowering [7, 18] (Chapter 43). That genotypes 6 and 9 also contribute large and positively suggests that, similarly, each is a photoperiod-sensitive genotype. Their sensitivity to photoperiod was unknown when this work was conducted, but independent tests conducted at the International Center for Tropical Agriculture at Palmira, Colombia (J. W. White, personal communication, 1993) verify classification of genotypes 6 and especially 9 as photoperiod sensitive, as shown in Table 1. That genotype 8 causes a near-zero contribution to the G × E interaction effect on the days to maturity, while 6 and 9 cause the largest positive contribution, suggests a difference between these photoperiod-sensitive genotypes in maturity genes of other classes. For example, in addition to showing that genotype 8 differs from genotype 5 by a single gene for photoperiod sensitivity, Mmopi's data [19] (see Chapter 45) show that genotype 5 carries a nonphotoperiod gene that causes it to flower earlier, regardless of both daylength and T, than all the other genotypes in this study except genotype 7 (Figure 3A).

For all traits analyzed (Figures 2 and 3), genotype 7 responded to T like the photoperiod-insensitive genotypes, in agreement with the previously published AMMI analysis of the days to flowering of the 10 genotypes of this study (Figure 8 of Ref. 7). In the legend of that paper's Figure 1, genotype 7 was described as photoperiod sensitive. The trials at the International Center for Tropical Agriculture (personal communication) indicate that genotype 7 is, in fact, photoperiod insensitive, as indicated by all the AMMI analyses within this paper and that in Figure 8 of Ref. 7.

V. DISCUSSION

A. Shifts of Paradigm Suggested by Whole System Research

1. Some Physiological-Genetic Bases for Shifts of Paradigm

PHOTOPERIOD GENE CONTROL OVER PARTITIONING. Summerfield and Roberts [21] use the term f for days to flowering. They explain that f results from the rate of development to flowering, not vice versa. Wallace et al. [11,12] (Chapter 43) show this rate (measured as $1/f$) to be high (relatively few days to flowering) if there is predominant partitioning of the available photosynthate to the reproductive organs, but low (many days to flowering) if most of the photosynthate is partitioned competitively toward continued growth of more vegetative organs. It is shown that under both short and long daylength a photoperiod-insensitive genotype inherently allows physiological dominance of partitioning the photosynthate toward rapid growth of flower buds, pods, and seeds. Some correlated consequences are rapid growth of the reproductive organs, which results in low f (high $1/f$) and days to maturity plus both a high rate of accumulation of yield and a high harvest index. Partitioning photosynthate to the reproductive growth deprives the vegetative growth of photosynthate, competitively, thereby causing the correlated consequences of reducing the rate of growth and development of additional nodes and leaves, axillary branches, and shoot elongation.

For bean growing under long daylength, the photoperiod-sensitive genotype causes physiological dominance for partitioning of the photosynthate toward continued growth and development of more nodes, leaves, branches, and elongation of shoots and branches [11,12]. This enlarges the plant biomass and leaf area, which progressively increases the capacity for photosynthesis. It competitively draws photosynthate away from continued growth of the earliest initiated flower buds, pods, and seeds, thereby reducing (and often aborting) their growth, with attendant reduction of the rate of accumulation of yield. Under short daylength the photoperiod-sensitive gene(s) are inactive, so the response is like the insensitive genotype under both short and long daylength.

Partitioning a quantitatively largest proportion of the photosynthate toward initiation and continued growth of flower buds is not the initial requirement. What is necessary is stronger sink activity of the cells and tissues that become the flower buds, pods, and seeds [7]. The level of sink activity is their gain in biomass, measured as the grams/gram of tissue/unit of time.

Insensitive and sensitive bean genotypes initiate their earliest buds simultaneously whether in short or long daylength [10,11]. After bud initiation, in long daylength, it is smaller sink activity of the buds of the sensitive genotype that causes the larger f (smaller $1/f$) than for the insensitive genotype (7), in agreement with Summerfield and Roberts [21]. After flowering this physiological dominance of partitioning toward the vegetative rather than reproductive organs of the sensitive genotype continues, if the daylength is long, so additional consequences are a low rate of growth of the reproductive (yield) organs both per day of seedfill and per day of plant growth, consequently with more days to maturity, a lower harvest index, and a lower rate of accumulation of yield per day. The low level for all these traits expressed by the sensitive genotype in long daylength reverses to a high level in short daylength, because the short daylength prevents the photoperiod gene activity, thereby allowing the photosynthate to be partitioned predominantly toward continued growth of the reproductive (yield) organs [11,12].

The three preceding paragraphs describe the physiological-genetic model of Wallace et al. [11,12] for explaining the genotype × daylength × temperature interaction effects on days to flowering and maturity. All the effects by T reported in this chapter on the rate of yield accumulation, the yield, and the days to maturity of the 10 genotypes support and further elucidate that model. We interpret the common optimum T near 22°C for all these traits to be that T which is high enough to facilitate good vegetative growth and development without also facilitating sufficient photoperiod gene activity to excessively restrict (negatively control, see Chapter 41) the partitioning of photosynthate toward continued growth and development of the reproductive organs. Evidence presented below indicates that such restriction increases synergistically with each of a more photoperiod-delaying daylength, a higher T and greater genotypic sensitivity to photoperiod. The synergistic increases are largest [7] after T rises above its optimum for the rate of accumulation of the yield, which is near, if not identical to, the optimum for the rate of development to flowering.

EFFECTS OF VEGETATIVE GROWTH ON YIELD. Crop yield accrues at some average rate over the duration of growth—the time the genotype uses (needs) to develop to harvest maturity. Shown in this chapter is the fact that when all environments have the same daylength, this phenological time span is established by the genotype in interaction with T, and the predominant control (80%) is by T. All 10 genotypes give their highest yield per unit land area at 15.7°C, but all accumulate yield at a higher rate at an intermediately higher T. The highest rate is at 18.0°C for the three photoperiod-sensitive genotypes but at 23.1°C for the seven insensitive genotypes. Thus the highest yield of every genotype is partially dependent on its longer duration of growth at 15.7°C and is also partially dependent on both the duration and rate of seedfill. Additionally, the size of the vegetative structure achieved by the time of seedfill, principally its leaf area (photosynthetic capacity), will partially determine the rate of seedfill (rate of accumulation of yield). All of these controls are accompanied also by the photoperiod gene control over the proportions of the available photosynthate partitioned competitively to support growth of the reproductive organs (accumulation of the yield) versus growth of the vegetative organs.

The photoperiod-insensitive genotypes yield more than the sensitive genotypes at T's immediately higher than about 22°C, the T's at which the insensitive genotypes flower earlier than the sensitive genotypes and also have the higher rates of accumulation of yield. The sensitive genotypes yield more than the insensitive ones at T's immediately below about 22°C, the T's at which they have higher rates of accumulation of yield than the insensitive genotypes and also flower earlier. The lower optimal T for rate of accumulation of yield by the photoperiod-sensitive genotypes than by the insensitive genotypes repeats published evidence for the same 10 genotypes plus additional evidence for 68 cultivars [7]. All this evidence indicates that photoperiod-sensitive genotypes flower in fewest days at a lower T_o than the T_o that causes the fewest days to flowering of photoperiod-insensitive genotypes. Graphic display of this for nine genotypes is shown in Figure 2 of Chapter 43.

2. Some Previously Described Shifts of Paradigm

Wallace et al. [11] list 24 traits in addition to f that are controlled by photoperiod gene activity. These pleiotropic controls by a single photoperiod gene over at least 25 traits are all explainable by competitive partitioning of the photosynthate between the reproductive growth versus continuation of the vegetative growth. That is, the correlated changes in the levels of all 25 pleiotropic traits are interpretable as a consequence of photoperiod gene control over partitioning of the available photosynthate between the reproductive and vegetative organs. Across the 10 genotypes compared in this chapter, a T_o value near 22°C results in the highest rate of accumulation of yield per day to harvest maturity (Figures 2B and E and 3E), the highest average rate of development to flowering (Figures 2G and 3A), the fewest days to harvest maturity (Figures 2H and K and 3F), and the fewest days for completion of seedfill (Figures 2I and 3C). A common biological cause is clearly suggested.

Wallace et al. [12] have described shifts from eight paradigms about photoperiod and T

effects that are suggested by our basic whole system research that has compared the yield of each genotype with the penultimate and therefore major physiological genetic components of the biological system that accumulates crop yield. Additional shifts of paradigm are suggested below.

3. Current Paradigm for Mathematical Modeling of Days to Flowering

Summerfield et al. [20] state: "The rate of progress to flowering is given by a linear response plane of the form

$$\frac{1}{f} = a' + b'T + c'P$$

where [*f* is the days to flowering, $1/f$ is the rate of development to flowering,] *T* is the mean diurnal temperature (°C), *P* is the photoperiod (h d^{-1}), and *a'*, *b'*, and *c'* are genotype-specific constants, of which *c'* has a negative value in short day plants such as rice. The equation applies within wide and unambiguous values of temperature and photoperiod, namely where $T_b < T < T_o$, where *T* is the temperature at any time, T_b is the base temperature (at and below which there is no progress toward flowering), and T_o is the optimum temperature (at which *f* is minimum, i.e., $1/f$ is maximum), and where $P_c < P < P_{ce}$."

Prior to the text cited, Summerfield and Roberts [20] (compare Ref. 21) define P_c as the critical photoperiod below which the genotype does not respond to daylength; *P* is the daylength, and P_{ce} is the daylength above which *f* is not delayed additionally. In the citation above, we inserted the bracketed material, which appeared prior to the text cited.

Summerfield et al. [20] model $1/f$ rather than *f* because they find $1/f$ to respond linearly to both daylength and temperature, until *T* rises above the T_o for flowering of the genotype. Their T_o for *f* is the same as our optimum *T* for *f*. In this chapter the T_o for *f* is shown to equal or be very near the T_o for the rate of accumulation of yield.

Our prior interpretations [7,11,12] and those in this chapter agree fully with a further statement by Summerfield et al. [20]. They indicate that large photoperiod × temperature interaction effects on *f* occur only when *T* is supraoptimal, that is, when $T > T_o$. They explain that they do not model the range $T > T_o$ because the photoperiod × temperature interaction eliminates the linear response to *P* and *T*. Thus, using reductive research, their model considers only that half of the A-shaped response of $1/f$ to *T* that remains below the T_o value for flowering.

Based on the data of this chapter and other recent reports [7,8], we suggest that the complete model for the days to flowering is

$$\frac{1}{f} = a' + b'T + c'P + d'P^*e'T$$

where *d* is the contribution to the G × E interaction effect by the photoperiod (daylength) and *e* is the contribution to this interaction effect by the temperature.

4. Suggested Paradigm Shifts for Mathematical Modeling of Days to Flowering

Bean is a short-day plant like the rice discussed by Summerfield et al. [21]. For the bean data in this chapter, across 14.7, 18.0, to 23.1°C ($T < T_o$), the correlation between *T* and *f* is 1.00 for nine of the genotypes and 0.99 for the tenth. For $1/f$ the correlations of 1.00 and 0.99 occur with the same frequency. Therefore, these field plantings at different *T* (elevations) do not support the greater linearity for $1/f$ than for *f* that Summerfield et al. [20] and Summerfield and Roberts [21] report based on many controlled-environment (and therefore reductive) experiments with many crops. The bean data in this chapter and Wallace et al. [7] do support linear response of 1/days to flowering across $T < T_o$. Data for flowering in Chapter 43 (Figures 2 and 3, both originally from Ref. 7) show additionally, however, that extension of the daylength causes minimal delay in flowering as long as $T < T_o$. These data provide the following enhanced understanding of the daylength $\times$ temperature interaction results because each higher T greatly enlarges the delay of f that the long daylength causes through the photoperiod gene activity. That is, each higher T amplifies the photoperiod gene activity that is facilitated by the long daylength.

The range $T < T_o < T$ results in a U-shaped curve for the days to flowering of every one of the 10 genotypes of this study (Figure 3A). This is an A-shaped response curve for the rate of development to flowering ($1/f$). An explanation of the U-shaped curve follows.

Each higher T decreases the days required to develop a node (see Figure 3 of Chapter 43); that is, the higher T increases the rate of vegetative development (i.e., decreases the number of days required to develop an additional node on the plant). The same higher T increases the node to flowering, however (see Figure 3 of Chapter 43). That is, the higher T decreases the rate of reproductive development by amplifying any photoperiod-gene-caused delay of the node to flowering (Chapter 43). Interpretation of that Figure 3 (Chapter 43) suggests that a consequence from the fewer days needed to develop a node due to each higher T is a tendency toward smaller f (larger $1/f$), but the same higher T also amplifies the photoperiod gene activity to cause a simultaneous tendency toward larger f (smaller $1/f$) by enlarging the photoperiod-gene-caused delay in the node to flowering.

The interpretation above, derived from Figure 3 of Chapter 43, suggests the following shortcoming for the modeling of f by Summerfield et al. [20] and Summerfield and Roberts [21]. Their linear modeling of only $T_b < T < T_o$ is modeling of only the half of the A-shaped response curve of $1/f$ for which the effect on f through the effect of T on $1/$days per node (the rate of vegetative growth and development) causes a larger change of f than the opposite change of f caused by the effect of T through the photoperiod gene activity. Therefore, because their model ignores that half of the A-shaped response of $1/f$ to T for which $T_o < T$, it ignores all the combinations of a genotype and an environment for which the photoperiod gene activity causes a larger change in f than the opposite change caused by the same T through its effect on $1/$days needed to develop an additional node.

Associated interpretations are presented in Wallace et al. [7], based on Figure 3 of Chapter 43 and Figure 3A of this chapter. One interpretation is that T_o is that T at which the two simultaneous but opposite effects by T on f (one effect through the rate of vegetative development and the opposite effect through the rate of reproductive development) are exactly equal so that each cancels the other. A further interpretation is that across $T < T_o$ the effect by T limits without preventing all photoperiod gene activity. In contrast, across $T_o < T$ the delay by photoperiod gene activity becomes the larger of the two opposite effects by T on f. Therefore, whenever T is and remains either higher or lower than T_o, the directly observable delay of f due to a change of T is due to that part of the larger effect on f that is not canceled by the smaller but opposite and simultaneous effect on f.

Our interpretation [7,8] (Chapter 43) agrees with those of Summerfield et al. [20] and Summerfield and Roberts [21] that the largest photoperiod $\times$ temperature interaction effects on days to flowering occur when $T_o < T$. There are no photoperiod $\times$ temperature interaction effects in the data of this chapter, however, because the same latitude and day of planting (see Section III) gave an identical daylength for the different T of each different elevation. Therefore, all the interaction effects on f presented within this chapter are genotype $\times$ temperature interaction effects. Actual daylength $\times$ temperature interaction effects on f are discussed in Chapter 43 and Ref. 7.

An interpretation beyond the insight which Summerfield and colleagues base their modeling on is that it is $T > T_o$ versus $T < T_o$ that is the major cause of the overall G $\times$ E interaction effect (see Figures 2 and 3 of Chapter 43). To illustrate further, in this paper up to 51% of the variation in rate of accumulation of yield (Figure 3E and H) and 27% the variation in yield (Figure 3D) is due to G × T interaction, in comparison with only 1% for the days to maturity, 2% for the days of seedfill, and 18% for the days to flowering. Illustrating the G × T effect on the yield, the highest rates of yield accumulation (its T_o) are near 18.0°C for the three photoperiod-sensitive genotypes but are near 23.1°C for the seven insensitive genotypes. Across all 10 genotypes this results in an average T_o near 22°C. The highest yield for all 10 occurs at 15.7°C, however.

5. Suggested Paradigm Shifts Relative to Growing Degree Days

The simultaneous but opposite effects on f by T described above indicate shifts of paradigm also relative to the concept of growing degree days (GGDs). The theoretical paradigm is that the GDD effect on f is constant across all Ts, until T is truly supraoptimal. We interpret supraoptimal as when the high T causes many enzymatic and physiological processes to function poorly or not at all (see discussion of Q_{10} effects on gene activity in Chapters 41 and 43). Summerfield et al. [20] use supraoptimal simply as $T > T_o$, and they apply the theoretical paradigm for GDD to model the near-linear left-half of the A-shaped response of $1/f$ to T (i.e., to model that part of the overall response for which $T < T_o$). From their admission [21], however, the response of $1/f$ to both T and P (the photoperiod) is near linear only if the $T < T_o$. Our interpretations above indicate that this T must be low enough that its amplification of the photoperiod gene activity does not cause a larger delay of f by delaying the node to flowering than the change of f the same T causes toward earlier flowering by decreasing the days used to develop each node. But this smaller effect on f through the photoperiod gene activity occurs only when $T < T_o$ (Figure 2 of Chapter 43) [22]. Because every T that is higher than T_o delays the flowering most strongly and synergistically for a photoperiod sensitive genotype, the paradigm that the effect of GDDs is linear and causes earlier flowering across all T's is not applicable to all development toward flowering. In agreement with this, Hodges [22] cites studies for several crops for which the relationship between GDDs and the development to flowering is not linear. The interpretations in this chapter suggest that the paradigm of a linear effect on f by GDDs is relatively accurate for both photoperiod-insensitive and photoperiod-sensitive genotypes if all T's $< T_o$, but this paradigm is 180° out of phase after T is high enough that $T_o < T$ because, then, each higher T delays the node and days to flowering by amplifying the photoperiod gene activity, and this delay of f is always larger than the simultaneous effect toward earlier f that each higher T will cause through the fewer days needed to develop a node.

Of the growing T's compared in this chapter, the only T at which bean yield is not high enough for commercial production is the 28.3°C at 50 m elevation. The observed f and successful commercial bean production across the range $T < T_o < T$ (14.7, 18.0, 23.1, and 25.4°C, with T_o being about 22°C) shows that a GDD does not cause the same advance in development to f at all growing T's. Further, $T > T_o$ is not supraoptimal in the sense that the overall biology continues to degrade. The interpretation that each higher T diverted the available photosynthate away from the reproductive (yield) organs and toward the vegetative organs suggests that the largest aerial biomass was accumulated at the 28.3°C environment. Unfortunately, the aerial biomass was not measured at any of the five T's.

A recent example of the nonlinear effect of GDDs on f has been reported for wheat, for which f is delayed by short daylength. Rawson and Richards [23] state: "In day-degree terms, heading was delayed by high temperature in all isolines under all photoperiods. In four of the isolines, the delay in heading due to high temperatures increased as the photoperiod shortened." This agrees fully with the larger delay of f of bean by long daylength combined with high T.

6. Suggested Paradigm Shifts for Modeling of Photoperiod × Temperature Effects on Yield

Grimm et al. [24] state that growth and yield models depend on good prediction of phenological events such as *f*. They describe the best of the multiple approaches they compared for modeling *f*. The estimates are based on both *T* and daylength. They state that the *f* derived by the model is for incorporation into the much more inclusive and complex model that simulates growth and yield under various soil, water, and management conditions. Their next sentence follows: "The dry matter partitioning in this model is controlled to a large extent by the timing of vegetative and reproductive stages." This assumption for the model is the reverse of the cause versus effect we infer throughout this chapter, as derived from our whole system research. Wallace et al. [10] present other examples of this usual application of the paradigm that the development is causal of the pattern of partitioning, in comparison with our reversed interpretation that the partitioning is causal of the development.

Mathematical modeling of the growth, development, and yield of different environments is a mode of whole system research. It will fall short, however, if the many parameters it applies toward estimating the daily gains in growth, development, and yield are all based on reductive research. A small error for many such parameters, especially for multiplicative effects, will result in large modeling errors. Estimation of such parameters and crop modeling will advance more rapidly and accurately when it incorporates the broadened insight derivable through the basic whole system research described in Chapters 41 to 45 of this book and in Wallace et al. [10], where its application toward more rapid breeding for higher yield is discussed.

B. Need for Complete Yield System Analysis

Understanding the effects of temperature and photoperiod on crop yield requires analysis of more than the phenological traits days to flowering and maturity and of the rate of development to flowering and the rates of accumulation of the yield. Our failure in this study to measure the aerial biomass accumulated on each yield trial plot results in an inability to evaluate the main and interaction effects by the genotypes and temperatures on two of the three major physiological-genetic components of the process of accumulating crop yield [10–12] (Chapter 41). The two that could not be evaluated are the net accumulated aerial biomass and the harvest index, the latter being the ultimate measure of partitioning between the reproductive (yield) organs and the other aerial organs. Our reductive focus on phenological traits in 1983 as these yield data were collected resulted in measurement of only the third major component, the days to harvest maturity, which is the time required to complete both the net accumulation of the biomass plus its partitioning between the reproductive (yield) and vegetative organs. Failure to measure the aerial biomass results in inability to assess the effect of the biomass on the yield, the effect on this biomass that is due to the different genotypes, the effect due to the different temperatures, and the effect due to the fully integrated measure of the partitioning of the biomass, which is quantifiable by the harvest index.

The measures of the biomass will reflect the leaf area and leaf area index (capacity for photosynthesis). Like the yield, the biomass accumulates at an average rate across a measurable time. Both yield and total aerial biomass are altered, competitively, by any change of the proportion of the accrued photosynthate that is partitioned to the vegetative organs versus to the reproductive (yield) organs [11]. Wallace et al. [10] describe how measurement of the days to flowering, days to maturity, the yield, and the aerial biomass of ongoing cultivar/yield trials, followed by calculation of the rates of biomass and yield accumulation and of the harvest index, followed by subsequent AMMI analysis of all these components of the process of accumulating yield, can inexpensively improve the efficiency of breeding for higher crop yield.

VI. SUMMARY AND CONCLUSIONS

To compare the responses to temperature (*T*) of the days to flowering with the associated changes in the yield, 10 tropical bean genotypes were grown in five tropical fields at different elevations; the average *T*s were 15.7, 18.0, 23.1, 24.5, and 28.3°C. For all *T*'s the daylength during growth varied from 13.0 to 13.5 and back to 13.0 h. Days to flowering, growing degree days to flowering, days to maturity, growing degree days to maturity, and yield were measured. Days of seedfill, rate of yield accumulation per day and per growing degree day, and rate per day and growing degree day of seedfill were calculated. Three photoperiod-sensitive genotypes expressed higher rates of yield accumulation at both 15.7 and 18.0°C than the rate for each of seven insensitive genotypes, but these higher rates versus lower rates were reversed at 23.1 and 25.4°C. Similarly, for most comparisons between just two photoperiod-sensitive or photoperiod-insensitive genotypes, the genotype with the higher rate of yield accumulation at 15.7 and 18.0°C had the lower rate at 23.1 and 25.4°C. From these highest rates, the rates of all 10 genotypes decreased progressively as *T* was raised to 28.3°C. Across the 10 genotypes grown at 15.7, 18.0, and 23.1 to 24.5°C, the G × E interaction described accounted for about 50% of the variation in rate of accumulation of yield per day to harvest maturity. Yield of all genotypes was near zero at 28.3°C and increased with each lower-growing *T* to the highest yield per unit land area at 15.7°C. Lower rates of accumulation of yield at 15.7 than 18.0°C, however, suggest that the yield is reduced additionally by further lowering of the mean *T*.

The results agree with the following conclusions from the model we developed previously to explain the variation in the days to flowering across genotypes, daylengths, and temperatures.

1. The time to flowering of bean is controlled by the low to moderate to high sensitivity to photoperiod of the genotype.
2. Further control over the days to flowering results from amplification by each higher temperature of the photoperiod gene activity.
3. The expressed level of this photoperiod activity is established jointly, by the genotype in interaction with the daylength and the temperature. The results support the following further concepts of the model developed previously.
4. The photoperiod gene activity controls the proportion of the photosynthate that is partitioned to the reproductive organs (to the yield) rather than being competitively partitioned to the continued growth of more vegetative organs.
5. The larger the proportion of the photosynthate partitioned to the reproductive (yield) organs, the faster their rate of growth.
6. For bean, the preflowering growth rate of the reproductive organs (the growth rate of the flower buds) controls the days to flowering.
7. After flowering, the same photoperiod gene activity controls the rate of growth of the seeds of bean, which is also control over the rate of accumulation of the yield.

In this study, in addition to the yield, only the days to flowering and days to maturity were measured. These days are two phenological traits (i.e., they vary with the climate). The net accumulated aerial biomass was not measured. Therefore, only one of the three major physiological genetic components of the process of accumulating crop yield is quantifiable. This, one of the measured phenological traits, is the days used (needed) by each cultivar to develop to harvest maturity; it is the time duration required to complete both of the other major components, both being physiological processes. Additional direct measurement of just the aerial biomass on each yield trial plot, which is the endpoint of the process of accumulating the net accumulated aerial biomass and is the second major physiological-genetic component of the overall process of accumulating the yield, would have facilitated calculation of the harvest index, which is the endpoint of the partitioning between the reproductive and vegetative organs, and is the second process and third major physiological-genetic component of yield accumulation. The harvest index quantifies the endpoint of the process of partitioning the net accumulation of aerial biomass between the reproductive (yield) organs and the nonyield (vegetative) organs of the plant. Thus the additional measurements of the aerial biomass in any yield trial facilitates a complete (see Chapter 41) yield system analysis.

The genotype × temperature interaction effect caused 50% of the variation in rate of accumulation of yield, 27% of the variation in yield, 2% of the variation in the days to flowering, and 1% of the variation in the days to harvest maturity. Quantification of these G × E effects using AMMI statistical analysis demonstrated the following additional benefits. First, AMMI quantifies the magnitude of positive or negative contribution to the G × E interaction effect by each genotype and by each environment. Second, the product of the contribution by a genotype multiplied by the contribution by an environment quantifies the G × E effect on the level of the trait (the realized yield, or rate of yield accumulation, or the days to flowering or maturity) of each treatment. Each treatment is the combination of one genotype and one environment (one temperature in this study). Third, the relative magnitude of the genetic and the environmental and the G × E interaction controls over the yield and the phenological traits are all interpretable from the pattern of change of the G × E interaction effect across the genotypes plus across the temperatures (the environments). These patterns are provided by the result of the AMMI analysis.

The economically important question to be answered by a yield trial is always: Which genotypes give the highest yield in which environment? Biologically and scientifically important questions that can be answered by yield system analysis of multiple yield trials followed by AMMI analysis are: What physiological-genetic attributes allow one cultivar to yield more than another? Why and how does the G × E interaction account for as much as 50% of the rate of the accumulation of the yield?

The results of the AMMI analysis reinforce the concept that in bean and other crops, the genotype × temperature interaction effect on both the days to flowering and maturity and on the yield all arise through direct control by the genotype for sensitivity to photoperiod and through modulation by the temperature of this gene activity. The photoperiod gene activity controls the rate of accumulation of the yield. These interpretations agree with prior evidence that the photoperiod gene activity controls the proportion of the photosynthate that is partitioned to the reproductive (yield) organs, which controls the rate of growth of these organs (which is also the rate of accumulation of the yield), which controls the days to flowering and to maturity, which control the adaptation of the cultivar to the duration of the growing season, which further controls the yield.

ACKNOWLEDGMENTS

This research was supported by Hatch Project 487 and the USAID-funded Bean/Cowpea Collaborative Research Support Program. This research is a contribution to the WR-150 project.

REFERENCES

1. R. A. Richards, *Field Crops Res.*, 26: 141 (1991).
2. D. H. Wallace, J. L. Ozbun, and H. M. Munger, *Adv. Agron.*, 24: 97 (1972).
3. D. H. Wallace, M. M. Peet, and J. L. Ozbun, in *CO_2 Metabolism and Plant Productivity* (R. H. Burris and C. C. Black, eds.), University Park Press, Baltimore, p. 43 (1976).

4. D. H. Wallace, *Well-Being of Mankind and Genetics*, Vol. I, Book 2, *Proceedings of the 14th International Congress on Genetics*, MIR Publishers, Moscow, pp. 306–317 (1980).
5. D. H. Wallace and R. W. Zobel, in *Handbook of Crop Productivity* (M. Rechicgl, Jr., ed.), CRC Press, Boca Raton, Fla., p. 137 (1982).
6. D. H. Wallace, in *Horticulture: New Technologies and Applications* (J. Prakash and R. L. M. Pierik, eds.), Kluwer Academic Publishers, Dordrecht, The Netherlands, p. 1 (1991).
7. D. H. Wallace, P. A. Gniffke, P. N. Masaya, and R. W. Zobel, *J. Am. Soc. Hortic. Sci.*, *116*: 534 (1991).
8. D. H. Wallace and G. A. Enriquez, *J. Am. Soc. Hortic. Sci.*, *105*: 583 (1980).
9. D. H. Wallace, *Plant Breed. Rev.*, *3*: 21 (1985).
10. D. H. Wallace, J. P. Baudoin, J. Beaver, D. P. Coyne, D. E. Halseth, P. N. Masaya, H. M. Munger, J. R. Myers, M. Silbernagel, K. S. Yourstone, and R. W. Zobel, *Theor. Appl. Genet.*, *86*: 27 (1993).
11. D. H. Wallace, K. S. Yourstone, P. N. Masaya, and R. W. Zobel, *Theor. Appl. Genet.*, *86*: 6 (1993).
12. D. H. Wallace, R. W. Zobel, and K. S. Yourstone, *Theor. Appl. Genet.*, *86*: 17 (1993).
13. H. G. Gauch, Jr., *Statistical Analysis of Regional Yield Trials: AMMI Analysis of Factorial Designs*, Elsevier, New York (1993).
14. R. W. Zobel, M. J. Wright, and H. Gauch, *Agron J.*, *80*: 388 (1988).
15. R. W. Zobel, *Adv. Soil Sci.*, *19*: 17 (1992).
16. A. F. H. Muhammad, Ph.D. dissertation, Cornell University, Ithaca, N.Y. (1983).
17. H. G. Gauch and R. W. Zobel, *Theor. Appl. Genet.*, *79*: 753 (1990).
18. K. S. Yourstone, Ph.D. thesis, Cornell University, Ithaca, N.Y. (1988).
19. S. G. Mmopi, Ph.D. thesis, Cornell University, Ithaca, N.Y. (1992).
20. R. J. Summerfield, S. T. Collinson, R. H. Ellis, E. H. Roberts, and F. W. T. Penning de Vries, *Ann. Bot.*, *69*: 101 (1992).
21. R. J. Summerfield and E. H. Roberts, *World Crops: Cool Season Food Legumes* (R. J. Summerfield, ed.), Kluwer Academic Publishers, Norwell, Mass., pp. 911–922 (1988).
22. T. Hodges, ed., *Predicting Crop Phenology*, CRC Press, Boca Raton, Fla. (1991).
23. H. M. Rawson and R. A. Richards, *Field Crops Res.*, *32*: 181 (1993).
24. S. S. Grimm, J. W. Jones, K. J. Boote, and J. D. Hesketh, *Crop Sci.*, *33*: 137 (1993).

45

Control of Days to Flowering of Bean (*Phaseolus vulgaris* L.) by Interaction of a Photoperiod Gene and a Nonphotoperiod Gene

Seja Mmopi and Rafael Rodríguez
Instituto de Ciencia y Tecnologia Agricolas, Guatemala City, Guatemala

Donald H. Wallace
Cornell University, Ithaca, New York

Porfirio N. Masaya
Interamerican Institute for Cooperation for Agriculture, San Jose, Costa Rica

Richard W. Zobel
Agricultural Research Service, U.S. Department of Agriculture, Ithaca, New York

I. INTRODUCTION

Based on evidence that a few photoperiod genes control partitioning of the aerial biomass between the reproductive and vegetative organs, Wallace et al. [1–3] (Chapter 41) proposed that the proportion of the photosynthate partitioned to the reproductive (yield) organs rather than to continued growth of the vegetative organs is controlled by just a few photoperiod and other maturity genes. They suggest that this control by a few photoperiod and other maturity genes gives high heritability to days to flowering, days to maturity, and harvest index. As a consequence, selection among a few thousand segregates has a relatively high probability of identifying some segregates that breed true for a higher harvest index. On the contrary, because total biomass is controlled by these few maturity genes plus most of the additional thousands of genes of the plant, its heritability is extremely low. Therefore, finding a segregate that breeds true for higher biomass requires selection among millions of segregates. Wallace et al. [1–3] (Chapter 41) suggest that this large difference in number of genes and consequent heritability explains why selection for higher yield during the last century has raised the yield of self-pollinated seed crops by enlarging the proportion of the aerial biomass that is partitioned to the reproductive (yield) organs, but with little or no gain in the total plant biomass. For agronomic reasons also, there were often deliberate attempts to select for more compact plants. In this chapter we present research that supports the concept that the major genetic control over days to flowering is by only a few photoperiod genes plus a few nonphotoperiod genes.

II. MATERIALS AND METHODS

A. Materials

For each of four Guatemalan black-seeded bean cultivars (Rabia de Gato, Ju-80, Quetzal, and San Martin), all with indeterminate growth habit, it was shown (Figure 3 of Chapter 43) that progressively higher growing temperatures decrease the days needed to develop a node, but simultaneously delay the node to flowering. Rabia de Gato and Ju-80 are both photoperiod insensitive, while Quetzal is moderately photoperiod sensitive and San Martin is highly photoperiod sensitive (J. W. White, personal communication).

The days to flowering of Rabia de Gato, San Martin, Ju-80, and Quetzal, respectively, were shown previously [4] to be 53, 52, 68, and 54 days at a mean growing temperature of 15.7°C; 41, 47, 58, and 54 days at 18.0°C; 31, 37, 39, and 39 days at 23.1°C; and 33, 45, 40, and 43 days at 28.3°C. Why is it that San Martin flowers as early as Rabia de Gato at 15.7°C but flowers latest of all at 28.3°C? Also, why does Rabia de Gato flower earlier than the others in all four environments? This study was designed to study such questions.

B. Methods

F_1 seeds of all six possible crosses among Rabia de Gato, Ju-80, Quetzal, and San Martin, including reciprocal crosses, were planted to produce F_2 seed. A single seed was planted from each plant of the F_2 generation. Single-seed descent continued through plants *of* the F_7 generation. The F_8 seed was increased to assure sufficient *seed* for the multiple plantings of this research. The generation advances began and were completed at the 23°C elevation with 13.3 h daylength in Guatemala.

Single-seed-descent advance is used because each F_8 generation progeny has 98.5% probability of being homozygous for each gene for which the parents of the cross differ. Also, each progeny has 50% probability of possessing the allele carried by each parent. For a gene that controls the days to flowering, therefore, half of the F_8 progenies are expected to flower early, like one parent, and the other half to flower later, like the second parent.

During May 1989, all F_8 progenies of all six crosses were planted in Guatemalan fields at elevations with mean daily temperatures near 19 and 29°C [5]. Hereafter, these are designated as the low-temperature (LT) and high-temperature (HT) environments. The 23°C elevation used in the previous studies [5] was not used, because its near-optimum temperature for flowering causes the smallest difference in days to flowering between the parents (Figure 1 of Ref. 5); this limits the merit of 23°C for studying the inheritance of the days to flowering. The daylength was 13.0 h in May, 13.3 in June, and 13.3 h in August. Four replications of a randomized complete block design were planted at both elevations. For each replicate, the days to first flowering was recorded for five plants. The mean of the five was used as the observed days for the replicate.

Each of the same F_8 progenies was also planted on June 29, 1989 in a summertime greenhouse at Ithaca, New York. The mean temperature was 26°C and the initial daylength was near 16 h. This is the long-daylength (LD) environment. Each F_8 progeny was represented by three plants in separate 15-cm pots. The soil medium was peat moss and vermiculite plus fertilizers. About half of the F_8 progenies from each cross were planted again in a wintertime greenhouse (January 1, 1990); this is the short-daylength (SD) environment. Supplemental light from 6:00 A.M. to 6:00 P.M. with 1000-W high-pressure sodium lamps provided a 12-h daylength. Light intensity at plant level was 50 W/m^2. Both day and night temperatures were 20°C. This January 1, 1990 planting included only half of the F_8 progenies of each cross, because other wintertime use of the greenhouse left insufficient space for all of them.

The days to flowering were analyzed with the AMMI statistical procedure [6]. Because of the unequal number of replications in the field and greenhouse, the AMMI analysis was applied to the average days to flowering of all observations within each of the four environments. The

AMMI estimation maximization routine [6,7] was used to estimate the days to flowering for each F_8 progeny that was not grown in the SD environment.

III. RESULTS AND DISCUSSION

A. General for All Crosses

A sufficient number of F_8 progenies to measure reliably the distribution of the days to flowering due to the genetic and environmental controls was obtained from five of the six possible crosses among the four parental cultivars. From these five crosses, one general result is lack of any detectable difference in days to flowering due to which of the two parents was the male or the female parent [4]. Therefore, all comparisons of the days to flowering in this chapter assume that the reciprocal crosses are identical. Another general result is the occurrence among the F_8 progenies of each of the five crosses of a bimodal distribution of the days to flowering within either two or three of the four environments. Still another general result is the presence within all five crosses of a G × E interaction effect for which the genetic control is best interpreted as many background genes with individually small effects on days to flowering.

B. One Cross Demonstrates Major Genetic Control over the Days to Flowering by One Single Photoperiod Gene

1. The Cross Rabia de Gato (Photoperiod Insensitive) × San Martin (Photoperiod Sensitive)

FREQUENCY DISTRIBUTIONS IN THE FOUR ENVIRONMENTS. In the LD environment, distribution of the days to first flower of the 119 F_8 progenies was divided about equally (63:59) between early and late flowering (Figure 1A). This bimodal distribution suggests that the 119 progenies represent an early homozygous genotype and a late homozygous genotype. Self-pollination to the F_8 generation provides 98.5% probability of homozygosity for every gene. The nearly equal number of progenies with early and late genotypes indicates control by a single gene. Each F_8 progeny within the early genotype class, as differentiated in the LD environment, is marked black within the distribution of the days to flower for each of the four environments (Figure 1A to D).

In both SD (Figure 1B) and LT (Figure 1C) bimodal early:late distribution of the days to flowering is not expressed; rather, the days to first flower of the early and late genotypes occur across a short, common time span. For the 58 of 119 F_8 progenies grown in this SD environment, 51 had first flowers across 6 days beginning at 26 days after planting; seven, including both the early and late genotypes, extend the total span to just 15 days (to 40 days after planting). (See Section II.B for the reason that only about half of all the progenies were grown in the SD environment.)

In the LD environment, all 63 progenies of the early genotype also opened a first flower across a span of only 7 days. In contrast, the span for progenies of the late genotype is 21 days. Thus the total span is 28 days across all 119 F_8 progenies. In the LT environment, 114 F_8 progenies opened a first flower across 9 days and five extend the total span to 15 days. In the HT environment (Figure 1D) the total span is 47 days, but the distribution approaches the same early:late distribution as under LD (Figure 1A). However, in this HT environment, some progenies of the early genotype flower relatively later than in LD, and a larger number of progenies with late genotype are shifted toward relatively earlier flowering. The result is a larger number with intermediate days to flowering and a skewed monomodal distribution (Figure 1D).

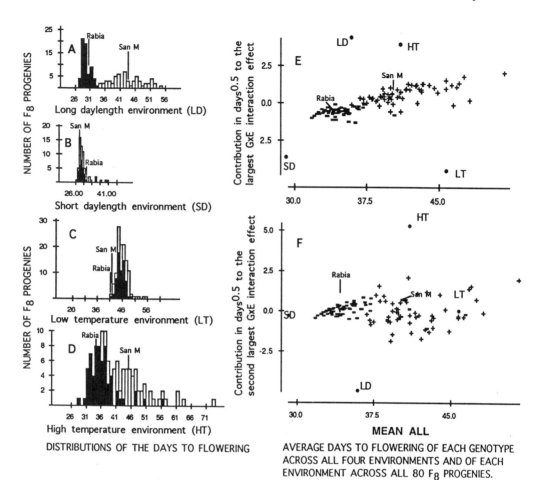

DISTRIBUTIONS OF THE DAYS TO FLOWERING

AVERAGE DAYS TO FLOWERING OF EACH GENOTYPE
ACROSS ALL FOUR ENVIRONMENTS AND OF EACH
ENVIRONMENT ACROSS ALL 80 F_8 PROGENIES.

Figure 1 Distributions of the days to flowering of 119 F_8 progenies of the cross between the photoperiod-insensitive cultivar Rabia de Gato and the highly photoperiod-sensitive cultivar San Martin when grown in long-daylength (A), short-daylength (B), low-temperature (C), and high-temperature (D) environments. Graphs E and F are biplots showing (1) the average days to flowering across the four environments of each F_8 progeny plotted against the contribution to the largest (E) and second-largest (F) G × E interaction effect on these days by the same progeny, and (2) the average days to flowering across all 119 F_8 progenies of each of the four environments plotted against the contribution by that environment to the largest (E) and second-largest (F) G × E interaction effect on these days. The origin of the differential marking of the early- and late-flowering F_8 progenies is within the LD environment (graph A), and each F_8 progeny is marked the same in graphs B, C, and D. In graphs E and F all of the early-flowering progenies are indicated with the symbol −, and all late-flowering progenies with the symbol +.

In the LD environment, the average of 30 days to first flowering for the early parent is centered within the narrow peak of the early F_8 progenies, and the 43-day average of the late parent is within the range of the late F_8 progenies (Figure 1A); such centralities of the parents also occur within the HT environment (Figure 1D). The average days to flowering of the early and late homozygous F_8 genotypes are 36 and 46 days. In the SD environment the early and late parents had first flowers at 30 and 29 days, respectively, while their days in LT were 41 and 42. In summary, in all four environments, the F_8 progenies of the early and late genotype express, respectively, the same relatively early or late flowering as the early-flowering photoperoid-insensitive parent or the late-flowering photoperiod-sensitive parent.

INTERPRETATION OF THE NUMBER OF MAJOR GENES. Chi-square analysis of 63 early:56 late F_8 progenies in the LD environment does not reject a hypothesized ratio of 1 early:1 late ($\chi^2 = 0.41$). This ratio, plus duplication in all four environments of the same response as the early photoperiod-insensitive parent Rabia de Gato or the late photoperiod-sensitive parent San Martin, indicate major control of the days to flowering by a single photoperiod gene.

DIFFERENTIAL VARIABILITIES OF THE EARLY AND LATE GENOTYPES. Within the LD environment, all plants of the F_8 progenies with the early genotype opened their first flower across days 27 to 33 days after planting, a span of just 7 days (Figure 1A). First flowers on progenies of the late genotype begin at 34 days and continue to 55 days (Figure 1A), a 22-day span.

The threefold larger span for the late genotype repeats the previous observation with F_8 progenies from the cross between the early-flowering photoperiod-insensitive cultivar Redkloud *and* the late-flowering photoperiod-sensitive cultivar Redkote [1]. These are an early and a late cultivar bred for the temperate climate of New York. In that cross, gene *Ppd* and its allele *ppd*, respectively, express the photoperiod-sensitive response (delay of flowering by an LD environment) and the photoperiod-insensitive response (without delay by LD). In that cross, in the LD environment, homozygous *Ppd* causes a fourfold larger span of the later (delayed) days to first flowering than the span of the earlier days for *ppd*. Under the SD environment, flowering of *Ppd* is early, with a narrow span of days like *ppd* expressed under any daylength. Wallace et al. [1] have discussed the bases for this.

THE ACTIVITY OF GENE *Ppd* IS MODULATED BY TEMPERATURE IN ADDITION TO DAYLENGTH. Mmopi [4] crossed San Martin with Redkloud (genotype *ppd*) and backcrossed the F_1 to both parents. The F_1, F_2, and backcross to each parent were grown in the summertime greenhouse during 1990. The backcross to the insensitive parent exactly fit the expected 1:1 ratio (10:10), and the ratio for both the F_1 and backcross to the sensitive parent was the expected 0 early:4 late. The ranges of days to flowering of both the early parent and the early segregates of the F_2, and of the backcross to the early insensitive parent, did not overlap with the range of days of the late sensitive parent, or of the late F_2 segregates, or of the known *Ppd* genotype (Redkote).

The collective data from these crosses suggest that Rabia de Gato carries the insensitive allele *ppd*, or an allele not easily differentiated from the *ppd* carried by Redkloud. The data imply also that San Martin carries a photoperiod-sensitive allele that is either *Ppd* or is not easily differentiated from *Ppd* by the techniques used.

The delay of flowering caused by both the LD (Figure 1A) and HT (Figure 1D) environments, and lack of delay in both the SD (Figure 1B) and LT (Figure 1C) environments, repeat prior evidence of control over the activity of *Ppd* by both daylength and temperature [5,8].

STATISTICAL QUANTIFICATION OF THE PROPORTIONAL CAUSES OF THE VARIATION IN THE DAYS TO FLOWERING. The AMMI analysis indicates that 25% of the total variation in days to flowering is accounted for by differences in genotype among the 119 F_8 progenies; differences among the four environments account for 53%; and G × E interaction accounts for 22%. The largest of two significant ($p < 0.0001$) sources of the G × E interaction effect accounts for 60.5% of the 22% of total variation in days to flowering that is caused by the G × E interaction. The second-largest source accounts for 35.7% of that 22%. Thus two statistically significant physiological-genetic sources of a G × E interaction effect on the days to flowering account for 96% of the total G × E interaction effect, which is for 21% (96% of 22%) of the total variation of the days to flowering.

INTERPRETATION OF THE LARGEST SOURCE OF G × E INTERACTION. Most of the 63 F_8 progenies of the earlier genotype *ppd* have a smaller average days to flowering (main effect due to their genotype) across all four environments than that of most of the averages (main effects) of the 56 F_8 progenies with the late genotype *Ppd* (Figure 1E). The range of averages for *ppd* is 31.9 to 40.8 days, compared with 35.9 to 48.3 days for *Ppd* (Figure 1E). Across all progenies with the early genotype *ppd*, the average is 34.8 days to flowering, in contrast to the 41.6-day average across all the F_8 progenies of *Ppd*.

Fifty-nine of the 63 progenies with *ppd* contribute negatively to the largest source of the G × E interaction effect on the days to flowering (Figure 1E). This range of contributions is −1.37 to 0.56 $days^{0.5}$, and the average is −0.56 $day^{0.5}$. On the contrary, all but eight of the 56 progenies with genotype *Ppd* contribute positively; this range is −0.60 to 2.02 $days^{0.5}$ and its average is 0.63 $day^{0.5}$.

The LD and HT environments contribute heavily and positively to this largest G × E interaction effect, while the SD and LT environments contribute heavily and negatively (Figure 1E). Therefore, both the LD and the HT environments interact positively with the late-flowering genotype *Ppd* to delay its flowering (i.e., to cause photoperiod gene activity), but both environments interact negatively with the early-flowering genotype *ppd* to strengthen its gene action toward earlier flowering. (Readers should consult Chapters 42 and 43 if they do not understand that both positive and negative contributions to the G × E interaction effect can result in both positive and negative G × E interaction effects.)

That the contribution to the G × E interaction effect on the days to flowering is positive for most F_8 progenies having the photoperiod-sensitive genotype *Ppd* while the contribution by most progenies with the insensitive genotype *ppd* is negative indicates that the *Ppd* and *ppd* genotypes cause a G × E interaction effect of opposite sign in all environments. That is, whenever the G × E interaction effect within a given environment increases the days to flowering of genotype *Ppd*, the G × E effect by *ppd* is a decrease of the days, and vice versa.

The F_8 progenies of the *ppd* and *Ppd* genotypes appear to respond alike in one way (Figure 1E): that for both genotypes, the more positive a progeny's contribution to the G × E interaction effect, the larger that progeny's average days to flowering across all four environments. The correlation between the contribution and the average days is 0.47 for progenies of *ppd* and 0.68 for *Ppd*.

INTERPRETATION OF THE SECOND SOURCE OF G × E INTERACTION. Contributions to the second source of G × E interaction by both the SD and LT environments are near zero (−0.3 and 0.3 $day^{0.5}$) (Figure 1F) rather than large and negative (−3.7 and −4.6 $days^{0.5}$) like their contributions to the largest source of G × E interaction (Figure 1E). The near-zero contributions by both SD and LT indicate that these two environments cause little G × E interaction. That the two homozygous genotypes both flower simultaneously in both the SD and LT environment (Figure 1B and C) explains why these environments cause little of this second-largest G × E interaction effect; both prevent the photoperiod activity of *Ppd*, thereby causing this genotype to flower early, like *ppd*.

The largest negative contribution is by the LD environment (−5.0 $days^{0.5}$), while the largest positive contribution is by the HT environment (5.3 $days^{0.5}$). Alleles *ppd* versus *Ppd* cause little if any of this second source of the G × E interaction. This is shown by 19 (34%) of the 56 homozygous F_8 progenies having the sensitive allele *Ppd* contributing positively to this second source while 37 (66%) contribute negatively, and is reinforced by 23 (37%) of the progenies of *ppd* contributing negatively and 40 (63%) contributing positively. The average contribution by progenies with *Ppd* is –0.19 day$^{0.5}$, while for *ppd* it is 0.17 day$^{0.5}$. Thus many contributions by both *ppd* and *Ppd* are near zero. Only 49% (31 progenies) of genotype *Ppd* and 17% (11 progenies) of *ppd* contribute with magnitude greater than ±0.5 day$^{0.5}$; in contrast, 57% and 65% cause contributions of ±0.5 day$^{0.5}$ or larger to the largest source of the G × E interaction.

Of the progenies with contribution larger than ±0.5 day$^{0.5}$, 19 of *Ppd* contribute negatively and 12 positively, while one of *ppd* contributes negatively and eight positively. A positive contribution to this second source of G × E interaction indicates a positive G × E interaction effect (later than average days to flowering) whenever the F$_8$ progeny is grown in the HT environment, but a negative interaction effect (relatively earlier flowering) when grown in the LD environment.

For the same progeny and environment, the G × E interaction effect due to the second physiological genetic source is additive to the negative or positive G × E interaction effect due to the largest source. Thus the two G × E interaction effects account jointly for 96% of the total G × E interaction effect on the days to flowering.

We interpreted (above from Figure 1A to D) the skewed monomodal distribution of the days to flowering in the HT environment (Figure 1D) to result because some F$_8$ progenies of genotype *ppd* shift toward relatively later flowering, while a larger number of genotype *Ppd* shift toward relatively earlier flowering than their average in the LD environment. Here we interpret the second source of G × E interaction to be the cause of this shift from an indisputable bimodal distribution of the days to flowering in the LD environment (Figure 1A) to a skewed monomodal distribution in the HT environment (Figure 1D).

Control of a G × E Interaction Effect by the Genetic Background   The only major genetic control in this cross is by one gene: the photoperiod gene shown to be genotype *Ppd*. That about one-third of the F$_8$ progenies of both this gene's homozygous photoperiod-insensitive genotype *ppd* and its homozygous photoperiod-sensitive genotype *Ppd* contribute to the second source of G × E interaction with sign opposite that of the other two-thirds (Figure 1F) indicates that the genetic control for this source is by other than the photoperiod gene. We consider here also that in all four additional crosses, the HT environment causes a large positive contribution to one of the statistically significant G × E interaction effects, for which the LT environment causes a large negative contribution. Thus high and low temperature cause the opposite change—toward earlier or later flowering. Therefore, because the major genetic control is by only one gene, but this gene is not causal of this pattern of G × E interaction, we interpret this second-largest cause of a G × E interaction effect on the days to flowering to be controlled by the environmental factor of high temperature versus lower temperature. We interpret the causal gene actions to be small effects by many background genes on the days to flowering, with the temperature-modulating positive or negative effects on the days to flowering by each of these many background genes. This interpretation will be reinforced as the results for each of the four additional crosses are presented.

C.   Four Crosses Show Major Genetic Control over the Days to Flowering by a Nonphotoperiod Gene

1. The Cross Rabia de Gato (Photoperiod Insensitive) × Quetzal (Moderately Photoperiod Sensitive)

FREQUENCY DISTRIBUTIONS INDICATE MAJOR CONTROL BY A SINGLE NONPHOTOPERIOD GENE. The distribution of the days to flowering of the F$_8$ progenies of the cross Rabia de Gato × Quetzal is bimodal in the LD (Figure 2A) SD (Figure 2B) and LT (Figure 2C) environments. This indicates that the major control is not by a photoperiod gene, since the early and late genotypes are differentiated only in the noninductive SD

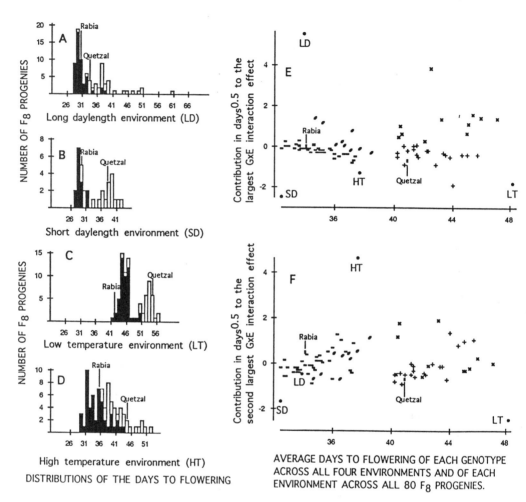

Figure 2 Distributions of the days to flowering of 80 F8 progenies of the cross between the photoperiod-insensitive cultivar Rabia de Gato and the moderately photoperiod-sensitive cultivar Quetzal when grown in long-daylength (A), short-daylength (B), low-temperature (C), and high-temperature (D) environments. Graphs E and F are biplots showing (1) the average days to flowering across the four environments of each F8 progeny plotted against the contribution to the largest (E) and second-largest (F) G × E interaction effect on these days by the same progeny, and (2) the average days to flowering across all 80 F8 progenies of each of the four environments plotted against the contribution by that environment to the largest (E) and second-largest (F) G × E interaction effect on these days. The origin of the differential marking of the early- and late-flowering F8 progenies is according to their nonoverlapping averages within the biplot of the largest G × E interaction effect (graph E). Within the distributions (graphs A, B, C, D) each of the same F8 progenies is designated as early or late. In graphs E and F all of the early-flowering progenies are indicated with the symbol – or /, and all late-flowering progenies with the symbol + or ×.

environment. Rather, control is by a single nonphotoperiod gene. The early and late genotypes of the nonphotoperiod gene are expressed within the HT environment (Figure 2D), but this is apparent only when the two genotypes are differentially marked. In the HT environment (Figure 2D), most of the early genotypes again flower early, but this is masked by a large number of the late genotypes flowering relatively earlier plus a few flowering relatively later.

THE LARGEST G × E INTERACTION EFFECT INDICATES MAJOR CONTROL BY A NONPHOTOPERIOD GENE PLUS PARTIAL CONTROL BY A PHOTOPERIOD GENE. The early versus late distributions of the days to flowering are obvious in the LD (Figure 2A), SD (Figure 2B), and LT (Figure 2C) environments. Rather than being based on one or another of these three distributions, all differential marking of the early and late nonphotoperiod gene in Figure 2 is based on the AMMI-determined nonoverlapping averages (Figure 2E). The largest G × E interaction effect shows the 50 F_8 progenies with the early genotype to have average days to flowering across all four environments, ranging from 32.3 to 38.4 days (Figure 2E). The 30 progenies with the late genotype have averages between 40.3 and 47.1 days. Because there is no overlap between these early and late averages (Figure 2E), the differential markings of the early versus late genotypes of the nonphotoperiod gene are based on them, as shown in Figure 2E. Flowering of the parent Rabia de Gato is with the early genotype, and that of the parent Quetzal is with the late group.

Within the largest G × E interaction effect, nine of the 50 early-flowering genotype (those in Figure 2E represented by the symbol /) and 10 of the 30 late-flowering genotypes (those in Figure 2E represented by the symbol ×) have larger positive contributions than the rest of the F_8 progenies of the corresponding early and late genotypes of the nonphotoperiod gene. The contribution by the LD environment is also positive. Therefore, in the LD environment each of these nine and 10 progenies causes a large and positive G × E interaction effect. That is, the LD delays the days to flowering beyond the progeny's average across all four environments. Expression of a photoperiod response is reinforced by the SD environment, causing the largest negative G × E interaction effect on the days to flowering of these nine and 10 F_8 progenies. A photoperiod response is expected for this cross because, independently, the parent Rabia de Gato has been shown to be photoperiod insensitive, while Quetzal is moderately sensitive (J. W. White, personal communication).

In summary, both the bimodal distributions (Figure 2A to D) and the largest G × E interaction effect (Figure 2E) show the major control over the days to flowering of the cross Rabia de Gato × Quetzal to be by a nonphotoperiod gene. The largest G × E interaction effect shows, additionally, that a subgroup within both the early and late genotypes responds to the LD environment with a delay of flowering. This indicates that photoperiod gene activity provides a second major control over the days to flowering.

G × E INTERACTION CAUSED BY THE GENETIC BACKGROUND. Within the second-largest G × E interaction effect on the days to flowering of the 80 F_8 progenies of the cross Rabia de Gato × Quetzal (Figure 2F), 23 of the 50 early-flowering F_8 progenies (Figure 2A to D) interact positively with the HT environment to delay their flowering beyond their average (Figure 2F). Eleven of the 30 late-flowering F_8 progenies also interact positively. All of these progenies interact negatively with the SD, LD, and LT environments to decrease the days to flowering of the progeny, whereas these environments interact positively with the remaining 25 and 19 F_8 progenies of the early and late genotypes of the nonphotoperiod gene. This same pattern of a G × E interaction effect will be shown to occur, with minor modifications, within three of the five crosses, while major modifications of it occur in the two crosses that involve the most photoperiod gene activity. As discussed for each of the two previous crosses, we interpret this G × E interaction pattern to result from modulation by high versus lower temperature of the effects on the days to flowering by differences in the many background genes.

2. The Cross Rabia de Gato (Insensitive) × Ju-80 (Insensitive)

FREQUENCY DISTRIBUTIONS INDICATE MAJOR CONTROL BY ONLY A NONPHOTOPERIOD GENE. The 84 F_8 progenies are marked as divided between an early-

and a late-flowering genotype within the frequency distribution of the days to flowering of the LD (Figure 3A) and SD (Figure 3B) environments and (but only with insight provided by the differential markings) the LT environment (Figure 3C). The differential marking of the early versus late genotypes of the nonphotoperiod gene is based on average days to flowering of the progenies in the biplot of the AMMI analysis; the early genotype includes four progenies that would be in the late genotype if the basis were the histogram distributions in the LD environment (Figure 3A). Within the HT environment, enough of the progenies of the early-flowering genotype are shifted toward later flowering by the HT environment, while a larger proportion of the late genotype is shifted toward earlier flowering, that the major control by a single nonphotoperiod gene is masked (Figure 3D). The same cause of masked gene action was described previously for the cross with major control by only a photoperiod gene (Figure 1D),

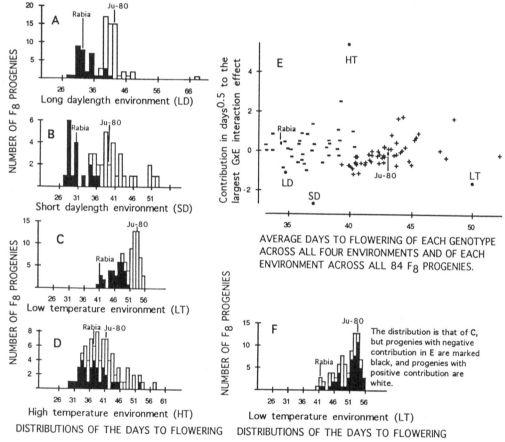

Figure 3 Distributions of the days to flowering of 84 F$_8$ progenies of the cross between the two photoperiod-insensitive cultivars Rabia de Gato and Ju-80 when grown in long-daylength (A), short-daylength (B), low-temperature (C), and high-temperature (D) environments. Graph E is a biplot showing the average days to flowering across the four environments of each F$_8$ progeny plotted against the contribution to the largest (E) and second-largest (F) G × E interaction effect on these days by the same progeny; there is only one statistically significant G × E interaction effect on these days. The origin of the differential marking of the early and late flowering F$_8$ progenies is according to their relatively distinct distributions within the biplot of the largest G × E interaction effect (graph E). Graph F is the distribution of graph D, but distinguishes between progenies with a positive versus a negative contribution to the G × E interaction effect.

and for the cross of the photoperiod insensitive Rabia de Gato × the moderately photoperiod-sensitive Quetzal (Figure 2D). This same masking of the control by a major gene also occurs for the two crosses yet to be discussed.

Independent testing (J. W. White, personal communication) has indicated that both Rabia de Gato and Ju-80 are photoperiod insensitive. In agreement, the bimodal distribution in both the LD (Figure 3A) and SD (Figure 3B) environments, with backup by the biplot of the AMMI analysis (Figure 3E), indicates lack of photoperiod gene activity.

The larger proportion of the 37 F_8 progenies of the early genotype occur near the beginning of flowering for this genotype in each of the LD (Figure 3A), SD (Figure 3B), and HT (Figure 3D) environments. On the contrary, in the LT environment (Figure 3C) there is gradual increase from flowering of few to many progenies on successive days. Such a gradual increase also occurs for the 47 progenies of the late-flowering genotype of this nonphotoperiod gene. Such a consistent increase in the number of progenies that flower on successive days does not occur in any environment for any other of the five crosses.

PROPORTIONAL CAUSES OF THE VARIATION. The AMMI analysis shows 26% of the overall variation in days to flowering to be caused by difference among the genotypes, 61% by difference among the environments, and 13% to be due to the G × E interaction effect on the days. There is but one statistically significant physiological genetic source of the G × E interaction effect on the days to flowering. It accounts for 63% of the 13% of the total variation that is due to the G × E interaction.

INTERPRETATION OF THE SINGLE SOURCE OF G × E INTERACTION. Both the SD and LD environments contribute negatively to the G × E interaction effect on the days to flowering (Figure 3E). This reinforces the lack of a photoperiod response indicated by the distributions of the days to flowering (Figure 3A to D). The LT environment also causes a negative contribution of similar magnitude. On the contrary, the HT environment contributes positively. Therefore, this G × E interaction effect on the days to flowering is due to a factor(s) of the HT environment that is at a different level in all three other environments. We interpret this single G × E interaction effect to be due to the highest of all temperatures being within the HT environment.

EFFECTS OF HIGH VERSUS LOW TEMPERATURE ON DAYS TO FLOWERING. For the distribution shown in Figure 3F, all F_8 progenies that contribute negatively to the single statistically significant G × E interaction effect (Figure 3E) have been marked. This distribution (Figure 3F) shows that the LT environment interacts similarly with both the early and late genotypes of the nonphotoperiod gene. Both genotypes show delay of flowering by the LT environment of the F_8 progenies, which have intermediate to latest flowering in the histogram for the LT environment (Figure 3C). The SD and LD environments contribute negatively, as does the LT environment (Figure 3E). Therefore, these three environments all delay the flowering of some of the progenies belonging to both the early and late genotypes of the nonphotoperiod gene, which indicates that genetic control is not by the contrasting genotypes of the nonphotoperiod gene. Although not illustrated with a histogram, the negative contribution of these genotypes multiplied by the positive 5.38 days$^{0.5}$ of the HT environment indicates a negative G × E interaction effect. The resulting change toward earlier flowering in the HT environment of these progenies with negative contribution is about 3.3 times larger than their delay of flowering in the LT environment. This is indicated in Figure 3E by the fact that the magnitude of the positive contribution by the HT environment is 3.3-fold larger than the magnitude of the negative contribution by the LT environment.

G × E Interaction Due to the Genetic Background The pattern of this single G × E interaction effect on the days to flowering (Figure 3E) is nearly identical to that of the second-largest G × E effect (Figure 2F) of the cross Rabia de Gato × Quetzal. We again interpret the pattern as being due to limited control over the days to flowering by each of many background genes, plus the modulation by high versus low temperature of the effect of each of these genes on the days to flowering. The early versus late genotypes of the nonphotoperiod gene do not interact differently within this G × E interaction effect.

3. The Cross Ju-80 (Insensitive) × San Martin (Sensitive)

FREQUENCY DISTRIBUTIONS IN SD AND LT ENVIRONMENTS INDICATE MAJOR CONTROL BY A SINGLE NONPHOTOPERIOD GENE. The frequency distributions in the SD and LT environments (Figure 4) are like those of the cross Rabia de Gato × Ju-80 (Figure 3) and the cross Rabia de Gato × Quetzal (Figure 2) in that they are bimodal in both the SD and LT environments. The early and late genotypes are differentiated most clearly in the LT environment (Figure 4C). Expression of both genotypes in the SD environment indicates a nonphotoperiod gene. Within both the LD (Figure 4A) and HT (Figure 4D) environments, the distributions of the early and late genotypes of the nonphotoperiod gene are unlike the distributions within these environments for any of the three crosses discussed previously (Figures 1 to 3). For this cross (Ju-80 × San Martin), in both the LD and HT environments, the time span of flowering of the early genotype overlaps fully with the span of the late-flowering genotype.

PROPORTIONAL CAUSES OF THE VARIATION. The differences in genotype among the F_8 progenies cause 23% of the total variation in the days to flowering, differences among the environments cause 57%, and the G × E interaction causes 20%. The largest physiological source of the G × E interaction effect on the days to flowering causes 63% of the total variation due to G × E interaction. The second largest source is causal of 31%. Thus two statistically significant sources account for 94% of the total G × E interaction effect on the days to flowering of the 86 F_8 progenies.

PATTERN OF THE LARGEST SOURCE OF G × E INTERACTION. The largest source of the G × E interaction effect on the days to flowering has the same pattern of contributions by the environments (Figure 4E) as that of the second-largest source within the cross Rabia de Gato × Quetzal (Figure 2F), which is the same as the single statistically significant source of G × E interaction of the cross Rabia de Gato × Ju-80 (Figure 3E). The pattern is that the HT environment causes the largest positive contribution to the G × E interaction effect on the days to flowering, while the SD, LD, and LT environments all cause nearly equal negative contributions with magnitude about 30% as large as that of the HT environment.

Further, for the cross Rabia de Gato × San Martin, its second-largest source of G × E interaction (Figure 1F) is a modification of the environmental contributions (Figure 4F) that occur in this cross plus the two additional crosses mentioned in the preceding paragraph (Figures 2F and 3E). The modifications for Rabia de Gato × San Martin are (Figure 1F): the contribution by the SD environment is near zero while still being negative, and the contribution by the LT environment is also near zero, although positive. The pattern of the largest source for the fifth cross (Quetzal × San Martin) will be shown to be yet another modification (Figure 5E).

G × E Interaction Caused by the Genetic Background   The pattern of the G × E interaction that occurs in all five crosses involves several features. The first is that this pattern may be either the largest or the second-largest G × E interaction effect. A second is that its largest positive contribution is always due to the HT environment. The third is that the magnitude of this contribution by the HT environment is severalfold larger than the contributions by any of the other environments. For three of the five crosses, all three other environments cause a relatively large negative contribution (Figures 2F, 3E, and 4F). Two modifications of this G × E interaction pattern are that one (the LD environment, Figure 5E) or two environments (the

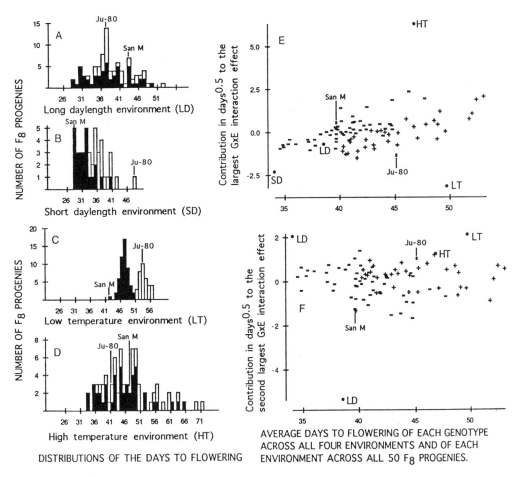

Figure 4 Distributions of the days to flowering of 88 F_8 progenies of the cross between the photoperiod-insensitive cultivar Ju-80 and the moderately photoperiod-sensitive cultivar San Martin when grown in long-daylength (A), short-daylength (B), low-temperature (C), and high-temperature (D) environments. Graphs E and F are biplots showing (1) the average days to flowering across all four environments for each F_8 progeny plotted against the contribution to the largest (E) and second-largest (F) G × E interaction effect on these days by the same progeny, and (2) the average days to flowering across all 88 F_8 progenies of each of the four environments plotted against the contribution by that environment to the largest (E) and second-largest (F) G × E interaction effect on these days. The origin of the differential marking of the early- and late-flowering F_8 progenies is according to their distribution within the low-temperature environment (graph C). In graphs E and F all of the early-flowering progenies are indicated with the symbol −, and all late-flowering progenies with the symbol +.

LT and SD environments, Figure 1F) cause an intermediate contribution. A fourth feature is that the F_8 progenies of the early and late genotypes of the nonphotoperiod gene (Figures 2F, 3E, 4E, and 5E) have low-to-high averages across all environments. These averages may be equal (Figures 1F, 2F, 3F, 4E, and 5E), but their points on the biplot for the progenies of the early and late genotypes remain separated by differences in their contributions to the G × E interaction effect. A fifth feature is that an increasingly positive contribution by the individual F_8 progenies within each of the two genotypes with the major control is correlated positively with the average days to flowering of the F_8 progeny (Figures 1F, 2F, 3E, 4E, and 5E). Again, we interpret these features of this pattern of G × E interaction to result from modulations by high versus lower temperature of the small effects due to each of many background genes on the days to flowering.

PATTERN OF THE SECOND-LARGEST SOURCE OF G × E INTERACTION. The LD environment contributes negatively to the second-largest physiological-genetic cause of a G × E interaction effect on the days to flowering of the F_8 progenies from the cross Ju-80 × San Martin (Figure 4F), and its magnitude is the largest (-5.36 days$^{0.5}$) of all the contributions by an environment. Each of the SD, LT, and HT environments, respectively, causes a lower magnitude of positive contributions (2.05, 2.11, and 1.21 days$^{0.5}$). Rather than large positive correlation (as occurs for the pattern that is common to all of the crosses, including its modifications), the correlations between the contributions and the average days to flowering are -0.04 across all 81 F_8 progenies, -0.41 across just the 52 progenies of the early genotype of the nonphotoperiod gene, and -0.17 across the 34 progenies of the late genotype.

As presented numerically in the preceding paragraph, the magnitude of the negative contribution by the LD environment is 2.5-fold larger than the magnitude of the largest positive contribution, which is by the LT environment, with the next largest being by the SD environment. Part of the F_8 progenies of both the early and late genotypes of the nonphotoperiod gene contribute positively, while part of both contribute negatively (Figure 4F). Those progenies that contribute negatively interact positively with the LD environment (i.e. the LD environment delays their flowering, which is a photoperiod response). Verifying a photoperiod response is that most of the progenies that flower latest within the LD environment contribute negatively to this second-largest G × E interaction effect on the days to flowering.

4. The Cross Quetzal (Moderately Photoperiod Sensitive) × San Martin (Strongly Photoperiod Sensitive)

FREQUENCY DISTRIBUTIONS INDICATE MAJOR CONTROL BY A SINGLE NONPHOTOPERIOD GENE. Bimodal distribution of the days to flowering of the 50 F_8 progenies occurs in both the SD (Figure 5B) and LT (Figure 5C) environments. This indicates an early and a late flowering phenotype, with the major control being by a nonphotoperiod gene. The days to flowering of this early and late genotype overlap almost completely in both the LD (Figure 5A) and HT (Figure 5D) environments. The distributions show no evidence of a response to photoperiod, even though the parent Quetzal is known to be moderately sensitive to photoperiod and San Martin to be highly sensitive.

PROPORTIONAL CAUSES OF THE VARIATION. Differences in genotype are causal of 27% of the total variation in the days to flowering, differences in the environments are causal of 51%, and the G × E interaction is causal of the remaining 22%. The largest physiological-genetic source of the G × E interaction accounts for 63% of the total G × E interaction effect, and the second largest accounts for 31%. Together, the two sources account for 94% of the G × E interaction, which is for 21% of the total variation in the days to flowering.

INTERPRETATION OF THE LARGEST SOURCE OF G × E INTERACTION. Both of the two statistically significant sources of G × E interaction (Figure 5E and F) indicate delay of the days to flowering by the LD environment. This does demonstrate the control expected over these days to flowering by the photoperiod gene activity that should be inherited from both parents.

The biplot for the largest G × E interaction effect (Figure 5E) is the second modification of that G × E pattern, which is common to all five crosses (see Section III.3.b). The first is the second-largest G × E interaction effect when the only major control is by photoperiod gene activity (Figure 1F). Like the distributions of the days to flowering (Figure 5A to D), for this

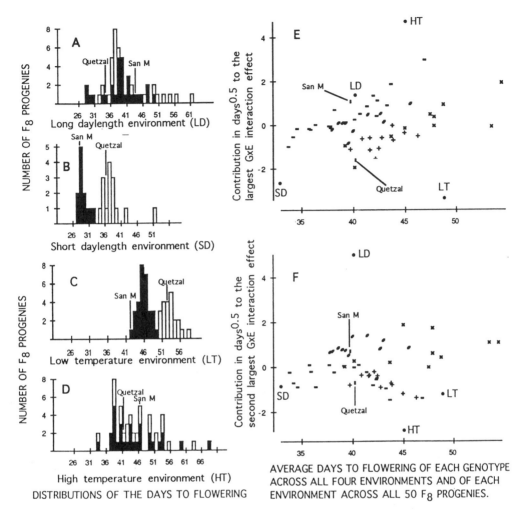

DISTRIBUTIONS OF THE DAYS TO FLOWERING

AVERAGE DAYS TO FLOWERING OF EACH GENOTYPE
ACROSS ALL FOUR ENVIRONMENTS AND OF EACH
ENVIRONMENT ACROSS ALL 50 F_8 PROGENIES.

Figure 5 Distributions of the days to flowering of 50 F_8 progenies of the cross between the moderately photoperiod-sensitive cultivar Quetzal and the highly photoperiod-sensitive cultivar San Martin when grown in long-daylength (A), short-daylength (B), low-temperature (C), and high-temperature (D) environments. Graphs E and F are biplots showing (1) the average days to flowering across all four environments for each F_8 progeny plotted against the contribution to the largest (E) and second-largest (F) G × E interaction effect on these days by the same progeny, and (2) the average days to flowering across all 88 F_8 progenies of each of the four environments plotted against the contribution by that environment to the largest (E) and second-largest (F) G × E interaction effect on these days. The origin of the differential marking of the early- and late-flowering F_8 progenies is according to their distribution within the low-temperature environment (graph C). In graphs E and F all of the early-flowering progenies are indicated with the symbol − or /, and all late-flowering progenies with the symbol + or ×.

cross (Figure 5E) as for the one with control by only a photoperiod gene (Figure 1F), the modified pattern (Figure 5E) also shows that the G × E effect is opposite for half of the progenies of both the early and late genotypes of the gene with the major control. For this cross, the contribution to this largest G × E interaction effect by the HT environment of 4.68 days$^{0.5}$ is 3.5-fold larger than the positive contribution (1.35 days$^{0.5}$) by the LD environment (Figure 5E). Thus the major control by the nonphotoperiod gene and the photoperiod gene cause a G × E interaction effect that results in the same delay or else earlier flowering of each F_8 progeny, but the positive G × E effect in addition to that of the photoperiod gene is 3.5-fold larger than the effect by the photoperiod gene activity. Any delay of flowering by the LD environment is a photoperiod response, which is reinforced by the following: Within this largest G × E interaction effect, both the SD and LT environments cause the opposite of the G × E interaction effect (earlier or later flowering) caused by the LD environment (Figure 5E).

In summary, the largest G × E interaction effect (Figure 5E) of this cross, between cultivars with different levels of sensitivity to photoperiod, shows the major (largest) control to be by a nonphotoperiod gene and shows a lower level of simultaneous major control by a photoperiod gene. Further, the much larger effect by the HT environment is a modification of a pattern of G × E interaction that occurs regardless of whether major control is by only the photoperiod gene (Figure 1F) or is by only the nonphotoperiod gene (Figure 3E).

INTERPRETATION OF THE SECOND-LARGEST G × E INTERACTION EFFECT ON THE DAYS TO FLOWERING. The contribution to the second-largest physiological genetic cause of a G × E interaction effect by the LD environment is large and positive (5.0 days$^{0.5}$), while contribution by the SD environment is negative (–0.89 day$^{0.5}$). This indicates that as for the largest cause of the G × E interaction, the second largest also involves photoperiod gene activity. The contribution by the HT environment is negative (–2.84 days$^{0.5}$) (Figure 5F) rather than the largest positive contribution as it is for the largest physiological-genetic cause of the G × E interaction (Figure 5E).

The Largest Major Genetic Control Is by the Nonphotoperiod Gene   The average days to flowering plotted on the *x*-axis of the second-largest G × E interaction effect is the same as that of the first. Therefore, from conclusions about the largest cause, one interpretation within the second-largest G × E interaction effect is that its largest main effect by a genetic factor (the largest effect on the average days to flowering of the progeny) depends on whether the F_8 progeny has inherited the early or the late genotype of the nonphotoperiod gene.

The Second Level of Major Control Is by the Photoperiod Gene   A second interpretation is that the nonphotoperiod gene interacts with the photoperiod gene, as also shown within the largest G × E interaction. This interpretation is based on the following evidence for a low and a high level of photoperiod gene activity within both the early and late genotypes of the nonphotoperiod gene activity.

Of the 29 progenies that in theory must be homozygous for the early nonphotoperiod gene, 11 have positive contributions of larger magnitude than the other 18 (Figure 5F), which suggests that they have the homozygous highly photoperiod-sensitive genotype of the parent San Martin. The 18 progenies that cause the more negative contributions to the G × E interaction are interpreted as having inherited the homozygous genotype that has less sensitivity to photoperiod of the parent Quetzal. These 18 progenies are identified in Figure 5F by the symbol –. Of the 18, some with intermediate averages cause positive contributions, with the largest being 0.07 day$^{0.5}$; progenies with the lowest averages contribute negatively (magnitude as large as –0.91 day$^{0.5}$), and those with the highest averages cause larger negative contributions (as large as –1.50 days$^{0.5}$). Thus, across the range of average days to flowering of these 18 F_8 progenies, the association between changes in contribution to the G × E interaction and the average days across all four environments results in an arch-shaped curve on the biplot (Figure 5F).

Of the 11 progenies with the larger positive contributions, which are symbolized by / in Figure 5F, the six with the smaller averages cause positive contributions between 0.52 and 0.84 day$^{0.5}$, while contributions of the five, all but one with still larger averages, range from 0.79 to 1.36 days$^{0.5}$ (Figure 5F). All 11 of these F_8 progenies are interpreted as being homozygous for the early genotype of the nonphotoperiod gene, with the additional interpretation that their larger magnitude of positive contribution to the G × E interaction (delay of flowering by the LD environment) indicates homozygosity also for the highly photoperiod-sensitive genotype of the parent San Martin. An interaction effect not seen before is that the contributions change toward a less positive (more negative) level as the average of the progeny is larger. This indicates a smaller delay of flowering by the HT environment. We interpret this as occurring because, on average, the high temperature causes earlier flowering of both genotypes of the non-photoperiod gene (Figure 3E and F), and of the less sensitive genotype (allele) of the photoperiod gene (Figure 1E), but causes later flowering of the sensitive genotype of the photoperiod gene (Figure 1E).

As within the early nonphotoperiod genotype, the biplot (Figure 5F) of the second-largest physiological genetic cause of a G × E interaction effect also subdivides the 20 homozygous F_8 progenies of its late genotype among at least three groups. One group of 10 progenies all contribute negatively, and of these the progeny with the lowest average (39.7 days) contributes -0.79 $day^{0.5}$ to the G × E interaction, one progeny has an intermediate average of 42.1 days, and its contribution is -0.20 $day^{0.5}$, and the progeny with the highest average (46.3 days) contributes -1.38 $days^{0.5}$. Thus for these progenies that are homozygous for the late genotype of the nonphotoperiod gene and are interpreted as being homozygous also for the lower level of sensitivity to photoperiod inherited from Quetzal, the association between the average days to flowering and the contribution to the G × E interaction results in an arch-shaped curve that is like the curve already described for the early nonphotoperiod genotype when it is also homozygous for this lower level of sensitivity to photoperiod.

Suggestions for Control by Modifying Genes   The several distinct magnitudes and positions on the biplot (Figure 5F) of the contributions by the 10 F_8 progenies interpreted as having the late genotype of the nonphotoperiod gene suggest two and possibly three subgroups (Figure 5F, symbolized by ×). Of the 10, for a subgroup of five that includes both the lowest averages and the positive contributions of lesser magnitude, the arch-shaped curve of the relationship between the average and the contribution to the G × E interaction effect is again apparent. For the four with the largest average days to flowering and the largest of all positive contributions, which leaves but two progenies in a seemingly third and intermediate grouping, the contribution changes toward a less positive magnitude as the average becomes larger. This is just as described previously for the early genotype nonphotoperiod gene when also homozygous for the sensitive genotype of the photoperiod gene.

IV.   OVERALL INTERPRETATIONS OF THE G × E INTERACTION EFFECTS

A.   Control by Two Major Genes

The interpretations presented above for the largest (Figure 5E) and second-largest (Figure 5F) G × E interaction effect of the cross Quetzal × San Martin are the simplest possible based on segregation of two major genes: a nonphotoperiod gene and a photoperiod gene, and the single-seed descent of each to a statistically probable level of 98.5% homozygosity. The bimodal frequency distribution between the early- and late-flowering genotypes of the nonphotoperiod gene are evident in the SD (Figure 5B) and LT (Figure 5C) environments, because the photoperiod gene expresses little, if any, activity in these environments (Figure 1E). On the contrary, in the LD and HT environments, activity of the photoperiod gene is expressed. Therefore, control over days to flowering involves the larger control by the nonphotoperiod gene interacting with a smaller but also major (i.e., relatively large and simultaneous) control over these days by the photoperiod gene (Figure 1).

B. Control by Minor Genes

Evidence for control by minor genes is limited. A larger number of F_8 progenies would facilitate better evaluation of the suggested control by minor genes. The evidence for minor control is the two or possibly three levels of positive contribution to the G × E interaction which attend control by the homozygous highly photoperiod-sensitive genotype of the photoperiod gene (Figure 5F; Section III.4.d) and either the early or late homozygote of the nonphotoperiod gene.

C. Control by the Biological System as a Whole

Here we ask: When the only major genetic control is by the photoperiod gene (Figure 5), why does quantitative variation of the days to flowering occur across the homozygous F_8 progenies of the early-flowering photoperiod-insensitive genotype? Similarly, why does quantitative variation of the contributions to G × E interaction occur also? Why do these same quantitative variations occur across the homozygous late-flowering photoperiod-sensitive genotype? Why are the flowering phenotypes controlled by these major genes not differentiated qualitatively? The same questions also apply to both the early and late genotypes of the nonphotoperiod gene.

For both the early- and late-flowering genotypes of the photoperiod gene when that gene is the sole major genetic control (Figure 1A), and for both the early and late genotypes of the nonphotoperiod gene with sole major control (Figure 3E), there is a G × E interaction effect for which the days to flowering of part but not all of the progenies of each genotype are quantitatively increased by high temperature but the days are quantitatively decreased for the other part.

Temperature can be expected to modulate the activity of every gene. It could be that much, even most, of the quantitative variation of the days to flowering is due to modulation by temperature of many background genes, which in comparison with the major controls by a few genes, cause small (even less than minor) control over the days to flowering. If this is so, the overall genotype of each F_8 progeny, which will vary with all the background genes that differ between the two parents, will differ from one F_8 progeny to another. This is the apparent situation. If this is the case, the quantitative variation of the days to flowering that results from these many genes with small effects and from their many small G × E interactions with the many factors of the overall environment account for much of the quantitative variation of the days to flowering. Control by many genes with a small effect is the classical explanation for a quantitatively variable trait.

In summary, we suggest that the positive correlation between the contribution to the G × E interaction by the F_8 progenies and their average days to flowering, which occurs whether the only major genetic control is by the photoperiod gene or by the nonphotoperiod gene, results from small controls by each of many background genes and the many positive and negative modulations of their different gene actions by the temperature.

V. TOOLS THAT FACILITATE ANALYSIS OF THE PHYSIOLOGICAL GENETICS OF QUANTITATIVE TRAITS

A. Initial Knowledge About a Quantitative Trait

This research began with the following knowledge [4] about the quantitative variation of the days to flowering of four Guatemalan bean (*Phaseolus vulgaris* L.) cultivars. Cultivar Rabia de Gato flowers earlier than the other three at low (18°C), intermediate (23°C), and high (28°C) temperature. San Martin flowers near the same number of days after planting as Rabia de Gato at 18°C, later than Rabia de Gato but earlier than Quetzal and Ju-80 at the intermediate 23°C, but far later that the other three cultivars at 28°C. Flowering of both Ju-80 and Quetzal is extremely late at 18°C, with Ju-80 being the latest, but both change to being intermediate among the four cultivars at 28°C.

The smallest overall differences in days to flowering among the four cultivars occurs at the intermediate temperature of 23°C (Figure 1). Also, 23°C causes the fewest days to flowering for each cultivar [4,5] (Chapters 43 and 44). Therefore, 23°C is certainly near, if not the actual optimum temperature for rapid development to flowering of all four cultivars, as shown for three of them in Chapter 44.

As this research began, the known insight about the genetic and environmental controls was knowledge that San Martin is a highly photoperiod-sensitive genotype and that photoperiod gene activity is amplified as the temperature rises [5,8] (compare Chapters 43 and 44). It was also known that Rabia de Gato is a photoperiod-insensitive genotype.

B.   Factors That Constrain Quantification of the Genetic, Environmental, and G $\times$ E Interaction Controls over Quantitative Variation

Our first attempts [1] to elucidate genetic control over the days to flowering of bean demonstrated [1] that the genotype $\times$ environment (G $\times$ E) interaction effect on these days is large, which constrains the elucidation of inheritance. A further constraint is not knowing the expected or exact genotype of any progeny among the early (F_2 and F_3)-generation segregates. A further constraint is that a heterozygous genotype can reverse its phenotypic dominance between environments. Another constraint is the inability to repeat field environments from year to year. The consequent G $\times$ E interaction effects result in a seemingly different inheritance because of changes of the phenotypic ratio among the segregates [1]. Often, this negates intended verification of segregation ratios and inheritance pattern using F_3 progeny and backcross progeny tests. The following tools alleviate these constraints.

C.   Tools for Elucidation of the Major Genetic, Environmental, and G $\times$ E Interaction Controls over Quantitatively Variable Traits

1. Single-Seed Descent of the Segregates from Crosses Until Homozygosity Is Assured

Selfing of each F_2 plant derived from a cross, coupled with single-seed descent from each F_2 plant through the F_7 generation, provides F_8-generation seed. For each gene for which the two parents differ, every F_8-generation progeny has 98.5% probability of being homozygous for the allele of the genes contributed to the cross by one parent or of being homozygous for the allele of the other parent. Single-seed descent from the F_2 generation simultaneously establishes 50% probability that each F_8 progeny possesses the allele (genotype) of each parent, and 100% probability of having the allele of one parent or the other parent. The homozygosity of genotype expected for each F_8 progeny, and its 50% probability of being the genotype (and expressing the phenotype) of each parent for any major genetic control of the trait, result in an expected bimodal distribution across all the F_8 progenies for the trait in environments that cause a large difference in the expressed level of the trait. If a gene exerts the only major control, the distribution can be expected to have a 1:1 ratio if the number of F_8 progenies is adequate, if the G $\times$ E effect within the given environment does not prevent the activity of the gene and alter the phenotype of some of the progenies, and if there is no gene $\times$ gene interaction due to simultaneous control of the trait by other gene(s).

2. Comparison of Frequency Distribution of the Quantitative Levels Expressed in Each Environment to Differentiate the Major Genetic Controls

If a trait is actually controlled by a single major gene, then, seemingly, any G × E interaction effect on the expressed quantitative level of the trait would arise because one environment allows the gene action that enlarges the trait, while another environment reduces or prevents (modulates) that gene action. If two or more genes control the trait, part of the G × E effect will arise from interaction between the simultaneous and synergistic (or simultaneous but competitive effects) of the genes. Yet another part of this G × E interaction will arise from the different effect(s) by each environment on the level of activity of each of the multiple genes of the genetic background.

Histogram presentation of the quantitative distributions of the trait within multiple environments, selected because they modulate one extreme or the other of the gene activity, will present a bimodal distribution within any environment for which one gene provides a sufficiently large proportion of the control over the quantitative trait. The bimodal distribution will not show the expected 1:1 ratio, however, even with a large number of F_8 progenies, if either another gene causes smaller but also major control, and/or if G × E interaction effects additionally alter the level of the trait.

3. AMMI Model of Statistical Analysis

The AMMI model quantifies the proportion of the total variation of the quantitative trait that results from the variation among the tested genotypes, from the variation among the environments, and from the total G × E interaction effect on the trait. The facilitating merit of AMMI analysis is that in addition to these three additive effects, it quantifies the negative or positive contribution to the total G × E interaction effect that is due to each F_8 progeny and the contribution that is due to each environment. The latter two contributions are multiplicative effects; their product equals the G × E interaction effect for the referenced genotype grown in the referenced environment.

4. Biplot Presentation of the Results from AMMI Analysis

TOTAL NUMBER OF G × E INTERACTION EFFECTS. The total number of G × E interaction effects on the expressed level of a quantitative trait is no fewer than the number of genotypes multiplied by the number of environments. The AMMI analysis considers this product to be the number of treatments, in order to facilitate its quantification of the individual G × E interaction effects. For example, for one AMMI analysis of this study (Figure 1E), this number of treatments is the 119 homozygous F_8 progenies times the four environments. The product indicates at least 476 separate G × E interaction effects on the expressed days to flowering. The total of a G × E effect on the days to flowering is the sum of the individual G × E effects for each of the 476 combinations of one F_8 progeny grown in one environment. The facilitating merit of AMMI analysis is that the relative magnitude of each of the 476 separate G × E interaction effects is readily apparent from the 123 contributions to the G × E interaction effect on each biplot (Figure 1E) that presents a statistically significant G × E effect identified by the AMMI analysis. The 123 points on the biplot facilitate far more rapid comprehension of the relative magnitude of the 476 individual G × E effects than does direct comparison of all 476 individual G × E effects that collectively constitute the largest G × E interaction effect. A second biplot (Figure 1F), with another 123 points, facilitates rapid comparison of the relative magnitudes of the 476 individual G × E effects that constitute the second-largest physiological genetic source of a G × E interaction effect.

SIMPLIFIED COMPARISONS AMONG THE G × E INTERACTION EFFECTS. The graph discussed as an example (Figure 1E) is called a biplot because it displays the relative contribution to the referenced G × E interaction for each of the 119 individual F_8 progenies and simultaneously displays the relative contribution by each of the four environments. The 123 points facilitate rapid comparisons of all of the 476 separate G × E effects because each G × E effect is the product of the contribution by the genotype multiplied by the contribution by the environment. The ability to compare the relative G × E effects more rapidly than by direct comparison of all 476 combinations of one F_8 progeny and one environment is illustrated next.

The contributions by a progeny that are near zero [these attend the F_8 progenies with the central positions on the y-axis of the biplot (Figure 1E)], when multiplied by the always larger negative or larger positive contribution of each environment, indicate a G × E interaction effect that is near zero. When these F_8 progenies are grown in any environment, the resulting G × E interaction effect is near zero. That is, the days to flowering of the referenced genotype when grown in the referenced environment is near the average days to flowering of that genotype across all the environments. On the biplot, this average of each F_8 progeny and of each environment is plotted against the contribution to the G × E by the referenced genotype or the referenced environment.

The contributions by the F_8 progenies with the larger positive contributions (Figure 1E) multiplied by the always larger positive contributions by the LD and HT environments indicate a large positive G × E effect. The large positive G × E effect indicates that the short-day and high-temperature environments both delay the days to flowering of the F_8 progeny. Multiplying the large positive contributions of the short-day and long-day environments times the contributions of the F_8 progenies that cause negative contributions of the largest magnitude indicates a negative G × E effect, thereby showing that the long-day and high-temperature environments cause the days to flowering of the referenced F_8 progeny to be even earlier than its comparatively small (early) average days across all four environments.

The advantages of presenting AMMI results in a biplot are that (a) interpretation of the G × E interaction effect is achievable with far fewer points on a graph than for direct comparison of the larger number of individual G × E interaction effects, and (b) interpretation is more rapid because all contributions near zero for either the genotypes (the F_8 progenies in the illustration described) or the environments result in a minimal G × E effect. Therefore, these can be ignored, while the interpreting scientist concentrates on the environments and genotypes that cause the largest G × E interaction effects. The biplot display of the contributions shows both which progenies and which environments cause opposite G × E interaction effects (increases or decreases of the days to flowering) plus the relative magnitude of these effects.

VI. SUMMARY AND CONCLUSIONS

Four tools collectively facilitated elucidation of the major genetic and environmental controls over a quantitatively variable trait: the days to flowering of bean. The tools are (a) homozygous segregates from crosses among cultivars that express different levels of the trait, (b) frequency distributions of the levels of the quantitative trait that are expressed by the F_8 progenies within contrasting environments which modulate large quantitative differences in the expressed level of the trait, (c) statistical analysis using the additive main effects and multiplicative interaction (AMMI) model to quantify the genotype × environment interaction effects on the expressed quantitative level of the trait, and (d) presentation of the results of the AMMI analysis in a biplot. The biplot facilitates rapid visual comparison of the main effects by the genotypes, the main effect by the environments, and most important, of the genotype × environment (G × E)-caused variations of the trait. Previous statistical models have not adequately quantified these complex G × E interaction effects.

The days to flowering was recorded for each of a large number of F_8-generation progenies from each possible cross among four parental cultivars. The F_8-generation progenies, which were developed by single-seed descent from the F_2-generation plants, were used because each progeny has a 50% probability of inheriting the genotype of each parent for any gene for which the parents possess different alleles, plus a 98.5% probability of being homozygous for that gene. The progenies were grown in four environments: two with a large contrast in temperature and two with a large contrast in daylength.

Application of all four tools reveals that one photoperiod gene and one nonphotoperiod gene each provides major genetic control over the days to flowering. For this group of cultivars, the nonphotoperiod gene implements the larger control (most research reports refer to a nonphotoperiod gene as an earliness gene). The pattern of the G × E interaction suggests small detectable controls by a few minor genes.

Bimodal distributions of the days to flowering of the progenies that are homozygous for both—the late genotype of the nonphotoperiod and the photoperiod-sensitive genotype—display obvious control by only the nonphotoperiod gene, and only within the short-daylength and low-temperature environments, both of which prevent photoperiod gene activity. In the long-daylength and high-temperature environments, the distribution is monomodal; the major genetic controls interact so that both are masked. The resulting monomodal distribution suggests quantitative control by many genes. Even when not evidenced in the frequency distributions of the days to flowering, control by the photoperiod gene is demonstrated in the biplot graphs that present the G × E interaction effects quantified by the AMMI analysis.

For both the early and late genotypes of the nonphotoperiod gene, high temperature causes earlier flowering of about half of the F_8 progenies but causes later flowering of the other half of both, and low temperature causes the opposite change for each F_8 progeny. Further, the same occurs within both the early (photoperiod-insensitive) and late (photoperiod-sensitive) genotypes of the photoperiod gene. That the same temperature causes early flowering of some and later flowering of other of the F_8 progenies that are homozygous for both the early and late genotypes of both the nonphotoperiod gene and the photoperiod gene indicates that this G × E interaction effect is controlled by neither of these genes. Nevertheless, this is the largest G × E interaction pattern on the days to flowering whenever the nonphotoperiod gene is present in the population of F_8 progenies, and it is the second-largest G × E interaction effect even when the only major control is by the photoperiod gene alone.

We suggest that the occurrence of this pattern of G × E interaction in the results of all five crosses is because the many differences in background genes each interact with the temperature to cause small effects on the days to flowering. We propose also that this G × E interaction, due to many background genes, is causal of much of the quantitative variation in the days to flowering. It is reasonable for many background genes to be involved, since time to flowering is strongly controlled by the competitive partitioning of the photosynthate between growth of the reproductive organs and competitive growth of the vegetative organs (compare Chapters 41, 43, and 44). The evidence suggests that both the photoperiod gene and the nonphotoperiod gene cause essentially qualitative effects on the days to flowering, but the photoperiod gene causes large G × E interaction effects, which add to the quantitative variation of the days to flowering. The nonphotoperiod gene causes minimal if any G × E interaction effects, leaving the genetic background as causal of the largest G × E interaction effect on the days to flowering.

ACKNOWLEDGMENTS

This research was supported by Hatch Project 487 and the USAID-funded Bean/Cowpea Collaborative Research Support Program. This research is a contribution to the WR-150 project.

REFERENCES

1. D. H. Wallace, K. S. Yourstone, P. N. Masaya, and R. W. Zobel, *Theor. Appl. Genet.*, *86*: 6 (1993).
2. D. H. Wallace, R. W. Zobel, and K. S. Yourstone, *Theor. Appl. Genet. 86*: 17 (1993).
3. D. H. Wallace, J. P. Baudoin, J. Beaver, D. P. Coyne, D. E. Halseth, P. N. Masay, H. M. Munger, J. R. Myers, M. Silbernagel, K. S. Yourstone, and R. W. Zobel, *Theor. Appl. Genet., 86*: 27 (1993).
4. M. M. Mmopi, Ph.D. thesis, Cornell University, Ithaca, N.Y. (1992).
5. D. H. Wallace, P. A. Gniffke, P. N. Masaya, and R. W. Zobel, *J. Am. Soc. Hortic. Sci.*, *116*: 534 (1991).
6. H. G. Gauch, Jr., *Statistical Analysis of Regional Yield Trials: AMMI Analysis of Factorial Designs*, Elsevier, New York (1993).
7. H. G. Gauch and R. W. Zobel, *Theor. Appl. Genet.*, *79*: 753 (1990).
8. D. H. Wallace and G. A. Enriquez, *J. Am. Soc. Hortic. Sci.*, *105*: 583 (1980).

46

Sustainable Primary Production—Green Crop Fractionation: Effects of Species, Growth Conditions, and Physiological Development on Fractionation Products

Rolf Carlsson

Institute of Plant Biology, Lund University, Lund, Sweden

I. INTRODUCTION

The major aim of early agriculture was to produce more and better food. However, for the last decade, the production of industrial raw material has also been advocated. These two global demands for food and industrial raw materials from agricultural lands prescibe optimal utilization of every potential source of plants and lands [1–5].

Early agriculture had a very limited impact on ecological systems. Modern agriculture, on the other hand, has a heavy impact on the ecological environment, due to the use of large amounts of inorganic fertilizers, irrigation, pesticides, and heavy agricultural machinery. This results in a reduced plant diversity, destruction of soil, pollution, gradual intoxication of the entire planet, and an increased energy cost per unit of food and industrial raw material produced. Thus this type of "traditional" agriculture definitely limits what it is presently possible to obtain from agricultural lands. The alternative is a better ecologically adapted agriculture based on the primary productivity of photosynthesis in green plants. Such an agriculture will be based on green crops, new plant species, and a renewed and expanded use of nitrogen-fixing plants.

Some of the ecological advantages of green plants are their continuous soil cover, thus reducing erosion by water and wind, and a reduced use of pesticides. A selection of nitrogen-fixing green plants will enrich the soil, as will the use of brown juice from green plants (see below).

Primary production by photosynthesis in green plants means that more than 20 tons of dry matter and 3 tons of protein per hectare can be produced in temperate climates, and 80 tons of dry matter and 6 tons of protein, in tropical areas [6]. Better access to solar energy and use of C4 plants (e.g., tropical grasses and Amaranthaceae species) represent more efficient photosynthesis [6].

II. GREEN CROP FRACTIONATION

Green crops are now used primarily as a forage and as a source of leafy vegetables. Very little of the green crop's potential for production of food and industrial raw materials is presently utilized. However, green crops can be used for manufacturing several products simultaneously by a process called green crop fractionation or wet fractionation of green crops.

Green crop fractionation is in progress in more than 70 countries, the majority of which are in the tropics [7]. Several hundreds of temperate and tropical species have been investigated for green crop fractionation [8–10]. Reviews on green crop fractionation and production of leaf nutrient concentrate (synonyms: leaf protein concentrate, "leaf protein" in the British literature, and leaf concentrate) have been given by Pirie [12,13], Costes [14], Wilkins [15], Graham and Telek [8], and in the proceedings of the 1st, 2nd, and 3rd Leaf Protein Research Conferences [9–11]. In this chapter the term *leaf nutrient concentrate* (LNC) is used. *Leaf protein* means protein in plant shoots or leaves.

In the first step of green crop fractionation, a fiber-enriched, pressed crop and an expressed, green juice are produced (Figure 1). In a second step the green juice can be fractionated by heat, acid, and centrifugation into a leaf nutrient concentrate and a brown, deproteinized juice.

The LNC consists mainly of chloroplastic and other organelle membranes plus denatured soluble proteins. The brown juice, based on vacuolar substances, is enriched in water-soluble constituents as sugars, organic acids, low-molecular-weight nitrogen substances, glycoside phytochemicals, and mineral ions (e.g., potassium, sodium, chloride, and nitrate). Apart from other uses (see below), the brown juice is utilized as a biofertilizer to recirculate minerals. This reduces the need to add inorganic fertilizers to the soil and reduces the salinity of soils.

The approximate compositions of the LNC and brown juice are given in Tables 1 and 2. By further fractionation of the proteins extracted from the green juice, a chloroplast membrane, lipid-enriched protein concentrate, and two concentrates of soluble white proteins, Rubisco protein, and other pooled soluble proteins, FI protein and FII proteins, respectively, can be obtained.

In this chapter we emphasize the effects of species, plant anatomy, growth conditions, and physiological development of plants on the yield, composition, and nutritive of LNC, FI, and FII protein concentrates. Pressed crops and brown juices are dealt with to a minor extent.

III. PLANT SPECIES FOR GREEN CROP FRACTIONATION

Among species investigated that are most suitable for LNC production are those with a high dry matter and protein production, and those that give a high yield of extracted, high-quality leaf protein as LNC. Such species often belong to such plant families as Amaranthaceae, Chenopodiaceae, Cruciferae, Cucurbitaceae, Leguminosae, and Solanaceae [7].

Centrospermae species such as the Chenopodiacae and Amaranthaceae plant families are known as leafy vegetables, spinaches, or pseudocereals in temperate or tropical countries. Also, many Chenopodiaceae species are grown in saline areas as forage plants [16]. The most favorable species for green crop fractionation belong to the plant genera *Amaranthus* (tropical spinaches, pseudocereals), *Atriplex* (saltbushes), *Beta* (leafy beets), *Celosia* (tropical spinaches), *Chenopodium* (pseudocereals, "spinaches"), *Kochia* (saltbushes), and *Spinacia* (spinach). These highly productive species can give leaf protein yields between 1000 and 1500 kg/ha.

Less productive Centrospermae species, although still used for green crop fractionation, are species of the families Aizoaceae (*Aptenia*, *Carpobrotus*, and *Mesembryanthemum* genera), Basellaceae (*Basella* and *Boussingaultia* genera), Chenopodiaceae (*Hablitzia*, *Rhagodia*, *Salsola*, and *Suaeda* genera), Portulacaceae (*Portulaca* species), and Tetragoniacea (*Tetragonia*

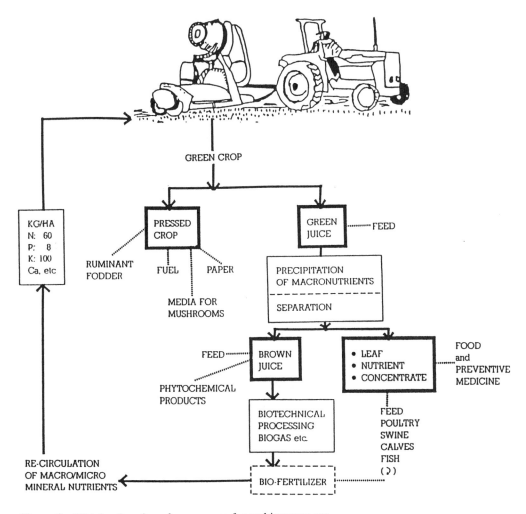

Figure 1 Wet fractionation of green crops for multipurpose use.

species) [16,17]. Many other Centrospermae species from saline or dry areas may be worth investigating by green crop fractionation.

Cruciferae species, such as those of *Brassica* (cabbages, kale, mustard, rape, turnip) and *Raphanus* (radish), give yields similar to those of the highly productive Centrospermae species mentioned above. Solanaceae species used for production of LNC are those of *Solanum* (potato, etc.) and *Nicotiana* (tobacco) [18]. *Nicotiana* species are investigated especially for their high yields of pure white leaf protein concentrates, especially of crystalline ribulose 1,5-biphosphate carboxylase (Rubisco) protein [19]. Cucurbitaceae species (cucumber, etc.) also produce high-quality LNC. However, the yield per hectare has not been investigated extensively.

Several Leguminosae species are being investigated for green crop fraction, as the crops are known to farmers and they fix nitrogen from the air. Suitable for the production of LNC are *Medicago sativa* (lucerne, alfalfa) and species of the genera *Lupinus, Phaseolus, Pisum, Trifolium*, and *Vicia* in temperate climates, and species of the tropical genera *Canavalia, Clitoria, Desmodium, Psophocarpus*, and *Vigna*. The yields of extractable leaf protein per

Table 1 Contents of Some Nutrients of Leaf Nutrient
Concentrates (per 100 g of Dry Matter)

True protein (g)	50–60
Essential amino acids (g/100 g protein)	
Lysine	5.6–7.3
Cysteine	1.2–1.7
Methionine	2.2–2.7
Lipids (g)	10–25
Fatty acids (g/100 g fatty acids)	
Palmitic acid	13–34
Oleic acid	12–46
Linoleic acid	4–27
Linolenic acid	10–50
β-Carotene (mg)	45–150
Carotenol (xanthophyll) (mg)	100–300
Starch (g)	2–5
Monosaccharides (g)	1–2
Crude fiber (g)	0.1–2
B vitamins (mg)	16–22
Tocopherol (mg)	15
Phylloquinone (mg)	2–8
Choline (mg)	220–260
Ash (g)	5–10
Calcium (mg)	400–800
Iron (mg)	40–70
Phophorus (mg)	240–570

Source: Data from Ref. 120.

hectare vary from 1000 to 3000 kg per year. Although no nitrogen fertilizer is necessary, both potassium and phosphorus fertilizers are needed for exploitation of the full potential of these and other promising Leguminosae species.

From other plant families one need only mention individual species, such as *Helianthus annuus* (sunflower) and *H. tuberosus* (Jerusalum artichoke), *Urtica dioica* (stinging nettle), green temperate cereals, and *Tithonia rotundifolia* (*T. tagetiflora*).

The species mentioned above are easy to wet fractionate (synonym: green crop fractionate) and give good-quality LNC. When nitrogen fertilized, many tropical grasses also give high yields of dry matter and protein. One such species used for LNC and fiber production is *Pennisetum purpureum* (elephant or Napier grass), which can yield 3000 kg of extracted leaf protein per hectare. However, the energy consumption for green crop fractionation of such C4 grasses is relatively high due to the high content of cell wall constituents. The energy to disintegrate *Zea mays* (C4 species) to get 1 kg of LNC is 10 times the amount needed to disintegrate *Chenopodium quinoa* (C3 species).

Other sources of species for green biomass production and green crop fractionation are those growing in marshes and inundated deltas, and aquatic plants. Such plants have an advantage: the ability to clean environmentally polluted, wet areas, and they can be used for the production of useful raw materials for paper pulps. Aquatic species worthy of mention include *Eichhornia crassipes*, *Ipomoea aquatica*, *Nasturtium officinale*, *Nymphea* spp., *Pistia*

Table 2 Composition of Brown Deproteinized Juice[a]

	Medicago sativa		Lolium perenne	
	Mean	Range	Mean	Range
BOD (mg O$_2$/L)	15,000	14,800–15,800	15,650	15,200–16,500
Dry matter	5.7	3.9–8.0	5.1	3.6–7.8
pH	5.4	4.6–5.8	5.7	5.0–6.3
	% DM		% DM	
TN	2.5	2.0–2.9	2.5	1.2–3.1
NPN	2.0	1.6–2.3	2.3	0.9–3.0
NO$_3$-N	0.1	0.03–0.20	1.3	0.48–1.78
KJ/g	15.2	14.3–16.1	12.8	10.3–15.2
WSC	31.6	16.5–39.8	36.4	31.3–40.9
Calcium	3.52	2.66–4.19	2.20	1.42–2.87
Magnesium	0.50	0.38–0.68	0.44	0.22–0.60
Sodium	0.66	0.24–1.43	1.51	0.77–3.10
Potassium	6.22	4.26–8.01	7.17	0.40–12.12
Phosphorus	0.41	0.24–0.60	0.47	0.19–0.69
	ppm		ppm	
Iron	56	33–85	59	48–83
Manganese	80	59–103	121	105–165
Zinc	75	50–102	90	68–118
	g amino acid/16 g N		g amino acid/16 g N	
Lysine	2.9		2.8	
Methionine	0.3		0.3	

Source: Ref. 15.

[a]BOD, biological oxygen demand; TN, total nitrogen; NPN, nonprotein nitrogen; KJ/L, energy value; WSC, water-soluble carbohydrate; DM, dry matter.

stratoides, *Polygonum* spp., and wetland plants such as *Arundo donax*, *Phragmites australis* (*P. communis*), *Typha latifolia*, and some *Atriplex* spp. [20].

IV. PLANT SELECTION AND BREEDING OF PLANTS FOR GREEN CROP FRACTIONATION

Preliminary studies to select cultivars, provenances, breeding lines (strains), and even individual plants from established agricultural plants as well as from promising, underexploited plants have been carried out. This was done to increase the yield and quality of LNC from plant shoots (see Refs. 20 and 21 with multiple references; additions noted separately): cultivars of *Brassica* spp. [22], *Helianthus annuus*, *H. debilis*, *Lolium multiflorum* [23,24], *Medicago sativa* [25–27], and *Solanum tuberosum* (potato). The provenances of *Amaranthus*, *Atriplex*, and *Chenopodium* species have also been studied, as well as strains of *M. sativa* and individual plants of *M. sativa* and *Trifolium pratense*. Soluble leaf proteins such as the FI and FII proteins, which are base materials for white leaf protein concentrates, have also been subjects for breeding and genetics studies for both legumes and green cereals [21].

V. PLANT ANATOMY AND GREEN CROP FRACTIONATION

Differences in yield and composition of LNC can be due to plant morphology and plant anatomy of a species, apart from effects of the plants' chemical compositions and the processing effects on the plant (see below). Plant shoots can be divided into leaves and stems. A large proportion of leaves yields more LNC per hectare than is yielded by a small proportion, provided that the shoot biomass production is similar. The stems also contain more fiber cells and xylem tissue cells than the leaf contains. A large amount of such "fiber" cells can reduce the protein extractability (see below).

Plant spacing or plant density affects the proportion of leaves as well as the green plant maturity. Denser plant cultivations produce longer stems, due to the effects of the composition of the light that penetrates the plant canopy—a known phytochrome effect. Also, the stem/leaf ratio will increase. Larger spacings cause the development of vegetative lateral shoots.

Most of the plant protein in a shoot is derived primarily from cell constituents of leaves such as chloroplasts, although other cytoplasmic organells and constituents, including the cytoskeleton, contribute. Very little plant protein is found in cell walls of green biomass. When green biomass harvested at a late stage, flower and seed proteins will be included in LNC, which affects LNC amino acid composition (Table 3).

A. Differences Due to Photosynthetic Systems

Leaf anatomy, especially as related to the plant anatomy of species with different types of CO_2 fixation (C3, C4 or CAM), greatly affects the content of protein in a leaf, from which LNC can be produced. C3 and C4 plants have different nitrogen needs for optimal photosynthetic efficiency [28]. The higher nitrogen efficiency of C4 plants is reflected in the low protein content of grasses [28], but not necessarily in dicotyledon plants such as *Atriplex* and *Amaranthus* species [28–30].

C3 leaves have both palisade and spongy parenchyma cell layers with pentose phosphate CO_2 fixation (Rubisco protein) in their chloroplasts, whereas this type of fixation is restricted to bundle-sheath cell chloroplasts in C4 leaves (28). The latter, of course, limits the amount of LNC that can be produced from these leaves, especially the amount of FI protein concentrate.

Leaf protein fractionation has shown that differences in protein content between C3 and C4 plants were due to lack of Rubisco protein/FI protein in normal C4 leaf mesophyll cells [28,31]. The contents of soluble protein and FI protein is relatively high in C3 grasses (33 to 48% and 29 to 43% of total protein, respectively) compared to C4 grasses (26 to 30% and 11 to 16%, respectively) [31]. By gel fitration a further coarse fractionation of soluble chloroplast proteins in a C4 plant was made [32]. The C3 dicotyledon plants contain up to 50% FI protein of the soluble leaf proteins [33].

Although C4 plants under normal, tropical conditions have a low protein content relative to C3 plants, they respond very well to an increase in nitrogen fertilizer by producing large amounts of biomass and protein [7,28]. Young tropical C4 grasses can contain 20% protein in their dry matter [28,34,35], while the protein content of mature plants is around 5% of the dry matter.

CAM species, often succulent plants, may have fewer or no leaves, a small amount of chloroplasts in their cells, and their chloroplasts are not very well developed [36]. Thus CAM

Table 3 Protein Amino Acid Composition of Flower Head, Leaf, Stem, Seed, and Leaf Nutrient Concentrate from Whole Plant Shoots[a] (g amino acid/16 g N)

Amino acid	Sample type				
	Flower	Leaf	Stem	Seed	LNC
Amaranthus cruentus					
Lysine	6.3	6.2	5.6	5.5	7.0
Threonine	4.4	4.6	4.8	3.6	5.0
Cysteine	1.7	1.3	2.0	2.2	1.6
Methionine	1.7	2.0	1.6	2.3	2.4
Valine	5.7	6.0	6.2	4.6	6.8
Isoleucine	4.9	5.2	5.1	4.0	5.8
Leucine	6.9	8.3	7.9	5.6	9.0
Tyrosine	3.7	3.5	2.7	3.2	4.4
Phenylalanine	4.5	4.7	4.4	4.0	5.8
Histidine	2.0	1.9	2.0	2.6	2.4
Proline	4.2	4.5	4.4	3.9	4.5
Arginine	4.9	4.8	4.2	8.7	6.0
Aspartic acid	9.3	9.1	9.8	8.3	10.0
Serine	5.6	4.7	5.0	5.9	5.6
Glutamic acid	9.8	9.9	11.0	15.5	10.8
Glycine	6.3	5.6	6.1	7.2	5.6
Alanine	4.7	5.3	5.6	3.7	5.8
N distribution in pland (%)	59	30	11	—	—
Chenopodium quinoa					
Lysine	7.4	6.7	5.3	5.5	6.9
Threonine	5.5	5.4	4.7	3.5	5.4
Cysteine	2.0	1.7	2.3	2.0	1.4
Methionine	2.5	2.6	2.6	2.3	2.5
Valine	7.2	6.9	6.1	4.7	7.0
Isoleucine	6.0	5.7	4.7	3.9	6.0
Leucine	9.5	9.5	7.5	6.1	9.8
Tyrosine	5.5	4.1	2.9	2.9	4.9
Phenylalanine	5.8	5.7	4.3	3.8	6.4
Histidine	2.7	2.5	1.9	2.8	2.6
Proline	5.6	5.2	4.5	3.6	5.2
Arginine	5.4	6.4	4.7	9.5	6.3
Aspartic acid	11.2	10.3	9.6	7.7	10.1
Serine	6.7	5.5	5.0	4.2	5.5
Glutamic acid	10.5	10.7	10.6	14.2	11.0
Glycine	6.8	6.1	5.7	5.4	5.6
Alanine	6.3	6.3	5.8	4.2	6.3
N distribution in plant (%)	56	33	11	—	—

Table 3 (*Continued*)

Amino acid	Flower	Leaf	Stem	Seed	LNC
			Sample type		
		Helianthus annuus			
Lysine	5.6	6.5	5.5	3.2	7.2
Threonine	5.2	5.8	5.0	3.2	5.5
Cysteine	1.1	1.3	1.3	1.7	1.6
Methionine	2.4	2.9	1.8	1.6	2.7
Valine	6.7	7.2	6.2	5.0	7.1
Isoleucine	5.2	5.5	4.8	4.7	6.1
Leucine	8.3	9.8	7.4	6.4	10.0
Tyrosine	3.3	4.0	2.0	2.4	4.9
Phenylalanine	4.5	5.6	4.3	4.5	6.4
Histidine	2.2	2.1	1.7	2.2	2.8
Proline	4.4	5.1	4.4	—	5.3
Arginine	5.2	5.2	4.3	8.7	6.4
Aspartic acid	10.8	10.5	9.5	—	10.2
Serine	5.2	4.8	4.7	—	5.5
Glutamic acid	10.6	11.3	10.3	—	11.2
Glycine	5.4	6.3	5.3	—	5.6
Alanine	5.7	6.6	5.5	—	6.4
N distribution in plant (%)	40	33	27	—	—

Source: Data from Ref. 103.

[a]LNC/leaf nutrient concentrate is washed with water. S.D. is 0.1 for all essential amino acids but lysine, isoleucine, and leucine (S.D. = 0.2). S.D. = 0.3 for other amino acids.

plants normally are not considered useful for LNC production. However, *Pereskia aculeata*, a bushlike Cactacea plant, has been investigated for LNC production [37,38].

B. C3 Plants

C3 plants are normally best for production of LNC, due to the high content of protein-rich chloroplasts in their leaves, and since the proportion of leaves with large lamina can be very high. C3 plants can have leaves of two or more layers of palisade parenchyma cells [39], which are most rich in chloroplasts. The number of chloroplasts per cell can vary from 50 upward. The higher the irradiance, the more layers of palisade cells are developed. The spongy parenchyma cells in the leaves contains fewer chloroplasts [39]. Shade leaves in a tree canopy or leaves of a shade plant may not have a palisade layer, and are relatively thin. Such leaves are a poor source of LNC production.

The amount of protein in a leaf depends on the plant's growth and development stage. The amount of protein in a leaf is related to leaf weight, area, and thickness. In the leaf the protein content is determined by the number of different mesophyll layers, intercellular space volume (cell densities), cell numbers, cell length and width, and the number of chloroplasts per cell.

These factors, in turn, are dependent on the position of the leaves (upper/lower ones; solar/shade ones), availability of water and fertilizers (especially nitrogen), and other factors.

The aforementioned factors have been studied for *M. sativa* leaves by Addy et al. [40] to determine optimal cell rupture and leaf protein extraction. This study showed that early in a season, the protein content of leaves and extraction yields increased between the leaf ages of 7 and 17 days, and for a late season, the same period appeared between days 28 and 38. Similar results for grasses and mixed herbage swards are the background for repeated cuts during a season, where each cut follows after a regrowth of 3 to 5 weeks (see below). The result of several cuts is to maximize leaf protein production in the dry matter per hectare.

For young leaves, the FI protein (the dominant chloroplast protein) content is gradually increased following chloroplast development. When leaf senescence starts, the lower leaves reduce their protein content faster than the upper leaves. This is a consequence of the rapid breakdown of FI protein in the chloroplast. Applications of cytokinins are known to delay FI protein breakdown and other effects of senescence.

As the palisade cells are more slender and brittle at the peak period, and as such palisade cell walls are more amenable to rupture, the maximum protein extraction occurs at the same peak period [40]. Another consequence of cell wall conditions (turgidity) is that during rapid water loss, as during harvest, the walls lose their rigidity, with the effect that cell rupture and protein extraction is reduced [40]. Irrigation that affects the cell turgor can increase leaf protein extraction yields.

VI. EFFECTS OF FIBERS AND CELL WALLS IN A PRESS CAKE ON EXTRACTION OF LEAF PROTEIN

Apart from the physical strength of the cell wall and its rupture for possible release of protein, the plant fibers and cell walls form a thick filter (ultrafilter) during expression of the green juice from a disintegrated or pulped plant material. The effect of this filter on protein extraction, in turn, depends on the thickness of a layer and its structural carbohydrate composition.

Such a fibrous press layer separates off large fractions of cell membranes, large protein molecules, and especially, denatured proteins. Denaturation occurs because of oxidized phenolics, quinones, which react with proteins, and release of acid constituents, organic acids and phenolics, from the vacuole. The latter reactions lower the pH of the green juice in relation to the relative high cytoplasmic and chloroplast pH.

Furthermore, a thick fibrous layer absorbs water from solutions, such as the green juice, and the proteins originally released are retained, unless thorough water washing of the press cake is performed. Plants with pectin-rich, or alga with alginate-rich cell wall constituents further increase the absorption of green juice in the pressed material. By chemical treatment of the pulped material with calcium salts the water-holding capacity of alga pulp and press cake can be reduced.

To avoid problems with cell walls due to the thickness of a fibrous press layer, a final layer thickness of about 2 cm is often recommended. Alternatively, continuous centrifuges can be used for green juice separation from thinner fibrous layers. However, too much fine fiber may separate out with the green juice. This causes the LNC to have a low protein content, which should be avoided.

VII. EFFECTS OF PLANT DEVELOPMENT STAGES ON PROTEIN EXTRACTION

Plant morphology, anatomy, cell structure, cell protein composition, and the biochemical production of secondary substances in the cell vary according to physiological plant development. Generally, secondary plant substances accumulate during plant growth and development. All these factors affect leaf protein extractability.

Secondary substances such as easyily oxidizable phenolic substances can reduce protein extractability due to the denaturation of proteins during plant pulping [41]. The same effect is produced by organic acids from the cell vacuole. On the other hand, saponins of Chenopodiaceae species and other species (e.g., *Medicago sativa*) may explain a high leaf protein extraction ratio [17]. The saponins break up membranes, and can thus facilitate protein release from the cell. Chenopodiaceae species also contain phenolics that do not easily oxidize. This can prevent quinone formation and its denaturation of protein.

The period for optimal protein extraction and maximum yield of extractable protein per hectare can fall at different times during a growth season. Factors affecting this period for different species and cultivars can be determined by the leaf/stem ratio and the leaf development stage. Cell wall composition and thickness, and the chemical composition of secondary substances in a cell, vary with the development stage of the plant and its leaves. Thus there are profound effects of plant physiological development when the optimal period for extraction of protein from the plant or its leaves appears.

Agronomic aspects of this have been studied by many scientists (e.g., Refs. 42 to 47). In practice, the peak period for leaf protein extraction is often at a prebloom stage, although some species (e.g., *Chenopodium quinoa*) can have increased leaf protein extraction yields after that period [44]. The maximum yield of leaf protein per hectare can therefore be obtained by cuts of regrown plants, where each regrowth of the plants has reached the physiological stage of prebloom. On the other hand, a maximum yield of carbohydrate in crop dry matter is given when only one harvest per season is made.

A. Factors Affecting Leaf Protein Extraction

Major factors correlated with leaf protein extraction ratios are the true protein content of plants or their leaves and their dry matter content [48,49]. The latter is correlated with the increase in cell wall thickness due to secondary growth, the proportion of stem in relation to leaf, the loss of lower leaves as the plant ages, and the loss of protein from lower leaves during the senescence.

Apart from pressing effects, the method for plant and cell disruption also affects the leaf protein extraction yields [50,51], which can modify the optimum period for leaf protein extraction. The energy used for plant material disintegration also varies from 8 to 220 kJ/kg processed crop due to the type of disintegrator [51]. This, of course, affects the heat development during processing, resulting in possible protein denaturation, and subsequently, the ratio of extratable protein will be reduced. The energy needed for pulping a crop for a final production of 1 kg of LNC from the juice released can vary from 0.6 kWh (brittle and slender *Chenopodium quinoa* plant cells, a C3 species) to 5.5 kWh (tough and fibrous *Zea mays*, a C4 species) (N. W. Pirie, personal communication, Rothamsted Experimental Station, Harpenden, Hertfordshire, England).

For crops that become too fibrous at the time for harvest (e.g., older plants), an increase in the pH of the extraction medium can alleviate the protein extraction. However, a high pH during protein extraction also extracts pectins, which lowers the protein content of the LNC. The plant physiological development is affected by the season for crop cultivation as well as by the geographic location. These effects can be reflected in the crop and LNC composition and uses [52,53].

B. Practical Approaches for Best Harvesting Stage

For a farmer the leaf protein extraction yields per hectare are essential for establishing good economic results. Therefore, various standard formulas for expressing yields of extractable leaf protein per hectare have been developed for standard disintegration and pressing techniques [48,49]. These formulas include plant protein content, dry matter content, plant age, plant type, growth, cultivation conditions, and growth seasons of the year.

The wet-fractionation techniques used for green crops depend on the use of required products from the green crop. Dewatering of a crop, a low-protein extraction technique, is used to faciliate green crop drying. An exhaustive protein extraction technique can serve other purposes, including production of LNC for feed and a pressed crop for silage, or production of LNC for feed/food, bioenergy (biogas, ethanol, solid fuel), phytochemicals, and a pressed crop for paper pulp production or biologically better tobacco products [18,54–56]. Well-irrigated plants can give an increased protein extraction ratio, and the same effect can be obtained by the addition of water during processing (pulping and pressing) [49].

Attainment of a specific stage of crop maturity may not be necessary for multipurpose use of products by green crop fractionation. In such cases the leaves can be separated from the stem for LNC production while the stem is used for production of fibrous products, or the tops of the plants are cut for LNC production [57] while the lower parts with fewer leaves are harvested for fiber products.

VIII. EFFECTS OF NITROGEN FERTILIZATION ON YIELD OF EXTRACTABLE PROTEIN PER HECTARE

In general, 16 elements are needed for complete plant growth and development, including its propagation. Nitrogen plays a major role for production of plant proteins for membranes, cytoskeletons, enzymes, and storage proteins. A plant's vegetative growth is often limited by its access to nitrogen. For green biomass studies, several nitrogen fertilizer trials have been performed to estimate effects on the yield of dry matter and protein per hectare. Also, for green crop fractionation the effects on extractability of leaf protein of fertilizer applications can be essential. Plants of different species and cultivars respond differently to the amount of nitrogen applied to the soil. Access to enough water in connection to nitrogen fertilization is essential.

A. Experiments Under Controlled Conditions in a Greenhouse

Under controlled conditions in a greenhouse, several species with a high protein productivity were studied by Lexander et al. [30]. Most species responded well to nitrogen fertilization (equivalent to 280, 560, and 840 kg N/ha) by increased true protein content and dry matter production. Amaranthaceae and Chenopodiaceae species, *Urtica dioica*, and *Helianthus annuus* responded especially well, whereas most grasses and legumes gave moderately positive responses. The former species, except for *H. annuus*, are nitrophilous species. Most species investigated showed no effect of increased fertilization on the protein extraction ratio (water was added during the extraction process).

B. Experiments Under Normal Field Conditions

Under field conditions relatively little nitrogen is given to the soil, at most between 100 and 300 kg N/ha for green crops. An increase in nitrogen fertilization at those levels increases the total dry matter and protein production per hectare [49,58–60]. The extraction ratio of protein was fairly unaffected by the fertilizer levels used [49,58–60].

C. Experiments with High Nitrogen Fertilizer Levels

The wish to maximize yields of extractable leaf protein has meant that up to 1000 kg N/ha has been applied, often as split top dressings. Again, an increase in dry matter and protein production was noted ([44]: 200 to 565 kg N/ha; [61]: 200 to 1000 kg N/ha; [62]: 240 to 480 kg N/ha;

[63]: 150 to 880 kg N/ha), as well as a fairly constant protein extraction ratio when the crop was harvested at the same age [44,62,63]. Intensive nitrogen fertilizer experiments in a greenhouse, where the equivalent of up to 4000 kg N/ha was applied, showed that green temperate cereals could cope with up to 1200 kg N/ha, while a nitrophilous *Chenopodium quinoa* could tolerate the highest level [17].

The increase in protein production by increased nitrogen fertilizer levels is based on increases in the crop protein content of the dry matter, in delayed senecense of otherwise yellowing or wilting base leaves, and in some cases in the development of new lateral shoots. The cells of plants richly supplied with nitrogen fertilizer are often very brittle, which can also enhance cell rupture and protein extraction. Unfortunately, no studies on chloroplast development in leaves and their brittleness were undertaken. For most plants, the increased nitrogen content of the plant was due to overproduction of soluble organic nitrogen compounds and accumulation of high nitrate contents in the leaf cell vacuoles [17].

IX. EFFECTS OF PHYTOCHEMICALS ON LEAF NUTRIENT CONCENTRATE/LNC PRODUCTION AND QUALITY

Yields of leaf protein and LNC extracted from the green juice, and LNC quality (estimated as its chemical composition and nutritive value determined in vitro and in vivo), can be influenced by a plant's age and physiological stage [20,64–67]. Apart from the normal organic substances in a plant, plants of different orders, families, species, and cultivars contain specific secondary compounds and other organic substances. Such compounds have been used to classify these plants [68]. These substances tend to accumulate in the plant as it gets older and during plant physiological development.

Several of the secondary plant substances, vacuolar stored nitrate and oxalic acid, have a negative influence on LNC quality, as they reduce its nutritive value. Such substances are called antinutritive substances. A general description of plant-specific secondary substances has been given by Liener [69]. Examples of such antinutritive substances are nitrogen-containing substances such as antiproteases, especially antitrypsins, phytohemagglutinins (lectins), toxic oligopeptides, amino acids (e.g., mimosine), fabatoxins, nitrate, and nitrite (the latter may turn into cancerogenic nitrosoamines). Several antinutritive glycosides stored in the vacuole for biological defenses include phenolics/tannins, saponins, plant estrogens, cyanoglycocides, alkaloids, and tioglycosides/goitrogenic substances. Many species, such as Chenopodicaeae species, accumulate oxalic acid in the vacuoles. These and other toxic or antinutritive substances are mixed into the green juice for production of LNC. Thus they can contaminate the final LNC and reduce its original high quality.

A. Presence of Antinutritive Substances in LNC

The presence of secondary substances in LNC have been demonstrated: for example, trypsin inhibitors [70,71], phytohemagglutins [71], phenolics/tannins [71–77] saponins [17,71,75,78,79], plant estrogens [80–83], cyanoglycosides [84–86], alkaloids [87–90], tioglycosides [22], and antinutritive substances such as oxalic acid and nitrate [75,89]. Generally, small amounts of antinutritive substances were found in LNCs compared to the amounts occurring in the original plant material. White leaf protein concentrates, especially, have low concentrations of secondary substances and other antinutritive substances [19,46,91–94].

In some cases, substances (e.g., mimosine or cyanoglycocides [74]) could not be detected in the LNC, although it is unlikely that they could be totally absent in it. Also, although solanines were present in the LNCs, there was no correlation with the amounts and the in vivo nutritive values [87,88]. In a study by the present author, nicotine-containing LNCs that reduced the nutritive value of LNC and killed rats [89,90] did not affect pigs fed the same LNC (unpublished results). Despite extremely low nicotine contents, tobacco FI protein isolates showed weak mutagenicity using the Ames test and three in vivo tests [94].

B. Positive Effects of Protein Concentrate Constituents

Quite a few secondary substances [95] and other LNC constituents can add positive dietary values when using LNC and leaf protein isolates: for example, saponins (hypocholesterolemic effect [78,96]), tioglycosides (anticarcinogenic effect [97]), β-carotene (anticarcinogenic effect [98,99]; antidiarrhea effect and benificial against pulmonary diseases [100]), vitamin E (anticarcinogenic [101]), and pure proteins as such (hypocholesterolemic effect [102]).

Secondary glycoside substances accumulate in the brown juice. For tobacco brown juice, this has been used as an advantage; the juice was utilized by the author as a biological pesticide to kill aphids and butterfly larvae on vegetable crops and cereals (unpublished results).

X. PLANT PHYSIOLOGICAL DEVELOPMENT AND LNC QUALITY

The quality of protein in LNC has been shown to be affected by the plant's growth stage at harvest. The amino acid composition of LNC has been studied [52,64,103,104], and the in vitro protein digestibility has been investigated [46,105]. The in vivo values of proteins of LNC were studied as to biological value, true digestibility, and net protein utilization, using rats [65,66,67] (Table 4).

The content of carotenoids, including β-carotene, and other pigments in LNC has been determined in relation to plant physiological development [24,53,106]. The content of

Table 4 Effects of Plant Physiological Development on Net Protein Utilization of Leaf Nutrient Concentrates of *Atriplex hortensis* and *Chenopodium quinoa*: A Rat Assay[a]

| | Physiological stage | | | |
	Vegetative	Bud	Flower	Seed
A. hortensis				
BV	58	68	58	52
TD	82	87	87	84
NPU	48	59	50	43
C. quinoa				
BV	60	—	65	61
TD	90	—	91	87
NPU	54	—	58	53
Casein control				
BV	69			
TD	100			
NPU	69			

Source: Data from Ref. 67.

[a]BV, biological value; TD, true digestibility; NPU, net protein utilization.

b-carotene in LNC was highest at the vegetative preflowering stage. It was also noted that more pigments were found in plants from a cooler growth period at a stage selected for plant harvest [53]. A LNC generally contains 10 to 20 times more β-carotene than is in the original plant material (unpublished results of the present author).

The general conclusion is that an optimum quality of LNC (protein and β-carotene) can be obtained by vegetative plants at a preflowering stage. Also, it should be noted that protein amino acids of plant parts other than leaves might affect the protein amino acid composition of LNC [103] (Table 3). Some of the qualitative differences in the protein of LNC could be ascribed to the saponins and phenolics [62]. The palatability of the LNC diet could have been affected by saponins [107,108]. The brown juice (whey) with saponins from *Medicago sativa* in a rat diet has also given an initial reduction of growth in rats [109]. However, the rats recovered after 3 weeks on the diet.

A negative correlation has been found between the contents of phenolics in LNC and its nutritive value ([67]: NPU; [110]: PER). Part of the negative correlation was due to the destruction of sulfur amino acids and lysine [66,67]. Other secondary substances, contaminating a LNC, can give similar negative responses (e.g., reduced feed intakes, reduced weight gains, or reduced nitrogen balances). Secondary substances can be washed off or neutralized [17,67,85,111,112], however, giving a LNC or white leaf protein concentrate with the same nutritive value as casein. Addition of sulfite or other reducing compounds is one way to conteract phenolic oxidation. This may prevent destruction of lysine and sulfur amino acids that are limiting in rat assays [67,110,111] (Table 5).

XI. PLANT SPECIES GIVING GOOD-QUALITY LNC

The physiological stage at plant harvest, as well as bad processing conditions, can reduce the nutritive value of protein in LNC. However, in cases where LNCs were produced from leafy plants at a vegetative stage, a qualitative comparison of LNC from different species may be justified (Tables 5 and 6).

LNCs with a high nutritive value have been found for species of *Amaranthus*, *Brassica*, *Chenopodium*, temperate cereals (wheat, rye, barley, and oats), and for legumes such as *Medicago sativa* and *Lupinus* species, compared to LNCs with lower nutritive values from species of *Leucaena*, *Manihot*, *Melilotus*, *Trifolium*, and *Vicia* [17,65,66,74,113–119]. The LNCs of some species of *Nicotiana* [89] and *Solanum tuberosum* [65,87,88] have fair nutritive values. Such results are due partially to differences in the composition and concentration of secondary plant substances of species and cultivars that contaminate LNC.

A positive effect of increased fertilizer levels on in vivo evaluation of green fraction LNC or *C. quinoa* was noted [63]. This was perhaps due to an enhanced level of protein in the LNC, based on an increased amount of soluble proteins from the fertilized plants [46]. White, soluble protein with higher protein contents than LNCs have higher nutritive value than that of whole LNC or green fraction LNC [19,111].

XII. WHITE LEAF PROTEIN PRODUCTS

White leaf protein products can be concentrates of soluble leaf proteins such as FI and FII proteins from the chloroplast and cytoplasm, concentrates of organic solvent precipitated proteins from the green juice, depigmented LNC (including membrane-bound proteins), and protein-fiber concentrates of depigmented leaves. These products have been described extensively [19]. Apart from better protein quality compared than that of an LNC, the white proteins have better functional properties, which make them desirable for the food industry [19].

Table 5 Protein Efficiency Ratio and Lysine, Cysteine, and Methionine Content of Protein in Leaf Nutrient Concentrates: Effects of Species and Addition of Reducing Sulfite During Processing[a]

Species and treatments	PER of LNC		Amino acids (washed LNC)		
	Nonwashed	Washed	Lys	Cys	Met
Amaranthus caudatus					
Water	1.6	1.8	—	—	—
Water + SO₃	2.0	1.9	—	—	—
A. hypocondriacus					
Water	—	1.8	6.5	1.2	2.6
Water + SO₃	—	1.9	6.5	1.2	2.4
A. retroflexus					
Water	—	1.3	—	—	—
Water + SO₃	—	1.9	—	—	—
Atriplex hortensis					
Water	1.3	1.5	6.6	1.1	2.7
Water + SO₃	1.8	2.0	7.0	1.3	2.7
Brassica hirta					
Water	—	2.1	7.3	1.5	2.8
Water + SO₃	—	2.1	6.9	1.3	2.8
Brassica napus					
Water	1.8	2.1	6.7	1.3	2.7
Water + SO₃	1.9	2.1	6.7	1.3	2.4
Chenopodium quinoa					
Water	2.0	2.0	6.9	1.4	2.5
Wate + SO₃	1.9	1.9	7.2	1.6	2.7
Helianthus annuus					
Water	—	1.6	—	—	—
Water + SO₃	—	1.9	6.7	1.3	2.6
Lagenaria siceraria					
Water + SO₃	—	1.7	6.7	1.2	2.5
Luffa acutangula					
Water + SO₃	—	2.0	6.9	1.3	2.6
Medicago sativa					
Water	—	1.5	6.2	1.2	2.1
Water + SO₃	—	2.0	6.2	1.1	2.3
Sorgum sudanense					
Water + SO₃	—	1.9	6.6	1.2	2.3
Casein	—	2.5	—	—	—

Source: Ref. 110.

[a]PER, protein efficiency ratio in 3-week rat assays; LNC, leaf nutrient concentrate; Lys, lysine; Cys, cysteine; Met, methionine; SO₃ = 0.5% sulfite of pulped material at pH 8.5; washed LNC: twice washed, 4 parts of distilled water + 1 part of wet LNC.

Table 6 Nutritive Value In Vivo of Leaf Nutrient
Concentrates in Rat Assays: TD, BV, and NPU[a]

Latin name	Parameter		
	TD	BV	NPU
Chenopodiaceae			
Atriplex hortensis [17]	95	82	78
Beta vulgaris [65]	79	67	53
B. vulgaris cv. cicla [116]	72	48	35
Chenopodium quinoa [17]	98	77	76
Cruciferae			
Brassica napus [65]	85	78	66
B. napus [116]	83	67	55
B. oleracera cv. acephala [114]	86	77	66
B. oleracea cv. acephala [116]	89	60	55
Tropaeolum majus [65]	91	84	75
Sinapis alba [65]	82	74	60
Graminae			
Dactylis glomerata [112]	—	—	31–32
Hordeium vulgare [65]	84	81	68
Lolium multiflorum [112]	—	—	29–30
Secale cereale [65]	77	76	58
S. cereale [116]	93	69	64
Triticum aestivum [65]	82	80	66
T. aestivum [116]	91	67	61
Leguminosae			
Lupinus sp. [65]	76	83	63
Medicago sativa [87]	82	59	48
M. sativa [112]	—	—	38–49
M. sativa [114]	86	77	66
M. sativa [116]	84	59	50
Melilotus officinalis [65]	81	38	31
Pisum sativum [65]	81	65	53
Trifolium fragiferum [65]	77	62	48
T. pratense [65]	71	54	38
Vicia sativa [65]	81	58	47
V. sativa [116]	82	53	43
Solanaceae			
Nicotiana glutinosa cv. T5 [89]	91	64	58
N. rustica cv. T33 [89]	98	61	60
N. tabacum mixed cv. [89]	90	60	58
Solanum tuberosum [65]	81	73	60
S. tuberosum [116]	76	60	46

[a]Source of data compiled is given in brackets. TD, true digestibility;
BV, biological value; NPU, net protein utilization. Leaf nutrient
concentrates (LNC) with NPU values higher than 60 are considered
good-quality LNC. Casein NPU values range from 70 to 84.

The first class of white protein products are of major practical interest. Relatively large quantities of soluble protein for the production of concentrates of FI and FII proteins can be obtained from plants of species of the families Chenopodiaceae, Cruciferae, Leguminosea, and Solanaceae. Crystalline, soluble protein isolates of Rubisco (FI protein) have been obtained from species of the genera *Nicotiana* and *Solanum* (Solanaceae) and *Spinacia* (Chenopodiaceae) [19].

Studies of FI and FII protein isolates, especially of the extremely pure, chrystalline protein Rubisco, have been undertaken primarily in Italy, the United States, Sweden, and France. The Rubisco isolates are *in essence* free of contaminating substances and are of major interest for dietary medicine.

XIII. SUMMARY AND CONCLUSIONS

The major aims of agriculture today are to produce more and better foods and industrial raw material from sustainable crops. Green crops are primary producers of raw materials. The primary production by photosynthesis of green crops gives maximum production per hectare and season. Green crops make optimal use of solar energy for raw material production compared to the use of grain crops.

Cultivation of green crops has several ecological advantages, such as a continuous cover of the soil and subsequently, diminishing erosion by wind and water. Use of nitrogen-fixing green legumes reduces the need for nitrogen fertilization, at the same time enriching the soil. The use of green crops means that almost any of the approximately 300,000 higher plant taxa can be used. Thus agriculture will not be limited by cultivation of the 12 major staple crops that produce grains. Furthermore, green crop cultivation is a way to save biodiversity.

The primary production of green crops, mainly C3 species in temperate climates, can yield more than 20 tons of dry matter and 4 tons of protein per hectare per year. The use of C4 species in the tropics can give dry matter yields exceeding a production of 80 tons/ha per year and protein yields of up to 6 tons.

In the first step, green crop fractionation (wet fractionation of green crops) produces a fiber-enriched pressed crop and a nutrient-rich green expressed juice. The large organic molecules, primarily proteins and lipophilic substances, of the green juice can be separated as a leaf nutrient concentrate (LNC), leaving a brown, deproteinized juice enriched in sugars and soluble mineral nutrients. The proteins of the green juice can be further fractionated into a green, chloroplast membrane–enriched LNC, plus FI protein (Rubisco protein) and FII protein as white leaf protein concentrates/isolates.

Each fractionation product can have specific uses (Figure 1). The fiber fraction can be used as ruminant fodder, solid fuel, paper pulp, biologically better tobacco products, or for the cultivation of mushrooms. The brown juice is used as a biofertilizer as such or in fermentation to produce biogas or organic chemicals, and can be used as a biologial pesticide or a source of pharmacologically active products.

Most animals, including primates, can get most of their essential nutrients, apart from energy constituents, from fresh green leaves. Therefore, dark green leaves are recommended for a healthy diet. Because of the bulkiness of leaves and their high water content, LNC production from the leaves or whole green plants are recommened as a food supplement, especially for young children or mothers with infants [12,120–122]. On the other hand, larger-scale green crop fractionation (up to 120 tons processed per hour) is used commercially for production of LNC for nonruminant feeds [10,11].

Even if LNC is a better food from a total nutrition point of view, white leaf protein concentrates have been interesting for the food industry as a protein supplement due to their better functional properties. The FI protein that can be obtained in a pure, crystalline form has been studied for use in medical diets.

The most suitable species for production of LNC and white protein concentrates are those belonging to the plant families Amaranthaceae, Cruciferea, Chenopodiaceae, Solanaceae, and selected Graminae and Leguminoseae species. Such species have vegetative, green plants from which a high yield of extractable leaf protein for good-quality LNC can be produced.

The plant morphology and anatomy, as well as the occurrence of secondary plant substances and other antinutritive substances, determines whether or not a plant species is suitable for LNC production or for production of other wet-fractionation products. The plant physiological development and age of the vegetative plant determine the yield of extractable leaf protein, the nutritive quality of LNC, and the yield of white leaf protein concentrates. The highest yield and best quality of LNC are derived from vegetative plants at the prebloom stage.

During plant development and maturation the proportion of dry matter ("fiber cells" and cell walls) of a plant increases and the protein content diminishes. This reduces the yield of extractable leaf protein per kilogram of fresh weight, but as the total dry matter production per hectare increases, the total yield of extractable leaf protein for LNC production increases to an optimum at the prebloom stage. These tendencies have introduced the practice of repeated cuts of regrown vegetative plants during a cultivation season for production of maximum of protein in the dry matter per hectare.

Nitrogen fertilization of nonlegume plants has a profoundly positive effect on leaf protein production in the plant, and on the yield of LNC, under proper irrigation conditions, provided that enough water is available. The effect on the plant of this fertilization is an increase in its protein content due to a variety of factors in cell metabolism, delay of senescence, and production of new vegetative lateral shoots.

The continuous buildup of secondary substances and other antinutritive substances in the plant during plant development and aging can reduce the yield of extractable leaf protein and the quality of the LNC. On the other hand, positive effects have been shown for secondary substances and others in LNC and protein concentrates

Thus selected plant species, grown for green crop fractionation, can give maximum yield of food, feed, and industrial raw materials if harvested during an optimum period. The yield and quality of LNC and white leaf protein products is greatly affected by the plant's development during a growth season.

REFERENCES

1. I. Hedberg, ed., "Systematic Botany: A Key Science for Tropical Research and Documentation," *Symb. Bot. Ups.*, *28*(3) (1988).
2. G. E. Wickens, H. Haq, and P. Day, eds., *New Crops for Food and Industry*, Chapman & Hall, London (1989).
3. M. Pessarakli, ed., *Handbook of Plant and Crop Stress*, Marcel Dekker, New York (1993).
4. National Research Council, *Underexploited Tropical Plants with Promising Economical Value*, National Academy Press, Washington, D.C. (1975).
5. National Research Council, *Saline Agriculture: Salt-Tolerant Plants for Developing Countries*, National Academy Press, Washington, D.C. (1990).
6. R. Carlsson, "An Ecologically Better Adapted Agriculture: Wet-fractionation of Biomass of Green Crops, Macro-alga, and Tuber Crops", *Recent Advances in Leaf Protein Research: Proceedings of the 2nd International Conference on Leaf Protein Research* (I. Tasaki, ed.), Faculty of Agriculture, Nagoya University, Togo-cho, Aichi-ken, Japan, pp. 19–23 (1985).
7. R. Carlsson, in *New Crops for Food and Industry* (G. E. Wickens, N. Haq, and P. Day, eds.), Chapman & Hall, London, pp. 101–107 (1989).

8. L. Telek and H. D. Graham, eds., *Leaf Protein Concentrates*, AVI Publishing, Westport, Conn. (1983).

9. N. Singh, ed., *Progress in Leaf Protein Research: Proceedings of the First International Conference on Leaf Protein Research*, Today and Tomorrow's Printers and Publishers, New Delhi (1984).

10. I. Tasaki, ed., *Recent Advances in Leaf Protein Research: Proceedings of the 2nd International Conference on Leaf Protein Research* (I. Tasaki, ed.), Faculty of Agriculture, Nagoya University, Togo-cho, Aichi-ken, Japan (1985).

11. P. Fantozzi, ed., *Proceedings of the 3rd International Conference on Leaf Protein Research* (P. Fantozzi, ed.), *Ital. J. Food Sci.* (special issue), Chiriotti Editori, Pinerolo, Italy (1989).

12. N. W. Pirie, *Leaf Protein: Its Agronomy, Preparation, Quality and Use*, IBP Handbook 20, Blackwell Scientific Publications, Oxford (1971).

13. N. W. Pirie, *Leaf Protein and Its By-Products in Human and Animal Nutrition*, 2nd ed., Cambridge University Press, Cambridge (1987).

14. C. Costes, ed., *Protéines Foliares et Alimentation*, Gauthier-Villars, Bordas, Paris (1981).

15. R. J. Wilkins, ed., *Green Crop Fractionation*, Occasional Symposium 9 British Grasslands Society, Grassland Research Institute, Hurley, Maidenhead, England (1977).

16. R. Carlsson, in *Handbook of Plant and Crop Stress* (M. Pessarakli, ed.), Marcel Dekker, New York, pp. 543–558 (1993).

17. R. Carlsson, Ph.D. dissertation, Lund University, Sweden, /LUNBDS/ (NBFB-1004)/1-8 (1975).

18. R. Carlsson, "New Tobacco Products and Phytochemicals from Selected, Field-Cultivated *Nicotiana* species," *Tobacco Protein Utilization Perspectives: Proceedings of a Seminar*, Report EUR 11923 EN (EC) (P. Fantozzi, ed.), Perugia, Italy, pp. 54–61 (1989).

19. R. Carlsson, "White Leaf Protein Products for Human Consumption: A Global Review of Plant Material and Processing Methods," *Tobacco Protein Utilization Perspectives: Proceedings of Round Table Meeting* (P. Fantozzi, ed.), Salsòmaggiore Terme, Italy, Oct. 1985, C.N.R., I.P.R.A., Rome, pp. 125–146 (1986).

20. R. Carlsson, "Trends for Future Applications of Wet-Fractionation of Green Crops," *EC Seminar: Forage Protein Conservation and Utilization* (T. W. Griffiths and M. F. Maguire, eds.), Agricultural Institute, Dunsinea Research Centre, Castleknock, Dublin, Ireland, pp. 57–82 (1982).

21. R. Carlsson, "Plant Selection Among Temperate Species for Production of Leaf Protein Concentrates," *Quality Aspects in Fodder Crops Breeding: Eucarpia Fodder Crop Section Meeting*, Radzikow, Poland, Sept. 1978, Biuletyn 135/1979, Institutu Hodowli i Aklimatyza cji Roslin, Blonie k/Warszawy, pp. 239–263 (1989).

22. C. K. Lyon, P. F. Knowles, and G. O. Kohler, *J. Sci. Food Agric., 34*: 849 (1983).

23. M. Bubicz and Z. Koter, *Rocz. Nauk Roln. Ser. B, 100*: 84 (1980).

24. M. Bubicz, B. Baraniak, G. Macik-Baranska, and B. Kota, *Pamiet. Pulaski, 72*: 91 (1980).

25. M. Bubicz, A. Jelinowska, and J. Majewski, *Pamiet. Pulawski 78*: 139 (1982).

26. M. Bubicz and A. Jelinowska, *Rocz. Nauk Roln. Ser. A, 105*(4): 41 (1983).

27. H. Ostrowski-Meissner and G. Falconer, "Selecting Lucerne (*Medicago sativa*) Cultivars for Commercial Crop Processing and Fractionation in Mid-west NSW, Australia," *Proceedings of the 4th International Conference on Leaf Protein Research* (H. T. Ostrowski-Meissner, ed.), Massey University, Palmerston North, New Zealand, Paper 19 (1993).

28. R. H. Brown, *Crop Sci., 18*: 93 (1978).

29. B. Koch, M. Kota, and I. M. Horwath, *Agrobotanika, 9*: 131 (1967).

30. K. Lexander, R. Carlsson, V. Schalen, A. Simonsson, and T. Lundborg, *Ann. Appl. Biol., 66*: 193 (1970).

31. P. Pheloung and C. J. Brady, *J. Sci. Food Agric., 30*: 246 (1979).

32. M. L. Fishman and D. Burdick, *J. Agric. Food Chem., 25*: 1122 (1977).

33. S. C. Huber, T. C. Hall, and G. E. Edwards, *Plant Physiol., 57*: 730 (1976).

34. L. Telek and F. W. Martin, in *Leaf Protein Concentrates* (L. Telek and H. D. Graham, eds.), AVI Publishing, Westport, Conn., pp. 81–116 (1983).

35. R. Carlsson, L. Jokl, and G. Amorim, *Nutr. Rep. Int., 30*: 323 (1984).

36. M. Kluge and I. P. Ting, in *Ecological Studies* (W. D. Billings, F. Golley, O. L. Lange, and J. S. Olson, eds.), Springer-Verlag, Berlin, Vol. 30, pp. 1–209 (1978).

37. M. de Souza Dayrell and E. Cardillo Vieira, *Nutr. Rep. Int.*, *15*, 529 (1977).
38. M. de Souza Dayrell and E. Cardillo Vieira, *Nutr. Rep. Int.*, *15*: 539 (1977).
39. J. T. O. Kirk and R. A. E. Tilney-Bassett, *The Plastids*, Elsevier/North-Holland Biomedical Press, Amsterdam (1978).
40. T. O. Addy, L. F. Whitney, and C. S. Chen, in *Leaf Protein Concentrates* (L. Telek and H. D. Graham, eds.), AVI Publishing, Westport, Conn., pp. 490–507 (1983).
41. W. S. Pierpoint, in *Leaf Protein Concentrates* (L. Telek and H. D. Graham, eds.), AVI Publishing, Westport, Conn., pp. 235–267 (1983).
42. D. B. Arkcoll, *Leaf Protein: Its Agronomy, Preparation, Quality, and Use* (N. W. Pirie, ed.), IBP Handbook 20, Blackwell Scientific Publications, Oxford, pp. 9–18 (1971).
43. W. R. Kehr, R. L. Ogden, and L. D. Satterlee, *Agron. J.*, *71*: 272 (1979).
44. R. Carlsson, *Acta Agric. Scand.*, *30*: 418 (1980).
45. G. P. Srivastava and A. K. Singh, "Effects of Extractants on Leaf Protein Yields from Different Cuttings of Berseem (*Trifolium alexandrinum* L.)," *Recent Advances in Leaf Protein Research: Proceedings of 2nd International Conference on Leaf Protein Research* (I. Tasaki, ed.), Faculty of Agriculture, Nagoya University, Togo-cho, Aichi-ken, Japan, pp. 254–255 (1985).
46. H. T. Ostrowski-Meissner, R. Carlsson, and D. R. McKenzie, *J. Food Qual.*, *7*: 27 (1984).
47. H. T. Ostrowski-Meissner, C. J. Pearson, and L. M. Shields, "The Effect of the Physiological Stage of Lucerne on the Degree of Protein Recovery from the Crop Used in the Wet-Fractionation Process," *Proceedings of the 4th International Conference on Leaf Protein Research* (H. T. Ostrowski-Meissner, ed.), Massey University, Palmerston North, New Zealand, Paper 21 (1993).
48. S. B. Heath, in *Plant Proteins* (G. Northon ed.), Butterworth, Boston, pp. 171–189 (1978).
49. H. T. Ostrowski-Meissner, in *Leaf Protein Concentrates* (L. Telek and H. D. Graham, eds.), AVI Publishing, Westport, Conn., pp. 562–600 (1983).
50. P. A. Carroad, H. Anaya-Serrano, R. H. Edwards, and G. O. Kohler, *J. Food Sci.*, *46*: 383 (1981).
51. H. W. Ream, N. A. Jorgensen, R. G. Koegel, and H. D. Bruhn, in *Leaf Protein Concentrates* (L. Telek and H. D. Graham, eds.), AVI Publishing, Westport, Conn., pp. 437–466 (1983).
52. H. T. Ostrowski-Meissner, C. J. Pearson, and L. M. Shields, "The Amino Acid Profiles of Extracts from Lucerne as Affected by Its Physiological Stage and Season of the Year at the Time of Harvest for Wet-Fractionation," *Proceedings of the 4th International Conference on Leaf Protein Research* (H. T. Ostrowski-Meissner, ed.), Massey University, Palmerston North, New Zealand, Paper 44 (1993).
53. H. T. Ostrowski-Meissner, "Profiles of Pigments in Leaf Extracts from Lucerne as a Factor of Its Physiological Stage and Geographic Location," *Proceedings of the 4th International Conference on Leaf Protein Research* (H. T. Ostowski-Meissner, ed.), Massey University, Palmerston North, New Zealand, Paper 45 (1993).
54. T. C. Tso and S. D. Kung, in *Leaf Protein Concentrates* (L. Telek and H. D. Graham, eds.), AVI Publishing, Westport, Conn., pp. 117–132 (1983).
55. R. Carlsson, "Wet Crop Fractionation," *Proceedings of Dri-Crops '89, 4th International Green Crop Drying Congress*, Cambridge, Agra Europe Ltd., London, pp. 108–112 (1989).
56. B. Holm-Christensen, "The Dehydration Plant as a Producer for the Cellulose Industry," *Proceedings of Dri-Crops '89, 4th International Green Crop Drying Congress*, Cambridge, Agra Europe Ltd., London, pp. 91–94 (1989).
57. H. T. Ostrowski-Meissner, G. Falconer, and L. M. Shields, "Increasing the Efficiency of Wet Crop Fractionation by the Two-Stage Harvest of the Lucerne Crop," *Proceedings of the 4th International Conference on Leaf Protein Research* (H. T. Ostrowski-Meissner, ed.), Massey University, Palmerston North, New Zealand, Paper 20 (1993).
58. D. B. Arkcoll and G. N. Festenstein, *J. Sci. Food Agric.*, *22*, 49 (1971).
59. H. G. Dakore, G. S. Reddy, and A. M. Mungikar, "The Yields of Extracted Leaf Protein from Maize and Sorghum Under the Influence of Fertilizers," *Recent Advances in Leaf Protein Research: Proceedings of the 2nd International Conference on Leaf Protein Research* (I. Tasaki, ed.), Faculty of Agriculture, Nagoya University, Togo-cho, Aichi-ken, Japan, pp. 204–206 (1985).
60. D. K. Bagchi, "Studies of the Effect of Nitrogen and Boron on Root and Extracted Leaf Protein Yield of Beet Root," *Advances in Leaf Protein Research: Proceedings of the 2nd International*

Conference on Leaf Protein Research (I. Tasaki, ed.), Faculty of Agriculture, Nagoya University, Togo-cho, Aichi-ken, Japan, pp. 227–228 (1985).

61. S. B. Heath and M. W. King, "The Production of Crops for Green Crop Fractionation", *Green Crop Fractionation*, Occasional Symposium 9 British Grasslands Society, Grassland Research Institute, Hurley, Maidenhead, England, pp. 9–21 (1977).

62. P. Hanczakowski and R. Lutynska, *Rocz. Nauk Zootech.*, *3*(1): 143 (1976).

63. R. Carlsson, P. Hanczakowski, and T. Kaptur, *Anim. Feed Sci. Technol.*, *11*: 239 (1984).

64. A. M. Smith and A. H. Agiza, *J. Sci. Food Agric.*, *2*: 503 (1951).

65. K. M. Henry and J. E. Ford, *J. Sci. Food Agric.*, *16*: 425 (1965).

66. R. Carlsson and E. M. W. Clarke, *Qual. Plant. Plant Foods Human Nutr.*, *33*: 127 (1983).

67. R. Carlsson, "Effects of Species, Plant Physiological Development Stages, Growth Conditions, and Process Conditions on the Nutritive Value of Leaf Protein Concentrates," *Progress in Leaf Protein Research*: *Proceedings of the First International Conference on Leaf Protein Research* (N. Singh, ed.), Today and Tomorrow's Printers and Publishers, New Delhi, pp. 201–208 (1984).

68. R. Hegnauer, *Chemotaxonomie der Pflanzen*, Vols. 2 to 6, Birkhauser Verlag, Basel, (1963, 1964, 1966, 1969, 1973).

69. I. E. Liener, *Toxic Constituents of Plant Foodstuffs*, Academic Press, New York (1969).

70. C. Humphries, *J. Sci. Food Agric.*, *31*: 1225 (1980).

71. L. Jokl, R. Carlsson, and R. C. Santos, "Effects of Processing Conditions on the Chemical Composition and Nutritive Value of Leaf Protein Concentrates from Tropical Legumes and from Leaves of Forest Trees," *Progress in Leaf Protein Research*: *Proceedings of the First International Conference on Leaf Protein Research* (N. Singh, ed.), Today and Tomorrow's Printers and Publishers, New Delhi, pp. 221–232 (1984).

72. K. Igarashi, Y. Sakamoto, and T. Yasui, *J. Agric. Chem. Soc. Jpn.*, *50*: 67 (1976).

73. B. Monties and J. C. Rambourgh, *Bull. Lias. 8 Groupe Polyphenol.*, Narbonne, France (1978).

74. P. R. Cheeke, L. Telek, R. Carlsson, and J. Evans, *Nutr. Rep. Int.*, *22*: 717 (1980).

75. P. R. Cheeke, R. Carlsson, and G. O. Kohler, *Can. J. Anim. Sci.*, *61*: 199 (1981).

76. N. L. Lahiry and L. D. Satterlee, *J. Food Sci.*, *40*: 1326 (1975).

77. J. Zarnowski, J. Okonski, S. Gwiazda, P. Lorek, and Z. Elzbieziak, *Rocz. Nauk Zootech.*, *25* (87): 257 (1987).

78. A. L. Livinston, B. E. Knuckles, R. E. Edwards, D. de Fremery, R. E. Miller, and G. O. Kohler, *J. Agric. Food Chem.*, *27*: 362 (1979).

79. M. Hegsted and H. Linkswiler, *J. Sci. Food Agric.*, *31*: 777 (1980).

80. R. G. Glencross, G. N. Festenstein, and H. G. C. King, *J. Sci. Food Agric.*, *23*: 371 (1972).

81. B. E. Knuckles, D. de Fremery, and G. O. Kohler, *J. Agric. Food Chem.*, *24*: 1177 (1976).

82. K. Igarashi and T. Yasui, *J. Agric. Chem. Jpn.*, *52*: 241 (1978).

83. J. C. Rambourgh and B. Monties, *Ann. Technol. Agric.*, *29*: 463 (1980).

84. C. S. Balasundaram, R. Chandramni, P. Muthuswamy, and K. K. Krishamoorty, *Indian J. Nutr. Diet.*, *13*: 11 (1976).

85. R. Carlsson and L. Telek, unpublished report for Mayaguez Institute for Tropical Agriculture, Mayaguez, Puerto Rico (1978).

86. M. Nandakumaran, C. R. Anathasubramanam, and P. A. Devasia, *Kerala J. Vet. Sci.*, *9*: 221 (1978).

87. P. Hanczakowski and M. Makuch, *Potato Res.*, *23*: 1 (1980).

88. P. Hanczakowski, B. Skraba, and M. Mlodkowski, *Amin. Feed Sci. Technol.*, *6*: 413 (1981).

89. R. Carlsson and P. Hanczakowski, "Leaf Nutrient Concentrate from *Nicotiana* Species: Nutritive Value *in Vivo* and Antinutritive Substances," *Proceedings of the 3rd International Conference on Leaf Protein Research* (P. Fantozzi, ed.), *Ital. J. Food Sci.* (special issue), Chiriotti Editori, Pinerolo, Italy, pp. 333–337 (1989).

90. P. Hanczakowski and B. Skraba, *Anim. Feed Sci. Technol.*, *24*: 151 (1989).

91. A. E. Harmuth-Hoene and J. F. Diehl, "Nutritive Value of Tobacco Protein Fractions," *Tobacco Protein Utilization Perspectives*: *Proceedings of Round Table Meeting* (P. Fantozzi, ed.), Salsomaggiore Terme, Italy, Oct. 1985, C.N.R., I.P.R.A., Rome, pp. 117–124 (1986).

92. M. Bacinelli and G. F. Montedoro, "Impiego di alcuni additivi nell'abbattimento di sostanze

inquinanti naturali delle frazioni proteiche de tabacco," *Tobacco Protein Utilization Perspectives*: *Proceedings of a Seminar*, Report EUR 11923 EN (EC) (P. Fantozzi, ed.), Perugia, Italy, pp. 34–36 (1989).

93. D. Lathia, B. Ledwig, and S. Kruchten, "Tobacco a Healthy Alternative: Nutritive Quality of Different Tobacco Protein Concentrates," *Tobacco Protein Perspectives: Proceedings of a Seminar*, Report 11923 EN (EC) (P. Fantozzi, ed.), Perugia, Italy, pp. 62–66 (1989).

94. H. W. Renner and R. Munzner, "Genotoxicity Testing of Tobacco Protein: A Safety Evaluation," *Proceedings of the 3rd International Conference on Leaf Protein Research* (P. Fantozzi, ed.), *Ital. J. Food Sci.* (special issue), Chiriotti Editori, Pinerolo, Italy, pp. 253–258 (1989).

95. J. Raloff, *Sci. News*, *133*(25): 397 (1988).

96. H. Ueda and M. Oshima, *Anim. Sci. Technol. Jpn.*, *63*: 1032 (1992).

97. L. W. Wattenberg and W. D. Loub, *Cancer Res.*, *40*: 2820 (1978).

98. Anonymous, *Lancet*, *2*: 325 (1984).

99. G. W. Burton and K. U. Ingold, *Science*, *224*: 569 (1984).

100. *Find Your Feet*, personal communication (1991).

101. R. M. McDonald, P. E. Donnelly, R. A. Mills, and S. R. Vaughan, "High Value Products from Lucerne: A New Zealand Perspective," *Recent Advances in Leaf Protein Research: Proceedings of the 2nd International Conference on Leaf Protein Research* (I. Tasaki, ed.), Faculty of Agriculture, Nagoya University, Togo-cho, Aichi-ken, Japan, pp. 117–178 (1985).

102. T. Horigome, Y. S. Cho, and M. Oshima, *Ital. J. Food Sci.*, *4*: 227 (1990).

103. R. Carlsson and G. O Kohler, "Composition of Leaf Protein Concentrates: Contributions of Nitrogenous Substances from Different Plant Shoot Parts to Whole Leaf Protein Concentrates of *Amaranthus cruentus*, *A. hypocondriacus*, *Chenopodium quinoa*, and *Helianthus annuus*," Report for Western Regional Research Center, USDA, Berkeley, Calif. (1978).

104. A. D. Correa, F. S. Espindola, T. N. Tanaka, and R. Piau, Jr., "Influence of the Age of Plants Producing Leaf Protein Concentrate," *Proceedings of the 3rd International Conference on Leaf Protein Research* (P. Fantozzi, ed.), *Ital. J. Food Sci.* (special issue), Chiriotti Editori, Pinerolo, Italy, pp. 444–446 (1989).

105. A. D. Correa, F. S. Espindola, T. N. Tanaka, R. Piau, Jr., L. Jokl, and A. Loures, "Amino Acid Composition, *in Vitro* Protein Digestibility, and Some Antiphysiological Factors of Leaf Protein Concentrates Obtained by Various Methods from *Cajanus cajan*," *Proceedings of the 4th International Conference on Leaf Protein Research* (H. T. Ostrowski-Meissner, ed.), Massey University, Palmerston North, New Zealand, Paper 47 (1993).

106. M. Bubicz and A. Jelinowska, "Carotenoid Content in Protein Concentrates in Relationship to Variety and Growth Phase of Alfalfa," *Proceedings of the 3rd International Conference on Leaf Protein Research* (P. Fantozzi, ed.), *Ital. J. Food Sci.* (special issue), Chiriotti Editori, Pinerolo, Italy, p. 503 (1989).

107. M. W. Pedersen, O. O. Anderson, J. C. Street, L. C. Wang, and R. Baker, *Poult., Sci.*, *51*: 458 (1972).

108. P. R. Cheeke, J. H. Kinzell, and M. W. Pedersen, *J. Anim Sci.*, *46*: 476 (1977).

109. E. L. Hove, E. Lohrey, M. K. Urs, and R. M. Allison, *Br. J. Nutr.*, *31*: 147 (1974).

110. R. Carlsson, M. R. Gumbmann, D. de Fremery, and G. O Kohler, "Quality of Leaf Protein Concentrates and Fibre Residues from Various Species, Grown in a Hot Temperate Climate," Report for Western Regional Research Center, USDA, Berkeley, Calif. (1978).

111. E. M. Bickoff, A. N. Booth, D. de Fremery, R. H. Edwards, B. E. Knuckles, R. E. Miller, R. M. Saunders, and G. O. Kohler, in *Protein Nutrition Quality of Foods and Feeds*, Part 2 (M. Friedman, ed.), Marcel Dekker, New York (1975).

112. A. A. Woodham, D. B. Arkcoll, R. A. Karmali, and E. M. W. Clarke, "The Effect of Processing Conditions on the Nutritive Value of Protein Concentrates Prepared from Green Leaves," *Proceedings of the 4th International Congress on Food Science and Technology*, Madrid, Vol. I (1976).

113. A. A. Woodham, *Proc. Nutr. Soc.*, *24*, xxiv (1965).

114. B. H. Subba Rau, K. V. R. Ramana, and N. Singh, *J. Sci. Food Agric.*, *23*: 233 (1972).

115. M. Tao, M. Boulet, G. J. Brisson, K. H. Huang, R. R. Riel, and J. P. Julien, *Can. Inst. Food Sci. Technol. J.*, *5*; 50 (1972).

116. P. Hanczakowski, *Rocz. Nauk Roln. Ser. B*, 97: *85* (1975).

117. S. K. Munshi, D. S. Wagle, and V. K. Tharpar, *J. Food Sci. Technol. (India)*, *12*: 23 (1975).

118. T. Horigome, *Jpn. J. Zootech. Sci.*, *48*: 267 (1977).

119. T. Lundborg, Ph.D. dissertation, Lund University, Sweden,/LUNDBS/(NBFB-1003)/1-12 (1979).

120. R. Carlsson, *Var Foda*, *6–7*: 270 (1983).

121. N. W. Pirie, "Maximising the Amount of β-Carotene in Leaf Protein," *Proceedings of the 3rd International Conference on Leaf Protein Research* (P. Fantozzi, ed.), *Ital. J. Food Sci.* (special issue), Chiriotti Editori, Pinerolo, Italy, pp. 416–419 (1989).

122. *Find Your Feet*, personal communication (1993).

Index